2023 China Semiconductor Technology International Conference (CSTIC 2023)

Shanghai, China
26-27 June 2023

IEEE Catalog Number: CFP2360Y-POD
ISBN: 979-8-3503-1101-3

**Copyright © 2023 by the Institute of Electrical and Electronics Engineers, Inc.
All Rights Reserved**

Copyright and Reprint Permissions: Abstracting is permitted with credit to the source. Libraries are permitted to photocopy beyond the limit of U.S. copyright law for private use of patrons those articles in this volume that carry a code at the bottom of the first page, provided the per-copy fee indicated in the code is paid through Copyright Clearance Center, 222 Rosewood Drive, Danvers, MA 01923.

For other copying, reprint or republication permission, write to IEEE Copyrights Manager, IEEE Service Center, 445 Hoes Lane, Piscataway, NJ 08854. All rights reserved.

*** This is a print representation of what appears in the IEEE Digital Library. Some format issues inherent in the e-media version may also appear in this print version.

IEEE Catalog Number: CFP2360Y-POD
ISBN (Print-On-Demand):   979-8-3503-1101-3
ISBN (Online):   979-8-3503-1100-6

Additional Copies of This Publication Are Available From:

Curran Associates, Inc
57 Morehouse Lane
Red Hook, NY 12571 USA
Phone: (845) 758-0400
Fax: (845) 758-2633
E-mail: curran@proceedings.com
Web: www.proceedings.com

TABLE OF CONTENTS

Research on Vt Window Improvement Process of 2T Sonos Embedded Flash..1
  Xiaokang Li, Shipu Li, Jun Qian

An Effective Method to Minimize the Difference Between Spot and Ribbon Beam in S/D Region
by Carbon Co-Implantation 3
  Long Feng, Zhiqiang Xiao, Jiaxing Xiao, Zhirui Li, Haitao Yan

Study of Breakdown Voltage Improvement of High-Voltage PLDMOS 6
  Wenting Duan, Donghua Liu, Haiyang Ling, Ying Cai, Feng Jin, Wensheng Qian

A Compact Sawtooth Wave Generator Based on Novel Z^2-FET Device..9
  Hui Xie, Yingxin Chen, J. Wan

Study of the Formation of Copper Void Defect and Process Optimization for Reduction in Dual
Damascene Process 12
  Hongliang Zhu, Shuhuai Jia, Dejing Ma, Fengjiao Wang

Study on N-Type MOS Capacitor in 55NM CMOS 16
  Hongliang Zhu, Xin Zhou, Dejing Ma, Haoqi Zheng, Tianfu Zhang

The Study on the Optimization and Electrical Behavior of NEDMOS Transistors for High-Voltage
Power Applications..18
  Chongkai Du, Chenchen Qiu, Jun Qian, Chang Sun

Fabrication and Characterization of a Novel Embedded Mirror Gate Sonos 22
  Ning Wang, Kegang Zhang

Investigation of a New Disturb Effect in the Aggressively Scaled Dual-Bit/Cell Split-Gate Floating-
Gate Flash Cell..25
  Yintong Zhang, Zhaozhao Xu, Dylan Zhou, Alan Shen, Fredric Liu, Ziquan Fang, Donghua
  Liu, Gordon Li, Wensheng Qian

Important Process Parameter and Its Sensitivity Check by Virtual Fabrication: Channel Hole Profile
Impact on Advanced 3D NAND Structure 29
  Qingpeng Wang, Pengfei Lyu, Lifei Sun, Yu De Chen, Cheng Li, Jacky Huang, Benjamin
  Vincent, Joseph Ervin

$HF_XZR_{1-X}O_2$ Ferroelectric Thin Film Grain Size Tuning Via Annealing Ramp Rate Achieving
Endurance >10^9 Cycles, $2P_R$ of 40.6μC/CM^2, Write Voltage Down to 1.5 V, and Switching Speed of
30 NS..32
  Zhixiong Li, Bing Zhou, Jiawei Xu, Shuaihang Xu, Jun Lan, Quanzhou Zhu, Yingjie Zhu, Jie
  Li, Xuewei Feng, Mei Shen, Feichi Zhou, Longyang Lin, Yida Li

FinFet Source/Drain Parasitic Resistance Optimization by TCAD Simulation..35
  Tongtong Luan, Xinqi Liu, Yu Gu, Xufeng Kou

Simulation Study of Gate-All-Around Nanosheet Devices Based on SOI Structure 38
  Yangyang Hu, Tianxiang Zhao, Mengmeng Yang, Jianhua Zhang, Kailin Ren

Leakage Reduction of GAA Stacked SI Nanosheet CMOS Transistors and 6T-SRAM Cell Via
Spacer Bottom Footing Optimization..41
  Jiaxin Yao, Xuexiang Zhang, Lei Cao, Junjie Li, Na Zhou, Qingkun Li, Yanzhao Wei, Yanna
  Luo, Jun Luo, Qingzhu Zhang, Huaxiang Yin

Investigation of Electrical Characteristics on Morphotropic Phase Boundary of $Hf_{1-x}Zr_xO_2$ for Dynamic Random Access Memories.. 45

Kun Zhong, Huaxiang Yin, Zhaohao Zhang, Fan Zhang

High Endurance Sonos Technology Improved by Design & Process Optimization... 48

Pingsheng Zhou, Kegang Zhang, Gordon Li, Hualun Chen, Xiang Yao, Weiran Kong

Investigation of the Doping Profile for Ion Implants and Rapid Annealing in Silicon Via an Improved Method.. 52

Zeqi Zha, Zhenhui Wang, Ya Wang

Experimental Investigation of Ultra-Low Temperature LA_2O_3/HFO_2 Bi-Layer Dipole-First Process Using PVD Method for Advanced IC Technology 55

Yanzhao Wei, Jiaxin Yao, Renren Xu, Qingzhu Zhang, Huaxiang Yin

Large Area CVD Mos_2 Memristor Suitable for Neuromorphic Applications... 58

Muhammad Zaheer, Tariq Aziz, Jun Lan, Quanzhou Zhu, Wenhui Wang, Mei Shen, Feichi Zhou, Longyang Lin, Xuewei Feng, Yida Li

Novel Channel-On-Fin (COF) IGZO-TFTS with Ultra-Scaled Back Gate Length of 23 NM 61

Shangbo Yang, Gaobo Xu, Gangping Yan, Zhiyu Song, Guoliang Tian, Yanna Luo, Yinan Yan, Lianlian Li, Huaxiang Yin

Investigation of Vertical Channel IGZO-TFT Based on PVD-IGZO 64

Zhiyu Song, Gaobo Xu, Gangping Yan, Shangbo Yang, Yanna Luo, Guoliang Tian, Yinan Yan, Huaxiang Yin

A Compact Model of Non-Volatile Ferroelectric Tunnel Fet with Ambipolarity for in-Memory-Computing Based Edge AI 67

Hanyong Shao, Jin Luo, Zhiyuan Fu, Qianqian Huang, Ru Huang

Design of Ferroelectric FET-Based Capacitive-Coupling Computing-In-Memory for Binary Neural Networks 71

Boyi Fu, Jin Luo, Weikai Xu, Qianqian Huang, Ru Huang

Investigation of Synergic Hydrogen Mitigation Technique for Top-Gate A-IGZO Thin-Film Transistors 75

Gangping Yan, Zhiyu Song, Haoqing Xu, Shangbo Yang, Chuqiao Niu, Guoliang Tian, Yanna Luo, Luoyun Zhang, Yunjiao Bao, Gaobo Xu, Huaxiang Yin

Characterization of Field Cycling Fatigue in HfZrOx Ferroelectric Capacitors... 78

Puyang Cai, Zhiwei Liu, Tianxiang Zhu, Zhigang Ji, Runsheng Wang, Ru Huang

Promoting Chip Probing Test Yield by Simple ISSG and Global Wet Process 81

Jingsong Peng, Kegang Zhang, Ning Wang, Yang Ding, Pingsheng Zhou, Yin Yin

Reliability Performance of Novel Tunneling Field Effect Transistors Based on Foundry Platform.................. 84

Yukun Tang, Qianqian Huang, Kaifeng Wang, Yongqin Wu, Hongyan Han, Ye Ren, Weihai Bu, Junhua Liu, Zhigang Ji, Ru Huang

Influence of Interfacial Layers and High-K Post Dielectric Annealing on the Characteristics of MOS Devices 87

Guanqiao Sang, Qingzhu Zhang, Huaxiang Yin, Junfeng Li, Xulei Qin

Optimized Wafer Edge Condition in Lithographic Process for Peeling Defect Reduction 90

Shanshan Chen, Hunglin Chen, Yin Long, Kai Wang

Silicide Profile Optimization on Active Area in 4XNM ETOX nor Flash Memory 92
    Hualun Chen, Yuxin Tong, Xiangyu Qi, Songhan Duan, Botong Liu, Chaoran Zhang, Lin Gu

Improvement of Standby Current Failure by Device Optimization on 4XNM ETOX NOR-Flash
Memory 95
    Hualun Chen, Zhuangzhuang Wang, Lin Gu, Yihang Du, Chun Yao, Xiaodong Mu, Wan Song

An On-Chip Superconducting Quantum Transponder 99
    Rutian Huang, Xinyu Wu, Xiao Geng, Jianshe Liu, Wei Chen

Effects of Floating Gate Profile on Cell Characteristics of 4XNM FG-First ETOX nor Flash
Memory 102
    Hualun Chen, Yihang Du, Lin Gu, Zhuangzhuang Wang, Chun Yao, Zhaozhao Xu

Improved Environmental Stability of N-Type Polymer Field-Effect Transistors Using Nickel
Contact Electrode 105
    Yuan Liu, Quanhua Chen, Rujun Zhu, Jinxiu Cao, Yong Xu

A New Method to Calculate Loading Effect in Embedded Flash 109
    Fangce Sun

A New Method to Improve Split Gate Flash Erase and Endurance 112
    Fangce Sun

Design and Simulation of a Superconducting Switch Based on Weakly Damped Superconducting
Quantum Interference Devices 115
    Xinyu Wu, Rutian Huang, Jianshe Liu, Wei Chen

Technologies for Superior Reliability in SiC Power Devices 118
    Min-Hwa Chi

The Study on Reducing Bit-Line Parasitic Capacitance in Advanced DRAM 122
    Yexiao Yu, Hong Ma, Zhongming Liu, Shaoyou Xiong, Dan Wang, Yang Zhang, Yi Yang

Enhancement of Pattern Depth in Plasmonic Lithography for Practical Application 125
    Dandan Han, Yayi Wei

Cyclegan-Based Mask Diffraction Model 128
    Jiaxiang Zhuo, Dongyong Xu, Yijiang Shen

Illumination Optimization for the Beol Dtco with 45 Degree Local Interconnection 132
    Xianhe Liu, Muzi Han, Yanli Li, Qi Wang, Qiang Wu

Process and Tool Monitor and Diagnosis Based on Overlay Data and Modeling 135
    Yi Tong, Libin Zhang, Yayi Wei, Tianchun Ye, Yun Wang

A Multi-Step Sraf to Improve Process Windows in Metal Layer 138
    Wei Wei, Lu Zhu, Dan Wang, Xiaoyan Sun, Yue Wang, Yueyu Zhang, Jianzhong Liu, Shiri Yu

Recent Progress of EUV Resist Development for Improving Chemical Stochastic 142
    Toru Fujimori

The Reduction of Yield Loss and Contact Overlay Shift by Optimizing the Process Profile of Pre-
Layer Process Integration 145
    Zhejun Liu, Wei Lu

The Analysis of Optical Critical Dimension (OCD) Signal Strength Between 5 Nm FinFET and 3 Nm Complementary FET (CFET) Vertical Gate Stacks 147
Qi Wang, Qiang Wu, Xianhe Liu, Yanli Li

Study On-Product Overlay Improvement for Immersion Lithography 150
Guoping Liu, Yinsheng Yu, Chi Zhang, Yuhui Li, Wei Cao, Qin Yuan, Hongwen Zhaoa

The Possibility of Using 193 NM Immersion Lithography Process for 5 NM Logic Design Rules 153
Qiang Wu, Yanli Li, Xianhe Liu, Qi Wang

Line-End Roundness and Voids Improvement of BEOL Metal Layer 157
Mudan Wang, Tiancheng Tu, Hui Zhao, Shirui Yu

OPC Correction Method Based on Corner to Corner Structure... 160
Qiguang Zhou, Dan Wang, Yueyu Zhang, Shirui Yu

Study on Inter-Layer Overlay of Stitching Lithography Technology 162
Hongmin Liu, Changcheng Gao, Qiongtao Wu

A Negative-Tone Photosensitive Epoxy Material.. 165
Ke Bai, April Tang, Jiangtao Mou, Jinfu Zheng

Tungsten/Silicon Oxide/Titanium Nitride Stack Etching 168
Jie Luo, Haochang Lyu, Linjie Hou, Baodong Han, Hongbo Sun, Chao Zhao

The Investigation of CF3I for High-Aspect-Ratio Cryogenic Dielectric Etch 172
Jianqiu Hou, Vina Xu, Kai Zhang, Ziyang Wu

SiARC Residue Reduction Methods with Minimizing Profile Change for Mask Patterning.......................... 176
Xingxing. Xu, Hexin Zhou, Quanbao. Li, Jian. Huang

Distortion Control When Etching DRAM Metal Contact 179
Jianqiu Hou, Yao Sun, Hui Xue, Ya Zhou, Hao Li, Zhiwen Luan, Zijian Chen, Zengwen Hu

Study on the Effect of Water Spraying Mode on the N Content of Wafer Surface After SC1 Cleaning in Light Doping Process.. 183
Jinlei Wang, Fenglin Guan, Mingguang Hang, Lili Jia, Fang Li, Xinhua Cheng

A Technical Optimization of Waferless Auto Clean for Aluminum Etcher 186
Li Qi, Qiang-Qiang Sang, Xing-Jun Yao, Li-Tian Xu, Jian-Kun Zhang, Li-Song Hu, Yi-Chang Liu, Chen Chen

Trimming of Silicon Nitride Hard Mask Using Cyclic Deposition and Etch Process 188
Li-Tian Xu, Pei Mei, Xing-Jun Yao, Chen Chen, Jia-Yun Zhang, Alan Zhang, Xiao-Peng Wu, Wu-Hao Han, Hong-Bo Lin, Nichole Zhang

Enabling Plasma Etch Solution for GaN Technology.. 190
Zoe Wang, Chunxiang Guo, Jian Liu, Yingxiong Feng, Lulu Guan, Kangning Xu, Qiao Huang, Lu Chen, Kaidong Xu

The Research of Special Gate Morphology Adjustment and Its Influence on Electrical Properties................. 193
Junjie Pan, Kai Qian, Lian Lu, Quanbo Li, Jun Huang, Yu Zhang

Study of Spacer Etching with PR Approach 196
Yuhao Yang, Siyuan Che, Xiangguo Meng, Lian Lu, Quanbo Li, Jun Huang, Yu Zhang

Silicon Surface Roughness Improvement During Plasma Etch... 198
Guang Yang, Li Zeng, Haiyun Zhu, Jing Wang, Zhongwei Jiang

Effect of Process Gas on Side Wall Angle in Silicon Trench Etching 200
*Yiming Ma, Guang Yang, Litian Xu, Zhongwei Jiang, Jing Wang, Qifei Wang, Donghan Wang,
Yongjie Zhou*

Investigation of CD Precise Control in Pitch Doubling Flow for Memory Industry 202
Zhao Liu, Baodong Han

The Plasma Etching of the Deep Hole Structure in Silicon with the Mixed Gas of SF_6, HBr and O_2.............. 206
Qifei Wang, Yiming Ma, Zhongwei Jiang, Jing Wang, Haiyun Zhu

Investigation of High Aspect Ratio Amorphous Carbon Etching in NAND Flash Memory 209
Li Zeng, Guang Yang, Zhongwei Jiang, Jing Wang, Li-Tian Xu

Study of Photo-Resist (PR) Strip Rate with High Temperature Pedestal for AL Patterning Process 212
Cheng Tian, Li-Tian Xu, Li-Song Hu, Xue-Hua Wang

Self-Limited Etching of Silicon Nitride Using Cyclic Process with CH2F2 Chemistry 215
Xue-Hua Wang, Li-Tian Xu, Tian Cheng

Strategy for Line Width Roughness (LWR) Reduction in Carbon Mandrel Patterning................................... 217
Yichang Liu, Li Qi, Litian Xu, Lianfu Zhao, Xingjun Yao, Zihan Zhang

Redistribution Layer Aluminum Advanced Etching Process Development 219
Xing-Jun Yao, Li-Tian Xu, Li Qi, Qiang-Qiang Sang, Li-Song Hu, Chen-Chen, Yi-Chang Liu

Deep Trench (DT) Etching Process for Power MOS Device 221
Chen Chen, Li-Tian Xu, Xing-Jun Yao

Technical Difficulties and Optimization Methods of Nmos Share CT in Contact Hole Etching..................... 223
Peng Zhu, Renhui Xu, Wentao Fu, Lei Sun, Yu Bao

Investigations of NAND Flash Device Cycling Performance Improvement Via FG Carbon-Doped
Polysilicon and Channel Corner Rounding 226
Lifeng Liu, Jun Wang, Xinruo Su

Optimization of SADP Process for Defect Reduction in Planar 2D NAND Flash.. 231
Lifeng Liu, Jun Wang, Yinan Ma, Zhenchao Sui, Yue Li

Control of Deposition in Cyclic Deposition/Etch Process... 236
Zhang Zihan, Wang Jing, Xu Litian

A Novel Method to Eliminate Bond Pad Crystal.. 238
*Fanshun Meng, Jiajin Wang, Yunbo Chen, Qiang Rui, Zhongkui Chen, Kuo-Jung Chen, Yi Liu,
Fengyang Li, Chao Sun*

Carbon Hard Mask Opening Process Development with Novel Sidewall Passivation in Memory
Manufacturing 241
Meng-Jiao Zhu, Li-Tian Xu, Jing Wang, Li Zeng, Zi-Han Zhang

The Optimization of Pin Hole Defect in High Resistance Process... 243
Lunan Zhu, Shan Huang, Xiaofeng Qu, Lei Sun, Quanbo Li, Tianpeng Guan, Yu Zhang

Growth and Reduction of Tiny Particle Defects in Selective SiGe Epitaxy S/D Devices 246
Zhiqiang Xiao, Cunzhe He, Dongliang Gao, Jiaxing Xiao, Haitao Yan, Zhenchao Sui, Xin Zhang

Technical Challenges in MRAM Fabrication 249
Y. H. Wang, X. L. Yang, Y. Tao, Y. H. Sun, Q. J. Guo, F. T. Meng, G. C. Han

Some Key Modifications of Theory Required to Understand the Leakage Current Mechanisms for Ferroelectric HfZro Capacitors Used in Microelectronics 253
W. S. Lau

Pulsed DC Parameters (Reverse Voltage, Duty Cycle, Pulsed Frequency) on Film Quality in Reactive Sputtered Aluminum Nitride Films 256
Wei-Yu Zhou, Xue-Li Tseng, Ning-Hsiu Yuan, Hsiao-Han Lo, Peter J. Wang, Ming-Yu Jiang, Yiin-Kuen Fuh, Tomi T. Li

Electroplating Process Improvement on Post-CMP Dishing Profile 262
Wenbo Wu, Tong Lei, Zhijun Zhu, Zhenhua Hu, Yushan Chi

Effect of Sub-Atmospheric Chemical Vapor Deposition SIO2 Film Deposition Process on Surface Chemistry Sensitivity 265
Jianan Wei, Xue Liu, Dapeng Ruan

Gas Distribution Effect on Thermal ALD AlN Film Thickness Non-Uniformity 269
Xiaomeng Liu, Qihui Zhang, Hao Deng

Substrate Effect on Thermal ALD AlN Film Growth Rate 271
Xiaomeng Liu, Tiantian Liu, Wenyi Liu, Xinyu Zhang, Hao Deng

Lau's Unified Schottky-Poole-Frenkel Theory with Asymmetric Distortion by Electron Charge Trapping Proposed to Explain the Current-Voltage Characteristics of High-K Metal-Insulator-Metal Capacitors 274
W. S. Lau

Cell Structure and Process Integration of a Novel 2T0C Technology for High-Density Dram Application 277
Zheng-Yong Zhu, Bok-Moon Kang, Jing Zhang, Xin-Lv Duan, Jin-Juan Xiang, Guan-Hua Yang, Di Geng, Wang Dan, Xie-Shuai Wu, Ming-Xu Liu, Gui-Lei Wang, Chao Zhao

The Study of Slip Defects in Furnace High Temperature Process 281
Sun Yan, Wei Simeng, Xie Yuanxiang

The Study of Silicon Nitride Films Deposited in Batch ALD System 288
Shiyao Cheng, Wei Kuai, Wenxu Duan, Xinyang Wang, Xiaomeng Liu, Shuo Cheng, Yuanxiang Xie, Xiaoping Shi

Influence of Ion Implantation on Void Defect Formation in Epitaxially Grown Silicon 292
Zeqi Zha, Zhenhui Wang, Ya Wang

The Effect of Sige Siconi Pre-Clean Time on Planner Logic Device Performance Study 295
Xuechun Zhang, Weichi Cheng, Li Ning, Jingang Wang

Mechanical Properties of Flip-Chip Bonding Structures for Micro-Led Devices: Cu-Cu Bonding with Passivation Layer and Indium Bumps Bonding 299
Kefeng Wang, Zehua Chen, Xiaoxiao Ji, Luqiao Yin, Xiuzhen Lu, Jianhua Zhang

A Novel Method to Optimize Sige Profile Using Co-Implantation 303
  Zhiqiang Xiao, Long Feng, Cunzhe He, Jiaxing Xiao, Dongliang Gao, Mingying Liu

A Study of Parasitic Capacitance Using Different Bit Line Spacer Integration Schemes in Advanced
DRAM 306
  Dempsey Deng, Qingpeng Wang, Yujia Zhong, Yu De Chen, Jacky Huang

Improvement of Sige Relaxation by a New Clamping Film Deposition Process Method 309
  Zhiqiang Xiao, Cunzhe He, Dongliang Gao, Haitao Yan, Zhenchao Sui, Xin Zhang

Some Methods to Reduce Micro Scratch Defect for Via Contact Tungsten Chemical Mechanical
Planarization Process 312
  Zhijie Zhang, Le Ning, Hongdi Wang, Zhiyang Liang

Study on the Mechanism of CMP Induced W Seam at Advanced Technology Node 317
  Shaojia Zhu, Yurong Que, Feng Shi, Mingfei Yu, Jian Zhang, Jingxun Fang, Yu Zhang

Study on the Mechamism of SIN Residue for ILD0CMP 319
  Yurong Que, Xing Ma, Jian Zhang, Hu Li, Jingxu Fang, Yu Zhang

Slurry System Establishment and Optimization for Advanced Cobalt Interconnect Metallization 322
  Lifei Zhang, Tongqing Wang, Xinchun Lu

Pattern Loading Improvement for CU CMP Process 325
  Lei Zhang, Yu Yang, Jian Zhang, Jingxun Fang, Yu Zhang

A Fem Model of Micro-Galvanic Corrosion Evolution at RU/CU Interface in H2O2 CMP Solution 328
  Shuo Gao, Qinhua Miao, Boyu Wen, Jie Cheng

Effects of Process to Material Removal in CMP: Modelling and Experiments 333
  Yanming Ren, Yiran Liu, Zijun Guan, Lei Zhu, Yuanda Gao, Wenjie Yu, Weimin Li

Improving 300mm Si Wafer Planarization Process with a Wholistic Approach 337
  Zijun Guan, Yiran Liu, Jiaming Fan, Yuanda Gao, Lei Zhu, Wenjie Yu, Weimin Li

Research Progress and Challenges of Chemical Mechanical Polishing for Silicon Carbide Wafer 340
  Lijuan Zhang

Research on the Dispersion Stability and Polishing Performance of Ceria Slurry 343
  Min Liu, Baoguo Zhang, Shitong Liu, Dexing Cui, Wenhao Xian, Pengfei Wu, Ye Wang

Study on the Slurry for Chemical Mechanical Polishing of Sapphire Wafer 346
  Wenhao Xian, Baoguo Zhang, Liu Min, Dexing Cui, Pengfei Wu, Ye Wang

Pad Surface Variation and Its Effect on SiO_2 Removal Rate in Ceria-Based CMP Slurry 349
  Chenchen Yang, Yu Yao, Enghoe Tan

Impact of Slurry for Dishing Reduction During CU CMP 352
  Yu Yao, Chenchen Yang, Enghoe Tan

Effect of Abrasive on the CMP Performance of C-Plane (0001) GAN Flim 354
  Jianghao Liu, Xinhuan Niu, Ni Zhan, Yida Zou, Yebo Zhu

Analysis of the Adsorption and Passivation Mechanism of JFCE on Copper Surface in Alkaline
CMP Slurry 357
  Ni Zhan, Xinhuan Niu, Yinchan Zhang, Fu Luo, Han Yan

Effect of Surfactants on CMP Properties of M-Plane Sapphire... 360
  Yida Zou, Xinhuan Niu, Ziyang Hou, Minghui Qu, Ni Zhan, Jianghao Liu

Methods for Fin Etching Profile Maintaining and Measurement 363
  Yun Xu, Tong Wu, Fairy Chen

Investigation of Different Gate Bias on PMOS HCI Performance 367
  Lei Li, Canny Chen, Atman Zhao

An Efficient Tool for Generating Test Program to Save Marginal Fail Chips 370
  Hanyan Chen

Research on TDDB Physical Mechanism of 28HKMG Mosfet 374
  Ting Wan, Hao Jiang, Yueqin Zhu, Ke Zhou

Reservoir Effect Study on Electro-Migration Behavior of ALCU Interconnects 377
  Jizhou Li, Kitty Wang, Weihai Fan

A Universal Auto Test Program Generation on Advantest V93000 ATE Platform 380
  Xin Song, Yefang Wang, Hanyan Chen

Innovation Test Technology for Ultra-High-Speed ADC on ATE 385
  Yanyan Chang, Tianyu Chen, Jiaying Xiang, Yichen Xiao

Computer Vision Technology Supported Rapid DRAM Capacitor Analyzing System Based on
TEM Image... 388
  Zhi-Yuan Gui, Chang Xu, Han Yan, Zhi-Yu Li

A Novel Model-Matching Based Scratch Tool Tracing System 391
  Shi-Qiang He, Yan-Qiu Zhang, Xiao-Lei Zhang, Chic-Kuo Fang

Faster AU-AL IMC Growth Under Chlorine Environment 394
  Liao Jinzhi Lois, Wang Bisheng, Zhang Xi, Hua Younan, Li Xiaomin

The Verification of TDDB Acceleration Model in Ultrathin Gate Dielectric 398
  Wen Ying, Canny Chen, Atman Zhao

The Design-Based Inspection Strategy for CU Void Defects Reduction 401
  Xingdi Zhang, Hunglin Chen, Yin Long, Kai Wang

Anomaly Detection of Non-Normal Distribution Wafer Acceptance Test Data Via GMM-Based
Method 404
  Junjun Zhuang, Yong Wang, Guiyun Mao, Xu Chen, Yansheng Wang, Zhengying Wei

Impact of Interface Trap Density on the Endurance of HFO_2/SI FEFETS 408
  Jiaqi Zheng, Yue Peng, Yanbin Yang, Dawei Gao, Rui Zhang, Genquan Han

Reliability Research on Micro Bump and C4bump in Large-Size 2.5D FCBGA 411
  Xiang Li, Zhuqiu Wang, Xiao He, Dan Yang, Na Mei

Design and Optimization of RC Triggered MV-NMOS for 28NM CMOS Technology ESD
Protection 415
  Jia Zhu, Lanying Wei, Yang Li, Jun Wu, Kun Wang, Wei Chen

Reliability Analysis of Metal Thermal Interface Materials for Ultra-Large Size FCLGA Package 418
  Keqing Ouyang, Zhuolun Wu, Zhuqiu Wang, Weilun Wang, Dan Yang, Na Mei

Impact of Interface Traps Generation on Flicker Noise Degradation in SI pMOSFETs 421
Yi Jiang, Luping Wang, Yanbin Yang, Dawei Gao, Rui Zhang

Research on Hot Carrier Injection Optimization of 28HKMG Technology 425
Weiwei Ma, Yang Li, Ran Huang, Yamin Cao, Wei Zhou

Application of Picosecond Ultrasonic Technology for CMOS Image Sensors 428
Johnny Mu, Kaixing Song, Johnny Jin, Cheolkyu Kim, Yaodong Huang, Hong Hong

Neutron Irradiation Induced Carrier Removal and Deep-Level Traps in N-Gan Schottky Barrier
Diodes.. 432
Jin Sui, Jiaxiang Chen, Haolan Qu, Ruohan Zhang, Min Zhu, Xing Lu, Xinbo Zou

Metavit-Trans: A Framework for Mixed-Type Defect Detection of Wafers with Vision Transformer
Combined with Meta-Learning and Transfer Learning 435
Junfeng Zhao, Lixin Tang

Lithography Hotspot Detection Based on Transfer Learning with High Resolution Networks 438
Hongzhe Wang, Lixin Tang

A Methdology for Testing Scan Chain with Diagnostic Enhanced Structure 441
Keqing Ouyang, Minqiang Peng, Shuai Wang, Guohua Zhou, Kai Wang

An End-To-End Detection Approach for Micropipe Defect of SIC Wafers Via Fusing Multiple
Hierarchical Features.. 444
W. X. Shi, T. G. Zhao, J. W. Zhang

A Novel Method to Achieve High Efficient Iteration of MBIST Pattern 447
Minqiang Peng, Keqing Ouyang, Feilong Pan, Guohua Zhou, Lei Chen

Calibration of Pitch Standards of SEM for Semiconductor Dimension Metrology Application 450
Wei Li, Yang Qu, Yushu Shi

Ultra-Wideband (UWB) Test Solution on V93000... 453
Kevin Yan, Daniel Sun

Applications of Picosecond Laser Acoustics to Power Semiconductor Device: IGBT and MOSFET............ 458
Johnny Dai, Cheolkyu Kim, Priya Mukundhan

RC-Triggered Silicon Controlled Rectifier-based ESD Clamp with Fast Transient Reaction........................ 461
Lingran Pan, Wenwen Zhang, Da-Wei Lai, Yidan Liang, Feijun Zheng

Virtual Metrology Modeling for CVD Film Thickness with Lasso-Gaussian Process Regression 464
*Shijia Yan, Cong Luo, Sen Wang, Shenglan Ding, Lei Li, Juan Ai, Qiang Sheng, Qing Xia, Zhi
Li, Qilin Chen, Shilin Li, Hongwei Dai, Yuting Zhong*

A Real-Time Detection Method for Wafer Probe Reference Die Shift... 468
Deguang Zheng, Kuan Lu, Bo Zhong, Shuxin Liu, Xiaofeng Liang

Novel Localization Approaches in Metal-Insulator-Metal Structure Failure Analysis..................................... 472
Lvye Fang, Hongtao Qian, Qinqin Yu

The Improvement Study of UTS CIS Bevel Peeling Defect Based on the Application of SEM API.............. 475
Xianghua Hu, Guangzhi He, Jingfeng Wang, Qiliang Ni

General Chip Digital Data Obtaining Solution on Ate 478
Steve Xie

An Efficient Protocol Framework Solution on V93000 482
Jun Chen, Xin Song, Yanfen Fang

Ultra-High-throughput Inline Probe Metrology and Inspection on EUV Resist 485
Andrew Humphries, John Cossins, Lei Feng

Study on E-Beam Induced Deposition with Gas Injection System 490
Fan Zhang, Yun Xu, Hongtao Qian

A Simulation Study on the Thermal Effectiveness of Graphene-Based Films in Intelligent Power
Modules 494
Jie Bao, Juan Hu, Yunyan Zhou, Yuan Xu

Effects of Different Catalysts on Epoxy Molding Compound 497
*Yangyang Duan, Wei Tan, Xingming Cheng, Lanxia Li, Hongjie Liu, Dandan Fan, Lingling
Liu, Xiaojuan Jiang, Liang Cui, Xingzhi Cui*

Rough Nickel PPF for Mold Adhesion Improvment 499
Wei-Gang Wu, Tsz-Chun Lo, Ka-Kiu So, Fai-Lung Ting, Maria Rzeznik

Rough Silver for Improved Lead-Frame Reliability 503
Fai-Lung Ting, Ka-Kiu So, Tsz-Chun Lo, Wei-Gang Wu, Maria Rzeznik

Printable Copper Sintering Paste for High-Power Die-Attach Application 507
Li Ma, Hongyun Li, Min Yao, Fen Chen, Xuelian Han, Yan Liu

Integrating High Frequency Radar Chip Using Laminated Substrate Transitions for System-In-
Package Design 511
Zhiqiang Fang, Boping Wu

Electromagnetic Interference Shielding Solution for System-In-Package 514
Lihong Liu, Jiongjiong Gu, Boping Wu

A Composite Photodector with Wide Dynamic Range and Small Area for Dynamic Vision Sensor
Application 518
Yaping Chen, Xiaona Zhu, Shaofeng Yu

Improve the Breakdown Voltage of Silicon Pixel Sensor with Optimized Multi-Guard Rings 521
Peng Sun, Gaobo Xu, Jianyu Fu, Mingzheng Ding, Yinan Yan, Luoyun Zhang, Huaxiang Yin

Study on Improvement of Dark Count Rate for Silicon Photomultiplier 524
Xing Chen, Zhigao Wang

Process Optimization and Performance Improvement of CMOS Microbolometer with a Salicided
Polysilicon Thermistor 527
Jiang Lan, Haolan Ma, Yaozu Guo, Ke Wang, Feng Yan, Yiming Liao, Xiaoli Ji

Investigation of Vertically Stacked Horizontal Gate-All-Around SI Nanosheet Ion Sensitive Field
Effect Transistor for Detection of C-Reactive Protein 530
*Yang Liu, Qingzhu Zhang, Junjie Li, Cinan Wu, Lei Cao, Yanna Luo, Zhaohao Zhang, Shuhua
Wei, Qianhui Wei, Jiaxin Yao, Jiawei Hu, Meiyan Qin, Enxu Liu, Yanchu Han, Lianlian Li,
Yinglu Li, Tao Yang, Na Zhou, Jianfeng Gao, Junfeng Li*

Monolithic 3D Integration of Dendritic Neural Network with Memristive Synapse, Dendrite and
Soma on Si CMOS 533
*Tingyu Li, Jianshi Tang, Junhao Chen, Xinyi Li, Han Zhao, Yue Xi, Wen Sun, Yijun Li,
Qingtian Zhang, Bin Gao, He Qian, Huaqiang Wu*

Simulation Investigation on the Characteristics of Gan-Based Multi-Quantum Wells Micro-Leds 536
Pengfei Ye, Youshan Gui, Yue Li, Ding Chen, Jinghao Yu, Yi Tong, Haixia Da

Near-Infrared Sensitivity Enhancement of CMOS Image Sensor with Germanium on Silicon
Structure 539
Hui Chen, Chenchen Qiu, Zhengying Wei, Chang Sun, Jun Qian, Yufei Peng

Differential Evolution with Multivariate Gaussian Sampling for Sensor Arrangement 542
Kuiling Du, Gang Tang

A 2A 4MHz Dual-Phase ZDS Hysteretic DC-DC Buck Converter with Peak Efficiency Above 90% 545
Yanye Chen, Quan Sun, Changyou Men, Lenian He

Improve Sparse Implicit Projection Via Incomplete Cholesky Factorization 548
Yang Yang, Fan Yang, Xuan Zeng

High Efficient Automatic Power/Ground Layout Routing Algorithm for Analog ICS 551
Jiaxin Zuo, Fei Li, Jing Wan

Implementing Boolean Function by Ternary Content Addressable Memory with Approximate
Match 554
Jian Shi, Weikang Qian

Verification of 100Gb/s Data-Rate Transceiving Through Silicon-Photonic Module in an FPGA
Platform 557
Xuhui Liu, Chun-Zhang Chen, Xiaoli Fang, Liang Wang, Quan Pan, Hanming Wu

Artificial Neural Network Compact Modeling Methodology for Complementary Field Effect
Transistor 560
Ouwen Tao, Xiaona Zhu, Yage Zhao, Rongzheng Ding, Shaofeng Yu, Ye Lu

A 14.7mW 4Gb/s/lane Wireless Through Silicon Interface for Memory Cube Exploiting 16-QAM
and Magnetic Resonance 563
Chonghui Sun, Rushuo Tao, Kun Yang, Xuhui Liu, C.-Z. Chen, Xiaolei. Zhu

A Hardware Accelerator for Standard Convolution and Depthwise Convolution 566
Fubang An, Wei Cao, Xuegong Zhou, Lingli Wang

A Multi-Layer Stacked 3-D SRAM System Based on Wireless Transceiver Using Inductively
Coupled Interface in 22-NM CMOS 569
Kun Yang, Chonghui Sun, Rushuo Tao, Jiannan Guo, Cheng Yang, D. Ma, Xiaolei Zhu

An Adaptive Controlled Chip-Level Wireless Power Transfer System with DPID Controller for
Wireless 3-D Stacked Chips 572
Rushuo Tao, Chonghui Sun, Kun Yang, Cheng Yang, Jiannan Guo, Xiaolei Zhu

An Improved Noise Canceling Sturdy 2-1 MASH Sigma-Delta Modulator with Multi-Bit SAR
Quantizer 575
Tengteng Mu, Lianxi Liu

Post-Training Quantization Or Quantization-aware Training? that is the Question 578
Xiaotian Zhao, Ruge Xu, Xinfei Guo

A Front-End for 1.5GSPS 12Bit Pipelined ADC 581
Xiuheng Wu, Xuan Guo, Fangyuan Xu, Zeyu Li, Hanbo Jia, Xinyu Liu

Logic Circuit Simulation Based on Semi-Tensor Product 584
Ruibing Zhang, Hongyang Pan, Zhufei Chu

CirSAT: An Efficient Circuit-Based SAT Solver Via Fanout-driven Decision Heuristic 587
Kunmei Hu, Zhufei Chu

Fast NoC Router Latency Estimation Using Machine Learning 590
Yang Li, Pingqiang Zhou

Lutplace: An Improved Lookup Table-Based Placement for Routability 593
Yihang Qiu, Yan Xing, Shuting Cai, Xingquan Li, Xiaoming Xiong

AcArm: A Novel Semiconductor Wafer Handling Robot 596
Donglin Chen, Lixin Tang, Dehong Cong, Jingchao Qiao

Efficient Partitioning and Communication Scheme-Based Distributed Edge Computing to
Accelerate Deep Neural Network 599
Xudong Lu, Cheng Zhuo

A Hybrid Training Framework for Speeding Up the Inference Process of Spiking Neural Networks 602
Ziwen Li, Yu Ma, Pingqiang Zhou

Attention-Based Mechanism for Technology Mapping Optimization 605
Zhaohui Yang, Yinshui Xia, Mengke Wang, Chenghao Yang, Xiaojing Zha

An Efficient ATPG Technology Based on Time Division Multiplexing Method 608
Minqiang Peng, Keqing Ouyang, Lunmao Zhou, Guohua Zhou

Learning-Based Performance and Power Model for Processor Microsecond DVFS 611
Yingtao Shen, An Zou

A High-Sensitivity and Large-Dynamic Range Readout Circuit for Polysilicon-Based
Microbolometer 614
Wei Zhu, Ke Wang, Yaozu Guo, Sheng Xu, Feng Yan, Yiming Liao, Xiaoli Ji

A Scalable and Configurable Low-Power Mixed Signal Neuromorphic Accelerators for Spiking
Neural Network 617
Yekuan Chen, Yiqi Meng, Yiling Chen, Xiaolei Zhu

RLCkt: An Analog Circuit Automatic Sizing Sage Based on Reinforcement Learning 620
Wangge Zuo, Lingge Liu, Fei Li, Yifei Huang, Liqian Zhang, Jing Wan

Convolutional Neural Networks on the Edge: A Comparison Between FPGA and GPU 624
Yichen Wei, Siyi Gong, Hongfei Mei, Longxing Shi, Xinfei Guo

Logic Optimization Sequence Tuning Based on Policy Search Deep Reinforcement Learning 627
Yu Jin, Haijiao Huang, Wenzhe Ye, Xuebing Zhang

Agile Full-Chip Sign-Off in the Post-Moore Era 630
Xiao Dong, Songyu Sun, Zhengrui Chen, Jianyi Yang, Cheng Zhuo

A 2-D Multi-Dielectric Capacitance Solver Based on Floating Random Walk Method 635
Jiahao Xu, Yibin Zhang, Shenghan Gao, Jiecheng Huang, Ming Yang, Wenjian Yu

Correlation Analysis Between Defect Scanning and Machine Components 638
Ming Guo

Essential Steps to Enable Analyzing Effective Resistance of ESD Paths-PG Routing Network Pruning and Resistance Contribution by Layer 640
Frank Feng, Abner Huang, Joe Huang, Dawson Chiou, Jeff Byrd, Nicholas Palmer, Charles McFalls, Akhil A. Gore

An 18-Bit 2MSPS SAR ADC with Double Passive Noise-Shaping Calibration 644
Xiao-Wei Zhang, Jian-Xiong Xi, Tao Wang, Le-Nian He

Design and Simulation of a PFM-PWM Hybrid Controller for DCDC Converter with CLLC Topology 647
Hai Liu, Lenian He, Quan Sun, Changyou Men

Design of an 8-Channel 12Bits 1MSPS SAR ADC 650
Zhengxue Shi, Quan Sun, Changyou Men, Lenian He

Author Index

China Semiconductor Technology International Conference 2023 (CSTIC 2023)

Editors:

Cor Claeys
KU Leuven
Leuven, Belgium

Hanming Wu
Zhejiang University
Hangzhou, China

Ru Huang
Southeast University
Najing, China

Beichao Zhang
HFC Semiconductors
Hangzhou, China

Xiaowei Li
ICT, CAS
Beijing, China

Steve X. Liang
JCET Semiconductor (Shaoxing) Co. Ltd
Shaoxing, China

Qinghuang Lin
LAM Research Corporation
Fremont, CA, USA

Hsiang-Lan Lung
Macronix International Ltd
Hsinchu, Taiwan, China

Linyong (Leo) Pang
D2S
San Jose, California, USA

Weikang Qian
University of Michigan-Shanghai Jiao Tong
University
Shanghai, China

Xinping Qu
Fudan University
Shanghai, China

Xiaoping Shi
Naura Microelectronics
Beijing, China

Ying Zhang

Santa Clara, CA, USA

PREFACE

This issue contains a selection of the accepted papers presented at China Semiconductor Technology International Conference 2023 (CSTIC 2023), June 26-27, 2023 in Shanghai, China. After reviewing a selection of the presentations have been considered for publication in IEEE Xplore.

CSTIC is the largest and the most comprehensive annual industrial semiconductor technology conference in China. It aims to provide a platform for executives, managers, engineers and researchers from around the world to exchange the latest developments in semiconductor technology and manufacturing and related fields. It also offers an opportunity for those who are interested in investing and collaboration opportunities in the semiconductor industry in Asia, particularly in China.

CSTIC covers all the aspects of semiconductor technology and manufacturing, including circuit design, system integration, devices, materials, patterning (lithography and etching), processes, integration, testing, reliability, device physics and manufacturing as well as emerging semiconductor technologies, including clean energy such as light emitting diodes (LEDs), III-V semiconductors, sensors and micro-electromechanical systems (MEMS).

CSTIC 2023, organized by Semiconductor Equipment and Material International (SEMI) and The Integrated circuit Materials Industry Technology Innovation Alliance (ICMTIA) and technically sponsored by the IEEE Electron Devices Society relies on a long time tradition, which started in 2001. The original International Semiconductor Technology Conference (ISTC) merged in 2009 to become CSTIC, aiming for a broad international representation and increased paper submissions from around the world. For CSTIC 2023 the papers came from all major semiconductor manufacturing regions in the world, including China, France, Japan, Korea, Singapore, Taiwan, United Kingdom and the United States of America. About 171 papers have been selected for oral presentations and approximate 191 papers for poster presentations after careful reviews by the conference organizing committee.

In total 203 papers are included in these Proceedings after peer reviews. They represent a snapshot of the recent developments in semiconductor technology and manufacturing in the world. In particular, they offer a glimpse into the state-of-the-art of semiconductor technology and manufacturing in China. These papers are divided into nine (9) chapters according to the nine symposia of CSTIC 2023:

- Device Engineering and Memory Technology

- Lithography and Patterning

- Dry & Wet Etch and Cleaning

- Thin Film, Plating and Process Integration

- Chemical-Mechanical Polishing (CMP) and Post-CMP Cleaning

- Metrology, Reliability and Testing

- Packaging and Assembly

- MEMS, Sensors and Emerging Semiconductor Technologies

- Design and Automation of Circuits and Systems

These Proceedings are very valuable to engineers and researchers in the fast-moving and growing semiconductor industry. It will give readers a clear understanding of the status of semiconductor technology and manufacturing in China. Furthermore, it will also serve as a useful reference for those who are interested in nanofabrication, micro- and nano-fluidics, micro- and nano-photonics, organic electronics, bio-chips, light emitting diodes (LEDs) and other clean energy technologies.

We thank the invited speakers and the authors, particularly the conference plenary speakers, Prof. Albert Fert, Research Director Unité Mixte de Physique, Nobel Laureate 2007, Prof. Ming Liu, Academician, CAS, Fudan University, Shanghai, Dr Yang Pan, Vice President, Lam Research, Jinrong Zhao, Chairman Executive Committee NAURA Technology, Beijing, and Dr. Yalin Xiang, Senior Vice President. KLA Corporation, San Jose, USA for their valuable contributions to CSTIC 2023. We also thank the more than 120 organizing committee members, particularly the symposium chairs, for their dedication and hard work to help improve the quality and to broaden the reach of CSTIC. These committee members are experts in their respective fields of semiconductor technology and are from well-known companies or prestigious institutions. They all have demanding day jobs, yet they have volunteered to help organizing this conference and to critically review papers presented in these Proceedings. Their contributions were crucial for the success of the conference. We are also indebted to the financial support from the sponsors of CSTIC 2023. Finally, we extend our sincere thanks to SEMI for their tireless efforts and their meticulous organizational skills to help organize CSTIC 2023 and to assemble and publish these CSTIC 2023 proceedings.

Hanming Wu, General Chair CSTIC 2023
Zhejiang University, Hangzhou, China

Cor Claeys, Co-Chair, CSTIC 2023
KU Leuven, Leuven, Belgium

CSTIC 2023 Organizing Committee

June 2023, Shanghai, China

Table of Contents

Preface

Chapter I - Device Engineering and Memory Technology

Design of Ferroelectric FET-Based Capacitive-Coupling Computing-In-Memory For Binary Neural Networks
Boyi Fu[1], Jin Luo[1], Weikai Xu[1], Qianqian Huang[1,2,3] and Ru Huang[1,2,3]
[1]*School of Integrated Circuits, Peking University, Beijing, China*
[2]*Beijing Advanced Innovation Center for Integrated Circuits, Beijing, China*
[3]*Chinese Institute for Brain Research, Beijing, China*

$Hf_xZr_{1-x}O_2$ Ferroelectric Thin Film Grain Size Tuning via Annealing Ramp Rate Achieving Endurance >10^9 cycles, $2P_r$ of 40.6 $\mu C/cm^2$, Write Voltage Down to 1.5 V, and Switching Speed of 30 ns
Zhixiong Li[1,2], Bing Zhou[1], Jiawei Xu[1], Shuaihang Xu[1], Jun Lan[1], Quanzhou Zhu[1], Yingjie Zhu[1], Jie Li[1], Xuewei Feng[3], Mei Shen[1], Feichi Zhou[1], Longyang Lin[1] and Yida Li[1]
[1]*Southern University of Science and Technology, Shenzhen, China*
[2]*Shenzhen Longsys Electronics Co., Ltd, Shenzhen China*
[3]*Shanghai Jiao Tong University, Shanghai, China*

Large Area CVD MoS_2 Memristor Suitable for Neuromorphic Applications
Muhammad Zaheer[1], Tariq Aziz[1], Jun Lan[1], Quanzhou Zhu[1], Wenhui Wang[1], Mei Shen[1], Feichi Zhou[1], Longyang Lin[1], Xuewei Feng[2] and Yida Li[1]
[1]*Southern University of Science and Technology, Shenzhen, China*
[2]*Shanghai Jiao Tong University, Shanghai, China*

The Study on Reducing Bit-Line Parasitic Capacitance in Advanced DRAM
Yexiao Yu, Hong Ma, Zhongming Liu, Shaoyou Xiong, Dan Wang, Yang Zhang and Yi Yang
Changxin Memory Technology, United Process Development, Hefei, China

Technologies for Superior Reliability in SiC Power Devices
Min-hwa Chi
Micro-Nano Technology College, Qingdao University, Qingdao, Shandong, China
GTA Semiconductor Co, Ltd, Shanghai, China

Leakage Reduction of GAA Stacked Si Nanosheet CMOS Transistors and 6T-SRAM Cell Via Spacer Bottom Footing Optimization
Jiaxin Yao[1,], Xuexiang Zhang[1,3†], Lei Cao[1,3], Junjie Li[1], Na Zhou[1], Qingkun Li[1,3], Yanzhao Wei[1,3], Yanna Luo[1,3], Jun Luo[1,3], Qingzhu Zhang[1,2] and Huaxiang Yin[1,2]
[1]*Integrated Circuit Advanced Process R&D Center, Institute Of Microelectronics of Chinese Academy of Sciences, Beijing, China*
[2]*key Laboratory of Microelectronics Devices and Integrated Technology, Institute of Microelectronics of Chinese Academy Sciences, Beijing, China*
[3]*University of Chinese Academy of Sciences, Beijing, China*

Experimental Investigation of Ultra-Low Temperature La$_2$O$_3$/HfO$_2$ Bi-Layer Dipole-First Process Using PVD Method for Advanced IC Technology

Yanzhao Wei[1,3], Jiaxin Yao[1], Renren Xu[1,3], Qingzhu Zhang[1,2] and Huaxiang Yin[1,2,3]

[1]Integrated Circuit Advances Process R&D Center, Institute of Microelectronics, Chinese Academy of Sciences, Beijing, China
[2]Key Laboratory of Microelectronic Devices and Integrated Technology, Institute of Microelectronics, Chinese Academy of Sciences, Beijing, China
[3]School of Integrated Circuits, University of Chinese Academy of Sciences, Beijing, China

A Compact Sawtooth Wave Generator Based on Novel Z²-Fet Device

Hui Xie, Yingxin Chen and Jing Wan

School of Information Science and Engineering, Fudan University, Shanghai, China

Fabrication and Characterization of a Novel Embedded Mirror Gate SONOS

Ning Wang and Kegang Zhang

Shanghai Huahong Grace Semiconductor Manufacturing Corporation, Shanghai, China

High Endurance SONOS Technology Improved by Design & Process Optimization

Pingsheng zhou, Kegang Zhang, Gordon Li, Hualun Chen, Xiang Yao and Weiran Kong

Shanghai Hua Hong Semiconductor Manufacturing Corporation, Shanghai, China

Promoting Chip Probing Test Yield by Simple ISSG and Global Wet Process

Jingsong Peng, Kegang Zhang, Ning Wang, Yang Ding, Pingsheng Zhou and Yin Yin

Shanghai Huahong Grace Semiconductor Manufacturing Corporation, Shanghai, China

An On-Chip Superconducting Quantum Transponder

Rutian Huang[1,2], Xinyu Wu[1,2], Xiao Geng[1,2], Jianshe Liu[1,2] and Wei Chen[1,2,3]

[1]Laboratory of Superconducting Quantum Information Processing,
School of Integrated Circuits, Tsinghua University, Beijing, China
[2]Beijing Innovation Center for Future Chips, Tsinghua University, Beijing, China
[3]Beijing National Research Center for Information Science and Technology, Beijing, China

Design and Simulation of a Superconducting Switch Based on Weakly Damped Superconducting Quantum Interference Devices

Xinyu Wu, Rutian Huang, Jianshe Liu and Wei Chen

School of Integrated Circuits, Tsinghua University, Beijing, China

Research on Vt Window Improvement Process of 2T SONOS Embedded Flash

Xiaokang Li, Shipu Li and Jun Qian

Shanghai Huali Microelectronics Corporation, Shanghai, China

An Effective Method to Minimize the Difference Between Spot and Ribbon Beam in S/D Region by Carbon Co-Implantation

Long Feng[1], Zhiqiang Xiao[1,2], Jiaxing Xiao[1], Zhirui Li[1] and Haitao Yan[1]

[1]Semiconductor Manufacturing North China (Beijing) Corp. Beijing, China
[2]School of Integrated Circuits, Tsinghua University, Beijing, China

Study of Breakdown Voltage Improvement of High-Voltage PLDMOS
Wenting Duan, Donghua Liu, Haiyang Ling, Ying Cai, Feng Jin and Wensheng Qian
HuaHong Grace Semiconductor Manufacturing Corporation, Shanghai, China

Study of the Formation of Copper Void Defect and Process Optimization for Reduction in Dual Damascene Process
Hongliang Zhu, Shuhuai Jia, Dejing Ma and Fengjiao Wang
Semiconductor Manufacturing International Corporation(SMIC), Beijing, China

Study on N-Type MOS Capacitor in 55nm CMOS
Hongliang Zhu, Xin Zhou, Dejing Ma, Haoqi Zheng and Tianfu Zhang
Semiconductor Manufacturing International Corporation(SMIC), Beijing, China

The Study on the Optimization and Electrical Behavior of NEDMOS Transistors for High-Voltage Power Applications
Chongkai Du, Chenchen Qiu, Jun Qian and Chang Sun
Shanghai Huali Microelectronics Corporation, Shanghai, China

Investigation of a New Disturb Effect in the Aggressively Scaled Dual-Bit/Cell Split-Gate Floating-Gate Flash Cell
Yintong Zhang[1], Zhaozhao Xu[1], Dylan Zhou[2], Alan Shen[2], Fredric Liu[2], Ziquan Fang[2], Donghua Liu[2], Gordon Li[2] and Wensheng Qian[2]
[1]Huahong Semiconductor (Wuxi) Limited, Wuxi, China
[2]Shanghai Huahong Grace Semiconductor Manufacturing Corporation, Shanghai, China

Important Process Parameter and its Sensitivity Check by Virtual Fabrication: Channel Hole Profile Impact on Advanced 3D Nand Structure
Qingpeng Wang, Pengfei Lyu, Lifei Sun, Yu De Chen, Cheng Li, Jacky Huang, Benjamin Vincent and Joseph Ervin
Coventor, Inc., A Lam Research Company, Shanghai, China

FinFET Source/Drain Parasitic Resistance Optimization by TCAD Simulation
Tongtong Luan[1], Xinqi Liu[2], Yu Gu[1] and Xufeng Kou[1]
[1]School of Information Science and Technology, ShanghaiTech University, Shanghai, China
[2]School of Physical Science and Technology, ShanghaiTech University, Shanghai, China

Simulation Study of Gate-All-Around Nanosheet Devices Based on SOI Structure
Yangyang Hu[1], Tianxiang Zhao[1], Mengmeng Yang[2], Jianhua Zhang[1] and Kailin Ren[1]
[1]School of Microelectronics, Shanghai University, Shanghai, China
[2]School of Computing, Shanghai University, Shanghai, China

Investigation of Electrical Characteristics on Morphotropic Phase Boundary of $Hf_{1X}Zr_xO_2$ for Dynamic Random Access Memories
Kun Zhong[1,2,4], Huaxiang Yin[1,2,4], Zhaohao Zhang[1,2,4] and Fan Zhang[1,2,3]
[1]Key Laboratory of Microelectronics Devices and Integrated Technology
[2]Institute of Microelectronics of Chinese Academy of Sciences, Beijing, China
[3]Xidian University Key Laboratory of Wide Bandgap Semiconductor Materials
[4]School of Integrated Circuits, University of Chinese Academy of Sciences, Beijing, China

Investigation of the Doping Profile for Ion Implants and Rapid Annealing in Silicon Via an Improved Method
Zeqi Zha, Zhenhui Wang and Ya Wang
Semiconductor Manufacturing International (Beijing) Corporation, Beijing, China

Novel Channel-on-Fin (COF) IGZO-TFTs with Ultra-Scaled Back Gate Length of 23nm
Shangbo Yang[1,2], Gaobo Xu[1,2], Gangping Yan[1,2], Zhiyu Song[1,2], Guoliang Tian[1,2], Yanna Luo[1,2], Yinan Yan[1,2], Lianlian Li[1,2] and Huaxiang Yin[1,2]
[1]Institute of Microelectronics, Chinese Academy of Sciences, Beijing, China
[2]University of Chinese Academy of Sciences, Beijing, China

Investigation of Vertical Channel IGZO-TFT Based on PVD-IGZO
Zhiyu Song [1,2], Gaobo Xu [1,2], Gangping Yan, Shangbo Yang [1,2], Yanna Luo [1,2], Guoliang Tian [1,2], Yinan Yan [1,2] and Huaxiang Yin [1,2]
[1]Institute of Microelectronics, Chinese Academy of Sciences, Beijing, China
[2]University of Chinese Academy of Sciences, Beijing, China

A Compact Model of Non-Volatile Ferroelectric Tunnel FET with Ambipolarity for In-Memory-Computing Based Edge AI
Hanyong Shao[1], Jin Luo[1], Zhiyuan Fu[1], Qianqian Huang[1,2,3] and Ru Huang[1,2,3]
[1]School of Integrated Circuits, Peking University, Beijing, China
[2]Beijing Advanced Innovation Center for Integrated Circuits, Beijing, China
[3]Chinese Institute for Brain Research, Beijing, China

Investigation of Synergic Hydrogen Mitigation Technique for Top-Gate A-IGZO Thin-Film Transistors
Gangping Yan[1,2,3], Zhiyu Song[1,2], Haoqing Xu[1,2], Shangbo Yang[1,2], Chuqiao Niu[1,2], Guoliang Tian[1,2], Yanna Luo[1,2], Luoyun Zhang[1,2], Yunjiao Bao[1,2], Gaobo Xu[1,2] and Huaxiang Yin[1,2]
[1]Key Laboratory of Microelectronics Devices and Integrated Technology, Institute of Microelectronics of Chinese Academy of Sciences, Beijing, China
[2]University of Chinese Academy of Sciences, Beijing, China
[3]Beijing Superstring Academy of Memory Technology, Beijing, China

Characterization of Field Cycling Fatigue In HfZrOx Ferroelectric Capacitors
Puyang Cai[1], Zhiwei Liu[2], Tianxiang Zhu[1], Zhigang Ji[2], Runsheng Wang[1] and Ru Huang[1]
[1] School of Integrated Circuits, Peking University, Beijing, China
[2]Departure of Micro/Nano Electronics, Shanghai Jiao Tong University, Shanghai, China

Reliability Performance of Novel Tunneling Field Effect Transistors Based on Foundry Platform
Yukun Tang[1], Qianqian Huang[2,4], Kaifeng Wang[2], Yongqin Wu[3], Hongyan Han[3], Ye Ren[3], Weihai Bu[3], Junhua Liu[1,2,4], Zhigang Ji[1] and Ru Huang[1,2,4]
[1]National Key Laboratory of Science and Technology on Micro/Nano Fabrication, Shanghai Jiao Tong University, Shanghai, China
[2]School of Integrated Circuits, Peking University, Beijing, China
[3]Semiconductor Technology Innovation Center (Beijing), Beijing, China
[4]Beijing Advanced Innovation Center for Integrated Circuits, Beijing, China

Influence of Interfacial Layers and High-k Post Dielectric Annealing on the Characteristics of MOS Devices
Guanqiao Sang[1,3], Qingzhu Zhang[1,2], Huaxiang Yin[1,2], Junfeng Li[1] and Xulei Qin[3]
[1]*Key Laboratory of Microelectronics Devices and Integrated Technology, Institute of Microelectronics of Chinese Academy of Sciences, Beijing, China*
[2]*University of Chinese Academy of Sciences, Beijing, China*
[3]*ChangChun University of Science and Technology, Changchun, China*

Optimized Wafer Edge Condition in Lithographic Process for Peeling Defect Reduction
Shanshan Chen, Hunglin Chen, Yin Long and Kai Wang
Shanghai Huali Integrated Circuit Corporation, Shanghai, China

Silicide Profile Optimization on Active Area in 4xnm ETOX NOR Flash Memory
Hualun Chen, Yuxin Tong, Xiangyu Qi, Songhan Duan, Botong Liu, Chaoran Zhang and Lin Gu
Huahong Semiconductor (Wuxi) Limited, Wuxi, China

Improvement of Standby Current Failure by Device Optimization on 4xnm ETOX NOR-Flash Memory
Hualun Chen, Zhuangzhuang Wang, Lin Gu , Yihang Du , Chun Yao, Xiaodong Mu and Wan Song
Wuxi Huahong Grace Semiconductor Manufacturing Corporation, Wuxi, China

Effects of Floating Gate Profile on Cell Characteristics of 4xnm FG-First ETOX NOR-Flash Memory
Hualun Chen, Yihang Du, Lin Gu, Zhuangzhuang Wang, Chun Yao and Zhaozhao
Wuxi Huahong Grace Semiconductor Manufacturing Corporation, Wuxi, China

Improved Environmental Stability of N-type Polymer Field-Effect Transistors Using Nickel Contact Electrode
Yuan Liu[1], Quanhua Chen[1], Rujun Zhu[1], Jinxiu Cao[1] and Yong Xu[1,2]
[1]*College of Integrated Circuit Science and Engineering, Nanjing University of Posts and Telecommunications, Nanjing, China.*
[2]*Guangdong Greater Bay Area Institute of Integrated Circuit and System, Guangzhou, China*

A New Method to Calculate Loading Effect in Embedded Flash
Fangce Sun
Department of Process Integration, Shanghai Huahong Grace Semiconductor Manufacturing Corporation, Shanghai, China

A New Method to Improve Split Gate Flash Erase and Endurance
Fangce Sun
Department of Process Integration, Shanghai Huahong Grace Semiconductor Manufacturing Corporation, Shanghai, China

Chapter II – Lithography and Patterning

Recent Progress of EUV Resist Development for Improving Chemical Stochastic**
Toru Fujimori
Electronic Materials Research Laboratories, FUJIFILM Corporation, Shizuoka, Japan

CycleGAN-Based Mask Diffraction Model*
Jiaxiang Zhuo, Dongyong Xu and Yijiang Shen
[1]School of Automation, Guangdong University of Technology
[2]Mega Education Center South, Guangzhou, China

Illumination Optimization for the BEOL DTCO with 45-Degree Local Interconnection
Xianhe Liu,[12], Muzi Han[1], Yanli Li[12], Qi Wang[1,2] and Qiang Wu[1,2]
[1]School of Microelectronics, Fudan University, Shanghai, China
[2]National Integrated Circuit Innovation Center, Shanghai, China

Process and Tool Monitor and Diagnosis Based on Overlay Data and Modeling
Yi Tong[1], Libin Zhang[1,2,3,4], Yayi Wei[1,2,3,4], Tianchun Ye[1,2,3] and Yun Wang[1,2]
[1]Guangdong Greater Bay Area Institute of Integrated Circuit and System, Guangzhou, China
[2]Institute of Microelectronics of Chinese Academy of Sciences (IMECAS), Beijing, China
[3]School of Integrated Circuits, University of Chinese Academy of Sciences, Beijing, China
[4]Nanjing Chengxin Institute of IC Technology, Nanjing, China

The Analysis of Optical Critical Dimension (OCD) Signal Strength Between 5 nm FinFET and 3 nm Complementary FET (CFET) Vertical Gate Stacks
Qi Wang[1,2], Qiang Wu[1,2], Xianhe Liu[1,2] and Yanli Li[1,2]
[1]School of Micro-Electronics, Fudan University, Shanghai, China,
[2]National Integrated Circuit Innovation Center, Shanghai, China

The Possibility of Using 193 nm Immersion Lithography Process for 5 nm Logic Design Rules**
Qiang Wu[1,2], Yanli Li[1,2], Xianhe Liu[1,2] and Qi Wang[1,2]
[1]School of Microelectronics, Fudan University, Shanghai, China
[2]National Integrated Circuit Innovation Center, Shanghai, China

Enhancement of Pattern Depth in Plasmonic Lithography for Practical Application
Dandan Han and Yayi Wei
University of Chinese Academy of Sciences, School of Integrated Circuits, Beijing, China

Study on Inter-Layer Overlay of Stitching Lithography Technology
Hongmin Liu, Changcheng Gao and QiongTao Wu
Semiconductor Manufacturing International Corporation, Beijing, China

A Negative-Tone Photosensitive Epoxy Material
Bai Ke, April Tang, Jiangtao Mou and Jinfu Zheng
Shandong Shengquan New Materials Co. Ltd, Jinan City, Shandong Province, China

A Multi-Step SRAF to Improve Process Windows in Metal Layer
Wei Wei, Lu Zhu, Dan Wang, Xiaoyan Sun, Yue Wang, Yueyu Zhang, Jianzhong Liu and Shiri Yu
Shanghai Huali Integrated Circuit Corporation, Shanghai, China

The Reduction of Yield Loss and Contact Overlay Shift by Optimizing the Process Profile of Pre-Layer Process Integration
Zhejun Liu and Wei Lu
Shanghai Huali Integrated Circuit Corporation, Shanghai, China

Study On-Product Overlay Improvement for Immersion Lithography
Guoping Liu, Yinsheng Yu, Chi Zhang, Yuhui Li, Wei Cao, QinYuan and Hongwen Zhao
Shanghai Huali Integrated Circuit Corporation, Shanghai, China

Line-End Roundness and Voids Improvement of BEOL Metal Layer
Mudan Wang, Tiancheng Tu, Hui Zhao and Shirui Yu
Technology Development, Shanghai Huali Integrated Circuit Corporation, Shanghai, China

OPC Correction Method Based on Corner to Corner Structure
Qiguang Zhou, Dan Wang, Yueyu Zhang, Shirui Yu
Shanghai Huali Integrated Circuit Corporation, Shanghai, China

Chapter III – Dry & Wet Etch and Cleaning

Trimming of Silicon Nitride Hard Mask Using Cyclic Deposition and Etch Process*
Li-Tian Xu[1], Pei Mei[1], Xing-Jun Yao[1], Chen Chen[1], Jia-Yun Zhang[2], Alan Zhang[2], Xiao-Peng Wu[2], Wu-Hao Han[2], Hong-Bo Lin[2] and Nichole Zhang[2]
[1]*Beijing NAURA Microelectronics Equipment Co. Ltd, Beijing, China*
[2]*ChangXin Memory Technologies, Inc. Ltd, China*

The Investigation of CF3I for High-Aspect-Ratio Cryogenic Dielectric Etch
Jianqiu Hou, Vina Xu, Kai Zhang and Ziyang Wu
Advanced Micro-Fabrication Equipment Inc., Shanghai, China

SIARC Residue Reduction Methods with Minimizing Profile Change for Mask Patterning
Xingxing Xu, Hexin Zhou, Quanbao Li and Jian Huang
Lam Research Service Co., Ltd, Shanghai, China

Investigations of NAND Flash Device Cycling Performance Improvement Via FG Carbon-Doped Polysilicon and Channel Corner Rounding
Lifeng Liu[1], Jun Wang[1,2] and Xinruo Su[2]
[1]*School of Integrated Circuit, Peking University, Beijing, China*
[2]*Semiconductor Manufacturing North China, Beijing, China*

Enabling Plasma Etch Solution for GaN Technology
Zoe Wang, Chunxiang Guo, Jian Liu, Yingxiong Feng, Lulu Guan, Kangning Xu, Qiao Huang, Lu Chen and Kaidong Xu
Jiangsu Leuven Instruments Co., Ltd., Xuzhou, Jiangsu, China

Effect of Process Gas on Side Wall Angle in Silicon Trench Etching
Yiming Ma, Guang Yang, Litian Xu, Zhongwei Jiang, Jing Wang, Donghan Wang, Qifei Wang and Yongjie Zhou
Beijing NAURA Microelectronics Equipment Co., Ltd., Beijing, China

Optimization of SADP Process for Defect Reduction in Planar 2D NAND Flash
Lifeng Liu[1], Jun Wang[1,2], Yinan Ma[2], Zhenchao Sui[1,2] and Yue Li[2]
[1]*School of Integrated Circuit, Peking University, Beijing, China*
[2]*Semiconductor Manufacturing North China, Beijing, China*

Distortion Control when Etching DRAM Metal Contact
Jianqiu Hou[1], Yao Sun[2], Hui Xue[2], Ya Zhou[1], Hao Li[2], Zhiwen Luan[1], Zijian Chen[2] and Zengwen Hu[1]
[1]*Advanced Micro-Fabrication Equipment Inc. Shanghai, China*
[2]*Chang Xin Memory Technologies Corporation, Hefei, China*

The Plasma Etching of the Deep Hole Structure in Silicon with the Mixed Gas of SF_6, HBr and O_2
Qifei Wang[1], Yiming Ma, Zhongwei Jiang[1], Jing Wang[1] and Haiyun Zhu[1]
ETCH II BU, Beijing NAURA Microelectronics Equipment Co. Ltd., Beijing, China

Investigation of CD Precise Control in Pitch Doubling Flow for Memory Industry
Zhao Liu and Baodong Han
Beijing Superstring Academy of Memory Technology, Beijing, China

Tungsten/Silicon Oxide/Titanium Nitride Stack Etching
Jie Luo, Haochang Lyu, Linjie Hou, Baodong Han, Hongbo Sun and Chao Zhao
Beijing Superstring Academy of Memory Technology, Beijing, China

Study on the Effect of Water Spraying Mode on the N Content of Wafer Surface after SC1 Cleaning in Light Doping Process
Jinlei Wang, Fenglin Guan, Mingguang Hang, Lili Jia, Fang Li and Xinhua Cheng
Shanghai Huali Integrated Circuit Corporation, Shanghai, China

A Technical Optimization of Waferless Auto Clean for Aluminum Etcher
Li Qi, Qiang-Qiang Sang, Xing-Jun Yao, Li-Tian Xu, Jian-Kun Zhang, Li-Song Hu, Yi-Chang Liu and Chen Chen
Beijing NAURA Microelectronics Equipment Co. Ltd, Beijing, China

The Research of Special Gate Morphology Adjustment and its Influence on Electrical Properties
Junjie Pan, Kai Qian, Lian Lu, Quanbo Li, Jun Huang and Yu Zhang
Shanghai Huali Integrated Circuit Corporation, Shanghai, China

Study of Spacer Etching with PR Approach
Yuhao Yang, Siyuan Che, Xiangguo Meng, Lian Lu, Quanbo Li, Jun Huang and Yu Zhang
Shanghai Huali Integrated Circuit Corporation, Shanghai, China

Silicon Surface Roughness Improvement during Plasma Etch
Guang Yang, Li Zeng, Haiyun Zhu, Jing Wang and Zhongwei Jiang
Beijing NAURA Microelectronics Equipment Co. Ltd, Beijing, China

Investigation of High Aspect Ratio Amorphous Carbon Etching in NAND Flash Memory
Li Zeng, Guang Yang, Zhongwei Jiang, Jing Wang and Li-Tian Xu
Beijing NAURA Microelectronics Equipment Co. Ltd, Beijing, China

Study of Photo-Resist (PR) Strip Rate with High Temperature Pedestal for Al Patterning Process
Cheng Tian, Li-Tian Xu, Li-Song Hu and Xue-Hua Wang
Beijing NAURA Microelectronics Equipment Co. Ltd, Beijing, China

Self-Limited Etching of Silicon Nitride Using Cyclic Process with CH_2F_2 Chemistry
Xue-hua Wang, Li-Tian Xu and Tian Cheng
Beijing NAURA Microelectronics Equipment Co. Ltd, Beijing, China

Strategy for Line Width Roughness (LWR) Reduction in Carbon Mandrel Patterning
Yichang Li, Li Qi, Litian Xu, Lianfu Zhao, Xingjun Yao and Zihan Zhang
Beijing NAURA Microelectronics Equipment Co. Ltd, Beijing, China

Redistribution Layer Aluminum Advanced Etching Process Development
Xing-Jun Yao[1], Li-Tian Xu[1], Li Qi[1], Qianq-qiang Sang[1], Li-song Hu[2], Chen-Chen[2] and Yi-chang Lu[2]
[1]Beijing NAURA Microelectronics Equipment Co. Ltd, Beijing, China
[2]Beijing NAURA Microelectronics Equipment Co. Ltd, Hefei, China

Deep Trench (DT) Etching Process for Power MOS Device
Chen Chen, Li-Tian Xu and Xing-Jun Yao
Beijing NAURA Microelectronics Equipment Co. Ltd, Beijing, China

Control of Deposition in Cyclic Deposition/Etch Process
Zihan Zhang, Jing Wang and Litian Xu
Beijing NAURA Microelectronics Equipment Co., Ltd. Beijing, China

A Novel Method to Eliminate Bond Pad Crystal
Fanshun Meng[1], Jiajin Wang[1], YunBo Chen[1], Qiang Rui[1], Zhongkui Chen[1], Kuo-Jung Chen[2], Yi Liu[2], Fengyang Li[2], Chao Sun[2]
[1]CanSemi Technology Inc., Guangzhou City, Guangdong Province, China
[2]Advanced Micro-Fabrication Equipment Inc. (AMEC), Shanghai, China

Carbon Hard Mask Opening Process Development with Novel Sidewall Passivation in Memory Manufacturing
Meng-Jiao Zhu, Li-Tian Xu, Jing Wang, Li Zeng, Zi-Han Zhang
Beijing NAURA Microelectronics Equipment Co. Ltd, Beijing, China

The Optimization of Pin Hole Defect in High Resistance Process
Lunan Zhu, Shan Huang, Xiaofeng Qu, Lei Sun, Quanbo Li, Tianpeng Guan and Yu Zhang
Shanghai Huali Integrated Circuit Corporation, Shanghai, China

Technical Difficulties and Optimization Methods of NMOS Share CT in Contact Hole Etching
Peng Zhu, Renhui Xu, Wentao Fu, Lei Sun and Yu Bao
Shanghai Huali Integrated Circuit Corporation(HLIC) Shanghai, China

Chapter IV – Thin Film, Plating and Process Integration

Technical Challenges in MRAM Fabrication**
Y. H. Wang[1], X. L. Yang[1], Y. Tao[1], Y. H. Sun[1], Q. J. Guo[1], F. T. Meng[1] and G. C. Han[1,2]
[1]*Zhejiang Chituo Technology Co., Ltd, Hangzhou, China*
[2]*Key Lab. of Spintronics Mater, Devices and Systems of Zhejiang Prov., Hangzhou, China*

Cell Structure and Process Integration of a Novel 2T0C Technology for High-Density DRAM Application*
Zheng-Yong Zhu[1], Bok-Moon Kang[1], Jing Zhang[1], Xin-Lv Duan[2], Jin-Juan Xiang[1], Guan-Hua Yang[2], Di Geng[2], Wang Dan[1], Xie-Shuai Wu[1], Ming-Xu Liu[1], Gui-Lei Wang[1] and Chao Zhao[1]
[1]*Beijing Superstring Academy of Memory Technology, Beijing, China*
[2]*Key Laboratory of Microelectronics Devices and Integrated Technology, Institute of Microelectronics, Chinese Academy of Sciences, Beijing, China*

The Study of Slip Defects in Furnace High Temperature Process*
Yan Sun, Simeng Wei and Yuanxiang Xie
Beijing NAURA Microelectronics Equipment Co., Ltd, Beijing, China

Improvement of SiGe Relaxation by a New Clamping Film Deposition Process Method
Zhiqiang Xiao[1,2], Cunzhe He[1], Dongliang Gao[1], Haitao Yan[1], Zhenchao Sui[1] and Xin Zhang[1]
[1]*Semiconductor Manufacturing North China (Beijing) Corp., Beijing, China*
[2]*School of Integrated Circuits, Tsinghua University, Beijing, Chin*

A Novel Method to Optimize SiGe Profile Using Co-Implantation
Zhiqiang Xiao[1,2], Long Feng[1], Cunzhe He[1], Jiaxing Xiao[1], Dongliang Gao[1] and Mingying Liu[1]
[1]*Semiconductor Manufacturing North China (Beijing) Corp. Beijing, China*
[2]*School of Integrated Circuits, Tsinghua University, Beijing, China*

A Study of Parasitic Capacitance Using Different Bit Line Spacer Integration Schemes in Advanced DRAM
Dempsey Deng, Qingpeng Wang, Yujia Zhong, Yu De Chen and Jacky Huang
Coventor Inc., A Lam Research Company, Shanghai, China

The Effect of SiGe SiCoNi Pre-Clean Time on Planner Logic Device Performance Study
Xuechun Zhang[1], Weichi Cheng[2], Li Ning[1] and Jingang Wang[1]
[1]*Semiconductor Manufacturing North China (Beijing) Corporation, Beijing , China*
[2]*Semiconductor Technology Innovation Center (Beijing) Corporation, Beijing , China*

Growth and Reduction of Tiny Particle Defects in Selective SiGe Epitaxy S/D Devices
Zhiqiang Xiao[1,2], Cunzhe He[1*], Dongliang Gao[1], Jiaxing Xiao[1], Haitao Yan[1], Zhenchao Sui[1] and Xin Zhang[1]
[1]*Semiconductor Manufacturing North China (Beijing) Corp., Beijing, China*
[2]*School of Integrated Circuits, Tsinghua University, Beijing, China*

Influence of Ion Implantation on Void Defect Formation in Epitaxially Grown Silicon
Zeqi Zha, Zhenhui Wang and Ya Wang
Semiconductor Manufacturing International (Beijing) Corporation, Beijing, China

Some Key Modifications of Theory Required to Understand the Leakage Current Mechanisms for Ferroelectric HfZrO Capacitors Used in Microelectronics
W.S. Lau
Nanyang Technological University (Retired), School of EEE, Singapore

Lau's Unified Schottky-Poole-Frenkel Theory with Asymmetric Distortion by Electron Charge Trapping Proposed to Explain the Current-Voltage Characteristics of High-k Metal-Insulator-Metal Capacitors
W.S. Lau
Nanyang Technological University (Retired), School of EEE, Singapore

Pulsed DC Parameters (Reverse Voltage, Duty Cycle, Pulsed Frequency) on Film Quality in Reactive Sputtered Aluminum Nitride Films
Wei-Yu Zhou[1], Xue-Li Tseng[1], Ning-Hsiu Yuan[2], Hsiao-Han Lo[2], Peter J. Wang[2], Ming-yu Jiang[2], Yiin-kuen Fuh[1] and Tomi T. Li[1]
[1]*Department of Mechanical Engineering, National Central University*
[2]*Delta Electronics, Inc., Taoyuan City, Taiwan, China*

The Study of Silicon Nitride Films Deposited in Batch ALD System
Shiyao Cheng, Wei Kuai, Wenxu Duan, Xinyang Wang, Xiaomeng Liu, Shuo Cheng, Yuanxiang Xie and Xiaoping Shi
Beijing NAURA Microelectronics Equipment Co., Ltd, Beijing, China

Mechanical Properties of Flip-Chip Bonding Structures for Micro-LED Devices: Cu-Cu Bonding with Passivation Layer and Indium Bumps Bonding

Kefeng Wang[1], Zehua Chen[1], Xiaoxiao Ji[2,3], Luqiao Yin[1,2], Xiuzhen Lu[1] and Jianhua Zhang[1,2]
[1]*School of microelectronics, Shanghai University, Shanghai, , China*
[2]*Key Laboratory of Advanced Display and System Applications, Shanghai University, Ministry of Education, Shanghai, , China*
[3]*School of Mechatronic Engineering and Automation, Shanghai University, Shanghai, China*

Electroplating Process Improvement on Post-CMP Dishing Profile
Wenbo Wu, Tong Lei, Zhijun Zhu, Zhenhua Hu and Yushan Chi
Lam Research Service Co., Ltd, Shanghai, China

Effect of Sub-Atmospheric Chemical Vapor Deposition SiO_2 Film Deposition Process on Surface Chemistry Sensitivity
JiananWei, XueLiu and Dapeng Ruan
Piotech Technology Co., Ltd., Shenyang, Liaoning Province, China

Gas Distribution Effect on Thermal ALD AlN Film Thickness Non-uniformity
Xiaomeng Liu, Qihui Zhang and Hao Deng
Piotech (Shanghai) Inc., Shanghai, China

Substrate Effect on Thermal ALD AlN Film Growth Rate
Xiaomeng Liu, Tiantian Liu, Wenyi Liu, Xinyu Zhang and Hao Deng
Piotech (Shanghai) Inc., Shanghai, China

Chapter V – CMP and Post-CMP Cleaning

Improving 300mm Si Wafer Planarization Process with a Wholistic Approach*
Zijun Guan[1], Yiran Liu[1], Jiaming Fan[1], Yuanda Gao[1], Lei Zhu[1,2], Wenjie Yu[1,2] and Weimin Li[1,2]
[1]*Shanghai Institute of IC Materials Co., Ltd., Shanghai, China*
[2]*Shanghai Institute of Microsystem and Information Technology, Chinese Academy of Sciences, Shanghai, China*

Study on the Mechanism of SIN Residue Improvement for ILD0 CMP
Yurong Que, Xing Ma, Jian Zhang, Hu Li, Jingxu Fang and Yu Zhang
Advanced Module Technology Dept., Shanghai Huali Integrated Circuit Corp., Shanghai, China

Some Methods to Reduce Micro Scratch Defect for Via Contact Tungsten Chemical Mechanical Planarization Process
ZhiJie Zhang, Le Ning, HongDi Wang and ZhiYang Liang
Semiconductor Manufacturing North China(Beijing) Corporation, Beijing, China

A FEM Model of Micro-Galvanic Corrosion Evolution at Ru/Cu Interface in H_2O_2 CMP Solution
Shuo Gao, Qinhua Miao, Boyu Wen and Jie Cheng
School of Mechanical Electronic and Information Engineering, China University of Mining & Technology-Beijing, Beijing, China

Slurry System Establishment and Optimization for Advanced Cobalt Interconnect Metallization
Lifei Zhang[1,2], Tongqing Wang[1,2] and Xinchun Lu[1,2]
[1]*Hwatsing Technology Co., Ltd, Tianjin, China*
[2]*State Key Laboratory of Tribology in Advanced Equipment, Tsinghua University, Beijing, China*

Analysis of the Adsorption and Passivation Mechanism of JFCE on Copper Surface in Alkaline CMP Slurry
Ni Zhan[1,2], Xinhuan Niu[1,2], Yinchan Zhang[1,2], Fu Luo[1,2] and Han Yan[1,2]
[1]School of Electronics and Information Engineering, Hebei University of Technology, Tianjin, China
[2]Tianjin Key Laboratory of Electronic Materials and Devices, Tianjin, China

Research on the Dispersion Stability and Polishing Performance of Ceria Slurry
Min Liu[1,2], Baoguo Zhang[1,2], Shitong Liu[1,2], Dexing Cui[1,2], Wenhao Xian[1,2], Pengfei Wu[1,2] and Ye Wang[1,2]
[1]*School of Electronics and Information Engineering, Hebei University of Technology Tianjin, China*
[2]*Tianjin Key Laboratory of Electronic Materials and Devices, Tianjin, China*

Effect of Abrasive on the CMP Performance of C-Plane (0001) GaN Flim
Jianghao Liu[1,2], Xinhuan Niu[1,2], Ni Zhan[1,2], Yida Zou[1,2] and Yebo Zhu[1,2]
[1]*School of Electronics and Information Engineering, Hebei University of Technology, Tianjin, China*
[2]*Tianjin Key Laboratory of Electronic Materials and Devices, Tianjin, China*

Effect of Surfactants on CMP Properties of M-Plane Sapphire
Yida Zou[1,2], Xinhuan Niu[1,2], Ziyang Hou[1,2], Minghui Qu[1,2], Ni Zhan[1,2] and Jianghao Liu[1,2]
[1]*School of Electronics and Information Engineering, Hebei University of Technology, Tianjin, China*
[2]*Tianjin Key Laboratory of Electronic Materials and Devices, Tianjin, China*

Study on the Slurry for Chemical Mechanical Polishing of Sapphire Wafer
Wenhao Xian[1,2], Baoguo Zhang[1,2*], Liu Min[1,2], Dexing Cui[1,2], Pengfei Wu[1,2] and Ye Wang[1,2]
[1]*School of Electronics and Information Engineering, Hebei University of Technology Tianjin, China*
[2]*Tianjin Key Laboratory of Electronic Materials and Devices, Tianjin, China*

Pad Surface Variation and Its Effect on SiO_2 Removal Rate in Ceria-Based CMP Slurry
Chenchen Yang, Yu Yao and EngHoe Tan
Semiconductor Manufacturing Beijing Corporation, Beijing, China

Study on the Mechanism of CMP Induced W Seam at Advanced Technology Node
Shaojia Zhu, Yurong Que, Feng Shi, Mingfei Yu, Jian Zhang, Jingxun Fang and Yu Zhang
Advanced Module Technology Development, Shanghai Huali Integrated Circuit Corporation, Shanghai, China

Pattern Loading Improvement for Cu CMP Process
Lei Zhang, Yu Yang, Jian Zhang, Jingxun Fang and Yu Zhang
Advanced Module Technology Development, Shanghai Huali Integrated Circuit Corporation, Shanghai, China

Effects of Process to Material Removal in CMP: Modelling and Experiments
Yanming Ren[1,2,3], Yiran Liu[2], Zijun Guan[2], Lei Zhu[2,3], Yuanda Gao[2], Wenjie Yu[2,3] and Weimin Li[2,3]
[1]*School of Materials and Chemistry, University of Shanghai for Science and Technology, Shanghai, China*
[2]*Shanghai Institute of IC Materials Co., Ltd, Shanghai, China*
[3]*Shanghai Institute of Microsystem and Information Technology, Chinese Academy of Sciences, Shanghai, China*

Impact of Slurry for Dishing Reduction During Cu CMP
Yu Yao, Chenchen Yang, and EngHoe Tan
Semiconductor Manufacturing Beijing Corporation, Beijing, China

Research Progress and Challenges of Chemical Mechanical Polishing for Silicon Carbide Wafter
Lijuan Zhang
Shanghai XinQian Semiconductor Co. Ltd., Shanghai, China

Chapter VI – Metrology, Reliability and Testing

Innovation Test Technology for Ultra-High-Speed ADC on ATE
Yanyan.Chang, Tianyu.Chen, Jiaying.Xiang and Yichen.Xiao
SA, Advantest (China) Co., Ltd, Shanghai, China

Ultra-Wideband (UWB) Test Solution on V93000
Kevin Yan and Daniel Sun
Business Development & Center of Expertise, Advantest (China) Co., Ltd, Shanghai, China

General Chip Digital Data Obtaining Solution on ATE
Steve Xie
Advantest, Shanghai, China

An Efficient Protocol Framework Solution on V93000
Jun Chen, Xin Song and Yanfen Fang
Advantest, Shanghai, China

A Novel Model-Matching Based Scratch Tool Tracing System
Shi-Qiang He, Yan-Qiu Zhang, Xiao-Lei Zhang and Chic-Kuo Fang
Fujian Jinhua Integrated Circuit Co., Ltd, Jinjiang, Quzhou, Fujian, China

Faster Au-Al IMC Growth under Chlorine Environment
Liao Jinzhi Lois, Wang Bisheng, Zhang Xi, Hua Younan[2] and Li Xiaomin[3]
[1]*WinTech Nano-Technology Services Pte. Ltd., Singapore*
[2]*Huawei Technologies Co Ltd, Bantian Huawei Base, , Shenzhen, China*
[5]*Wintech-Nano (Suzhou) Co., Ltd., Suzhou, China*

The Design-Based Inspection Strategy for Cu Void Defects Reduction
Xingdi Zhang, Hunglin Chen, Yin Long and Kai Wang
Shanghai Huali Integrated Circuit Corporation, Shanghai, China

Reliability Research on Micro Bump and c4bump in Large-Size 2.5D FCBGA
Xiang Li[1,2], Zhuqiu Wang[2], Xiao He[2], Dan Yang[2] and Na Mei[1]
[1]*State Key Laboratory of Mobile Network and Mobile Multimedia Technology, Shenzhen, China*
[2]*Department of Reliability Engineering, Sanechips, Shenzhen, China*

Reliability Analysis of Metal Thermal Interface Materials for Ultra-Large Size FCLGA Package
Keqing Ouyang[2], Zhuolun Wu[1.2*], Zhuqiu Wang[2], Weilun Wang[3], Dan Yang[2] and Na Mei[1]
[1]*State Key Laboratory of Mobile Network and Mobile Multimedia Technology, Shenzhen, China*
[2]*Department of Reliability Engineering, Sanechips Technology Co., Ltd., Shenzhen, China*
[3]*Department of Packaging and Testing Engineering, Sanechips Technology Co., Ltd., Shenzhen, China*

An Efficient Tool for Generating Test Program to Save Marginal Fail Chips
Hanyan Chen
Advantest, Shanghai, China

A Universal Auto Test Program Generation on Advantest V93000 ATE Platform
Xin Song, Yefang Wang and Hanyan Chen
Advantest (China) Co. Ltd, Shanghai, China

A Methodology for Testing Scan Chain with Diagnostic Enhanced Structure
Keqing Ouyang[1,2*], Minqiang Peng[1,2], Shuai Wang[1,2], Guohua Zhou[1,2] and Kai Wang[1,2]
[1]*State Key Laboratory of Mobile Network and Mobile Multimedia Technology, Shenzhen, Guangdong, China*
[2]*Dept.of back-end design, Sanechips Technology Co., Ltd., Shenzhen, Guangdong, China*

A Novel Method to Achieve High Efficient Iteration of MBIST Pattern
Minqiang Peng[1,2], Keqing Ouyang[1,2*], Feilong Pan[1,2], Guohua Zhou[1,2] and Lei Chen[1,2]
[1]*State Key Laboratory of Mobile Network and Mobile Multimedia Technology, Shenzhen, Guangdong, China*
[2]*Sanechips Technology Co., Ltd, Shenzhen, Guangdong, China*

Impact of Interface Trap Density on the Endurance of HFO$_2$/Si FeFETS
Jiaqi Zheng[1], Yue Peng[2], Yanbin Yang[3], Dawei Gao[1], Rui Zhang[1] and Genquan Han[2]
[1]*School of Micro- and Nano-Electronics, Zhejiang University, Hangzhou, China*
[2]*School of Microelectronics, Xidian University, Xi'an, China*
[3]*Institute of Zhejiang Intelligence Lab, Chengdu, China*

Impact of Interface Traps Generation on Flicker Noise Degradation in Si pMOSFETs
Yi Jiang[1], Luping Wang[1], Yanbin Yang[2], Dawei Gao[1] and Rui Zhang[1]
[1]*School of Micro- and Nano-Electronics, Zhejiang University, Hangzhou, China*
[2]*Institute of Zhejiang Intelligence Lab, Chengdu, China*

Research on Hot Carrier Injection Optimization of 28HKMG Technology
Weiwei Ma, Yang Li, Ran Huang, Yamin Cao and Wei Zhou
Shanghai Huali Integrated Circuit Corporation Shanghai, China

Applications of Picosecond Laser Acoustics to Power Semiconductor Device: IGBT and MOSFET
Johnny Dai[1], Cheolkyu Kim[2] and Priya Mukundhan[1]
[1]*Onto Innovation, Budd Lake, New Jersey, USA*
[2]*Onto Innovation, Gyunggi-do, Korea*

RC-Triggered Silicon Controlled Rectifier-based ESD Clamp with Fast Transient Reaction
Lingran Pan[1], Wenwen Zhang[2], Da-Wei Lai[2], Yidan Liang[1] and Feijun Zheng[1]
[1]*Zhejiang University, Hangzhou, China*
[2]*Pride Silicon Technology Co., Ltd, Hangzhou, China*

Ultra-high-Throughput Inline Probe Metrology and Inspection on EUV Resist*
Andrew Humphries, John Cossins and Lei Feng
Infinitesima Limited, Abingdon, Oxfordshire, United Kingdom

Computer Vision Technology Supported Rapid DRAM Capacitor Analyzing System Based on TEM Image
Zhi-Yuan Gui, Chang Xu, Han Yan and Zhi-Yu Li
Fujian Jinhua Integrated Circuit Co., Ltd., Jinjiang, Quanzhou, China

Calibration of Pitch Standards of SEM for Semiconductor Dimension Metrology Application
Wei Li[1], Yang Qu[2] and Yushu Shi
[1]*Division of Center for Advanced Measurement Science，National Institute of Metrology, Beijing, China*
[2]*Institute of Microelectronics, Chinese Academy of Sciences (CAS), Beijing, China*

Virtual Metrology Modeling for CVD Film Thickness with Lasso-Gaussian Process Regression
Shijia Yan, Cong Luo, Sen Wang, Shenglan Ding, Lei Li, Juan Ai, Qiang Sheng, Qing Xia, Zhi Li, Qilin Cheng, Shilin Li, Hongwei Dai and Yuting Zhong
Wuhan Xinxin Semiconductor Manufacturing Co., Ltd.(XMC), Wuhan, China

The Improvement Study of UTS CIS Bevel Peeling Defect Based on the Application of SEM API
Xianghua Hu, Guangzhi He, Jingfeng Wang and Qiliang Ni
Shanghai Huali Microelectronics Corporation, Shanghai , China

Study on E-Beam Induced Deposition with Gas Injection System
Fan Zhang, Yun Xu and Hongtao Qian
Semiconductor Manufacturing International (Shanghai) Corporation, Shanghai, China

Methods for Fin Etching Profile Maintaining and Measurement
Yun Xu, Tong Wu and Fairy Chen
Semiconductor Manufacturing International (Shanghai) Corporation Failure Analysis Laboratory, Shanghai, China

Investigation of Different Gate Bias on PMOS HCI Performance
Lei Li, Canny Chen and Atman Zhao
Semiconductor Manufacturing International (Shanghai) Corp., Shanghai, China

Research on TDDB Physical Mechanism of 28HKMG MOSFET
Ting Wan, Hao Jiang, Yueqin Zhu and Ke Zhou
Shanghai Huali Integrated Circuit Corporation, Shanghai, China

Reservoir Effect Study on Electro-Migration Behavior of AlCu Interconnects
Jizhou Li, Kitty Wang and Weihai Fan
Semiconductor Manufacturing International (Shanghai) Corp., Shanghai, China

The Verification of TDDB Acceleration Model in Ultrathin Gate Dielectric
Wen Ying, Canny Chen and Atman Zhao
Q&R, Semiconductor Manufacturing International Corporation (SMIC) Shanghai, China

Anomaly Detection of Non-Normal Distribution Wafer Acceptance Test Data Via GMM-Based Method
Junjun Zhuang, Yong Wang, Guiyun Mao, Xu Chen, Yansheng Wang and Zhengying Wei
Shanghai Huali Microelectronics Corporation, Shanghai, China

Design and Optimization of RC Triggered MV-NMOS for 28nm CMOS Technology ESD Protection
Jia Zhu, Lanying Wei, Yang Li, Jun Wu, Kun Wang and Wei Chen
Semiconductor Manufacturing International Corporation, Beijing, China

Application of Picosecond Ultrasonic Technology for CMOS Image Sensors
Johnny Mu[1], Kaixing Song[3], Johnny Jin[1], Cheolkyu Kim[2], Yaodong Huang[3] and Hong Hong[1]
[1]Onto Innovation, Shanghai, China
[2]Onto Innovation, Sungnam-si, Gyunggi-do, Korea
3Galaxycore, Shanhai, China

Neutron Irradiation Induced Carrier Removal and Deep-Level Traps in N-GaN Schottky Barrier Diodes
Jin Sui[1,2,3], Jiaxiang Chen[1,2,3], Haolan Qu[1,2,3], Ruohan Zhang[1,5], Min Zhu[1,2,3], Xing Lu[4] and Xinbo Zou[1,5]
[1]SIST, ShanghaiTech University, Shanghai, China
[2]Shanghai Institute of Microsystem and Information Technology, CAS, Shanghai, China
[3]School of Microelectronics, University of Chinese Academy of Science, Beijing, China
[4]School of Electronics and Information Technology, Sun Yat-sen Univ., Guangzhou, China
[5]Shanghai Engineering Center of Energy Efficient and Custom AI IC, Shanghai, China

MetaViT-Trans: A Framework for Mixed-Type Defect Detection of Wafers with Vision Transformer Combined with Meta-Learning and Transfer Learning
Junfeng Zhao[1,2] and Lixin Tang[1]
[1]National Frontiers Science Center for Industrial Intelligence and Systems Optimization, Northeastern University, Shenyang, China
[2]Key Laboratory of Data Analytics and Optimization for Smart Industry (Northeastern University), Ministry of Education, Shenyang, China

Lithography Hotspot Detection Based on Transfer Learning with High Resolution Networks
Hongzhe Wang[1,2] and Lixin Tang[1]
[1]National Frontiers Science Center for Industrial Intelligence and Systems Optimization, Northeastern University, Shenyang, , China.
[2]The Key Laboratory of Data Analytics and Optimization for Smart Industry (Northeastern University), Ministry of Education, Shenyang, China

An End-to-End Detection Approach for Micropipe Defect of SiC Wafers Via Fusing Multiple Hierarchical Features
W. X. Shi, T. G. Zhao and J. W. Zhang
State Key Laboratory of Precision Measurement, Department of Precision Instruments, Tsinghua University, Beijing, China

A Real-Time Detection Method for Wafer Probe Reference Die Shift
Deguang Zheng, Kuan Lu, Bo Zhong, Shuxin Liu And Xiaofeng Liang
Wafer Test Department, NXP Semiconductors, Tianjin, China

Novel Localization Approaches in Metal –Insulator-Metal Structure Failure Analysis
Lvye Fang, Hongtao Qian and Qinqin Yu
Semiconductor Manufacturing International (Shanghai) Corp., Shanghai, China

Chapter VII – Packaging and Assembly

Integrating High Frequency Radar Chip Using Laminated Substrate Transitions for System-in-Package Design
Zhiqiang Fang and Boping Wu
JCET Group Co., Ltd., Shanghai, China

Electromagnetic Interference Shielding Solution for System-in-Package
Lihong Liu [1], Jiongjiong Gu [1] and Boping Wu[2]
[1]JCET Group Co., Ltd., Jiangyin, China
[2]JCET Group Co., Ltd., Shanghai, China

Printable Copper Sintering Paste for High-Power Die-Attach Application
Li Ma, Hongyun Li, Dr. Min Yao, Fen Chen, Xuelian Han and Yan Liu
Indium Corporation (Suzhou) Suzhou, Jiangsu, China

Rough Nickel PPF for Mold Adhesion Improvement
Wei-Gang Wu[1], Tsz-Chun Lo[1], Ka-Kiu So[1], Fai-Lung Ting[1] and Maria Rzeznik[2]
[1]Dupont, On Lok Tsuen, Fanling, Hong Kong SAR, China
[2]Dupont, Marlborough, Massachusetts, USA

Rough Silver for Improved Lead-Frame Reliability
Fai-Lung Ting[1],Ka-Kiu So[1],Tsz-Chun Lo[1], Wei-Gang Wu[1] and Maria Rzeznik[2]
[1]Dupont, On Lok Tsuen, Fanling, Hong Kong SAR, China
[2]Dupont, Marlborough, Massachusetts, USA

A Simulation Study on the Thermal Effectiveness of Graphene-Based Films in Intelligent Power Modules
Jie Bao, Juan Hu, Yunyan Zhou Yuan Xu
Engineering Technology Research Center of Intelligent Microsystem of Anhui Province, Huangshan University, Huangshan, China

Effects of Different Catalysts on Epoxy Molding Compound
Yangyang Duan, Wei Tan, Xingming Cheng, Lanxia Li, Hongjie Liu, Dandan Fan, Lingling Liu, Xiaojuan Jiang, Liang Cui and Xingzhi Cui
Jiangsu Hua Hai Cheng Ke Advanced Material Co. Ltd., Lianyungang, China

Chapter VIII – MEMS, Sensors and Emerging Semiconductor Technologies

Monolithic 3D Integration of Dendritic Neural Network with Memristive Synapse, Dendrite and Soma on Si CMOS
Tingyu Li, Jianshi Tang, Junhao Chen, Xinyi Li, Han Zhao, Yue Xi, Wen Sun, Yijun Li, Qingtian Zhang, Bin Gao, He Qian and Huaqiang Wu
School of Integrated Circuits, Beijing Innovation Center for Future Chips (ICFC), BNRist, Tsinghua University, Beijing, China

Differential Evolution with Multivariate Gaussian Sampling for Sensor Arrangement
Kuiling Du[1,2] and Gang Tang[3]
[1]National Frontiers Science Center for Industrial Intelligence and Systems Optimization, Northeastern University, Shenyang, China.
[2]Key Laboratory of Data Analytics and Optimization for Smart Industry(Northeastern University), Ministry of Education, Shenyang, China.
[3]China Construction Fifth Engineering Bureau Ltd, Changsha, Hunan, China

Near-Infrared Sensitivity Enhancement of CMOS Image Sensor with Germanium on Silicon Structure
Hui Chen, Chenchen Qiu, Zhengying Wei, Chang Sun, Jun Qian and Yufei Peng
Shanghai Huali Microelectronics Corporation, Shanghai, China

A Composite Photodector with Wide Dynamic Range and Small Area for Dynamic Vision Sensor Application
Yaping Chen, Xiaona Zhu and Shaofeng Yu
School of Microelectronics, Fudan University, Shanghai, China

Improve the Breakdown Voltage of Silicon Pixel Sensor with Optimized Multi-Guard Rings
Peng Sun[1,2], Gaobo Xu[1,2], Jianyu Fu[2], Mingzheng Ding[1], Yinan Yan[1,2], Luoyun Zhang[1,2] and Huaxiang Yin[1,2]
[1]Integrated Circuit Advanced Process R&D Center, Institute of Microelectronics, Chinese Academy of Sciences, Beijing, China
[2]University of Chinese Academy of Sciences, Beijing, China

Study on Improvement of Dark Count Rate for Silicon Photomultiplier
Xing Chen and Zhigao Wang
Semiconductor Manufacturing International Corp., Beijing, China

Process Optimization and Performance Improvement of CMOS Microbolometer with a Salicided Polysilicon Thermistor
Jiang Lan[1], Haolan Ma[1], Yaozu Guo[1], Ke Wang[1], Feng Yan[1], Yiming Liao[2] and Xiaoli Ji[1]
[1]*School of the Electronic Science and Engineering, Nanjing University, Nanjing, China*
[2]*School of Electronic and Optical Engineering, Nanjing University of Science and Technology, Nanjing, China*

Investigation of Vertically Stacked Horizontal Gate-All-Around Si Nanosheet Ion Sensitive Field Effect Transistor for Detection of C-Reactive Protein
Yang Liu[1,2], Qingzhu Zhang[1], Junjie Li[1*], Cinan Wu[2*], Lei Cao[1], Yanna Luo[1], Zhaohao Zhang[1], Shuhua Wei[3], Qianhui Wei[4], Jiaxin Yao[1], Jiawei Hu[1,3], Meiyan Qin[1], Enxu Liu[1], Yanchu Han[1,2], LianLian Li[1], YingLu Li[1,3], Tao Yang[1], Na Zhou[1], Jianfeng Gao[1] and Junfeng Li[1]
[1]*Key Laboratory of Microelectronic Devices and Integration Technology, Institute of Microelectronics, Chinese Academy of Sciences, Beijing, China*
[2]*College of Big Data and Information Engineering, Guizhou University Guiyang, China*
[3]*School of Information Science and Technology, North China University of Technology, Beijing, China*
[4]*State Key Laboratory of Advanced Materials for Smart Sensing GRINM Group Co. Ltd., Beijing, China*

Simulation Investigation on the Characteristics of GaN-Based Multi-Quantum Wells Micro-LEDs
Pengfei Ye[1], Youshan Gui[2], Yue Li[1], Ding Chen[1], Jinghao Yu[1], Yi Tong[2] and Haixia Da[1]
[1]*College of Electronic and Optical Engineering & College of Flexible Electronics (Future Technology), Nanjing University of Posts and Telecommunications, Nanjing, China.*
[2]*College of Integrated Circuit Science and Engineering, Nanjing University of Posts and Telecommunications, Nanjing, China*

Chapter IX – Design and Automation of Circuit and Systems

Logic Circuit Simulation Based on Semi-Tensor Product
Ruibing Zhang, Hongyang Pan and Zhufei Chu
EECS, Ningbo University, Ningbo, China

CirSAT: An Efficient Circuit-based SAT Solver via Fanout-driven Decision Heuristic
Kunmei Hu and Zhufei Chu
EECS, Ningbo University, Ningbo, China

RLCkt: An Analog Circuit Automatic Sizing Sage Based on Reinforcement Learning
Wangge Zuo[1], Lingge Liu[1], Fei Li[2], Yifei Huang[2], Liqian Zhang[2], Jing Wan[1]
[1]*School of Information Science and Engineering, State key lab of ASIC and System, Fudan University, Shanghai, China*
[2]*Suzhou Foohu Technology Co., Ltd., Shanghai, China*

Efficient Partitioning and Communication Scheme-Based Distributed Edge Computing to Accelerate Deep Neural Network
Xudong Lu and Cheng Zhuo
Zhejiang University, Hangzhou, China

Post-training Quantization or Quantization-aware Training? That is the Question
Xiaotian Zhao, Ruge Xu and Xinfei Guo
University of Michigan -Shanghai Jiao Tong University Joint Institute, Shanghai Jiao Tong University, Shanghai, China

Agile Full-Chip Sign-Off in the Post-Moore Era**
Xiao Dong[1], Songyu Sun[1], Zhengrui Chen[1], Jianyi Yang[2] and Cheng Zhuo[2]
[1]College of Information Science & Electronic Engineering, Zhejiang University, Hangzhou, China
[2]School of Micro-Nano Electronics, Zhejiang University, Hangzhou, China

Learning-Based Performance and Power Model for Processor Microsecond DVFS
Yingtao Shen and An Zou
UM-SJTU JI, Shanghai Jiao Tong University, Shanghai, China

Fast NoC Router Latency Estimation Using Machine Learning
Yang Li and Pingqiang Zhou
School of Information Science and Technology, ShanghaiTech University, Shanghai, China

Artificial Neural Network Compact Modeling Methodology for Complementary Field Effect Transistor
Ouwen Tao[1], Xiaona Zhu[2*], Yage Zhao[2], Rongzheng Ding[2], Shaofeng Yu[2] and Ye Lu[1]
[1]School of Information Science and Technology, Fudan University, Shanghai, China
[2]School of Microelectronics, Fudan University, Shanghai, China

A Hybrid Training Framework for Speeding Up the Inference Process of Spiking Neural Networks
Ziwen Li, Yu Ma and Pingqiang Zhou
School of Information Science and Technology, ShanghaiTech University, Shanghai, China

An 18-Bit 2MSPS SAR ADC with Double Passive Noise-Shaping Calibration
Xiao-Wei Zhang, Jian-Xiong Xi, Tao Wang and Le-Nian He
School of Micro-Nano Electronics, Zhejiang University, Hangzhou, China

Design of an 8-Channel 12bits 1MSPS SAR ADC
Zhengxue Shi[1], Quan Sun[2], Changyou Men[2] and Lenian He[1]
[1]School of Micro-Nano Electronics, Zhejiang University, Hangzhou, China
[2] Hangzhou Vango Technology Inc., Hangzhou, China

A 2A 4MHz Dual-Phase ZDS Hysteretic DC-DC Buck Converter with Peak Efficiency Above 90%
Yanye Chen[1], Quan Sun[2], Changyou Men[2] and Lenian He[1, 2]
[1]School of Micro-Nano Electronics, Zhejiang University, Hangzhou, China
[2]Hangzhou Vango Technology Inc., Hangzhou, China

Correlation Analysis Between Defect Scanning and Machine Components
Ming Guo
Shanghai Huali Integrated Circuit Corporation, Shanghai, China

Design and Simulation of a PFM-PWM Hybrid Controller for DCDC Converter with CLLC Topology
Hai Liu[1], Lenian He[1], Quan Sun[2] and Changyou Men[2]
[1]School of Micro-Nano Electronics, Zhejiang University, Hangzhou, China
[2]Hangzhou Vango Technology Inc., Hangzhou, China

Improve Sparse Implicit Projection via Incomplete Cholesky Factorization
Yang Yang, Fan Yang and Xuan Zeng
[1]School of Microelectronics, Fudan University, Shanghai, China

High Efficient Automatic Power/Ground Layout Routing Algorithm for Analog ICs
Jiaxin Zuo[1], Fei Li[2] and Jing Wan[1]
[1]State key lab of ASIC and System, School of Information Science and Engineering, Fudan University, Shanghai, China
[2]Suzhou Foohu Technology Co., Ltd., China

Implementing Boolean Function by Ternary Content Addressable Memory with Approximate Match
Jian Shi[1] and Weikang Qian[1,2]
[1]University of Michigan-SJTU Joint Institute, Shanghai, China
[2]MoE Key Lab of AI, Shanghai Jiao Tong University, China

Verification of 100Gb/s Data-Rate Transceiving through Silicon-Photonic Module in an FPGA Platform
Xuhui Liu [1], Chun-Zhang Chen [1,2], Xiaoli Fang [1], Liang Wang [1], Quan Pan [1,3] and Hanming Wu [1,2]
[1] Peng Cheng Laboratory, Shenzhen, China
[2] School of Micro-Nano Electronics, Zhejiang University, Hangzhou, China
[3] School of Microelectronics, Southern University of Science and Technology, Shenzhen, China

A 14.7mw 4Gb/s/Lane Wireless Through Silicon Interface for Memory Cube Exploiting 16-QAM and Magnetic Resonance
Chonghui Sun[1], Rushuo Tao[1], Kun Yang[1], Xuhui Liu[2], C.-Z. Chen[2] and Xiaolei Zhu[1]
[1]School of Micro-Nano Electronics, Zhejiang University, Hangzhou, China
[2] Peng Cheng Laboratory, Shenzhen, China

A Hardware Accelerator for Standard Convolution And Depthwise Convolution
Fubang An[1], Wei Cao[2], Xuegong Zhou[2] and Lingli Wang[1]
[1]School of Microelectronics, Fudan University, Shanghai, China
[2]Institute of Big data, Fudan University, Shanghai, China

A Multi-Layer Stacked 3-D SRAM System Based on Wireless Transceiver Using Inductively Coupled Interface in 22-nm CMOS
Kun Yang[1], Chonghui Sun[1], Rushuo Tao[1], Jiannan Guo[4], Cheng Yang[4], D.Ma[2,3] and Xiaolei Zhu[1,3]

[1]School of Micro-Nano Electronics, Zhejiang University, Hangzhou, China
[2]School of Computer Science And Technology, Zhejiang University, Hangzhou, China
[3]Zhejiang Lab, Hangzhou, China
[4]JCET Group Co., Ltd, Shanghai, China

An Adaptive Controlled Chip-Level Wireless Power Transfer System with DPID Controller for Wireless 3-D Stacked Chips

Rushuo Tao[1], Chonghui Sun[1], Kun Yang[1], Cheng Yang[2], Jiannan Guo[2] and Xiaolei Zhu
[1]School of Micro-Nano Electronics, Zhejiang University, Hangzhou, China
[2]JCET Group Co., Ltd, Shanghai, China

An Improved Noise Canceling Study 2-1 Mash Sigma-Delta Modulator with Multi-Bit SAR Quantizer

Tengteng Mu and Lianxi Liu
School of Microelectronics, Xidian University, Xi 'an, China

A Front-End for 1.5GSPS 12bit Pipelined ADC

Xiuheng Wu[1,2], Xuan Guo[1], Fangyuan Xu[1,2], Zeyu Li[1,2], Hanbo Jia[1] and Xinyu Liu[1]
[1]Institute of Microelectronics of The Chinese Academy of Sciences, Beijing, China
[2]School of Integrated Circuits, University of Chinese Academy of Sciences, Beijing, China

LUTPLACE: An Improved Lookup Table-Based Placement for Routability

Yihang Qiu[1,2], Yan Xing[1], Shuting Cai[1], Xingquan Li[2] and Xiaoming Xiong[1]
[1]School of Integrated Circuits, Guangdong University of Technology, Guangzhou, China
[2]Peng Cheng Laboratory, Shenzhen, China

AcArm: A Novel Semiconductor Wafer Handling Robot

Donglin Chen[1,2], Lixin Tang[1], Dehong Cong[1,2] and Jingchao Qiao[1,3]
[1]National Frontiers Science Center for Industrial Intelligence and Systems Optimization, Northeastern University, Shenyang, China
[2]Key Laboratory of Data Analytics and Optimization for Smart Industry (Northeastern University), Ministry of Education, Shenyang, China
[3]Liaoning Engineering Laboratory of Data Analytics and Optimization for Smart Industry, Shenyang, China

Attention-Based Mechanism for Technology Mapping Optimization

Zhaohui Yang, Yinshui Xia, Mengke Wang, Chenghao Yang and Xiaojing Zha
School of Ningbo University, Ningbo, China

An Efficient ATPG Technology Based on Time Division Multiplexing Method

Minqiang Peng[1,2], Keqing Ouyang[1,2], Lunmao Zhou[1,2] and Guohua Zhou[1,2]
[1]State Key Laboratory of Mobile Network and Mobile Multimedia Technology
[2]Sanechips Technology Co., Ltd, Shenzhen, Guangdong, China

A High-Sensitivity and Large-Dynamic Range Readout Circuit for Polysilicon-Based Microbolometer

Wei Zhu[1], Ke Wang[1], Yaozu Guo[1], Sheng Xu[1], Feng Yan[1], Yiming Liao[2] and Xiaoli Ji[1]
[1]School of the Electronic Science and Engineering, Nanjing University, Nanjing, China
[2]School of Electronic and Optical Engineering, Nanjing University of Science and Technology, Nanjing, China

A Scalable and Configurable Low-Power Mixed Signal Neuromorphic Accelerators for Spiking Neural Network
Yekuan Chen[1], YiQi Meng[1], Yiling Chen[1] and Xiaolei Zhu[1,2]
[1]*School of Micro-Nano Electronics, Zhejiang University, Hangzhou, China*
[2]*Zhejiang Lab, Hangzhou, China*

Logic Optimization Sequence Tuning Based on Policy Search Deep Reinforcement Learning
Yu Jin, Haijiao Huang, Wenzhe Ye and Xuebing Zhang
Beijing University of Chemical Technology, Beijing, China

A 2-D Multi-Dielectric Capacitance Solver Based on Floating Random Walk Method
Jiahao Xu, Yibin Zhang, Shenghan Gao, Jiecheng Huang, Ming Yang and Wenjian Yu
Department of Computer Science and Technology, Tsinghua University, Beijing, China

Essential Steps to Enable Analyzing Effective Resistance of ESD Paths-PG Routing Network Pruning and Resistance Contribution by Layer
Frank Feng, Abner Huang, Joe Huang, Dawson Chiou, Jeff Byrd, Nicholas Palmer, Charles McFalls, and Akhil A. Gore
Synopsys Inc. & Synopsys Taiwan Co., Ltd, Mountain View, CA,USA

Convolutional Neural Networks on the Edge: a Comparison between FPGA and GPU
Yichen Wei, Siyi Gong, Hongfei Mei, Longxing Shi and Xinfei Guo*
University of Michigan – Shanghai Jiao Tong University Joint Institute Shanghai Jiao Tong University, Shanghai, China

RESEARCH ON VT WINDOW IMPROVEMENT PROCESS OF 2T SONOS EMBEDDED FLASH

Xiaokang Li, Shipu Li [*], and Jun Qian
Shanghai Huali Microelectronics Corporation, Shanghai 201203, China
*Corresponding Author's Email: lixiaokang@hlmc.cn

ABSTRACT

In this work, a strong correlation between ONO interlayer film thickness and Flash window is analyzed based on SONOS (Silicon-Oxide-Nitride-Oxide-Silicon) embedded flash with 2T structure. Also, the thickness of interlayer film Si-ON between the bottom tunneling oxide (TUNOX) layer and the storage layer has a decisive effect on Flash window. The storage layer has good effect on the capability of data retention after 85° baking. The ONO film layer is grown by continuous thermal oxidation by PEDIFF and the growth in furnace tubes requires a silicon source (DCS) to adjust the bottom TUNOX thickness because of the mechanism of thermal deposition for bottom oxide layer. The flat band voltage shift in the programmed device is measured at high temperature to observe the thermal excitation of electrons from the nitride traps in retention mode. The trap energy distribution was extracted using the charge decay rates and WAT parameters related to cell can be significantly improved by adjusting the thickness of TUNOX and BTMSIN, of which window is increased by 21%. By regulating the thickness of TOPOX and TOPSIN, the performance of VTE after baking can be increase by 8%.

INTRODUCTION

Huali 55nm eFlash platform is developed by embedding flash storage structure on 55LP standard process platform. Fig.1 shows the standard 2T storage unit of SONOS cell, consisting of a CG and a SG. As seen from the partial enlargement of CG ONO film structure, the structure of CG is consisted of poly silicon at the top, middle oxide/nitride/oxide layer and silicon substrate at the bottom. Top oxide acts as a barrier which prevents electrons from entering poly structure. The bottom oxide acts as tunnel oxide. For the ONO layer is the actual storage layer, the thickness and film quality of ONO are on high requirements and requiring special stricter control. SONOS eFlash stands for Silicon Oxide Nitride Oxide Silicon embedded flash.

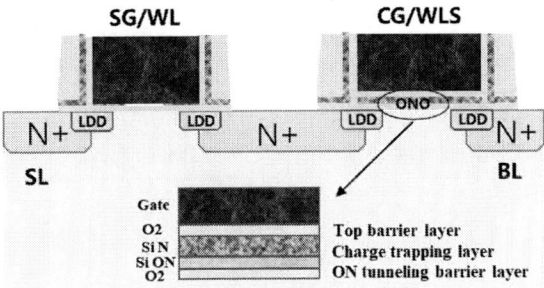

Figure 1: (a) Storage cell of 2T structure

RESULTS AND DISCUSSION

Figure.2(a) shows the structural optimization of baseline (BSL) process flow which inducing the Wetting process to regulate the thickness of TOPOX and TOPSIN in the HTO Remove step. As shown in Figure 2(b), the chip window after baking is increased by 8%.

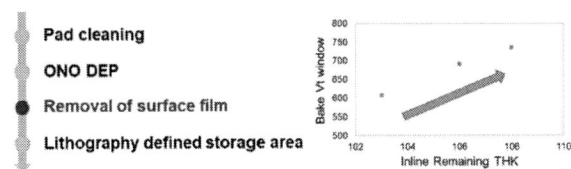

Figure 2. （a）Baseline process flow (b) HTO Remove condition change result for Chip Window

The window improvement mechanism after baking is analyzed further. Adjusting the cleaning time of HTO remove process is actually equivalent to adjusting the thickness of TOPSIN. In Flash memory field, high temperature induces the loss of electrons easily, and the thickness of trap layer can effectively adjust the electron trapping ability. Therefore, the electron capture ability of the trap layer is significantly improved in the condition of high temperature, which can effectively curb the loss of electrons. Similarly, a shortcoming of Eflash is the long erasure time. When the thickness of the trap layer reaches the limit, flash erase time will also be lost. Compared with the thickness of BSL, the trapping layer distribution was extracted using the charge decay rates and the thickness of Top Nitride film is greatly increased. Therefore, the electronic storage capacity of product is improved effectively and the overall Cell Vt window is also improved.

In ONO loop of BSL process, ONO film deposition

process of the CG of the 2T structure in the Cell area is the key layer for flash window. Based on the existing BSL condition, the combination of different thickness of TUNOX and BTMSIN is implemented. Three different conditions are matched to BSL conditions respectively and the involving machines could suffer little impact for only adjusting the new deposition condition. When it comes to this process, an operation of wet process and RTA process can be omitted, which also has a large capacity release for the actual FAB production. In addition, the size of Cell area is small, so the process of film deposition in Diffusion will significantly impact on the actual thickness effect.

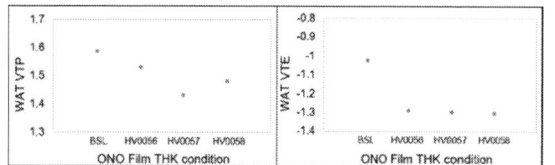

Figure 3. TUN OX and Bottom Nitride condition change result for WAT

The improvement of WAT window can be obviously seen by comparing the relevant WAT parameters of Cell in Figure 3. Vtp parameter is the Vt when the Cell is written. After the window improvement process, we can see the mean value of Vtp downgrades from 1.58 V to 0.98 V comparing with the BSL condition using the WAT window improvement maximum condition (HV0056). Although the mean value of Vtp dropped slightly by 3%, the overall Cell window increases from initial 2.61 V to 2.82V with consideration of the improvement of Vtpi. Vtpi is the Vt of the inhibit bit when the Cell is written. After the window improvement process, the mean value of Idspi is increased by 26%. Idspi and Idsp are another two WAT parameters which can characterize the Cell. Idspi is the current when the Cell is written. After the window improvement process, the mean value of Idspi is increased by 24%. Idsp is the leakage when the Cell is written. It can be seen that after process optimization, the Vt window of the entire Cell area is significantly improved.

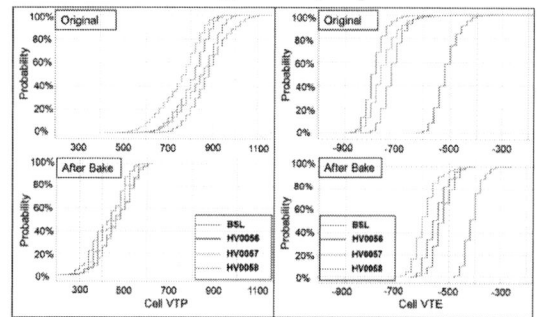

Figure 4. The comparison of Chip Vt and after Bake Vt window for baseline and new condition

Figure 4 shows the comparison of Chip Vt and Vt window after baking between the baseline and new condition, which can fully support that the difference of ONO film thickness can greatly influence the Cell Vt. Compared with BSL, new condition shows the great increase of Erase Vt, which is similar to the Vt after 85° baking, and the Cell VT is obviously improved. As shown in P charts, we can easily see the longer tailing when Cell is erased, suspecting the strong correlation with SG DIDL leakage. Similar to WAT relation between VTP and VTPI, original VTP is smaller than BSL condition. It can be easily found that the mean value of VTP increase from 0.42 V to 0.46 V, and VTE increase from -0.41 V to -0.54 V after baking, thereby the Vt window after baking is increased by 20%.

CONCLUSION

Firstly, ONO film profile which is the core storage structure is optimized. Based on the HTO remove process, ONO film profile of Top nitride thickness is optimized so that the effective area of storage structure is greatly expanded. Secondly, the thickness of tunneling oxide and bottom nitride can significantly influence cell Vt. Combined with the above process optimization conditions, it can be seen that the WAT parameters related to Cell are significantly improved and the overall window is significantly improved. Vtp mean value is increased by 9% after baking and VTE mean value is increased by 34%. To avoid backside tuning issue, increase the thickness of TOPSIN to improve the cycling and reduce the thickness of TOPOX (ISSG) to reduce the equivalent oxide thickness. In addition, improvement experiments are conducted on ONO film and other process, as well as the overall process window confirmation experiments are in progress.

REFERENCES

[1] Liqun Dong* , Zhenghong Liu, *IEEE TRANSACTIONS ON ELECTRON DEVICES*, VOL. 61, NO. 11, NOVEMBER 2014

[2] Zhenghong Liu*, Liqun Dong, *IEEE JOURNAL OF SOLID-STATE CIRCUITS,* VOL. 49, NO. 8, AUGUST 2014

[3] Jae-Young Sung, *IEEE TRANSACTIONS ON ELECTRON DEVICES* , vol. 48, no. 5, pp. 1302–1309, May 2021.

[4] Sung-Kun Park, *IEEE TRANSACTIONS ON ELECTRON DEVICES,* VOL. 67, NO. 3, 922-928, MARCH 2020,

[5] Sung-Kun Park, *IEEE TRANSACTIONS ON ELECTRON DEVICES*, VOL. 68, NO. 4, 1585-1592, APRIL 2021.

[6] Zhaozhao Xu , *Microelectronics Reliability* VOL. 84, 157-162(2018).

AN EFFECTIVE METHOD TO MINIMIZE THE DIFFERENCE BETWEEN SPOT AND RIBBON BEAM IN S/D REGION BY CARBON CO-IMPLANTATION

Long Feng[1], Zhiqiang Xiao[1,2*], Jiaxing Xiao[1], Zhirui Li[1], Haitao Yan[1]*

[1]Semiconductor Manufacturing North China (Beijing) Corp. Beijing, China

[2]School of Integrated Circuits, Tsinghua University, Beijing, China

*Corresponding Author's Email: ZhiQiang_Xiao@smics.com; Frank_Feng@smics.com

ABSTRACT

The device matching of different implanters with single-wafer ribbon beam to batch-wafer spot beam is difficult due to their significant differences in beam shape and scanning architecture, which lead to diverse damage accumulation and final amorphous layer thickness. Especially for Source/Drain (S/D) doping with high concentration and high mechanical stress, implant condition with single wafer ribbon beam may suffer severe dislocation defect while not with the batch type spot beam. In this paper, by modulating the dose rate or damage process on the single-wafer implanter, a series of experiment were performed to match the batch-wafer spot beam dislocation performance. Carbon co-implantation was demonstrated as the most effective way to avoid dislocation formation, while changing beam current, using cold-implantation or using Ge co-implantation has no positive effect on the silicon defect relieving.

Keywords—ion implantation; matching; dislocation; dose rate; carbon implant

INTRODUCTION

With the critical dimension continuous shrinking and moving to larger diameter substrates, a switch from multi-wafer to single-wafer high current implanters occurred around the 65 nm process node due to the precise dosage and angle control of single-wafer implanter [1]. However, when the single-wafer new architecture was incorporated into existing processes, device matching to existing batch-wafer spot beam implanter becomes a big problem. The significant differences of the single-wafer tool in beam shape and scanning architecture can affect beam density reaching the wafer and dwell time, which result in different damage and silicon defect properties.

Especially, in the source and drain (S/D) area, the challenge of the two architecture matching become even more tremendous. Firstly, the high dose S/D doping easily creates severe damage to the silicon lattice and lead to high density end of range (EOR) silicon defect. The EOR defect may agglomerate to silicon dislocation in the post-anneal process and result in enhanced dopant diffusion and junction leakage [3]. Secondly, the increased density of shallow trench isolation (STI) associated with the novel films lead to high stress, which has a big effect on defect nucleation during solid phase epitaxy growth (SPEG) of amorphous silicon induced by implantation. Thus in the S/D implantation, by affecting the amorphous layer thickness and silicon defect generation, the dislocation ratio fluctuates sharply with the dose rate variation [4]. For the two type implanters, one-dimensional mechanical scan coupled with single-wafer ribbon beam usually has more serious dislocation defect than the two-dimensional scan coupled with batch-wafer spot beam (Figure 1). Since the silicon dislocations severely degrade the device performance, it is urgent to develop an effect route to optimize the single-wafer ribbon beam process and alleviate the dislocations.

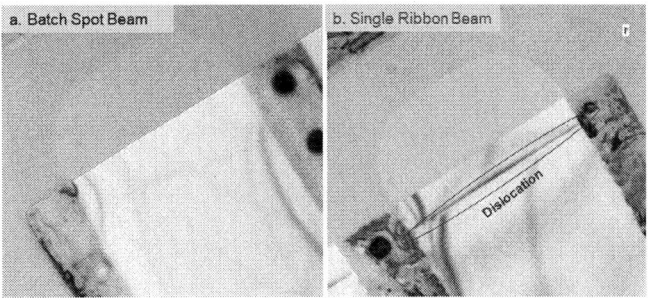

Figure 1. TEM cross-section of post S/D implantation anneal: a) no obvious dislocation observed for the batch spot beam implanter; b) severe AA dislocation observed for the single ribbon beam implanter.

Numerous studies [1,4,5] reported the dose rate relationships of the different type implanters and its impact on the device performance. However, the methods in those literature cannot eliminate the dislocations in single-wafer implanter and match the batch-wafer performance in Figure 1. In the present study, to match the two different type implanters in the S/D implantation, a series of attempts were performed, such as tuning ion beam parameters, cold implantation or co-implantation. Thermal-wave (TW) and Transmission Electron Microscopy (TEM) were employed to the damage and amorphous layer characterization. The final TEM analysis and electrical properties represented by junction leakage related yield loss show that carbon co-implantation was the most effective way to avoid dislocation and match the two type implanters.

EXPERIMENTS AND RESULTS

Due to the difference in the way that the ion beam is formed and transported to the wafer, as well as differences in the dynamics of the wafer motion under the beam [1], there are other differences between the two architectures besides dose rate difference, like dose and angle non-uniformity difference. To exclude those factors effect on the final device difference, we firstly check the dose and angle difference of the two architecture. The dose split (+/-3%)and angle split (+/-2°) result shows little difference, which indicate that the dose or

979-8-3503-1101-3/23 $31.00 © 2023 IEEE

angle non-uniformity difference has little contribution to the final electrical differences.

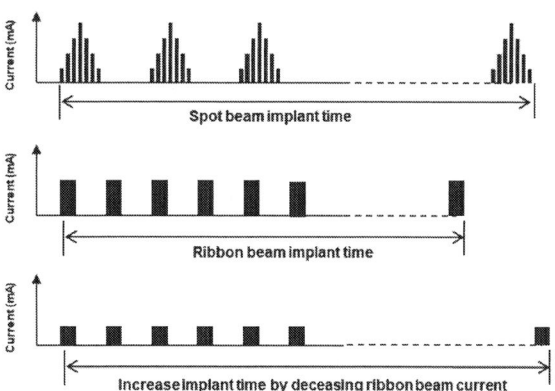

Figure 2 Illustration of spot beam and ribbon beam dose rate

For the major dose rate difference of the two architecture, as shown in Figure 2, with the same beam current, though a ribbon beam system has much lower beam density due to the higher beam size, its dwell time is much longer than the spot beam system. This results in less relaxation period during the time of implant than the spot beam system and leads to different damage accumulation and thermal profile for the two architecture. Generally, for a given species and beam current with a high dose implantation, due to the self-anneal and consequently implant damage relaxed effect of the spot beam system, implantation with ribbon beam creates more silicon damage and thicker amorphous layer. Since the relation between the residue defect and dose rate was still in dispute, combining with high compression stress from STI in S/D region, the dislocation formation process was complicated [6-8]. In order to elucidate the relationship of the dislocation and eliminate the dislocations in ribbon beam implanter, a series of experiment was designed as shown in Table 1.

For the baseline group, the S/D implantation was performed with Arsenic, 30KeV, 5E15, Tilt 0, 6mA on a spot beam Quantum and ribbon beam Trident high current implanter respectively. Since the major difference of the two architecture comes from dose rate, the first group experiments attempt to decrease the ribbon beam current from 6mA to 1mA or 0.1mA in order to prolong the relaxed period, which can approach the spot beam dose rate performance. In practice, the exactly matching through simply recipe optimization or beam tuning was hard to be achieved due to the intrinsic beam scan dynamic and beam shape difference of the two architecture. As the dislocation mostly derived from the micro defect in the amorphous/crystallization (a/c) interface, cold implantation, which is deemed as an ideal method to form perfect amorphous layer and sharp a/c interface with lower defect density, was introduced for the second group split. For the third and fourth split, co-implantations were incorporated for the modification of the defect in a/c interface. The Ge implantation was also used in the fourth split to magnify the amorphization process with a thicker and more perfect amorphous layer. The carbon is thought as a good candidate for residual EOR defect elimination because it easily traps excess interstitials by the Carbon-Si pair.

TABLE 1 EXPERIMENT CONDITIONS FOR DISLOCATION REDUCTION

No.	Condition	Tool Type	Split Condition
0	Baseline	Batch spot beam	As, 30KeV, 5E15,Tilt 0,6mA
		Single ribbon beam	As, 30KeV, 5E15,Tilt 0,6mA
1	Beam current	Single ribbon beam	As, 30KeV, 5E15,Tilt 0,1mA
			As, 30KeV, 5E15,Tilt 0,0.1mA
2	Cold Implant	Single ribbon beam	As, 30KeV, 5E15,Tilt 0,6mA,-100°C
3	Add Ge IMP	Single ribbon beam	Ge30KeV, 5E14,Tilt 0,3mA
			As, 30KeV, 5E15,Tilt 0,6mA
4	Add Carbon IMP	Single ribbon beam	C6KeV, 5E14,Tilt 0,3mA
			As, 30KeV, 5E15,Tilt 0,6mA

The split result represented by the normalized leakage yield loss was shown in Figure 3. For the baseline group, as mentioned above, the split with ribbon beam suffers serious AA dislocation and accordingly has about 20X higher yield loss than the spot beam split. As for the beam current split, previous study [6] reported that with the increasing dose rate the amorphous layer thickness increase linearly while the a/c interface thickness (EOR region) keep constant. After annealing, the EOR defect density decreases with the increasing dose rate, which indicates that lower dose rate implantation may be much easier to create dislocation than the higher one. However, the above mechanism was not applicable for this case. As the first split result shown in Figure 3, with decreasing beam current (dose rate), the yield loss can be decreased with better dislocation performance. Especially for the lowest beam current (0.1mA) split, the yield loss can decrease 50% than the ribbon beam baseline. However, this yield loss still much higher than the spot beam baseline. Moreover, the beam current 0.1mA is near the lower current limit of the tool capability and is not applicable for mass production due to its low throughput.

The second group split result shows that with -100°C cold ribbon beam condition, the yield loss increased about 30% than the ribbon beam baseline, which indicated that the thicker amorphous layer with perfect a/c interface may induced much worse dislocation in this case. This guess can be further proved by the third group Ge co-implantation split. For the Ge co-implantation split with thickest amorphous layer, the yield loss is the worst within all the experiment conditions.

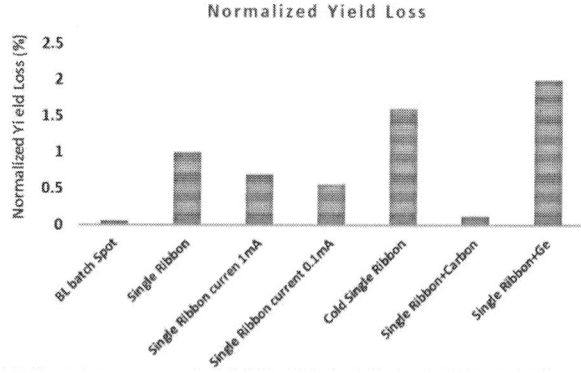

Figure 3 Normalized yield loss with different implant conditions

The best yield performance comes from the third split with carbon co-implantation. As shown in Figure 3, its yield performance is comparable with baseline spot beam result, which indicates that carbon can suppress the dislocation formation effectively in the S/D implant. The incorporation of carbon paves an indirect but available way for dislocation matching between ribbon and spot beam matching.

Based on the phenomenon discussed above, we supposed two possible mechanisms for the dislocation formation in the S/D implantation with high concentration doping. For the first one, the dislocation formation appears to depend critically on the total number of Si atoms displaced during implantation [9]. The more damage and amorphous to the silicon, the more displaced silicon atoms. These displaced atoms provide the mobile Si interstitials which agglomerate to form dislocations. With the spot beam or low dose rate implantation, the anneal effect and consequently implant damage relaxed can decrease the displaced silicon atom, while with the carbon co-implantation the silicon interstitials can be getter by the carbon-interstitial pairs. Another possible mechanism may relate to the stress induced dislocation. Silicon crystals compress when a high concentration of arsenic impurities is electrically active [10]. With the degree of crystallization of the arsenic doped silicon increases, the stress of the crystal becomes higher. With stress above critical level, it will give way to misfit dislocations. Thus the thicker amorphous layer and more perfect the a/c interface, the more dislocations with the crystal. For the carbon co-implantation split, the tensile stress can be obtained from substitutional carbon atoms, which relaxed the lattice compression and accordingly the dislocation is avoided.

CONCLUSIONS

The device matching of single-wafer ribbon beam to batch-wafer spot beam in the S/D implantation of arsenic impurities was investigated. A series of attempts were performed to modify the dislocation performance of single-wafer ribbon beam and match the yield performance of the batch-wafer spot beam. The carbon co-implantation was proved as the most effective method to modify the dislocation of ribbon beam, while the traditional cold implantation or Ge co-implantation modification with thicker amorphous layer and better EOR defect resulted worse dislocations. Based on the result of the splits, it is supposed that the emergence of the dislocations in this case may come from agglomeration of the super-saturated silicon interstitials caused by the implant damage or the crystal misfit induced by over-critical compression stress from the substitutional arsenic atoms and the STI.

REFERENCES

[1] M. S. Ameen, M. A. Harris, C. Huynh and R. N. Reece, "Dose Rate Effects: the Impact of Beam Dynamics on Materials Issues and Device Performance", AIP Conference Proceedings 1066, 30, 2008.

[2] P. Ferreira et al., "Elimination of Stress Induced Silicon Defects in Very High-Density SRAM Structures", 31st European Solid-State Device Research Conference, September 2001

[3] B. Colombeau; B. Guo et al, "Advanced CMOS devices: Challenges and implant solutions.", Phys. Status Solidi A 211,No.1 101-108, 2014.

[4] Y. L. Chin et al., "Study on the influence of implant dose rate and amorphization for advanced device characterization," 2014 20th International Conference on Ion Implantation Technology (IIT), pp. 1-2, 2014.

[5] K. Shim et al. "Impact of Dose Rate Effects and Damage Engineering on Device Performance", AIP Conference Proceedings 866, 137 ,2006

[6] Lance S. Robertson, Aaron Lilak, Mark E. Law, and Kevin S. Jones. "The effect of dose rate on interstitial release from the end-of-range implant damage region in silicon", Appl. Phys. Lett. 71, 3105-3107 (1997)

[7] R. Simonton, J. Shi, T. Boden, P. Maillot, and L. Larson, Mater. Res. Soc. Symp. Proc. 316, 153 (1994).

[8] J. R. Liefting, J. S. Custer, R. J. Schreutelkamp, and F. W. Saris, Mater.Sci. Eng. B 15, 173 (1992).

[9] F W Saris et. "Avoiding dislocations in ion implanted silicon" Microelectronic Engineering 19, 357-362(1992).

[10] G. S. Cargill, J. Angilello, and K. L. Kavanagh, Phys. Rev. Lett. 61, 1748(1988).

[11] Victor Moroz et, Applied Physics Letters 87, 051908 (2005)

Study of Breakdown Voltage Improvement of High-VoltagePLDMOS

Wenting Duan, Donghua Liu, Haiyang Ling, Ying Cai, Feng Jin ,Wensheng Qian

HuaHong Grace Semiconductor Manufacturing Corporation, Shanghai 201206, China

*Corresponding Author's Email: Wenting.Duan@hhgrace.com

ABSTRACT

In this paper, the method of high-voltage PLDMOS breakdown voltage (BV) improving is studied, in which high-voltage NLDMOS BV is not sacrificed. The method is increasing high-voltage PLDMOS drift region dose of high energy implant and decreasing that of middle energy implant, but the PNP punch-through BV of high-voltage PLDMOS is suffered in this method. Then both BV and PNP punch-through BV are considered, finally, the high-voltage PLDMOS which both BV and PNP punch-through BV meet target is got in TCAD simulation.

Key Words—High-voltage PLDMOS; Breakdown Voltage; PNP punch-through BV

INTRODUCTION

LDMOS device is widely applied as the switch in power management circuits. The high-voltage PLDMOS in this paper is applied on 80V-100V in high-voltage BCD process platform whose market prospects in the field of power and power distribution will become very broad, such as LED lighting, high efficiency motor drive, AC/DC conversion, etc [1][2]. The breakdown voltage (BV) of this device is very important and the concentration of PLDMOS drift region is a key factor in determining the breakdown voltage and conduction characteristics of the device [3][4]. The drift region of low concentration and deep depth is a common method employed in high-voltage LDMOS.

DEVICE STRUCTURE

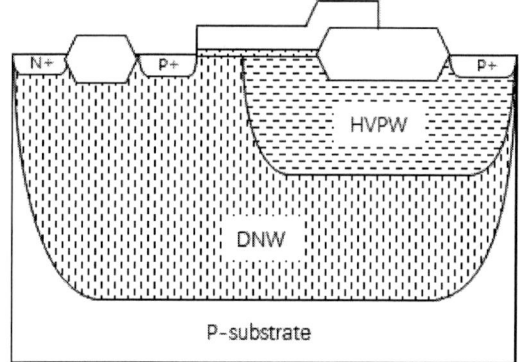

Figure 1: PLDMOS device structure in this paper

The high-voltage PLDMOS device cross-section is schematically shown as Fig.1, the PLDMOS drift region is formed by high-voltage P-type well (HVPW), and the deep N-type well (DNW) is used to isolate P-substrate and

be channel area. The DNW is much deeper than HVPW. Low concentration of HVPW is required for high BV of HVPW/DNW junction and PLDMOS. The effective drift region concentration is the compensation doping of HVPW and DNW, which is even lower than that of DNW. The DNW cannot be deplete completely, which cause PLDMOS BV is lower than target. The DNW is the drift region of high-voltage NLDMOS, the HVPW is the channel area of high-voltage NLDMOS. the PLDMOS BV improvement should be considered with NLDMOS BV.

RESULTS AND DISCUSSION

For DNW fully depletion under HVPW in PLDMOS, HVPW doping concentration increasing is attempted. There are mainly three parts of HVPW implantation: low energy implantation, middle energy implantation and high energy implantation. PLDMOS BV is increased with HVPW doping concentration increasing, and the impact to BV is more as the implantation energy is higher, and the BV has a peak value, which is shown in Fig. 2. The high energy implantation is the most effective to DNW depletion, the depletion region(grey area) comparison is shown in Fig. 3.

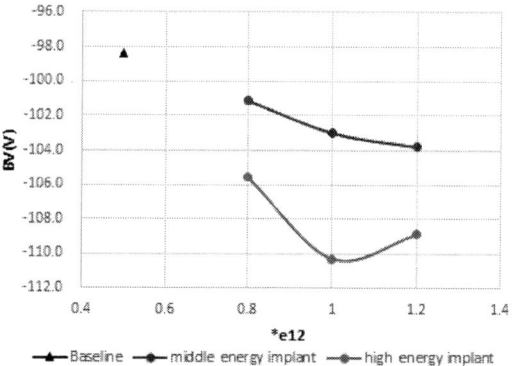

Figure 2: PLDMOS BV trend vs HVPW middle energy and high energy implantation

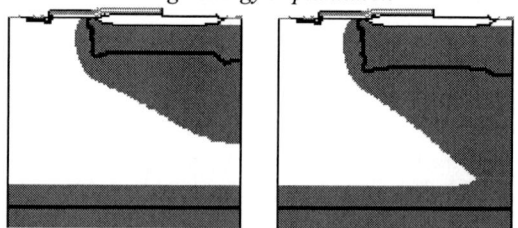

a. baseline   b. HVPW high energy implant 1e12
Figure 3: PLDMOS depletion region comparison

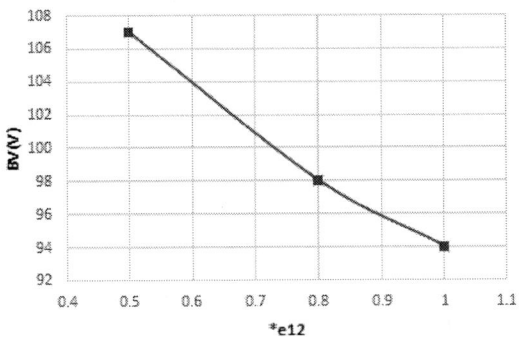

Figure 4: NLDMOS BV trend vs HVPW high energy implantation

The BV of high-voltage NLDMOS with HVPW as channel area decrease with HVPW high energy concentration increasing, as shown in Fig. 4. HVPW total concentration need to be constant to keep NLDMOS BV stable. So the method to improve PLDMOS BV that the high energy implantation concentration increasing and middle energy implantation concentration decreasing is suggested. Because the high energy implantation concentration of HVPW is more sensitive to PLDMOS BV.

The PLDMOS BV increase to meet target by increasing the high energy implantation concentration and decreasing middle energy implantation concentration of HVPW. But the PNP punch-through BV of PLDMOS decrease obviously due to high energy implantation concentration increasing, as shown in Fig. 5, the PNP punch-through BV happened at 80V before the device BV which is 115V. The parasitic PNP (HVPW/DNW/ P-sub) is triggered.

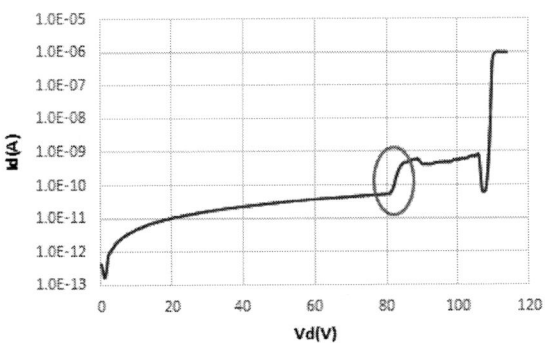

Figure5: PLDMOS BV curve

The shallower HVPW junction depth and lighter HVPW bottom concentration is helpful for the PNP punch-through BV increasing. To balance the PNP punch-through BV and device BV, the energy and concentration of high energy implantation of HVPW are reduced, and the concentration of middle energy

implantation is increased slightly, then the PNP punch-through BV is increased with the PLDMOS device BV dropping little. Finally, the PNP punch-through BV and device BV both meet target. The longitudinal doping profile of PLDMOS drift region is shown in Fig. 6. The concentration of low energy implantation is the highest for low on resistance and NLDMOS threshold voltage, the concentration of middle energy implantation is the lowest for drift region HVPW depletion completely and higher PNP punch-through BV, the concentration of high energy implantation is higher than middle energy implantation for higher device BV by DNW under HVPW depletion completely.

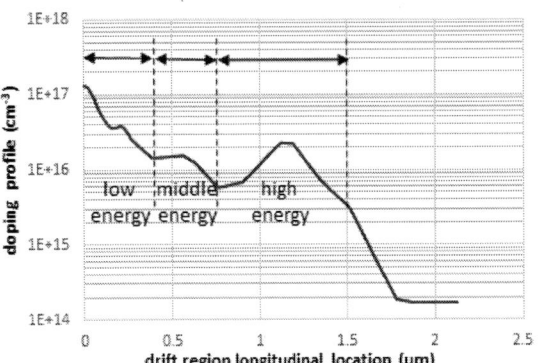

Figure6: longitudinal doping profile of PLDMOS drift region

With the optimization of HVPW doping profile, the high voltage PLDMOS PNP punch-through BV and device BV both meet target, and the high voltage NLDMOS BV is also considered.

CONCLUSION

In this paper, the high voltage PLDMOS is carefully designed by drift region doping profile optimization, and gets competitive BV performance, including the PNP punch-through BV and device BV. The high voltage NLDMOS BV is also considered as the channel area shared with PLDMOS. The high energy implantation of drift region is the most critical factor for both the PNP punch-through BV and device BV of high voltage PLDMOS. Finally, the high voltage PLDMOS device BV is improved by 10V at least and both device BV and PNP punch-through BV meet target in TCAD simulation.

The method of the high voltage PLDMOS BV optimization can be also used in other high voltage device with different process.

ACKNOWLEDGEMENTS

The authors would like to thank all members of device group and high voltage group of HHgrace for their great support on this work.

REFERENCES

[1] T. R. Efland, Chin-Yu Tsai and S. Pendharkar, "Lateral thinking about power devices (LDMOS)," International Electron Devices Meeting 1998. Technical Digest (Cat. No.98CH36217), San Francisco, CA, USA, 1998, pp. 679-682.

[2] He J , Xi X , Chan M , et al. Linearly graded doping drift region: a novel lateral voltage-sustaining layer used for improvement of RESURF LDMOS transistor performances[J]. Semiconductor Science & Technology, 2002, 17(7):721-728.

[3] Shaari S , Hanim A R , Mardiana B , et al. Modeling And Analysis Of Lateral Doping Region Translation Variation On Optical Modulator Performance[J]. AIP Conference Proceedings, 2010, 1325(1):297-300.

[4] Yang K , Guo Y , Pan D Z , et al. A Novel Variation of Lateral Doping Technique in SOI LDMOS With Circular Layout[J]. IEEE Transactions on Electron Devices, 2018, PP(4):1-6.

A COMPACT SAWTOOTH WAVE GENERATOR BASED ON NOVEL Z²-FET DEVICE

Hui Xie, Yingxin Chen, and J. Wan*

School of Information Science and Engineering, Fudan University, Shanghai, China

Email: jingwan@fudan.edu.cn

ABSTRACT

A sawtooth wave generator with compact circuit and high integration density is realized by integrating the switching and hysteresis comparison functions into a novel semiconductor device, Z²-FET (Zero subthreshold swing and Zero impact ionization FET). The frequency and amplitude of the sawtooth wave can be flexibly controlled by the gate voltage. The circuit is further fabricated with CMOS-compatible process. The measurement indicates excellent sawtooth wave generator performance which agrees with TCAD simulation.

INTRODUCTION

Sawtooth signal generator is widely used in modern electronic equipment. Its waveform is characterized by frequency, amplitude, linearity and other parameters, which greatly affect its application. From the initial use of Boot-strap and Miller circuit to utilize linear capacitor charge-discharge principle [1], researchers have proposed different sawtooth generators for different applications [2-4]. Takai N [3] combined two triangular waves to achieve a sawtooth wave with low descent time which reduces the errors in duty ratio when it is applied as a component of switching regulator. Li X [4] utilizes AlGaN/GaN

MIS-HEMT(Metal-Insulator-Semiconductor High Electron Mobility Transistor) as the switching device in sawtooth wave generator to get a high peak-to-peak voltage which is benefit to its use on power converter applications.

The generation of analog sawtooth waves can be roughly divided into two types: the integrations of input square wave and the linear charging-discharging of capacitor. The latter has attracted much attention due to its excellent linearity. It generally consists of a current source, switches, and a hysteresis comparison module. However, the existing problem is that the switches with large subthreshold swing deteriorates its linearity, while the additional comparison module used to define V_{peak} and V_{valley} will increase the circuit complexity, which is detrimental to large-scale integration.

Z²-FET [5] is an ideal switching device with both gate-controlled hysteresis and zero subthreshold swing. Its essence is a field-effect doped thyristor formed by gate bias. It has found applications in memories, ESD (electrostatic discharge) protection, photodetection, and

biosensing fields [5-8]. A single Z²-FET can realize both high-speed switching and hysteresis comparison functions. The sawtooth wave generator module with Z²-FET as the core has a compact circuit size and can control the output amplitude and frequency flexibly with its gate control.

SAWTOOTH GENERATOR CIRCUIT

The Synopsys Sentaurus TCAD simulation is implemented to analyze the proposed circuit. Doping induced bandgap narrowing, Philips unified model, electric field and doping-dependent mobility model, SRH thermal generation-recombination model with doping-dependent carrier lifetime are employed as physical models in the simulation. A channel acceptor doped N-type Z²-FET is used, and the sawtooth generator composed of Z²-FET is schematically shown in Fig. 1(a).

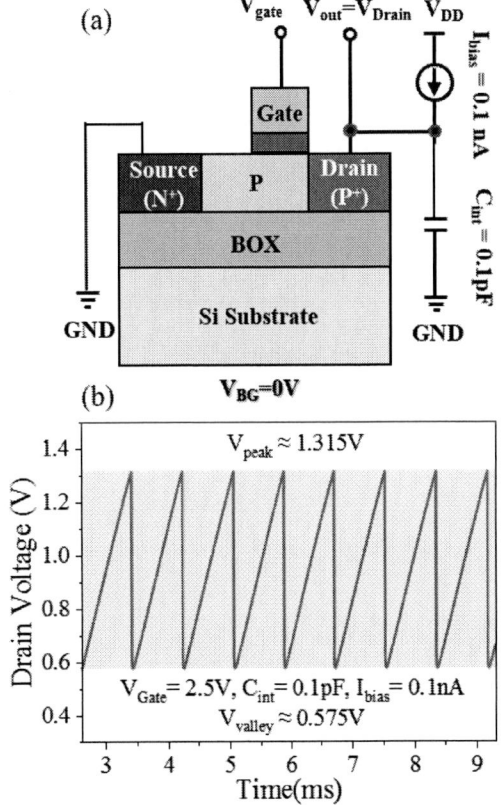

Fig. 1. (a)The circuit diagram of sawtooth wave generator with Z²-FET. The device parameters of

Z^2-FET are: $L_G=L_{IN}=0.2\mu m$, $T_{ox}=5nm$, $T_{si}=28nm$, $T_{box}=145nm$, $N_A(P^+)=10^{20}cm^{-3}$, $N_D(N^+)=10^{20}cm^{-3}$, $N_A(P)=10^{17}cm^{-3}$. (b)The generated sawtooth wave.

The current source charges the capacitor at a fixed rate and increases the voltage on the capacitor (C_{int}) linearly. When the capacitor voltage (C_{int}) reaches turn-on voltage (V_{ON}), Z^2-FET conducts, and C_{int} discharges rapidly ($I_{ON}\gg I_{bias}$). When the voltage on the C_{int} decreases to V_{OFF}, the device is turned off and C_{int} is charged linearly again. This eventually generates a sawtooth wave. According to the working principle, the circuit parameters and period of the sawtooth wave are related by:

$$T = \frac{C_{int}}{I_{bias}}(V_{on} - V_{off}), \quad I_{bias} \gg I_{off} \quad (1)$$

As the output wave show in Fig. 1(b), the average peak voltage of the wave is $V_{peak}=V_{ON}=1.31V$, the average valley voltage $V_{valley}=V_{OFF}=0.57V$, and the average period is 822µs indicating a frequency of 1216HZ. The rising stage of the sawtooth wave shows an excellent linearity, with an average linearity=0.85%. Its average rise time $t_r\approx724\mu s$, and average fall time $t_f\approx0.45\mu s$, indicating a sharp drop.

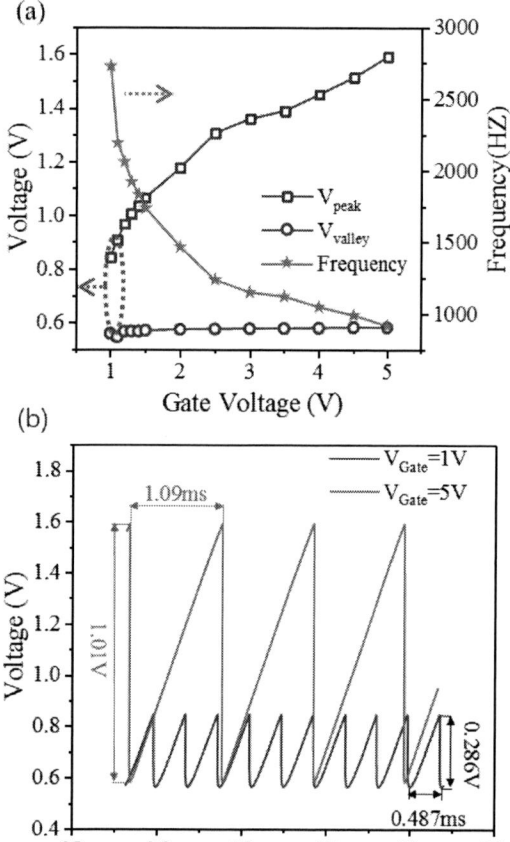

(a)

(b)

Fig. 2. (a)The change of output V_{peak}, V_{valley} and Frequency when only gate voltage changed.(b)The output waveform of the circuit at $V_G=1V$ and $5V$

Gate Voltage Control

The frequency and amplitude of output wave can be flexibly adjusted by changing the gate voltage (V_G). The V_G directly changes the turn-on voltage of the Z^2-FET by modulating the barrier height in L_G. This indirectly affects the output frequency through different charging time required to charge the capacitor to turn-on voltage. Fig. 2(a) shows the characteristic trend of the output wave when the gate voltage is changed. The minimum allowable gate voltage $V_{Gmin}\approx1V$, and the corresponding maximum frequency $f_{max} \approx 2.7kHz$, and minimum amplitude $A_{min}\approx0.3V$. With the increase of V_G, the valley voltage is almost constant and the peak voltage increases. The increase of peak voltage decreases the frequency consequently, since the charging rate is constant. With $V_G=5V$, the V_{peak} increases to 1.6V with the frequency down to 914Hz

Fabrication and Measurement

The sawtooth wave generator with Z^2-FET is further fabricated with CMOS-compatible process in our clean room. The Z^2-FET and the capacitor are integrated in the same chip. The fabrication starts with a SOI substrate which has $T_{si}=28nm$ and $T_{box}=145nm$. The active region is defined by photolithography followed by wet etching in TMAH (Tetra-Methyl Ammonium Hydroxide) solution. After, photolithography and ion implantation steps are used to define the anode (P^+) and cathode (N^+) regions. The gate oxide is 20nm HfO_2 high-k dielectrics deposited by atomic layer deposition (ALD). The metal gate is eventually formed by photolithography and deposition of 20nm Ti/80nm Au with electron beam evaporation (EBE). The top view of the device is shown in Fig. 3(a). The device has gate length of 3um,intrinsic length of 3µm and width of 10µm, Ti/Au metal pads (100µm×100µm) are formed in the source and drain and used as contacts. The big drain pad also forms a capacitor with the back gate through the buried oxide, which is needed in the sawtooth wave generator circuit.

The fabricated circuit is measured with a transistor analyzer. Unlike the p-type Z^2-FET in TCAD simulation, the fabricated Z^2-FET is a n-type device. Thus, all ports are oppositely biased compared to the p-type device. The V_G is biased with a constant voltage of -3V. Instead of doping the channel as in the TCAD simulation, the back gate is biased with $V_{BG} = 4V$ to induce the required carrier injection barriers in the channel of the fabricated device. A constant current source of -0.5µA is applied on the drain port. The drain voltage is recorded as the output. As shown in Fig. 3(b), the output wave at the drain port shows a sawtooth shape, perfectly reproducing the results from the TCAD simulation. The constant current source charges the capacitor in the drain contact pad, which increases the V_{out} linearly. As V_{out} reaches V_{ON}, the Z²-FET is turned on followed by rapid discharge of the

(a)

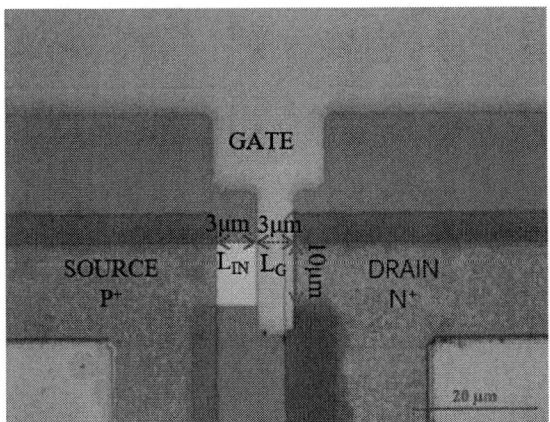

(b)

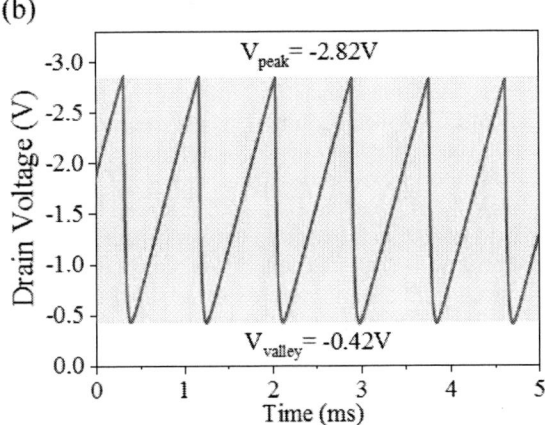

Fig. 3 (a) The top view of Z²-FET. (b) The experimental output sawtooth waveform

capacitor through Z²-FET. The measured sawtooth wave has period of 855µs, frequency of 1170HZ, V_{valley}=-0.42V and V_{peak}=-2.82V.

CONCLUSION

The sawtooth wave generator proposed in this paper uses novel semiconductor device Z²-FET as the core in order to realize small size and high integration density. TCAD simulation has been conducted to study the sawtooth wave generation mechanism. The circuit has been further fabricated with CMOS-compatible process. The measurement results show excellent sawtooth waveform agreeing with the TCAD simulation well. The proposed sawtooth wave generator has compact size and high integration density. The frequency and amplitude of the sawtooth wave can be flexibly tuned by its gate voltage. Thus, it is expected to be used in applications

such as BIST(Built-In Self-Test) where the circuit area is desired to be small enough.

ACKNOWLEDGEMENTS

This work was supported by the National Key R&D Program of China under Grant 2021YFA120050 and Grant 2018YFB2202800, Shanghai Science and Technology Commission "explorer project" under Grant 21TS1401300. Part of the sample fabrication was conducted at Fudan Nano-fabrication Lab.

REFERENCE

[1] K. Sarma and V. Mathew, "A Linear Sawtooth Generator," *IETE Journal of Research,* vol. 16, no. 2, pp. 134-139, 1970.

[2] M. T. Abuelma'Atti and M. K. Alabsi, "Current-controlled sawtooth generator," *Active and passive electronic components,* vol. 27, no. 3, pp. 155-159, 2004.

[3] N. Takai and Y. Fujimura, "Sawtooth generator using two triangular waves," in *2008 51st Midwest symposium on circuits and systems,* 2008: IEEE, pp. 706-709.

[4] X. Li, M. Cui, and W. Liu, "A full GaN-integrated sawtooth generator based on enhancement-mode AlGaN/GaN MIS-HEMT for GaN power converters," in *2019 International Conference on IC Design and Technology (ICICDT),* 2019: IEEE, pp. 1-3.

[5] J. Wan, C. Le Royer, A. Zaslavsky, and S. Cristoloveanu, "A systematic study of the sharp-switching Z2-FET device: From mechanism to modeling and compact memory applications," *Solid-State Electronics,* vol. 90, pp. 2-11, 2013.

[6] Y. Solaro *et al.,* "Z2-FET: A promising FDSOI device for ESD protection," *Solid-state electronics,* vol. 97, pp. 23-29, 2014.

[7] J. Liu *et al.,* "Dynamic coupling effect in Z 2-FET and its application for photodetection," *IEEE Journal of the Electron Devices Society,* vol. 7, pp. 846-854, 2019.

[8] S. Cristoloveanu *et al.,* "A Review of Sharp-Switching Band-Modulation Devices," *Micromachines,* vol. 12, no. 12, p. 1540, 2021.

STUDY OF THE FORMATION OF COPPER VOID DEFECT AND PROCESS OPTIMIZATION FOR REDUCTION IN DUAL DAMASCENE PROCESS

Hongliang Zhu[1]*, Shuhuai Jia[1], Dejing Ma[1] and Fengjiao Wang[1]

[1]Semiconductor Manufacturing International Corporation(SMIC), Beijing 100176, China

*Corresponding Author's Email: Randy_Zhu@smics.com

ABSTRACT

Copper void defect issue is becoming more sensitive for the quality of backend interconnect metals with the downscaling of the semiconductor devices. There are many reasons for the formation of copper void. In this paper, the q-time between ECP to CMP, and the influence of different ECP additives, different ECP anneal times and different CMP slurry were studied to reduce copper void defect. With the q-time between ECP to CMP longer, the Cu void performance is worse for the self-annealing effect. And the copper void defect performance is obviously different when different ECP additives were used. As the purity is higher of Cu ECP additive, the copper void performance is better. In addition, the ECP annealing time longer, the defect condition is becoming worse. These conclusions are effective to reduce the copper void defect.

INTRODUCTION

With copper metal line wide application in back end of line in semiconductor fabrication. The quality of copper film become more and more important [1-2]. But with the downscaling of semiconductor devices, the defect of copper film become worse, especially the copper void defect [3~6]. The copper void will be detected as a missing defect after chemical mechanical polishing (CMP). In the damascene structure, electrochemical plating (ECP) was used to fill the trench. Normally there is a ECP annealing process after copper bulk ECP process. After ECP process, the Cu grain size is ultra-fine [7]. With the ECP annealing process or self-annealing effect, copper grain size will grow up from surface to the internal. And with the grain size grow up, copper film resistance and hardness decreased gradually [8]. As a result, the CMP remove rate changes. After the copper film deposition, there are stress formation for copper deposition in the low-k material [9]. At the same time, vacancy occurs and migration along the gradient of stress. When the grain size grows up, stress migration and vacancy gather lead the formation of void. This problem will lead to yield loss and metal line reliability quality such as electro migration(EM) and stress migration(SM). So the optimization of the copper void become a more and more urgent thing.

Further for copper interconnect metallization process and dual damascene process, barrier/copper seed, cooper bulk deposition, and copper CMP process were completed to form the copper metal interconnect. In this paper ECP and CMP process were investigated. The different condition for ECP and ECP annealing, and CMP process were implemented to check the impact on the copper void defect performance. At the same time, we study the influence of the q-time of ECP to CMP to copper void defect.

EXPERIMENTAL

300mm silicon wafers were used for the different conditions experiment on ECP to CMP stage. For damascene structure, metal trench was etched as the first step, then TaN/Ta barrier layer and Cu seed layer were deposited with physical vapor deposition in this paper. Then series of wafers run with different conditions for ECP process, ECP annealing and CMP process. In this paper, the q-time between ECP to CMP, and different process conditions such as different ECP additives, different ECP annealing times and different slurries were study to the impact of copper void defect. The different ECP additives were adopted to get better gap fill capability and better film characteristic. And different ECP anneal times were detected to the influence of copper void defect performance. At last, the copper void defect performance with acidic and alkaline slurry were studied. At the same time, after CMP process, all experiment conditions were inspected the copper void defect performance.

RESULT AND DISCUSSION

All the defect maps are present as random fail, there is no special pattern as the Fig1, the void defect were mainly located at the thinner metal line and SRAM area. For the thinner metal line which metal trench were thinner, it requires higher gap fill ability. And for SRAM area, there are higher metal density and thinner metal lines, so it also requires higher gap fill ability. As a result, in these areas, more copper void defects were detected.

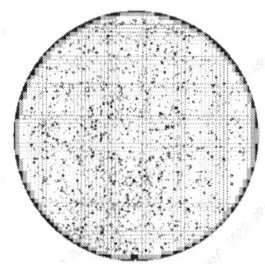 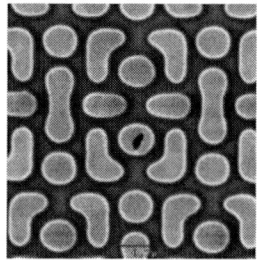

(a) defect map   *(b) SRAM defect image*

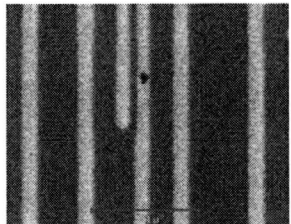

(c) thinner metal void defect
Fig1: Copper void defect map and image

Also the TEM result were examined for the experiment result. As the Fig2, from the TEM and cross-section sketch map, the void was form in the center of the trench. And the TaN/Ta layer was deposited well. Copper loss was found in the center of the trench. So the void defect was principally suspect to relate with ECP process and CMP process.

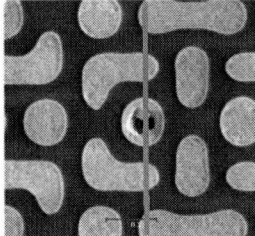

(a) copper void image (b) void cross section sketch map
Fig2: Copper void defect image and sketch map

Firstly, the Bright field inspection result as the Fig3 and the defect counts as the Fig4. When the q-time is short with 12h, defect counts is lower. But as long as the q-time longer to 24h，defect performance obviously become worse. And when the q-time longer to 48h, defect become very serious. So with the q-time between ECP to CMP becomes longer for all experiment conditions, the copper void defect becomes more and more worse. The cooper void defect different results with q-time conditions mainly because the self-annealing effect after copper ECP process. There are vacancy and residual stress exist in copper film. With the growth of grain size and the residual stress release, the vacancies concentrate along the gradient of stress. The self-annealing effect can continue with long time about 200hs. Even if there is an annealing process after ECP, the self-annealing effect still exist. So in this experiment, with the q-time between ECP to CMP longer, the vacancies concentrate more adequately. So after CMP process, the vacancy concentrates to form void defect.

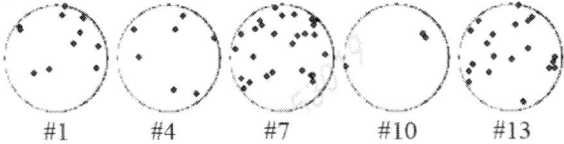

(a) ECP to CMP q-time 12h defect map

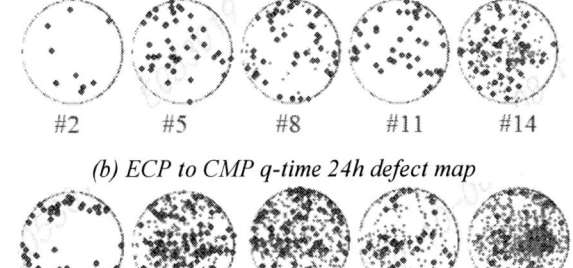

(b) ECP to CMP q-time 24h defect map

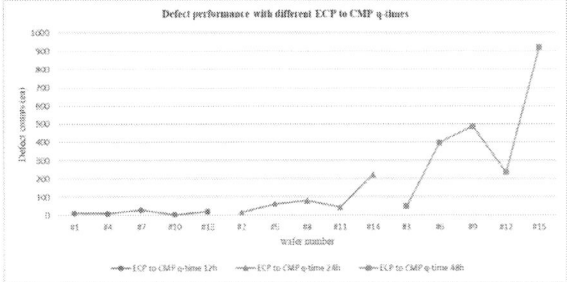

(c) ECP to CMP q-time 48h defect map
Fig3: The copper void defect performance with ECP to CMP q-time change for different experiment conditions

Fig4: The copper void defect counts with ECP to CMP q-time change for different experiment conditions

Secondly, with the ECP process split experiment, there are two kinds additives were used. The different conditions defect performance as the Fig5. The units 1 and 4 q-time from ECP to CMP is short, units 2 and 5 q-time little longer, and units 3 and 6 q-time is more longer. When the q-time of ECP to CMP is short, for the void defect, there is no different. But with q-time longer that the self-annealing effect will release. For the additive of ECP condition1, the copper void defect is better than condition2 obviously when the q-time were 24h and 48h. So the result of ECP additive condition1 is better. For the reason, the additive of condition1 has better gap fill ability especially for thinner metal trench. On the one hand, the ability accelerates the deposition of copper bulk exceed the additive of condition2. So the cooper bulk can be filled faster and decrease the hole in bulk in the trench. On the other hand, the purity of condition1 is higher than condition2. With the film impurity higher, the film defect and vacancy will be more, then concentration phenomena will be more obvious, so the void defect after CMP process will be worse.

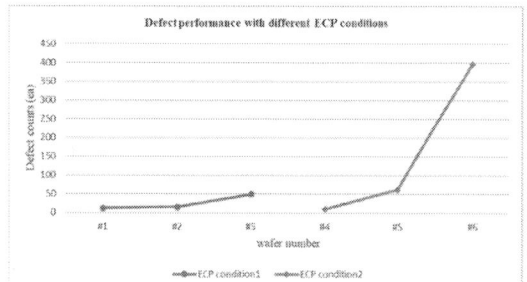

Fig5: The copper void defect counts with ECP to CMP q-time change for different ECP additive conditions

For another experiment, in this part, three different conditions were adopted with different ECP annealing times. The result as Fig6, for all conditions, with q-time longer, the defect become worse, and with annealing time longer, the defect performance become worse.

On the one hand, after compare these three conditions, when the ECP to CMP q-time become longer from 12hrs to 24hrs and 48hrs, all conditions defect become more and more worse, especially for the condition3. And the reason for defect becoming worse with q-time become longer has been expound in the first part.

On the other hand, when the q-time is short of 12hrs, there is no obvious different for three conditions. When the q-time is short, all conditions performance is comparable and the defect counts were low. This is because the self-annealing effect not sufficient with short q-time, and the concentration of vacancy is not obvious. Copper void defect become little worse for condition1 and condition2, when q-time longer to 24hrs. But for condition3, the defect become worse obviously. Then the q-time is to 48hrs, all conditions defect count become higher and defect performance become worse, especially for condition3. But condition1 is better, condition3 the worst. That means with the ECP annealing time longer, the defect become worse. When q-time longer, ECP annealing time longer, the defect performance is worse. Because the annealing time was longer than others and the grain growth and self-annealing effect were obviously, and the vacancy concentrates more quickly and obviously. And the void formation is ongoing. It was worst when the q-time longer to 48hrs. the defect counts become higher than other conditions for condition3.

After ECP process, the copper grain size is ultra-fine, and the RS is very high. So the ECP annealing process is needed to make the grain grow up and improve film property. And if the annealing time short, that means the grain growth on sufficient, and then the copper grain will growth up spontaneously with self-annealing effect. And that will lead to other issue such as the hillock defect and electro migration(EM) or stress migration(SM) test fail. While if the ECP annealing time too long, the grain size will become bigger, and this will lead to the vacancy to

concentrate more quickly and sufficient. At the same time, with the ECP annealing time longer, the remain stress will release more completely. Along the stress gradient, the vacancy will concentrate to the trench. Because for the over-hang effect, the trench is the stress higher location, so the vacancy will concentrate to the trench more and more. So with the ECP annealing time longer, the trench will gather more vacancy to form void defect.

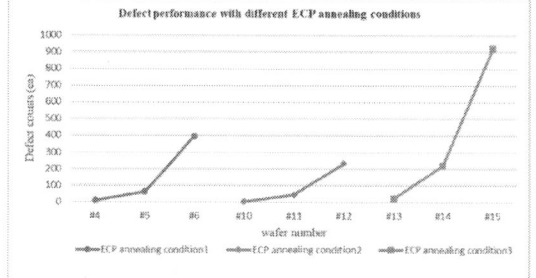

Fig6: The copper void defect counts with ECP to CMP q-time change for different ECP annealing conditions

For the last experiment, to find the influence of Cu CMP process to copper void, different Cu CMP conditions with acidic and alkaline slurries were studied to the copper void defect performance. The defect results with different Cu CMP process as the Fig7. Firstly, same phenomenon was found in this experiment. With the q-time longer, all conditions defect become worse. But for two conditions, there is no obviously different for all q-time conditions. So this means that the different Cu CMP process conditions have no obvious influence on copper void defect.

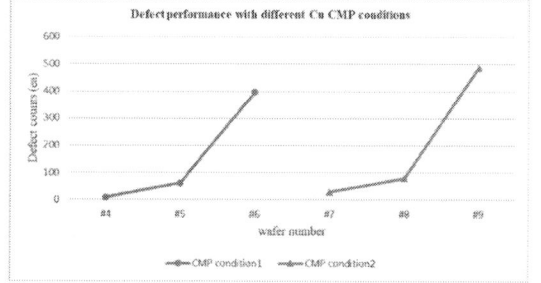

Fig7: The copper void defect counts with ECP to CMP q-time change for different Cu CMP conditions

CONCLUSION

Copper void defect become more and more serious in BEOL process with the downscaling of semiconductor devices, it may lead to yield loss or reliability failures. In this paper, we do a series of experiments to study the influence conditions of ECP and CMP process to the formation of copper void. The experiments results can help understand the formation of copper void and these conclusions are effective to improve the copper void performance. Firstly, with the ECP to CMP q-time longer, the copper void defect become more and more worse. Secondly, for different ECP annealing time, the defect is

979-8-3503-1101-3/23 $31.00 © 2023 IEEE   14

better with the time shorter, but this may lead to other defect and reliability issue. And the defect is worst with ECP annealing time longer. So the ECP annealing time need select one appropriate time. Thirdly, different CMP slurries have no obvious influence on void defect.

ACKNOWLEDGEMENTS

The authors would like to acknowledge the SMIC Logic team members. Without their sincere assistance and support and analysis and idea, it could not be succeeded.

REFERENCES

[1] T. C. Huang, C. H. Yao, W. K. Wan, C. C. Hsia and M. S. Liang, "Numerical modeling and characterization of the stress migration behavior upon various 90 nanometer Cu/Low k interconnects," Proceedings of the IEEE 2003 International Interconnect Technology Conference (Cat. No.03TH8695), Burlingame, CA, USA, 2003, pp. 207-209.

[2] Y. K. Lim et al., "Novel dielectric slots in Cu interconnects for suppressing stress-induced void failure," IEEE International Electron Devices Meeting, 2005. IEDM Technical Digest., Washington, DC, USA, 2005, pp. 179-182.

[3] Y. A. Wahab, A. F. Ahmad and Z. Awang, "Queue Time Impact on Defectivity at Post Copper Barrier Seed, Electrochemical Plating, Anneals and Chemical Mechanical Polishing," 2006 IEEE International Conference on Semiconductor Electronics, Kuala Lumpur, Malaysia, 2006, pp. 938-943.

[4] R. L. de Orio and S. Selberherr, "Formation and movement of voids in copper interconnect structures," 2012 IEEE 11th International Conference on Solid-State and Integrated Circuit Technology, Xi'an, China, 2012, pp. 1-4.

[5] J. Tseng et al., "Embedded Metal Voids Detection to Improve Copper Metallization for Advanced Interconnect," 2018 IEEE International Interconnect Technology Conference (IITC), Santa Clara, CA, USA, 2018, pp. 169-171.

[6] Y. Cao et al., "The influence of anneal condition on copper film property in ECP process," 2016 China Semiconductor Technology International Conference (CSTIC), Shanghai, China, 2016, pp. 1-3.

[7] R. Huang, W. Robl, T. Detzel and H. Ceric, "Modeling of stress evolution of electroplated Cu films during self-annealing," 2010 IEEE International Reliability Physics Symposium, Anaheim, CA, USA, 2010, pp. 911-917.

[8] R. Huang, W. Robl, H. Ceric, T. Detzel and G. Dehm, "Stress, Sheet Resistance, and Microstructure Evolution of Electroplated Cu Films During Self-Annealing," in IEEE Transactions on Device and Materials Reliability, vol. 10, no. 1, pp. 47-54, March 2010.

[9] C. Huang, A. Juan and K. C. Su, "Stress Induced Voiding Behavior of Electroplated Copper Thin Films in Highly Scaled Cu/low-k interconnects," 2020 IEEE International Reliability Physics Symposium (IRPS), Dallas, TX, USA, 2020, pp. 1-3.

STUDY ON N-TYPE MOS CAPACITOR IN 55NM CMOS

HongLiang Zhu[1]*, Xin Zhou[1], DeJing Ma[1], HaoQi Zheng[1], and TianFu Zhang[1]

[1]Semiconductor Manufacturing International Corporation(SMIC), Beijing 100176, China

*Corresponding Author's Email: Randy_Zhu@smics.com

ABSTRACT

This work performs fundamental physical measurements test and electrical measurements test on a N-Type metal oxide semiconductor capacitor (MOSCAP). Experimental results show that the thermally grown oxide is thicker when the doping energy decrease with the fixed doping dosage; the oxide is thicker when the doping dosage increase with the fixed doping energy. When the capacitor is larger, the range (which stands for the voltage dependence) is also larger, and the requirement for capacitor and the range is a tradeoff (the smaller range is the better). Various gate oxide conditions have also been examined, and the result shows that as the gate oxide shrinking down, the break down voltage appear some abnormal, which suspect the interface quality is not so excellent and the corner rounding for the gate oxide is worse.

INTRODUCTION

In the development of the metal oxide semiconductor field effect transistors (MOSFETs), scaling down the size is the main way to enhance the performance [1]. The capacitor should have a large capacitance per unit area and a high linearity. In order to get a high capacitor of MOSCAP, several ways were used to increase the capacitor, such as impurity doping adjustment and oxide thickness reduction. The most effective way to tune the capacitor of MOSCAP is changing the oxide thickness. But there is a limit to reduce the oxide thickness. So it is necessary to study the correlation between the doping and the capacitor, such as the doping element, energy and dosage.

In this work, we examine the influence of different doping element, different doping energy and different doping dosage, on the n-MOSCAP. In term of the doping, the oxide thickness reduction also has been arranged.

EXPERIMENTAL

In our experiment, we use the traditional 55nm CMOS process flow, and add one additional phosphorus Ion-Implantation to form the N type MOSCAP. To make the split condition clear and easy to analysis, we do the experiment with the following parameters as variables:

Fixed energy

Keep the thermal condition and other Ion-Implantation is all same except the doping dosage for MOSCAP.

Fixed dosage

Keep the thermal condition and other Ion-Implantation is all same except the doping energy for MOSCAP.

Fixed MOS capacitor

To form the same MOS capacitor, we use various gate oxide thicknesses with the different doping condition to make the same thermally grown oxide thickness.

RESULT AND DISCUSSION

Firstly, different doping energy shows the different thermally grown oxide thickness (the thermal condition and other Ion-Implantation is all same except the implantation condition for MOSCAP). As shown in Fig. 1, the thermally grown oxide is thicker when the doping energy decrease with the fixed doping dosage. When the energy fixed, the oxide is thicker with the doping dosage increase.

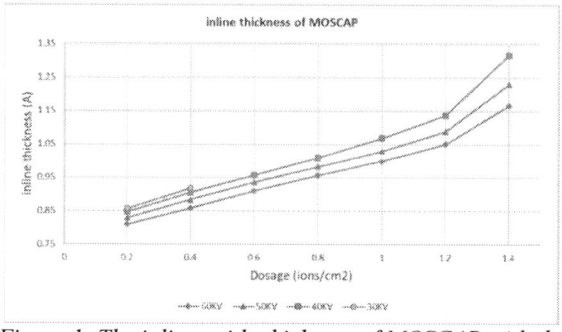

Figure 1: The inline oxide thickness of MOSCAP with the different IMP energy and dosage

When the energy fixed, the oxide is thicker with the doping dosage increase. We can get the dosage sensitivity when energy fixed, as shown in Fig. 2. When the IMP energy fixed, the inline thickness of MOSCAP is thicker about 1.7A with the dosage bigger.

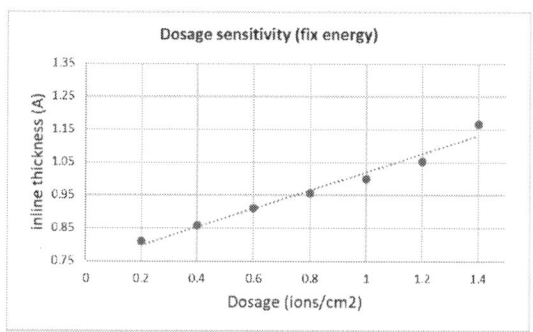

Figure 2: The inline oxide thickness of MOSCAP with the

different IMP dosage (fixed energy 60K)

When the dosage fixed, the oxide is thinner with the doping energy increase. We can get the energy sensitivity when dosage fixed, as shown in Fig. 3. When the IMP dosage fixed, the inline thickness of MOSCAP is thinner about 1.8A with the energy bigger.

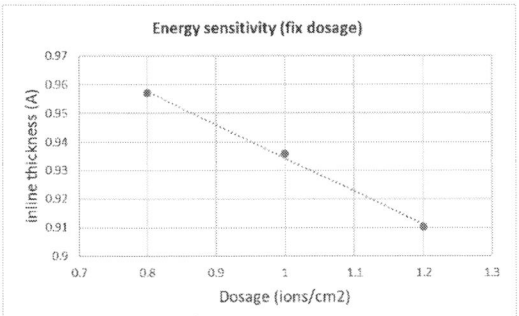

Figure 3: The inline oxide thickness of MOSCAP with the different IMP energy (fixed dosage 6E14)

MOS capacitors use the gate-to-bulk capacitance of a standard MOSFET feature with a high capacitance per unit area but a strong voltage dependence. The strong voltage dependence caused by different charge distributions in the accumulation, inversion, and depletion regions. And can be reduced by a series or parallel compensation technique then achieve a high linearity capacitor [3].

The MOS capacitors are tested by a constant voltage operating on the GATE separate in the inversion mode (COXI) and accumulation mode (COXA).

We calculate the range of capacitor by using COXI/COXA. And from the result shown in Fig. 4, we can get the conclusion that when the capacitor is larger, the range is also larger. But the requirement for capacitor and the range is a tradeoff (the smaller range is the better). So we need to get a balance.

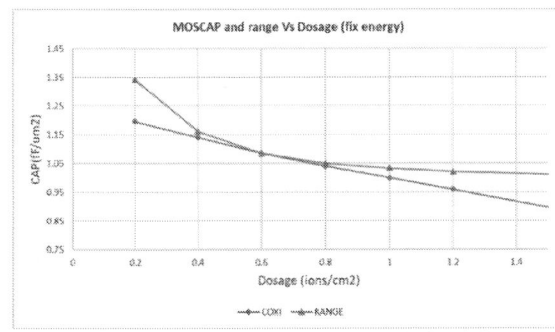

Figure 4: The MOSCAP and range with the different IMP dosage with fixed energy

Various gate oxide thicknesses have been examined

with the different doping condition to make the same MOS capacitor. And then we test the break down voltage of the gate oxide for the MOSCAP to confirm the oxide quality. The result shows that the break down voltage falling to 0V on some sites when the oxide shrinking down, and the map of the fail sites is random.

Suspect the interface quality is not so excellent. The oxide quality will be damaged if too much element into oxide layer. We do TEM to check the gate oxide profile, and the image shows that oxide corner rounding of the thinner condition is obvious worse than the baseline gate oxide condition.

CONCLUSION

In summary, through the analysis, the split and the result, it is clear that the thermally grown oxide is thicker when the doping energy decrease with the fixed doping dosage; the oxide is thicker when the doping dosage increase with the fixed doping energy. When the capacitor is larger, the range (which stands for the voltage dependence) is also larger. And as the gate oxide shrinking down, the break down voltage appears some abnormal, which suspect the interface quality is not so excellent and the corner rounding for the gate oxide is worse.

ACKNOWLEDGEMENTS

The authors would like to acknowledge the SMIC Logic team members. Without their sincere assistance and support and analysis and idea, it could not be succeeded.

REFERENCES

[1] FU-YUAN JIN. Abnormal Positive Bias Temperature Instability Induced by Dipole Doped N-Type MOSCAP, *2019*, ELECTRON DEVICES SOCIETY, VOLUME 7, 2019, pp. 897-901.

[2] Kyle M.Bothe. Capacitance Modeling and Characterization of Planar MOSCAP Devices for Wideband-Gap Semiconductors With High-k Dielectrics, 2012, IEEE TRANSACTIONS ON ELECCTRON devices, VOL. 59, NO. 10, pp. 2662-2666

[3] Thomas Tille. Design of Low-Voltage MOSFET-Only ΣΔ Modulators in Standard Digital CMOS Technology, 2004, IEEE TRANSACTIONS ON CIRCUITS AND SYSTEMS, VOL, 51, NO. 1, pp. 96-109

The Study on the Optimization and Electrical Behavior of NEDMOS Transistors for High-Voltage Power Applications

Chongkai Du*, Chenchen Qiu1, Jun Qian1, Chang Sun1

1Shanghai Huali Microelectronics Corporation, No. 568 Gaosi Road

Shanghai 201620, People's Republic of China

*Corresponding Author's Email: duchongkai@hlmc.cn

Abstract

For the application of high voltage (>40V) power supply, the optimization of electrical characteristics of extended drain metal oxide semiconductor field-effect transistors (EDMOS) is systematically studied. The key dimensions and doping conditions of 40V N-channel EDMOS (NEDMOS) are optimized through practical tests, computer optimization program, TCAD simulation and silicon experiment to achieve the target of electrical characteristics.

Keywords—NEDMOS; Electrical characteristics ; Device optimization ; TCAD;

INTRODUCTION

With the development of consumer electronics applications, more and more attention is paid to the research of power integrated circuit devices[1]. N-channel extended drain mos(NEDMOS) ,which has high voltage and high current driving ability, excellent compatible with CMOS technology, is mainly used in power management applications of various working voltages, providing excellent performance with low cost [2].With the improvement of lithography accuracy, the process node of NEDMOS is gradually increased from 130nm to 90nm, and then to bellow 55nm [3]. The miniaturization of device dimensions while maintaining a high breakdown voltage poses a severe challenge to the structure design and process optimization of NEDMOS [4].

In this paper, we investigate the high voltage (40V) behavior of NEDMOS. In order to combine accuracy and robustness, the computer code developed internally is used to construct an analysis mode and optimize the device dimensions and doping conditions [5].The correlation between device dimensions, doping conditions and electrical characteristics is studied. The optimum dimension and doping condition is obtained by computer optimization program, and is verified by TCAD simulation and silicon experiment.

EXPERIMENTAL DESIGN

The schematic of the NEDMOS is shown in Fig. 1. The device is implemented in a BCD process platform. Deep n-well isolates the device from p-type substrate. N-drift is high resistance area which can withstand high voltage. The channel is the region of p-body under poly, which is generated by large angle injection. Compare with traditional NLDMOS, source and bulk of NEDMOS are isolated by STI.

The main electrical characteristics of NEDMOS selected as the experimental objectives include the drain-to-source breakdown voltage(BVDS), the saturation current(Idsat), the specific ON-resistance(Rsp-on) and the threshold-voltage(Vtgm). The device dimension from A to I as shown in Fig. 1 and shallow doping conditions of p-body and n-drift are selected as the experimental variables. The values of each experimental variable are set reasonably. The device electrical data is obtained through silicon experiment.

In order to analyze the multi-objective and multi-variable optimization problem more accurately, a multi-dimensional model is built based on genetic algorithm training by using the internally developed software. The optimum dimension and doping condition is explored through automatic combination of machine learning. The prediction accuracy is evaluated by the TCAD process simulation analysis and silicon experiment.

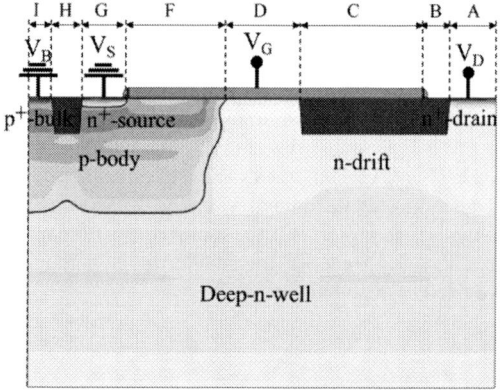

Figure 1 The cross-section of the n-channel EDMOS

RESULTS AND DISCUSSIONS

The absolute importance coefficient of the experimental variables is shown in Figure 2. The importance coefficient is automatically calculated through computer programs based on genetic algorithms and multi-objective optimization. The larger the absolute value of the coefficient of influence of the experimental variable, the more significant it is to the electrical characteristics. The device dimensions B, C, D, F, the p-body energy and dosage is more significant. The relationship between these significant variables and electrical characteristics is discussed in detail as below.

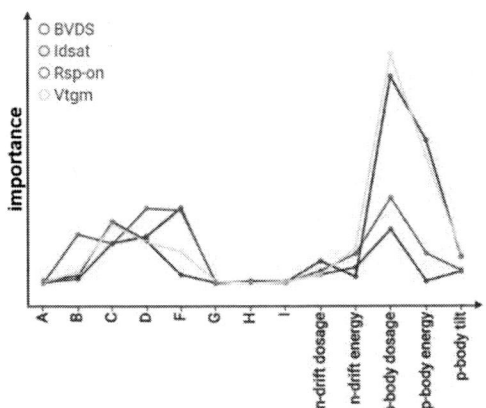

Figure 2 The absolute importance coefficient of the experimental variables on the electrical characteristics

Figure 3 shows the correlation between dimension B and the electrical characteristics. The increase of dimension B is little significant on BVDS, Idsat and Vtgm with shift about 1.8~4.5%. The increase of dimension B is more significant on Rsp-on with about 19.4% increase. The length of the n-drift area extension poly with high resistance increases, and the device total length increases lead to a large increase in Rsp-on.

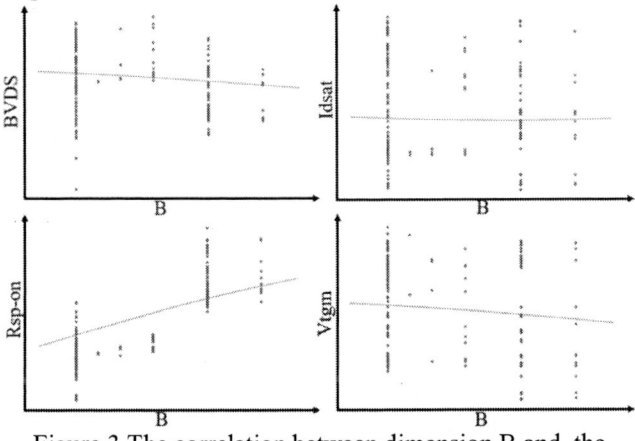

Figure 3 The correlation between dimension B and the electrical characteristics

Figure 4 shows the correlation between dimension C and the electrical characteristics. With the increase of size C, BVDS improve about 24.4%, Rsp-on increase about 15.8%, Idsat and Vtgm increases about 16~23.3%. With the length of the n-drift area overlap poly and STI with high resistance increases, the n-drift voltage sharing ratio increase, while the weak breakdown areas voltage sharing ratio decrease. However, high resistance area and the device total length increases lead to a large increase in Rsp-on .

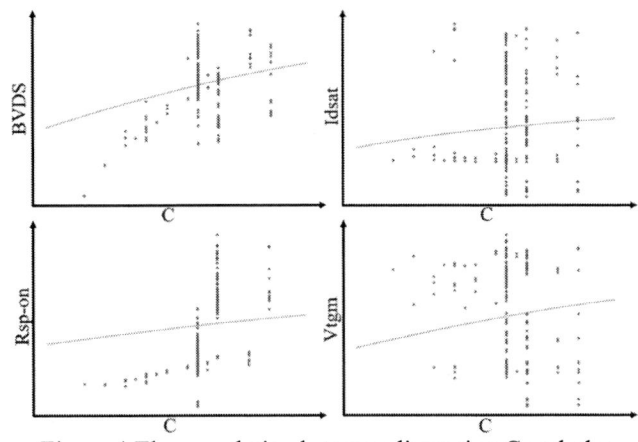

Figure 4 The correlation between dimension C and the electrical characteristics

Figure 5 shows the correlation between dimension D and the electrical characteristics. With the increase of dimension D, BVDS decrease about 16.9%, Rsp-on increase about 29.6%, Vtgm and Idsat increases about 16.3~18.7%. The dimension D increases may affect the voltage sharing, the reason leading to BVDS decrease needs further study. The length of the high resistance n-drift increases, and the device total length increases leading to a large increase in Rsp-on.

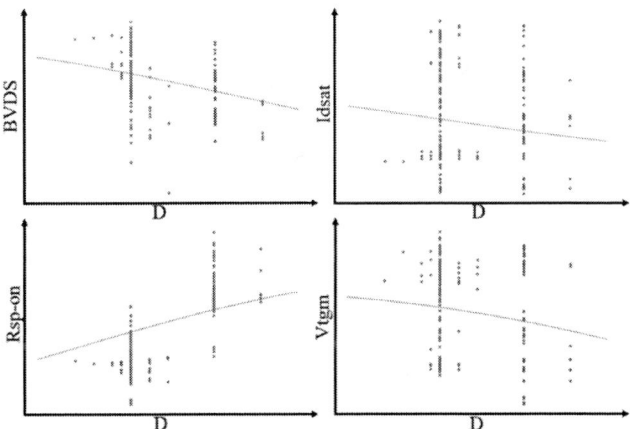

Figure 5 The correlation between dimension D and the electrical characteristics

Figure 6 shows the correlation between dimension F and the electrical characteristics.With the increase of dimension F, BVDS shift about 3%, Idsat decrease about 29.7%, Rsp-on increase about 28.7%, Vtgm increase about 12.7%.The increase of channel length F makes the channel is more difficult to conduct which leading to high Vtgm. The channel resistance increase leading to low Idsat and high Rsp-on.

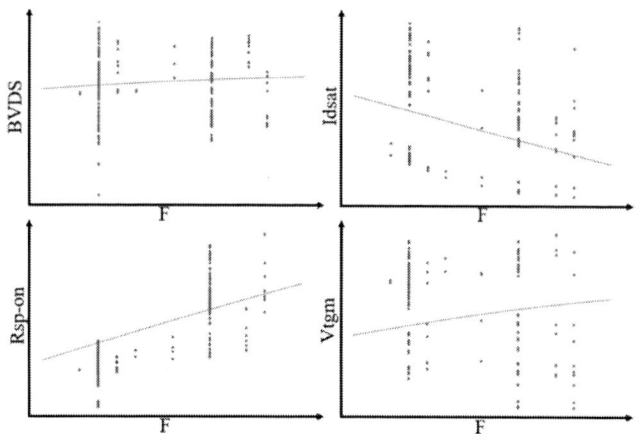

Figure 6 The correlation between dimension F and the electrical characteristics

Figure 7 shows the correlation between p-body dosage and the electrical characteristics. With the increase of p-body dosage, BVDS and Rsp-on increases about 21.2%~33.6%, Idsat decrease about 81.5%, Vtgm increase about 90.2%. Channel doping concentration affects the position of fermi-level in the semiconductor. If the heavier the channel P-type doping, the closer the Fermi level is to the valence band and the farther away from the intrinsic Fermi level, the more difficult it is for the surface to invert, the higher the voltage drop is needed for the semiconductor surface depletion zone to achieve the inversion, so increasing the channel doping results in an increase in the Vtgm. Channel inverse carrier reduction leading to the decrease of Idsat and the increase of BVDS and Rsp-on.

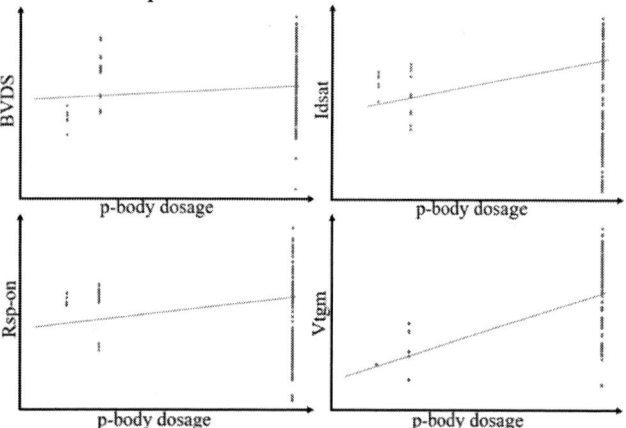

Figure 7 The correlation between p-body dosage and the electrical characteristics

Figure 8 shows the correlation between p-body energy and the electrical characteristics. With the increase of p-body energy, BVDS and Rsp-on shift about 0.7%~11.9%, Idsat increase about 56.4%, Vtgm decrease about 49.8%. With channel doping depth increase, channel surface doping concentration decreases which leading to the decrease of Vtgm and the increase of Idsat.

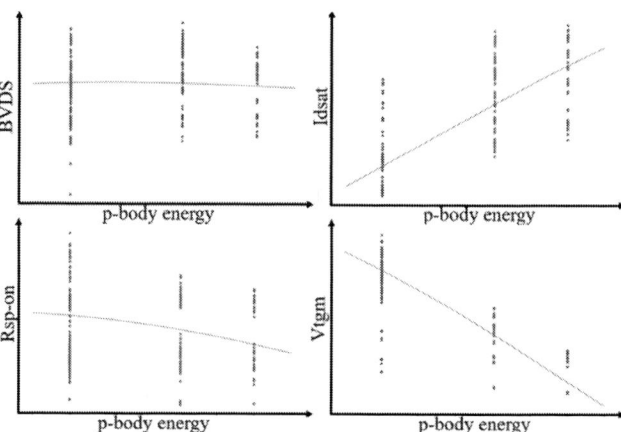

Figure 8 The correlation between p-body energy and the electrical characteristics

According to the silicon experiment split，much more values is assignd to the variables of the computer multi-dimensional model, the electrical characteristics prediction values are auto obtained. Compare with the electrical characteristics target and range, the optimum variable group is selected. From the TCAD simulation, the I_{DS}-V_{GS} and I_{DS}-V_{DS} curve of the optimum dimension and doping condition is shown in Figure9. The electrical characteristics obtained by TCAD simulation is excellently coincide with the computer multi-dimensional model prediction.

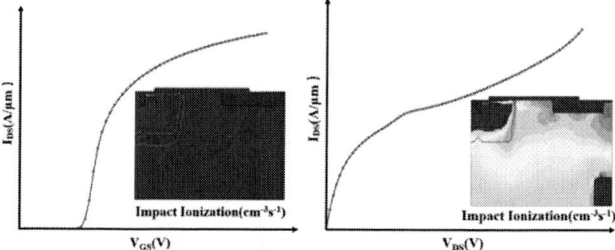

Figure 9 the TCAD simulation I_{DS}-V_{GS} and I_{DS}-V_{DS} curve of the optimum dimension and doping condition

The optimum dimension and doping condition is verified in silicon experiment. The silicon experiment data is well coincide with the prediction values obtained by computer multi-dimensional model，with about 6.2%~23.9% decrease of the BVDS, Idsat and Rsp-on and about 25.8% increase of Vtgm. The silicon experiment data of the optimum NEDMOS meets the target of electrical characteristics.

CONCLUSION

In this work, the high voltage about 40V behavior of NEDMOS is studied, and the main achievements are as follows: the dimensions and shallow doping condition of NEDMOS is optimized by the computer multi-dimensional mode, and are verified by TCAD simulation and the silicon experiment. Based on the computer multi-dimensional mode, the correlation between device dimensions, doping conditions and electrical characteristics is studied. The device dimensions B, C, D, F, the p-body energy and dosage is more significant. By the increase of device dimensions B, C, D and F, the high resistance area and the device total length increase leading to the increase of Rsp-on. The dimension C increase leading to n-drift voltage sharing ratio increase, and the BVDS is improved. With the increase of size F, the channel is more difficult to conduct which leading to high Vtgm. The high dosage and shallow energy p-body doping leading to channel inverse carrier reduction , which result in the decrease of Idsat and the increase of BVDS and Rsp-on.

REFERENCES

[1] C. Hung W.,F. Tu Y.,C. Chang T.etc., Abnormal On-Current Degradation Under Non-Conductive Stress in Contact Field Plate Lateral Double-Diffused Metal-Oxide-Semiconductor Transistor With 0.13-µm Bipolar-CMOS-DMOS Technology, IEEE Electron Device Letters, 2022, 43（5）, 769-772

[2] Lu L.,Lin F.,Ma S.etc., Hot-Carrier-Induced Reliability for Lateral DMOS Transistors With Split-STI Structures, IEEE Journal of the Electron Devices Society, 2021, 9, 1188-1193

[3] Lu L.,Ye R.,Liu S.etc., Hot-Carrier-Induced Reliability Concerns for Lateral DMOS Transistors with Split-STI Structures, 2021, 1-3

[4] Ma J.,Zhang L.,Zhu J.etc., Silicon-on-Insulator Lateral DMOS With Potential Modulation Plates and Multiple Deep-Oxide Trenches, IEEE Transactions on Electron Devices, 2021, 68（10）, 5073-5077

[5] Coyne E.,Geary S.,Brannick A.etc., Parasitic NPN and PNP Latch-Up Within a Single DMOS for High Voltage Reliability, IEEE Transactions on Electron Devices, 2020, 67（8）, 3291-3297

FABRICATION AND CHARACTERIZATION OF A NOVEL EMBEDDED MIRROR GATE SONOS

Ning Wang*, Kegang Zhang

Shanghai Huahong Grace Semiconductor Manufacturing Corporation, Shanghai 201203, China

*Corresponding Author's Email: Ningning.Wang@hhgrace.com

ABSTRACT

A novel embedded mirror gate SONOS is fabricated and its electrical performance is observed. Compared with traditional 2T SONOS, the mirror gate SONOS with two split-gate SONOS symmetrically distributed on sides of a selective gate can save over 25% of cell area. The fabrication of the embedded mirror gate SONOS is quite simple, and no additional mask is added as compared to 2T cells. Dedicated source line and shared source line array structure of the mirror gate SONOS are investigated. The E/P curve of both arrays reveal good Vt window. Furthermore, the shared source line array applied in a differential IP passes the functional test in a wide operating temperature of -40~85 °C.

INTRODUCTION

With the explosively development of industrial and consumer electronics, the market for non-volatile memories (NVM) for data and code storage are rapidly growing these years [1-2]. Large memory capacity, fast erase/program speed, long endurance, high reliability, wide operating temperature, as well as low cost are urgently needed for NVMs [3]. Among the various NVMs, Silicon-oxide-nitride-oxide-silicon (SONOS) memory products perfectly meet all the needs and are widely used in commercial [4-5].

The commonly used 2-Transistor (2T) SONOS device is composed of one memory gate (MG) transistor and one selective gate (SG) transistor, as shown in Fig.1(a). The SG is a regular NMOS, while the MG is identical to the SG only with the gate dielectric changed to an oxide-nitride-oxide (ONO) layer. This special structure of 2T SONOS gives it merit of easy fabrication and low cost, especially in embedded memories, such as bank cards and electricity/water/gas meters [6]. However, the size shrinking is a challenging problem for 2T SONOS.

Here we report a novel embedded mirror gate SONOS device, which can save over 25% of the cell size by facial self-aligned fabrication as comparing to 2T SONOS. In the mirror gate device, two SONOS MGs symmetrically distribute on sides of one SG, and the three gates share the same channel, as illustrated in Fig. 1(b). The fabrication is quite simple and only two mask is needed in cell loop. Dedicated source line and shared source line array structures of the mirror gate SONOS, as well as their pros and cons are investigated. Both two arrays show good

erase/program performance.

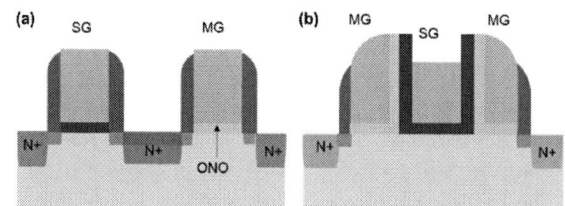

Fig.1. Cross-sectional schematics of (a) 2T SONOS and (b) mirror gate SONOS.

FABRICATION AND STRUCTURE

Fig.2 shows the main fabrication process steps of the embedded mirror gate SONOS. The SG in memory cell and N/P MOS in logic circuits use the same gate oxide and gate poly. The SG and logic NMOS share the same P-well. As shown in Fig.2(a), the deep N well (DNW) and N/P well are generated first, followed by the growth of SG gate oxide, SG poly, and a silicon-nitride (SiN). Then the SG gate is defined by lithography, and formed by the following dry etch, while the logic part is still covered by SiN, as shown in Fig.2(b). Next, thin tetra-ethoxy-silane (TEOS) oxide is deposited and etched to form an oxide spacer between the SG and latter MG, as presented in Fig.2(c). Here the MG Vt implant is implanted as the SiN acting as hard mask. The ONO dielectric layer and MG poly are deposited along the SG gate, as illustrated in Fig.2(d). In Fig.2(e), the MG poly and ONO are self-aligned etched to form the MG gate. The gate length of MG is defined by the thickness of the MG poly. Only two masks are used in the cell loop. One is to define the DNW, and the other is to define the SG gate. Then logic gates are formed and source and drain (N+/P+) are implanted, as presented by Fig.2(f). This embedded SONOS is fabricated at 90-nm technology node.

Since the MG gate is deposited along the SG gate without cutting, the mirrored MG gates are connected in one page, as illustrated in Fig. 3(a). During cell operation, the voltage is simultaneously forced on the mirrored two MGs. This leads to unique operation of mirror gate SONOS.

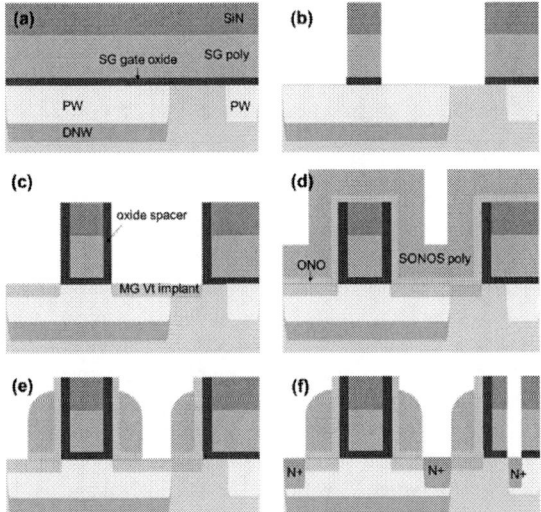

Fig.2 Main fabrication process steps of the embedded mirror gate SONOS. (a) DNW, N/PW formation, and deposition of SG oxide, SG poly and SiN, (b) SG photo and etch, (c) Oxide spacer formation and MG Vt implant, (d) ONO and SONOS poly deposition, (e) SONOS poly and ONO etch, (f) logic gate etch and N+/P+ implant.

Two types of mirror gate SONOS array are investigated, as show in Fig.3 (b) and (c). Both arrays are programmed and erased by page. After erase, all the data is "0". In program, the programmed data can be "1" or "0". Fig.3(b) shows the dedicated source line (DSL) array, in which each column has one bit line (BL) and one line (SL). The MGs in one page connect to one SONOS word line (WLS), while the SGs connect to another word line (WL). Table 1 shows the operation table of the DSL array. Vpos is a positive value ranging from 7V to 12V. Vp0 is a inhibit value about 3~6V. The specific value of Vpos and Vp0 depend on the thickness and composition of ONO dielectric. Vpwr is a positive value larger than the Vth of SG. Mostly the Vpwr can be 0.8~3V. The Vlim is about 0.6~1.6V. For erasing, the SONOS PW is forced Vpos while SONOS gate is 0V. Holes are injected to the ONO dielectric so that the mirrored MGs are all erased to "0". As for programming, the SONOS PW is 0V, the selected WLS is Vpos, and unselected WLS is 0V. For the units P "1", their BL and SL are 0V, and the Vpos gate voltage injects electrons into ONO, saving data "1". For the units P "0", their BL and SL are forced an inhibit voltage Vp0, which reduces the gate-channel voltage to "Vpos-Vp0", unable to make electrons tunneling and keeping "0". This also means the mirrored two MGs in one unit are programed "1/1" or "0/0" simultaneously. In read, the selected WL is forced Vpwr, and the BL is Vlim. The cell status can be read by the current of BL. If Ibl is larger than the reference current, the cell is "0" ("0/0"), otherwise, the cell is "1" ("1/1").

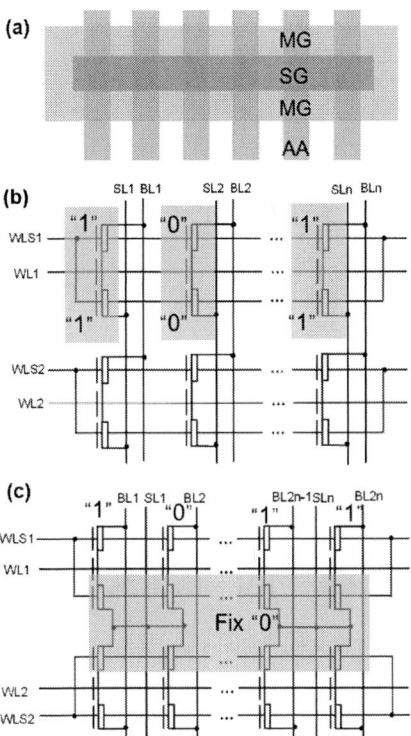

Fig.3 (a) Layout of the mirror gate SONOS. Array structure of the (b) DSL and (c) SSL mirror gate SONOS.

Table 1. Operation table of the DSL mirror gate SONOS

operate	cell	Vwl	Vwls	Vbl	Vsl	Vpw
Erase	select	0V	0V	float	float	Vpos
	unselect	0V	Vpos	float	float	
Program	P "1"	0V	Vpos	0V	0V	0V
	P "0"	0V	Vpos	Vp0	Vp0	
	unselect	0V	0V	0/Vp0	0/Vp0	
Read	select	Vpwr	0V	Vlim	0V	0V
	unselect	0V	0V	Vlim	0V	

Table 2. Operation table of the SSL mirror gate SONOS

operate	cell	Vwl	Vwls	Vbl	Vsl	Vpw
Erase	select	0V	0V	float	float	Vpos
	unselect	0V	Vpos	float	float	
Program	P "1"	0V	Vpos	0V	Vp0	0V
	P "0"	0V	Vpos	Vp0	Vp0	
	unselect	0V	0V	0/Vp0	Vp0	
Read	select	Vpwr	0V	Vlim	0V	0V
	unselect	0V	0V	Vlim	0V	

Fig. 3(c) illustrates the array of shared source line (SSL) mirror gate SONOS. In SSL, each two adjacent columns share one SL. This can reduce cell size compared to DSL. The operation table of SSL is mostly same as DSL, as shown in Table 2. The difference is in program, the selected SL is forced inhibit Vp0, while the BL voltage can be Vp0 (P "0") or 0V (P "1"). This means that the MGs near the common SL are always fixing "0", and the

979-8-3503-1101-3/23 $31.00 © 2023 IEEE

MGs away from the SL can be programmed to "1" or "0". As for reading, the "0/1" refers to "1", while the "0/0" refers to "0".

RESULTS AND DISCUSSION

The cross-sectional Transmission Electron Microscope (TEM) photo of the mirror gate SONOS is shown in Fig.4. The as prepared cell exhibits perfect symmetrical structure. The MGs and SG share one channel on horizon, and are isolated by an oxide spacer and ONO dielectric on vertical. Both the tops of MG and SG are deposited with silicide for better connection with contacts.

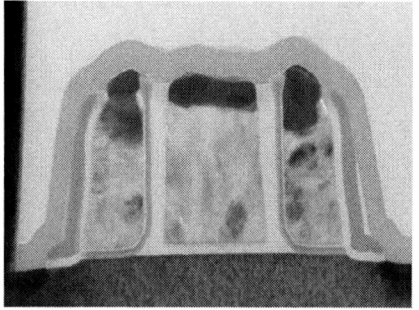

Fig. 4. TEM photo of the proposed mirror gate SONOS.

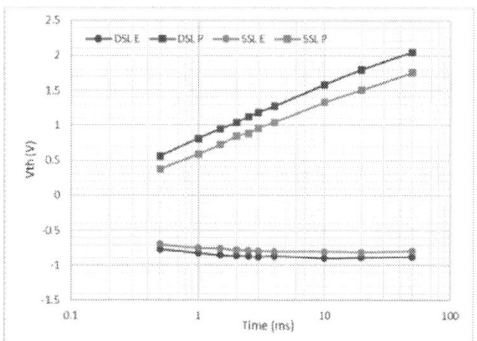

Fig. 5 E/P curves of the DSL and SSL mirror gate SONOS array under 8V E/P voltage.

The electrical performance of the DSL and SSL arrays are both investigated. Fig. 5 demonstrates the erase/program (E/P) curve of the two arrays under 8V E/P voltage. The abscissa is the erase or program time. The ordinate is the threshold voltage of the SONOS. The Vtp (threshold voltage in program state) curve of DSL is obviously higher than that of SSL, and this superiority of DSL in Vte (threshold voltage in erase state) is smaller. The "1" status of SSL is actually "1/0" and that of DSL is "1/1". The "0" status of both array is "0/0". In practical, both large Vt window and fast E/P speed are needed. Overall consideration, we choose erase time as 2.5ms and program time as 1.5ms. Under this condition, the DSL mirror gate SONOS shows a Vt window of 1.82V and the window for SSL is 1.51V. Both the Vt values are enough for application. Researchers can choose different arrays for different application considerations. Considering smaller cell size, we apply the SSL array in a 128Kb differential IP. In the IP functional test, specific test data are successfully erased, programmed and read at 25, 85 and -40 °C, respectively.

CONCLUSION

Fabrication, array structure, and electrical performance of a novel embedded mirror gate SONOS is demonstrated in this report. Applying simple self-aligned technology, no additional mask added, the mirror gate SONOS can save over 25% of the cell size. This small size is ascribed to the removal of common source between SG and MG, as well as the thinner MG defined by self-align etch. Array structures and operation table of DSL and SSL are investigated. The SSL can save cell size while the DSL can achieve a larger Vt window. At 8V E/P voltage, erase 2.5ms, and program 1.5ms, the DSL and SSL show Vt window of 1.82V and 1.51V, respectively. The DSL array employed in a differential IP passes the functional test in a wide operating temperature of -40~85 °C.

REFERENCES

[1] Ramkumar K, Prabhakar V, Keshavarzi A, et al. SONOS memories: Advances in materials and devices[J]. MRS Advances, 2017, 2(4): 209-221.

[2] Zhao C, Zhao C Z, Taylor S, et al. Review on non-volatile memory with high-k dielectrics: Flash for generation beyond 32 nm[J]. Materials, 2014, 7(7): 5117-5145.

[3] Charge-trapping non-volatile memories[M]. Cham, Switzerland: Springer, 2015.

[4] Xu Z, Liu D, Xiong W, et al. Investigation and three implementations for low power self-aligned 1.5-T SONOS flash device[C]//2018 China Semiconductor Technology International Conference (CSTIC). IEEE, 2018: 1-5.

[5] Hu J, Xu Z, Zhang K, et al. Improvement of Cell's Performance for Low Power Self-Aligned Split-Gate SONOS Memory Device[C]//2019 China Semiconductor Technology International Conference (CSTIC). IEEE, 2019: 1-4.

[6] Xu Z, Liu D, Hu J, et al. Device scaling considerations for sub-90-nm 2-bit/cell split-gate flash memory cell[J]. Solid-State Electronics, 2019, 152: 46-52.

INVESTIGATION OF A NEW DISTURB EFFECT IN THE AGGRESSIVELY SCALED DUAL-BIT/CELL SPLIT-GATE FLOATING-GATE FLASH CELL

Yintong Zhang[1,], Zhaozhao Xu[1,*], Dylan Zhou[2], Alan Shen[2], Fredric Liu[2], Ziquan Fang[2], Donghua Liu[2], Gordon Li[2], Wensheng Qian[2]*

[1]Huahong Semiconductor (Wuxi) Limited, Wuxi 214029, China
[2]Shanghai Huahong Grace Semiconductor Manufacturing Corporation, Shanghai 201203, China
*Corresponding Author's E-mail: Zhaozhao.xu@hhgrace.com ; Yintong.zhang@hhgrace.com

ABSTRACT

A new disturb effect had been experimentally shown in an aggressively scaled triple self-aligned split-gate floating-gate (FG) NOR-type Dual-bit/cell (NORD) flash array. The mechanism of this new disturb was demonstrated and studied by a series of test experiments and the assistant of technology computer-aided-design (TCAD) simulation. It was revealed that non-suppressed disturb effect was ascribed by the hot electron injection, which originated from the strong electronic field and high current density in the substrate under the edge of disturbed FG. Based on the simulated result, we proposed a method of self-aligned etching shallow trench contact on bit line (BL) to suppress this new disturb effect. It was illustrated that the current density in the substrate under the edge of disturbed FG was effectively reduced, which fundamentally decreased the possibility of injection.

Key Words—NOR-type Dual-bit/cell (NORD); Split-gate; Hot electron injection; Program disturb; Source-side injection

INTRODUCTION

Nonvolatile flash memory has become widely used in various integrated circuit for consumer-level and automotive-level electronic products.[1] Among these, split-gate flash cells, using source side injection (SSI) for programming, are becoming one of the dominate devices due to its high program efficiency.[2] However, with the shrink of cell pitch, one concern is the unintentional soft programming of non-selected bit during the program operation.[3,4] Considering the reliability of programming and erasing cells is the most basic requirement for flash products, adressing the disturb issue is the key to develop high performance aggressively scaled flash memory.[5,6]

In this article, a new disturb effect in dual-bit/cell split-gate floating-gate flash array was presented. A detailed investigation into the mechanism driving the degraded performance was carried out by a series of test experiments and Sentaurus TCAD tools from Synopsys technology. It was revealed that non-suppressed disturb effect was ascribed by the hot electron injection. Process improvement solutions were proposed of self-aligned etching shallow trench contact on BL and the expected effect was performed by TCAD simulation. The results illustrated that the current density in the substrate under the edge of disturbed FG was effectively reduced, which fundamentally decrease the amount of hot electrons and the possibility of injection.

DEVICE AND DISTURB CHARACTERISTICS

The cell structure of the split-gate floating-gate flash in BL direction is shown in Fig. 1, which contains one sharing select-gate (SG) [word line (WL)] and two physically separated bits. The cell utilizes SSI programming and poly-to-poly Fowler-Nordheim (FN) tunneling erasing.

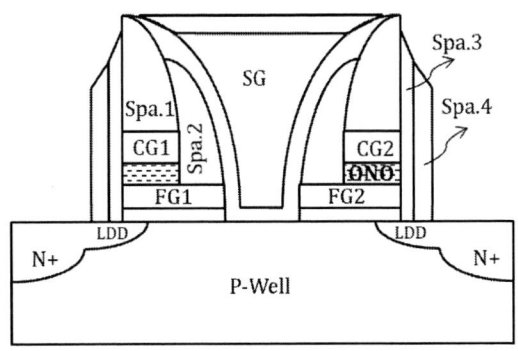

Figure 1. Schematic cross-sectional drawings of our NORD memory cell.

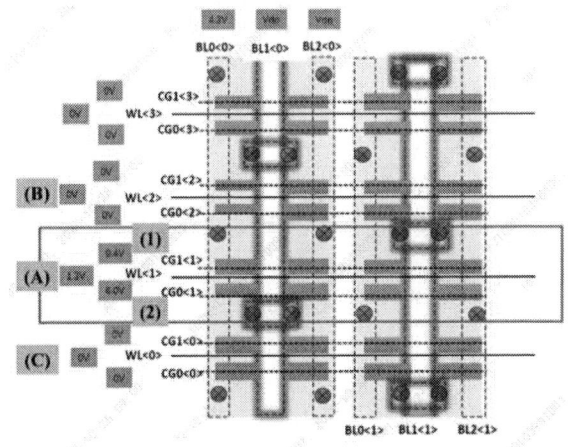

Figure 2. Array architecture and program biases schematic of the NORD flash array (Bit-1 of Cell-A is the selected bit for programming).

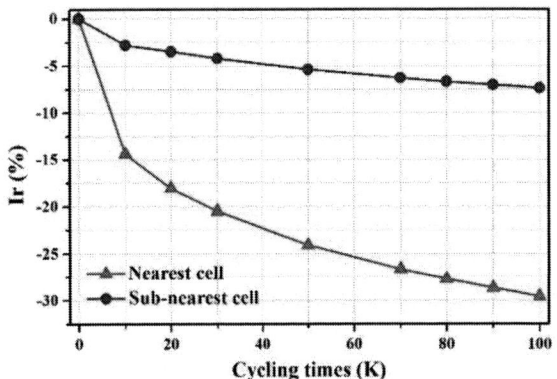

Figure 3. The decay degree of read current of erase state for the nearest bit and the sub-nearest bit in cycling tests.

As exhibited in Fig. 2, the cells were implemented in a NOR-type array, where the N^+ BL junctions sharing one contact with adjacent cell were connected as a cell-string. The bias conditions for programming operation are also illustrated in Fig. 2. To generate program current for programming the selected bit (bit-1 of Cell-A), a high voltage of 4.3 V and low voltage of V_{dp} were applied on BL-0 and BL-1, respectively. When performing program/erase cycling test, there was a new disturb effect that the nearest sector (Cell-B) which shares the same high voltage bias on BL with the selected one suffered disturb. In details, as shown in Fig. 3, with the cycling times accumulating, the read current density (Ir) of erase-state decreased continuously. After 100K times of cycling, the Ir of the nearest bit had a decay by about 30%, while that of the sub-nearest bit (Cell-C in Fig. 2) was only about 5%.

RESULTS AND DISCUSSIONS

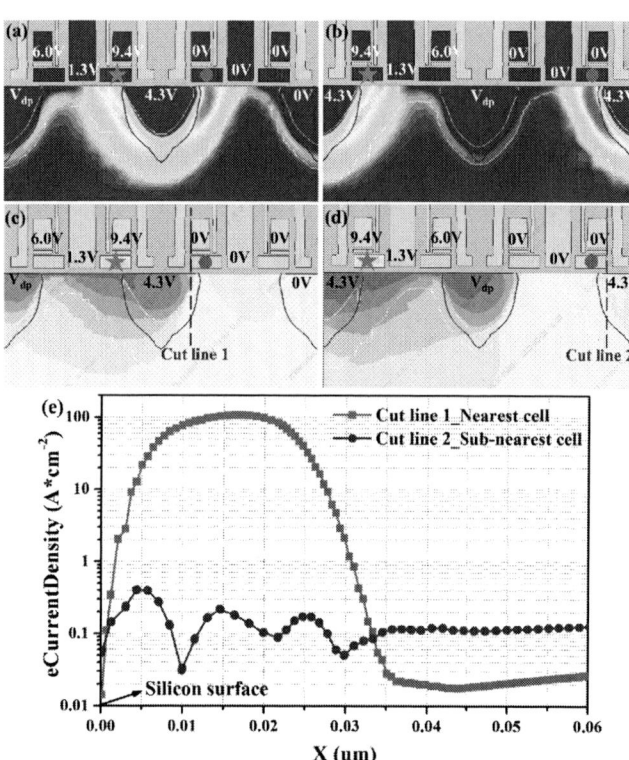

Figure 4. Simulated distribution of (a, b) electron temperature and (c, d) electron current density under program bias conditions for the selected bit (pink star) and its (a, c) nearest bit or (b, d) sub-nearest bit (red circle). (e) Electron current density along the cut lines for the nearest bit and sub-nearest bit.

To identify the program disturb mechanism, device simulations were firstly performed to understand electron states at various locations in the channel of the selected cell and its adjacent cell. Fig. 4 a illustrates the electron temperature distribution under programming conditions, where the selected bit and the nearest bit were marked by pink star and red circle, respectively. It's obvious that there is a piece of red area near

the BL junction under the FG of the nearest bit due to the high electric field, which indicates the energy of the electrons in this region is high[7]. A similar situation also can be observed in the sub-nearest bit as shown in Fig. 4 b. Additionally, the distributions of the electron current density in the substrate under programming condictions are exhibited in Fig. 4 c and d, which shows significant difference between the nearest bit and the sub-nearest bit. The electron current density along the cut lines were plotted in Fig. 4 e. It is clear that the electron current density in the substrate under the FG edge of the nearest bit is two orders of magnitude higher than that of the sub-nearest bit, which indicates there are more hot electrons in that region of the nearest bit. Therefore, the most likely cause of the disturb issue is the injection of unwanted hot electrons in the substrate into the nearest FG [8].

Base on the above simulation results, a series of test experiments were performed to evaluate the effect of BL voltage (V_{BL}) and programming current (I_{dp}) on the disturb. Fig. 5 indicates that the decrease of V_{BL} can mitigate the disturb characteristics due to the reduction of electrons energy[9]. What's more, Fig. 6 displays the disturb characteristic with different I_{dp} conditions. Tuning the I_{dp} to lower values also can weaken the degree of disturb, which can be explained by low current density and low possibility for electrons injection into FG[10].

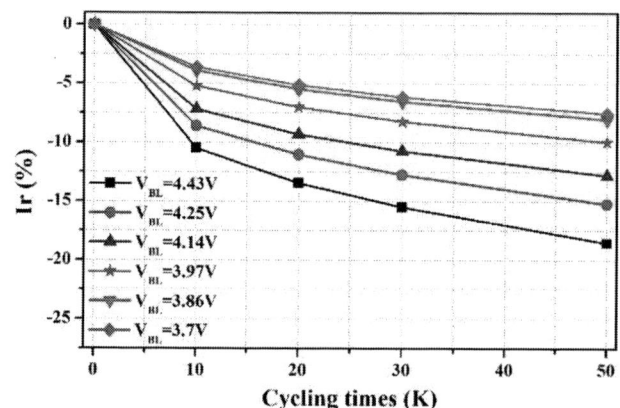

Figure 5. The decay degree of read current of erase state for the nearest bit in cycling tests under different BL bias conditions.

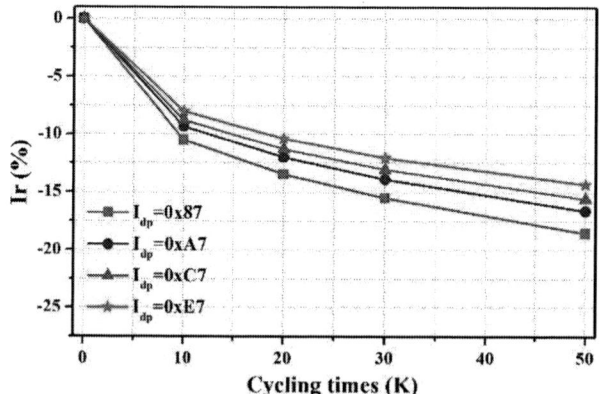

Figure 6. The decay degree of read current of erase state for the nearest bit in cycling tests under different program current condition. (The applied I_{dp} drecreases gradually from 0x87 to 0xE7.)

To further understand the kinetics of the disturb

mechanism, the disturb characteristics with negative voltage on the CG of nearest bit were investigated as comparison. The simulated distribution of electron temperature are exhibited in Fig. 7. With the increase of the negative voltage on CG (V_{CG}), the energy of the electrons in the substrate under the FG of the disturbed bit was significantly weakened. Additionally, test results are shown in Fig. 8. It can be seen that the disturb can be effectively confined since the V_{CG} is larger than -4V. These results supports the argument that the disturb mechanism was driven by hot electron injection, since a negative voltage on CG provide a vertical electric field and thus slows down the hot electron injection mechanism [11,12].

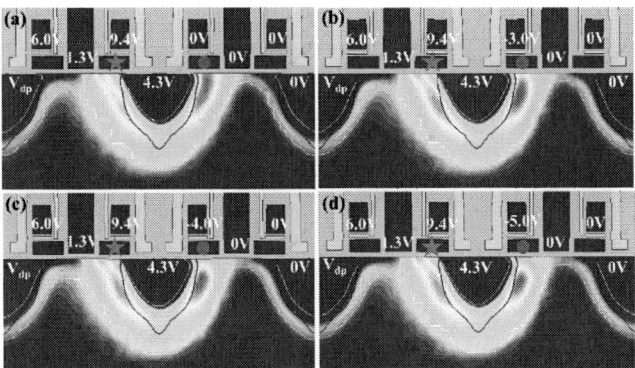

Figure 7. Simulated distribution of electron temperature under program bias conditions and the CG of the nearest bit biased on (a) 0.0V, (b) -3.0V, (c) -4.0V, (c) and (d) -5.0V. The selected bit is marked by pink star and its nearest bit is marked by red circle.

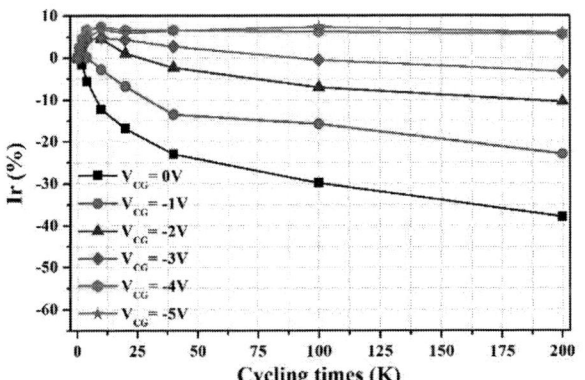

Figure 8. The decay degree of read current of erase state for the nearest bit in cycling tests under different CG bias conditions.

Therefore, combining the above simulation and test results, the program disturb mechanism of the nearest bit can be summarized as following. The program current from source side wasn't totally and timely collected by the high voltage on the drain side. The residual electrons were scattered by the lattice and accelerated by the electronic field in the depletion region of drain side[13]. Some of those electrons had a certain probability directly toward FG of the nearest bit. If they possessed sufficient energy to overcome the energy barrier, the injection of hot electrons into the FG would happen and caused disturb[14]. Therefore, the disturb effects can be attributed to two necessary reasons. One is the strong electronic field in the depletion region of BL, which provides energy for electrons. The other is high current density in substrate under the edge of

the nearest FG, which guarantees enough amount of electrons that fundamentally increase the possibility of injection.

Since the high voltage on BL have an influence on the program efficiency, which offers lateral electric field for SSI programming, decreasing the current density under the edge of the nearest FG is an alternative method to improve the disturb characteristic. As shown in Fig.9 a, a shallow trench was formed by self-aligned etch on the surface of N$^+$ BL and then silicide was formed on the surface of the trench as BL contact. The trench contact added area and depth of low resistance region, which effectively improve the ability of electron collection of BL. As displayed in Fig. 9 b, after the formation of trench contact, the current density under the edge of the nearest FG reduces by two orders of magnitude, which can fundamentally decrease the possibility of injection and significantly improve the anti-disturbance characteristic of the flash array.

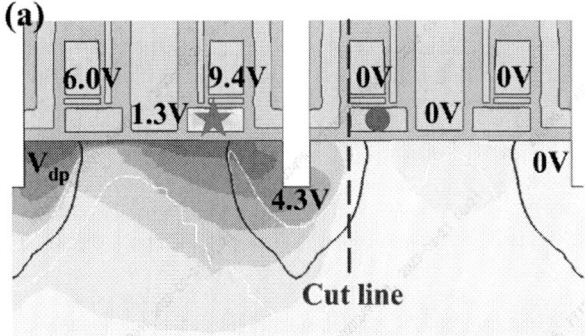

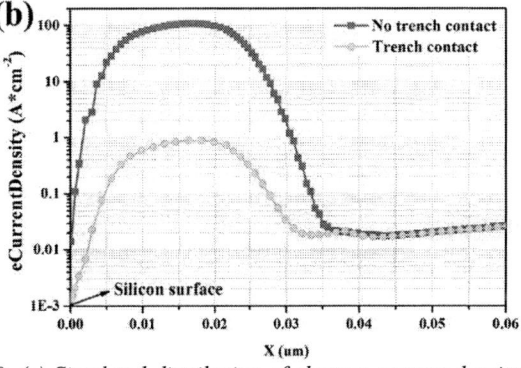

Figure 9. (a) Simulated distribution of electron current density after forming BL trench contact under program bias conditions for the selected bit (pink star) and its nearest bit (red circle). (b) Electron current density along the cut lines for baseline cell and the trench contact cell.

CONCLUSION

A new disturb effect was observed in NORD flash array when applying program-erase cycling test. A detailed investigation into the mechanism was performed by test experiments and TCAD simulation. The results suggested that the disturb was caused by the injection of unwanted hot electrons into the nearest bit, which were originated from the strong electronic field and high current density in the substrate under the edge of disturbed FG. Process improvement solutions were proposed of self-aligned etching shallow trench on the surface of N$^+$ BL and the expected effect of the trench contact was performed by TCAD simulation. The results illustrated that the current density in the substrate under the edge of disturbed

FG was effectively reduced, which fundamentally decreased the amount of hot electrons and the possibility of injection.

REFERENCE

[1] L. Fang, J. Gu, B. Zhang, W. R. Kong and S. C. Zou, "A Highly Reliable 2-Bits/Cell Split-Gate Flash Memory Cell With a New Program-Disturbs Immune Array Configuration", IEEE Transactions On Electron Devices, vol. 61, no. 7, pp. 2350-2356, 2014.

[2] T. Xu, Z. Cao, G. Han, H. Chen and H. Wang, "A Method to Solve Reverse Tunneling Disturb Issue for SuperFlash @ Memory".

[3] A. Datta, R. Asnani and S. Mahapatra, "A Novel Gate-Assisted Reverse-Read Scheme to Control Bit Coupling and Read Disturb for Multibit/Cell Operation in Deeply Scaled Split-Gate SONOS Flash EEPROM Cells", IEEE Electron Device Letters, vol. 30, no. 8, pp.885-887, 2009.

[4] H. H. Wang, C. W. Hung, H. H. Kuo, T. Yang, J. Huang, C. J. Hwang, Y. T. Lin, T. C. Ong and L. C. Tran, "A Novel Program Disturb Mechanism Through Erase Gate in a 110nm Sidewall Split-Gate Flash Memory Cell", pp. 43-47, 2007.

[5] T. H. Hsu, H. T. Lue, P. K. Hsu, T. H. Yeh, P. Y. Du, G. R. Lee, C. J. Chiu, K. C. Wang, and C. Y. Lu, "A Vertical Split-Gate Flash Memory Featuring High-Speed Source-Side Injection Programming, Read Disturb Free, and 100K Endurance for Embedded Flash (eFlash) Scaling and Computing-In-Memory (CIM)", IEEE International Electron Devices Meeting, no. 20, pp. 111-114, 2020.

[6] C. Bukethal, G. Tempel, R. Strenz and J. Power, "Analysis and Optimization of Program Disturb in Split-gate Cells using Source Side Injection and Impact on Further Cell Size Reduction", IEEE, 2013.

[7] V. Markov, K. Korablev, A. Kotov, X. Liu, Y. B. Jia, T. N. Dang, and A. Levi, "Charge-Gain Program Disturb Mechanism in Split-Gate Flash Memory Cell", IEEE IIRW Final Report, pp. 43-47, 2007.

[8] Y. Cai, X. Zhang and R. Huang, "Counter-Lightly-Doped-Drain (C-LDD) Structure for Multi-Level Cell (MLC) NOR Flash Memory Free of Drain Disturb", IEEE, pp. 207-210, 2011.

[9] Y. Cai, P. Tang, S. Qin and R. Huang, "Investigation of Source Potential Impacts On Drain Disturb in Nanoscale Flash Memory".

[10] Y. H. Wang, Y. S. Tsair, A. C. Kang, W. T. Chu, E. Chen, J.R. Shih, H.W. Chin and K. Wu, "Novel Cycling-induced Program Disturb of Split Gate Flash Memory", IEEE 45th Annual International Reliability Physics Symposium Phoenix, pp. 558-563, 2007.

[11] H. C. Sung, T. F. Lei, T. H. Hsu, Y. C. Kao, Y. T. Lin, and C. S. Wang, "Novel Program Versus Disturb Window Characterization for Split-Gate Flash Cell", IEEE Electron Device Letters, vol. 26, no. 3, pp. 194-196, 2005.

[12] V. Markov, and A. Kotov, "Program Disturb Induced by Interface-Trap-Assisted Field and Thermal Electron Emission in the Channel of Split-Gate Memory Cell", IEEE Transactions On Device And Materials Reliability, vol. 14, no. 2, pp. 672-680, 2014.

[13] C. Dunn, J. MacPeak, S. Bo, B. Kirkpatrick, B. Horning, T. Grider, C. O'Brien, S. H. Barna, A. Vigil, J. Nafziger, L. Preiss, K. DeShields, V. Markov, J. Kim, N. Do and A. Kotov, "Program Disturb Mechanism in Embedded SuperFlash Technology".

[14] H. Chen, Z. Xu, W. Xiong, J. Zhang, X. Xu, H. Wang, Y. Dang, J. Wang, W. Song, T. Tian, D. Liu, W. Qian and W. Kong, "Fabrication and optimization of aggressively scaled Dual-Bit/Cell Split-Gate Floating-Gate flash memory cell in 55-nm node technology", Solid-State Electronics, vol. 194, pp. 108316, 2022.

IMPORTANT PROCESS PARAMETER AND ITS SENSITIVITY CHECK BY VIRTUAL FABRICATION: CHANNEL HOLE PROFILE IMPACT ON ADVANCED 3D NAND STRUCTURE

Qingpeng Wang*, Pengfei Lyu, Lifei Sun, Yu De Chen, Cheng Li, Jacky Huang, Benjamin Vincent and Joseph Ervin

Coventor, Inc., A Lam Research Company, Shanghai, China
*Corresponding Author's Email: Qingpeng.Wang@lamresearch.com

ABSTRACT

A virtual DOE-based process sensitivity check was performed for two tiers of channel holes in a 3D NAND device. The channel hole tilt distance, twist angle, and their sensitivities to the visible area in silicon-oxide-nitride-oxide (SONO) punch process were analyzed. The results show that controlling the upper tilt distance is more important for offering a larger visible area. Also, a negative tilt distance difference and a closer twist angle between the upper and lower channel holes offer a good choice for providing larger visible punch area.

INTRODUCTION

To enhance memory cell density, 3D NAND devices with hundreds of layer pairs and multiple tiers are the mainstream in advanced flash memory development [1,2]. Due to its high aspect ratio, the channel hole (CH) often shows a bowing profile, smaller bottom CD, tilt distance, twist angle, and other non-ideal profile in the etch process. The inter-step impact between different tiers (upper tier and lower tier) can be dominated by these non-ideal profiles. Below, Figure 1(a) shows the upper channel hole profile, (b) the top area, (c) the middle/bow area, and (d) the bottom/joint area [3].

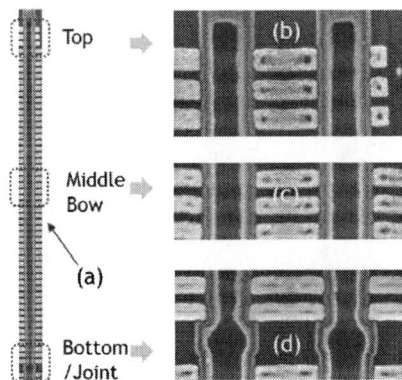

Figure 1: a typical channel hole profile in a 2 tiers 96L 3D NAND structure (Courtesy from TechInsights).

Performing channel hole profile splits on a silicon wafer is very expensive, and even impossible, due to the indirect and complicated link between process parameters and structure profile. Under such circumstances, virtual fabrication using SEMulator3D® offers an inexpensive and accurate way to clearly understand the inter-step profile correlations [4]. In this paper, we will use a simple channel hole punch through example to demonstrate how virtual fabrication can help detect the process margin and provide guidance for inline process spec control.

PROBLEM DESCRIPTION AND FLOW SETUP

In a two-tier 3D NAND structure, the upper and lower channel hole profile can be different, and this combination of different profiles leads to different top-down visible areas. The visible area is the key metric to determine whether the bottom SONO layer can be punched through to make sure the bit cells connect to the common source line. Here, we build two tiers of 64 oxide-nitride stacks and create the upper and lower channel holes by user-defined profiles, then use virtual metrology to measure the visible area of the structure as a metric of the SONO punch process window.

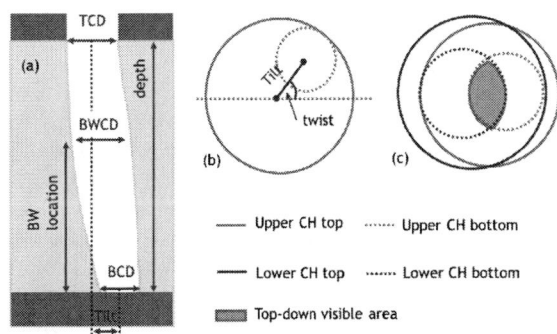

Figure 2: Schematic of a typical channel hole profile, with (a) cross sectional view, (b) top view, (c) overlap and visible area.

Figure 2 (a) shows a cross-sectional view of the channel hole profile with the major profile variables identified: top CD (TCD), bottom CD (BCD), bow CD (BWCD), and bow location (BW location). Figure 2 (b) shows a top view of the channel hole, showing the tilt distance and twist angle. Figure 2 (c) shows the overlap of the upper and lower channel holes; the overlapping visible area, which is a metric of the SONO punch process window, is shown in blue. Figure 3 shows the CH profile generated from the SEMulator3D® software. The top CD, bottom CD, bow CD, bow location, tilt distance and twist angle can be defined by the user in the model. Figure 3 (h) and (i) show the top and cross-sectional view of the two-tier channel structure.

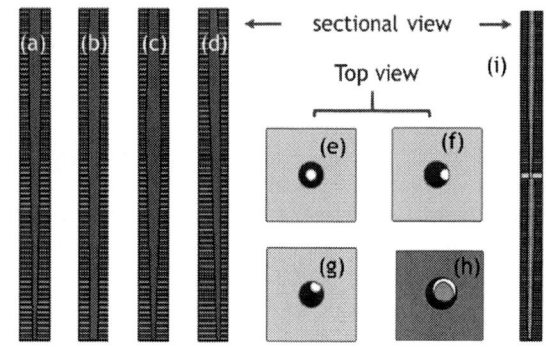

Figure 3: Different channel profiles showing (a) baseline, (b) small top CD and large bottom CD, (c) larger bow CD and lower bow location, (d) tilt, (e) baseline top view, (f) top view with tilt, (g) top view with tilt and twist, (h) top view with both tiers of the visible area, (i) two tiers of CH.

DOE DESIGN AND REGRESSION ANALYSIS

In the SONO punch process, the visible punch area can be impacted by top CD, bottom CD, tilt distance, twist angle, bow CD, bow CD location, etc. But which ones are most important? What are their sensitivities? What are their proper specification ranges? These are problems that need careful study, which can be done using a virtual DOE through the SEMulator3D® analytics feature. For simplicity, we fixed the top CD, bottom CD, bow CD, and bow location, and only varied the tilt distance and twist angle of both the upper and lower CH. Then we selected the visible area for output metrology data to demonstrate the virtual DOE functions. Table 1 shows the DOE range for each input factor. The software performed 500 Monte Carlo DOE runs using these settings.

TABLE 1: DOE VARIABLES AND SPLIT RANGE

Item	Min (nm)	Max (nm)	Range (nm)
TCD	80	80	0
BCD	40	40	0
BWCD	120	120	0
BW Location	2500	2500	0

Item	Min (nm)	Max (nm)	Range (nm)
Tilt Distance	0	30	30
Twist Angle	0°	360°	360°

Based on the virtual DOE results, the combined visible area with respect to each input variable was investigated first. It is obvious that a small lower and upper tilt distance results in a larger visible area (Figure 4 (a) (b) (c)). The maximum visible area can be obtained when the upper twist angle and lower twist angle are close to each other (Figure 4 (d)). These results make sense, confirming the correct logic of the virtual experiments.

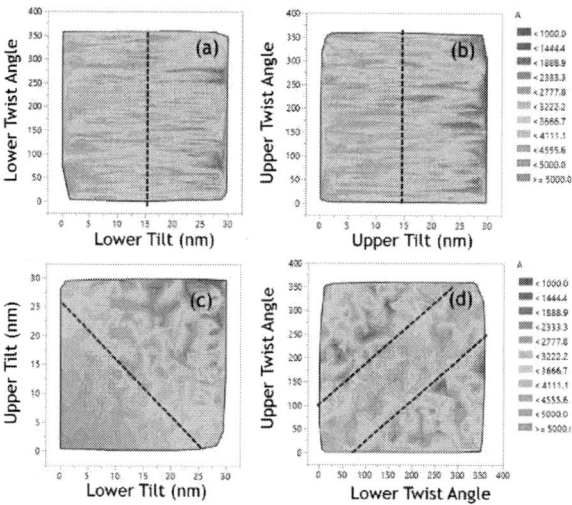

Figure 4: Visible area contour with respect to tilt distances and twist angles.

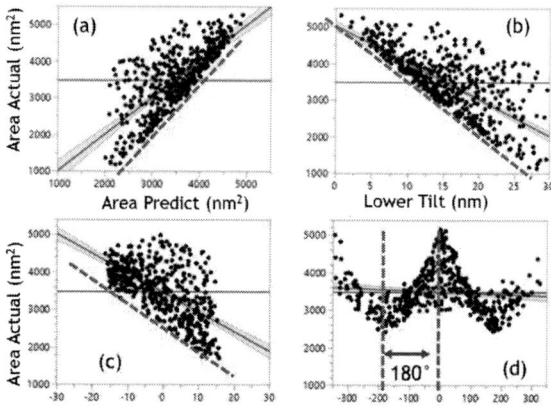

Figure 5: Visible area leverage plot with respect to tilt distance and twist angles.

To gain insight into the relationship between the

visible area and the input variables, we created three new independent variables: lower tilt distance, delta tilt distance (the difference between upper tilt and lower tilt), and delta twist angle (the difference between the upper and lower twist angles). Then regression analysis was performed to generate leverage plots that show the relationship between these variables and the resulting visible area. Figure 5 (a) shows the predicted visible area vs. the actual visible area. Figure 5 also shows the visible area vs. the lower tilt distance (b), delta tilt distance (c), and delta twist angle (d).

In the worst case, the visible area will decrease about 150 nm^2 when the lower tilt distance increase 1 nm. Also, the visible area will decrease 70 nm^2 when the delta tilt distance increase 1 nm. The analysis shows that if the lower tilt is unavoidable, letting the upper tilt grow smaller than the lower tilt will result in a larger visible area. (In other words, upper tilt distance control is more important than lower tilt distance control.)

The visible area has no correlation with a single twist angle, but is strongly correlated with differences in the upper and lower twist angle (delta twist angle). The closer the twist angles are close to each other, the larger the visible area will be. If twisting is unavoidable, ensuring twist in the same direction can mitigate the effects. In a real fabrication process, this may require using the same etch tool, chamber, and recipe for etching the lower and upper channel holes.

In this virtual DOE, only four variables were used to demonstrate the capabilities of virtual fabrication. More variables like the CDs can be added to the model to conduct a more complicated DOE for this case and offer a more well-rounded study. This methodology can also be used as a feed forward control system. Suppose the bottom tier is already made on a Si wafer; then the twist, tilt, and CD information of the bottom tier can be measured. Based on the metrology distribution of the bottom tier, virtual fabrication can help determine the corresponding specifications of the top tier to help produce a two-tier structure with a larger visible area.

CONCLUSION

In this paper, two tiers of a channel hole structure in 3D NAND were studied to demonstrate the powerful capability of virtual fabrication. The four factors incorporated into the model were the tilt distances and the twist angles of the lower and upper channel holes. These were selected as the DOE input variables, and a Monte Carlo analysis with 500 runs was used to study their sensitivities to the visible punch area and offer guidance for channel profile control. The results show that controlling the upper tilt distance is more important than controlling the lower tilt distance. If tilt and twist are unavoidable, a negative difference in tilt distances and similar twist angles for the upper and lower channel holes

can still provide a relatively large visible punch area. More variables, like the CDs, can be added to conduct a more thorough analysis, and this model can be combined with Si data to form a feed forward process control system to enlarge the process window.

REFERENCES

[1] Micheloni, R., et al. "Architectural and integration options for 3D NAND flash memories." Computers 6.3 (2017): 27.
[2] Parat, K., and A. Goda. "Scaling trends in NAND flash." 2018 IEEE International Electron Devices Meeting (IEDM). IEEE, 2018.
[3] TechInsights Toshiba 96-Layer 3D NAND tear down report.
[4] http://www.coventor.com/products/semulator3d

HF$_X$ZR$_{1-X}$O$_2$ FERROELECTRIC THIN FILM GRAIN SIZE TUNING VIA ANNEALING RAMP RATE ACHIEVING ENDURANCE >10^9 CYCLES, 2P$_R$ OF 40.6 µC/CM^2, WRITE VOLTAGE DOWN TO 1.5 V, AND SWITCHING SPEED OF 30 NS

Zhixiong Li[1,2#], Bing Zhou[1#], Jiawei Xu[1], Shuaihang Xu[1], Jun Lan[1], Quanzhou Zhu[1], Yingjie Zhu[1], Jie Li[1], Xuewei Feng[3], Mei Shen[1], Feichi Zhou[1], Longyang Lin[1], Yida Li[1]*
[1]Southern University of Science and Technology, Shenzhen 518055, China
[2]Shenzhen Longsys Electronics Co., Ltd, Shenzhen 518000, China
[3]Shanghai Jiao Tong University, Shanghai, China 200240
[#]Equal contributions, [*]Corresponding Author's Email: liyd3@sustech.edu.cn

ABSTRACT

The effect of annealing temperature ramp rate on the grain growth in Hf$_x$Zr$_{1-x}$O (HZO) ferroelectric thin film is investigated. Using X-Ray Diffraction (XRD) characterization, we find that the ferroelectric phase grain size and uniformity can be improved when the temperature ramp rate is decreased. When fabricated in a MIM capacitor structure, the device can achieve 2P$_r$ of ~40.6 µC/cm^2 better crystalline uniformity as compared to the other devices. Consequently, endurance exceeding 10^9 cycles, 30 ns switching speed, polarization voltages down to 1.5V, and multi-states were demonstrated, paving the path for its use in high performance storage and computing applications.

Keywords—Annealing Ramp Rate; Ferroelectric; HZO

INTRODUCTION

Ferroelectric memory demonstrates unique electrical properties, holding tremendous potential in the field of non-volatile storage and in-memory computing. [1-2] The recent demonstration of improved performance of Hf$_x$Zr$_{1-x}$O$_2$ (HZO) ferroelectric thin film, its versatility to be integrated in various device architectures such as capacitor and field-effect-transistors, and CMOS compatibility have generated increasing research interests in this area. [3] However, there are challenges in achieving a reasonably small ferroelectric phase grain size with good uniformity, and good interfacial properties, so as to enhance the device performance and reliability. [2]

HZO ferroelectric films are typically formed via atomic layer deposition (ALD), followed by a post deposition annealing step, aiding the transformation of the initial non-ferroelectric amorphous phase into the orthorhombic ferroelectric phase. [4] The annealing step provides a tuning knob to optimize the performance of HZO ferroelectric thin film, by affecting the growth dynamics of the ferroelectric phase grain. Various work investigating the effect of annealing on HZO ferroelectric thin film are mostly focused on parameters such as annealing temperature, time, and atmosphere. [5-7] On the other hand, the annealing ramp rate, a tuning knob that is widely used in CMOS industry is another promising approach. Hwang et al has reported on the modulation of dielectric constant and grain size of Al-doped HfO$_2$ thin film using a combination of fast temperature ramp +

cooling process with improved endurance. However, the temperature studied is high (800℃) and is not suitable for BEOL integration. [8] Apart from this, studies on the annealing ramp rate on HZO ferroelectric phase control has been limited.

In this work, we investigated the effect of annealing ramp rate using rapid thermal process (RTP) system on HZO ferroelectric thin film. Using X-Ray Diffraction (XRD) and Metal-Insulator-Metal (MIM) capacitor structure, the resulting grain size and electrical performance of the HZO thin film respectively were characterized. It is found that a larger temperature ramp rate results in a larger grain size but with poorer crystalline uniformity. Consequently, the slowest temperature ramp rate of 5℃/s shows the highest remanent polarization (2P$_r$) of ~40.6 µC/cm^2 better crystalline uniformity as compared to the other devices. Consequently, endurance exceeding 10^9 cycles, 30 ns switching speed, down to 1.5 V polarization voltages and multi-states operations.

EXPERIMENT

The process flow and schematic of the ferroelectric MIM capacitor is shown in Figure 1. The bottom electrode consisting of Ti/Pt (5/20 nm) was first defined using lithography and deposited using e-beam evaporator on a SiO$_2$ on Si substrate. Subsequently, the HZO layer was deposited using atomic layer deposition (ALD) by alternating single HfO$_2$ and ZrO$_2$ layers for a total of 100 cycles at 200℃; the Hf and Zr precursors used were TDMAH and TDMAZ respectively, while H$_2$O was used as the oxygen source.

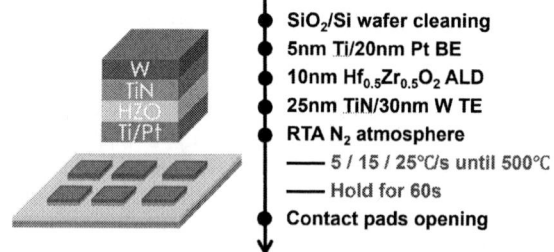

Figure 1: Process flow and schematic of the ferroelectric MIM capacitor. Devices with different post-metal annealing conditions as indicated are fabricated.

The deposited HZO film has a thickness of ~10 nm measured by ellipsometer. Thereafter, the top electrode consisting of TiN/W (25/30 nm) were deposited using sputtering. Finally, a post metal anneal was used performed to form the ferroelectric phase in the HZO thin film. Three different annealing temperature ramp rates were used – 5 ℃/s, 15 ℃/s, and 25 ℃/s, to a final temperature of 500℃, and held for a total time of 60 s. In all cases, the devices were annealed under N_2 atmosphere. MIM capacitors of sizes 20 x 20 μm² were fabricated. The electrical characteristics were measured using Keysight B1500A semiconductor parameter analyzer.

RESULTS AND DISCUSSION

Figure 2(a) shows the XRD spectra obtained of the annealed HZO films for the three different conditions as indicated. The orthorhombic phase peaks are located at 30.6 and 35.2 degrees. Using Scherrer's formula, the calculated grain size and crystallinity uniformity (FWHM) of each film (over 3 data points) is plotted out in Figure 2(b). It is seen that a smaller grain size (~10%) is resulted from a slower temperature ramp rate, and larger degree of crystallinity uniformity. The smaller grain size and higher degree of uniformity could be beneficial in a reduced device-to-device variation and control the V_{th} variation when integrated in a FEFET. [1]

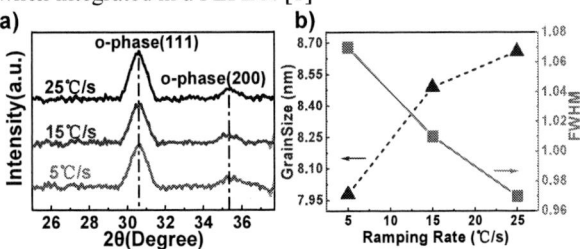

Figure 2: (a) XRD Spectra of the different HZO film annealed under different conditions as indicated, and (b) orthorhombic grain size calculated from (a) using Scherrer's formula as a function of annealing ramp rate.

Figure 3(a) shows the C-V curve of the fabricated capacitors. A typical butterfly curve is seen for all the different devices. However, we do observe a minor asymmetric C-V of higher and lower capacitance states (HCS/LCS) at DC 0V in all cases. The asymmetry can be attributed to excess oxygen vacancies at the bottom electrode interface. This induces the domain wall pinning effect to a down-polarized state even after positive sweep, resulting in more charges in the HCS as shown in the schematic in Figure 3(b). [9] Figure 3(c) shows the P-V curves measured of the different devices within the range of -4 V to 4 V. The $2P_r$ values as a function of the temperature ramp rate are extracted and plotted out in Figure 3(d). It is seen that a slower annealing temperature ramp rate favors a larger dielectric constant, with a higher degree of polarization.

In the subsequent discussion, we will focus on devices' performance annealed at the lowest annealing temperature ramp rate, i.e., 5°C/s. Figures 4(a) shows the polarization current and voltage vs time of the device, with

voltage waveforms used shown in the figure inset. In determining the switching speed, we used a pulse amplitude of 3 V. The polarization current (switching pulse current – non-switching pulse current shown in waveform in Figure 4(a)) is observed to increase when the applied voltage exceeds 1.5 V, indicating the minimum operating voltage of our device; Thereafter, we define the switching speed of our device at applied 3 V to be the time delay between when the maximum write voltage is applied to when the maximum polarization current is reached. A switching speed of 30 ns is extracted, noting that the sampling rate of the current measurement is 10ns/sample.

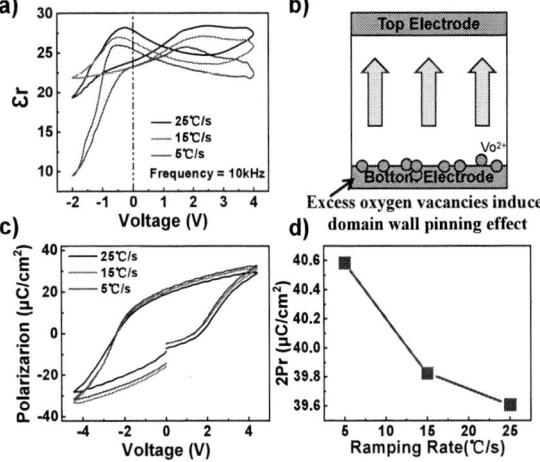

Figure 3: (a) Asymmetric small-signal C-V shows a memory window at DC 0 V with 10 kHz 100 mV AC small signals applied, (b) schematic illustration of excess oxygen vacancies inducing domain wall pinning, (c) P-V curves measured of the different devices within the range of -4 V to 4 V, and (d) extracted $2P_r$ and $2P_s$ values from (c) as a function of temperature ramp rate.

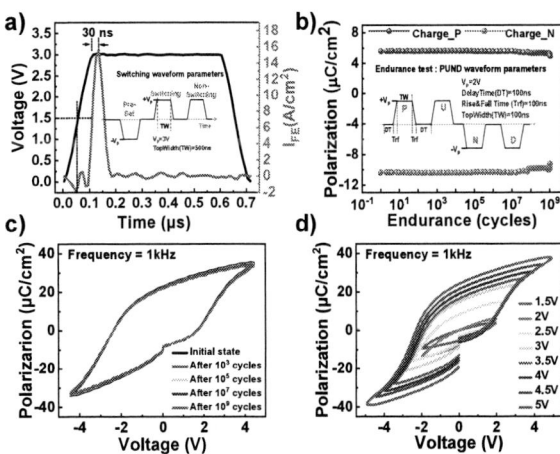

Figure 4: (a) Switching speed, (b) endurance, (c) P-V curves at different number of operating cycles, and (d) multi-states of the ferroelectric capacitor using different write voltages down to 1.5 V

In Figure 4(b), we measured the endurance of the device up till 10^9 cycles without any observable

979-8-3503-1101-3/23 $31.00 © 2023 IEEE

degradation, further supported by Figure 4(c) where the P-V curves at different number of operating cycles are shown. In Figure 4(d), we show the multi polarization states of the device by performing DC voltages sweep over different voltage range, down to a write voltage of 1.5 V. Table I benchmarks our fabricated ferroelectric capacitors with related works that studied the effect of various annealing approaches in tuning HZO ferroelectric thin film performance. Our proposed annealing approach using annealing temperature ramp rate modulation results in devices showing excellent overall performance including $2P_r$, switching speed, and endurance.

Table I: Benchmark of the performance of related HZO ferroelectric capacitors

Device type	2Pr (μC/cm^2)	Endurance (cycles)	Switching Speed (ns)
TiN/HZO/TiN[5]	31.8	>10^8	-
TiN/HZO/TiN[6]	38	>4x10^7	-
TiN/HZO/TiN[10]	49	-	66.5
Pt/HZO/TiN[11]	30	>10^8	-
This work (5°C/s)	40.6	>10^9	30

CONCLUSION

In summary, we have presented on the study of HZO thin-film grain size tuning via annealing temperature ramp rate approach. The ferroelectric grain size and uniformity is found to be affected by the ramp rate. It was found that a lower temperature ramp rate (5°C/s used in this work) results in a smaller grain size but higher degree of crystallinity uniformity, as characterized by XRD. We further show that ferroelectric capacitors fabricated with the smallest annealing temperature ramp rate can achieve an endurance exceeding 10^9 cycles, write voltages down to 1.5 V, fast-switching speed of 30 ns, and multi-states operation. Our results pave an alternative process tuning knob for further optimization of HZO ferroelectric thin film grain size and uniformity for high performance ferroelectric memory.

ACKNOWLEDGEMENTS

This work was supported by the National Natural Science Foundation of China (Grant No. 62174074, 62274081, 62104091, 52273246), Young Innovative Talent Project Research Program (Grant No. 2021KQNCX077), Zhujiang Young Talent Program (Grant No. 2021QN02X362) and the Shenzhen Fundamental Research Program (Grant No. JCYJ20220530115014032, JCYJ20220530115204009, JCYJ20190809143419448). We would also like to acknowledge the Core Research Facilities (CRF) at SUSTech for the facilities used, and the technical support provided by the staff and engineers at the CRF.

References

[1] R. Yang, "In-memory computing with ferroelectrics", Nature Electronics, vol. 3, pp. 237–238, May 2020.

[2] Asif Islam Khan, Ali Keshavarzi, and Suman Datta, "The future of ferroelectric transistor technology", Nature Electronics, vol. 3, pp. 588–597, October 2020.

[3] T. Mikolajick, U. Schroeder, and S. Slesazeck, "The Past, the Present, and the Future of Ferroelectric Memories", IEEE Transactions on Electron Devices, vol. 67, no. 4, pp. 1434-1443, April 2020.

[4] A. Chernikova, M. Kozodaev, A. Markeev, Yu. Matveev, D. Negrov, and O. Orlov, "Confinement-free annealing induced ferroelectricity in $Hf_{0.5}Zr_{0.5}O_2$ thin films", Microelectronic Engineering, vol.147, pp. 15-18, 2015.

[5] Y. Choi, C. Han, J. Shin, S. Moon, J. Min, H. Park, D. Eom, J. Lee, and C. Shin, "Impact of Chamber/Annealing Temperature on the Endurance Characteristic of Zr:HfO$_2$ Ferroelectric Capacitor", Sensors, vol.22, pp. 4087, 2022.

[6] D. Lehninger, R. Olivo, T. Ali, M. Lederer, T. Kämpfe, C. Mart, K. Biedermann, K. Kühnel, L. Roy, M. Kalkani, and K.Seidel, "Back-End-of-Line Compatible Low-Temperature Furnace Anneal for Ferroelectric Hafnium Zirconium Oxide Formation", Phys. Status Solidi A, 217: 1900840, 2020.

[7] Si Joon Kim, Jaidah Mohan, Harrison Sejoon Kim, Su Min Hwang, Namhun Kim, Yong Chan Jung, Akshay Sahota et al., "A Comprehensive Study on the Effect of TiN Top and Bottom Electrodes on Atomic Layer Deposited Ferroelectric Hf(0.5)Zr(0.5)O(2)Thin Films", Materials, vol. 13, no. 13, July 2020.

[8] Junghyeon Hwang, Minki Kim, Minhyun Jung, Taeho Kim, Youngin Goh, Yongsun Lee, and Sanghun Jeon, "Relatively Low-k Ferroelectric Nonvolatile Memory Using Fast Ramping Fast Cooling Annealing Process", IEEE Transactions on Electron Devices, vol. 69, no. 6, pp. 3439-3445, June 2022.

[9] Yuan-Chun Luo, Jae Hur, Tzu-Han Wang, Anni Lu, Shaolan Li, Asif Islam Khan, and Shimeng Yu, "Experimental Demonstration of Non-volatile Capacitive Crossbar Array for In-memory Computing", IEEE International Electron Devices Meeting (IEDM), pp. 1-4, 2021.

[10] Batzorig Buyantogtokh, Venkateswarlu Gaddam, and Sanghun Jeona, "Effect of high pressure anneal on switching dynamics of ferroelectric hafnium zirconium oxide capacitors", Journal of Applied Physics, vol. 129, no. 244106, June 2021.

[11] Yeriaron Kim, Jiyong Woo, Solyee Im, Yeseul Lee, Jeong Hun Kim, Jong-Pil Im, Dongwoo Suh, Sang MoYang, Sung-MinYoon, and Seung Eon Moon, "Optimized annealing conditions to enhance stability of polarization in sputtered HfZrO$_x$ layers for non-volatile memory applications", Science Direct, vol. 20, Issue 12, December 2020.

FINFET SOURCE/DRAIN PARASITIC RESISTANCE OPTIMIZATION BY TCAD SIMULATION

Tongtong Luan[1], Xinqi Liu[2], Yu Gu[1], and Xufeng Kou[1]*

[1]School of Information Science and Technology, ShanghaiTech University,
Shanghai 201210, China

[2]School of Physical Science and Technology, ShanghaiTech University, Shanghai 201210, China

*Corresponding Author's Email: kouxf@shanghaitech.edu.cn

ABSTRACT

This paper reports the strategies to reduce parasitic resistance for advanced FinFETs based on TCAD. The conductance integration method is applied to extract the channel resistance and parasitic resistance components. It is found that the resistance under the spacer (R_{spa}) accounts for approximately 50% of the Source/Drain (S/D) parasitic resistance, so reducing R_{spa} is more valuable for lowering parasitic resistance. We optimize the device structure and process after analyzing their impact on individual components of the parasitic resistance. Our simulation results show that the R_{spa} can be effectively reduced by increasing trench depth, decreasing the spacer thickness, decreasing proximity, and decreasing fin width. Furthermore, we evaluate the DC/AC performance of the device and propose the device structure with optimized process parameters for high device performance.

INTRODUCTION

For advanced FinFETs, the proportion of parasitic S/D resistance to the total device resistance increases with the device scaling. As the R_{ext}/R_{on} ratio increases, the effective current (I_{eff}) decreases, indicating worse device performance [1]. Therefore, reducing parasitic resistance is imperative for advanced FinFETs. To reduce parasitic resistance, we propose to subdivide the parasitic resistance and conduct an analysis of the process and structure parameters that affect each individual component of the parasitic resistance. The conductance integration method can accurately extract channel resistance and each parasitic resistance component with realistic and complex FinFET structure [2]. Currently, several approaches have been proposed to reduce parasitic resistance, including in-situ doping and laser spike annealing [1,3]. However, further optimization of device structure and process parameters is necessary for effective reduction of parasitic resistance and improvement of device DC/AC performance.

In this paper, the channel resistance and the parasitic resistance components will be extracted based on TCAD, including spacer resistance, contact resistance and S/D bulk resistance. Given that R_{spa} constitutes a significant fraction of the parasitic resistance, we will conduct a thorough analysis of the device structure and process parameters that affect R_{spa}, such as trench depth, spacer

thickness, proximity, and fin width. According to the DC/AC performance of the FinFET, we propose guidelines for optimizing the device structure and process parameter to reduce R_{spa} and improve the device performance.

METHOD

In this paper, a N-type FinFET simulation deck was developed by Sentaurus TCAD tools [4]. As shown in Figure 1, the relationship between the components of channel resistance and the individual components of S/D parasitic resistance is expressed as:

$$R_{on} = R_{ch} + R_{ext}, \qquad (1)$$

$$R_{ext} = R_{spa} + R_{sd} + R_c, \qquad (2)$$

where R_{on} is the total resistance, R_{ch} is channel resistance, and R_{spa} is the resistance under the spacer, which is between the channel and S/D epi. According to the measured data, R_c is set as the reference value 1000 Ω in TCAD, and R_{sd} is the total resistance in S/D epi. R_{ch} and R_{spa} can be determined by using the conductance integration method. Hence, using Eqs.(1) and Eqs.(2), R_{ext} and R_{sd} can be determined.

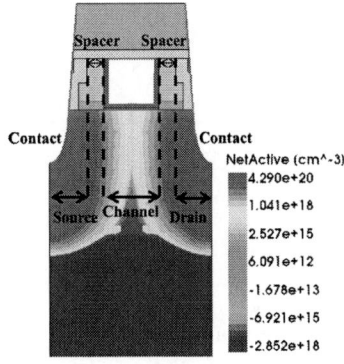

Figure 1: FinFET doping profile with resistance division

Historically, the parasitic S/D resistance has been extracted using the R_{on} versus the reciprocal of the overdrive voltage $1/(V_g-V_t)$ technique [5]. As shown in Figure 2, the curve of R_{on} and the individual components

979-8-3503-1101-3/23 $31.00 © 2023 IEEE

of R_{on} against $1/(V_g-V_t)$. Wherein, the device operates in the linear region (usually $V_d = 0.05V$), the total resistance is plotted against $1/(V_g-V_t)$ with TCAD are shown in black, and R_{ext} is determined by extrapolating to the value of R_{on} corresponding to $1/(V_g-V_t) = 0$. There is little difference between the parasitic resistance value extracted by the conductance integration method and the parasitic resistance value extracted by the R_{on} versus $1/(V_g-V_t)$ technique. Furthermore, solely the channel component of the resistance displays variation with respect to $1/(V_g-V_t)$, a trend that aligns with that of R_{on}. In contrast, the individual components of parasitic resistance remain largely unaffected by $1/(V_g-V_t)$, which agrees with the ideal expectations. The conductance integration method can not only extract the parasitic resistance accurately, but also subdivide the parasitic resistance into R_{spa}, R_{sd} and R_c, which provides effective aid for analyzing and optimizing the parasitic resistance.

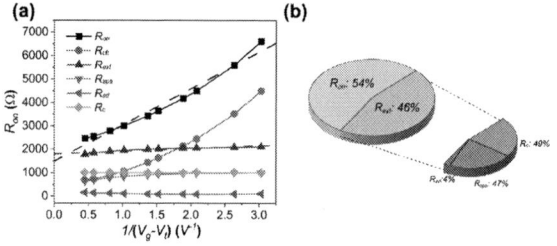

Figure 2: (a) The extraction of resistance (b) The breakdown of FinFET total resistance

As shown in Figure 2(b), the parasitic resistance accounts for nearly half of the total resistance, so it is imperative to reduce the parasitic resistance to improve the device performance. Specifically, R_{spa} constitutes a substantial fraction of the parasitic resistance, thus reducing R_{spa} is necessary to effectively reduce the parasitic resistance.

ANALYSIS AND OPTIMIZATION

In this section, the effects of changing the trench depth, the proximity, the spacer thickness, and the fin width on the channel resistance and the individual components of parasitic resistance are analyzed and investigated for the FinFET structure. Additionally, a comprehensive analysis of the DC and AC performance of the device is conducted to determine the optimal process range. In this paper, the AC performance of the device is characterized using C_{gd0}, where smaller capacitance corresponds to better AC performance. Under a unified benchmark of leakage current (I_{off}), the ratio of saturation current (I_{dsat}) to the target saturation current (I_{dsat_target}) is employed to characterize the DC performance of the device. A larger ratio indicates better DC performance, as the I_{dsat} is larger

for the same leakage current.

Trench depth

The trend of the variation of each resistance component with the trench depth is shown in Figure 3(a). A rise in trench depth can significantly decrease R_{ext}, which comprises of both R_{spa} and R_{sd}, while having a negligible effect on R_{ch}. The trench is used for epi to grow, and its depth determines the volume of epi. Increasing the trench depth results in an increase in the epi size and an enhancement in the diffusion from S/D. This leads to more carriers diffusing into the spacer region, but the carrier influence on the spacer region is limited due to the fact that only the volume of epi increases without changing the epi doping concentration. At the same time, R_{sd} is significantly reduced due to the increase in epi volume. Hence, a deeper trench is preferable for reducing the parasitic resistance. Nonetheless, excessively increasing the trench depth can lead to a reduction in the control capability of the fin, an increase in the leakage current, a negative impact on the DC characteristics of the device, and an increase of the capacitance. Therefore, it is important to note that the trench depth cannot be increased indefinitely, and to consider both the DC and AC performance of the device comprehensively. Figure 3(b) shows the variation curve of the DC performance and C_{gd0} of the device as the trench depth changes from 51nm to 60nm. It can be observed that the device performance is optimal in the part circled in red in the figure, suggesting that a trench depth of around 56nm is the best process point.

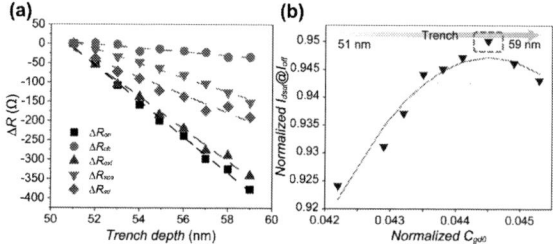

Figure 3: (a) Variation of each resistance component with the trench depth (b) Normalized $I_{dast}@I_{off}$ v.s. Normalized C_{gd0} with the trench depth varies from 51nm to 59nm

Proximity

The proximity refers to the distance of the gap between the channel and the epi. As shown in Figure 4, the proximity needs to be set appropriately to have a proper diffusion distance. On the one hand, if the proximity is too small, the impurity scattering in the channel region increases, resulting in declined channel mobility and decreased SCEs, such as higher I_{off} and higher drain-induced-barrier-lowering (DIBL), leading to the deterioration of the DC performance. On the other hand, if the proximity is too large, the carrier concentration that diffuses into the spacer region will decrease, causing an increase of R_{spa} and R_{sd}, while R_{ch} is almost unaffected by the vary of proximity. As shown in Figure 4(b), the region circled in red exhibits higher performance and smaller capacitance, indicating that the optimal process range of the proximity is between 8 nm to 8.5 nm.

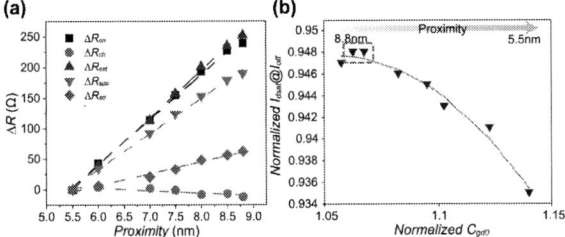

Figure 4: (a) Variation of each resistance component with the proximity (b) Normalized $I_{dast}@I_{off}$ v.s. Normalized C_{gd0} with the proximity varies from 8.8nm to 5.5nm

Spacer thickness

The spacer material used in this paper is SiOCN. As shown in Figure 5(a), the spacer thickness (T_{SiOCN}) has a significant impact on parasitic resistance. Upon reducing the spacer thickness from 10nm to 6.5nm, R_{spa} exhibited a decline of approximately 1.5% per nm. The decrease of the spacer thickness will lead to a decrease of proximity and an increase of epi volume increase, thereby resulting in increasing the carrier diffusion to the spacer region significantly. R_{spa} decreases greatly due to the decrease of carrier concentration in spacer region. And R_{sd} is also decreased due to the increase in epi volume, while R_{ch} slightly increases due to the enhanced impurity scattering in the channel region. As shown in Figure 5(b), decreasing the spacer thickness can effectively improve the device DC performance, but it also reduces the distance between the epi facet and poly, leading to the increase of C_{gd0}. Upon careful consideration of tradeoffs, the optimal spacer thickness is found to be around 7 nm.

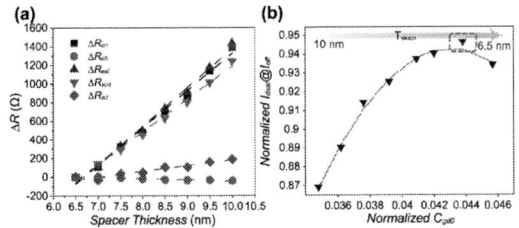

Figure 5: (a) Variation of each resistance component with the spacer thickness (b) Normalized $I_{dast}@I_{off}$ v.s. Normalized C_{gd0} with the spacer thickness varies from 10nm to 6.5nm

Fin width

As shown in Figure 6, reducing the fin width can effectively improve the electrostatic performance, thereby alleviating leakage currents and improving DC

performance of nanoscale devices. However, the narrow fin structure can lead to the degradation of mobility in channel region and increase the parasitic resistances, including R_{spa} and R_{sd}. Taking all factors into consideration, the optimal range for fin width is 5.1 nm - 5.3 nm.

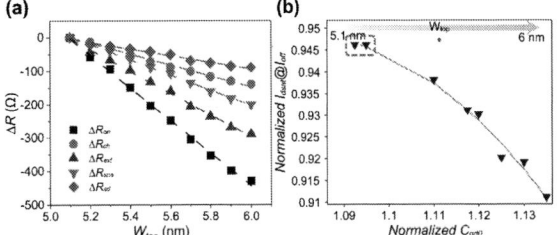

Figure 6: (a) Variation of each resistance component with the fin width (b) Normalized $I_{dast}@I_{off}$ v.s. Normalized C_{gd0} with fin width varies from 5.1nm to 6nm

CONCLUSION

In this paper, the conduction integration method is used to extract channel resistance and individual parasitic resistance. The result indicates that R_{spa} accounts for a considerable fraction in S/D parasitic resistance, and the R_{spa} can be effectively reduced by increasing trench depth, decreasing the spacer thickness, decreasing proximity, and decreasing fin width. Moreover, the effects of varying the trench depth, spacer thickness, proximity, and fin width on device performance were analyzed in terms of both DC and AC. The optimized device structure consists of a trench depth of approximately 56nm, a proximity range of 8nm-8.5nm, a spacer thickness of approximately 7nm, and a fin width of approximately 5.1nm-5.3nm.

ACKNOWLEDGEMENTS

This work was partially supported by Talent Laboratory, ShanghaiTech University.

REFERENCES

[1] Wu, Heng, et al. *IEEE International Electron Devices Meeting (IEDM)*. IEEE, 2018, pp.35-4.
[2] Narayanan, Sudarshan, et al. *Solid-State Electronics*, vol.123, 2016, pp. 44-50.
[3] Khaja, Fareen Adeni. *MRS Advances* 4.48, pp. 2559-2576.
[4] Sentaurus user manuals, P-2019.03.
[5] Chang, Yang-Hua, and Yao-Jen Liu. *IEEE ICSICT*. IEEE, 2010, pp. 1910-1912.

SIMULATION STUDY OF GATE-ALL-AROUND NANOSHEET DEVICES BASED ON SOI STRUCTURE

*Yangyang Hu[1], Tianxiang Zhao[1], Mengmeng Yang[2], and Jianhua Zhang[1], Kailin Ren[1] ***
[1]School of Microelectronics, Shanghai University, Shanghai 200444, China
[2]School of Computing, Shanghai University, Shanghai 200444, China
*Corresponding Author's Email: renkailin@shu.edu.cn

ABSTRACT

Owing to the excellent controllability of the Gate-All-About Nanosheet (GAANS) device, it's the most promising technology to achieve 3 nm technology node. However, there are few reports on electrical performances of GAANS devices on SOI substrates. In this article, the influences of device structural parameters on electrical characteristics are investigated by TCAD simulations. It's concluded that with the increase of channel thickness and interface trap density, the sub-threshold swing and the peak transconductance increases, owing to the increase in effective channel width and the degradation in carrier mobility at the interface. Nevertheless, SOI substrates are effective in suppressing gate leakage.

INTRODUCTION

From the 22 nm technology node, Fin-FET have become mainstream devices. With the continuous scale of CPP and cell height, the Fin-FET node was changed from 10 nm to 5 nm [1]. When reaching a technological node of 5 nm or more, two key problems arise which limit the scalability of Fin-FET structures. One is that the short channel effect in Fin-FET becomes serious; the other one is to further reduce the effective width caused by fins, which is difficult to compensate by increasing the fin height [2]. It has been proven that GAANS have excellent gate controllability and scalability, and have the best electrostatic control ability, the ability to control short channel effects[3]. Consequently, they are considered to be the most promising devices for scaling and extending the length and spacing of gates beyond the limits of Fin-FET. In addition, the structure of vertically stacked nanosheets allows GAA FET to achieve greater drive current, greater efficient channel width and higher DC performance while reducing unit size [4].

In preview works, the influence of various device structure parameters on the electrical characteristics of GAA nanosheet FET have been investigated, such as parasitic channel height[4], work function and channel doping concentration[5]. At the same time, there are also simulation studies on the influence of radiation effects on the DC and RF characteristics of GAA Nanosheet FET[6] and Effects of total ionization dose (TID) and electrostatic discharge on GAA Nanosheet FET[7]. The influence of the geometry of GAA Nanosheet FET and the thermal conductivity of various materials on the self-heating

effect[8]. However, few articles have examined the influence of interface trap density and channel thickness on the GAANS FET device based on the SOI structure by simulation.

In this paper, the influences of channel thickness and interface trap density on the performance of an SOI based device with the gate length of 15 nm, the nanosheet width of 10 nm, and the thickness of 5 nm are investigated by TCAD simulations. This will help eliminate the gate leak problem caused by the reduction in gate length.

TCAD SIMULATION APPROACH

The GAANS FETs mentioned in this document refer to an N-type MOSFET. Fig. 1 shows a transversal view of a GAANS device in two different directions. Correspondingly to Fig. 1, Table 1 shows some key structural parameters of the GAANS device, where the gate length is 15 nm, and the width, thickness, and spacing of the nanosheet are respectively 10 nm, 5 nm, and 9 nm. A GAANS device with a single fin and two nanosheets have been investigated by TCAD simulations with Silvaco. Fig. 2 (a) illustrates a simulated 3D structure diagram of the apparatus, and Fig. 2 (b) shows a transverse view of the X-Z plane of the 3D structure of the apparatus. Fig. 2 demonstrates that a three-gate channel is formed on the substrate, often called a parasitic channel.

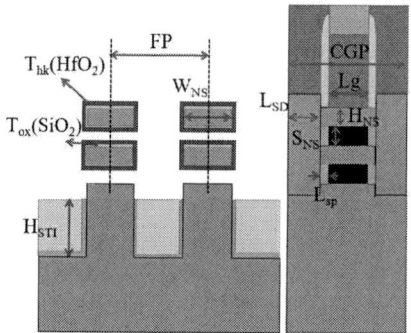

Figure 1: Schematic of cross section of the GAANS device

The simulation model a low field mobility model that characterizes carrier mobility, a Shockley-Read-Hall (SRH) composite model that characterizes phonon transitions, an Auger composite model that characterizes Auger transitions, and a Band-to-Band tunneling model that considers tunneling effects are added. The default parameters for Si are used for all four models.

TABLE I. KEY STRUCTURE PARAMETERS OF THE GAA NANOSHEET DEVICE

Parameters	Value
Contact Gate Pitch (CGP) (nm)	45
Fin Pitch (FP) (nm)	21
STI Height (HSTI) (nm)	50
Gate Lenth (Lg) (nm)	15
Source/Drain (LSD) (nm)	9
Spacer Length (Lsp) (nm)	6
Nanosheet Thickness (TNS) (nm)	5
Nanosheet Width (WNS) (nm)	10
Nanosheet Spacer (SNS) (nm)	9
High-k (HfO$_2$) Thickness (Thk) (nm)	1
Oxide (SiO$_2$) Thickness (Tox) (nm)	1

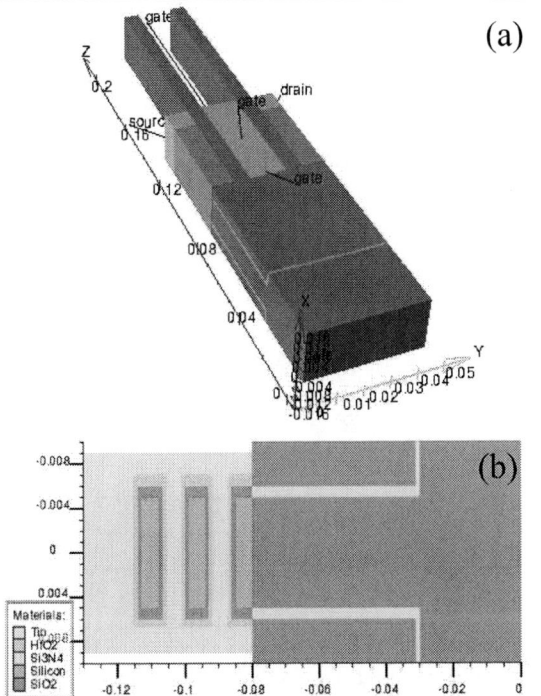

Figure 2: (a) 3D schematics of GAANS in the simulations; (b) Cross section of GAANS in X-Z plane

RESULTS AND DISCUSSION

Impact of channel thickness on the performance of GAA device

The channel thickness influences the actual channel width of the device. Therefore, while keeping the other size parameters unchanged, the electrical characteristics of GAANS at a thickness of 3 nm, 5 nm, 7 nm and 9 nm are investigated. Fig. 3 (a) observes the characteristic transfer curves for various nanosheet thicknesses, and Fig.

3 (b) observes the transconductance change with the gate voltage. It can be seen that as the channel thickness increases, the effective channel width increases, the threshold voltage and I_{on} both increase, and the control ability of the gate to the drain current increases.

Impact of interface trap density on the performance of GAA device

Due to the fact that SiO$_2$ surrounds all four surfaces of the GAANS device, the trap effect at the Si and SiO$_2$ interfaces have a significant impact on carrier transport. Here, the impact of interface trap densities on device characteristics have been discussed, at various trap density of 1.0×10^{11} cm^{-2}, 1.0×10^{12} cm^{-2}, 7.0×10^{12} cm^{-2}, and 1.0×10^{13} cm^{-2}.

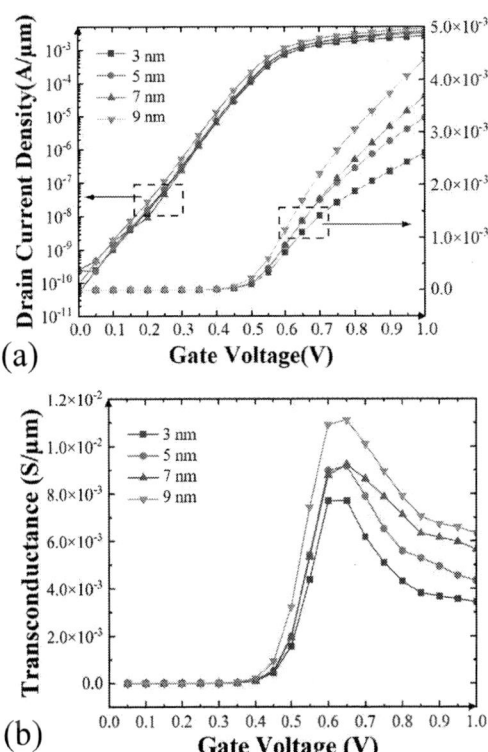

Figure 3: (a) transfer characteristic and (b) transconductance at different channel thickness values

Fig. 4 (a) illustrates the transfer characteristic curves for different interface trap densities, and Fig. 4 (b) illustrates the variation of transconductance with gate voltage. As shown, as the density of the interface traps increases, the threshold voltage increases. The reason is that the interface traps the electrons with a negative charge, which depletes the electrons in the lower channel. Therefore, a larger gate voltage is required to open the channel. When the interface trap density is 7.0×10^{12} cm^{-2}, the gate voltage can no longer linearly modulate the drain current, possibly because the carrier mobility near the trap surface decreases as the trap density increases.

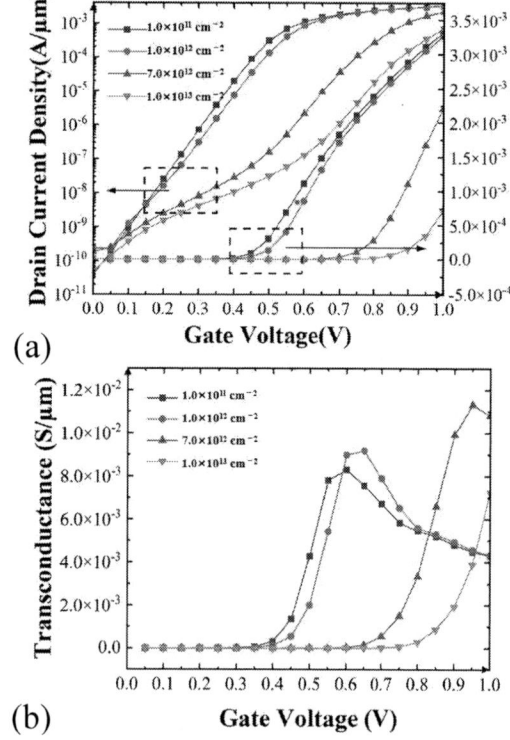

(a)

(b)

Figure 4: (a) transfer characteristic and (b) transconductance at various interface trap density values

Fig. 5 notes the sub-threshold swing under different conditions. It is found that with the increase of interface trap density, the sub-threshold swing increases. Although the channel thickness has changed, it is not enough to cause a significant change in the depletion region capacitance, so there is little difference in their sub-threshold swing.

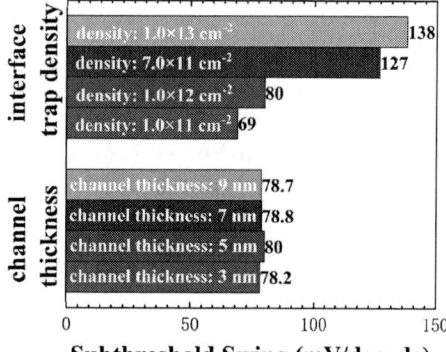

Subthreshold Swing (mV/decade)

Figure 5: sub-threshold swing

CONCLUSIONS

In this paper, the SOI based GAANS FET with different channel thicknesses and different interface trap densities have been investigated. The results show that with the increase of the effective channel width, the

control ability of the gate to the drain current increases, and both the turn-on current and the peak transconductance value increase. In contrast, the electrical characteristics of the GAANS device with a thickness of 9 nm are the best. The increase in the interface trap density increases the threshold voltage. When the interface trap density is too large, the gate voltage cannot linearly modulate the drain current, so the interface trap density should be controlled to be as small as possible.

ACKNOWLEDGEMENTS

This work was supported by the National Natural Science Foundation of China under Grant 62204150 and the Shanghai Natural Science Foundation under Grant 23ZR1422500.

REFERENCES

[1] J. -S. Yoon et al., "Performance, Power, and Area of Standard Cells in Sub 3 nm Node Using Buried Power Rail," in IEEE Transactions on Electron Devices, vol. 69, no. 3, pp. 894-899, March 2022.

[2] J. Ryckaert et al., "Enabling Sub-5nm CMOS Technology Scaling Thinner and Taller!," 2019 IEEE International Electron Devices Meeting (IEDM), San Francisco, CA, USA, 2019, pp. 29.4.1-29.4.4.

[3] R. Ritzenthaler et al., "Vertically Stacked Gate-All-Around Si Nanowire CMOS Transistors with Reduced Vertical Nanowires Separation, New Work Function Metal Gate Solutions, and DC/AC Performance Optimization," 2018 IEEE International Electron Devices Meeting (IEDM), 2018, pp. 21.5.1-21.5.4.

[4] Y. Choi et al., "Simulation of the effect of parasitic channel height on characteristics of stacked gate-all-around nanosheet FET," Solid-State Electronics, vol. 164, p. 107686, Feb. 2020.

[5] Y. Liu et al., "Vertical nanowire/nanosheet FETs with a horizontal channel for threshold voltage modulation," Journal of Semiconductors, vol. 43, no. 1, p. 014101, Jan. 2022.

[6] Y. Ma et al., "The influences of radiation effects on DC/RF performances of $L_g = 22$ nm gate-all-around nanosheet field-effect transistor, " Semiconductor Science and Technology, vol. 37, no. 3, p. 035010, Jan. 2022.

[7] J. Lee and M. Kang, " TID Circuit Simulation in Nanowire FETs and Nanosheet FETs," Electronics, vol. 10, no. 8, p. 956, Apr. 2021.

[8] L. Cai, W. Chen, G. Du, J. Kang, X. Zhang and X. Liu, "Investigation of self-heating effect on stacked nanosheet GAA transistors," 2018 International Symposium on VLSI Technology, Systems and Application (VLSI-TSA), Hsinchu, Taiwan, 2018, pp. 1-2.

LEAKAGE REDUCTION OF GAA STACKED SI NANOSHEET CMOS TRANSISTORS AND 6T-SRAM CELL VIA SPACER BOTTOM FOOTING OPTIMIZATION

Jiaxin Yao[1,†], Xuexiang Zhang[1,3†], Lei Cao[1,3], Junjie Li[1], Na Zhou[1], Qingkun Li[1,3], Yanzhao Wei[1,3], Yanna Luo[1,3], Jun Luo[1,3], Qingzhu Zhang[1,2,]*, and Huaxiang Yin[1,2,]*

[1] Integrated Circuit Advanced Process R&D Center, Institute of Microelectronics of Chinese Academy of Sciences, Beijing 100029, China

[2] Key Laboratory of Microelectronics Devices and Integrated Technology, Institute of Microelectronics of Chinese Academy of Sciences, Beijing 100029, China

[3] University of Chinese Academy of Sciences, Beijing 100049, China

*Corresponding Authors: yinhuaxiang@ime.ac.cn, zhangqingzhu@ime.ac.cn, [†]Equal contribution

ABSTRACT

In this work, the significant leakage reduction approach is proposed and investigated by critical spacer bottom footing (SBF) optimization for gate-all-around (GAA) stacked Si nanosheet (SiNS) transistors. The fabricated GAA stacked SiNS CMOS transistors and standard 6T static random-access memory (SRAM) cell have achieved reduced static-state leakage current by an order of magnitude and improved read static noise margin (RSNM) by +12.67% due to SBF optimization. The performance boosting origins from improving the gate and source/drain overlap doping profile via SBF optimization, and thus suppressing the band to band tunneling leakage current correspondingly. The proposed leakage reduction approach can inspire and broaden the ultralow power performance application for the state-of-the-art GAA nanosheet technology.

Keywords—GAA CMOS; Nanosheet; Leakage; SRAM-cell; Spacer bottom footing

INTRODUCTION

For the state-of-the-art GAA nanosheet technology, leakage reduction is crucial to realize low power and low cost application at the aggressive nanoscale dimension. At present, leakage reduction of gate-all-around (GAA) stacked Si nanosheet (SiNS) transistors is focused on the reducing leakage of sub-fin region via (1) novel device structure, (2) ground plane doping optimization. On the one hand, for device structure optimization, bottom dielectric isolation [1-3] and narrow sub-fin [4] schemes are successfully proposed to achieve low off-state current, which need very complicated integration efforts. On the other hand, for both novel isolation structure and ground plane doping [5], the leakage is only reduced sub-fin leakage. However, the reducing channel intrinsic leakage remain quite issues of overlap region between gate and source/drain area. In this work, the significant leakage reduction approach is proposed and investigated by

critical spacer bottom footing (SBF) optimization for GAA stacked SiNS CMOS transistors. The SBF impact on the off-state leakage and device/circuit performance will be investigated in the fabricated CMOS transistors and 6T-SRAM cell. The proposed leakage reduction approach is simple and effective, which can inspire and broaden the ultralow power performance application for the GAA nanosheet technology.

METHOD AND EXPERIMENT

GAA SiNS CMOS transistors and 6T-SRAM cell is started with the 8-inch P<100> bulk silicon wafer. As shown in **Fig. 1(a)**, the fabrication process is followed with previous reports and similar to general GAA process [5,6] with selective SiGe etch for channel release in replace metal gate (RMG) module. In order to control overlap region between gate and S/D, the critical SBF approach is designed and manipulated by controlling spacer over-etch ratio for GAA SiNS CMOS devices. Therefore, the corresponding lateral bottom spacer width is broadened nearest the first channel after the poly-gate formation shown in **Fig. 1(b)**. After NS channel selective release, HfO_2 and dual-metal work function layer are deposited, patterned and selective removed for CMOS RMG module. The CMOS devices for 6T static random-access memory (SRAM) cell demos are fabricated after the following modules of CMOS HKMG formation and backend interconnect processes.

SRAM is often used to directly exchange data with CPU registers, and its high-frequency operation makes its stability extremely important. The stability of SRAM is mainly limited by read static noise margin (RSNM) which is more vulnerable than write static noise margin (WSNM) and hold static noise margin (HSNM). In addition, for better system performance, the proportion of SRAM array in system on chips (SoCs) is increasing, and its static-state leakage current has become a problem that cannot be ignored. We tested RSNM and static-state leakage current for SRAM cells. The RSNM is defined as the maximum amount of noise voltage that can be tolerated by the cell

979-8-3503-1101-3/23 $31.00 © 2023 IEEE

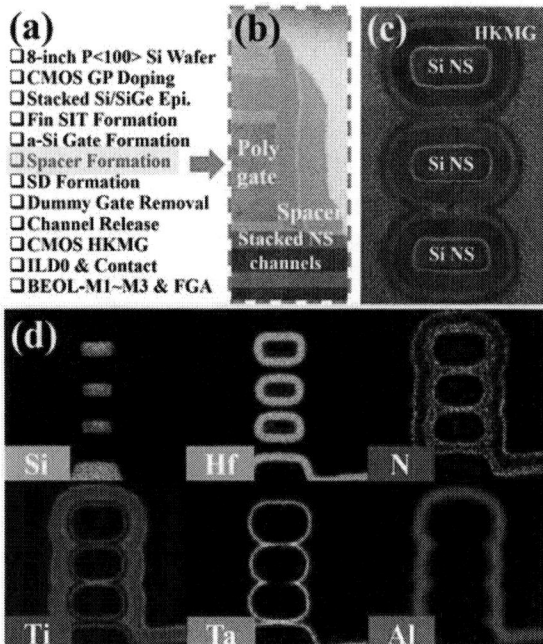

Fig.1 (a) The fabrication process flow of GAA SiNS CMOS transistors and 6T-SRAM cells; (b) Cross-sectional TEM results (along the Fin direction) for the critical spacer bottom footing optimization for CMOS transistors and 6T-SRAM leakage reduction; (c) Cross-sectional TEM results (across the channel direction) of the fabricated stacked GAA SiNS and (d) EDS mapping for Si, Hf, N, Ti, Ta. and Al distributions for CMOS HKMG structure in GAA SiNS pFETs.

while still maintaining the correct data and extracted by the maximum square definition method. Static-state leakage current is obtained by measuring the GND terminal current when the access (AX) transistors is closed and the bit lines are connected to VDD.

RESULTS AND DISCUSSIONS

The fabricated GAA SiNS CMOS transistor is shown in **Fig. 1(c)**. the cross-sectional TEM results (along the Fin direction) are demonstrated for the critical spacer bottom footing optimization for GAA CMOS transistors and 6T-SRAM cell leakage reduction. The corresponding EDS mapping of Si, Hf, N, Ti, Ta. and Al distributions for HKMG filling is depicted in **Fig.1 (d)**, including HfO_2, TiN, TaN, TiAlC layers. Three silicon nanosheet channel is vertically stacked and the conformal HKMG is deposited around SiNS for CMOS devices.

The IdVg characteristics for GAA SiNS CMOS transistors are shown in **Fig. 2(a)** and **2(b)**, respectively. Compared to the PMOS devices without SBF optimization, the off-state leakage is decreased for IdVg curves at Vdlin and Vdsat conditions. The subthreshold slope (SS) is

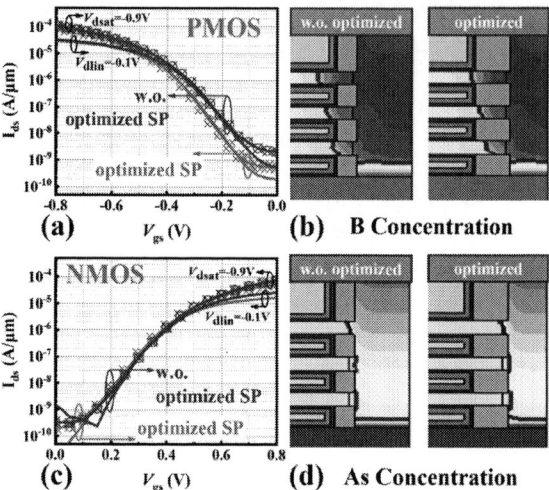

Fig. 2 Comparisons of device performance for GAA SiNS CMOS devices with SBF optimization. (a) PMOS IdVg characteristics, (b) comparison of gate and source/drain overlap boron doping profile for PMOS with and without the optimized SBF; (c) NMOS IdVg characteristics, (d) comparison of gate and source/drain overlap arsenic doping profile for NMOS with and without the optimized SBF.

optimized from 76.39 mV/dec to 71.15 mV/dec, which is accounted by doping profile change with the spacer bottom optimization and decreasing the lateral Boron diffusion into channel (**Fig. 2(b)**). For NMOS, the off-state leakage is suppressed, and SBF has nearly no impact on NMOS SS (~76 mV/dec), due to less Arsenic lateral diffusion into channel (**Fig. 2(d)**).

SBF impacts on Vtsat of GAA SiNS CMOS devices are also investigated. For NMOS, the statistical Vtsat (median) is lightly increased from 0.428V to 0.457 mV by SBF optimization approach (**Fig. 3(a)**). For PMOS, the Vtsat (median) is stable at 0.404V, while the standard deviation of Vtsat is greatly decreased by SBF optimization approach (**Fig. 3(b)**). The Ion-Ioff mapping for GAA stacked SiNS CMOS devices are shown in Fig. 3(c) and 3(d). The off-state leakage is decreased by a order of magnitude both for PMOS and NMOS devices (PMOS not shown here). The driving current of PMOS demonstrates no obvious degradation. NMOS driving current is lightly decreased due to the increase of overlap region and the corresponding lateral resistance.

SBF impact on the circuit performance is investigated on the fabricated GAA SiNS 6T-SRAM cell. The schematic of the standard 6T-SRAM cell and the fabricated circuit die is shown in Fig. 4(a). The 6T-SRAM cell consists of two pull-down (PD) NMOSFETs and two pull-up (PU) PMOSFETs to form two interlocked CMOS inverters, and two access (AX) transistors are controlled

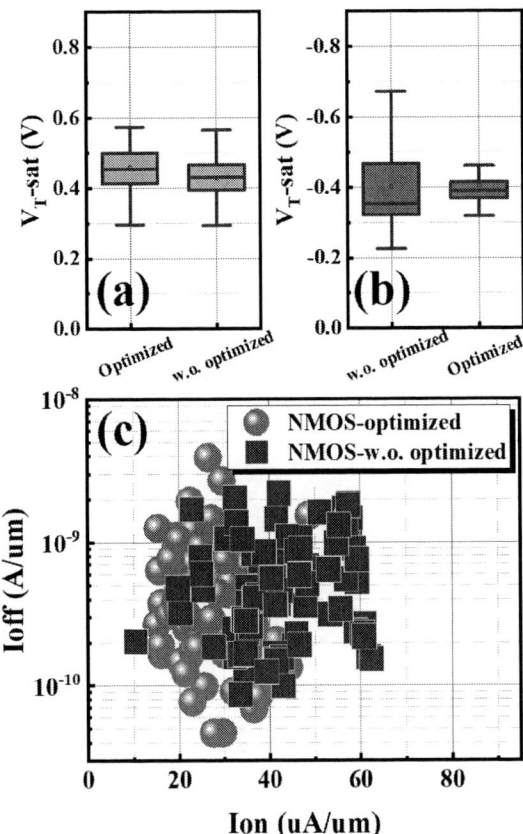

Fig. 3 SBF optimization impact on GAA SiNS (a) NMOS V_T, (b) PMOS V_T, and (c) Device leakage reduction and driving performance.

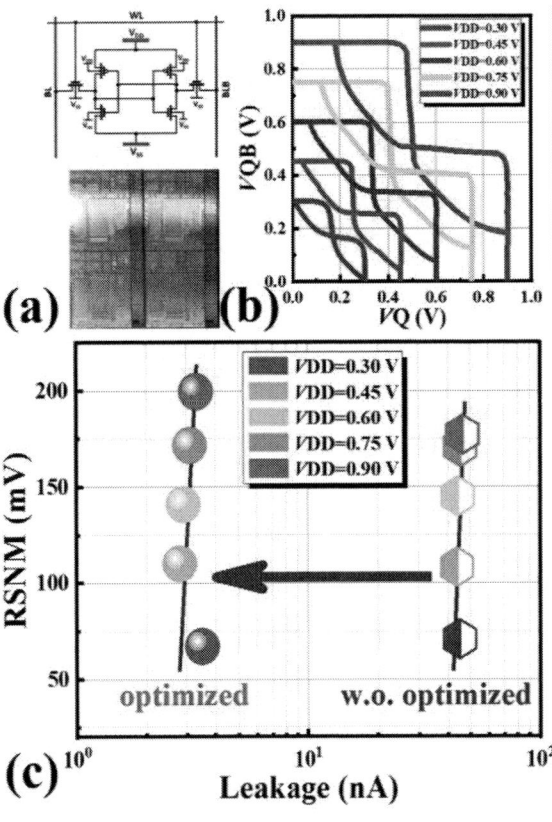

Fig. 4 (a) The schematic of 6T-SRAM cell, and the fabricated SRAM die, (b) the measured RSNM for fabricated 6T-SRAM cell with GAA stacked SiNS CMOS transistors for various operation voltages; (c) leakage and RSNM improvements for 6T-SRAM cell at different operation voltages by SBF optimization.

by word line (WL) and connected to complementary bit lines (BLs). Butterfly curves under various operation voltages is shown in **Fig. 4(b)** for fabricated 6T-SRAM cell with GAA stacked SiNS CMOS transistors. The measured RSNM (Read Static Noise Margin) is extracted from the maximum inscribed square of each butterfly curve in read states, as shown in **Fig. 4(c)**. 6T-SRAM cell have achieved reduced static-state leakage current by an order of magnitude and improved read static noise margin (RSNM) by +12.67% at 0.9 V operating voltage due to SBF optimization.

The SRAM cell mainly has two types of leakage paths. One is that VDD flows through the PU PMOSFET, the PD NMOSFET, and finally flows to GND. The second is that BL flows through the AX NMOSFET, the PD NMOSFET, and finally reaches GND. There are always off-state transistors on both paths, so the leakage current of SRAM cell is determined by the current allowed by the off-state transistors. Before SBF optimization, the off-state leakage of PFET is an order of magnitude larger than that of NFET. The leakage current of SRAM cell mainly comes from the

path that VDD flows through the off-state PU PMOSFET, the on-state PD NMOSFET, and finally reaches GND. After SBF optimization, the off-state leakage of PFET drops to the same order of magnitude as that of NFET, and the leakage current on the path of SRAM cell also drops significantly.

There are two main factors that affect the RSNM of the SRAM cell. One is the ratio of the AX NMOSFET to the PD NMOSFET (PD ratio), which has not changed before and after SBF optimization. The second is the flipping threshold of the inverter that makes up the SRAM cell. After SBF optimization, the driving ability of the PMOSFET is slightly stronger than that of the NMOSFET, which makes the flipping threshold of the inverter close to VDD/2, and finally increases the RSNM.

CONCLUSION

In this work, the significant leakage reduction approach is proposed and investigated by critical spacer bottom footing (SBF) optimization for GAA stacked SiNS transistors. The fabricated GAA stacked SiNS CMOS devices and standard 6T-SRAM cell have achieved reduced static-state leakage current by an order of magnitude and improved read static noise margin (RSNM) due to SBF optimization. The novel SBF optimization approach has great potential of leakage reduction for the advanced GAA nanosheet technology.

ACKNOWLEDGEMENTS

This work is supported by Strategic Priority Research Program of Chinese Academy of Sciences (Grant No. XDA0330300), Beijing Natural Science Foundation program (Grant No. 4224096), and Beijing nova program (Grant No. Z201100006820084). The authors would like to thank all staff at the Integrated Circuit Advanced Process Center, IMECAS, for their support in the fabrication on CMOS pilot line.

REFERENCES

[1] N. Loubet, et al., Stacked nanosheet gate-all-around transistor to enable scaling beyond FinFET,*2017 Symposium on VLSI Technology*, Kyoto, Jun. 5-8, 2017, pp. T230-T231.

[2] J. Zhang, et al., Full Bottom Dielectric Isolation to Enable Stacked Nanosheet Transistor for Low Power and High Performance Applications, 2019 IEEE International Electron Devices Meeting (IEDM), San Francisco, CA, Dec. 7-11, pp. 1-6-1-11-6-4.

[3] L. Cao, et al., Bottom Dielectric Isolation to Suppress Sub-Fin Parasitic Channel of Vertically-Stacked Horizontal Gate-All-Around Si Nanosheets Devices, *2022 China Semiconductor Technology International Conference (CSTIC)*, Shanghai, Jun. 20-21, 2022, pp.1-3.

[4] J. Gu, et al., Narrow Sub-Fin Technique for Suppressing Parasitic-Channel Effect in Stacked Nanosheet Transistors, *IEEE Journal of the Electron Devices Society*, 2021, vol.10, pp. 35-39.

[5] Q. Zhang, et al., Optimization of Structure and Electrical Characteristics for Four-Layer Vertically-Stacked Horizontal Gate-All-Around Si Nanosheets Devices, Nanomaterials, 2021, vol.10(3), pp. 646-1-646-16.

[6] J. Yao, et al., Record 7(N)+7(P) Multiple VTs Demonstration on GAA Si Nanosheet n/pFETs using WFM-Less Direct Interfacial La/Al-Dipole Technique, *2022 International Electron Devices Meeting (IEDM)*, San Francisco, CA, Dec. 3-7, pp. 811-814.

INVESTIGATION OF ELECTRICAL CHARACTERISTICS ON MORPHOTROPIC PHASE BOUNDARY OF $HF_{1-x}ZR_xO_2$ FOR DYNAMIC RANDOM ACCESS MEMORIES

Kun Zhong[1,2,4], Huaxiang Yin[1,2,4], Zhaohao Zhang[1,2,4*], Fan Zhang[1,2,3]*

[1]Key Laboratory of Microelectronics Devices and Integrated Technology
[2]Institute of Microelectronics of Chinese Academy of Sciences, Beijing 100029, China
[3]Xidian University Key Laboratory of Wide Bandgap Semiconductor Materials
[4]School of Integrated Circuits, University of Chinese Academy of Sciences, Beijing 100029, China
*Corresponding author's E-mail: zhangzhaohao@ime.ac.cn, yinhuaxiang@ime.ac.cn;

ABSTRACT

This paper investigates the properties of different Zr compositions in $Hf_{1-x}Zr_xO_2$ films. Due to the morphotropic phase boundary (MPB) between the orthorhombic ferroelectric phase and the tetragonal anti-ferroelectric phase, a maximum κ value of 55 was obtained for 9 nm $Hf_{0.4}Zr_{0.6}O_2$ films. Furthermore, the sufficiently low leakage current of the capacitors was confirmed. Finally, a promising switching endurance of up to 5×10^9 cycles was also performed, which becomes one of the driving forces for next-generation dynamic random access memories (DRAM) cell capacitor dielectric materials.

INTRODUCTION

For further scaling of DRAM, smaller dielectric film thicknesses or equivalent oxide thicknesses will soon become necessary. Numerous high-κ dielectric materials such as rutile-structured TiO_2 ($\kappa \approx 100\text{-}140$), Al-doped TiO_2 ($\kappa \approx 60\text{-}100$), and $SrTiO_3$ ($\kappa \approx 150\text{-}200$) have been reported [1-4]. However, on the one hand, the leakage current increases rapidly as the film thickness decreases and therefore does not reduce the EOT as expected. On the other hand, these films are incompatible with conventional TiN electrodes primarily due to the interfacial reaction that creates the low-κ interfacial layer and the inadequate crystallization caused by the lack of a lattice match.

The newly discovered morphological phase boundary (MPB) between the ferroelectric orthorhombic and tetragonal phases in the HfO_2-ZrO_2 solid solution may hold promise for further reducing EOT with TiN as an electrode material in DRAM technology. The deposition processes of both materials (HfO_2 and ZrO_2) are highly matured in the semiconductor fabrication process, where the HfO_2 is a base material for the high-k gate dielectric layer in the logic transistor, and the ZrO_2 is the capacitor dielectric. Moreover, because of its rather large band gap (>5.5eV), the leakage current through TiN/HZO/TiN capacitors can be sufficiently low to meet the requirement e for a DRAM capacitor[5].

In this paper, we performed extensive electrical characterizing on a thin film of $Hf_{1-x}Zr_xO_2$ with 9 nm thickness. By varying zirconium levels (Hf: Zr=5:5, 4:6, 3:7, 2:8, 1:9) of the films, the MPB between the o-phase and t-phase is investigated based on TiN/HZO/TiN capacitors (metal-ferroelectric-metal, MFM). Finally, the endurance of the capacitors is measured to meet the requirements of advanced dynamic random access memory (DRAM) capacitors.

Device Fabrication.

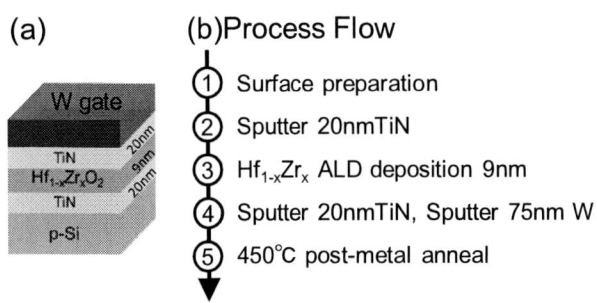

(a)   (b)Process Flow
① Surface preparation
② Sputter 20nmTiN
③ $Hf_{1-x}Zr_x$ ALD deposition 9nm
④ Sputter 20nmTiN, Sputter 75nm W
⑤ 450℃ post-metal anneal

Figure 1: The schematic of the MFM Capacitor(a) and (b) Process flow for the fabrication of the capacitor with Hf1-xZrxO2 as the dielectric.

Figure 1(a) shows the schematic of the MFM Capacitor. The process flow for fabricating the capacitor with $Hf_{1-x}Zr_xO_2$ as the dielectric is shown in Figure 1(b). 8 inches p-type Si wafers with 8-12 Ω•cm were utilized as a substrate. Next, the 20-nm-thick TiN electrodes were deposited using DC reactive sputtering onto the Si substrates. Then, the 9 nm thick $Hf_{1-x}Zr_xO_2$ film was deposited alternately, introducing Hf and Zr organic precursors. Subsequently, HZO films were deposited on the Si substrate with TiN electrodes. After the tungsten (W,75nm)/TiN(20nm) top electrode, which was deposited via DC sputtering (TiN contacted the HZO film), post-metallization annealing was performed for 25 seconds at 450℃ in an N_2 atmosphere via rapid thermal annealing to crystallize the films. The variation of Zr composition in HZO films is achieved by changing the ratio of Hf and Zr growth cycles.

979-8-3503-1101-3/23 $31.00 © 2023 IEEE

Electrical Measurements.

P-V curves, leakage current density-electric field (J-V), and capacitance-voltage (C-V) were acquired with a ferroelectric analyzer (TF Analyzer 3000, aixACCT Systems GmbH).

RESULTS AND DISCUSSION

We investigated the characteristics of the 9 nm MFM capacitors with varied Zr concentrations (Hf: Zr=5:5, 4:6, 3:7, 2:8, 1:9). As Figure 2 shows, the P-E curve gradually changes from the FE phase to the AFE phase as the Zr ratio rises from 0.5 to 0.9. At Hf: Zr = 0.4:0.6 and 0.3:0.7, the FE and AFE phases can coexist in the films, indicating the possibility that the MPB phenomenon exists in these films.

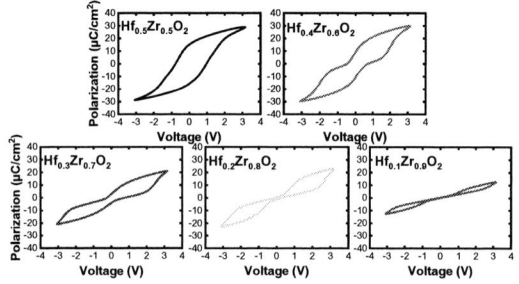
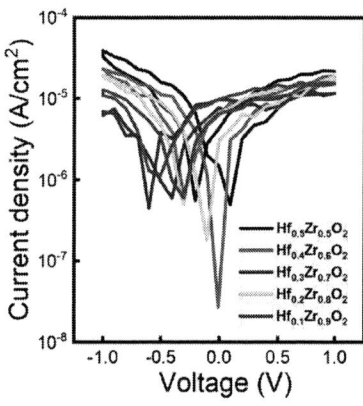

Figure 2: P-V curves of the 9-nm-thick Hf1-xZrxO₂ films.

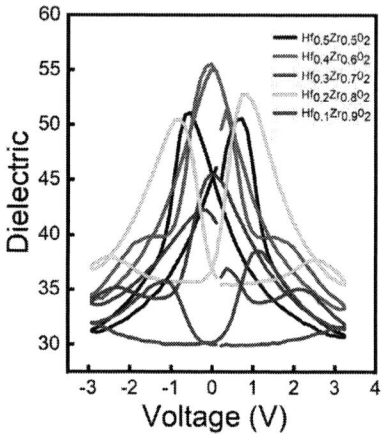

Figure 4: κ-V curves of the 9nm-thick Hf1-xZrxO₂ films

The κ-V curves are derived from the capacitance-voltage (C-V) measurements. Figure 4 shows that $Hf_{0.4}Zr_{0.6}O_2$ has the highest dielectric constant among the zirconium-containing capacitances. This extreme value can be reached close to the approximate zero fields, demonstrating the presence of MPB in the $Hf_{1-x}Zr_xO_2$ system.

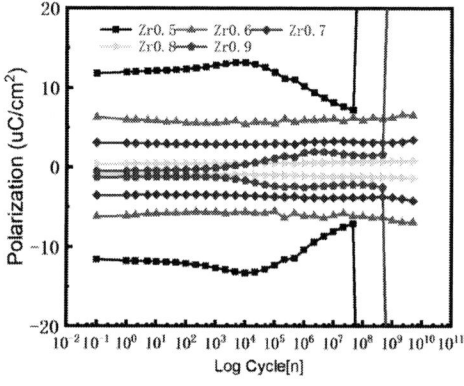

Figure 5: Polarization as a function of cycling number

The endurance characteristics of the MPB films are also performed. Figure 5 shows that the 9 nm thick $Hf_{0.5}Zr_{0.5}O_2$ films can withstand less than 1×10^8 cycles, and the $Hf_{0.1}Zr_{0.9}O_2$ films can endure less than 1×10^9 cycles, while the 9-nm-thick $Hf_{0.4}Zr_{0.6}O_2$ films can resist a high operation voltage of 2.4 V for more than 5×10^9 cycles, holding promise to be one of the driving forces for future ultra-low power transistors and ultra-high density memories.

Figure 3: J-V curves of the 9nm-thick Hf1-xZrxO₂ films

Subsequently, we further analyzed the electrical properties of the films. The leakage current density-electric field (J-V) and constant-electric field (κ-V) on the $Hf_{1-x}Zr_xO_2$ films are shown in Figure 3 and Figure 4, respectively. Figure 3 illustrates that the low current density of these films is sufficient to meet the capacitance requirement in DRAM technology.

CONCLUSION

TABLE I.

$Hf_{1-x}Zr_xO_2$	The Properties Of Different Zr Compositions In $Hf_{1-x}Zr_xO_2$ Films		
	Maximum κ Value	*Current Density (A/cm²)*	*Endurance (cycle)*
$Hf_{0.5}Zr_{0.5}O_2$	51	3.89×10^{-5}	4.6×10^{7}
$Hf_{0.4}Zr_{0.6}O_2$	55	2.30×10^{-5}	$>5\times10^{9}$
$Hf_{0.3}Zr_{0.7}O_2$	45	6.11×10^{-6}	$>5\times10^{9}$
$Hf_{0.2}Zr_{0.8}O_2$	52	1.93×10^{-5}	$>5\times10^{9}$
$Hf_{0.1}Zr_{0.9}O_2$	38	1.27×10^{-5}	4.6×10^{8}

The variation of the HfO_2-ZrO_2 solid solution as a function of the Hf/Zr ratio (Hf: Zr=5:5, 4:6, 3:7, 2:8, 1:9) is summarized in the experimental observations, as shown in Table I. The $Hf_{0.4}Zr_{0.6}O_2$ was discovered to have 55 peak values. Moreover, the maximum value of the dielectric constant appears near the zero electric fields, indicating the presence of the MPB phenomena. The leakage current via TiN/HZO/TiN (MFM) capacitors can also be suitably low, and additional durability testing demonstrated the suitability of this MPB for DRAM capacitor applications.

ACKNOWLEDGEMENTS

Thanks to all the teachers and students of the research group, as well as the scholars who provided consultation and help.

REFERENCES

[1] Kim, S. K.; Popovici, M. Future of Dynamic Random-Access Memory as Main Memory. MRS Bull. 2018, 43, 334−339.

[2] Hwang, C. S. Prospective of Semiconductor Memory Devices: from Memory System to Materials. Adv. Electron. Mater. 2015, 1,1400056.

[3] Jeon, W.; Yoo, S.; Kim, H. K.; Lee, W.; An, C. H.; Chung, M. J.; Cho, C. J.; Kim, S. K.; Hwang, C. S. Evaluating the Top Electrode Material for Achieving an Equivalent Oxide Thickness Smaller than 0.4 nm from an Al-Doped TiO2 Film. ACS Appl. Mater. Interfaces 2014, 6, 21632−21637.

[4] Popovici, M.; Swerts, J.; Redolfi, A.; Kaczer, B.; Aoulaiche, M; Radu, I.; Clima, S.; Everaert, J.-L.; Van Elshocht, S.; Jurczak, M. Low Leakage Ru-Strontium Titanate-Ru Metal-Insulator-Metal Capacitors for Sub-20 nm Technology Node in Dynamic Random Access Memory. Appl. Phys. Lett. 2014, 104, 082908.

[5] Park, M. H.; Kim, H. J.; Kim, Y. J.; Lee, Y. H.; Moon, T.; Kim,K. D.; Hyun, S. D.; Hwang, C. S. Study on the Size Effect in Hf0.5Zr0.5O2 Films Thinner than 8 nm before and after Wake-up Field Cycling. Appl. Phys. Lett. 2015, 107, 192907.

HIGH ENDURANCE SONOS TECHNOLOGY IMPROVED BY DESIGN &PROCESS OPTIMIZATION

Pingsheng zhou*, Kegang Zhang, Gordon Li, Hualun Chen, Xiang Yao, Weiran Kong
Shanghai Hua Hong Semiconductor Manufacturing Corporation, Shanghai 201203, China
*Corresponding author. Tel.: +86-18801928579. E-mail addresses: Pingsheng.zhou@hhgrace.com

BIOGRAPHY

Pingsheng Zhou now is a senior technology engineer of Shanghai Hua Hong Semiconductor Manufacturing Corporation. His research interest focuses on embedded Non-volatile memory.

ABSTRACT

In this paper, we present a design and process optimization to improve SONOS cell endurance cycle. Generally, the endurance cycle can be effectively increased by reducing the voltage. But with the voltage dropping, large V_T margin loss will be found. In this case, we design a differential cell which can enlarge cell V_T window. In addition, we optimize the process condition to get the ideal V_T window of the differential cell. Finally, the endurance cycle of SONOS IP(512Kb) is improved from 500K to 25M cycles at 85 degrees.

Keywords—SONOS memory; high endurance; SONOS cell V_T window, reliability

INTRODUCTION

SONOS memory has the advantages of high density, high endurance for programing cycle, etc., which is electrically erasable programmable read-only memory (EEPROM) [1]. The most popular trapping type of NVM device was observed in SONOS memory. Low-power, low programming voltage and small size were natural attributes included in the SONOS memory. The main characteristic of the SONOS is the application of the tunneling oxide-nitride-blocking oxide (ONO) instead of the common oxide[2]. This paper evaluates all the wafer-level reliability that occurs in our cell arrays, focusing mainly on the study of endurance.

The 2-transistor SONOS cell which is composed of a memory transistor and a select transistor is fabricated at deep submicron technology node in our design. For these devices, long-term safe use is one of the most important features. However, in most EEPROM technologies, the normal processes used to write and read cells can cause stress, which eventually reduces the properties of the memory [3]. As device sizes shrink, it becomes increasingly challenging to maintain the durability of smaller SONOS. As shown in Figure1, the V_{TE} (threshold voltage of erase) of the weakest die after 500K endurance cycles in standard cell is close to 0,which means that while the number of cycles continues to increase, the cell would fail. The endurance cycle can be effectively increased by reducing the voltage, as shown in Figure 2. After operation voltage dropping, the V_{TE} window drop slop becomes smaller, which means as the endurance cycles increasing, V_{TE} window is enlarged at the end of lift. But for the V_{TP} (threshold voltage of program) is shifted down obviously, the entire window of SONOS becomes 200mV smaller in the beginning.

In this work, we try to improve a SONOS device V_T window by design optimization based on the voltage reduction. Firstly, we use a differential SONOS cell structure instead of stand cell, and the V_T window difference is shown in Figure 3. The V_T window in standard cell depends on the minimum value of a and b(min(a,b)). The alphabet of a stands for V_T of post programming, and letter b stands for V_T of post erasing. But the V_T window for differential cell is a+b, as shown in Figure 3, which means it can greatly improves cell window.the advantage is the differential cell can get larger V_T window, faster operation speed, more endurance cycles.

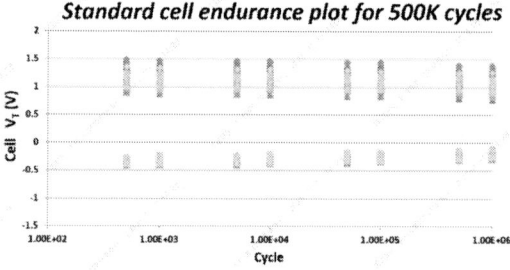

Fig.1: The standard cell endurance plot for 500K cycles.

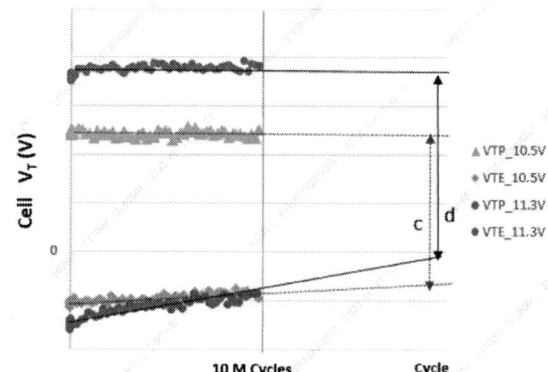

Fig.2: The end of life V_T window distribution.

979-8-3503-1101-3/23 $31.00 © 2023 IEEE

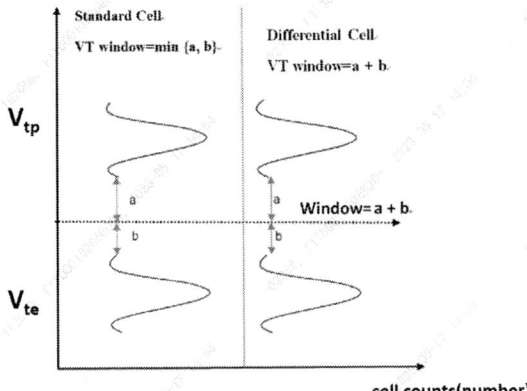

Fig.3: V_T window difference between standard cell and differential cell

Device and Measurement Details

1. cell structure

As shown in Fig.4 left figure, this differential structure comprises two memory transistors and two select transistors both located in a P-well and isolated from each other by a field oxide or a shallow trench isolation (STI). The pair of transistors each comprises a gate, a source, and a drain region. The Fig.4 right figure depicts the differential cell circuit diagram. The drains are connected together in the same column as a bit line, and the gates are connected together in the same row to form a word line. The SONOS gate is called WLS(word line SONOS). The sources of the select transistors are connected together in the same row [4]. Here, Fower-Nordheim (FN) or direct tunneling is formed at high electrical field. When the electric field between the substrate and the gate polysilicon is removed, the tunnel prevents the captured electrons from being lost into the substrate [5]. Besides, Process fluctuations can cause window inconsistencies in wafer-to-wafer or within wafer in standard cell, which can be well improved in the differential cell

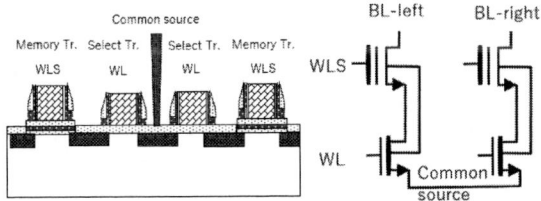

Fig.4: the figure in the left depicts cross-sectional drawing of differential cell. The figure in the right depicts schematic circuit diagram.

2. operation modes

There are 3 modes (program, erase and read) for traditional SONOS cell. For the erase operation, negative bias applies to gate, source floating, while positive applies to drain and bulk. The electrons are pulled back into the channel through the screen oxide and the channel is open. The read current, called I_{DSE} (read current after erasing), is always about uA(10^{-6}A) level. The program operation can be divided into 2 kinds: program "1" and program "0". The program "1" operation can be understood as below: positive bias applies to gate, source floating, and the drain and bulk are input by negative bias. Actually the electrons are pulled into the nitride via screen oxide and the channel is switched off at the same time. The I_{DSP} which means the read current after programming, is about nA(10^{-9}A) level. The program "0" can be imaged as a special status, which applies positive bias to gate, source floating, and a light positive bias is forced to drain and bulk. At the operation condition, the SONOS can't be program successfully, which called the program inhibit status, with the read current about uA(10^{-6}A) level.

The above introduction is all about the traditional SONOS cell, which is different with our new design, called differential cell. The differential cell have two cell, left and right cell. As shown in table1, for the post program status, the two cell have two different status: "1" or "0". For the (left cell, right cell) matrix, (1,0) stands for the entire cell Program "1" status, and (0,1) stands for the entire cell Program "0" status. The left and right cells states can be judged by the sense current. In this design, the differential cell status can be identified if the left and right cell read current have enough margin to distinguish. By the method, it obviously improve the entire cell margin. For erase mode, it only has one status, which means that two cells pull electrons back into channel simultaneously.

Table.1: the differential cell operation modes.

Status	Left Bit	Right Bit
erase	0	0
Program"0"	0	1
Program"1"	1	0

Process Optimization results

Due to design constraints, there is a minimum detectable point about the current, called sense current, which means the current can't be identified if it is lower than the sense current. Considering this cause, the SONOS window should be calculated from erase state current to sense current. But the post-program current is so small that the window causes some waste from the post-program current to the sense current.

In this paper, we present a smart process to improve SONOS cell V_T window through SONOS cell V_T shift. By adjusting the concentration of N-doped in the SONOS channel, the current in the channel can be changed and the whole window of SONOS can be shifted..

We arrange different split experiment group (BL~BL-6) to check the process change effect. As shown in Fig.5, the SONOS initial V_T was adjusted by process change so that the whole window is panned to V_{TE}

979-8-3503-1101-3/23 $31.00 © 2023 IEEE

direction. The erase current will be bigger and more advantageous in this way, and the current under the program will also be bigger. Fig.6 depicts yield loss under different process split. After 250 degrees 34 hours bake (equivalent to 293 years at 85 degrees), the residual value of SONOS V_{TP}/V_{TE} which reflects the data retention ability after SONOS cell trimming needed to be checked. The result illustrates that BL-2 split yield loss is minimal, which means it's data retention ability is the best. If the process change causes the V_{TP} to continue to decline, the yield loss will be increases.

Fig.7 chip probed test map depicts that BL-2 yield is the highest and red yield loss is the lowest. This process change greatly improves the yield and this data proves that yield loss becomes lower from BL (baseline condition) to BL-2, but yield loss becomes higher from BL-3 to BL-6. This suggests that BL-2 is the inflection point for yield loss and reliability is the best under this condition.

Fig.8 shows the SONOS single cell V_T window changing trend after constantly erasing and programming at 85 degrees. During 10M cycles at 85 degrees, and baking at 125 degrees about 10 hours after every 500K cycles, V_T window gets about 200mV smaller, but not a single die fails.

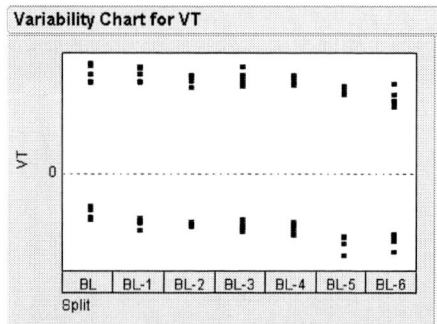

Fig.5: SONOS cell V_T under different process split

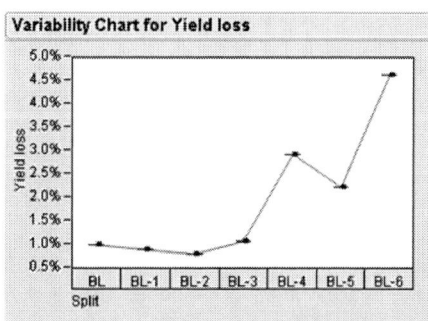

Fig.6: Yield loss under different process split.

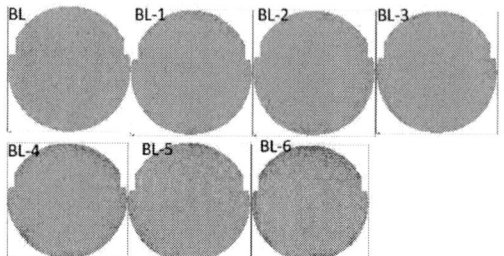

Fig.7: the figure illustrates chip probed test map with different experiment.

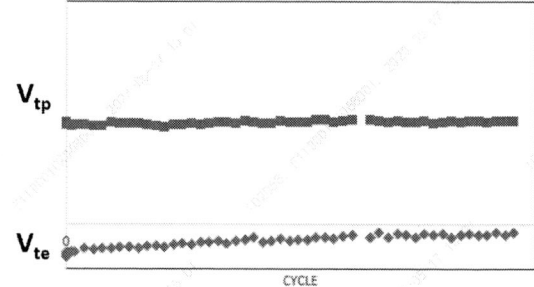

Fig.8: the figure shows the SONOS single cell V_T window changing trend after constantly erasing and programming at 85 degrees.

Reliability of optimized device

Based on the above experimental data and analysis, we found that the endurance of differential cell becomes better after V_T window shifting. The threshold voltage window decreases when the cycles of programming and erasing increase. The anti-erasing ability of SONOS memory is primarily judged by the programming/erasing threshold voltage, i.e. by functional verification failure [6].

First, we verified SONOS cell capability after process changing, with Table.2 showing passing results for samples taken through different erase and program cycles. After CP(chip probe) trimming, erase and program are cyclic operated at different modes (bulk/chip/page), different patterns (00~FF) and different voltages (1.7V~5.5V) at 85 degrees. Reading V_{TP} and V_{TE} value after every 500K cycles, then reading V_{TP} and V_{TE} again after baking for 10 hours at 125 degrees. If no samples fail, keep erasing and program cycles. The results tell us none of the check samples failed after repeating the above loop to 25M cycles. This good result is greatly ahead of the similar products in the existing market. Then we want to know how many cycles the differential cell's endurance fail. There are total of 3 fail dies until 59M erasing and programming cycles at 85 degrees.

The above data is used to calculate SONOS cell failure instance time, which is FIT. As shown in Table 3, suppose one cycle per hour, under the confidence coefficient of 90%, FIT sis only 0.99 after 59M cycles. Suppose one cycle per minute, FIT is only 59.6 after 59M cycles.

Table.2: different cycles pass/fail result

Test Items	Test Condition	Fail Criteria	spec.	Results fail/total sample
Endurance & PCHTDR(post cycling high temperature data retention)	Total 25M cycle @85℃ Bake @125℃/10hours after every 5M cycle	Data verify	0/77	0/114
Endurance	Total 30M cycle @ 85℃			1/114
	Total 60M cycles @85℃			3/114

Table.3: FIT calculation result

Sample	Confidence coefficient	Cycle	Test Time (hrs)	Failure	FIT
114	90%	59M	983333 hrs （if 1min/cycle）	3	59.6
			59M hrs （if 1hr/cycle）	3	0.99

Conclusions

We described 95nm high yielding, high reliability SONOS differential cell which fabricated on a 300mm manufacturing line. The differential cell will enlarge a SONOS device V_T window cell based on the voltage reduction. Experimental process implant adjustments in the SONOS region changed the SONOS cell erase and program V_T values, thereby expanding the existing window. From the experimental yield, it can be seen that the differential cell V_T window shift is around 100mV, which is the highest yield under BL-2 condition and the lowest yield loss after baking. Checking single cell value at the same time，during 1000K cycles at 85 degrees, and baking at 125 degrees about 10 hours after every 500K cycles, V_T window gets about 200mV smaller, but not a single die fails. Next, verify IP reliability, cell endurance performance is improved from 100K to 59M cycles. While cell data retention keeps no degradation, and devices reliability shows no suffer.

Acknowledgments

We acknowledge contributions of many dedicated Hua Hong Semiconductor Manufacturing Corporation colleagues, especially the Production testing development team and reliability verification team.

References

[1] M. H. White and D. Adams, "Low-Voltage SONOS Nonvolatile Semiconductor Mem-ories (NVSMs)," GOMAC 2000.

[2] K. Ramkumar, "Materials and device relialility in SONOS memories," Charge-Trapping Non-Volatile Memories, pp. 1-54: Springer, 2017.

[3] J. E. Brewer and M. Gill, Nonvolatile Memory Technologies with Emphasis on Flash, 2008.

[4] Jan M. Rabaey, Anantha Chandrakasan, Borivoje, Nikolic, " Digital Integrated Circuits: A Design Perspective, Second Edition." 2003, pp. 289-295

[5] K. Ramkumar, "Materials and device relialility in SONOS memories," Charge-Trapping Non-Volatile Memories, pp. 1-54: Springer, 2017

[6] J. H. Choi, C. G. Yu, and J. T. Park, "Nanowire width dependence of data retention and endurance characteristics in nanowire SONOS flash memory," Microelectronics Reliability, vol. 64, pp. 215-219, 2016.

INVESTIGATION OF THE DOPING PROFILE FOR ION IMPLANTS AND RAPID ANNEALING IN SILICON VIA AN IMPROVED METHOD

Zeqi Zha[1]*, Zhenhui Wang[1]*, Ya Wang[1]

[1] Semiconductor manufacturing International (Beijing) Corporation, Beijing 100176, China
*Corresponding Author's Email: zhazeqi@163.com, zhenhui_wang@smics.com

ABSTRACT

Nowadays, applications of ion implantation and rapid annealing in semiconductor fabrication are widespread. The doping profile in silicon determines the depth and lateral spread of p-n junctions. An optimal combination of ion implantation and annealing conditions for tuning the electrical properties of device usually takes a lot of time and money. In this paper, we present a cheap and efficient method for investigation the implanted impurity distribution on Sentaurus TCAD tools. The simulated boron/phosphorus/arsenic doping profiles are in good agreement with SIMS measurements for a wide range of implant energies and annealing temperature. In addition, the simulation of resistance in different depths is also investigated.

INTRODUCTION

Due to the precision dose control, the ion implantation is widely used in the semiconductor manufacturing, especially for foundry fab facilities. Together with rapid thermal annealing (RTA), it was possible to combine higher conductivities of doped layers and shallow p-n junctions, so that better optimized devices can be designed and manufactured [1]. However, this poses a challenge for the selection of an optimized implant/RTA recipe that meets the different requirements of devices. The doping profile in crystalline silicon can be controlled by energy of ion beam and the annealing temperature. Suitable ion energy and annealing temperature are often achieved over long experimental cycles. TCAD simulation tools offer researchers the ability to study electrical and structural properties of device before manufacturing. In this paper, TCAD was used to investigate the doping profile of ion implants in silicon. The simulated doping profiles are in good agreement with Secondary Ion Mass Spectroscopy (SIMS) and spreading resistance probe (SRP) measurements. In addition, the correlation between the beam energy and the depth of peak concentration after annealing process was also investigated.

EXPERIMENTAL METHODS

Due to the high sensitivity and depth resolution of SIMS, it was applied to study the doping profile for a wide range of implant energies. In this paper, the dual-Pearson distribution was used to simulate the doping profile. The dual-Pearson function has been shown to be effective in simulating the distribution of doped ions obtained by ion implantation. It uses the sum of two Pearson functions which describe the non-channeling and channeling components in implanted silicon dopant profile. The dual-Pearson function can be expressed as below:

$$f_{dual-Pearson}(x) = ratio \cdot f_{non-channel}(x) + (1-ratio) \cdot f_{channel}(x) \qquad (1)$$

where the $f_{non-channel}(x)$ and $f_{channel}(x)$ are Pearson function. The Pearson distribution function is defined in a differential form as follows:

$$\frac{df(x)}{dx} = \frac{(x-a)f(x)}{b_0 + b_1 x + b_2 x^2} \qquad (2)$$

in which a, b_0, b_1 and b_2 are constants.

The doping profile of crystal silicon wafer after annealing at 1100°C 20s was also characterized by SIMS. The Charged-Pair model in TCAD was used to simulate the annealed doping profile.

The spreading resistance probe (SRP) measurement is also used to analyze the doping distribution after annealing. The SRP measurement is the method of choice for producing accurate resistivity profiles in the vertical direction. The resistivity is determined by the carrier concentration and the mobility, both of which are related to the dopant concentration. The resistivity was calculated by following form:

$$\rho(x) = 1/q(n(x) \cdot \mu_n + p(x) \cdot \mu_p) \qquad (3)$$

By building a simple one-dimensional PN junction model, the TCAD software can be used to calculate the carrier concentration ($n(x)$, $p(x)$) and mobility (μ_n, μ_p) for a given ion implantation and annealing condition.

RESULTS AND DISCUSSION

As implanted ions enter the silicon, they give up their energy to the lattice atoms and eventually come to rest at certain depths in the silicon. Figure 1a shows a typical doping profile obtained by ion implantation. The doping profile consists of two parts, a non-channeling area and a channeling area (the tail of doping profile). Blood et al. have shown that the tail in implanted profiles is mainly due to atoms scattering into channels rather than an interstitial diffusion mechanism at the implantation temperature [2].

979-8-3503-1101-3/23 $31.00 © 2023 IEEE

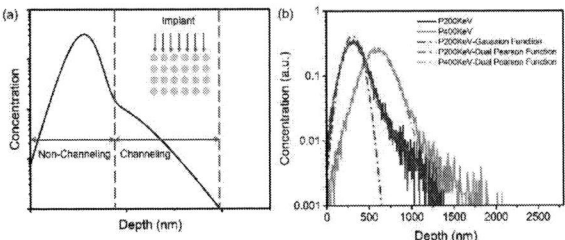

Figure 1: (a) Schematic Diagram of channel and non-channel regions in doping profile obtained by ion implantation. (b) Comparison of different simulated doping profiles for 200 keV Phosphorus implant and comparison of the measured doping profile for 200 keV/400 keV. Experimental SIMS data is shown solid lines, TCAD simulation as dash-dotted lines.

Figure 1b shows the experimental measured and simulated doping profiles for 200 keV/ 400 keV phosphorus implants in crystalline silicon. The data in Figure 1b show that the phosphorus 200 keV doping profile and phosphorus 400 keV doping profile exhibit distinct channeling properties at implantation depths of ~750 nm and ~1180 nm, respectively. Due to the asymmetry of the doping profile, the simulated curve based on the Gaussian function can't accurately describe the concentration of doping atoms (see Figure 1b, red dashed line). Moreover, a dual-Pearson approach to modelling ion-implanted phosphorus concentration profiles in crystalline silicon is shown in Figure 1b. The better result shows that dual-Pearson function is effective in simulating doping profile.

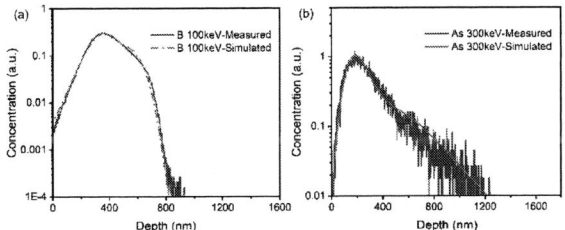

Figure 2: Doping profiles (by SIMS and TCAD) for 100 keV Boron ion (a) and 300 keV Arsenic ion (b) implanted into crystalline silicon; Experimental SIMS data is shown solid lines, simulation curves as dash-dotted lines.

The doping profiles of the boron and arsenic ion in crystalline silicon were also simulated by the dual-Pearson function. Figure 2a and Figure 2b show the simulated doping profiles of the boron at 100 keV and arsenic at 300 keV, respectively. It can be seen that there is good agreement between the doping profile simulated by dual-

Pearson function for boron/arsenic and the experimental doping file. Moreover, the depth of 300 keV arsenic ion implantation is shallower than that of 100 keV boron ion implantation, indicating that the implantation depth do not only depend on the beam energy but also on other factors like the species.

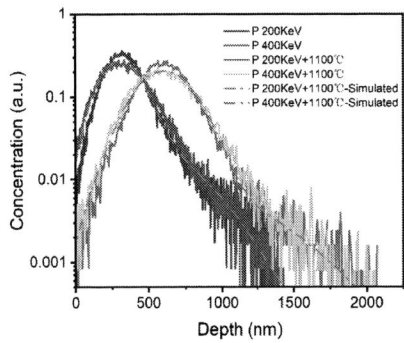

Figure 3: Unannealed and annealed doping profile (by SIMS) for 200 keV Phosphorus ion and 400 keV phosphorus ion implanted in crystalline silicon. Experimental SIMS data is shown solid lines, simulation curves as dash-dotted lines.

Figure 3 shows doping profiles of 200 keV/400 keV phosphorus implants annealed by RTA at 1100°C 20s. During RTA, the implanted damage anneals and simultaneously phosphorus diffuses deeper into the silicon [3]. In the SIMS data of the 400 keV phosphorus ions, it can be seen that the phosphorus concentration at the depth of ~1000 nm is higher than before RTA. This indicates that the diffusion of phosphorus ions during annealing changes the doping profile. And the position of peak concentration depth gradually moved ~30 nm and ~ 25 nm towards the silicon surface for 200 keV/400 keV phosphorus, respectively. The simulation data calculated by the TCAD Charged-Pair model are in agreement with the annealed doping profile (see Figure 3 dash-dotted lines). After electrical activation of implanted impurities, the concentration of doping ions can also be characterized by SRP measurement. As shown in Figure 4a and Figure 4b, the experimental and simulated doping profiles show that the position of peak concentration depth measured by SIMS is not significant different from the position of lowest resistivity depth measured by SRP. Therefore, under the condition of RTP 1100°C 20s, the peak electrically active concentration depth can be obtained from SIMS measurement after annealing.

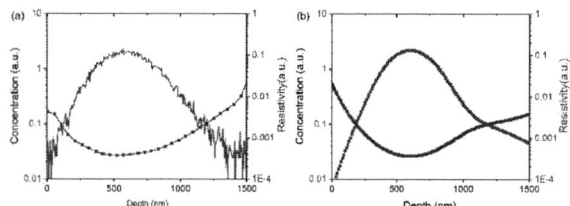

Figure 4: (a) Experimental SIMS / SRP data and (b) the corresponding simulated doping profile / resistivity profile for 400 keV Phosphorus ion implanted in crystalline silicon.

The implanted depths for boron, phosphorus and arsenic in silicon after annealing 1100°C 20s with different energies (< 800 keV) were also simulated (see Figure 5). According to the data in Figure 5, the depth and energy of ion implantation are positively correlated. Moreover, it is obvious that the implantation depth of heavy atoms is less than that of light atoms under the same conditions.

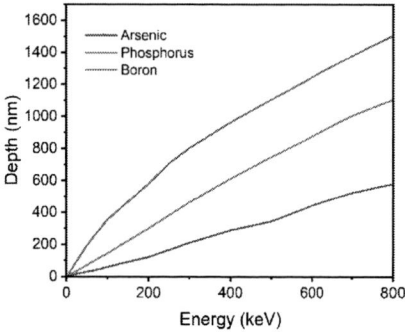

Figure 5: Depth in silicon of Boron, Phosphorus and Arsenic ion for different energies, under 1100°C 20s annealing conditions.

CONCLUSIONS

In this paper, the simulated doping profile by TCAD was used for investigation the doping profile of ion implants in silicon. Good agreement was found between the simulated boron/phosphorus/arsenic doping profiles and experimental SIMS data. Under typical RTA condition 1100°C 20s, the correlation between the beam energy and the depth of peak concentration after annealing process for different species was also investigated. It can help researchers to quickly identify the implantation depth. This cheap and efficient simulated method is very practical for finding optimal implant/RTA conditions.

REFERENCES

[1] Lerch. phys. stat. sol. (a), vol.158, 1996, pp. 117-136.

[2] P. Blood, G. Diurnally and MA. Wilkins. J. Appl. Phys., vol. 45, 1974, pp. 5123-5128.

[3] Hill Chris. Nuclear Instruments and Methods in Physics Research Section B: Beam Interactions with Materials and Atoms, vol.19, 1987, pp. 348-358.

EXPERIMENTAL INVESTIGATION OF ULTRA-LOW TEMPERATURE LA₂O₃/HFO₂ BI-LAYER DIPOLE-FIRST PROCESS USING PVD METHOD FOR ADVANCED IC TECHNOLOGY

Yanzhao Wei[1,3], Jiaxin Yao[1], Renren Xu[1,3], Qingzhu Zhang[1,2], and Huaxiang Yin[1,2,3*]*

[1] Integrated Circuit Advanced Process R&D Center, Institute of Microelectronics, Chinese Academy of Sciences, Beijing 100029, China

[2] Key Laboratory of Microelectronic Devices and Integrated Technology, Institute of Microelectronics, Chinese Academy of Sciences, Beijing 100029, China

[3] School of Integrated Circuits, University of Chinese Academy of Sciences, Beijing 100049, China

*Corresponding Author's Email: yaojiaxin@ime.ac.cn; yinhuaxiang@ime.ac.cn

ABSTRACT

In this paper, a La_2O_3/HfO_2 bi-layer dipole-first (DF) process is proposed and investigated by ultra-low temperature PVD dielectric laminates to achieve lower gate effective work function (EWF) for monolithic 3D-IC (M3D) application. The impacts of ultra-low temperature La-dipole on EWF modulation and interfacial properties are comprehensively investigated. It is found that the flat-band voltage (V_{FB}) negatively shifts 60 mV with sub-1nm La_2O_3 thickness, which provides an effective way to meet the require of Si conduction band-edge EWF modulation. Furthermore, the electron trap/detrap densities (Not) and interfacial trap densities (Dit) are suppressed by La_2O_3/HfO_2 bi-layer DF process to improve device performance. These results exhibit a promising bi-layer DF process in low thermal integration for advanced IC technology.

Keywords—Dipole-first; fine range V_{FB} modulation; low temperature; magnetron sputtering; La-dipole; La_2O_3

INTRODUCTION

To meet the needs of high performance and low consumption requirements of transistors in high integration density circuits like monolithic 3D-IC (M3D), a low thermal budget and volume-less multiple threshold voltage (multi-Vt) technique is in need [1-2]. Dipole engineering is a promising approach for threshold voltage fine/wide modulation with less space using, and La_2O_3 is the mainstream flat-band voltage (V_{FB}) shifter for NMOSFETs as it provides Si conduction band-edge effective work function (EWF) by La-dipole formation [3]. For La-dipole formation, depositing La_2O_3 directly on SiO_2 IL as dipole-first (DF) process and depositing La_2O_3 on HfO_2 high-k layer with drive-in anneal as dipole-last (DL) process is the commonly utilized method. However, the wide range V_{FB} negative modulation obtained by DF process is superfluous for the need of fine range Vt modulation, and thermal budget in DL process for fine range V_{FB} negative modulation is too high to be compatible with M3D application [3-5].

In this paper, we have experimentally investigated a

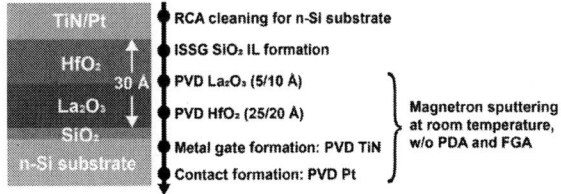

Figure 1: Key flow and schematic of the ultra-low temperature process gate stack. The thermal budget of HKMG structure is at room temperature.

La_2O_3/HfO_2 bi-layer dipole-first (DF) approach via ultra-low temperature PVD process. Less than 100 mV V_{FB} fine range negative modulation was achieved and precisely controlled by sub-1nm ultra-thin La_2O_3 thickness. Meantime, the electron trap/detrap densities (Not) and interfacial trap densities (Dit) were suppressed by La_2O_3/HfO_2 bi-layer DF process to improve device performance.

EXPERIMENT AND FABRICATION

Fig.1 shows the key flow and schematic of high-k metal gate (HKMG) structure with La_2O_3/HfO_2 laminated dielectric used for this study. First, an n-type Si (100) wafer with resistivity of 2-5 Ω·cm was used as the substrate. After etching native oxide on wafer by buffered oxide etchant (BOE), a 1.3 nm SiO_2 interface layer (IL) was grown by *in-situ* steam generation (ISSG) process. For the La_2O_3/HfO_2 laminated dielectric high-k layer, a room temperature magnetron sputtering process was utilized. 5 Å and 10 Å La_2O_3 was respectively deposited on IL firstly. To keep the physical thickness of high-k laminated layer at 3 nm, 25 Å and 20 Å HfO_2 layer was sputtered on La_2O_3 respectively. As a comparison, the high-k dielectric of reference sample is a pure 3-nm HfO_2 layer without La_2O_3. Then, a 2-nm TiN was sputtered on La_2O_3/HfO_2 laminated dielectric high-k layer as the metal gate. At last, a 100-nm gap-filling metal of Pt was sputtered. The TiN and Pt sputtering process temperature is at room temperature. No post deposition annealing

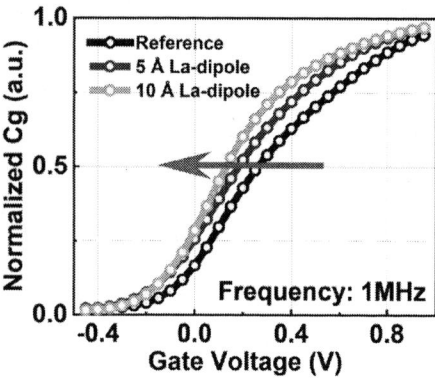

Figure 2: C-V curves of different thickness La-dipole devices and reference device via room temperature PVD.

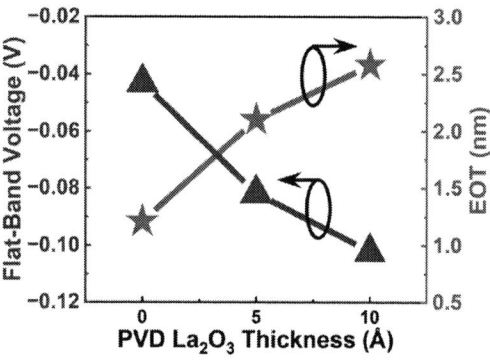

Figure 3: Flat-band voltage and EOT value variations of devices with PVD La_2O_3 in a 3-nm high-k stack.

(PDA) process and forming gas annealing (FGA) process were utilized. For device testing, the capacitance-voltage (C-V) characteristics were measured by Keithley 4200 semiconductor parameter analyzer in the air ambient at room temperature.

RESULTS AND DISCUSSIONS

The C-V curves of devices with different thickness of PVD La_2O_3 are showed in Fig.2. The C-V curves were measured at 1 MHz frequency from inversion region to accumulation region. As a comparison to the reference device, the device with PVD La_2O_3 in high-k stack shows an obvious negative shift on C-V curves. With the thickness of La_2O_3 increasing from 5 Å to 10 Å, the C-V curve keeps negative shifting. The extracted V_{FB} value from C-V curves are presented in Fig.3. V_{FB} value of devices with PVD La_2O_3 thickness of 0 Å, 5 Å, 10 Å is -0.04 V, -0.08 V, and -0.10 V, respectively. A maximum

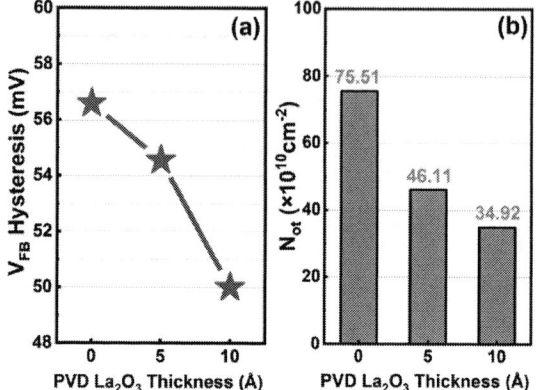

Figure 4: The variation of (a) voltage hysteresis and (b) Not of devices with PVD La_2O_3 in a 3-nm high-k stack.

of 60 mV V_{FB} negative shift was obtained by 10 Å PVD La_2O_3 in high-k stack, which can be seen as a fine range V_{FB} modulation. The trend of V_{FB} negative shifting with the thickness of PVD La_2O_3 increasing is basically linear, which shows that the precise fine range V_{FB} negative modulation is achieved by changing PVD La-dipole layer thickness under 1 nm.

The extracted equivalent oxide thickness (EOT) value from C-V curves is also showed in Fig.3. The EOT value of devices with PVD La_2O_3 thickness of 0 Å, 5 Å, 10 Å is 1.21 nm, 2.10 nm, and 2.58 nm, respectively. As a comparison to the reference device, an around 1-nm EOT penalty was obtained by sputtered La_2O_3 in high-k stack. The reason for EOT increasing can be considered as the low-k LaSiOx formation at La_2O_3/SiO_2 interface [6]. Although the thermal budget of HKMG sputtering is ultra-low, La_2O_3 is easily to react with SiO_2 [6], especially a sufficient 1.3-nm SiO_2 IL is presented. According to our prior work [7], EOT value and interface charge density (Nss) would cause small range of V_{FB} shifts. Nevertheless, the dominant origin of V_{FB} shifts in this experiment is La-dipole in high-k layer.

Considering the defect affected by La_2O_3 in gate stack bulk and at interface, the trap/detrap electrons density (Not) and interface trap density (Dit) were investigated by the same C-V test method as our prior work [7]. Fig.4(a) shows the V_{FB} hysteresis value extracted from dual-sweep C-V curves of devices with different thickness of PVD La_2O_3 in gate stack. With the La_2O_3 thickness increasing from 0 Å to 10 Å, the V_{FB} hysteresis is decreased from 56.6 mV to 50.0 mV. In Fig.4(b), the Not value calculated by V_{FB} hysteresis is decreased from 75.51×10^{10} cm^{-2} to 34.92×10^{10} cm^{-2} at the same time, indicating that 53.8 % trap/detrap electron density reducing is achieved by La_2O_3 in gate stack. This result can be considered both by oxygen vacancy passivation effect of La_2O_3 and thickness of HfO_2 with more oxygen vacancies decreasing [8].

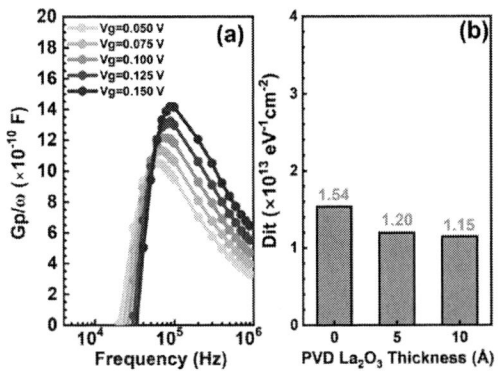

Figure 5: The variation of (a) Gp/ω with frequency at different gate voltage of device with 5 Å PVD La$_2$O$_3$ and (b) Dit values.

For the interface quality, the conductance method with multi-frequency C-V measurement from 1 KHz to 1 MHz was applied, and the measured parallel conductance/frequency (Gp/ω) curve at different gate voltage is showed in Fig.5(a). Fig.5(b) presents the calculated Dit value of devices with different thickness of PVD La$_2$O$_3$ in gate stack. With the La$_2$O$_3$ thickness increasing from 0 Å to 10 Å, the Dit is decreased from 1.54×10^{13} eV^{-1} cm^{-2} to 1.15×10^{13} eV^{-1} cm^{-2}, indicating an obvious interface trap passivation effect by La$_2$O$_3$ in gate stack [9]. Both Not and Dit decreasing exhibits an effective ability on defect suppression in gate stack bulk and at interface by ultra-low temperature magnetron sputtering La$_2$O$_3$ in gate stack.

CONCLUSION

In this paper, a La$_2$O$_3$/HfO$_2$ bi-layer DF process is proposed and comprehensively investigated by room temperature magnetron sputtering dielectric laminates. A 60 mV V$_{FB}$ fine range negative modulation is achieved by 10 Å PVD La$_2$O$_3$, and can be precisely controlled by La$_2$O$_3$ thickness in sub-1nm. Furthermore, the Not and Dit are suppressed by La$_2$O$_3$/HfO$_2$ bi-layer DF process to improve interface trap and electron trap/detrap density. These results exhibit a promising bi-layer DF process in low thermal integration for advanced IC technology.

ACKNOWLEDGEMENTS

This work is supported by Beijing Natural Science Foundation program (Grant No. 4224096), Strategic Priority Research Program of the Chinese Academy of Sciences (Grant No. XDA0330302), and Beijing nova program (Grant No. Z201100006820084). The authors would like to thank all staff at the Integrated Circuit Advanced Process Center, IMECAS, for their support in the fabrication on 8-inch CMOS pilot line.

REFERENCES

[1] M. -J. Yu, R. -P. Lin, Y. -H. Chang and T. -H. Hou, "High-Voltage Amorphous InGaZnO TFT With Al2O3 High- k Dielectric for Low-Temperature Monolithic 3-D Integration," in *IEEE Transactions on Electron Devices*, vol. 63, no. 10, pp. 3944-3949, Oct. 2016.

[2] J. Yao et al., "Record 7(N)+7(P) Multiple VTs Demonstration on GAA Si Nanosheet n/pFETs using WFM-Less Direct Interfacial La/Al-Dipole Technique," 2022 International Electron Devices Meeting (IEDM), San Francisco, CA, USA, 2022, pp. 34.2.1-34.2.4.

[3] K. Kakushima et al., "Origin of flat band voltage shift in HfO2 gate dielectric with La2O3 insertion," Solid-State Electronics, vol. 52, no. 9, pp. 1280-1284, 2008.

[4] B. Lee, S. R. Novak, D. J. Lichtenwalner, X. Yang and V. Misra, "Investigation of the Origin of V$_T$/V$_{FB}$ Modulation by La$_2$O$_3$ Capping Layer Approaches for NMOS Application: Role of La Diffusion, Effect of Host High-k Layer, and Interface Properties," in IEEE Transactions on Electron Devices, vol. 58, no. 9, pp. 3106-3115, Sept. 2011.

[5] H. Arimura et al., "Dipole-First Gate Stack as a Scalable and Thermal Budget Flexible Multi-Vt Solution for Nanosheet/CFET Devices," 2021 IEEE International Electron Devices Meeting (IEDM), San Francisco, CA, USA, 2021, pp. 13.5.1-13.5.4.

[6] L. N. Liu, W. M. Tang, and P. T. Lai, "Advances in La-Based High-k Dielectrics for MOS Applications," Coatings, vol. 9, no. 4.

[7] R. Xu et al., "Experimental Investigation of Ultrathin Al$_2$ O$_3$ Ex-Situ Interfacial Doping Strategy on Laminated HKMG Stacks via ALD," in IEEE Transactions on Electron Devices, vol. 69, no. 4, pp. 1964-1971, April 2022.

[8] E. Nadimi et al., "Interaction of oxygen vacancies and lanthanum in Hf-based high-k dielectrics: an ab initio investigation," Journal of Physics: Condensed Matter, vol. 23, no. 36, p. 365502, 2011/08/25 2011.

[9] K. Kakushima et al., "Interface and electrical properties of La-silicate for direct contact of high-k with silicon," Solid-State Electronics, vol. 54, no. 7, pp. 715-719, 2010.

LARGE AREA CVD MOS$_2$ MEMRISTOR SUITABLE FOR NEUROMORPHIC APPLICATIONS

Muhammad Zaheer[1], Tariq Aziz[1], Jun Lan[1], Quanzhou Zhu[1], Wenhui Wang[1], Mei Shen[1], Feichi Zhou[1], Longyang Lin[1], Xuewei Feng[2*], and Yida Li[1*]

[1]Southern University of Science and Technology, Shenzhen 518055, China
[2]Shanghai Jiao Tong University, Shanghai 200240, China
*Corresponding Author's Email: fengxw@sjtu.edu.cn, liyd3@sustech.edu.cn

ABSTRACT

Two-dimensional materials (2DMs) are promising candidates for future electronics due to their atomic thickness and excellent electronic properties. 2DMs based memristors possess unique properties to achieve low operating voltages and efficient switching. In this work, we report on CVD grown, 5-layers MoS$_2$ memristors based crossbar array via transfer approach. The MoS$_2$ memristors show forming free, bipolar resistive switching, operating voltages of 0.7 V, and analog states. The resistive switching mechanism of the MoS$_2$ memristors are revealed to be space charge limited conduction (SCLC) and conductive filament operation mechanism via I-V analysis. Our results pave the way towards enabling large-area 2DMs memristor crossbar array in future neuromorphic applications.

INTRODUCTION

Among all 2D materials, Molybdenum disulfide (MoS$_2$) has attracted tremendous research interest due to their excellent electrical features, tunable bandgap and excellent chemical properties. MoS$_2$, atomic layered transition-metal dichalcogenides (TMDs), has been used in various applications such as photodetectors, energy storage, tribological fields and electronics. Specifically, the possibility of bandgap modulation via layers number engineering has attracted this class of materials much research interests in logic and memory devices. Of which, the use of MoS$_2$ as memristors is promising due to its unique electrical properties. However, much work is still needed to improve the performance of MoS$_2$-based memristor devices, such as research on robustness, stability, and set voltage tunning.

Most of the recently reported MoS$_2$ memristors rely on the mechanical exfoliation method, however this approach encounters scalability, uniformity and size control problems when fabricating large-scale crossbar arrays. In order to solve these problems, various approaches have been adopted such as chemical vapor deposition (CVD), atomic layer deposition (ALD), and pulse laser deposition (PLD). Among these, CVD is a commonly used method to produce high-quality and large area MoS$_2$ films. To date, three terminal memtransistors are widely investigated using CVD grown polycrystalline MoS$_2$, however the set voltage is very high. For instance, Sangwan et al., reported gate-tunable MoS$_2$ memtransistor with a set voltage up to 80 V [1]. Leveraging on the top gate configuration and high-k gate dielectric, Wang et al., reduces the set voltage to 10 V [2], but still far from meeting the low-power requirement of in-memory computing. To the best of our knowledge three terminal memtransistor devices usually have switching voltage much larger than two terminal memristors, therefore they are not ideal choices for use in large scale circuitry, instead, two terminal memristor is a better candidate. Kalita et al., fabricated Gr/MoS$_2$ memristor by CVD method having Ni top and Gr as bottom electrode, the switching voltage of the device was about 2.8 to 2.9 V [3]. However, the above-mentioned reports still show high operating/switching voltages. In this aspect, a comprehensive study of CVD grown MoS$_2$ for low operating voltage/switching voltages has yet to be conducted.

Hence in this work, we report on a large area memristor crossbar array based on 5-layer (few layers) CVD grown MoS$_2$ [4, 5]. The entire process is fully compatible with current CMOS technology. The resultant few layer memristor devices exhibit forming free bipolar memristor behavior with low operating voltage of 1 V to -1 V, analog switching behavior, and retention exceeding 100 hours at 85℃. The switching mechanism is then elucidated via I-V analysis to be due to a combination of space charge limited conduction (SCLC) and conductive filament.

EXPERIMENT

The optical image and process flow of the memristor crossbar array (10×10) are shown in Figure 1 (a) and (b) respectively.

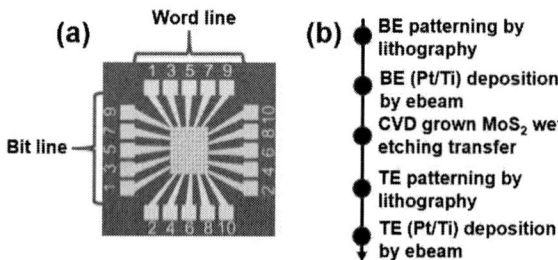

Figure 1:(a) Optical image of crossbar array taken by optical microscope, (b) process flow of large area MoS$_2$ memristor crossbar array.

The memristor crossbar is of a MIM structure where the switching layer or intermediate layer is sandwiched between top electrodes (TE) and bottom electrodes (BE). Both the TE (Pt/Ti) and BE (Pt/Ti) are deposited using an e-beam evaporation method where Pt has 20 nm and Ti has 5 nm thickness. The switching layer is made of MoS_2 which is transferred by PMMA assisted transferred method [5] from sapphire substrate, which was grown by CVD method. The peak difference of A_{1g} and E^1_{2g} is 24.2 cm^{-1}, which reveals that the MoS_2 film has a 5-layer structure [6]. Moreover, Raman spectra from two distinct areas reveals that the MoS_2 film is uniform as shown in Figure 2. The fabricated 10×10 memristor array has a single cell size of 5×5 μm^2. Individual devices are selected by applying biases to the appropriate bit line and word line. The electrical characteristics were measured using Keysight B1500A semiconductor parameter analyzer and Everbeing C-4 probe station with a temperature-controlled chuck. Positive sweep was applied to the top electrode, whereas the bottom electrode was constantly grounded during all measurements.

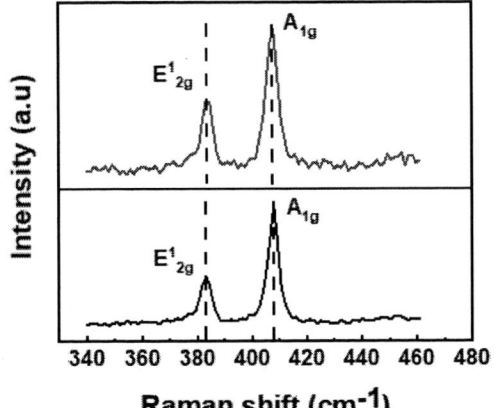

Figure 2: Raman spectra of CVD grown MoS_2 after transfer to crossbar arrays shows A_{1g} and the E^1_{2g} peak of five layers.

RESULTS AND DISCUSSION

Figure 3(a) shows the semi-log electrical current-voltage features of MoS_2 device under ambient conditions where voltage bias varies from 1 to -1 V with 1 mA compliance current. A typical bipolar memristor behavior with an HRS/LRS ratio of ~10 is seen, where HRS is high resistive state and LRS is low resistive state. In addition, no electroforming process was observed for our fabricated devices. Figure 3(b) shows the same device retention characterization over 100 hours at 85℃ at a reading voltage of 0.2 V. Excellent stability of the LRS and HRS, with no signs of degradation are seen. Figure 3(c) shows the multi-states characterizations of the device with current compliance of 100/300/1000 µA during the set process. Figure 3(d) plots the LRS of the device as a

function of the different current compliance at 0.2 reading voltage, showing that good conductance linearity can be achieved within the range of current compliance of 100 – 500 µA; beyond 500 µA, the LRS saturates.

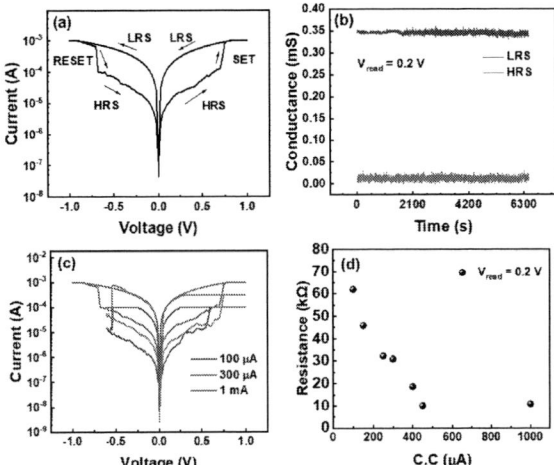

Figure 3: Semi-logarithmic I-V characterization of MoS_2 memristor (a) Electrical characterization under +1 V to -1 V and 1 mA compliance current, (b) Retention characterization for over 100 hours at 0.2 reading voltage. (c) I-V curve under compliance current of 100/300/1000 µA, and (d) Multiple resistive states under various compliance current at 0.2 reading voltage.

In order to investigate the conductance mechanism of the memristor device, the semi-logarithmic I-V curve of Figure 3 (a) was further explored using double logarithmic curve fitting as shown in figure 4(a) and 4(b). Initially under positive electrical bias the fitting slope was ~0.98 showing ohmic behavior, corresponding to HRS state. As the sweeping voltage increases the fitting slope ~1.86 shows child's law behavior. Further, analysis suggests trap-controlled space charge limited conduction (SCLC) accounts for state switching behavior. This behavior is reported by many research groups [7-9] and may account for the rapid changes in current. Then further increase in sweep voltages shows sharp increased tendency which leads to state changes i.e., from HRS to LRS state (slope ~0.99) and finally device switch to SET state, hence SCLC dominates the conduction mechanism in the positive region. When the sweep voltage direction changes 0 to -1.0 V, the curve fitting has slope of ~1 which depicts ohmic behavior, this evident conductive filament behavior [10]. As the applied voltage further increases state change occurs and devices switch to HRS state (slope fitting ~1). Hence under the negative voltage regime, the Pt atoms under the electric bias may oxidize and migrate to the bottom electrode. This shows diffusion of the conductive filament and the resistance switching occurs i.e., from LRS to HRS. So, the conductive filament behavior dominates in the negative region. Table I shows the comparison of various related 2DM memory devices with the MoS_2 memristor presented in this work, showing the lowest switching voltage with potential for large-scale implementation.

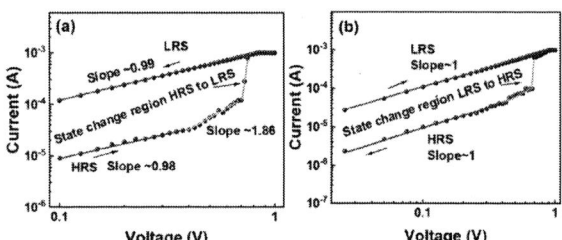

Figure 4: Double log fitting illustration at (a) set and (b) reset process shows SCLC and conductive filament behavior.

TABLE I: PERFORMANCE PARAMTER BENCHMARK WITH VARIOUS REPORTED WORK

Switching Layer	Switching voltage (V)	Electrodes	Growth method	Device structure
Gr/MoS₂ [3]	2.8-2.9 V	Ni (TE) Gr (BE)	CVD	Memristor
1L-MoS₂ [11]	5-9 V	N⁺⁺ - Si bottom gate	CVD	Transistor
1L-MoS₂ [12]	3.5-8.3 V	Ti (TE), Si (BE)	CVD	Gate-tunable memristor
1L-MoS₂ [1]	Threshold voltage: 20 V(HRS), 10 V LRS)	Ti/Au (TE), Si (BE)	CVD	Mem-transistor
5L-MoS₂ This work	0.7 V	Pt/Ti	CVD	memristor

CONCLUSION

We reported on MoS_2 memristor devices prepared by CVD grown and PMMA assisted transfer method with forming free, low operating voltage (0.7 V set and -0.6 V reset), with good retention. Raman spectra shows successful transfer of 5-layer MoS_2 on desired (10 × 10) high density crossbar arrays. Curve fitting shows SCLC behavior mechanism dominates in the positive bias regime and during the negative bias region conductive filament-based behavior dominates. Our devices show route to the high-density CMOS compatible MoS_2 based memristor devices for emerging memory applications.

ACKNOWLEDGEMENTS

This work was supported by the National Natural Science Foundation of China (Grant No. 62174074, 52273246), Young Innovative Talent Project Research Program (Grant No. 2021KQNCX077), Guangdong Provincial Department of Education Innovation Team Program (2021KCXTD012) and the Shenzhen Fundamental Research Program (Grant No. JCYJ20220530115014032, JCYJ20220530115204009, JCYJ20190809143419448). We would also like to acknowledge the Core Research Facilities (CRF) at SUSTech for the facilities used, and the technical support provided by the staff and engineers at the CRF.

REFERENCES

[1] V. K. Sangwan *et al.*, "Multi-terminal memtransistors from polycrystalline monolayer molybdenum disulfide," *Nature,* vol. 554, no. 7693, pp. 500-504, 2018.

[2] L. Wang *et al.*, "Artificial synapses based on multiterminal memtransistors for neuromorphic application," *Advanced Functional Materials,* vol. 29, no. 25, p. 1901106, 2019.

[3] H. Kalita *et al.*, "Artificial neuron using vertical MoS_2/graphene threshold switching memristors," *Scientific reports,* vol. 9, no. 1, pp. 1-8, 2019.

[4] H. Yu *et al.*, "Wafer-scale growth and transfer of highly-oriented monolayer MoS_2 continuous films," *ACS nano,* vol. 11, no. 12, pp. 12001-12007, 2017.

[5] X. Li and H. Zhu, "Two-dimensional MoS_2: Properties, preparation, and applications," *Journal of Materiomics,* vol. 1, no. 1, pp. 33-44, 2015.

[6] J. Tao *et al.*, "Growth of wafer-scale MoS_2 monolayer by magnetron sputtering," *Nanoscale,* vol. 7, no. 6, pp. 2497-2503, 2015.

[7] J. Chen *et al.*, "High-performance memristor based on MoS_2 for reliable biological synapse emulation," *Materials Today Communications,* vol. 32, p. 103957, 2022.

[8] W. Wang *et al.*, "MoS_2 memristor with photoresistive switching," *Scientific Reports,* vol. 6, no. 1, pp. 1-11, 2016.

[9] J.-W. Lee and W.-J. Cho, "Fabrication of resistive switching memory based on solution processed PMMA-HfO_x blended thin films," *Semiconductor Science and Technology,* vol. 32, no. 2, p. 025009, 2017.

[10] J. Lan *et al.*, "Zinc-Alloyed HFO₂ Synaptic RRAM with Operating Voltage and Switching Energy Enhancement," in *2022 China Semiconductor Technology International Conference (CSTIC),* 2022: IEEE, pp. 1-3.

[11] G. He *et al.*, "Thermally assisted nonvolatile memory in monolayer MoS_2 transistors," *Nano letters,* vol. 16, no. 10, pp. 6445-6451, 2016.

[12] V. K. Sangwan *et al.*, "Gate-tunable memristive phenomena mediated by grain boundaries in single-layer MoS_2," *Nature nanotechnology,* vol. 10, no. 5, pp. 403-406, 2015.

NOVEL CHANNEL-ON-FIN (COF) IGZO-TFTS WITH ULTRA-SCALED BACK GATE LENGTH OF 23 NM

Shangbo Yang[1,2], Gaobo Xu[1,2*], Gangping Yan[1,2], Zhiyu Song[1,2], Guoliang Tian[1,2], Yanna Luo[1,2], Yinan Yan[1,2], Lianlian Li[1,2] and Huaxiang Yin[1,2]

1Institute of Microelectronics, Chinese Academy of Sciences, Beijing 100029, China
2 University of Chinese Academy of Sciences, Beijing 100049, China
*Corresponding Author's Email: xugaobo@ime.ac.cn

ABSTRACT

In this paper, we showed a novel Channel-On-Fin (COF) InGaZnO4 thin-film transistors (IGZO-TFTs) with ultra-scaled gate. The 23-nm practical gate length in the back-gate device was realized on P+ silicon substrate by the traditional spacer pattern shift technology. The devices obtained outstanding electrical characteristics, e.g., high on/off current ratio of 108, promising subthreshold swing (SS) of 98.3 mV/decade, and slight Drain-Induced Barrier Lowering (DIBL) of 11mV/V. In addition, the influence of gate insulator thickness was also investigated.

Keywords—IGZO; short channel effect; channel on fin (COF); thin film transistor (TFT);

INTRODUCTION

The amorphous IGZO-TFT has great potential for the future development of highly integrated high-performance memory due to its ultra-low turn-off leakage current, large switching current ratio, low process temperature, and Back End of Line (BEOL) process compatibility [1-2]. There are recent implementations and understandings of IGZO nanoscale devices, but theses short-channel devices were defined based on source-drain spacing [3-5]. Due to the limitation of lithographic linewidth, the preparation of devices with practical short gate length (L) is still lacking, which is very unfavorable to the study of short channel effect of IGZO TFTs.

In this paper, we propose and fabricate a novel Channel-On-Fin (COF) IGZO TFTs based on spacer pattern shift technology. An ultra-scaled gate length of 23nm in COF devices is achieved. The influence of gate isolator thickness on the device performance is evaluated. The proposed device paves the way for the investigation of short channel effect in the IGZO TFTs.

EXPERIMENT

The process flow and structure of COF IGZO-TFT are shown in Fig.1(a) and (b), respectively. First of all, BF_3 ions were implanted into the Si substrate, and the rapid thermal annealing was carried out to active the ions, enhancing the conductivity of bottom Si. Then 23-nm Si-Fin was formed as the back-gate electrode using spacer pattern shift technology, where the Scanning Electron Microscope (SEM) picture is shown in Fig.1(c). The SiO_2 isolation layer was deposited and polished to expose

Si-Fin back-gate, followed by the deposition of HfO_2 as the gate insulator using Atomic Layer Deposition (ALD) at 300°C with Tetrakis (Ethylmethylamino) hafnium as precursor and water vapor as oxygen source. The 15-nm HfO_2 was deposited for Device COF-1, while the 6-nm gate dielectric was fabricated for Device COF-2. The 20 nm a-IGZO was deposited using magnetron sputtering (target material In: Ga: Zn= 1:1: 1), where the ratio of Ar and O2 was 30:3 sccm. The IGZO was patterned by the wet etching of the dilute nitric acid. Finally, the 60 nm Mo was sputtered and patterned using dry etching as the source/drain electrode, as shown in Fig. 1(d). The detail of two kinds of IGZO devices is listed in Table I.

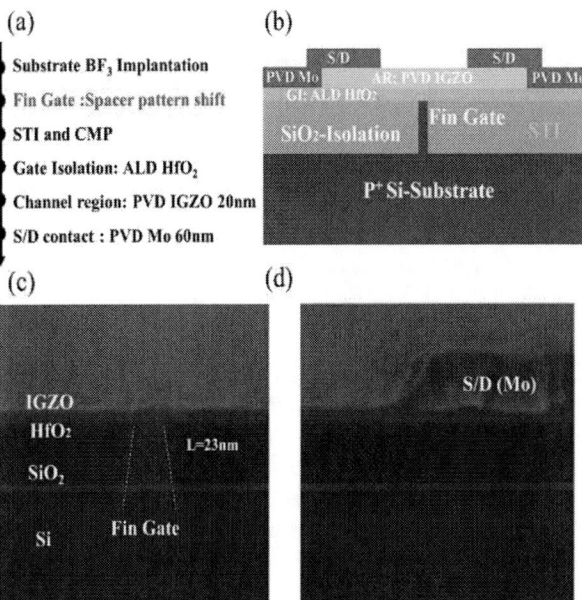

Fig 1. (a) The process flow of COF IGZO-TFT. (b) Structure diagram of the COF IGZO-TFT device. (c) SEM cross-section and (d) Mo source /drain electrode of the IGZO TFTs.

TABLE I.   THIN FILM GROWTH CONDITIONS OF DEVICES

Device	Thickness of Oxide	Thickness of IGZO
COF-1	15nm ALD HfO_2	20nm PVD IGZO
COF-2	6nm ALD HfO_2	20nm PVD IGZO

Fig.2. shows the transfer characteristic curve of Device COF-1 with L of 23 nm under the condition of V_{DS} = 1 V. The device obtained a high on/off current ratio (I_{on}/I_{off}) of 108. The threshold voltage (Vth) is about -0.75 V (defined at I_{DS} = 10 pA), and the Ion (defined as I_D at Vth + 2 V) is 321.9 nA, where the subthreshold swing (SS) is about 236.2 mV/decade.

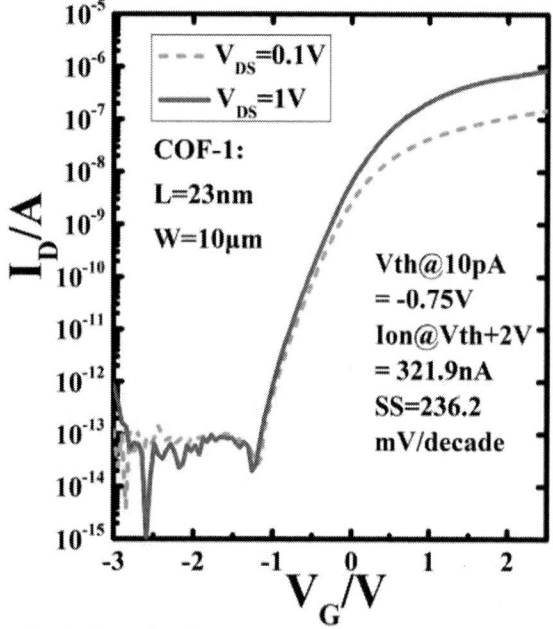

Fig 2. Transfer characteristic curves of Device COF-1 with a 15-nm HfO_2 gate isolator.

RESULTS AND DISCUSSION

The transfer characteristic curves of COF-2 device is shown in Fig. 3(a). After the HfO_2 thickness is reduced from 15nm to 6nm, the device still achieves a decent Ion/off ratio of 108 at V_{DS} = 1 V. The Vth, I_{on}, and SS are about -0.39 V, 686.0 nA, and about 98.3 mV/decade, respectively. The optimization of SS may be caused by the reduction of oxide layer thickness and electrostatic control enhancement of the gate. Fig. 3(b) shows the corresponding output characteristics of Device COF-2, demonstrating that the maximum driving current reaches 2.4µA under the gate voltage of 3V. The COF IGZO-TFTs all show good ohmic behavior since the output curves are not crowded in the linear region. These results suggest that the proposed COF IGZO-TFTs with a gate length of 23 nm meets the demands of practical short channel devices.

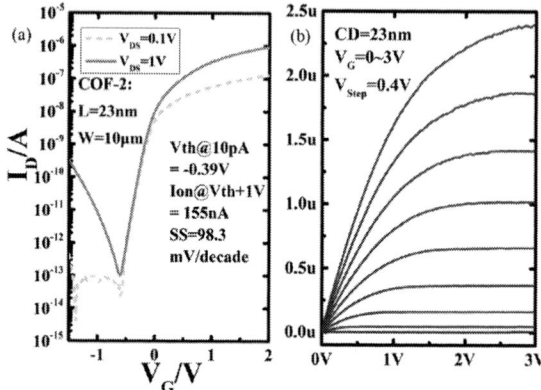

Fig 3. (a) Transfer characteristic curves and (b) output characteristics of Device COF-2 with a 6-nm HfO_2 gate isolation.

TABLE II.   DEVICES PERFORMANCE SUMMARY

W/L 10µm/23nm	COF-1	COF-2
SS(mV/dec)	236.2	98.3
V_{th} (V)	-0.75	-0.39
I_{on} (nA)	321.9	686.0
DIBL(mV/V)	66	11
I_{on}/I_{off}	10^8	10^8

Table 2 summarizes the device key parameters of COF-1 and COF-2. Comparing COF-1 and COF-2 devices, as HfO_2 thickness decreased from 10nm to 6nm, SS is reduced due to the increase of gate capacitance [4], and the threshold voltage also drifts forward due to the improvement of gate electrostatic control capability. As the subthreshold swing is smaller, COF-2 has higher I_{on}. The parasitic resistance caused by the source drain spacing is larger in the present device, resulting in hindrance to improvement of device performance. COF-1 and COF-2 have 66mV/V and 11 mV/V excellent DIBL characteristics respectively, indicating that reducing gate dielectric thickness can suppress DIBL effect [5]. It is worth mentioning that Device COF-2 shows Gate-Induced Drain Leakage (GIDL) effect, which may increase the static power consumption.

CONCLUSION

In this work, A novel COF IGZO-TFTs with an ultra-scaled gate was demonstrated. The device gate length of 23 nm was successfully realized based on the spacer pattern shift process technology, and the good device performances were achieved by reducing the thickness of the gate isolation layer. This work provided a feasible approach for the realization and SCE research of nanoscale IGZO TFTs.

ACKNOWLEDGEMENTS

This work was supported in part by the High-Density 3D IGZO DRAM Projects of the Beijing Superstring Academy of Memory Technology under Grant SAMT-ZK-KT-22030102.

REFERENCES

[1] Nomura, K., Ohta, H., Takagi, A. et al. Room-temperature fabrication of transparent flexible thin-film transistors using amorphous oxide semiconductors. Nature 432, 488–492 (2004).

[2] K. Han, et al, "Top-Gate Short Channel Amorphous Indium-Gallium-Zinc-Oxide Thin Film Transistors with Sub-1.2 nm Equivalent Oxide Thickness," 2021 5th IEEE Electron Devices Technology & Manufacturing Conference (EDTM), 2021,

[3] C. Wang et al., "Extremely Scaled Bottom Gate a-IGZO Transistors Using a Novel Patterning Technique Achieving Record High Gm of 479.5 μS/μm (VDS of 1 V) and fT of 18.3 GHz (VDS of 3 V)," 2022 IEEE Symposium on VLSI Technology and Circuits (VLSI Technology and Circuits), 2022, pp. 294-295.

[4] K. Chen et al., "Scaling Dual-Gate Ultra-thin a-IGZO FET to 30 nm Channel Length with Record-high Gm,max of 559 μS/μm at VDS=1 V, Record-low DIBL of 10 mV/V and Nearly Ideal SS of 63 mV/dec," 2022 IEEE Symposium on VLSI Technology and Circuits (VLSI Technology and Circuits), 2022, pp. 298-299,

[5] K. Han, et al, "High Field Temperature-Independent Field-Effect Mobility of Amorphous Indium–Gallium–Zinc Oxide Thin-Film Transistors: Understanding the Importance of Equivalent-Oxide-Thickness Downscaling," in IEEE Transactions on Electron Devices, vol. 68, no. 1, pp. 118-124, Jan. 2021.

INVESTIGATION OF VERTICAL CHANNEL IGZO-TFT BASED ON PVD-IGZO

Zhiyu Song [1, 2], Gaobo Xu [1,2*], Gangping Yan, Shangbo Yang [1, 2], Yanna Luo [1, 2], Guoliang Tian [1, 2], Yinan Yan [1, 2] and Huaxiang Yin [1, 2]

[1] Institute of Microelectronics, Chinese Academy of Sciences, Beijing 100029, China
[2] University of Chinese Academy of Sciences, Beijing 100049, China
*Corresponding Author's Email: xugaobo@ime.ac.cn

ABSTRACT

In this paper, a step type vertical channel In-Ga-Zn-O thin-film transistor (IGZO-VTFT) with two independent gates on either side of the steps and inverters consist of the VTFTs have been investigated towards high density integration with higher efficiency and lower cost. The TFTs have on/off current ratio over 1×10^8 and SS of 118 mV/dec. The reversal source-drain characteristics is also investigated. The inverter based on the VTFTs have gain of 6.1.

Keywords—IGZO-TFT, vertical channel transistor, PVD IGZO, inverter.

INTRODUCTION

Due to its low-temperature fabrication and back end of line (BEOL) compatible, IGZO-TFT has great potential for high-density monolithic 3D integration[1, 2]. As a more compact transistor structure, the vertical channel IGZO TFT (IGZO-VTFT) can further improve the integrated density at the same gate length. Previous reports of IGZO-VTFT focus on the impacts of atomic layer deposition (ALD) IGZO, demonstrating a promising prospect in scaling devices[3-5]. As a widely used deposition method, the physical vapor deposition (PVD) process has a faster growth rate and lower cost. However, the research on PVD IGZO-VTFT with high performance and reliability is still lacking, and the behaviors of its logic application like inverters remains unknown.

In this paper, Step type PVD IGZO VTFT with low thermal budget (< 400 °C) has been investigated. By annealing at 350°C, the influence of shallow donor H introduced by the ALD gate dielectric process is removed, and the threshold voltage of the device is optimized. The IGZO-VTFTs achieve on/off current ratio over 1×10^8 and subthreshold swing (SS) of 118 mV/dec. The reversal source-drain characteristics is also investigated. Benefitted from the good device performance, the inverter with a decent gain of 6.1 is employed.

EXPERIMENT

Fig.1(a) and (b) show the schematic diagram of the structure and process of the devices, respectively.

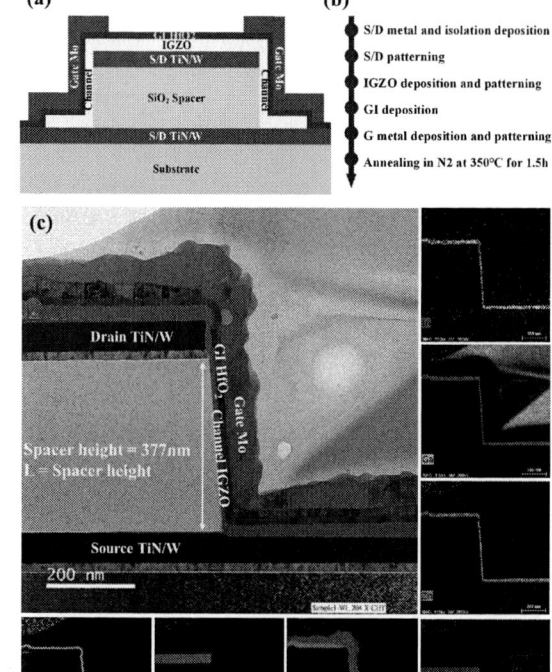

Figure.1 (a) Process flow of device fabrication; (b) schematic diagram of the device structure; (c) TEM image and EDS map for several elements of the devices in one side of the step.

For the IGZO-VTFT, magnetron sputtered 60-nm molybdenum as the source electrode. Next, 400 nm plasma enhanced chemical vapor deposition (PECVD) SiO_2 was deposited as inter-layer dielectric (ILD). Then magnetron sputtered 60-nm molybdenum was deposited as the drain electrode, and patterned down to the source electrode. Afterward, the IGZO was deposited by magnetron sputtering using a ceramic target (In: Ga: Zn = 1:1:1 mol%) at room temperature as active area, where the flow rate ratio of argon : oxygen during sputtering was 50:5 sccm. The active area was patterned using wet etch by dilute nitric acid. ALD 15nm HfO_2 as the gate dielectric. Finally, 60nm molybdenum was deposited and patterned by dry etching as the gate electrode. The fabricated devices are annealed in 350°C in N_2 atmosphere for 1.5

hours. Transmission electron microscope (TEM) image for cross-sectional view and Energy Dispersive Spectroscopy (EDS) map are shown in Fig.1(c). The gate length of the devices is 377nm, which is the thickness of the ILD. The IGZO on the side wall is 13-nm thick and covers evenly, and the elements have been not diffused. The electrical characteristics of the devices were measured using Agilent 4156C.

RESULT AND DISCUSSION

Figure. 2 (a) and (b) show the transfer and output curves of the IGZO-VTFTs before and after annealing. Detailed performance parameters of the IGZO-VTFT after annealing are listed in Table 1. The as-fabricated devices have negative V_{th} due to the large amount of hydrogen donors introduced by the ALD process of the gate insulator [6]. After 350°C annealing for 1.5h, extra H atoms escapes from the IGZO film, causing the device threshold voltage to drift forward to 0.42V. The devices obtain SS of 118 mV/dec, leakage current less than 10^{-14}A and on/off current ratio over 10^8. The output characteristic curve indicates drive current reaches 5μA when $V_G=V_D=3V$. The output curves under small drain voltage exhibit a rectification characteristic, indicating the source/drain contact are Schottky contact.

Table. 1 Detail parameters of performance for the VTFTs after annealing

W/L(μm)	20/0.4
on/off ratio	$>10^8$
SS (mV/dec)	118
V_{th} (V)	0.42
I_{off} (A/μm)	<1f
I_{on} (μA/μm) (I=Vth+1V)	0.1

Due to the asymmetry of IGZO-VTFT structure and the different damage of the top and bottom electrodes of the steps during the process, VTFT has the characteristics of asymmetric source-drain performance. The transfer characteristic curves of the top and bottom source-drain electrodes after reversal and their Drain Induced Barrier Lowering(DIBL) effect are discussed. After reversing the source-drain, the on-state current of the VTFT decreases, the off-state leakage current increases, and the SS deteriorates, as shown in Figure 3(a). This may be due to the fact that the plasma is more severely damaged to the bottom electrode during etching and dry degumming, resulting in an additional Schottky barrier created by IGZO contact with the bottom electrode. Moreover, the film near the contact between the bottom electrode and IGZO is thick, which made the gate control ability is weakly, resulting in the performance degradation of the device after source-drain reversal.

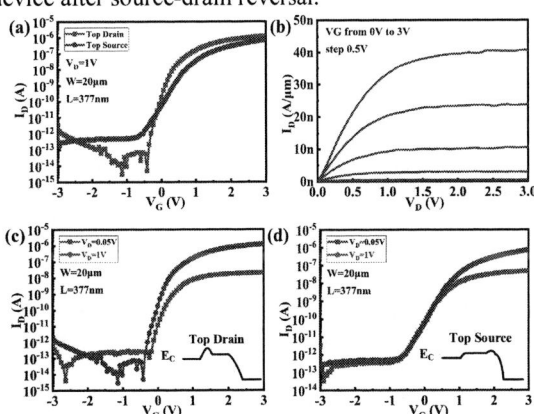

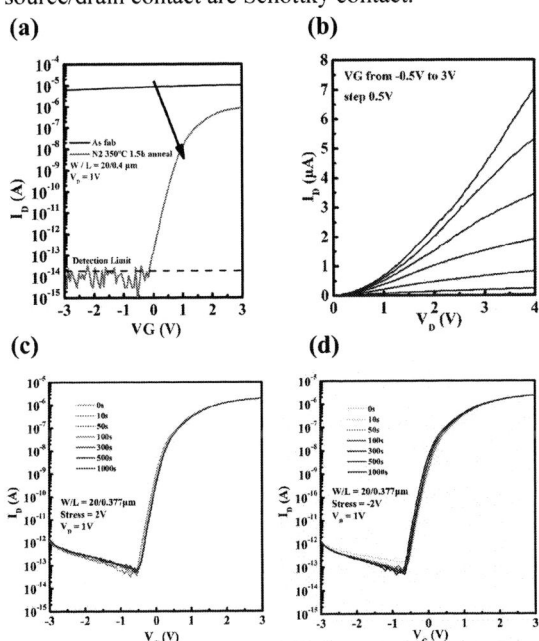

Figure.2 (a) transfer curve of the IGZO-VTFT; (b) output curve of the IGZO-VTFT; (c) PBS of VTFTs; (d) NBS of VTFTs.

Figure.2 (c) and (d) shows the transfer curves with the positive and negative bias stress. The results show that under the bias stress of +2/-2V, the threshold voltage drift after 1000s is about +67mV/-78mV.

Figure.3 (a) Transfer curve of the IGZO-VTFT with before and after reversal Source-Drain; (b) output curve of the IGZO-VTFT with reversal Source-Drain; (c) the DIBL of the VTFTs before reverse the Source-Drain; (d) the DIBL of the VTFTs after reverse the Source-Drain.

Figure 3(b) shows the output characteristic curve after source-drain reversal, and its drive current is 40nA/μm at $V_G=V_D=3V$. When the drain voltage is small, the linearity of the drain current curve is good, and there is no obvious crowding phenomenon, which proves that the top electrode of the VTFT is in good contact. Figure 3(c) and (d) show the DIBL characteristics before and after source-drain reversal, with DIBL significantly decreasing after source-drain reversal. After the source-drain reversal, the bottom electrode is the drain, and the additional barrier compensates for the lower barrier caused by the drain voltage, making it have a smaller DIBL.

The schematic diagram of Pseudo-CMOS inverter design circuit is shown in Figure 4(a). The inverter structure consisting of above VTFTs is shown in Figure 4(b). Compared to planar devices, vertical devices have a smaller cell area at the same gate length, enabling higher integration density. The VTC curve is shown in Figure 4 (c), when the VDD = 0.5V, 1V, 1.5V and 2V. The V_{IN} is from -0.5V to 1V. As Figure 4(d) shows, the gain of the inverter is 6.1 when the VDD at 2V.

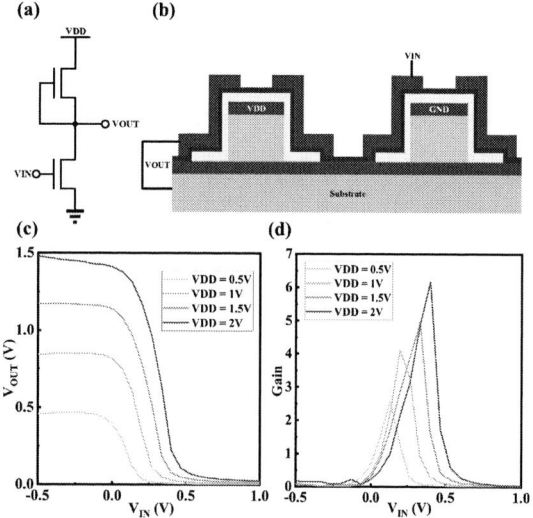

Figure.4 (a) schematic diagram of the inverter circuit; (b) schematic diagram of the inverter structure; (c) the VTC curve of the inverter. (d)Gain of the inverter.

CONCLUSION

In this work, the performance, reliability and simple logic units based on step type vertical channel IGZO-TFTs with low thermal budget (<400°C) have been investigated. The VTFTs achieved on/off current ratio over 1×10^8 and subthreshold swing (SS) of 118 mV/dec. The inverter composed of the VTFTs obtain gain of 6.1. It provides ideas for high density monolithic 3D integration with higher efficiency and lower cost.

ACKNOWLEDGEMENTS

This work was supported in part by the High-Density 3D IGZO DRAM Projects of the Beijing Superstring Academy of Memory Technology under Grant SAMT-ZK-KT-22030102.

REFERENCES

[1] A. Belmonte *et al.*, "Capacitor-less, Long-Retention (>400s) DRAM Cell Paving the Way towards Low-Power and High-Density Monolithic 3D DRAM," presented at the 2020 IEEE International Electron Devices Meeting (IEDM), 2020.

[2] X. Duan *et al.*, "Novel Vertical Channel-All-Around(CAA) IGZO FETs for 2T0C DRAM with High Density beyond 4F2 by Monolithic Stacking," in *2021 IEEE International Electron Devices Meeting (IEDM)*, 11-16 Dec. 2021 2021, pp. 10.5.1-10.5.4.

[3] S.-N. Choi and S.-M. Yoon, "Implementation of In-Ga-Zn-O Thin-Film Transistors with Vertical Channel Structures Designed with Atomic-Layer Deposition and Silicon Spacer Steps," *ELECTRONIC MATERIALS LETTERS,* Article vol. 17, no. 6, pp. 485-492, 2021 NOV 2021.

[4] Y.-M. Kim, H.-B. Kang, G.-H. Kim, C.-S. Hwang, and S.-M. Yoon, "Improvement in Device Performance of Vertical Thin-Film Transistors Using Atomic Layer Deposited IGZO Channel and Polyimide Spacer," *IEEE Electron Device Letters,* vol. 38, no. 10, pp. 1387-1389, 2017.

[5] H.-J. Ryoo *et al.*, "Device Characterization of Nanoscale Vertical-Channel Transistors Implemented with a Mesa-Shaped SiO2 Spacer and an In–Ga–Zn–O Active Channel," *ACS Applied Electronic Materials,* vol. 3, no. 9, pp. 4189-4196, 2021.

[6] H. Y. Noh, J. Kim, J.-S. Kim, M.-J. Lee, and H.-J. Lee, "Role of Hydrogen in Active Layer of Oxide-Semiconductor-Based Thin Film Transistors," *CRYSTALS,* Article vol. 9, no. 2, 2019 FEB 2019, Art no. 75.

A COMPACT MODEL OF NON-VOLATILE FERROELECTRIC TUNNEL FET WITH AMBIPOLARITY FOR IN-MEMORY-COMPUTING BASED EDGE AI

Hanyong Shao[1], Jin Luo[1], Zhiyuan Fu[1], Qianqian Huang[1,2,3], and Ru Huang[1,2,3*]*

[1]School of Integrated Circuits, Peking University, Beijing 100871, China
[2]Beijing Advanced Innovation Center for Integrated Circuits, Beijing 100871, China
[3]Chinese Institute for Brain Research, Beijing 102206, China.
*Corresponding Author's Email: hqq@pku.edu.cn; ruhuang@pku.edu.cn

ABSTRACT

The novel ferroelectric tunnel FET (FeTFET) with ambipolarity has attracted much attention for area- and energy-efficient edge AI applications. To facilitate the circuit simulation based on the new device, in this work, a compact FeTFET model is established through calculating the ambipolar current based on surface potential and tunneling paths of both the source and drain junctions under all-range of gate bias, and the ferroelectric model based on dynamic Preisach model is also included in the gate stack with non-volatility. Based on the proposed model, the FeTFET demonstrates XNOR operation in one device, showing its high applicability for further FeTFET-based circuit design and simulation.

INTRODUCTION

Tunnel FET (TFET) with band-to-band tunneling (BTBT) mechanism has attracted much attention as one of the most promising candidates for ultralow-power applications due to its steep subthreshold swing [1]. Moreover, by taking advantage of the ambipolar BTBT behavior of TFET [2] and combining CMOS-compatible hafnium oxide based ferroelectric (FE) gate stack, the linearly inseparable problem of XNOR-like logic can be implemented in one single ferroelectric tunnel FET, enabling content addressable memory (CAM) [3] and encryption-embedded multiply-accumulate operation [4] for non-volatile in-memory-computing (nvCIM) based edge AI with potential high area- and energy-efficiency.

To evaluate the performance of FeTFET-based nvCIM circuit, the compact model of FeTFET with both ambipolar behavior and ferroelectric behavior is necessary. However, most TFET-based models were mainly aimed at the logic applications where the ambipolarity behavior is not preferred [5][6], and thus the tunneling current under negative gate-to-source bias (V_{gs}) were not considered for the n-type device model.

In this paper, based on our previously developed analytical current and capacitance model of TFET under positive gate bias ([7-9]), we further establish the compact model of FeTFET for the first time. Based on the analysis of the symmetry of surface potential in tunneling junction of drain and source regions, the physical model of

ambipolar BTBT current under all-range gate bias (both positive and negative V_G & V_{gs}) is established. Combining with dynamic Preisach model for non-volatile modulation of FE gate stack [10], the compact model of FeTFET is developed in HSPICE, enabling simulation analysis for novel FeTFET-based nvCIM applications.

MODEL DERIVATION AND ANALYSIS

The modeling method of FeTFET device is shown in Fig. 1(a), and the FE layer is stacked on the gate of the TFET (n-type is considered in this paper as illustrated). By modeling the surface potential in both source and drain regions under all-range V_{gs}, the ambipolar BTBT current is obtained with analytical tunneling paths, along with source/drain to gate capacitance of TFET structure. Further connecting the dynamic Preisach model based FE layer to the gate of TFET, the FeTFET model is established.

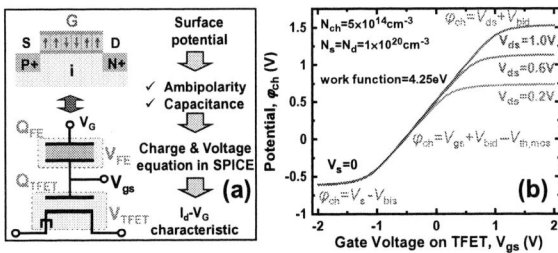

Figure 1: (a)Device structure, equivalent circuit diagram and the modeling method for FeTFET in this work. (b)Modeling results of channel surface potential far from source and drain regions at different gate and drain bias.

The surface potential model for all-range gate bias

The surface potential $\varphi_{ch}(V_{gs}, V_{ds})$ of the FeTFET channel far away from the source and drain under non-destructive readout mode, which is modulated by both gate and drain bias, is first calculated for further modeling lateral surface potential $\varphi_{sf}(x)$ and energy band structure of tunnel junction. Based on our previous $\varphi_{ch}(V_{gs}, V_{ds})$ model under only positive V_{gs} [8], the modulation of φ_{ch} by source under negative gate bias is calculated and the analytical $\varphi_{ch}(V_{gs}, V_{ds})$ under all-range

V_{gs} and V_{ds} is obtained by utilizing a nested iteration of a smooth join function (Eq. 1, all the equations are listed at the end of this paper) to ensure the continuation and derivability. Fig. 1(b) shows the model results of $\varphi_{ch}(V_{gs}, V_{ds})$. It can be seen that when the V_{gs} is relatively large, φ_{ch} is modulated by the source or drain potential, and when the V_{gs} is relatively small, it is linearly related to the gate bias.

Furthermore, φ_{ch} is used as one of the boundary conditions to solve the Poisson equation of the lateral surface potential at both source and drain tunnel junctions. Utilizing the parabolic approximation and Gaussian box method [8][9], the surface potential of source region under all-range V_{gs} can be obtained firstly (Eq. 2-3). According to the symmetry of surface potential in tunnel junction of drain and source regions, independent coordinates at these regions are set respectively in Fig. 2(a). An equivalent gate voltage parameter $V_{gs}^{mid} = (V_{ds} + V_s)/2$ is introduced to simplify the calculation of the potential at the drain junction (Eq. 4-5), with the premise of satisfying the boundary conditions and approximately consisting with the differential equations. Due to the condition of surface potential continuity, the width of depletion region (L_{d1}, L_{d2}) can be solved as Eq. 6. Therefore, the surface potential model under all-range V_{gs} for ambipolar BTBT in FeTFET is established and the modeling results are shown in Fig. 2(b)(c), which indicates dual-modulation effects of $\varphi_{sf}(x)$ with different gate and drain bias [7].

Ambipolar BTBT current modeling

Based on $\varphi_{sf}(x)$, the energy band of source and drain junctions can be derived (Fig. 3a & b) and the tunneling width W_t is obtained for further calculation of BTBT current. Considering the tunneling of holes at the drain junction in Fig. 3(b), the lateral distance corresponding to $\Delta\varphi = E_g/q$ is $W_{t,D}$, which is solved based on surface potential model (Eq. 2, 5) and expressed as Eq. 7. The beginning and ending point of the tunneling window (green shaded area) at the drain junction with minimum and maximum tunneling paths can be obtained (Eq. 8-10). Modeling of tunneling path at source junction is similar to the above discussion, which is given by Eq. 11-13.

Furthermore, the BTBT current is calculated based on the integral of tunneling generation rate G_{tun} in Kane model [11] (Eq. 14). The distribution of G_{tun} in the tunneling windows of both source and drain regions are considered. For x-direction parallel to the channel, the generation rate is dominated by the exponential term of W_t, and thus G_{tun} in the tunneling window will decay exponentially from $x = x_{begin}^{min}$ (corresponding to the shortest tunneling width) to the left and right sides until the tunneling window closes, that is, $G_{tun}(x) \approx G_{tun}^{max} \cdot \exp(-|x - x_{begin}^{min}|/\lambda_t)$. For y-direction, the G_{tun} decreases with depth exponentially as $G_{tun}(y) \approx G_{tun}^{max} \cdot \exp(-y/t_{eff})$, where t_{eff} is effective channel thickness. Finally, the integral of the ambipolar BTBT current can be expressed as Eq. 15. The simulated results based on our proposed model are shown in Fig. 3(c)(d). There is an off-state platform when V_{ds} is small since the tunneling windows in both source and drain junctions are not open (Fig. 3c). When V_{ds} increases, the output characteristic curve will saturate at V_{ds}=1.2V as given in Fig. 3d.

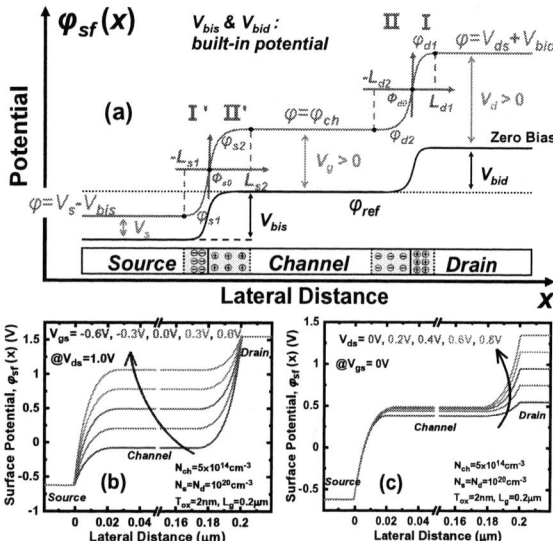

Figure 2: (a) Schematic diagram of the lateral surface potential $\varphi_{sf}(x)$ of the device and the depletion region near the tunnel junctions under the given gate and drain bias, two independent coordinates are established.
(b) Modeling results of $\varphi_{sf}(x)$ while drain bias stays at 1V as gate bias changes between -0.6V and 0.6V.
(c) Modeling results of $\varphi_{sf}(x)$ while gate bias stays at 0V as drain bias changes from 0V to 0.8V.

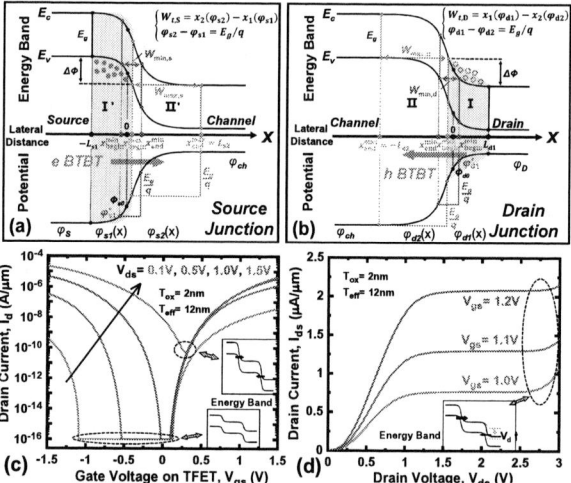

Figure 3: Energy band, surface potential, and tunneling paths at the (a) source- and (b) drain-channel junctions. Modeling results and energy band of the (c) transfer and (d) output curves without considering the FE switching.

FeTFET model with ambipolarity and non-volatility

The FeTFET device is constructed by connecting the FE capacitor to the gate of the TFET in series (Fig. 1a). Based on the proposed surface potential model, the effects of ambipolarity on inversion/depletion layer charge and gate capacitance model are also reconsidered [9]. Further combining the proposed analytical ambipolar TFET model with dynamic Preisach model based FE capacitor in HSPICE [9], the FeTFET with both ambipolarity and ferroelectric non-volatility is established.

The results of FeTFET model are shown in Fig. 4. When applying gate voltage sweeping with relatively low amplitude, the FeTFET is operated in the non-destructive readout mode without hysteresis in I_{ds}-V_G curve (Fig. 4a) since the voltage drop on the FE layer is not enough for FE polarization switching. When applying relatively high V_G (Fig. 4b) that can make the FE switching, the I_{ds}-V_G curve shows typical FE hysteresis with ambipolarity.

By applying different programming pulses and then measuring I_{ds}-V_G in non-destructive readout mode, the threshold voltage of FeTFET can be programmed to different states, representing different weights as shown in Fig. 4c. Only if input V_G level and stored weight are the same, the FeTFET will have a relatively high current I_H, otherwise, it is in the off-state with low current I_L (red), indicating the XNOR-operator (Fig. 4d).

CONCLUSION

In this paper, based on the surface potential model in tunneling junction of drain and source regions, a compact model of FeTFET is established with ambipolar BTBT current under both positive and negative V_G and non-volatile modulation of FE gate stack. The XNOR operation in one FeTFET device is also simulated based on proposed model, showing its high applicability for further FeTFET-based circuit design and simulation for edge AI application.

ACKNOWLEDGEMENTS

This work was supported by National Key R&D Program of China (2018YFB2202801), NSFC (61927901), Beijing SAMT Project (SAMT-BD-KT-22030101), 111 Project (B18001), and Tencent Foundation through the Xplore Prize.

REFERENCES

[1] Q. Huang et al., *IEDM*, pp. 187-190, 2012.
[2] X. Li et al., *DRC*, pp. 101-102, 2019.
[3] J. Luo et al., *VLSI*, pp. 226-227, 2022.
[4] J. Luo et al., *IEDM*, pp. 36.5.1-36.5.4, 2022.
[5] P. Xu et al., *IEEE TED*, 64(12), 5242-5248, 2017.
[6] Y. Guan et al., *IEEE TED*, 65(11), 5213-5217, 2018.
[7] C. Wu et al., *IEEE TED*, 61(8), 2690-2696, 2014.
[8] C. Wang et al., *SCI CHINA INF SCI*, 58(2), 1-8, 2015.
[9] J. Wang et al., *J. Appl. Phys*, 116(9), 094501, 2014.
[10] Z. Fu et al., *CSTIC*, pp. 1-4, 2020.

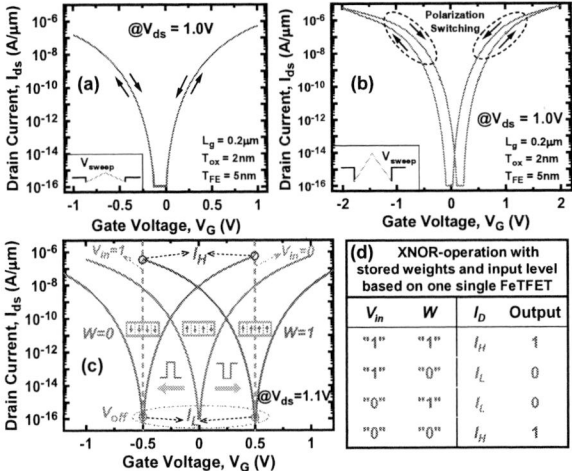

Figure 4: Modeling results of I_{ds}-V_G curve with relatively (a) low and (b) high sweep amplitude showing hysteresis. (c) I_{ds}-V_G curves after applying different programming pulses and the implement schematic of XNOR-operation; (d) Truth table of FeTFET based XNOR-operation.

A. Surface Potential Model (ψ_{ch} & ψ_{sf})

$$\psi(x) = \frac{1}{2}\left[\varphi_1(x) + \varphi_2(x) \pm \sqrt{[\varphi_1(x) - \varphi_2(x)]^2 + \delta^2}\right] \quad (1)$$

$$\varphi_{d1}(x) = -\frac{qN_d}{2\varepsilon_{Si}}(x - L_{d1})^2 + (V_{ds} + V_{bid}) \quad (2)$$

$$\varphi_{d2}^*(x) = \underbrace{\left[\varphi_{ch} - (V_{gs} - V_{fb} - \phi_0)\right]\cosh\left(\frac{x + L_{d2}}{\lambda}\right)}_{f(V_{gs})} + (V_{gs} - V_{fb} - \phi_0) \quad (3)$$

$$f'(V_{gs}) = -f(2V_{gs}^{mid} - V_{gs}) = -\varphi'_{ch} + (V'_{gs} - V_{fb} - \phi_0) \quad \text{in which } V'_{gs} = 2V_{gs}^{mid} - V_{gs} \atop \text{and } \varphi'_{ch} = \varphi_{ch}(2V_{gs}^{mid} - V_{gs}) \quad (4)$$

$$\varphi_{d2}^*(x) = \underbrace{\left[-\varphi'_{ch} + (V'_{gs} - V_{fb} - \phi_0)\right]\cosh\left(\frac{x + L_{d2}}{\lambda}\right)}_{f'(V_{gs})} + \varphi_{ch} + \varphi'_{ch} - (V'_{gs} - V_{fb} - \phi_0) \quad (5)$$

$$L_{d1} = \sqrt{\frac{-\Phi_{d0} + (V_{ds} + V_{bid})}{qN_d/2\varepsilon_{Si}}} \qquad L_{d2} = \lambda \cdot \cosh^{-1}\left[\frac{\Phi_{d0} - (\varphi_{ch} + \varphi'_{ch} - (V'_{gs} - V_{fb} - \phi_0))}{-\varphi'_{ch} + (V'_{gs} - V_{fb} - \phi_0)}\right]$$

$$\Phi_{d0} = \sqrt{\left[-\varphi'_{ch} + (V'_{gs} - V_{fb} - \phi_0)\right]^2 - 2\phi_d\left[(V_{ds} + V_{bid}) - (\varphi_{ch} + \varphi'_{ch} - (V'_{gs} - V_{fb} - \phi_0))\right] + \phi_d^2} \quad (6)$$
$$-\phi_d + (\varphi_{ch} + \varphi'_{ch} - (V'_{gs} - V_{fb} - \phi_0))$$

in which $\lambda^2 = t_{ox}t_{eff}\varepsilon_{Si}/\varepsilon_{ox}$ $\phi_0 = qN_{ch}\lambda^2/\varepsilon_{Si}$ $\phi_d = qN_d\lambda^2/\varepsilon_{Si}$

B. Tunneling Window Model (W_{tun} & x_{tun})

$$W_{t,D}(\varphi_{d1}) = \left[L_{d1} - \sqrt{\frac{-\varphi_{d1} + (V_{ds} + V_{bid})}{qN_d/2\varepsilon_{Si}}}\right] - \left\{\lambda \cdot \cosh^{-1}\left[\frac{\varphi_{d1} - E_g/q - (\varphi_{ch} + \varphi'_{ch} - (V'_{gs} - V_{fb} - \phi_0))}{-\varphi'_{ch} + (V'_{gs} - V_{fb} - \phi_0)}\right] - L_{d2}\right\} \quad (7)$$

$$\varphi_{d1}^* = \sqrt{\left[-\varphi'_{ch} + (V'_{gs} - V_{fb} - \phi_0)\right]^2 - 2\phi_d\left[(V_{ds} + V_{bid}) - (\varphi_{ch} + \varphi'_{ch} - (V'_{gs} - V_{fb} - \phi_0) + E_g/q)\right] + \phi_d^2} \quad (8)$$
$$-\phi_d + (\varphi_{ch} + \varphi'_{ch} - (V'_{gs} - V_{fb} - \phi_0) + E_g/q)$$

Drain tunnel

$$\begin{cases} x_{begin}^{min} = L_{d1} - \sqrt{\frac{-\varphi_{d1}^* + (V_{ds} + V_{bid})}{qN_d/2\varepsilon_{Si}}} \\ x_{end}^{min} = \lambda \cdot \cosh^{-1}\left[\frac{\varphi_{d1}^* - E_g/q - (\varphi_{ch} + \varphi'_{ch} - (V'_{gs} - V_{fb} - \phi_0))}{-\varphi'_{ch} + (V'_{gs} - V_{fb} - \phi_0)}\right] - L_{d2} \end{cases} \quad (9) \qquad x_{begin}^{max} = \begin{cases} L_{d1} - \sqrt{\frac{-(\varphi_{ch} + E_g/q) + (V_{ds} + V_{bid})}{qN_d/2\varepsilon_{Si}}} & \text{when } \Phi_{d0} - \varphi_{ch} < E_g/q \\ \lambda \cdot \cosh^{-1}\left[\frac{\varphi_{ch} + E_g/q - (\varphi_{ch} + \varphi'_{ch} - (V'_{gs} - V_{fb} - \phi_0))}{-\varphi'_{ch} + (V'_{gs} - V_{fb} - \phi_0)}\right] - L_{d2} & \text{when } \Phi_{d0} \geq E_g/q \end{cases} \quad (10)$$

$$\varphi_{s1}^* = \sqrt{\left[\varphi_{ch} - (V_{gs} - V_{fb} - \phi_0)\right]^2 - 2\phi_s[(V_{gs} - V_{fb} - \phi_0 - E_g/q) - (V_s - V_{bis})] + \phi_s^2} + \phi_s + (V_{gs} - V_{fb} - \phi_0 - E_g/q) \quad (11)$$

Source tunnel

$$\begin{cases} x_{begin}^{min} = \sqrt{\frac{\varphi_{s1}^* - (V_s - V_{bis})}{qN_s/2\varepsilon_{Si}}} - L_{s1} \\ x_{end}^{min} = L_{s2} - \lambda \cdot \cosh^{-1}\left[\frac{\varphi_{s1} + E_g/q - (V_{gs} - V_{fb} - \phi_0)}{\varphi_{ch} - (V_{gs} - V_{fb} - \phi_0)}\right] \end{cases} \quad (12) \qquad x_{begin}^{max} = \begin{cases} \sqrt{\frac{(\varphi_{ch} - E_g/q) - (V_s - V_{bis})}{qN_s/2\varepsilon_{Si}}} - L_{s1} & \text{when } \varphi_{ch} - \Phi_{s0} < E_g/q \\ L_{s2} - \lambda \cdot \cosh^{-1}\left[\frac{(\varphi_{ch} - E_g/q) - (V_{gs} - V_{fb} - \phi_0)}{\varphi_{ch} - (V_{gs} - V_{fb} - \phi_0)}\right] & \text{when } \varphi_{ch} - \Phi_{s0} \geq E_g/q \end{cases} \quad (13)$$

C. Tunneling Current Model (G_{tun} & I_{tun})

$$G_{tun} = A_{kane} \cdot \frac{\xi^2}{\sqrt{E_g/q}} \cdot \exp\left(\frac{-B_{kane}(E_g/q)^{3/2}}{\xi}\right) = A_{kane} \cdot (E_g/q)^{3/2} \cdot \frac{1}{W_t^2} \cdot \exp\left(-\frac{W_t}{\lambda_t}\right) \quad \text{in which } \xi = \frac{E_g/q}{W_t} \quad (14)$$

$$I_{tun} = q \cdot \iiint G_{tun}(x,y,z)\,dxdydz = q \cdot G_{tun}^{max} \cdot \int_{x_{open}}^{x_{close}} \exp\left(-\frac{|x - x_{begin}^{min}|}{t_{eff}}\right)dx \cdot \int_0^{t_{eff}} \exp\left(-\frac{y}{t_{eff}}\right)dy \cdot \int_0^{W_g} dz = q \cdot G_{tun}^{max} \cdot \lambda_t \left[2 - \exp\left(-\frac{|x_{begin}^{max} - x_{begin}^{min}|}{t_{eff}}\right) - \exp\left(-\frac{|x_{open} - x_{begin}^{min}|}{t_{eff}}\right)\right] \cdot t_{eff}(1 - e^{-1}) \cdot W_g \quad (15)$$

[11] E. O. Kane, *J. Appl. Phys.*, 32(1), 83-91, 1961.

DESIGN OF FERROELECTRIC FET-BASED CAPACITIVE-COUPLING COMPUTING-IN-MEMORY FOR BINARY NEURAL NETWORKS

Boyi Fu[1], Jin Luo[1], Weikai Xu[1], Qianqian Huang[1,2,3]* and Ru Huang[1,2,3]*

[1]School of Integrated Circuits, Peking University, Beijing 100871, China

[2] Beijing Advanced Innovation Center for Integrated Circuits, Beijing 100871, China

[3]Chinese Institute for Brain Research, Beijing 102206, China

Corresponding Author's Email: hqq@pku.edu.cn, ruhuang@pku.edu.cn

ABSTRACT

Ferroelectric FETs (FeFETs) have been applied to computing-in-memories (CIMs) for binary neural networks (BNNs) because of the advantages of ultra-low energy cost, while suffering from large output variation or write disturb. This work proposes a novel FeFET-based CIM for BNNs, reducing the output variation by exploiting charge-domain scheme and avoiding write disturb by exploiting additional access transistors. Simulation results show that compared with conventional FeFET-based current-domain design, the proposed design can improve accuracy by 20% on MNIST by reducing the output variation. Compared with previous SRAM-based and RRAM-based CIM for BNNs, the proposed design can reduce energy cost by 54% and 98%, showing its great potential for edge AI applications.

INTRODUCTION

Binary neural networks (BNNs) have attracted much attention for embedded applications because of their light-weight and high speed [1]. Recently, various studies have been proposed to implement BNNs with computing-in-memory (CIM) architecture, which can reduce the energy cost and delay caused by data transmission. Compared with SRAM-based CIMs for BNN [2], non-volatile memories (NVMs) based CIMs for BNN have the potential for higher density and lower energy consumption, such as RRAM [3] and PCRAM [4]. Among NVMs, ferroelectric FETs (FeFETs) have attracted significant attention for its high on/off ratio and low write power [5]. Based on FeFET, current-domain CIM for BNN proposed in [5] computes and senses in current mode, suffering from high output variation caused by the large variation of FeFET on-state resistance. In order to solve the variation problem, FeFET-based charge-domain CIM for BNN proposed in [6] computes and senses in charge or voltage mode. This is because the output variation of charge-domain CIM mainly depends on the capacitance, whose variation is much lower than FeFET on-state resistance. However, it still suffers from write disturb issue.

In this work, we propose a novel FeFET-based CIM design for BNN, including a 4-transistor-1-capacitor (4T1C) cell and a capacitive-coupling array. All the performance, precision accuracy and array-level energy

cost of our proposed CIM are evaluated by simulation. This design can avoid write disturb and improve the accuracy on MNIST from less than 80% to 93% by reducing output variation. Besides, the proposed array reduces the energy cost by 54% and 98% compared with SRAM-based CIM and RRAM-based CIM respectively.

DESIGN AND OPERATIONS OF FEFET-BASED CAPACITIVE-COUPLING CIM

Design of the capacitive-coupling cell for BNN

BNNs binarize the weights and activations [1] to be +1 and -1, and thus the multiplication and accumulation (MAC) operation is simplified to be XNOR operation and bit-counting. Although BNNs are simplified, they have shown acceptable accuracy in image dataset classification tasks like MNIST and CIFAR-10.

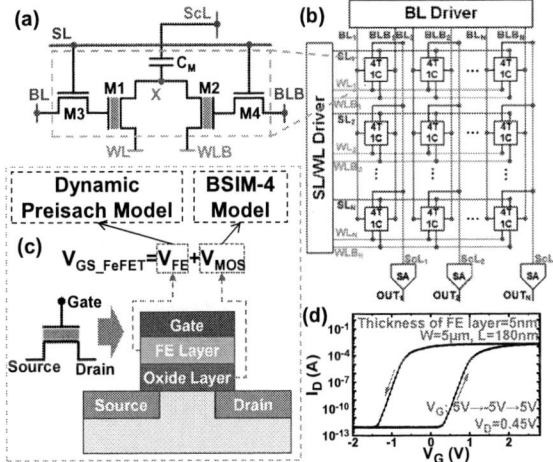

Figure 1: (a)(b) Cell and array structure of the proposed design. (c) The modelling framework of FeFET. (d) I_D-V_G hysteresis curve of FeFET.

For FeFET with a ferroelectric layer in the gate stack, as the polarization of FE layer can be set to different state, FeFETs can be set to two states by applying different gate voltage, and one state is of high threshold voltage (V_T) and the other is of low V_T. The cell of the proposed design in this work for XNOR operation is composed of two FeFETs (M1 and M2), two access transistors (M3 and M4) and a capacitor (C_M) as shown in Fig. 1(a), which can both store weights and perform XNOR operation. The capacitive-coupling array structure for BNNs is shown in Fig. 1(b). The two FeFETs are programmed to be in different states (Low V_T and High V_T) to store weights. The state, in which M1 is set to low V_T and M2 is set to high V_T, represents weight "+1", and the opposite state represents weight "-1". The word lines WL and WLB are complementary when computing, representing input activations. To analyze the properties of the design, the FeFET adopted in this work is modelled based on the dynamic Preisach model for ferroelectric layer [7] and BSIM-4 model for MOSFET, as shown in Fig. 1(c), and all the simulations are carried out in HSPICE tools. Fig. 1(d) shows the FeFET I_D-V_G curves, which show hysteresis behavior.

Write Operation

TABLE I shows the voltage setup for programing. Write operation of the array is carried out row by row, and when programming a certain row, the SL of the selected row is set to VDD, with others set to GND. Therefore, M3 and M4 in the cells of the selected row can transmit V_{BL} and V_{BLB}, and M3 and M4 in the cells of the unselected row can cut off V_{BL} and V_{BLB} to avoid write disturb.

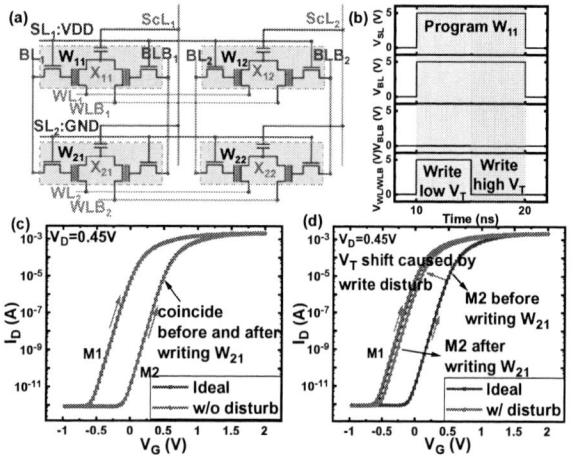

Figure 2: (a) Schematic of array programming. (b) Input voltage of programming. (c)(d) The I_D-V_G curves of FeFETs without and with write disturb.

The write operation has two phases. In phase one, the cell of high V_T is programmed and in phase two, the cell of low V_T is programmed, similar to the methods in [6]. Take programming weight "+1" to the cell W_{11} in the 2×2 array as an example (see Fig. 2(a)). SL_1 is set to VDD and SL_2 is set to GND in the operation, and V_{BL1}/V_{BLB1} is biased to V_w/GND for setting M1 of low V_T and M2 of high V_T. In phase one, V_{WL} and V_{WLB} are set to V_w, and thus $V_{GS1}=0$, $V_{GS2}=-V_w$. Therefore, M2 is programmed to the state of high V_T. In phase two, V_{WL} and V_{WLB} are set to GND, and thus $V_{GS1}=V_w$, $V_{GS2}=0$. Therefore, M1 is programmed to the state of low V_T. The waveform of the

input voltage is shown in Fig. 2(b), and the I_D-V_G curves of M1 and M2 in W_{11} is shown in Fig. 2(c)(d).

TABLE I. WRITE SCHEME FOR THE ARRAY

Weight	Target state	Phase	V_{ScL}	V_{BL}	V_{BLB}	V_{SL}(selected)	V_{SL}(unselected)	V_{WL}	V_{WLB}
+1	M1:Low V_T	1	GND	V_w	GND	VDD	GND	V_w	V_w
	M2:High V_T	2						GND	GND
-1	M1:High V_T	1	GND	GND	V_w	VDD	GND	V_w	V_w
	M2:Low V_T	2						GND	GND

TABLE II. XNOR LOGIC IN ONE CELL

Stored weight		Input		Output	
Value	$V_T(M1)/V_T(M2)$	*Value*	V_{WL}/V_{WLB}	*Value*	V_X
"+1"	Low/High	"+1"	V_R/GND	"+1"	V_R
"+1"	Low/High	"-1"	GND/V_R	"-1"	GND
"-1"	High/Low	"+1"	V_R/GND	"-1"	GND
"-1"	High/Low	"-1"	GND/V_R	"+1"	V_R

To validate the write-disturb-resistance of this design, cell W_{21} is programmed "-1" after W_{11} has stored weight "+1". In this design, the I_D-V_G curves of M1 and M2 in W11 after programming W21 coincide with the curves before, as shown in Fig. 2(c). However, if both the SLs are all set to VDD and access transistors can all transmit V_{BL} and V_{BLB}, which is equivalent to lack of access transistors, the I_D-V_G curves of M2 in W_{11} after programming W_{21} in Fig. 2(d) shows significant V_T shift, indicating severe write disturb.

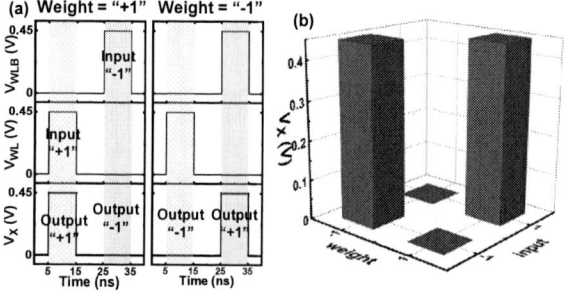

Figure 3: (a) Input and output voltage waveform diagram of XNOR operation in the cell (V_R=0.45V). (b) Output voltage of different weight and input.

XNOR Operation for BNN

In this stage, all the SLs are set to VDD to transmit V_{BL} and V_{BLB}, and BLs and BLBs are all set to GND to guarantee the gate voltage of each FeFET is set to GND. Initially, V_{ScL} is set to GND and then floating, and V_{WL}/V_{WLB} are all set to GND. For each cell, since either M1 or M2 has already been set to low V_T, V_X is also GND. Then V_{WL} and V_{WLB} of each row are set complementary. Here we define $V_{WL}/V_{WLB}=V_R$/GND as input "+1" and $V_{WL}/V_{WLB}=$GND/V_R as input "-1".

V_X can be derived from V_{WL} and V_{WLB} divided by the

979-8-3503-1101-3/23 $31.00 © 2023 IEEE

drain-source resistance of M1 and M2 respectively. Because of the high on-off ratio of FeFETs and the complementary state of M1 and M2, the V_X value can be approximated as V_R or GND. TABLE II shows the XNOR logic in one cell, and Fig. 3(b) shows the output voltage of different weight and input. Fig. 3(a) shows the waveform diagram of the input and output voltage of the XNOR operation in our proposed design.

FeFET-based Capacitive-coupling Array for MAC of BNN

The operational process for MAC can be divided into two phases. In phase one, ScL is pre-charged to GND and then floating. In the second phase, XNOR operation is performed in each cell, as shown in Fig. 4(a). V_X in each cell represents the output of the cell, and V_{ScL} represents the result of MAC performed in the column.

Assuming that there are N cells in one column, among which M cells have the output of $V_X=V_R$, and the others have the output of $V_X=$GND. Fig. 4(b) shows the equivalent circuit diagram, and for ideally C_M in every cell is identical, V_{ScL} can be expressed as follows:

$$V_{ScL} = \frac{\sum_{i=1}^{N} V_{Xi} \cdot C_{Mi}}{\sum_{i=1}^{N} C_{Mi}} = \frac{\sum_{i=1}^{N} V_{Xi}}{N} = \frac{M V_R}{N}.$$

Fig. 4(c) shows the ideal MAC result when N is set to 128.

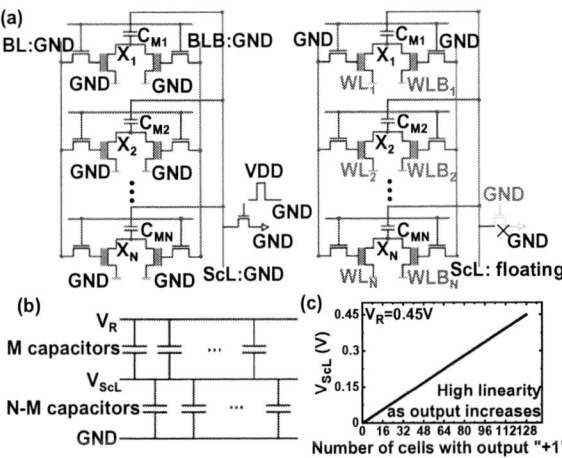

Figure 4: (a) MAC operation of the array. (b) Equivalent ScL charging capacitance. (c) Output of MAC performed by the array.

EVALUATION OF THE ACCURACY AND ENERGY BASED ON THE PROPOSED DESIGN

The influence of the variation on the proposed array for classification applications is further evaluated, using the variation of C_M in [6] and FeFET on-state resistance in [8]. The network we build on MNIST test set is shown in Fig. 5(a), and the result is shown in Fig. 5(b). According to the result, although the accuracy of the network reduces

when the variation increases, this work improves the accuracy compared with the design in [5].

The energy cost in a single MAC of the array is also evaluated compared with SRAM-based design [2] and RRAM-based design [3]. In the simulation, the size of array is set to be 128×128. The SAs are from [9]. The results are shown in Fig. 5(c). It is shown that this work can save more energy than that in [2], because in charge-domain works, energy is mainly consumed by capacitance charging, and the equivalent capacitor of capacitive-coupling scheme is only about 1/3 of that of charge-sharing scheme.

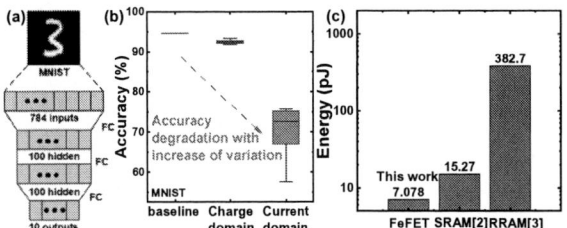

Figure 5: (a) Network structure in this work. (b) Accuracy based on the proposed array with variation. (c) Energy cost of MAC operations. C_M is set to 1.2fF [2]. We use $R_{on}=10K\Omega$ and $R_{off}=1M\Omega$ for RRAMs. Evaluations are carried out at 100MHz.

CONCLUSION

We have proposed a novel FeFET-based CIM design for BNNs, with each cell composed of 2 FeFETs, 2 transistors and 1 capacitor. Simulation results demonstrate that our design can avoid write disturb compared with the previous FeFET-based charge-domain CIM and reduce the output variation compared with current-domain CIM. The proposed design can largely reduce energy consumption compared with SRAM-based design, showing its great potential for edge AI applications.

ACKNOWLEDGEMENTS

This work was supported by National Key R&D Program of China (2018YFB2202801), NSFC (61927901), Beijing SAMT Project (SAMT-BD-KT-22030101), 111 Project (B18001), and Tencent Foundation through the Xplore Prize.

REFERENCES

[1] M. Courbariaux et al., arXiv:1602.02830.
[2] H. Valavi, et al., *JSSC*, June 2019, pp. 1789-1799.
[3] X. Sun et al., *DATE*, 2018, pp. 1423-1428.
[4] J. V et al., *Nature communications*, 2020, pp. 11: 1-13.
[5] X. Chen et al., *DATE*, 2018, pp. 1205-1210.
[6] G. Yin et al., *T-CAS2*, July 2021, pp. 2262-2266.
[7] Z. Fu et al., *CSTIC*, 2020, pp. 1-4.
[8] T. Soliman et al., *IEDM*, 2020, pp. 29–2.

[9] S. Yu, et al., *CICC*, 2020, pp. 1-4.

INVESTIGATION OF SYNERGIC HYDROGEN MITIGATION TECHNIQUE FOR TOP-GATE A-IGZO THIN-FILM TRANSISTORS

Gangping Yan,[1,2,3] Zhiyu Song,[1,2] Haoqing Xu,[1,2] Shangbo Yang,[1,2] Chuqiao Niu,[1,2] Guoliang Tian,[1,2] Yanna Luo,[1,2] Luoyun Zhang,[1,2] Yunjiao Bao,[1,2] Gaobo Xu,[1,2,] and Huaxiang Yin[1,2]*

[1] Key Laboratory of Microelectronics Devices and Integrated Technology, Institute of Microelectronics of Chinese Academy of Sciences, Beijing 100029, China

[2] University of Chinese Academy of Sciences, Beijing 100049, China

[3] Beijing Superstring Academy of Memory Technology, Beijing 100176, China

*Corresponding Author's Email: xugaobo@ime.ac.cn

ABSTRACT

In this work, a simple hydrogen mitigation method for the top-gate amorphous In-Ga-Zn-O thin-film transistor (a-IGZO TFT) is proposed. The influences of synergic modulations between post-dielectric thermal treatment and Al2O3 buffer layer acting as oxygen-blockers are investigated to address hydrogen issues. Using the proposed optimization approach, the self-aligned channel annealing is successfully performed to realize > 104x increase in on-current of a-IGZO TFTs. These results provide useful guidance for high-performance a-IGZO devices in monolithic three-dimensional integration.

INTRODUCTION

The top-gate amorphous In-Ga-Zn-O (a-IGZO) thin-film transistor (TFT) has been extensively investigated in three-dimensional technology due to its good uniform, ultra-low leakage, excellent subthreshold swing (SS), and low-temperature process compatibility [1-5]. However, the a-IGZO TFT suffers from hydrogen problems, which may deteriorate the threshold voltage (V_{TH}) in devices. Previous works investigate the hydrogen mitigation methods such as O_2 annealing and using an oxygen-tunnel structure, but these approaches are limited by the degraded source and drain (S/D) contact and complicated fabrication process [5-6].

Herein, a simple hydrogen mitigation method is proposed by evaluating the synergic modulations between post-dielectric annealing and the Al_2O_3 buffer layer acting as oxygen blockers. Using the simple approach, the self-aligned channel annealing is successfully performed to eliminate the effect of hydrogen and maintain good S/D contact, where a >10^4x increase in on-current (I_{ON}) of a-IGZO TFTs is achieved.

EXPERIMENT

An a-IGZO TFT with traditional top gate structures was fabricated on the substrate, as shown in Fig. 1(a). Fig. 1(b) depicts the summary of the IGZO transistor integration. First, a conventional SiO_2 buffer layer was deposited through the plasma-enhanced chemical vapor

deposition (PECVD) for Devices A and B, where Device C contained the Al_2O_3 buffer layer. Next, a 20 nm-thick IGZO was sputtered at room temperature (RT) using a radio frequency (RF)-type sputtering system with a target that had the atomic ratio of In:Ga:Zn = 1:1:1. The a-IGZO thin film was patterned by dilute HNO_3 wet etchant (HNO_3/H_2O = 20:1). Subsequently, the S/D contact electrode of 60 nm-thick molybdenum (Mo) was sputtered, followed by the dry etching process to patterning. The channel width (W) and length (L) were 5 and 4 µm, respectively. After the SiO_2 deposition as the gate insulator (GI) by PECVD, the annealing in air ambient was carried out at RT for Device A and at 250 °C for Devices B and C. Finally, a 50-nm thick aluminum (Al) was sputtered and patterned as the gate electrode. Fig. 1(c) is the scanning electron microscope (SEM) image, in which each layer is observed clearly. Fig. 1(d) shows the top view of the staggered IGZO TFT, where each part is defined well. The multiple sets of devices are listed in Table I.

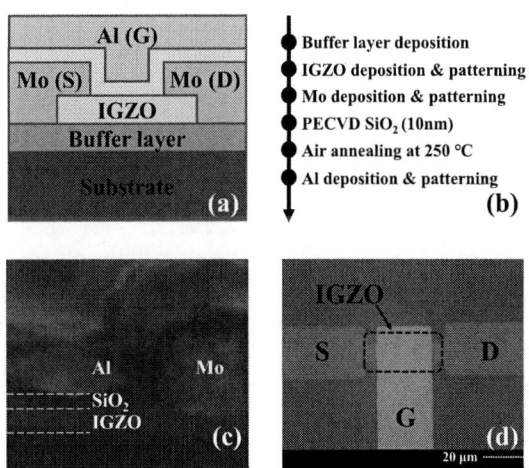

Figure 1: (a) Schematic for top-gate a-IGZO TFTs with SiO2/Al2O3 buffer layer. (b) Summary of the top-gate IGZO transistor integration. (c) SEM image and (d) optical image of the top gate device.

TABLE I.   SUMMARY OF MULTIPLE SETS OF DEVICES

Device	Buffer Layer	Annealing Temperature
A	SiO_2	Room Temperature
B	SiO_2	250 °C
C	Al_2O_3	250 °C

RESULTS AND DISCUSSIONS

Fig. 2(a) illustrates the transfer characteristics of varying a-IGZO TFTs with a fixed drain voltage (V_{DS}) of 1 V. The gate voltage (V_{GS}) is swept from -4 V to 4 V, where the gate current (I_{GS}) of different devices is low, meaning the breakdown of the GI does not occur. It is obtained that the as-fabricated device with SiO_2 buffer layer (Device A) is in the normally on state at a high drain current (I_{DS}) level of 4×10^{-4} A. This is caused by the introduction of hydrogen into the a-IGZO channel during the GI deposition process. Hydrogen generally acts as a shallow donor, which causes an increase in the Fermi level of the a-IGZO thin film and offers free carriers. Once the a-IGZO film is doped by hydrogen donors, the conductivity of the channel enlarges. To address the effect of hydrogen, annealing is conducted to restrain hydrogen donors in a-IGZO (Device B). However, the maximum I_{DS} of 9 nA in Device B is insufficient to be operated in integrated circuits, as shown in Fig. 2(b). As the post-dielectric annealing in air ambient is performed, the oxygen atom can permeate into the a-IGZO thin film to passivate the hydrogen and oxygen vacancy, leading the device to have switching characteristics. Whereas, the permeation in the SiO_2 buffer layer can severely deteriorate the S/D contacts, as depicted in Fig. 3(a), which dominates the inferior performance of Device B.

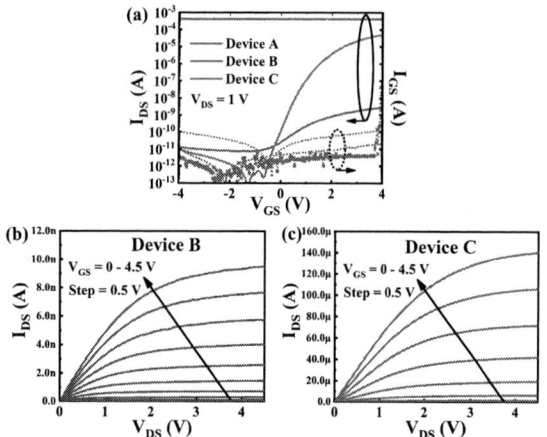

Figure 2: (a) Transfer characteristics of a-IGZO devices with a fixed V_{DS} of 1 V. Output characteristics of devices (b) B and (c) C, where the V_{GS} is swept from 0 V to 4.5 V with a step of 0.5 V.

For solving this obstacle, the a-IGZO device with an Al_2O_3 buffer layer is proposed (Device C). The Al_2O_3 layer is expected to act as an oxygen blocker, which prevents oxygen from migration, as shown in Fig. 3(b). Fig. 2(c) depicts the output characteristics of Device C, where the V_{GS} is swept from 0 V to 4.5 V with a step of 0.5 V. Benefitted from the Al_2O_3 buffer layer, the self-aligned channel annealing is successfully realized. Hence, compared to Device B, Device C exhibits a 10^4x increase in I_{DS}, and the high current reaches about 140 μA. In addition, the improved SS of 171.5 mV/dec in Device C is also obtained, showing a relatively steep transfer curve. The significant performance improvement of Device C manifests that the fabrication Al_2O_3 buffer layer and conducting post-dielectric annealing in a-IGZO TFTs are effective and simple ways to address the hydrogen problems.

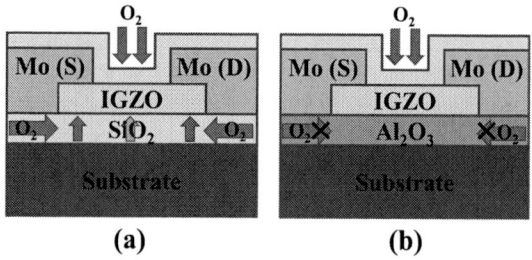

Figure 3: (a) Schematics of the top-gate IGZO devices with (a) SiO_2 and (b) Al_2O_3 buffer layer, highlighting the O_2 role during the annealing in air-ambient.

Due to the introduction of high-κ Al_2O_3 as the suffer layer on the silicon substrate, the plausibility of double-gated operation of Device C is evaluated, where the substrate can be operated as the back gate. Fig. 4 illustrates the transfer characteristics of Device C with the swept top gate or back gate from -4 V to 4 V and fixed V_D of 1 V. The V_{TH} of the back gate device is larger than that of the top gate one. This is probably caused by the difference of electron concentration between the top and back channel. During etching the S/D regions by the reactive ion etching process, the concentration of indium (In) at the top channel increases because relatively weak In-O bonds are preferentially broken by ion bombardment [7], resulting in the formation of the more conductive top channel and thus the more negative V_{TH} of the top-gated device. Such mechanism also explains the increased off-state and on-state current of the device using the top gate. It is noting that the SS of the back-gated device is larger than that of the top gate transistor. Owing to the Al_2O_3 buffer layer on the silicon substrate, few oxygens penetrate into the back channel to passivate the interface defects, so that the trapping and de-trapping electrons at the back channel is more severe than that at the top channel. The excellent electrostatic control of the back gate electrode also yields the V_{TH} modulation of the top gate device, enabling the proposed device a better adaption in different application conditions.

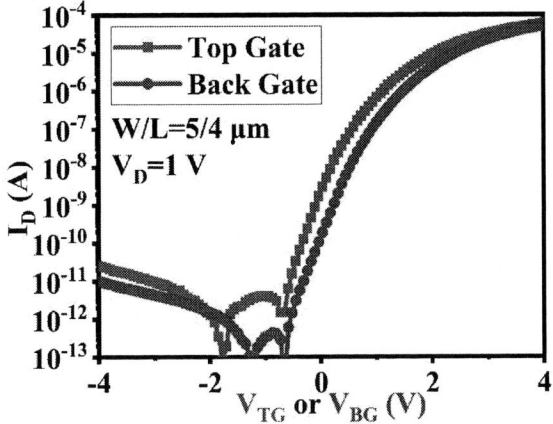

Figure 4: Typical transfer characteristics of Device C under top-gated or back-gated operation, where the gate voltage is swept from -4 V to 4 V with a fixed V_D of 1 V. The gate width is 5 μm and the channel length is 4 μm.

CONCLUSION

This article discusses the synergic modulations between post-dielectric annealing and the Al_2O_3 buffer layer acting as oxygen blockers. With respect to the as-fabricated device, the current remains at a high level due to the introduction of excess hydrogen donors into the channel. Annealing in an O-rich environment is a useful approach to alleviate the effect of hydrogen, but it seriously degrades the source and drain contact, resulting in inferior electrical properties in a-IGZO devices with a conventional SiO_2 buffer layer. It is obtained that the device with Al_2O_3 exhibits remarkable performance enhancement due to the self-aligned channel annealing. The double-gated operation of the proposed device is also evaluated. This synergic and simple hydrogen mitigation technique provides useful guidance for high-performance top gate a-IGZO devices in monolithic three-dimensional integration.

REFERENCES

[1] Zhao Z, Gomez J, Ye H, Imani M, Yin X, Deng S, Melanson B, Zhang J, Gong X, Abusleme A, Datta S. Computational Associative Memory Based on Monolithically Integrated Metal-Oxide Thin Film Transistors for Update-Frequent Search Applications. In2021 IEEE International Electron Devices Meeting (IEDM) 2021 Dec 11 (pp. 37-6). IEEE.

[2] Chand U, Fang Z, Chun-Kuei C, Luo Y, Veluri H, Sivan M, Feng LJ, Tsai SH, Wang X, Chakraborty S, Zamburg E. 2-kbit Array of 3-D Monolithically-stacked IGZO FETs with Low SS-64mV/dec, Ultra-low-leakage, Competitive μ-57 cm 2/Vs Performance and Novel nMOS-Only Circuit Demonstration. In2021 Symposium on VLSI Technology 2021 Jun 13 (pp. 1-2). IEEE.

[3] Yan G, Yang H, Liu W, Zhou N, Hu Y, Shi Y, Gao J, Tian G, Zhang Y, Fan L, Wang G. Mechanism Analysis of Ultralow Leakage and Abnormal Instability in InGaZnO Thin-Film Transistor Toward DRAM. IEEE Transactions on Electron Devices. 2022 Mar 25;69(5):2417-22.

[4] Wang, C., Kumar, A., Han, K., Sun, C., Xu, H., Zhang, J., ... & Gong, X. (2022, June). Extremely Scaled Bottom Gate a-IGZO Transistors Using a Novel Patterning Technique Achieving Record High G m of 479.5 μS/μm (V DS of 1 V) and f T of 18.3 GHz (V DS of 3 V). In 2022 IEEE Symposium on VLSI Technology and Circuits (VLSI Technology and Circuits) (pp. 294-295). IEEE.

[5] Belmonte A, Oh H, Rassoul N, Donadio GL, Mitard J, Dekkers H, Delhougne R, Subhechha S, Chasin A, van Setten MJ, Kljucar L. Capacitor-less, long-retention (> 400s) DRAM cell paving the way towards low-power and high-density monolithic 3D DRAM. In2020 IEEE International Electron Devices Meeting (IEDM) 2020 Dec 12 (pp. 28-2). IEEE.

[6] Belmonte A, Oh H, Subhechha S, Rassoul N, Hody H, Dekkers H, Delhougne R, Ricotti L, Banerjee K, Chasin A, van Setten MJ. Tailoring IGZO-TFT architecture for capacitorless DRAM, demonstrating> 10 3 s retention,> 10 11 cycles endurance and L g scalability down to 14nm. In2021 IEEE International Electron Devices Meeting (IEDM) 2021 Dec 11 (pp. 10-6). IEEE.

[7] Park J , Kim S , Kim C , et al. High-performance amorphous gallium indium zinc oxide thin-film transistors through N2O plasma passivation[J]. Applied Physics Letters, 2008.

CHARACTERIZATION OF FIELD CYCLING FATIGUE IN HfZrO$_x$ FERROELECTRIC CAPACITORS

Puyang Cai[1], Zhiwei Liu[2], Tianxiang Zhu[1], Zhigang Ji[2,], Runsheng Wang[1,*], and Ru Huang[1]*
[1] School of Integrated Circuits, Peking University, Beijing 100871, China
[2] Departure of Micro/Nano Electronics, Shanghai Jiao Tong University, Shanghai 200240, China
*Corresponding Author's Email: r.wang@pku.edu.cn; zhigangji@sjtu.edu.cn

ABSTRACT

In this paper, the fatigue behavior of HfZrO$_x$ (HZO) ferroelectric (FE) capacitor is thoroughly characterized. By proposing an empirical model applicable to the entire process of fatigue, we found that the fatigue effect is dominated by the process of charge migration to non-switching regions and charge exchange with electrodes to form the localized built-in field and cause domain-pinning.

INTRODUCTION

In recent years, Hf-based ferroelectric (FE) material has been regarded as the candidate for high-density nonvolatile memory devices due to its CMOS compatibility, scalability, high speed, and low operation voltage [1]. However, the endurance problem poses significant challenges for their practical adoption for mass production. One of its endurance problems is fatigue effect, which refers to the degradation of remnant polarization (P_r) under the bipolar field cycles. The main cause of fatigue is generally attributed to pinning of seed domains or domain walls by charged defects. New defect generation, defect movement and the process of charge trapping/de-trapping are possible origins of these charged defects [2]-[3], but a consensus on the dominant factor has not been reached. Moreover, current fatigue models of ferroelectrics are only applicable to certain stages of the decay process [4] or have shown inadequate fitting to the fatigue trend [5]-[6]. A comprehensive model that can accurately describe the entire fatigue process is still missing. Developing a quantitative fatigue model can aid in understanding the underlying physical mechanism and predicting the degradation of ferroelectrics during operation.

This paper presents a thorough investigation of the fatigue effect in HfZrO$_x$ (HZO) ferroelectric capacitors, including the development of a quantitative model that accurately describes the entire fatigue process. The model is applied to analyze fatigue at different temperatures, voltage amplitudes, and operation frequencies in detail, and the dominant factors for fatigue are revealed.

DEVICE FABRICATION

The capacitors used in this work are TiN/Zr:HfO$_2$/TiN (MFM) structures. Its fabrication process starts with the magnetron sputtering onto the SiO$_2$/Si wafer to form a 60 nm thick TiN bottom electrode. Secondly, 12 nm HZO thin film with Hf:Zr ratio of 1:1 was deposited by atomic layer deposition (ALD) at 250°C. Thirdly, photolithography was conducted to form patterned electrodes, and then 60 nm top electrode TiN was fabricated by magnetron sputtering. Finally, the as-fabricated HZO film was crystallized at 450°C for 1 minute in N$_2$ ambient.

CHARACTERIZATION AND MODELING OF FATIGUE

The pulse sequences to examine the switching kinetics and endurance of the MFM capacitor are shown in Fig. 1(a) and (b), respectively. To exclude the leakage current, positive-up-negative-down (PUND) pulse sequences were applied to measure the FE component.

A typical FE polarization charge density-time relationship is shown in Fig. 1(c), from which the relationship of switched polarization and the corresponding operation frequency of square wave can be extracted. As shown in the inset of Fig. 1(c), a switching time distribution tail can be found between 100kHz and 500kHz. The evolution of P_r with bipolar stress cycling at

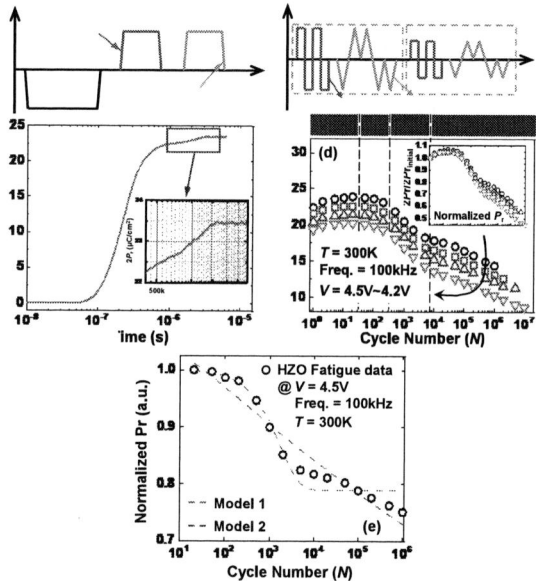

Figure. 1: The polarization-time relationship: (a) voltage setup and (c) experimental data. Inset: enlarged view of the box section. The switching time is transformed to the corresponding operation frequency. The evolution of P_r under different switching voltages: (b) voltage setup and (d) experimental data. Inset: the evolution of $Pr/Pr_{initial}$ with cycling. (e) Fatigue data fit by previous models, which are not applicable to the entire process of fatigue.

979-8-3503-1101-3/23 $31.00 © 2023 IEEE

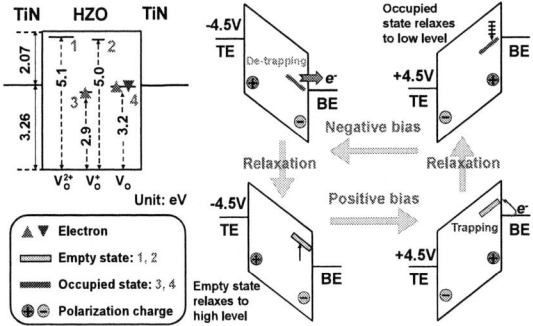

Figure 2: Energy level of oxygen vacancy in 0, +1 and +2 charge states [7], and the band diagram of the trapping and de-trapping process.

different voltage amplitudes are summarized in Fig. 1(d), which can be divided into four stages. Stage I is referred to as wake-up, where the P_r is increasing. Subsequently, the fatigue process starts with a reduction of P_r. Based on different degradation rates, this process can be further divided into three stages: stage II (slow), III (logarithmic), and IV (saturated). Our results indicate that lower pulse amplitudes lead to higher fatigue rates, which can be further confirmed by the normalized P_r values in the inset of Fig. 1(d). As shown in Fig. 1(e), fatigue models proposed for traditional FE materials fail in describing the fatigue data of HZO.

Defect generation, defect movement, and the trapping/de-trapping of charges are potential factors to induce charged defects and cause domain-pinning. However, if the first two ones are dominant factors, the fatigue rate should be faster at higher voltage, which is not the case in our experiments. Oxygen vacancy (V_O) is well recognized as the type of trap that can impact the FE properties during field cycling [7]. As the commonly used TiN electrode in the MFM capacitor is oxygen reactive, it is often partially oxidized during the ALD process and causes the aggregation of V_O at the interface. According to the formation energy given by DFT calculations [7], V_O is stable in the neutral or positively charged state in FE HZO. Hence, we assumed that FE domains are pinned by V_O^{2+} at the top or bottom interface and will be de-pinned upon the injection of electrons.

Charge redistribution within the ferroelectric film and charge exchange between the electrodes are two factors that can result in the charging of V_O and domain-pinning, which can help explain the lower fatigue rate at higher voltage amplitudes. Despite most domains are switched at the high electric field of ~3.5MV/cm, a distribution tail of switching field still exists, and some domains remain non-switched, as shown in Fig. 1(d). Thus, akin to the split-up effect [8], charges are redistributed to the non-switching region to form a localized built-in field with a direction that aligns with the FE polarization and cause the domain-pinning. This localized field will cause the FE domains in

the adjacent switching region to align with the same direction, thereby enlarging the non-switching region. With smaller voltage, more non-switched regions exist, and the redistribution effect becomes more pronounced, resulting in a higher fatigue rate.

Another perspective to explain the fatigue trend is the charge exchange with electrodes. The band diagram in Fig. 2 shows an example of defects near the bottom electrode, while defects near the top electrode follow a symmetric process. When a negative bias is applied to the top electrode, electrons are de-trapped, and V_O tends to become positively charged, pinning the seed domains at the interface. When the polarity is changed, electrons tend to be trapped into the V_O^{2+} near the bottom electrode, and FE domains are de-pinned. Therefore, if bipolar stress is applied, trapping and de-trapping will occur simultaneously, and the defects will finally come to an equilibrium occupancy, resulting in a decrease in $2P_r$. Based on these analyses, we proposed an empirical model to describe the process of fatigue, as follows:

$$R(N) = \frac{2P_r(N)}{2P_{rmax}} = -Aexp(-BN^{-m}) + 1 \quad (1)$$

$$\lim_{N \to 0} R(N) = 1 \quad (2)$$

$$\lim_{N \to \infty} R(N) = 1 - A \quad (3)$$

where R is the reliability function, representing the normalized P_r, A is a fitting parameter that decides the fatigue rate and ratio of pinned domains in the final state, and B and m are fixed parameters related to the energy and spatial distribution of defect. Next, fatigue behavior at different temperatures, voltages and frequencies is characterized and studied in detail with the proposed model.

The dependence of fatigue on temperature is first characterized. As shown in Fig. 3(a), the voltage amplitude and frequency are fixed at 4.5V and 100kHz, and our model can well reproduce the entire fatigue process as the temperature is varied from 275K to 350K. The fatigue rate increases with higher temperature, and $2P_r$ degrades to a lower final value. Since fatigue is a thermally activated process, it follows Arrhenius law:

$$A(T) \propto \exp(-E_a/k_B T) \quad (4)$$

where E_a is the activation energy, k_B is the Boltzmann constant, and T is temperature. As shown in Fig. 3(b), Ln(A) exhibits a good linear relationship with $1/T$, and E_a is extracted as 32meV, which is consistent to the calculated results in other studies [4]. Since the activation energy for defect generation (2eV [9]) and movement (0.7eV [10]) are both much higher than 32meV, they should not be the primary factor responsible for fatigue. Instead, this low E_a is not far from the phonon energy $\hbar\omega_0$ [11], indicating that P_r degradation is caused by the charged defects induced by electron de-trapping.

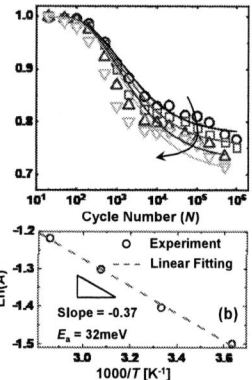

Figure. 3. (a) Experiment fatigue data and (b) extracted fatigue rate A at different temperatures.

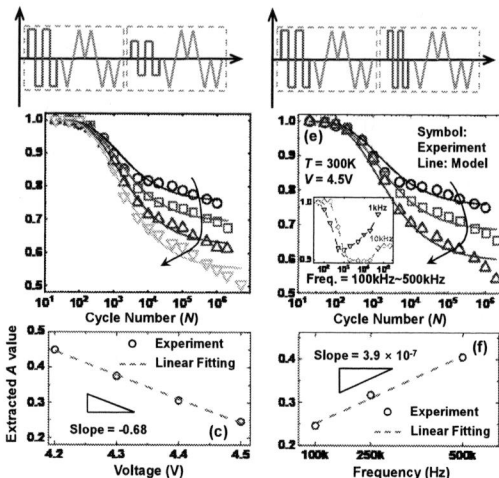

Figure. 4. (a) Voltage setup, (b) experiment fatigue data and (c) extracted fatigue rate A at different voltages. (d) Voltage setup, (e) experiment fatigue data and (f) extracted fatigue rate A at different frequencies.

Next, we analyze the fatigue data obtained at different voltage pulse amplitudes using the proposed model. Fig. 4(a) illustrates the voltage setup, and Fig. 4(b) shows the results. We observe that fatigue accelerates gradually as the voltage is lowered from 4.5V to 4.2V, and our new model can describe all the fatigue curves well. As shown in Fig. 4(c), the values of A we extract are linearly related to voltage, implying that fatigue rate is regulated by voltage amplitude through band bending. Since the defect energy level of empty state is higher than the occupied state [7], trapping at a positive bias is much harder than de-trapping at a negative bias (Fig. 2). With higher voltage amplitude, electron traps at higher energy levels can be pulled down closer to the Fermi level. This increases the possibility for electrons to be trapped and more domains to be de-pinned, resulting in a slower fatigue rate.

Finally, the effect of frequency on fatigue is investigated. Fig. 4(d)-(e) show the voltage setup and fatigue results from 100kHz to 500kHz, where our model can fit the fatigue curves well. As shown in Fig. 4(f), the extracted fatigue rate A follows a linear relationship with frequency within this specific range. At higher frequencies, electron trapping becomes more difficult because of the larger capture time constant induced by the higher energy level of the empty state. Therefore, de-trapping dominates and the fatigue rate increases. However, when the frequency drops below 10kHz, only very weak fatigue happens, which can even recover after a certain number of cycles, as shown in the inset of Fig. 4(e). This recovery can be attributed to the fact that a more complete switching and charge injection from the electrode can occur at low frequency, which reduces the non-switched region and inhibits the formation of charged V_O^{2+}.

Based on the analysis above, we can deduce that charge redistribution within the ferroelectric thin film and charge exchange with electrodes are dominant factors for P_r fatigue, instead of defect generation. Although new defects may be generated upon repeated field cycling, they would not affect the FE properties if they are in a neutral state. It is important to notice that breakdown is more likely to occur under higher voltage, which is different from the voltage dependence of fatigue. Therefore, the operating conditions should be chosen carefully to optimize the endurance of ferroelectric capacitor.

CONCLUSION

In this work, we have conducted a comprehensive characterization of fatigue for HZO ferroelectric thin film. Our proposed empirical model has successfully captured the entire fatigue process, revealing that the fatigue rate is influenced by charge redistribution within the ferroelectric film and the charge exchange with the electrode, leading to the creation of charged defects and domain-pinning. Moreover, different voltage dependence of fatigue and breakdown implies that to find an appropriate operating condition is crucial for practical application.

ACKNOWLEDGEMENTS

This work was supported by NSFC (61927901, 62125401), and the 111 Projects (B18001).

REFERENCES

[1] M. Sung et al., *IEDM*, pp. 33.3.1-33.3.4, 2021
[2] M. Pešic et al., *AFM*, pp. 4601-4612, 2016
[3] F. P. G. Fengler et al., *AEM*, 4, 1700547, 2018
[4] J. W. Adkins et al., *APL*, 117, 142902, 2020
[5] X. J. Lou et al., *PRB*, 75, 2007
[6] I. K. Yoo et al., *phys. stat. sol.*, (a) 133, 565, 1992
[7] Damir R. Islamov et al., *Acta Mater.*, pp. 47-55, 2019
[8] T. Schenk, et al., *ACS AMI*, 7, 36, 20224, 2015
[9] S. Bradley et al., *Phys. Rev. Applied*, 4, 064008 2015
[10] N. Capron, et al., *APL*. 91, 192905, 2007
[11] L. Vandelli, et al., *IEEE TED*, pp. 2878-2887, 2011

PROMOTING CHIP PROBING TEST YIELD BY SIMPLE ISSG AND GLOBAL WET PROCESS

Jingsong Peng [1], Kegang Zhang[1], Ning Wang[1], Yang Ding[1], Pingsheng Zhou[1], and Yin Yin[1]*

[1] Shanghai Huahong Grace Semiconductor Manufacturing Corporation, Shanghai 201203, China

*Corresponding Author's Email: Jingsong.Peng@hhgrace.com

ABSTRACT

STI has become a widely used technique for the SONOS flash memory. Large STI step height caused by process variation, however, may generate STI divot and following ONO and polysilicon residues, leading to low yield of chip probing test. Lowering the STI may avoid the severe divot while the narrow width effect will be enhanced resulting in V_T rolling off and more leakage. In this paper, a promoted fabrication process is demonstrated involved in ISSG and global wet etch. The nitride will be retained after ONO etch to be a screen layer for well implant and further etched by a global wet process after PR tripping. The remaining HTO can act as a hard mask and be regenerated by ISSG. Thus, the original screen pad oxide can be removed before ONO deposition by a wet process, during which the STI divot will be significantly flattened to eliminate ONO residues. More importantly, the final step height will be well retained instead of only lowering the STI. This promoted process distinctly solve the yield loss by STI divot, which is potential for mass production of various SONOS flash memories.

INTRODUCTION

Shallow trench isolation (STI) technique is of importance for the CMOS process technology, such as the silicon-oxide-nitride-oxide-silicon (SONOS) flash memory. The isolation for early immature CMOS process mainly involved in the p-n junction, which may suffer the latch-up and poor integration [1, 2]. The arising of the STI technique exactly meets the requirement of powerful isolation for CMOS due to its negligible field encroachment, preferable planarity, elimination of latch-up and low junction capacitance [3]. To solve the problem caused by the sharp top corner in STI, where the crowded electrical fields and potential defects will emerge, corner rounding is the most widely used method [3-4].

However, the issue of oxide divot arises while the device scaling down, leading to some residues of tunneling oxide-nitride-blocking oxide (ONO) film and following polysilicon residues at the STI edges. These residues will act as some potential bridges between word line (WL) and source line (SL), leading to the low yield of chip probing test, which becomes a more critical problem especially with a higher STI step height. Thus, the variation of STI step height during fabrication process may cause the ONO or polysilicon residues, leading to the function failure of SONO flash memory. Adding etch time

to lower the step height during STI forming process can be effective to eliminate the ONO residues. The following wet etch on pad oxide, however, should further decrease the step height. As a result, the narrow width effect will be enhanced, which the threshold voltage (V_T) of narrow devices will roll off and the leakage will increase [5].

Herein, to solve this problem, we developed a simple fabrication process involved in the in-situ stream generation (ISSG) technique helped by global wet etch. During this promoted process, the wet etch on pad oxide was set before ONO deposition to flatten the severe STI divot, well solving the problem of ONO and polysilicon residues on STI top edges. After CMOS p-well or n-well was defined, the ONO etch will only remove the high temperature oxide (HTO), the original blocking oxide of ONO film. The remaining nitride will serve as a screen layer for well implant and removed via the following global wet after photo resist (PR) stripping. During the global wet process, the HTO at the SONOS p-well play the role of a hard mask to protect beneath nitride. The ISSG technique was utilized to re-oxidize the nitride to obtain new blocking oxide of ONO. The resultant chip probing test yield can be improved by almost 1.5 times due to the elimination of polysilicon residues, indicating a simple and high-yield fabrication process for various SONOS memories.

EXPERIMENT DETAILS

SONOS cells in this work were mainly fabricated based on 95-nm technology node. 2-transisor (2T) SONOS was adopted with a select gate transistor (SG) as well as a memory SONOS transistor. After the STI trench was obtained, the undercutting of pad oxide was etched to facilitate the corner rounding. Then, the liner oxidation was utilized to round the top corner, and furthermore, to eliminate the damage of trench side wall during dry etch process [3]. The corner rounding, giving rise to the undercutting below nitride layer, however, will hinder the following oxide deposition leaving a void near top corner arose. Finally, a severe divot was formed at STI edges because the void can be only dipped into a small amount of etch solution. After the STI was completed, an ONO film was deposited. The server divot, however, obstructed the etch process of the curled ONO film in the divot, leading to some ONO residues illustrated as Figure 1a and b. the remaining pad oxide will serve as a screen layer during well implant. The next etch process of polysilicon may also result in residues in the divot at the STI edges.

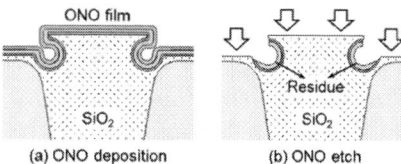

Figure 1: The forming process of ONO residues.

To remove the ONO residues, our promoted fabrication process involved in ISSG helped by global wet process. Before ONO deposition, the diluted HF solution was used to remove pad oxide, at the same time, the severe divot will be flattened (Figure 2a) avoiding obstruction on ONO etch. After the ONO deposition, the dry etch on entire ONO during original NMOS and PMOS process was replaced by one-step dry or wet each of only HTO layer. The leaving nitride acted as the screen layer during well implant instead of original pad oxide. The global wet process after PR stripping is shown in Figure 2b. The nitride on NMOS/PMOS wells was firstly removed by H_3PO_4 solution. Simultaneously, the HTO on the p-well of SONOS served as a hard mask to prevent etch on beneath nitride. Secondly, the global wet process was involved in oxide remove via diluted HF solution both for NMOS/PMOS and SONOS wells. Finally, the blocking oxide of ONO was obtained from ISSG oxidation of nitride.

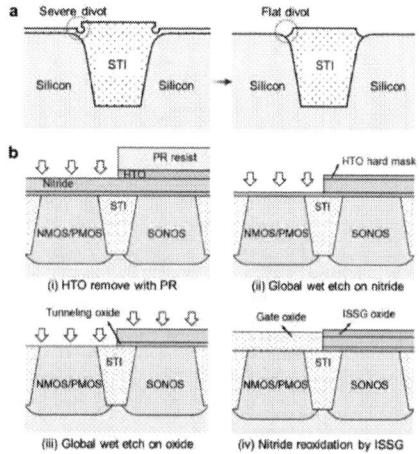

Figure 2: (a) wet etch on pad oxide to flatten the divot at STI edges. (b) Global wet process after PR stripping and ISSG reoxidation.

RESULTS AND DISCUSSION

The TEM of STI in original process after ONO deposition are shown in Figure 3a. The original STI with a higher step height shows a severe divot and the ONO film demonstrates a curled morphology. Thus, the ONO film cannot be removed completely leading to the residues containing silicon nitride (inserted figure in 3a). From the SEM image of top view for cell array after ONO etch, the ONO residues on the edges of STI is clearly found to be widespread over this region (Figure 3b).

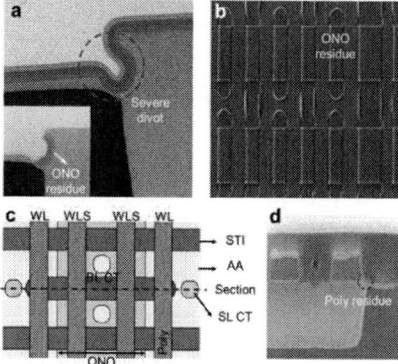

Figure 3: (a) STI from original process after and before (insertion) ONO etch. (b) Cell array region after ONO etch. (c) Layout of the cell array of 2T SONOS. (d) The section for the cells.

The layout of 2T SONOS is shown in Figure 3c. The SEM result (Figure 3d) shows a clear polysilicon residue which is very close to the SL contact (CT). This polysilicon residue will act as a bridge between the SL CT and WL polysilicon. Since the erase function involves in the hole injection process from the well to silicon nitride trap layer [6, 7], a V_{POS} is applied to the bit line (BL) and the p-well while a V_{NEG} is applied to the WLS to trigger the hole injection. The polysilicon bridge, however, connects the WL and SL, lowering the V_{POS}, decreasing the erase voltage and leading to the erase failure.

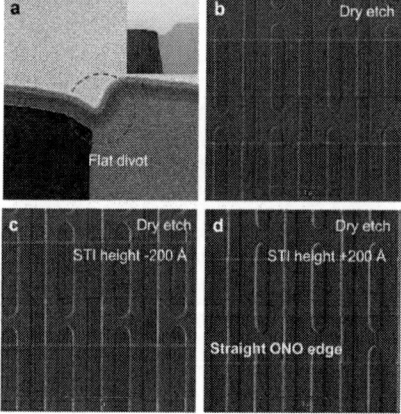

Figure 4: (a) STI from promoted process after and before (insertion) ONO etch. Cell array region with baseline (b), -200 Å (c), +200 Å (d) STI height.

Owing to the global wet process helped by ISSG technique, the nitride can be retained before well implant, which can act as a screen layer to reduce the effect of ion channeling and protect the silicon substrate [8]. Thus, the pad oxide is not required before well implant, which can be removed before ONO deposition. After removing pad oxide by diluted HF solution, the STI was significantly lowered. And more importantly, the edges demonstrating a sharp angle caused by severe divot will be removed leaving a flattened divot. As shown in Figure 4a, this flat divot averted the curled ONO film, and next, eliminated the residues after ONO etch (inserted figure in 4a). The SEM image of top view for cell array after ONO etch (Figure 4b) shows all clean edges of STI, indicating the fully elimination of ONO residues.

We also conducted the experiments about STIs with different step height by promoted process. The results shows that both lowered STI by 200 Å and heightened STI by 200 Å eliminated the ONO residues, leading to all clean edges (Figure 4c and d). The process margin was distinctly enlarged to ensuring high yield. This is due to the adjustment on STI profile by diluted HF solution after STI has been accomplished. It is worth emphasizing that the final STI step height of original and promoted process is close to each other. The STI from original process will possess a higher step height than that from promoted process before ONO deposition. The promoted process, however, skips the following wet etch on pad oxide. Thus, the promoted process actually well maintains the original STI step height, instead of simply lowering STI. Considering that the lower STI may generate serious narrow width effect, resulting in declining V_T and rising leakage of narrow devices [5], the promoted process well balances the STI height and profile.

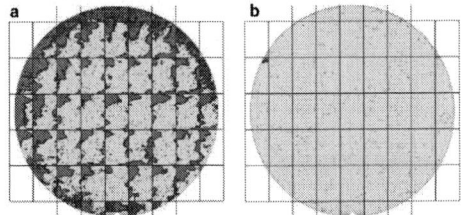

Figure 5: Chip probing test maps of the wafers from original (a) and promoted (b) fabrication process.

Figure 5 shows the chip probing test map from original and promoted process. The yield map demonstrates a clear by-shot type (Figure 5a). The top left corners suffer massive yield loss, where the dies mainly failed at program "0". This yield loss mainly stems from the weak erase process caused by polysilicon residues. Furthermore, the bridge between WL and SL CT caused by polysilicon residues will be even worse with a poor photolithography overlay shift of contact, leading to the by-shop pattern. The chip probing test from promoted process, on the contrary, demonstrated an enhanced yield which is more than 1.5 times that from original process. And the by-shot pattern is disappeared (Figure 5b). The elimination of polysilicon residues during promoted process contributes a lot to the improvement on yield.

CONCLUSION

A promoted fabrication process of SONOS memory involved in global wet process and ISSG technique was demonstrated to eliminate the ONO and polysilicon residues in this paper. This promoted process utilized the nitride from ONO film to be the screen layer for well implant and the original screen pad oxide can be therefore removed before ONO deposition. Simultaneously, the severe divot at the STI edges can be flattened during wet etch on pad oxide, eliminating ONO residues and boosting the yield of chip probing test. What' more, the adjustment on STI profile will not significantly lower the final STI step height, avoiding enhanced narrow width effect. This promoted fabrication process, which is simple and high-yield, should be also suitable for the fabrication of some other SONOS flash memories.

REFERENCES

[1] G. J. Hu, *IEEE Transactions on Electron Devices,* vol. 31, 1984, pp. 62-67.

[2] B. Gregory and B. Shafer, *IEEE Transactions on Nuclear Science,* vol. 20, 1973, pp. 293-299.

[3] M. Nandakumar, A. Chatterjee, S. Sridhar, K. Joyner, M. Rodder, and I.-C. Chen, *International Electron Devices Meeting 1998. Technical Digest (Cat. No. 98CH36217)*, San Francisco, December 6-9, 1998, pp. 133-136.

[4] H. Y. Chiu, Y. K. Fang, T. H. Chou, Y. T. Chiang, and C. I. Lin, *Semiconductor Science and Technology,* vol. 22, 2007, pp. 1157-1160.

[5] Y. Kim, S. Sridhar, and A. Chatterjee, *IEEE Electron device letters,* vol. 23, 2002, pp. 600-602.

[6] J. Bu and M. H. White, *Solid-State Electronics,* vol. 45, 2001, pp. 113-120.

[7] N. C. Leng, A. J. C. Har, and O. Y. W. Tat, *2012 4th Asia Symposium on Quality Electronic Design (ASQED)*, Penang, July 10-11, 2012, pp. 128-132.

[8] C. Park, K. Klein, A. Tasch, R. Simonton, and G. Lux, *International Electron Devices Meeting 1991 [Technical Digest]*, Washington, December 8-11, 1991, pp. 67-70.

RELIABILITY PERFORMANCE OF NOVEL TUNNELING FIELD EFFECT TRANSISTORS BASED ON FOUNDRY PLATFORM

Yukun Tang[1], Qianqian Huang[2,4], Kaifeng Wang[2], Yongqin Wu[3], Hongyan Han[3], Ye Ren[3], Weihai Bu[3], Junhua Liu[1,2,4], Zhigang Ji[1*] and Ru Huang[1,2,4*]*

[1]National Key Laboratory of Science and Technology on Micro/Nano Fabrication, Shanghai Jiao Tong University, Shanghai 200240, China
[2]School of Integrated Circuits, Peking University, Beijing 100871, China
[3]Semiconductor Technology Innovation Center (Beijing), Beijing 100176, China
[4]Beijing Advanced Innovation Center for Integrated Circuits, Beijing 100871, China
*Corresponding Author's Email: hqq@pku.edu.cn; zhigangji@sjtu.edu.cn; ruhuang@pku.edu.cn

ABSTRACT

In this work, the hot carrier injection (HCI) and negative bias temperature instability (NBTI) of a novel tunneling field effect transistor (TFET) manufactured by standard CMOS baseline platform is experimentally demonstrated. Results show that the reliability of the novel dopant segregated TFET (DS-TFET) is superior to the conventional MOSFET, and the reliability behavior of DS-TFET shows the similar temperature dependence to that of MOSFET. The proposed DS-TFET exhibits both excellent device performance and reliability, showing its great potential for ultra-low-power applications.

INTRODUCTION

Tunneling field effect transistor (TFET) [1, 2] with band-to-band tunneling (BTBT) mechanism has attracted much attention for ultra-low-power applications due to its low leakage current, sub-60 mV/dec sub-threshold swing, and high process compatibility. Conventional Si TFETs suffer from low on-state current (I_{ON}) and ambipolar effects. Recently, a novel Si TFET with dopant segregation technology (DS-TFET) based on CMOS baseline platform has been manufactured, experimentally demonstrating the I_{ON} enhancement of 3 decades and record high I_{ON}/I_{OFF} ratio of 10^7 among TFETs by industry-manufacturers [3]. Nevertheless, besides device performance, the device reliability is also an essential aspect for high-volume production. There are few studies on the reliability of TFETs reported, and most of the research are based on the TFET devices fabricated by non-standard processes [4-6]. In our previous work, the conventional Si TFETs using 0.13 μm CMOS baseline platform has been manufactured, and the time-dependent dielectric breakdown (TDDB) and positive bias temperature instability (PBTI) have been investigated, showing that conventional TFETs have a different degradation mechanism and worse reliability compared with MOSFETs [7]. In this work, we systematically investigated the hot carrier injection (HCI) and negative bias temperature instability (NBTI) of the novel DS-TFETs based on the 55 nm CMOS baseline platform.

DEVICE AND EXPERIMENT

The novel n-type and p-type DS-TFET devices are fabricated simultaneously with standard CMOS based on foundry platform [3, 8].

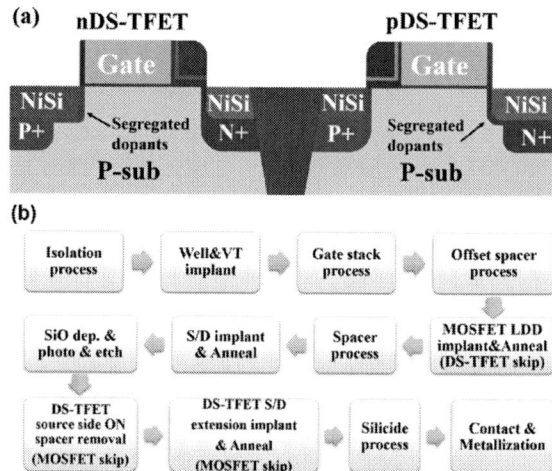

Figure 1: (a) The schematic structure of DS-TFETs. (b) Process flow for DS-TFET and MOSFET.

The device structure is shown in Fig.1(a). The device has an asymmetric sidewall structure. A self-aligned segregation source junction is designed. NiSi is formed at the source side to obtain a pocket layer at the interface of NiSi/Si along the direction of the channel at the gate edge in order to obtain a larger electric field for a higher BTBT generation rate. Fig.1(b) shows the specific process flow.

Both DS-TFETs and standard MOSFETs with 55-nm gate length and 1-μm gate width are measured for the reliability comparison. The DC-IV (measure - stress - measure) method is used for testing by Keysight 4082F Semiconductor Tester. The threshold voltage (V_{th}) of DS-TFET is obtained by constant current method.

RESULTS AND DISCUSSION

HCI for n-type DS-TFET and MOSFET

In the standard 55 nm CMOS foundry platform, the

979-8-3503-1101-3/23 $31.00 © 2023 IEEE

worst HCI stress condition of MOSFET is $V_g=V_d$, and the other ports are grounded. DS-TFET adopts the same stress condition as MOSFET for comparison. The HCI degradation curves of nDS-TFET and nMOSFET are shown in Fig.2(a). The test temperature was at room temperature.

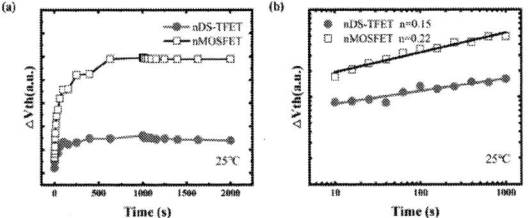

Figure 2: (a) V_{th} degradation of nDS-TFET and nMOSFET versus HCI stress time (0-1000 s) and recovery time (1000-2000 s). (b) Log-Log plot of HCI degradation in nDS-TFET and nMOSFET.

It indicates that the degradation of nDS-TFET is much smaller than that of nMOSFET at the same stress voltage, which is different from the conclusions of some pervious reports [4]. The recovery of nMOSFET is almost negligible, while the recovery of nDS-TFET is about 14% during the recovery stage. Log-Log plot of HCI degradation is shown in Fig.2(b). It is demonstrated that both nDS-TFET and nMOSFET conform to the law of $\Delta V_{th}=At^n$, and the n value of nDS-TFET and nMOSFET is 0.15 and 0.22 respectively.

For conventional nTFET, the P+ source region at the gate-source overlap will not be completely inverted to N+ when HC stress is applied, which means the potential drop of the gate dielectric at the gate-source overlap region is larger than other regions and the corresponding region of nMOSFET obviously. Furthermore, this peak electric field can cause the worse HCI degradation in conventional TFET compared with MOSFET [4, 7]. Nevertheless, in this work, the peak electric at source region can be effectively reduced and eliminated in nDS-TFET by dopant segregation technology. Moreover, the HCI degradation is caused by the interface state at SiO_2/Si interface or the traps in gate oxide layer generating from high-energy carriers by impact ionization. Besides the impact of electric field, the generation of high-energy carriers is also related to the density of carriers, and the on-current of nDS-TFET by the standard process is much smaller than that of nMOSFET. Therefore, the lower carrier density of nDS-TFET makes the generation rate of high-energy carriers much smaller for better reliability.

The HCI is almost irreversible damage, which can be illustrated by the recovery results of the nMOSFET. However, the recovery of nDS-TFET is relatively larger because the damage in nDS-TFET is concentrated in the narrow region generated by tunneling, while the damage of nMOSFET is in the whole channel region. The

difference in the value of slope n means that the defects generated by HCI are not identical in two devices, and the breaking ratio of Si-H and Si-O bonds in nDS-TFET is larger than that in nMOSFET [9].

HCI for p-type DS-TFET and MOSFET

The HCI stress condition of p-type devices is $V_g=V_d$ similarly with other ports grounded. The test temperature is 125℃. As shown in Fig. 3, the degradation of the pDS-TFET is also smaller than that of the pMOSFET under the same HC stress voltage, while the difference of degradation is not as large as that of the n-type device. The V_{th} of pDS-TFET and pMOSFET recovered by 13% and 6% respectively in 1000s recovery stage. The degradation conforms to the law of $\Delta V_{th}=At^n$, and the n values of pDS-TFET and pMOSFET are 0.33 and 0.21, respectively.

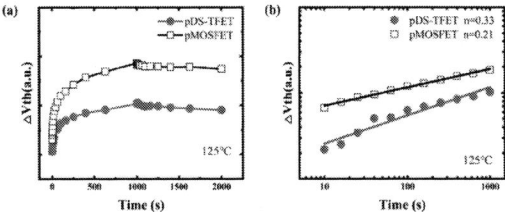

Figure 3: (a) V_{th} degradation of pDS-TFET and pMOSFET versus HCI stress time (0-1000 s) and recovery time (1000-2000 s). (b) Log-Log plot of HCI degradation in pDS-TFET and pMOSFET.

The main difference between n-type and p-type devices is that the carriers contribute to HCI degradation are mainly hot holes. Comparing with hot electrons, hot holes have smaller mobility and larger effective mass. Thus, they are not as easy to obtain high energy as hot electrons. In addition, the effect of carrier mobility in pMOSFET under HC stress is far stronger than that of pDS-TFET, and thus the reduction of carrier mobility makes the difference value of HCI degradation between pMOSFET and pDS-TFET smaller than that of n-type devices. As for the discrepancy in the degradation of pDS-TFET and pMOSFET, like the n-type devices, it is also determined by the weak source peak electric field by the dopant segregation and the lower carrier density of the pDS-TFET altogether.

NBTI for p-type DS-TFET and MOSFET

The NBTI stress condition of p-type devices is the constant V_g with other ports grounded. The test temperature is 125℃. As shown in Fig.4, it is obvious that the degradation behavior of the p-type device under NBT stress is significantly different from that of HC stress. The V_{th} degradation of pDS-TFET and pMOSFET after 1000s stress is almost the same. The V_{th} of pDS-TFET and pMOSFET recovered by 11% and 29% respectively in 1000s recovery stage. The degradation also conforms to the law of $\Delta V_{th}=At^n$, and the n values of pDS-TFET and pMOSFET are 0.25 and 0.22, respectively.

When p-type devices are under NBT stress, holes will be injected from the channel to the gate oxide layer, resulting in the generation of interface traps and oxide layer charges, resulting that the device parameter such as V_{th} and saturation drain current will be drifted. Unlike HCI degradation, NBTI mainly occurs at Si/SiO_2 interface in the whole channel and the overlap of the source/drain and gate. Therefore, the damage is almost uniform across the channel in MOSFETs. However, conventional TFETs usually have larger degradation than MOSFETs due to the source peak vertical electric field. Benefiting from the relatively uniform vertical electric field, the degradation of DS-TFET under NBT stress is almost the same as MOSFET.

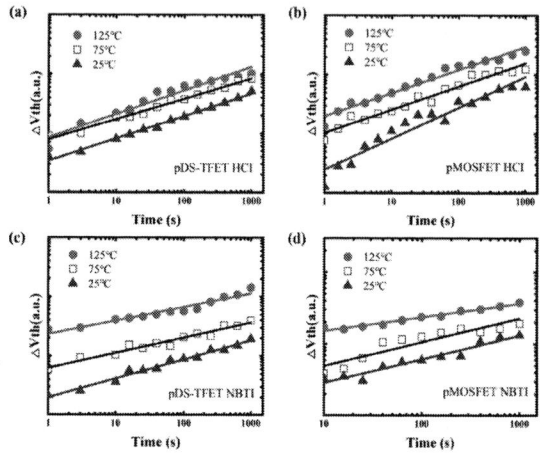

Figure 5: Log-Log plot of (a) HCI degradation in pDS-TFET, (b) HCI degradation in pMOSFET, (c) NBTI degradation in pDS-TFET, and (d) NBTI degradation in pMOSFET at different temperature.

CONCLUSIONS

This paper explores the device-level reliability of DS-TFET and MOSFET based on the standard 55 nm foundry platform. For both n- and p-type devices, the reliability performance of DS-TFETs are better than MOSFETs under HC stress conditions, which can be resulted from the designed dopant segregated source without peak vertical electric field and the lower carrier density. The degradation behavior of DS-TFETs and MOSFETs are almost the same under NBT stress condition, and show the similar temperature dependence. The proposed DS-TFET exhibits both excellent device performance and reliability, showing its great potential for ultra-low-power applications in the future.

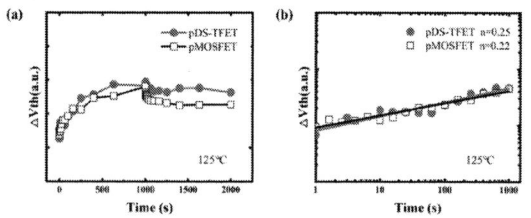

Figure 4: (a) V_{th} degradation of pDS-TFET and pMOSFET versus NBTI stress time (0-1000 s) and recovery time (1000-2000 s). (b) Log-Log plot of NBTI degradation in pDS-TFET and pMOSFET.

ACKNOWLEDGEMENTS

This work was supported by National Key R&D Program of China (2018YFB2202801), NSFC (61927901), Beijing SAMT Project (SAMT–BD–KT-22030101), 111 Project (B18001), and Tencent Foundation through the Xplore Prize.

Effect of temperature on the reliability

The dependence of DS-TFET reliability impacted by temperature is further investigated. Fig.5(a) and Fig.5(b) show the HCI degradation of pDS-TFET and pMOSFET at different temperature respectively. It is demonstrated that the V_{th} degradation of both DS-TFET and MOSFET is stronger with the increasing temperature. In addition, the DS-TFET and MOSFET both conform to the law of $\Delta V_{th}=At^n$, and the slope n values of the same device are similar at different temperature. In deep submicron devices, the dominant mechanism of hot carrier generation is electron-electron scattering rather than mobility effects. The carrier concentration and carrier energy will increase with the growing temperature. Therefore, the HCI degradation of the two kinds of devices is more serious with the increase of temperature.

Fig.5(c) and Fig.5(d) show the NBTI degradation of pDS-TFET and pMOSFET at different temperature respectively. Same as HCI degradation, the NBTI degradation of both pDS-TFET and pMOSFET are stronger with the increasing temperature. It can be attributed that the main cause of the NBTI effect is the breaking of the Si-H bond, which is a thermally activated process and is very sensitive to temperature, and thus the increase in temperature makes the Si-H bond more easily broken.

REFERENCES

[1] Q. Huang, et al. *IEDM*, pp. 8.5.1-8.5.4, 2012.

[2] G. Dewey, et al. *IEDM*, pp. 33.6.1-33.6.4, 2011.

[3] K. Wang, et al. *IEEE ESSDERC*, pp. 360-363, 2022.

[4] G. Jiao, et al. *IEDM*, pp. 1-4, 2009.

[5] G. Han, et al. *VLSI-TSA*, pp. 1-2, 2012.

[6] W. Mizubayashi, et al. *IEDM*, pp. 14.3.1-14.3.4, 2015.

[7] Q. Huang, et al. *IEDM*, pp. 31.5.1-31.5.4, 2016.

[8] Y. Li, et al. *ICSICT*, pp. 1-3, 2020.

[9] S. Mahapatra, et al. *IEEE TED*, vol. 53, no. 7, pp. 1583-1592, 2006.

INFLUENCE OF INTERFACIAL LAYERS AND HIGH-k POST DIELECTRIC ANNEALING ON THE CHARACTERISTICS OF MOS DEVICES

Guanqiao Sang[1,3], Qingzhu Zhang[1,2], Huaxiang Yin[1,2], Junfeng Li[1], Xulei Qin[3*]*

[1]Key Laboratory of Microelectronics Devices and Integrated Technology, Institute of Microelectronics of Chinese Academy of Sciences, Beijing 100029, China

[2]University of Chinese Academy of Sciences, Beijing 100049, China

[3]ChangChun University of Science and Technology, Changchun 130022, China

[*]Corresponding Author's Email: qxl@cust.edu.cn, zhangqingzhu@ime.ac.cn

ABSTRACT

In this paper, investigation of the effects of Post Dielectric Annealing (PDA) and Interfacial Layers (IL) on the characteristics of Metal-Oxide-Semiconductor (MOS) devices. Based on the traditional MOS fabrication process, the PDA method provides a solution to the problem of large interface trap density (D_{it}) after direct deposition of HfO_2 medium, and successfully reduces D_{it} by about 78%. At the same time, compare the interface state and electrical characteristics of various IL layer deposition methods. The results showed that the Atomic Layer Deposition (ALD) method reduced D_{it} by nearly 60% over the thermal oxidation method. And the positive shift of the flat band voltage (V_{fb}) was increased by about 150 m. In addition, analyze the relationship between the change of V_{fb} and the internal charge of the dielectric layer is analyzed, it is found that annealing and IL layer dielectric deposition methods change V_{fb} by affecting the interface defect charge (Q_{it}) and oxide defect charge (Q_{ot}). In this work, the results provide a guidance for the interface optimization of advanced devices.

Keywords—MOS devices、IL、ALD、V_{fb}、D_{it}、HK PDA、Q_{ot}、Q_{it}

INTRODUCTION

With the scaling of the MOSFETs, dielectric based on high-k materials have been widely adopted. Among all evaluated high-k materials, HfO_2 based dielectrics are considered to be one of the most promising gate materials due to their high dielectric constant, relatively large band gap, and good thermal stability with Si [1]. Before high-k layers, high-quality SiO_2 or SiO_xN_y interface layer (IL) is required to reduce the lattice miss-match induced mobility and reliability degradation [2]. However, the intrinsic high oxygen vacancy concentration in HfO_2 based gate dielectrics is easy to induce threshold voltage hysteresis, leakage current, and flat-band voltage shifting that all degrade the performance of transistors. Among them, high-k PDA is a practicable and effective way [3]. When CMOS technology moves into the GAA era, DI O_3 or ALD SiO_2 is often used to obtain thinner and high-quality IL layers, while reducing the thermal budget [4]. In this

work, we carried out and investigated the effects of two different forming methods for IL (ALD and thermal oxidation) and HK PDA on properties of the MOS devices. The results showed that IL layer fabricated by ALD process with a PDA after HfO_2 deposition obtained the best C-V characteristics.

EXPERIMENTS

The MOS devices fabrication process and the scheme of the devices are shown in Fig.1 (a) and (b), respectively. In order to investigate the effects of different SiO_2 layer deposition methods on the interface state characteristics, and the effects of HfO_2 annealing on the interface and electrical characteristics of the device. Alternatively, the SiO_2 with an oxide layer thickness of 1nm was grown by rapid thermal oxidation (RTO) annealing and ALD. In order to observe the quality of each dielectric layer of the MOS device, the TEM characterization of the sample was shown in Fig.1 (c).

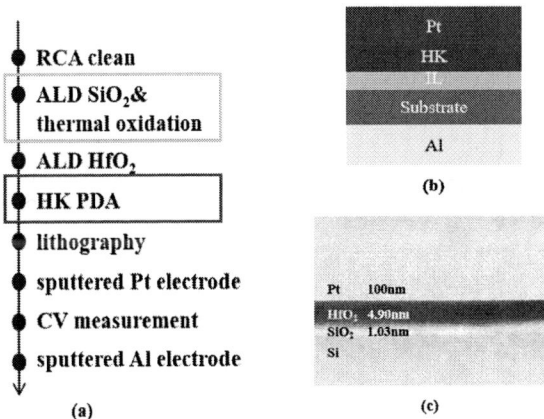

Fig. 1: (a) MOS devices fabrication process (The yellow part is the IL and the purple part is the HK annealing). (b) Schematic of MOS capacitor. (c) The cross-sectional TEM image of deposition HfO_2 (4.90 nm)/ALD SiO_2 (1.03 nm)/Si. The active area of this capacitor is $100 \times 100 \ \mu m^2$.

RESULTS AND DISCUSSIONS

Fig.2 (a) shows the CV curves of the NMOS capacitance at a frequency of 1MHz. In order to characterize the interface quality, the hysteresis characteristics of three groups were measured (see Fig.2 (b)).

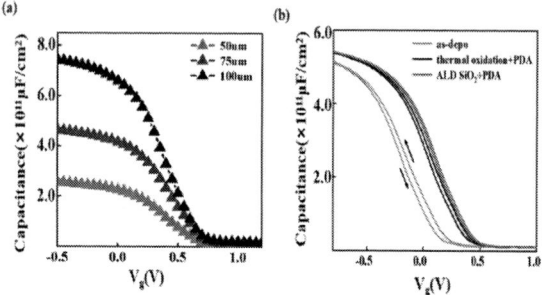

Fig. 2: (a) CV curves with three different capacitor sizes. (b) Dual-Sweep CV curves for the three samples (from inversion to accumulation and accumulation to inversion region).

Fig.3 (a) show the total amount of defective charges in the SiO_2 layer in three groups of samples (as-depo, ALD SiO_2 + PDA, thermal oxidation + PDA), respectively. As can be seen from the image, there are the largest densities of defect charge for the device without HK PDA, and the lowest densities of for that of the ALD SiO_2 after PDA process. The results indicates that the PDA process can effectively reduce the density of defects and vacancies in IL. At the same time, it was observed that the defects of SiO_2 grown by ALD were lower than those of thermal oxidation after PDA process. The results indicate the PDA is the most important factor for reduce the bulk IL layer defects, which is more important than that of the growth method of the IL layer.

To investigate the effect of HK PDA, ALD SiO_2 and thermal oxidation growth SiO_2 on the quality of MOS capacitor interface, we measured the multi-frequency C-V curves using conductance method. The D_{it} of the three samples was calculated after the measurement, as shown in Fig.3 (b). The interface trap density after HK PDA decreases by about 78%. The result of the significantly reduction of the interface states density is consistent with previous reports. Moreover, the D_{it} of ALD growing SiO_2 samples is also lower than that of thermal oxidation, which indicates that the ALD process induces lower interface defect charge (Q_{it}), lattice defects and lattice damage at the interface. This result is contrary to the conclusion of Hiller D et al., which has generally concluded that the interface state density of the SiO_2 layer grown by thermal oxidation is generally considered to be better [5]. It maybe because the precursors we use to prepare the oxide layer with ALD do not contain H_2O, and most of the precursors of ALD SiO_2 have H_2O, which leads to the involvement of –OH group. The –OH group causes partial atomic polarization

in the layer, and changes in interface properties and permittivity have been considered to be related to hydroxyl content [6]. Another reason may be because the thickness of the film layer reported in this paper is only 1nm, and the ultra-thin film thickness and lower thermal oxidation temperature are also the reasons.

Fig.3 (c) shows the EOT values of multiple groups of MOS before and after the sputtering back electrode. When the bottom was sputtered the electrode, the EOT decreased by about 1 nm, which is consistent with the theoretical value. The attached image in Fig.3 (d) shows the V_{fb} values of multiple groups of different sizes capacitances under the three conditions actually measured, while the V_{fb} conditions normalized are shown in the main figure. It is not difficult to see that the V_{fb} increases significantly after annealing, that's up nearly 40% and offset of about 150 m, however, the different SiO_2 deposition modes had little effect on V_{fb}.

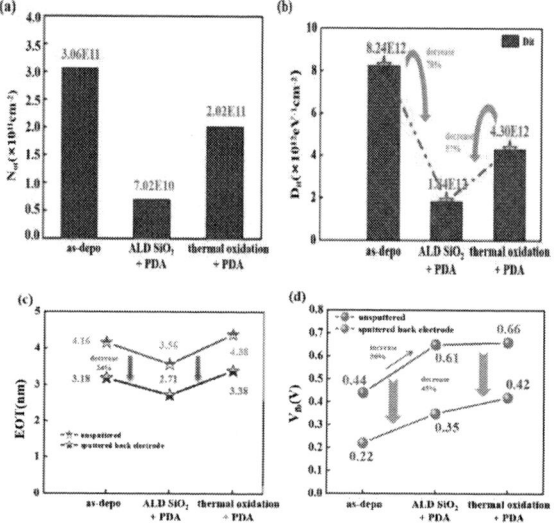

Fig. 3: Extracted N_{ot}, EOT, V_{fb}, and D_{it} results for three samples. (a) Trap/detrap electrons density (Q_{ot}) variations with difference samples. (b) Comparison of different interface trap density (D_{it}) of three groups of samples. (c) EOT change curve. (d) Flat-band voltage (V_{fb}) change curve.

In the analysis of V_{fb} changes, it can be seen that the flat-band voltage is closely related to the charge in the SiO_2 layer. The charges connected to the SiO_2 are shown in Fig.4 (a), then combined with the flat-band voltage formula, the HK annealed sample has a small Q_{it}, thus with a larger V_{fb}. For the same annealed ALD and thermally oxidized SiO_2 samples, it can be seen from Fig.3 (b) that for the growth method of the oxide layer, the Q_{it} of the ALD method is smaller than that of the thermal oxidation method, so it can be inferred that the flat-band voltage of the ALD method is larger. However, fig.3 (d) shows that the flat-band voltage of the ALD and thermal oxidative growth IL layers is basically the same, and even the voltage of the thermal oxidation method is larger. Fig. 4 (b) analyzes the reason of V_{fb} shift from the perspective of energy band. Since the value of V_{fb} is measured between the depletion area and the accumulation area in the CV curves, during the transition from the positive to negative gate pressure, the carrier drifting from the substrate to the gate is captured by the interface trap charge, resulting in less holes in the accumulation area. While the samples with HK PDA have small Q_{it} and less trapped carriers, the flat-band voltage is large. Similarly, the oxide defect charge (Q_{ot}) also captures the carriers, thus causing the difference in the flat band voltage between sample 2 and sample 3.

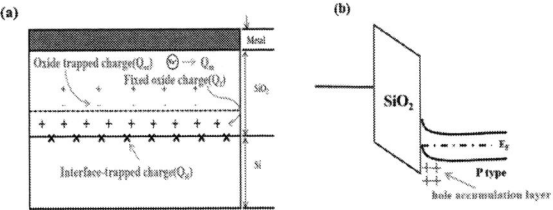

Fig. 4: (a) The charges connected to the SiO_2 (including the Q_{it} and Q_{ot}). (b) Capacitor energy banding in the accumulation zone.

CONCLUSION

Based on the conventional MOS capacitance fabrication process, the PDA method successfully reduced the density of interface states by 78%. Compared the effect of ALD and thermal oxidation deposition of 1nm SiO_2, there are nearly 60% (D_{it}) for the SiO_2 prepared by ALD method compared with that of the thermal oxidation. In addition, the HK annealing would affect V_{fb}. Meanwhile, we systematically analyzed the change pattern of the flat band voltage in the three sets of samples from the perspective of internal charge, the results found that the annealing can affect the Q_{it}, and the different IL layer deposition process can affect the Q_{ot} in SiO_2, which causes the shift of V_{fb}. The results provide a good reference for the interface optimization for advanced devices.

ACKNOWLEDGEMENTS

This work was supported in part by the Strategic Pilot Project of the Chinese Academy of Sciences-Class A, under Grant XDA0330302; in part by the Youth Innovation Promotion Association, Chinese Academy of Sciences, under Grant Y9YQ01R004; and in part by the National Natural Science Foundation of China under Grant 91964202 and Grant 61904194.

REFERENCES

[1] L. Kang, et al. "Single-layer thin HfO$_2$ gate dielectric with n$^+$-polysilicon gate," 2000 Symposium on VLSl Technology Digest of Technical Papers, 2000, pp. 44 - 45.

[2] M. L. Green, et al. "Nucleation and growth of atomic layer deposited HfO$_2$ gate dielectric layers on chemical oxide (Si-O-H) and thermaloxide (SiO$_2$ or Si-O-N) underlayers," Journal of Applied Physics, 2002, pp.7168-7174.

[3] He, Yonggen, et al. "Investigation of different post HK annealing impact on HK film property and device performance." 2014 20th International Conference on I$_{on}$ Implantation Technology (IIT), 2014, pp.1-4.

[4] Radamson, et al. "State of the art and future perspectives in adv-anced CMOS technology." Nanomaterials, 2020, pp. 155-164.

[5] D. Hiller, et al. "Low temperature silicon dioxide by thermal atomic layer deposition: Investigation of material properties." Journal of Applied Physics 107, 2010, pp. 122-132.

[6] B. Fowler, et al. "Relationships between the material properties of silicon oxide films deposited by electron cyclotron resonance chemical vapor deposition." Journal of Vacuum Science & Technology B, 1994, pp. 441-448.

Optimized Wafer Edge Condition in Lithographic Process For Peeling Defect Reduction

Shanshan Chen[1], Hunglin Chen[1], Yin Long[1], and Kai Wang[1]
[1]Shanghai Huali Integrated Circuit Corporation, Shanghai 201314, China
*Corresponding Author's Email: chenshanshan@hlmc.cn

ABSTRACT

As technology develops and transistor size keeps shrinking, it is critical to define high-resolution patterns in integrated chip manufacturing by optimizing bevel rinse condition of lithography processes. The presence of multiple organic film stacks requires corresponding complicated etching processes and may lead to unwanted damages in the post removal. The wafer edge and bevel just need to be protected since wafer processing issues can be induced by bevel profile, and the etching could change its shape while the wafer bevel is usually naked to plasma. This paper presented an un-health rinse condition at the bevel area that leaked to the striping of hard mask oxide and became the peeling source. The oxide film cannot be totally removed at the etch process and is easy to peel off at the bevel area. With step-by-step investigation, the peeling source occurred in photo step and is enhanced after etch process. A series of experiments were conducted to find out the optimized bevel rinse condition. With the solution, the defect chart showed that the peeling source defect level dropped down obviously.

KEYWORDS

Peeling Defect; Edge Bead Remove (EBR); Bevel Rinse Condition.

INTRODUCTION

For photolithography spin coating process, there are several basic requirements to large-scale chip production, including organic photoresist material, process parameter window, standard control, less defect and so on. Photolithography is a process by transferring the geometric pattern design from the mask to the wafer [1, 2].

Nowadays there are two typical lithographic schemes, one is layer 1, layer 2 and layer 3 tri layer scheme, which is widely picked up in advanced node. The other is layer 1 and layer 2 Bi-layer scheme, which is usually picked up in wide line technology. Therefore, lithographic technique depends on photosensitivity materials is introduced into a design of polymer structure. Edge cleaning is a key factor as defect around this area can spread across the wafer and cause yield loss [1].

During the photo process, while the organic compound strip, wafer edge need to be clear in the end. Exposure more time to high temperature during the degumming process is conducive to the interaction of photoresist organic materials, for example layer 1 and layer 2, which lead to the organics more difficult to remove with plasma or wet cleaning. In order to maintain the stable operation of fab, the edge bead removal (EBR) area need to be optimized, for example, which should be narrow and the width should be controlled at a certain level [1, 2].

In this paper, research was carried out to solve problem existing at wafer edge. The formation of peeling defect is related to edge EBR process from step by step data. Quite often an organic layer such as layer 3 in between the layer 2 and film layers is also needed as an intermittent etching mask since directly using layer 1 and layer 2 as the mask can be very difficult. The un-covered layer would lead to form a redundant structure at wafer edge, and was easy to cause the hard mask oxide to bulge and form a source of peeling source in the subsequent process. The peeling defect could be greatly reduced even disappeared while optimization of the bevel rinse condition, and ultimately existing problem was solved.

RESULTS AND DISCUSSION

The architecture of poly loop integration is presented. An Oxide layer and hard mask are deposited, which is situated in oxide film below. In addition, layer 1, layer 2 and layer 3 tri layer locate above the thin film.

For lithography process, any defect formed in etch steps might probably cause yield loss. Figure 1 shows peeling defects post etch process; Fig.1 (a) display the distribution of failed chips, which is 12 o'clock wafer edge signature map; (b) is the peeling scheme.

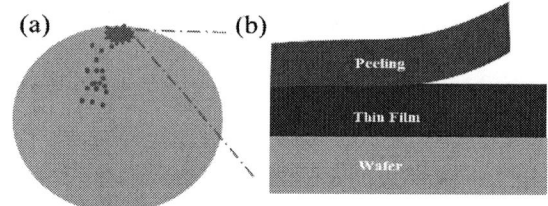

Fig.1. (a) Peeling defect wafer map signature; (b) Peeling Defect Scheme.

To clarify the root cause of wafer edge peeling defect formation, serial experiments were executed step by step, with the purpose of checking the defect source. Fig.2 shows peeling defect OM image of wafer edge. An abnormal area is found at the edge which is different from the other areas, and defects occur after etch process. This suggests that the source of wafer edge peeling source might come from the photo, and the hard mask layer stripping may be one of the sources of wafer edge peeling defect, which is relate to the wafer edge rinse condition [3,4].

Fig.2. Step by step defect source for OM image.

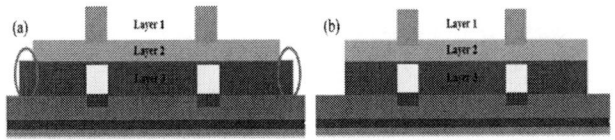

Fig.3. Peeling source' fail mode diagram (a) film stack without EBR,(b) film stack with EBR

Fig.3 shows the possible formation mechanism of peeling source. It can be seen from the Fig.3 (a) that there is an abnormal layer 3 spin coating on the top of under layer. In photography process, the weakness of film adhesion would cause the formation of some abnormal residue while different film stacking together. Because the different bevel rinse condition of layer 2 and layer 3, which lead to a partial etch area, and in the subsequent process form the peeling defect source, and lifted to the surface of wafer. This suggest that film layer stripping may be one of the sources of wafer edge peeling defect. Because the hard mask layer will be partly left on the surface while the layer 3 coverage at wafer extreme edge, which may cause the risk of peeling defect. Furthermore, in order to found out optical bevel rinse condition, EBR of CIP1, CIP2 and CIP3 experiment was carried out. While changing the bevel rinse condition of layer 3 the same as layer 2, the peeling source problem can be solved, as shown in Table 1 and Fig.3 (b). Therefore, the peeling defect could be greatly reduced even disappeared while optimization of the bevel rinse condition, and ultimately existing problem was solved, as shown in Fig.4. [5, 6]

Layer	PR	EBR split			
Photo Coating	Layer3	BSL	CIP1	CIP2	CIP3
Photo	Layer2				
	Layer1				

Table1 EBR split for CIP remove process names with layer 1, 2, 3.

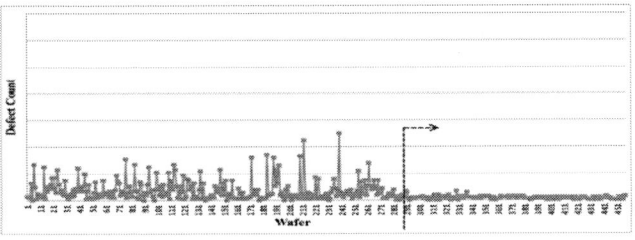

Fig.4 Trend Chart after actions for peeling defects.

CONCLUSION

In this investigation, experiences showed that optimization of EBR condition have taken over excellent chemistries developed by photoresist for microlithography. Serial experiments were executed step by step, in order to check the peeling source and future optimize the EBR recipe condition. Inadequate edge EBR has a significant contribution to this class of defect. Ultimately, we come to a conclusion that the layer 3 bevel rinse condition will produce abnormal pattern structure and lead to the strip of hard mask oxide, which will aggravate the peeling defect at the subsequence etch process. By future improving the rinse condition at wafer edge, the peeling defect can be greatly reduced even disappear. Therefore, the wafer edge defect yield loss can be reduced while optimizing the peeling defect.

ACKNOWLEDGEMENTS

The authors would like to thank our YE teams for their support and FA department for the supports of TEM/EDS analysis, also Litho team for their technical discussion

REFERENCE

[1] X.Q. Liu, B.Q. Liu, "Investigation of Removing Standing Wave Effect During Litho Process," CSTIC, 2022.

[2] Z. Wang, K. Feng, "Research of Lithograph Process of Polyimide Photo Resist for Passivation Thick Films," CSTIC, 2018.

[3] Seung Hwa Hyun, Doo-Hyun Cho, "Vision based EBR Metrology for Edge Bead Removal Optimization," IPFA 2020.

[4] Remy. Charavel, et al., "Wafer Bevel Protection During Deep Reactive Ion Etching," IEEE Transactions on Semiconductor Manufacturing, 2011, Volume: 24, Issue: 2.P358-365.

[5] M. Boumerzoug, "Optimized BARC films and etch byproduct removal for wafer edge defectivity reduction," ASMC (2014).

SILICIDE PROFILE OPTIMIZATION ON ACTIVE AREA IN 4XNM ETOX NOR FLASH MEMORY

Hualun Chen [1], Yuxin Tong[1]*, Xiangyu Qi[1], Songhan Duan[1], Botong Liu[1], Chaoran Zhang[1], and Lin Gu[1]

[1] Huahong Semiconductor (Wuxi) Limited, Wuxi 214000, China
* Corresponding Author's Email: yuxin.tong@hhgrace.com

ABSTRACT

Silicide overgrowth along active area corner results in undesired charge accumulation and electric field distribution in silicide, especially in advanced tech-node. An optimized silicide profile on 4Xnm ETOX (EPROM tunnel oxide) Nor flash has been obtained with the decrease of self-aligned-silicide aligned block (SAB) dry etch time and thick NiPt film deposition. The better anti-breakdown capability of the adjusted device is attributed to the uniform distribution of electric field in the silicide. This study presents an effective method to boost breakdown resistance for advanced technology nodes devices.

INTRODUCTION

In order to realize the reduction of contact resistance and the RC delay associated with the silicon gate, Ni-based silicide has been widely applied in 55nm node Nor flash and below. In general, the interface between silicide and silicon is smooth, expect the undesired pipping effect, which can be eliminated by doping Pt in Ni alloy [1]. However, due to the corner rounding of active area and influence from subsequent dry or wet etch process, the silicide overgrowth along the boundary of active area is inevitable. These silicide growth behavior may lead to unexpected leakage or breakdown. The active area size of 4Xnm Nor flash is less than 60nm with a high corner rounding proportion, results in an arch-bridge-like shaped silicide. Since the narrow pitch of Nor flash device, this abnormal silicide profile exhibits great impact on the breakdown performance of the device. Herein, we demonstrate an optimized silicide profile to reduce leakage current from contact (CT) to bulk and optimize electric field distribution in NiSi alloy.

EXPERIMENTAL

The process flows of Nor flash fabrication are shown in Fig.1. The SAB Dry etch time and NiPt deposition thickness experiments were carried out respectively. In the two sets of splits, the anneal time and other parameters kept constant.

Notably, NiPt in all samples was excessive and won't react completely during the first annealing process. Subsequently, the remaining NiPt alloy was removed through metal strip to avoid effecting the phase transformation ($Ni_2Si \rightarrow NiSi$ [2]) during the second anneal process.

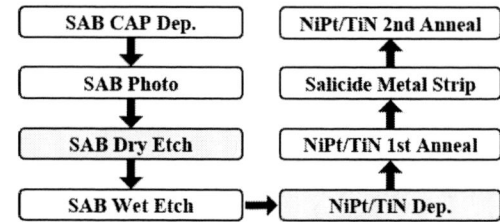

Figure 1: Experimental flow chart.

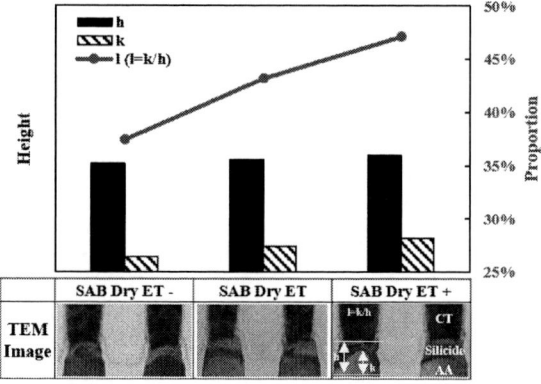

Figure 2: TEM image and related data of samples with different SAB dry etch time.

RESULTS AND DISCUSSION

As presented in Fig. 2, it can be observed that when increasing SAB dry etch time, the thickness of NiSi alloy along active area corner (record as h) increases correspondingly, In addition, the silicide thickness difference between active area center and corner was evaluated by l ratio, where l is silicide arch height to total height ratio. The l ratio significantly increases with dry etch time, indicating silicide was intended to growth along the active area corner since more active area exposed under longer dry etch process. Fig. 3 displays the bridge mapping (bit line to bit line, abbreviated as BL to BL) of SAB dry etch time splits, which shows that the bit line bridge parameter of the sample with increasing etch time is lower than that of other samples at wafer edge region; it can be reasonably explained that the concentration of electric field at the tips of silicide bottom leads to very high leakage between CT and bulk.

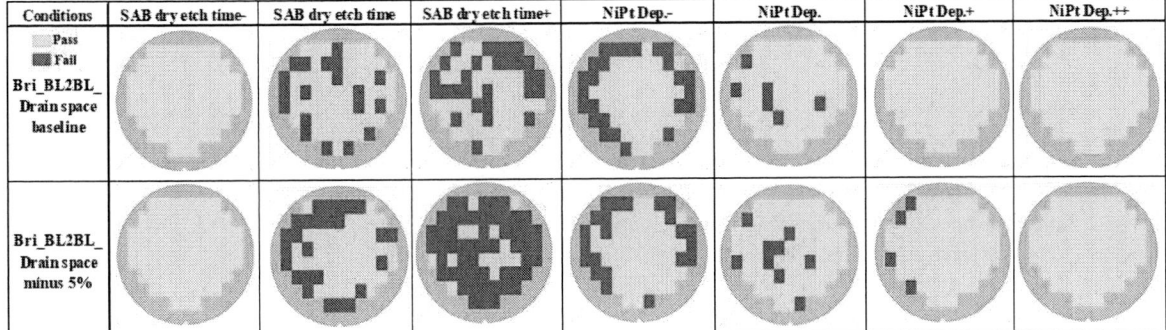

Fig.3 BL to BL bridge mapping of samples with different SAB dry etch time and NiPt deposition thickness.

What's more, the same conclusion can also be obtained from the curves as shown in Fig.4-6; the distribution of bit line stressed voltage becomes worse as SAB dry etch time increases (Fig. 4); the resistance of sample B is lower than that of sample A because of larger cross section silicide area (Fig. 5). In addition, sample A with sharper silicide bottom tips has much higher low field leakage current than sample B, further indicating a worse leakage performance.

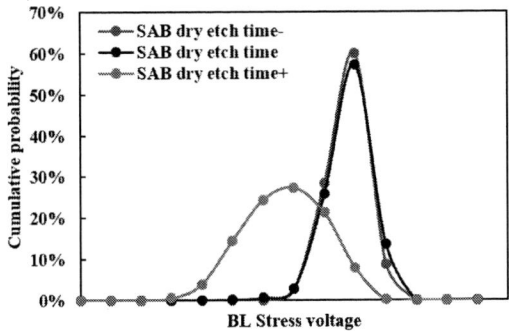

Figure 4: BL stress curve by SAB dry etch splits.

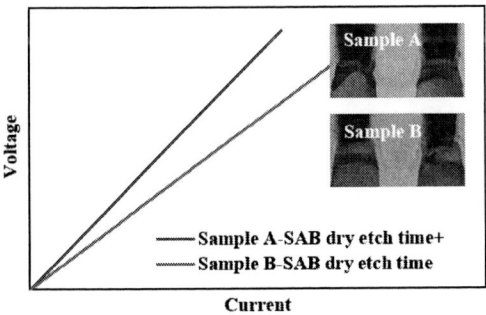

Figure 5: Active area resistance curve of samples with different SAB dry etch splits (sample A and B correspond to the samples of SAB dry etch time+ and SAB dry etch time respectively in Fig.4).

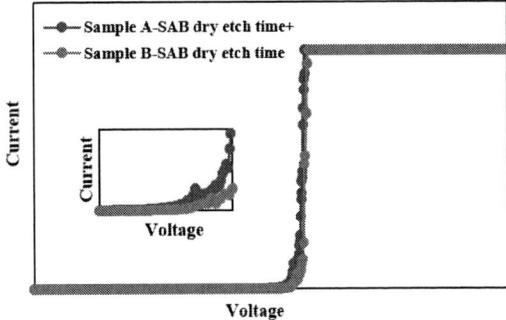

Figure 6: I-V curve of samples with different SAB dry etch splits.

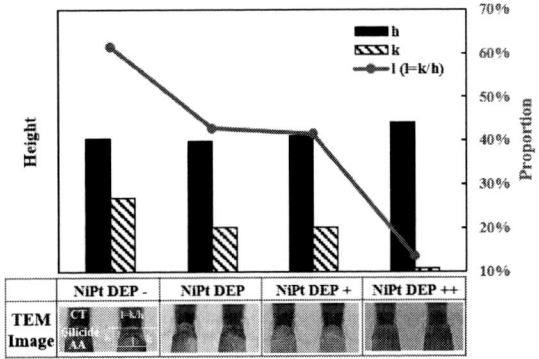

Figure 7: TEM image and related data of samples with different NiPt deposition thickness.

Considering the deterioration of electric performance of arch-bridge-like silicide profile, further improvement was conducted through increasing NiPt deposition thickness. TEM images of samples with different NiPt deposition thickness are exhibited in Fig. 7, with thick NiPt film deposition, the silicide thickness difference ratio l significantly decreased, meanwhile, the interface between silicide and silicon is no longer arch-bridge-like profile. The BL breakdown voltage goes back to pass level (Fig. 3), which can be attributed to the uniform distribution of electric field in the silicide.

CONCLUSION

The optimized silicide profile was achieved by reduction of SAB dry etch time and thick NiPt film deposition on 4Xnm ETOX Nor flash, which significantly enhanced BL breakdown voltage as a result of uniform electric field distribution in silicide. The present work provides effective ways and guidelines of silicide profile optimization on advanced tech-node devices that need further improving-anti-breakdown capability.

REFERENCES

[1] T. Sonehara, A Hokazono, et al. *IEEE T. Electron Dev.*, vol. 1, 2011, pp. 3778-3786.

[2] H. Iwai, T. Ohguro and S.-i. Ohmi. Microelectron Eng., vol. 1, 2002, pp. 157-169.

IMPROVEMENT OF STANDBY CURRENT FAILURE BY DEVICE OPTIMIZATION ON 4XNM ETOX NOR-FLASH MEMORY

Hualun Chen[1], Zhuangzhuang Wang[1]*, Lin Gu[1], Yihang Du[1], Chun Yao[1], Xiaodong Mu[1], Wan Song[1]

[1]Wuxi Huahong Grace Semiconductor Manufacturing Corporation, Wuxi 214028, China

*Corresponding Author's E-mail Addresses: Zhuangzhuang.Wang@hhgrace.com

ABSTRACT

The self-aligned STI technique is used for the fabrication of ETOX (Erase through oxide) NOR-flash as the feature size scaling to 4Xnm node owing to the advantages in improving the poly gap-filling window and eliminating the floating gate poly residue. In this paper, the failure mode of standby current was characterized by the WAT test, junction stain analysis and TCAD simulation. The poor performance of well/well isolation is attributed to the inordinate diffusion of p-type implanted ion because of the additional thermal budget. At last, the failure ratio of standby current and isolation BV performance can see significant boost by optimizing the HVP-well filed implantation condition determined by the simulation and experiment results.

INTRODUCTION

NOR-flash memory has been widely used in consumer and automotive electronics, industrial production and other application areas due to its higher read/write speed and good compatibility with CMOS [1, 2]. Recently, the increasing demand for high density, low cost and low power memory devices continuously drives the aggressive scaling for the nor flash memory. However, for traditional standard ETOX NOR-flash structure, there are enormous challenges in the course of cell scaling down, such as the balance of floating gate (FG) and control gate (CG) gap-filling window, and the poor performance of poly residue resulting from the FG CMP (chemical mechanical polishing), etc.. According to the study [3], the self-aligned STI (SA-STI) process is one of the most available techniques for the fabrication of advanced nor flash benefitting by its enhancement of gap-filling

capacity and resolution of FG poly residue. However, the standby leakage current is a big challenge in our research of 4Xnm ETOX NOR flash memory with SA-STI technique.

In this paper, the root cause of standby current failure in 4Xnm SA-STI NOR-flash memory was investigated by the isolation and junction bridge voltage (BV) test, junction stain analysis and TCAD simulation. After optimizing the isolation of well/well, the fail bin related standby current of 4Xnm NOR-flash was significantly improved.

DEVICE FABRICATION

Fig.1 illustrates that the process sequence of the aforementioned traditional (a) and the self-aligned STI technique ETOX NOR- flash. For the traditional process flow, as shown in Fig.1(a), AA/STI pattern is defined after nitride deposition, and the trench is subsequently refilled with CVD oxide followed by a densification anneal. Analogously, the CMP anneal with slightly lower temperature performed after the STI CMP, which aims to release stress and repair polish damage. After removing the pad nitride, the channel implantation and the high voltage N-well (HVN) and P-well implantation are executed sequentially. The floating gate formation by poly CMP process after the tunneling oxide and floating gate deposition. It is worth noting that the formation of HV N-well and P-well prior to the growth of tunneling oxide and the deposition of floating gate polysilicon and nitride. The AA/STI pattern is defined over the nitride/FG polysilicon/tunnel oxide/substrate stack and then refilled with CVD oxide followed by a densification anneal. On the one hand, the application of the SA-STI technique could eliminate the poly residue ascribable to the change of FG pattern definition mode, from poly CMP to one-step etching. On the other hand, attributing to the preposition of HV well formation the two anneal steps, STI oxide anneal and STI CMP anneal, would introduce the additional thermal budget during the ion activation of HV well, which might impact the cell device performance.

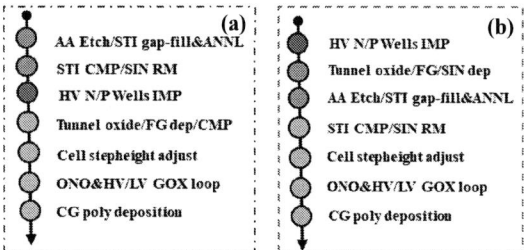

*Fig.1: Illustration of **(a)** traditional ETOX Nor-flash and **(b)** self-aligned STI technique ETOX Nor-flash*

RESULTS AND DISCUSSION

Fig.2 (a) shows that the chip prober (CP) yield map of SA-STI, exhibiting that the failure ratio of standby current with SA-STI structure is nearly 100%. To contrast the electrical properties of HVPW field implantation baseline condition with that of self-aligned STI structure, the standby current curves of point of reference (POR) and SA-STI sample are tested and the results are shown in Fig.2(b). It is obvious that the standby current performance of SA-STI is worse than that of the POR.

Generally speaking, the poor performance of standby current is often blamed on the leakage current of peripheral transistors. For this reason, the well/well isolation BV and plus/well junction BV performances of various MOSFET are measured.

Schematic of isolation and junction BV test key with a finger structure are shown in Fig. 3. In process testing, the voltage is recorded as the bridge voltage at which the leakage current reaches a target value with 0~20V sweep voltage. According to the BV test data which are shown in Fig.2(b), both the LNW and HNW to HPW isolation BV of SA-STI sample are obviously lower than that of the POR, however, there are no significant differences in the N+ to HPW junction BV of the two samples. Based on the

reason above, it is possible that the key to the failure of N-well to HVP-well isolation lies in the isolation between wells rather than that between N+ and HVP-well.

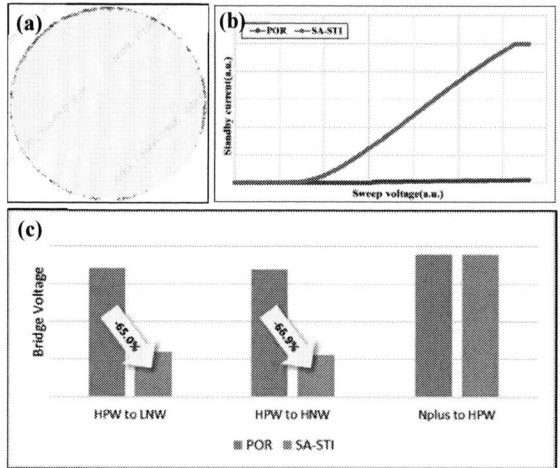

*Fig.2: **(a)** Chip Probe yield map of SA-STI sample and **(b)** Standby current and (c) BV performance of POR and SA-STI samples*

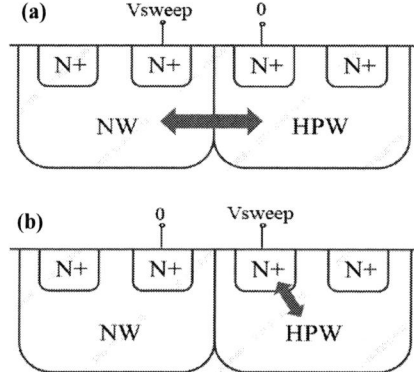

*Fig.3: Schematic of **(a)** N-Well to HVP-Well isolation BV and **(b)** N+ to HVP-Well junction BV test condition*

Junction stain method as a common instrument is widely applied in dopants inspection of MOSFET, on account of its advantages of low cost, large area, two-dimensional profile [4]. It works as follows: the N+/P+ dopant regions as the cathode/anode electrodes and the CuSO4/HF aqueous solution as the electrolyte compose the electrolytic cell, and in the electrolysis process, the ion doping profile is exposed by the selective deposition of copper ion on the cathode/anode electrodes [5]. Fig.4

illustrates the HVNMOS junction stain profile of POR and SA-STI samples, significantly, the junction profile of SA-STI structure shows more diffusion compared with the POR under the same N-plus and Lightly Doped Drain (LDD) implantation condition.

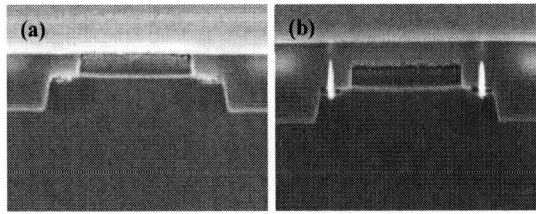

Fig.4: Junction stain analysis of (a) POR and (b) SA-STI

Based on the analysis results of BV test and junction stain, the poor isolation performance of NW to HPW and long diffusion distance of N-plus and LDD implantation can be explained by the lower positive type boron ion concentration in the P-well, which may result from the severe diffusion in the additional thermal budget process of SA-STI technique.

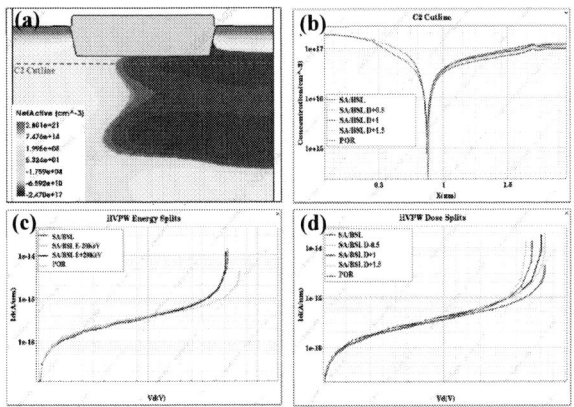

Fig.5: TCAD simulation of (a) ion diffusion profile in isolation testkey structure, (b) Ion concentration at C2 cutline, Isolation BV by (c) energy split and (d) dosage

Fig.5 (a) shows the TCAD simulation results of ion diffusion profile in isolation structure. Based on series of HPW field implantation dosage and two structure of POR and SA-STI, the ion concentration is simulated and the curves along C2 cutline are shown in Fig. 5(b). The distribution of boron ion concentration in SA-STI sample

is comparable to POR when the dosage changes to D+1.5. Fig. 5(c) and Fig.5 (d) illustrate the Id-Vd simulation curves of series energy and dosage split in HPW field implantation, there are no obvious impact by energy split when keep the same implantation dosage. Moreover, the Id-Vd is more sensitive to the dosage of field implantation and the bridge voltage matches the POR performance when the dosage is D+1.5.

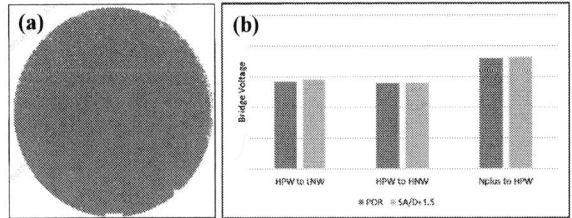

Fig.6 (a) Chip Probe yield map of SA-STI with HPW field implant dosage+1.5 and (b) BV performance of POR and SA-STI with HPW field implant dosage+1.5

The map of chip probe yield in SA-STI with HPW field implantation dosage +1.5 condition is displayed in Fig. 6(a). The failure ratio of standby current(less than 3%) with dosage +1.5 condition shows an obvious improvement compared with SA-STI baseline condition, and both the LNW and HNW to HPW isolation BV of dosage+1.5 are comparable to the POR as shown in Fig.6(b).

CONCLUSION

In this work, the failure mode of standby current in 4Xnm SA-STI NOR-flash memory was clarified. It was shown that the standby current failure and the dramatic decrease of isolation BV can be attributed to the inordinate diffusion of implanted p-type ion on account of the additional thermal budget. Meanwhile, the standby current failure and isolation BV performance have significantly improved by the dosage+1.5 of HPW field implantation condition.

REFERENCE

[1] C. Y. Lu, K. Y. Hsieh, and R. Liu. *Microelectronic Engineering*, vol. 86, 2008, pp. 283–286.

[2] Y. H. Song, J. Y. Lee, S. E. Lee, and J. H. Park. Jpn. *J. Appl. Phys.*, vol. 46, 2007, pp. 5067–5070.

[3] M. Kondo, T. Nakauchi, S. Lto, N. Aoki, and H. Ishiuchi. *2006 International Conference on Simulation of Semiconductor Processes and Devices*. IEEE

[4] Y. S. Chuang, C. A. Huang, J.H. Hsu and Y. J. Wu. *2017 19th Electronics Packaging Technology Conference*. IEEE

[5] B. Jin, W. L. Ma. *Semiconductor Inspection &Testing Technologies*, vol. 37, 2012, pp. 0078-0083.

AN ON-CHIP SUPERCONDUCTING QUANTUM TRANSPONDER

Rutian Huang[1,2], Xinyu Wu[1,2], Xiao Geng[1,2], Jianshe Liu[1,2], and Wei Chen[1,2,3]*

[1] Laboratory of Superconducting Quantum Information Processing,
School of Integrated Circuits, Tsinghua University, Beijing, 100084, China
[2] Beijing Innovation Center for Future Chips, Tsinghua University, Beijing, 100084, China
[3] Beijing National Research Center for Information Science and Technology, Beijing, 100084, China
*Corresponding Author's Email: weichen@tsinghua.edu.cn

ABSTRACT

Transmitting signals from one port to different ports at low temperature is necessary in superconducting quantum test system. An on-chip superconducting quantum transponder (SQT) is designed. The SQT has the function of routing signals from one port to multiple ports. Routing signals from one port to different ports can be realized through the impedance match between SQUIDs and output ports. In addition, due to the non-reciprocity of SQT, the isolation between sources and receivers can be achieved. The dual function of "transmit" and "isolate" may play an important role in superconducting quantum computing.

INTRODUCTION

With the increasing complexity of quantum integrated circuits, there are increasing demands for the processing computing power of nodes [1][2], so the transponder is necessary. However, when transmitting signals in the superconducting quantum test system, the input and output ports cannot be isolated, which will inevitably lead to the interference between input and output information [3][4]. In order to amplify quantum signals, reflective amplifiers such as Josephson parametric amplifiers (JPAs) or impedance-matched parametric amplifiers (IMPAs) are usually used in superconducting quantum computing test system [5][6]. In order to prevent the reflection of signals, the nonreciprocity device is needed [7]-[9]. Therefore, a signal transponder, which can transmit signals from one port to different ports at low temperatures and separate input from output ports, is necessary.

To meet above requirements, a superconducting quantum transponder (SQT) which has the function of "transmit" and "isolate" is designed. The SQT adjusts the impedance between the bias line and the output port. When the impedance match is realized at one port, the input signal is output from this port corresponding to the impedance matching unit. When the impedance mismatch occurs, the input signal continues to be transmitted to the next port of the forwarder along the circular direction until the port of the impedance match is encountered, and finally the information is transmitted out. It can transmit signals from one port to different ports, and isolate the signal source and receiver. The SQT is an on-chip device with nonreciprocity and can be integrated with superconducting circuits as well as reflective amplifiers such as JPA or IMPA to rout the amplified information.

PRINCIPLE AND ANALYSIS

As shown in Fig.1, the SQT is a multi-port device, taking the 6-port as an example, which consists of a circulator [5], four SQUIDs, four capacitors and four bias lines. The SQT adjusts the equivalent inductance of the SQUID by adjusting bias line to change its impedance. That is to say, the magnetic flux through SQUID is adjusted by bias line, and then the inductance of SQUID is changed. The impedance of the output port is also changed simultaneously. So, the impedance of the output port can be regulated by the bias line. The signal is output from impedance matching port, and the port with impedance mismatch has no signal output. The signal routed from one port to different ports can be realized.

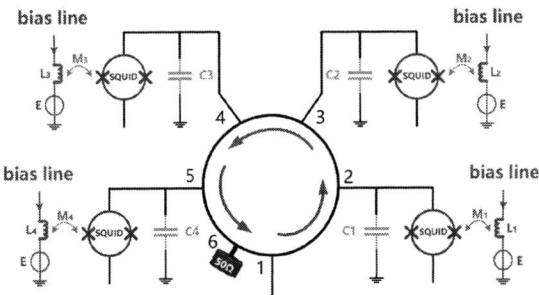

Figure 1: Schematic of a superconducting quantum transponder.

Similarly, taking the five-port SQT as an example, port 1 is connected to the input signal, and port 2, 3, 4 and 5 are connected to a SQUID respectively, port 6 is impedance matching with 50 Ω. The scattering matrix between the six ports is as follows:

$$
\begin{bmatrix}
b_{1,\text{out}}(t) \\
b_{2,\text{out}}(t) \\
b_{3,\text{out}}(t) \\
b_{4,\text{out}}(t) \\
b_{5,\text{out}}(t) \\
b_{6,\text{out}}(t)
\end{bmatrix}
=
\begin{bmatrix}
S_{11} & S_{12} & S_{13} & S_{14} & S_{15} & S_{16} \\
S_{21} & S_{22} & S_{23} & S_{24} & S_{25} & S_{26} \\
S_{31} & S_{32} & S_{33} & S_{34} & S_{35} & S_{36} \\
S_{41} & S_{42} & S_{43} & S_{44} & S_{45} & S_{46} \\
S_{51} & S_{52} & S_{53} & S_{54} & S_{55} & S_{56} \\
S_{61} & S_{62} & S_{63} & S_{64} & S_{65} & S_{66}
\end{bmatrix}
\begin{bmatrix}
a_{1,\text{in}}(t) \\
a_{2,\text{in}}(t) \\
a_{3,\text{in}}(t) \\
a_{4,\text{in}}(t) \\
a_{5,\text{in}}(t) \\
a_{6,\text{in}}(t)
\end{bmatrix}
\quad (1)
$$

Here, $a_{i,\text{in}}(t)$ and $b_{j,\text{out}}(t)$ represent input and output port parameters, respectively. S_{ij} (i, j = 1, 2, 3, 4, 5) indicates the transmission characteristic parameters from port j to port i.

The ideal scattering parameter matrix of SQT yields:

$$\begin{bmatrix} b_{1,out}(t) \\ b_{2,out}(t) \\ b_{3,out}(t) \\ b_{4,out}(t) \\ b_{5,out}(t) \\ b_{6,out}(t) \end{bmatrix} = \begin{bmatrix} 0 & 0 & 0 & 0 & 0 & 0 \\ S_{21} & 0 & 0 & 0 & 0 & 0 \\ S_{31} & 0 & 0 & 0 & 0 & 0 \\ S_{41} & 0 & 0 & 0 & 0 & 0 \\ S_{51} & 0 & 0 & 0 & 0 & 0 \\ 0 & 0 & 0 & 0 & 0 & 0 \end{bmatrix} \begin{bmatrix} a_{1,in}(t) \\ a_{2,in}(t) \\ a_{3,in}(t) \\ a_{4,in}(t) \\ a_{5,in}(t) \\ a_{6,in}(t) \end{bmatrix} \quad (2)$$

In the ideal scattering parameter matrix, all parameters except S_{i1} (i = 2, 3, 4, 5) are 0, that is, there is no crosstalk between different ports. It is noticeable that the S_{61} is 0 due to the port 6 is impedance matching with 50Ω. That is, if all ports are impedance mismatching, the signal will be finally routed to port 6. The transmission characteristics and crosstalk between different ports could be obtained by the simulation scattering parameter. Therefore, the equivalent inductance of the SQUID can be adjusted by bias line so that the signal outputs from the desired port.

A SQUID is a circular structure which two Josephson junctions are connected in parallel, and the critical current of SQUID can be modulated by a flux passing through the ring.

According to the Josephson effect [10], the relationship between the current of Josephson junction and the phase difference between two electrodes, and the relationship between the voltage of Josephson junction and the phase difference between two electrodes are respectively:

$$\begin{cases} I = I_c \sin \varphi , \\ V = \dfrac{\hbar}{2e} \dfrac{d\varphi}{dt} = \dfrac{\Phi_0}{2\pi} \dfrac{d\varphi}{dt} . \end{cases} \quad (3)$$

with

$$\Phi_0 = \frac{\hbar}{2e} . \quad (4)$$

And the inductance L_{JJ} of Josephson junction can be obtained as follows:

$$L_{JJ} = \frac{\Phi_0}{2\pi} \frac{1}{I_C \cos \varphi} , \quad (5)$$

while, L_{JJ} is also an equivalent nonlinear inductance of Josephson junction. Thus, the equivalent nonlinear inductance of SQUID is:

$$L_{SQUID}(\Phi_{ext}, I_{bias}) = \frac{\Phi_0}{4\pi \sqrt{\left[I_C \cos\left(\pi \frac{\Phi_{ext}}{\Phi_0}\right)\right]^2 - \frac{I_{bias}^2}{4}}} \quad (6)$$

where Φ_{ext} is the magnetic flux through SQUID, Φ_0 is flux quantum, the critical currents of both Josephson junctions are all I_c, and I_{bias} is the current of bias line [10]. So, by adjusting the current of bias line can modulate the inductance of SQUID, the impedance of the port is also

tunable. That is, the current of bias line can adjust the critical current of SQUID. Fig.2 (a) illustrates the circuit of bias line and SQUID. L_{bias} is the inductance of bias line, L_{SQUID} is the inductance of SQUID. L_{bias} and L_{SQUID} can be mutual inductance with each other, and M is the coefficient of mutual inductance between L_{bias} and L_{SQUID}. By adopting this bias line and SQUID circuit and using the WRspice@ software, the relationship of the current of bias line (I_{bias}) and the critical current of SQUID (I_{SQUID}) can be simulated. The simulation result is displayed in Fig.2 (b).

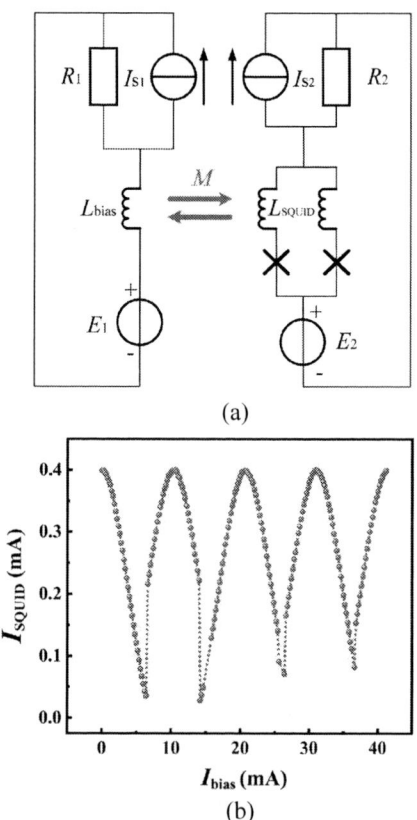

(a)

(b)

Figure 2: The relationship of the current of bias line (I_{bias}) and the critical current of SQUID (I_{SQUID}).

Fig.2 displays the relationship of the current of bias line (I_{bias}) and the critical current of SQUID (I_{SQUID}). The simulation result is consistent with the theoretical Equation (7) [10].

$$I_{SQUID} = \frac{4\pi M I_{bias} \sqrt{\left[I_C \cos\left(\pi \frac{M I_{bias}}{\Phi_0}\right)\right]^2 - \frac{I_{bias}^2}{4}}}{\Phi_0} , \quad (7)$$

where Φ_0 is the reduced flux quantum $\hbar/2e$. M is the coefficient of mutual inductance between L_{bias} and L_{SQUID}.

The SQT is simulated, and the transmission characteristics and crosstalk between different ports are obtained by the simulation scattering parameter. The simulation results are shown in Fig.3. It can be seen that

979-8-3503-1101-3/23 $31.00 © 2023 IEEE

the peak value of S_{41} is -6 dB, and those of S_{21}, S_{31} and S_{51} are -100 dB, which means that the signal can be transmitted from port 1 to port 4, while the signal is not output from the other ports. Therefore, the equivalent inductance of the SQUID can be adjusted by bias line so that the signal outputs from the desired port.

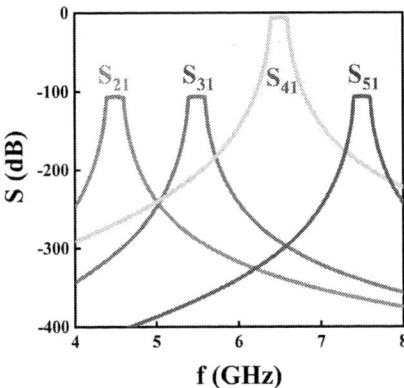

Figure 3: S parameter of superconducting quantum transponder.

The SQT can also be integrated with reflective amplifiers such as JPA or IMPA to output the amplified signal to the desired port. At the same time, the SQT can avoid the signal reflected from the output port.

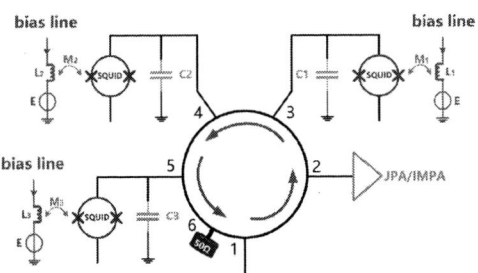

Figure 4: Diagram of the SQT integrated with JPA/IMPA.

Fig.4 shows the diagram of the SQT integrated with JPA/IMPA. The signal is input from port 1, output from port 2, and enters the circulator again after being amplified by JPA/IMPA. The output signal is determined by the impedance match between port and SQUID. This integration method can effectively reduce the cooling burden of dilution refrigerator.

CONCLUSION

In superconducting quantum test system, transmitting signals from one port to different ports at low temperature is necessary. The SQT can transmit signals to any channel based on demand without crosstalk. The SQT can not only realize the function of transmitting signals from one port to different ports by adjusting the equivalent inductance of

SQUID through bias line, but also play the role of isolation between load and quantum processor. It eliminates the destructive influence on the quantum processor when the superconducting quantum computing test system fails, and ensures the safety of the quantum processor.

ACKNOWLEDGEMENTS

We acknowledge Yingshan Zhang, Yunfan Shi, Changhao Zhao, Kaiyong He, Qing Yu, Yongcheng He, Liangliang Yang and Mingjun Cheng for calculating and valuable discussions. This work is partially supported by the National Natural Science Foundation of China (Grant Nos. 11653001, 11653004, and 60836001), and key R&D program of Guangdong province (Grant No. 2019B010143002).

REFERENCES

[1] Barends R, Kelly J, Megrant A, et al. Coherent Josephson Qubit Suitable for Scalable Quantum Integrated Circuits [J]. *Physical Review Letter*, 2013, 111: 080502.

[2] Arute F, Arya K, Babbush R, et al. Quantum supremacy using a programmable superconducting processor [J]. *Nature*, 2019, 574:505-510.

[3] Wu Y, Bao W, Cao S, et al. Strong Quantum Computational Advantage Using a Superconducting Quantum Processor [J]. *Physical Review Letter*, 2021, 127: 180501.

[4] Gong M, Wang S, Zha C, et al. Quantum walks on a programmable two-dimensional 62-qubit superconducting processor. *Science* 2021;372: 948–52.

[5] Chapman B J, Rosenthal E I, Kerckhoff J, et al. Widely Tunable On-Chip Microwave Circulator for Superconducting Quantum Circuits [J]. *Physical Review X*, 2017, 7(4).

[6] Abdo, Baleegh et al. "On-chip single-pump interferometric Josephson isolator for quantum measurements." *arXiv: Quantum Physics* (2020): n. pag.

[7] Krantz P, Kjaergaard M, Yan F, et al. A Quantum Engineer's Guide to Superconducting Qubits [J]. *Applied Physics Reviews*, 2019, 6(2):021318.

[8] J. Kelly, R. Barends, A. G. Fowler, A. Megrant, E. Jerey, T. C. White, D. Sank, J. Y. Mutus, B. Campbell, Y. Chen et al., State preservation by repetitive error detection in a superconducting quantum circuit, *Nature*, vol. 519, no. 7541, pp. 66-69, 2015.

[9] Gu X, Kockum A F, Miranowicz A, et al. Microwave photonics with superconducting quantum circuits[J]. *Physics Reports*, 2017, 718: 1-102.

[10] Zhang Yingshan. Study of Key Devices and Techniques for Scalable Superconducting Quantum Computing Systems. *Tsinghua University*, 2019.

EFFECTS OF FLOATING GATE PROFILE ON CELL CHARACTERISTICS OF 4XNM FG-FIRST ETOX NOR FLASH MEMORY

Hualun Chen[1], Yihang Du[1]*, Lin Gu[1], Zhuangzhuang Wang[1], Chun Yao[1], Zhaozhao Xu[1]
[1]Wuxi Huahong Grace Semiconductor Manufacturing Corporation, Wuxi 214028, China
*Corresponding Author's E-mail Addresses: *Yihang Du@hhgrace.com*

ABSTRACT

The FG-first technique is proposed in order to evade the typical process challenges as the NOR-flash memory scaling down to below 4Xnm generation. In this letter, the effect of floating gate top profile on the flash cell characteristics are investigated, the characteristics of arch-FG structure and planar-FG structure are trialed by the simulation and experiment. The planar-FG structure features better floating gate profile uniformity, larger cell memory window, tighter threshold voltage distribution width and higher chip prober yield, which can greatly enhance the process window and device window for the 4Xnm NOR-flash memory.

INTRODUCTION

Continuous scaling down of the feature size has led to a lot of significant challenges on manufacturing process of the floating gate-based (FG) NOR flash memory. Traditionally, the self-aligned FG (FG-last technique) is used for fabricating the typical standard ETOX (Erase through oxide) NOR-flash structure, the benefit of this method is a lager gate coupling ratio in the nearly fully covered structure, thus achieving better charge loss tolerance and significantly better process window [1, 2]. However, the poor poly chemical mechanical polishing (CMP) performance and control gate (CG) gap fill capability are the main scaling limitation. Specifically, poly dishing and oxide erosion induced by CMP pattern density effect would create undesirable circuit performance degradation. In order to avoid the above-mentioned process challenges, the FG-first technique is proposed, where the floating gate is defined by the pattering and etching after the stacked film (tunnel oxide/floating gate/nitride) deposition. According to the study of literatures [3], the FG-first process is one of the most available techniques to reduce the dispersion the threshold voltage due to the electric field of the active area corner. Therefore, the FG-last cell structure has been replaced by a simpler FG-first structure, which elucidates the side effects of poly CMP and gain the control gate gap-filling process window. In the previously proposed FG-first structure, the effect of floating gate top profile on the cell

characteristics is seldom studied. In this letter, the effect of arch FG top profile (arch-FG) and planar FG top profile (planar-FG) structure on the cell characteristics has been investigated.

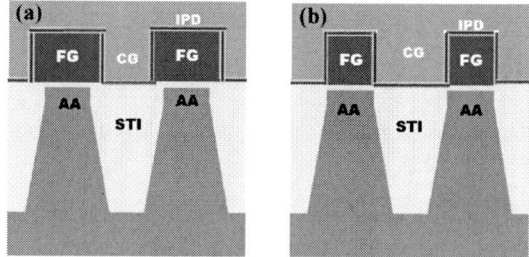

Fig.1 Schematic structure of (a) FG-last and (b) FG-first technique

DEVICE FABRICATION

The experimental Flash cell and chip were fabricated by a 4Xnm ETOX NOR process with an FG-first structure. After growing the tunneling oxide and depositing the phosphorus-doped floating gated, FG-first processes were executed and then STI isolation was completed. It is noteworthy that the FG profile on the FG-first structure was controlled by the wet etch on the dry-wet combination etching process. After the cell oxide recess adjustment, the inter-poly dielectric (IPD) and CG were deposited sequentially. Finally, the middle-end of line (MEoL) and back-end of line (BEoL) processes were performed to accomplish the 4Xnm NOR-flash memory product, aforementioned manufacturing process as shown in Fig.2.

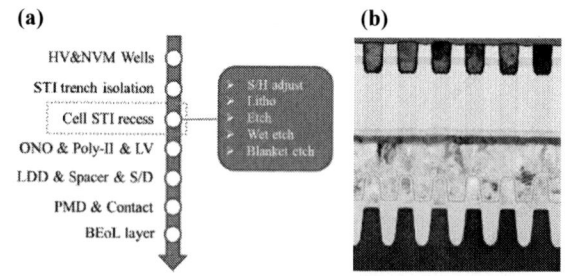

Fig.2 (a) Process flowchart description and (b) Cell TEM cross section of FG-first technique

RESULTS&DISCUSSION

Fig. 3 shows the electric field profile of arch-FG and planar-FG top profile obtained by a device simulation of the NOR-flash cell with cell design rule of 4Xnm. The control gate voltage of 10V and floating gate voltage of 6V were forced on the program process, while the control gate voltage of 10V and floating gate voltage of 2V were forced on the erase process. The simulation results show that the electric field becomes the maximum on the FG top center on the arch-FG structure, while for the planar-FG structure, the maximum electric field occurs at FG top edges. According to the simulation results, the coupling ratio of planar-FG structure is higher than arch-FG structure.

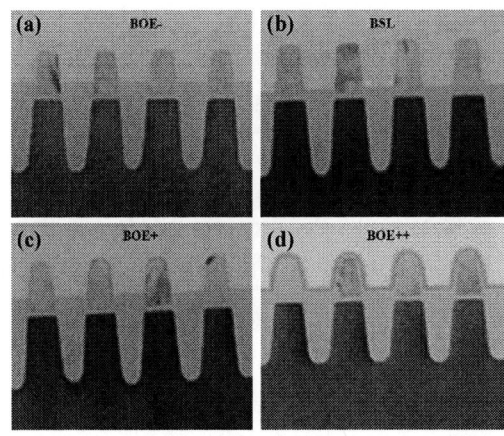

Fig.4 *FG top profiles with different BOE and dry etching time*

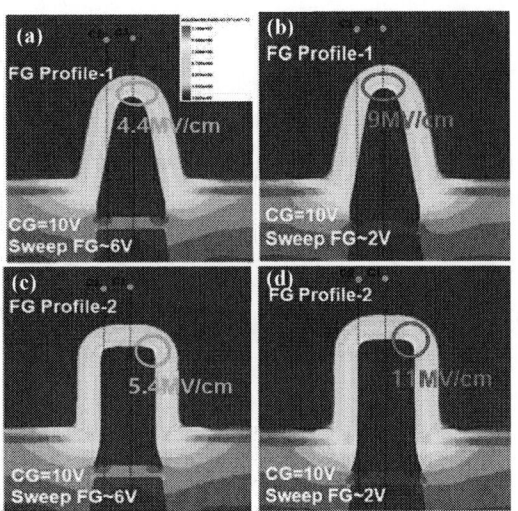

Fig.3 *Simulation results for an electric field profile of a FG-first ETOX NOR cell in a design rule of 4Xnm. (a) Program and (b) erase status of arch-FG structure, (c) program and (d) erase status of planar-FG structure*

In Fig.4, the change of FG top profile with the etching time of buffered oxide etchant (BOE) solution is illustrated. More specifically, in order to obtain the profile as shown in Fig.4, the BOE wet etching combining with other etching process is utilized. It is well known that the Si-O compound is generated easily at grain boundaries of polysilicon because of their low solubility in silicon crystal [4], which results that the Si-O bonds are well segregated into the grain boundaries [5]. Consequently, the etch rate of poly-SiO$_2$ film on the FG-first structure can be effected by the BOE wet etching and the dry etching process sequentially, which leading to the rounding-shape FG profile.

The single cell 4Xnm NOR-flash cell characterization of planar-FG structure and arch-FG structure was presented as Fig. 5. The Id (Vg) curves showing the memory window between the programmed and erased status, the memory window of planar-FG structure is larger than arch-FG structure, as shown in Fig. 5(a), which attributes to the higher threshold voltage (Vt) of planar-FG structure. Fig.5(b) shows the impact of FG top profile on the FG-first process, in comparison to the arch-FG structure, the on-state and off-state's channel current of planar-FG structure is slightly lower. The higher threshold voltage and lower channel current performance fit well with the simulation result. In order to further verify the coupling ratio of arch-FG profile structure and planar-FG structure, the coupling ratio was measured from the test-key monitor parameters, the results as shown in Fig.5(c). By comparison with planar-FG structure samples, the arch-FG structure samples show lower coupling ratio. Notice that the coupling ratio range of planar-FG structure samples is narrow than arch-FG structure, which is due to the poor uniformity of FG top profile during the cell oxide recess etching process (as presented in Fig.4(d)). The natural threshold voltage distribution in one page of the NOR-flash samples with arch- and planar-FG structure under specific programmed voltage, as shown on Fig.5(d), the natural threshold voltage distribution width of planar-FG structure can be tightened as compared to the arch-FG structure. The poor uniformity of top FG profile and the lower gate coupling ratio widens the natural threshold voltage distribution of arch-FG structure. Simultaneously, as reduction of FG volume from planar-FG to arch-FG structure, the stored electrons decrease and the sensitivity of electron loss becomes severe, leading to lower programmed threshold voltage.

979-8-3503-1101-3/23 $31.00 © 2023 IEEE

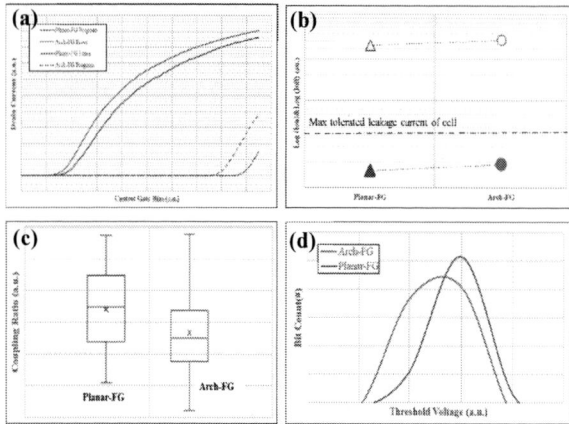

Fig.5 The cell characteristics of arch-FG and planar-FG structure. (a) Cell Id (Vg) curves in write and erase states, (b) the current of on and off status, (c) coupling ratio and (d) one page natural Vt distribution

The chip prober (CP) yield and main fail bins of planar-FG structure and arch-FG structure are presented in Fig.6. It is shown that the yield of planar-FG structure is higer than arch-FG structure, while the main fail bins relevant to cell characteristics are lower than arch-FG structure. The arch-FG strcture samples display lower coupling ratio and worse FG profile uniformity, which lead to the higher fail bins of cell characteristics due to lower program voltage and higher volatility for the flash cell.

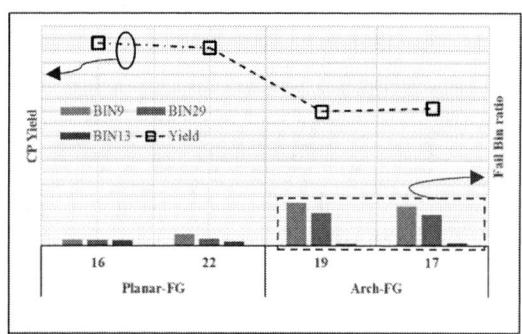

Fig.6 The yield and main fail bin of planar-FG and arch-FG structure

The high temparature data reliablity (HTDRB) of planar-FG and arch-FG structure are measured by checking ICKB Vt distribution after wafer baking at 250 °C for 24 hours. Compared with planar-FG structure samples, the widen post-bake Vt distribution of arch-FG strcture samples indicates the worse HTDRB performance, which is related to the wose electric field distribution of arch-FG structure. As mentioned previously, the arch-FG structure samples exhibit the worse floating gate profile uniformity, which result in the widen Vt distribution. What is more important, the roughess of FG

wrap around profile can be the leakage path of the stored electrons in the floating gate, giving rise to the lower post-bake Vt, as shown in Fig.7.

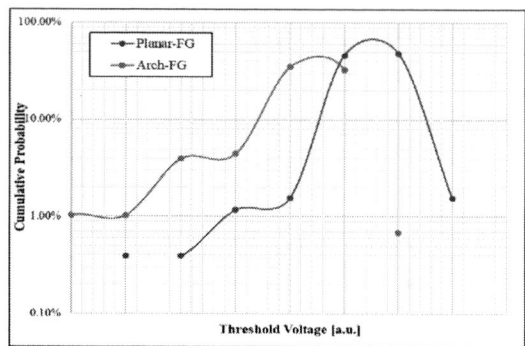

Fig.7 The HTDRB Vt distribution of planar-FG and arch-FG structure

CONCLUSION

In this work, the FG profile of different cell STI recess etching by adjusting BOE etching and subsequent dry etching was evaluated, the electrical characteristics, yield and HTDRB of planar-FG and arch-FG structure at 4Xnm NOR-flash was investigated. The FG profile uniformity, memory window, coupling ratio, natural Vt distribution width, the yield and HTDRB can be significantly improved by the planar-FG structure.

REFERENCE

[1] M. Park, C.-S. Lee, S.-H. Hur, K. Kim, and W.-S. Lee, "The effect of field oxide recess on cell VTH distribution of NAND flash cell arrays," IEEE electron Device Lett., vol. 29, no. 9, pp. 1050–1052, Sep. 2008.

[2] Z. –S. Wang, Y. –J. Lee, R. Yang, Y. –C. Li, H. –H. Chen and C. J. Lin, "A New Recess Method for SA-STI NAND Flash Memory" EEE electron Device Letters, vol. 33, no. 6, pp. 896-898, June 2012.

[3] M. Kondo, T. Nakauchi, S. Lto, N. Aoki and H. Ishiuchi, "Simulation of NOR-Flash Memory Cells Focusing on Channel Effects on VTH Dispersion," 2006 International Conference on Simulation of Semiconductor Processes and Devices. IEEE.

[4] M. Hamasaki, T. Adachi, S. Wakayama, and M. Kikuchi, Crystallographic study of semi-insulating polycrystalline silicon (SIPOS) doped with oxygen atoms," J. Appl. Phys., vol.49, p. 3987, 1978.

[5] J. H. Jeon., J. S. Yoo, "A novel method for a smooth interface at poly-SiO_x/SiO_2 by employing selective etching," IEEE Electron Devices Letters, 21(4), 152-154, 2000.

IMPROVED ENVIRONMENTAL STABILITY OF N-TYPE POLYMER FIELD-EFFECT TRANSISTORS USING NICKEL CONTACT ELECTRODE

Yuan Liu[1], Quanhua Chen[1], Rujun Zhu[1], Jinxiu Cao[1], and Yong Xu[1,2]*

[1]College of Integrated Circuit Science and Engineering, Nanjing University of Posts and Telecommunications, Nanjing 210023, China.

[2]Guangdong Greater Bay Area Institute of Integrated Circuit and System, Guangzhou 510535, China

*Corresponding Author's Email: xuyong@njupt.edu.cn

ABSTRACT

N-type organic field-effect transistors have been challenged by poor environmental stability. In this work, we found that by using nickel (Ni) instead of the commonly used gold (Au) as contact electrodes, N-type polymer transistors showed much improved stability upon annealing in nitrogen and exposure to air. Specifically, pronounced ambipolar conduction was observed for Au-contacted devices, whereas Ni-contacted devices maintained good unipolar properties and superior stability. These findings are critical for developing stable n-type polymer transistors to make complementary polymer circuits at low cost.

INTRODUCTION

In recent years, organic field-effect transistors (OFETs) have been used in many areas [1]. Among them, flexible integrated circuits, transparent display, sensor, and storage demonstrate promising potential [2, 3], owing to their unique advantages of being suitable for large areas and great flexibility [4]. However, as one of the cornerstones of future flexible electronics, OFETs have not yet been fully developed for practical applications. This is because in complementary circuits, both of P-type and N-type devices require high performance and stability. Compared to the relatively matured P-type counterparts, N-type polymer field-effect transistors are hard to achieve high performance and stability simultaneously.

A main issue arises from the contacts. In general, the contact between metal and organic semiconductors is Schottky-like with a high energetic barrier (e.g.,> 0.2 eV) [5], which limits charge injection and degrades device performance. For top-gate OFETs, although the dielectric layer can serve as a protective layer to protect the underlying organic semiconductor film from physical and chemical attracts, the transistor itself inevitably absorbs moisture from air. Those water and oxygen, equivalent to P dopants, can affect device performance and stability, in particular for N-type devices [6].

Up to now, gold has been the most commonly used source and drain electrodes in N-type polymer OFETs, thanks to its excellent conductivity and good stability. On the other hand, the high cost of gold and its high work function hinder its application for mass production and high performance. In order to achieve better electron injection, many methods have been attempted [7], e.g., doping and adding interlayers [8, 9]. In fact, the most effective way is to use low-work-function metals as source/drain electrodes. From the perspective of the energy band diagram, most N-type organic semiconductors have the LUMO (Lowest Unoccupied Molecular Orbital) of 3-4 eV, while the work function of gold is about 5.1 eV, resulting in a large Schottky barrier after forming a gold-organic semiconductor contact. If metals with low work functions are used as contact electrodes (such as calcium, potassium), the stability of the devices is poor due to the extreme instability of these metals in air.

In this study, we used nickel (Ni) instead of the widely used gold (Au) as the source/drain electrodes for n-type polymer OFET. The polymeric semiconductor was N2200 (N (2-octodecyl) naphthalene-1,4,5,8bis) (2,6-dialkyl) - 2,6-5,50 - (2,29-dithiophene). Our experimental results showed that Ni-contacted devices outperformed the Au-contacted rivals. In particular, we observed much improved environmental stability of the Ni-contacted devices when annealed in N_2 and exposed to air. By contrast, the Au-contacted devices degraded significantly and exhibited quickly increasing ambipolar characteristics.

DEVICE FABRICATION AND CHARACTERIZATIONS

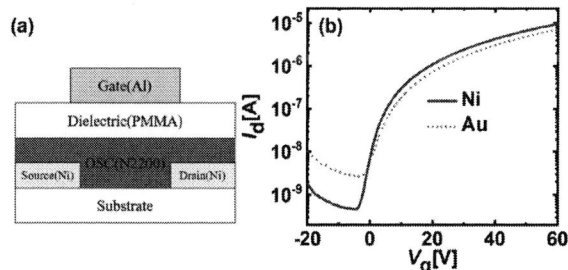

Figure 1: (a) Device structure. (b). Saturation transfer characteristics of N2200 OFETs with Au and Ni source/drain electrodes (at V_d = 60 V).

The devices were fabricated on glass substrate with top-gate, bottom-contact (TG-BC) structure; see Figure 1 (a). The substrates were cleaned twice in alcohol and DI ultrasonically for 20 min each, and then dried at 100 ℃ in air for 20 min. Next, 50 nm Ni or 5/50 nm Ni/Au was evaporated in vacuum (3×10^{-4} Pa) onto the substrate through shadow mask for source/drain (S/D) electrodes, forming a uniform channel width (W) of 1200 μm and diverse channel lengths (L) of 100-500 μm. Then, N2200 (7 mg/ml, dissolved in 1,2-dichlorobenzene; purchased from Ruixun, Mw>100 K) was spinning coated (first 500 rpm for 5 s and then 2000 rpm for 60 s) and annealed at 200 ℃ for 20 min. Afterwards, PMMA (100 mg/ml polymethyl methacrylate, soluble in n-butyl acrylate) was spinning coated onto the N2200 layer at 2000 rpm for 60 s. Annealing was performed in a nitrogen-filled glove box at 80 ℃ overnight. Finally, 100 nm-thick aluminum (Al) gate electrodes were evaporated using shadow mask. Electrical characterizations were carried out in air by using a probe station connected with semiconductor parameter analyzer (Keysight B1500A).

EXPERIMENTAL RESULTS

As shown in Fig. 1(b), the N2200 OFETs with Au and Ni contacts exhibit very similar saturation transfer characteristics, where the threshold voltage is about 0 V and the drain current reaches virtually the same magnitude. Although the performance improvement is limited, the cost of fabrication is considerably reduced, providing a path for future large-scale application.

To probe stability, we first investigated the effect of thermal annealing. Thermal annealing is routinely implemented in OFET fabrication to evaporate the residual solvent and to condense the deposited film. Yet, impurities (e.g., oxygen and water) are still inevitable. They electronically manifest themselves as traps or tail states in the bandgap, degrading performance and stability. For a top-gate device, the final annealing applies to the gate dielectric, which means that, traps can be reintroduced during the subsequent deposition of the gate electrode. Therefore, we annealed device after all fabrications.

Polymers are porous in nature. Oxygen and water can easily penetrate to behave as p-dopants, detrimental to N-FET. As seen in Fig. 2, before annealing, the Au-contacted OFETs suffer from ambipolar conduction. Anneal at 80 ℃ for 12 h eliminates this effect, in line with the literature where annealing fixed traps. The mobility extracted by the Y function method (YFM) [10] and the threshold voltage extracted in the saturation regime indicate that mobility slightly increases from 0.17 to 0.18 cm^2/Vs and threshold voltage decreases from 1.92 to 0.89 V. Owing to the transition from ambipolar to unipolar conduction, the subthreshold swing reduces from 5.67 to 5.21 V/dec, and the on–off ratio rises from 794 to 2551. This is because the thermal annealing evaporated the absorbates and turned the devices back into unipolar. In contrast, annealing did not play an important role on Ni-contacted OFETs that remains fairly good unipolar properties.

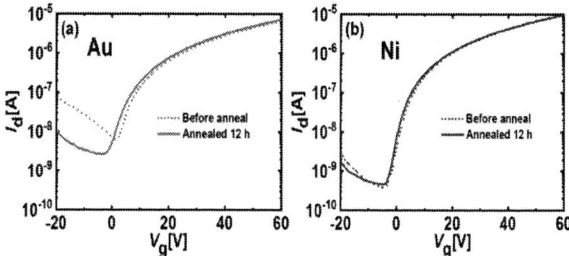

Figure 2: Annealing test. Saturation transfer characteristics of N2200 OFETs with Au (a) and Ni (b) contacts before and after annealing in N_2, where $V_d = 60$ V and $L = 100$ μm.

To compare the stability of the two devices, we placed the two devices in air for 5 days. Figure 3(a) shows the transfer characteristic curves of the two devices after exposure to air. The relevant electrical parameters are also listed in Table 1. After five days of air exposure, the threshold voltage of the Au-contacted device increased from 1.08 V to 17.31 V, the mobility decreased from 0.11

Table 1: Comparison of performance parameters of two representative devices before and after exposure to air.

Device conditions	Threshold voltage(V)	Mobility (cm^2/V·s)	Subthreshold swing (V/dec)	On/off ratio
Au (no exposure)	1.08	0.11	1.92	1.58×10^6
Au (exposure 5 days)	17.31	0.03	4.65	2.07×10^3
Ni (no exposure)	0.16	0.12	2.20	3.54×10^5
Ni (exposure 5 days)	9.27	0.10	3.60	2.47×10^6

cm^2/Vs to 0.03 cm^2/Vs, the subthreshold swing increased from 1.92 V/dec to 4.65 V/dec, and the on/off ratio decreased from 1.58×10^6 down to 2.07×10^3. In addition, pronounced ambipolar characteristics arose from that device. Notably, the Ni-contacted device parameters remained unipolar properties with much less degradation.

To understand the differences in performance and stability, the contact resistance is extracted by using the modified transfer-length method (M-TLM), as shown in Figure 3 (b) for the above two sets of devices. It can be seen that, at V_g-V_t=45 V, the contact resistance is 3.34×10^5 Ω.cm and 1.13×10^5 Ω.cm, for the Au- and Ni-contacted devices, respectively. This result indicates much contact resistance with much less gate-voltage dependency by using Ni instead of Au as the contact electrodes.

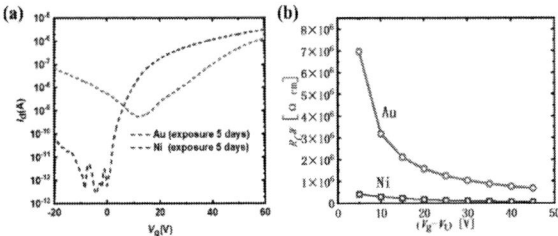

Figure 3: (a) Comparison of the saturation transfer characteristic curves of Ni and Au-contacts OFET after 5 days of exposure to air. (b) Contact resistance extracted by using M-TLM at different V_g.

To further explore the underlying mechanism, we take the device physics of the metal-semiconductor contact into consideration, and the energy-band diagram is illustrated in Fig. 4. Since only the contact metal changes in this work, we mainly focus on charge injection. After five days of exposure to air, the polymeric semiconductor film inevitably absorbs water and oxygen, which were reported to be equivalent to P dopants negatively degrading the device performance of N-type OFETs. The significant inter-diffusion between gold and N2200 leads to high-density gap states, resulting in much reduced injection barriers for both of electrons and holes. More importantly, the diffusion-induced defects accommodate H_2O and O_2, which is the origin of air instability. So, degradation can be amplified when exposed to air. However, the nickel-contact devices did not change much. Nickel has a lower work function, i.e., smaller barrier to electron injection. On the other hand, compare to gold, nickel is more easily oxidized in air to form a nickel oxide film, which is an intrinsic broad-band-gap semiconductor with good thermal and chemical stability. This ultra-thin, stable oxide layer acts as a diffusion barrier layer to prevent the inter-diffusion of metals and polymers, avoiding the formation of contact defects. Therefore, the superior stability of Ni-contacted devices mainly benefits

from the optimization of metal-semiconductor contact.

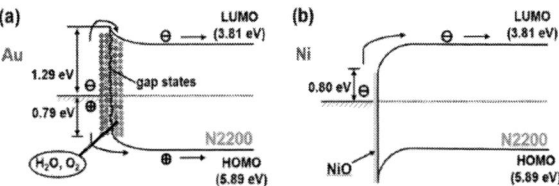

Figure 4: Schematic energy band diagram of N2200 with contacts using Au (a) and Ni (b), where the LUMO and HOMO represent the lowest unoccupied molecular orbital and the highest occupied molecular orbital, respectively.

In the end, after all above experiments, we performed thermal annealing again. The devices were annealed in N_2-filed glove box at 80 ℃ for 12 hours. Interestingly, their electrical properties almost recovered, as shown in Figure 5. This result indicates that the observed degradation does originate from H_2O and O_2, which can be eliminated by thermal annealing.

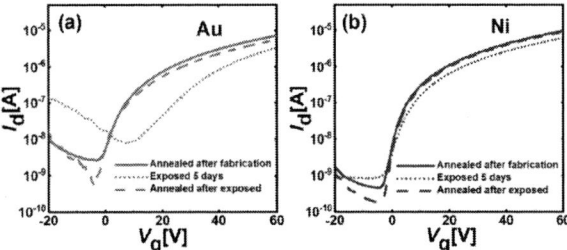

Figure5: Annealing test. Saturation transfer characteristics of N2200 OFETs with Au (a) and Ni (b) contacts under three conditions: Annealed after fabrication; Exposed 5 days; Annealed after exposed, where V_d = 60 V and L = 100 µm

CONCLUSION

In summary, we used Ni instead of the widely used Au as the source/drain electrode in N-type polymer N2200 OFETs. Owing to the lower contact resistance, the Ni-contacted device slightly outperformed the Au-contact rival. In particular, we found much improved stability for Ni-contacted device upon annealing and exposure to air. The ultrathin, stable oxide interlayer of nickel oxide is inferred to prevent inter-diffusion between metal and polymer and thus avoid defect formation at the contact. Therefore, Ni/N2200 contacts and the composed devices can resist the effects of water and oxygen when exposed to air. In addition to its lower price and good adhesion to the substrate, Ni is promising for use of contact metal to making high-performance, highly stable N-type polymer transistors.

REFERENCES

[1] H. Sirringhaus, *Adv. Mater.*, vol. 26,2014,pp. 1319-1335.

[2] A. Bilgaiyan, S. I. Cho, M. Abiko, K. Watanabe, and M. Mizukami, *Sci. Rep.*, vol. 11,2021,pp. 11710.

[3] J. W. Borchert, R. T. Weitz, S. Ludwigs, and H. Klauk, *Adv Mater*, vol. 34,2022,pp. e2104075.

[4] S.-W. Liu, C.-C. Lee, H.-L. Tai, J.-M. Wen, J.-H. Lee, and C.-T. Chen, *ACS Appl. Mater. Inter.,* vol. 2,2010,pp. 2282-2288.

[5] Y. Xu, H. Sun, and Y.-Y. Noh, *IEEE Trans. Electron Devices*, vol. 64,2017,pp. 1932-1943.

[6] M. S. A. Abdou, F. P. Orfino, Y. Son, and S. Holdcroft, *J. Am. Chem. Soc.*, vol. 119,1997,pp. 4518-4524.

[7] L. Bürgi, T. J. Richards, R. H. Friend, and H. Sirringhaus, *J. Appl. Phys.*, vol. 94,2003,pp. 6129-6137.

[8] T. He, M. Stolte, Y. Wang, R. Renner, and C. D. Frisbie, *Nat. Mater*, vol. 20,2021,pp. 1532-1538.

[9] F. Huang, D. Lu, Y. Ji, Y. Xu, and J. Chu, *IEEE Trans. Electron Devices*, vol. PP,2021,pp. 1-7.

[10] Y. Xu, T. Minari, K. Tsukagoshi, J. A. Chroboczek, and G. Ghibaudo, *J. Appl. Phys.*, vol. 107,2010,pp. 7.

A NEW METHOD TO CALCULATE LOADING EFFECT IN EMBEDDED FLASH

Fangce Sun[1*]

[1] Department of Process Integration, Shanghai Huahong Grace Semiconductor Manufacturing Corporation, Shanghai 201203, China

*Corresponding Author's Email: Fangce.Sun@HHGrace.com

ABSTRACT

Loading effect is a very common phenomenon in IC manufacture process, such as etch process, diffusion process and thin film process, etc. In many cases, process recipe is designed to ensure enough process window to cover different level of loading effect, which is originated from various end customer applications which lead to diverse layout. In some cases, process recipe should be chosen precisely base on specific loading effect to gain best process window and product yield due to rigid device performance requirement. In this paper, the difficulty of loading effect calculation in embedded flash is introduced and a new method of calculation is discussed.

INTRODUCTION

Floating gate tip plays critical role in split gate flash program、 erase operation, by influencing the flash coupling voltage and electric field during program and erase operation, to determine the flash performance and reliability. Therefore, it is crucial and meaningful to well control the floating gate tip's height and sharpness.

In split gate flash process flow, the final floating gate tip is formed by a self-aligned etch process, refer Fig.1. It is blanket etch process to etch the useless floating gate material, using the oxide which is named of floating gate spacer as etch hard mask. It is a key process related with the floating gate tip and its etch loading effect can bring huge performance difference among various products even using the exactly same process flow.

In general, the loading effect will have a strong correlation with products' one layer mask pattern density. Then it can make up the loading effect by tuning the etch process base on the loading effect correlation. Several papers have already studied this topic [1-4].

It works quite well in non-embedded flash applications in traditional way, but it doesn't work in embedded flash. It shows no correlation between the floating gate layer pattern density and remain oxide thickness after self-aligned etch, refer Fig.2. It is because a large proportion of floating gate is etched during active etch process, refer Fig.3. Therefore it should be taken in account when calculating the loading effect of the self-aligned etch process.

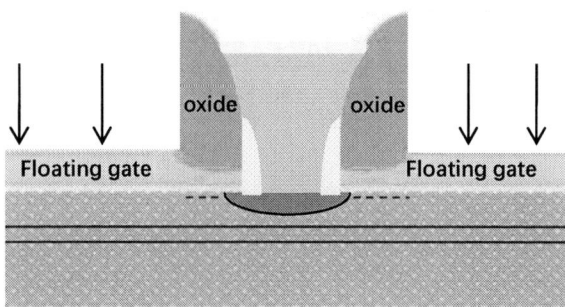

Figure 1: A self-aligned etch process to form the floating gate tip.

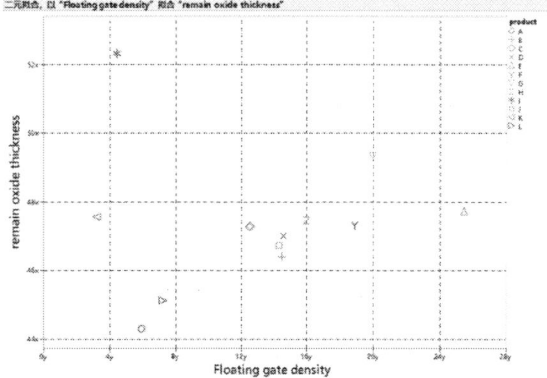

Figure 2: Floating gate pattern density and remain oxide thickness correlation after self-aligned etch .

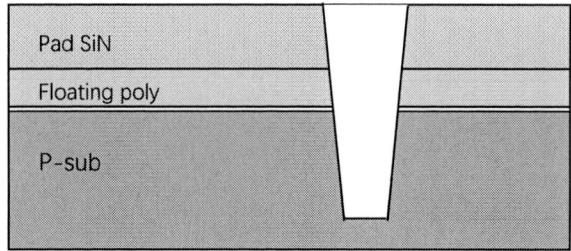

Figure 3: A large proportion of floating gate is etched during active etch process.

EXPERIMENT

There are two typical flash cell types, refer Fig.4. One structure has oxidation process after source line poly deposition. This oxide film above source poly can be a hard-mask during the self-aligned etch process. Therefore, the related region should be excluded when computing the loading effect of the self-aligned etch process.

In contrast, for the structure without source poly oxidation, the source poly is exposed during the self-aligned etch, and its portion should be taken in account.

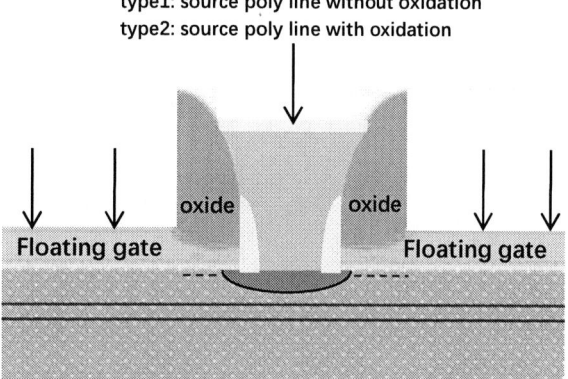

Figure 4: Two types of flash cell structure for self-aligned etch : source line with& without oxidation

TEST RESULTS

After taking account the floating gate loss during active etch and source poly line contribution for different types of flash cell structure to get the overall density for the self-aligned etch, both types of flash cell will have strong correlation between overall pattern density and remain oxide, refer Fig.5 and Fig.6.

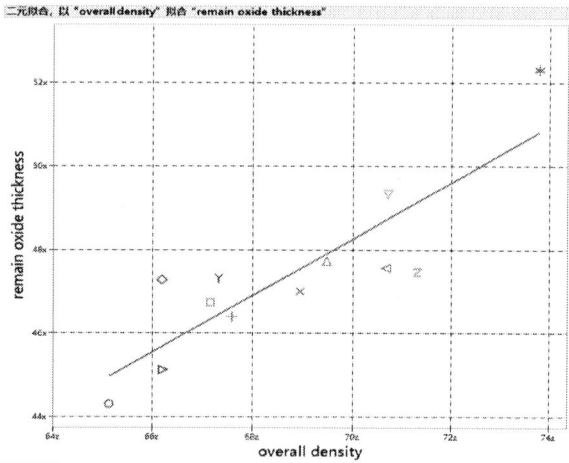

Figure 5: overall density and remain oxide thickness after self-aligned etch correlation for type1 flash cell.

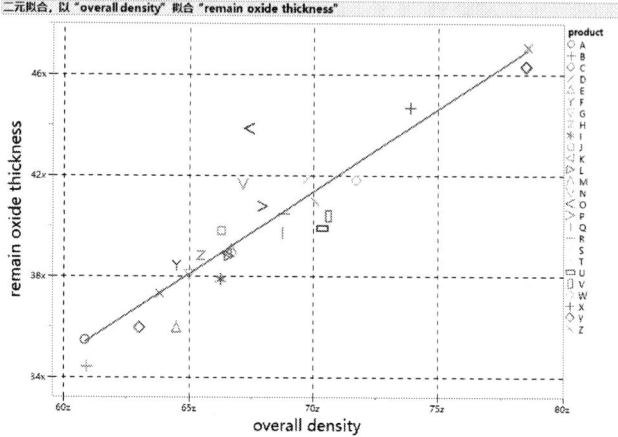

Figure 6: overall density and remain oxide thickness after self-aligned etch correlation for type2 flash cell.

DISCUSSION

Self-aligned etch process will etch the floating gate material exposed in the air, but the STI region floating gate material has already been consumed during active etch process. The flash cell array region which has oxide spacer, will protect the floating gate not to be etched, therefore this portion of floating gate should be excluded. In this section, two types of flash cell structure will be discussed separately, the source poly with or without oxidation will impact the etching loading because source line poly will be etched during the self-aligned etch process too for the flash cell without source poly oxidation.

In the 1st case, let us discuss the source poly with oxidation flash cell. The exposed floating gate material equals the active layer pattern density minors floating gate layer pattern density multiply a factor A. The factor A is determined by a given flash cell array dimension, it equals flash cell array ACT width divided by ACT pitch. Therefore it is a constant instead of a variable. In the flash cell region, not all the floating gate material is protected during etch process, because a portion of the floating gate in the flash cell region has already been consumed in the active etch process too. The overall density equation should be as below shows (ACT means the active layer and FG means the floating gate layer):

Overall density (real etch clear ratio)= ACT density-FG density*factor A(=Flash cell array: ACT width/ACT pitch)

In the 2nd case, for source poly without oxidation flash cell, the source line poly will be etched too. The related equation will be slightly changed as below shows, adding the source poly portion.

Overall density (real etch clear ratio)= ACT density-FG density*factor A(=Flash cell array: ACT width/ACT pitch)+ source poly density

Source line density varies during the self-aligned etch. In the beginning of the etch process, the source poly density should be equal as the FG density. And in the end of the etch process, the source poly density will be smaller, varies from dimension A to dimension B, refer Fig.7. In the end of the process, the source poly density should be FG density* (B/A). Then the average source poly density will be the average of the beginning and end of the etch process:

Source poly density=FG density*[(1+B/A)/2]

Then the equation is changed as:

Overall density (real etch clear ratio)= ACT density-FG density*factor A(=Flash cell array: ACT width/ACT pitch)+ FG density*[(1+B/A)/2]

The B/A should be a constant when the self-aligned etch recipe is fixed although this constant may have a little variation from process variation and product difference. And this data should be easily got by flash cell SEM cross section image measurement.

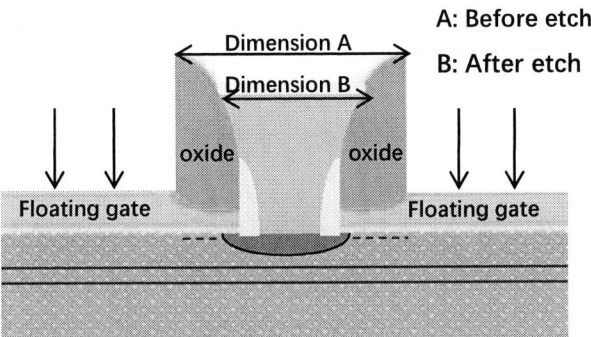

Figure 7: Source line density varies during self-aligned etch for the flash cell without oxidation

CONCLUSION

1, There is no correlation between the floating gate layer pattern density and remain oxide thickness after self-aligned etch in embedded flash. Self-aligned etch loading effect cannot be calculated from single layer pattern density information.
2, To compute the loading effect of self-aligned etch, flash cell layout dimension and active layer pattern density should be taken in account too, besides the floating gate layer pattern density.
3, Overall density formula is different for the flash cell with and without source line poly oxidation.

REFERENCES

[1] Dong-You Choi et al. Reduction of loading effects with the sufficient vertical profile for deep trench silicon etching by using decoupled plasma sources. Materials Processing Technology, 2009, pp.5818-5829.
[2] C.Hedlund et al. Microloading effect in reactive ion etching, American Vacuum Society, 1994, pp.1962-1965
[3] Guang Yang et al. Optimization of Shallow Trench Isolation CD micro loading, CSTIC, 2002, 1-3.
[4] H.Jansen et al. The black silicon method II: the effect of mask material and loading on the reactive ion etching of deep silicon trenches, Microelectron.Eng, 1995, pp.475-480

A NEW METHOD TO IMPROVE SPLIT GATE FLASH ERASE AND ENDURANCE

Fangce Sun[1]*

[1] Department of Process Integration, Shanghai Huahong Grace Semiconductor Manufacturing
Corporation, Shanghai 201203, China
*Corresponding Author's Email: Fangce.Sun@HHGrace.com

ABSTRACT

Split gate flash floating gate tip's sharpness is critical for flash erase function and endurance performance. To form sharper floating gate tip, more etch time of an isotropic etching process is required and this requirement leads to thicker floating gate thickness. However, the floating gate deposition thickness is limited in traditional way due to STI HDP void concern. In this paper, a new method to increase 150Å floating gate thickness by using a CeO_2 base slurry STI CMP is introduced and the related experiment result is discussed.

INTRODUCTION

There are several papers discussed the split gate super flash or EEPROM erase and endurance [1-6].

Split gate flash erase and endurance performance can be improved by enhancing floating gate tip's sharpness.

In split gate flash process flow, the floating gate tip depends on an isotropic etch process, refer Fig.1. More etch time will have sharper floating gate tip. Floating gate material is a material for electron storage, therefore its thickness has a low spec to ensure data storage. The more isotropic etch time requires thicker floating gate deposition thickness.

With technology develops, the floating gate deposition thickness faces its limitation. STI HDP process has 3:1 aspect ratio limitation (3 means the trench step height, and 1 for STI width), refer Fig.2. If the aspect ratio is bigger than 3, there will be STI HDP void defect risk. Therefore, the pad nitride thickness plus floating gate thickness has a limitation as upper spec.

Pad nitride thickness has its limitation due to STI CMP requirement. Too thin pad nitride film thickness will have STI CMP induced scratch defect.

In traditional way, split gate flash erase and endurance improvement is limited by the floating gate deposition thickness.

A novel way helps remove the above limitation, by through using a CeO_2 base slurry STI CMP which has an additive to protect the nitride film and much higher nitride to oxide selectivity, refer to Fig.3.

With the new slurry's help, the pad nitride thickness can be reduced and floating gate thickness can be increased.

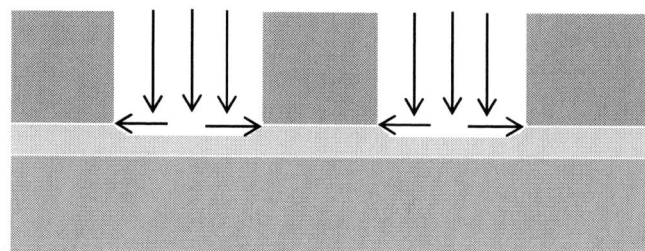

Figure 1:An isotropic etch process to form the floating gate tip.

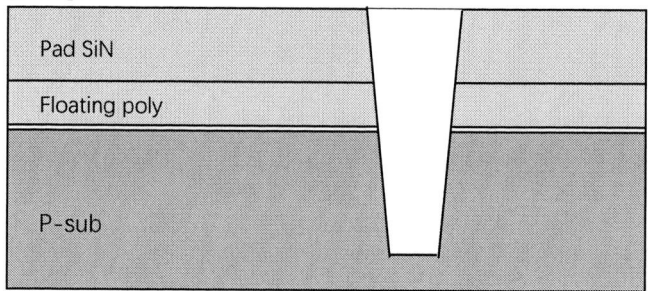

Figure 2: Pad nitride thickness plus floating gate poly thickness has a limitation as STI HDP aspect ratio requires.

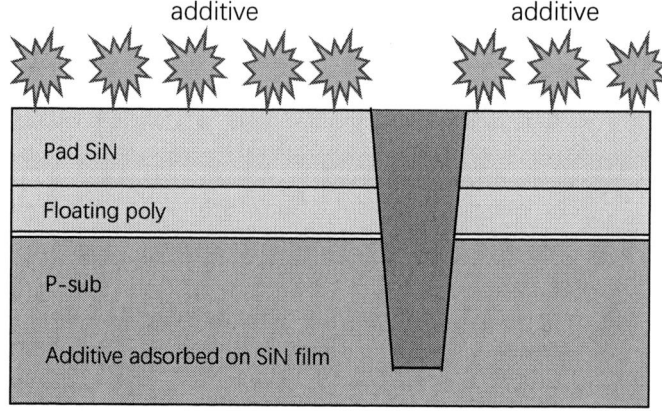

Figure 3: CeO_2 base slurry with an additive: protect nitride film, low nitride polish rate

EXPERIMENT

The floating gate deposition thickness is increased about 150Å while pad nitride thickness is reduced about 300Å to prevent STI HDP void defect. STI etch recipe should be fine-tuned to keep the STI trench depth match with baseline performance. STI CMP changes from SiO_2 base slurry whose nitride to oxide selectivity is about 3 to CeO_2 base slurry whose nitride to oxide selectivity is about 26. The former nitride loss of STI CMP is about 200 Å while the latter's nitride loss is only 25Å.

On account of the increased floating gate thickness, the isotropic etching time can be increased 30s (5Å per second).

The SEM cross section pictures of the POR process and new process are compared and two kinds of full loop process wafers' yield is compared by using the same CP test program and their endurance data is collected too.

TEST RESULTS

The floating gate tip becomes much sharper of the new process, compared with POR process, refer Fig.4 and Fig.5. The new process product yield is also much improved, refer Fig.6. The POR process wafer can only pass 20K endurance cycling while the new process wafer can pass 100K endurance.

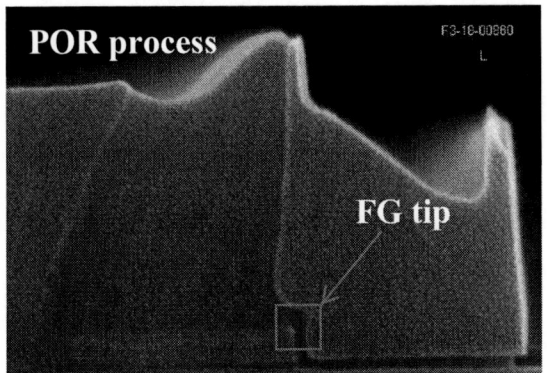

Figure 4: POR process SEM cross section after flash cell process finished

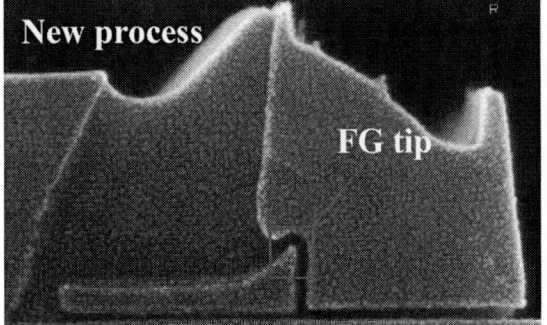

Figure 5: New process SEM cross section after flash cell process finished

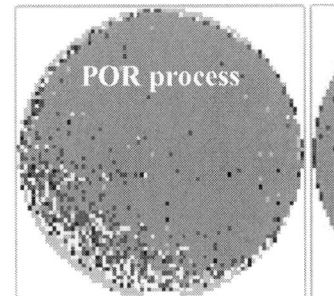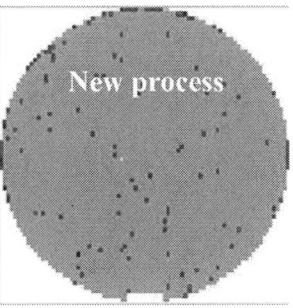

Figure 6: POR& New process CP yield, POR wafer yield is 74%, major failure bin is erase related fail; New process wafer yield is 97%, almost 0% yield loss related with erase failure.

DISCUSSION

Split gate flash yield is related with flash erase, program performance and its program inhibit performance to prevent program disturb, etc. And it has strong dependency with flash cell dimension. Its reliability such as data retention and endurance has strong correlation with flash structure too. When flash cell's one dimension is changed to meet specific goal, the side effect should be well evaluated.

Split gate flash erase performance depends on the voltage bias between the word line and floating gate, tunnel oxide thickness and floating gate tip's sharpness. To improve the flash erase, enlarge the voltage bias, reduce the tunnel oxide thickness and enhance floating gate tip's sharpness are three ways to meet the goal. Enlarge the voltage bias will bring more stress for the tunnel oxide during endurance cycling. Then it will lead to poor endurance performance. Reduce the tunnel oxide thickness may have data retention failure risk. Enhance floating gate tip's sharpness can both benefit flash erase and endurance performance.

Although it should be taken in account that reverse tunneling program disturb may be one side effect of enhancing floating gate tip's sharpness, this kind of side effect could always be eliminated because the program operation voltage on the source line is about 8V, coupling to floating gate voltage is about 5V. Therefore the reverse tunneling will hardly happen under this low word line and floating gate voltage bias.

There are several ways to enhance the floating gate tip's sharpness because several processes will contribute the sharpness of the floating gate tip. But among all the processes, the isotropic etch process should be the most fundamental process because this process forms the floating gate tip and later processes only trim the floating gate tip base on its result.

More etch time will lead to sharper floating gate tip in the isotropic etch process. But if we don't want to impact the flash cell data retention, we cannot simply to add etch time without changing the floating gate deposition thickness.

It should be no problem to have a thick floating gate film at 0.18um node because of the bigger ACT pitch size, compared with more advanced technology node, such as 0.13um or 0.11um. The ACT pitch size limits the STI trench depth, and it combines the effect of STI HDP gap filling capability to limit the pad nitride plus floating gate total thickness.

To increase the floating gate thickness, there are several ways:

1, the process uses shallower STI trench depth combines thicker floating gate thickness, then the device leakage should be well controlled.

2, introduce the HARP process (High Aspect Ratio Process) to replace the STI HDP process to overcome the aspect ratio limitation, then there will be extra HARP equipment cost.

3, introduce the CeO_2 base slurry STI CMP to overcome the STI CMP POR process low nitride to oxide selectivity. In this way, the STI HDP aspect ratio limitation is still existed. But by replacing the STI CMP slurry to enlarge the nitride to oxide selectivity, the pad nitride thickness can be reduced then the floating gate thickness has room to be increased.

The experiment shows positive results for both the SEM cross section and wafer yield performance. It proves it is a new and practical way to improve flash erase and reliability.

CONCLUSION

1, Enhancing floating gate tip's sharpness can both benefit flash erase and endurance performance. The fundamental way to enhance the floating gate tip is to increase the isotropic etch time and it requires thicker floating gate deposition thickness.

2, In traditional way, there is limitation to increase the floating gate thickness due to STI HDP aspect ratio requirement.

3, By replacing the STI CMP slurry as CeO_2 base to enlarge the nitride to oxide selectivity, the pad nitride thickness can be reduced then the floating gate thickness has room to be increased.

REFERENCES

[1] Xian Liu et al. Endurance Characteristics of Super Flash Memory, ICSICT, 2006, pp.763-765
[2] Wen-Ting Chu et al. Shrinkable triple self-aligned field-enhanced split-gate flash memory, IEEE tans. On Electron Devices, 2004 pp.1667-1671
[3] A.Chimenton et al. Overerase phenomena: an insight into flash memory reliability, Proc. of IEEE, 2003, pp.617-626

[4] T.I. Wu et al. Characterization of split gate flash memory endurance degradation mechanism, Proc. of 11th IPFA, 2004, pp.115-117
[5] S. Bhattacharya et al. Improved performance and reliability of split gate source-side injected Flash memory cells, IEDM Tech. Dig., 1996, pp. 339–342.
[6] K.-C. Huang et al. The impacts of control gate voltage on the cycling endurance of split gate Flash memory, IEEE Electron Device Lett., 2000, pp. 359–361

DESIGN AND SIMULATION OF A SUPERCONDUCTING SWITCH BASED ON WEAKLY DAMPED SUPERCONDUCTING QUANTUM INTERFERENCE DEVICES

Xinyu Wu[1], Rutian Huang[1], Jianshe Liu[1] and Wei Chen[1]*

[1]School of Integrated Circuits, Tsinghua University, Beijing, 100084, China
[*]Corresponding Author's Email: weichen@tsinghua.edu.cn

ABSTRACT

Superconducting switches are important components in multiplexers based on superconducting quantum interference devices (SQUIDs). In time-division multiplexing (TDM) circuits, the superconducting switches are mainly used to activate and control input SQUIDs. The superconducting switch has the advantages of low power consumption and easy control. Here, a superconducting switch based on weakly damped SQUIDs is proposed. The dynamic resistance of the superconducting switch can be increased by appropriately adjusting the Stewart-McCumber parameter β_c of the Josephson junction. This design is expected to reduce the number of SQUIDs in arrays and simplify the fabrication processes of switches.

INTRODUCTION

Superconducting quantum interference devices (SQUIDs) are considered to be suitable for the readout of superconducting transition-edge sensors (TESs) since SQUIDs have low input impedance, low noise, and low power dissipation. The TES coupled to the SQUID has good single-pixel performance. With the extension of TES arrays, the thermal load and wiring complexity present challenges for the cryogenic readout. Therefore, it is essential to develop SQUID-based multiplexers in which multiple TESs share one readout line. At present, different multiplexers have been proposed, including time-division multiplexing (TDM) [1], code-division multiplexing (CDM) [2], frequency-division multiplexing (FDM) [3], and microwave SQUID multiplexing (MW-Mux) [4].

TDM is a mature architecture with low readout noise and low crosstalk. In each pixel of TDM, a switch is required to control the corresponding input SQUID. TDM circuits utilizing room-temperature switches have complex architectures and high-power dissipation. Superconducting switches operating at cryogenic temperatures are preferred. At present, superconducting switches mainly include two forms, namely, Zappe-style interferometers [5] and SQUID-based switches [6].

In order to ensure that the input SQUID can be activated and generate a large output voltage swing, the dynamic resistance of the superconducting switch needs to be much larger than that of the input SQUID. Increasing the number of SQUIDs can increase the dynamic resistance of the superconducting switch, but this requires high consistency of SQUIDs in the array. Here, the design and simulation of a superconducting switch based on weakly damped SQUIDs are reported. The simulation results show the dynamic resistance of the superconducting switch can be increased by adjusting the Stewart-McCumber parameter β_c of the Josephson junction properly.

PRINCIPLE AND ANALYSIS

Figure 1 shows the schematic of a single pixel of TDM circuit. The superconducting switch (red dashed box) is connected in parallel with an input SQUID (SQ) and a resistor R. The input SQUID is used to measure TES signals. The superconducting switch consists of low-inductance SQUIDs, and it is usually in the form of a series array. The screen parameter β_L of the low-inductance SQUID is extremely small, and thus its critical current is low at half-integer flux quanta. The dynamic resistance of the superconducting switch increases with the number of SQUIDs.

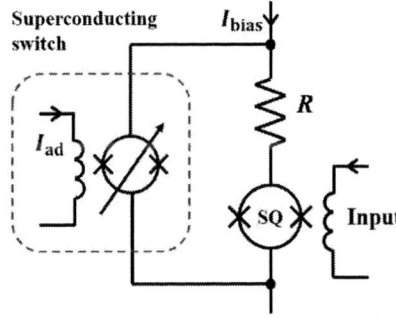

Figure 1: Schematic of a single pixel in the TDM circuit

The bias current I_{bias} is designed to be slightly larger than the current I_{sq} flowing through SQ and smaller than the critical current I_c of the superconducting switch. The switch is in the superconducting state at integer flux quanta. When a current I_{ad} that can generate a flux of half-integer flux quanta is applied to the control line, the switch is in the resistive state, and a large fraction of I_{bias} flows into the input SQUID. If the dynamic resistance of the switch is not large enough, the current flowing through the input SQUID may fail to activate it. Therefore, the dynamic resistance of the switch needs to be much larger than that of the input SQUID.

Figure 2 shows the schematic of a part of the superconducting switch based on weakly damped SQUIDs. In each unit of the switch, SQUID is a first-order parallel gradiometer that is formed by two Ketchen-type SQUIDs. The first-order parallel gradiometer reduces the total inductance of the SQUID loop and minimizes the interference of external magnetic field. The Stewart-McCumber parameter of Josephson junction is designed to be about 1 ($\beta_c \approx 1$), allowing the SQUID to achieve a large dynamic resistance while keeping the non-hysteretic characteristics. The superconducting switch proposed here contains a total of 16 such units and is fabricated by the standard Niobium process. To reduce the hot electron effect, each shunt resistor in the superconducting switch is connected to a colling fin.

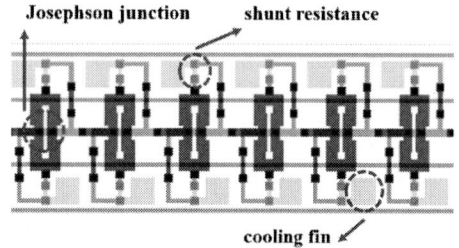

Figure 2: Schematic of a part of the superconducting switch based on weakly damped SQUIDs

RESULTS AND DISCUSSION

The inductance of the first-order parallel gradiometer is extracted by using InductEX and the simulation circuits of this device are set up by using WRspice. Figure 3(a) shows the flux-voltage characteristics of the switch. It can be seen that the control current of 1.32 mA corresponds to one flux quantum (Φ_0). Figure 3(b) is the current-voltage characteristics of the switch at different control currents. When a control current of 660 μA is applied to the control line, the switch shows resistive behavior, with a normal resistance of about 150 Ω. With the same number of SQUIDs in the series array, the dynamic resistance of superconducting switches based on weakly damped SQUIDs is larger than that of traditional superconducting switches.

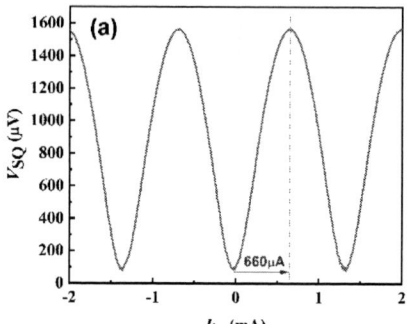

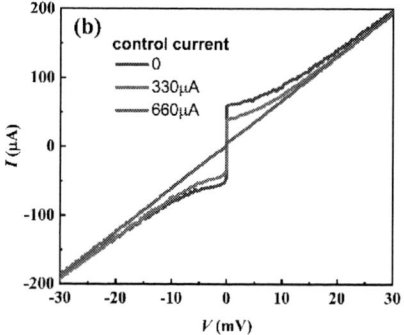

Figure 3(a): Flux-voltage characteristics of the switch; (b): Current-voltage characteristics of the switch at different control currents

To verify the function of the superconducting switch, a 4-channel TDM circuit is simulated by WRspice. The control current is sequentially applied to each channel to activate the corresponding input SQUID. Figure 4(a) shows the response of the input SQUID to the input signal in a single channel. Figure 4(b) shows the total response of the 4-channel TDM circuit at different control currents. When the control current of 660 μA is applied to one control line, the switch gets into the resistive state and the voltage of the input SQUID responds to the input current signal. Simulations show that the switch can be used to control the input SQUID of the TDM circuit.

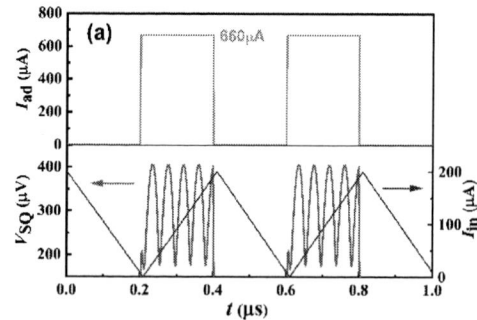

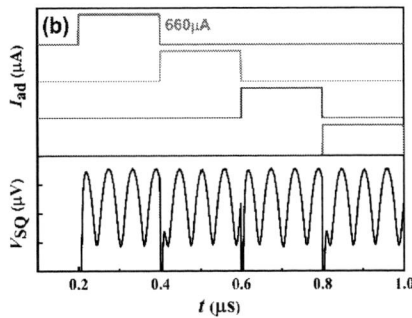

Figure 4(a): Top: the control current in a single channel. Bottom: the response of the input SQUID to the input current signal. (b): Top: the control current in different channels. Bottom: the total voltage response of the 4-channel TDM circuit.

CONCLUSION

In this paper, a superconducting switch based on weakly damped SQUIDs is designed and simulated. The device structure of superconducting switch is based on the low-inductance SQUID and series array. The dynamic resistance of the switch can be increased without hysteresis by increasing the parameter β_c appropriately. The superconducting switch has low power dissipation and is compatible with semiconductor fabrication processes. In addition, compared with the conventional superconducting switch, this switch requires less SQUIDs when the normal resistance is roughly the same, which reduces the occupied chip area and the difficulty of device fabrication.

ACKNOWLEDGEMENTS

We acknowledge Changhao Zhao, Qing Yu, Yongcheng He, Kaiyong He, Xiao Geng, Gengting Dai, Liangliang Yang and Mingjun Cheng for valuable discussions. This work is supported by the National Natural Science Foundation of China (Grant Nos. 11653001, 11653004 and 60836001).

REFERENCES

[1] W. B. Doriese, K. M. Morgan, D. A. Bennett, E. V. Dension, C. P. Fitzgerald, J. W. Fowler, J. D. Gard, J. P. Hays-Wehle, G. C. Hilton, K. D. Irwin, Y. I. Joe, J. A. B. Mates, G. C. O'Neil, C.D. Reintsema, N. O. Robbins, D. R. Schmidt, D. S. Swetz, H. Tatsuno, L. R. Vale and J. N. Ullom. *J. Low Temp. Phys.*, vol. 184, 2016, pp. 389-395.

[2] K. M. Morgan, B.K. Alpert, D. A. Bennett, E. V. Denison, W. B. Doriese, J. W. Fowler, J. D. Gard, G. C. Hilton, K. D. Irwin, Y. I. Joe, G. C. O'Neil, C. D. Reintsema, D. R. Schmidt, J. N. Ullom and D. S. Swetz. *Appl. Phys. Lett.*, vol. 109, 2016, 112604.

[3] K. Sakai, Y. Takei, R. Yamamoto, N. Y. Yamasaki, K. Mitsuda, M. Hidaka, S. Nagasawa, S. Kohjiro and T. Miyazaki. *J. Low Temp. Phys.*, vol. 176, 2014, pp. 400-407.

[4] B. Dober, Z. Ahmed, K. Arnold, D. T. Becker, D. A. Bennett, J. A. Connors, A. Cukierman, J. M. D'Ewart, S. M. Duff, J. E. Dusatko, J. C. Frisch, J. D. Gard, S. W. Henderson, R. Herbst, G. C. Hilton, J. Hubmayr, Y. Li, J. A. B. Mates, H. McCarrick, C. D. Reintsema, M. Silva-Feaver, L. Ruckman, J. N. Ullom, L. R. Vale, D. D. Van Winkle, J. Vasquez, Y. Wang, E. Young, C. Yu and K. Zheng. *Appl. Phys. Lett.*, vol. 118, 2021, 062601.

[5] H. H. Zappe. *IEEE Trans. Magn.*, vol. 13, 1977, pp. 41-47.

[6] J. Beyer and D. Drung. *Supercond. Sci. Technol.*, vol. 21, 2008, 105022.

Technologies for Superior Reliability in SiC Power Devices

Min-hwa Chi

Micro-Nano Technology College, Qindao University, Qindao, Shandong, China 266071

GTA Semiconductor Co, Ltd，Shanghai，China

Email: 18017378580@qq.com

ABSTRACT

Power devices based on SiC material have demonstrated superior performance than Si based power devices in high breakdown field, high power density, lower switching loss, and high operation temperatures, etc. It is also perceived as the most promising power devices for future applications in electric vehicles and clean energy systems. Thus, in addition to its superior performance (than Si power devices), it shall have further superior reliability and stability in applications. We review state-of-art technologies for achieving high reliability in SiC devices in gate oxide, trench profile, internal field minimization, current loading equalization, and minimization of electro-migration and stress migration, etc.

Keywords—SiC Power Devices, Reliability, Electric vehicles.

INTRODUCTION

Wide bandgap (WBG) materials, e.g. Si-Carbide (SiC) and Gallium nitride (GaN), are widely adopted with great promise for future high-power applications as related to their intrinsic advantages [1-2] compared to Si (Fig. 1). These new materials (SiC and GaN) with wide band-gap have a high breakdown electric field and thermal conductivities compared with Si, thus they can result in power devices with superior performance in high voltage/power capability, higher switching frequency, better thermal conduction, higher junction temperature, and better radiation hardness. Thus, the integration of power devices and circuits can achieve better performance and improved efficiency with less cooling needed. Currently, SiC based power MOSFET is forth coming and practically able to replace all the current Si-based devices. Power devices based on wide band-gap materials are increasingly widespread for high-power applications.

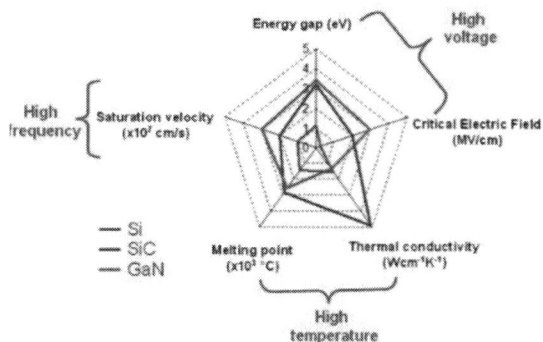

Fig. 1: A comparison of wide bandgap materials to Si [1-2].

The application range of Si, GaN, and SiC devices is illustrated in Fig. 2. The switching loss of SiC MOSFET is relatively lower as its switching speed is faster, so it can be used in high-frequency conditions. In comparison, the switching loss of Si-based IGBT is higher due to its inherent slower switching speed related to its bipolar nature. Compared with Si IGBT, the electrical conduction loss of SiC MOSFET is lower as related to higher doping concentration in the drift region. The smaller depletion width in SiC MOSFET also reduces the ON-state resistance, leading to less electrical conduction loss. In addition, SiC MOSFET has much lower output capacitances and gate charge; this leads to switch at much higher dv/dt and di/dt. High switching speed enables low switching loss and high switching frequency, which can improve the power density and efficiency of power module. Moreover, switching loss of Si IGBT will increase significantly at higher operational temperature, while switching loss of SiC MOSFET has little variation over temperature. Therefore, SiC MOSFET possesses superior advantages in middle-high voltage applications over competing Si counterparts.

Material	Bandgap (eV)	Mobility (cm²/V*s)	Permittivity	Vsat (cm/s)	Critical field (V/cm)
Si	1.1	1400	11.8	1×10^7	3×10^5
GaAs	1.42	8500	12.8	2×10^7	4×10^5
4H-SiC	3.23	260	9.7	2×10^7	2.9×10^6
GaN (Bulk)	3.4	900	9	2.5×10^7	3.3×10^6
GaN (HEMT)	3.4	1800	9	2.5×10^7	3.3×10^6

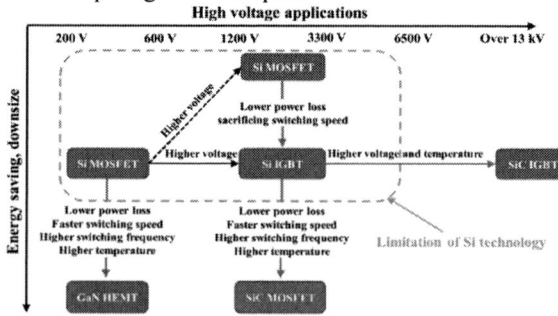

Fig.2: Application range of Si, GaN, and SiC power devices [3]

Wide-bandgap power devices in electric vehicles [4]

Recent progresses of wide band-gap (WBG) power semiconductor devices and applications in electrical vehicles (EV) is a main trend as illustrated in Fig. 3 [4]. For SiC devices, the power ratings of commercial devices can achieve up to 3.3kV and 800A for motor drive in EV. For GaN devices, the device power rating is lower (e.g. less than 900V and 150A) with fast switching beyond MHz range as suitable for on-board DC/DC converters and chargers. Both SiC and GaN device characteristics are still new to designers, thus device characterization is an important area. Under fast WBG device switching transients, the parasitic inductances and overall thermal resistance in device packaging and integration scheme with power electronics in converters are to be minimized. In respect to their applications for e-mobility, the objectives include the increase of power efficiency and density, and the reduction of EMI of on-board power electronics systems. To achieve these objectives, techniques of soft switching, multifunction integration of power converters, and new power module structures are important.

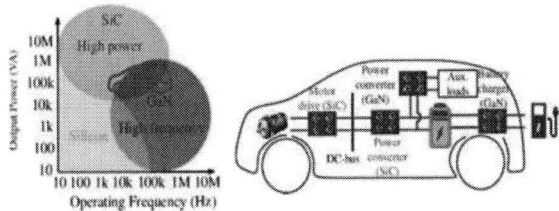

Fig. 3. Wide band-gap (WBG) device applications in electrical systems and EV [4]. The parasitic inductances and overall thermal resistance in device packaging and integration schemes with power electronics shall be minimized for fast WBG device switching transients.

SiC integrated circuits [5]

SiC is the only wide bandgap (WBG) material system as already demonstrated capability of integration with digital circuits as illustrated in Fig. 4 [5]. However, SiC MOS devices have main obstacle to overcome in the areas of low inversion-channel mobility and high interface trap-density at the SiC-oxide interface. Among the 3 popular SiC poly-topes (3C, 4H, 6H), 3C-SiC is known to have the highest-quality interface to SiO_2 (i.e. achieving D_{it} < 10^{11}/cm²/eV and mobilities in the range of 200cm²/Vs with dry oxidation), but its lower bandgap and defective bulk epitaxial quality limit its commercial applications. 4H–SiC is preferred for power devices due to its higher bulk mobility and slightly higher bandgap, though 6H-SiC shows higher MOS inversion channel mobility. CMOS has been demonstrated on both of 4H–SiC and 6H–SiC polytopes.

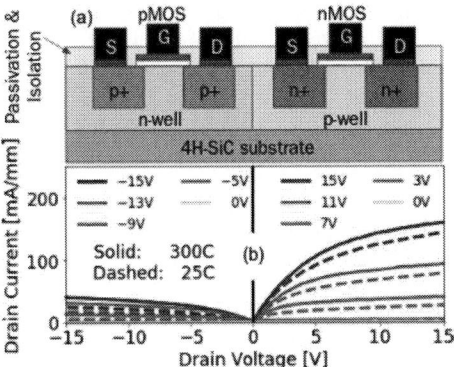

Fig. 4. (a) Technology cross section of Raytheon's HiTSiC process. (b) I–V characteristics of a 1.2-μm pMOS and nMOS at 25 °C and 300 °C) [5]

RELIABILITY TECHNOLOGY FOR SiC IN FEOL

Gate-ox process with "H2-Ox-NO" 3-steps [6]:

The 3-step process (Fig. 5), i.e. H_2 etching, oxide deposition, and annealing in NO, can yield higher drain current while keeping normally-off characteristics. The MOSFET fabricated with the "H₂-CVD-NO" process exhibited a higher mobility (~72–80 cm²/Vs) at gate voltage of 5–10 V, as a 2-fold improvement in channel mobility (vs standard process of "Ox+NO") as well as process window, failure analysis, and yield enhancement, etc.

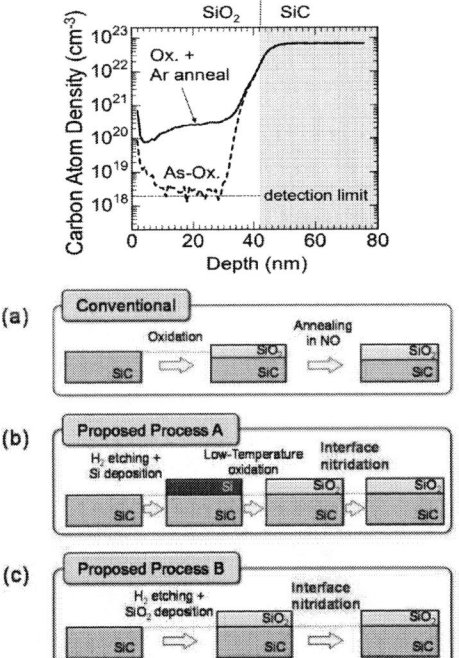

Fig. 5. The profile of C atom density in thermal oxide (SiO_2)/SiC samples increases after anneal (Ar/1300ºC). The enhanced gate-oxide formation in SiC MOSFETs is shown [6]

Gate-ox process with Si implanted surface [7]

The performance in 1.2KV 4H-SiC VDMOS can be improved by using a Si implanted surface (~10kV, ~10^{14} – 10^{15} cm^{-2}, 600ºC) as in Fig. 6 [7]. The Si implantation forms a slightly Si-rich layer on the top surface for reducing the creation of carbon clusters during thermal oxidation. Firstly, the devices with Si implantation on the surface shows a decrease of carbon atom percentage from XPS and EDX analysis, an improvement of sub-Vt slope (SS) and mobility in n-MOSFETs, and a lower interface state density near the conduction band edge from n-MOS capacitors. Secondly, high-voltage VDMOS with Si implantation on the surface shows an improvement in SS, drain current (I_D), breakdown voltage, and electrical safe operating area (SOA). Finally, the reliability including positive and negative bias temperature instability (PBTI/NBTI) and Ron stability under high-voltage pulses are also shown improvement.

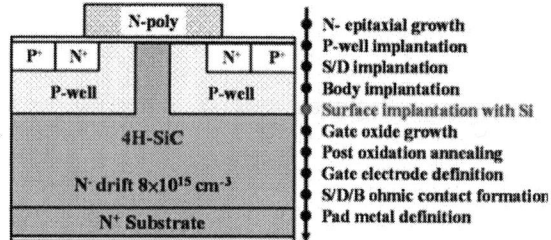

Fig. 6: Schematic of 4H-SiC power MOSFETs with brief process flow. A surface implantation using Si at 600ºC is used for better quality of gate oxide [7]

High-k dielectric with metal or poly-gate [8-9]

Despite the recent progresses in SiC power MOSFET technology and its commercialization, the defective MOS interface still impacts the full potential of SiC devices and CMOS integrated circuits. Recent results using high-k gate stack technology (Fig. 7) shows significant reduction of the density of interface states (Dit) along with superior stability of threshold voltage (Vt) for low voltage SiC power MOSFETs. The findings indicate virtually no Vt-shift during characterization with respect to various starting gate voltage and ramp. Furthermore, the dynamic switching results also show virtually no Vt shift for $V_{GS,start}$ > -12V. The low defect density of the high-k/SiC dielectric interface has a remarkable improvement not only on the performance, but also in terms of Vt hysteresis, extrinsic defects, and extrapolated lifetime at operating voltage.

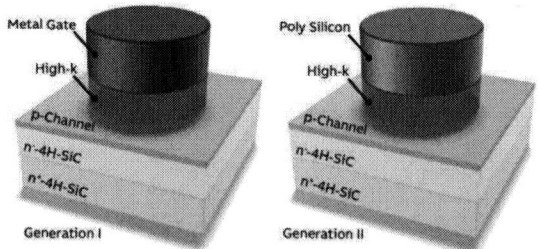

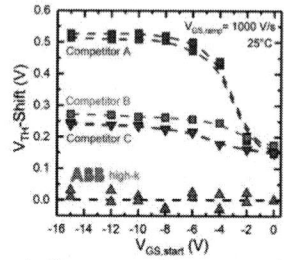

Fig. 7: Schematic illustration of the gate stack for ABB's high-k with poly-gate or metal gate. Room temperature Vt-shift as function of $V_{GS,start}$ (1.2kV MOSFETs). A constant gate voltage ramp of 1000V/s is used for these measurements. [8-9]

Trench profile for higher breakdown voltage [10]

We found a new method to eliminate the bottom thinning effect by forming an outward notch shape (with characteristic angle α) at the trench bottom corner by tuning etching parameters as in Fig. 8. The simulation of the new notch-shape trench MOSFET showed that BV is ~788V (@turn-off), and V_{CE}=2.73V (@turn-on) with lower field around the notch of the trench bottom. More simulation data with α as a parameter are also shown. The notch angle is proposed to be <80º for the best BV and process capability. The mechanism of the BV enhancement by the outward notch shape at bottom corners is discussed here. After turn-on to on-state, the higher electric field around the sharper poly-gate can induce more accumulation electrons in the drift region adjacent to the notch corner, thus, Ron is reduced slightly (though the notch corner occupied a little bit n-drift region). The net effect appears a small effect on Ron. After turn-off to off-state, the breakdown voltage increased. Firstly, the electric field in the depletion region is more evenly distributed, and therefore improving the BV. Secondly, the Si-oxide inside the notch shape corner may be thicker, and more resistant to breakdown than Si based devices. The sharper corner at the notch does not contribute to larger electrical field during off-state as the field around the notch is related to the ionized dopants in the depletion layer outside the notch.

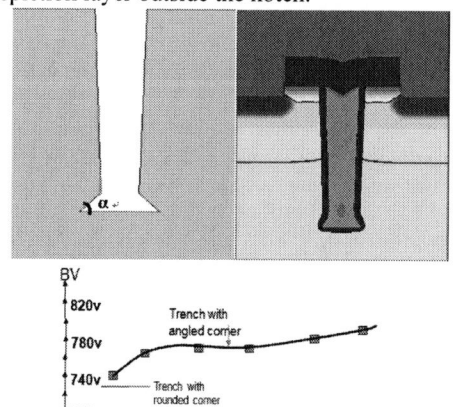

Fig. 8: New trench etching with outward notch shape at the bottom corner and final structure after doped poly-Si filling and planarization. The simulation shows that lower electric field intensity (i.e. higher breakdown voltage) at off-state around the notch shape of bottom corner.

RELIABILITY TECHNOLOGY FOR SiC IN BEOL AND PACKAGING

Ag sintering technology [11]

Pressure-assisted sintering is a reliable sintering approach for bonding power devices for high-temp applications. However, the pressure-assisted sintering can cause a decrease in the yield and efficiency for bonding multiple chips. To achieve robust sintered Ag attachment efficiently under significantly low sintering pressure, new heating sources (e.g., electric-current, selective laser, microwave, etc.) can be combined using pressure for low-temp sinter-joining of Ag particles in the future. The decrease in sintering pressure is essential to reduce the cost of the special sintering equipment. The pressure-less sintering (Fig. 9) has inherent advantages for sinter-bonding multiple chips targeting for high-density integration of a power module. However, pre-defects should be controlled strictly for acceptable reliability. Furthermore, it is important to control precisely the warpage of substrates to ensure good wettability and inter-diffusion.

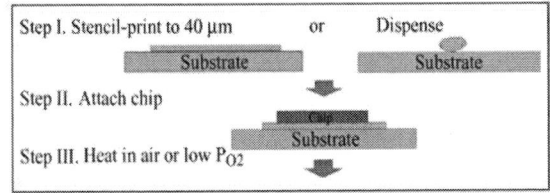

Fig. 9: A pressure-less Ag sintering process. [11]

Double-sided cooling in packaging [12]

SiC devices are cable of switching at higher speed with increased frequency. Need to develop new packages with low parasitic inductance (from Al wires) and high-efficiency cooling for SiC-based power packaging or modules. Superior properties of SiC devices cannot be fully utilized if it is used simply as a direct replacement of conventional Si devices. A state-of-art SiC high-power inverter module is shown in Fig. 10 using new materials for high temperature over 220°C, e.g. the flip-chip bonding for source and gate interconnections using Bi–Ag solder paste (with solidus temp ~260°C and liquidus temp ~360°C) for much high thermal stability than conventional Pb-free solder. In order to increase the wettability of solder paste, an electroless Ni immersion Au process on the Al bonding pads on the MOSFET surface was used. Cu clips were attached using Ag sintering materials as drain interconnection. Thermal dissipation was also enhanced by twice vs conventional single-side cooling power module.

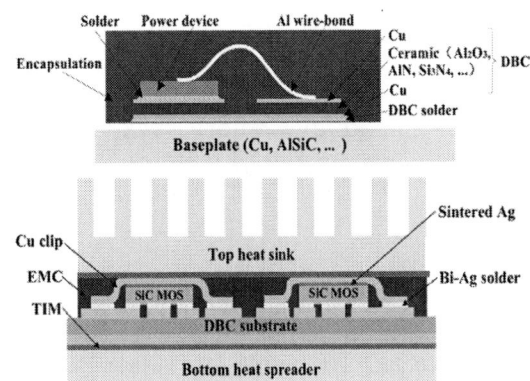

Fig. 10: Conventional packaging (upper) and a state-of-art SiC packaging with double-sided cooling structure [12].

CONCLUSIONS

Power devices based on SiC material have demonstrated superior performance than Si based power devices as the most promising for future applications in electric vehicles. Thus, in addition to its superior performance, it shall have superior reliability and stability in applications. The state-of-art technologies for achieving superior reliability in SiC devices in gate oxide, trench profile, internal field minimization, current loading equalization, and minimization of electro- and stress-migration are reviewed.

REFERENCES

[1] L. Spaziani and L. Lu, "Silicon, GaN and SiC_ there is room for all: an application space overview of device considerations", ISPSD , p.8, 2018. [2] F. Roccaforte, et.al., "Processing issues in SiC and GaN power devices technology - the cases of 4H-SiC planar MOSFET and recessed hybrid GaN MISHEMT", CAS, p.7, 2018. [3] F. Hou, et.al., "Review of Packaging Schemes for Power Modul", IEEE J. of emerging and selected topics in power electronics, v.8, No.1, p.223, 2020. [4] Thang V. Do, et.al., "Reviewing of Using Wide-bandgap Power Semiconductor Devices in Electric Vehicle Systems: from Component to System", VPPC, 2020. [5] S. J. Bader, et.al., "Prospects for wide bandgap and ultrawide bandgap CMOS Devices", IEEE ED, p.4040, 2020. [6] T. Kimoto, et.al., "Physics and Innovative technologies in SiC power devices", IEDM, p.761, 2021. [7] J.-W. Hu,, et.al., "1.2 kV 4H-SiC VDMOSFETs with Si-implanted surface: performance enhancement and reliability evaluation", ISPSD, paper#SiC-P2, p.211, 2021. [8] S. Wirths , et.al., "Vertical power SiC MOSFETs with high-k gate dielectrics and superior Vt stability", ISPSD，p226，2020. [9] S. Wirths, et.al., "Study of 1.2kV high-k SiC power MOSFETS under harsh repetitive switching conditions", ISPSD, p.107, 2021. **(10)** M. Chi, J. Liu, P. Li, "Impact of topology of trench gate bottom corner for power MOSFET and IGBT ", poster paper #1-9, CSTIC, 2022. [11] H. Yan, et.al., "Brief review of Ag sinter-bonding processing for packaging high-temperature power devices", CJEE, V.6, No.3, p.25, 2020. [12] F. Hou, et.al., "Review of packaging schemes for power module", 2020. IEEE J. Emerging and selected topics in power electronics, V.8, No.1, p.223, 2020.

THE STUDY ON REDUCING BIT-LINE PARASITIC CAPACITANCE IN ADVANCED DRAM

Yexiao Yu[1]*, Hong Ma[1], Zhongming Liu[1], Shaoyou Xiong[2], Dan Wang[2], Yang Zhang[2], Yi Yang[2]
[1]Changxin Memory Technology, United Process Development, Hefei, China
[2]Changxin Memory Technology, United Process Development, Hefei, China
*Corresponding Author's Email: Yexiao.Yu@cxmt.com

ABSTRACT

In this paper, 3D TCAD (Technology Computer Aided Design) process and device simulation methods are used to explore the relationship between CBL (parasitic capacitance of bit line) and structure. The results show that BL spacer thickness and material property are critical to CBL. We suggest that BL profile should be vertical, which can give more room to BL spacer, then CBL will be reduced. Through process simulation, when BL TB ratio (top CD: bottom CD) increase from 0.6 to 0.9, the thickness of BL spacer increases by 1nm, and CBL decreases by 2.7aF/cell.

Keywords—TCAD, parasitic capacitance, BL spacer, BL profile, TB ratio, simulation

INTRODUCTION

As DRAM shrinking for higher density and speed, CBL plays a key role of achieving sense margin in the scaling [1]. Based on reverse engineering data from TechInsight, advanced DRAM manufacturers have developed different CBL reduction roadmaps. The CBL reduction way of S company is BL airgap spacer structure; H company uses BL metal levelling with low k spacer solution; For M company, BL low k spacer is also needed in the absence of air spacers, meanwhile, BL metal film thinning is used to reduce CBL.

In WAT testing, CBL is mainly composed of four parts: BL to NC, BL to BW, BL to BL, BL to substrate capacitance (formula 1) [2], among which CBL_NC is the main index affecting CBL, with a contribution rate of 80%~90%.

$$CBL = CBL\_NC + CBL\_BL + CBL\_BW + CBL\_sub \quad (1)$$

Combine CBL_NC define formula (2), there is a strong correlation between CBL_NC and BL structure parameters. Series of BL structure parameters were listed in Table 1. Then, we modelled DRAM array structure for further study (see Fig.1).

$$CBL\_NC = \varepsilon \frac{BL\ metal\ height\ \times BL\ length}{BL\ spacer\ thickness} \quad (2)$$

In this work, the sensitivity of CBL to structure was modelled and analyzed. It is found that BL spacer thickness and material property are the most sensitive indexes to CBL. Since BL design pitch has been determined (formula 3), the dynamic relationship between BL CD, NC CD and BL spacer thickness must be considered when designing the new CBL reduction scheme. In this work, a new approach for CBL reduction was put up: BL PNR vertical profile → NC CD enlarge → BL to NC distance enlarge → thicker BL spacer→ CBL reduction (see Fig.2). It will give value guide for manufactures.

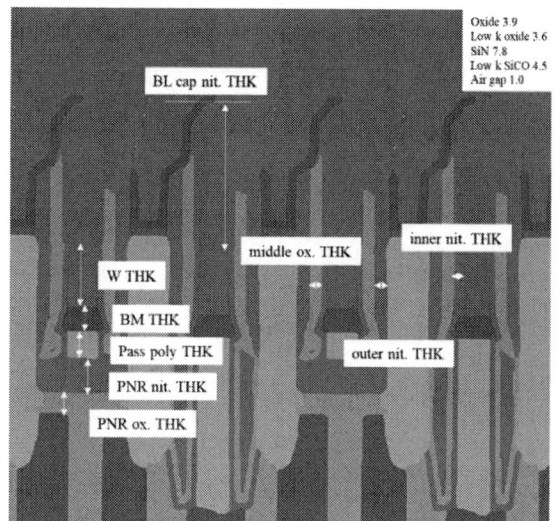

Fig.1.DRAM array structure and CBL related process parameters

Table 1 BL key structure parameters

Vertical	Structure Parameters	Horizontal	Structure Parameters
BL Metal Height	BL W THK	**BL Spacer Thickness**	Middle oxide THK
	BL BM THK		Outer nitride THK
	Pass BL Poly THK	**BL Spacer Dielectric**	Low k inner nitride
	PNR nitride THK		Middle oxide
	PNR oxide THK		Air gap

$$BL\ design\ pitch = BL\ CD + 2 \times BL\ spacer + NC\ CD \quad (3)$$

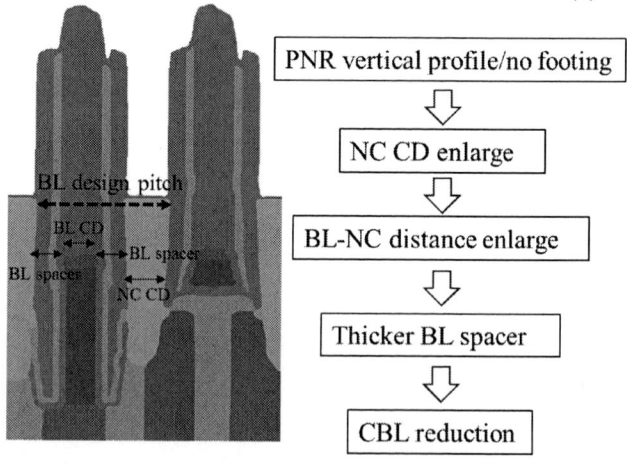

Fig.2.CBL reduction process design scheme

BL PROFILE INVESTIGATION AND MODELLING

In a 1y/1z DRAM reverse engineering report from TechInsights, different BL profiles were found among leading DRAM manufactures (see Fig.3). In process manufacturing, BL process conditions always affect BL profile, and even CBL performance. To evaluate the BL process condition effects on CBL performance, a short loop DRAM process flow from active area (AA) to capacitor landing pad (M0) was built up (see Fig.4). CBL device model was extract form DRAM structure.

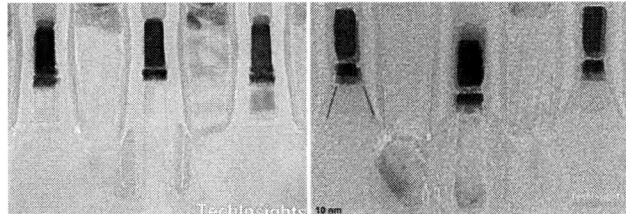

Fig.3.Section view of the BL profiles of 1y/1z DRAM from different manufactures (Courtesy: TechInsight)

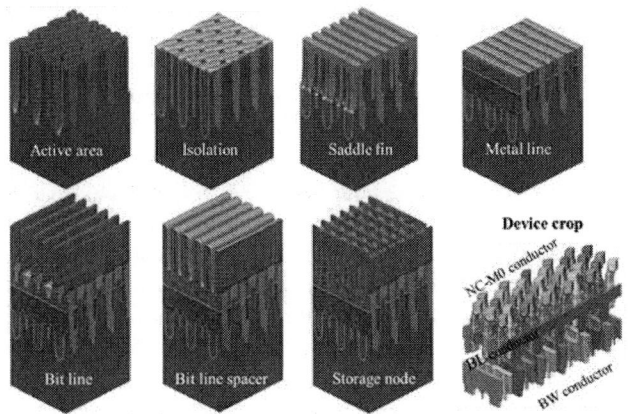

Fig.4 DRAM process flow from active area (AA) to capacitor landing pad (M0) and CBL device model

BL PROCESS CONDITION EFFECTS ON CBL SIMULATION

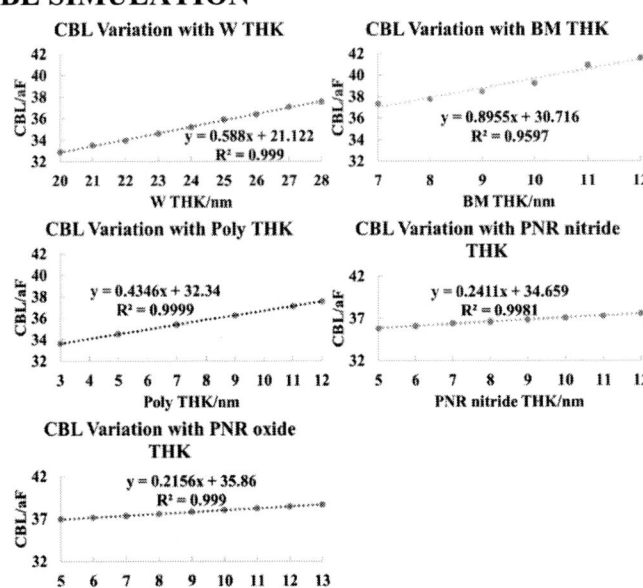

Fig.5.BL metal and PNR height effects on CBL

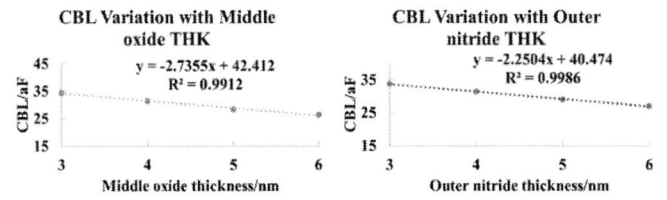

Fig.6.BL spacer film thickness effects on CBL

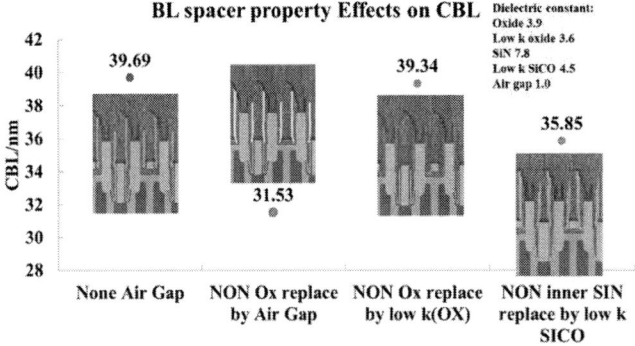

Fig.7.BL spacer film property effects on CBL

On the basis of the previously defined virtual model (see Fig.4), we conducted a series of experiments to summarize the relationship between CBL and process conditions (see Fig.5 & 6). Different process conditions have different contributions to CBL, and the order is as follows: BL spacer>BL metal > BL poly > BL PNR film. In addition, it is found that CBL can be reduced significantly by inner nitride replaced by low k oxide and middle oxide replaced by air gap (see Fig.7). We summarized above results in Fig.8. The conclusion is changing spacer material properties and thickness can make CBL reduce significantly.

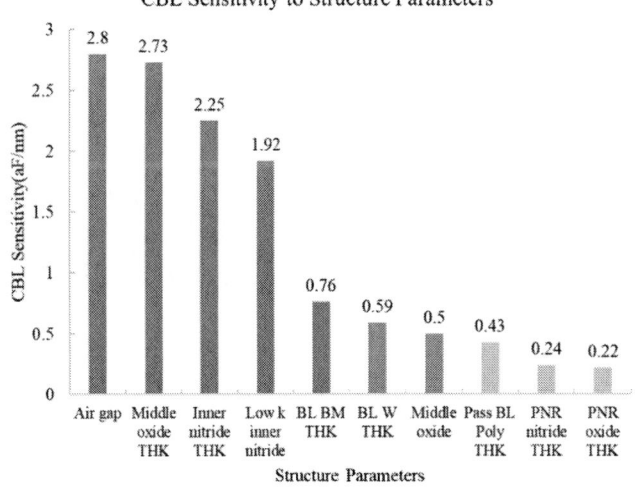

Fig.8.CBL sensitivity orders to BL process

However only increase BL spacer thickness is not reasonable, which will cause smaller NC CD and even induce DVC issue[3]. In the latest CBL reduction process design scheme (see Fig.2), BL etch process make a crucial part, which can bring more room for BL spacer and make CBL reduced indirectly.

BL PROFILE OPTIMIZATION FOR CBL

REDUCTION

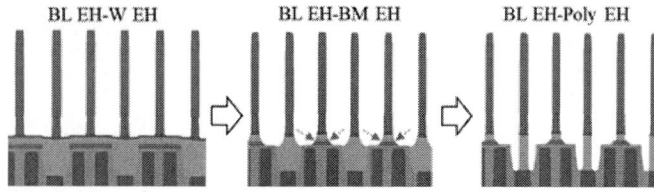

Fig.9.BL etch process movie

In the BL etch process (see Fig.9), BM metal is very difficult to trim vertically. In manufacturing, BM etching less induce BL PNR tapper profile, which is unfriendly to NC to AA contact; When BM over etching, although BM profile can make vertical enough, it is easy for the dielectric damage under BM film and BL to AA leakage [4]. In the below process model, we use BL BT ratio to describe BL profile briefly, then do BL BT ratio split for process window (see Fig.10).

$$PNR\ BT\ Ratio = \frac{Bottom\ CD}{Top\ CD} \quad (4)$$

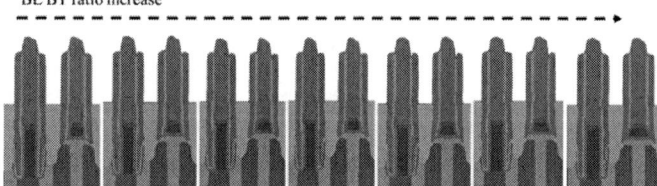

Fig.10.BL PNR profile BT ratio defines (A) and NC-AA 3D contact area(B&C)

In this study, we split series of BL BT ratio experiments (see Fig.11). Try to find the correlation between BL BT ratio and BL spacer thickness. When we make BL BT ratio reach from 0.6 to 0.9, we can make BL spacer increase from 3 to 4, indicating the method can reduce CBL about 2.7aF/cell.

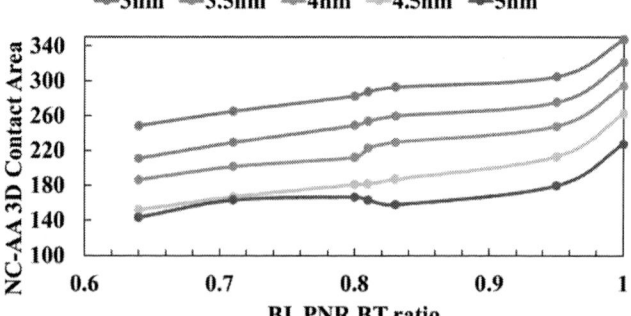

Fig.11.BL profile BT ratio split

Pass BL poly BT ratio effects on DVC

Fig.12.BL PNR BT ratio correlation with NC-AA 3D contact area at different BL spacer thickness

CONCLUSION

In this paper, BL process condition effects on CBL were modelled and analyzed. The CBL sensitivity order is: BL spacer(property & thickness) > BL metal thickness > BL poly thickness > BL PNR thickness. Therefore, we provide a new scheme to reduce CBL: vertical BL PNR profile give more room to BL spacer thickness, then increase BL spacer thickness and achieve CBL reduction. Through NC-AA 3D contact process model, the results shows CBL was reduced by 2.7aF/cell when PNR BT ratio improve from 0.6 to 0.9. It is a valuable ways to solve sense margin probelem and yield improvement in manufactures.

REFERENCES

[1] Q. Han, M. Cai, B. Wu and K. Cao, "A DTCO approach on DRAM bit line capacitance and sensing margin improvement," 2020 IEEE 15th International Conference on Solid-State & Integrated Circuit Technology (ICSICT), 2020, pp. 1-3.

[2] X. Liu, J. Jeon, B. Wu and M. Cao, "Simulation Studies about the NON Spacer Effects on the DRAM Access Transistor Performance," 2020 IEEE 15th International Conference on Solid-State & Integrated Circuit Technology (ICSICT), 2020, pp. 1-3.

[3] Y. Yu, Z. Liu, G. Xiong and H. Ma, "Evaluating the Impact of Process Variations on Storage Node Contact Area Under Different Etching Models," 2022 54th International Conference on Solid States Devices and Materials (SSDM), 2022, J-10-03.

[4] Y. Yu, Z. Liu, J. Cui, Z. Kong, G. Xiong and H. Ma, "The Study of Bit Line to Storage Node Contact Leakage in Advanced DRAM," 2022 IEEE 5th International Conference on Electronics Technology (ICET), 2022, pp. 118-122.

ENHANCEMENT OF PATTERN DEPTH IN PLASMONIC LITHOGRAPHY FOR PRACTICAL APPLICATION

Dandan Han and Yayi Wei*

University of Chinese Academy of Sciences, School of Integrated Circuits, Beijing, 100049, China,
*Corresponding Author's Email: weiyayi@ime.ac.cn

ABSTRACT

A hybrid plasmonic waveguide (HPW) patterning system, which is composed of the plasmonic BNA-PR layer-silver reflector, has been proposed to solve the shallow pattern issue in the plasmonic lithography. Theoretical calculations show that line array patterns in sub-20 nm-feature can be achieved with high aspect ratios and pattern uniformity, mainly due to the forward and reflected waves can remove the exponential decay and allow a higher wave vector to participate the patterning process. Eventually, the general criteria of the proposed MIM waveguide patterning system is also discussed, which can broaden the application of the plasmonic lithography in nanoscale fabrication.

INTRODUCTION

Photolithography as the essential technology of micro and nano manufacturing, has been the driving force for the development of the semiconductor industry and is also the important reason why integrated circuits developed following the Moore's law [1-2]. Unfortunately, the further advance of photolithography resolution is limited by the diffraction of light waves. According to the Fourier analysis theory, the diffraction limit of photolithography is principally because the subwavelength information is carried by the evanescent waves, and the evanescent waves only existing in the near-field [3]. Thus, the pattern resolution of photolithography may also can be enhanced to deep-subwavelength scale by utilizing the component of evanescent waves. Plasmonic lithography using a bowtie nanoaperture (BNA), which was recently studied, offers a maskless, low cost, and high throughput approach for large-scale patterning [4-6]. With state-of-the-art plasmonic lithography tool, the pattern resolution has been demonstrated to reach sub-10 nm-feature size by utilizing the surface plasmon waves (SPWs) at the interface between BNA and photoresist (PR). The SPWs are evanescent waves, which induced by free electron oscillations at the exit of plasmonic BNA and has a wavelength much smaller than that of the wave in free space, hence, the degree of confinement itself can directly improve the pattern resolution [7]. The exposure system of plasmonic lithography was originally composed of a plasmonic BNA and a PR layer to stimulate the SPWs, despite the plasmonic BNA can provide field confinement in deep subwavelength, the SPWs decay exponentially in the PR layer generally cause a rapid loss of the high spatial frequency contents. The exponential decaying characteristic significantly limits the pattern depth for plasmonic lithography, experimental results demonstrated that the pattern with sub-100nm feature size in PR exhibited a poor profile depth of less than 10 nm [5, 8]. Increasing the aspect ratio and achieving high pattern uniformity are critical to plasmonic lithography for practical applications.

In this work, a hybrid plasmonic waveguide mode is proposed to overcome the shallow pattern depth issue of plasmonic lithography. The resolution, depth, and uniformity of nano-patterns are the key indicators of the exposure quality. Hence, maintaining a uniform and strong near-field enhancement effect in the PR layer is of significant importance to increase the pattern depth.

MODEL

Characteristics analysis of the hybrid plasmonic waveguide mode

In order to study the field confinement ability in PR layer, a finite differential time domain (FDTD, ansys lumerical v2021) calculation is performed to obtain the electronic field distribution in the plasmonic lithography systems. The geometries and dimensions of the plasmonic lithography system is shown in Figure 1a. A transverse magnetic (TM) polarized plane wave is illuminated on the plasmonic BNA, and the wavelength is 365 nm. At this illumination wavelength, The permittivities of Al, Ag, and PR are,, and, respectively. The outline dimension of BNA in an Al film is assumed to be 150 nm x 150 nm, and the ridge-gap size is set as 20 nm. Figure 1b indicates that the SPWs reflected by the Ag reflector film is reasonably coupled with the SPWs excited by the plasmonic BNA/PR interface, which greatly compensates the intensity attenuation in PR layer and thus leads to the photoenergy distribution in the PR layer is consistent with the two interfaces. This is mainly due to the asymmetry electromagnetic mode excitation in the HPW structure, which can enhance the resonance coupling between the generated SPWs at the top and bottom of PR layer and the plasmonic waveguide. Significantly, the uniform distribution of the field intensity of SPWs in the PR layer improves the pattern depth, and ensures high resolution and uniformity of patterns. Therefore, the findings demonstrate that the HPW structure is useful for improving the aspect ratio of plasmonic lithography.

979-8-3503-1101-3/23 $31.00 © 2023 IEEE

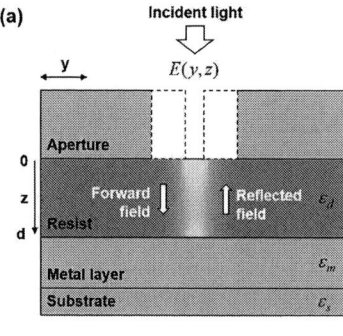

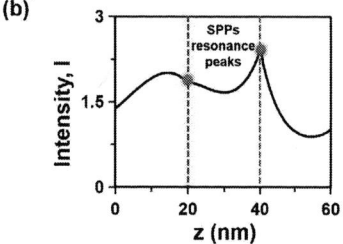

Figure 1: (a) Schematic configuration of the HPW structure in plasmonic lithography system including a scanning plasmonic BNA, PR, metal-reflector layer, and Si substrate. (b) HPW structure, Red and blue dotted lines imply the interfaces of plasmonic BNA/PR and PR/metal reflector respectively, two red circles mark the generated SPWs resonance peaks.

Achievable resolution in plasmonic lithography

Typically, the achievable resolution of plasmonic lithography is decided by the transmitted field intensityalong the z-direction of the PR layer, and the imaging contrast, where z is the distance from the exit of the plasmonic BNA, r is the radial coordinate, and is the peak value of the intensity in the plane [9]. By calculating the resolution value at a different gap size where the imaging contrast equals 0.1 [5, 9], we can plot the achievable resolution as a function of gap size for the hybrid plasmonic waveguide structure, as shown in Figure 2. Obviously, smaller than 10 nm resolution can be realized by the hybrid plasmonic waveguide structure as the gap size decreases from 20 nm to 6 nm.

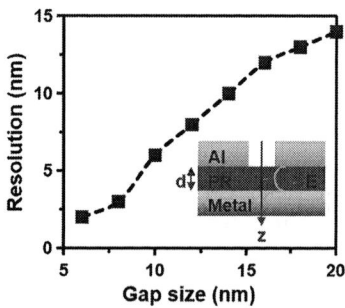

Figure 2: The achievable resolution of the HPW structure

RESULTS AND DISCUSSION
The generated line array pattern profile by HPW structure

We investigated the performance of the HPW in plasmonic lithography by evaluating the pattern quality of line array patterns. The electric field intensity distribution in the PR layer for a line array pattern is depicted in Figure 3a. The field distribution shows that the SPWs are excited at the top and bottom surfaces of PR layer. The forward- and backward-propagating SPWs in the PR layer interfere with each other, and consequently pattern profiles with high uniformity and high contrast are clearly observed in the PR.

Figure 3b shows the field intensity distribution at the top, middle, and bottom lines in the PR layer corresponding to the plasmonic lithography with HPW structure. It is worth to note that the half-pitch (HP) resolutions of these 3 curves in HPW structure are almost identical, which are 20.5 nm, 21.5 nm, and 20.5 nm, respectively. In addition, the image contrasts defined by of these 3 lines are 0.98, 0.93, and 0.96, respectively. Interestingly, the maximum field intensity in the PR layer becomes higher than the incident light, which further confirms the field enhancement effects by the strong coupling strategy between SPWs and plasmonic waveguide. The high field intensity and image contrast in the PR layer ensure the exposure stability in the patterning process.

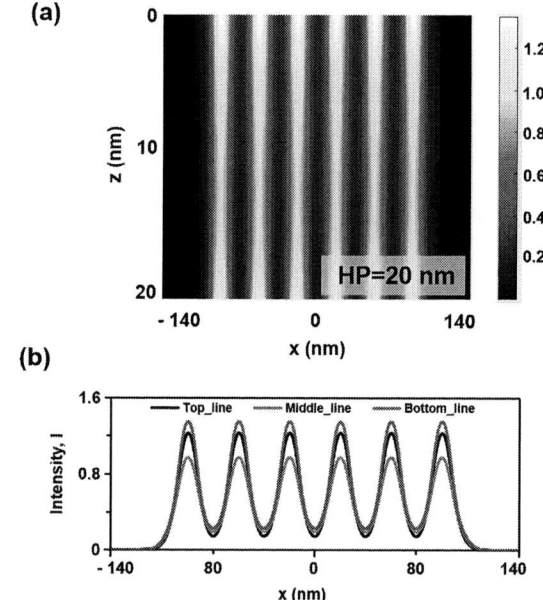

Figure 3: (a) Electric field distribution in the PR layer corresponding to the HPW structure. (b) Electric field intensities at 3 positions of z=0 nm, z=10 nm and z=20 nm in the PR layer corresponding to the HPW structure.

Spatial frequency filtering in HPW structure

To gain a better insight of the nature of pattern depth enhancement and the spatial frequency selection in plasmonic lithography with HPW structure, we quantitatively analyze the normalized intensity distribution in the HPW mode. For a comparison, the normalized intensity maps in the PR layer of the HPW patterning system are given in Figure 4a for PR with a 20 nm thickness and in Figure 4b for PR with a 40 nm thickness. The normalized intensity at the center of the plasmonic BNA along the z-direction in the PR layer corresponding to the Figure 4a and b are plotted in Figure 4c. Clearly the electric field intensity in the HPW structure can be significantly enhanced, and it is sensitive to the PR thickness. To achieve high aspect ratio feature, the effective index n_{eff} of the HPW can be adjusted to satisfy the phase matching condition, by optimizing the thickness of PR [10, 11]. These results verify the spatial frequency selection characteristic of the HPW structure, and indicate that such a HPW mode design in plasmonic lithography can be used to expose a PR of different thickness, and thereby generating patterns in PR with the desired aspect ratio features, which is useful for practical applications .

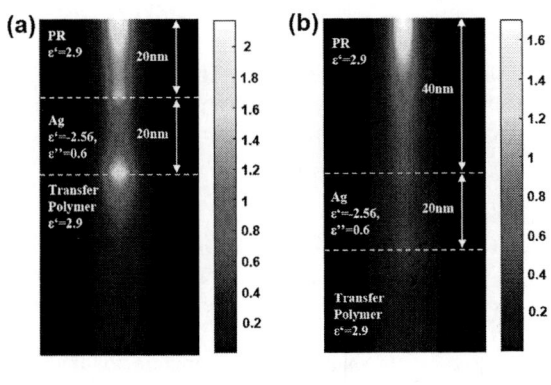

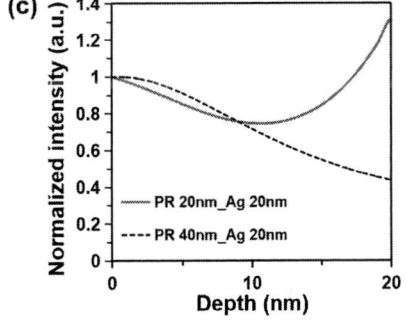

Figure 4: (a) Normalized intensity map in the PR layer of the plasmonic lithography with HPW structure corresponding to the PR thickness d_1=20 nm, and (b) d_1=40 nm.(c) The intensity distribution as a function of exposure depth.

CONCLUSION

In summary, we have numerically demonstrated that the half-pitch resolution of the generated patterns by plasmonic lithography can go beyond 10 nm with high aspect ratio. This is achieved by employing a HPW structure through antisymmetric coupled SPWs in PR layer. The HPW structure is based on the spatial frequency filtering characteristics of evanescent waves, so that high-k can be selected , which results in the generated pattern with high quality. Hybrid plasmonic waveguide mode and evanescent waves amplification are numerically proved to contribute greatly to the patterning process in plasmonic lithography, with considerably improve the aspect ratio. Furthermore, we expect that the configuration of HPW structure will be helpful in extending the practical application of the plasmonic lithography.

ACKNOWLEDGEMENTS

This work is supported by University of Chinese Academy of Sciences (Grant# 118900M032)

REFERENCES

[1] T. Ito and S. Okazaki. *Nature,* vol. 406, 2000, pp. 1027–1031.

[2] R. Garcia, A. W. Knoll, and E. Riedo. *Nat. Nanotechnol.,* vol. 9, 2014, pp. 577–587.

[3] G. Tallents, E. Wagenaars, and G. Per. *Nat. Photonics,* vol. 4, 2010, pp. 809–811.

[4] W. Srituravanich, N. Fang, C. Sun, Q. Luo, and X. Zhang. *Nano Lett.,* vol. 4, 2004, pp.1085–1088.

[5] S. Kim, H. Jung, Y. Kim, J. Jang, and J. W. Hahn. *Adv. Mater.,* vol. 24, 2012, OP337–OP344.

[6] P. Gao, M. Pu, X. Ma, X. Li, Y. Guo, C. Wang, Z. Zhao, and X. Luo. *Nanoscale,* vol. 12, 2020, pp. 2415–2421.

[7] D. Han, C. Park, S. Oh, H. Jung, and J. W. Hahn. *Nanophotonics,* vol. 8, 2019, pp. 879–888.

[8] P. Gao, M. Pu, X. Ma, X. Li, Y. Guo, C. Wang, Z. Zhao, and X. Luo. *Nanoscale,* vol. 12, 2020, pp. 2415–2421.

[9] Y. Chen, J. Qin, J. Chen, L. Zhang, C. Ma, J. Chu, X. Xu, and L. Wang. *Nanotechnology,* vol. 28, 2017, pp. 055302.

[10] X. Chen, F. Yang, C. Zhang, J. Zhou, and L. J. Guo. *ACS Nano,* vol. 10, 2016, pp. 4039–4045.

[11] C. Wang, P. Gao, Z. Zhao, N. Yao, Y. Wang, L. Liu, K. Liu, and X. Luo. *Opt. Express,* vol. 21, 2013, pp. 20683–20691.

CYCLEGAN-BASED MASK DIFFRACTION MODEL

Jiaxiang Zhuo, Dongyong Xu and Yijiang Shen*

[1]School of Automation, Guangdong University of Technology

[2] Mega Education Center South, Guangzhou, 510006, P. R. China

*(86) 18665399614, yjshen@gdut.edu.cn

ABSTRACT

Modeling of photomask nearfield in extreme ultra-violet (EUV) lithography is one of the fundamental tasks for process modeling and physical verification. Rigorous electromagnetic field (EMF) simulation tools are too slow for large area applications, necessitating approximations for computational lithography applications. We duly propose, in this paper, a fast learning-based mask diffraction model with cycle generative adversarial network (CycleGAN) emphasis for EUV mask nearfield simulation. The network parameters are trained in a supervised manner. The proposed CycleGAN network learns the forward and inverse mapping between the photomask and the nearfield in a supervised manner by incorporating cycle consistency loss combinatorically with generative loss functions in both ways. We do not discriminate the learning of the real part and imaginary part of the nearfield by training them simultaneously. Investigation of the weighting the forward and inverse generative loss is also conducted. Simulation results merit fast and accurate inference from photomask to the nearfield and vice versa. It is also shown that different weights, in general, do not undermine the overall performance.

Keywords: near-field; deep learning; Cycle generative adversarial network (CycleGAN); photomask; rigorous electromagnetic field (EMF) simulation

INTRODUCTION
Paper Length

Computational lithography is a core technology in very large-scale integration (VLSI) [1] manufacturing, and its continuous improvement has extended the lifecycle of Moore's Law [2]. However, as semi-conductor devices continue to shrink in size, diffraction effects become more pronounced, particularly in the extreme ultra-violet (EUV) range, rendering the thin mask model (TMM) defined by the Hopkins assumption inadequate. As a result, mask near-field computation has become increasingly important in both forward and inverse lithographic imaging.

Rigorous electromagnetic field (EMF) solvers for Maxwell equations, such as the finite difference time domain (FDTD) [3] method and rigorous coupled wave analysis (RCWA) [4], are not practical for full-chip simulation due to their computational complexity. To address these issues, the rapid development of machine learning has encouraged practitioners to explore design-for-manufacturing (DFM) applications.

For instance, some machine learning techniques use fully convolutional networks (FCN) [5] or generative adversarial networks (GAN) [6] to achieve fast near-field calculation. These networks enable end-to-end and pixel-to-pixel data transformation, but they only achieve the forward mapping from the photo-mask to the near-field and learn the real and imaginary parts separately through training different networks. Consequently, such generative networks are not suitable for reverse mapping.

In this paper, we take a new approach to address mask near-field computation inspired by the recent development of the cycle consistent adversarial network (CycleGAN) [7]. Drawing inspiration from machine translation dual learning [8], CycleGAN employs reciprocal translators, denoted as G and F, which are both bijections and inverses of each other. To train G and F simultaneously, a cycle consistent loss is added to encourage $G(F(x)) \approx x$ and $F(G(y)) \approx y$, where x and y represent the photo-mask and the near-field, respectively. The CycleGAN network is capable of achieving the forward and inverse mapping between the photo-mask and the near-field. Moreover, the network can generate multiple mask near-fields simultaneously, which improves computation speed and accuracy.

This paper is structured as follows. First, we introduce the CycleGAN framework used to calculate mappings between the photomask and the nearfield. Then, we present the simulation results that verify the merits of the proposed CycleGAN method and compare it with the FCN method [9]. Lastly, some conclusion remarks are provided.

NETWORK FRAMEWORK AND DETAIL
Data Preparation

The mask pattern data includes rectangle arrays and contact arrays, which have corresponding near-fields that are rigorously calculated using waveguide-based electromagnetic field simulator in the Sentaurus TM Lithography software. The rectangle arrays in the Y coordinate have a CD of 14nm, while the ones in the X coordinate have CD values of 14nm and 28nm. The pitches of the rectangle arrays fall in the range of 40nm to 100nm. Fig. 1 provides examples of rectangle arrays. A numerical aperture (NA) of 0.33 is used for the lithography system, with coherent illumination for simplicity. The thick-mask is represented by a pixel-based binary matrix, where absorber regions

979-8-3503-1101-3/23 $31.00 © 2023 IEEE

and non-absorber regions are re- presented by pixels of 1 and 0, respectively.

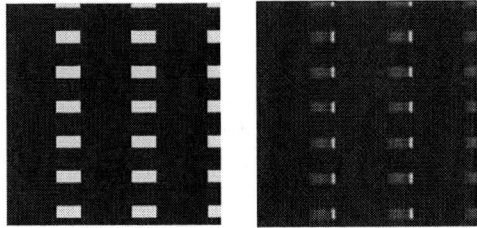

Figure 1: An example of the input mask pattern for training and the associated near field calculated by EMF method.

DNF Modeling

The DNF (diffraction near-field) [10] of a thick mask is calculated as follows:

$$E_{DNF} = E_{UV} \times E_{IN} = \begin{bmatrix} E_{XX} & E_{XY} \\ E_{YX} & E_{YY} \end{bmatrix} \times E_{IN}, \quad (1)$$

where E_{IN} is the incident electric field to the mask, and E_{UV} is the complex-valued diffraction matrices where U = X or Y and V = X or Y represent the polarizations as

$$E_{UV} = E_{(Real)UV} + iE_{(Im\ ag)UV}, \quad (2)$$

with $E_{(Real)UV}$ and $E_{(Imag)UV}$ are the real and imagery parts of the complex-valued near-field.

Network Framework

CycleGAN [7] is composed of a pair of generators and a pair of discriminators. The forward mapping from the photo-mask to the near-field by Generator A2B is shown in Fig. 2, while the inverse mapping from the near-field to the photo-mask by Generator B2A is shown in Fig. 3. By training the two generative networks, we can achieve mutual mapping between the photo-mask and the near-field.

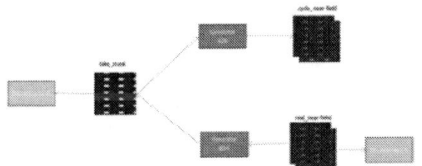

Figure 2: the forward mapping from the photo-mask to the near-field by Generator A2B.

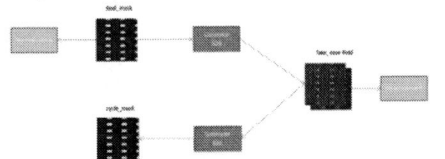

Figure 3: the inverse mapping from the near-field to the photo-mask by Generator B2A.

CycleGAN's generator typically comprises down-sampling and up-sampling layers, along with

Residual blocks. The former extracts and interprets input data, while the latter recovers pixel-wise information. In near-field calculation, the down-sampling path captures mask pattern geometric features, with the up-sampling path recovering pixel-based diffraction matrices. Residual blocks convert input/output characteristics, with depth adjusted by altering block count based on conversion complexity.

The output channel of the last network layer in the forward generator is 8 with tanh function, while the backward generator uses the softmax function and has an output channel of only 2. Such difference is induced by a backward mapping as a classifier and the forward mapping as a regression network. The discriminator is comprised of 4 convolution layers whose stride is set 2 for pooling effect.

Both mapping functions have same adversarial losses. For the mapping function G: X(mask) → Y(near field) and its discriminator D_{NF}, we give the objective as

$$L_{GAN}(G, D_Y, X, Y) = E_{y \sim p_{data}(y)}\left[log\,D_y\,(y)\right] + E_{x \sim p_{data}(x)}\left[log\,(1 - D_y(G(x)))\right], \quad (3)$$

in which G tries to generate near-field G(x) that look similar to near-field from domain Y by minimizing this objective against an adversary D_y that tries to maximize it, i.e., $\min_G \max_{D_Y} L_{GAN}(G, D_Y, X, Y)$. An adversarial loss for the inverse mapping F: Y(near field) → X(mask) and its related discriminator D_X is designed similarly as $\min_F \max_{D_X} L_{GAN}(F, D_X, Y, X)$.

Relying solely on adversarial losses cannot ensure an accurate learning from an individual input mask to a desired output near-field. To further constrain the space of potential mapping functions, we define the mapping to be cycle-consistent: for each mask in the domain X, the image translation cycle should be capable of returning the mask back to its original form or x → G(x) → F(G(x)) ≈ x. Similarly, for each near-field from domain Y, G and F should also satisfy backward cycle consistency y → G(y) → F(G(y)) ≈ y. A cyclic consistency loss is given as:

$$L_{cyc}(G, F) = E_{x \sim p_{data}(x)}[\|F(G(x)) - x\|_1] + E_{y \sim p_{data}(y)}[\|G(F(y)) - y\|_1]. \quad (4)$$

By incorporating Eq. (4) into the loss function

$$L(G, F, D_x, D_y) = L_{GAN}(G, D_y, X, Y) + L_{GAN}(F, D_x, Y, X) + \lambda L_{cyc}(G, F), \quad (5)$$

where λ controls the relative importance of the two objectives. We finally aim to solve:

$$G^*, F^* = arg \min_{G,F} \max_{D_x,D_y} L(G, F, D_X, D_Y). \quad (6)$$

Training Details

The implementation of the networks is performed using PyTorch on a Windows PC equipped with an Intel(R) Core(TM) i7-6850K CPU and two NVIDIA GeForce

GTX 1080 Ti (11G) GPUs. The batch size is set to 2, and the optimizer uses Adam [11], with the learning rate initially set to 0.0002 and then decreases as the number of epochs increases. We also normalize the real and imaginary parts to a range of $[-1,1]$.

EXPERIMENTAL RESULT

The metric to measure the accuracy are different for the forward and backward networks. For the forward network, we use the mean square error (MSE) loss proposed in Lin's paper as the validation criterion. However, for the backward network, as it is a segmentation network, we use the common validation function mean intersection over union (MIOU).

Fig. 4 illustrates the convergence curves of the loss for both the generator and discriminator in cycleGAN. The curves show fast convergence, and the loss values of both the generator and discriminator reach convergence after a few iterations.

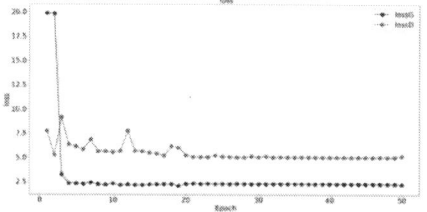

Figure 4: loss of generator and discriminator in 50 epoch.

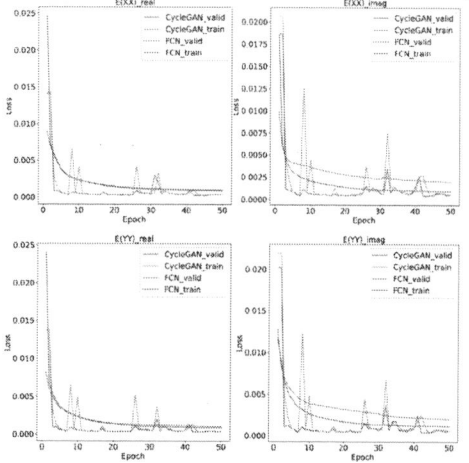

Figure 5: Convergence of near-field generation for cycleGAN and FCN.

Fig. 5 shows the convergence curves of the mean square error (MSE) [12] loss for the forward generator G_AB of the CycleGAN model and FCN during 50 epochs of training. It can be seen that CycleGAN produces some oscillations due to the adversarial nature of GANs, but the trained model achieves much better results than FCN.

Fig. 6 and Fig. 7 show the generated results of the

near-field for contact arrays and rectangle arrays mask respectively. It can be seen that CycleGAN produces better results especially in terms of contour features and details. In CycleGAN, the generated masks are cyclically regenerated to minimize the adversarial consistency loss. With the contour as the strongest mask features, the loss function is minimized by backpropagation for better forward capability to focus on the contour features in the mask vicinity. Additionally, the superior performance in capturing details by coupling the real and imaginary fields with polarizations, allows for information learning in between with more prominent results.

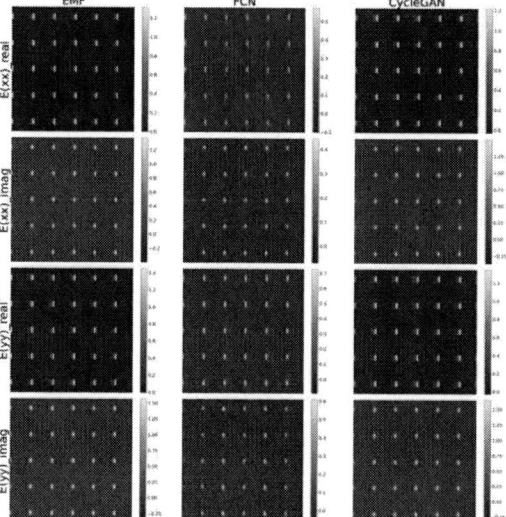

Figure 6: Near-field generation for contact arrays with cycleGAN and FCN.

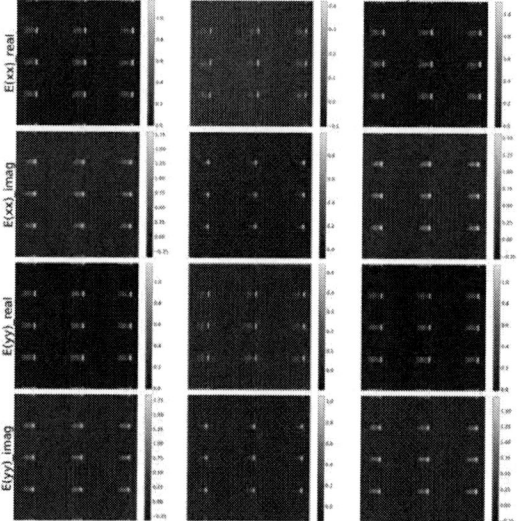

Figure 7: near-field generation for rectangle arrays with cycleGAN and FCN.

Fig. 8 shows the validation curves of mean intersection over union (MIOU) for generating masks from the near-field using both FCN and CycleGAN. It can also be seen that the trained model of CycleGAN achieves slightly better results than FCN.

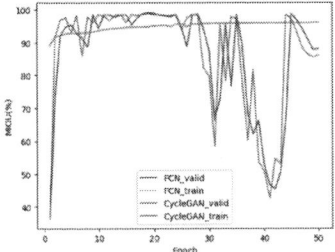

Figure 8: MIOU of mask generation for cycleGAN and FCN.

Fig. 9 shows the generated masks from near-field masks using both FCN and CycleGAN. It can be observed that CycleGAN produces masks with slightly better detail compared to FCN.

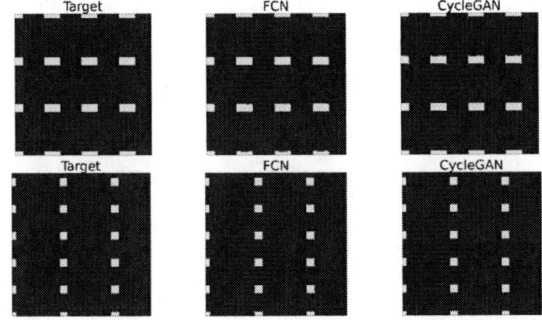

Figure 9: Mask generation with cycleGAN and FCN.

It is duly noted from the quantitative comparison in Table 1 and Table 2 that the proposed CycleGAN based

TABLE I.   MSE LOSS VALUE COMPARSION

Method	MSE Loss Value			
	$E_{\mathrm{real}}(XX)$	$E_{imag}(XX)$	$E_{\mathrm{real}}(YY)$	$E_{imag}(YY)$
CycleGAN	2.75E-4	3.93E-4	3.75E-4	4.93E-4
FCN	7.36E-3	8.97E-3	6.42E-3	8.17E-3

TABLE II.   MIOU COMPARSION

Method	MIOU
	Near-Field to Mask
CycleGAN	99.15%
FCN	97.13%

TABLE III.   RUNTIME COMPARSION

Method	Run Time (s)	
	Mask to Near-Field	Near-Field to Mask
CycleGAN	0.00917	0.00987
FCN	0.01102	0.00329

network outperforms the FCN model [9] in terms of both MSE loss and MIOU. The runtime comparison in Table 3

shows that the computation of mask near-field is faster than the FCN model, only on the forward mapping.

CONCLUSION

In this article, we propose a fast mask near-field calculation method based on the cycleGAN model. The generation performance and computational speed of the cycleGAN model outperform four separate FCN networks because the cycle-consistent loss and multiple near-fields are coupled with more information about the correlation between masks and near-field. Additionally, the cycleGAN's reverse network can decouple the original ILT from spatial images to masks, making it more versatile.

ACKNOWLEDGEMENTS

This paper is partially supported by Natural Science Foundation of China (62174037), Natural Science Foundation of Guangdong Province, China (2020A1515010633, 2021A1515012000). The authors would also like to thank the support from the Institute of Microelectronics, the Chinese Academy of Sciences for providing rigorously computed mask near-fields.

REFERENCES

[1] W. Oldham, S. Nandgaonkar, A. Neureuther, et al. IEEE Transactions on Electron Devices, vol. 26, 1979, pp. 717-722.

[2] R. Schaller. IEEE Spectrum, vol. 34, 1997, pp. 52-59.

[3] S. Burger, R. Köhle, L. Zschiedrich, W. Gao, F. Schimdt, R. März and C. Nölscher. 25th annual BACUS Symposium on photomask technology, Monterey, 2005, pp. 599216.

[4] P. Schiavone, G. Granet and J. Robic. Microelectronic Eng., vol. 57, 2001, pp. 497-503.

[5] J. Lin, L. Dong, T. Fan, X. Ma, R. Chen and Y. Wei. IEEE International Workshop on Advanced Patterning Solutions, 2020, pp. 1-4.

[6] I. Goodfellow, J. Pouget-Abadie, M. Mirza, B. Xu, D. Warde-Farley, S. Ozari, A. Courville and Y. Bengio. Proceedings of NIPS, 2014, pp. 2672-2080.

[7] JY. Zhu, T. Park, P. Isola and A. Efros. Proceedings of the IEEE I. Conf. Comp. Vis., 2017, pp. 2223-2232.

[8] D. He, Y. Xia, T. Qin, L. Wang, N. Yu, T. Liu and W. Ma. Proceedings of NIPS, 2016, pp. 820-828.

[9] J. Lin, L. Dong, T. Fan, X. Ma, R. Chen and Y. Wei.IEEE International Workshop on Advanced Patterning Solutions, 2020, pp. 1-4.

[10] Y. Chengzhen, M. Xu and Z. Junbi. International Workshop on Advanced Patterning Solutions, 2022, pp. 1-4.

[11] D. Kingma and J. Ba. arXiv e-prints, 2014, pp. arXiv:1412.6980.

[12] R. Fisher. JR Statistical Society, vol. 85.

ILLUMINATION OPTIMIZATION FOR THE BEOL DTCO WITH 45 DEGREE LOCAL INTERCONNECTION

Xianhe Liu[12]*, Muzi Han[1], Yanli Li[12], Qi Wang[12], Qiang Wu[12]

[1]School of Microelectronics, Fudan University, No. 825 Zhangheng Road, Pudong New Area, Shanghai 201203, China
[2]National Integrated Circuit Innovation Center, Pudong New Area, Shanghai 201203, China
*Corresponding Author's Email: xianheliu@fudan.edu.cn

ABSTRACT

In previous work we have shown a few typical patterns containing 45-degree local interconnection in the BEOL layers under 7 nm logic design rules. Our simulation result has demonstrated that the 45-degree design can save mask area up to 20%. As illumination source optimization has been widely applied in lithography in advanced technology nodes for improving process window, in this work, we will report a study on the subject that, with source optimization, how the SO may affect the process window in the patterns containing the 45-degree design. Some typical patterns using conventional design rules will be presented first, then the 45-degree interconnection design will be fully investigated under illumination source optimization.

Keywords—source illumination optimization, process window, 45-degree interconnection, minimum area

INTRODUCTION

In advanced logical nodes for integrated circuit manufacturing, with the chip feature size continuously scaling down, maintaining the process window, ie. EL (Exposure Latitude), MEF (Mask Error Factor) and DoF (Depth of Focus) becomes more and more challenging. Source mask co-optimization (SMO) technique has been proven a practical method to improve process window and image fidelity by a synergy effect from jointly optimizing mask and source.[1-4] With optimized pixilated free-form illumination, a better process window may be achieved compared to the conventional sources such as annular, dipole and C-Quad.[5] In SMO method thousands of combinations of sources and masks for a designed pattern will be evaluated, we input the main factor CD (Critical Dimension), EL (Exposure Latitude), MEF (Mask Error Factor) different weightings in the merit cost function, so as to give a considerable optimization to the utmost.[6, 7]

SMO has been widely applied in lithography nowadays in favor of improving the process window in advanced nodes.[8, 9] Various typical mask patterns, such as dense or isolated lines or contacts, and more complicated 2D patterns with corner rounding, has been analyzed under SMO which showed an overall ~8-10% process window gain in the average.[10] In our previous work, we have presented some Z-like 2D patterns containing 45-degree local interconnection in the BEOL

layers under 7 nm logic design rules [11], which brings to a 3-20% mask area saving after careful mask optimization (MO). Based on which, we conducted a study in this paper by applying SO on the Z-like patterns with 45-degree connecting lines, in order to analyze how the SO may affect the process window in these untraditional designed patterns, compared with the conventional 2D typical patterns.

RESULT AND DISCUTION

The typical patterns for 28 nm node were reported in Wu's work[10]. For the 90 nm pitch Line/Space (L/S) dense trench, a dipole SO'ed illumination source in the X direction was occurred, whereas a C-Quad source for Tip-to-Tip (TtT) pattern. In the first case the EL values are significantly increased, while in the latter the process window gain is less evident.

In this study, the effect of SMO for various 2D typical features were analyzed near the extreme 193 nm immersion lithography condition, with a pitch of 80 nm and targeting Critical Dimension (CD) of 40 nm. With the aid of multiple patterning SALELE (Self-Aligned Litho-Etch Litho-Etch), 20 nm CD metal connecting lines can be obtained for the 7 nm logic node chip manufacturing. The Aerial simulator CF-Litho was used firstly for the mask pre-optimization using a 110° dipole illumination source, followed by the global source optimization in a self-developed program. We will discuss in the first step the L/S and TtT typical patterns. Next the Z-like features with one or three 45-degree connecting lines were investigated using the SO program.

SO for L/S and TtT typical 2D patterns

The Source Optimization (SO) result for L/S pattern (80 nm pitch) was shown in the Figure 1. During the mask pre-optimization, the mask CD was set to 37 nm for fabricating 40 nm target CD. Negative Tone Development (NTD) was applied without taking in account of Photo Decomposable Base (PDB), the appropriated EL was obtained ~10%. After SO, the optimized source remained as dipole, yet the angle became narrower for enhancing the X-direction pattern. The EL values were considerably increase to ~19%, meanwhile the MEF value was decreased by 50%.

However, for the TtT pattern, the SO effect was less

significant. Compared with the initial source dipole 110° in Y-direction, serval pixelate dots were added in the X-direction and near the center change in the SO'ed source, which helps to enhance the tip-to-tip spacing as shown in the figure 2, the EL for the linecut 2 was increased a bit to ~10%, with a smaller CD ~ 76 nm and a smaller MEF<5. In general, the SMO has shown an apparent benefit for the typical rectilinear 2D patterns in 7 nm node. In the next section, the result of Z-like features with 45-degree connecting lines will be presented.

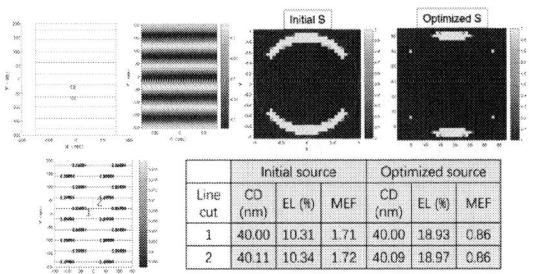

Figure 1: Simulated Line/Space(L/S) 2D patterns, contour and aerial image. CD and EL values under initial (dipole 110°) and optimized illumination sources were listed in the table.

Line cut	Initial source			Optimized source		
	CD (nm)	EL (%)	MEF	CD (nm)	EL (%)	MEF
1	40.00	10.31	1.71	40.00	18.93	0.86
2	40.11	10.34	1.72	40.09	18.97	0.86

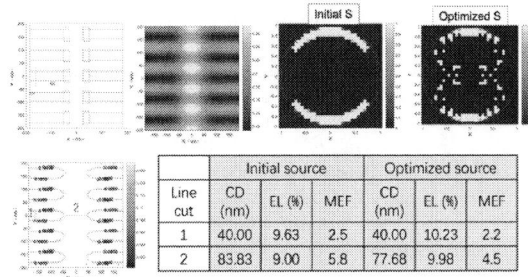

Figure 2: Simulated Tip-to-Tip (TtT) 2D patterns, contour and aerial image. CD and EL values under initial (dipole 110°) and optimized illumination sources were listed in the table.

Line cut	Initial source			Optimized source		
	CD (nm)	EL (%)	MEF	CD (nm)	EL (%)	MEF
1	40.00	9.63	2.5	40.00	10.23	2.2
2	83.83	9.00	5.8	77.68	9.98	4.5

SO for Z-like feature with 45-degree connection

In our previous work [11], the slanted 45-degree connection has been shown to make a great advantage for mask area saving in the 7 nm advanced logical node. The mask area can reduce to 3%-20% using the 45-degree connection, depending on the number of the Z-like turns.

Firstly, we investigated the pattern with one Z-like turn. As illustrated in the Figure 3, the 45-degree connecting line is created via a succession of 3 aligning rectangles. As the linewidth correlating to the electrical current flow should be perpendicular to the contour line, the CD and EL value for the slanted connecting line need to be remeasured with an appropriated slanted linecut instead of the horizontal one, as shown in the Figure 3 (linecut 2). The accurate CD and EL in the position linecut

2 is ~ 87 nm and 15%, while in the linecut 3 the EL is ~8%, hardly reaching the minimal process window (> 9% for all linecuts). The optimized result using SO was presented in the Figure 4. Less pronounced change occurred in the optimized illumination source compared with the initial one. The connecting line in the linecut 2 shrank a bit to 80 nm CD with the EL decrease a bit to ~13%, meanwhile in the linecut 3 EL was improved to ~10%, satisfying the minimal process window. Besides the MEF for linecuts 1-3 were improved where the optimized values are all≤ 10.

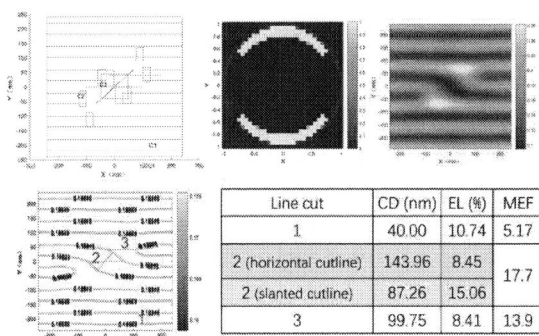

Figure 3: Simulated 2D patterns with one Z-like shape, contour and aerial image. CD and EL values under illumination sources (dipole 110°) were listed in the table. The red line refers to the perpendicular line cut, which brings to an accurate measurement of CD and EL in the position linecut 2.

Line cut	CD (nm)	EL (%)	MEF
1	40.00	10.74	5.17
2 (horizontal cutline)	143.96	8.45	17.7
2 (slanted cutline)	87.26	15.06	
3	99.75	8.41	13.9

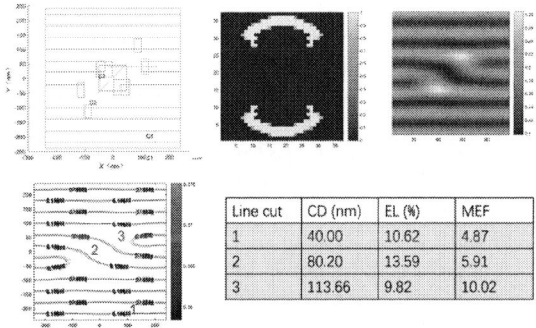

Figure 4: Simulated 2D patterns with one Z-like shape, contour and aerial image. CD and EL values under the optimized illumination source were listed in the table.

Line cut	CD (nm)	EL (%)	MEF
1	40.00	10.62	4.87
2	80.20	13.59	5.91
3	113.66	9.82	10.02

Finally, we investigated the 45-degree pattern with three dense Z-line turns. In our previous work, this type of design may help to save ~ 20% mask area. As shown in the figure 5, under the initial illumination source of 110°, the 40 nm dense lines (linecut 1) possessed a limited EL value ~10%; meanwhile for the slanted connecting lines/spacings (linecuts 2-4) and the line-end spacing (linecut 5), the EL values were ~9-11%. Under the SO'ed source, the EL in linecut 1 showed a remarkable increase to 12%, while the other linecuts maintain the acceptable EL value ~9-11%. The MEF for all linecuts were slightly decreased as well. Therefore, the SMO is an important technique which may gain process window to varying degrees for both rectilinear and 45-degree connecting Z-like patterns. However, it is worth noticing that for these Z-like patterns the mask optimization (MO) effect is predominant, as it brings to a considerable mask area saving, while the SO makes rather slight gain for the process window.

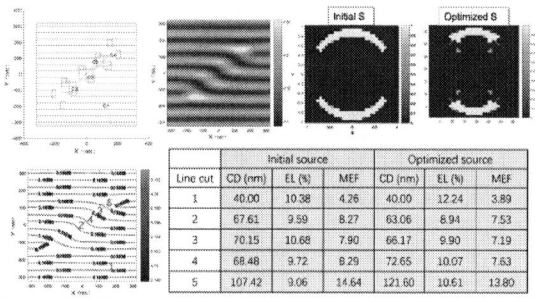

Figure 5: Simulated 2D patterns with three Z-like shapes, contour and aerial image. CD and EL values under initial (dipole 110°) and optimized illumination sources were listed in the table.

CONCLUSION

In this paper we discuss the effect of the SMO for different 2D patterns in 7 nm advance logic node. Simulation for dense Line/Space and Tip-to-Tip patterns, as well as Z-like features with 45-degree connection were analyzed under 193 nm immersion resolution limit condition (80 nm pitch, 40 nm CD) using NTD to meet enough process window (e.g., EL ~9-11% in all line cuts). The Source optimization (SO) result shows that after optimization the SO'ed free-form sources basically are closed to the dipole source in Y-direction, with some small change depending on the targeting patterns. The rectilinear (L/S or TtT) and slanted (Z-like) features can all be improved under SO. Dense L/S pattern shows the most significant improvement in EL, while for the other patterns the process window gain may be less evident (~ 1%). In particular, for the dense Z-like slanted connecting lines, the mask optimization (MO) produces a more important effect on mask area saving than the SO in process window gain.

ACKNOWLEDGEMENTS

The authors of this paper thank School of Microelectronics Fudan University and National Integrated Circuit Innovation Center for the support of this work.

REFERENCES

[1] E. R. Alan et al., "Optimum mask and source patterns to print a given shape," in Proc.SPIE, 2001, vol. 4346, pp. 486-502, doi: 10.1117/12.435748. [Online]. Available: https://doi.org/10.1117/12.435748

[2] T. Kehan et al., "Benefits and trade-offs of global source optimization in optical lithography," in Proc.SPIE, 2009, vol. 7274, p. 72740C, doi: 10.1117/12.814305. [Online]. Available: https://doi.org/10.1117/12.814305

[3] D. Yunfei et al., "Considerations in source-mask optimization for logic applications," in Proc.SPIE, 2010, vol. 7640, p. 76401J, doi: 10.1117/12.848865. [Online]. Available: https://doi.org/10.1117/12.848865

[4] M. David et al., "Demonstrating the benefits of source-mask optimization and enabling technologies through experiment and simulations," in Proc.SPIE, 2010, vol. 7640, p. 764006, doi: 10.1117/12.846716. [Online]. Available: https://doi.org/10.1117/12.846716

[5] Z. Jörg et al., "Generation of arbitrary freeform source shapes using advanced illumination systems in high-NA immersion scanners," in Proc.SPIE, 2010, vol. 7640, p. 764005, doi: 10.1117/12.847282. [Online]. Available: https://doi.org/10.1117/12.847282

[6] Y. Kazuyuki, N. Seiji, T. Kazuhiro, and U. Takayuki, "Challenges for low-k1 lithography in logic devices by source mask co-optimization," in Proc.SPIE, 2010, vol. 7640, p. 76401K, doi: 10.1117/12.846263. [Online]. Available: https://doi.org/10.1117/12.846263

[7] M. Thomas et al., "Simultaneous source-mask optimization: a numerical combining method," in Proc.SPIE, 2010, vol. 7823, p. 78233X, doi: 10.1117/12.865965. [Online]. Available: https://doi.org/10.1117/12.865965

[8] Z. Lulu et al., "Lithographic exposure latitude aware source and mask optimization," in Proc.SPIE, 2021, vol. 12073, p. 1207302, doi: 10.1117/12.2604844. [Online]. Available: https://doi.org/10.1117/12.2604844

[9] G. Xuejia, L. Yanqiu, D. Lisong, and L. Lihui, "Co-optimization of the mask, process, and lithography-tool parameters to extend the process window," Journal of Micro/Nanolithography, MEMS, and MOEMS, vol. 13, no. 1, p. 013015, 3/1 2014, doi: 10.1117/1.JMM.13.1.013015.

[10] Q. Wu, Y. Li, X. Liu, X. Zhu, S. Yu, and W. Zhang, "Considerations in the Setting up of Industry Standards for Photolithography Process, Historical Perspectives, Methodologies, and Outlook," in 2022 China Semiconductor Technology International Conference (CSTIC), 20-21 June 2022 2022, pp. 1-7, doi: 10.1109/CSTIC55103.2022.9856921.

[11] X. Liu, Y. Li, and Q. Wu, "A Study of Improved Design Rules Through Allowing 45-Degree Metal Lines," in 2022 China Semiconductor Technology International Conference (CSTIC), 20-21 June 2022 2022, pp. 1-4, doi: 10.1109/CSTIC55103.2022.9856712.

PROCESS AND TOOL MONITOR AND DIAGNOSIS BASED ON OVERLAY DATA AND MODELING

Yi Tong[1], Libin Zhang[1,2,3,4,*], Yayi Wei[1,2,3,4,**], Tianchun Ye[1,2,3] and Yun Wang[1,2]

[1] Guangdong Greater Bay Area Institute of Integrated Circuit and System, Guangzhou 510535, China
[2] Institute of Microelectronics of Chinese Academy of Sciences (IMECAS), Beijing 100029, China
[3] School of Integrated Circuits, University of Chinese Academy of Sciences, Beijing 100049, China
[4] Nanjing Chengxin Institute of IC Technology, Nanjing 211899 China
*Corresponding Author's Email: zhanglibin@ime.ac.cn
**Corresponding Author's Email: weiyayi@ime.ac.cn

ABSTRACT

The overlay plays an important role in chip manufacturing for all technology nodes, but controlling it becomes increasingly difficult when the process node is less than 7nm or the number of 3D layers exceeds 100. The primary challenge in overlay control arises from tool and process variation, which can have a significant impact on chip performance and yield. This paper presents a systematic review and analysis based on overlay data and modeling to monitor process and tool variations by exploring and discussing four critical issues related to overlay control: the overlay mark selection method, over-fitting diagnosis method, scanner tool monitoring method, and process grouping method. The selected methods and models can be applied in various areas, including process development, tool monitoring, feedback model selection, and lot grouping.

INTRODUCTION

During advanced lithography patterning development, overlay is one of the most three key targets – resolution, overlay, and throughput -- associated with the high-volume manufacturing. The resolution could be improved using a shorter wavelength, larger NA and smaller k1, or the computational lithography technologies and multi-patterning technologies, according to the R=k1*λ/NA. However, it is hard to shrink the overlay dramatically as it associates with all tools and processes. Five main reasons -- tools, metrology, reticle, process and feedback control model -- are associated with the overlay performance. A lot of effects are proposed to improve the overlay performance through increasing the self-diagnosis and stability performance of scanner and other tools. However, the process variation is now the main overlay onset and should be carefully studied and controlled. Some published papers have shown the difficulty of overlay feedback during process variation, in both logic process control and 3D NAND overlay after etching processes [1-5].

An overlay budget tree is shown schematically in Fig. 1. Although five different budgets are separate lists, they are closely interconnected each other. For example, the heating of reticle occurs during the lithography process and is limited by applying a scanner controlling model. The primary metrology issue occurs with process-induced overlay mark damage. And the over-fitting happens when the number and placement of overlay marks or fields are selected incorrectly.

Detailed description of overlay budget is quite complicated. So, we focus on the following questions related to the metrology, process and feedback, and will answer four questions. (1) How do the engineers choose overlay mark to balance the advanced process requirements and measurement burden? During metrology, engineers want to measure the overlay mark as fewer as possible. However, the feedback model accuracy needs to enlarge the number as much as possible. (2) How to evaluate the selected model's accuracy? Over-fitting should be avoided and the process-robustness should be as larger as possible. (3) How to analyze the process or tool variation in the R2R feedback system? (4) What is the recommended tool matching model for the overlay controlling.

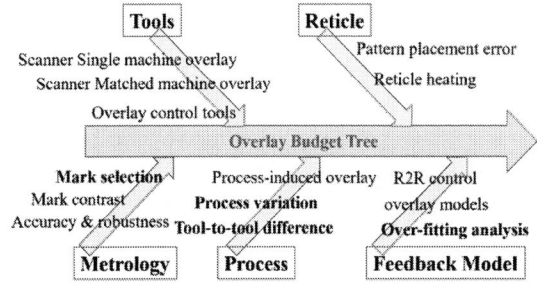

Figure 1: Overlay budget tree schematic

In this paper, we will give a systematic analysis of the overlay origin and control strategies. Four mathematical analysis methods or models are given to answer the above four questions. And we hope some of the views and models could be applied in the advanced manufacturing to improve the yield controlling and problem diagnostic.

METHODS
Overlay mark and model selection method

The arrangement of overlay mark is essential for metrology and feedback in industry, both inter-field and intra-field are used to describe the overlay performance across wafer and within every exposure field. For HVM, it should balance the number of overlay mark and the feedback precision. That is, it is desirable to carefully design mark's location and minimize the numbers to match the process requirements. Besides, for a fixed mark number, as is shown in Fig. 2, one could have a better overlay performance by choosing the proper feedback parameters. Detailed selection method could be found in the patent [6]

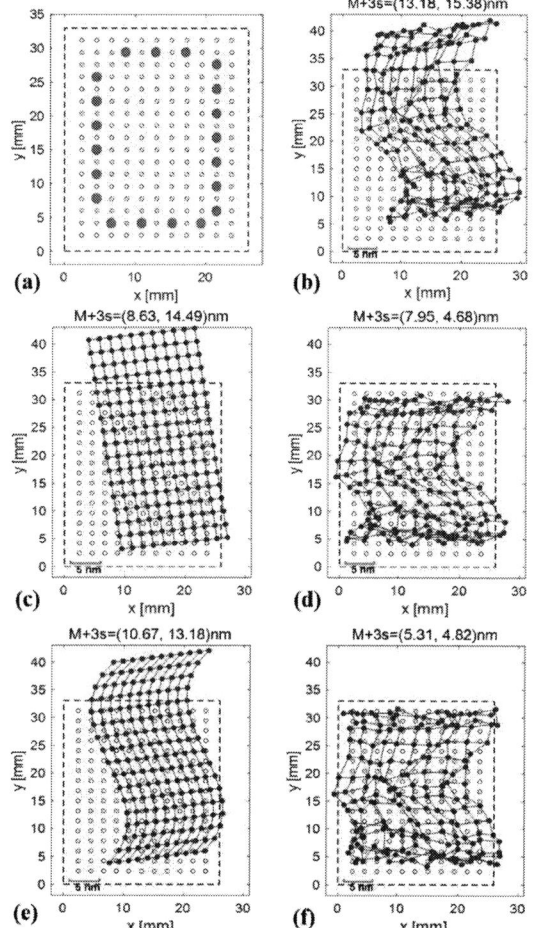

Figure 2: Comparison of 6 linear feedback model and selected best feedback model by fixing 6 parameters. (a) initial overlay data and selected marks. (b) initial overlay map. (c) fitting map of linear model. (d) residual map of linear model. (e) fitting map of selected best 6 parameter model. (f) residual map of selected best 6 parameter model.

As an example to describe the method for choosing a better feedback model, an intra-field overlay raw data are used. Reference method is linear method with 6 parameters. Overlay residual after fitting is about (7.95, 4.68) nm in X and Y direction, respectively. By comparison, applying the method we have proposed, a better overlay feedback model with 6 parameters and higher order parameters are selected, and the overlay residual is reduced more than 30% in X direction.

The result demonstrates that the common used method may be not the best model for overlay control. We recommend to give a systematic of the overlay performance and find the best fitting parameters. Thus, the mark number could not increase and the overlay performance could be maintained or improved.

Over-fitting diagnosis method

Over-fitting usually occurs when the number of overlay marks is less than the minimum requirement of the feedback model, or the distribution is severely uneven. For example, when some overlay marks are damaged during process or the fields are in the wafer edge. Therefore, the feedback model should be carefully checked and tuned to avoid the over-fitting.

To solve such problem, it is suggested to give a view on the fitting maps, both for interfiled and intra-field. One example is shown in Figure 3 with different fitting maps of inter-field and intra-field model, which show that the overlay fitting model has a larger fitting performance in edge positions [7].

During applying such model, the overlay mark selection in the field and across wafer should be according to the data selection methods, and Nyquist sampling is the most suggested sampling method for inter and intra field feedback.

Besides, some methods are used in the feedback model to avoid the overfitting. One is to reduce the fitting order when the number of overlay mark is less than the requirements. The other is to narrow the parameter fitting range as describe in ref. [7].

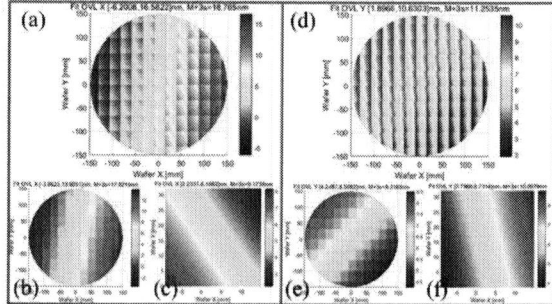

Figure 3: Overlay fitting maps in X and Y direction with inter-field and intra-field fitting respectively. (a) Inter and Intra field fitting map in X direction; (b) inter-field fitting map in X direction; (c) intra-field fitting map in X direction; (d)-(f) the fitting maps in Y direction.

Zernike-CPE method for tool monitoring

Process variation mostly comes from etching, CMP and annealing. For the advanced logic and memory manufacturing, it is difficult to control the overlay within the overlay scope without considering tool difference. CPE feedback model is normally used for better overlay control. However, it needs lots of overlay metrology data. So it is necessary to reduce the metrology data with CPE model applied when there is process variation or tool matching.

We have proposed a method by introducing Zernike model (Zernike-CPE model). That is, after establishing a basic CPE model, only several exposure fields are needed to be measured. Then Zernike model is applied to find the difference between the metrology lot to the reference one. After transferring the Zernike model to CPE model, one could get the updated CPE model considering the process variation. The detailed comparison could be found in the published paper [8] and a simple flow could be seen in Fig. 4. It also demonstrates that the overlay residual is similar to the CPE-only model for the Zernike order larger than 10, as seen in Fig. 5, which is suitable for the HVM application.

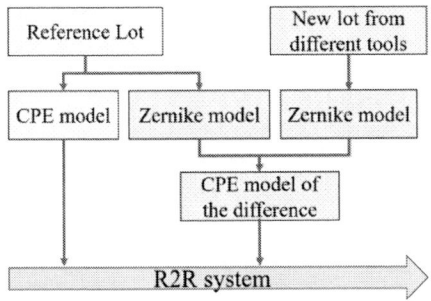

Figure 4: Proposed Zernike-CPE model for tool matching.

Tool grouping and monitoring method

The benefit for using Zernike-CPE method is that it could be used to monitor the process or tool variation and give a grouping analysis. If we know the tool difference, we could add such difference to the R2R system for controlling. However, it is important to check such tool difference is stable or time variation.

The steps for applying Zernike-CPE model for lot grouping is as follows:

Step1: calculate the Zernike parameters of the lots from different tools.

Step2: calculate the difference between the above lots to the reference one.

Step3: give a distance between different lots by calculating the Zernike fitting maps.

Step4: plot the cluster matrix to show the tool difference.

The cluster matrix or cluster tree could be used for the process or tool variation monitoring.

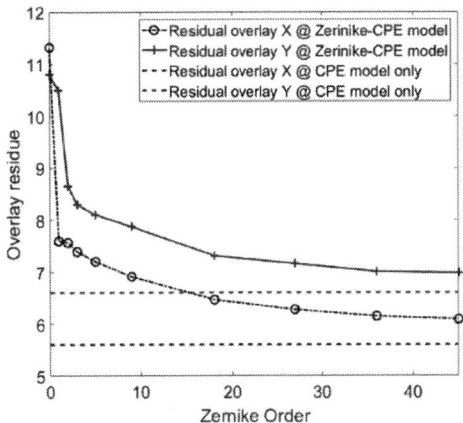

Figure 5: Overlay residual vs Zernike order. [8]

CONCLUSIONS

In this paper, we give an analysis on the overlay budget, propose and summarize four different methods to deal with the most difficult problems faced in advanced lithography. Different models and methods are proposed to deal with the mark selection, overfitting, process variation and tool monitor. We also build a software to integrate these models and methods to help the industry for the better overlay control during R&D.

ACKNOWLEDGEMENTS

This work is supported by Guangdong Province Research and Development Program in Key Fields (2021B0101280002), Guangzhou City Research and Development Program in Key Fields (No. 202103020001), University of Chinese Academy of Sciences (No. 118900M032) and China Fundamental Research Funds for the Central Universities (No. E2ET3801). We also thank the support of Key Laboratory of Microelectronics Devices & Integrated Technology of IMECAS.

REFERENCES

[1] J. Mulkens, M. Kubis, W. Tel, et al.. *Proc. SPIE*, vol. 10585, 2018, pp. 105851L.

[2] Y. Chen, S. Lin, K. Chen, C. Ke, and T. Gau, *Proc. SPIE*, vol. 8681, 2013, pp. 86812P.

[3] D. Boef, and J. Arie. *Surface Topography Metrology & Properties*, vol. 4(2) 2016, pp. 023001.

[4] X. Jian, Q. Long, Q. Chen, H. Zhi, Y. Wang, Z. Yang, and Z. Mao, *CSTIC*. 2017, pp. 7919769.

[5] Y. Feng, P. Xuan, D. Wu, et al., *Proc. SPIE*, vol. 11611, 2021 pp. 116110V.

[6] Y. Tong, Y. Wei, and L. Zhang, patent: 202310088349.4.

[7] E. Ma, L. Zhang, Y. Feng, L. Ma, S. Zhang, and Y. Wei, *J. Vac. Sci. Technol. B*, vol. 40(4), 2022, pp. 042601.

[8] L. Zhang, Y. Feng, Z. Song, S. Yang, and Y. Wei, *J. Vac. Sci. Technol. B*, vol. 40(6), 2022, pp. 062604.

A MULTI-STEP SRAF TO IMPROVE PROCESS WINDOWS IN METAL LAYER

Wei Wei*, Lu Zhu, Dan Wang, Xiaoyan Sun, Yue Wang, Yueyu Zhang, Jianzhong Liu, Shiri Yu

Shanghai Huali Integrated Circuit Corporation, Shanghai, China

*Corresponding Author's Email: weiwei@hlmc.cn

ABSTRACT

In the metal layer of the integrated circuit, the layout has dense lines with high density and isolated lines with low density. The T-opening is in an isolated position in the layout. However, dense lines have mismatched process windows with isolated lines in the metal layer, lowering the process window with T-openings and analogous structures. Sub-Resolution Assist Feature (SRAF) is often used to improve the contrast, depth of field, and process redundancy of the lithography process to improve the process window of the layout in integrated circuits. Due to the different density features between the upper and lower (longitudinal) distribution of the horizontal pattern in the T-shaped opening, the diffraction spectrum distributions around the horizontal main pattern generate high discrepancy. As a result, the diffraction spectrum difference will lead to the mismatch of the process windows and reduce the process window of the layout. Here, we introduce a multi-step SRAF addition method evenly to reduce the contrast of graphics with different sizes, then improve the difficulty of the photolithography process.

Keywords—SRAF; Multi-step addition; Diffraction Spectrum; Process Window

INTRODUCTION

The loss of DOF in dark-field layers such as via, contact, and metal poses a challenge for inverse lithography in the IC manufacturing process. In advanced nodes, the regular design rule increases the design complexity on contact and metal trenches patterning [1]. In consideration of the increasing design and manufactural difficulty, Sub-Resolution Assist Feature (SRAF) such as OPC/SRAF insertion approaches are applied to produce the desired wafer patterns. As yet, SRAF has become a primary method for advanced node manufacture. Metal layers of the back end of the line (BEOL) are the most critical part of the advanced node. The intricate layout has both high-density lines and low-density isolated lines in metal layers. Hence, dense and isolated line patterns in the layout have inconsistent process windows, increasing the difficulty to mask manufacture. In order to improve the process window of complex metal layer patterns, the SRAF technology is often used to improve the visibility of the lithography process graphics depth of field and the redundancy technology. Besides, it is necessary that the feasible tuning of the placement of SRAF will be of benefit to the manufacturing processes to acquire good control of the DOF, but also the CD (critical dimension) uniformity and PV-band (process variation band).

Due to mask shop limitations, the rule-based SRAF allows complete control of the OPC flow and is mask shop compliant. This approach can be tricky for critical pattern configurations. Previous research has shown that patterns presenting the "T" or "L" shapes have a high risk of SRAF printability [1]. The rule-based SRAF method could achieve acceptable accuracy for simple and regular target patterns effectively, but it should not apply to complex layouts [3]. In this study, due to the discrepancy of critical dimensions in complex metal layer patterns such as T-opening shape and multidimensional patterns, and the diffraction spectrum distribution difference of the patterns will lead to the mismatch of the process window, reducing the process window of the layout, and improving the difficulty of the photolithography process. The regular rule-based SRAF generation around the complex BEOL metal-layer layout will cause a high PV-band for the photolithography.

A multiple-step rule-based SRAF addition method is proposed to optimize metal layer pattern contracts. We demonstrate that some modifications of the SRAF insertion step, width and position bring an obvious DOF increment of above 33%, more precise ADI (After Developing Inspection) CD and lesser PV-band.

SRAF INSERTION

This study focuses on the tuning of SRAF placement between dense and no-dense metal layer patterns using a simple but effective rule-based SRAF method. As we know, for isolated patterns, there is sufficient space between the designed pattern to place SRAF and have the correct assist. However, in semi-dense and dense configurations, it is difficult concerning the insufficient space for SRAF insertion. Here, we introduce a multiple-step SRAF method to fill the inhomogeneous space rationally. As seen in Figure 1, two types of metal layer layouts are selected for SRAF configuration verification. The space between patterns for SRAF insertion is inhomogeneous. In Figure 1 (a), the targeted horizontal metal layer pattern is surrounded by four vertical patterns. The target CD of the horizontal pattern is 68 nm, and the CD of the surrounding vertical patterns is 136 nm. In Figure 2, the larger CD (126 nm) pattern located in the central position is surrounded by smaller CD (63 nm) patterns and the space between patterns has not been unevenly distributed throughout the layout. Hence the two typical kinds of layouts are selected for the SRAF configuration study.

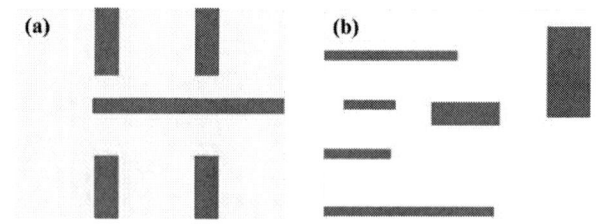

Figure 1. The (a) T-opening sharp pattern and (b) multidimensional pattern before OPC and SRAF insertion.

There are three critical parameters when SRAF generation: width, space of SRAF and the space between SRAF and main patterns. In the regular rule-based SRAF method, SRAF patterns were generated and placed around the main patterns at once following the empirical formula derived by the inverse lithography technique. Nevertheless, owing to the uneven distribution of patterns in the layout, the inserted SRAF cannot change the layout's spatial distribution in a continuous region, resulting subsequent lithograph process didn't achieve the desired effect.

In the three-step cycle SARF insertion rule, the first inserted SRAF patterns were treated as the main patterns, and the second SRAF patterns were placed around the former. Then, the first and second SARF patterns were treated as the main to generate the new third SRAF. After SRAF insertion, we can get four mask patterns from Figure 1, which are input for the OPC process. The treated patterns will be as target layers. Then, the OPC process is handled using Calibre nmOPC and output layouts as shown in Figure 2. Figure 2 shows the one-step SARF insertion of (a) T-opening and (c) multidimensional post-OPC patterns. The SARF insertion rule determined that the SARF pattern must be generated around the objective pattern with a fixed size. In this SRAF rule, only when the objective pattern CD is smaller than 80 nm did the SRAF appear. As seen in Figures 2(c), and (d), the patterns in layout became more well distributed, which is useful to improve the lithograph process.

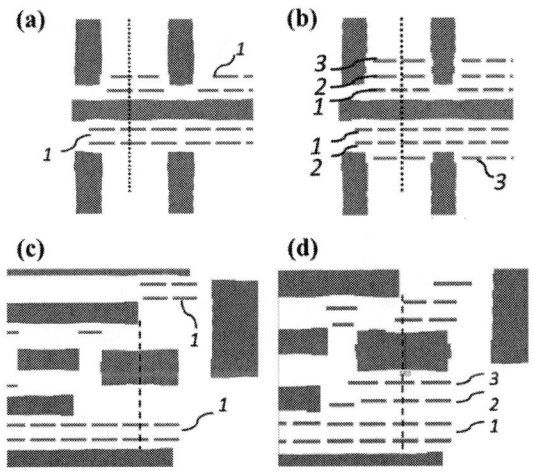

Figure 2. One-step SRAF insertion of (a) T-opening and (c) multidimensional post-OPC patterns. Three-step cycle SARF insertion of (b) T-opening and (d) multidimensional post-OPC patterns.

SIMULATED AND WAFER RESULTS

PV-band is the difference between the maximum and minimum simulated CD contour derived by variable energy dose and focus [3]. Figure 3 shows the OPC result and calculated PV-band of two different SRAF-generating methods. The simulated condition is set as focus +/- 35 nm and dose +/- 4%. As shown in Figure 3, using the three-step cycle SARF insertion method, the PV-band decreased 18% in T-opening patterns and 39% in multidimensional patterns.

Foucs: +/- 35 nm
Dose : +/- 4%

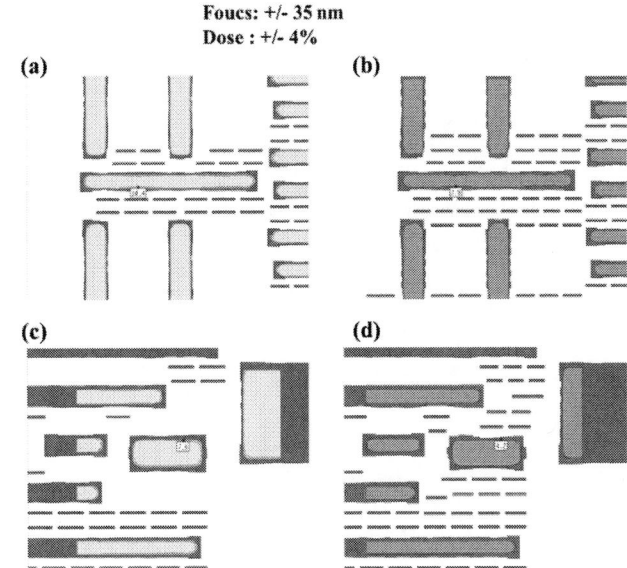

Figure 3. The process window comparison between these two SRAF methods. We generate a PV-band using Calibre RET-Flow tool based on the OPC model. The PV-band result shows one-step SARF insertion of (a) T-opening (PV-band) and (c) multidimensional (PV-band) post-OPC patterns. Three-step cycle SARF insertion of (b) T-opening PV-band and (d) multidimensional post-OPC patterns.

Meanwhile, the corresponding diffraction spectra simulated by the process window (PW) model are shown in Figure 4. The light intensity data is collected from dashed line positions in Figure 2. The light intensity vs. cutline distance curve demonstrates that three-step SRAF insertion patterns possess better image contrast than one-step SRAF insertion patterns in the nominal PW condition (focus = 0 nm, dose = 0%). In particular, the enhancement of intensity contrast at the pattern's edge performs more significantly. We could get ADI CD near the target more easily according to the three-step SRAF insertion method.

We obtain ADI results from a FEM (focus energy matrix) wafer. The SEM images of layouts in Figure 2 are shown in Figure 5 at the nominal PW condition. The ADI CD in T-opening post-OPC patterns generated by (a) the One-step SRAF insertion is 65.4 nm and (b) the three-step SRAF insertion method is 66.9 nm. The target CD is 68.0 nm. Similarly, the ADI CD in multidimensional post-OPC patterns generated by (C) the One-step SRAF insertion method is 124.6 nm and (b) the three-step SRAF insertion method is 121.0 nm. The target CD is 126.0 nm.

(a)

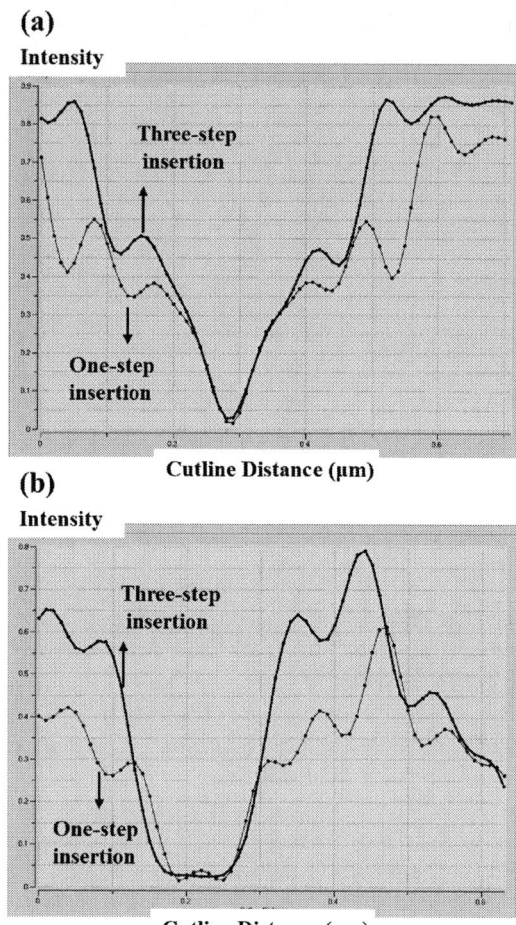

(b)

Figure 4. The corresponding diffraction spectra of designed layouts: the light intensity vs. cutline distance around selected patterns of (a) T-opening and (b) multidimensional post-OPC patterns. The data is collected from dashed line positions in Figure 2, respectively.

The above ADI data show that all the CD in the nominal condition is in spec (about +/- 10% of the target value). Finally, the DOF values are calculated from the FEM. In the T-opening layout, the DOF is 45 nm for a one-step SRAF insertion post-OPC pattern and 60 nm for a three-step SRAF insertion post-OPC pattern. In the multidimensional layout, the DOF is 60 nm for a one-step SRAF insertion and 90 nm for a three-step SRAF insertion.

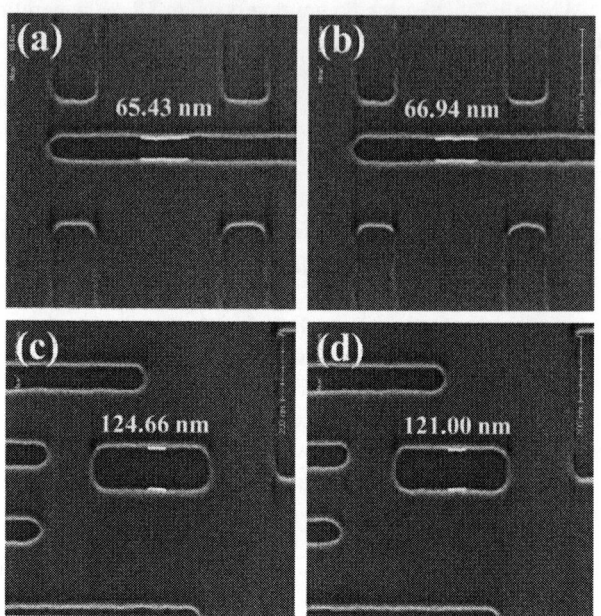

Figure 5. The SEM images of ADI results: One-step SRAF insertion of (a) T-opening and (c) multidimensional post-OPC patterns. Three-step cycle SARF insertion of (b) T-opening and (d) multidimensional post-OPC patterns.

CONCLUSION

A three-step insertion approach has been developed to optimize regular rule-based SRAF placement for advanced nodes. This method has demonstrated the capability to improve 33% and 50% DOF for T-opening and multidimensional layouts in the metal layer. The model simulation also illustrates this method can produce a smaller PV-band and high contrast at the pattern edge, which is effective for optimizing the photolithography process.

ACKNOWLEDGMENTS

I am very grateful for many valuable discussions with Honglan Zhu, and Pan Zhang. Especially, I greatly appreciate my colleague Chenhua Yuan for his useful suggestions and data collected.

REFERENCES

[1] V. Farys, F. Robert, C. Martinelli, Y. Trouiller, F. Sundermann, C. Gardin, J. Planchot, G. Kerrien, F. Vautrin, M. Saied, E. Yesilada, F. Foussadier, A. Villaret, L. Perraud, B. Vandewalle, J. C. Le Denmat, and M. K. Top "Study of SRAF placement for contact at 45 nm and 32 nm node", *Proc. SPIE 6924, Optical Microlithography XXI*, 69242Z (12 March 2008); https://doi.org/10.1117/12.774091

[2] M. B. Alawieh, Y. Lin, Z. Zhang, M. Li, Q. Huang and D. Z. Pan, "GAN-SRAF: Subresolution Assist Feature Generation Using Generative Adversarial Networks," in *IEEE Transactions on Computer-Aided Design of Integrated Circuits and Systems*, vol. 40, no. 2, pp. 373-385, Feb. 2021, doi: 10.1109/TCAD.2020.2995338.

[3] Q. Yanhui, L. Baoxuan, W. Dan, C. Yanpeng and Y. Shirui, "Enlarge Process Window of BSI in DTI Loop: A Novel OPC Approach to Add Sraf," *2020 China Semiconductor Technology International Conference (CSTIC)*, Shanghai,

China, 2020, pp. 1-3, doi: 10.1109/CSTIC49141.2020.9282428.

RECENT PROGRESS OF EUV RESIST DEVELOPMENT FOR IMPROVING CHEMICAL STOCHASTIC

Toru Fujimori

Electronic Materials Research Laboratories

FUJIFILM Corporation

4000 Kawashiri, Yoshida-Cho, Haibara-Gun, Shizuoka, 421-0396, Japan

toru.fujimori@fujifilm.com

ABSTRACT

In 2019, extreme ultraviolet (EUV) lithography has been applied to high volume manufacturing (HVM). With recent rapid progress on the source power improvement, and EUV lithography development including photoresist materials has been achieved HVM requirements. However, the performance of EUV resist materials are still not enough for the expected HVM requirements. The critical point is the stochastic issues, which will be become 'defectivity', like a nano-bridge or a nano-pinching. 2 (two) major stochastic issues, which are photon stochastic and chemical stochastic, were observed in the lithography steps. And the improvement status of each stochastic issues including lithographic results are described.

INTRODUCTION

In 2019, finally, EUV lithography has been applied to HVM for preparing advanced semiconductor devices, like 5 nm technology node and beyond. That was very important year for EUV enthusiasts and semiconductor industry. Because it takes for a long time, more than 30 years, to study EUV lithography for realizing HVM. In fact, the 1st paper of EUV lithography was published in 1986 by prof. Kinoshita's group [1]. However, it had been so difficult to realizing EUV lithography for a long time. According to the closing remarks in EUVL symposium in 2017, the key factors for EUV lithography realization were light source, mask defectivity and inspection, and photo resist materials. The most critical issue was light source until 2016 [2]. With recent rapid progress on source power improvement [3], photo resist materials and processes explorations are more and more accelerated to achieve HVM requirements. A key factor for the realization of EUV lithography is the choice of EUV resist materials that is capable to resolving below 15 nm half pitch with high sensitivity [4]. Unfortunately, the performance of EUV resist is still NOT enough for the true HVM requirements, even by using the qualified EUV resist materials for the 1st generation. One critical issue is 'stochastic issues', which will be become 'defectivity' [5]. The understanding of the status of the stochastic issues of EUV lithography and their improvement procedure were described in this paper.

RESULTS AND DISCUSSION

Classification of the stochastic effect

In the past, speaking of the stochastic issue of EUV lithography was basically considered from poor photon number from EUV light source, which means 'photon shot noise'. It was still critical concerning point of the stochastic issue, even with recent progress on source power improvement. However, the stochastic issue is not only from them but also from EUV materials and processes, called 'Chemical stochastic' [6]. The 'Chemical stochastic' means caused from resist materials and processes for lithography, materials uniformity in the film, catch the photon efficiency, reactive uniformity in the film, and dissolving behavior with the developer (Figure 1).

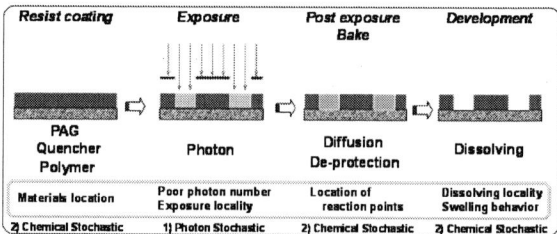

Figure 1: The analyzing summary of stochastic factors in lithography

Photon stochastic (Photon shot noise)

Photon stochastic, photon shot noise effect, on EUV lithography was well-known issue and various studies have emphasized its impact on LWR performance [7-9]. As well as LWR, the photon shot noise effect on defectivity must be considered to obtain pattern quality satisfying HVM requirements. 'Organic high EUV absorption materials' was designed and synthesized to be able to catch photon more efficiency from the light source (Fig. 2).

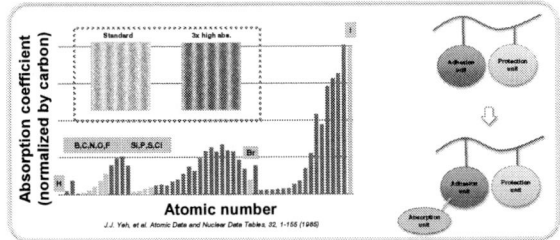

Figure 2: The design of 'Organic high EUV absorption materials.

High absorption resist, 'Organic high EUV absorption materials', showed 20% dose reduction with keeping its LWR value. Besides that, nano-bridge is clearly decrease on the high absorption resist (Fig. 3) [10].

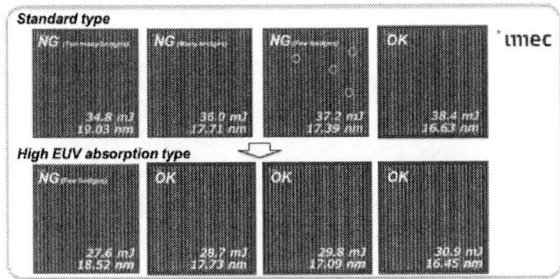

Figure 3: The lithographic performance of 'Organic high EUV absorption materials.

Chemical stochastic (Materials location randomness)

The other hand, 'Chemical stochastic', the materials location randomness, could be observed in the film. A traditional CAR material is mainly composed from polymers, PAGs (Photo Acid Generator) and quenchers, which locate variously in the film. The location locality of the materials in the film induced solubility randomness during the development process, which observed worse LWR and defectivity, like nano-bridge and nano-pinching (Fig. 4).

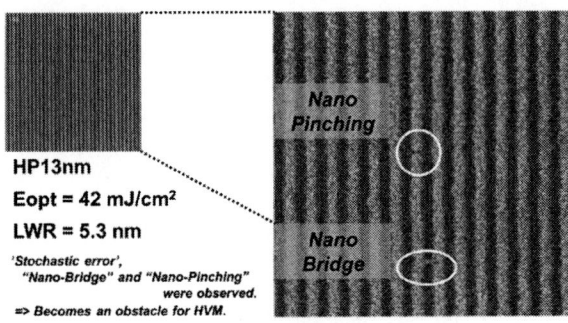

Figure 4: Example of the stochastic errors, Nano-pinching and Nano-bridge.

One of the famous methods to reduce the 'Chemical stochastic' is higher PAG loading formulation. The simulation study and the experimental study were reported (Fig. 5) (Fig. 6) [6].

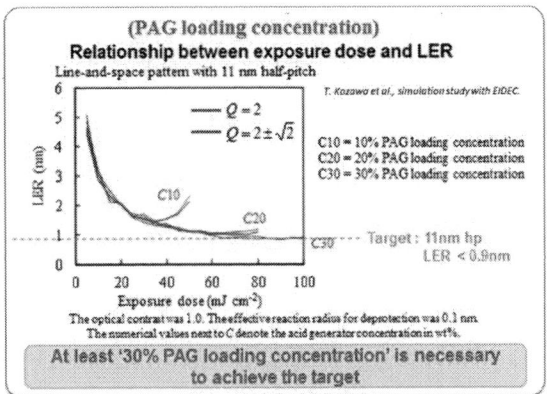

Figure 5: The simulation study of PAG loading effects.

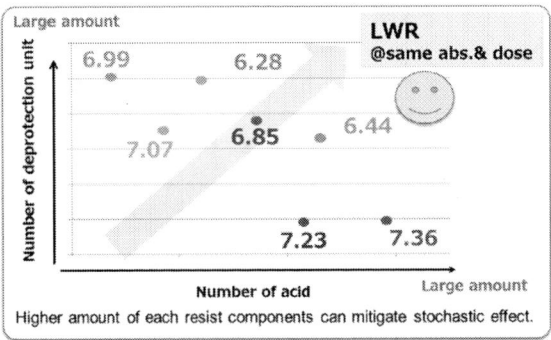

Figure 6: The experimental results of PAG loading effects and protecting groups loading effects for LWR.

However, even introduced higher loading technology, it still remained the fluctuation, like unexpected repulsion, interaction and aggregation. To reduce such an unexpected behavior and expected to control of the materials location in the films, 'The novel functionalized materials' was designed and synthesized to introduce connection unit, interaction, or bounded.

'The novel functionalized materials' were designed and synthesized to introduce connection unit, interaction unit and bounded to approach the uniform film preparation. 'The novel functionalized materials' observed excellent lithographic performance on 13 nm half pitch without pinching and bridging (Fig. 7) [11-13]

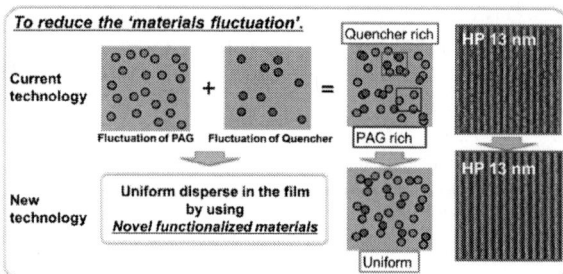

Figure 7: The concept and lithographic results of novel functionalized materials.

Chemical stochastic (Dissolving randomness)

Also, the dissolving stochastic with developer, a kind of 'Chemical Stochastic', could be observed during the development process due to their swelling behavior. One of the key items to improve them was Negative-tone imaging (NTI, using organic solvent-based developer) process (Fig. 8).

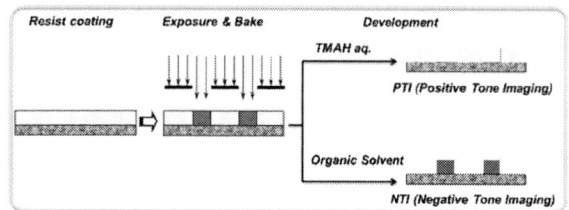

Figure 8: The process flow of the lithography, PTI (Positive-tone imaging) and NTI (Negative-tone imaging).

NTI process provided lower swelling and smoother dissolving behavior than Positive-tone imaging (PTI) process, which reduced the dissolving stochastic. Therefore, negative-tone imaging with EUV exposure (EUV-NTI) has huge advantages for the performance, especially for improving LWR, which will be expected to resolve RLS trade off. Also, NTI system has already introduced to manufacturing with ArF exposure. So, it seems not so difficult to use EUV-NTI for manufacturing.

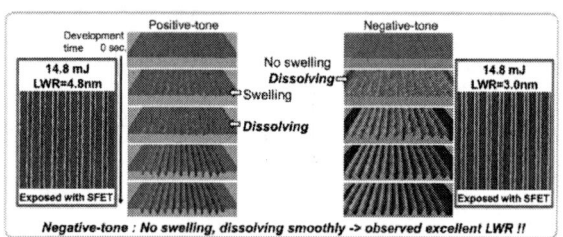

Figure 9: The dissolution behavior comparison between PTI and NTI by using in-situ high speed AFM.

Comparison of dissolving behavior between positive-tone and negative-ton by using in-situ high speed AFM, NTI process observed less swelling than PTI process. The lithographic performance of NTI process observed 60% better LWR performance than PTI process under the same exposure dose owing to their smooth-dissolving behavior and non-swelling properties (Fig. 9).

CONCLUSION

This study hereby showed excellent improving the lithographic performance due to reduce the stochastic issue, 'Photon stochastic' and 'Chemical stochastic'. 'Organic EUV high absorption materials', 'The novel functionalized materials', and 'Negative-tone imaging with EUV exposure (EUV-NTI)' were very effective to reduce the stochastic effect. The most critical issue is how to apply each item. These technologies were expected to apply the real EUV lithography HVM

REFERENCES

[1] H. Kinoshita, et.al., *The Japan Society of Applied Physics*, **28-ZF-15** (1986).

[2] P. Naulleau and P. Gargini, *International Symposium on Extreme Ultraviolet Lithography, closing remarks,* (2017).

[3] A. A. Schafgans, D. J. Brown, I. V. Fomenkov, Y. Tao, M. Purvis, S. I. Rokitski, G. O. Vaschenko, R. J. Rafac, D. C. Brandt, *Proc. SPIE 10143, 1014311 (2017)*.

[4] T. Fujimori, *International Microprocesses and Nanotechnology Conference, the 34th, MNC* (2021).

[5] P. De Bisschop, J. Van de Kerkhove, J. Mailfert, A. Vaglio Pret, J. Biafore, *Proc. SPIE*, **9048** (2014) 904809.

[6] T. Fujimori, *International Symposium on extreme ultraviolet lithography,* **11147** (2019).

[7] R. L. Brainard, P. Trefonas, J. H. Lammers, C. A. Cutler, J. F. Mackevich, A. Trefonas, and S. A. Robertson, *Proc. SPIE*, **5374** (2004).

[8] J. M. Hutchinson, *Proc. SPIE*, **3331** (1998).

[9] S. Bhattarai, W. Chao, S. Aloni, A. R. Neureuther, P. Naulleau, *Proc. SPIE*, **9422** (2015) 942209.

[10] H. Furutani, M. Shirakawa, W. Nihashi, K. Sakita, H. Oka, M. Fujita, T. Omatsu, T. Tsuchihashi, N. Fujimaki, and T. Fujimori, *J. Photopolym. Sci. Technol. (2018)*.

[11] T. Fujimori, *International symposium on extreme ultraviolet lithography* (2018).

[12] T. Fujimori, *China Semiconductor Technology International Conference (CSTIC)*, (2020-2022).

[13] T. Fujimori, *International Workshop on Advanced Patterning Solutions (IWAPS)*, (2020-2022).

The Reduction of Yield Loss and Contact Overlay Shift by Optimizing the Process Profile of Pre-Layer Process Integration

Zhejun Liu*, and Wei Lu

Shanghai Huali Integrated Circuit Corporation
Shanghai, China
*+86-151-2101-5906, liuzhejun@hlmc.cn

Abstract

This paper presents the end-of-the-line yield failure related to contact (CT) overlay deviation, and we studied a failure mode to minimize CT overlay deviation. By changing the profile of underneath layers, the thickness jumps up at the wafer edge was reduced, resulting in improved CT Photo ERO and CT OVL and final yield were improved.

Keywords—Contact; Overlay; ERO; Profile Improvement.

Introduction

With the development of chip manufacturing process and the reduction of critical dimensions, the alignment of key steps in the process plays an increasingly important role. For example, CT (Contact) is an important part for connecting FEOL MOS FET and BEOL metal lines. Its overlay (OVL) to Poly or AA determines whether the MOS FET can be effectively driven by the external input. [1-3] When the Poly size is small enough, overlay deviation of a few nanometers may result in serious device and yield failure. Although currently there are various means (BSE, LIS, etc.) to compensate CT OVL to improve the precision and accuracy of CT OVL, there are too many factors affecting CT OVL, including the profile of the pre-layer and photolithography conditions, and some small variables may make CT OVL shift. These need special attention in the process of manufacturing. In this paper, we present a case of yield failure caused by CT OVL deviation. It was found that the NG wafers all have the problem of large CT Photo ERO (Edge Roll Off), and the root cause was that the profile after the ILD Dep and CMP in the front layer was high in the wafer edge, resulting in the lithography overlay deviation. For this reason, we improved the profile of the front layer and resulting in improved CT overlay and subsequent yield.

As shown in FIG. 1.a, the CP stack map showed that the yield failure was mainly concentrated in the Wafer Edge area, especially in the 3 o 'clock and 9 o 'clock directions. We backtracked to the inline investigation and found that this CP failure map matched the OVL map distribution from inline CT to poly. The OVL at 3 o 'clock was larger than the wafer center, while the OVL at 9 o 'clock was smaller. Normally, the closer the CT OVL value is to 0, the more accurate the

alignment is, and no deviation occurs, as shown in CT OVL map by wafer center. However, in this case, CT OVL of wafer edge showed obvious shift, especially the OVL shift at 3 o 'clock, which would lead to serious functional failure. FIG. 1.b showed that serious CT overlay deviation had occurred in

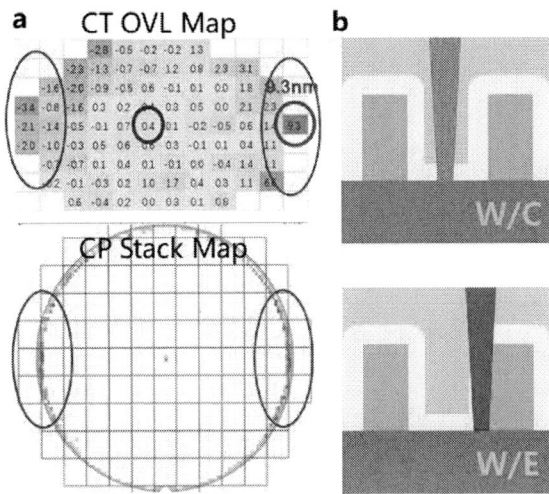

wafer edge area. The CT that should theoretically land in the AA area lands at the boundary between AA and Poly due to the OVL shift, which may cause AA to Poly short, which further leads to device and yield failure.

Figure 1. a. The CT OVL Map & CP Stack Map of NG Wafer; b. The schematic diagram of CT OVL at wafer center and wafer edge.

Figure 2. a. The Correlation between CP Mbist Fail Rate and CT to Poly OVL; b. The Correlation between CT OVL @ 3 o'clock and the ERO value by wafer.

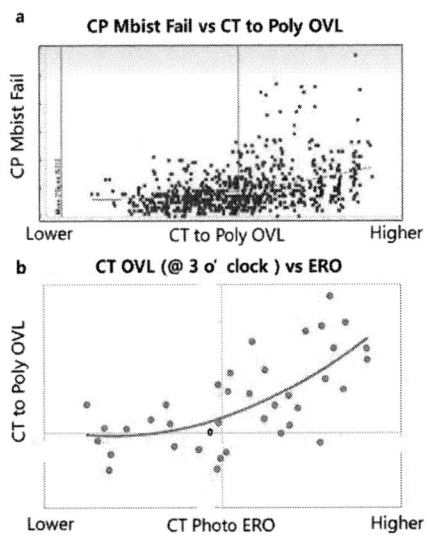

979-8-3503-1101-3/23 $31.00 © 2023 IEEE

Figure 2.a shows the correlation between CP Mbist Fail and CT to Poly OVL. It can be seen that the larger the shift of OVL between CT and Poly is, the more serious CP Mbist Fail will be, which is consistent with our failure model. ERO (Edge Roll Off) is a measurement parameter used to characterize the flatness of wafer surface before lithography process.The larger the value, the larger the fluctuation of the wafer surface and the more uneven the wafer.Figure 2.b shows a positive correlation between CT OVL and ERO, that is, with the increase of ERO value of wafer, CT to Poly OVL

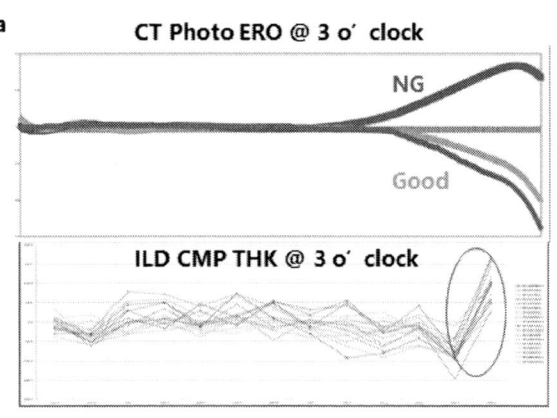

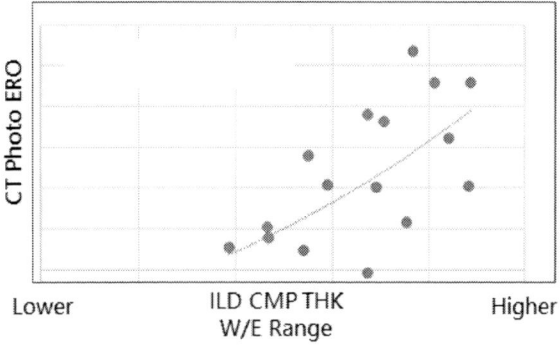

shift becomes more serious.

Figure 3. a. The CT ERO Value and ILD CMP THK Profile by radius of NG and Good Wafer; b. The correlation between CT ERO value and ILD CMP THK range at wafer edge.

Figure 4. Schematic diagram of the effect of the profile of pre-layer ILD CMP on CT Photo process.

Through investigating the radius chart of CT ERO (FIG. 3.a), we found that the ERO value of NG wafer jumped significantly at the 3 o'clock of wafer edge, while the ERO of good wafer at wafer edge was low. Further investigation of inline found that the reason for the high ERO at wafer edge

was that the CMP thickness of the pre-layer ILD had a significant jump at 3 o'clock, which led to ERO jump. Figure 3.b also illustrates this correlation well.

FIG. 4 shows the influence of the thickness profile of pre-layer ILD CMP on the CT Photo process. It can be seen that when the thickness of wafer edge is thicker, the photoresist of CT Photo will inherit the profile of the pre-layer, thus affecting the photolithography process, resulting in the shift of CT OVL and abnormal CT hole

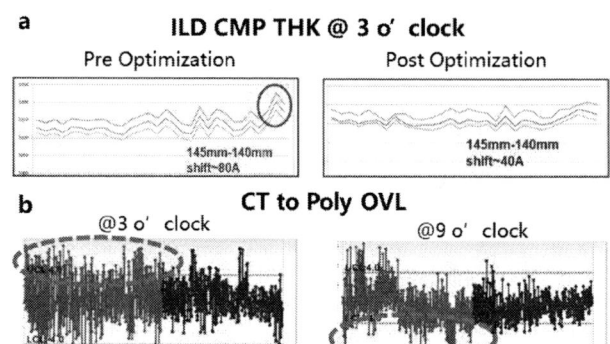

opening.

Figure 5. a. The profile of pre-layer ILD CMP at 3 o'clock of pre/post profile optimization; b. The CT to Poly OVL improvement at 3 & 9 o'clock after pre-layer profile optimization.

Therefore, we optimized the pre-layer profile and effectively reduced the thickness at wafer edge by improving ILD Dep and CMP processes. As shown in FIG. 5.a, the range of wafer edge decreased by about 50% after profile optimization. FIG. 5.b shows the influence of profile optimization on CT to Poly OVL. OVL jump at 3 o'clock and OVL drop at 9 o'clock before optimization are both significantly improved and have a positive effect on yield.

In conclusion, we studied a failure mode in which the yield failed due to CT overlay deviation. By improving the profile of the pre-layer, the thickness jump at wafer edge was reduced, resulting in improved CT Photo ERO and CT OVL and reduced yield loss.

References

[1] G. Deng, J. Hao and Q. Wu, "Investigation into PR profile representation through method of OVL focus subtraction based on a case of Overlay AEI-ADI offset on contact layer of advanced technology node," 2016 China Semiconductor Technology International Conference (CSTIC), 2016, pp. 1-5, doi: 10.1109/CSTIC.2016.7463984.

[2] K.T. Turner, S. Veeraraghavan and J.K. Sinha, "Predicting distortions and overlay errors due to wafer deformation during chucking on lithography scanners", Journal of Micro/Nanolith. MEMS MOEMS, vol. 8, no. 043015, 2010.

[3] C. M. Ke et al., "Overlay similarity: a new overlay index for metrology tool and scanner overlay fingerprint methodology", Proc. SPIE, vol. 7272, pp. 72720E, 2009.

The Analysis of Optical Critical Dimension (OCD) Signal Strength Between 5 nm FinFET and 3 nm Complementary FET (CFET) Vertical Gate Stacks

Qi Wang[1,2]*, Qiang Wu[1,2], Xianhe Liu[1,2], Yanli Li[1,2]*

[1]School of Micro-Electronics, Fudan University, No. 825 Zhangheng Road, Shanghai 200433, PR China,
[2]National Integrated Circuit Innovation Center, Pudong New Area, Shanghai 201203, China
*Corresponding Author's Email : *wangqi_fd@fudan.edu.cn, li_yanli@fudan.edu.cn*

ABSTRACT

The use of Optical Critical Dimension (OCD) technique to characterize the optical properties of periodic semiconductor structures, has achieved great success in semiconductor manufacturing over the past decades. OCD is an indirect and nondestructive optical metrology, in which physical properties of a sample is obtained through optical scattering spectra and modeling simulation. As in-line measurement of three dimensional (3D) structures is more and more critical in semiconductor manufacturing, the improvement of the accuracy and sensitivity for the latest 3D architectures has become inevitable. This article will provide an analysis of OCD Signal Strength for both 5 nm FinFET and 3 nm Complementary FET (CFET) architectures, and provide some insight and advise for OCD modeling in advanced technology nodes.

Key words—Optical scatterometry, Optical critical dimension (OCD), Complementary FET (CFET), FinFET.

1. INTRODUCTION

When the integrated circuit manufacturing enters into the 3 nm technology node, Complementary FET (CFET) architectures, by utilizing shared gate for vertically stacked nanosheet n-FET and p-FET, have achieved notable reduction on cell areas [1]. Owing to multi-channels' gate-all-around design and shared utilization of gate area, CFET structure's better electrostatic gate control and device performance has shown great application potential [2].

However, such complicated multi-stacked structures pose significant challenges not only for precision etch, CMP and EPI processes, but also for metrology such as optical scatterometry. The optical critical dimension (OCD) metrology, owing to its fast, non-destructive, and in-line-compatible features, has proven to be an effective tool to detect 3D periodic structure's geometry at several-nanometer scale [3]. As the penetration depths of visible light in the nanostructures are around 20 nm to 60 nm, the contrast between different materials can be very weak, especially for those "deeply-buried" layers. For instance, the inner spacers are formed by Atomic Layer Deposition (ALD) of spacer material on the recessed pockets of the SiGe super-lattice. These inner spacers insulate the metal gates and the source/drain terminals, which are instrumental to the gate reliability and parasitic capacitance [4]. Therefore, accurate control and measurement on inner spacers' dimension uniformity is crucial to the manufacturing.

In this paper, we will perform a study on the critical dimensions in CFET structures by using optical scatterometry simulation based on our verified Rigorous Coupled Wave Analysis (RCWA) algorithm. We will compare 3 nm CFET structure's signal strength with 5 nm FinFET structure to obtain the signal loss due to dimension shrinkage and complexity increase. By tuning detection parameters, such as angle-of-incidence, illumination wavelength, we will report our findings on optimized parameter settings for measurement, and give recommendations for building OCD models of CFET structures.

2. SIMULATIONS AND SETUPS

In general, the modeling of OCD are based on the change of amplitude and phase from reflected light. There are many methods originated from the numerical solutions of Maxwell's equations for the optical scatterometry modeling of light-nanostructure interaction, e.g. the finite element method, the boundary element method, the finite-difference time-domain method and the the rigorous coupled-wave analysis (RCWA). The main disparities are the numerical techniques adopted in solving the governing equations and boundary conditions. Here we chosen RCWA to extract profile dimensions due to its good convergence and relatively easy implementation in codes [5].

The RCWA technique uses a set of regression analysis to reach optimum structural parameter configuration for the given sample's optical spectra. For the implementation of RCWA modeling, the nano-structure's geometry information should first be stratified along the vertical (Z) direction into multiple layers and build the device's dielectric constant grid at each layer. Then, relative permittivity of each layer is expanded into Fourier series, and through the match of the expansion coefficients at various boundaries of the layers, we can get the transmitted and reflected field amplitude and phase [6]. The illumination angle can be altered by tuning the incidence polar angle θ and azimuthal angle φ [7].

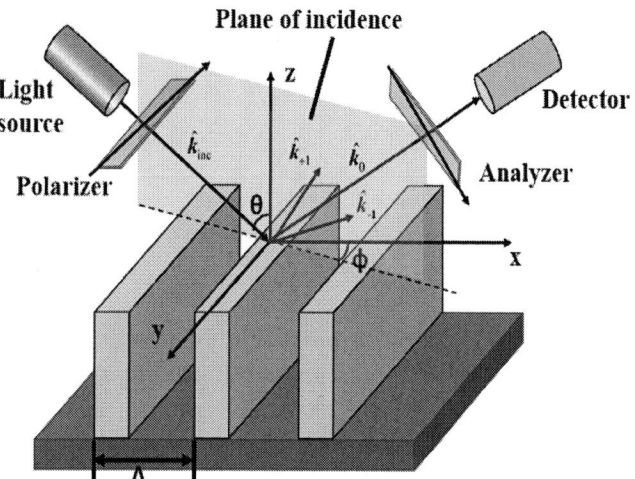

Figure. 1. Schematic of optical scatterometry for periodic structures with spectroscopic and angular functions.

We have successfully verified our model based on binary gratings structures, in which the simulation outputs fit well with the experimental results [8]. We built a typical FinFET structure depicted in Figures 2(a)-2(b). The pitch between unit cells is 48 nm. The metal gate structure has an 28 nm length, 30 nm width, and 55 nm height. From edge to core, each layer has the thickness of 2.5, 3.5, 1, 1, 2, 1.5 nm for TiN, TiAl, TaN, TiN, HfO_2, SiOCN in Figure 1 (b). The spacers around the gate are made of Si_3N_4 with 5 nm thickness. we stratify the FinFET with 8 slices along Z-axis to represent its geometry.

We use single CFET gate stack and one Fin structure as the basic unit cell. The pitch between cells also choose 48 nm. The total gate length is 32 nm, in which the core has 16 nm length and the gate oxide (mainly HfO_2) has 3 nm thickness. PMOS and NMOS cores are made of TiN and TiAl, which are both surrounded by the gate oxide. The inner spacers are made of Si_3N_4 with 5 nm thickness. The source & drain of PMOS and NMOS are made of SiGe and SiP with 16 nm length. Figures 2(c)-2(d) show both the x-z and x-y cross-sections of the CFET unit cell. we stratify the CFET with 38 slices along Z-axis to approximately represent its geometry. The Total height of CFET is 231 nm. During the simulation, a grid size of 2 nm is chosen, as it can both maintain the accuracy and efficiency of modeling.

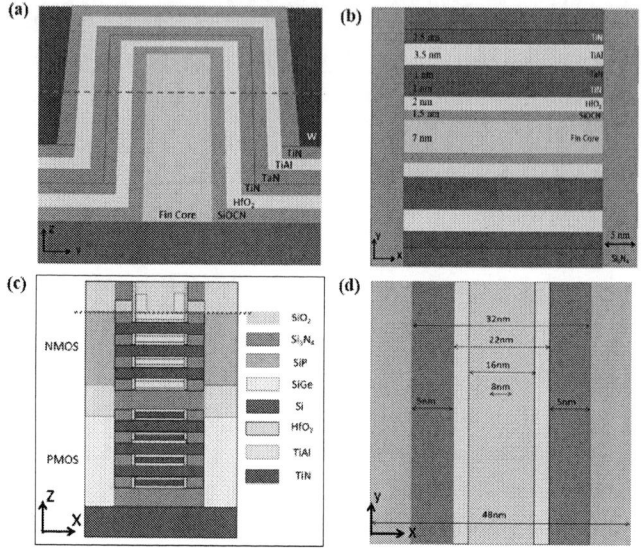

Figure 2. Diagram of FinFET and CFET model structure. (a) y-z cross-section of FinFET.. (b) x-z cross-section of FinFET at red line layer.(c) x-z cross-section of CFET. (d) x-y cross-section of CFET at red line layer.

3. RESULTS AND DISCUSSION

To analyze Optical Critical Dimension (OCD) signal strength, we first find the most sensitive incident azimuthal angle φ and polar angle θ sets for each of the CFET and FinFET structures. The outputs in the form of R_p, R_s is Fresnel reflection coefficients of p-polarized and s-polarized light, respectively. We will also get two ellipsometric parameters which is quantified by amplitude ratio Ψ and phase difference D.

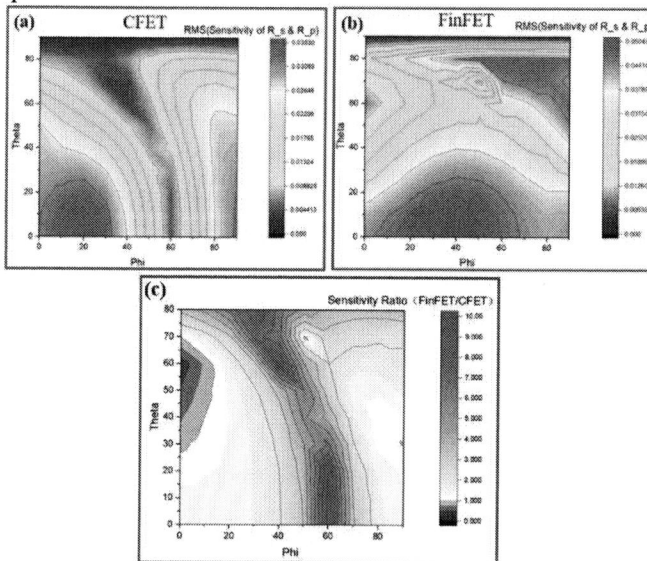

Figure. 3. Schematic of angular sensitivities of (a) CFET (b) FinFET with 2nm inner spacer variation at 300 nm wavelength, (c) Sensitivity Ratio of FinFET/CFET

Figures 3(a)-3(b) show the sensitivities of inner spacer variation at 300 nm wavelength for both CFET and FinFET structures, which indicates that when polar angle $\theta = 0$ (normal incidence), the inner spacer of both CFET and FinFET has high sensitivity for 2 nm spacer changing. Besides normal incidence (polar angle $\theta = 0$), the oblique incidence condition of polar angle $\theta = 70\sim80°$ and azimuthal angle $\varphi = 60\sim80°$ also has high sensitivity for FinFET. Such angular settings will break the the symmetry along the plane of incidence of light, and provide extra information which will helps in the decorrelation of simulation variables associated with feature dimensions, thus enable determination of a more accurate structural optical model.

Then, we will compare the signal strengths for both kinds of FET. From Figure 3(c), we can learn that FinFET has higher signal strength than CFET at almost all illumination conditions. It may indicate that, although the pitch of the CFET doesn't shrink to more tight design rule (e.g. 48 nm to 36 nm), the dimension complexity and stacking height of CFET may have caused the reduction of signal strengths comparing to FinFET structure.

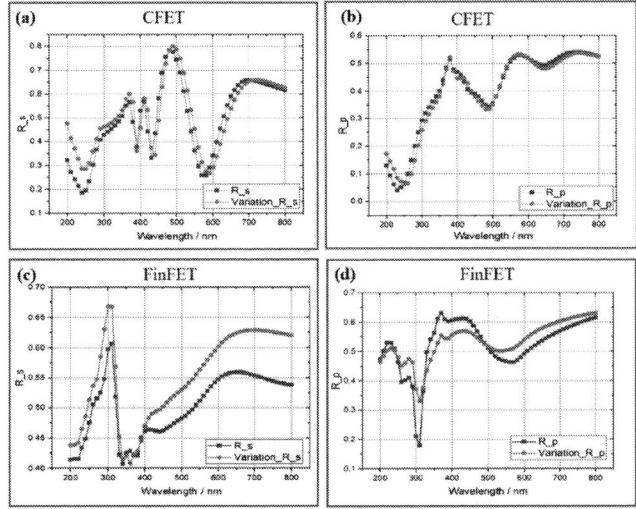

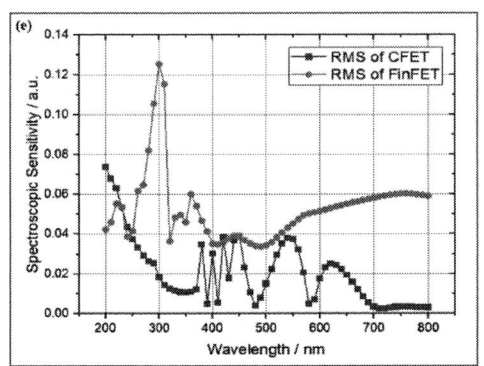

Figure 4. Schematic of (a-b) CFET's R_s & R_p spectroscopic reflectance and (c-d) FinFET's R_s & R_p spectroscopic reflectance (e) Spectroscopic Sensitivity with inner spacer variation in the form of RMS(Root Mean Square).

Apart from the above angular analysis, spectroscopic scatterometer uses dispersed wavelength light to obtain the spectral information with the zeroth-order diffracted beam in Figures 5(a)-5(d). By introducing 2 nm spacer variation, both CFET and FinFET has notable magnitude and peak position

shifts for s- or p- polarized light beams. By collecting these shifts and gathering them in Figure 5(e) in Root Mean Square (RMS) form, it has shown that the spectroscopic sensitivity for CFET is lower than for FinFET during 200-800 nm on the whole. This reduction of signal strengths means that more vertically stacked layers in CFET increase the height of the structure and the background signal, thus will pose more detection sensitivity challenge in OCD metrology.

4. CONCLUSION

In this paper, we have demonstrated a modeling analysis of optical scatterometry to characterize subsurface features in advanced nanoscale structures. We built two sets of OCD simulations for both 5 nm FinFET and 3 nm CFET architectures. By fully optimizing angular and spectroscopic measurement conditions, our results show that the signal strength is still sufficient to sense the subsurface structural features in both 3D structures. That the scattering signal strength in FinFET over the most spectral range is stronger than in CFET may be due to CFET's complex and higher stack along the vertical direction. In summary, the signal reduction from the 5 nm FinFET to 3 nm CFET, according to our study, is around 15~20%, which is incremental. As a result, we believe that the OCD technique will still be an industrially viable method in inline monitoring for the critical dimensions in semiconductor manufacturing.

5. REFERENCES

[1] Vincent B, Boemmels J, Ryckaert J, et al. "A benchmark study of complementary-field effect transistor (CFET) process integration options done by virtual fabrication." IEEE Journal of the Electron Devices Society, 8: 668-673 (2020).

[2] Ryckaert J, Schuddinck P, Weckx P, et al. "The Complementary FET (CFET) for CMOS scaling beyond N3." 2018 IEEE Symposium on VLSI Technology. IEEE, 141-142(2018)

[3] N. G. Orji, M. Badaroglu, B. M. Barnes, C. Beitia, B. D. Bunday, U. Celano, R. J. Kline1, M. Neisser, Y. Obeng, and A. E. Vladar, "Metrology for the next generation of semiconductor devices", Nat Electron. s41928-018-0150-9 (2018)

[4] Madhulika Korde, Subhadeep Kal, Cheryl Alix, et al. "Nondestructive characterization of nanoscale subsurface features fabricated by selective etching of multilayered nanowire test structures using Mueller matrix spectroscopic ellipsometry based scatterometry", J. Vac. Sci. Technol. B 38, 024007 (2020)

[5] Moharam MG, Grann EB, Pommet DA, Gaylord TK, "Stable implementation of the rigorous coupled-wave analysis for surface-relief gratings: enhanced transmittance matrix approach". J Opt Soc Am A 12:1077–1086, (1995).

[6] Li L, "Use of Fourier series in the analysis of discontinuous periodic structures". J Opt Soc Am A 13:1870–1876, (1996).

[7] Huang H-T, Kong W, Terry FL Jr, "Normal-incidence spectroscopic ellipsometry for critical dimension monitoring". Appl Phys Lett 78:3983–3985, (2001).

[8] Wang Q.,et al. "Verified Optical Scatterometry Model for Line-Space and Metal-Gate Structures" ,2021 International Workshop on Advanced Patterning Solutions (IWAPS):10.1109 (2021)

STUDY ON-PRODUCT OVERLAY IMPROVEMENT FOR IMMERSION LITHOGRAPHY

Guoping Liu*, Yinsheng Yu, Chi Zhang, Yuhui Li, Wei Cao, QinYuan, Hongwen Zhao

Shanghai Huali Integrated Circuit Corporation, No.6, Liangteng Rd., Pudong District, Shanghai, China

*Corresponding Author's Email: liuguoping@hlmc.cn

ABSTRACT

This paper reported a novel alignment optimization method to improve on-product overlay (OPO) on an immersion lithography layer. Results show that the optimized alignment strategy makes the decorrected overlay improve 18.6% and 43.9% in the x/y directions, respectively. Budget breakdown shows the main residual overlay performance indicator (ROPI) variations from the common wafer 2nd and 3rd terms. Compared with the default alignment strategy, the optimized alignment strategy with high order wafer alignment (HOWA3) control solution brings much benefit in reducing lot-to-lot variation, which can increase OPO by ~36.4%/7.6% in x/y directions. The alignment optimization method is an effective method to improve OPO.

INTRODUCTION

The shrinking design rule means the increase of the number of process layers in devices, which results in an increase of control complexity of the on-product overlay (OPO). Overlay (OVL) are influenced by a variety of factors, both lithographic and non-lithographic. Photolithography includes alignment, measurement accuracy, run-2-run (R2R) control and other factors. Non-lithographic factors such as chemical mechanical process (CMP) and etch usually directly affect OPO. The accurate alignment process of scanner guarantees the overlay performance between previous and current layers [1].

Lot-to-lot (L2L) variation is a key factor affecting OPO improvement. Many researchers have reported improvement in OPO by overlay control strategy optimization. Bhattacharyya et al. introduced an accurate and robust overlay metrology for OPO improvement [2]. We have reported a combined feed forward and feedback control method to improve a back-end-of-line immersion layer [3]. However, an advanced OVL control method must ensure that its alignment is accurate and robustness. Few researches have focused on improving alignment, especially when using advanced lithography machines, there are more and more effective ways to improve alignment.

In this paper, we reported a novel alignment optimization method to improve OPO on an immersion lithography layer. Alignment strategy are optimized by exploring the proper color and sampling. The decorrected

OVL is utilized to represent the goodness of alignment. Budget breakdown method is used to reveal the contribution components of ROPI. Based on the overall analysis and optimization results, the final improvement effect of layer n on OPO is simulated. The alignment optimization method we reported is an effective method to improve OPO.

EXPERIMENTAL

A BEOL immersion layer n was used to study OPO improvement. All wafers with the diameter 300 mm were exposed on an ASML TWINSCAN immersion scanner with ORION alignment system. 4 kind of colors with x/y polarizations are studied. Image based overlay (IBO) was used to characterize overlay performance. The IBO measurement was performed on Archer 600. The inline alignment model is 6-par and the alignment sampling is the full entire shots. Overlay is corrected by intra-field high order process correction (iHOPC) model.

RESULTS AND DISCUSSION

POR alignment KPI check and alignment sampling optimization

In ASML scanner system, several key performance indicators (KPIs) are utilized to quantitatively characterize the fine wafer alignment (FIWA) signal quality for the ORION alignment system, such as the wafer quality (WQ), multiple correlation coefficient (MCC), residual overlay performance indicator (ROPI) and color-2-color (C2C) variation. All of the parameters will be collected and recorded during lots exposure. WQ is determined by the signal contrast from the alignment marks. It requires higher WQ values and small variations to cover the process variation. MCC shows how good is the signal fit, and it is determined by the signal-to-noise ratio and mark background. ROPI indicates how much residuals can impact OVL, so the smaller the ROPI, the better the alignment performance.

Fig. 1 shows the simulation alignment KPIs results of under 4 colors at x/y polarizations (Px/Py). The POR alignment strategy is Red1_Px, as indicated by the red dotted boxes in Fig. 1. The results of WQ, MCC and ROPI are presented in Fig. 1 (a), (b) and (c), respectively. Combined the WQ, MCC and ROPI values, three candidates including FIR1_Px, Green1_Py and NIR1_Px show clear better improvement than the POR, as indicated by the blue, green and black dotted boxes in Fig. 1.

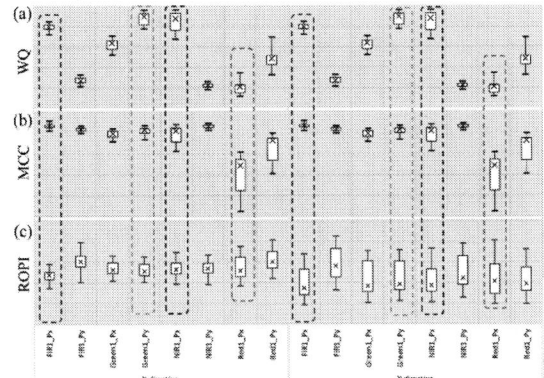

Figure 1: Comparison the alignment key performance under different colors and polarization: (a) WQ; (b) MCC; (c) ROPI

Different colors mean the different detection wavelength. Differences between colors indicate alignment marks asymmetry, which can be used as the alignment robustness indicator. Fig. 2 shows the ROPI color-2-color variation for the three candidates and POR alignment strategies. As presented in Fig. 2(a), the default alignment sampling is all full shots (the total number is 79). The shape and size of the bubble map represents the C2C difference. The wafer edge shots show relatively larger C2C variation than the center shots, as indicated by the red box in Fig. 2(a). The large C2C variation at wafer edge shots may introduce alignment uncertainty, and eventually affect the OPO. It is suggested to remove these bad shots to improve OPO, and the optimized alignment map is displayed in Fig. 2(b). We have checked that all the alignment KPIs exhibits better performance under the optimized new alignment map (NM) than that of default alignment map.

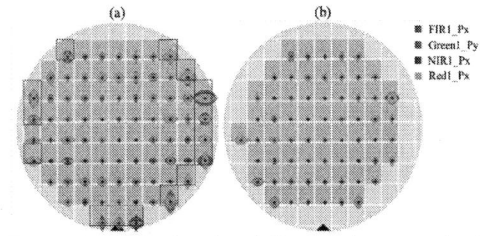

Figure 2: Color-2-color difference between the 3 alignment candidates and POR: (a) the default alignment sampling map; (b) the optimized new alignment sampling map. Note: (a) and (b) share the same scaling.

Decorrected OVL and ROPI budget breakdown

It is well known that the decorrected OVL can indirectly reflect the alignment quality. The smaller the decorrected OVL, the better the alignment. Fig. 3 shows the decorrected overlay trend chart under the 3 alignment candidates and POR using NM and POR map. There is no

clear difference among these alignment strategies in the x direction, and the wafer-2-wafer difference decreases after using the optimized alignment map. However, distinguished differences are seen in the y direction, especially when using the illuminations of Green1_Py and FIR1_Px. The mean+3sigma of all the test wafers are used to describe the decorrected OVL variation, as shown in Fig. 3(c). Compared with POR, the FIR1_Px_NM decreased by 18.6% and 43.9% in the x/y directions, respectively. Therefore, we will use the FIR1_Px_NM illumination for OPO optimization.

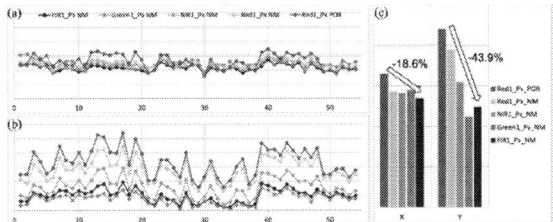

Figure 3: Decorrected overlay trend chart under the 4 alignment candidates using the new alignment map (NM) and POR: (a) Decorrected overlay x; (b) Decorrected overlay y; (c) Mean+3sigma of the decorrected overlay. Note: (a) and (b) share the same scaling.

Budget breakdown method is usually used to analyze the contribution of context types and model terms to the alignment or overlay. Fig. 4 shows the ROPI budget breakdown results matrix with context (rows) and model terms (columns). The context is composed of common parts, lot-2-lot (L2L) and wafer-2-wafer (W2W) variation. Model terms consist of translation x/y (Tx/Ty) and linear (magnification, rotation). Because the POR model is 6-par, their terms' contributions are consistent across common, L2L and W2W. The most contribution components to ROPI is common wafer 2nd and 3rd terms. L2L and W2W also see variations in wafer 2nd and 3rd terms, which can be corrected by high order wafer alignment model.

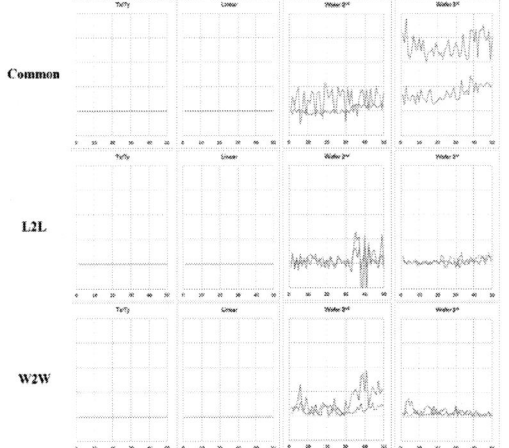

Figure 4: ROPI budget break down result (all charts share the same scaling)

979-8-3503-1101-3/23 $31.00 © 2023 IEEE   151

OPO improvement

As mentioned above, both the alignment strategy and sampling for better illumination are optimized, OVALIS is used to simulate OPO performance under the optimized alignment strategy. The exponentially weighted moving average (EWMA) method is used for OPO feedback (FB) control, and the exponent λ is 0.3. FIR1_Px_NM is believed to be more suitable, then it is used for improving OPO and decreasing variation. Fig. 5 shows the simulated OPO trend chart under the 6-par (W1) and HOWA3 (W3) alignment model. For the POR alignment strategy, namely Red1_Px_NM, HOWA3 can reduce OVL and its variation in the x direction, but increase OVL in the y direction, which may be related to different polarization directions with OVL. The same phenomenon exists with optimally aligned illumination FIR1_Px_NM. It is clear that FIR1_Px_NM W3 improves OPO in both x/y directions compared with Red1_Px_NM W1. When we convert it to the OPO mean+3sigma, FIR1_Px_NM W3 is 36.4%/7.6% lower than Red1_Px_NM W1 in the x/y directions, respectively, as depicted in Fig. 5(c).

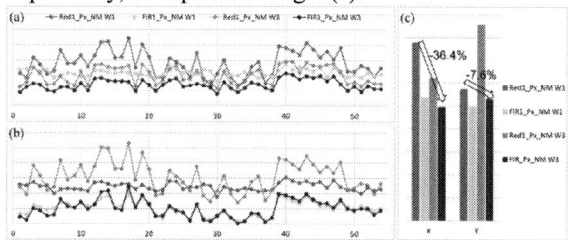

Figure 5: The simulated OPO trend chart under the 6-par and HOWA3 alignment model: (a) overlay x; (b) overlay y; (c) Mean+3sigma of the OPO. Note: (a) and (b) share the same scaling.

CONCLUSIONS

In this study, the alignment strategy and sampling are optimized to improve OPO on an immersion lithography layer. Results show that the default alignment strategy can be optimized by color change and sampling adjustment. The optimized alignment strategy makes the decorrected overlay improve 18.6% and 43.9% in the x/y directions, respectively. Budget breakdown shows the main ROPI variations from the common wafer 2nd and 3rd terms. Compared with the default alignment strategy, the optimized alignment strategy with high order wafer alignment (HOWA3) control solution brings much benefit in reducing lot-to-lot variation, which can increase OPO by ~36.4%/7.6% in x/y directions, respectively. The alignment optimization method is an effective method to improve OPO in lithography.

ACKNOWLEDGEMENTS

The authors would like to appreciate great supports from colleagues of lithography group of Advanced Modules Technology Development in Shanghai Huali Integrated Circuit Manufacture Corporation (HLMC). The authors also acknowledge many individuals from KLA for their valuable supports and fruitful discussions to this paper.

REFERENCES

[1] Lulu Lai, Rui Qian, Biqiu Liu, et al. *2020 China Semiconductor Technology International Conference (CSTIC)*, Shanghai, China, 2020, pp. 1-3.

[2] Kaustuve Bhattacharyya, Arie den Boef, Marc Noot, et al. *Proc. SPIE, Metrology, Inspection, and Process Control for Microlithography XXXI, 101450A*, 2017, doi: 10.1117/12.2257662

[3] Guoping Liu, Yinsheng Yu, Chi Zhang, et al. *International Workshop on Advanced Patterning Solutions (IWAPS)*, Beijing, China, 2022, pp. 11-13.

THE POSSIBILITY OF USING 193 NM IMMERSION LITHOGRAPHY PROCESS FOR 5 NM LOGIC DESIGN RULES

Qiang Wu*[12], Yanli Li*[12], Xianhe Liu[12], and Qi Wang[12]

[1]School of Microelectronics, Fudan University
No. 825 Zhangheng Road, 201203, Shanghai, China
[2]National Integrated Circuit Innovation Center
Pudong New Area, Shanghai 201203, China
*Corresponding Author's Email: wu_qiang@fudan.edu.cn

ABSTRACT

It is generally accepted that the Extreme Ultra-Violet (EUV) lithography has been used in large scale starting at the 5 nm logic technology node. Here we perform an analysis on the possibility of manufacturing the 5 nm design with 193 nm immersion lithography with multiple patterning techniques as it has aroused quite some interest in the industry. We will analyze process challenges from linewidth uniformities and overlay, as well as the challenges in etch process.

INTRODUCTION

Since typical Fin Pitches (FP) for the 7 nm and 5 nm designs, respectively, are 30 nm and 22.5~25 nm. Typical Contacted Poly Pitches (CPP) for the 7 nm and 5 nm designs, respectively, are 56 nm and 48~50 nm. Typical Minimum Metal Pitches (MMP), for the 7 nm and 5 nm designs, respectively, are 40 nm and 30~32 nm [1], as shown in Table I. For the Front-End-Of-the-Line (FEOL) fin and the gate layers, respectively, Self-Aligned-Quadruple-Patterning (SAQP) and Self-Aligned-Double-Patterning (SADP) with cuts are usually used. For the Back-End-Of-the-Line (BEOL) layers, Self-Aligned Litho-Etch Litho-Etch (SALELE, or SALE2) with cuts can be used with 193 nm immersion lithography for the 7 nm metal process. For the 5 nm's 30 nm pitch, with EUV, same approach can be applied. However, for the 193 nm immersion, it seems that such process has to be modified to include Litho-Etch Litho-Etch (LELE) processes which is susceptible to extra overlay error. For the via layers, extra hard mask layers need to be used with improved etch process. Similar challenges exist for the Middle-Of-the-Line (MOL). In this paper, we will analyze the process challenges based on a model SRAM cell with a FP of 22.5 nm, a CPP of 50 nm, and a MMP of 30 nm from a previous study [2].

TABLE I.　FIN, GATE, METAL, AND VIA BASIC DESIGN RULES AND LITHOGRAPHY METHODS FOR VARIOUS TECHNOLOGY NODES

Logic Tech Node	14 nm 2015	10 nm 2017	7 nm 2019	5 nm 2021	3 nm 2024	2.1 nm 2027	1.5 nm 2030	1.0 nm 2033
Fin Pitch (nm)	48 193i SADP	33 193i SAQP	27–30 193i SAQP	22.5–25 193i SAQP	20 (?) 193i SAQP	18 0.33NA EUV SADP	14 0.33NA EUV SAQP	14 0.33NA EUV SAQP
for Nano-plate/Nano-wires (nm)					21 193i SAQP	18 0.33NA EUV SADP	14 0.33NA EUV SAQP	14 0.33NA EUV SAQP
Gate Pitch (nm)	78/84–90 193i SP	66 193i SADP	54–58 193i SADP	48–50 193i SADP	36–42 193i SADP/SAQP	32 193i SAQP , 0.33NA EUV SADP	32 193i SAQP , 0.33NA EUV SADP	32 193i SAQP , 0.33NA EUV SADP
Metal Pitch (nm)	64 193i LE2	44 193i SALE2	40 193i SALE2/ 0.33NA EUVSP	30–32 0.33NA EUV SALE2	20–24 0.33NA EUV SALE2	14–18 0.55NA EUV SALE2	14 0.55NA EUV SALE2	14 0.55NA EUV SALE2
Via Pitch (nm)	64–80 193i LE2	44–62 193i LE2	40–56 193i LE3/ 0.33NA EUV SP	36–50 0.33NA EUV SP/LE2	30–36 0.33NA EUV LE2	25 0.55 NA EUV LE3/ EUV+DSA	20 0.55 NA EUV LE4/ EUV+DSA	20 0.55 NA EUV LE4/ EUV+DSA

With the introduction of the 0.33NA Extreme Ultra-Violet (EUV) lithography, the resolution has been increased from the 76 nm pitch to the 26 nm pitch, or nearly 3 times. However, due to the existence of the photon absorption stochastics, the minimum pitch for the EUV single exposure is around 36 nm, and for the logic devices, the minimum pitch for the 193 nm immersion seems to be around 80-90 nm. The improvement in lithographic resolution from the 193 nm immersion to the 0.33NA EUV is about 2.4×. From a previous paper, we have studied a model 5 nm logic process with the 0.33NA EUV lithography [2]. Here as follows, we will make an attempt to analyze the same 5 nm logic process with 193 nm immersion lithography and our analysis will be based on the SRAM cell.

A 5 NM SRAM CELL WITH EUV AND WITH 193 NM FLOWS (IF POSSIBLE)

As shown in Figure 1, a 6 track (6T) and 1 fin for each of the Pull Down (PD), Pass Gate (PG), and Pull Up (PU) SRAM cell can be made with 0.33NA EUV and 193 nm immersion lithography and major layers are depicted. nm immersion lithography and our analysis will be based on the SRAM cell, which has been described by a previous study [2].

979-8-3503-1101-3/23 $31.00 © 2023 IEEE

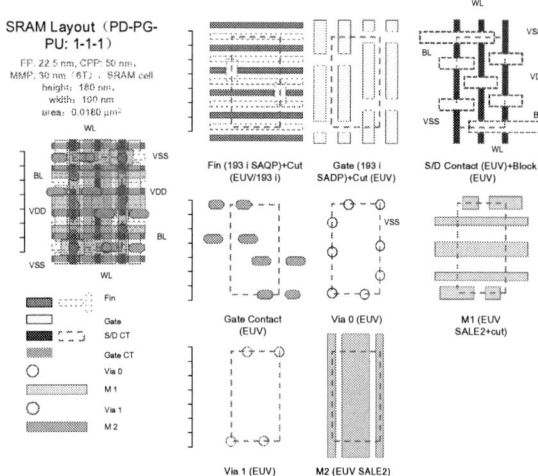

Figure 1: Schematic diagrams showing top down patterns for various layers of a 5 nm SRAM cell under the availability of 0.33NA EUV lithography.

With 193 nm immersion lithography only, we have tentatively designed a process with increased number of exposures, as described in Figure 2. Here we only present the scheme based on imaging resolution of the 193 nm immersion lithography, more detailed analysis must be carried out for pattern shape difference between the 193 nm process and the 0.33 NA EUV process since the EUV process can provide smaller corner rounding which may contribute to better shape enclosure between the connecting layers. Here we have provide a simple example for the gate contact, which has changed from more rectangular shape in Figure 1 to the round shape in Figure 2, this shape change may cause potential short risk along the gate length direction (Y direction) between the shared contacts connecting the source/drain (S/D) and gate.

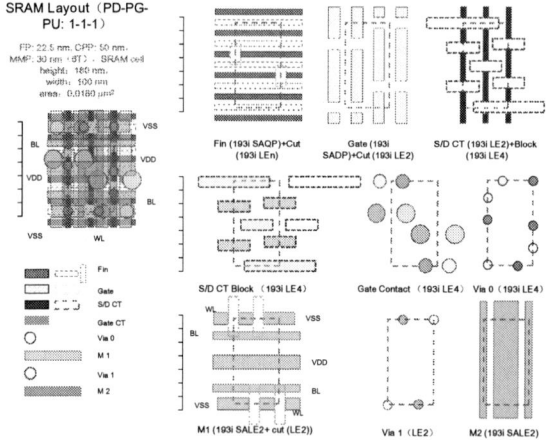

Figure 2: Tentative schematic diagrams showing top down patterns for various layers of a 5 nm SRAM cell

CHALLENGES

Here we list some major process challenges with the change of 0.33NA EUV lithography to the 193 nm immersion lithography.

Cut Process

In logic integrated circuit manufacturing, starting around the time when the 65 nm process is developed, cut process has been adopted to increase the patterning density. Figure 3 shows a schematic of S/D block (or cut) mask for the SRAM cell with different colored rectangles representing individual split masks under 193 nm immersion lithography. Due to the left-right and up-down symmetries of the SRAM cell to cell layout and due to the more difficult mask split scheme in Litho-Etch (LE), Litho-Etch multiple patterning process, repeating periods are 4 cell along the X direction and 2 cell along the Y direction. For photolithography, what we are focused on is the ability to project the designed patterns with maximal fidelity. For the cut process, due to the existence of minimum imaging area dictated by the design rule, or the limit of photolithography, a 2.5 square must be observed. A square is defined by the critical dimension (CD) squared. For example, the CD for a 193 nm lithography with Negative Tone Developing (NTD) process is around 37~50 nm. If we take 45 nm for example, the minimum area is $45 \times 45 \times 2.5 = 5062.5$ nm^2, or a bar with a width of 45 nm and a length of 112.5 nm.

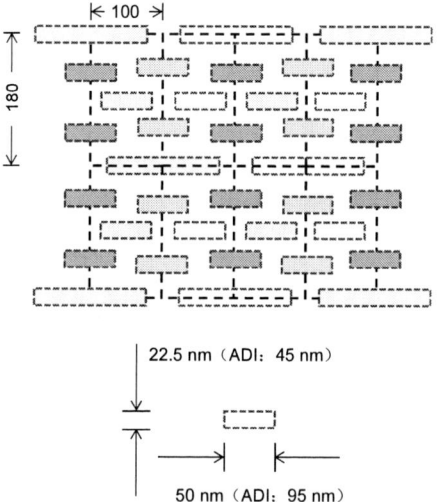

Figure 3: A tentative schematic diagram showing top down patterns for various split colors of the S/D contact block mask with 193 nm immersion lithography and the smallest cut shape.

In Figure 3, the minimum mask dimension is 22.5 nm × 50 nm. In the plasma etch process, the etch bias along the length direction is usually 2 times of that of the width direction. Since the minimum lithography area is around 2.5 squares, or 5062.5 nm^2, a same area rectangle with 45 nm ×112.5 nm, if used as the target of lithography (After Developing Inspection, ADI), will become 22.5 nm × 67.5 nm after etch (After Etch Inspection, AEI), which is the minimum area that can be produced by the 193 nm immersion lithography with NTD process and regular plasma etch process. However, it is still larger than the shape shown in Figure 3! To make the 22.5 nm × 50 nm will violate the design rule and will reduce the process window of current lithography process.

Contact Process

As pointed out early, the second challenge is with gate contact. The change of more rectangular shape in Figure 1 to the round shape shown in Figure 4 due to the resolution limit of 193 nm imaging may cause potential shortage between neighboring conductors. Besides, due to the minimum ADI CD under the 193 nm immersion lithography is around 55 nm with NTD (65 nm with PTD), the minimum AEI CD is around 30 nm. An AEI CD smaller than 30 nm may introduce residual or defectivity. Comparing the ADI CD of 25 nm under the above mentioned EUV process, there is increased risk of shortage.

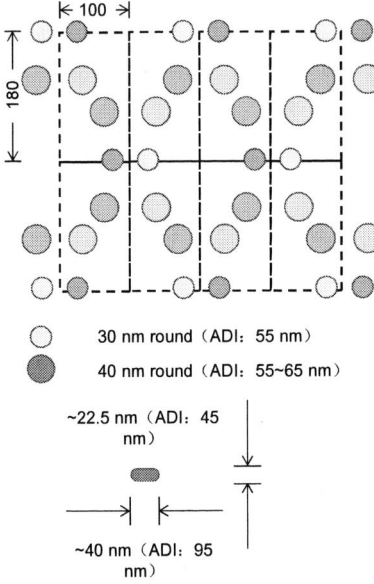

○ 30 nm round（ADI：55 nm）

● 40 nm round（ADI：55~65 nm）

Figure 4: A tentative schematic diagram showing top down patterns for various split colors of the Gate contact mask with 193 nm immersion lithography. Original more rectangular shared gate contact is included for comparison.

Via Process

The via process is similar to the gate contact process.

The challenge is for the etch process to shrink an ADI CD of 55 nm to about 16 nm under a MMP of 30 nm. For the SRAM cell, however, it seems that a Via 0 CD of 20~25 nm can be accepted and the minimum pitch for the metal 1 seems to be 45 nm, as shown in Figures 5 and 6. However, even an AEI via CD of 25 nm is challenging for an ADI CD of 55 nm. A easy to think solution may be to use 2 Hard Mask (HM) and to divide the via shrink process into 2 steps.

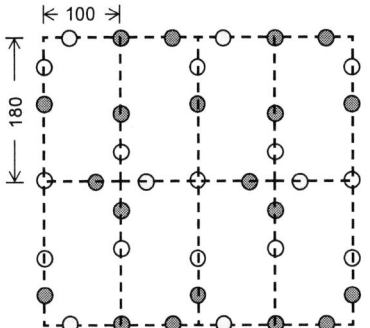

Figure 5: A tentative schematic diagram showing top down patterns for various split colors of the Via 0 mask with 193 nm immersion lithography.

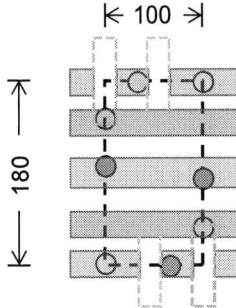

Figure 6: A tentative schematic diagram showing top down patterns for various split colors of the M1 mask with 193 nm immersion lithography.

Metal Process and Overlay

A metal 1 pattern for the SRAM has been shown in Figure 6, which basically adopts a Self-Aligned Litho Etch Litho Etch (SALELE, or SALE2) process with cut process, in which the cut process adopts a Litho Etch Litho Etch (LELE, or LE2) process. As discussed, this metal process can adopt 193 nm immersion SALELE with LELE cuts with the minimum pitch of 90 nm for each single exposure. But for the general metal process where the minimum pitch is 30~32 nm, should 193 nm immersion is used, the self alignment will at least be partially broken down since at least a triple patterning is needed.

As shown in Figures 7(a)-7(e), if the 193 nm immersion lithography is to be adopted for the metal process with a MMP of 30~32 nm, the original self alignment will be partially broken down with at least one space between the metal lines not self aligned, some slight overlay may cause short or reliability issues. A analysis between the SALELE process with the simple LELE with no self alignment is discussed in Ref. [3], where a 2.75 nm Edge Placement Error (EPE) may be added between the non-self-aligned metal LEs instead of just 0.56 nm from SALELE process, about 2.2 nm (3σ) increase!

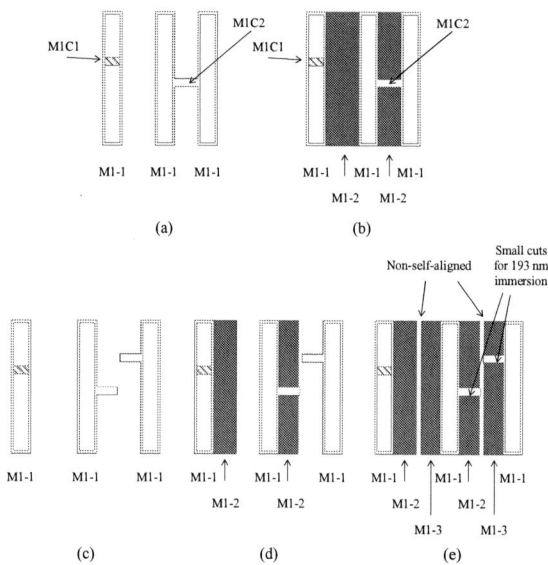

Figure 7: A schematic diagram showing top down patterns for various split colors of the M1 mask under triple patterning scheme with 193 nm immersion lithography with (a),(b) showing the original SALE2 patterning scheme, (c)-(e) showing the issues with the LE3 patterning under the same M1-1 mask.

Besides, according to the Figure 7, for the M1-2 and M1-3 cuts, the maximum AEI dimension is 60~64 nm with no overlay or length window (length, assuming a MMP of 30~32 nm) by 25 nm (width). If ±10% length variation plus 2~3 nm 3σ overlay window is considered, the length of the cut may be at 48~51 nm. As we discussed earlier, if we use NTD, we can have a minimum ADI linewidth of 45 nm in width and around 110 nm in length, which corresponds to a 25 × 70 nm AEI dimension. The maximum length of 48~51 nm will be very difficult to achieve after etch, because it will violate the minimum area rule for the photolithography substantially! An easy way to solve this may be similar to that for the via layers, which is to use 2 hard masks to divide the cut shrink process into 2 steps.

A more detailed study may result in the recommendation to relax the design rules by a few

nanometers (~5-7 nm) to avail extra budget in EPE and to relieve the etch process complexity.

SUMMARY

We have presented a study of the 5 nm logic photolithography process with only 193 nm immersion lithography. A few key challenges have been identified, such as the possible of circuit shortage with round shared contacts, the difficult etch process with very large etch bias between the ADI and AEI linewidth targets for the via and cut processes, the extra EPE between non self-aligned metal lines. Finally, we believe that some relaxation of the design rules and the use of 2 hard mask to relieve the difficulty from the large etch bias for the contact/cut/via processes may help.

ACKNOWLEDGEMENTS

The authors of this paper would like to thank School of Microelectronics Fudan University and National Integrated Circuit Innovation Center (NICIC) for the support of this work.

REFERENCES

[1] National Integrated Circuit Innovation Center, "2019 China's Integrated Circuit Development Roadmap".

[2] Qiang Wu, Yanli Li, et al., "A Study of Image Contrast, Stochastic Defectivity, and Optical Proximity Effect in EUV Photolithographic Process under Typical 5 nm Logic Design Rules", *Proc. CSTIC2020, IEEE Xplore.*

[3] Qiang Wu, Yanli Li, Xianhe Liu, Xiaona Zhu, Shaofeng Yu, "A CDU Budget and Process Window Study with EUV Lithography for 3 nm CFET Logic Processes and an Outlook for Future Generations", *Proc. ICSICT2022, IEEE Xplore.*

LINE-END ROUNDNESS AND VOIDS IMPROVEMENT OF BEOL METAL LAYER

Mudan Wang*, Tiancheng Tu, Hui Zhao, Shirui Yu

Technology Development, Shanghai Huali Integrated Circuit Corporation, Shanghai, China

*Corresponding Author's Email: wangmudan@hlmc.cn

ABSTRACT

Cu Voids are a common defect frequently occurred in BEOL metal layer of any technology. According to the characteristic of orientation, location and formation mechanism, it can be divided into several types such as line end voids, line voids, island voids, seam voids and so on. Increase of line end roundness can improve the Cu fill performance and decrease the occurrence of line end voids. In this paper, methods to enhance the line end roundness are demonstrated from several OPC aspects. An effective verify approach to detect the non-round line end is also discussed. Combining effective OPC approach and verify method can avoid more voids and rapidly improve BEOL yield.

INTRODUCTION

With the continuous shrink of the semiconductor chip dimension, the wavelength of lithography has been much larger than the pattern critical dimension (CD), which caused fatal problems to light diffraction and interference. It is commonly called optical proximity effect (OPE) and it can finally result in pattern distortion at wafer level. Common pattern distortion type includes critical dimension offset, line bridging, line-end shortening/shrinking and corner rounding. Small line-end CD and roundness increase the difficulty for Cu filling and thus increase the probability of voids happening.

Cu plating process (ECP) is a critical step in BEOL processes. With the development of integrated circuit, Cu gap filling in more and more thin trench became a big challenge. The Cu void can lead to metal open which can cause end-of-line (EOL) yield failures. This is especially critical when the voids occur in the vicinity of via connecting to the previous metal layer which could create an open or a reliability concern. Cu void can be caused by many factors such as Cu plating process, Cu barrier/seed process, pre layer CD and so on [1-4]. The CD of line end is smaller than the adjacent one-dimensional (1D) CD due to the existence of OPE. It will make the Cu gap filling at line end more difficult than the 1D parts and more easily generate line end voids (LEVs).

In this paper, we propose an optical proximity correction (OPC) approach to improve the line-end roundness and voids by enlarging the line-end CD. In addition, a valid OPC verify approach is applied to characterize and measure the line-end CD and roundness degree. Combining valid CD enlarging technique and verify approach make line-end voids defect decrease from 2000 to 10 and then improve the Cu gap filling performance.

LINE-END ROUNDNESS CHARACTERI-ZATION AND IMPROVEMENT

From the beginning of 40nm node, model-based OPC correction has been widely used to enhance resolution and regarded as a popular technique to obtain mask pattern. An illustration was plotted to describe the process of model_based OPC, as shown in Figure 1. Original input design, the target designer desired, was divided into many fragments with certain length. A compact model was used to simulate the target and corresponding contour was formed. The difference between the simulated contour and the desired target around pattern edge, also called Edge Placement Error (EPE) is detected by the sites placed along pattern edge [5]. The correction recipe will receive the EPE feedback and then guide the edge segment movement to minimize the difference. The process of EPE feedback and then segment movement will continually iterate and until the EPE results converge reasonably well. The correction mode applies equally to the line end. However, the EPE of line end usually is difficult to detect precisely due to the amount and orientation of sites placed. Just relying on MBOPC is not enough. A more convenient and controllable method is required to improve the line end performance.

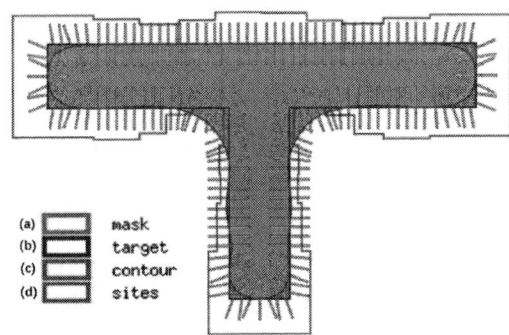

Figure 1. MB OPC operation flow: (a) mask formed by series of edge segments' movement; (b) the input target of designer desired; (c) simulated contour of current mask shape; (d) sites placed along pattern edge.

979-8-3503-1101-3/23 $31.00 © 2023 IEEE

LINE-END ROUNDNESS CHARACTER-IZATION APPROACH

Verify approach is very important to check whether the result is ok and whether it meets the tape-out criteria. Two new verify approaches were proposed to calculate and characterize the line-end CD and roundness as shown in Figure 2. The first way is to measure the CD of locations close to line end 30nm and 50nm respectively. Location 50nm is a position at the transition from 1D to 2D where contour begins to curve. CD at location 30nm can directly reflect the roundness of line-end tip. The second way is to calculate the area ratio of the line-end contour area vs. curve target area around line-end 30nm region. The more closely the value approximate to 1, the more round the line end is.

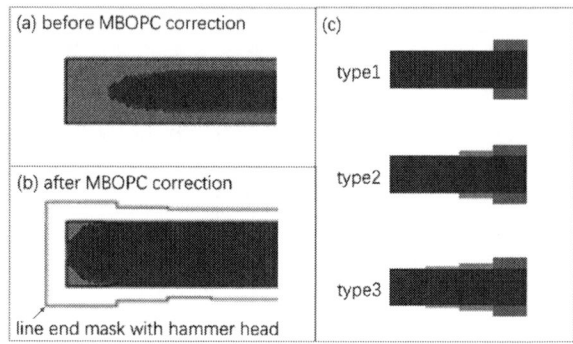

Figure 3. illustration of hammer type: (a) line end before MBOPC correction; (b) line end after MBOPC correction; (c) different hammer head types.

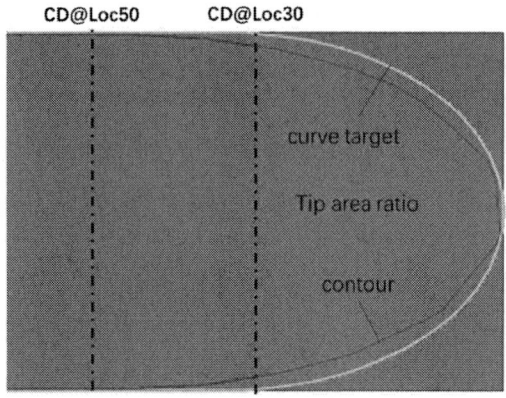

Figure 2. Two verify methods of line-end roundness: (a) CD at location 30nm and 50nm; (b) tip area ratio of contour vs. curve target of line-end 30 nm region.

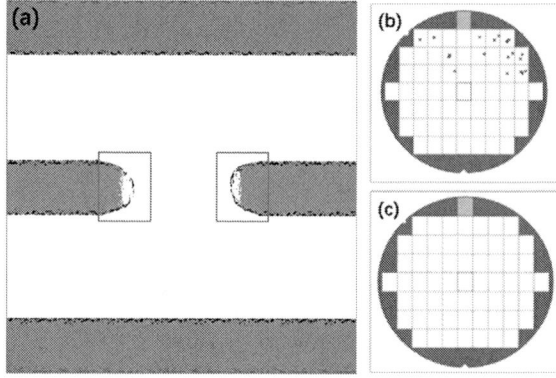

Figure 4. Line-end voids improvement before and after adding hammer. (a) SEM image of LEVs; (b) void scan before adding hammer; (c) void scan after adding hammer.

LINE-END ROUNDNESS IMPROVEMENT BY OPC APPROACH

For the line-end CD shrink and shorten issue, a commonly used OPC technique to correct the line-end mask is to add hammer head around line end. Figure 4 shows the schematic diagram of hammer head. The shorten and shrink issue of line end was greatly improved after MBOPC correcting by adding hammer head at the line end, as shown in Figure 3(a) and 3(b). This inspired us to add hammer head at the line end of target layer to enlarge line-end CD and roundness. Three type hammer head was designed as shown in Figure 4(c). Hammer head was added at the first three fragments region adjacent to line end. The final hammer shape can be designed and adjusted by changing the width and gradient of neighbor two hammer head.

Ten kind of hammer head structures were designed and the difference is the width of hammer head located at the first three fragments, respectively. Table 1 summarized the line-end performance parameters using above two verify approaches. The result demonstrates that type1 hammer head with only one fragment show relatively smaller line end CD and tip area ratio compared with type2 and type3 hammer head. Generally speaking, the larger the hammer head width is, the more round the line end is. Among the 10 samples, sample08 and sample10 with the largest hammer width display distinct advantage not only the CD at loc30 and loc50 but also the tip area ratio. LEVs count scan map of Figure 4 shows that LEVs were effectively cleaned up with the help of line end hammer and the Cu plating properties were improved in the help of enlarged line-end CD and roundness. It can be demonstrated that adding hammer at line end can improve the line-end CD and roundness by suitably designing hammer structure.

TABLE 1. LINE-END ROUNDNESS PERFORMANCE OF DIFFERENT HAMMER HEAD TYPE

hammer width	CD@Loc30	CD@Loc50	tip_area_ratio
0.0-0.0-0.0	40.3	41.5	88.11%
0.5-0.0-0.0	40.1	41.4	88.82%
1.0-0.0-0.0	40.7	41.5	89.22%
1.5-0.0-0.0	40.1	41.5	89.17%
2.0-0.0-0.0	40.5	41.5	89.10%
1.0-0.5-0.0	40.6	42.5	90.27%
1.5-1.0-0.0	40.6	41.6	90.39%
2.0-1.5-0.0	41.4	42.6	91.44%
1.5-1.0-0.5	40.7	41.9	90.10%
2.0-1.5-1.0	41.4	42.6	91.76%

CONCLUSION

In this paper, line-end CD and roundness were improved by adding hammer head around the line end of target layer. The enlarged line-end CD and tip area ratio can effectively improve line-end Cu gap filling performance and thus decrease LEVS count. Two kind of verify methods were also proposed to help monitor and detect those weak patterns with small CD and sharp line end, which can avoid LEVs happening and causing electrical failure. Combining effective OPC approach and verify method can avoid more voids and rapidly improve BEOL yield.

REFERENCES

[1] Weiye He, Beichao Zhang, Jian Kang et al., "The contributions of barrier resputter for BEOL integration", *ECS Transactions*, vol. 44, no. 1, pp. 487-492, 2012.

[2] Yu Bao, Xuezhen Jing, Jingjing Tan et al., "Optimization of Metallization Processes for 28-nm-node Low-k /Cu Multilevel Interconnects", ECS Transactions, vol. 44, no. 1, pp. 477-480, 2012.

[3] Xuezhen Jing, Jingjing Tan, Jiquan Liu, 32/28NM BEOL CD GAP-FILL CHALLENGES FOR METAL FILM, CSTIC, 2015.

[4] Xingdi Zhang, Hunglin Chen, Yin Long, Kai Wang, THE INSPECTION SOLUTIONS OF 3BAR STRUCTURE CU VOID IN BEOL ADVANCED SEMICONDUCTOR PROCESS, CSTIC, 2020.

[5]Yongqiang Hou, QiangWu,Optical Proximity Correction Methodology and Limitations, CSTIC, 2021.

OPC correction method based on corner to corner structure

Qiguang Zhou*, Dan Wang, Yueyu Zhang, Shirui Yu

Shanghai Huali Integrated Circuit Corporation

Shanghai, China

*Corresponding Author's Email: zhouqiguang@hlmc.cn

Abstract

In the OPC correction method for corner to corner structure, it is considered that the distance between the photoresist boundary line (PR line) and its own graph and the distance between the adjacent graph of the previous layer should be kept at a certain distance. The structure of the light mask designed in this paper is characterized by concave in the middle and convex on adjacent sides, which meets the rule of the mask production, so as to reduce the risk of photoresist peeling and improve the yield of silicon wafer chip. This paper also analyzes the optimization of different cutting angles and obtains the method of the optimization size of cutting angles.

Key words: OPC correction method; corner to corner structure; mask production rule; photoresist peeling; Angle cutting

Background

With the development of semiconductor technology, OPC has become an important member in the whole process. As the node process becomes more advanced, the graphics of the design layout become more complex. Especially the corner to corner structure (as shown in Figure 1). In the corner to corner structure, the distance between the corner to corner structure of the graph of the local layer (B as shown in Figure 1) should be satisfied, and the graph of the local layer should be enveloped by other graphs of the internal layer (A as shown in Figure 1). Therefore, this corner to corner structure has little space to be used for graph correction. As we all know, the exposed shape of the square Angle figure will become circular, which will lead to the distance A of the inner figure does not meet the technical requirements. Graphics are complex and space is small, which requires OPC engineers to use special methods to solve these problems.

The traditional way is to the Angle of graphic grows a cornerite structure (as shown in figure 2), the cornerite graph (with shadow), although this structure can guarantee the photoresist line to internal graphics distance far enough, but there is no guarantee that the distance between the corner to corner structure far enough outside, and the risk of violation of mask making rules (figure 2 a star location).

In this paper, the Angle cutting method is used to solve the problem of this kind of corner to corner structure. The influence

of different Angle cutting methods on OPC correction is also analyzed.

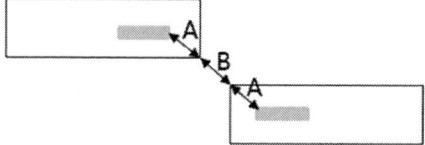

FIG. 1 corner to corner structure

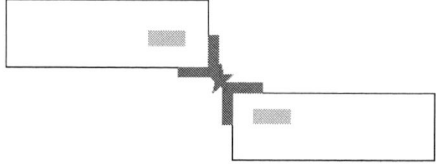

FIG. 2 Traditional OPC correction method

Technical solution

In the OPC correction of corner to corner structure, the distance between the current layer and the previous layer should be considered. In the limited corner to corner structure space, the operation for OPC correction is really difficult, because the rule of the mask production should also be considered, that is, the minimum distance between the figures should not be too small. This paper designs a graphic structure characterized by concave in the middle and convex on adjacent sides (Figure 3). This method is different from the traditional corner OPC modified figure (Figure 2), and meets the rule of the mask production.

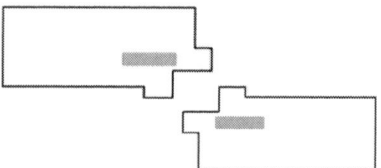

FIG. 3 Design drawing of special light mask

Result analysis

The special light mask structure with concave middle and convex sides is simulated, and the effect Figure 4 is obtained. The purple-shaded figure is the front layer; The blue-shaded figure is the light mask of the layer; The white curve is the photoresist profile (PR line). As can be seen in FIG. 4, when the corner to corner structure distance of the layer pattern is about 80 nm, including the distance to the front layer is about 20 nm, which is in line with the OPC correction expectation, and when the forward and opposite distance of the layer mask pattern is 60 nm, it meets the requirements for making the mask. Figure 5 is the figure after exposure on the silicon wafer. It can be seen that the shape of the photoresist conforms to the OPC simulation results.

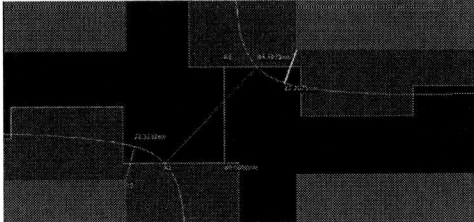

FIG. 4 Effect of specially designed light mask

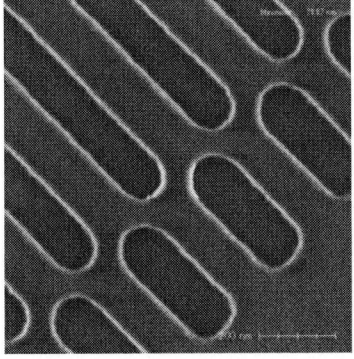

FIG 5. The pattern after exposure on silicon wafer

Optimal experiment

The cutting Angle of different sizes is cut out in the corner of the target pattern, as shown in FIG. 6. The specific method is that the spacing P between the small missing angles dug out in the corner to corner structure is larger or smaller than the minimum spacing value made by the light mask. The simulation results show that when the spacing P is smaller than the minimum spacing value of the mask (as shown in FIG. 7), the yellow diagonal line is the target graph of OPC, and the cut Angle size P is set to 40 nm, which is smaller than the 60 nm rule of the mask production), the repaired mask (blue graph) is retreated by about 10 nm (as shown by the red arrow). And the OPC correction result is not good (the corner to corner structure spacing of the photoresist line is about 50 nm). The reason for the analysis is that the rules of light mask fabrication are taken into account when OPC is operating. In the process of each OPC operation iteration, the edge of corner cutting will be receded for the rule of mask fabrication, so the OPC calculation efficiency will be lost.

In this paper, OPC simulation experiments are also done for different shapes of cutting angles (as shown in Figure 8). The red dashed box is the square cutting Angle with side length of 60 nm. The results show that the mask repaired by the square Angle cut is still stuck by the mask making rules (the distance between the two corners of the mask is 40 nm), and the OPC correction result is not good (the distance between the photoresist lines is about 60 nm).

Therefore, when we do Angle cutting, we should ensure that the distance P of the corner to corner structure is larger than the minimum spacing value of the mask. Also consider the distance from the interior of the current layer to the front layer.

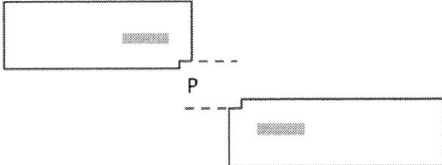

FIG. 6 Optimization analysis of different cutting angles

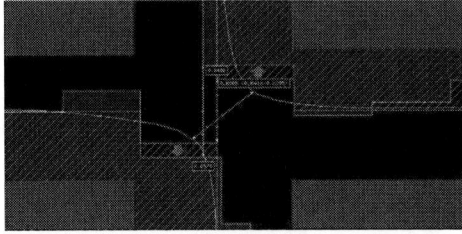

FIG. 7 The cut Angle size is smaller than that of the rule of mask production

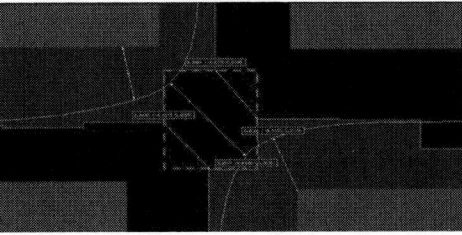

FIG. 8 Influence of square cut Angle on OPC result

Acknowlegement

First of all, I would like to thank Shirui Yu for his valuable advice and guidance in the design and testing of this paper, and Yueyu Zhang, Yanpeng Chen and Dan Wang for their help in the writing of this paper. Finally, I would like to thank all OPC colleagues for their support.

STUDY ON INTER-LAYER OVERLAY OF STITCHING LITHOGRAPHY TECHNOLOGY

Hongmin Liu[1], Changcheng Gao[1], QiongTao Wu[1]*

[1] Semiconductor Manufacturing International Corp., Beijing, China

*Corresponding Author's Email: HongMin_Liu@smics.com

ABSTRACT

The interlayer overlay shift during semiconductor process may lead to a series of serious failure problems including short circuit, open circuit, and graphics failure. Furthermore, existing compensatory techniques are incapable of completely controlling the OVL shift induced by the stitching lithography process for large-area devices. In this paper, inline overlay reveal that the displacement of overlay marks on stitching mask significantly affects the process feedback and interlayer alignment. Therefore, a design rule of overlay mark dedicated to stitching reticle, which states the arrangement of overlay marks in the same exposure image, was proposed to improve the alignment for stitching lithography process.

INTRODUCTION

The overlap of lithography layers is critical in semiconductor manufacturing processes as it serves the purpose of build devices with specific application requirements and functions by stacking up through multiple layers. Aside from the lithography machine's precision, the main causes of inter-layer overlay (OVL) shift are wafer deformation induced by film deposition, thermal budget, CMP, and other operations, but it can be compensated by setting compensation items in the exposure procedure.

With the growing demand for large-array CCD/CIS image sensors in application such as medical imaging and industrial inspection, the size of products has exceeded the maximum field size (~33mm) of the scanner during the exposure process [1-2]. In order to solve the problem that the traditional lithography technology is incapable of preparing large-array productions, many experts developed stitching lithography technology. However，stitching multiple images within single reticle may result out-of-control interlayer overlay. In this paper, a design rule is proposed and used to compensate overlay shift by examining the principle of stitching influence on photography.

COMPENSATION FOR OVERLAY SHIFT

A. stitching lithography technology

The stitching lithography technology was schematically shown in Fig.1, when the size of products exceeds the lithography machine scanner to exposure, it is necessary to distinct images inside a single reticle based on their functions, such as pixel array area and peripheral circuit of image sensors. As demonstrated in "Fig.1", the B image was exposed several times due to the repetitive design, whereas the A image was only exposed twice. Eventually, a complete sensor consisting of two B images and one A image is formed on the wafer.

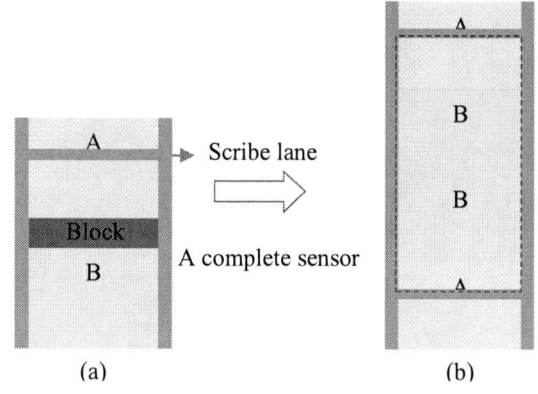

Fig.1: (a) A reticle sample for 1D stitching process; (b) A complete sensor with two B images and one A image

B. Inter-layer overlay adjustment

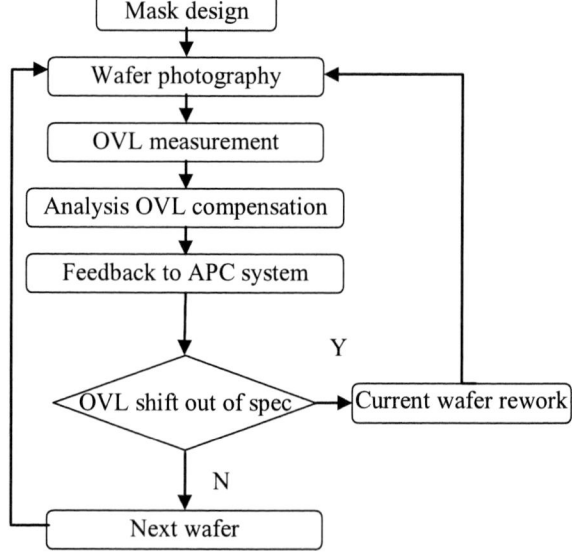

Fig.2: The APC system feedback overlay compensation to ensure manufacture stability.

There are two traditional approaches for increasing the tolerance of interlayer alignment. The first method is to extend the critical dimension of the pattern in layout design to ensure that the process shift is tolerable, which is mainly used for processes with nodes larger than 65nm. Another technique, which is usually utilized in sub-65 nm node processes to ensure manufacture stability, is to configure an APC system to feedback the compensation value to optimize pattern shift for next wafer, as shown in "Fig.2".

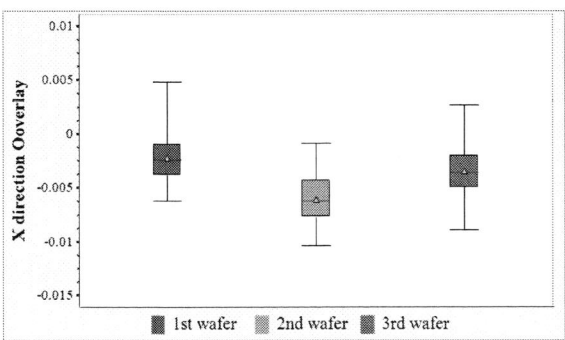

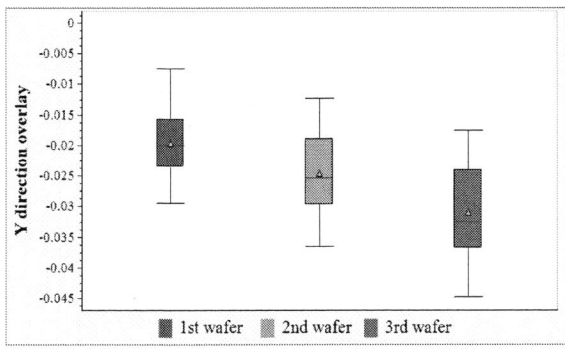

Fig.3: The X and Y direction overlays of 1D stitching photography after feedback from the APC system.

The inline overlay shift of traditional photography, which is controlled by the APC feedback system, is often less than 15 nm (using 293 nm lithography). However, various exposure images during single reticle will inevitably result in alignment changes due to splicing, causing the pattern of the 1D stitching process shifts larger and larger after several repetitions, as shown in "Fig.3".

ANALYSIS OF 1D STITCHING OVERLAY SHIFT

The dimension between the actual exposure image and that calculated compensation does not match, which is why the APC system cannot control the interlayer alignment of 1D stitching process. "Fig.4" illustrates a complete sensor in blue dashed box using the expansion

coefficient in the compensation as an example. When OVL measurements with the complete sensor reveal the shrink state, an expansion coefficient can be obtained as compensation and feedback for the next lithography. Yet, in this scenario, the compensation tendency is the inverse of the actual demand, resulting in poor inter-layer alignment after APC feedback.

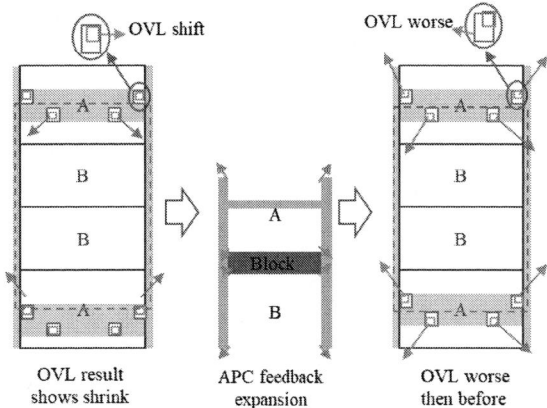

Fig.4: APC system failure when controlling inter-layer overlay of 1D stitching photography due to image splicing.

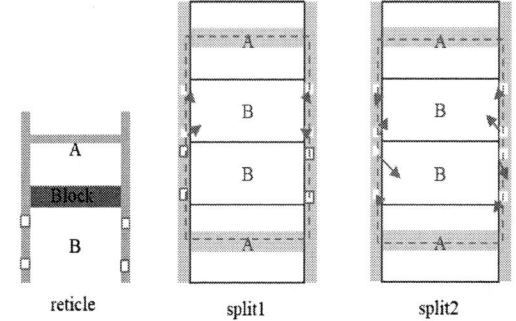

Fig.5: Two split verifications for designing OVL mark positions on the stitching reticle.

Based on the above-mentioned analysis, the displacement in reticle design was proposed to disperse the OVL mark reasonably within one of the different exposure images. As show in Fig.5, the yellow OVL mark of split 1 and split 2 designs were tested individually. The results show that the maximum OVL residue is respectively about 3.5nm @ split 1 and 9.8 nm @ split 2 as shown in Fig.6, which demonstrates that split 1 has a smaller pattern shift and is more appropriate for maintaining the stability of inline process. As a result, to effectively restrain the influence of image splicing on the inter-layer alignment, the design rule of positioning the OVL mark for measurement within the same exposure image must be observed.

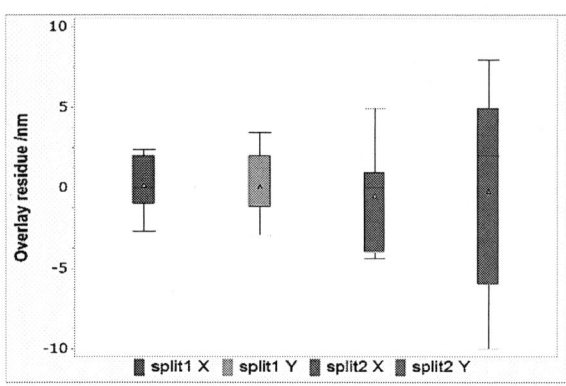

Fig.6 The inherent poly residue in the 1D stitching lithography after APC compensation.

CONCLUSION

In this paper, to compensate overlay shift and achieve stability of inter-layer alignment for stitching lithography technology, the design rule for overlay mark on stitching reticle at sub-65nm node process was proposed, which clearly specifies that measured OVL marks must be placed within the same exposure image.

REFERENCES

[1] D. Durini, F. Matheis, C. Nitta, W. Brockherde and B. J. Hosticka, "Large full-well capacity stitched CMOS image sensor for high temperature applications," *2010 Proceedings of ESSCIRC*, 2010, pp. 130-133.

[2] Yu B , Jia W , Zhou C , et al. Grating imaging scanning lithography[J]. *Chinese Optics Letters*, 2013.

A NEGATIVE-TONE PHOTOSENSITIVE EPOXY MATERIAL

Ke Bai*, April Tang, Jiangtao Mou, Jinfu Zheng
*Shandong Shengquan New Materials Co. Ltd
Jinan City, Shandong Province, China
*Corresponding Author's Email: bai.ke@shengquan.com

ABSTRACT

A negative-tone photosensitive epoxy material has been developed for MEMS applications at Shandong Shengquan New Materials Co. Ltd. It has excellent film properties, including high tensile strength, low modulus, excellent high temperature stability, chemical resistance, high transparency, and high aspect ratio Photolithography, mechanical, thermal properties, chemical resistance and transmittance of the photosensitive epoxy material is described in this paper.

INTRODUCTION

MEMS has been identified as one of the most promising technologies for the 21st Century and has the potential to revolutionize both industrial and consumer products by combining silicon-based microelectronics with micro- machining technology. Its techniques and microsystem-based devices have the potential to dramatically affect of all of our lives and the way we live.

Lithography technology is widely applied in MEMS manufacturing. There are 3 different exposure method: contact exposure, proximity exposure, projection exposure. High-performance dielectric material is normally designed for micro-electro mechanical systems (MEMS) and package process [1]. The photosensitive materials applied in MEMS manufacturing requires excellent lithography performance and excellent comprehensive properties, it is also suitable for many other applications, such as sensors and packaging, semiconductor device manufacturing and packaging, and micro-machining. In addition to its main application in MEMS and packaging, photosensitive epoxy material can also be applied as adhesive layer materials and structural support materials that require convex-concave patterns [2] , such as displays/micro arrays [3] /microfluidic [4] channels/structural materials for printer heads; due to its excellent patterning capability, Photosensitive epoxy material is also becoming the main material for many research institutions and engineers to make models and prototyping [5] . In short, Photosensitive epoxy material materials are still one of the most commonly used negative photosensitive materials in many nano fabs and cleanrooms.

By application, the MEMS market is segmented into consumer electronics, automotive, industrial, aerospace & denfense, healthcare, telecommunication and others. The consumer electronics segment acquired the largest share in 2021, and expected to grow at a significant CAGR from 2022 to 2031 for MEMS market size. With the development of MEMS industrial, we believe that there will a stable increasing in photosensitive epoxy material.

In this paper, we presented a new negative-tone photosensitive epoxy material developed at Shandong Shengquan New Materials Co. Ltd. This commercially available product shows excellent material properties along with outstanding photolithography property.

METHODOLOGY

Materials

Substrate: 6 inches bare Silicon wafer, orientation 100, resistivity 0.001-100,000 ohm cm.

The SQED photoresist (Shandong Shengquan New Materials Co. LTD, Jinan City, Shandong Province, China) was chosen as the structured layer in this study.

Edge beads removal: SQED-E002 (Shandong Shengquan New Materials Co. LTD, Jinan City, Shandong Province, China)

Developer: SQED-D002 (Shandong Shengquan New Materials Co. LTD, Jinan City, Shandong Province, China)

Rinser: SQED-R002 (Shandong Shengquan New Materials Co. LTD, Jinan City, Shandong Province, China)

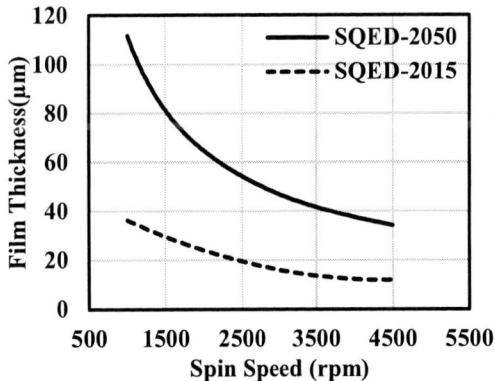

Fig. 1: Spin Speed Curves for SQED-2000 Series

The coating conditions (rotation speed/time) determine the thickness of the film. In order to obtain different film thicknesses, the current SQED-2000 series include two products, SQED-2050 and SQED-2015, which can cover a film thickness range from 10 microns to 120 microns. Figure 1 shows the relationship between the recommended coating speed and film thickness of SQED-2050 and SQED-2015 (spin/speed Curve).

Photolithography Process

The process flow of the photosensitive epoxy material includes the following steps: priming, coating, soft bake, exposure, post exposure bake, development, and hard bake as shown in Fig. 2.

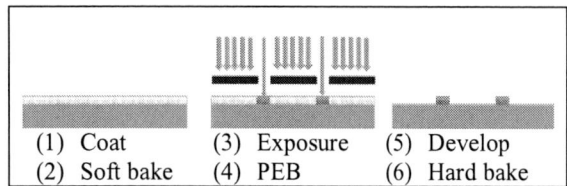

| (1) Coat | (3) Exposure | (5) Develop |
| (2) Soft bake | (4) PEB | (6) Hard bake |

Fig. 2: Photolithograph process

The photoinitiator in the photoresist absorbs photons and undergoes a photochemical reaction to generate a strong acid, which acts as an acid catalyst to promote the occurrence of crosslinking reactions during the baking process. Only the photoresist in the exposed areas contains strong acid, while the unexposed areas do not have such strong acid. In the subsequent intermediate baking process, the exposed area is catalyzed by strong acid, and the molecules are cross-linked. The crosslinking reaction proceeds in a chain, and each epoxy group can react with other epoxy groups in the same molecule or in different molecules. As mentioned earlier, each epoxy group has an pre-connected of another epoxy groups, and then a dense cross-linked network is formed after extended cross-linking. This network is insoluble in the developer solution. In the unexposed area, the photoresist is not cross-linked and is dissolved in the developer, so the reverse pattern of the mask plate is formed after development.

Characterization

After hard baking, the film is fully crosslinked. Then the film is stripped from 6 inches bare silicon wafer. Film properties, like mechanical properties, thermal properties, are investigated by below apparatus.

The TMA experiments were conducted using TMA equipment (Q400, TA Instruments, USA). The coefficient of thermal expansion (CTE) can be determined by dimension change – temperature curve.

The DMA experiments were conducted using DMA equipment (DMA GABO, NETZSCH, Germany). The glass-transition temperature (Tg) can be determined by tan δ.

The tensile experiments were conducted using tensile equipment (Dage 4000, Nordson, UK). Elongation to break, Youngs modulus, Tensile Strength can be determined by distance force curve.

The TGA experiments were conducted using TGA equipment (Q50, TA Instruments, USA). Weight loss were conducted in oxygen and nitrogen atmosphere.

Chemical resistance is conducted after hard bake on 6 inches silicon wafer when photosensitive epoxy material fully cross-linked. Film thickness is tested before and after chemical resistance test with thickness measurement system (ST 5000, K-MAC, Korea). Film appearance is visually inspected with microscope (OLYMPUS, MX63L, Japan) before and after chemical resistance test.

Transmittance experiments were conducted using ultraviolet spectrophotometer (UV 1900, SHIMAZDU, Japan).

RESULTS AND DISCUSSION
Photolithography

Figure 5 shows the photolithography property of the photosensitive epoxy material, which was obtained by a contact aligner with cut-off filter at a exposure dose 300mJ/cm2 (i-line @ 365nm wavelength). Strong agitation was applied for development.

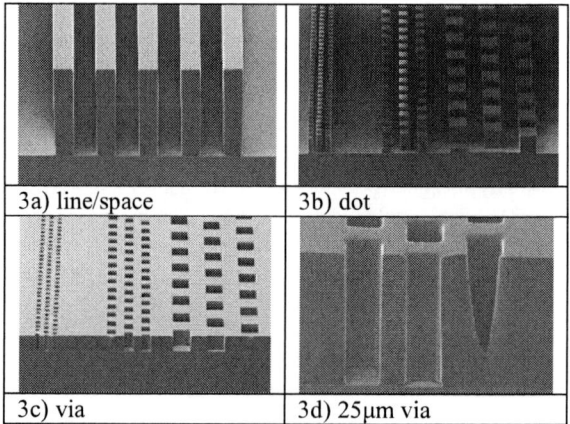

Fig.3: Photolithography of the negative tone photosensitive epoxy material after final cure at 150°C.
3a) Line/space structure;
3b) Dot structures;
3c) Via structures.
3d) 25µm via structure.

As SEM images shown in the Fig.3, line/space, dot, via structures, 25µm via are clearly developed for the films with thickness of 100µm (after final cure@150°C) in Fig 3a, 3b, 3c, 3d respectively. After development, the substrates were clear without any scum. An uniform and no cracking pattern was obtained. Pattern profile was almost vertical to the substrate. The epoxy photoresist also showed high resolution and aspect ratio, it could get 15µm line / space and 25µm via structure.

Mechanical / Thermal Properties

As shown in Table I, the negative tone photosensitive epoxy material serials material demonstrates outstanding performance in all major material properties and therefore indicates a great potential in many applications. In addition, the current material is developed from a versatile material designing platform and many film properties can be adjusted by changing resin formation and/or changing formulation additives.

TABLE I
FILM PROPERTIES

Film spec	Test result	Unit
Youngs Modulus	2.3	GPa
Tensile Strength	99.4	MPa
CTE	50.3	ppm/°C
Elongation	14	%
T_g	237.1	°C
5% weight loss (N2)	373.1	°C
5% weight loss (Air)	337.0	°C

Chemical Resistance

Table II shows the chemical resistance of the photosensitive epoxy material with 10um film thickness coated on 6 inches silicon wafer. Thickness and visual inspection were conducted before and after chemical resistance test. Film loss refers to thickness percentage change before and after chemical resistance test.

As shown in Table II, the photosensitive epoxy material shows no apparent film appearance and film loss change in organic solvents, acid aqueous, alkaline aqueous. In addition, the photosensitive epoxy material shows slight swelling in acetone, which is typical characteristic of negative tone photoresists.

TABLE II CHEMICAL RESISTANCE

Chemical	Condition	Film Appearance	Film Loss
IPA	RT,30min	No change	<1%
Ethanol	RT,30min	No change	<1%
NMP	RT,30min	No change	<1%
Acetone	RT,30min	Slight swelling	<1%
1%HF	RT,30min	No change	<1%
2.38TMAH	RT,30min	No change	<1%
DMSO	RT,30min	No change	<1%
30%H_2O_2	RT,30min	No change	<1%
10%H_2SO_4	RT,30min	No change	<1%
10%KOH	RT,30min	No change	<1%

Transmittance

Transmittance verse wavelength of SQED-2000 series photoresist with 10um thickness film coated on glass wafer. The film was cured at 200°C for 30min on hot plate. As shown in Fig. 4, the photosensitive epoxy material shows high transmittance on visible wavelength region. The transmittance is 98.5% and 100% at 426 nm and 500nm wavelength respectively.

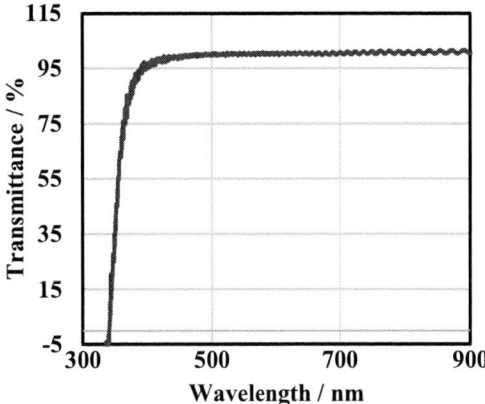

Fig. 4: Transmittance verse wavelength curve

The negative-tone photosensitive epoxy material has excellent film properties, including high tensile strength, low modulus, excellent high temperature stability, chemical resistance, high transparency, and high aspect ratio, while retaining low temperature crosslink properties. With low Young modulus, the negative tone material is ideal choice for MEMS / nano structure.

CONCLUSION

A new negative tone photosensitive epoxy material is presented in this paper. The material shows excellent mechanical, thermal properties, chemical resistance, high transparency for MEMS and packaging applications. Processing of the material is straightforward and compatible to most semiconductor fabrication processes.

REFERENCES

[1] Yang, L., et al., Review on stationary phases and coating methods of MEMs gas chromatography columns. Reviews in Analytical Chemistry, 2020. **39**: p. 247-259.

[2] Fujita, M. and M. Hibi, Stamper, method of forming a concave/convex pattern, and method of manufacturing an information recording medium. 2007.

[3] Yu, L., et al., High-performance UV-curable epoxy resin-based microarray and microfluidic immunoassay devices. Biosensors & bioelectronics, 2009. **24**: p. 2997-3002.

[4] Olmos Carreño, C., et al., Epoxy resin mold and PDMS microfluidic devices through photopolymer flexographic printing plate. Sensors and Actuators B: Chemical, 2019. **288**.

[5] Jenkins, G., Rapid prototyping of PDMS devices using SU-8 lithography. Methods Mol Biol, 2013. **949**: p. 153-68.

TUNGSTEN/SILICON OXIDE/TITANIUM NITRIDE STACK ETCHING

Jie Luo, Haochang Lyu, Linjie Hou, Baodong Han, Hongbo Sun, and Chao Zhao[*]

Beijing Superstring Academy of Memory Technology, Beijing 100176, China

*Corresponding Author's Email: chao.zhao@bjsamt.org.cn

ABSTRACT

The periodic stacked film etching is a key enabler of 3D device manufacturing. This work demonstrates a silicon oxide/tungsten/titanium nitride periodic stack etching in one etching step. Because of the chemical reaction mechanism differences for various materials, a wave-like sidewall is always observed after one step etching. Sub-100nm holes on single and periodic stacks are etched with Cl2/NF3/C4F6/Ar gases. By tailoring the gas ratio and bias power, passivation distribution and ion bombardment strength are optimized and a smooth sidewall with an excellent vertical profile is achieved.

INTRODUCTION

Challenges, induced by quantum effect, are pending to be solved when advanced devices are scaled down to sub-ten-nanometer era. 3D integration is emerging as a solution to raise the integration level without the scaling induced quantum effect [1-3]. To date, silicon oxide/poly silicon periodic stack, silicon oxide/silicon nitride periodic stack, silicon/silicon-germanium periodic stack and gate stack etching are developed and reported [4-7]. With those researches, 3D NAND and 3D nano-wire devices are able to be transferred from lab to fab with manufactured and commercialized process flow. Recently, a novel scheme for 3D DRAM has been presented，which is strongly dependent on a metal/dielectric periodic stack [3]. And stacking film etching is a critical process which has notable direct correlation with device performance during the 3D DRAM fabrication [8-10]. Therefore, in this work, silicon oxide/tungsten/titanium nitride (SiO2/W/TiN) periodic stack etching is studied. And a tunable and reliable etching solution is proposed for 3D DRAM research and manufacture.

To etch the stack, etching gases need to be selected accordingly. Table I. shows some gases which are reported to be used as the main etching gases for the target materials. For W and TiN etching, a NF3/Cl2 mixing gas is a good alternative since it includes F and Cl to remove W and TiN by forming gas by-product of WF6 and TiCl4. Thus, NF3/Cl2/N2 [12] is preferred. And Ar is used to replace N2 to promote TiN etching rate by suppressing N's concentration [13]. At the same time, C4F6 is added for SiO2 etching [15]. Therefore, in this work, NF3/Cl2/C4F6/Ar is determined as the mixing etching gas.

TABLE I.   REPORTED ETCHING GASES FOR W, TiN AND SiO2.

Material	Gas	Reference
W	SF6/O2	[11]
	NF3/Cl2/N2	[12]
TiN	CHF3/Ar	[13]
	Cl2/Ar	[13]
	BCl3/Ar	[13]
SiO2	C4F8/Ar/O2	[14]
	C4F6/Ar	[15]
	C4F8/Ar	[15]

EXPERIMENT

An inductively coupled plasma (ICP) etching chamber, Applied Materials-Centris Sym3 Y, is used for the stack etching. The SiO2/W/TiN periodic stack is prepared by plasma enhanced chemical vapor deposition and atomic layer deposition technics. The SiO2, W and TiN layers are stacked as Figure 1. 60nm holes are patterned by the 193nm DUV immersion lithography system. And the carbon is used as etching hard mask during the etching process.

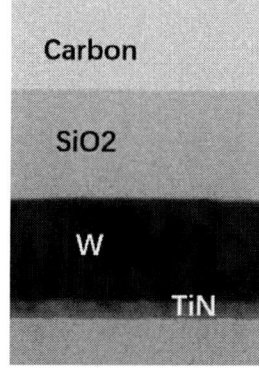

Figure 1: A SiO2/W/TiN stack with carbon hard mask.

RESULTS AND DISCUSSION

Optical Emission Spectrum (OES)

A single SiO2/W/TiN stack is etched firstly. The gas flow rates of NF3/Cl2/C4F6/Ar are set as a:b:c:d., The elements and chemical bonds can be identified according to the results from optical emission spectrum. CN, N2, CCl, CO, CF and SiF2 which are considered as the components of etching by-products are marked in the spectrum in Figure 2.

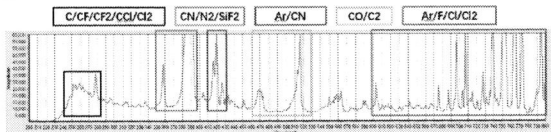

Figure 2: An optical emission spectrum (200nm~800nm) of SiO2/W/TiN stack etching. Etch gases are NF3/Cl2/C4F6/Ar.

Vertical profile - Cl2 and NF3

To understand the sidewall passivation mechanism, sample #1/2/3 are prepared with different NF3/Cl2 flow rate, a:0, a:b and 2a:0. The split table is shown in Table II. To estimate the SiO2 etch rate, the ARC layer, SiON (removed during etching process) removal rate is calculated (1). The thickness of SiON is designed as znm and the removing time is identified by SiF2 signal (395.5nm) which is shown in Table II.

$$SiON\ etch\ rate = \frac{Thickness_{SiON}(znm)}{Delay\ Time_{Signal\ Transforming}} \qquad (1)$$

Comparing sample #1/2/3 in table II, it is found that SiON etch rate raises as Cl2 or NF3 increasing, which can be explained that the involvement of Cl2 or NF3 contributes to removing the surface passivation and promoting etching process. And it is also confirmed by the different profiles for various material species in table III. The addition of more Cl2 and NF3 will facilitates the formation of vertical profiles transformed from the taper profiles. And comparing sample #2 and #3, both Cl2 and NF3 work for SiO2 profile optimization. However, Cl2 removes the sidewall passivation on both SiO2 and W, NF3 only removes the sidewall passivation on SiO2, not the one on W. In other word, NF3 can only clean the polymer on SiO2 or SiON, but Cl2 removes the polymer on both SiO2/SiON and W.

TABLE II. SiON ETCH RATE UNDER DIFFERENT FLOW RATE..

Sample #	NF3/Cl2/C4F6/Ar Gas Flow Rate	SiF2 signal and SiON Etch Rate
1	a:0:c:d	2ynm/min
2	a:b:c:d	3ynm/min
3	2a:0:c:d	3ynm/min

TABLE III. BY CHANGING THE CL2 AND NF3 GAS FLOW RATE, SIO2 AND W PRESENT DIFFERENT SIDEWALL PROFILES.

Sample#	TEM
1	
2	
3	

The passivation polymer components are analyzed with the OES data at Figure 2. The OES curve shows strong signals of CN and CF. Based on that, a C-based polymer, CxNyFz, is considered as the sidewall passivation polymer on SiO2 and W sidewalls. Focusing on the cleaning gases, NF3 and Cl2, NF3 introduces N element which is a component of CxNyFz, it is not a perfect choice for N based polymer removal. But for Cl2, it does not introduce any other polymer related elements, (2).

$$C_xN_yF_z + Cl_2 = CCl_x + CNCl + NF_3 + NCl_3(N_2 + Cl_2) \ (2)$$

That means Cl2 helps to clean the passivation up and continue the etching reaction.

NF3 can only help with participations, such as the oxygen in SiO2, (3).

$$C_xN_yF_z + NF_3 + SiO_2 = SiF_4 + CO + NO \qquad (3)$$

Smooth Ox and W Interface – Ion Bombardment

2MHz bias power is identified as a factor which affects ion bombardment strength and ion angle distribution during etching process. Sample #2/4/5 are prepared with the same gas ratio, NF3/Cl2/C4F6/Ar, a:b:c:d, but different bias powers, h/j/k, h<j<k. The results are shown in Table IV.

TABLE IV. *DIFFERENT BIAS POWERS PROVIDE DIFFERENT SiO2 AND W INTERFACE. IT CAN BE UNDERSTOOD BY ION BENDING EFFECT INDUCED HIGH BOMBARDMENT AT THE SiO2 AND W INTERFACE.*

Sample #	2M Bias Power	TEM Images
2	h	
4	j	
5	k	

With different bias power, SiO2 and W's interface are different. h bias power delivers a discontinued interface - a SiO2 recessing on the W top surface. However, when the bias power reaches j, the recessing amount is less. This recessing can be understood by an ion bending effect which is induced by the W layer. As a conductive metal, W would like to attract and remove positive ions during etching process. Because of the attraction, ions are bended towards SiO2 and W's interface and the SiO2 recessing is observed. In other words, there is an additional local electric field at the interface of SiO2 and W because of the material difference. Under this field, ions are bent and bombard the SiO2/W interface. Then, it boosts SiO2 etch rate [16] and causes the SiO2 recessing.

In this work, the SiO2 recessing is alleviated by a high bias power. As the bias power increasing, ion bombardment energy is enhanced [17]. And a high-energy ion, j bias power and above, is strong enough to avoid the bending by the additional local electric field. All in all, a high bias power etching process delivers a smooth profile.

Periodic Stacked Film Etching

Periodic stacked SiO2 and W films are etched with similar etching conditions above. h bias is used for sample a, Figure 3(a) By increasing bias power from h to j, etching profile becomes vertical but carbon is sacrificed too much during etching process, shown in Figure 3(b). Figure 3(c) delivers a longer etching time result under h bias power. Although the 5 cycles periodic stacked SiO2 and W film is etched, the profile is too taper. All in all, the periodic stacked SiO2 and W film etching with carbon hard mask is limited by the SiO2/W and carbon selectivity.

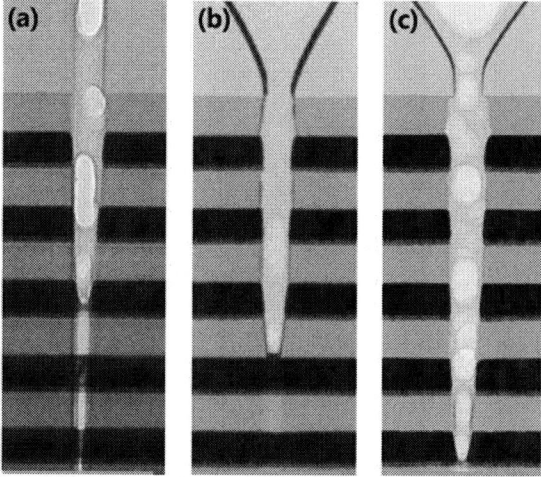

Figure 3: 300s etching under (a) h bias power and (b) j bias power. Considering the carbon hard mask loss during etch process, (c) long etching time with h bias power is used, finally.

CONCLUSION

In this work, we demonstrate a sub-100nm hole etching with a carbon hard mask on a SiO2/W/TiN periodic film stack. A NF3/Cl2/C4F6/Ar mixing gas is used and holes are formed in one etching step. Sidewall passivation on SiO2 and W surfaces are changed by removing carbon (forming CClx) and nitrogen (forming CNCl) from the surfaces, simultaneously. And a sweet point is found, a vertical profile is achieved. At the meantime, ion bombardment is discovered as a major factor for discontinued SiO2/W interface. In the end, a 5-cycles periodic stacking film is etched which delivers a promising start for high-aspect-ratio SiO2/W/TiN etch researching.

REFERENCES

[1] D. Y. Chen et al., 2009 IEEE International Electron Devices Meeting (IEDM), 2009, pp. 1-4.
[2] Q. Huo et al., IEEE Electron Device Letters, vol. 41, no. 3, pp. 497-500.
[3] K. Huang et al., 2022 IEEE Symposium on VLSI Technology and Circuits, pp. 296-297.

[4] Z. Yang et al., ASMC 2013 SEMI Advanced Semiconductor Manufacturing Conference, 2013, pp. 24-26.

[5] M. -K. Jeong, H. -I. Kwon and J. -H. Lee, 2009 IEEE International Memory Workshop, 2009, pp. 1-2.

[6] V. Paraschiv, W. Boullart, E. Altamirano-Sánchez, Microelectronic Engineering, Volume 105, 2013, Pages 60-64。

[7] Shin, Myoung Hun, et al., Japanese journal of applied physics 44.7S (2005): 5811.

[8] L. Petti et al., IEEE Electron Device Letters, vol. 36, no. 5, pp. 475-477.

[9] Y. Kim, H. Kang, G. Kim, C. Hwang and S. Yoon, IEEE Electron Device Letters, vol. 38, no. 10, pp. 1387-1389.

[10] H. Kim, J. Yang, G. Kim and S. Yoon, 2018 25th International Workshop on Active-Matrix Flatpanel Displays and Devices (AM-FPD), 2018, pp. 1-4.

[11] J. Berthold, C. Wieczorek, Applied Surface Science, Volume 38, Issues 1–4, 1989, Pages 506-516.

[12] Zaleski M, Simpson L, Guerry A. Electrochem. Soc. Proc. 98-4, 1998: 203-209.

[13] Tonotani, J., et al., Journal of Vacuum Science & Technology B: Microelectronics and Nanometer Structures Processing, Measurement, and Phenomena 21.5 (2003): 2163-2168.

[14] Matsui, Miyako, Tetsuya Tatsumi, and Makoto Sekine. Journal of Vacuum Science & Technology A: Vacuum, Surfaces, and Films 19.5 (2001): 2089-2096.

[15] Li X, Hua X, Ling L, et al., Journal of Vacuum Science & Technology A: Vacuum, Surfaces, and Films, 2002, 20(6): 2052-2061.

[16] Schaepkens, Marc and Gottlieb S. Oehrlein. Journal of The Electrochemical Society 148 (2001).

[17] Liu, J., G. L. Huppert, and H. H. Sawin. Journal of applied physics 68.8 (1990): 3916-3934.

THE INVESTIGATION OF CF3I FOR HIGH-ASPECT-RATIO CRYOGENIC DIELECTRIC ETCH

Jianqiu Hou[1], Vina Xu[1], Kai Zhang[1], and Ziyang Wu*

[1]Advanced Micro-Fabrication Equipment Inc. Headquarter. Shanghai 201201, China

*Corresponding Author's Email: ziyangwu@amecnsh.com

ABSTRACT

Considering the high surface adsorption capacity in cryogenic condition, we can achieve high etch rate, good bottom circularity and less risk of chamber arcing for high-aspect-ratio etch, by using C1 gases and lower power. However, the strategy will cause top layer blowing out and sharp etch front. We here explore a specific gas, CF3I, to deposit on side-wall. It can significantly decrease bowing while results in the reduced etch rate and mask. By tuning source and bias power, we can further enlarge etch front and mitigate the negative effect of CF3I.

INTRODUCTION

The high-aspect-ratio (HAR) dielectric etch is a critical and very challenging process when fabricating advanced memory device. Channel hole etch is a typical HAR etch [1]. The film stack before etching is composed by multiple pairs of silicon oxide (OX) and silicon nitride (SiN) layers (Figure 1). After etching, the sacrificial SiN will be removed and the channel hole will be filled with metal. It works as a control gate in the whole device. The aspect ratio of channel hole is usually more than 30 while the critical dimension (CD) should be controlled after etching.

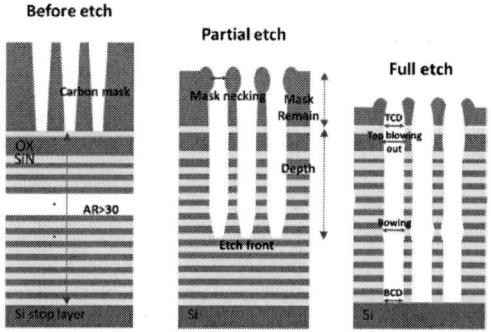

Figure 1: The schematic of channel hole etching

We use a capacitive coupled plasma (CCP) etch tool for HAR dielectric etch. The high ion energy is helpful to keep good anisotropy even in very deep holes. The RF power, produced by two generators, can form the plasma between upper and lower electrodes. The reactive ions in plasma will be accelerated by the bias voltage in plasma sheath, bombard the wafer surface to remove layers thus forming the target HAR patterns. We here use two RF generators with >40 MHz and <2 MHz frequency, respectively, to decoupled control ion density and ion

energy. The higher power of high frequency will increase the collision probability of gas molecules, resulting in higher ion and radical density. The higher power of low frequency will increase the bias voltage between plasma and wafer surface, leading to higher ion bombardment.

For channel hole etch, we should avoid bottom under etch meanwhile CD excessive enlargement. To address the challenge of anisotropy, the traditional strategy is using C4 gases, such as C4F6 and C4F8, to form a large amount of polymers. They can deposit on mask and side-wall due to the high molecule weight and adhesion coefficient. However, excessive polymer will increase the risk of mask blocking and bottom under etch. The schematic of the polymer deposition is shown in Figure 2. On the one hand, the bottom residual polymer is difficult to be pumped out. On the other hand, by the continual bombardment of ions, the etch front will be positively charged [2]. It will repel the reactive ions, resulting in reduced ER and bottom distortion [3].

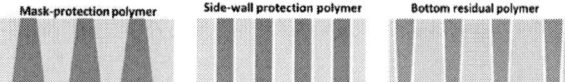

Figure 2: The schematic of polymer deposition

To enlarge the process window between small CD and blocking or under etch, we should increase the low frequency power i.e. bias power as much as possible. The strategy is not efficient because most energy are wasted on forming and removing polymer without etching target materials. Besides, too much polymer and ultra-high power will result in a challenge of chamber stability and reliability. Excessive polymer always deposits on chamber wall, surface of parts and wafer backside, leading to high probability of arcing with high power. Although frequent offline clean can remove the polymer, it interrupts the continual production thus increases overall cost.

Therefore, we choose the other strategy called cryogenic etch [4]. Lower wafer temperature (<-20°C) will significantly increase the adsorption coefficient of etchant and polymer, leading to higher ER and deposition rate. In that case, C1 gases, such as CF4, CHF3 and CH2F2, are enough to form polymers, which can deposit on side-wall and mask. The polymers produced by C4 gases and C1 gases are much different on molecule weight and structure [5]. For example, the chain growth of C4F6 is based on radical chain polymerization due to the unsaturated bond. Under the plasma excitation, it is readily to form a radical then connect with another C4F6 monomer to form a longer chain radical. By the repetitive reaction between chain radical and monomer, the chain is quickly extended and finally forms a high-molecule-weight polymer. Considering the two double bonds i.e. two reactive sites in one C_4F_6 monomer, the network structure will be also established by crosslinking. For C1 gases, the chain growth is based on radical coupled polymerization. Single radicals can be combined to each other, and then form dimers, tetramer step by step. Compared with the radical chain polymerization, the reaction rate much decreases with chain length due to steric hindrance, resulting in the less molecule weight and incompact polymer.

At room temperature, the incompact polymer formed by C1 gases cannot properly deposit on side-wall and mask. However, when we decrease the temperature to -60°C, the adhesion coefficient will be much increased, leading to the improved protection of mask and side-wall. Compared with the performance of C4 gases at room temperature, the bottom polymer from C1 gases can be easily removed by lower bias power. Besides, the surface concentration of etchant is also increased at lower temperature, resulting in much higher ER. Consequently, using the strategy of C1 gases at low temperature can produce the polymer that can deposit on side-wall and mask with low risk of bottom under etch. Therefore, the very high power is not necessary, resulting in the lower risk of arcing (Figure 3).

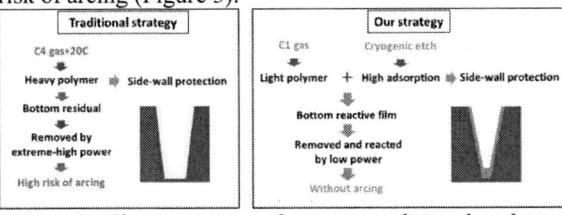

Figure 3: The comparison between traditional etch and cryogenic etch.

However, there are still some disadvantages in cryogenic etch. The main side-effect is top layer blowing out, caused by lacking top-layer-protection polymer or byproduct. To address the challenge, we provide a specific halogen-C1 gas, CF3I, to form protective byproduct on top layer, as shown in Figure 4. The structure that can passivate OX surface is identified as SiOIx[5]. By using CF3I, we can obviously mitigate the blowing out of top layer. Based on this concept, an appropriate cryogenic condition is proposed and optimized by tuning power, flow and pressure on CCP etch tool. We found that, when the ratio of bias power to source power is 2, the profile becomes straight. Besides, the flow ratio of CF3I to main gases, such as CF4, CHF3 and CH2F2, should be minor than 1/20. Too much CF3I will results under etch due to the iodine contained byproduct. In that case, if we increase O2 or bias power to open them, the mask will be much

consumed. A pressure range from 20mT to 25mT is appropriate because higher pressure usually makes a sharp bottom while lower pressure decreases ER. According to our experimental result and experience, other halogen contained gases, such as HBr, CH2Br2, Cl2, CH2Cl2 and so on, are also promising to be applied in cryogenic etch.

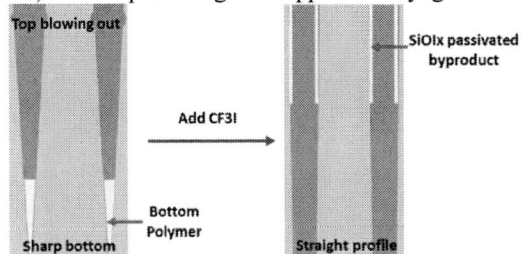

Figure 4: The top-layer protection by CF3I.

EXPERIMENTAL

All etching experiments were conducted in a capacitive coupled plasma (CCP) etch tool with two radio frequency (RF) generators (60 MHz source and 400 kHz bias power). The etch tool was AMEC HD RIE plus. The cryogenic chiller was supplied by Beijing Instrument Industry Group Co., Ltd. The CF3I gases was purchased from Beijing Yuji Science & Technology Co., Ltd.

To evaluate the dielectric film ER, we used the "blanket wafer", a 300mm bare Si wafer coated by ~1μm OX or ~600nm SiN film. A KLA SCD device was used to measure the thickness of OX or SiN film on 47 points by radius before and after etch. The blanket ER can be calculated by the change of film thickness and etch time.

To avoid wasting resources, we used a small chip sample that cut from whole wafer for experiment. The 2cm x 3cm chip has already been pre-coated with designed film stack for following etching process. We then pasted the chip on a carrier Si wafer in AMEC lab.

Scanning Electron Microscopy (SEM), Hitachi 4800, in AMEC lab, was applied to observe the cross-section and top view of samples. To observe the bottom circularity, we immersed post-etch coupon in 1% HF solution for 5-10 min then used ultrasonic to clean before SEM analysis.

Transmission Electron Microscope (TEM) and Energy Dispersive X-Ray (EDX) Spectroscopy analysis were supported by MSSCORPS CO., LTD.

RESULT AND DISCUSSION

Cryogenic etch process baseline

To figure out the advantages of CF3I, we should firstly choose a recipe as the baseline (BSL) in cryogenic etch. Based on our previous results of experiments, we decide to use CHF3, CH2F2 and O2 to etch channel hole. The ratio of CHF3 to CH2F2 is around 1. The ratio of bias power to source power is 1 and total power is less than 5kW. The pressure is 20mT. The recipe can both achieve high in-hole ER, more than 300nm/min, and high selectivity of dielectric materials to mask, around 6. However, the profile exhibits some disadvantages including top layer blowing out and sharp bottom etch front, as shown in Figure 5a. We plan to add CF3I to optimize the profile with less ER drop and sufficient mask selectivity in the next step.

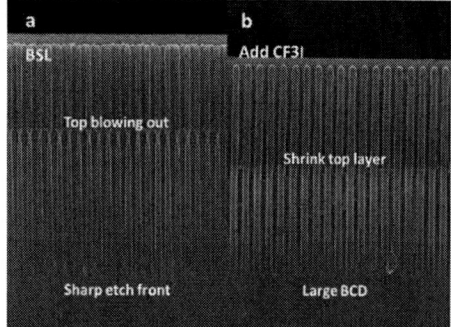

Figure 5: The SEM cross-section of post etch profile (a) without and (b) with CF3I.

Blanket etch rate

We studied some basic properties of CF3I, such as ER and OX to SiN ER selectivity (O/N sel.), on blanket wafers. We used the recipe containing CF3I, Ar and O2. The result is shown in Table I. We found that the OX ER of CF3I is much higher than SiN ER both at 20°C and -60°C. It indicates that CF3I can passivate the dielectric surface and more oxygen radicals can mitigate it. Besides, the OX ER at 20°C is similar with that at -60°C. The SiN ER is decreased by 22% when reducing the temperature to -60°C. It can be ascribed to the stronger deposition with decreased temperature.

TABLE I. The blanket OX and SiN ER at 20 °C and -60 °C

Temperature	20 °C	-60 °C
OX ER (nm/min)	307.9	305.3
SiN ER (nm/min)	32.6	25.4
O/N sel.	9.4	12.0

To evaluate the effect of CF3I on BSL recipe, we also study the BSL ER on OX and SiN blanket wafers with and without CF3I, as shown in Table II. When we add 10 sccm CF3I, the OX and SiN ER are decreased by 4% and 5%, respectively. The result shows that CF3I may deposit polymer/byproduct in cryogenic etch. However, when we increase the ratio of bias power to source power from 1 to 4 and increase the total power from 5kW to 7.5kW, CF3I can increase the OX and SiN ER by 2% and 6%, respectively. It indicates that we can control the deposition of iodine contained byproduct by tuning RF power.

TABLE II. The blanket OX and SiN ER change with power and CF3I

Condition	Low power	+CF3I	High power	+CF3I
OX ER (nm/min)	242.5	232.3	447.2	457.4
SiN ER (nm/min)	341.8	327.0	606.3	640.8

Top layer and in-hole etch rate

According to the previous ER, we predict that the byproduct of CF3I can deposit on the side-wall and bottom in deep holes, resulting in smaller TCD and lower ER. The deposition will be aggravated at lower temperature. More O2 and higher bias power can tune the profile and mitigate the side-effect of CF3I on in-hole ER and profile. To demonstrate them, we compare the TCD and in-hole ER of two chips etched by BSL recipe and CF3I contained recipe, respectively. The etching depth and TCD are measured by the SEM cross-section images. When adding CF3I, the in-hole ER is decreased from 372 nm/min to 118 nm/min and TCD is decreased by 23%. Besides, the mask shape and thickness are maintained. The result indicates that the byproduct of CF3I probably deposits on the side wall and bottom but not mask. To figure out the composition of byproduct, we analyze the elements at etch front by Energy Dispersive X-Ray spectroscopy. The result shows that the percentage by weight of iodine is around 7.4 wt%. The rest elements are 54.2 wt% silicon and 38.4 wt% oxygen. Although the top layer blowing out is obviously mitigated, as shown in Figure 5b, the negative effect on in-hole ER is not acceptable. When we increase O2, the in-hole ER is increased to 329 nm/min and TCD is enlarged by 20 nm. However, more O2 will accelerate the consumption of mask. The trade-off effect is hard to balance. When we increase the bias power from 1.5kW to 4.5kW, the in-hole ER is increased to 206 nm/min while TCD and mask consumption are not changed. When we further increase the bias power to 6kW, the increase of ER is not obvious. We suspect that 6kW power is so high that may increase wafer temperature thus decrease the reactive layer formed by C1 gases. All the data of in-hole ER and TCD change are recorded in Table III. Consequently, to increase in-hole ER with less side-effect, tuning bias power is the better choice.

TABLE III. The change of in-hole ER, TCD and mask remain with CF3I, O2 and bias power.

Condition	BSL	+CF3I	+O2	4500W	6000W
In-hole ER (nm/min)	372	118	273	206	201
TCD (nm)	87.3	67.5	75.3	68.8	77.0
Mask (nm)	1265	1263	1116	1272	1184

Straight profile and bottom circularity

The ideal profile of channel hole should be a straight cylinder with better bottom circularity. Because C1 gases are more likely to form incompact polymer, the profile usually shows bowing and small bottom CD (BCD). The ratio of bias to source power is effective to shrink bowing and enlarge BCD. We test four pairs of source and bias power, including 1.5kW/1.5kW, 1.5kW/4kW, 2kW/4kW and 2.5kW/4kW. The profiles, observed by SEM, are shown in Figure 6. The profile is tapper when the power is 1.5kW/1.5kW. As the bias power is increased to 4kW, BCD is enlarged thus profile shows more straight. It can be ascribed that the bottom residual are readily to be removed by higher power. Besides, as we increase the source power to 2.5kW, bowing can be decreased by 15%, however, mask closure and ER drop are appeared due to the excessive polymer deposition by high density radical. The results indicate that the best source/bias power ratio should be 2kW/4kW.

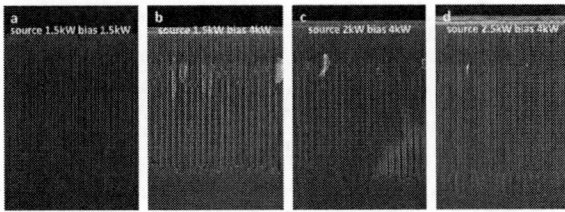

Figure 6: The SEM cross-section of post etch profile with various power ratio

In the end of etch, all channel holes will land on a Si stop layer. To check the bottom circularity, we immerse the chip into 10% HF aqueous solution then clean it with ultrasonic to remove all the OX and SiN layers. We can directly observe the shape of etch front by SEM top view, as shown in Figure 7. We found that the chip etched by C1 gases and cryogenic condition shows good bottom circularity. It is better than the result etched by C4 gases at room temperature. Generally, the degree of distortion is correlation with the none-uniform deposition of polymer at the interlayer between carbon mask and OX film [6]. Because we do not use dense polymer in cryogenic strategy, the negative effect of deposition might be mitigated.

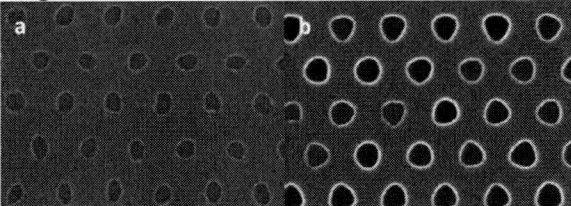

Figure 7: The SEM top views of bottom circularity resulted from (a) C4 gases and high power at room temperature, (b) C1 gases and low power at -60ºC.

CONCLUSIONS

To obtain the straight profile, high in-hole ER, appropriate mask selectivity and good bottom circularity in cryogenic HAR etch, we use CF3I, as an additive, in the C1-gases and low-power recipe. CF3I can produce some $SiOI_x$ byproduct to passivate side-wall thus mitigate the top layer blowing out. By further tuning source and bias power, we can enlarge BCD with less effect on the drop of mask selectivity and in-hole ER. Due to the less polymer in cryogenic etching, we also observe the good bottom circularity. Compared with the traditional condition of C4 gases and high power at room temperature, the cryogenic recipe shows less risk of arcing. We think the experience of CF3I can also be applied in other halogen contained gases.

ACKNOWLEDGEMENTS

The authors acknowledge the helpful collaboration with the colleagues in the companies. The authors are also thankful to the support of cryogenic chiller from Beijing Instrument Industry Group Co., Ltd. The authors thank the providence of specific gas from Beijing Yuji Science & Technology Co., Ltd.

REFERENCES

[1] S. PARK, J. Lee, J. Jang, J. K. Lim, H. Kim, J. J. Shim, M. Yu, J. Kang, S. J. Ahn, and J. Song, *Symposium on VLSI Technology*, 2021, pp. 1-2.

[2] C. Han, Z. Wu, C. Yang, L. Xie, B. Xu, L. Liu, Z. Yin, L. Jin, and Z. Huo, *Semiconductor Science and Technology*. vol. 35, 2020, pp. 045003.

[3] A. Rezvanov, A. V. Miakonkikh, A. S. Vishnevskiy, K. V. Rudenko and M. R. Baklanov, *Journal of Vacuum Science & Technology B*. vol. 35, 2017, pp. 021204.

[4] T. Iwase, M. Matsui, K. Yokogawa, T. Arase and M. Mori, *Journal of applied physics*. vol. 55, 2017, pp. 06HB02.

[5] W. W. Stoffels, E. Stoffels, and K. Tachibana, *Journal of Vacuum Science & Technology A*. vol. 16, 1998, pp. 87.

[6] S. Huang, S. Shim, S. Nam and M. Kushner. *Journal of Vacuum Science & Technology A*. vol. 38, 2020, pp. 1-11.

SIARC RESIDUE REDUCTION METHODS WITH MINIMIZING PROFILE CHANGE FOR MASK PATTERNING

[1]Xingxing. Xu, [1] Hexin Zhou, [1] Quanbao. Li, [1]Jian. Huang

[1]Lam Research Service Co., Ltd, Shanghai 201203, China

*Corresponding Author's Email: Xingxing. Xu@lamresearch.com

ABSTRACT

Tri-layer mask scheme including photoresist (PR), Si-based anti-reflection coating (SiARC) and spin-on carbon (SOC) is applied to transfer PR pattern and shrink the critical dimension (CD) to the underneath patterning layer (PL). However, due to the presence of different pitches, SiARC residues were often observed on the remained pattering layer after the etch and the wet clean, especially on the patterns with the larger pitches. However, the trade-offs between the SiARC residue removal efficiency and the profile control bring big challenges for finding proper solution that only fully removes the residues but minimizes profile change. In this study, the reason for the formation of SiARC residues is illustrated. Moreover, we show how to balance these trade-offs to eliminate the SiARC residues but also to minimize the profile changes.

Keywords – Tri-layer patterning, SiARC residues, residue reduction, profile control

Introduction

With the device scaling down, the requirements on the critical dimension and the profile become more and more strict. In the advanced technology nodes, tri-layer mask scheme including photoresist (PR), Si-based anti-reflection coating (SiARC) and spin-on carbon (SOC) is applied to meet the strict requirements. [1-3] The high selectivity of PR to SiARC enables more precise lithography by thinner PR and the rich polymers during SiARC etch enables CD shrinkage. Meanwhile, the balance between SiARC and SOC is critical for the desired profile of the underneath patterning layer (PL).

On the other hand, residue and defect control is of crucial importance to the yield. However, the more complicated structures, e.g., different pitches, iso/dense pattern, long/short trench, result in narrow process window for controlling defects. Therefore, how to reduce residues and defects and meanwhile to keep profile of all the complicated structures during etch is current big issues.

In this study, we show how to differentiate the SiARC residue during etching with PR/SiARC/SOC tri-layer mask and illustrate its formation scheme. Moreover, three different SiARC residue removal approaches according to the integration compatibility are introduced and their trade-offs with profile control are discussed. Finally, we show how to balance these trade-offs to minimize the profile changes while removing the SiARC residues.

Results and Discussions

Tri-layer patterning scheme for etch

Figure 1 shows a general scheme of the tri-layer etch process. Post SiARC etch (figure 1b), the CD of PR is transferred to SiARC and the CD can be shrunk if heavy polymer presented in this step. The SiARC is then as the mask for SOC etch. Post SOC etch, there is usually SiARC remaining (Figure 1c). The BT step is used to break through the top layer of the substate and the etch amount of BT determines the top CD of the substrate. Usually, the remaining SiARC is removed in this step (Figure 1d). The SOC is the mask for the following main etch and over etch (ME and OE) steps. The SOC profile is critical for the underneath substrate. Finally, the SOC is stripped in the ash step.

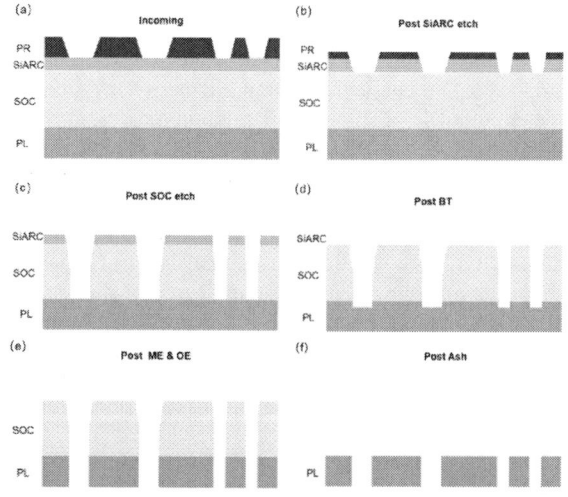

Figure 1. Tri-layer mask etch scheme.

SiARC residue identification and formation scheme

In the etch process with tri-layer patterning scheme, the balance of the etch of the different layers is a big issue. Especially, when different pitches are present, how to balance the selectivity and the etch amount between layers are big challenges. If the process is not in a good balance, the residues and defects occur.

Figure 2 shows the top view and the cross-section view of the SiARC residues found in the larger pitches. This type of residue is usually found in the middle of lines due to the unbalance of the etch amount between the small and the large pitches.

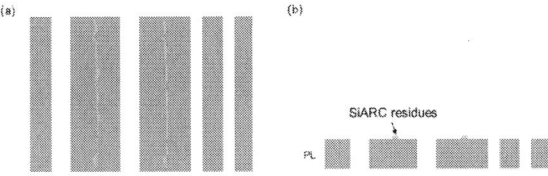

Figure 2. SiARC residue identification. (a) top view (b) cross section view.

As illustrated before, the SiARC is the mask for the SOC etch. During the SOC etch, the SiARC losses to some extend and the remaining SiARC is supposed to be fully removed post BT. However, during either SOC etch or BT step, the ion bombard is usually controlled to keep the selectivity. The chemical etch based on the radicals plays an important role.

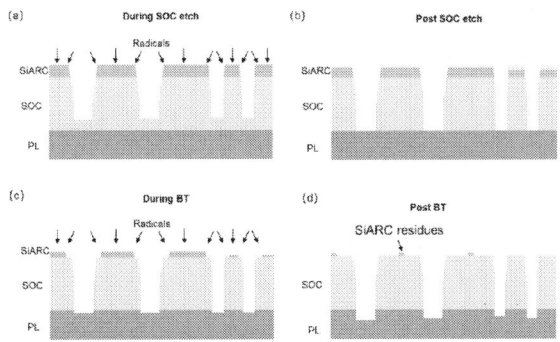

Figure 3. The SiARC residue formation due to different lateral etch amount between the large and small patterns.

Chemical etch is isotropic and meanwhile the larger pattern requires higher lateral etch amount. As seen in Figure 3, when the SiARC on the smaller pattern has been fully removed, SiARC residues on the larger pattern remain. In some applications, if the main etch step for the underneath PL has high selectivity to SiARC or the heavy byproducts during the PL etch covers the residues, these remaining SiARC residues may be difficult to remove.

SiARC residue removal approaches and trade-offs with profile
From above, we can know the SiARC residues mainly result from two aspects: 1) the pattern loading during SiARC removal; 2) the low sensitivity of the PL etch steps to SiARC. Importantly, due to the integration issue, the PL etch steps need to limit the use of fluorine. With

the limits, we tested three SiARC residues removal approaches, i.e., (a) stronger ash bias, (b) stronger BT with higher source power and bias power, and (c) lower SOC to SiARC selectivity during SOC main etch (ME). In the following, we will discuss their SiARC residue removal performance and their influence on the profile.

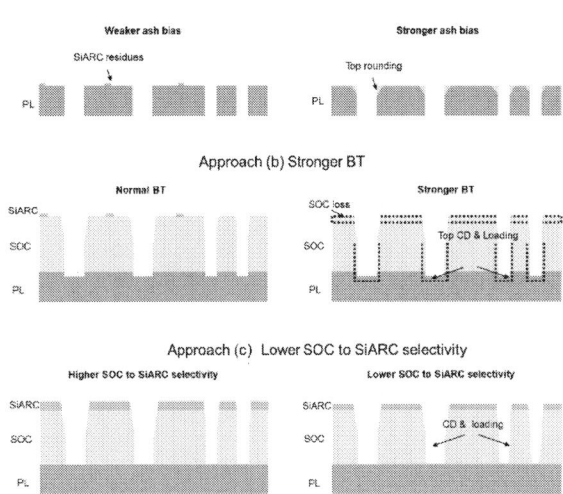

Figure 4. The trade-offs between the residue reduction approaches and the profile control. Approach (a) stronger ash bias (b) stronger BT (c) Lower SOC to SiARC selectivity during SOC step.

The approach (a) is to add the bias voltage in the ash step. However, if the bias voltage is low, e.g ~ 50V, the SiARC residues cannot be fully removed even with longer ash time. Higher bias, e.g., 100V, can fully remove the residues, while top rounding profile was observed. Different ash step OE amount were tested. We found that it was difficult to remove all the residues without top damage. The reasons accounting for this could be: I) Too much SiARC residues require large Ash OE; II) During the Ash OE, the substrate is exposed to the complex chemistry in the chamber and the high ion bombardment sputters the top corner; III) the energy required to break the bonds in SiARC (SiOC) residues is high, comparable to that break the PL molecular bonds.

The approach (b) is to use stronger BT by higher source power and higher bias voltage. We tested two conditions: 1) stronger BT but keeping the time to enlarge BT etch amount (EA); 2) stronger BT with reducing the time to keep the same EA. The 1st condition with larger BT EA can fully remove the residues. However, it has a big influence on the PL profile, e.g., the top CD enlarged, and the profile tapered. The reason for this may be the inefficient SOC height to maintain the substrate CD. Thicker SOC may help maintain the profile with stronger

BT but the attention to the through pitch loading should be paid. The 2nd condition with the same EA has acceptable influence on the profile while the residue cannot be fully removed.

The approach (c) is to lower the selectivity between SOC and SiARC during SOC etch to decrease the remained SOC thickness. The lower selectivity between the SiARC and SOC was realized by lower O2 amount during SOC etch. This approach also obviously improved the residue performance. The profile change is also controlled. However, since the O2 etch is isotropic and, the SOC etch is high aspect ratio etch, change O2 amount may result in change in the iso/dense loading.

Table 1. The SiARC residue removal performance of the three approaches and their influence on the profile and loading, EA is the abbreviation for etch amount.

Approaches		Residue remain	PL profile change	Influence on loading
Baseline		Worse	Baseline	Baseline
(a) Stronger ash bias	50V	Medium	Little	No
	100V	Free	Big, top rounding	No
(b) Stronger BT	Larger EA	Free	Big, Top CD	Medium
	BSL EA	Slightly	Little	Little
(c) Lower SOC to SiARC selectivity	Less O2	Slightly	No	Big

The SiARC residue removal performance of the three approaches and their trade-offs on the profile and the pitch loading are listed in table 1. After comparing the three approaches, we can see all the three approaches can improve the SiARC residue performance, but it is difficult to modify only one step to remove all the SiRAC residues without changing profile. Therefore, combining the different residue removal approaches may minimize the profile change and remove all the residues. Our results show that the combination of any two approaches can remove all the SiARC residues. Regarding the top profile control, we do not suggest that combining the stronger ash bias with the stronger BT since they could increase the risk of top damage. Lower SOC to SiARC selectivity has little influence on the substrate profile but it has the biggest influence on the through pitch loading. Therefore, it dominates the loading after combining with either stronger ash bias or stronger BT. In the real practice, the appreciate combinations are determined by

the process window, i.e., for the changed profile or the loading, which one is more difficult to be tuned back.

In our case, the through pith loading can be tuned back through modification of SiARC etch process. Therefore, we finally chose to combine the lower SOC to SiARC selectivity with stronger ash bias (50V).

Conclusion

In this study, the tri-layer mask patterning scheme with PR/SiARC/SOC is introduced. We illustrate the formation scheme of SiARC residues and show the method to identify it. three different SiARC residue removal approaches, i.e., stronger ash bias, stronger break through (BT) and lower SOC to SiARC selectivity during SOC ME, and their trade-offs with the physical profile control on underneath layer are discussed. We found the combination of lower SOC to SiARC selectivity with either stronger BT or stronger ash bias has smallest influence on substrate profile. Mitigation with SiARC modification can cover the CD and through pitch loading issue.

References

[1] Yayi Wei, Martin Glodde, Hakeem Yusuff, Margaret Lawson, Sang Yil Chang, Kwang Sub Yoon, Chung-Hsi Wu, Mark Kelling, "Performance of tri-layer process required for 22 nm and beyond," Proc. SPIE 7972, Advances in Resist Materials and Processing Technology XXVIII, 79722L (15 April 2011)

[2] Owe-Yang, Dah Chung, Toshiharu Yano, Takafumi Ueda, Motoaki Iwabuchi, Tsutomu Ogihara and Shozo Shirai. "Development of high-performance tri-layer material." SPIE Advanced Lithography (2008).

[3] Ashim Dutta, Jennifer Church, Joe Lee, Brendan O'Brien, Luciana Meli, Chi Chun Liu, Saumya Sharma, Karen Petrillo, Cody Murray, Eric Liu, Katie Lutker-Lee, Qiaowei Lou, Chris Cole, Angélique Raley, Akiteru Ko, Subhadeep Kal, Jake Kaminsky, Aelan Mosden, Henan Zhang, Shan Hu, Lior Huli, Naoki Shibata, Dave Hetzer, Chia- Angélique Raley, Akiteru Ko, Subhadeep Kal, Jake Kaminsky, Aelan Mosden, Henan Zhang, Shan Hu, Lior Huli, Naoki Shibata, Dave Hetzer, Chia-Yun (Sharon) Hsieh, "Strategies for aggressive scaling of EUV multi-patterning to sub-20 nm features," Proc. SPIE 11323, Extreme Ultraviolet (EUV) Lithography XI, 113230V (6 April 2020);

Distortion Control when Etching DRAM Metal Contact

Jianqiu Hou[1], Yao Sun[2], Hui Xue[2], Ya Zhou[1], Hao Li[2], Zhiwen Luan[1], Zijian Chen[2]* and Zengwen Hu[1]*

[1]Advanced Micro-Fabrication Equipment Inc. Shanghai 201201, China
[2]Chang Xin Memory Technologies Corporation, Hefei 230093, China
*Corresponding Author's Email: Zijian.Chen@cxmt.com, zengwenhu@amecnsh.com.

ABSTRACT

Contact etch is a typical high-aspect-ratio etch, which highly requires less distortion at the bottom. We suspect that the non-uniform polymer deposition on the mask necking and the random charging on the side-wall will both result in distortion. The former mainly contributes to circularity. The later will cause the random direction at the bottom of hole. Based on the mechanism, we investigate some gases to improve the distortion by carbon mask modification and dielectric side-wall discharging.

INTRODUCTION

The high-aspect-ratio (HAR) etch is very challenging in fabricating advanced memory devices, such as 3D NAND and DRAM [1]. Contact is usually a cylinder holes composed of silicon oxide (OX). The aspect ratio is generally higher than 20. After etching, tungsten (W) will be filled into it for electronic transmission.

Figure 1 shows a typical tri-layer structure of contact etching. At first, we should etch a dielectric anti-reflective coating (DARC) layer with a photo resist (PR) mask. The elements in DARC include Si, O and N thus we use CF chemistry to etch it. Because PR is too soft to maintain initial circularity of holes under ions bombardment, we usually avoid using bias power and lower pressure when etching DARC.

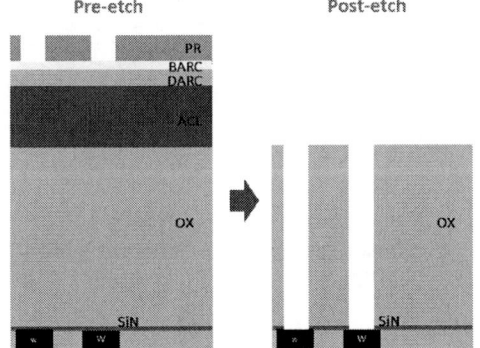

Figure 1: A typical try-layer structure of contact, pre and post etching

At next step, we use the patterned DARC as the mask to etch an amorphous carbon layer (ACL). The aspect ratio of ACL holes close to 10, implying the demand of strong side-wall protection during etching. COS is a good candidate. It can combine with O2 to achieve an anisotropic carbon etching.

Finally, the patterned ACL film can work as the mask of etching OX layer. The step is called as main etch (ME). It is very challenging because we should not only keep the circularity but also enhance the twisting of holes. Bottom distortion will increase the difficulty of the following filling step, leading to the negative effects on electrical performance of wafers.

To address the challenge, most previous researches focus on the mechanism of OX distortion in plasma etch [2-4]. Distortion refers to the worse circularity during etch. Besides, the variation between two adjacent holes can also be identified as distortion. Generally, the distortion can be ascribed to the limitation of etchant diffusion in deep and small holes [5]. The incident ions can positively charge side-wall and etch front, leading to the increased charge repulsive force with etching time [6]. In that case, reactive ions are difficult to transport to bottom, leading to the non-uniform isotropic etching [7].

The traditional solution to improve distortion is power pulsing. In this mode, high and low power can be cycled with very high frequency. Etching will be occurred at high level while discharging and excluding byproduct will be occurred at low level. Therefore, the etching resistance is reduced and the efficiency of power is increased. With the sufficient etchant, the local uniformity will be improved.

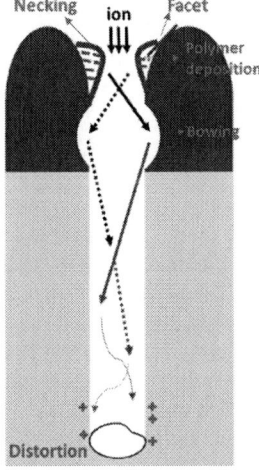

Figure 2: The evolution of distortion from mask to bottom

We find a specific metallic halogen compound that can much improve bottom circularity in ME step. We suspect that the discharging function of the gas can improve the local uniformity of OX etch. However, the risk of under etch is simultaneously increased because metal-byproducts are difficult to be pumped out thus deposit at the etch front.

Except the negative effects from OX surface, we found that the interlayer between ACL mask and OX film is readily to be deformed during plasma etch. The root cause might be the non-uniform deposition of polymer on carbon surface, as shown in Figure 2. Adding N2 and CH2F2 can improve it, leading to the enhanced bottom circularity of holes.

EXPERIMENTAL

All etching experiments were conducted in a capacitive coupled plasma (CCP) etch tool with two radio frequency (RF) generators (>40 MHz source and <2 MHz bias power). The etch tool was AMEC HD RIE plus.

To avoid wasting resources, we used a small chip sample that cut from whole wafer for experiment. The 2cm x 3cm chip has already been pre-coated with designed film stack for following etching process. We then pasted the chip on a carrier Si wafer in AMEC lab.

Scanning Electron Microscopy (SEM), Hitachi 4800, in AMEC lab, was applied to observe the cross-section, the top view and elemental analysis of samples. To observe the bottom circularity, we immersed post-etch coupon in dilute HF aqueous solution for 5-10 min then used ultrasonic to clean before SEM analysis.

Transmission Electron Microscope (TEM) and Energy Dispersive X-Ray (EDX) Spectroscopy analysis were supported by MSSCORPS CO., LTD.

We used a plasma damage monitoring system (WT-2000, SEMI-lab) to measure wafer surface charge pre and post plasma etching.

RESULT AND DISCUSSION

By step check distortion formation

Figure 3 shows the evolution of distortion in different steps by SEM analysis. The distortion becomes most severe in ACL etching. On the one hand, the soft material is easy to be damaged under plasma. On the other hand, the carbon surface shows deep affinity with carbon polymer. Therefore, the precise pattern transfer becomes more difficult in ACL mask open. The top circularity of holes will be improved after ME step when etching OX film. However, the bottom circularity is still bad. The distortion at the bottom may be caused by the limitation of etchant diffusion. Based on the result, we should improve distortion both on ACL and ME step by modifying carbon mask profile and discharging OX sidewall.

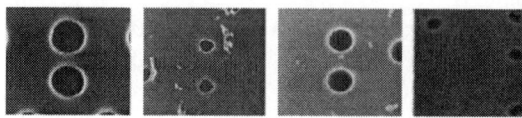

Figure 3: The SEM images of distortion evolution

Metal-halogen compound discharging effect

Adding a small amount of metal-halogen compound, in ME step is the most effective method to improve bottom distortion when etching OX holes. As shown in Figure 4, the circularity is significantly increased when increasing the specific gas flow. Besides, the orientation of surrounding holes also becomes more consistence.

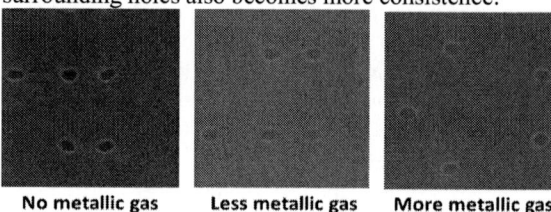

Figure 4: The SEM images of bottom distortion with increased metallic gas flow

We think the better distortion can be ascribed to the discharging function of metallic byproduct. To prove it, we use a plasma damage monitoring system to check the remaining charge on wafer surface. After HAR plasma etch, the charges are increased. When adding the metallic halogen gas in plasma, surface charge is significantly decreased, implying the discharging effect. The result is shown in Figure 5.

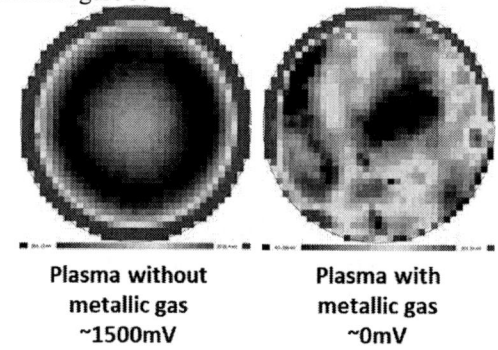

Figure 5: The wafer surface remaining charge after plamsa treatment with and without metallic gas.

We also find some side effects caused by metallic gas. For example, when using a metallic-gas as an addictive, the under etch is more likely to be occurred, as shown in Figure 6. It can be ascribed that the metal byproducts prefer to deposit on etch front, which is demonstrated by EDX analysis. Decreasing chamber pressure can solve the problem meanwhile maintain the advantages on distortion caused by metal discharging.

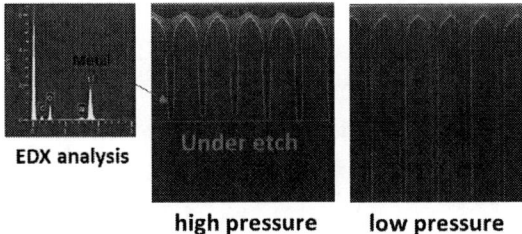

Figure 6. *Contact under etch caused by metal byproduct.*

Besides, when we increase metallic gas flow to larger amount, the chamber condition is not stable and cannot be recovered by in-situ clean. We think too much metallic byproduct will deposit on chamber parts surface, leading to the higher risk of plasma leak.

N2 modifying mask necking

Mask profile will highly influence the dielectric layer in patterning transfer. When etching a deep hole, the corner of the mask will become rounding caused by ion bombardment. The polymers prefer to deposit on the outstretched facet, leading to the mask necking. We observe that the polymer deposition on mask necking is usually non-uniform i.e. asymmetric, as shown in Figure 4. Because the molecule weight distribution of polymers is very wide by radical or coupling polymerization, the deposition on mask cannot be equal to every hole. The random deposition can also be ascribed to the limitation of radical concentration and diffusion. Consequently, the asymmetric polymer deposition on mask necking will result in the distortion, twisting, bowing and the variation between adjacent holes. Besides, the mask necking is more likely to form in short axis direction, leading to the decrease of holes x/y ratio i.e. circularity.

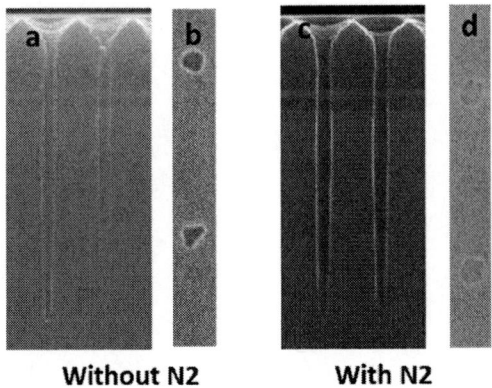

Figure 7. *The cross-section and bottom-dimple of contact etch (a, b) without and (c, d) with N2.*

N2 is a good candidate to mitigate the formation of mask necking due to the less bombardment than Ar and the capacity of removing carbon like polymer in ME step. Figure 7 shows the comparison of profile and distortion with or without adding N2. The mask necking is obviously mitigated with N2, resulting in the better bottom circularity when landing on the stop layer.

CH2F2 improving mask profile

In addition to N2, we find that CH2F2 can also modify mask profile both in ACL and ME step, as shown in Figure 8. When opening carbon mask, we usually use COS to protect carbon side-wall. However, a top-necking and middle-bowing profile is readily to be formed caused by the deposition of S-byproduct on the corner facet. In that case, a small amount of CH2F2 can mitigate necking and bowing by forming volatile H2S with S-byproduct. The enhanced mask profile can further increase the circularity in the following pattern transfer steps, as shown in Figure 8a. For the same reason, we can also use CH2F2 in ME step. The mask necking is improved, leading to the better bottom circularity, as shown in Figure 8b.

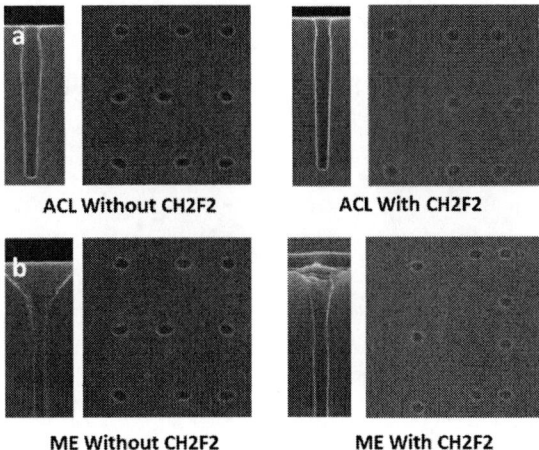

Figure 8. *The carbon mask cross-section and OX layer bottom-dimple of contact etch without and with CH2F2 in (a) ACL step or (b) ME step.*

CONCLUSIONS

Distortion is very challenging in HAR etch. It is related with non-uniform polymer deposition and positive charging on material surface. By studying the evolution of contact distortion step by step, we found that both OX side-wall charging and carbon mask profile can influence the distortion. To address the challenging, we investigate three kinds of gases, including metallic gases for OX side-wall discharging, N2 and CH2F2 for mask necking. All three gases are effective to improve the bottom circularity thus decrease the variation between adjacent holes. We will do more detail researches on mechanism and apply the strategies to other very HAR etching.

ACKNOWLEDGEMENTS

The authors acknowledge the helpful collaboration with the colleagues in the companies. The authors are thankful to the TEM analysis from MSSCORPS CO., LTD.

REFERENCES

[1] Y. Kim, S. Lee, T. Jung, B. Lee, N. Kwak and S. Park, *Proc. SPIE.* vol. 9428, 2015, pp. 942806.

[2] J. Lee, I. Jang, S. Lee, C. Kim and S. Moon, *Journal of the Electrochemical Society.* Vol. 157, 2010, pp. 142-146.

[3] J. Kim, S. Lee, S. Cho and G. Yeom, *Journal of Vacuum Science & Technology A: Vacuum, Surfaces and Films.* vol. 33, 2015, pp. 021303.

[4] M. Miyake, N. Negishi, M. Izawa, K. Yokogawa, M. Oyama and T. Kanekiyo, *Japanese Journal of Applied Physics.* vol. 48, 2009, pp. 8HE01.

[5] S. Huang, S. Shim, S. Nam and M. Kushner, *Journal of Vacuum Science & Technology A.* vol. 38, 2020, pp. 023001.

[6] A. Shibkov, M. Abatchev, H. Kang and M. Lee, *Electronics Letters,* vol. 32, 1996, pp. 890-891.

[7] D. Kim, E. Hudson, D. Cooperberg, E. Edelberg and M. Srinivasan, *Thin Solid Films.* vol. 515, 2007, pp. 4874-4878.

STUDY ON THE EFFECT OF WATER SPRAYING MODE ON THE N CONTENT OF WAFER SURFACE AFTER SC1 CLEANING IN LIGHT DOPING PROCESS

Jinlei Wang*; Fenglin Guan; Mingguang Hang; Lili Jia; Fang Li; Xinhua Cheng
Shanghai Huali Integrated Circuit Corporation, Shanghai, China
*Corresponding Author's Email: wangjinlei@hlmc.cn

ABSTRACT

In the lightly doping process, it was found that the N content on wafer surface increased after wet cleaning process. In this paper, XPS was used to characterize the N content on wafer, and the effect of different CO_2 water spraying modes on the N content of the wafer after SC1 cleaning was studied. The results show that the CO_2 water spraying by fix mode is not enough to clean the center of the wafer, which leads to high N content in the area. Compared with fix mode, the CO_2 water spray can directly cover the wafer center by scan mode, which can enhance the cleaning ability of CO_2 water to the wafer center and effectively reduce the N content in the area after the wet process.

Keywords—Lightly doping process; Wet cleaning process; N content

INTRODUCTION

With the development of integrated circuits, the size of devices is continuously reduced in proportion to improve the performance and density of devices per unit area. As the feature size of devices is reduced to submicron and deep submicron, the hot carrier effect will occur near the drain, which will lead to the degradation of the performance of the device and affect their reliability [1,2]. In the process flow of integrated circuits, a certain width light doping region is formed between the drain and the channel by a process of light doping ion implantation before the side wall formation (LDD). LDD process can reduce the peak electric field near the drain to weaken the hot carrier effect [3]. The region of ion implantation is defined by PR development in LDD process. After ion implantation, PR was removed by ashing and wet clean. The chemical used in wet clean is SC1 (standard clean 1, a mixture of ammonia, hydrogen peroxide and water). It is found that the N content increased after LDD wet clean process. However, it was found that the content of N element increased on the wafer surface after LDD wet cleaning process. There are some researches show that excessive N content on the wafer surface will cause photoresist "poisoning", which will affect the accuracy of the pattern defined by the lithography process, and reduce the performance of the device [4,5].

In this paper, the LDD wet cleaning process is simulated by bare wafer (non-pattern wafer). X-ray photoelectron spectroscopy (XPS) was used to characterize the N content on the wafer surface before and after wet process. The effect of different spray modes of CO_2 water on the N content of wafer surface after SC1 in LDD process.

EXPERIMENT

Bare wafer was used to simulate the LDD wet cleaning process in this paper. The equipment used in the wet process is a single cleaning machine (SU3200, DNS, Japan). The chemical used is standard clean 1 (SC1, a the volume ratio of 1:2:50 $NH_4OH:H_2O_2:H_2O$) and CO_2 water. In the process, SC1 is used to clean the wafer surface firstly. And then CO_2 water is sprayed onto the surface. Finally, the wafer is dried by rotating it at high speed. The N content was characterized on the wafer surface before and after the wet process by XPS.

RESULT

During the integrated circuit manufacturing process, LDD loop requires repeated lithography, implantation, asher, and wet cleaning processes to meet the different electrical requirements of devices. In practical production, it was found that the SRAM 114 ISB of CVTP area was larger in the wafer center after once rework, and the yield was lower, as shown in Fig. 1(a). After twice reworks, the reduction of wafer center yield was intensified (as shown in Fig. 1(b)). Subsequently, XPS was used to characterize the oxide pad of CVTP, and the results are shown in Fig. 2. Compared with pure oxide, the oxide pad exhibits an obvious N-peak, and the intensity of N-peak is obviously different in different regions. The intensity of N-peak gradually decreases from center to edge of the wafer. Sites (0,0), (2,0), (4,0), (6,0) represent the positions on the wafer from center to edge. It indicates that the N content of CVTP oxide pad gradually decreases from center to edge, as shown in Fig. 2(b).

979-8-3503-1101-3/23 $31.00 © 2023 IEEE

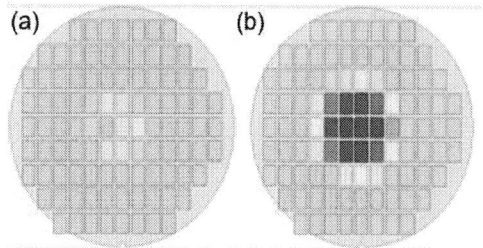

Figure 1: The yield loss map of wafer after rework.

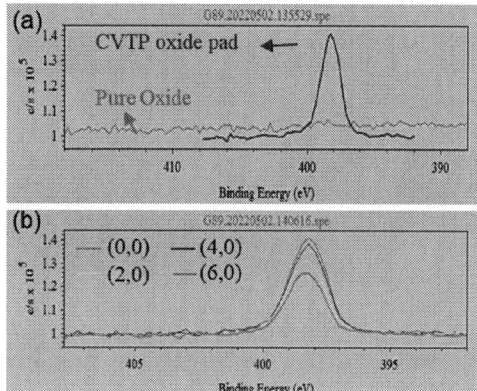

Figure 2: XPS analytical spectrum of the wafer surface.

In order to solve the problem of high N content of CVTP Oxide Pad in wafer center, the wet cleaning recipe was optimized. The CO_2 water spray is fix mode after SC1 in original recipe, and it is sprayed at a point near the wafer center. After recipe modification, the CO_2 water spray mode is changed from fix to scan mode. During CO_2 water spraying process, the spray nozzle moves continuously on the wafer surface. Then, the wet cleaning process was evaluated by bare wafer, and the N content was characterized by XPS before and after process.

After the wet process with different spray mode, the N content increment at different regions on the wafer surface is shown in Table 1. The result shows that the N content increment in the center is obviously higher than that in the middle and edge region after wet process using fix mode. This is consistent with the results of product wafer (shown in Fig. 2(b)). In the optimization recipe, the spray mode is changed to scan mode. Compared with fix mode, the N content increment at middle and edge has no obvious change, but the N content increment at center of wafer becomes negative. This indicates that the N content of wafer center can be significantly reduced by changing the spray mode from fix to scan mode.

TABLE I.   THE N CONTENT INCREMENT OF DIFFERENT REGIONS AFTER WET PROCESS USING FIX MODE OR SCAN MODE (ATOMIC %)

	fix mode		scan mode	
	wafer1	wafer2	wafer3	wafer4
center	0.2326	0.1830	-0.0980	-0.6224
middle	0.1083	-0.0077	-0.0697	-0.1189
edge	-0.0878	-0.0766	-0.1098	-0.0703

DISCUSSION

In practical production, it was found that the SRAM 114 ISB of CVTP area was larger in the wafer center after rework, and the yield was lower. The results of XPS analytical spectrum show that the N content of CVTP oxide pad in the region with lower yield is higher. It was found that the N element may be introduced by the wet cleaning chemical SC1. We simulated the wet process by bare wafer. The result shows that the N content of wafer center increased after wet process. This is consistent with the results of product wafer (shown in Fig. 2(b)). Furthermore, it can result in insufficient CO_2 water cleaning of wafer center. CO_2 water is sprayed at a fixed point 22 mm from wafer center, and then covered wafer surface by rotation. This may be the reason that the cleaning ability to wafer center is insufficient. A small amount of residual N element in SC1 is absorbed by the natural oxide layer, causing the N content in wafer center to increase.

In the optimized recipe, the spray mode is changed from Fix to scan mode. Fig. 3 shows the scan mode diagram. After the SC1 step is completed, the CO_2 water is transmitted along the arm to the nozzle. The nozzle moves continuously on the wafer surface between -97mm and 53mm (radius) and CO_2 water is sprayed out through nozzle. At the same time CO_2 water covers the wafer surface by rotating at high speed. During the CO_2 water step, the chemical can directly cover the wafer center. It enhances the cleaning ability to wafer center. SC1 residue is removed completely, and the N content in wafer center is reduced after wet process.

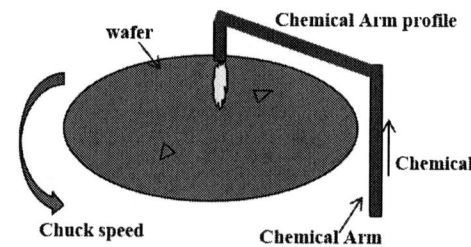

Figure 3: Diagram of water scan mode.

CONCLUSION

In LDD wet cleaning process, the CO_2 water cleaning ability is insufficient for wafer center by using fix mode after SC1, and the oxide pad adsorbs the N element, resulting in a high N content in the area. After the wet process recipe is optimized, the spray mode is changed from fix to scan mode. Under the control of the arm, CO_2 water can directly cover the wafer center. It enhances the water cleaning ability to wafer center, and reduce the N content in the area.

REFERENCES

[1]. D. Saha, D. Varghese and S. Mahapatra. Role of anode hole injection and valence band hole tunneling on interface trap generation during hot carrier injection stress. IEEE Electron Device Letters, 2006, 27(7): 585-587.

[2]. Y. Qian, Y. Gao and A. K. Shukla et al. Modeling of Hot Carrier Injection on Gate-Induced Drain Leakage in PDSOI nMOSFET. International Conference on Integrated Circuits, Technologies and Applications (ICTA), 2021, 239-240.

[3]. D. T. Wen. Integrated Circuit Manufacturing Process and Engineering Application. China Machine Press, 2018.

[4] Z. F. Gan, Z. B. Mao. A method to reduce photoresist poisoning. Chinese Patent: CN106019876A, 20161012.

[5] X. Wei. Method of Reducing photoresist poisoning by Dual Graphics Process. China Patent: CN113394080A, 20210914.

A Technical Optimization of Waferless Auto Clean for Aluminum Etcher

Li Qi[1]*, Qiang-Qiang Sang, Xing-Jun Yao, Li-Tian Xu, Jian-Kun Zhang, Li-Song Hu, Yi-Chang Liu, Chen Chen

[1] Beijing NAURA Microelectronics Equipment Co. Ltd
No. 8 Wenchang Avenue, Beijing Economic-Technological Development Area. Beijing City, China

*Corresponding Author's Email: qili01@naura.com

Abstract

Aluminum etching by-products accumulate in the inner wall of the process chamber, the failure to remove such etching products would lead to the shift of chamber condition. In this regard, the paper has developed an efficient waferless auto clean (WAC) to facilitate the by-product removal of aluminum etching chamber. It is found that clean efficiency can be improved by the optimization of cleaning step sequence, the introduction of CH4 in the Cl2 cleaning step and the introduction of Cl2 in the O2 cleaning step

Keywords — aluminum etcher, waferless auto clean, deposition removal, plasma etch

Introduction

By-products(including photoresists, etching precursors, and etching products) formation can be found in the plasma etching process, which will be deposited on the etch chamber walls. Such deposits not only cause process drift, but also introduce particle contamination that affect final yields[1]. Therefore, wet cleaning needs to be periodically applied on the chamber, which is very time consuming, plasma wafer-less auto clean (WAC) after each wafer is commonly used to clean the chamber, thus extending the Mean Time Between Clean (MTBC)[1,2].

For aluminum etcher, chlorine (Cl2) is commonly regarded as the prime candidate for etching gas in WAC, and the main reactions are as follows[3]:

$$Al+ (x)Cl \rightarrow AlCl_x \uparrow$$

Incomplete removal of the aluminum-containing products will deposit on the chamber's inner wall or around the electrode. When such by-products accumulate to a certain level, it will cause process shift, power loading abnormalities, and particle contamination, further shortening the MTBC.

This paper aims to develop a WAC technology, more efficient approach to realize the by-product removal from the aluminum etching cavity, thereby extending the machine mass production MTBC, improving equipment utilization, and reducing manufacturing costs.

Experiment

The equipment NAURA NMC612G etcher was adopted in this experiment.

Two WAC recipes in Table 1 and Table 2 has be developed in this study. The high-pressure(HP) step is adopted to clean the upper part of the chamber, low-pressure(LP) step to clean the lower part of the chamber. Oxygen-contained gas is used to clean organic by-products, and chlorine-contained gas to clean metal by-products containing aluminum[4].

In order to compare the cleaning ability of WAC-1 and WAC-2 on the aluminum etching chamber. The same aluminum product was run in the same chamber, and check chamber condition at the same RF hours(150hrs).

Table.1 WAC-1 Recipe

Step Name	Pressure	Gas
O(HP)	High	O2
Cl(HP)	High	Cl2
O(LP)	Low	O2
Cl(LP)	Low	Cl2

Table.2 WAC-2 Recipe

Step Name	Pressure	Gas
Cl(HP)	High	Cl2, CH4
O(HP)	High	O2, Cl2
O(LP)	Low	O2, Cl2
Cl(LP)	Low	Cl2

Result and discussion

Fig1 shows the picture of chamber condition with different WAC. It can be seen that there are obvious contaminants on the chamber surface according to the WAC-1, while the chamber walls are perfectly clean according to the WAC-2, indicating that the WAC-2 is of an extremely efficient cleaning ability.

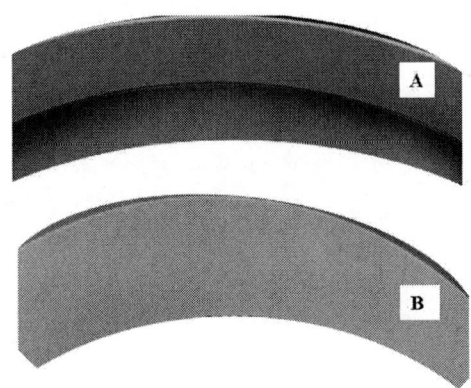

Fig.1 Chamber condition of different WAC at same RF hours: (A) WAC-1, (B)WAC-2

This mainly attributes to the following modification of WAC-2:

1. The gas sequence change from step1 to 4: WAC-1 adopts the cleaning sequence of "O → Cl→O→Cl", while WAC2 is modified to " Cl→O→O→Cl". This is due to the fact that large amounts of aluminum-containing by-products remain in the chamber after etching aluminum, which can easily react with oxygen-active groups to generate Al-O compounds that cannot be easily removed. The replacement of O with Cl in the first step enables the chlorine active group to react with the Al in the by-products to form $AlCl_x$ which can be discharge from the chamber, greatly improving the cleaning efficiency.

2. The introduction of small volumes of CH4 to Cl2-base clean step. H radical will be generated when CH4 plasma excitation, On the one side, H can react with OH and other groups to reduce the polymer; on the other hand, H has a reducing effect and can make AlO be reduced to Al, which requires less efforts to be removed; finally, CH4 can also react with Cl2 to generate HCl, which enables the improvement on the removal rate of Al-containing deposits such as Al and AlO.

$$H+ OH -> H_2O$$

$$H+AlO -> Al+OH$$

$$CH4+Cl2 -> CH3Cl+HCl$$

3. The introduction of small volumes of Cl2 to the O-based clean step. The addition of Cl2 prevents the reaction of O and Al to form Al-O compounds that are difficult to remove while ensuring that O2 plasma remove the organic byproducts.

Conclusion

Three ways as follows were investigated in this study: 1. Optimize the Clean step sequence; 2. Introduce small volumes of CH4 to the Cl-based cleaning gas step; 3. Introduce small volumes of Cl2 to the O-based cleaning gas step. The cleaning efficiency for the aluminum etching polymer is enhanced to ensure a clean chamber, thus extending the MTBC and improving the tool utilization.

References

[1] Xu, S., Sun, Z., Qian, X., Holland, J., & Podlesnik, D. (2001). Characteristics and mechanism of etch process sensitivity to chamber surface condition. Journal of Vacuum Science & Technology B: Microelectronics and Nanometer Structures, 19(1), 166. [2] Cunge, G., Pelissier, B., Joubert, O., Ramos, R., & Maurice, C. (2005). New chamber walls conditioning and cleaning strategies to improve the stability of plasma processes. Plasma Sources Science and Technology, 14(3), 599–609.
[3] Qiang, F., Huang, C. L., Hendrianto, J., Song, J., Lv, M., Wang, K., & Shi, C. (2011). Clean Mode Al Etch Process Development for Defect Reduction.
[4] Winters, H. F. (1985). Etch products from the reaction on Cl2 with Al(100) and Cu(100) and XeF2 with W(111) and Nb. Journal of Vacuum Science & Technology B: Microelectronics and Nanometer Structures, 3(1), 9.

TRIMMING OF SILICON NITRIDE HARD MASK USING CYCLIC DEPOSITION AND ETCH PROCESS

Li-Tian Xu[1]*, Pei Mei[1], Xing-Jun Yao[1], Chen Chen[1], Jia-Yun Zhang[2], Alan Zhang[2], Xiao-Peng Wu[2], Wu-Hao Han[2], Hong-Bo Lin[2] and Nichole Zhang[2]

[1] Beijing NAURA Microelectronics Equipment Co. Ltd

[2] ChangXin Memory Technologies, Inc. Ltd

*Corresponding Author's Email: xulitian@naura.com

ABSTRACT

Silicon nitride (SiN) hard mask (HM) has been investigated to achieve sub-100nm of high aspect ratio (HAR) pattern. In this study, SiN HM trimming technique is developed by mean of cyclic reactive ion etching (RIE) for the vertical profile required in advanced DRAM memory fabrication. Furthermore, the SiN CD trim rate is revealed as ~2 nm per cycle.

INTRODUCTION

Memory technology faces different challenges to reduce the capacitor cell area. The bitline (BL) structure is a SiN/W/WN/TiN/WSi/poly-Si stack in 32nm mobile DRAM [1]. The top silicon nitride (SiN) film as hard mask (HM) is implemented not only to define the critical dimension (CD) of BL pattern but also to play an isolation for the neighboring storage node contact (SNC) [2, 3]. SiN HM has been adopted to achieve sub-100nm of high aspect ratio (HAR) pattern [4]. In this study, SiN HM trimming technique is developed by mean of cyclic reactive ion etching (RIE) process for the vertical BL profile in DRAM memory fabrication.

EXPERIMENT

Fig.1 illustrates a schematic SiN HM manufacturing process on NAURA NMC612E high-density inductive coupled plasma (ICP) etcher. The film scheme started with lithography defined BL pattern (Fig.1 (a)). Then, a carbon mask is patterned via oxygen based plasma (Fig.1 (b)). Next, a selective SiN etch process is executed for SiN HM open and stopping on BL metal layer (Fig.1 (c)). Lastly, an in-situ SiN trimming process is carried out to meet the CD target and produce the vertical SiN HM profile (Fig.1 (d)).

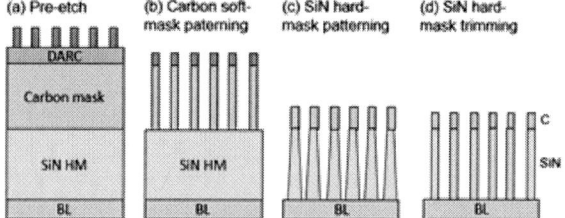

Fig.1 SiN HM manufacturing process flow: (a) pre-etch with incoming BL pattern, (b) DARC and carbon masks etching, (c) SiN HM etching, and (d) SiN HM lateral etch for CD trimming.

As for SiN trimming, we take advantage of the cyclic two sequential steps to deposit a non-conformal thicker film on the top part of SiN HM, followed by an etch step for the sidewall pullback. The key factor of success (KFS) is to create a vertical profile of SiN HM.

RESULT AND DISCUSSION

A. Traditional one-step trim process

Traditional SiN trim process used one-step SF_6-based chemistry. Compared to pre-trim profile as in Fig.2 (a), this dry trim process intended to shrink the top part of SiN HM and induce a tapered profile as Fig.2 (b). Increasing the trim rate by using 12mT instead of 5mT may further worsen the SiN profile slope as Fig.2 (c). It will cancel out the downstream process margins for BL CD control and memory storage node (SN) leakage performance.

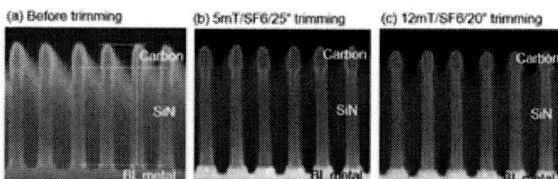

Fig.2 Traditional one-step SiN HM trim: (a) Before trimming, (b) trimmed by 5mT/SF6/25sec, (c) trimmed by 12mT/ SF6/ 20sec.

B. Cyclic trim with C_xF_y deposition

Fig.3 demonstrates the non-conformal C_xF_y deposition on SiN HM surface. This C_xF_y film selectively deposits a thicker layer on top area of SiN sidewall than the bottom area. In calculation, the deposition rate is about 6A/sec.

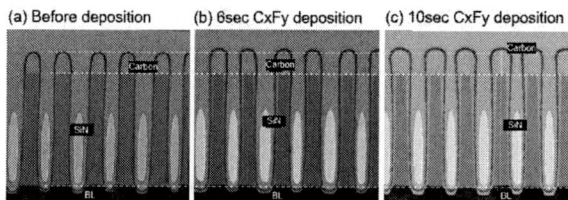

Fig.3 Cross-section images of C_xF_y deposition layer on SiN HM: (a) before deposition, (b) post 6sec deposition, (c) post 10sec deposition.

979-8-3503-1101-3/23 $31.00 © 2023 IEEE

Cyclic trim process with alternated C_xF_y deposition and NF_3-based dry etch can form vertical SiN sidewall. But the C_xF_y polymer may accumulate in profile bottom area and result in SiN footing as Fig.4 (a) with 50Wb in etch step. As the etch bias power increased to 70Wb, little SiN footing still existed and SiN notch showed up under the carbon mask as Fig.4 (b). As the etch bias power further increased to 100Wb, Fig.4 (c) showed significant SiN top notch issue due to the ion scattering bombardment effect, even though the footing profile was addressed. Therefore, the process window is narrow because of the trade-off between bottom footing and top notch issues.

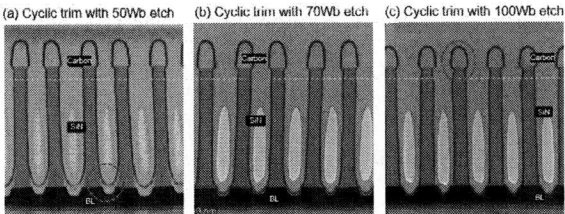

Fig.4 Cross-section images of SiN HM profiles after 3 cycles trim with 6sec C_xF_y deposition and 27sec etch, depending on different etch bias powers: (a) 50Wb, (b) 70Wb, and (c) 100Wb.

C. *Cyclic trim with SiO_x deposition*

In term of SiO_x deposition, Fig.5 (b) shows the quasi-atomic layer deposition (ALD) approach which formed 90A SiO_x layer on carbon mask, compared to the baseline without deposition Fig.5 (a). Besides, Fig.5 (c) shows 60A SiO_x deposition layer by using $SiCl_4/O_2$ plasma within 15sec deposition time. Both of the deposition methods trended to grow the SiO_x film on the surface of carbon mask rather than the sidewall of SiN HM. It did not provide enough passivation on the sidewall of SiN HM.

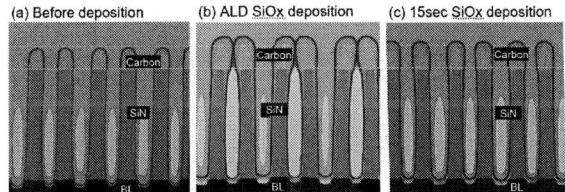

Fig.5 Cross-section images of SiO_x deposition layer on SiN HM: (a) before deposition, (b) 3cycles ALD SiO_x deposition with 30A/cycle deposition rate, (c) 15sec SiO_x deposition with 4A/sec deposition rate.

After the cyclic trim processes with alternated SiO_x deposition and F-based dry etch, Fig.6 (a~c) indicated that the SiN HM profiles were still tapered, respectively, even with the different dry etch conditions of SF_6/12sec, NF_3/15sec and CF_4/15sec. It also proves that SiO_x film almost grow on the surface of carbon mask, not on the SiN sidewall.

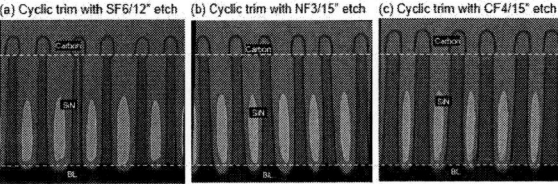

Fig.6 Cross-section images of SiN HM profiles after 3 cycles trim with ALD SiO_x deposition and the 3 different dry etch conditions: (a) SF_6/12", (b) NF_3/15", and (c) CF_4/15".

D. *Cyclic trim with SiN_x deposition*

Fig.7 (b) demonstrates more vertical SiN HM profile through the cyclic trim process with an alternated novel SiN_x deposition and NF_3-based chemistry dry etch, compared to pre-trim profile of Fig.7 (a). Moreover, using higher bias power in etch step as shown in Fig.7 (c) can help to eliminate the bowing of SiN HM profile. It is notable that the thickness of carbon HM almost does not be consumed during this cyclic trim process.

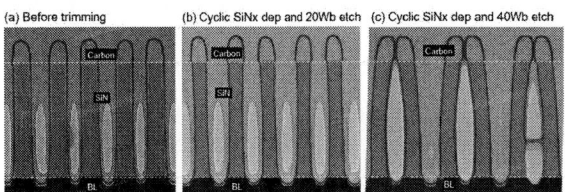

Fig.7 Cross-section images of SiN HM profiles: (a) before trim, (b) after 3 cycles trim with SiN_x deposition and NF_3/20Wb etch, and (c) after 3 cycles trim with SiN_x deposition and NF_3/40Wb etch.

In calculation, the CD trim rates for SiN HM were 2.2 and 1.6 nm/cycle, respectively, in the cases of 20Wb and 40Wb etch bias power. It also reveals that the CD of SiN HM is tunable via cycle number of the cyclic trim process.

CONCLUSION

We evaluated the three different SiN HM trim methods. Cyclic trim process with the novel SiNx deposition step and NF3-based etch step shows the benefits of not only tunable AEI CD but also vertical SiN HM profile. This technique can meet the requirements for the DRAM BL manufacturing in the advanced memory application.

REFERENCES

[1] Dick James, "Recent Advances in Memory Technology", ASMC 2013, 386 - 395, (2013)

[2] Jaeho Won, et al., "Failure Analysis of Bit Line to SNC Leakage Fail in 2x nm DRAM Using Nano-Probing Technique", ISTFA 2014,

[3] Yexiao Yu, et al., "The Study of Bit Line to Storage Node Contact Leakage in Advanced DRAM", 2022 IEEE 5th International Conference on Electronics Technology (ICET)

[4] V. Jovanović, et al., "Sub-100 nm silicon-nitride hard-mask for high aspect-ratio silicon fins", (2012)

Enabling Plasma Etch Solution for GaN Technology

Zoe Wang[1], Chunxiang Guo[1*], Jian Liu[1], Yingxiong Feng[1], Lulu Guan[1], Kangning Xu[1], Qiao Huang[1], Lu Chen[1], Kaidong Xu[1]

[1] Jiangsu Leuven Instruments Co., Ltd., Xuzhou, Jiangsu, China

Phone: +86 516-86995099 Email: chunxiang.guo@leuven-instruments.com

ABSTRACT

High breakdown voltages, high electron mobility and saturation velocity of gallium nitride (GaN) make it suitable for high frequency, high voltage, high temperature and high efficiency applications. Plasma dry etch is one of the key device manufacturing processes which played an important role in advanced GaN technology development. Partnered with several GaN device research institutes, IDMs and foundries, Leuven Instruments has been continuously accelerating tool hardware/function design cycle to fulfill the device manufacturing requirement, improving tool robustness, enhancing tool process capabilities. We've successfully developed a full GaN dry etching solutions from well-tuned anisotropic profile and smooth etching sidewall/surface control to minimize or even eliminate surface damage by low power & ALE technique. Leuven Instruments could provide an integrated plasma dry etching solutions to our customer in support various GaN technology development. *Keywords—Inductively Coupled Plasma (ICP) etching; GaN Technology; Atomic Layer Etch (ALE); Low damage.*

INTRODUCTION

As a wide bandgap material, gallium nitride (GaN) demonstrates superior physical and electrical properties, such as high dielectric strength, high operating temperature, high current density, high switching speed, and low on-resistance. Such attributes offer significant advantages in mid-power high radio frequency 5G applications, as well as power supplies and motor controls for industrial, commercial, and even more stringent automotive applications. The high-volume production of various GaN electronic devices, from Schottky barrier diode (SBD) to high electron mobility transistor (HEMT), all require meticulous device design and continuous process/equipment improvement to cost-down and stay competitive in the market. Coordination between design houses, chip makers, and equipment vendors is the key enabler for GaN technology to succeed in various markets [1].

As one of the key device manufacturing process technology, plasma dry etching delivers precise device patterning, with well-tuned anisotropic etch control, smooth sidewall and surface profile, equal etch rate for different crystallographic orientations, as well as ability to compensate for etch issues related to defects, uniformity, or quality imperfections. Furthermore, device manufacturing requires high etch selectivity or at least equal etch rate etching for different materials, minimal damage to the device active region, elimination of the electron trapping in the semiconductor, high efficiency, power density, speed, and gain, high etch rate for better mass production throughput, high uniformity for high yield. These imposed many challenges to device design and manufacturing, which require more advanced plasma dry etching equipment and techniques to support new technology development [2-3].

Leuven Instruments partnered with several GaN device research institutes, IDMs and foundries and worked extensively on specific plasma dry etch process development to optimize device performance and yield. Our expertise in plasma dry etching equipment and hardware provides reliable support for new advanced GaN device manufacturing companies. In this paper, we provide integrated etching process solutions for GaN etching with damage-free surfaces, vertical etch profile, high selectivity, improved uniformity, and high yield.

EXPERIMENTS

Samples used for this study were provided by Sinovio Semiconductor Technology (Dongguan) Co., Ltd. Etching and deposition processes were performed with inductively coupled plasma etching system (Pishow® A) and plasma enhanced chemical vapor deposition (Shale® A) from Leuven Instruments Co., Ltd. The ICP system was designed and optimized for superior low bias power process as low as 1-2 W. The reflected power can be controlled to within 3%, as proved by extensive fast-switch plasma on-off ignition marathon test. Scanning electron microscopy (SEM) and atomic force microscopy (AFM) are used in this study to quantify/characterize the etched profiles and the surface roughness/damage.

RESULT AND DISCUSSION

Pre-treatment

At clean room atmosphere, the GaN epitaxial layer tends to generate a non-uniform thin gallium oxide (Ga_2O_3) surface oxidation layer. This Ga_2O_3 layer would act as a micro mask during the dry etching process, resulting in pits or pillars, which severely affects the electrical performance and reliability of GaN devices.

To avoid defects caused by this surface natural oxidation, the material was treated with oxygen plasma to obtain a uniform oxidized layer on the GaN surface [4]. Subsequently, the oxidized layer can be selectively removed by Lewis acidic BCl_3 plasma [5].

Fast etch and soft etching

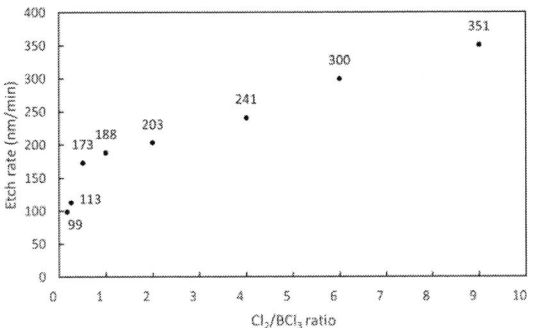

Figure 1. Etch rates as a function of Cl_2/BCl_3 ratio with constant total gas flow.

Cl_2 and BCl_3 are generally used for GaN fast etching since the boiling point of $GaCl_3$ (201℃ at 1 atm) [6] is low. In this study, a gas mixture of Cl_2 and BCl_3 (volumetric 4:1) was introduced to produce anisotropic and smooth profile. The process pressure was fixed at 10 mTorr. High process pressures would reduce mean free path of plasma ions, which further decreases etch rates and lead to tapered profile. Etch rates increases with the percentage of Cl_2 if the total gas flow is kept constant, as shown in Figure 1. However, further increasing the Cl_2 ratio would lead to sub-trench, since the passivation protection from the BCl3 was reduced, therefore the increased etch rate leads to undesired sub-trenching at the bottom corner. When the flow of BCl_3 was higher than Cl_2, etch rates dramatically decreased, we attributed this to the decreased chloride ion flux and increased influence of boron-containing-polymer passivation.

The ICP source power and the bias power are essential parameters that significantly affect the etching process results. Increased ICP power resulted in higher chloride ion flux and higher etch rates. Further increasing ICP power to higher than 800 W would not lead to further etch rates increases, and on the contrary, this might result in a slight decrease in etch rate. Increasing bias power could lead to an increase in the etch rates and give rise to more vertical profiles. The selectivity to photoresist reduced significantly with high bias power and led to a dual angle sidewall profile, as the physical bombardment is too aggressive at such high bias settings.

To achieve high selectivity between GaN and AlGaN during device fabrication, we developed an oxygen containing soft etching system. We introduced oxygen to form a layer of Al_2O_3 thin film as a passivation layer when AlGaN was exposed to the plasma to slow down the AlGaN etching. The AlGaN initial loss can be suppressed from 2 nm to 0.8 nm by introducing bias pulsing and the selectivity between GaN and AlGaN can reach as high as 20:1.

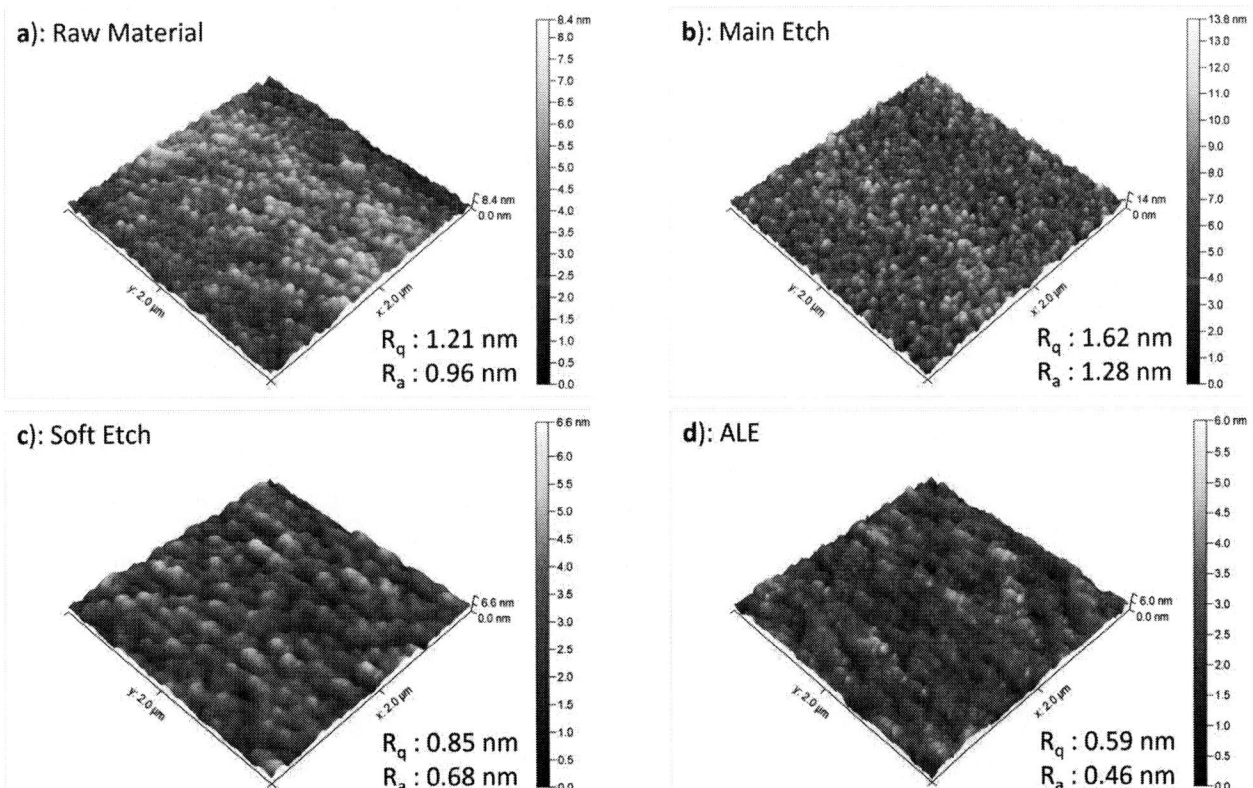

Figure 2. Roughness of GaN surface, data collected with blanket GaN wafer: **a)** Raw material, $R_q = 1.21$ nm, $R_a = 0.96$ nm; **b)** after main etch, $R_q = 1.62$ nm, $R_a = 1.28$ nm; **c)** after soft etch, $R_q = 0.85$ nm, $R_a = 0.68$ nm; **d)** after 5 loops of ALE, $R_q = 0.59$ nm, $R_a = 0.49$ nm.

Under optimized process conditions, the GaN etch rates are typically ~100 nm/min for fast etch and ~5 nm/min for soft etch. The final profile is around 80° after fast etch and soft etch. Compared to the raw material (R_q = 1.21 nm), the surface roughness slightly increased after the fast etch (R_q = 1.62 nm) and decreased after soft etch (R_q = 0.85 nm) (Figure 2. a-c).

Atomic layer etching

Atomic layer etching offers atomic level etching precision and ultra-low damage manufacturing process by removing material layer-by-layer[7]. Typically, an ALE loop includes two successively controlled steps that begins with the self-limiting surface modification, followed by the removal of the modified layer, as depicted in Figure 3.

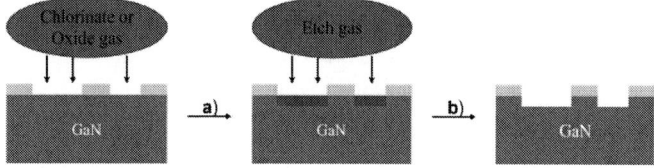

Figure 3. mechanism of ALE etching: **a)** modification step, **b)** modified layer removal.

The performance and mechanism of both O_2-BCl_3[8] and BCl_3-Ar GaN ALE reactions have been extensively studied at Leuven Instruments. In the O_2-BCl_3 system, both the modification and etching steps are chemical based reactions. The GaN surface is oxidized by O_2 plasma to give a Ga_2O_3 layer of 2.30 nm, which is then removed by Lewis acidic BCl_3 plasma. For the BCl_3-Ar system, the surface modification is based on chlorination by BCl_3 plasma which offers a 0.4 nm thick $GaCl_3$ layer. Compared to O_2-BCl_3, the removal of the chlorinated layer depends on the bombardment of low bias Ar plasma. In both cases, the GaN etch depths increased monotonically as the ALE loops increased, which indicated that the ALE reactions were atomic-level self-limited and the etch rates per cycle were stable. Interestingly, the etched GaN surface was getting smoother after ALE process, demonstrating ALE having superior ability to repair or even eliminate surface damages (Figure 2, d). These studies provide a deep insight into the ALE mechanism of GaN material and the essential role of the ALE technique in fabricating reliable and high performance GaN devices.

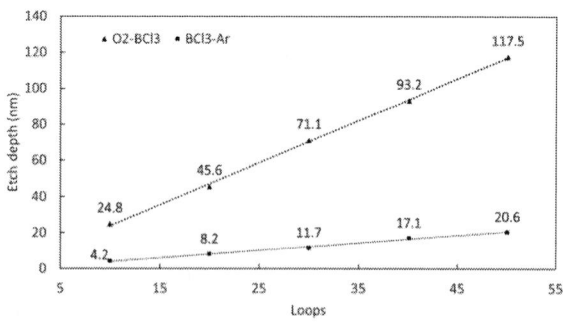

Figure 3. Etch depths as function of ALE loops for O_2-BCl_3 and

BCl_3-Ar system within the self-limiting window.

CONCLUSION

This article introduces a full plasma dry etch solution for GaN technology. Surface treatment, fast etch and soft etch are used to produce vertical profiles, low-damage surface, and high selectivity to AlGaN. Two ALE systems have been developed and studied in detail which can help achieve controllable etching rates, smooth etched surfaces, and nearly zero surface damage. These innovations by Leuven Instruments Pishow® A plasma dry etching equipment provide integrated process solutions for GaN device fabrication and enable our customers to further advance their chip design.

REFERENCE

[1]   K. H. Teo, Y. Zhang, N. Chowdhury, et al. J. Appl. Phys. **2021**, 130, 160902.

[2]   D. Marcon, B. De Jaeger, S. Halder, et.al. IEEE Transactions on Semiconductor Manufacturing, **2013**, 26-3, 361.

[3]   D. S. Rawal, H. K. Malik, V. R. Agarwal, *et al.* IEEE Transactions on plasma science, **2012**, 40, 2211.

[4]   Shul R, McClellan G, Casalnuovo S, et al., Inductively coupled plasma etching of GaN, Applied physics letters, **1996**, 69(8), 1119-1121.

[5]   Chiu H-C, Yang C-W, Chen C-H, et al., Characterization of enhancement-mode AlGaN/GaN high electron mobility transistor using N2O plasma oxidation technology, Applied Physics Letters, **2011**, 99(15), 153508.

[6]   Tripathy, S., Ramam, A., Chua, S. J., Pan, J. S., & Huan, A.. Characterization of inductively coupled plasma etched surface of GaN using Cl2/BCl3 chemistry. Journal of Vacuum Science & Technology A: Vacuum, Surfaces, and Films, **2001**, 19(5), 2522-2532.

[7]   Guan, L., Li, X., Che, D., Xu, K., Zhuang, S. Plasma atomic layer etching of GaN/AlGaN materials and application: An overview. Journal of Semiconductors, **2022**, 243(11), 113101.

[8]   LIN, Yuan, et al. A novel digital etch technique for p-GaN Gate HEMT. In: 2018 IEEE International Conference on Semiconductor Electronics (ICSE). IEEE, **2018**, 121-123.

THE RESEARCH OF SPECIAL GATE MORPHOLOGY ADJUSTMENT AND ITS INFLUENCE ON ELECTRICAL PROPERTIES

Junjie Pan*, Kai Qian, Lian Lu, Quanbo Li, Jun Huang, Yu Zhang

Shanghai Huali Integrated Circuit Corporation

Shanghai City, China

*Corresponding Author's Email: panjunjie@hlmc.cn

ABSTRACT

In the key nodes of semiconductor process manufacturing, gate characteristic size, line width uniformity, side wall profile, line width roughness and other characteristics are strictly controlled process parameters. The gate CD of polysilicon directly affects the electrical performance of CMOS devices, and the gate side profile is directly related to the gate performance. In current mainstream Gate processes, Gate profiles are typically "Vertical" side walls. In this paper, the gate side wall profile is successfully change from "Vertical" to "reverse tapered" shape by adjusting the etching process parameters such as ME (Main Etch), SL (Soft Landing), OE (Over Etch), which is greatly beneficial to improve the Dummy-gate removal and subsequent metal filling process window, meanwhile achieves a 2%-3% improvement in the AC Electrical performance of the Device.

Keywords—ICP Etch, Polymer Gas, Gate Profile, Electrical Performance

INTRODUCTION

As CMOS Technologies enter 28nm node and beyond, in order to improve current drive capability and reduce short channel effect, the thickness of gate oxide and poly CD are gradually reduced. To overcome these problems, materials with high dielectric constant (K) are widely used as gate dielectric layer instead of silicon oxide layer [1]. At the same time, due to the constraint of Gate exhaustion, high dielectric materials are often paired with Metal gates. Under the current 28nm process node, the etching process of the gate is facing severe challenges.

In HK gate etch process, poly etch consists of three steps: ME (Main etch) / SL (Soft landing) /OE (Over etch). As a gas with a high selection ratio, the CF_4 removes the main film in the ME stage and etch the Lighter polymer profile of the Gate. HBr/O_2 as the main gas in the SL, it has a good selection ratio for gate oxide (>30:1) and Captures the signal of poly thin film descent as the etch end point in the SL. OE step is the removal of the poly remain film after ME and SL and control the profile at the bottom of poly [2, 3]. OE amount needs to be controlled well because too much OE amount leads to upper Necking, too little leads to Footing of the poly, in addition, excessive OE amount can also easily cause substrate Pitting, which will directly lead to the decline of the saturation current of the device.

Poly profile is one of the most critical parameters in the gate etching process. Conventional Poly profiles include Vertical, Bowling, Tapered and Reverse Tapered. In this paper, the effects of advanced etching parameters of 28nm on different profile of Gate were investigated under the conditions of plasma source etching. A reverse tapered profile of HK Gate is obtained by optimizing etching parameters such as Polymer gas ratio in SL, OE amount and ME, which greatly beneficial to improve DPR (Dummy Poly Removal) process, and broadens the process window of subsequent steps such as metal filling, meanwhile greatly improve the AC performance of the device.

EXPERIMENT

All the dry etching experiments were performed in the same types of 300mm commercial etcher of ICP system with tunable ESC (Electrostatic Chuck) and TCCT (Transformer Coupled Capacitive Tuning). In the poly etching process, the O_2 ratio of SL is set as A, A-10% and A-20%, respectively. After etching, the poly cross-section poly profile after wet cleaning was observed under TEM; Under the same experimental conditions, OE amount was set to B, B+10% and B+20%, respectively. After etching, the poly was observed under TEM for CD data analysis. Under the same experimental conditions, the ME process was divided into ME1 and ME2, and the etching process was carried out by adjusting the SF_6 gas ratio in ME2. After etching, the poly cross-section poly profile after wet cleaning was observed under TEM. Cross section profiles of gate were imaged via TEM (Transmission Electron Microscope). EPD system detected and monitored the etch endpoint was used in all experiments. Table 1 is the experimental design table, in which A and B are the benchmark values.

Experimental group	Etch parameter	Value
1	SL polymer gas / O_2	A
		A-10%
		A-20%
2	OE amount	B%
		B+10%
		B+20%
3	ME	C
	ME1+ME2	C_1+C_2

Tab.1 Design group in etch experimental

RESULT AND DISCUSSION

Impact of Polymer gas of SL and OE amount on poly Profile

The vertical direction of Poly electrode was set as X direction. Fig.1 shows the profile TEM of poly etching in X direction under different etching conditions. Fig.1 I and II show the gate of vertical profile obtained at O_2 flow rate A. It can be seen that with the decrease of Polymer Gas (O_2), the profile of HK gate gradually changed from vertical to reverse tapered, and an ideal reverse tapered gate profile was obtained. This is due to in SL etching stage, the side wall of gate will produce passivated layer. Such as SiOxFy and other polymers will remain adsorbed around the side wall of HK Gate to form a certain protection for the side wall [4]. Polymer Gas (O_2) is the source of passivated layer. Figure (III) and (IV) are profile of the bottom silicon substrate when O_2 ratio is A-10% and A-20%, respectively. Under the relatively high O2 ratio, the silicon substrate is flat, but with the reduction of the O2 ,it can be seen that the bottom of the silicon has been damage, which is mainly for the existence of O2 can not only improve the overall rate of the etching process, and can have a high selection ratio for gate oxide.

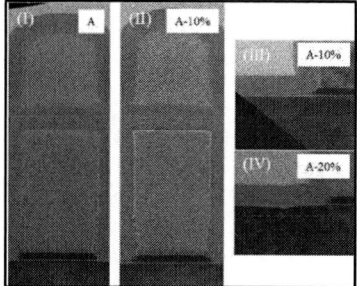

Fig.1 Poly profile of different O2 ratio in the SL, (I) O2: A; (II) O2: A-10 %;(III) O2: A-10 %;(IV) O2: A-20%

Fig.2 shows the influence of different etch OE amount on Middle CD-Bottom CD after gate etch under the same experimental conditions，and the target of the MCD-BCD is b. As OE Amount changes, so does the profile at the bottom of the gate. When OE amount is B, the etched profile is close to Vertical. Increasing the OE amount can improve the gate bottom footing and increase the lateral etching ability, which the poly profile closer to the target under the OE amount of B+10%. As the OE amount continues increase, although the poly MCD-BCD also approaches the target, but the gate TEM show the excessive OE amount causes necking which is not expected to see.

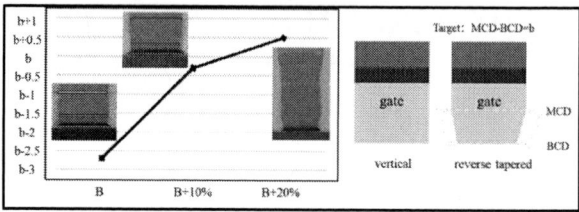

Fig.2 Poly profile of different OE amount in the OE

Effect of ME (Main Etch) on Gate Profile

ME is the key part of poly etching, and about 80% of the poly were removed on ME. Fig.3 is the poly profile obtained after dividing the ME into ME1 and ME2 (increased SF_6 gas) in Gate etching stage. According to the Fig.3, the ideal inverted trapezoidal gate can also be obtained after adjusting the ME, and it can be seen that the transition from vertical to reverse tapered becomes smoother. This is due to the isotropy of SF_6 in the ME stage. The etching reaction principle of highly active F radical with Si and SiO_2 is as follows[5]:

$$SF_6 \rightarrow F; Si \text{ and } 4F \rightarrow SiF_4 \uparrow$$
$$4F \text{ and } SiO_2 \rightarrow SiF_4 \text{ and } O_2 \uparrow$$

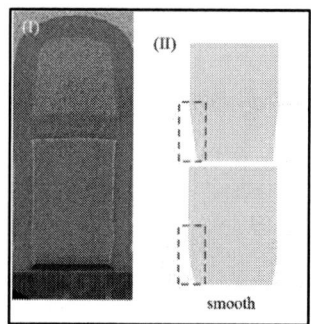

Fig.3 The poly profile TEM of the gate after ME, (I) ME1+ME2; (II) Schematic diagram

Electrical Performance of HK gate (Reverse tapered Profile)

Gate Profile has a great relationship with the electrical properties of CMOS device. Fig.4 shows the WAT results of different poly profile. As can be seen from the figure, The FREQ-AC of a Gate with a Reverse Tapered profile performed 2-3% better than the Vertical profile. This indicates that Gate of reverse tapered profile obtained by adjusting etching parameters can further improve the chip stability and better electrical performance.

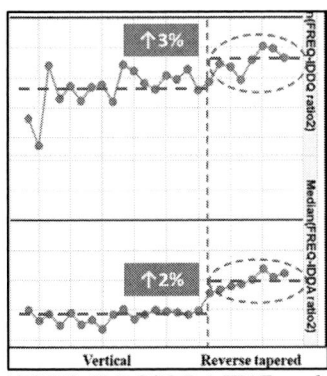

Fig.4 FREQ-IDDQ/IDDA WAT analysis

CONCLUSIONS

In this paper, we study the effect of etching parameters including Polymer Gas Ratio, OE amount and ME on gate profile. The study found that in gate etching process, with the decrease of polymer Gas Ratio and the increase of OE amount, the bottom morphology of gate gradually changed from Vertical to reverse tapered. At the same time, by adjusting the SF6 gas ratio of ME, a smoother gate reverse tapered profile can be obtained to achieve accurate regulation of the gate profile. An ideal reverse tapered profile can be achieved with the most suitable etch parameters. In WAT AC performance tests, gate of reverse Tapered profile improved electrical performance by about 2-3% compared to Vertical profile.

ACKNOWLEDGMENTS

Thanks for all the authors for great work and collaboration. Thanks for FA (failure analysis) for TEM support.

REFERENCES

[1] N. Yamagishi et al, 2003 Dry Process International Symposium, (2003) 105.

[2] Mike Barnes et al., Inductively coupled plasma source with controllable power deposition, US Patent 6, 507, 155.

[3] J. Kedzierski, International Electron Devices Meeting. Technical Digest (Cat. No.01CH37224), Washington, DC, USA, 2001, pp. 19.5.1-19.5.4. S. Junking, Y. HeeJoung, S.YeolMun, Transactions on Semiconductor Manufacturing, 20, 150(2007).

[4] Liang, C. W., Chen, M. T., Jenq, J. S., Lien, W. Y. etc. A 28nm poly/SiON CMOS technology for low-power SoC applications, 2011 Symposium on VLSI Technology - Digest of Technical Papers > 38 - 39 Proceedings Article March 29, 2013.

[5]. Zhang Y. Low Temperature Plasma Etching Control through Ion Energy Angular Distributions in and 3-Dimensional Profile Simulation. University of Michigan, 2015.

STUDY OF SPACER ETCHING WITH PR APPROACH

Yuhao Yang*; Siyuan Che; Xiangguo Meng; Lian Lu; Quanbo Li; Jun Huang; Yu Zhang
Shanghai Huali Integrated Circuit Corporation
Shanghai City, China
*Corresponding Author's Email: yangyuhao@hlmc.cn

ABSTRACT

Over the past decade, advances in CMOS FEOL process technology nodes have made each generation of technology to achieve higher integration density and performance. However, the gate oxide thickness will lead to an exponential increase of leakage current which is usually caused by Lightly-doped drain (LDD) process. Furthermore, this appearance makes the some region insufficient to meet the electrical performance requirements. In order to solve the leakage issue caused by LDD, we develop a new spacer etch with Photo Resist (PR) approach, but this approach will cause some problems such as profile, width and uniformity issue. The formation mechanism of profile and improvement of uniformity are analyzed in this paper. The new approach shows better performance in the spacer etch process with PR. This paper provides an effective strategy for spacer etch process in the PR environment.

Keywords— Spacer etch, PR, Profile, Uniformity

INTRODUCTION

In the past decade, with the advancement of silicon CMOS FEOL process technology nodes, each generation of technology has higher integration density and performance [1-2]. However, the reduction of threshold voltage (Vt) and gate oxide layer thickness also leads to an exponential increase in leakage current, which becomes a major contributor to the total power and design challenges [2-4].

However, due to the introduction of some regions, metal ions will invade OX in the subsequent LDD process, resulting in grid leakage [5-7]. It makes this region insufficient to meet the electrical performance requirements. In order to solve the leakage problem of this region, this paper protects the region by introducing PR mask into spacer etching. Paper studies the morphology formation mechanism and uniformity improvement in PR environment, and provides a reasonable and effective approach for spacer etching with PR.

EXPERIMENT

In order to solve the problem of electric leakage caused by the introduction of ion implantation in some regions, a part of SIN is retained on the region OX to protect the region, so that the final electrical property is improved. Therefore, PR mask is added to this region, as shown in Fig.1. However, due to the introduction of PR, the

spacer etching is in a more complex environment. This is a huge challenge for spacer etching..

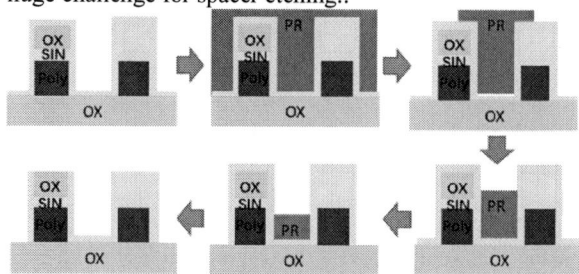

Fig.1. Spacer etching schematic with PR

RESULT AND DISCUSSION

Fig.2 (a) and (b) shows the topography of logic area with or without PR. It can be seen that in an environment with PR the width increases by about 10% compared with BSL, and the shoulder is shown to be more smooth and the footing is smaller. It is speculated that PR etch back process before conventional spacer etching results in the formation of SIxOnCyNz polymer on the surface of the original spacer, as shown in Fig.2 (c). The formation of this polymer changes the original spacer film, resulting in a decrease in the etch rate. The changes of SINME etch rate can also be clearly seen from its Curve, as shown in Fig.3. Moreover, the presence of the polymer has a good protective effect on the shoulder position. However, the conventional morphology makes plasma scatter and refract, which results in the formation of footing.

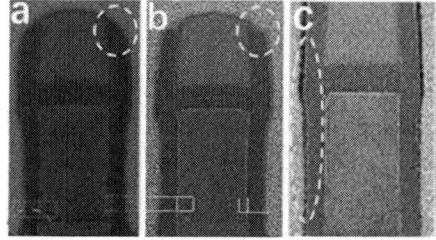

Fig.2. (a) NFET morphology in regular spacer etching
(b) NFET morphology in spacer etching with PR
(c) NFET morphology after PR etch back process

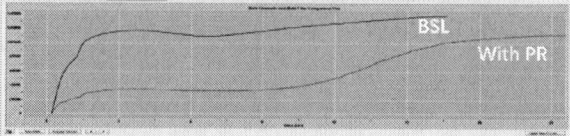

Fig.3. SINME etching curve with and without PR environment

Fig.4 shows the uniformity map of width in an environment with or without PR. It show that in an environment with PR, the uniformity of width becomes much worse than the width without PR. However, uniformity did not improve with regular TCCT and ESC and etc. It is presumed that under the influence of pump in the chamber, uneven SIxOnCyNz polymer appeared in the center and edge of wafer. Regulation by conventional methods had little effect on this polymer, resulting in changes in uniformity.

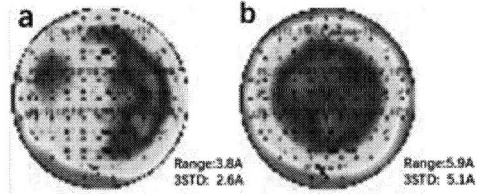

Fig.4. (a) The uniformity of width in an environment without PR

(b) The uniformity of width in an environment with PR

According to the above analysis, spacer etching in an environment with PR will generate an uneven layer of SIxOnCyNz on the surface of spacer walls due to the adoption of O2 in the PR etch back process, as shown in Fig.5. However, this polymer layer cannot be uniformly removed under the subsequent CH3F gas in over etching process, so the width and uniformity will be greatly changed.

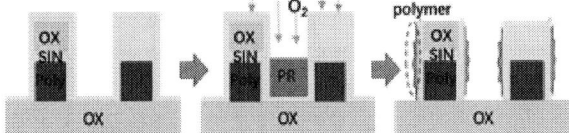

Fig.5. Schematic diagram of SixOnCyNz formation mechanism

Subsequently, the surface polymer was removed with the added Trim step to adjust width and achieve better uniformity. The Fig.6 shows a curve after improvement. It can be seen that with added trim, spacer etching did not have a flat phase and the rate increased obviously. Tab.1 shows the range and 3sigma of width under different conditions. The introduction of Trim into spacer etching with PR significantly improves wafer uniformity. And electrical properties eventually is improved as well.

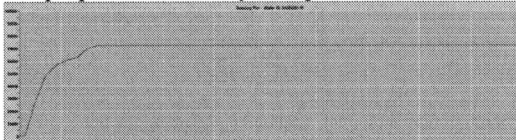

Fig.6. SINME etching curve with trim step

CONCLUSIONS

To solve the problem of electrical change in some area, spacer etching process with PR was studied. The results showed that the introduction of PR into spacer etching resulted in uneven SIxOnCyNz polymer formation on the original spacer walls. This resulted in changes of etching rate and uniformity. The polymer on the surface of spacer needs to be removed by introducing trim to achieve the same performance and performance as spacer etching without PR. It provides an effective method for the preparation of stable devices in some area.

ACKNOWLEDGMENTS

Thanks for all the authors for great work and collaboration. Thanks for FA (failure analysis) for TEM support.

REFERENCES

[1] K. Roy, et. al. Leakage Current Mechanisms and Leakage Reduction Techniques in Deep-Submicron CMOS Circuits, Proceeding of IEEE, vol. 91, pp. 305–327, Feb 2003.

[2] M.H. CHEN, et al. The study of leakage current performance improvement in high voltage technology. Proceeding of IEEE, pp. 158-160, 2012.

[3] Andrew P. Dumlao, Raminda U. Madurawe, Thomas McFarlane. Gate-aided drain to filed breakdown of high voltage NMOS devices. Proceeding of IEEE, pp. 841-844, 1990..

[4] H. Bostpn and K. Academic, High Voltage Devices and Circuits in Standard CMOS Technologies, Springer, R1, 1998.

[5] V. Parthasarathy, R. Zhu, V. Khemka, T. Roggenbauer, A. Bose, P. Hui, et al., A 0.25μm CMOS based 70V smart power technology with deep trench for high-voltage isolation, IEDM '02., pp. 459–462, 2002.

[6] Kyeonglan Rho. Effects of LDD Spacer Etches on Spacer Widths, Subsequent Oxide Growths and Yield Enhancement. Proceeding of IEEE, pp. 210-211.Aug 2002.

[7] Federico Faccio, et al. Influence of LDD Spacers and H+ Transport on the Total-Ionizing-Dose Response of 65-nm MOSFETs Irradiated to Ultrahigh Doses. Proceeding of IEEE, pp. 164-174, Oct 2017.

Silicon surface roughness improvement during plasma etch

Guang Yang*, Li Zeng, Haiyun Zhu, Jing Wang, Zhongwei Jiang

Beijing NAURA Microelectronics Equipment Co. Ltd
No. 8 Wenchang Avenue, Beijing Economic-Technological Development Area. Beijing City, China

*Corresponding Author's Email: yangguang@naura.com

Abstract

To reduce etch depth loading in shallow trench isolation and deep trench isolation etch, excess oxygen is added to slow down the etch rate in large opening area. The approach often causes surface roughness problem in large open area. This paper studies the surface roughness mitigation in STI etch with Cl_2/O_2 plasma through a comparison experiments conducted with a NAURA poly etcher. Results show that high BRF/SRF power ratio with low pressure could reduce surface roughness. However it cannot resolve the issue completely. By using synchronous pulsing plasma, we can significantly reduce open area surface and micro-trenching compare with traditional continuous wave (CW) plasmas.

Keywords—Surface roughness, passivation, silicon etch, pulsed plasma.

Introduction

As the device dimensions continue to scale down few nanometer nodes, chip electrical performance poses ever more strict requirements for plasma etching, such as precise control of the feature profile, critical dimension (CD), and surface roughness formed during plasma etching [1-3]. The precision control of Si etching with Cl- and Br-based chemistry is indispensable for the fabrication of transistor shallow trench isolation [4, 5]. The surface roughness and micro-trenching at the feature bottom affects the uniformity of bottom surface, which is responsible for a recess and damage in substrates, and in turn, also leads to the variability in transistor performance.

A larger number of studies are developed to understand the mechanisms responsible for the formation and evolution of surface roughness during plasma etching of blank (or planar) Si substrates. The mechanisms include geometrical shadowing [6], surface reemission of etchants [7], and effects of etch inhibitors [8]. Up to now, the understanding has still not been fully established.

This paper we investigate the influence of Si surface roughness during plasma etching in Cl-based plasmas using ICP tool. By varying SRF and BRF power and gas pressure, Si surface roughness phenomenon is improved, but microtrench issue still hard to eliminate. Based on above situation, advanced synchronized pulsed plasma were applied, and Si surface become smoothed compared with traditional continuous wave (CW) plasmas.

Experiment

The experimental etch in this paper were completed on one commercial inductively coupled plasma (ICP) etcher (NAURA) from China. The plasma was excited with two RF coils (13.56 MHz) to improve the ions flux uniformity. We selected silicon oxide as mask, Cl_2-based gas considered as main etchant. In this paper, there were three main power coupling settings. The first one is both source and bias power in continuous wave plasmas. The second one is to keep the source in CW RF mode and the substrate bias in pulsed mode. The third one is fully synchronized source and bias pulsing. Cross section profiles were imaged via Transmission Electron Microscope (TEM).

Result and discussion

As shown in Fig. 1, surface roughness is assumed to occur when the difference in depth between the feature top and that at the center of the trench, and microtrench is assumed to occur when the difference in depth between the feature bottom adjacent to sidewalls and that at the center of the trench.

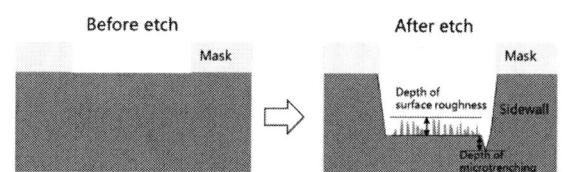

Fig 1. Illustration of surface roughness and microtrench.

A. Effects of SRF/BRF power

Fig. 2 shows TEM micrographs of Si feature bottom roughness with different SRF power. The surface roughness is more significant appear with increased SRF power, while the microtrench at the feature bottom is being less significant. Note the higher SRF power result in lower dc self-bias voltage observed in experiments, we get the conclusion that there is relation between dc self-bias and surface roughness in some degree.

Fig. 3 shows TEM images of the feature bottom roughness that etched with high and low bias RF voltage. It can be seen that bottom surface become smooth at high voltage. The above results indicate that as the bias power or ion energy is increased, the roughness of etched Si surfaces decreases. This phenomenon is attributed to the effects of a small amount of etch products/byproducts during plasma process.

979-8-3503-1101-3/23 $31.00 © 2023 IEEE

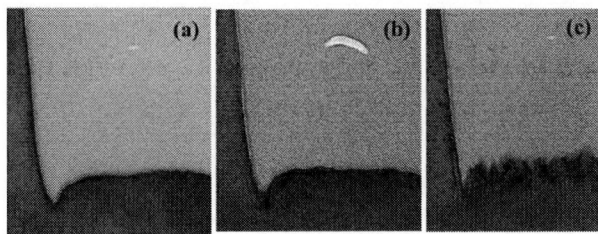

Fig.2 Trends of the etched bottom surface of Si with different SRF power: (a) SRF=X-100; (b) SRF=X; (c) SRF=X+100

These byproducts stick or deposit preferentially on the top of the feature, forming surface passivation layers (or micromask) and inhibiting etching thereon. Higher bias power may attack the passivation or residual polymer left on the bottom surface after the etch process. As a result, less byproduct of open area lead to smooth surface. However, the microtrench phenomenon almost unchanged.

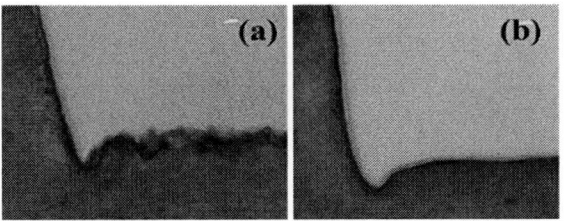

Fig.3 Bottom surface images (a) low bias RF voltage and (b) high bias RF voltage

B. Effects of gas pressure

Different gas pressures is tested as shown in Fig. 4. It can be seen that roughness depth decrease significantly with the decreasing of gas pressure. On the other hand, microtrench depth exhibits only a slight decrease with low gas pressure. These results indicate under appropriate low pressure range, microtrench depth is very low, and surface roughness issue is suppressed. The pressure in a plasma processing chamber lower, the mean free path of ion longer. Likewise, ion bombardment energy and ion direction stronger, which can affect rate and profile in etch processes.

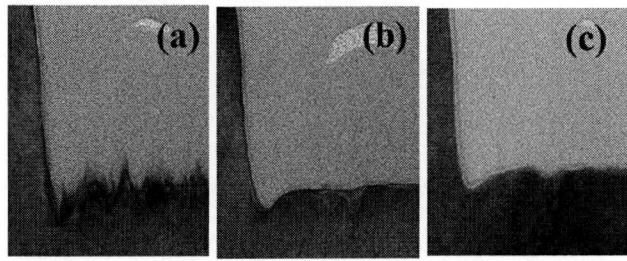

Fig.4 Trends of the etched bottom surface of Si with different gas pressures: (a) Pa=P+20 mtorr; (b) Pb=P mtorr; (c) Pc=P-20 mtorr

C. Effects of pulsed plasma mode

Fig. 5 compare bottom surface roughness performance in continuous wave (CW) bias RF mode, pulsed bias mode and synchronous pulsed mode that keep ICP source in CW RF mode. It can be seen in Fig. 5a, surface roughness is worse in CW bias RF mode. By using bias pulsed mode (Fig.5b),

surface roughness has been well controlled compared to Fig.5a. The reason may due to pulsed plasma reduce the byproduct deposition in plasma etch.

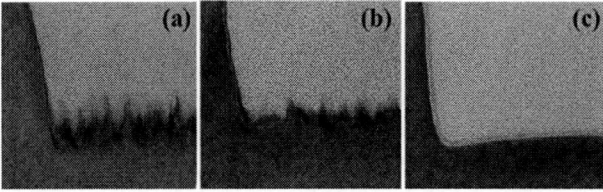

Fig.5 Bottom surface improvement images in (a) CW plasma mode (b) pulsed bias mode and (c) synchronous pulsed mode

To obtain good smooth performance, we test synchronous pulsing, as shown in Fig.5c. Synchronous pulsing significantly improve bottom roughness, and greatly decrease the depth of microtrench. This is mainly because that etch by-product deposition is totally same in the open area under synchronous pulsed mode, lower-energy positive ions and negative ions are recombined during the after-glow phase (power off period), the chemistry of the discharge is decreased, thus, the difference of etchant amount built up on bottom is decreased. Such polymer deposition is uniform at etched area. In addition, plasma bombardment became weak during power on period, which leads to better microtrench performance [9].

Conclusion

To reduce etch surface roughness and microtrench during silicon etch. We have investigated the possibility parameters like, SRF/BRF power, gas pressure and bias RF mode. The experiments showed that with the high BRF/SRF power ratio, the surface roughness is improved, meanwhile, this phenomenon is reduced under appropriate low pressure range. However it cannot resolve the issue completely. Therefore, different bias etch mode is introduced, and it is demonstrated that synchronous pulsing plasma could largely improve both surface performance and microtrench issue. So synchronous pulsing is definitely one of the most promising techniques method to achieve smooth bottom surface in the future.

References

[1] G. S. Oehrlein, R. J. Phaneuf, and D. B. Graves, J. Vac. Sci. Technol., A 29, 010801 (2011).

[2] E. Gogolides, V. Constantoudis, G. Kokkoris, D. Kontziampasis, K. Tsougeni, G. Boulousis, M. Vlachopoulou, and A. Tserepi, J. Phys. D: Appl. Phys. 44, 174021 (2011).

[3] K. J. Kanarik, T. Lill, E. A. Hudson, S. Sriraman, S. Tan, J. Marks, V. Vahedi, and R. A. Gottscho, J. Vac. Sci. Technol., A 33, 020802 (2015).

[4] H. Abe, M. Yoneda, and N. Fujiwara, Jpn. J. Appl. Phys. 47, 1435 (2008).

[5] V. M. Donnelly and A. Kornblit, J. Vac. Sci. Technol. A 31, 050825 (2013)

[6] P. Brault, P. Dumas, and F. Salvan, J. Phys. Condens. Matter 10, L27(1998).

[7] E. Zakka, V. Constantoudis, and E. Gogolides, IEEE Trans. Plasma Sci. 35, 1359 (2007).

[8] E. Gogolides, V. Constantoudis, G. Kokkoris, D. Kontziampasis, K.Tsougeni, G. Boulousis, M. Vlachopoulou, and A. Tserepi, J. Phys. D:Appl. Phys. 44, 174021 (2011).

[9] F.Y. Xiao, Q. H. Han, H.Y. Zhang, SPIE, 10149 (2017)

Effect of process gas on Side Wall Angle in Silicon Trench Etching

Yiming Ma *, Guang Yang, Litian Xu, Zhongwei Jiang, Jing Wang, Qifei Wang, Donghan Wang,

Yongjie Zhou

Beijing NAURA Microelectronics Equipment Co., Ltd.
Beijing, china
*Corresponding Author's Email: mayiming@naura.com

Abstract

The side wall angle is the main parameter to measure the profile of the side wall of trench. In this paper, HBr, SF_6, O_2, CF_4 are used as the main etching gases, and the sidewall morphology is adjusted under the condition of optimizing the etching sequence of mixed gases, smooth and continuous sidewall morphology is obtained.

Keywords: isotropic; etch; side wall; trench; morphology

Introduction

Trench isolation is the main method for device isolation in large scale integrated circuit manufacturing, which is widely used in the isolation process of 0.25μm and below semiconductor technology nodes [1-3]. With the increasing requirements for semiconductor chips, the size of the chip is getting smaller and smaller, and the size of the trench isolation structure is also decreasing. Under the condition of constant depth, the depth-width ratio used for trench isolation structure increases gradually, so the etching process becomes more and more difficult [4,5].

Scholars at home and abroad have studied this phenomenon and made improvements in process parameters and device structure. However, its research on process parameters is relatively simple, and no breakthrough progress has been achieved. Although the improvement of device structure is effective, it greatly increases the complexity of manufacturing, and cannot be extended to all micro devices.

In this paper, the process parameters affecting the trench morphology of inductively coupled plasma (ICP) etching are analyzed. By optimizing the etching sequence of mixed gas, a method to optimize the discontinuity of the side wall of the isolation trench is proposed, and it is applied to the trench isolation process, and good results have been achieved.

Experimental details

In the experiment, a silicon wafer with a diameter of 300mm (12 inches) was used, and SiO_2 and Si_3N_4 were successively epitaxial on the silicon substrate as hard masks. The ICP equipment of NAURA was used for etching. The RF power was coupled into the plasma source through the inductor coil, and the etching gas was ionized under the joint action of high-frequency magnetic field and induced electric field to generate high-density plasma. The gas flow rate is controlled by the gas flow controller, and the plasma generation and bias generation are respectively generated by independent RF sources. In this experiment, two methods are adopted to verify the effect of the experiment, the first method adopts a one-step method, using HBr, SF_6, O_2 and He gases to etch to the target depth. The second method adopts two-step method. When depth-width ratio is less than or equal to 5, the first step is performed, and HBr, SF_6, O_2 and He are used to the target depth; when the depth-width ratio is greater than 5, the second step is performed, and HBr, CF_4, O_2 and CL_2 are used to etch to the target depth. The process conditions of the one-step method are as follows: the etching gases include HBr, O_2 and He, and the etching time is 90s.

The process conditions used in the two-step method are as follows: (1) the etching condition used in the first step, when the trench depth-width ratio is less than or equal to 5. The etching gas includes HBr, SF_6, O_2 and He, and the etching time is 40s. (2) the process conditions adopted in the second step are used when the trench depth-width ratio is greater than or equal to 5. HBr is the main etching gas, and the etching gas contains HBr, CL_2, CF_4, He, and the etching time is 50s.

Results and discussions

Figure1 and figure 2 show the trench sidewall morphology obtained under one-step and two-step etching conditions respectively. Figure 1 shows the etching morphology obtained by one-step method. The side wall of the trench is discontinuous. The angle of the side wall of the upper part is 86 degrees, and the angle of the side wall of the lower part is 80 degrees. There are sharp corner at the junction, which will lead to electric leakage.

The vertical sidewall morphology is mainly determined by alternating periods of passivation and etching. If the passivation is excessive, there will be more etching above and less etching below the trapezoidal morphology, which will lead to the stop of the etching seriously. If the etching excessive, serious etching will occur on a certain part of the silicon substrate, and in severe cases, etching break will occur on a certain part of the silicon substrate. Only when the etching and passivation reach a balance, can the side wall form a continuous and smooth shape close to 90 degrees.

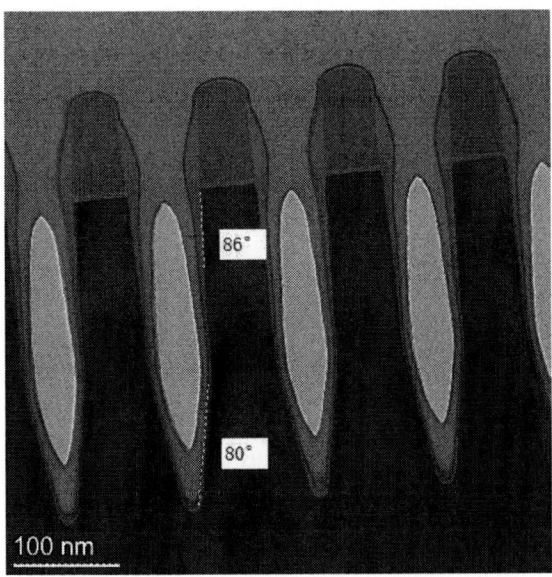

Figure 1. Trench side wall morphology after one-step etching

It can be seen from figure 1 that when the depth to width ratio is greater than or equal to 5, due to the enrichment of etching by-products on the side wall, the etching will have an inflection point here, the angle of the side wall of the trench below the inflection point starts to decrease, and the side wall of the trench above the inflection point is discontinuous. The difference between figure 2 and figure 1 is that the combination of SF_6 and O_2 is replaced by CF_4 and CL_2.

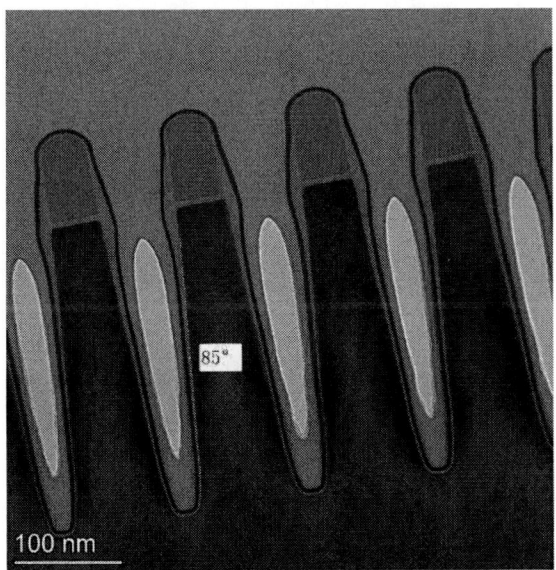

Figure 2. Trench side wall morphology after one-step etching

HBr, SF_6 and O_2 can form a large number of barrier particles in the etching process, that is, these gases simultaneously provide etching particles, energy particles and precursor particles forming the barrier layer during the etching process, and these precursor particles are adsorbed to the side wall surface layer to form a barrier layer [6,7]. The particles obtained after the ionization of SF_6 can also polymer protection sidewalls with the etching reactants. The barrier layers prevent the etching ions from contacting the silicon matrix, resulting in less and less density of active groups reaching the side wall, and the etch speed is much smaller than passivation speed, thus creating a corner in this part. CF_4 can remove a part of the barrier layer (such as SiO_2), and can remove a part of the barrier layer generated in the first step of etching [8,9]. CL_2 mainly plays an etching role, reducing the ability of etching gas to form a barrier layer. Therefore, replacing SF_6 and O_2 with CL_2 and CF_4 can solve the problem of corner when the side wall depth-width ratio is greater than 5 due to the heavy side wall protection.

Conclusions

The discontinuity of trench sidewall can be solved by two-step etching. First, use the combination of HBr, SF_6 and O_2 to etch the smooth sidewall morphology, and then use the combined gas of HBr, CL_2 and CF_4 to solve the sidewall discontinuity caused by the thick barrier layer, providing technological support for the subsequent silicon trench etching.

References

[1] Wang F , Liu S , Yin X , et al. Shallow trench isolation structure design for through silicon vias stress reduction[J]. International journal of numerical modelling: Electronic networks, devices and fields, 2022(4):35.

[2] Mica I , Roffarelloa P M , Dutartre D . Impact of the Substrate Specifications on the Extended Defects Induced by the Deep Trench Isolation [J]. ECS transactions, 2021(4):102.

[3] Wang F , Liu S , Yin X , et al. Shallow trench isolation structure design for through silicon vias stress reduction[J]. International journal of numerical modelling: Electronic networks, devices and fields, 2022(4):35.

[4] Lang J . Atomic-scale sidewall passivation for micro LED devices [J]. SID International Symposium: Digest of Technology Papers, 2022(3):53.

[5] Hudson E A . Technique to deposit sidewall passivation for high aspect ratio cylinder etch. 2019.

[6] Miwa K , Mukai T . Influences of reaction products on etch rates and line widths in a poly-Si/oxide etching process using HBr/O-2 based inductively coupled plasma.

[7] Sriraman S , Panagopoulos T , Paterson A , et al. Feature Scale Model of Shallow Trench Isolation (STI) Etch in HBr Plasma and Comparison with Experiments[J].

[8] Kim J H , Cho S W , Park C J , et al. Angular dependences of SiO2 etch rates at different bias voltages in CF4, C2F6, and C4F8 plasmas[J]. Thin Solid Films: An International Journal on the Science and Technology of Thin and Thick Films, 2017(Sep.1):637.

[9] Levko D , Raja L . Optimization of silicon etch rate in a CF4/Ar/O2 inductively coupled plasma [J]. Journal of vacuum science and technology, B. Nanotechnology & microelectronics: materials, processing, measurement, & phenomena: =JVST B, 2022(3):40.

INVESTIGATION OF CD PRECISE CONTROL IN PITCH DOUBLING FLOW FOR MEMORY INDUSTRY

Zhao Liu, Baodong Han
Superstring Academy of Memory Technology Beijing, China
*zhao.liu@bjsamt.org.cn

ABSTRACT

The conventional pitch doubling process aims to achieve smaller pitch size beyond photo patterning limitation cross technology node for both DRAM and NAND industry. With the shrinkage of chip size, the structure's critical dimension (CD) is more significant to enable the circuitry to function properly. Firstly, this research starts with a brief investigation of the traditional pitch doubling process from analytical study on the origins of CD variation. Secondly, we study the conventional CD control mechanism, such as run-to-run feedback tuning, SPC (Statistical Process Control) basic knowledge and point out its limitations with detailed experimental analysis. Lastly, a novel CD control concept of feed-forward tuning instead of feedback tuning is proposed. And this paper also describes a method that is how to build up the mathematical model for optimal CD control as well as its implementation in industry. Such novel tuning mechanism achieved precise CD control with a dramatic improvement in process capability (process capability index Cpk) with 32% CD variation reduction and 1.4% Yield gain.

INTRODUCTION

In memory industry, Pitch refers to the total length of a line (printable feature) and the adjacent space. Half of the length of minimum pitch fabricated on the chip active area remarks the technology node for the semiconductor industry. With the shrinkage on the chip size, the pitch length approaches to its limitation [1]. Lithographic printing capability is typically considered as the bottle neck for technology node revolution in memory industry. ASML extreme ultraviolet lithography (EUV) system receives overwhelming attention globally. It is published that ASML's most advanced lithography system TWINSCAN NXE:3600D enables EUV volume production at 3nm logic nodes. Self-aligned double patterning (SADP) [2][3] process is another technique to achieve smaller pitch length beyond lithographic patterning limitation. This paper demonstrates a deep study on conventional pitch doubling scheme and identifies inline Critical dimension CD control challenges. Following with the novel tuning method development to finally achieve consistent good inline results after implementation.

MAIN BODY

SADP Integration scheme.

Conventional pitch doubling process starts with a Photo resist patterning followed by either resist trim or hard mask (HM) etch to form a mandrel over an underlying layer. (Figure 1) The underlying layer functions as an etch stopping layer (ESL) for downstream spacer etch, normally Silicon Oxynitride (SiON) or Dielectric anti-reflective coating (DARC) are preferred due to their high selectivity to oxide during spacer etch. After 1st mandrel patterning is established, it proceeds to atomic layer deposition (ALD) for oxide spacer deposition at room temperature. Next comes to oxide spacer etch to remove the oxide on the horizontal surface but keep the sidewall spacer. After the wet strip for the mandrel removal, the patterned oxide spacer serves as a new mask and will transfer to the next layer. Furthermore, self-aligned quadruple patterning (SAQP) is simply repeating the spacer deposition to spacer strip scheme.

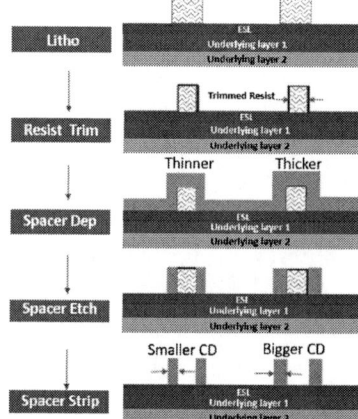

Figure 1: Pitch doubling scheme

The pitch doubling process is well employed to achieve a smaller bitline/wordline profile in memory industry. As it is known that CD is a measurable parameter to quantify pitch length. As shown in Figure 2, after mandrel removal, the spacer line (Line 1) is formed. Line 1 CD is defined by oxide spacer film thickness during the ALD process. The thicker spacer deposition, the bigger line 1 CD is achieved post spacer strip. During the pattern transfer etch, line 1 acts as a new mask and underlayer 1 is etched to form line 2. Both line 1 CD and pattern transfer etch process will affect line 2 CD, for example, normally CHxFx/HBr (polymerizing gas) is applied during pattern transfer etch process to tune the CD to bigger size.

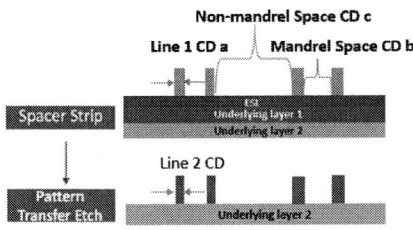

Figure 2: Line & Space CD

For Space CD, the mandrel Space CD is controlled by upstream resist CD and non-mandrel Space CD is impacted by both line CD and mandrel Space CD. The simplification of all CD parameters follows the below formula.

$$2a + b + c = Constant. \qquad (1)$$

Inline CD control mechanism

Statistical Process Control (SPC) is a well-known statistical method to monitor the stability and performance of a process or production. Enormous studies and in-depth research have been done for its application in industry [4][5]. The most popular SPC tool is the control chart to record real-time data for monitoring process behavior and trigger a series of reactions when an event occurs. Figure 3 is a typical control chart with 3 main components Upper Control Limit (UCL), Control Limit (CL or Target) and Lower Control Limit (LCL).

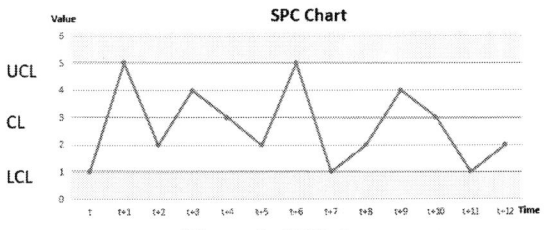

Figure 3: SPC chart

When talking about process tuning, simply means maintaining the measurement parameters as closer to target as possible. The deviation to target or so-called variation is inevitable, but process engineers still put continuous effort to achieve progressively smaller variation. Current industrial process tuning is mainly based on Running Process Adjustment (RPA), a feedback tuning loop, which is widely adopted in Lithography, Etching and other modules. For example, as mentioned before, the mandrel Space CD is mainly affected by incoming resist CD. After each run of the litho patterning step, the After the Develop Inspection (ADI) CD data

enters into the SPC chart. The deviation from the data to target is calculated and feeds back to litho system to adjust the dose for the next run. If the data fails into OOC (out-of-control) yellow region, the wafer will track into rework route to strip the photoresist on top and re-pattern again.

Feedback tuning mechanism limitation---Lagging effect

The frequency of such a feedback tuning scheme is highly dependent on measurement (sampling) rate. The normal practice is to assign different sampling rates to different measurement steps based on their criticality which may vary from time to time in consideration of production capability. This paper takes Carbon etch as pattern transfer etch for illustration purpose. HBr is the typical etching (tuning knob) gas, while higher HBr injection increases final line CD. When there is a small drift in Spacer Deposition, that is a lot with thicker Oxide (ALD) film proceeds to downstream step. When the lot is not sampled post Spacer Strip, but measured after Carbon etch. Its thicker incoming Oxide film results in bigger Line 1 CD (estimated) and transfers to bigger Line 2 CD (measured). Carbon etch feedback tuning mechanism is triggered and the tuning knob HBr gas flow will be wrongly adjusted lower as the RPA tuning scheme is designed for its own step (Carbon etch) process control not in compensation for upstream (Spacer Deposition) drift. If next lot with similar thicker Oxide Spacer deposition gets measured directly after Spacer Strip, bigger Line 1 CD will further tune down HBr even lower. Mismatch sampling rate and feedback tuning mechanism postpone CD recovery. Such a lagging effect makes Carbon etch RPA false adjustment and causes low accuracy in CD tuning.

NOVEL TUNING METHOD DEVELOPMENT

ALD film thickness variation and low Line 1 CD sampling rate is the main contributor to the lagging effect. For film deposition step, the process control is usually based on monitoring the blanket wafer, an indirect measurement to indicate film thickness on a real structure. The fundamental fix is to find an inline film thickness indicator with a higher sampling rate. Spacer Oxide etch employs endpoint etching detection mode to cater for incoming film variation. The thicker ALD film the longer etch time. Plot the Linear regression between Line 1 CD and ALD film and it demonstrates below correlation with R^2 around 0.9 conceptually as shown in Figure 4. Using endpoint etch time to predict Line 1 CD to realize a 100% sampling rate with high accuracy assuming the Spacer etch itself is stable.

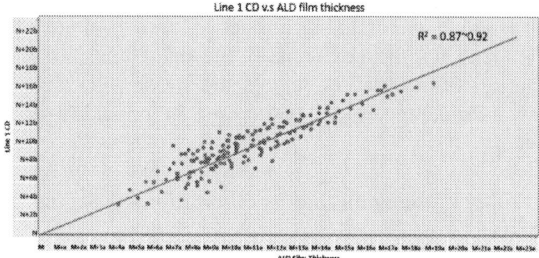

Figure 4: Line 1 CD v.s ALD film thickness

Scenario 1 No litho patterning between Spacer Strip to pattern transfer etch

To simplify the illustration, we still use Carbon etch as the example. Design one Special Work Request (SWR) lot to collect Line 1 CD, run HBr gas skew at Carbon etch and measure Line 2 CD to determine the tuning slope (Figure 5). After the tuning slope (α) is defined, make use of endpoint time from Spacer Oxide etch to model Line 1 CD1 for every lot and adjust HBr (tuning knob) gas flow to compensate for the individual lot. The new tuning scheme follows the below formula.

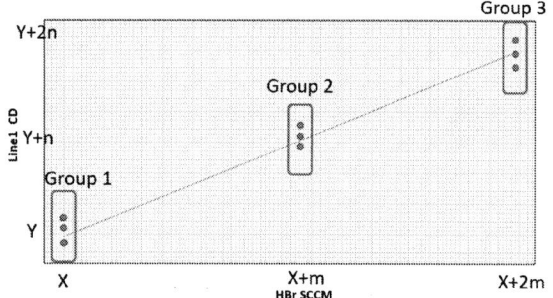

Figure 5: Line 1 CD skew from different Group with tuning slope n/m= α

Novel tuning = Feedforward(FF)+Feedback(FB)

$$FF = -(Line\ 1\ CD1* -Target)\frac{i}{\alpha} \qquad (2)$$

The FB portion in the above formula follows its current practice, the only change is to add Feedforward tuning. Line 1 CD1* indicates the calculated (modeled) CD1 value from Spacer Oxide etch endpoint time, while if the wafer is sampled for Line 1 CD1, the real measurement data will be substituted in the formula. The tuning adjustment coefficient i normally chose from 30% ~70%.

Scenario 2 At least one litho patterning between Spacer Strip to pattern transfer etch

For such a situation (Figure 6), since the CD1 & CD2 size differs (CD1 is relatively smaller from SADP), due to the Aspect-ratio dependent etching (ARDE) effect [6], the tuning slope (HBr tuning knob) on CD1 is different from CD2. Firstly, follow scenario 1 SWR to collect CD2

tuning slope (β). Secondly, design another SWR lot with different Line 1 CD2 skew to run same HBr gas at Carbon etch and collect Line 2 CD2 to determine the correlation γ . (Line 2 CD2 vs. Line 1 CD2 at same HBr tuning knob) Lastly, with all the above information the novel tuning mechanism follows the below formula.

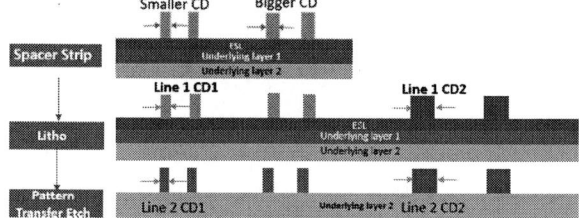

Figure 6: CD2 patterning after Spacer Strip

Novel tuning = Feedforward(FF)+Feedback(FB)

$$FF = \frac{-(Line\ 1\ CD1* -Target)*\frac{\beta i}{\alpha\gamma}}{Dose} \qquad (3)$$

$$FB = \frac{(Line\ 1\ CD2 - Target)}{Dose} \qquad (4)$$

The FB portion for litho tuning in above formula follows its current practice, here it shows a conceptual FB controller for illustration purposes. The novel tuning mechanism for litho consists of two parts. FB controller built to faster detect any drift in Litho itself. FF controller enables dose to be adjusted for each lot according to it calculated Line 1 CD1, similar to scenario 1 but this time the focus is on CD2. Meaning to purposely offset litho dose to make the final result on-target.

RESULT AND CONCLUSION

Such novel tuning method mitigates the lagging effect due to low sampling rate with more accuracy in CD control. Below table 1 shows the result of implementation on scenario 2. The actual CD measurement data is omitted due to confidentiality reasons. The average CD variation reduction is 32% and Product B achieve Yield improvement around 1.4%, considered as a huge gain for mature Yield product.

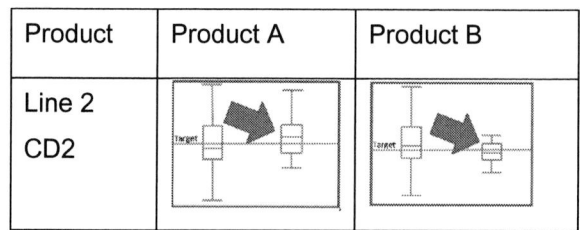

Table 1 Inline CD performance with new tuning method

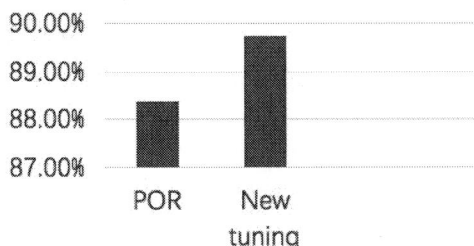

Figure 7: Product B Yield performance with new tuning (89.7% v.s 88.3%)

ACKNOWLEDGMENTS

The author would like to extend her sincere thanks and appreciation to Guilei, Ares for their valuable suggestions and support.

REFERENCES

[1] Yaegashi, H. , et al. *Novel approaches to implement the self-aligned spacer double-patterning process toward 11-nm node and beyond."* Proceedings of SPIE - The International Society for Optical Engineering 7972.1(2011).

[2] Bencher C . *Self-Aligned Double Patterning[J].* SemiconductorInternational,2008,31(10):37-38,40,42,45

[3] Mohanty, N. , et al. *"Challenges and mitigation strategies for resist trim etch in resist-mandrel based SAQP integration scheme."* Proceedings of SPIE - The International Society for Optical Engineering 9428(2015).

[4] Oakland, J. S. . "Statistical Process Control." John Wiley & Sons Inc New York Ny 14.2(2014):20–29.

[5] Spanos, C. J. . "Statistical process control in semiconductor manufacturing." Proc IEEE 80.6(1992):819-830.

[6] Chung C K, *Geometrical pattern effect on silicon deep etching by an inductively coupled plasma system.* J. Micromech. Microeng,2004.

The Plasma Etching of the Deep Hole Structure in Silicon with the Mixed Gas of SF_6, HBr and O_2

Qifei Wang1*, Yiming Ma[1], Zhongwei Jiang[1], Jing Wang[1], Haiyun Zhu[1]

[1]ETCH II BU, Beijing NAURA Microelectronics Equipment Co. Ltd., Beijing, China

*Corresponding Author's Email: wangqifei@naura.com

Abstract

We studied the inductively coupled plasma (ICP) etching of deep silicon hole with the etchant gas of SF_6/HBr/O_2. We mainly researched the influence of HBr flow and bias voltage during the plasma etching process. Based on the profile of Si hole, we found that increasing the HBr flow could obviously enhance the etching rate within a proper range and the bias voltage could control the roughness of the sidewall. Through balancing the HBr flow and bias voltage, we could realize the vertical and smooth sidewall of the Si hole, which could be applied to other plasma etching of Si hole requiring high quality profile.

Keywords — plasma etching, silicon hole, roughness

Introduction

Fluorine-based plasma has been widely applied into the plasma etching of silicon with high aspect ratio structure including Si trench, Si hole and so on, which could be used in the fabrication of memory devices and microelectromechanical system (MEMS) components [1~3]. Former researches have revealed that the oxygen could suppress the lateral etching during the plasma etching process in the mixed gas of SF_6/O_2 [4~6]. Further studies exhibit that O_2 could help to generate oxide passivation layers on the sidewall, which largely decreases the etching rate on the vertical direction. To further improve the anisotropic etching property, HBr could also be added into the SF_6/O_2 without decreasing the etching rate. The addition of HBr could decrease the F-to-O ratio in the plasma, which could enhance the sidewall passivation and suppress the undercut.

However, with O_2 concentration increasing, the etching process could not be able to continue and the etching stop phenomenon could appear. We found that the addition of HBr could enhance the etching process, which shown that HBr could not only decrease the F-to-O ratio but also enhance the ion bombardment in etching process [7]. The improvement on the ion bombardment could weaken the passivation layer and the sidewall roughness could be worse. In this study, we explored the influence of HBr on the ion bombardment and also resolved the roughness of sidewall by changing the RF-bias voltage, which further revealed that the roughness was determined by the ion energy during the etching process.

Experiment setup

The plasma etching processes were performed in the inductively coupled plasma reactor (NAURA 612C). The plasma was maintained with the 13.56 MHz RF power source and the electrostatic chuck (ESC) was also powered by the second RF power at 13.56 MHz. Detailed parameters were listed in tables on the manuscript.

Results and discussion

The profile of Si holes etched with different amount of HBr flow were shown in Fig. 1, and detailed process parameters were listed in the Table 1. When the HBr flow increased from 20 sccm to 60 sccm, the etching rate became faster, which increased from 10.1 nm/s to 27.9 nm/s. In particular, the phenomenon of etching stop appeared when the HBr flow was under 60 sccm. However, when the HBr flow increased from 60 sccm to 100 sccm, the etching rate dropped a little from 27.9 nm/s to 26.0 nm/s. According to the variation of

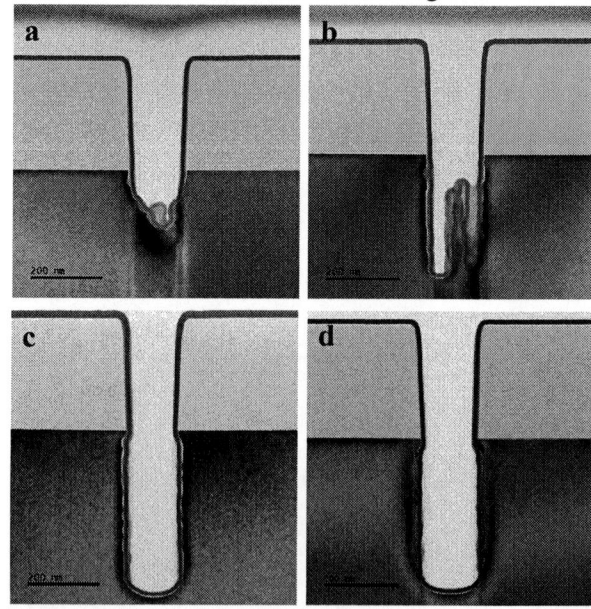

Fig. 1. The TEM cross section images of Si hole etching in SF6/HBr/O2 plasma with different HBr flow. Detailed process parameters are listed in Table 1. (a) 20 sccm HBr, (b) 40 sccm HBr, (c) 60 sccm HBr, (d) 100 sccm HBr.

TABLE 1

DETAILED PARAMETERS IN PLASMA ETCHING

Process parameter	a	b	c	d
Pressure (mTorr)	40	40	40	40
Source RF (W)	1000	1000	1000	1000
RF-bias voltage (V)	-350	-350	-350	-350
SF_6 (sccm)	50	50	50	50
HBr (sccm)	20	40	60	100
O_2 (sccm)	80	80	80	80
He (sccm)	200	200	200	200
ESC temperature (^{o}C)	25	25	25	25
Process time (s)	16	16	16	16

etching rate, we assumed that the etching rate of Si was limited by two factor, which included the rate of removing surface passivation layer and concentration of fluorine radicals. When the gas flow of HBr was below 60 sccm, the etching process of Si was ion-limited. More HBr could improve the ability of removing the passivation layer on the bottom of S i hole, which was benefit of the reaction between F radicals andSi. The appearance of etching stop could be related to the influence of electrons accumulated on the surface of sidewall. These electrons could attract the incident ions and change the initial direction, which made the ion density increased near the sidewall. Due to the lower density of incident ions, the passivation layer could not be effectively removed on the center of Si hole bottom and led to the etching stop (Fig. 1b).When concentration of F radicals was enough for the etching process, the etching rate increased rapidly with HBr

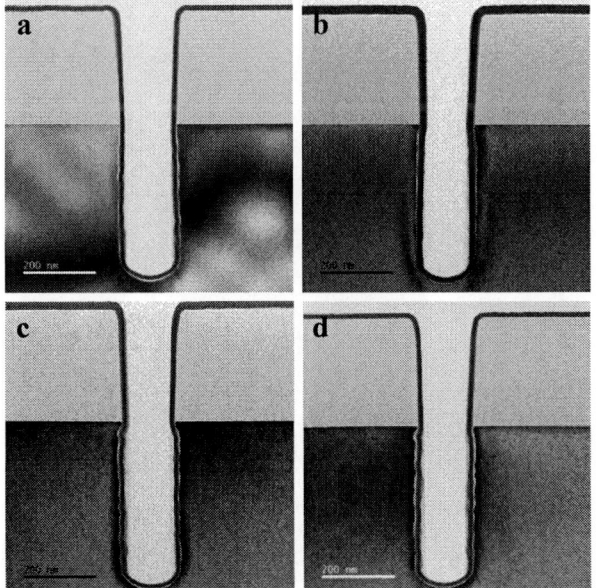

Fig. 2. The TEM cross section images of Si hole etching in SF6/HBr/O2 plasma with different RF-bias voltages. Detailed process parameters are listed in Table 2. (a) -250V, (b) -300V, (c) -350V, (d) -400V.

flow rising and the passivation layer could be completely removed when the HBr flow reached 60 sccm. With further increasing of HBr flow, the limitation of etching rate could turn to the concentration of F radicals. Due to the H could suppress the concentration of F radicals, the etching rate began to slow down slightly with the HBr flow increasing from 60 sccm to 100 sccm. Based on the profile of Si holes etched with different HBr flow, we could conclude that the concentration of HBr is an essential factor during the etching process with the etchant of $SF_6/HBr/O_2$. On the one hand, the HBr could enhance the ability of removing the passivation layer. On the other hand, the excess of HBr could bring too much H into the etching process, which could decrease the concentration of F radicals in the etching process. Finally, we determined that the proper gas flow of HBr could be 60 sccm in this work.

TABLE 2

DETAILED PARAMETERS IN PLASMA ETCHING

Process parameter	a	b	c	d
Pressure (mTorr)	40	40	40	40
Source RF (W)	1000	1000	1000	1000
RF-bias voltage (V)	-250	-300	-350	-400
SF_6 (sccm)	50	50	50	50
HBr (sccm)	60	60	60	60
O_2 (sccm)	80	80	80	80
He (sccm)	200	200	200	200
ESC temperature (^{o}C)	25	25	25	25
Process time (s)	16	16	16	16

Although we could achieve the vertical sidewall in the Si hole, there was the obvious roughness appearing on the surface of sidewall etched with $SF_6/HBr/O_2$. In Fig. 2, we exhibited the influence of the RF-bias voltage on the sidewall roughness during the plasma etching process. With the RF-bias voltage increasing, the etching rate did not changed obviously for the depth of Si hole was almost the same, which is consistent with the former assumption that the etching process was limited by the concentration of F radical. Further increasing the RF-bias voltage could not enhance the etching rate for the RF-bias voltage mainly influenced the ion energy. Referring to the sidewall roughness, the sidewall could be smoother under the voltage of -250 V and -300 V. Higher voltage could lead to more severe roughness during the etching process with -350 V and -400 V. The origin of roughness was that the sidewall protection was not enough. And the isotropic etching of F radicals could cause the lateral etching when the sidewall Si was exposed, which could lead to the roughness profile. The damage of sidewall passivation layer could mainly come from the diffraction of positive ions that were accelerated in the sheath and bombarded both the sidewall and bottom of the Si hole. The lower voltage led to lower energy of ions but larger ion incident angles, which increased the ions densities bombarding the sidewall and induced the roughness and larger sidewall angle in Fig. 2a. On the contrary, the higher RF-bias voltage could shrink the ion incident angle and decreased the amount of ions reaching the sidewall. However, the energy of diffracted ions could increase with the voltage rising and the ability of removing the sidewall passivation layer could also be enhanced, which could make worse roughness with the same sidewall angle (Fig. 2b~2d). In this work, when the RF-bias voltage was -300V, we could realize the balance of the ion energy and ion incident angle and achieve the vertical profile of Si hole and smoother sidewall.

Summary and conclusions

In the plasma etching of Si deep hole with $SF_6/HBr/O_2$ gas etchant, the HBr flow could play an important role in controlling the etching rate and profile. Through increasing the HBr flow, the property of etching process could convert from the ion-limited process to radical-limited process by enhancing the ability of removing the passivation layer. Besides, the HBr could also decrease the F radical concentration with gas flow rising, which could suppress the etching rate when the HBr flow is too high. For the sidewall protection, the key point is to achieve the balance of ion energy and ion incident angle. The ion with high energy could easily damage the sidewall passivation layer and the larger ion incident angle could lead to the fact that more ions could bombard the sidewall and make worse sidewall profile. The proper balance of ion energy and ion incident angle could be benefit for realizing the smooth sidewall profile. In this work, we also propose the assumption how the reaction occurring during the plasma etching process, which could help to understand the theory of Si hole plasma etching.

References

[1]   A. A. Ayón and R. Braff, "Characterization of a Time Multiplexed Inductively Coupled Plasma Etcher", Journal of The Electrochemical Society, vol. 146, pp. 339, 1999.

[2] G. Craciun and M. A. Blauw, "Temperature influence on etching deep holes with SF6/O2 cryogenic plasma", Journal of Micromechanics and Microengineering, vol. 12, pp. 390, 2002.

[3] D. L. Flamm and V. M. Donnelly, "The reaction of fluorine atoms with silicon", Journal of Applied Physics, vol. 52, pp. 3633-3639, 1981.

[4] S. Gomez and R. J. Belen, "Etching of high aspect ratio features in Si using SF6/ O2/ HBr and SF6/ O2/ Cl2 plasma", Journal of Vacuum Science & Technology A, vol. 23, pp. 1592-1597, 2005.

[5]   V. K. Singh and E. S. G. Shaqfeh, "Simulation of profile evolution in silicon reactive ion etching with re‐emission and surface diffusion", Journal of Vacuum Science & Technology B: Microelectronics and Nanometer Structures Processing, Measurement, and Phenomena, vol. 10, pp. 1091-1104, 1992.

[6] W. C. Tian and J. W. Weigold, "Comparison of Cl2 and F-based dry etching for high aspect ratio Si microstructures etched with an inductively coupled plasma source", Journal of Vacuum Science & Technology B: Microelectronics and Nanometer Structures Processing, Measurement, and Phenomena, vol. 18, pp. 1890-1896, 2000.

[7] H. F. Winters, J. W. Coburn, "Surface science aspects of etching reactions", Surface Science Reports, vol. 14, pp. 162-269, 1992.

Investigation of High Aspect Ratio Amorphous Carbon Etching in NAND Flash Memory

Li Zeng[1]*, Guang Yang[1], Zhongwei Jiang[1], Jing Wang[1], Li-Tian Xu[1]

[1] Beijing NAURA Microelectronics Equipment Co. Ltd
No. 8 Wenchang Avenue, Beijing Economic-Technological Development Area. Beijing City, China
*Corresponding Author's Email: zengli01@naura.com

Abstract

The amorphous carbon layer (ACL), used as the hard mask for the etching of deep hole in NAND flash memory, was etched using O_2/SO_2 and pulsed dual-frequency inductively coupled plasma. The oxide mask was gradually collapsed and distorted during the ACL etching due to the high bias power, which can cause poor uniformity and even etching stops. In this paper, a method is proposed to remove the mask above the hole and enlarge the mask space CD, which is good for byproduct to diffuse out and reactants to reach the etch front and then good uniformity and bottom critical dimension controllability are obtained.

Keywords —High aspect ratio; ACL etching; NAND flash memory; CD controllability; Uniformity

Introduction

NAND flash memory has been one of the most successful storage technologies in the past few decades, and there are efforts to scale down the effective cell size [1]. Therefore, the integrated three-dimensional architectures of NAND flash memory is emerged to overcome the scaling limit of planar NAND flash arrays [2]. The etching process in three-dimensional NAND flash mainly include the etching of staircase, channel hole, trench and contact hole. In the process of channel hole etching, the mask should be thick and have good rigidity, which would not be depleted during the channel hole etching process due to its high aspect ratio. Among the various hard mask materials, amorphous carbon layer (ACL) has been widely used due to the high etch selectivity over a photoresist and Si-based materials, easy deposition, and easy removal after the dry etch process [3].

Recently, plasma techniques with dual-frequency and high bias power have been widely used for the high aspect ratio etching of different materials. Meeting the standard of the critical dimension and maintaining good uniformity of the high aspect ratio process is challenging and critical. In this study, two kinds of methods are proposed to improve the profile and uniformity of high aspect ratio hole mask etching process for NAND flash memory. The mask distorted and blocked the hole, which could slow down or stop the byproducts getting out the hole. The bottom CD of the hole increased significant by adding BT step to reopen the mask during the ACL etching, which is especially significant and helpful for the mask hole etching process.

Experiment

The film stacks of channel hole hard mask etch process before etch and after etch are illustrated in Fig.1. There are six kinds of film layers：

(1) The photo resist, Silicon bottom anti-reflective coating(HM1) and organic under layer(HM2) are called the tri-layer structure;
(2) The HM3 layer can be oxide and oxide/SiOC, which is used as mask during the ACL etching;
(3) The amorphous carbon layer is the main etching material for this process，which was etched using pulsed dual-frequency inductive coupled plasma. The pulsing mode is used in this process to increase the selectivity of etching material to mask, which is by turning on and off the radio frequency power periodically at a certain frequency during the etching. We use four steps to etch ACL layer: ME1, ME2, ME3 and OE with ramped source/bias power and different gas ratio, in which OE is carried out to check oxide recess and uniformity.
(4) Oxide is the stop layer for ACL layer etching.

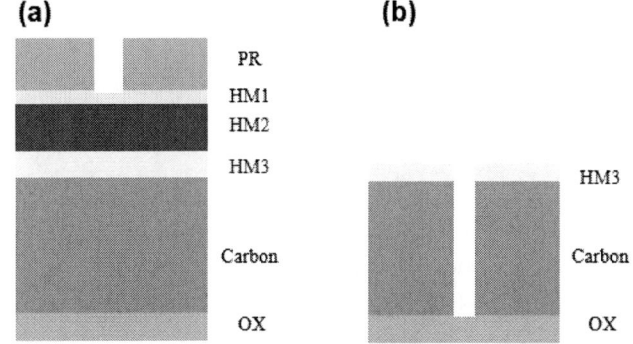

Fig.1 Film stack of deep hole hard mask etch process(a)before etch(b) after etch.

Result and discussion

As is seen in Fig.2, the profile of oxide mask is gradually collapsed and distorted during the ACL etching. The profile of the mask is still square when etch to a quarter of the ACL depth; and then the top of the mask becomes rounded and expands into the hole when etched to half of the ACL depth; the mask is completed distorted after full etch of the ACL layer. The by-products are left at the bottom of the hole because the hole is partly blocked by distorted oxide mask, which can cause poor uniformity and even etching stops.

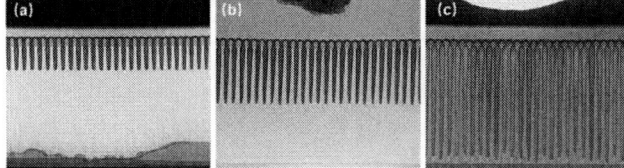

Fig.2 TEM images of mask profile for different etch depth of carbon layer:(a)1/4, (b)1/2 and (c) full etch.

TEM images and EDS mapping were carried out to test the chemical composition of the mask and substances at the bottom of the hole, which are shown in figure 3. The mask mainly contains Si and O elements while substances at the bottom of the hole mainly contains C and S elements. We proposed that the upper part of the mask was bombarded and deposited down at the mask below under high bias power and thus expands into the hole. The byproduct is hard to get out of the hole and less plasma can get in the hole, which may cause poor uniformity and even etching stops.

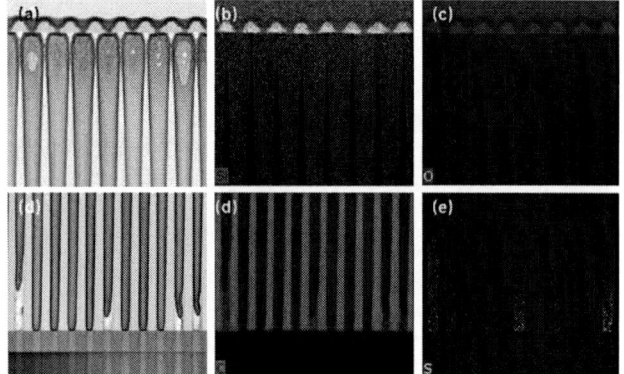

Fig. 3 (a)TEM ,(b) and (c) EDS mapping of mask ; (d)TEM ,(e) and (f) EDS mapping of substances at the bottom of hole.

In order to get the byproduct out of the hole rapidly and timely, we add BT steps between ME steps to remove the mask above the hole during the ACL etching, which uses CF4 and O2 as the main etching gas.

TEM results with different bias power of ME1, ME2, ME3 and OE steps is shown in figure4.different bias power used in result of figure4a and figure4b. The etching depth decreased when bias power increased. The space CD of mask is too small and etching stop happened when increasing the bias power. Which are consistent with the analysis that the upper part of the mask was bombarded and deposited down at the mask below under high bias power and thus expands into the hole. The byproducts accumulated at the bottom of the hole due to the mask space CD shrunk and caused etching stops.

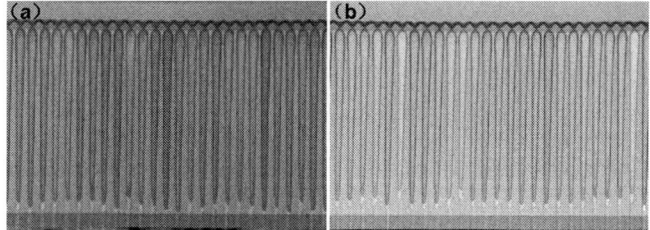

Fig.4 TEM images of (a) before and (b) after adding BT1 step between carbon etch.

We add BT1 step between ME2 and ME3 to remove the mask which extended to and blocked the hole. The main etching gas used are CF4/Ar/O2. To remove the hole blocks thoroughly, high pressure and continuous wave mode are used in BT1 step. The detailed morphology characterization of the hole was conducted by using TEM (figure 5a). The space CD increased and the byproduct at the bottom of the hole was nearly removed. But the bottom CD and uniformity of the hole need to be further improved.

Based on the above analysis, we further add three BT2 steps among ME1, ME2, ME3 and OE to remove the mask above the hole, which uses CF4/O2 as the main etching gas and decrease the bombardment effect for more oxide mask remaining. The detailed conditions for BT2 step, we decreased the pressure and use pulsing wave mode to keep enough mask remaining. As is

illustrated in figure 5c, byproducts at the bottom are almost removed and the bottom CD of the hole are enlarged due to the reopened mask. Finally, we increase the bias power and which is marked as BT3 step. The space CD increased when adding three BT3 steps between ME1, ME2, ME3 and OE. The bias power in BT3 step increased to obtain enough bombardment energy and etch the substance produced during ACL etching completely. The result is shown in figure 5d, the critical dimension of the hole is significantly enlarged and the uniformity is well improved.

Fig.5 TEM images of (a)before and(b) after adding 1 BT step and (c,d) after adding 3 BT steps between carbon etch.

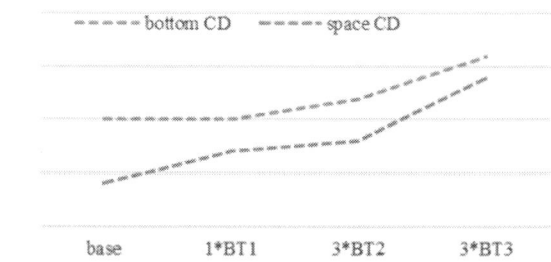

Fig.6 plots of mask space CD and hole bottom CD of different conditions.

The trend plots of mask space CD and hole bottom CD of different conditions are illustrated in figure6. Adding BT steps between ME and OE steps is efficient for mask hole bottom CD control and uniformity improve. As is shown in figure6, by adding three BT3 steps among ME1, ME2, ME3 and OE steps, the bottom CD of the hole could increase a lot, which is especially significant and helpful for the mask hole etching process.

Conclusion

In this paper, adding BT steps between ME and OE steps are proposed to improve the profile and uniformity of channel hole mask etching process for NAND flash memory. The mask distorted and blocked the hole, which could slow down or stop the byproducts getting out the hole. The bottom CD of the hole could increase by adding BT step to reopen the mask during the ACL etching.

References

[1] Christian Monzio Compagnoni, Akir a Goda, Alessandro S.

Spinelli, Peter Feeley, Andrea L. Lacaita, and Angelo Visconti, Reviewing the Evolution of the nand Flash Technology, Vol.105,1609-1633,2017.

[2] Geun Ho Lee , Sungmin Hwang , Junsu Yu and Hyungjin Kim , "Architecture and Process Integration Overview of 3D NAND Flash Technologies" Appl. Sci., Vol.11, 6703, 2021.

[3] Min Hwan Jeon, JinWoo Park, Deok Hyun Yun, Kyong Nam Kim, and Geun Young Yeom, "Etch Properties of Amorphous Carbon Material Using RF Pulsing in the O2/N2/CHF3 Plasma" Journal of Nanoscience and Nanotechnology, Vol. 15, 8577–8583, 2015.

STUDY OF PHOTO-RESIST (PR) STRIP RATE WITH HIGH TEMPERATURE PEDESTAL FOR AL PATTERNING PROCESS

Cheng Tian, Li-Tian Xu*, Li-Song Hu, Xue-Hua Wang

Beijing NAURA Microelectronics Equipment Co. Ltd

*Corresponding Author's Email: xulitian@naura.com

ABSTRACT

In this paper, we studied the influence of pedestal temperature and pressure on Photo-resist(PR) shrink firstly. The results show that the higher pressure, the faster PR shrink when pedestal temperature was >280℃, but the maximum shrink is >10,000Å. When the pressure and heating time are fixed, the higher pedestal temperature is, the greater PR shrink will be. In addition, we designed the corresponding orthogonal matrix tests, and obtained the influence trend of power, pressure and ashing gas on the etching rate and uniformity of PR. The results show that O_2 is the main ashing gas, and the smaller the proportion of O_2, the lower the etching rate. Power is positively correlated with etching rate, while pressure is negatively correlated with etching rate.

INTRODUCTION

Photo-resist(PR) removal is one of the most important processes in the semiconductor industry. plasma etching process can remove PR anisotropy, making the delineation of features possible, which is difficult to replicate in wet etching process [1,2]. Dry etching has been used more and more in PR removal process in recent years. In this paper, we discuss the effect of wafer pedestal temperature on the shrinkage of the PR and study the effect of pressure, power and ashing gas on the etch rate of the PR aim to optimize the metal line pattern in-situ PR strip process.

EXPERIMENT AND RESULTS

The general flow of Aluminum(Al) etching process is shown in Fig.1, We mainly studied the factors that affect PR strip process after metal etching in this flow.

The initial thickness of PR used in this experiment was about 30000Å. We use Nano film thickness detector to measure the thickness of evenly dispersed 49 points on PR wafers before and after the PR strip process, and the average value was taken as the shrink or etching amount of PR. In order to calculate the etch rate, we studied the change of PR etch amount with time under the conditions of Power on or Power off with 1:10 N_2/O_2 gas ratio and >280℃ wafer pedestal temperature. Before etching, PR was heated by a >280℃ pedestal for 15sec under a high pressure firstly, and the results were shown in Fig.1.

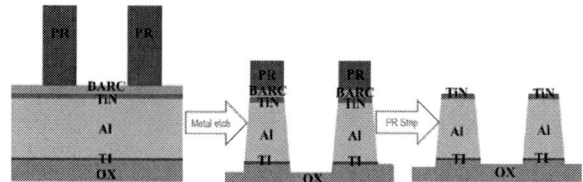

Fig.1.The general flow of Al etching process.

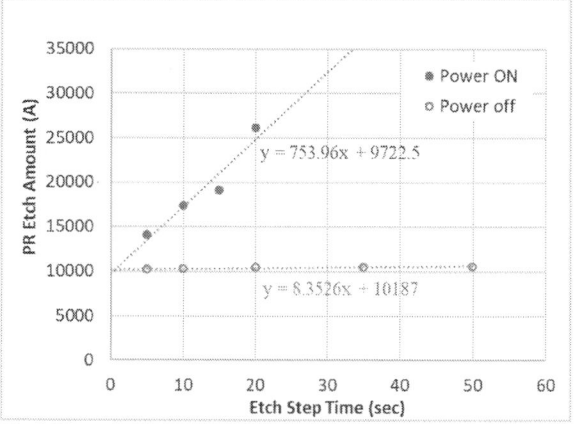

Fig.2.PR etch amount as a function of time.

From Fig.2, it is can be see that ER is 754.0Å/Sec when the power is turned on, and when the power is off, ER is 8.4 Å/Sec, which is hardly etched. In addition, we can know that heating will lead to PR shrink.

In order to further analyze the shrink properties of PR, we used >280℃ pedestal to heat PR Wafers for different time under the Low and High pressure respectively, and obtained the following data.

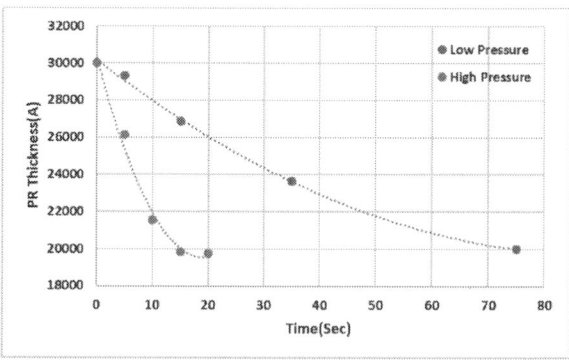

Fig.3.PR thickness as a function of heating time

979-8-3503-1101-3/23 $31.00 © 2023 IEEE

As shown in Fig.3, PR will gradually shrink with the increase of time by heating, and the shrink was faster on the higher pressure, and its maximum shrink is near 10,000Å.

In addition, we also studied the influence of pedestal temperature on PR shrink. As shown in Fig.4, when the pressure and heating time were fixed, the higher pedestal temperature was, the greater PR shrink would be, and the PR shrink would increase about 49Å for every 1℃ increase in temperature.

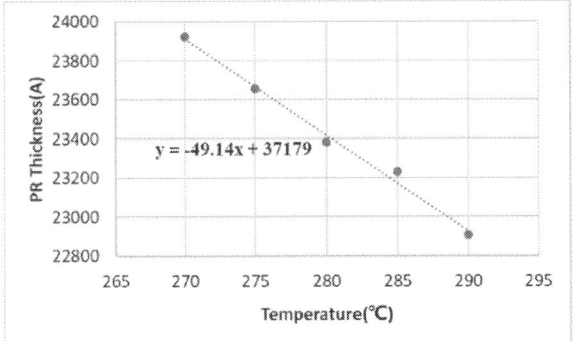

Fig.4.PR thickness as a function of pedestal temperature

In order to obtain the influence of pressure, power and flow of ashing gas on the etching rate and uniformity, The orthogonal matrix test with four factors and three levels was designed, as shown in Table 1. The four factors were pressure, power, flow of H_2O, and flow ratio of N_2 to O_2. The orthogonal table of $L9(3^4)$ was selected for a total of nine experiments. The pressure is Low, Middle, High; The power is Low, Middle, High; The H_2O flow were 0 sccm, low and high flow rate, respectively. The gas flow ratios of N_2 to O_2 are 1:10, 1:3 and 1:1.5, respectively. The JMP software was used to analyze the experimental results, and the trend diagram of the influence of pressure, power and gas flow on the etch rate and uniformity of PR was obtained.

Table.1. Orthogonal matrix test

Test number	Pressure (mT)	Power (W)	H₂O flow (sccm)	N₂/O₂ raio (sccm/sccm)	Etch rate (Å/Sec)	Uniformity (%)
1	Low	Low	0	1 : 10	589.7	19
2	Low	Middle	Low	1 : 3	549.7	24
3	Low	High	High	1 : 1.5	429.8	24
4	Middle	Low	Low	1 : 1.5	263.9	21
5	Middle	Middle	High	1 :10	291.0	25
6	Middle	High	0	1 : 3	565.6	25
7	High	Low	High	1 : 3	90.0	97
8	High	Middle	0	1 : 1.5	377.7	46
9	High	High	Low	1 :10	378.5	45

As can be seen from Fig.5, the higher pressure is, the lower etch rate is, and the uniformity decreases slightly at first. When the pressure exceeds Middle, the uniformity decreases significantly, so lower pressure is the most ideal pressure condition. The higher power, the greater etching rate and the better uniformity, so higher power is the best etching power; With the increase of H_2O flow, the etch rate decrease and the uniformity become worse. The larger N_2/O_2 ratio is, the lower the etch rate is, and the uniformity become worse first and then better. Therefore, the best gas ratio is N_2/O_2 at 1:10, However, the best ashing parameters will depend on the different process requirement, such as etch rate, uniformity, selectivity, by-products and corrosion.

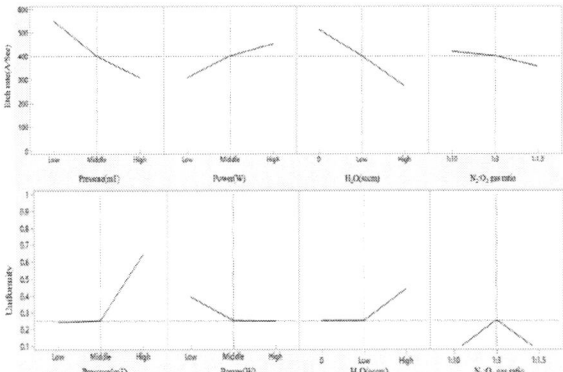

Fig.5.Effects of pressure, power, gas flow on etch rate and uniformity

DISCUSSION

With the increase of pressure, The decrease of ashing rate was due to the degree of dissociation decrease [1]. The higher power, can produce more oxygen active radical, accelerate the rate of chemical etching, so the greater power, the greater etch rate ; With the increase of the flow of H_2O and N_2, the proportion of O_2 in the gas mixture decreases, which leads to the decrease of the concentration of oxygen active radical in the plasma and the weakening of the chemical etching effect, so the etching rate decreases [3].

CONCLUSION

PR will shrink by heating, and its shrink rate is positively correlated with the pressure. pressure, power and gas flow have influence on PR etching rate and uniformity. With the increase of pressure, the etching rate decreases and the uniformity decreases. With the increase of power, the etching rate increases and the uniformity becomes better. In the ashing process of PR, O2 is the main etching gas, as the proportion of O2 decreases, the etch rate decreases.

The orthogonal matrix experiment can help to find the optimal process parameters systematically and efficiently.

REFERENCES

[1] Takagi K I, Ikeda A, Fujimura T, et al. Inductively coupled plasma application to the resist ashing[J]. Thin Solid Films, 2001, 386(2):160-164.

[2] Graves D B. Plasma processing[J]. IEEE Transactions on Plasma Science, 1994, 22(1):31-42.

[3] Han K H, Kang J G, Han S U, et al. Photo-resist ashing by atmospheric pressure glow discharge[J]. Current Applied Physics, 2007, 7(2):211-214.

SELF-LIMITED ETCHING OF SILICON NITRIDE USING CYCLIC PROCESS WITH CH2F2 CHEMISTRY

Xue-hua Wang, Li-Tian Xu*, Tian Cheng

Beijing NAURA Microelectronics Equipment Co. Ltd

*Corresponding Author's Email: xulitian@naura.com

ABSTRACT

In this paper, we used the alternating cycle of adsorption and desorption processes to carry out self-limited etching of silicon nitride, and explored the use of different hydrofluorocarbon (HFC) gases in the adsorption process in order to present self-limiting characteristics. Two gases as CHF_3 and CH_2F_2 are selected to conduct two sets of experiments respectively. The results show that CHF_3 produces etching effect on SiN, without deposition effect, and no self-limiting characteristics are presented. In contrast, the etching rate of SiN firstly increases and then decreases with the increase of CH_2F_2 step time. It is suitable for precise etching thickness control by the cycle number.

INTRODUCTION

Atomic Layer etching technology, which involves the removal of one or several monolayers of material at a time in multiple repetitive cycles, has the ability to achieve controlled, self-limiting etching of the material and is very popular in industry, especially in very thin film etching, compared to traditional plasma etching. Typical Atomic Layer etching processes involve surface modification and etched surface modification layers. Surface modification can be achieved by adsorption of gas[1-3], deposition of fluorocarbon polymer layers[4,5] or ion bombardment,6 and then etching away the surface modification layer by high-energy ion bombardment or spontaneous reaction with suitable gas/free radical. Ishii et al.[7] reported that hydrogen fluorocarbon (HFC) plasma was used to etch the atomic layer of silicon nitride. At first, a layer of fluorocarbon polymer was deposited on the surface for surface modification, and then an ion bombardment was carried out to selectively remove the modified layer.

Posseme et al. proposed an alternative method to achieve self-limiting etching of SiN by modifying the surface of SiN with hydrogen ions and then removing the modified layer with diluted hydrofluoric acid. Sherpa et al.8used this method to remove the modified layer using fluorine free radicals generated in the plasma, and the thickness of the removal material depends on the degree of surface modification (i.e. the ionic energy of hydrogen ions). As a result, the thickness of the etched material (a few nanometers per cycle) is significantly higher than that of the typical process. Recently, it has been demonstrated that self-limited etching of SiN can be adsorbed by CH3F gas, and Kim et al. attributed the surface modification of SiF to chemisorption of CH_3F on the surface of SiF. Analogs to other HFC gases, such as CH_2F_2, CHF_3, and CF_4, differences in surface interactions between different gases (e.g., H: F or C: F ratios in the gas, how they adsorb, and the chemistry of the modified surface layer) have some effect on self-limited etching of SiN.

In this paper, we used the alternating cycle of adsorption and desorption processes to carry out self-limited etching of silicon nitride, and explored the use of different hydrofluorocarbon (HFC) gases in the adsorption process in order to present self-limiting characteristics.

EXPERIMENT AND RESULTS

In the adsorption step, CHF_3 and CH_2F_2 were studied to explore which chemistry would present self-limiting characteristic for SiN etching. In the desorption step, Ar/O_2 plasma with 100Wb RF bias power and a constant step time is applied to the wafers. In the experiment, SiN film with a thickness of 1000A was used as the ER monitor wafers. The experimental conditions and SiN etch rate (ER) data are shown in Table 1.

Table.1. List of the self-limited etching experimental results

Test number	Pressure (mT)	SRF (W)	BRF (W)	Gas flow (sccm)	Time (s)	SiN ER (Å/Cycle)
1	Low	Middle	0	CHF_3	6sec	33
2	Low	Middle	0	CHF_3	12sec	56
3	Low	Middle	0	CHF_3	18sec	71
4	Low	Middle	0	CHF_3	24sec	85
5	Low	Middle	30	CH_2F_2	6sec	15
6	Low	Middle	30	CH_2F_2	12sec	32
7	Low	Middle	30	CH_2F_2	18sec	29
8	Low	Middle	30	CH_2F_2	24sec	23

Fig. 1 shows the relationship between the deposition time and the cyclic etch rate (ER). It can be seen that SiN ER increases with the increase of CHF_3 deposition time. There is no self-limiting characteristics because CHF_3 plasma produces etching effect on SiN.

In contract, SiN etching rate increases first and then decreases as CH_2F_2 deposition time increases. It appears the self-limiting characteristics on the etching of SiN. CH_2F_2 plasma can form a modification layer on SiN surface, but does not etch the SiN film. The ER trended down because of insufficient desorption time in Ar/O_2 bombardment step.

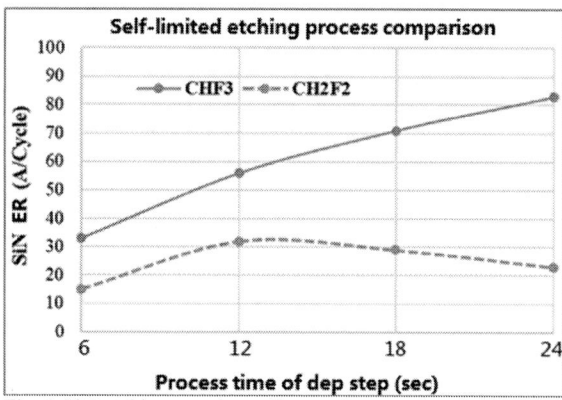

Fig. 1 SiN etching rate depending on dep step time in the cyclic process.

The etching of SiN with CH_2F_2 can show obvious self-limiting characteristics. The etching thickness of each cycle remains unchanged. Therefore, the etching depth monotonically increases with the increase of the cycle number as Fig.2

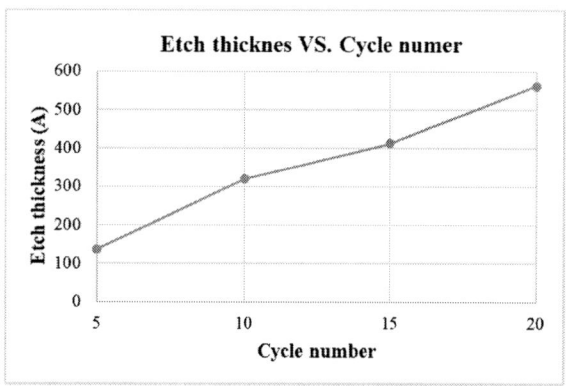

Fig. 2 The etching thickness of SiN varies with the cycle number

Eventually, we selected appropriate experimental conditions and applied them to the actual demo for SiN spacer and Si recess etch, where the SiN thickness is 29A. The result showed that self-limiting process can etch the SiN film precisely by the cycle number. SiN spacer is free of plasma damage on sidewall area.

CONCLUSION

Through experiments, it is found that SiN is etched by alternating absorption and desorption processes, different hydro-fluoride gases will produce different effects on SiN etching. When CH_2F_2 is used as deposition, the etching rate of SiN increases first and then decreases with the increase of CH_2F_2 deposition time, showing significant self-limited etching characteristics.

REFERENCES

[1] S. D. Athavale and D. J. Economou, J. Vac. Sci. Technol. B 14, 3702 (1996).

[2] T. Matsuura, J. Murota, Y. Sawada, and T. Ohmi, Appl. Phys. Lett. 63, 2803 (1993).

[3] W.-H. Kim, D. Sung, S. Oh, J. Woo, S. Lim, H. Lee, and S. F. Bent, J. Vac. Sci. Technol. A 36, 01B104 (2018).

[4] D. Metzler, C. Li, S. Engelmann, R. L. Bruce, E. A. Joseph, and G. S. Oehrlein, J. Vac. Sci. Technol. A 34, 01B101 (2016).

[5] C. Li, D. Metzler, C. S. Lai, E. A. Hudson, and G. S. Oehrlein, J. Vac. Sci.Technol. A 34, 041307 (2016).

[6] N. Posseme, O. Pollet, and S. Barnola, Appl. Phys. Lett. 105, 051605 (2014).

[7] Y. Ishii, K. Okuma, T. Saldana, K. Maeda, N. Negishi, and J. Manos, Jpn. J. Appl. Phys. 56, 06HB07 (2017).

[8] S. D. Sherpa and A. Ranjan, J. Vac. Sci. Technol. A 35, 01A102 (2017).

Strategy for line width roughness (LWR) reduction in carbon mandrel patterning

Yichang Liu, Li Qi, Litian Xu, Lianfu Zhao, Xingjun Yao, Zihan Zhang

[1] Beijing NAURA Microelectronics Equipment Co. Ltd
No. 8 Wenchang Avenue, Beijing Economic-Technological Development Area. Beijing City, China

*Corresponding Author's Email: liuyichang@naura.com

Abstract

Multi-patterning has been widely used in the semiconductor manufacturing industry to get narrow pitch at beyond 40nm technology node. Self-aligned double patterning (SADP) with a carbon pattern as the first mandrel shows significant process to meet the patterning pitch line/space features. The requirements for the Line Width Roughness (LWR) of mandrel reformation etch is extremely strict. This paper focuses on LWR reduction in carbon mandrel patterning etch. The effects of HBr curing organic mask and $C_XH_YF_Z$ deposition are investigated. LWR is optimized with HBR curing and $C_XH_YF_Z$ deposition.

Keywords: mandrel; etch; Line Width Roughness (LWR); Self-aligned double patterning (SADP)

Introduction

In order to pursue higher graphic density and smaller process nodes, multiple exposure techniques such as LELE (litho-etch-litho-etch) and SADP (self-aligned double patterning) have been developed on the basis of the common glue-expose - develop - etching process [1]. By etching the hard mask(HM), the first layer of pattern is transferred to the HM below it, and finally a pattern with twice the pattern density is obtained on the substrate. Fig.1 illustrates the schematic of SADP manufacturing process in NAURA NMC612E inductive coupled plasma (ICP) etcher. The incoming pattern as Fig.1(a) which is after lithography [2]. Using PR mask, the mandrel is etched via fluorocarbon and oxygen gas based plasma (Fig.1(b)). Next, the oxide spacer layer will deposit as next HM(Fig.1(c)). Then, the mandrel will be removed after spacer etch (Fig.1(d)). Finally, we etch the target film stack(Fig.1(e)). In this process, the requirements for the Line Width Roughness (LWR) of mandrel is extremely strict, which has a critical effect on the roughness of the target film stack [3]. This paper focus the first step as the mandrel etch. We aim to LWR reduction by HBr curing organic mask and $C_XH_YF_Z$ gas deposition effect in the mandrel, which is make important sense in IC fabrication.

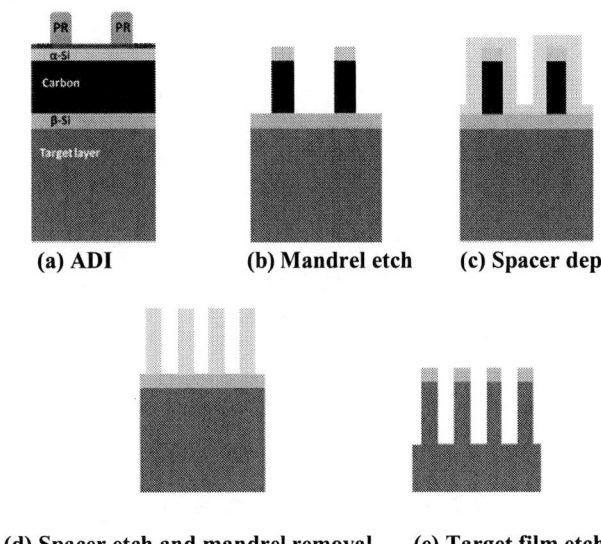

(a) ADI   (b) Mandrel etch   (c) Spacer dep

(d) Spacer etch and mandrel removal   (e) Target film etch

Fig.1 SADP process flow

Experiment

The equipment NAURA NMC612E inductive coupled plasma (ICP) etcher was adopted in this experiment. As for LWR reduction, the effects of different curing parameters on mandrel etch were studied. After that, the effects of different carbon-fluorine gas deposition processes on LWR reduction also studied. As for LWR reduction, we use the HBr gas to curing the organic mask, then the $C_XH_YF_Z$ gas was deposited in the mandrel, finally we get the target film stack. The result and discussion as follow.

Result and discussion

A. Different HBr time split effect on LWR reduction

We use the HBr and He gas body to make the LWR reduction. Fig. 2 shows the picture of examines the effect of different curing times on the LWR reduction. Compare to pre-etch as Fig. 2, It can be seen that it's obviously reduction on the LWR due to the HBr curing, and the curing time at 30s around is optimal. The current mechanism to explain that HBr can reduce LWR is the effect of vacuum ultraviolet(VUV) in HBr plasma. It can release photons to cure PR, as well as change the molecular structure of the PR surface, which can reduce the glass transition temperature (Tg) of the photoresist and the PR surface reflow, which is beneficial to the smoothing of the PR surface [4].

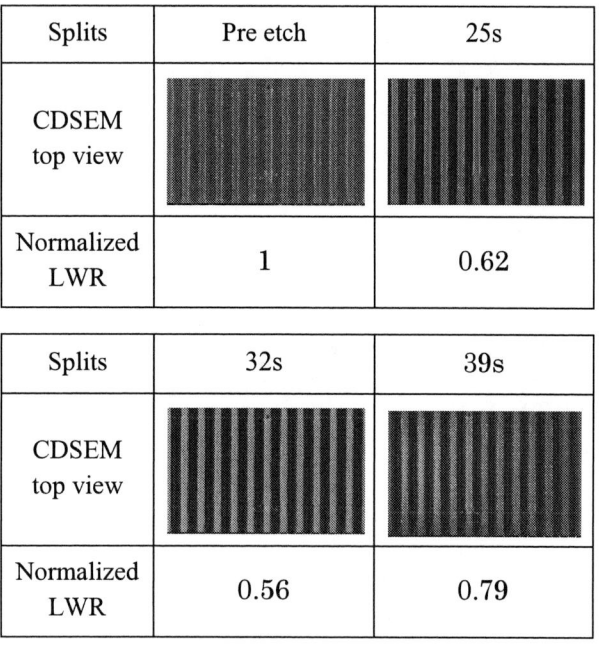

Splits	Pre etch	25s
CDSEM top view		
Normalized LWR	1	0.62

Splits	32s	39s
CDSEM top view		
Normalized LWR	0.56	0.79

Fig. 2 LWR reduction by HBr cure PR mask

B. Different $C_X H_Y F_Z$ gas deposition effect on LWR reduction

Aim to LWR reduction, we try to use the deposition gas of $C_X H_Y F_Z$ in the organic mandrel etch. As shown in Fig.3, the LWR of mandrel etch with CHF_3 gas is improved by one tenth of magnitude as compared to none dep. This is due to fluorocarbon gas can form polymers when it

Splits	Non Dep	CHF_3 Dep
CDSEM top view		
Normalized LWR	1	0.91

Splits	CH_2F_2 Dep	CH_3F Dep
CDSEM top view		
Normalized LWR	1.04	1.08

Fig. 3 LWR reduction by $C_X H_Y F_Z$ gas deposition

decomposes into plasma which can protect the side wall to a certain extent and optimize the LWR of mandrel. [5] However, when the gas deposition capacity used is too strong, it will lead to a slight increase in LWR. As well as deposited gas, such as CH_2F_2 and CH_3F, formed polymers on the surface of mandrel will make the critical dimension of mandrel to thicken, which has a negative impact on advanced manufacturing in semiconductor process.

Conclusion

In this study, we demonstrated HBr curing and fluorocarbon gas deposition to improve LWR in organic mandrel patterning etch. CHF_3 gas deposition is obviously in LWR reduction. In addition, the LWR can be significantly reduction by curing PR with HBr in a proper time, which have reference value for the development and optimization of other mandrel etching processes.

References

[1] Nihar Mohanty, et al., "Challenges and mitigation strategies for resist trim etch in resist mandrel based SAQP integration scheme", Proc. of SPIE Vol. 9428 94280G-6.

[2] Ling Li, et al., Investigation of surface modifications of 193 and 248 nm photoresist materials during low-pressure plasma etching. Oehrlein. J. Vac. Sci. Technol. B. vol. 22, 2004, pp. 2594-2603I

[3] Mimi Dai, et al., Efectively improving local critical dimension uniformity of small hole arrays by photo resist treatment. CSTIC 2022-3-56.

[4] Li Quanbo, et al., "Optimization and Development of 40 nm Gate Etching", China Integrated Circuit. 2015-4-191.

[5] Erhu Zheng, et al., "The Line Roughness Improvement in Self-aligned Multiple Patterning", ECS Transactions, 92 (1) 65-70 (2019).

Redistribution Layer aluminum advanced etching process development

Xing-Jun Yao[1]*, Li-Tian Xu[1], Li Qi[1], Qiang-qiang Sang [1], Li-song Hu[2], Chen-Chen[2], Yi-chang Liu[2]

[1] Beijing NAURA Microelectronics Equipment Co. Ltd

No. 8 Wenchang Avenue, Beijing Economic-Technological Development Area. Beijing City, China
No. 388, Xingye Avenue, Economic and Technological Development Area, Hefei City, China

*Corresponding Author's Email: yaoxingjun@naura.com

Abstract

Advanced Package continues to play a major role in the semiconductor industry. Redistribution Layer (RDL) is an important part of today's advanced packaging, it is the electrical extension and interconnection of X-Y plane. It changes its contact position through wafer level metal wiring process and bump process, and arranges pins to new, more spacious areas, so that IC can be applied to different packaging forms. In the process of RDL plasma etching process tuning, we overcome the bottom CD and OX recess uniformity Worse, double slope and other issues to successfully realize the mass production of RDL on etch machine ;

Keywords —Advanced Package, RDL, plasma etch, dry etch, OX recess, double slope.

Introduction

Recently, Fan-out Wafer Level Packaging (FOWLP) has been emerged as a promising technology to meet the ever increasing demands of the consumer electronic products[1]. The RDL interposer has generic structural advantages in interconnection integrity and bump joint reliability, which allows further scaling up of The package size for more complicated functional integration[2].

A new concept of Si-less redistribution layer (RDL) platform was proposed for server/HPC application with advantages of packaging cost reduction, warpage control, and enhanced reliability[3]. In the advanced packaged FIWLP and FOWLP, RDL is the most critical technology. In 2.5D IC integration, besides the TSV on the silicon substrate, RDL is also indispensable. The network is interconnected through RDL and distributed to different positions, so as to connect the Bump of the chip above the silicon substrate and the Bump below the substrate. In 3D IC integration, for up and down stacking generally, TSV can directly complete the electrical interconnection function for the same chip. If different types of chips are stacked up and down, they need to be re stacked through RDL. The new wiring layer aligns the upper and lower layers of chips to complete the electrical interconnection.

Although the film structure of the RDL process is not complex, many problems will be encountered in the etching process. At present, we have realized the mass production of RDL.

Experiment

Fig.1 illustrates the schematic film structure in the RDL etch process. There are five kinds of film layers as below list：
1) Soft Mask,
2) Mask1
3) Al
4) Mask2
5) OX stop layer
We use NMC612G ICP chamber for the etching.

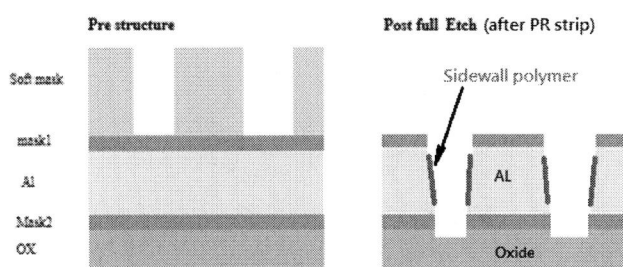

Fig.1 The film structure of RDL etch process.

In the process of RDL etching, many problems will be encountered, such as heavy polymers, the bottom CD and OX recess uniformity worse, and the double slope issue.

For the problem of heavy polymer, we further analyze the role of BCl3. The BCl3 decomposes into BClx atomic clusters and BCl3+ positive ions in plasma, which have three functions: ① BCl3+ positive ions have a large molecular weight and are an important source of physical bombardment to enhance the physical bombardment effect; ② BClx clusters can recombine with Cl atoms (BClx+1), consuming Cl atoms on the side wall surface, thus reducing side wall etching and improving the anisotropy of etching; ③ BCl3 is very easy to react with O and H ions. After preferential reaction, the O and H ions generated in the cavity and in the reaction process are taken away to reduce the possibility of Al etch termination and corrosion. And as to the function of Cl2, Cl2 etch Al is pure chemical etching, the product is AlCl3, the boiling point and melting point are very low, easy to sublimate. Therefore, when using Cl2 etch Al, the side wall must be protected; We overcame this problem by designing DOE experiments and increasing power. Figure 2 is a picture of RDL sidewall polymer after the soft mask removed. As shown in figure 3, the RDL sidewall polymer issue can be thoroughly solved by using low pressure and high bias power approaches in Al metal etch step.

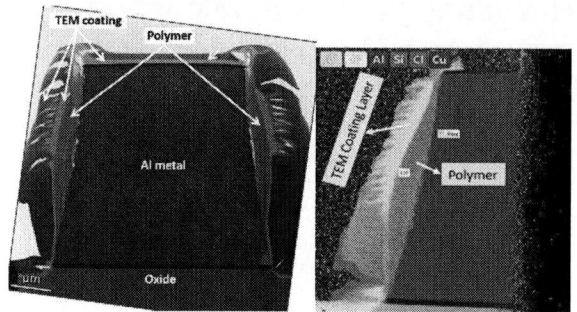

Fig.2　X-TEM and EDX analysis for RDL sidewall polymer after PR strip.

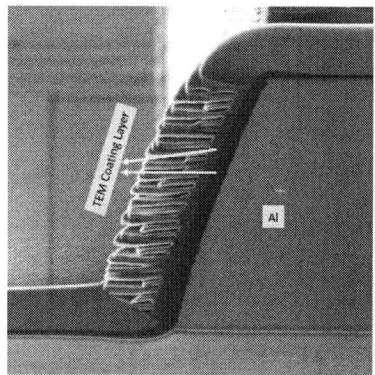

Fig.3　RDL sidewall polymer solved by the optimal Al metal etch process.

For the issue of RDL sidewall double slope, we found that power plays a key role in solving the issue of double slope through a large number of experiments and analysis of the experimental data collected in fab. By adjusting the air intake of BCl3 and controlling the amount of Cl2, the junction increases the BRF power to clear the polymer of the Al sidewall, we have solved the problem of double slope, so that the RDL sidewall will not have the problem of double slope again. The reasonable solution of this problem plays an important role in promoting mass production. Figure 4 is a picture of RDL sidewall double slope issue.

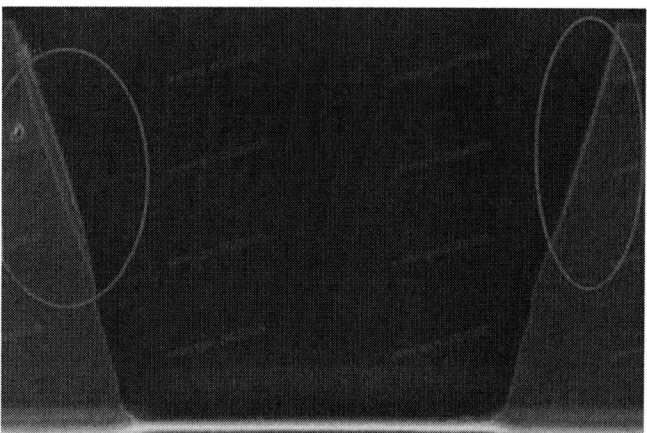

Fig.4　RDL sidewall double slope.

For bottom CD U% worse, edge Space CD larger than center and Ox recess U% worse , Edge Ox recess larger than center issue, after analyzing the data collected in the laboratory before, we found that we can improve the uniformity by regulating the intake mode and controlling the current. The temperature also plays a great role in improving the uniformity.

Result and discussion

In addition to solving the above representative problems, we also carried out some other thoughts and experiments, such as why not use the fluorine based gas etch: because the fluorine based gas etch Al will generate AlF3, which has a high melting point and is extremely difficult to volatilize, and will form a polymer that is difficult to remove, so the chlorine based gas etch Al is generally used instead of the fluorine based gas; Sources of polymer in Al etching process: ① plasma physically bombards PR and captures C to obtain polymer; ② BClx cluster side wall protection; What kinds of gas can quickly generate polymer to protect the side wall of Al: N2/CHF3/CH4. N2 will produce too many polymers in the etching process, which is easy to form trapezoidal sidewalls; CHF3 is not perfect for side wall protection; CH4 can achieve uniform sidewall protection and vertical profile.

Conclusion

The solution of the above problems has played a great role in promoting the mass production of RDL. At present, RDL is in stable mass production on the etch machine. At the same time, we are constantly exploring more advanced process methods and constantly optimizing our RDL process.

References

[1] Vempati Srinivasa Rao，"Development of High Density Fan Out Wafer Level Package (HD FOWLP) with Multi-layer Fine Pitch RDL for Mobile Applications"，2015

[2] Yi-Hang Lin "Multilayer RDL Interposer for Heterogeneous Device and Module Integration"，2008

[3] Kyoung-Lim Suk "Low Cost Si-Less RDL Interposer Package for High Performance Computing Applications"，2005

Deep trench (DT) etching process for power MOS device

Chen Chen[1]*, Li-Tian Xu[1], Xing-Jun Yao[1]

[1] Beijing NAURA Microelectronics Equipment Co. Ltd
No. 8 Wenchang Avenue, Beijing Economic-Technological Development Area. Beijing City, China

*Corresponding Author's Email: chenchen03@naura.com

Abstract

In this article, the effects of SF6/O2 containing plasmas were investigated to achieve bottom rounding of deep trench etching. External controllable parameters such as the ration of SF6/O2 gas flow were used to study the changes of Si etch rate, Si profile, and oxide selectivity. And this paper explained how to control the profile and the trench depth and the effect of temperature and pressure on Si top undercut.

Keywords — silicon, SF6/O2, bottom rounding, undercut, inductive coupled plasma, and NMC612D.

Introduction

The deep trench etching in single crystal silicon is useful for a wide range of applications, which include: metal-oxide semiconductor field effect transistor(MOSFET) circuits, diodes, and storage capacitors for memory cells [1]. Etching of deep trench is implemented not only to define the critical space dimension(CD), recess, but also to play bottom rounding [2]. Dry etching is the best candidate to etch 1000nm deep trench in silicon with vertical sidewalls. SF6/O2 gas mixtures in reactive ion etching plasma have been shown to yield high etch rates as well as bottom rounding. The process was optimized for etch rate ratio, etch rate, and trench sidewall angle by independently varying the etch parameters of temperature, pressure, SF6/O2 ratio [3,4].

Experiment

Fig.1 illustrates the schematic deep trench manufacturing process in NAURA NMC612D high-density inductive coupled plasma etcher. The film scheme started with lithography defined pattern Fig. 1 (a). A Si-O mask is formed in C4F6/O2 plasma as Fig.1 (b). PR striping process is executed as Fig.1 (c). Deep trench etching process is carried out in order to produce vertical Si profile and bottom rounding as shown in fig.1 (d).

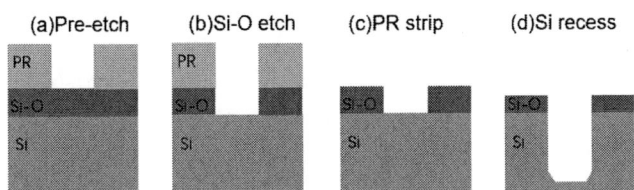

Fig.1 Deep trench etching manufacturing process flow: (a) pre-etch with incoming pattern, (b) Si-O etching, (c) PR striping, and (d) Si etching for vertical profile and bottom rounding.

Result and discussion

A. HBR-based plasma process

Traditional deep trench etching process with HBR-based step was used to check the bottom rounding. Compared to pre-rounding etching profile as Fig.2(a), the HBR-based process shows no effect for the Si bottom rounding as Fig.2(b). Decreasing the process pressure by using low pressure instead of high pressure also no obviously improved as Fig.2(c).

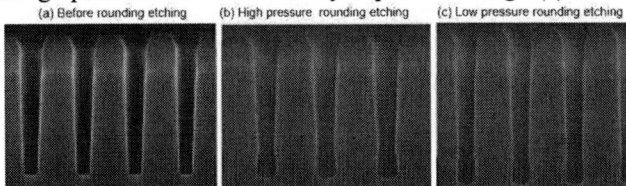

Fig.2 Traditional HBR-based process approach: (a) before rounding step, (b) post rounding step with high pressure, (c) post rounding step with low pressure.

B. One-step etching for Si and bottom rounding

Fig.3 shows the TEM of deep trench based on SF6/O2 plasma. Fluorine neutrals that react with silicon atoms are produced by SF6 dissociation. For this model, we have taken fluorine and oxygen atoms into account. It is well known that the addition of atomic oxygen in the total neutral flux oxidizes the SF6 radicals, increasing the fluorine concentration. The reactivity of atomic oxygen leads the formation of SiOxFy sites which form a passivation layer. This protective film is not removed from the sidewall, which is a necessary condition to obtain a bottom rounding trench profile. The concentration of oxygen and fluorine radicals in the plasma is critical for the bottom rounding profile.

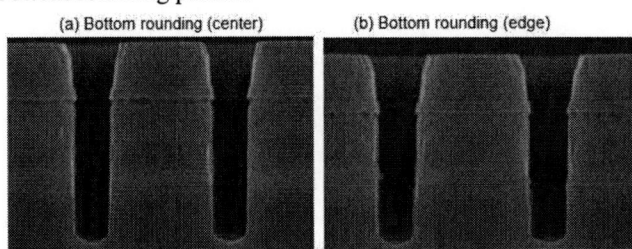

Fig.3 SF6/O2 based plasma for deep trench process: (a) Center area profile, (b) Edge area profile.

C. *Undercut improving with SiOx deposition*

The Si top undercut described in Fig.4(a). Undercut is caused by the diffraction of ions while entering the trench. Moreover, the undercut is difficult to completely remove. The effect of SiOx deposition before Si etching was observed for the optimized etch process mentioned. The results shown in Fig.4(b) shows the Quasi-atomic layer deposition (ALD) approach which formed the residue on the bottom of deep trench. Besides, Fig.5(c) shows the same phenomenon by using SiCl4/O2 plasma deposition process. As a result, deposition methods trend to grow the SiOx film on the surface of Silicon rather than the sidewall of Si-O mask. It does not provide better profile than the former.

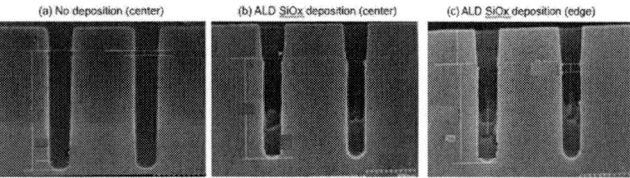

Fig.4 Cross-section images of SiOx deposition before Silicon etching: (a) No deposition, (b) ALD SiOx deposition (center), (c) ALD SiOx deposition (edge).

D. *Temperature optimization for undercut improving*

Fig.5 demonstrates deep trench etching profile through temperature optimization with Si etching step. The trench undercut disappearance with decrease of temperature in Si etching step. The Si etch rate is low because of the decrease of temperature. Owing to the changes of SF6/O2 ratio, temperature, pressure, high aspect ratio, the deep trench profile could be fabricated successfully.

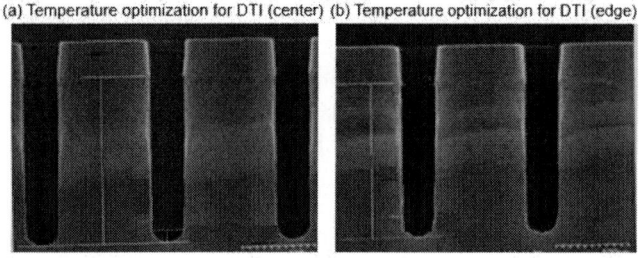

Fig.5(a) temperature optimization for DTI profile (center), and (b) temperature optimization for DTI profile (edge).

Conclusion

We have developed a new etch model under an SF6/O2 plasma mixture. This model takes the effects of plasma etching, etched species deposition and the passivation mechanisms into account. We have verified that low temperature is the principal mechanism for the undercut formation.

References

[1] R. J. Shul and S. J. Pearton, Handbook of Advanced Plasma Processing Techniques (Springer, Berlin, 2000).

[2] A. Grill, Cold Plasma in Materials Fabrication (IEEE, New York, 1994), p. 231.

[3] S. Tachi, K. Tsujimoto, S. Arai, and T. Kure, J. Vac. Sci. Technol. A 9, 796, (1991).

[4] S. Aachboun and P. Ranson, J. Vac. Sci. Technol. A 17, 2270, (1999).

TECHNICAL DIFFICULTIES AND OPTIMIZATION METHODS OF NMOS SHARE CT IN CONTACT HOLE ETCHING

Peng Zhu*, Renhui Xu, Wentao Fu, Lei Sun,Yu Bao

Shanghai Huali Integrated Circuit Corporation(HLIC) Shanghai 201314, China

*Corresponding Author's Email: zhupeng@hlmc.cn

ABSTRACT

As the last generation of single graphics technology representative, 22nm is favored by advanced semiconductor manufacturers because of its higher cost performance. However, its key size is reduced to the process limit, the process window is very small, which limits the mass production of many semiconductor manufacturers. The most likely and most difficult problem to solve NMOS share CT open in contact hole etching(CT-ET). In this paper, based on local CDU, the removal of by-products in etching and regulating pattern loading of large holes and small holes, continuous process improvement is carried out, and effective solutions are proposed. Finally, the CDU was reduced by 36%, the contact resistance of CT was reduced by 30%, the NMOS Share CT size was increased by 100Å while the square CT size remained the same, and the NMOS share CT open was basically eliminated, which greatly improved the product yield and meet the requirements of process integration and mass production.

Key Word: Contact etch，Pattern loading，BVC，NMOS share CT，By product，CD uniformity

INTRODUCTION

Because SRAM (Static Random Access Memeory) has high read and write speed, simple operation and does not need dynamic refresh, more and more applications in the network, server, digital products related chip[1, 2]. With the continuous advancement of the logic process, in order to meet the higher and higher process integration requirements on the 22/28nm technology, the importance of contact hole etching as a bridge connecting the front component and the rear interconnection is self-evident, and it is one of the most important key processe[3,4]. The reduction of the key size brings about a higher depth to width ratio, which brings a lot of difficult problems to the contact hole etching process, and is greatly related to the improvement of yield. There are a large number of SRAM components with dense graphics. In order to realize the local interconnection between Gate and Source and Drain, large holes, also known as Share CT, which fall on gate and S/D at the same time, are often introduced into contact hole etching (Figure 1b). In this paper, the 22nm contact hole process development, aiming at the NMOS Share CT open problem, the process has done the corresponding analysis and research, and put forward the corresponding countermeasures and solutions.

The CT-ETCH process involved in this paper is shown in Figure 1. All the experiments involved were completed on the Vigus-LK2 machine of Tokyo Electric Electronics (TEL), which is also the mainstream contact hole etching machine of 22/28 nm CT-ETCH at present.

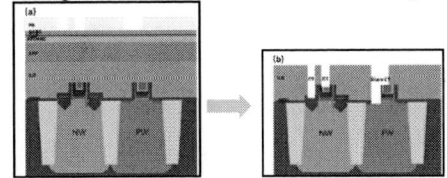

Fig 1: CT-ETCH process flow diagram:(a)Pre-CT ETCH;(b)Post CT ETCH

EXPERIMENT AND DISCUSSION

Technical Difficulties of NMOS Share CT in Contact Hole Etching

As shown in Figure 2, SRAM has two types share CT structures, one is PMOS Share CT(PSCT), the other is NMOS Share CT(NSCT), and they all fall on the metal grid. The difference is that the PSCT also falls in silicon germanide (SiGe), while the NSCT falls in the Active Area (AA). Because the SiGe is nearly 190A higher than AA, the part of NSCT falls in AA has a larger depth-to-width ratio, which greatly increases the difficulty of etching. Compared with 28nm, the key size of Gate at 22nm remains the same, but the space between gate is reduced by 7%, and the process size requirement of Contact etching at 22nm is reduced from 400 Å to 375 Å On the premise of ensuring sufficient process short window between CT and gate, It is also necessary to ensure the open process window of CT, especially the structure of NSCT with extremely high depth to width ratio. The traditional contact hole etching technology cannot meet the etching requirements.

Figure 3a shows the results of tungsten filling and grinding after contact hole etching followed by Ebeam detection. BVC results showed 21% defect die, resulting in a large yield loss, all of the defect occured in NSCT structures. BVC was even worse at the edge of the wafer than at the center. By comparing CD Map(FIG. 3b) and TEM results of fixed-point failure analysis, we found that CT open was caused by incomplete etching of NSCT in the AA region, and the main influencing factors were as follows:

1) During the etching process, NSCT will touch the metal gate, which will produce metal-containing by-products. These by-products are difficult to be removed which affects the etching rate at the bottom

of the SCT hole;

2) The CD map shows that the CD of edge of the wafer is smaller than the target size 375 Å by 20~30 Å.The CDU is worse, and the reduction of CD will increase the difficulty of removing the by-products;

3) The CD difference between two adjacent NSCT is 100 Å, which greatly reduces the CT open window. However suck a large gap in the size of locally adjacent holes is hardly caused by etching, indicating that the local CDU of Litho is worse under the current process condition;

4) Fig 3c shows that OVL is shift 55 Å to the left as a whole, which will further reduce the bottom CD of the left NSCT;

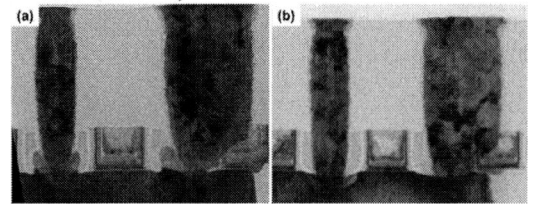

Fig 2: The difference of Share CT:(a)PMOS Share CT;(b) NMOS Share CT

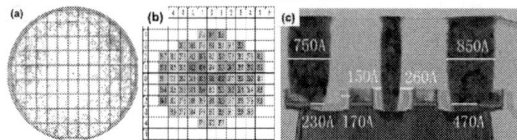

Fig 3: N-MOS Share CT open: (a) BVC; (b) CD Map;(c) TEM

Improvement of CT-ETCH CDU

Local CDU refers to the size difference between different holes in a small local area. Some small NSCT were prone to appear because of the worse local CDU. Therefore, it is urgent to improve the local CDU and increase the process window of CT open. At present, the 22nm CT film stack contains NFDARC+CAP OX, and the thickness was 200A+300A. The combined film has a reflectance of up to 3.2% when exposed to photolithography (Fig. 4a), As a result, serious bowling occurs in the optical resistance (PR) topography (Fig. 5b), which is not conducive to etching and further deteriorates the local CDU. When the thickness of NFDARC+CAP OX becomes 260A+50A, the simulation results are shown in Fig. 4d, e, f. the reflectance of the new thickness of the film can be greatly reduced from 3.2% to 0.1%. The simulation results of PR under this reflectance show that bowling can be effectively improved, with large upper part and small bottom part and smooth middle part.

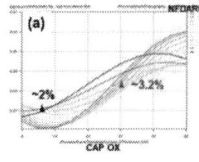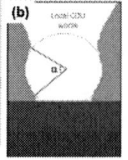

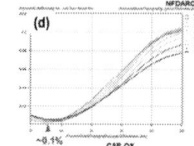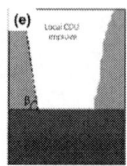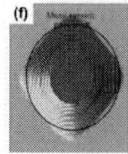

Fig 4: the PR profile of the different NFDARC+CAP thickness:(a)the reflectance of old film;(b)the PR profile simulation of the old film layer;(c)the PR plane simulation of the old film layer;(d)the reflectance of new film;(e)the PR profile simulation of the new film layer;(f)the PR plane simulation of the new film layer

In order to improve the local CDU, the thickness of NFDARC+CAP OX was changed from 200A+300A to 260A+50A, and then the CDU after CT etching was measured. As shown in Fig 5, the CD map of the old film thickness did not appear special map, the CDU was still as high as 42 Å, more than 10% of the average of CD 369 Å. This is the reason that the in-plane CDU cannot be further reduced due to worse local CDU. However, the CDU of the new film can be increased to 27 Å, which is 36% higher than that of the old film. At the same time, the Range of CD size in the plane is also greatly reduced from 88 Å to 39 Å, which can effectively avoid some abnormally small CD holes and increase the CT open process window.

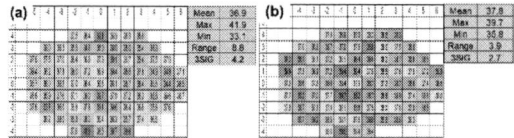

Fig 5: The CT AEI CDU of the different thickness of NFDARC+CAP:(a) old film;(b) new film

Improvement of Removal Effect of Etched by Products in CT-ETCH

At the 22/28nm technical node, the Gate material is changed to Metal Al or W, whick is called Metal Gate(MG). In this paper, MG is Al, and some heavy Al-containing by-products will be generated during the etching process. The by-products were difficult to be removed, so the by products produced during the process of NSCT etching will be attached to the side wall and bottom, greatly increasing the difficulty of etching and reducing the etching rate, resulting in a sharp decrease or even open at the bottom CD of NSCT. Therefore, in order to ensure the smooth etching of NSCT in AA area, the first task is to take measures to reduce the bottom by-product. Our current gas for etching SiN is CH3F, which is a by-product gas with a high selectivity ratio of SiN to Oxide, CO_2 is used in PET(Post Etch Treatment). Since CO_2 contains oxygen, there is a risk that metal will be oxidized, which will lead to high CT resistance. Firstly, PET is replaced with N2/H2 gas. The high reducibility of H2 ensures both removal of bottom by-products and the risk of oxidation of NiSi on AA surfaces. In order to achieve the removal of by-products in the process of SiN etching, the current SiN was divided into two steps and the PET was added in the middle, and the previous SiN+PET was changed into SiN+PET+SiN+PET. In this method, the by-products were removed when the SiN etching was half and then the SIN etching was carried out, which could effectively remove the bottom SiN. The result is shown in the Fig 6, The use of N2/H2 PET and SIN changed into two steps can effectively improve CT resistance, CT Rs on Ploy increased by about 10% from 130 Ω to 116 Ω, and the CT Rs on AA increased by about 30% from 114 Ω to 79 Ω.

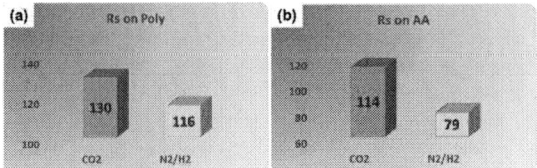

Fig 6: Influence of different gases of PET on CT resistance:(a);Rs on Poly(b)Rs on AA

The Improvement of Pattern Loading

The best solution to the problem of NSCT open is to make the CD of NSCT bigger and make sure that more of NSCT falls on AA, For the 22nm technical node, Poly space is reduced by 7% because poly CD remains unchanged. Considering the process window problem of short CT and MG, the overall CD cannot be enlarged. Therefore, a method is needed to keep the square CT CD and increase large holes CD, especially NSCT CD.

The Ploymer generated during the etching process attaches to the side wall to protect the side wall from etching, which lead to the shrink CD after Etching. That is to say, the more polymer generated during the etching process, the weaker ability of polymer clean, and the more serious the CD shrink. In this paper, the pattern loading between small holes and large holes is adjusted by using this feature, so as to reduce the shrink of NSCT CD and reduce polymer during the etching process of NSCT, we make use of the characteristics that the size of holes is larger and polymer in the holes is easier to be extracted. Choose to reduce the pressure of OX ME and SIN steps, wherein the pressure of OX ME is reduced from 100mt to 30mt, and the pressure of SIN is reduced from 20mt to 10mt. As shown in Figure 7, by adjusting the pressure of OX ME and SIN, while ensuring the small hole CD remains unchanged, The CD of NSCT can be increased by 100 Å, which greatly increases the open window of NSCT.

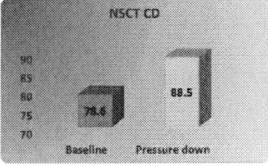

Fig 7: Influence of different pressure on NSCT

By changing the thickness of NFDARC+CAP,

splitting SIN into two steps, and changing PET into N2/H2 and using lower pressure during OX ME and SIN steps, these three main adjustment methods are combined as our CIP(Continue improve process), as shown in Figure 8. The results of BVC show that the number of NSCT BVC can be effectively reduced. DD% decreased from 21% to 0.04%, greatly increasing the yield. The BVC at the edge of the wafer is suspected to be caused by the worse OVL at the edge of wafer.

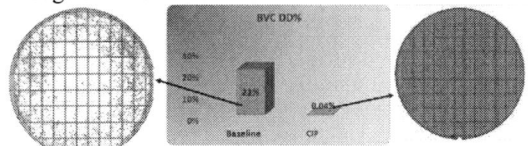

Fig 8: Influence of CIP on BVC

CONCLUSION

In this paper, the problem of NMOS Share CT open appeared in the development of 22nm CT-ETCH, and the CT process was optimized by improving local CDU, by-product and pattern loading. The problem of NSCT was greatly reduced. By changing the thickness of NFDARC+CAP OX to reduce the reflectivity, the profile of PR was optimized, the local CDU were reduced, and the situation of unusually small CD hole was avoided. by separating SIN+PET into SIN+PET+SIN+PET, and changing PET from CO2 to N2/H2, the by product of the etching process was reduced, and the contact resistance of CT was optimized. By reducing the pressure of OX ME and SIN, the pattern loading is improved, and the NSCT CD increases by nearly 100 Å while the square CD remains unchanged, NSCT BVC can achieve basic zero clearing after the combination of the three directions, meet the requirements of process integration, and avoid the yield loss caused by BVC.

ACKNOWLEDGMENT

The author sincerely thanks the Huali Advanced Module Research and Development dry etching module and process integration Department for their strong assistance, suggestions and beneficial discussions. At the same time, I would like to thank TEL for their help and cooperation.

REFERENCE

[1] X.P. Wang, et al, "Impact of Etching Chemistry and Sidewall Profile on Contact CD and Open performance in Advanced Logic Contact Etch", CSTIC, 2010.

[2] Thuy Tran-Qulun, et al, p.70, Advanced Semiconductor Manufacturing, 2004.

[3] X.P. Wang, et al, "Dry Etching Solutions to Contact Hole Profile Optimization for Advanced Logic Technologies", CSTIC, 2012.

[4] Jing-Yong Huang, "Contact Etch Schemes at Advanced Logic Technology Nodes", CSTIC, 2014.

INVESTIGATIONS OF NAND FLASH DEVICE CYCLING PERFORMANCE IMPROVEMENT VIA FG CARBON-DOPED POLYSILICON AND CHANNEL CORNER ROUNDING

Lifeng Liu[1], Jun Wang[1,2], and Xinruo Su[2]*

[1]School of Integrated Circuit, Peking University, Beijing 100871, China
[2] Semiconductor Manufacturing North China, Beijing, China
*Corresponding Author's Email: lfliu@pku.edu.cn

ABSTRACT

Program/Erase Cycling is the most important reliability index for NAND flash devices. There are many factors that can affect the cycling performance. Two major leakage models: floating gate to channel and bit line interference are investigated in this paper. Correspondingly, two process optimization schemes are proposed, including carbon-doped polysilicon floating gate and Active Area rounding corner. Cycling performance improvement has been demonstrated by ECC correction ratio reduction and break down voltage test after cycling.

Key Words: NAND, Program/Erase Cycling, Floating Gate, Carbon-doped Polysilicon, Corner Rounding.

INTRODUCTION

NAND flash memory has been widely used as a storage medium for digital system in past decades. The growth of mobile applications, such as smart phones and tablet computers make the demand of NAND flash memories increased significantly. The requirements with higher performance and storage capabilities drive NAND flash technologies towards the physical limits. The technology scaling down makes NAND flash memory Cell disturbance continued worse. In this case, with the rapid growth of storage density, bit error rate (BER) has also increased significantly, limits the development of NAND flash memory. Cell-to-cell interference and program disturbance induced degradation of program and erase (P/E) cycling endurance become a big reliability challenge of NAND flash. [1]. To evaluate the reliability performance of NAND Flash device, cycling performance test via error correction code (ECC) is proposed. The development of error correction code (ECC) in NAND Flash memory firstly concentrates on capacity improvement instead of performance. After cycling P/E, spare area usage is measured, higher spare area ratio indicates a higher fail ratio, which means worse cycling performance.

Random telegraph noise (RTN) and charge de-trapping (CD) play important roles in conventional planar NAND flash device reliability performance. They provide major contributions for the width of the threshold voltage (Vt) distribution after program and erase during the life of the memory device. [2] In this paper, floating gate (FG) consist of carbon-doped polysilicon and new scheme to achieve rounding corner on channel are adopted to reduce RTN and CD impact, and improved the cycling performance of 2X NAND Flash memory. The introduction of Carbon-doped poly, which is known as a wide bandgap semiconductor, with a larger conduction band offset (3.62eV) against SiO2 than polysilicon (3.2eV) results in an improved electron storage capability and leads to a lower hole tunneling current from the p-type substrate and improves data retention performance. [1] However, carbon-doped polysilicon FG and its' mechanism produces challenges to the etching process. In this paper, carbon-doped poly fabrication method is improved to achieve comparable floating gate profile. Cycling test is taken and the mechanism is also proposed. In the other side, corner rounding fabrication approach is also to achieve a better Active Area profile. Experiments demonstrated the improvement of break down voltage and program disturbance.

RESULTS AND DISCUSSION

Carbon-doped polysilicon Floating Gate fabrication and cycling Study

Silicon carbide (SiC-3C) has a larger conduction band offset against SiO2 than polysilicon does. Therefore, it is no surprising that the smaller valence band offset against SiO2 compared to polysilicon result in lower bias condition between p-type substrate and floating gate. [1] Carbon-doped polysilicon is considered as potential floating gate (FG) material to improve cycling performance of NAND Flash device. In this paper, a 5% carbon-doped polysilicon floating gate is fabricated by using Si2H6, SiH3 and C2H4 gas chemistry at 525℃ to form floating gate. And 950℃ N2 anneal was taken after deposition.

However, the introduction of carbon-doped polysilicon also exerts significant influence on FG vertical profile. In order to investigate the influence of FG doping condition, same FG etching recipe is performed to both without and with carbon-doped FG, which profiles are shown in Fig. 1. As illustrated in images, carbon-doped polysilicon FG shows a taper profile with double slope. Compared with FG without carbon-doped, a rough sidewall can also be observed, which indicated a more

979-8-3503-1101-3/23 $31.00 © 2023 IEEE

polymer rich process take place. Besides, a significant even-odd critical dimension bias occurs, which also indicated a different polymer behavior in different FG material etching process.

In order to achieve a vertical profile, carbon-fluorine base gas is employed in FG polysilicon etch process. During the etch process, with the consume of polysilicon and silicon oxide hard mask, gaseous SiF byproduct is formed and silicon and oxide is etched. Meanwhile, carbon atom which come from carbon fluorine base gas can react with native silicon oxide and hard mask oxide, which form CO_2/CO. However, carbon can also form a carbon base polymer deposit on the sidewall to protect silicon surface during polysilicon etch. With the introduction of carbon-doped polysilicon, the equilibrium of carbon atom is reset and carbon doped polysilicon offer more carbon atoms in process.

In polysilicon FG etch process, deposition of the carbon base polymer on the sidewall is driven by CF_2 radicals, which has a small sticking coefficient of ~0.004, [4] CF_2 base polymer tend to reach the bottom after reflecting multiple times on the sidewalls. After the introduction of carbon-doped polysilicon, deposition of polymer changes to driven by carbon. Carbon, which has a large sticking coefficient of 0.5, [4] tends to deposit on the upper part of the structure. After carbon deposit on the upper part of profile, the ion scattering rate is also increase and form a scattering ion flux which result in a relative taper profile. Moreover, carbon polymer continually deposits along the sidewall from upper part to the bottom, protect FG profile from etch. Carbon polymer in this process is no rich enough and finally formed a rough sidewall, in addition, the width of FG is also impacted by carbon based polymer, and induce even-odd phenomenon in double patterning process. [5]

![Figure 1 SEM images]

Figure 1: Post FG-ET profile of (a) polysilicon floating gate and (b) carbon-doped polysilicon in same fabrication recipe condition.

In order to compare the cycling performance of polysilicon and carbon-doped polysilicon, a comparable FG profile is in need. In this case, lower carbon-fluorine ratio chemical is applied in FG-ET and AA-ET to achieve a more vertical profile in carbon-doped polysilicon approach. ECC takes care of the failures during the life of the NAND flash device. Measurements of ECC after certain P/E cycles had been taken on carbon-doped polysilicon floating gate and normal polysilicon one, and

spare area usage was calculated to characterize the cycling performance. The cycling endurance of NAND Flash device can be evaluated by the spare area usage ratio. The usage of spare area started from ECC0, and the max spare area is ECC7 in our experiments with 8 bits' errors correction. Higher ECC usage ratio indicated higher bit error rate (BER) which result from worse cycling performance.

After adjust fabrication process, different doping conditions' floating gate NAND flash devices with stable Vt are designed to de-coupled other interference factors. We can consider the only difference between two conditions is floating gate material. The spare area usage conditions ratio after 10K P/E and 20K P/E cycles are shown in Fig. 2, respectively.

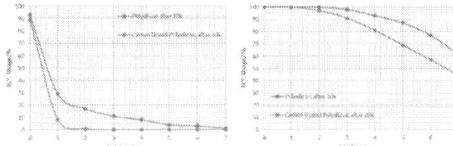

Figure 2: Each ECC spare area ratio usage ratio of carbon-doped polysilicon and normal polysilicon after 10K P/E cycles and 20K P/E cycles

As shown in Fig. 2, both carbon-doped polysilicon and normal polysilicon have higher ECC usage ratio after 20K P/E than that of 10K P/E . ECC0 to ECC3 is fully used after 20K P/E while only ECC0 is on relatively high level after 10K P/E. Moreover, the spare area usage ratio of carbon-doped polysilicon is lower than that of polysilicon in both 10K P/E cycles, which indicate a better cycling performance of carbon-doped polysilicon floating gate against normal polysilicon one. The same trend is also found even after 20K cycles. To explain the mechanism of cycling improvement result from carbon-doped polysilicon, microstructure investigation and a theatrical synthesis base on formation mechanism is proposed. Transmission electron microscopy is employed to demonstrate the difference. As shown in Fig. 3, TEM analysis shows a significant difference in grain size between carbon-doped poly and normal polysilicon. As shown in Fig. 3, carbon-doped polysilicon shows a smaller grain size (~19nm) while conventional polysilicon has much larger grain size of more than 300nm.

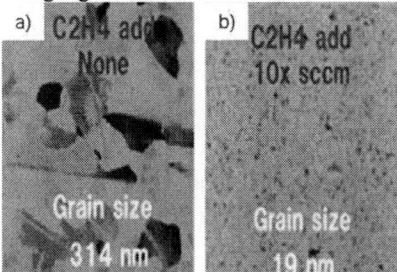

Figure 3: TEM images of a) polysilicon with a grain

size of ~300nm and b) carbon-doped polysilicon of ~20nm

TEM results imply that small grain size may result in a better cycling performance, two possible mechanisms are proposed.

The first mechanism is based on defect model. Defects in polysilicon, such as site or layer dislocation, can be easily generated during gate formation. These defects will trap electrons during program and erase operation. In fact, electrons have relatively high speed and energy, which will collide with sidewall of defects. As a result, the defects in polysilicon will become larger after collisions induced by program and erase actions, and these defects tend to attract more electrons and finally induce Vt shift and erase fail. After P/E cycles, electron is trapped in both floating gate and tunneling oxide, which will induce Vt and Verase shift and result in cycling regression. As we have controlled other parameters shift, here we only consider the influence of the floating gate.

However, carbon-doped polysilicon introduces an additional carbon doping process into the polysilicon layer. In previous studies, researchers reported that a SiC layer tend to formed at the surface with a depth of ~3nm after high dose of carbon doping into bulk silicon. The interfacial energy between Si and SiC is high enough to stop further carbon diffusion into silicon bulk. After the increasing of carbon dose, carbon atoms can transport along the polysilicon grain boundaries and suppress the polysilicon growth. Here carbon atoms play an important role in grain size and growth speed control. With slower growth speed, carbon-doped polysilicon with small grain size tends to have less defect, which results in less electron trapped possibility in program and erase operation. Smaller defect space will also limit the energy of trapped electrons, which leads to less lattice damage in same time period.

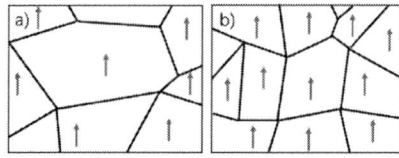

Figure 4: The FG inside electric field density Schematic a) polysilicon and b) carbon-doped polysilicon.

On the other hand, electric field control model is another side to demonstrate better cycling performance in carbon-doped polysilicon. Non-defect polysilicon grains can be considered as having same potential during program and erase. Thus, after program status, electrons are stored at grain boundaries. Considering this situation, the density of electric field that inside carbon-doped polysilicon (as shown in Fig.4b) is more than that of normal polysilicon (as shown in Fig. 4a). In both program and erase actions, varies potential is applied on floating gate. As the potential on floating gate is comparable, the electric field inside varies. Carbon-doped polysilicon, which has small grain size, have stronger electric field compared with normal polysilicon. This difference indicates a better control of electrons and enable to both program and erase more electrons during P/E cycles. As illustrated before, trapped electrons will enlarge defect area and induce more electrons trapping. Better electrons manipulation will also lead to a superior cycling performance.

AA profile tuning and rounding corner study

In 2X nm NAND flash technology generation, cell to cell coupling effects exert tremendous influence on reliability performance of NAND flash device. Three types of coupling are mainly considered, as shown in Fig. 5(a).

The first is the FG to FG coupling model. In cycling operation process, FG program status can be influenced by a nearby FG. This phenomenon can be attributed to an ununiformed EFH (Electric Field Height) and a small space between FGs. EFH, which is designed for enhanced coupling between FG and CG, will influence the electric field distribution during P/E cycling. Un-uniformed EFH leads to non-balance electric field distribution between FG, and result in unexpected Vt shift. Another physical cause to induce FG to FG coupling is considered as parasitic capacitance. The FG to FG capacitance across isolation layer becomes considerably larger as the isolation space becomes close to ~20nm.

Figure 5: a) coupling effect in NAND device structure and FG-channel corner rounding performance in b) BL condition and c) CIP Optimized condition.

The second is the CG to FG coupling model. This coupling ratio can be calculated via equivalent oxide thickness (EOT) of ONO layer between FG and CG. With the increasing of depletion layer, the EOT tends to increase and induces efficiency loss of operation Vt. FG and CG coupling is influenced by Phosphate concentration of gate deposition. In order to achieve optimized Vt distribution, an appropriate CG/FG phosphate concentration ratio has been required.

The most important coupling model is FG to channel edge coupling. The FG-Channel edge coupling mainly come from AA (Active Area) polysilicon damage and active area channel edge sharp corner. AA polysilicon plasma induced damage will induce defect in AA edge, these defects will trap electron during P/E cycles. Moreover, as mentioned in Carbon-doped polysilicon parts, electrons with high kinetic energy will enlarge defect area by collision during program and erase.

Besides, AA edge sharp corner also exerts significant influence on FG-channel edge coupling. A possible mechanism can be described as below. Firstly, sharp corner position has high stress compared with other positions. [6] During program and erase cycle in NAND Flash, an unexpected temperature variation tends to happen. In temperature shift process, high-stress positions have larger thermal expansion than low-stress ones. In this way, more defects are generated in high-stress zones at sharp corner of channel edge, and result in trapped electron accumulation at sharp corner after P/E cycles. Therefore, a higher program voltage is required after P/E cycles.

Moreover, with the reduction of NAND flash memory technology nodes, the parasitic capacitance caused by the size of the active region become more important and cannot be ignored [7], especially when the tip of the active region shows a sharp corner shape. Specifically, during the program operation of the memory unit, the channel electrons will cross the floating gate and directly enter the control gate, and finally the electrons will be reversed and lost from the contact connected to the control gate, and then the memory unit program operation fails, which seriously affects the device performance in the process of multiple erases and writes, and reduces the reliability of the device. [8]

However, program voltage shift is one side of the disadvantages of trapped electron. Besides, charged sharp corner also leads to a higher potential between channel top edge and FG bottom edge. Point discharge phenomenon tends to appear in sharp corner high potential condition result from trapped charge. High potential bias between channel and FG will induce unexpected leakage current, resulting in FG electrons condition shift. [9]

In order to fabricate AA with rounding corner, an optimized approach has been applied. In conventional approach, floating gate etch (FG-ET) is designed to etch stop on tunneling oxide layer. After FG-ET, AA-ET employed Chlorine and Bromine base gas for further process. Therefore, channel width is smaller than that of gate oxide. This structure makes it difficult for post treatment of rounding sharp corner, as shown in Fig. 5b). This phenomenon results from both recipe structure design and etching gas selection. In order to achieve a high aspect ratio channel, chlorine base gas with high power is employed. Chlorine gas, which is widely used as polysilicon etch, has high selectivity against dielectric material. This phenomenon results from bond strength difference. In this case, the bond strength of the Si-Cl bond and Si-Br bond are smaller than the Si-O bond, so that the polysilicon oxide etch rate becomes extremely low with Cl and Br. However, in AA-ET process, high selectivity leads to an undercut profile. This profile with undercut structure makes it difficult for sharp corner rounding in post process.

the endpoint of FG-ET landing on gate oxide. In order to meet the breakdown voltage of NAND Flash device, STI structure with a depth more than 2000A is in need, and horizontal etch result from high selectivity process can easily form a channel sharp corner during etch process.

In order to achieve a profile with corner rounding, we fabricate a corner rounding profile fabrication method via optimized approach in FG-ET step. When carbon exists in process chamber, Si-O bonds are weakening and broken through the formation of C-O bonds for the reason that the strength of the C-O bond is larger than the Si-O bonds. Because of this, a Gate oxide breakthrough step which used fluorine carbon base gas is added to punch through oxide in FG-ET. In this process, the endpoint of FG landing under oxide layer and a thicker top hard mask layer is in need for low selectivity step in FG-ET. This approach can form partial STI structure in FG-ET, and the corner of AA is exposed. In post AA-ET process, less horizontal etch is in need and a relatively rounding AA corner is formed. After AA-ET, a fully rounding corner can be formed after post treatments, such as high temperature oxidation (HTO) and related TEM data of this approach is shown as Fig. 5c).

To evaluate the cycling performance of new optimized rounding corner scheme, a test structure is designed and the breakdown voltage of BL (Bit line) to WL (word line), which could represent the coupling effect as well as the effect of parasitic capacitance between WL-Channel is tested in both sharp and rounding conditions. As shown in Fig. 7, breakdown voltage(VBK) performances between BL and WL in different cycle numbers are demonstrated.

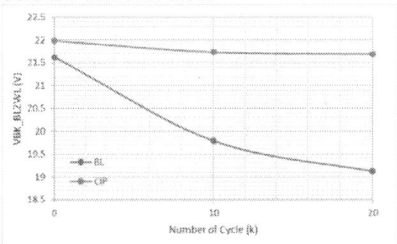

Figure 7: Breakdown Voltage between BL and WL performance of Baseline sharp approach and optimized rounding approach (CIP)

In baseline process condition, the breakdown voltage of WL to BL trend down to ~80% of its initial status after 20K cycling test, which denotes the coupling of Channel to WL (Control Gate) is getting stronger. This phenomenon indicated that the electrons are escaped and lost from the contact connected to the control gate after programming and erase cycles. Besides, the Range of voltage is also becoming worse, which indicated an ununiformed distribution within wafer in the baseline sharp corner process. However, the breakdown voltage of improved process condition has no obvious change after 10K and 20K cycles, respectively. and the tendency of range is stable compared with the baseline process condition. In another word, active area corner rounding reduces the coupling and parasitic capacitance effect between Channel and Control Gate and improving the Device cycling performance of 2X NAND Flash.

CONCLUSIONS

Carbon-doped polysilicon FG NAND Flash memory have been fabricated, and cycling performance have been evaluated by ECC spare area usage. In order to achieve comparable profile between polysilicon FG and carbon-doped one, FG etching process optimization is also taken. Experimental results demonstrate the improved cycling performance of carbon-doped polysilicon after 10K and 20K P/E cycles, respectively. Possible mechanisms of cycling improvement are proposed.

Besides, coupling effects of FG-channel are investigated. The mechanism of FG and channel coupling is discussed, a new FG/AA etching process scheme with rounding corner using breakthrough path is proposed and break down voltages between Bit line and Word line of baseline approach and new scheme are tested. BL-WL breakdown voltage show no descending after 10K and 20K cycles while baseline condition suffer breakdown when voltage reach ~80% of the initial status. Our physical and electrical characterization evidences exemplify the cycling and FG-channel coupling performance, which can be considered an effective approach to future NAND Flash process technology development.

REFERENCES

[1]. Jing Pu, Sun-Jung Kim, Seung-Hwan Lee .etc, IEEE Electron Device Letters. Vol. 29, No. 7 (2008)

[2]. LaiQian Luo, Kalya Shubhakar,Sen Mei. etc, IEEE Transactions on Device and Materials Reliability, Vol. 18, No. 1 (2018)

[3]. Huayong Hu, Peng Wang, Jianyao Liu, Hongwei Zhang, Qiang Wu, 2016 China Semiconductor Technology Internation Conference(CSTIC), 2016

[4]. N. Negishi, M. Izawa, K. Yokogawa, Y. Momonoi, T. Yoshida, K. Nakaune, H. Kawahara, M. Kojima, K, Tsujimoto and S. Tachi: Proc. Symp. Dry Process, 31,2000.

[5]. Kikuo Yamabe, Keitaro Imai, IEEE Electron Device Letters. Vol. 34, No. 8 (1987)

[6]. Mincheol Park, Keonsoo Kim, Jong-Ho Park and Jeong-Hyuck Choi, IEEE Electron Device Letters. Vol. 30, No. 2 (2009)

[7]. Zih-Song, Yajui Lee, Rex Yang and Chrong Jung Lin. Etc, IEEE Electron Device Letters. Vol. 33, No. 6 (2012)

[8]. Myounggon Kang, Joon-Sung Yang and Ik-Joon Chang, IEEE Electron Device Letters. Vol. 37, No. 3 (2016)

[9]. Carmine Miccoli, Giovanni M. Paolucci, Christian Monzio Compagnoni, Alessandro S Spinelli and Akira Goda, 2015 IEEE International Reliability Physics Symposium.

[10]. Y. S. Kim et al., Proc. IEEE Int. Rel. Phys. Symp. (IRPS), pp. 599-603, May 2010.

[11]. Hualun Chen, Ran Xu et al. 2022 China Semiconductor Technology International conference (CSTIC), 2022

[5] R. Nicole. *J. Name Stand. Abbrev.*, in press.

OPTIMIZATION OF SADP PROCESS FOR DEFECT REDUCTION IN PLANAR 2D NAND FLASH

Lifeng Liu[1], Jun Wang[1,2], Yinan Ma[2], Zhenchao Sui[1,2], Yue Li[2]*

[1]School of Integrated Circuit, Peking University, Beijing 100871, China
[2] Semiconductor Manufacturing North China, Beijing, China
*Corresponding Author's Email: lfliu@pku.edu.cn

ABSTRACT

With 2D NAND flash technology reached 24nm and beyond, tiny defects including bridge and pattern collapse have become the major yield killer defect in AA loop. In this work, the mechanism of tiny bridge and pattern collapse defect formation and improvement solutions are studied. For tiny bridge defect, the results showed that PR residue by periphery AA photo insufficient development in self-aligned double patterning process is the major cause of tiny bridge defect. Enhanced photo rinse & puddle process, SADP core/spacer line CD inline control and core etch profile optimization are all effective solutions to reduce tiny bridge defect. For pattern collapse defect, repeating structures with high aspect ratio are vulnerable by surface tension of drying liquid between patterns. Optimized wet cleaning sequence to ensure structure in continuous wetting status showed great improvement for pattern collapse defect reduction.

Index Terms—**Advanced NAND flash, Lithography, Plasma Etch, SADP, Tiny bridge defect, pattern collapse.**

INTRODUCTION

In 2D NAND flash, as the increasing request for cell density, critical dimension keep shrinking from 38nm to 24nm then even to 19nm. As integrated circuits line width continually shrinks, some critical dimensions are out of the conventional lithographic equipment limitation. SADP (Self-Aligned Double Patterning) technique successfully fills in the gap between 193nm immersion lithography and EUV lithography [1]. The complexity of SADP process and increasing cell density results in tiny defects like residue, bridge, line broken, pattern collapse and partial etch becoming the major yield killers [2]. Especially for tiny bridge, due to the SADP process multi steps, undetectable bridge defect at current stage may still impact on the following process [3]. Increasing aspect ratio by smaller feature size (floating gate width) [6] and certain STI depth to ensure isolation of NAND flash keep touching the limitation of current cleaning and drying method for pattern collapse defect. Drying liquid with low surface tension and high contact angle are essential to minimize the cohesive force impact on structure [7, 8].

In this paper, we analyzed the tiny bridge defect performance and yield impact in mass production

manufacturing, proposed several solutions to reduce tiny bridge defects and verified in production level, including optimization of lithography process, plasma etching process and integration flow. A new concept with continuous cleaning sequence is proposed to reduce pattern collapse defect, and significant improvement with larger aspect ratio are obtained.

EXPERIMENT

All the experiments and data analyses are performed on mass production NAND flash platform at 38nm or 24nm node. Relative SADP process schematic flow is shown in Figure 1. Several layers of films were firstly deposited as hard-mask (HM) stack. After cell area lithography/etch process, cell area core patterns were formed. Subsequently, spacer film deposition and 1-step spacer etch & core removal were performed. To define periphery AA pattern at the same STI etch step, an extra periphery AA photo step was inserted. Then, the final AA patterns were formed by silicon etching with cell area using spacer as hard-mask and periphery area using PR as mask. The inline defect inspection and review image were performed on KLA-Tencor defect scan tool and AMAT scanning electron microscope (SEM) review tool respectively. The SEM and transmission electron microscope (TEM) was applied for defect morphology and pattern profile observation. And to further understand the defect composition, a surface element analysis was performed using energy dispersive X-ray spectroscopy (EDS) on TEM samples

- SADP film deposition
- Cell area photo
- Core etch & clean
- Spacer deposition
- Spacer etch & clean
- Periphery area photo
- Cell/Peri. HM etch & clean
- Cell/Peri. STI etch & clean

Figure 1: Schematic process flow of conventional SADP technology used in NAND flash AA loop.

RESULTS AND DISCUSSION

Together with the shrinkage of line CD and increase of cell array density, defectivity becomes the top one yield killer in NAND flash manufacturing. Furthermore, due to the complex process of SADP, tiny defects which are difficult to be detected in the current step may appear after the subsequent etch or clean steps.

Typical tiny bridge and pattern collapse defect map on wafer and top view SEM image are as shown in Figure 2. From wafer dimension analysis, tiny bridge defects are all located in a wafer-centered spin map with radius about 100mm. From the SEM image post AA hard-mask etch step in Figure 2(b), two neighbored AA lines are observed to be slightly connected with a transparent, horizontal tiny line. These data indicated that a lithography development step with high rotation speed and specific nozzle dispense behavior is highly suspected as defect formation causes. While for pattern collapse, nearly all defect located at wafer top/down edge area with horizontal cell lines.

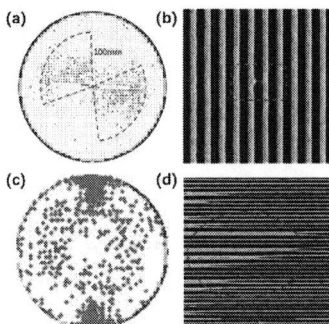

Figure 2: Tiny defect map on 300mm pattern wafer and top view SEM image, bridge defect (a, b) and pattern collapse (c, d).

These tiny defects would induce serious failure in CP yield test, bridge results in hard fail with silicon partial etch while pattern collapse always results in soft fail with only space shrinkage between two cell lines. As shown in Figure 3(a), direct CP yield loss and flash memory redundancy compensation composite map with more than 200pcs production wafers was observed as typical spin map, and shared the same radius of 100mm; while no significant special map can be found on a specific single wafer level. Electrical failure analysis (EFA) is performed by Bitmapping using MOSAID, and the major fail bin for this defect impacted die is double bit line failure: two neighbored bit lines are operated at the same time while they are supposed to be separated. Further physical failure analysis (PFA) results as shown in Figure 3(b & c) showed that at both on bit-line (BL) and word-line (WL) direction, tiny bridge between two neighbored AA lines was observed to be insufficient silicon etch.

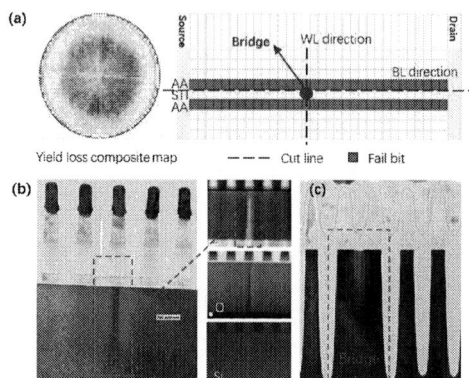

Figure 3: CP yield loss composite map and EFA analysis (a), and PFA results on BL direction (b) and WL direction (c).

Considering the SADP process flow, the major cause of tiny bridge defect can be concluded as shown in Figure 4: In periphery AA photo step, photo resist will be totally filled into cell array hard-mask space, decomposed with photo acid generation in the following light exposure and finally removed by lithography develop (DEV) process. If there is any factor caused insufficient PR development, tiny residues would remain in cell array space bottom or corner, and further block STI etch process leading to AA bridge defect formation.

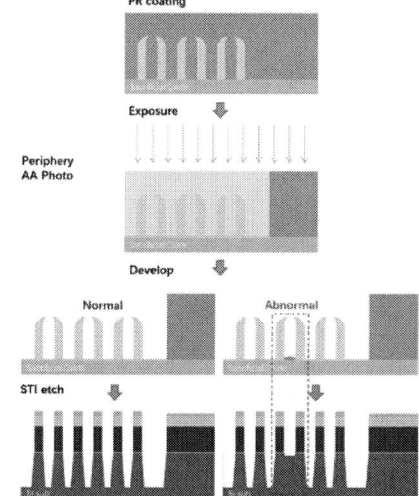

Figure 4: Mechanism illustration of tiny bridge defect.

Thus, tiny bridge defect performance and its failure model on yield loss is quite clear: tiny residue formed during AA-PH DEV step induced block etch and resulted in tiny bridge. Possible solutions to reduce this type of defect and improve CP yield could be: (a) Lithography DEV efficiency improvement; (b) Core/spacer inline control to enlarge DEV process window; (c) Optimized SADP approach with straight spacer profile. For pattern collapse defect, application with continuous wetting cleaning sequence is studied to reduce defectivity.

AA Photo DEV Rinse & Puddle Process Optimization

A typical lithography DEV process is described in Figure 5(a): a short step of DIW pre-wet is for better solvent wetting, then large amount of solvent is dispensed on wafer with PR after exposure, the decomposed PR tend to dissolve into solvent and flush away. Thus a certain long time of solvent puddle with extremely low spin speed is essential to ensure PR solubility. After that a DIW rinse step with large flow rate is performed to exchange solvent with DIW. Finally, a spin dry step with high rotation speed is used to make wafer surface dry.

Based on this sequence, we choose solvent puddle and DIW rinse step as major factors to evaluate tiny bridge defect dependency on process time. As a result, shown in Figure 5(b), increasing both puddle and rinse step time showed lower defect count than baseline condition. The contour chart was divided into 3 effective regions with different defect count. Region A is more close to current baseline condition by short puddle and rinse time, and further decreasing puddle and rinse time would make defect count higher as worse test. Region B refer to puddle time decreasing and rinse time increasing at the same time. As an extreme case with 0s puddle time and much longer rinse time, defect count is acceptable. But this could be a potential risk for different line space may have different response, line space related studies are still ongoing. Region C is the target area in this case with longer puddle and rinse time at the same time. As longer puddle/rinse time, PR residues are more easily to be removed after DEV process finished.

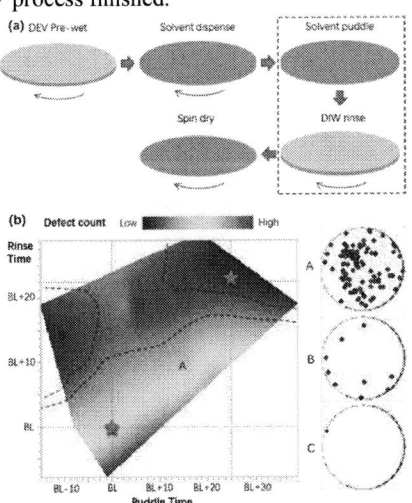

Figure 5: Lithography DEV process illustration (a)

and puddle & rinse time effect on tiny bridge defect performance (b).

Core/Spacer Etch CD Inline Control

Based on the current SADP run-path, an amorphous carbon (APF) core and silicon oxide spacer combination was chosen for easier process and lower cost [4]. However, a relative taper core profile was obtained due to the properties of APF (as shown in Figure 6a). Thus, a similar core space profile with smaller top CD and larger bottom CD after spacer etch and core removal would further enhance the limitation of following AA lithography process: photo resist residues especially at core space bottom corner area are difficult to remove in DEV step.

At this point of view, spacer inline CD shrink may be available to enlarge DEV process window. As shown in Figure 6b, a series of spacer CD split test are performed to confirm the influence on tiny bridge defect. Based on these results, spacer inline CD 1nm smaller could reduce tiny bridge defect by one order, and as spacer CD increase defect map with special spin pattern occurred.

Spacer CD shrink benefits on tiny bridge defect, but this benefit is not without limitation. In all our split test, to maintain pitch walking between lines transferred by SADP process, core CD and spacer oxide deposition thickness are varied at the same time. Each core CD referred to a best-fit spacer deposition thickness and resulted in a very narrow range of tunable spacer CD. Worse pitch walking with even/odd space difference larger than 1nm would also significantly impact on wafer defect performance and final yield. During the pattern transformation, smaller spacer CD will result in smaller AA CD, thus a narrow channel was formed with a larger threshold voltage in final electrical property test. Also due to the high aspect ratio in SADP STI etch process [5], smaller AA is more vulnerable to suffer pattern collapse in the following wet cleaning steps.

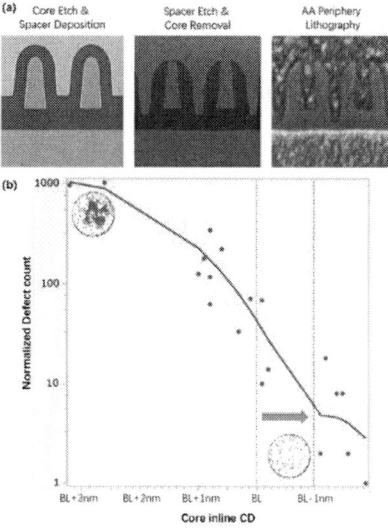

Figure 6: SADP process by step TEM/SEM image (a) and correlation between tiny bridge defect count and core etch inline CD (b).

Cap Core Approach for Profile Improvement

Spacer CD inline target optimization and force control is a possible method to reduce defect count, but cannot completely solve this issue. Aiming to minimize the process change, a novel cap core run-path with no film stack and film thickness change is proposed to optimize core profile and enlarge tiny bridge defect process window. As shown in Figure 7, a DARC layer with certain thickness is applied as HM in APF core etch. Only by core etch recipe DARC ME/OE and APF ME step process time decrease, an APF core with DARC cap profile can be obtained. Due to the sheltering effect and selectivity between DARC and APF, straighter APF profile with uniform top/bottom CD is achieved. In the following spacer etch process, extra ME step process time is needed to consume cap DARC, and a uniform core space profile is transferred after spacer etch. TEM images of different run-path showed the results above.

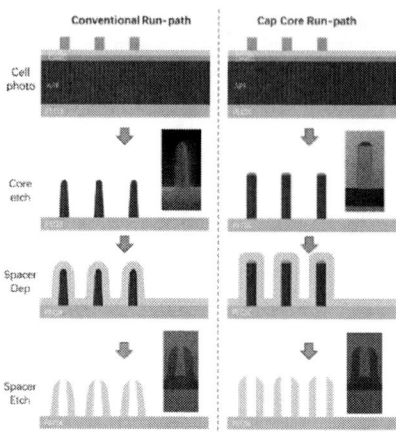

Figure 7: Comparison between conventional APF core/silicon oxide spacer scheme run-path and new Cap core run-path.

By this CIP run-path, uniform core space profile is considered to improve lithography rinse capability and reduce possible PR residue hidden at core space corner area. As a result, tiny bridge defect performance is outstanding even for core CD up to BL +2.7nm in Figure 8(a), Thus this Cap core run-path enlarged CD window especially for tiny bridge defect in SADP process, Furthermore, wafer redundancy usage in CP yield at 38nm NAND flash mass production level is summarized to compare long term baseline, normal run-path and Cap core run-path, as shown in Figure 8(b). Compared with long term baseline and normal run-path, Cap core run-path wafer showed an extremely low level performance in wafer level redundancy consumption, indicating the great improvement for optimization of core etch profile in tiny bridge defect reduction, and demonstrated a positive result for planar NAND key feature size scaling down.

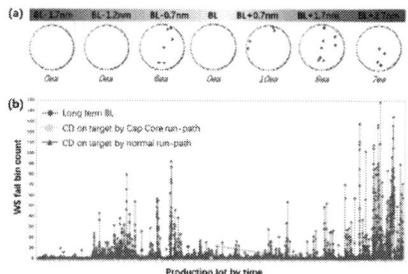

Figure 8: New Cap core run-path tiny bridge defect performance by core CD split (a) and production lot wafer redundancy usage comparison (b).

Continuous Cleaning Sequence to improve pattern collapse

Cohesive force of drying liquid on patterns used in wet process is the main cause for pattern collapse in HAR structures, and the key impact factors for structure deformation are surface tension, contact angle and aspect ratio [8]. Based on theoretical studies, drying liquid with lower surface tension and higher contact angle show benefit on cohesive force reduction. Currently, we already used single wafer clean equipment with hot IPA + nitrogen instead of DIW in drying step to balance cohesive force during fluid evaporated, and for baseline STI trench depth no pattern collapse is found, in Figure 9(a). However, for higher STI trench depth split to achieve better isolation performance, worse map pattern collapse is suffered when STI trench depth 300A higher than baseline.

A new method of continuous wet cleaning process is applied to improve high aspect ratio structure pattern collapse defect [9], as shown in Figure 9(b). Based on conventional wet clean recipe, several steps of overlapping chemical dispense/DI rinse is inserted to minimize wafer surface evaporating before drying step. Defect performance is comparable for baseline STI trench depth, while for STI trench depth +300A, continuous wet cleaning showed significant defect count decrease and better defect map in Figure 9(c). Thus, continuous wet cleaning is necessary for higher aspect ratio structure and may enlarge pattern collapse defect window for current structure.

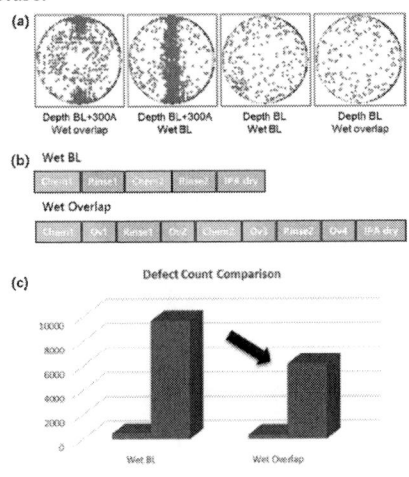

Figure 9: Pattern collapse defect map (a) and defect count (c) comparison, and schematic illustration for conventional cleaning sequence and continuous sequence (b).

CONCLUSION

In this paper, the mechanism of tiny bridge defect formation and improvement solutions in 2D NAND flash manufacturing are studied. The results showed that PR residue by periphery AA photo insufficient development in SADP process is the major cause of tiny bridge defect. Enhanced lithography rinse & puddle process, core/spacer line CD inline force control and core etch profile optimization are all effective solutions to reduce tiny bridge defect. And a novel continuous cleaning sequence could significantly improve pattern collapse defect by reducing cohesive force generated during cleaning and drying.

ACKNOWLEDGEMENTS

The authors gratefully acknowledge the support of NAND flash integration and yield enhancement team of SMNC for their contribution to this work.

REFERENCES

[1] C. Yi et al., "14nm Fin SADP Patterning Processes and Integration," *2020 China Semiconductor Technology International Conference (CSTIC)*, 2020, pp. 1-3, doi: 10.1109/CSTIC49141.2020.9282604.

[2] G. Ying, Y. Dai, J. Wang and H. Zheng, "Study on Lithography Defect Reduction for 19nm NAND SADP Process," *2021 International Workshop on Advanced Patterning Solutions (IWAPS)*, 2021, pp. 1-4, doi: 10.1109/IWAPS54037.2021.9671057.

[3] Q. Cao, Q. Ni and J. Song, "Tiny SADP Defect Detection and Reduction for 19nm NAND Flash Technology Semiconductor Manufacturing Engineering," *2022 China Semiconductor Technology International Conference (CSTIC)*, 2022, pp. 1-3, doi: 10.1109/CSTIC55103.2022. 9856819.

[4] H. Liu et al., "SNC SADP Spacer etch process development using carbon hard mask mandrel for sub advanced process DRAM," *2022 China Semiconductor Technology International Conference (CSTIC)*, 2022, pp. 1-3, doi: 10.1109/CSTIC55103.2022.9856784.

[5] V. V. Iyengar et al., "Collapse-free patterning of high aspect ratio silicon structures for 20nm NAND Flash technology," *2015 26th Annual SEMI Advanced Semiconductor Manufacturing Conference (ASMC)*, 2015, pp. 53-57, doi: 10.1109/ASMC.2015.7164450.

[6] C. Monzio Compagnoni, A. Goda, A. S. Spinelli, P. Feeley, A. L. Lacaita and A. Visconti, "Reviewing the Evolution of the NAND Flash Technology," in *Proceedings of the IEEE*, vol. 105, no. 9, pp. 1609-1633, Sept. 2017, doi: 10.1109/JPROC.2017.2665781.

[7] T. Tanaka et al., "Mechanism of Resist Pattern Collapse during Development Process", *Jpn. J. Appl. Phys.*, 1993, 32 pp 6059.

[8] G. Kim et al., "Effect of Drying Liquid on Stiction of High Aspect Ratio Structures", *Solid State Phenomena*, 2012, Vol. 187 pp 75-78.

[9] Patent in application, Y. Ma, X Liu, "A Method to Improve High Aspect Ratio Structure Pattern Collapse Defect", 2019.

CONTROL OF DEPOSITION IN CYCLIC DEPOSITION/ETCH PROCESS

Zhang Zihan, Wang Jing, Xu Litian
Beijing NAURA Microelectronics Equipment Co., Ltd.
Beijing, china
*Corresponding Author's Email: Zhangzihan@naura.com

ABSTRACT

Nowadays, Deposition/Etch cycle process is widely used in the process of transistor manufacturing. This kind of process can reduce consumed volume of sacrificial mask and improve selectivity. This paper concentrates on the critical parts of deposition: control deposition shape in ICP etch tools. We focused on the ways of pillar deposition by optimized recipe parameters, and got various dep-shape product.

INTRODUCTION

In the past decade, Moore's law has been used by semiconductor industry as predicative indictors of industry. In order to achieve higher development speed of semiconductor manufacturing, etch need more and more advance technologies are applied to process[1].

Cyclic Deposition/ Etch (CDE) is a high selectivity etch process. Comparing to traditional etch process, less mask is consumed in CDE process, even non-consumed sacrificial layer. CDE can complete critical HARC process and less plasma damage. In CDE series process are used in etch tool , the most critical part is controlled of deposition[2]. We have studied the experiments with different parameters (including polymer gas body, pressure, time, temperature, dilute gas volume, bias power) to inquiry deposition status.

EXPERIMENT

In the experiments, a silicon wafer with a diameter of 300 mm (12 inches) was used. The ICP equipment of NAURA was used for etching. The plasma was maintained with the 13.56 MHz RF power source and the electrostatic chuck (ESC) was also powered by the second RF power at 13.56 MHz.

RESULTS AND DISCUSSION

We used two kind of main polymer gas for deposit carbon-hydrate product on the pillars.
1. Methyl fluoride (CH3F), Fig. 1 shows this gas is the chemical compound contains one carbon atom, three hydrogen atoms and one fluorine atom. It is a flammable gas with no color and quite heavier as compared to air.
2. Hexafluoro-1,3-butadiene (C4F6) is a relatively new etch gas for the manufacturing of semiconductor devices, especially in critical etch processes that need high aspect ratio and selectivity. C4F6 have higher light resistance and silicon nitride selectivity than C4F8. The molecular structure C4F6 is in below Fig. 2.

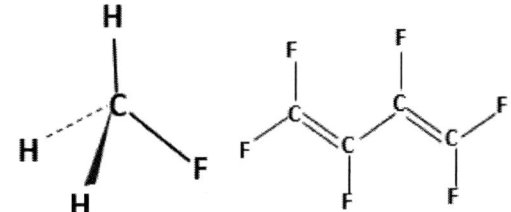

Figure 1: CH3F structure Figure 2: C4F6 structure

Table 1: Test A/B/C/D detail compared.

Test	Bias Power	He (sccm)	N2 (sccm)	Temperature (Deg)
A	0	Mild	0	Mild
B	0	Mild	0	Low
C	Low	Mild	0	Mild
D	0	Low	Mild	Mild

CH3F gas deposition results are compared in Fig. 3, similar recipes are used in the four tests: Mild source power, 50 sccm (standard cubic centimeter per minute) CH3F, mild He, and 15s RF on times. Detail variation factors showed in below Table 1.

Four TEM results are shown in Fig. 3. Test A deposit polymer height close to 80A, and the product have circular arc shape deposition. Test B reduce temperature, therefore polymer have a stronger physical absorptive capacity between sidewall and product [3]. B deposit product close 92A, this proved lower temperature lead to more polymer passivation. Adding low bias power for changing deposition shape result is shown in Test C. Comparing C to A profile, C have a more sharpen deposition from. An interesting phenomenon appears that low bias power lead to thicker carbon-hydrate on the pillar [4]. This reason may be caused by ion energy enhanced by bias power sheath so that deposition rate of radical and polymer cluster are influenced. some He is used instead of N2 in Test D, N2 lead to less deposition rate and uniformity deposition product.

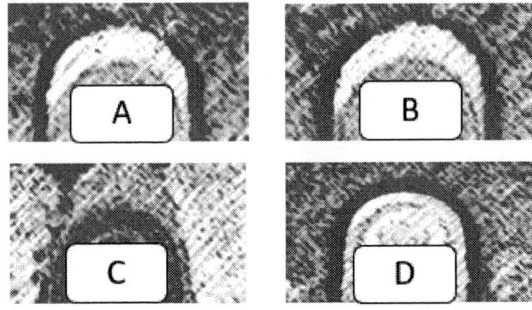

Figure3. CH3F gas all sort of depsition.

C4F6 gas deposition results are shown in Figure4, similar recipes are used in E/F tests: Mild source power, little O2, mild He dilute gas and 15s RF on times. Compared C4F6 to CH3F gas deposition effect, C4F6 is a heavy polymer gas, leading to more deposition product. As is shown in Test E TEM image, this polymer had higher adsorption and worse roughness surface. The byproducts of dissociated C4F6 are cyclic annular structure polymer, which is flocculence profile. Although deposition rate of C4F6 is higher, but soft byproduct is easily eliminated by etch. C4F6 and O2 gas ratio are adjusted in Test F for optimizing the quality of etch-byproduct.

We found the increasing O2 lead to more compact polymer layer in Test F TEM.

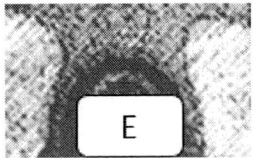 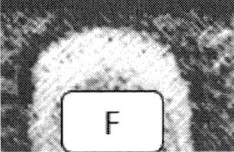

Figure4. C4F6 depsit shape TEM.

CONCLUSION

In this paper, we discussed the critical way of controlling deposition process. Two polymer gas experiment results proved the different gas create various deposition product. Diversified factors are used Test A/B/C/D to influence process. All of the above tests indicated that the systematic experiments provided many strategies of controlling deposition in Cyclic Deposition/ Etch. These results can be widely used in different CDE processes, and effectively optimized etch process. At the same time, it is helpful for tuning process recipe more logical and clear.

REFERENCES

[1] M. L. Kempsell, E. Hendrickx, A. Tritchkov, K. Sakajiri, K. Yasui, S. Yoshitake, et al, "Inverse lithography for 45-nm-node contact holes at 1.35 numerical aperture," J. Micro/Nanolith. MEMS MOEMS, vol. 8(4), pp. 043001, October, 2009.

[2] Fanton,La Greca, Jain, Prentice, Simiz, et al., "Process window optimizer for pattern based defect prediction on 28nm metal layer," Proc. SPIE 9778, (2016).

[3] Mike Barnes et al., Inductively coupled plasma source with controllable power deposition, US Patent 6, 507, 155.

[4] S. Panda, R. Ranade, and G.S. Mathad, "Etching high aspect ratio silicon trenches," J. Electrochem. Soc., 150, G612-G616 (2003).

A NOVEL METHOD TO ELIMINATE BOND PAD CRYSTAL

Fanshun Meng[1]*, Jiajin Wang[1], YunBo Chen[1], Qiang Rui[1], Zhongkui Chen[1]
Kuo-Jung Chen[2], Yi Liu[2], Fengyang Li[2], Chao Sun[2]
[1]CanSemi Technology Inc., Guangzhou City, Guangdong Province, China
[2]Advanced Micro-Fabrication Equipment Inc. (AMEC), Shanghai, China
*Corresponding Author's Email: fred.meng@cansemitech.com

ABSTRACT

Bond pad crystal defect is a kind of AlxOyFz crystal generated on aluminum bond pads, which would affect wire bonding and bump performance, resulting in lower bond pull and ball shear force, and cause discolored on bond pads. In this paper, to eliminate the crystal defect, simplified new methods were developed, including continue optimize fluoride-reducing recipe during etch process to remove polymer on backside of the wafer, and an unsealed teeth of front opening shipping box (FOSB) to suppress moisture gathering on the wafer surface, furthermore, a thin dense layer of oxidation film was formed by applying oxygen plasma treatment process to isolate underlayer from fluorine source and moisture. Because of these improvements, crystal defects on the special map of the wafer disappeared, and the defect-free storage time could be extended to 36 months at least.

Keywords: pad crystal; moisture; fluoride-reducing recipe; oxygen treatment; Front Opening Shipping Box (FOSB).

INTRODUCTION

Aluminum bond pad is a metal surface exposed on the wafer, which is formed after etching the passivation layer, and plays a vital role in integrated circuit industry. [1,2] Key factors of the formation of crystal including aluminum source from the metal layer itself, fluorine source from the polymer residue after passivation etching process and the storage enviroment, and so on. With the development and optimization of semiconductor production technology, and the critical dimension continue getting smaller, while the IC package precision requirements and the difficulty of controlling bonding quality and reliability are increasing obviously. [3] Considering the importance of bond pad, it is necessary to ensure that its surface is clean and defect free. However, it is observed in the integrated circuit industry that crystalline defects on the surface of the aluminum pad become progressively worse. [4,5] Thus, great efforts have been devoted to elimate the crystal defects, such as the control of the temperature and humidity of the storage enviroment, and the control of the fluorine content. [6,7]

Pani and co-workers analyzed the cause of the crystallization defects in detail, and then achieved 36 months of crystalline defect-free by reducing the surface fluorine level, that is, using a polymer free etch process or higher efficiency solvent clean to eliminates pad bonding with fluorocarbon polymer.[6] Kwon reported that after CF$_4$ etch, the addition of argon sputter step and 10min nitrogen purge in the FOUP and the fluorine concentration has dropped from 230 ng/L to 20 ng/L. [7]

However, with the increasing demand for power device, power management, fingerpint sensor chips, and so on, special processes have been transferred from 8-inch fab and widely applied in 12-inch fab. This technology evolution comes with higher chip performance requirement, especially for thicker passivation film, which needed more etching time and led to more polymer deposition and eventually more defect sources.

Meanwhile, althogh domestic semiconductor manufacturing equipments exhibit promising performance, the defect ratio is still higher than leading supplier. Thus, a new and reliable method is highly desirable for solving this bond pad defect.

Usually, pad crystal defects with random map (Fig. 1a), in this paper, spcial map crystal defects were deserved (Fig. 1b), located at wafer 3 & 9 o'clock. To eliminate this type of defect, we reported a new effective method using optimized fluoride-reducing recipe during passivation etch process to remove polymer on backside of wafer, and adopting a FOSB (Type-B) with unsealed teeth to avoid moisture accumulation on the wafer surface during the storage. Meanwhile, implement oxygen treatment as a precaution to isolate aluminum contacting with moisture and fluorine. After eliminating the three factors of crystal formation separately, the crystal defects on the aluminum bond pad were disappeared completely.

EXPERIMENT

The experimental sample was prepared by 12-inch wafers in the FAB environment with a standard CMOS logic flow. To characterize the crystal defects on the aluminum bond pad, we used the KLA (8825) to take optical photos on the wafer. The structure and element composition of crystal defects were analyzed by SEM (Scanning Electron Microscope) and EDX (Energy Dispersive X-ray spectroscopy) provided by KLA EDR (7XXX), respectively. TEM (Transmission electron microscopy, FEI Tecnai TF20, 200KV) was also conducted for wafer surface characterization of bond pad.

RESULTS AND DISCUSSION

As shown in Fig. 1b, crystal defects only occured at the wafer edge about 3 & 9 o'clock. It is observed that in Fig. 1c there are clusters of transparent crystal abnormalities at the pad surface under the optical microscope (OM). Meanwhile, SEM (Fig. 1d) shows the obvious crystal structure of defect, and EDX (Fig. 1e) analysis shows that the elements of defect include mainly aluminum and fluorine element. Aluminum element obviously from the metal layer itself, from day one, we already used low-fluorine content passivation etch recipe, if fluorine element from bond pad surface or sidewall, should be random map, as shown in Fig. 1a.

Thus, it is wondering why crystal defects only occured at 3 & 9 o'clock of the wafers? And where does the fluorine element from? Firstly, we carefully observed the edge of the wafers under OM and found that there was a circle of dendritic-like crystals at 3mm from the edge of the backside (Fig. 2a). EDX confirmed fluorine at the dendritic-like crystals (Fig. 2c).

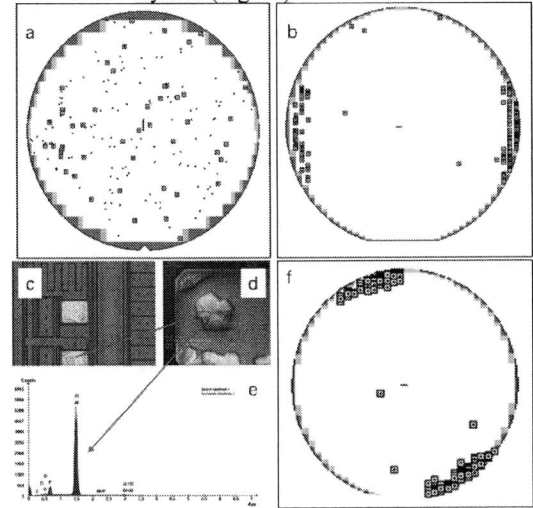

Figure 1: Crystal defect of radnom map(a), 3&9 o'clock special map (b), OM (c) and SEM (d) image and EDX (e) of crystal defect, 5&11 o'clock special map (f)

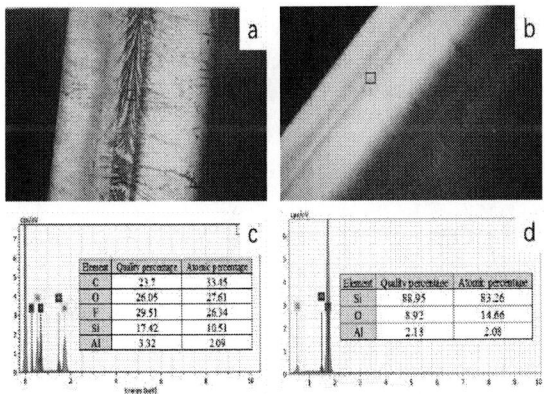

Figure 2: OM images and EDX analysis results of the backside edge of the wafer before (a,c) and after (b,d) the optimized recipe usage.

Considering the wafer manufacturing process, the dendritic-like crystals on the backside edge of the wafer are clearly shown polymer residues from the etching process. Usually, special gases, such as C_4F_8, C_4F_6, are added during the etching process to form polymers to protect the sidewall to achieve a good topography. However, too much polymer cannot be removed after the post etch solvent clean process, meanwhile, polymer residues will be deposited to the backside of wafer edge, which also exposed to plasma enviroment, during AMEC D-RIE etching chamber. To remove polymer residues from the backside of the wafer, fluorine-based gases are continuely reduced by 28% in the etching process recipe. Fig. 2b shows that dendritic-like crystals on the backside of the wafer disappear after the optimized recipe usage EDX analysis result (Fig. 2d) also manifests that there is no fluorine residue on the backside of the wafer.

Moreover, we found that the special map at 3&9 o'clock exactly match FOSB teeth, the position where wafer is placed in the FOSB. Then, we adjusted the notch of wafer in FOSB and found that the position of the crystal defect map also changed, as shown in Fig. 1f.

Figure 3: Type-A FOSB (a) and Type-B FOSB (b).

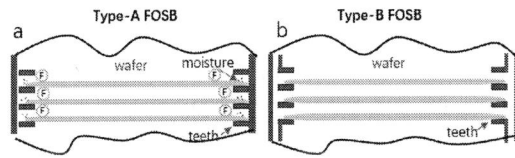

Figure 4: Schematic of the crystal defect formation Type-A FOSB (a) and Type-B FOSB (b).

After carefully observation of FOSB, it is found that the teeth supporting the wafer is a closed architecture (Fig. 3a), so that moisture in the air could be easily to gather here (Fig. 4a), and the volatile etching residual polymer containing fluorine on the wafer surface will also be blocked by the FOSB teeth and fell on the wafer surface. To improve this phenomenon, as shown in Fig. 3b, the teeth of another type FOSB (Type-B) with unsealed reconstruction can avoid moisture in the FOSB gathering here (Fig. 4b).

Considering the reliability concern of aluminum bond pad, a simple but effective preventive action is also implemented to further prevent crystal defect formation. In general, aluminum is an active element that is easily oxidized to form a thin layer of oxidation film, it could to prevent the internal aluminum further oxidation. However, the natural oxidization of aluminum is very time consuming, residual fluorine of the etch process can easily react with aluminum to generate crystalline defects during the oxidation process.

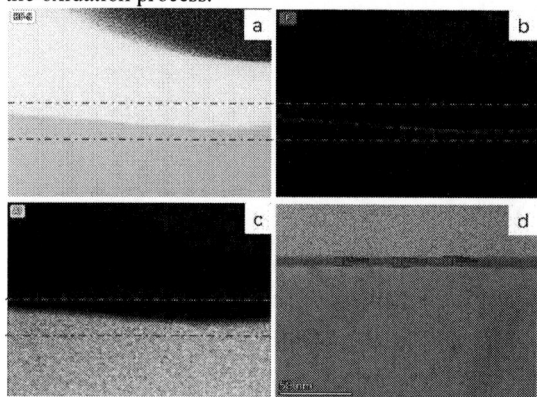

Figure 5: TEM (a) and the elemental mapping (b,c) images used baseline condition; TEM image (d) of aluminum bond pad surface by adding oxygen treatment.

It was found that there was the fluorine element gathering around 50 angstroms above the aluminum bond Pad surface without using optimized process (Fig. 5a, b, c). With the idea of oxidation isolation, by appling an oxygen treatment step after wafer manufacturing process of the solvent clean process to form a thin layer on aluminum pad surface immediately to prevent crystal defect formation. From Fig. 5d, about 60 angstroms uniform thin oxidation film formed on the surface of bond pad after oxygen treatment.

Based on above results, it is speculated that fluorine volatilizes from the wafer backside polymer residue, and then reacts with the moisture gathered around the FOSB teeth and the aluminum on the aluminum pad surface, finally result in crystal defects. According to the above analysis, the key elements forming a crystallization defect include the fluorine residue during the passivation layer etching process and the aluminum of metal pad under the humidity of the storage environment. Firstly, reducing fluorine-based gases usage in etching process eliminates polymer on the backside of the wafer. Secondly, Type-B FOSB, which hollowed out the teeth of supporting wafer was used to reduce the accumulation of moisture around the teeth. Finally, oxygen treatment as a precaution isolates aluminum contact with moisture and fluorine. Therefore, the worse test experiment at high temperature and high Humidity (Table 1) showed that crystal defect

did not appear even after 36months, as shown in Fig. 6.

TABLE I. *Experiment and corresponding shelf lifetime.*

Shelf Lifetime	Worse Test Experiment		
	Temperature	*Humidity*	*Time*
1 year	85°C	85%	32 hours
2 years	85°C	85%	64 hours
3 years	85°C	85%	96 hours

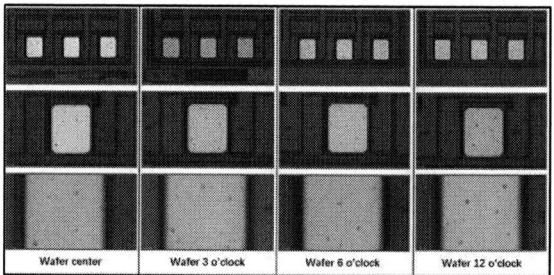

Figure 6: OM images of aluminum bond pad after 96 hrs worse test experiment.

CONCLUSIONS

Experimental results showed that fluoride-reducing recipe during passivation etching provides effective remove polymer on the backside of the wafer, Type-B FOSB indeed avoids the accumulation of moisture at the tooth, and a thin dense oxidation film is formed on the pad surface through oxygen treatment, which isolated aluminum to react with fluorine and moisture.

Benefited from these improvements, the crystal defects on the 3 & 9 o'clock of wafer edge disappeared, and the defect-free storage time extended to 36 months at least. This simplified but effective method provides one more solution to the wafer manufacturing inductry.

ACKNOWLEDGEMENTS

The authors gratefully acknowledge the CanSemi Integration, Quality, Etch, and AMEC Teams' support.

REFERENCES

[1] Y. N. Hua. *Semiconductor Electronics,2002. Proceedings*, pp. 177-181, 2002
[2] Y. N. Hua et al. *International Symposium for Testing and Failure Analysis*, pp. 495-504, 2002
[3] J. S. Chen et al. *ISMS 2001*, 297-299, 2001
[4] C.Y. Sun. *2020 China Semiconductor Technology International Conference*, 2020
[5] S-Ho. Ahn et al. *Proceedings 46th Electronic Components and Technology Conference*, pp. 107-112, 1996
[6] Pani et al. *J. Vac. Sci. Technol. B*,36,2166-2746, 2018
[7] Kwon et al. *Aerosol and Air Quality Research*, 17: 936–941, 2017

CARBON HARD MASK OPENING PROCESS DEVELOPMENT WITH NOVEL SIDEWALL PASSIVATION IN MEMORY MANUFACTURING

Meng-Jiao Zhu [1]*, Li-Tian Xu [1], Jing Wang [1], Li Zeng [1], Zi-Han Zhang [1]

[1] *Beijing NAURA Microelectronics Equipment Co. Ltd*

No. 8 Wenchang Avenue, Beijing Economic-Technological Development Area. Beijing City, China

*Corresponding Author's Email: zhumengjiao@naura.com

ABSTRACT

When the logic memory cells are scaled, the aspect ratio of the holes increases, and the profile control of these structures becomes increasingly challenging. The carbon hard mask serves as a sacrificial template to transfer its profile to the underneath layer. It is expected that the etched features have vertical profile to achieve better CD uniformity and less pattern roughness.

In this article, we report a study on the commonly carbon hard mask etching process with O2 and SO2 plasma. In order to achieve bowing free carbon profile, a novel sidewall passivation gas is added for the profile control during hole manufacturing in logic memory application.

Keywords — carbon hard mask, logic memory, silicon based deposition, bowing free.

INTRODUCTION

In semiconductor manufacturing processes, carbon hard masks are widely used, and the morphology of carbon hard masks has a significant impact on the performance of subsequent devices. [1]. The bowed profile is a bottleneck in the logic architecture, resulting in a decrease in the downstream process margins due to bridge or broken between the neighbor holes.

As is well known, sidewall bowing is caused by lateral etching deviating from ion bombardment, and the lack of sufficient protective layer on the sidewall can accelerate sidewall bending [2]. The ion trajectory may deviate from the surface normal due to collisions between ions and neutral ions in the plasma sheath, reflection of ions from the mask facet, and negative charge heaped up on the mask sidewall surface [3].

As the etching progresses, the spin on carbon (SOC) is eroded and tapered by ion sputtering, and then the ion scattering will tilt into the features that will produce bowing [4]. A possible solution to reduce bending is to reduce ion scattering or enhance sidewall protection at the bowing position.

In this paper, the division of SO2/O2 gas ratio during the etching process was studied, and different bias powers duty cycle and temperature changes were applied to control the sidewall bowing during the SOC profile etching process. In addition, we propose a silicon-based deposition to protect the SOC sidewalls in order to get rid of the bowing profile for the small diameter of holes during logic manufacturing.

EXPERIMENT

Experiments were performed using NAURA's etch tool high-density inductive coupled plasma (ICP) etcher to etch the SOC type of carbon hard mask. There are four kinds of film layers before etch as shown in Fig.1(a), such as dielectric hard mask-1, dielectric hard mask-2, SOC hard mask and bottom SiON as etch stop layer.

Fig.1 SOC HM manufacturing process: (a) pre-etch with HM-1 pattern, (b) post HM-2 etch, (c) post SOC HM open

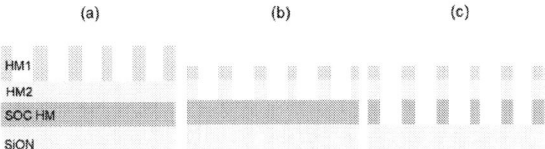

Four etch steps are used for the HM open process, such as BT step for native oxide etch, HM-2 step for HM-2 etch, SOC step for SOC HM open. In this experiment, polymer gas (e. g., HBr, CH3F, CH2F2, SO2, and a novel Si-based gas et al.) are needed to passivate the sidewall for carbon layer etch. After carbon etch, cross-sectional SEM or TEM was performed to check carbon profile. Key process qualification specifications include top CD, bottom CD and top dielectric mask remain. Bow-bottom CD is a measure of bowing performance: smaller bow-bottom CD represents better bowing control, due to a carbon profile that is more vertical.

RESULTS AND DISCUSSIONS

A. Passivation gas effect on SOC profile

The different deposition gases of CH3F, CH2F2, and Si-based gas are separately added in SOC etch for sidewall protection. As a result, the SOC profile with Si-based gas is the best for vertical profile compared to CH3F and CH2F2 as Fig.3. The mechanism might be that the passivation film is much robust during the SOC etch.

Fig.2 Cross-section TEM profiles comparison of post SOC HM open by adding different passivation protections: (a) CH3F, (b) CH2F2 and (c) Si-based gas.

B. Bias power pulsing effect on SOC profile

Bias power puling is used in this paper for high etching selectivity of SOC to dielectric mask. As is shown in Fig.3, when the duty cycle (DC) of pulsing increased from 20% to 40% and 60 %, the profile of SOC becomes less bowing and more vertical. It tends to be more anisotropic etch with higher pulsing duty cycle, and vertical profile are obtained.

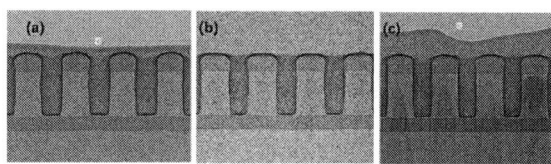

Fig.3 Cross-section TEM profiles comparison among the different pulsing duty cycle in SOC etch step: (a) 20% DC, (b) 40% DC, and (b) 60% DC.

C. Gas ratio effect on SOC profile

When O2 and SO2 are used to etch SOC layer, O2 plays as main etch gas. On the other hand, SO2 plays as sidewall passive gas because sulfur contained polymer formation in plasma. The O2 to SO2 gas ratio effect is illustrated in Fig.4. The sidewall is better protected and the bowing reduction is improved significantly with higher polymer gas ratio.

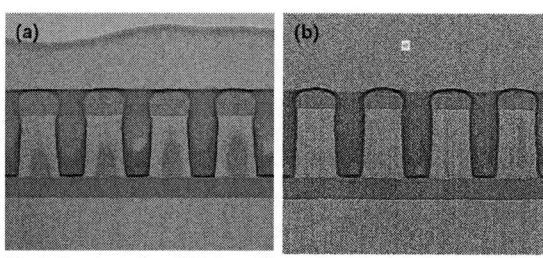

Fig.4 Cross-section TEM profiles comparison between (a) high O2/SO2 gas ratio and (b) low O2/SO2 gas ratio.

D. Temperature effect for SOC profile improvement

The profile of SOC becomes vertical at high temperature as Fig.5. It supports that the deposition species not only have low sticking probability but also conformably stick to the sidewall at high temperature.

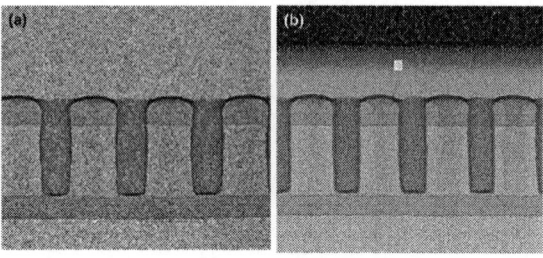

Fig.5 Cross-section TEM profiles comparison between (a) low temperature and (b) high temperature.

CONCLUSION

Naura etch tool high-density inductive coupled plasma (ICP) etcher has demonstrated excellent SOC profile control during the carbon hard mask open. These results may help to develop the related HM pattern transfer process for the device feature scaling in future.

REFERENCES

[1] Lee J K, Jang I Y, Lee S H, et al. Mechanism of sidewall necking and bowing in the plasma etching of high aspect-ratio contact holes[J]. Journal of The Electrochemical Society, 2010, 157(3): D142.

[2] Huang S, Shim S, Nam S K, et al. Pattern dependent profile distortion during plasma etching of high aspect ratio features in SiO2[J]. Journal of Vacuum Science & Technology A: Vacuum, Surfaces, and Films, 2020, 38(2): 023001.

[3] Shen M, Lill T, Hoang J, et al. Progress report on high aspect ratio patterning for memory devices[J]. Japanese Journal of Applied Physics, 2023.

[4] Iwase T, Kamaji Y, Kang S Y, et al. Progress and perspectives in dry processes for nanoscale feature fabrication: fine pattern transfer and high-aspect-ratio feature formation[J]. Japanese Journal of Applied Physics, 2019, 58(SE): SE0802.

THE OPTIMIZATION OF PIN HOLE DEFECT IN HIGH RESISTANCE PROCESS

Lunan Zhu*, Shan Huang, Xiaofeng Qu, Lei Sun, and Quanbo Li, Tianpeng Guan, Yu Zhang

[1] Shanghai Huali Integrated Circuit Corporation, Shanghai City, China

*Corresponding Author's Email: zhulunan@hlmc.cn

ABSTRACT

With the continuous shrinkage of the critical dimension and thinning of the film thickness in IC manufacture, the requirements of threshold voltage need advanced performance. However, limited by the structure and material, there is no further improvement of the Gate thickness. Nowadays, the high resistance process (HR) plays the role of separate-voltage which has been widely used to solve this problem. However, the existence of serious pin hole defect issue in traditional HR process leads to the underlying Co missing, which seriously impacts the device performance and limits application of HR process. In this paper, a method was proposed by reducing the SiN etching plasma bombardment capacity and adjusting the etching gas ratio based on dense SiN film, which can solve the issue of pin hole completely and further be applied in IC manufacture.

Keywords—High resistance etch; Pin hole; underlayer damage; defect scan

INTRODUCTION

In semiconductor process, the resistance structure formed by HR layer can divide voltage and limit current, so it is widely used in analog circuits. The traditional HR process adjusts resistance by ion doping of polysilicon (silicon oxide gate). With the development of integrated circuit technology, the line width CD (critical dimensions) continuously shrink, the leakage problem becomes increasingly prominent. In order to solve this problem, the introduction of high-K metal gate instead of silicon oxide gate. At the same time, it is impossible to use the traditional high resistance polysilicon to make HR. Therefore, high resistance TiN film for HR preparation was proposed.

However, New challenges arise in HR process. The defects of pin hole are easily appeared in HR etch open area, which would directly lead to underlayer damage, and further decrease the performance of device. Therefore, it is necessary to avoid those defects. In this paper, the influence factors of pin hole defects in HR etching process was analyzed in detail and the solution is given. The perfect morphology is obtained by adjusting gas ratio and plasma bombardment ability based on dense SiN film, which greatly beneficial to improve HR process.

EXPERIMENT

The HR dry etching experiments were performed by inductively coupled plasma (ICP) with Metal etcher tool from LAM Corporation. The morphology image was observed by scanning electron microscope (SEM) from Hitachi.

RESULT AND DISCUSSION

HR Etch Process Flow and Problems

The current HLMC HR etch process is shown in Figure 1. After HR photo process (Post HR PH), the composition of the experimental wafer film stack is Cap SiN\TiN \HM SiN\Barc\PR from bottom to top. At first, Barc is etched by CL_2 gas. C and F compositional gas flow is designed principally to etch SiN film with selectivity to TiN film. Then, residual PR and Barc were dry striped. Finally, it should be noted that the TiN film were removed by wet process.

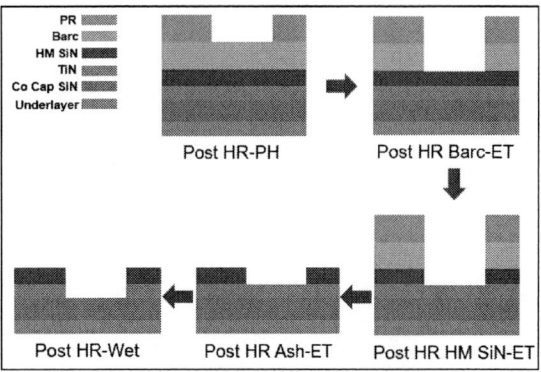

Figure 1: HR Etch Process flow

Figure 2: The pin hole defect in HR open area

However, after the HR process was completed, the serious pin hole defects were found in the HR etch open area, especially at the junction of the HR etch open area and the unopen area, as shown in Figure 2. It can cause damage to the underlayer, which will affect the performance of the device.

Causes Reasoning and Verification of Pin Hole Defects

In consideration of the pin hole defect and the process of HR etching, two hypotheses have been proposed. The first hypothesis is that the loose structure of SiN film leads to its poor blocking ability during etching. Let's analysis this idea in more detail as shown in Figure 3. Firstly, the holes of SiN film were transferred down to the TiN film under the etching bombardment of strong plasma CL_2 gas during barc etch. The next step of HM SiN etching, CxFy gas is used to further transfer the holes in the TiN film to Co Cap SiN. Finally, hydrofluoric acid wet etching removes TiN layers from open areas which leads to Co damaged at the bottom. On the other hand, the side wall is not straight in the etching process, the etching ions were sputter down along the side wall and caused the etching rate near the side wall to be faster, this phenomenon is called the micro trench effect. Therefore, there is a more serious pin hole at the junction of the HR etched open area and the unetched area. As figure 4 shows this.

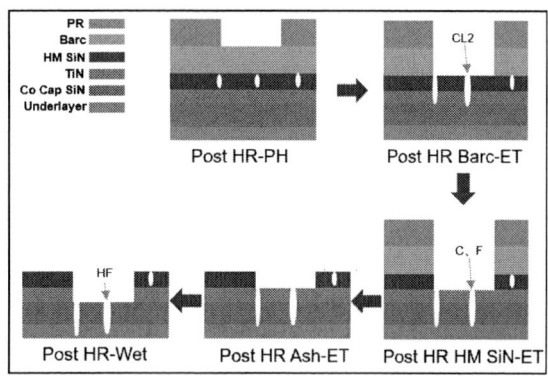

Figure 3: Hypothesis of pin hole defect

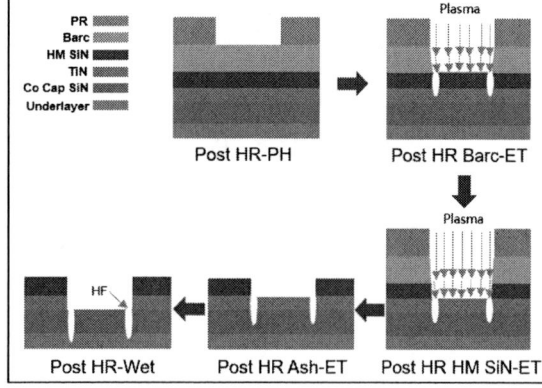

Figure 4: Hypothesis of pin hole defect

Solutions and Results for Improving Pin Hole Defects

According to the characteristics of HR-ET process conditions, the optimization is mainly carried out from three aspects. Firstly, it is increase the densification of HM SiN film to improve its etching resistance. Secondly, the holes can be lower transfer in HM SiN etching process by reducing the etching bombardment ability and time. Thirdly, the C/F gas ratio is increased to provide stronger polymer deposition for reduced the etching rate at the side wall corners. The experimental design as shown in table 1 based on the optimization scheme.

Table 1: Experiment design scheme

Item	1	2	3	4	5	6
Increase the density of HM SiN		V	V	V	V	V
Bais power-50%	V	V		V	V	V
Bais power-75%			V			
OE time-45%				V		
CF4/CHF3 5:1-> 2:1					V	
CF4/CHF3 5:1-> 5:7						V

Table 2: Experiment SEM image

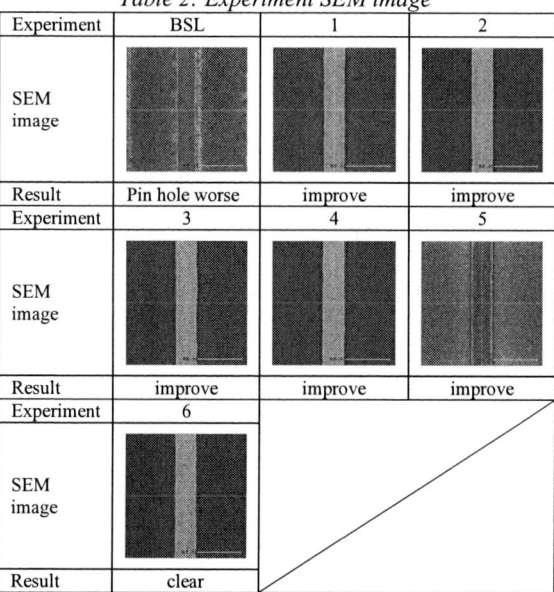

Experiment	BSL	1	2
SEM image			
Result	Pin hole worse	improve	improve
Experiment	3	4	5
SEM image			
Result	improve	improve	improve
Experiment	6		
SEM image			
Result	clear		

It can be seen that pin hole defects have been improved in all experiments through the analysis of experimental results and SEM graph data, The defect improvement is most obvious in experiment six, as shown in table 2. Therefore, it is necessary to integrate the above three optimization methods (such as: adjust the density of HM SiN, reduce the etching bombardment capacity, improve the C/F gas ratio) to solve the pin hole problem and establish a healthy HR etching process. At the same time, it should be noted that increasing the C/F gas ratio is one of the most effective regulation methods. However, it is increased the C/F ratio will bring more by-products, which will affect the Angle of the side wall and adversely affect the control of the device size. Therefore, the increase of the C/F ratio should be appropriately adjusted according to the process requirements.

CONCLUSION

In this paper, the solution of pin hole defects in high resistance etching is studied and analyzed in detail. We found the pin hole defects can be effectively solved by adjusting the density of SiN films, the etching bombardment ability and the etching gas C/F ratio. The SEM results was used to verify that the improved morphology meets the requirements. Also, This mechanism can be extended to pin hole defects in other etching processes.

ACKNOWLEDGEMENTS

The author sincerely thanks the supports and instructive suggestions from the colleagues in the dry etching module of Advanced Module Technology Development department and in the process integration of Technology Development II department.

REFERENCES

[1] H. Ekinci, N. M. S. Jahed, M. Soltani and B. Cui. IEEE Trans. Nanotechnol., vol 20, 2021, pp. 33-38.

[2] S. Senturia. *Proceedings of Transducers 2003*, Boston, June 8-12, 2003, pp. 10-15.

[3] R. Kaneko , 2009 International Workshop on Junction Technology, Kyoto, Japan, 2009, pp. 116-118.

[4] T. Tsuchiya, O. Tabata, J. Sakata and Y. Taga. *J. Microelectromech. Syst.*, vol. 7, 1998, pp. 106-113.

[5] YangKyu Choi, International Electron Devices Meeting. Technical Digest (Cat. No.01CH37224), Washington, DC, USA, 2001, pp. 19.1.1-19.1.4.

GROWTH AND REDUCTION OF TINY PARTICLE DEFECTS IN SELECTIVE SIGE EPITAXY S/D DEVICES

Zhiqiang Xiao[1,2]*, Cunzhe He[1]*, Dongliang Gao[1], Jiaxing Xiao[1], Haitao Yan[1], Zhenchao Sui[1], Xin Zhang[1].
[1]Semiconductor Manufacturing North China (Beijing) Corp.
Beijing, China
[2]School of Integrated Circuits, Tsinghua University, Beijing, China
*Corresponding Author's Email: ZhiQiang_Xiao@smics.com; Kevin_HCZ@smics.com

ABSTRACT

Selective SiGe epitaxy has been widely used as stressors in source/drain (S/D) regions of Metal–Oxide-Semiconductor Field Effect Transistor (MOSFET) to enhance hole mobility of channel. However, SiGe tiny particle defects in SRAM (Static Random Access Memory) area frequently results in yield loss in fab mass production. Tiny particles always grow both on top and sidewall of silicon nitride outside the poly gate. In this paper, the causes of tiny particles were analyzed and 5 effective solutions for defect reduction were provided..

INTRODUCTION

Tiny particle defects are the most common defects in SiGe epitaxy processes, which are always caused by poor deposition selectivity [1]. Tiny particle defects often appear in large flat pad regions and on the sidewall of silicon nitride outside polysilicon gate (Figure 1). The former generally does not have an impact on product yield, while the latter, commonly referred to as shoulder defects, does not kill yield when the particle number is small. However, when the quantity is sufficient, they can kill yield by failure mode of contact open or NiSi alloy missing [2,3]. Generally, the more advanced the technology, the lower the tolerance of particle defect quantity. For example, on the 28nm HKMG platform, particle defects exceeding 2000ea may have a yield loss, more than 400ea for 22nm.

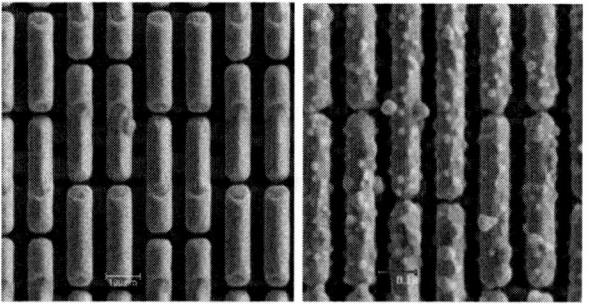

Figure1: Tiny particle defect after SiGe deposition

The number of tiny particle defects usually represents low or high deposition selectivity in epitaxy process. In general, epitaxy film is difficult to grow on silicon oxide or silicon nitride surfaces, and even if it grown, the growth rate is far lower than that on silicon surfaces. Generally, films grown on the surface of silicon oxide or silicon nitride are amorphous and easily removed by etching gas hydrogen chloride (HCL).

There are three main sources of tiny particle defects: 1. The polymer residues brought about by the previous etching steps can become nucleation points for particle defects; 2. The etching step may cause damage to the silicon nitride side walls, which can lead to the growth of particle defects in the exposed areas of the polysilicon gate [2]; 3. The etching/deposition selectivity of the epitaxial layer itself is insufficient, that is, the process recipes are not optimized enough. For sources 1 and 2, optimizing the etching process can be effectively addressed, while for sources 3, it can be solved through tuning of epitaxy process parameters. This work is focused on solving particle defects through epitaxy process optimization.

EXPERIMENTS

Bright field inspection (BFI) is used to check tiny particles defects after SiGe film deposition. KLA-Tencor tool is used to check thickness profile of SiGe films. TEM is used to check SiGe morphology and film thickness. Five different epitaxy methods are provided to remove tiny particle defects after S/D recess finished.

RESULTS AND DISCUSSION

Two methods are used to verify the sources of tiny particle (Figure 2): firstly, BFI defect scanning is performed after deposition of each layer, the defect map and review image show tiny particles appear only when L2 existed. In order to confirm this result, the second method is used. After each layer of thin film is deposited, a HCL flushing step is inserted, and the result is consistent with previous conclusion, only HCL flushing of post L2 deposition, can significantly remove the particle defects.

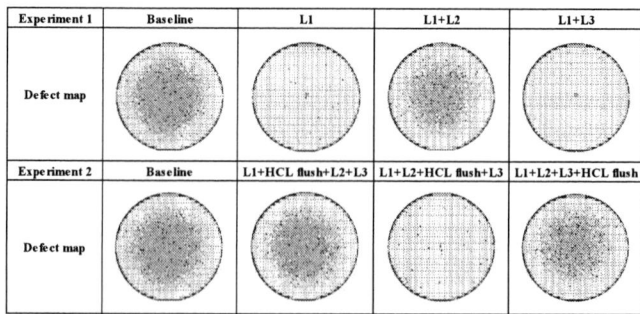

Figure 2: Experiment 1 (single layer check) and 2 (post HCL flushing) for confirm the tiny particle defects source

Post L1 deposition HCL flushing has no impact on defect count, while that of post L3 deposition can remove only a

979-8-3503-1101-3/23 $31.00 © 2023 IEEE

small number of particles. Therefore, particle defects occur during the growth of L2, and continuous deposition of L3 will make the particle become larger. Post L3 HCL flushing only removes the very small particles and has little effect on larger defects. Increasing the gas flow rate and extending the flushing time might completely remove tiny defects, but L3 or even L2 will be damaged. The etching/deposition ratio (E/D ratio) of L2 epitaxial layer is inappropriate and some optimization is needed to improve the defect condition.

High temperature L2

Different process temperature of L2 is used in this method. From Figure 3, it can be seen that increasing of L2 process temperature can effectively reduce the defect count, while decreasing that leads to an overload of defects, which contradicts our understanding. As the process temperature increases, the deposition rate of the thin film will become faster, making it easier to nucleation and grow on the surface of silicon oxide or silicon nitride, especially when the surface is relatively rough. We speculate that this may be related to the selective etching ability of HCL. At low temperatures, the etching ability of HCL is weak, and cannot remove particle defects completely during film growth. At high temperatures, the etching ability of HCL is strong, which can inhibit the growth of defects at the beginning.

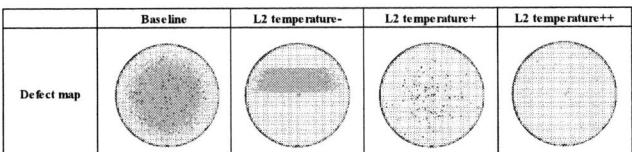

Figure 3: The impact of L2 temperature on SiGe defects

However, from the thickness curve of pattern wafer (Figure 4), we find that the thickness of wafer edge is lower than that of wafer center, indicating the temperature or the etching gas at the edge is lower than that at center area. But there are no obvious particles can be found at wafer edge, while most of the defects are located at wafer center, which is difficult to explain. We speculate that the etching ability of HCL at high temperatures dominates the removal of particle defects, while the distribution of HCL on the wafer surface dominates at low temperatures. In other words, HCL gas is mainly distributed at wafer edge, and can effectively remove defects located at this

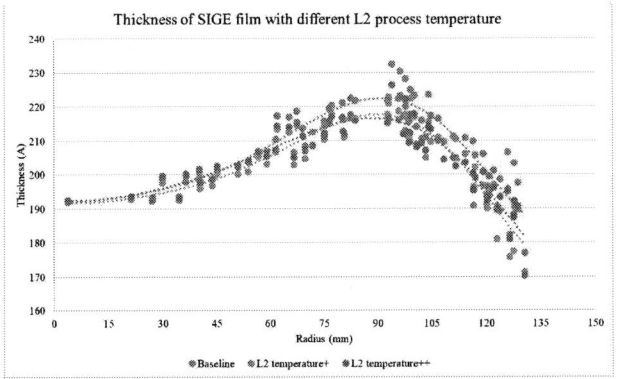

Figure 4: Thickness of SiGe film with different L2 temperature

area with sufficient amount at low temperatures, while only a small amount of HCL gas can reach wafer center, resulting in poor etching effect. At high temperatures, the etching ability of

HCL is greatly enhanced, and even a small amount of HCL in wafer center can remove most of the particle defects. And the increase of temperature enhances the thermal motion of HCL molecules, making it easier to arrive at the wafer center, enhancing the etching ability.

Carrier gas H2 flow rate enhancement

In the epitaxy process, the role of H2 is to mix with the precursor and inject it into reaction chamber. As mentioned earlier, HCL cannot easily reach the wafer central region at low temperatures, resulting in a large number of particle defects. Therefore, the second method is to increase the carrier gas flow rate, making it easier for HCL to enter the central region. Figure 5 shows that the number of defects in wafer center get much smaller when more carrier gas is involved. However, this method has side effects: increasing the carrier gas flow rate leads to a decrease in the concentration of precursors, which in turn leads to a reduction of film growth rate. Therefore, it is necessary to extend the time to compensate for the loss of thickness.

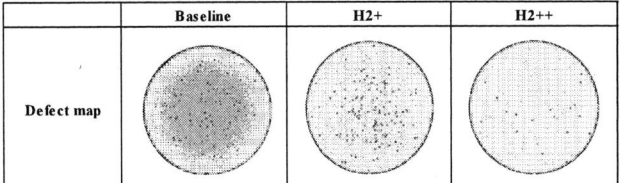

Figure 5: The impact of Carrier gas H2 flow rate on SiGe defects

Post deposition HCL flushing

After L2 deposition is completed and before L3 deposition, a HCL flushing step is introduced. This allows for the removal of particle defects immediately after they have grown. As can be seen from the previous Figure 1 Experiment 2, after adding HCL flushing, the number of particles drops to a level of less than 100ea. Compared with the active area (AA), poly gate and SIN side wall is much easier for HCL gas to arrive. Moreover, Si or SiGe film grown on SiO2 or SiN is amorphous, which is more easily etched than mono-crystalline Si or SiGe. According to the TEM images (Figure 6), this method has almost no damage to the S/D epitaxy films. The only potential problem is that it may increase the overall thermal budget due to adding a new step. This might have impact on the device performance and requires further study.

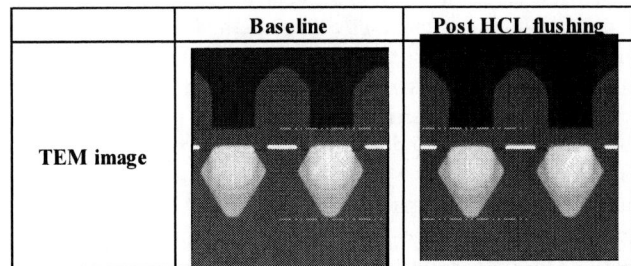

Figure 6: Post HCL flushing has no impact on pattern wafer thickness

Sacrificial layer introduction

The sacrificial layer is an evolutionary version of the previous method. From the TEM data mentioned above, it can be seen that HCL flushing does not damage the S/D epitaxy layer, but some micro-effects may not be detected by TEM. In order to reduce or avoid the impact of direct HCL flushing on the film, we introduce the sacrificial layer method. There are two schemes: Sacrificial SiGe, after L2, another layer of SiGe is grown, and then HCL flushing is used to remove both the particles and sacrificial layer; Sacrificial Si, before L3 deposition, a silicon film with several nanometers is first grown, and then both particles and sacrificial layer are removed before the real L3 film grows. The key of this method is to control the thickness of sacrificial layer and the etching conditions of HCL flushing. From the results in Figure 7, both methods could effectively remove particle defects. The potential risk remains that the thermal budget might cause a shift in electrical performance.

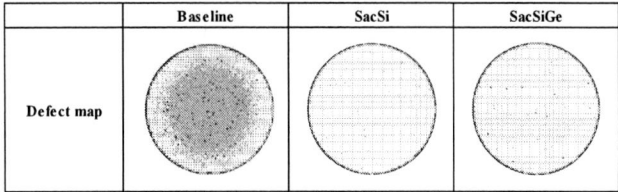

Figure 7: Sacrificial layer remove tiny particles effectively

E/D ratio optimization

Tiny particles are coming from the bad selectivity of film deposition and enhancement of the selectivity is the most direct method: reducing the growth rate and increasing the etch rate by tuning the precursor flow rate. Figure 8 shows when E/D ratio gets larger, the number of tiny particles will be reduced, but the film thickness will also get smaller, and much more time will be taking to finish process. And more HCL usage might impact the micro-loading of the epitaxy deposition. Combination with other previous methods can be a good solution.

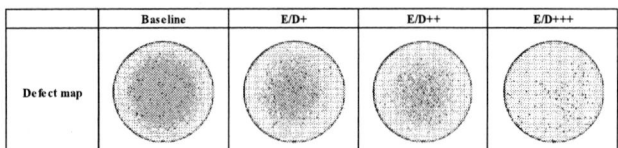

Figure 8 : E/D ratio impact on tiny particle defects

CONCLUSION

In this paper, five different methods are used to remove the SiGe tiny particle defects formed during L2 deposition. Among these methods, L2 high temperature, post L2 HCL flushing, H2 carrier gas enhancement and sacrificial film deposition are the most effective methods. Post L2 HCL flushing and sacrificial film deposition are just two remedial measures after defects have already formed, while L2 high temperature and H2 carrier gas enhancement are two methods for inhibiting the formation of defects during film growth. The introduction of new steps in the first two methods leads to longer process times and potential risk of WAT shift, which is not conducive to large-scale mass production of integrated circuits. In comparison, the latter two methods do not cause an increase in process time. From the experiment result, HCL etch ability and gas distribution might be the two dominant factors for tiny particle formation. Which method is better still depending on the results of electrical properties and yield test, which would be carefully studied in the future.

REFERENCES

[1] H. Tu, Y. He, Y. He, J. Liu, Jin, G. Cai, L. Yu and Y. Huang. "Investigation of embedded silicon germanium typical defect solution for advanced CMOS process," China Semiconductor Technology International Conference (CSTIC). IEEE. pp. 1-3, 2016.

[2] Q. Yan, C. Kun, H. Chen, L. Yin and W. Kai. "Defect Principle and Improvement of 28nm Germanium Silicon Epitaxial Growth Process," China Semiconductor Technology International Conference (CSTIC). IEEE. pp. 1-2. 2021.

[3] Q. Huang, Y. Chen, J. Hong, Q. Yan, J. Tan and H. Zhou. "SiGe Layers Defect of 28nm Node PMOSFETS In Advanced CMOS Technology," China Semiconductor Technology International Conference (CSTIC). IEEE. 2019. pp. 1-3.

TECHNICAL CHALLENGES IN MRAM FABRICATION

Y. H. Wang[1], X. L. Yang[1], Y. Tao[1], Y. H. Sun[1], Q. J. Guo[1], F. T. Meng[1], and G. C. Han[1,2]*

[1]Zhejiang Chituo Technology Co., Ltd, Hangzhou 311300, China

[2]Key Lab. of Spintronics Mater, Devices and Systems of Zhejiang Prov.，Hangzhou 311300, China

*Corresponding Author's Email: hguchang@hotmail.com

ABSTRACT

We have developed integration process for manufacturing spin-transfer torque (STT)-magnetic random access memory (MRAM). We have realized volume production of 64Mb chips and are ready to provide engineering samples with a capacity of up to 256Mb. In this paper, fundamental challenges in current CoFeB-and MgO-based STT-MRAM with perpendicular magnetic anisotropy (PMA) are highlighted in terms of materials and process aspects. The technical challenges in the key processes including chemical mechanical polishing (CMP), etching, and thin film deposition in MRAM fabrication are discussed in more details. Challenges and possible solutions related to device performance optimizations, scaling and reliability are also addressed.

INTRODUCTION

Spin torque transfer (STT) magnetic random access memory (MRAM) is one of the promising next generation memory technologies due to its nonvolatility, fast write-time, and easy integration with complementary metal-oxide-semiconductor (CMOS) circuitry[1]. The key element of a STT MRAM cell is a perpendicular magnetic tunnel junction (pMTJ) consisting of a free layer (FL) and a reference layer (RL) separated by an MgO barrier layer, which can provide high tunnel magnetoresistance (TMR) ratio, high thermal stability，high STT efficiency and better scalability. To realize the practical applications of STT-MRAM, MTJ elements must meet a set of performance requirements on switching current (Ic), read/write speed, write endurance, thermal stability, TMR ratio, and resistance-area (RA) product, as well as scalability. The challenge is that the improvement in these performance parameters cannot be realized independently. They need to be optimized with tradeoff for different applications. In addition, they are significantly affected by MTJ material and geometric parameters, and there are almost always some degree of parameter variations in the actual fabrication process. In this paper, we analyze the impact of manufacturing process on the performances of STT-MRAM chips and challenges they have to face for mass production.

CHALLENGES IN BIT CELL DESIGN

High density MRAM uses the 1-MTJ/1-transistor (T) memory cell structure with CMOS transistor as the access device. Fig.1 shows a typical structure of a STT MRAM

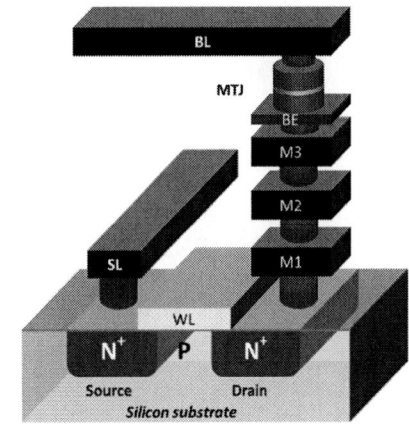

Figure 1. A typical MRAM cell structure

cell in this architecture. Generally, MTJ device is deposited at higher metal layers, usually between layers M4 and M5 or M3 and M4, as shown in Fig.1. Although the CMOS fabrication process might introduce defects which could lead to resistive open in the metal lines connecting the transistor source, gate and drain, there would not be an issue for a fab with mature CMOS fabrication process of the technique node. The challenge is mainly on performance power area (PPA) optimization for CMOS architecture design to provide sufficient driving current for MTJ switching while keep the low cell size for high density and low leakage for low power consumption. In addition, as MTJ device needs to switch between two states by STT currents, the CMOS should be able to provide bidirectional currents. As the CMOS is connected in series to MTJ device, its parasitic resistance will reduce the effective TMR of MTJ and decreases the read margins, the low working resistance is required for CMOS. In our MRAM products, proper MRAM bit-cell, array and chip structure is optimized through Design Technology Co-Optimization (DTCO). Transistor width and length, MTJ size and RA, process-induced device parameter variations and switching stochasticity are all included in spice model for bit-cell design and simulation. Then bit-cell numbers in one common bitline (BL) and parasitic resistance compensation are considered in MTJ array design and post-layout simulations for improved read margin. Furthermore, temperature compensation, sensing amplifier offset reduction and power supply optimization are applied in overall MRAM structure design to achieve excellent PPA goals.

979-8-3503-1101-3/23 $31.00 © 2023 IEEE

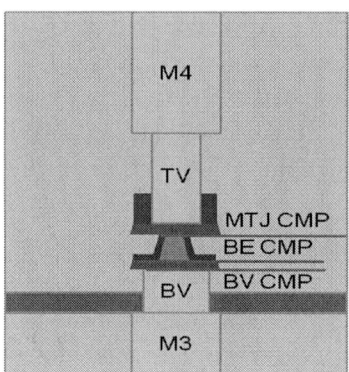

Fig.2. Critical CMP processes in MRAM fabrication

CHALLENGE IN CMP PROCESS

CMP is widely used for planarization in the manufacturing process of integrated circuit (IC). As shown in Fig.2, after the CMOS logic is fabricated, the wafer should be prepared for MTJ stack deposition by CMP process. It is risky to build the MTJ device on copper as the copper voids and hillocks can have severe effects on MTJ performance. As a result, MTJ of the conventional STT-MRAM is located off the axis of bottom electrode (BE) contact, resulting in a large cell size. Fig.2 shows an on-axis scheme which is usually used in today's MRAM technology. CMP is required to obtain ultra-smooth BE surface with a roughness of less than 0.2nm to apply the on-axis scheme, since the thickness of MgO tunnel barrier needs to be less than 1 nm for STT switching. In this on-axis scheme, two critical CMP processes are required for bottom via (BV) and bottom electrode (BE). In these CMP processes, both under-polishing and over-polishing of the surface can introduce defects. As indicated in Fig.2, BV is used to connect metal line (M3) to BE. CMP is employed to remove bulk Cu, which is deposited by an electroplating process, Cu barrier and dielectrics. Challenge in BV CMP process is to control a small dishing (<3nm) while obtain smooth Cu surface. Large dishing in BV induced by over-polishing causes Cu diffusion while under-polishing-induced rough surface affects MTJ performances. It is even more challenging for a wafer with both dense and isolate pattern areas. By optimizing CMP process and using special patterning design, we have successfully achieved a small dishing of 2-3nm, which can meet the requirement of the integration process without risk of Cu diffuse. To achieve ultra-smooth surface for MTJ stack, BE has to be deposited and polished on BV as shown in Fig.2. Different from conventional damascene barrier layer polishing process, BE CMP requires an opposite polishing selectivity for which there is no mature slurry in the market. For BE CMP, under-polishing may cause issues such as poor MTJ performances (low endurance, low TMR and so on) or even electrical shorts in some cases, while over-polishing could result in Cu diffusion or

residual slurry particles that are left behind. In order to overcome this issue, except the selection of suitable BE materials, more than one CMP process has been employed to control the BE thickness and the roughness. After optimized BE CMP process, the roughness of the BE is well below 0.2nm without any risk of Cu diffusion. The other key CMP process is the MTJ CMP which is used to open MTJ device for contact to the top electrode. In this CMP process, at least three different materials are involved: oxide, nitride and hard mask (HM). CMP must remove the nitride on the top of MTJ completely without over-polishing HM to affect or damage MTJ performances. Challenges in this CMP process are as follows: 1), MTJ pattern density is extremely low (less than 10%) in the wafer, it is difficult to detect HM signal and properly stop on it without much over-polishing; 2) as a suitable HM thickness is required by patterning process, the remaining HM after MTJ etching is very thin, there is no much window for CMP; 3) MTJ CMP relates to three different materials, as the polishing rate is different for different materials, the flattening of the whole wafer is difficult to control. In addition, a heavy oxidation of HM may be introduced in this MTJ CMP process as shown in Fig.3. A normal pre-sputtering process is not sufficient to clear the oxide on the top of MTJ, resulting in large serial resistance or even open. This issue was solved with a proper patterning design to eliminate the so-called antenna effect.

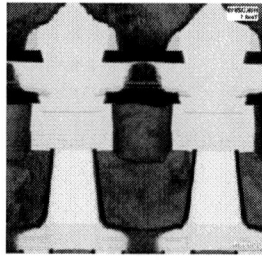

Fig.3. SEM picture showing HM oxidation induced by CMP

CHALLENGES IN MTJ ETCH PROCESS

MTJ etch is the fabrication step which is the most difficult to control. The target of such a process is to achieve a low short bit error rate (BER), high uniformity and low damage. Improper etching causes not only shorts due to sidewall re-deposition, but also degradation in MTJ performances due to damage and large variations in MTJ sizes. Challenges in MTJ etch come from several aspects. As MTJ materials are generally nonvolatile in normal process temperatures, an ion beam etching (IBE) is widely used for MTJ etching. This physical etching process inevitably induces re-deposition onto the etched surface, resulting in short circuit of MTJ. Therefore, a post-etch cleaning process has to be implemented to clear away the redeposited metals. A large cleaning angle is generally required to remove the metal deposited on the sidewall. However, as the density increases, the pitch between MTJs becomes smaller and smaller, limiting cleaning angle due to shadowing effect. In order to avoid the limitation by a small pitch, a special process flow with large over etch into oxide is developed to remove the redeposited metals. This process needs to use a bottom via with small and high aspect ratio, which is also a challenge. Further challenge comes from the fact that it is very difficult to find the redeposited metals on the sidewall of MgO even using Transmission Electron Microscope (TEM). On the other hand, even a few metallic atoms deposited on the MTJ sidewall can cause MTJ reliability issue, which is not allowed in MRAM product. An oxygen showering post-treatment process has been used to cure the etching damage[2]. However, excessive oxygen may also cause the oxidation of free layer, resulting in poor performance of MTJ device. Another challenge in IBE process is the sidewall damage, which may be produced by oxidation, ion implantation and stress introduced in MTJ etching and subsequent deposition of the protect layer. In addition, IBE process may also result in uniformity issues including variations in resistance, switching current/voltage, TMR and coercivity (H_c). Although MTJ pillar diameter (CD) is mainly determined by lithography and subsequent HM etching. Large resistance variation can be introduced by MTJ etching process. These variations are found to be related to etching angle, etching speed, HM profile, HM material structure, HM and MTJ thickness. All these parameters affect the short BER and MTJ performances (TMR, H_c and so on). It is very difficult to find an etching window to obtain all the best in these parameters. Proper compromise has to be exercised to achieve the good overall performances of MRAM product. Fig.4 shows an example of etching process optimization. By carefully controlling each step of MTJ etching to reduce etching damage, MTJ performance is much enhanced with TMR increase from 160% to 182%，and H_c from 1490 Oe to 2280 Oe while keeping the short BER to a level of less than 1ppm and an acceptable R variation.

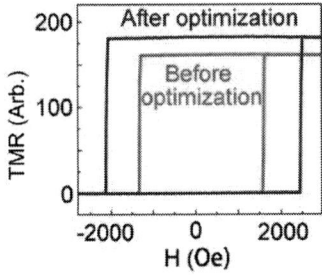

Fig.4. An example of MTJ etch process optimization

Fig.5 indicates the process development of MTJ

etching in the early stage. As shown in Fig.5, MTJ yield without any short increases from about 20% to higher than 99.9% (typically no short MTJ is detected in about 1000 MTJs). At the meantime, the average value of TMR increases from 65% of TMR measured in thin film level by Current in Plane Technology(CIPT)to 92%, achieving nearly no loss of TMR.

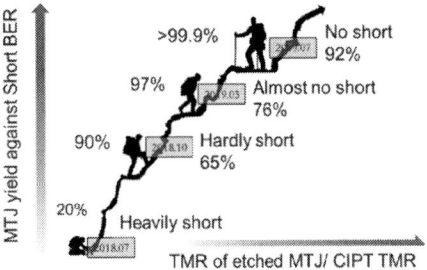

Fig.5. Process development of MTJ etching

CHALLENGES IN THIN FILM DEPOSITION

MTJ stack is the key for MRAM performance. MTJ stack consists of more than 30 ultrathin layers with atomic thickness. It is also a common practice to use so-called a dusting layer which is not even a continuous layer with nominal thickness less than one atomic layer to improve MTJ performances. The main issues that can arise during MTJ thin film deposition are: material intermixing, rough surface layers, and partial oxide of FL and RL at the interface with MgO barrier. These issues can lead to a wide variation in the cell resistance and switching current.

As shown in Fig.6, based on their functioning in the MTJ material structure, a MTJ is composed of seed layer, synthetic antiferromagnet (SAF), RL, MgO barrier, FL and capping layer, where each functioning layer consists of many ultrathin layers with different materials and thickness. The seed layer is used to induce strong fcc (111) texture for the (111) growth of SAF with sufficient PMA for pinning the magnetization of the RL. Other than serving as the seed for the textured growth of SAF, the seed layer needs to have extremely smooth surface. A rough surface can result in not only a low PMA of SAF , RL and FL, but also degradation in the breakdown. The rough surface is prone to element diffusion at high

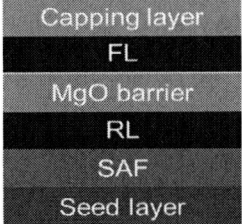

Fig.6. Main functioning structure of a MTJ

temperature (400°C) annealing process. Therefore, the seed material and deposition condition need to be carefully selected and optimized. It is common practice to use multilayered seed material structure with reduced thickness to prevent material diffusion at high temperature and provide the smooth surface.

The SAF structure is generally composed of two Co/Pt multilayers separated by an exchange coupling layer. It is essential for the SAF structure to provide strong PMA to fix the RL magnetization without increasing the roughness. It is a big challenge to control the interface roughness as each layer is of a thickness in order of atomic monolayer. The material intermixing is hardly eliminated, thus increasing the roughness. In addition, PMA of SAF comes from the interfaces between Co and Pt (or other materials), higher PMA requires more interfaces, which further increases the roughness. Thicker SAF structure is a burden for MTJ etching in current state of the art bottom pinned MTJ structure. As a result, careful process control needs to be implemented to obtain SAF with high PMA and low thickness.

Reference layer is one of the major magnetic layers with its magnetization fixed in one direction. One of the challenges for RL is that the underneath SAF is typically fcc crystalline structure, while RL should form bcc crystalline structure to obtain high TMR ratio. As a result, a transition layer is inserted between magnetic RL and SAF. This transition layer should be thin enough to avoid magnetic decoupling between RL and SAF while thick enough to function as structural transition from fcc to bcc.

MgO barrier is the key to achieve good performance of a MTJ device. As TMR is mainly determined by electron tunneling of $\Delta 1$ Bloch states [3], any conductance from other electron states will lower TMR. Therefore, the MgO barrier with perfect (001) crystalline structure and without any defect and impurity is desired. For MgO deposition, one big challenge comes from the imperfections at the two interface with RL and FL. Excessive oxygen may oxidize magnetic materials of FL and RL while insufficient oxygen results in the O vacancies, degrading the tunneling effect. In addition, as technology scales down, RA of MTJ has to be reduced to provide high endurance. Low RA needs ultrathin MgO layer, which brings along further challenge in the uniformity control. Finally, MTJ should be thermally stable in the annealing with a temperature up to 400 °C. Any materials diffusing into the MgO barrier cause the degradation of MTJ performances such as TMR and VBD.

Free layer determines the major functions of MRAM. Challenges in FL optimization come from that the performance parameters are heavily correlated and need to be optimized with tradeoff for different applications. For example, high PMA which is required for data retention may increase switching current and thus reduce endurance. As a result, PMA should be tuned through process and material engineering to meet the data retention requirement with the lowest Ic for writing endurance. PMA comes from the interfaces at FL/ MgO barrier and FL/Capping. The interfacial PMA is originated from the orbital hybridization of O2p-Fe3d and O2p-Co3d at the interfaces [4]. As a result, the performances of pMTJ is not only sensitive to the material and crystalline structure of FL, RL and MgO barrier, but also their interfaces, defects and impurity. For example, boron is required to achieve good crystalline structure. However, excessive boron at the interface and FL may also depress PMA and increase write error rate, respectively Further challenge in FL comes from the scaling requirement for high density application. As the data retention scales down as CD^2, FL with new materials and structure has to be developed. For example, more than two FL/MgO interfaces [5] are used for CD less than 30nm. When CD is further scaled down to below 10nm, shape anisotropy is probably needed to provide sufficient data retention, which brings along big challenges in thin film deposition and patterning process [6].

In order to enhance PMA and STT efficiency, capping layer consists of several different material layers in the state of the art MTJ. In this structure, the first capping layer deposited directly on FL is MgO. This MgO capping layer needs to be thick enough to provide good interface with FL and block the heavy metals from diffusion into FL to eliminate the pumping effect and thus enhance STT efficiency. But thick MgO increases MTJ resistance and decreases TMR. Therefore, a properly designed capping layer structure and careful deposition process control are required to enhance STT efficiency (thus endurance), TMR, and data retention.

SUMMARY

In conclusion, main challenges in MRAM chip fabrication are addressed in bit-cell structure design and manufacturing process. By solving all critical issues faced in the integration process, we have realized volume production of 64Mb chips and are ready to provide engineering samples with a capacity of up to 256Mb.

ACKNOWLEDGE

This work was supported by the National Key Research and Development Program of China (Grant No. 2020AAA009000).

REFERENCES

[1] J.C. Slonczewski, U.S.Patent 5,695,864 (Dec.9, 1997).

[2] J. Jeong and T. Endoh, Jpn. J. Appl. Phys. 56, 04CE09 (2017)

[3] W. H. Butler et al, Phys. Rev. B 63, 054416(2001)

[4] X. Yang et al, Phys. Rev. B, 84, 054401 (2011).

[5] K. Nishioka et al. VLSI 2019,T11-4

[6] K. Watanabe et al., Nat. Commun. 9, 663, 2018.

SOME KEY MODIFICATIONS OF THEORY REQUIRED TO UNDERSTAND THE LEAKAGE CURRENT MECHANISMS FOR FERROELECTRIC HfZrO CAPACITORS USED IN MICROELECTRONICS

W.S. Lau

Nanyang Technological University (Retired), School of EEE, Singapore 639798
E-mail: lauwaishing@yahoo.com.sg

ABSTRACT

Previously in 2020, the author proposed a new unified theory regarding the image force dielectric constant for Schottky emission leakage current in high-k dielectric. In this paper, the author will try to point out that this theory can be applied to ferroelectric hafnium zirconium oxide (HfZrO) capacitor structures.

INTRODUCTION

Capacitor structures based on high-k dielectric materials deposited by atomic layer deposition (ALD) have been used in integrated circuit (IC) technology. Recently, a special kind of high-k dielectric material known as ferroelectric materials have attracted worldwide interest. Hafnium zirconium oxide (HfZrO) is one of that family of materials. Ferroelectric material can be used for memory applications [1]. Recently, ferroelectric material has been demonstrated to provide negative capacitance for gate dielectric applications [2], resulting in significant improvement of the subthreshold swing of MOS transistors.

However, the leakage current vs. voltage (I-V) curve of ferroelectric HfZrO capacitor structures is still not well understood. In this paper, the author will try to explain the deeper reasons why it may be difficult to identify the correct leakage current mechanism.

THEORY

Scientists have been trying to use the theories of Schottky emission (SE) and Poole-Frenkel (P-F) emission theories to explain the leakage current of MIM capacitors. In the equations for SE and P-F, there is a parameter known as the "image force dielectric constant" ε_{if}. The SE image force dielectric constant ε_{if_SE} is usually considered to be equal to the PF image force dielectric constant ε_{if_PF}.

The current dominant theory (Theory X1) regarding the image force dielectric constant is that $\varepsilon_{if} = n^2$, where n is the refractive index of the insulator measured in the visible light range. The author would like to point out that Theory X1 is an imperfect theory. The author would like to point

out that there exists a theory (Theory X0) older than Theory X1; for example, in 1953, Krömer [3] pointed out that that $\varepsilon_{if} = 1$ for germanium. In 1964, Sze et al. [4] pointed out that that $\varepsilon_{if} = n^2$ for silicon according to their internal photoemission experiment for a metal film on silicon structure. After 1964, Theory X1 became the dominant theory whereas Theory X0 became forgotten.

Many scientists believe that the image force dielectric constant $\varepsilon_{if} = n^2$ for high-k dielectric materials and used this assumption to distinguish whether the leakage current mechanism is SE or PF. The author would like to name this approach as Theory and Practice X. Previously in 2020, the author would like to unify both Theory X0 and Theory X1 into a more generalized theory (Theory L) [5]. In this new Theory L, Theory X0 and Theory X1 are not mutually exclusive; instead, they are 2 parts of the same unified theory. Theory L includes (a) $\varepsilon_{if} = n^2$ approximately, (b) $n^2 > \varepsilon_{if} > 1$ and (c) $\varepsilon_{if} = 1$ approximately. For a high-k dielectric, the refractive index is usually about 2; Theory L points out that ε_{if} can have a value from slightly bigger than 4 to slightly smaller than 1. Fig. 1 shows the dielectric constant as a function of frequency. At the upper frequency range, there is a frequency range A with $\varepsilon = n^2$ and there is another frequency range B with $\varepsilon_{if} = 1$. Theory L just means the image force dielectric constant can correspond to the frequency range A + B. Internal photoemission spectroscopy of a metal film on ferroelectric HfZrO sometimes shows up experimental results that the image force dielectric constant is about 1. This is not acceptable for the old Theory X1 but acceptable for the new Theory L.

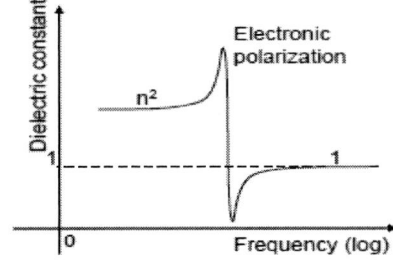

Fig. 1 The dielectric constant as a function of frequency at the higher frequency range showing that the image force dielectric constant can be smaller than 1.

After adopting the new Theory L, the method to distinguish SE and P-F has to be modified [5]. The author noticed that this can be done for MIM capacitors by examining the 2 curves of J vs $E^{1/2}$ plotted together for both polarities of bias voltage in the same figure instead of just examining 1 curve of J vs $E^{1/2}$ for only one polarity of bias voltage. The only key assumption is that $\varepsilon_{if\_SE\_top} = \varepsilon_{if\_SE\_bottom} = \varepsilon_{if\_PF}$. Theory and Practice X will not be used. For example, two different but parallel lines imply SE for both polarities.

EXPERIMENTAL SUPPORT

As discussed above, internal photoemission (IPE) is an important experimental technique to study the interface between a high-k dielectric and a metal or semiconductor [6]. The image force dielectric constant can be measured by IPE. The IPE barrier height can be measured as a function of the applied electric field. The IPE barrier height plotted as a function of the square root of the applied electric field yields a slope which can be used to calculate the image force dielectric constant. The author has analyzed existing experimental data published in the literature on HfZrO. For examples, Jenkins et al. presented their IPE results on HfZrO in 2021 [7]. They measured the IPE barrier as a function of the applied electric field but did not calculate the image force dielectric constant, as shown in Fig. 2. According to the supporting information from Jenkins et al. [8], only a minority of their experimental results are consistent with Theory X1. There exist some experimental data which show up some sort of "screening effect"; the IPE barrier height is insensitive to the change of the applied electric field. If the experiment data showing up "screening effect" are ignored, 2 cases can be observed: (a) $\varepsilon_{if} = n^2$ approximately, and (b) $\varepsilon_{if} = 1$ approximately. Case (a) is Theory X1 while Case (b) is actually Theory X0.

In 2015, Park et al. analyzed the leakage current mechanism of their ferroelectric HfZrO MIM capacitors [9]. They sticked to the Theory and Practice X and identified the leakage current mechanism as P-F. In 2020, Chen et al. analyzed the leakage current mechanism of their ferroelectric HfZrO MIM capacitors [10]. By ignoring Theory and Practice X without informing the readers, they identified the leakage current mechanism as SE. The author managed to extract data from a figure in Chen et al. and re-analyzed their data; the author found that the image force dielectric constant was about 1.3 for positive bias and about 1.5 for negative bias. If the Theory and Practice X is correct, the image force dielectric

constant should be about 4. In this paper, the author would like to point out that the experimental results of Park et al. can be considered as SE instead of P-F; the key point is the adoption of the author's new theory L. The Theory and Practice X should not be taken too seriously. If the leakage current was due to P-F as suggested by Park et al., the I-V characteristics should be symmetrical. In reality, the I-V characteristics were asymmetrical experimentally. Park et al. pointed out that the image force dielectric constant was found to be in the range of 3.8 -5.9 if the leakage current was due to P-F. The author suggests that the leakage current can be interpreted as SE with the image force dielectric constant in the range of 3.8/4 -5.9/4, i.e. 0.95-1.475.

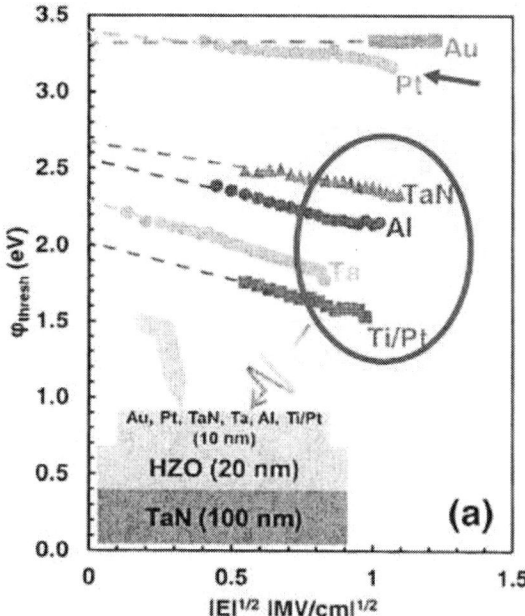

Fig. 2 The barrier height measured by internal photoemission as a function of the square root of the electric field for various top electrode materials on hafnium zirconium oxide (HfZrO) according to Jenkins et al. 2021 [7]. The image force dielectric constant calculated from the slope is about 1 for TaN, Al, Ta and Ti/Pt. (Red ellipse.) The image force dielectric constant calculated from the slope is consistent with Theory and Practice X only for Pt. (Blue arrow.) For Au, the barrier height appeared to be independent of the applied voltage.

CONCLUSION

Many scientists in the world had been influenced by Theory and Practice X. This approach is not totally wrong; it is not totally correct. For example, in 2011, Hikita et al. obtained IPE experimental data that $\varepsilon_{if} = 1$ approximately in their Au/Nb: $SrTiO_3$ Schottky junctions [11]. (Note: $SrTiO_3$ is also a ferroelectric material similar to HfZrO.) They tried to think out some theory to explain away their experimental observation. According to the author's new theory L, it is not necessary to think out some new theory to explain away the IPE results from Hikita et al. Moreover in 1994, Matsuhashi and Nishikawa [12] made a detailed report on $M/Ta_2O_5/n^+$-poly-Si capacitors (M= metal) with various metals as top electrode. The leakage current for negative voltage applied to the top electrode strongly depended on the metal used for the top electrode and the temperature dependence of the leakage current for negative voltage applied to the top electrode was strong; the leakage current mechanism seemed to be Schottky emission. However, the image force dielectric constant obtained experimentally was only 0.3-0.8. This experimental observation appeared to make Matsuhashi and Nishikawa puzzled, resulting in no conclusion regarding the leakage current mechanism. The author would like to point out that this problem can be easily solved by adopting the author's Theory L such that the measured image force dielectric constant with a value of 0.3-0.8 is actually not a problem at all.

Some scientists consider the image force dielectric constant with a value of about 1 is unphysical. This may not be the case. One possibility is that electrons can move much faster than expected such that the high-k dielectric cannot polarize fast enough. There are other possibilities. We can easily imagine that if there is a nano-sized hole in the bulk of the high-k dielectric, then the dielectric constant at the nano-sized hole will be about 1. There is no scattering in the nano-sized hole and electrons can move through it much more rapidly. It is commonly known that there are oxygen vacancies; there is a possibility that a nano-sized hole can be created because of the presence of oxygen vacancies. Similarly, we can easily imagine that if there is a nano-sized hole at the interface of metal and the high-k dielectric, then the dielectric constant at the nano-sized hole will be about 1. Another possibility can also be thought out. The electric field in the high-k dielectric may slightly penetrate into the metal such that a depletion layer exists in the metal at the interface. When a metal is depleted of free electrons, the dielectric constant is about 1. If the image force barrier lowering (Schottky effect) occurs in the metal depleted of free electrons out of the high-k dielectric, the image force dielectric constant can have a value about 1. There is much less scattering in the metal depleted of free electrons and electrons can move through it much more rapidly.

Previously in 2022, the author pointed out there are two special cases for a semiconductor Schottky diode [13]: (1) the image force dielectric constant can have a value equal to optical dielectric constant and (2)) the image force dielectric constant can have a value equal to 1 approximately. The same idea can also be applied to a high-k dielectric like HfZrO.

REFERENCES

[1] H.-G. Kim, D.-H. Hong, J.-H. Yoo and H.-C. Lee, "Effect of process temperature on density and electrical characteristics of $Hf_{0.5}Zr_{0.5}O_2$ thin films prepared by plasma enhanced atomic layer deposition," *Nanomaterials*, Vol. 12, no. 3, article no. 548, 2022.

[2] S. Salahuddin and S. Datta, "Use of negative capacitance to provide voltage amplification for low power nanoscale devices," *Nano Letters*, vol. 8, no. 2 (2008), pp. 405-410.

[3] H. Krömer, "Zur theorie des germaniumgleichrichters und des transistors," *Z. Physik*, vol. 134 (1953), pp. 435-449. (In German)

[4] S.M. Sze, C.R. Crowell and D. Kahng, "Photoelectric determination of the image force dielectric constant for hot electrons in Schottky diodes," *J. Appl. Phys.*, vol. 35, no. 8 (Aug. 1964), pp. 2534-2536.

[5] W.S. Lau, "Some key modifications of theory required to understand the leakage current mechanisms for MIM capacitors used in DRAM technology," *CSTIC 2020 (China Semiconductor Technology International Conference, Shanghai, 2020, IEEE)*, pp. 1-3, 2020.

[6] V.V. Afanas'ev and A. Stesmans, "Internal photoemission at interfaces of high-k insulators with semiconductors and metals," *J. Appl. Phys.*, vol. 102, no. 8 (2007), article no. 081301.

[7] M.A. Jenkins, K.E.K. Holden, S.W. Smith, M.T. Brumbach, M.D. Henry, C. Weiland, J.C. Woicik, S.T. Jaszewski, J.F. Ihlefeld and J.F. Conley Jr., "Determination of hafnium zirconium oxide interfacial band alignments using internal photoemission spectroscopy and X-ray photoelectron spectroscopy," *ACS Appl. Mater. Interfaces*, vol. 13 (2021), pp. 14634-14643.

[8] M.A. Jenkins, K.E.K. Holden, S.W. Smith, M.T. Brumbach, M.D. Henry, C. Weiland, J.C. Woicik, S.T. Jaszewski, J.F. Ihlefeld and J.F. Conley Jr., "Determination of hafnium zirconium oxide interfacial band alignments using internal photoemission spectroscopy and X-ray photoelectron spectroscopy," *ACS Appl. Mater. Interfaces*, vol. 13 (2021), supporting information.

[9] M.H. Park, H.J. Kim, Y. J. Kim, T. Moon, K.D. Kim, Y.H. Lee, S.D.Hyun and C.S. hwang, "Study on the internal field and conduction mechanism of atomic layer deposited ferroelectric $Hf_{0.5}Zr_{0.5}O_2$ thin films," *Journal of Materials Chemistry C*, vol. 3 (2015), pp. 6291-6300.

[10] W.-C. Chen, Y.-C. Zhang, P.-H. Chen, Y.-T. Tseng, C.-H. Wu, C.-C. Yang, P.-Y. Wu, Y.-F. Tan, S.-K. Lin, W.-C. Huang, H.-C. Huang, T.-M. Tsai and T.-C. Chang, "Investigation on the current conduction mechanism of $HfZrO_x$ ferroelectric memory," *J. Phys. D*, vol. 53 (2020), article no. 445110.

[11] Y. Hikita, M. Kawamura, C. Bell and H.Y. Heang, "Electric field penetration in Au/Nb: $SrTiO_3$ Schottky junctions probed by bias-dependent internal photoemission," *Appl. Phys. Lett.*, vol. 98 (2011) article no. 192103.

[12] H. Matsuhashi and S. Nishikawa, "Optimum electrode materials for Ta_2O_5 capacitors for high- and low-temperature processes," *Jpn. J. Appl. Phys.*, Part 1, vol. 33, no. 3A (March 1994), pp. 1293-1297.

[13] W.S. Lau, "Mechanism of reverse leakage current in Schottky diodes fabricated on large bandgap semiconductors like Ga_2O_3 and diamond Part II," *ECS Trans.*, vol. 108(6), pp. 39-56, 2022.

PULSED DC PARAMETERS (REVERSE VOLTAGE, DUTY CYCLE, PULSED FREQUENCY) ON FILM QUALITY IN REACTIVE SPUTTERED ALUMINUM NITRIDE FILMS

Wei-Yu Zhou[1], Xue-Li Tseng[1], Ning-Hsiu Yuan[2], Hsiao-Han Lo[2], Peter J. Wang[2], Ming-yu Jiang[2], Yiin-kuen Fuh[1]*, Tomi T. Li[1]

1 Department of Mechanical Engineering, National Central University
2 Delta Electronics, Inc.
Taoyuan City, Taiwan
Center Phone: +886-3-4227151#37313, and *Corresponding Author's Email: michaelfuh@gmail.com

ABSTRACT

Piezoelectric aluminum nitride (AlN) and scandium nitride (AlScN) thin films using reactive pulsed DC magnetron sputtering have shown the grown films on silicon (100) substrates without arcing during the deposition. Limited in-depth DC pulse studies have been done on the role played by variables like pulse frequency, duty cycle, and reverse voltage in the deposition process, despite the fact that many researchers have now demonstrated that pulsed DC magnetron sputtering can be used frequently to produce fully dense, defect-free films. Operating conditions were routinely altered, and the deposition technique was continuously updated. The goal of this study is to look at how the pulse parameters affected the deposition process and then how the pulse parameters and the film properties are correlated. Fifteen (15) distinct design of experiment (DOE) combinations by Box-Behnken experimental method on pulse parameters were executed for film deposition and characterized the films crystallinity, microstructures, and surface roughness to find out film properties in correlation with pulse parameters. This is the way that used for obtaining optimal pulse conditions is based on the subsequent response surface method (RSM) of DOE model.

Keywords—Reactive Pulsed DC magnetron sputtering; Aluminum nitride (AlN); Design of experiment (DOE); Pulse parameters

INTRODUCTION

Aluminum nitride exhibits unique physical and chemical characteristics, making it an attractive material for a variety of applications. Due to its excellent surface acoustic wave velocity, chemical stability and exceptional mechanical strength [1, 2], AlN has become a popular material for various applications including electronics, optoelectronics, and acoustics. In recent years, there has been a growing need for high-performance materials, particularly in the realm of electronic devices where there is a growing trend towards miniaturization and high-speed operation. Traditional materials are gradually unable to meet these requirements, and AlN has become an indispensable material today due to its excellent performance.

In this study, we utilized reactive magnetron sputtering with an asymmetric bipolar pulse power source to deposit AlN thin films on silicon (100) substrates. Reactive magnetron sputtering is capable of depositing high-quality and uniform AlN thin films at low temperatures [3, 4]. However, the arcing generated during the process can damage the film and make the plasma unstable [5], which affects the quality of the film. The power supply has the most direct impact on the plasma, so precise control of the pulse parameters (reverse voltage, pulse frequency, and duty cycle) is crucial. Box-Behnken design of experiments (DOE) was utilized to establish the experimental matrix and investigate the correlation between the surface roughness of films and the pulse parameters using RSM [6, 7]. To verify the successful deposition of the thin films, the experimental samples were measured using X-ray diffraction (XRD), field-emission scanning electron microscopy (FE-SEM) and atomic force microscopy (AFM). Typical thin film quality measurement results using XRD, FE-SEM, and AFM are shown in Fig. 1.

EXPERIMENTAL WORK

In this experiment, a reactive pulsed DC magnetron sputtering technique was employed to deposit AlN thin films on silicon (100) substrates. Three pulse parameters were modulated and 15 experiments with different parameter combinations were designed using the Box-Behnken method to investigate the correlation between the pulse parameters and the surface roughness of AlN thin films. Table I displays the configuration parameters of PVD process chamber, where the power, pressure, gas flow rate, and other fundamental chamber settings are fixed. The variables and their respective levels used in the experimental design are presented in detail in Table II. Each variable has three levels, including reverse voltage (A1), pulse frequency (A2), and duty cycle (A3). Each level of these three variables is divided into low, medium, and high values, represented as -1, 0, and +1, respectively. The experimental combinations are determined based on the setup presented in Table III. The experimental combinations for depositing AlN thin films with a process time of 30 minutes were determined by utilizing the fixed configuration outlined in Table I.

To ensure the quality of the thin films, we employed XRD (D8 Advance, Bruker US) which delivers Cu-Kα (λ = 1.5418 Å) radiation to detect the AlN (002) peak and to confirm the crystallinity and orientation of the AlN film. SEM was utilized to observe the microstructure of film surface and cross-section, providing crucial information on the morphology and

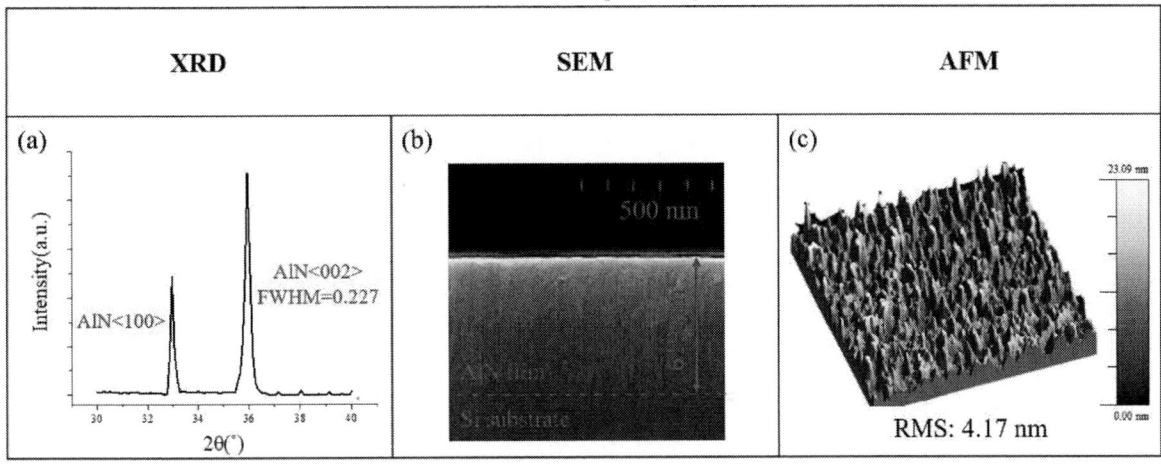

Fig. 1. Typical thin film quality measurement results of XRD, FE-SEM and AFM for S7.

crystallographic features of the film. Additionally, AFM (Nanoview 1000, China) was utilized to measure the average surface roughness of the samples. These analytical techniques are widely used in materials science research and have been proven to be effective in characterizing thin films.

RSM is a statistical analysis approach primarily utilized for experimental design and optimization, with the aim of investigating the influence of multiple independent variables on one or more response variables. RSM finds wide application in various fields such as manufacturing, chemistry, and industrial engineering. In this particular investigation, RSM was implemented through the utilization of multiple regression analysis and a quadratic polynomial equation model to optimize the performance of response variables to the maximum extent and effectively reducing experimental costs and saving time. The experimental procedure of this study consists of three steps. The first step involves defining the research variables and obtaining parameter combinations with different levels through experimental design to deposit AlN thin films. The second step utilizes the measured surface roughness and corresponding parameters of each group of AlN thin films to establish a model through polynomial regression. Finally, in the third step, 2D and 3D response surface plots are used for analysis to determine the optimal values of pulse parameters.

TABLE II. DESIGN OF EXPERIMENT (DOE): INTERACTING VARIABLES AND FACTOR LEVELS

Variables	Signs	Levels		
		-1	0	+1
Reverse voltage (V)	A_1	50	70	90
Pulse frequency (kHz)	A_2	100	175	250
Duty Cycle (%)	A_3	80	85	90
Roughness (nm)	B	Dependent variable		

TABLE III. RESULTS OF THE BOX-BEHNKEN DESIGN METHOD

Run	V_1	V_2	V_3	Run	V_1	V_2	V_3	Run	V_1	V_2	V_3
S1	0	-1	+1	*S6*	0	0	0	*S11*	0	-1	-1
S2	+1	+1	0	*S7*	0	+1	+1	*S12*	+1	0	-1
S3	-1	+1	0	*S8*	0	0	0	*S13*	0	0	0
S4	-1	-1	0	*S9*	-1	0	+1	*S14*	0	+1	-1
S5	+1	-1	0	*S10*	+1	0	+1	*S15*	-1	0	-1

TABLE I. CONFIGURATION OF THE PHYSICAL VAPOR DEPOSITION (PVD) CHAMBER PROCESS

Process chamber setup	Values	Units
Power	1200	W
Pressure	2	mTorr
Target-to-Substrate distance	80	mm
Deposition time	30	min
Substrate temperature	R.T.	°C
N_2	60	sccm
Total gas flow rate	60	sccm

RESULTS AND DISCUSSION

The crystallite structure and quality of AlN films were deduced by analyzing the positions of XRD peaks, and the findings are listed in Table IV, a typical example, see Fig. 1(a). Two distinct diffraction peaks were observed between 33° and 36° under different parameter combinations. By consulting the Joint Committee on Powder Diffraction Standards (JCPDS) database, we confirmed that the 33° diffraction peak corresponds to AlN (100) and the 36° peak to AlN (002), which verifies the successful deposition of the thin film. In addition, we calculated the FWHM of the 002 peak as an indicator of the thin film quality. The smallest value of 0.1° was listed for S3, indicating a sharper diffraction peak and a more regular crystal lattice structure with superior crystal quality. The studies and reviews [8] and [9] have documented comparable results.

To further verify the effective deposition of the AlN film on a silicon (100) substrate, cross-sectional images of the film were captured using a FE-SEM with a magnification of 20,000× for clear visualization. The thickness of the deposited film during the 30-minute process was found to be between 540-647 nm, see Table IV. It can be found that S11, S12, and S15 have lower deposition rates, which is attributed to their duty cycle of 80% (low level). Furthermore, the observation of the microscopic structure of the cross-sectional view of the film confirmed the quality of the AlN film, see Fig. 1(b), as the presence of compact columnar crystal structures was observed [10].

Due to the critical impact of surface roughness on the performance of AlN material, including potential changes in its optical, thermal, and mechanical properties. Specifically, in terms of optical performance, the different levels of scattering caused by surface roughness can affect the material's performance. Regarding thermal conductivity, changes in surface roughness can alter thermal impedance and thus impact the stability of heat dissipation performance. Additionally, surface defects may affect the strength of the material in terms of mechanical performance. This study aims to analyze and improve the surface roughness of thin films for better design and application. Fig. 2 shows the surface roughness (RMS) of thin film measured using AFM (tapping mode). According to the measurement data, the four groups of films (S3, S4, S9, and S15) with lower surface roughness in the experiment have a common feature, which is the use of a low-level reverse voltage of 50 V. When sputtering the insulating materials, reverse voltage plays a critical role in effectively preventing target surface insulation and reducing the possibility of arc damage to the film. High reverse voltage will reduce the quality of the film [2], so its value must be carefully controlled.

In this study, we employed Minitab 19 software to conduct multiple regression analysis on the variables (A1, A2, and A3) and the response variable (B). A second-order polynomial equation was utilized to elucidate the relationship between the pulse parameters and the surface roughness of AlN film, with the detailed equation presented in Eq.1. The results of the response surface regression analysis are summarized in Table V, revealing an R-square value of 85.91%, indicating that the model has a significant explanatory power. Specifically, the R-square value indicates the percentage of total variation in the regression model that can be explained, with a higher value indicating a better fit and a model that has a strong explanatory power. The p-value is used to indicate the significance of a factor, with values less than 0.05 typically considered significant [11, 12]. Among the quadratic terms, it can be seen that the A1*A1 has the most significant impact, with a p-value of 0.016.

$$
\begin{aligned}
B \text{ (roughness)} = & \; 157.8 + 0.235 * A_1 - 0.0037 * A_2 - 3.86 * A_3 \\
& - 0.002090 A_1 * A_1 + 0.000059 A_2 * A_2 + 0.02257 A_3 * A_3 \\
& + 0.000023 A_1 * A_2 + 0.00080 A_1 * A_3 - 0.000173 A_2 * A_3
\end{aligned}
\tag{1}
$$

Fig. 3 displays the contour plot and 3D surface plot based on the surface roughness of thin film. These plots utilize color and three-dimensional space to provide a clear visualization of the relationships between the variables and the response variables, facilitating the prediction of the optimal parameter combination. Through experimentation and measurement of various parameter combinations, a set of optimal parameters (reverse voltage: 50 V, pulse frequency: 145.455 kHz, and duty cycle: 85.1515%) was established through minimization of surface roughness. The predicted surface roughness was 1.606 nm RMS.

TABLE IV. THE FULL WIDTH AT HALF MAXIMUM (FWHM) AND FILM THICKNESS DATA IN EACH EXPERIMENTAL COMBINATION

Run	002 peak FWHM (°)	Thickness (nm)
S1	0.291	610
S2	0.234	647
S3	0.100	601
S4	0.333	577
S5	0.328	590
S6	0.368	626
S7	0.227	619
S8	0.322	570
S9	0.537	613
S10	0.338	600
S11	0.284	577
S12	0.456	560
S13	0.298	560
S14	0.373	606
S15	0.321	540

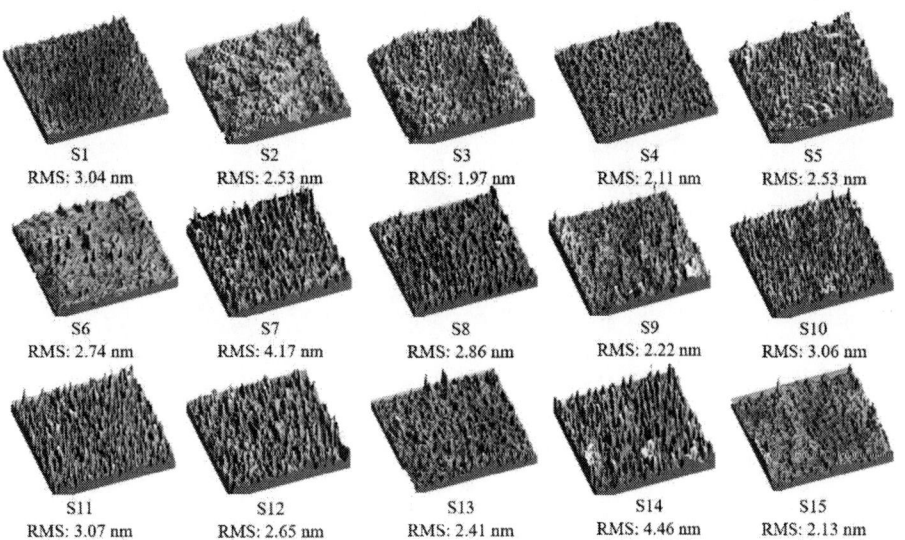

Fig. 2. AFM images of AlN films deposited on Si substrates are presented along with their corresponding RMS surface roughness values (S1-S15).

TABLE V.   A SUMMARY FROM RESPONSE SURFACE REGRESSION ANALYSIS

Summary of model

R-squared: 85.69%	R-squared (adj): 59.93%	R-squared (pred): N.D.

ANOVA

Source	DF[1]	Adj SS[2]	Adj MS[3]	F[4] Value	P[5] Value
Model	9	5.95718	0.66191	3.33	0.099
Linear	3	1.39655	0.46552	2.34	0.190
A_1	1	0.68445	0.68445	3.44	0.123
A_2	1	0.70805	0.70805	3.56	0.118
A_3	1	0.00405	0.00405	0.02	0.892
Square	3	4.51323	1.50441	7.56	0.026
A_1*A_1	1	2.57951	2.57951	12.96	0.016
A_2*A_2	1	0.41231	0.41231	2.07	0.210
A_3*A_3	1	1.17520	1.17520	5.91	0.059
2-Way Interaction	3	0.04740	0.01580	0.08	0.968
A_1*A_2	1	0.00490	0.00490	0.02	0.881
A_1*A_3	1	0.02560	0.02560	0.13	0.734
A_2*A_3	1	0.01690	0.01690	0.08	0.782
Error	5	0.99482	0.19896		
Lack of fit	3	0.98655	0.32885	79.56	0.012
Pure Error	2	0.00827	0.00413		
Total	14	6.95200			

1. DF: degree of freedom; 2. Adj SS: adjusted sum of squares; 3. Adj MS: adjusted mean square; 4. F: statistical test; 5. P: statistical value

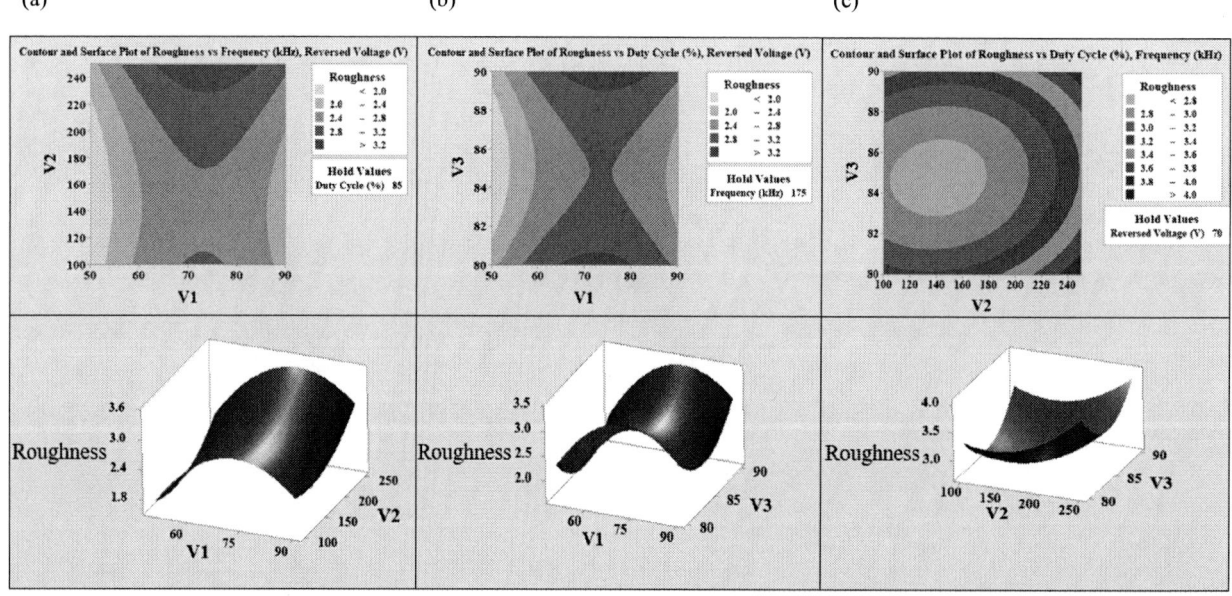

Fig. 3. *2D views based on the factors and levels, and 3D views that incorporate multiple factors simultaneously. 3D views be visualized in three different ways. (a) 3D view in A1*A2, the frequency against the reverse voltage while holding the third factor, A3 as a constant. (b) 3D view in A1*A3, the duty cycle against the reverse voltage while holding the second factor, A2 as a constant. (c) 3D view in A2*A3, the duty cycle against the frequency while holding the third factor, A1 as a constant.*

CONCLUSION

In this research, we deposited AlN thin films on silicon substrates, utilized Box-Behnken experimental design method and RSM to investigate how the power supply affected the deposition process and to determine the relationship between pulse parameters and thin film properties. By creating a quadratic polynomial equation to optimize the pulse parameters, we minimized the surface roughness of thin film. The optimal parameters predicted were a reverse voltage of 50 V, pulse frequency of 145.455 kHz, and duty cycle of 85.1515%, resulting in an RMS value of surface roughness of 1.606 nm.

ACKNOWLEDGEMENTS

The authors would like to acknowledge Delta Electronics, Inc., Taiwan, for funding and offering technical support in this project.

REFERENCES

[1] C. C. Wang, M. C. Chiu, M. H. Shiao, and F. S. Shieu, "Characterization of AlN thin films prepared by unbalanced magnetron sputtering," J. Electrochem. Soc. 151(10), F252, 2004

[2] J. S. Cherng, and D. S. Chang, "Effects of pulse parameters on the pulsed-DC reactive sputtering of AlN thin films," Vacuum, 84(5), 653-656, 2009

[3] D. Dallaeva, S. Țălu, S. Stach, P. Škvarada, P. Tománek, and L. Grmela, "AFM imaging and fractal analysis of surface roughness of AlN epilayers on sapphire substrates," Appl. Surf. Sci. 312, 81-86, 2014

[4] J. Vlček, A. D. Pajdarová, and J. Musil, "Pulsed dc magnetron discharges and their utilization in plasma surface engineering," Contrib. Plasma Phys. 44(5 - 6), 426-436, 2004

[5] H. C. Barshilia, B. Deepthi, and K. S. Rajam, "Growth and characterization of aluminum nitride coatings prepared by pulsed-direct current reactive unbalanced magnetron sputtering," Thin Solid Films, 516(12), 4168-4174, 2008

[6] J. Adamczyk, N. Horny, A. Tricoteaux, P. Y. Jouan, and M. Zadam, "On the use of response surface methodology to predict and interpret the preferred c-axis orientation of sputtered AlN thin films," Appl. Surf. Sci. 254(6), 1744-1750, 2008

[7] H. H. Lo, W. L. Chen, P. J. Wang, W. Lai, Y. K. Fuh, and T. T. Li, "Residual stress classification of pulsed DC reactive sputtered aluminum nitride film via large-scale data analysis of optical emission spectroscopy," J. Adv. Manuf. Technol. 119, 7449-7462, 2022

[8] A. Iqbal, G. Walker, L. Hold, A. Fernandes, A. Lacopi and F. Mohd-Yasin, "Sputtering of aluminium nitride (002) film on cubic silicon carbide on silicon (100) substrate: Influences of substrate temperature and deposition power," J. Mater. Sci: Materials in Electronics, 31, 239-248, 2020

[9] A. Iqbal and F. Mohd-Yasin, "Reactive sputtering of aluminum nitride (002) thin films for piezoelectric applications: A review," Sensors, 18(6), 1797,2018

[10] V. Brien, P. Miska, B. Bolle, and P. Pigeat, "Columnar growth of ALN by rf magnetron sputtering: Role of the {1 0 1 3} planes," J. Cryst. Growth, 307(1), 245-252, 2007

[11] S. Abid, R. Messadi, T. Hassine, H. Ben Daly, J. Soulestin, and M. F. Lacrampe, "Optimization of mechanical properties of printed acrylonitrile butadiene styrene using RSM design," J. Adv. Manuf. Technol. 100, 1363-1372, 2019

[12] I. Asiltürk, S. Neşeli, and M. A. Ince, "Optimisation of parameters affecting surface roughness of Co28Cr6Mo medical material during CNC lathe machining by using the Taguchi and RSM methods," Measurement, 78, 120-128, 2016

ELECTROPLATING PROCESS IMPROVEMENT ON POST-CMP DISHING PROFILE

Wenbo Wu[1*], Tong Lei, Zhijun Zhu, Zhenhua Hu, and Yushan Chi
[1]Lam Research Service Co., Ltd, Shanghai 200120, China
*Corresponding Author's Email: Claus.Wu@lamresearch.com

ABSTRACT

Copper damascene metallization has been widely utilized in integrated circuit fabrication to form metal interconnect structures through electrochemical plating (ECP) technology. When some large pads are patterned on incoming wafers, excessive dishing profiles are often observed after ECP and chemical mechanical polishing (CMP) process. Solutions to overcome overpolishing issues involve optimization of both CMP and ECP processes. For ECP process, leveler additives in plating solution are the major contributor to leveling the big pad. By adjusting the rotation speed of the wafer during plating steps, the mass transfer of large organic molecules near the wafer surface can be varied. In this work, it is observed that the total dishing defect count under optical microscope observation can be significantly reduced by decreasing the rotation speed during plating. It is expected that this study will be beneficial to copper electrochemical deposition for wafers with large Cu pads.

INTRODUCTION

As device sizes continue to shrink, copper damascene metallization has emerged as a substitute for traditional vacuum-deposited aluminum-based processes in silicon integrated circuits (ICs). This is due to its lower bulk resistivity and improved electromigration (EM) stability, as discussed in previous research [1]. As interconnect levels increase in back-end-of-line (BEOL) layers, feature geometries are also progressively becoming larger. Consequently, some large copper pads must be filled in the final global layer, also known as the ultra-thick metal (UTM) layer.

Electrochemical plating (ECP) is the most popular method for copper (Cu) deposition due to its low cost and high aspect ratio filling capabilities. Copper plating chemistries used in IC manufacturing typically require an acidic solution containing a small number of additive components, such as accelerators, suppressors, and levelers. The desired super-conformal plating characteristics are achieved in submicron inlaid features through a remarkable synergy created by using different additives in combination [2].

However, bottom-up filling becomes much more challenging when filling features are excessively large. After the CMP process, thickness differences known as dishing defects are typically observed, particularly in large-size features at the global interconnection layers, as shown in Figure 1. It is widely observed that the dishing profile deteriorates with increasing linewidth, leading to an increase in line resistance and current density [3]. Additionally, the resulting inferior local surface planarity complicates the next-level lithography process and metal line fabrication [4]. Minimizing dishing profiles can reduce the required ECP deposition film thickness, CMP polishing time, and slurry usage, etc.

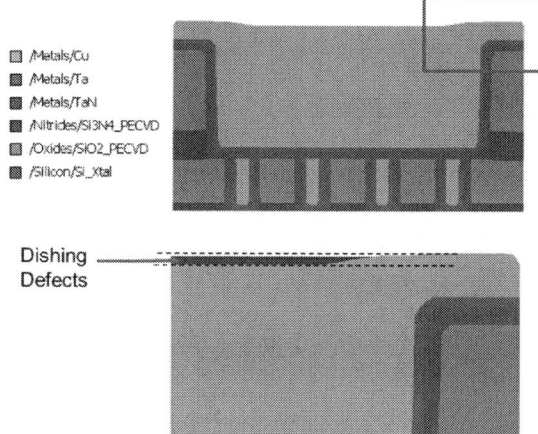

/Metals/Cu
/Metals/Ta
/Metals/TaN
/Nitrides/Si3N4_PECVD
/Oxides/SiO2_PECVD
/Silicon/Si_Xtal

Dishing Defects

Figure 1: (a) Cross section view of UTM pads after ECP copper deposition and CMP process and (b) illustrated dishing defect under SEM observation

Intensive research has been conducted to overcome dishing issues, including optimization of both CMP and ECP processes. When adjusting electrochemical plating conditions, leveler and suppressor effects are mainly focused on eliminating overplating by tuning plating temperatures and additive components. However, they can also have side effects on copper plating on dense features, necessitating a dedicated plating tool for UTM copper deposition or a compromise on thinner copper deposition performances.

Therefore, in this work, we present an alternative method for reducing dishing defects by tuning the rotation speed of wafers during the ECP process. The rotation speed can be adjusted through recipe parameter settings and assigned for each individual process of different metal layer deposition. This enables optimization of the UTM plating process without affecting the performance of any other metal layer deposition.

EXPERIMENTAL

The plating solution consisted of CuSO$_4$ (40 g/L), H$_2$SO$_4$ (10 g/L) and HCl (50 ppm). Bottom-up copper fill was achieved by the conventional damascene process where three additives were used, thus accelerator, suppressor, and leveler. Copper fill tests were done on SABRE® Excel copper plating tool that processed 300mm wafers. The patterned wafer was immersed in the solution and a specifically optimized current waveform was applied for optimal bottom-up Cu fill quality. The structure wafers used for filling evaluation had been patterned with large pads of the size larger than 50 μm x 50 μm and the depth greater than 3 μm. The typical stack for pattern structures was SiO$_2$/TaN/Ta/Cu. Post plating, wafers underwent spin-rinse-dry process.

RESULTS AND DISCUSSION

In this study, we assessed the performance of bottom-up filling for large copper pads using an optical microscope (OM). Discoloration defects were counted after multiple steps, including CMP, via redistribution, aluminum pad layer deposition, and passivation layer deposition, to assess the performance of the process. We investigated the impact of different plating rotation speeds on discolor defect counts, and the results are presented in Figure 2. The defect counts were normalized to the BSL performance. Our findings indicate that a decrease in the rotation speed led to fewer discolor defects under OM observation, indicating better bottom-up filling performance for large pads. By simply reducing the rotation speed, the optimization of dishing defect counts resulted in a 500-fold reduction compared to the original performance. The improved results were further confirmed by both top view and cross-sectional view observations under an optical microscope and after FIB cut. However, due to intellectual property concerns, we are unable to include these images in this paper.

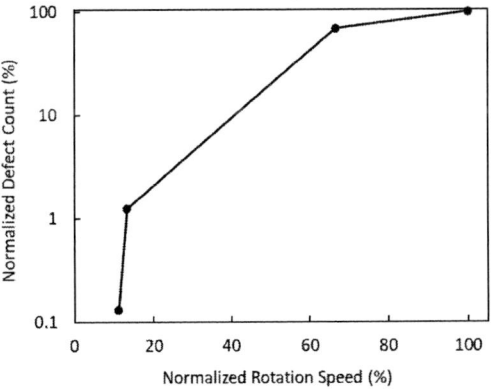

Figure 2: The total defect counts at different plating rotation speed

In addition to the improvement of the dishing profile, a change in the thickness profile of the plated copper film was observed when altering the rotation speed. As illustrated in Figure 3, a higher thickness was detected at the wafer edge when the rotation speed during plating decreased, while an insignificant variation was observed at the wafer center. This phenomenon may be attributed to the different mass transfer behaviors of the organic additives in the plating solution at altered rotation speeds, particularly the levelers with large molecular weights, which are sensitive to solution agitation. Further experiments on reaction parameters and various additive components are currently being conducted to gain a better understanding of the underlying mechanism.

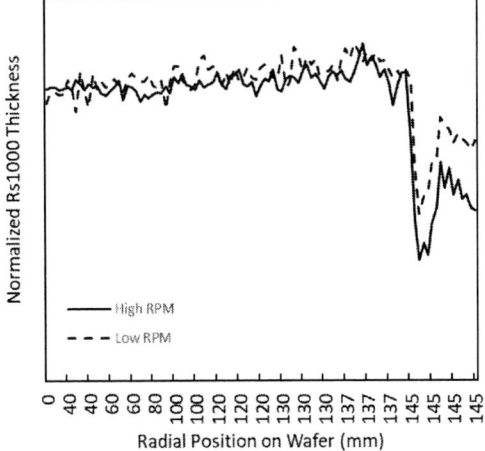

Figure 3: Thickness profiles for various rotation speeds

Of all three types of additives, leveler molecules are believed to be the most important for achieving a level final copper film. When filling dense arrays in local and intermediate interconnect layers, the large size of leveler molecules impedes their diffusion to the bottom of the feature. This results in the suppression of copper growth on the sidewalls and in the field, enabling bottom-up filling to occur.

Our results suggest that, in the case of copper deposition in UTM layers (as shown in Figure 4), the leveler effect operates differently than the common model for dense feature filling. Due to the significantly larger pad sizes, the diffusion limit of bulky leveler molecules becomes negligible, enabling them to easily diffuse to the bottom of the pads and restrain plating rate evenly over the feature. This results in more conformal plating during the ECP process, and ultimately leads to crater defects after CMP. To address this issue, it is beneficial to have more leveler in planar areas and less in trench areas to help eliminate craters. By understanding these mechanisms, we applied a reduction in the wafer rotation speed to hinder leveler spreading to the wafer trench, thus relieving dishing profiles in this case.

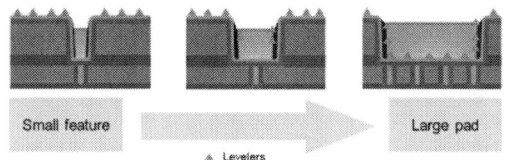

Figure 4: Leveler in the big pad restrains plating rate of bottom as the feature size increases

The convection in the plating cell is the primary driver of mass transfer of leveler reagents from bulk solution to the wafer surface. Considering the wafer as a rotating disk electrode (RDE), the mass-transfer coefficient (m_o) of levelers can be formulated as follows:

$$m_0 = 0.62 D_0^{\frac{2}{3}} \omega^{\frac{1}{2}} v^{-\frac{1}{6}} \qquad (1)$$

where D_o is the diffusion coefficient of leveler molecules, ω the angular velocity of the rotating electrode, and ν the kinematic viscosity of the plating solution. As the rotation speed increases, the mass transfer of leveler molecules from bulk solution to wafer surfaces also increases. Therefore, decreasing the rotation speed can improve the bottom-up filling performance. Similarly, temperature can be used as a tuning parameter to slow down the mass transfer of leveler molecules since the diffusion coefficient (D_o) is temperature dependent. We have experimentally verified that reducing the cell temperature can enhance the bottom-up filling performance of large pads. However, lower process temperatures can have side effects on the copper filling of small features in the local and intermediate interconnect layers. It can also be time-consuming and laborious to adjust the plating bath temperature for different metal layer depositions. Consequently, adjusting the rotation speed is a more convenient tuning knob for reducing the dishing effect.

CONCLUSION

In conclusion, this study proposes a model for dishing-shaped growth in large pad areas during copper electrolysis deposition, with the mass transfer of leveler additives in solution identified as the major contributor to this phenomenon. Our results demonstrate that the rotation speed during plating steps can serve as a tuning knob to improve the dishing defects. Specifically, we observed that a lower rotation speed can significantly reduce dishing-shaped growth in large pads, as evidenced by a decrease in the defect count under optical microscope scanning. Overall, these findings provide insights into the mechanisms behind dishing defects and offer practical guidance for improving the quality of copper electrodeposition in industrial applications.

ACKNOWLEDGEMENTS

The authors would like to thank Yu Lu and Caigan Chen for their kind help with this paper.

REFERENCES

[1] R. Havemann and J. Hutchby. *Proceedings of the IEEE*, vol. 89, 5, 2001, pp. 586-601.

[2] P. Vereecken, R. Binstead, H. Deligianni and P. Andricacos, *IBM Journal of Research and Development*, vol. 49, 1, 2005, pp. 3-18.

[3] V. Nguyen, P. Van Der Velden, R. Daamen, H. Van Kranenburg and P. Woerlee, *International Electron Devices Meeting*, 2000, pp. 499-502.

[4] Guanghui Fu and A. Chandra, *IEEE Transactions on Semiconductor Manufacturing*, vol. 16, 3, 2003, pp. 477-485.

EFFECT OF SUB-ATMOSPHERIC CHEMICAL VAPOR DEPOSITION SIO2 FILM DEPOSITION PROCESS ON SURFACE CHEMISTRY SENSITIVITY

JiananWei[1], XueLiu[1], DapengRuan[2]*

[1]Piotech, INC

[2] Shenyang City , Liaoning Province

*Corresponding Author's Email: jianan.wei@piotech.cn

ABSTRACT

Sub-atmospheric chemical vapor deposition (SACVD) oxide films have sensitivity when deposited on different surfaces. By adjusting the process, the sensitivity can be enhanced or reduced. In this paper, the SACVD chamber was used to deposit undoped silicate glass (USG) films, and the film thickness was tested with an ellipsometer. Taking the bare Si surface as a control, and using plasma enhanced-TEOS (PE-TEOS), thermal silicon oxide and a structure wafer as the experimental surface, the thicknesses of the films on different surfaces were compared under a specific process. The single variable test (SVT) such as TEOS concentration, temperature and pressure were also carried out, and the influence trend of each parameter on the surface sensitivity was tested. Ultimately, the sensitivity of the surface can be enhanced or reduced by adjusting the process.

Keywords—SACVD, USG film, surface sensitivity, process

INTRODUCTION

SiO2 film plays an important role in electronic devices [1]. In recent years, with the continuous reduction of the size of integrated circuits, more requirements have been placed on the performance of SiO2 films. Especially for high aspect ratio gap filling, high coverage of the film is required, applications such as interlayer dielectric (ILD) and shallow trench isolation (STI) films, etc. At present, USG films deposited by SACVD have good gap filling ability, SA USG films have been widely used in integrated circuit fabrication [2][3].

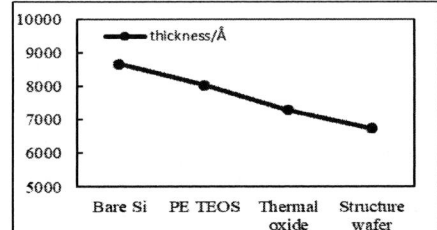

Fig1. USG film thickness under different surfaces

The USG films deposited by SACVD have sensitivity on different kinds of surfaces, and the film thickness is different on different surfaces. As shown in Fig 1. In other words, the deposition rate (DR) is different on different surfaces.

Studies have shown that the type of chemical bonds and electrostatic characteristics of the surface will affect the deposition rate of the film. Yuko Nishimoto, Noboru Tokumasu and other scholars found that when using atmospheric-pressure chemical vapor deposition (APCVD) to deposit SiO2 films on bare Si and thermal silicon oxide, respectively, the deposition rate on bare Si is significantly faster than that on thermal silicon oxide. The reaction is inhibited due to the presence of a large number of –Si–O–Si– on the thermal oxide surface. The same result was found on SACVD films [4]. K. Kwok and other scholars use SACVD to deposit SiO2. The deposition rates of SiO2 films on bare Si, thermal silicon oxide, plasma enhanced chemical vapor deposition TEOS (PECVD-TEOS), PECVD-SiH4 and PECVD TEOS-N2O were compared respectively, and it was found that the deposition rates on thermal silicon oxide and PECVD-TEOS were lower than other surfaces. The deposition rate is the slowest on thermal silicon oxide [5]. Since the surface of thermal silicon oxide has negative electrostatic characteristics, the reaction source is also negatively charged, so the repulsive effect inhibits the growth of the film.

The surface sensitivity of SA USG films is affected by the deposition process. In chip production, in order to match the requirements of process flow, sometimes the film thickness of the product needs to be adjusted. Especially in products with grooves, it is necessary to adjust the surface sensitivity to improve the gap filling performance. Therefore, we need to adjust the process to enhance or reduce surface sensitivity. In order to study the effect of process on surface sensitivity, PE-TEOS, thermal silicon oxide and a structural wafer were used as surfaces, and bare Si was used as a control. By adjusting the process conditions, the influence of process parameters on the surface sensitivity was studied and the reasons were analyzed

EXPERIMENT AND RESULTS

USG film was deposited on bare Si with a thickness of 8700A using SACVD 8inch chamber. The film process condition is shown in Table 1.

TABLE 1
PROCESS CONDITIONS

Temp. (℃)	Pressure (Torr)	TEOS(g. min⁻¹)	He /sccm	O3 / g. m⁻³
520-550	500-580	1.6-2	12000	160-200

The SVT data of temperature, pressure and TEOS concentration were collected. The process conditions and results are shown in Figure 2. Δthickness is the thickness on bare Si surface minus that on PE-TEOS surface. The larger the Δthickness is, the stronger the surface sensitivity is.

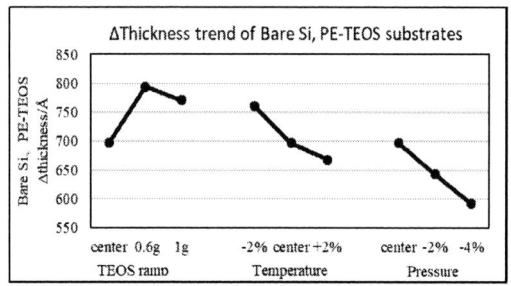

Fig. 2 Trend of bare Si and PE-TEOS surface process results -1

The experiment shows that the TEOS ramp of 0.6g/min has the maximum surface sensitivity. Therefore, taking this condition as the central condition, the SVT experiment was continued, as shown in Figure 3.

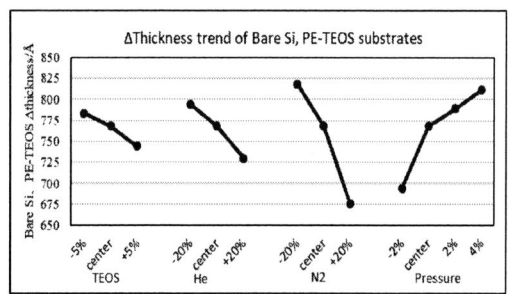

Fig. 3 Trend of bare Si and PE-TEOS surface process results -2

We continue to take Table 1 as the central condition to conduct film deposition on bare Si and thermal silicon oxide respectively, adjust the TEOS concentration, He flow rate, N2 flow rate and deposition pressure. The influence trend of the test process on the sensitivity of the surface is shown in Figure 4.

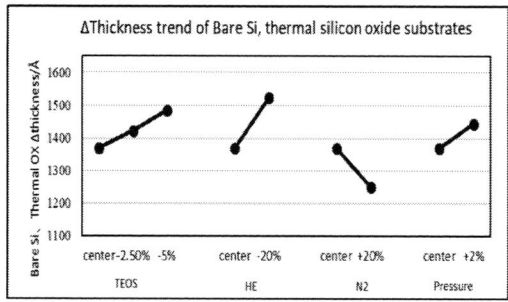

Fig. 4 Trend of bare Si and thermal silicon oxide surface process results

In the process of adjusting product wafer deposition process, fab have limited temperature, pressure and ozone to ensure film properties. According to the research results, the effects of TEOS concentration and total gas flow rate on the surface sensitivity of a structure wafer were studied. Thickness changes of the film on the grooves inside the structure were observed. The experimental conditions and results are shown in the following Table.

TABLE 2
WAFER SURFACE FILM THICKNESS RESULTS OF BARE SI AND A STRUCTURE

Process conditions	Thickness on bare Si/A	Thickness on structure wafer/A	ΔThickness /A
Center conditions	9637	7458	2179
Reduce gas flow by 10%	9616	7386	2230
Reduce TEOS concentration by 5%	9478	7204	2274
Reduce gas flow by 10% and TEOS concentration by 5%	9455	7058	2397

DISCUSSION OF EXPERIMENTAL RESULTS

Experimental results show that TEOS concentration, temperature, pressure, and gas flow rate during the reaction all affect the deposition rate of the surface. PE-TEOS and thermal silicon oxide have the same influence trend. On the structural wafer, TEOS The trends in the effect of concentration and gas flow on sensitivity are also consistent with the other two surfaces.

TEOS concentration

During the deposition of SA-USG thin films, TEOS was used as a reactant, and after being catalyzed by ozone at high temperature, silicon dioxide was formed on the surface. In general, the higher the concentration of TEOS, the faster the film deposition rate [6]. Comparing the experimental results, the surface sensitivity is enhanced when TEOS ramp and when the TEOS concentration decreases. The drop in deposition rate on PE-TEOS and thermal silicon oxide is greater than that on bare silicon. From the perspective of the reaction principle, TEOS is oxidized by the oxygen atoms generated by the decomposition of O3 to generate reaction precursors, which are generally considered to be silanol and SiO, etc. Silanol and the surface form a SiO2 film [7]. During the reaction of silanol with the surface, the type of chemical bonds on the surface will affect the rate of the reaction. Generally, silanol is easy to combine with -OH quickly to remove H2O form the SiO2, the reaction rate is very fast. A large number of -H bonds exist on the bare silicon surface, which are oxidized to generate -OH bonds after encountering O3, so the reaction rate is fast.

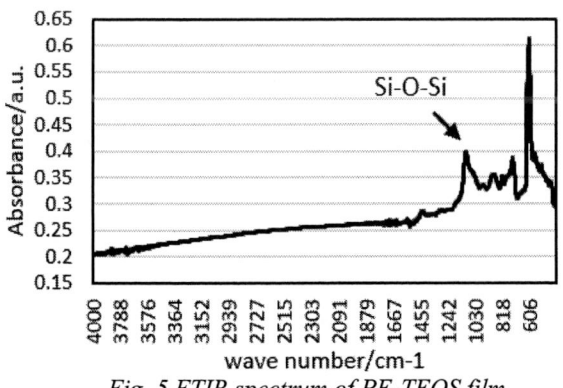

Fig. 5 FTIR spectrum of PE-TEOS film

As shown in Figure 5, on PE-TEOS and thermal silicon oxide surfaces, there are more –Si–O–Si– bonds, but less -H and -OH bonds, so the film deposition rate is slow, as shown in Figure 6 [4]. It can be seen that the deposition rate is affected by both the concentration of the reactant precursor and the number of -OH on the surface. When the number of -OH on the surface is large, the reaction rate is mainly affected by the concentration of the precursor. After the precursor is reduced, the binding site can still be found quickly, so the reduction of the reaction rate is small. When the number of -OH on the surface is small, the time for precursor to search for the binding site is the main factor affecting reaction rate. After the TEOS concentration decreased, the deposition rate decreased, and the time for the precursor to search for the binding site was further prolonged, so reaction rate decrease again. Lead to enhanced surface sensitivity.

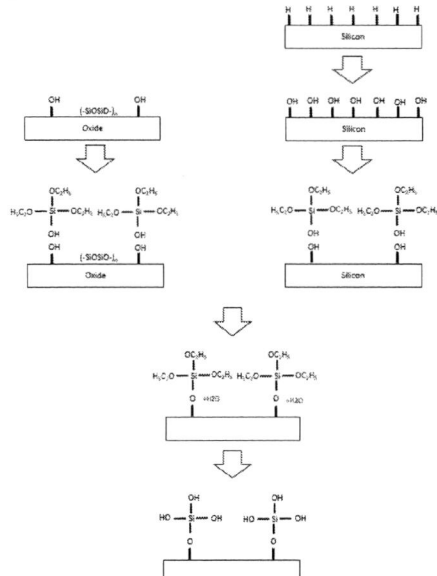

Fig. 6 Chemical bonds of bare Si and thermal silicon oxide surfaces

Temperature

As can be seen inFigure 2, the sensitivity of the films on PE-TEOS surfaces weakened after the temperature was increased. On the bare Si surface, the thickness decreases greatly, and the thickness of the PE-TEOS surface changes less, indicating that the temperature has little effect on the PE-TEOS surface. During the high temperature reaction, the gas phase reaction changes, and the reaction products are easy to diffuse outward, so that part of the generated SiO2 products diffuse to the surface of the wafer without forming a film [6]. The surface of the Bare Si surface is smooth, and the reaction products are easily taken away by the airflow after diffusing outward. On the PE-TEOS surface, due to the high surface roughness and good adhesion of the surface, the part of the reaction product diffused out is reduced, so the reduction in the film thickness is small.

Pressure

Figure 2 and Figure 3 show that the surface sensitivity increases slightly with the increase of pressure. This result is consistent with Vladislav's findings [11]. Kim and Gill proposed that when TEOS reacts with O3, the intermediate products produced include precursors that can form thin films and other by-products that cannot form thin films. These two reactions have a competitive relationship. The reaction rate is affected by the number of precursors in the intermediate, the more precursors, the faster the reaction rate [8][9]. In addition, scholars I. A. Shareef et al. proposed that the proportion of useless by-products in the reaction process increased after the pressure increased, and the film precursors became less and the concentration decreased [10]. Therefore, the effect of pressure and TEOS concentration on surface sensitivity is similar. The increase of pressure corresponds to the decrease of TEOS concentration and the increase of surface sensitivity.

Gas flow

From Figure 3 and Figure 4, it can be seen that the surface sensitivity decreases after the fluxes of He and N2 increase. When the pressure in the chamber is fixed, the gas flow rate increases, which leads to a faster flow rate of the gas in the chamber and a faster movement of the reaction precursor. Therefore, the binding site will be quickly found to form film. There are a large number of binding sites in bare Si, and the precursor can quickly find the binding site on the surface of the surface. When the speed of the precursor is accelerated, the improvement of the reaction rate is limited. On PE-TEOS and thermal silicon oxide surfaces, there are fewer binding sites for precursors. When the precursor moves faster, it will greatly reduce the time for the precursor to find binding sites and increase the deposition rate, lead to a decrease in surface sensitivity.

CONCLUSION

Using SACVD chamber, USG films were deposited on bare Si, PE-TEOS, thermal silicon oxide and structural wafer, and the influence of deposition process on surface sensitivity was studied. The result is as follows :

1）The higher the concentration of TEOS, the weaker the surface sensitivity.

2）The higher the temperature, the weaker the surface sensitivity.

3）The higher the pressure, the stronger the surface sensitivity.

4）The higher the gas flow rate, the weaker the surface sensitivity.

REFERENCES

[1] Wang Changlin, "Main chemical reactions and methods of silicon dioxide thin film deposition technology." Journal of Baotou Vocational and Technical College.2008.12

[2] Cary Ching, Harry Whiteselll, Shankar, Venkataraman. "Improving Electrical Performance Using SACVD Oxide Films." processing and manufacturing. 2008.6.26-29

[3] Zhang Miao, "Research on Optimization of borophosphorous Glass Process in 0.16um DRAM." Fudan University, 2009

[4] Yuko Nishimoto, Noboru Tokumasu, and Kazuo Maeda. "Thermal desorption spectra of SiO2 films deposited on Si and on thermal SiO2 by retreat hylortho silicate/O3 atmospheric-pressure chemical vapor deposition." 1999

[5] K. Kwok, E. Yieh, S. Robles, and B. C. Nguyen. "Surface Related Phenomena in Integrated PECVD/Ozone-TEOS SACVD Processes for Sub-Half Micron Gap Fill: Electrostatic Effects." 1994

[6] Li Juan, "Deposition of silica film by TEOS-O3 by CVD method ." Excellent master's degree theses in China Literature Database (Master), 2004, (04).

[7] K. Fujino, Y. NIshimoto, N. Tokumasu, K. Maeda, "Silicon Dioxide Deposition by Atmospheric Pressure and Low-Temperature CVD Using TEOS and Ozone." J Electrochem Soc, 1990, 137(9) 2883-2887.

[8] E. J. Kim and W. N. Gill, J. Electrochem. Soc. 141, 3462 ~1994.

[9] M. Young, The Technical Writer's Handbook. Mill Valley, CA: University Science, 1989.

[10] Shareef I A, "Role of gas phase reactions in subatmospheric chemical‐vapor deposition ozone/TEOS processes for oxide deposition". Journal of vacuum science & technology. B Microelectronics and nanometer structures: processing, measurement, and phenomena: an official journal of the American Vacuum Society, 1996, 14(2):772-788.

[11] Vladislav Y. Vassiliev and Jia-Zhen Zheng. "The investigation on the deposition kinetics of sub-atmospheric pressure glass film".1997.9.

[12] G Hummer , G Halter, M Grossl. "Calculated and measured flow conductance for butterfly valves". J. Vacuum, 1990, 2126-2128.

Gas Distribution Effect on Thermal ALD AlN Film Thickness Non-uniformity

Xiaomeng Liu[1]*, Qihui Zhang[1], and Hao Deng[1]

[1]Piotech (Shanghai) Inc., C13 1568 Feidu Road, Lingang New Area, China (Shanghai) Pilot Free Trade Zone

*Corresponding Author's Email: Xiaomeng.liu@sh.piotech.cn

ABSTRACT

Aluminum nitride film has been studied for various applications attributed to its high dielectric constants and large band gap. Atomic layer deposition method to obtain AlN thin film was applied in this report due to its precise control film thickness and excellent film uniformity. To figure out the effect of gas distribution on film property, measurements of thickness uniformity by ALD deposition system and gas distribution simulation by Ansys software were both employed to help us improving film thickness non-uniformity. With the refinement, gas distribution could be modified within reactant chamber and would make an improvement of film thickness uniformity.

INTRODUCTION

Aluminum nitride is inclined to be widely used in Surface acoustic wave (SAW), microwave amplifiers, high-frequency and high-power transistors and light emitting diodes ascribed to its high dielectric constants, good thermal conductivity and large band gap [1]. Among all the deposition methods to obtain AlN thin film, atomic layer deposition method (ALD) using trimethyl-aluminum(TMA) and ammonia stands out with surface self-limiting growth mechanism. With the trend of device miniaturization in semiconductor industry, ALD method is ramping up as an attractive technique due to its atomic level control of thin film deposition process [2].

Attributed to its accurate thickness control capability, high uniformity within wafer and excellent step coverage stability, ALD method is introduced into the MOSFET structure and to fabricate trench capacitors for DRAM. Most ALD process is related to binary reactions (precursor A and reactant B) similar to chemical vapor deposition (CVD). During ALD process, films are formed layer-by-layer in atomic scales in which the surface of substrate is exposed to the precursor and the reactant separately and alternately by gas injection and purge cycles [3]. Thin films are deposited on the substrate by chemical adsorption of surfactants. With precisely adjusted process conditions, such as deposition temperature, reaction chamber pressure, precursor/reactant pulse and purge flow rate and time, growth rate of thin film and film quality can be adjusted.

EXPERIMENTAL METHOD

The aluminum nitride depositions test was carried out with ALD system, PF-300T Altair dual station system, designed by Piotech Inc. AlN thin film deposition was performed on 12-inch wafers at 350C on susceptor heater and 2.4Torr of chamber pressure. The precursor, TMA (Trimethyl-Aluminum) was used as Al-source and the reactant, NH3 (ammonia) as N-source in the present work. Film thickness and uniformity of 49 points within wafer were measured by KLA Aleries 8500 system. Besides measurement of film deposition on wafer, we also employed engineering simulation software, Ansys fluent module, to simulate gas distribution within the whole reactant volume during deposition process.

RESULTS AND DISCUSSION

Besides the susceptor temperature and chamber pressure, AlN thin film can also be affected by gas distribution, since the film is deposited by chemisorption on surface during the precursor and reactant injection cycles. In the present report, the gas was injected-into reaction chamber. In our process, precursor vapor TMA was delivered by carrier nitrogen and pulsed into the reaction chamber within first ALD half-pulse process. After nitrogen purging the chamber, reactant NH3 was introduced second ALD half-cycle process. And the flow rate proportion of TMA carrier gas and NH3 was precisely controlled.

During this work, we were focused on gas distribution within reaction chamber. The tubes account ratio of precursor and reactant was taken as a variant. The film was deposited in the chamber, and film thickness was measured. To figure out gas distribution during process, two kinds of gas injecting into chamber were simulated. Film thickness results and gas flow simulation results (in longitudinal cross section) were shown in Fig.1(a) and (b), separately. In Fig.1 (a), film thickness map indicated that several local points at radius-98mm were over deposited and with higher thickness. Concerning to film thickness uniformity, it was calculated and indicated by non-uniformity as 10%. With obvious kinds of regional thickness center on the film thickness map, we found they were consistent with the injection hole position of precursor and reactant. Gas distribution within chamber was shown that large NH3 gas flow rate caused partial accumulation center upper wafer 1mm. When we came to the longitudinal cross section figure, the regional large NH3 gas flow rate accumulation points caused vortex gas flow cross the whole wafer.

979-8-3503-1101-3/23 $31.00 © 2023 IEEE

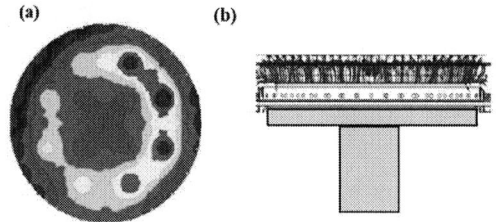

Fig. 1 Regional Design: (a) Film thickness deposition map with 49 points measured; (b) Gas flow distribution by Ansys simulation visual in longitudinal cross section.

To remove region highest thickness point within wafer, we re-designed total injection hole numbers and precursor and reactant ratio. After retrofit, gas flow rate was also re-simulated and film was deposited under same process conditions, including gas flow rate, heater temperature and pressure in all steps. The results were shown as in Fig. 2(a) and (b). With total injection holes increasing, gas distribution upper wafer was more uniform. The longitudinal cross section of gas distribution figure, the regional large NH3 gas flow rate accumulation points were removed significantly, and the vortex gas flow cross whole wafer was largely modified. Under new design, film thickness was more uniform, and partial thicker points were all eliminated. Film non-uniformity was enormously improved and decreased from 10% to 2% without any recipe tuning.

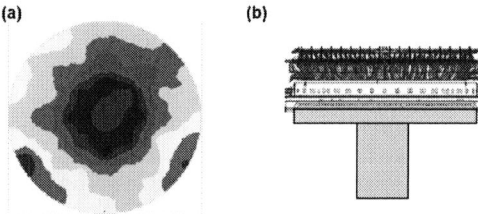

Fig. 2 New Design: (a) Film thickness deposition map with 49 points measured; (b) Gas flow distribution by Ansys simulation visual in longitudinal cross section.

CONCLUSION

The evaluation of a twin-chamber ALD system was performed with AlN film process analysis. The pattern of film thickness modification was carried out by new design by increasing total injection ratio. Employing Ansys simulation method, gas distribution helped us understanding the benefit of retrofit. Film non-uniformity was enormously decreased from 10% to 2% under same deposition process conditions. System and process refinement is in progress for further improvement.

ACKNOWLEDGEMENTS

I would like to give my heartfelt thanks to all the people who ever helped me in this paper and related works. My since and hearty thanks and appreciations to my supervisor in company. Many thanks to my group for generous supporting on this work.

REFERENCE

[1] S.C. Jain, M. Willander, J. Narayan, R. Van Overstraeten, J. Appl. Phys. 87 (2000) 965–1006.
[2] SUNTOLA T., *Thin Solid Films*, 1992, 216(1): 84-89.
[3] Richard W. Johnson, Adam Hultqvitst and Stacey F.Bent, Material Today,(2014) Vol.0.

Substrate Effect on Thermal ALD AlN Film Growth Rate

Xiaomeng Liu[1]*, Tiantian Liu[1], Wenyi Liu[1], Xinyu Zhang[1], and Hao Deng[1]

[1]Piotech (Shanghai) Inc., C13 1568 Feidu Road, Lingang New Area, China (Shanghai) Pilot Free Trade Zone

*Corresponding Author's Email: Xiaomeng.liu@sh.piotech.cn

ABSTRACT

Atomic layer deposition(ALD) method is widely employed to obtain thin films with accurate thickness. Since ALD process nucleation was verified to be island nucleation, i.e. V-M mode, growth rate could be enormously effected by substrates. In present report, AlN film was obtained by thermal ALD process, and growth rate per cycle of film on bare wafer with native oxide and well-grown silica film were measured by Optical equipment and TEM. We found that the uniform amorphous silica could significantly enhance growth rate than on bare wafer with native oxide attributed to higher surfactant group chemisorption on silica.

INTRODUCTION

Aluminum nitride film has been widely employed into integrated circuit as passivation layer and corrosion resistant layer attributed to its high thermal conductivity, high dielectric constants and high density [1]. Except metal-organic chemical vapor deposition (MOCVD), atomic layer deposition (ALD) is more preferred to be used for AlN film since it can result in high uniformity and conformity of film not only on 2D planar substrate but also in 3D structures. ALD film growing mode could be separated into F-M Mode, S-K Mode and V-M Mode. Based on the analysis of growth rate and growth time relationship in In2O3 [2] and ZnO [3] film, obtained by ALD process, island nucleation, i.e. V-M mode, is the preferred growth mode of ALD process in first tens of growing cycles which were verified by AFM and AES. As deposition continuing, small nucleation island could be merged into thin films and V-M growing mode would end only after film thickness reaches nanometer levels. And during the process of merging island structure into thin film, amorphous phase and polycrystalline structure would be formed due to large number of grain boundaries which is caused by the lattice mismatch and surface free energy difference, etc. [4].

Besides the growing mode, the surfactant group concentration also plays an important role in growth rate. A higher concentration of surfactant group could attract more targeting adduct on surface which would lead to higher growth rate of ALD process. This process would be finished gradually until saturation and film deposition is from hetero-epitaxy nucleation to homogeneous nucleation. Puuryunen [5] found that in the ALD Al2O3 deposition process, the surface concentration of hydroxide radical (OH) increase 6 times, the film composition ratio of methyl and Al ratio would be modified obviously. Therefore, the analysis of thin film growth rate in ALD process is quite important to obtain film with accurate thickness.

EXPERIMENTAL METHOD

In the present report, we focused on growth rate of aluminum nitride which was obtained by ALD system, PF-300T Altair dual station, designed by Piotech Inc. AlN film was deposited on 12-inch wafers at 350C on susceptor heater and 2.4Torr of chamber pressure. To figure out the substrate effect of heteroepitaxy nucleation and growing rate, AlN also deposited on bare silicon wafer with [001] orientation and amorphous silica substrate. A working as precursor TMA (Trimethyl-Aluminum), was introduced as Al source and NH3 (ammonia) as N-source. KLA Aleries 8500 system was employed to obtain film thickness and uniformity with 49 points within a wafer.

RESULTS AND DISCUSSION

In the present study, AlN films were deposited on bare Silicon wafers with [001] orientation, which has 8A native oxidation on surface. To determine the relationship of growth rate and growth cycle numbers, deposition thicknesses were measured after 5 cycles to 500 cycles.

In Fig. 1(a), film thickness with cycle number relationship is plotted. Film thickness is increasing linearly with cycle numbers. Film growth rate per cycle and cycle number correlation is also calculated with native oxide layer, as shown in Fig.1 (b). Growth rate per cycle, GPC showed a decreasing trend with cycle number for the first 30 cycles. Then growth rate per cycle began to increase to 1.86A/cycles till 500 cycles, and after that it became stabilized. But when native oxide thickness was substrate, growth rate per cycles showed a linearly relationship with cycle numbers, then it showed increasing with cycle numbers until 500 cycles to be saturated.

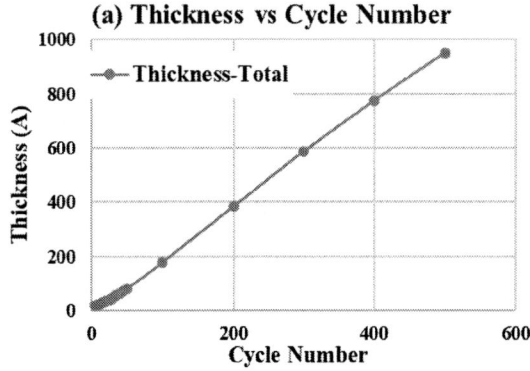

(a) Thickness vs Cycle Number

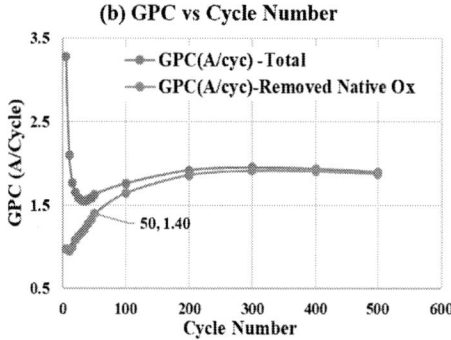

Fig. 1 ALD deposition on bare wafer with [100] orientation with native oxide (a) Relationship between thickness and cycle number; (b) Relationship between growth rate per cycle with cycle number.

In order to investigate the effect of silica thickness on the growth rate per cycles, AlN film was deposited on silicon oxide with a 200A of thickness by TEOS precursor plasma enhanced chemical vapor deposition (PECVD) method. AlN film was deposited on 200A silica with 16 cycles. As shown in Fig. 2(a), growth rate per cycles of AlN film on bare wafer and silica were compared using KLA Aleries 8500 system measured thickness. It can be seen from the figure that the growth rate of AlN film deposited by thermal ALD could be affected by the thickness of silica.

Growth rate of AlN film on 200A PECVD oxide was 20% higher than that of bare wafer with 8A native oxide. The thickness of AlN film deposited on 200A PECVD oxide was also verified by Transmission Electron Microscope(TEM) measurement and Energy Dispersive Spectroscopy(EDS) analysis. According to the test results, we found that the thickness of silica substrate has a significant effect on the growth rate.

As shown in TEM figure, AlN thin film deposited by thermal ALD process and silicon oxide deposited by PECVD were both in amorphous phase, while bare wafer was crystalline phase with [001] orientation. TEM results showed that 200A silica was much more uniform than 8A native oxide layer. The thinner native oxide layer was non-uniform because it was formed on strong crystalline oriented silicon. The hetero-interface growth of native oxide on silicon led closed packing atoms, which generated a large amount of residual thermal stress within native oxide layer.

Unlike the distorted native oxide layer, amorphous PECVD silicon oxide was more uniform and could provide more chemical adsorption sites on the reaction surface. The possibility of nucleation of AlN films in the first few cycles is enhanced by amorphous oxide. The growth rate of AlN film increased with the increase of chemical adsorption of adduct on silicon substrate. Therefore, the increased growth rate of AlN films could be attributed to the improved chemisorption efficiency of the precursor on amorphous silica.

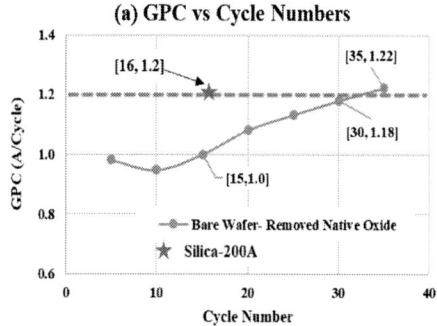

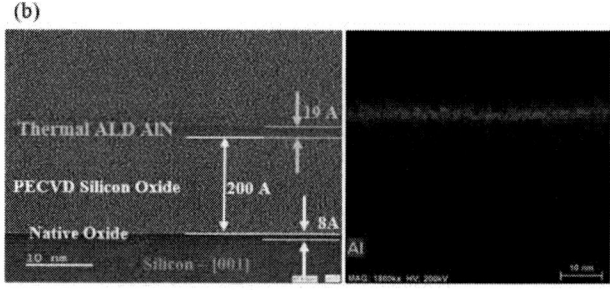

Fig. 2 ALD deposition (a) Relationship between growth rate per cycle with cycle number on bare wafer with native oxide thickness removed and on 200A PECVD Silicon Oxide; (b) Transmission Electron Microscope (TEM) and Energy Dispersive Spectroscopy (EDS) map of AlN thin film deposited on 200A PECVD Silicon Oxide.

CONCLUSION

AlN thin films were prepared thermal ALD method on bare wafers containing 8A native oxide layer and 200A PECVD silicon oxide film by. It was found that the growth rate of AlN film per cycle was linearly related to the number of deposition cycles on the bare wafer with thin native oxide. The periodic growth rate of AlN thin films was analyzed by optical measurement and transmission electron microscopy. It was found that amorphous silicon oxide films could significantly increase the growth rate of AlN thin films, which was attributed to the high chemisorption efficiency of its precursor.

979-8-3503-1101-3/23 $31.00 © 2023 IEEE

ACKNOWLEDGEMENTS

I would like to give my heartfelt thanks to all the people who ever helped me in this paper and related works. My since and hearty thanks and appreciations to my supervisor in company. Many thanks to my group for generous supporting on this work.

REFERENCE

[1] S.C. Jain, M. Willander, J. Narayan, R. Van Overstraeten, J. Appl. Phys. 87 (2000) 965–1006.

[2] ASIKAINEN T, RITALA M, LESKELA R, et al., Appl. Surf. Sci., 1996, 99(2): 91-98.

[3] LIM J, SHIN K, KIM H, et al., Thin Solid films, 2005, 475(1): 256-261.

[4] YOUSFI E B, FOUACHE J, LINCOT D., Appl. Surf. Sci., 2000, 153(4): 223-234.

[5] AHN C H, WOO C H, HWANG S Y, et al., Surface and Interface Analysis, 2010, 42(6/7): 955-958.

LAU's UNIFIED SCHOTTKY-POOLE-FRENKEL THEORY WITH ASYMMETRIC DISTORTION BY ELECTRON CHARGE TRAPPING PROPOSED TO EXPLAIN THE CURRENT-VOLTAGE CHARACTERISTICS OF HIGH-K METAL-INSULATOR-METAL CAPACITORS

W.S. Lau

Nanyang Technological University (Retired), School of EEE, Singapore 639798
E-mail: lauwaishing@yahoo.com.sg

ABSTRACT

In this paper, a unified Schottky-Poole-Frenkel theory with asymmetric electron charge trapping will be presented. The application of this theory to high-k MIM capacitors will be explained.

INTRODUCTION

The physics underlying the leakage current vs. voltage (I-V) characteristics of high-k MIM capacitors is not well understood even though intensive research has been done for about 60 years. It is expected that electron traps may have an important role. However, nobody has proposed a theory regarding how electron charge trapping can influence the I-V characteristics of high-k MIM capacitors. One of the leakage current mechanisms is the Schottky emission (S-E) mechanism. The author believes that the Schottky emission process can be distorted by charge trapping. Quite frequently, the I-V curves of high-k MIM capacitors are asymmetric such that the leakage current has a dependence on the polarity of the bias voltage. It can be speculated that the distortion by charge trapping also has a dependence on the polarity of the bias voltage. In this paper, the author will propose his new theory of Schottky emission with asymmetric distortion by charge trapping.

THEORY

The leakage current in high-k MIM capacitors can be due to the Schottky emission (S-E) mechanism or the Poole-Frenkel (P-F) mechanism. Previously in 2012, the author proposed a unified Schottky-Poole-Frenkel (S-P-F) theory including both mechanisms [1]. He also pointed out that there are 3 cases when log I is plotted against $V^{1/2}$ for both polarities of the bias voltage V, as shown in Fig. 1. Experimental results show that extra theory is still needed. The key point of this extra theory is that the high-k MIM capacitor can change during the I-V measurement process. People have been speculating about the effect of electron traps on the electrical properties of high-k capacitor structures. The author's theory is that in state-of-the-art

high-k capacitor structures, the high-k dielectric material has some deep donors and some deep electron traps. The deep donors in the bulk of the high-k dielectric are responsible for the P-F effect. The deep donors at the top and bottom interfaces can reduce the effective Schottky barrier ϕ_{Bn}. If the leakage current mechanism is the S-E mechanism or Fowler-Nordheim (F-N) tunneling, the interfacial deep donors can cause bigger leakage current due to the reduction of the effective Schottky barrier height. Then what is the effect of deep electron traps? The author would like to point out that the effect of deep electron traps is that ϕ_{Bn} increases during the I-V measurement process when the applied voltage increases.

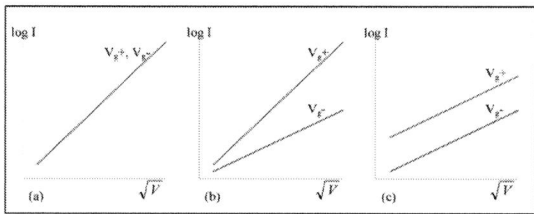

Fig. 1 3 diagrams of 2 curves of log I vs $V^{1/2}$ plotted together for both polarities of the bias voltage V. (a) P-F for both polarities, (b) P-F for positive bias and SE for negative bias and (c) SE for both polarities. The readers should note that the case of SE for both polarities tend to happen at relatively high voltage compared to the other two cases.

Hu et al. [2] pointed out that positive fixed charges in a high-k dielectric can cause a decrease of ϕ_{Bn}. By the same logic, the author believes that negative fixed charge in a high-k dielectric can cause an increase of ϕ_{Bn}. When an electron trap is empty, it is neutral; when an electron trap is occupied by an electron, it is negatively charged. Electron traps at the metal/high-k interface can capture electrons, resulting in negative charge trapping; negative charges at the interface repel negative electrons coming from the metal and effectively cause an increase of ϕ_{Bn}. Thus ϕ_{Bn} increases during the I-V measurement process when the applied voltage increases, resulting in a distortion of the I-V characteristics, as shown in Fig. 2. If the I-V characteristics have polarity asymmetry such that the leakage current for one polarity is much stronger than that for the other polarity, it can be imagined that the distortion of the I-V characteristics due to charge trapping by electron traps is significantly stronger for one polarity than for the other polarity, as shown in Fig. 3. This is the basis of the theory of "asymmetric distortion by charge trapping" proposed by the author.

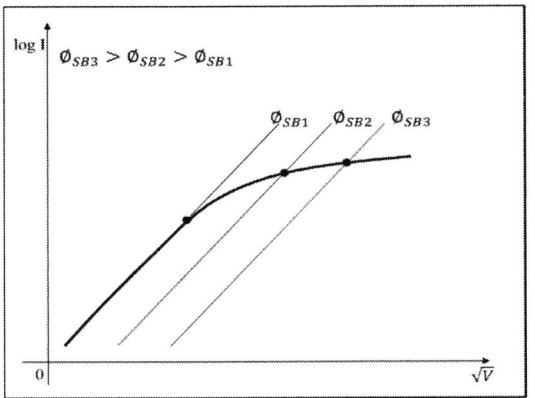

Fig. 2 This diagram shows that the plot of log I vs $V^{1/2}$ is a straight line at low voltage. When the voltage increases, the number of electrons injected from metal into high-k dielectric increases resulting in an increase of the effective Schottky barrier height. The plot of log I vs $V^{1/2}$ will be distorted from a straight line at low voltage into a curved line at high voltage.

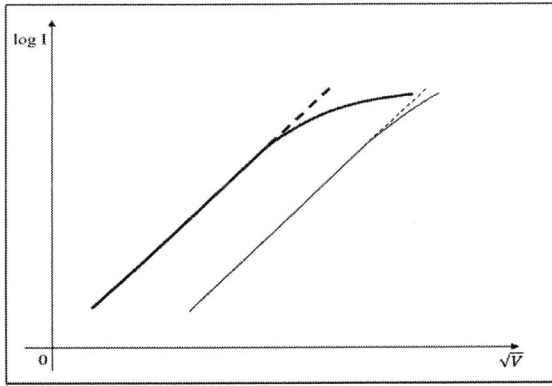

Fig. 3 This is a diagram of 2 curves of log I vs $V^{1/2}$ plotted together for both polarities of the bias voltage V. When the top and bottom electrodes have two different effective work functions, the log I vs $V^{1/2}$ plotted together for both polarities of the bias voltage V will show up as 2 straights lines. Electrons injected from metal can be trapped by

electron traps at the interface of the metal and high-k dielectric, resulting in distortion of the I-V characteristics. If the top and bottom interfaces are similar in terms of charging trapping capability, it can be expected that the upper line will be more distorted than the lower line. This is the situation of asymmetric distortion according to the author's theory.

EXPERIMENTAL SUPPORT

Previously in 2016, the author and co-workers have reported on the I-V characteristics of W/Ta$_2$O$_5$/W MIM capacitors [3]. Subsequently, the author proposed a new Ohmic-S-P-F theory in 2022; he pointed out the existence of O-SPF-1 and O-SPF-2 I-V characteristics [4]. In this paper, the author will point out that the I-V characteristics of W/Ta$_2$O$_5$/W MIM capacitors are consistent with the O-SPF-2 model. However, it appeared to the author that something extra is still required. In this paper, the author will propose a combination of the O-SPF-2 model together with his new theory of Schottky emission including asymmetric distortion by charge trapping.

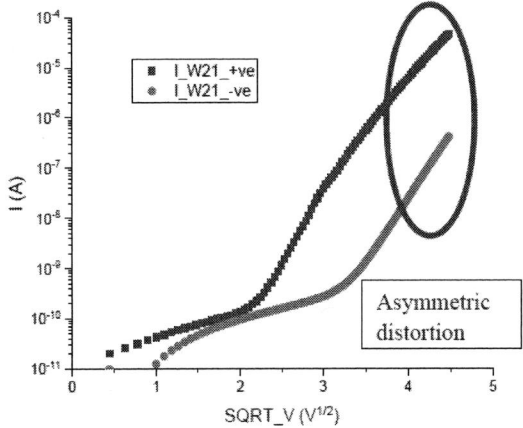

Fig. 4 This is a diagram of 2 curves of log I vs $V^{1/2}$ plotted together for both polarities of the bias voltage V for a W/Ta$_2$O$_5$/W MIM capacitor.

When the top and bottom electrodes have two different effective work functions, the log I vs $V^{1/2}$ plotted together for both polarities of the bias voltage V will show up as 2 straights lines, which are parallel to each other. Experimental results show 2 almost parallel lines. Electrons injected from metal can be trapped by electron traps at the interface of the metal and high-k dielectric, resulting in distortion of the I-V characteristics. If the top and bottom interfaces are similar in terms of charging trapping capability, it can be expected that the upper line will be more distorted than the lower line. This is the situation of asymmetric distortion according to the author's theory as shown in the region enclosed by an ellipse. If one of the interfaces is stronger than the other in terms of charging trapping capability, it is possible that the lower line will be more distorted than the upper line.

When there is distortion of the I-V characteristics due to chare trapping, the traditional method of curve fitting to find out the leakage current mechanism is difficult. This is because the validity of the theoretical equations of various leakage current mechanisms becomes doubtful due to the presence of charge trapping. Under such a situation, the approach of "pattern recognition" proposed by the author before [4] is easier to carry out in reality. Similarly, a combination of the O-SPF-2 model together with his new theory of Schottky emission including asymmetric distortion by charge trapping is possible.

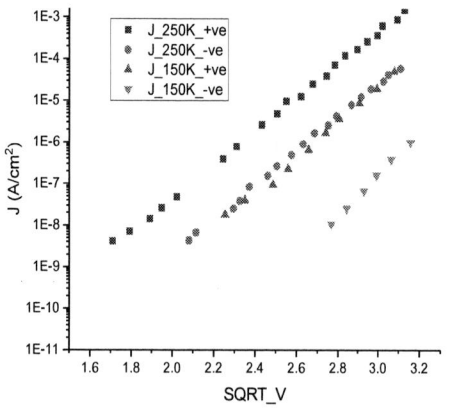

Fig. 5 This is a diagram of curves of log I vs V$^{1/2}$ plotted together for both polarities of the bias voltage V at two temperatures (250 K & 150 K) for a TiN/Ta$_2$O$_5$/TiN MIM capacitor. The asymmetric distortion at 150 K is noticeably stronger than that at 250 K.

Furthermore, the author extracted data from Deloffre et al. 2005 [5] and re-analyzed the data, resulting in Fig. 5. J is given by an equation like a constant x $[-\alpha(V)^{1/2}]$. For 250 K, the values of α are 9.18 and 9.42 for positive and negative bias voltage, respectively. For 150 K, the values of α are 9.83 and 11.8 for positive and negative bias voltage, respectively. The author noticed the asymmetric distortion has a temperature dependence. The asymmetric

distortion at 150 K is noticeably stronger than that at 250 K.

CONCLUSION

Previously, the author proposed the existence of O-SPF-1 and O-SPF-2 I-V characteristics for high-k MIM capacitors [4]. In this paper, the author extends his previous theory resulting in the proposal of the existence of O-SPF-1 and O-SPF-2 I-V characteristics with asymmetric distortion by charge trapping. Since the O-SPF-2 I-V characteristics can be observed more easily experimentally at room temperature than O-SPF-1, the author showed experimental evidence of the existence of O-SPF-2 I-V characteristics with asymmetric distortion by charge trapping instead of O-SPF-1. Furthermore, a variant of the O-SPF-2 type of I-V characteristics is the O-S (Ohmic-Schottky) type of I-V characteristics with the P-F portion negligible. The author noticed that the O-S type of I-V characteristics with asymmetric distortion by charge trapping is also commonly observed experimentally.

REFERENCES

[1] W.S. Lau, "An extended unified Schottky-Poole-Frenkel theory to explain the current-voltage characteristics of thin film metal-insulator-metal capacitors with examples for various high-k dielectric materials," *ECS J. Solid State Sci. Technol.*, vol. 1, pp. N139-N148, 2012.

[2] J. Hu, A. Nainani, Y. Sun, K.C. Saraswat and H.-S. P. Wong, "Impact of fixed charge on metal-insulator-semiconductor barrier height reduction", *Appl. Phys. Lett.*, vol. 99, article number 252104, 2011.

[3] D.Q. Yu, W.S. Lau, H. Wong, X. Feng, S. Dong and K.L. Pey, "The variation of the leakage current characteristics of W/Ta$_2$O$_5$/W MIM capacitors with the thickness of the bottom W electrode," *Microelectronics Reliability*, vol. 61, pp. 95-98, 2016.

[4] W.S. Lau, "The application of Lau's Schottky-Poole-Frenkel theory to distinguish leakage current mechanisms in high-k MIM capacitors by pattern recognition," *CSTIC 2022 (China Semiconductor Technology International Conference, Shanghai, 2022, IEEE)*, pp. 1-3, 2022.

[5] E. Deloffre, L. Montes, G. Ghibaudo, S. Bruyere, S. Blonkowski, S. Becu, M. Gros-Jean and S. Cremer, "Electrical properties in low temperature range of tantalum oxide dielectric MIM capacitors," *Microelectronics Reliability*, vol. 45, pp. 925-928, 2005.

CELL STRUCTURE AND PROCESS INTEGRATION OF A NOVEL 2T0C TECHNOLOGY FOR HIGH-DENSITY DRAM APPLICATION

Zheng-Yong Zhu[1]*, Bok-Moon Kang[1], Jing Zhang[1], Xin-Lv Duan[2], Jin-Juan Xiang[1], Guan-Hua Yang[2], Di Geng[2], Wang Dan[1], Xie-Shuai Wu[1], Ming-Xu Liu[1], Gui-Lei Wang[1] and Chao Zhao[1]

[1]Beijing Superstring Academy of Memory Technology, Beijing 100176, China
[2]Key Laboratory of Microelectronics Devices and Integrated Technology, Institute of Microelectronics, Chinese Academy of Sciences, Beijing, China
*Corresponding Author's Email: Zhengyong.Zhu@bjsamt.org.cn

ABSTRACT

A new DRAM 2T0C cell is introduced to resolve those special issues for traditional 2T0C DRAM. In this technology, the read transistor holds dual gates. The data is stored in one gate of read transistor, and the other gate is used to control read operation. By writing different-level voltages into storage gate, the read transistor will have different threshold voltages by using the other gate as control gate. Compared with 1T1C DRAM, read operation in this new technology is non-destructive and therefore no explicit capacitor is required. Low leakage is essential for write transistor to obtain long data retention time. A few cell structures of this new 2T0C technology are discussed for high-density DRAM application, and challenges of process integration is also analyzed.

INTRODUCTION

It is well known that the leakage of transistor is critical for DRAM products since it directly affects data retention time, refreshing frequency and power consumption. In addition to leakage, another concern for current 1T1C DRAM is the required large storage capacitor to build up detectable signal in digital line during sensing period, and it becomes one of main challenge to promote DRAM following 1T1C technology.

Recently Indium-Gallium-Zinc-Oxide (IGZO) transistor has attracted extensive study due to extremely low leakage [1], and it has been used for low-power display technology. It is expected that IGZO technology will improve current 1T1C DRAM, and furthermore make the 2T0C technology one promising route to future DRAM [2, 3]. Some remarkable progresses have been reported. For example, it has been demonstrated the gate length of IGZO transistor can be scaled down to less than 20 nm. In addition, a new compact $4F^2$ 2T0C cell structure was reported by using two channel-all-around (CAA) transistors, and it exhibits remarkable superiority due to high cell density, and simple process flow [4].

However, the traditional 2T0C technology has some inherent issues, including current sharing, IR-Drop of read word line (RWL) and so on. In latest year, we proposed a new DRAM 2T0C cell to resolve those issues [5, 6]. The main difference from traditional 2T0C is the read transistor in the new cell has dual gates. By writing different-level voltages into one gate of read transistor, which is used as storage node (SN), the read transistor will have different threshold voltages when the other gate works as control gate. In this paper, we will discuss the cell structure for this new 2T0C technology and process flow of integration. Challenges for future DRAM application is also analyzed briefly.

DUAL-GATE 2T0C CELL

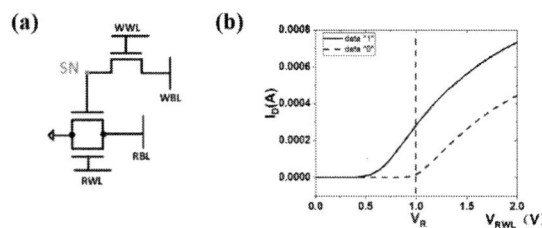

Figure 1: (a)Schematic of dual-gate 2T0C, (b) simulated I-V curves of read transistor for data "1" and "0".

Figure 1(a) presents the schematic of dual-gate 2T0C cell. It is seen that the cell is composed of two transistors, one for read operation and the other for write operation. To access data reliably, the read transistor is designed to have separate dual gates. The provided data in write bit line (WBL) is written into one gate of read transistor by turning on write transistor with write word line (WWL). The written data will be stored after write transistor is turned off. The storage node holds parasitic capacitance with other terminals, and it can serve data storage. The stored data can be kept for long time by refreshing operation. To reduce refreshing frequency, the write transistor with low leakage is preferred. The other gate of read transistor is connected to RWL and used to control read operation. In theory, the read transistor exhibits different threshold voltages (V_{TH}) when SN holds data "1" and "0", as shown in Fig. 1(b). Read transistor will become "ON" or "OFF", depending on stored data in SN, when proper voltage, V_R as seen in Fig. 1(b), is applied to the selected RWL. The stored data is thus read out through read bit line (RBL). For unselected rows, the associated read transistors are turned off by applying negative voltage to remaining RWLs, and thus there is no disturbing sneaky current as occurring in traditional 2T0C. Much more flexible control is achieved for our new dual-gate 2T0C DRAM. In addition, RWL is free of IR-drop issue.

CELL STRUCTURE

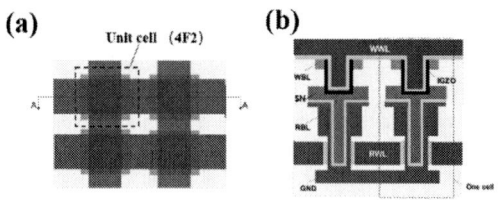

Figure 2: Schematic cell structure (a) top view and (b) cross section along A-A

There are a few embodiments for our dual-gate 2T0C cell. For example, planar or vertical structure can be implemented for both write and read transistors. In this paper we will focus on vertical structure, since it occupies less cell footprint as reported previously [4]. It is easy to extend the traditional 2T0C cell composed of two channel-all-around (CAA) transistors to our dual-gate 2T0C cell structure as seen Figs. 2 (a) and 2(b). It is seen that the read transistor has two individual gates. The internal surrounded gate is used as SN, and the outer surrounding one is part of RWL. One source/drain (S/D) terminal of read transistor is grounded (GND), and the other is connected to RBL. As discussed previously, write transistor is required to have low leakage, and special channel materials are necessary. IGZO material shows very good potential for this application. Finally, Fig. 2(a) shows comparable $4F^2$/cell density can be achieved as traditional 2T0C.

PROCESS FLOW

To manufacture the new 2T0C cell structure as shown in Figs. 2(a) and 2(b), the dummy RWL process is used for the dual-gate read transistor. The main process flow is shown in Fig. 3. Initially the substrate is prepared with one metal layer under a thin insulating layer, and it is grounded and used as source terminal of read transistor. After that, another sacrificial layer is grown and then patterned into the dummy gate for RWL as seen Fig. 3(a). It should be noted that this gate is along WL orientation, and its thickness determines the channel length of read transistor. The next step is to deposit another thin insulating layer. Both the above two thin insulating layers are used to isolate gate with source or drain. After that, one metal

layer is deposited and patterned along BL direction orthogonal to WL orientation, as seen Fig. 3(b). The following is to punch holes through to grounded metal layer to define read transistor as shown in Fig. 3(c). Then channel, gate dielectric and internal gate are then grown consecutively. To obtain conformal channel and gate dielectric on the side wall of hole and good gate filling, atomic layer deposition (ALD) is preferred. The bottom part of channel layer forms source contact with grounded metal layer. It should be noted that hole radius should be larger than total thickness of conformal channel and gate dielectric to allow space for internal gate, which is used as SN. The following patterning of internal gate and channel is performed to isolate storage nodes of adjacent cells as seen in Fig. 3(d).

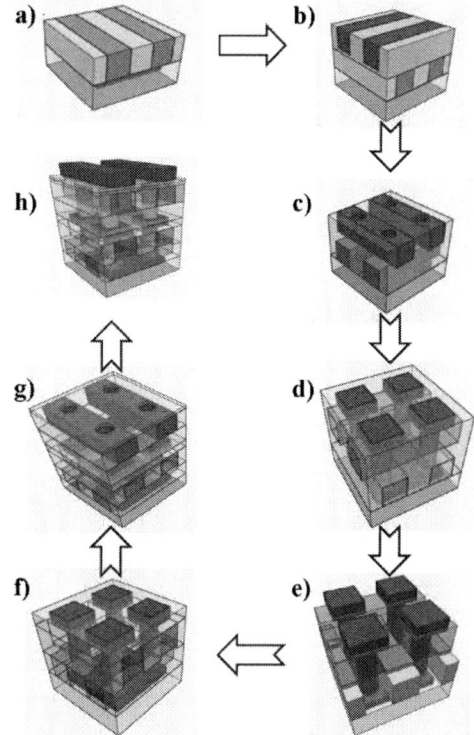

Figure 3: main process flows from (a) to (h) of new 2T0C cell using dummy RWL replacement technique.

The following step is to expose the dummy gate of read transistor by patterning and selectively etching the dielectric on top of the dummy gate. Following the well-known gate-last process in CMOS technology, then the dummy gate is removed by wet etch, and outer surface of channel is exposed, as shown in Fig. 3(e). Then gate dielectric and metal gate is grown by ALD in turn. The final schematic is shown as Fig. 3(f) when metal gate is recessed and planarization is completed with insulating dielectric filling. It is seen that the top surface of internal gate is exposed after planarization by chemical mechanical polishing (CMP). The remaining steps are to form write transistor, and most of them are the same as the process for read transistor except for dummy gate. One insulating layer is deposited, and its thickness define channel length of write transistor. Next, another metal layer is grown, and patterned as WBLs. Hole patterning for write transistor is completed in following step. The related etching process stops on metal layer of SN as seen in Fig. 3(g). The channel, gate dielectric and gate are grown consecutively following the previous flow for read transistor. Then the patterning for channel and gate is completed. After the deposition of insulating layer for planarization. The gate material is exposed by CMP. Finally, the metal layer is deposited and patterned for WWLs as shown in Fig. 3(h). Figure 4 shows the TEM image of CAA transistor with channel length less than 100 nm.

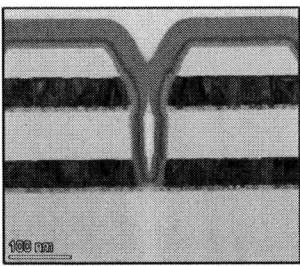

Figure 4: TEM of a CAA transistor structure

For 2T0C, extremely low leakage of write transistor is required to obtain a certain data retention time. One of available technology is IGZO transistor. For read transistor, one important factor is high current ratio of data "1" and "0" as shown in Fig. 1(b) to well distinguish the storing bit. Since there is always V_{th} variation among different transistors, and larger ΔV_{th} window is preferred. One way to improve this is to utilize thinner gate dielectric for internal gate as compared with that for RWL [6]. In addition, high read current for data "1" is also required for high-speed sensing process during read operation. Therefore, some other materials with high mobility can used for read transistor. For example, poly silicon or two-dimension materials usually have higher mobility than IGZO, and can be used for read transistor.

DISCUSSION

For 2T0C, non-destructive read operation is one of most remarkable difference compared with conventional 1T1C memory. Therefore, there is no severe constriction on the cell capacitance in theory. However, SN stability is still very critical to promise correct functionality and increase retention time. As considering above, one optimized structure is proposed as shown in Fig. 5. The major difference is to swap locations of RBL and GND metal layers. It is seen that GND metal layer is placed above RWL, and much more overlapping area between GND and SN is achieved but without other overheads. Since GND keeps constant all the time, instability of SN due to coupling effect by the fluctuation of other signals is much reduced because of the stabilizing of large capacitance between SN and GND. This benefit will be more significant for memory array, because coupling issue is more serious due to the coupling from adjacent cells. In addition to stability improvement, the retention time is also increased due to large total SN capacitance, and this will release the leakage restriction of write transistor or the frequency of refreshing operation. Another possible improvement of new structure is the reduction of capacitance of RBL. It is seen that overlapping area between RBL and SN can be smaller as compared previous 2T0C cell through the optimization of structure parameter. It is well known that the capacitance loading of BL in 1T1C memory is one of most important factors, since large enough ratio of cell capacitance (C_S) to total capacitance of each BL (C_{BL}) is required, and this will define maximum size of memory array. For 2T0C, the reduction of RBL capacitance loading also makes significance, since it helps to speed up read operation. \

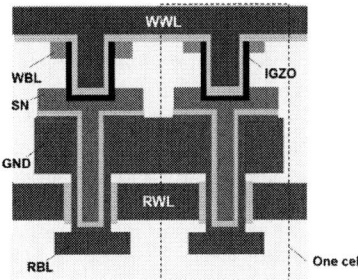

Figure 5: Schematic cell structure with improvement on SN stability

The proposal of 2T0C based on CAA transistors indeed can help to achieve $4F^2$ cell density and simplify process flow for integration. However, it also brings some disadvantage to the cell operation. For example, large capacitive loading of WBL and WWL is accompanied with the CAA write transistor, in which large overlapping capacitance is formed between WWL and WBL. Smaller capacitance loading of WWL is preferred to achieve good pulse signal with short rising and falling time, and it will help to reduce the required switch-on time of WWL. The WBL capacitive loading has direct relation with the quantity of charged or discharged charge during write/refresh/pre-charge operation, and it has severe impacts on the speed of corresponding operation. Figure 6 shows another structure of 2T0C cell, in which the write transistor holds gate-all-around structure. It can be inferred that the capacitance between WWL and WBL is much reduced, and that higher operation speed is achieved.

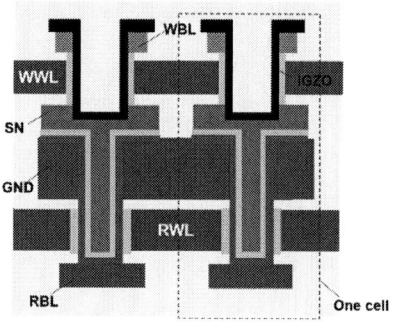

Figure 6: Schematic cell structure with improved structure of write transistor

As mentioned above, write transistor is required to hold extremely low leakage for both traditional and new 2T0C technology. Among all explored channel materials, IGZO is the most promising for this application until now. However, there are still some issues to address. Contact resistance is one outstanding one, especially for CAA transistor, since processing cannot be operated selectively on source and drain region without damaging channel. One solution was reported through inserting ITO between S/D metal and IGZO for planar device in latest year [7]. In similarity, it is expected that multiple stacking layers of S/D structure can be used to improve contact for CAA IGZO transistor.

Although the proposed vertical 2T0C structure with CAA transistors holds some superiority to planar 2T0C, 3D or multiple stacking of vertical cells is required to compete with 1T1C DRAM technology. For multiple stacking scheme, besides the increasing of manufacturing cost with the number of stacking layers, another challenge is the sensitivity of IGZO material to thermal budget, which is probably the biggest challenge to explore application for vertical CAA 2T0C technology. New 3D integration for both 1T1C and 2T0C is more promising way to promote DRAM technology.

CONCLUSIONS

A new dual-gate DRAM 2T0C cell is introduced to resolve those issues for traditional 2T0C. Based on $4F^2$ vertical 2T0C structure, dummy gate replacement is implemented to complete process integration. A few cell structures of this new 2T0C technology are discussed as

concerning to performance improvement. The contact issue and possible solution is addressed. Finally, challenges of 3D integration for this vertical 2T0C structure are analyzed. It is pointed out that new 3D integration for both 1T1C and 2T0C is required to promote DRAM technology.

ACKNOWLEDGEMENTS

This work was supported by the National Key R&D Program of China (Grant No.2022YFB3606900）

REFERENCES

[1] K. Kato, Y. Shionoiri, Y. Sekine, K. Furutani, T. Hatano, T. Aoki, M. Sasaki, H. Tomatsu, J. Koyama, and S. Yamazaki, Jpn. J. Appl. Phys., vol. 51, 2012, pp. 021201(1-7).
[2] A. Belmonte, H. Oh, N. Rassoul, G.L. Donadio, J. Mitard, H. Dekkers, R. Delhougne, S. Subhechha, A. Chasin, M. J. van Setten, L. Kljucar, M. Mao, H. Puliyalil, M. Pak, L. Teugels, D. Tsvetanova, K. Banerjee, L. Souriau, Z. Tokei, L. Goux, G. S. Kar, Tech. Dig. - Int. Electron Devices Meet., 2020, pp. 609-612.
[3] A. Belmonte*, H. Oh, S. Subhechha, N. Rassoul, H. Hody, H. Dekkers, R. Delhougne, L. Ricotti, K. Banerjee, A. Chasin, M. J. van Setten, H. Puliyalil, M. Pak, L. Teugels, D. Tsvetanova, K. Vandersmissen, S. Kundu, J. Heijlen, D. Batuk. J. Geypen, L. Goux, G. S. Kar, Tech. Dig. - Int. Electron Devices Meet., 2021, pp. 226-229.
[4] X. Duan, K. Huang, J. Feng, J. Niu, H. Qin, S. Yin, G. Jiao, D. Leonelli, X. Zhao, W. Jing, Z. Wang, Q. Chen, X. Chuai, C. Lu, W. Wang, G. Yang, D. Geng, L. Li and M. Liu, Tech. Dig. - Int. Electron Devices Meet., 2021, pp. 222-225.
[5] Z. Zhu, B. Kang, D. Wang, X. Wu, J. Son, Y. Yu, D. Xiao, J. Dai, G. Wang, A. Yoo, K. Cao and C. Zhao, ICSICT, 2022, pp. C6-5.
[6] W. Lu, Z. Zhu, K. Chen, M. Liu, B. Kang, X. Duan, J. Niu, F. Liao, D. Wang, X. Wu, J. Son, D. Xiao, G. Wang, A. Yoo., K. Cao, D. Geng., N. Lu, G. Yang, Z. Zhao, L. Li, and M. Liu, Tech. Dig. - Int. Electron Devices Meet., 2022, pp. 611-614.
[7] H. Qianlan, L. Qijun, Z. Shenwu, G. Chengru, L. Shiyuan, H. Ru and W. Yanqing, Tech. Dig. - Int. Electron Devices Meet., 2022, pp. 619-623.

THE STUDY OF SLIP DEFECTS IN FURNACE HIGH TEMPERATURE PROCESS

SUN Yan[1]*, WEI Simeng[1] *, XIE Yuanxiang[1]

Beijing NAURA Microelectronics Equipment Co., Ltd
[1]No. 8 Wenchang Avenue, Beijing E-Town, Beijing 100176, P. R. China
*Corresponding Author's Email: sunyan01@naura.com

ABSTRACT

As the feature size of device continues to shrink, the whole industry puts forward higher requirements for the performance and quality of silicon chips. Meanwhile, in order to reduce the cost of integrated circuit manufacturing, it has become a trend to use larger wafers (e.g. from 8 inches to 12 inches) for various silicon based devices.

For those high temperature processes treated in furnace, especially exceeding 1000℃, a larger wafer size leads to an increase in thermal stress caused by gravity mostly, which often results in the defects including slip dislocations during heat treatment in high temperature oxidation and anneal processes. The dimension and distribution of slip defects are critical for devices. In some cases the aggregation and severe defects may cause device failure. Therefore, it becomes much more important to find methods to control the defect level by optimization of certain process condition to reduce the occurrence of slips during heat treatments. For instance, it is very critical for the design of wafer support structure, which is related to higher temperature requirements of vertical oxidation diffusion equipment.

There are many literatures reporting on the generation and inhibition of silicon wafer slip defects. These papers mainly focus on the nailing effect of oxygen atoms, the influencing factors of gravity stress, the prediction of temperature rise and fall rate, and the theoretical calculation etc. But there are few summaries to analyze the correlations between them. Moreover, measurement techniques and early theories are needed to be updated iteratively. In this review, we describe the recent progresses in the field of silicon wafer slip and discuss the effect and following improve methods to control slip defects during various thermal processes. Meanwhile summarize the influencing factors of slip line defects, such as wafer strength, gravity-dependent stress and thermal stress, based on the formation mechanism of slip lines. Finally, we propose an optimization method to control slip defects for high-temperature process in vertical furnace.

INTRODUCTION

In order to reduce the cost of IC manufacturing and improve production efficiency, it has become a trend to use the large size wafers. The increase in wafer diameter leads to increased gravitational and thermal stresses, making it more susceptible to slip defects during high-temperature process. The occurrence of slip defects is harmful to semiconductor device, as it may cause short circuits or leakage. In addition, as the feature size of IC device continues to shrink, advanced processes become more sensitive to slip defects. Therefore, it is crucial to reduce wafer slip defects during thermal treatment.

During wafer thermal treatment, slip occurs when thermal stress and gravity-dependent stress exceeds the critical shear stress. In the past few decades, many studies on the generation and suppression of slip defects have been reported. Fujise et al. [1] investigated the relationship between slip defects and oxygen concentration in silicon wafers. As well as the locking effect by oxygen atoms of dislocations, it was found that strain rate also has an important role on the value of critical stress. Fischer et al. [2] theoretically calculated the generation conditions for slip in 200 mm silicon wafers in vertical and horizontal furnaces. Jurkschat et al. [3] proposed a method to calculate the critical shear stress of silicon wafers containing oxygen precipitation to prevent slip defects. Aghabekyan et al. [4] reported that slip defects first occur in the concave and convex areas of the wafer and increase with increasing processing temperature. Nilson et al. [13] introduced the factors that affect gravitational and thermal stresses in wafers and gave an example of the highest allowable temperature rise rate without slip defects, depending on wafer size, boat structure, and wafer vertical spacing.

There are many literatures reporting on the generation and inhibition of silicon wafer slip defects. These papers mainly focused on the nailing effect of oxygen atoms, the influencing factors of gravity stress, the prediction of temperature rise and fall rate, and the theoretical calculation etc. But there are few systematic studies on the suppression of slip line defects. In this review, we summarize the influencing factors of slip line defects, such as wafer strength, gravity-dependent stress and thermal stress, based on the formation mechanism of slip lines. Wafer strength is related to factors such as process temperature and doping concentration. Gravity stress is determined by the size and thickness of the wafer, the material and structure of boat. Thermal stress is related to the temperature rise and fall rate and wafer spacing. By studying the mechanism and factors influencing slip defects, this review has significant guidance for improving slip line defects in high-temperature process, especially in vertical furnaces.

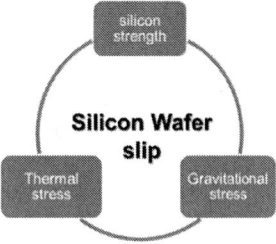

Figure 1: Effect of induced Slip during thermal process in silicon wafer

The Strength of Silicon Wafer

Wafer strength can be evaluated by the critical shear stress, which is an important factor affecting the plastic deformation of materials. [1] The higher the critical shear stress, the less likely slip defects will occur. Therefore, studying the influence of critical shear stress on wafers is of great significance for improving wafer strength. The critical shear stress strongly depends on temperature, which is also an important process parameter in heat treatment. In addition, the critical shear stress is also related to the concentration of oxygen and doped boron atoms in the wafer. [5-7]

The Effect of Temperature

Temperature is an important factor affecting the critical shear stress of silicon crystal, which is a physical quantity that truly reflects the initial yield stress. Usually, the critical shear stress also tends to decrease as the temperature increases.[1, 8-9] Hu reported the temperature dependence of the critical resolved shear stress in oxygen-free silicon determined in the temperature range 800~1050℃. To separate out the contribution of oxygen on the critical shear stress, this study used wafers of float-zone (FZ) silicon, containing < 1 × 1016 atoms/cm3 of oxygen. As shown as Fig. 2, the stress to vary only about 30% in the temperature range studied.

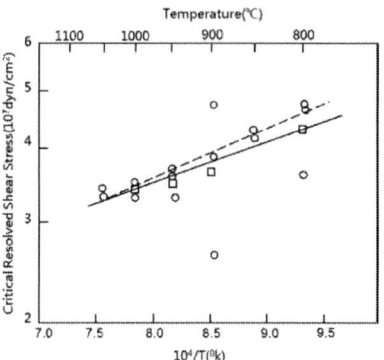

Figure 2: Critical resolved shear stress of dislocation movement in silicon function of temperature

Subsequently, Senkader et al. [8] studied the critical shear stress values of Czochralski (CZ) silicon wafers within a certain temperature range and provided the formula for calculating the critical shear stress as follows:

$$\tau_u = AC_o \exp(\frac{E_a}{kT})$$

Where A is a constant, T (K) is the test temperature, k is the Boltzmann constant (8.62 × 10–5 eV K−1), Co is the oxygen concentration. They showed the critical shear stress dependents on both temperature and oxygen concentration. However, these two factors are not independent, and the critical shear stress decreases as the temperature increases due to the decrease in oxygen concentration within the wafer at higher temperatures. The critical shear stress decreases with increasing temperature. Therefore, in order to improve the strength of the wafers, the process temperature should be lowered within the allowable range to reduce the occurrence of slip defects.

The Effect of Oxygen Concentration

Oxygen atoms in silicon wafers have a pinning effect, which can increase the critical shear stress and inhibit slip during thermal process. In 1977, Hu [10] first demonstrated the pinning effect of oxygen atoms in silicon wafers, which can increase the critical shear stress of silicon by up to four times under certain oxygen concentration conditions. This discovery made people realize that completely oxygen-free silicon wafers may not be suitable for complex integrated circuit process. Subsequently, Harada et al. [11] compared the indentation behavior of CZ and FZ Silicon crystals (Fig. 3), and confirmed that the indentation rosette extensions around the indentations made of the FZ silicon wafer are larger than those of the CZ silicon wafer. This is because the CZ silicon wafer contains a certain amount of oxygen atoms, mainly from the quartz crucible container used in the preparation process. While during the growth of the FZ silicon wafer, the molten silicon does not contact any object, resulting in higher single crystal purity and very low oxygen content. Thus, the floating-zone silicon wafer is prone to produce defects such as dislocations and slips during high-temperature process. Therefore, most IC process use CZ silicon wafers.

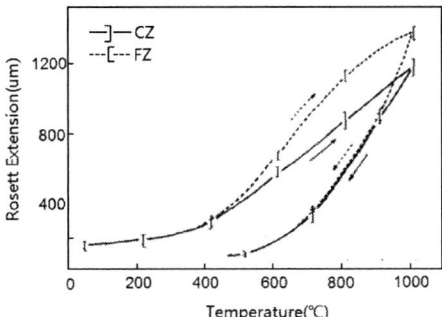

Figure 3: Rosette extensions plotted against the indentation temperature for the float-zone and Czochralski specimens subjected to a heat cycle up to 1000 ℃

Interstitial oxygen atoms can inhibit dislocations and reduce slip defects. However, at a certain process temperature, the oxygen concentration reaches saturation and forms oxygen precipitates. Proper amounts of oxygen precipitates can adsorb impurities and improve wafer strength. However, excessive oxygen precipitates can become sources of slip dislocations, and during thermal treatment, wafers are more prone to deformation, such as warpage. In 1980, Leroy and Plougonven [12] found that wafer strength is related to the quantity and form of oxygen precipitates. As shown in Figure 4, as the concentration of oxygen precipitates increases, the critical shear stress decreases. Furthermore, due to the influence of oxygen precipitation and the stress field around them, high oxygen precipitates are beneficial to the formation of defects in the center of the wafer, while low oxygen precipitates are beneficial to the formation of defects on the edge. Subsequently, many studies discussed the mechanism of slip dislocations caused by oxide precipitates [14-18]. During thermal treatment, as the concentration of oxygen precipitates increases, the oxygen atom concentration decreases, and the critical shear stress also decreases. Oxygen precipitates also become sources of slip dislocations, making slip more likely to occur. Therefore, it is necessary to control the concentration of oxygen precipitates to reduce slip.

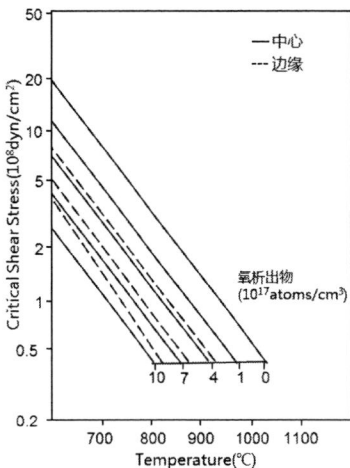

Figure 4: Critical shear stress at the center and at the edge of the wafer as a function of the temperature and the oxygen precipitated

The Effect of Boron

Silicon wafers must be doped with certain impurities to achieve the desired electrical properties. Doping the wafer with group III elements, such as boron, can form P-type wafers, while doping with group V elements such as phosphorus, arsenic, or antimony can form N-type wafers. Boron doping is the most commonly used, making P-type wafers more widely applied. It has been reported in the literature that boron doping will enhance the wafer strength. For instance, in 2000, Fukuda [19] reported the relationship between slip length and temperature for heavily boron-doped (P+ and P++) 300mm wafers (as shown in Figure 5), demonstrating that higher boron doping concentration leads to higher wafer strength and less susceptibility to slip.

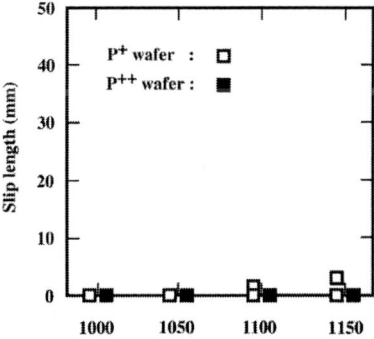

Figure 5: Temperature dependence of slip length in heavily boron-doped (P⁺ and P⁺⁺) 300 mm wafers

Gravity-related Stress

In order to improve production efficiency and save costs, the wafer diameter has been increased from 200mm to 300mm. As the wafer diameter increases, gravity-related stress also increases significantly, and slippage is more likely to occur under high temperature processing conditions. Therefore, it is necessary to suppress the occurrence of slippage in large-size wafers. In this paper, we will discuss the effects of wafer size and thickness on gravity-related stress, as well as methods for reducing gravity stress by optimizing boat material and structure.

Effects of Wafer Size and Thickness

As the wafer diameter increases, the slip defects become more severe in high-temperature because the self-gravity induced bending stress largely increases with wafer size. [19-22] Gravity stress is closely related to the size and thickness of the wafer. Fukuda [19] and YOO et al. [21] reported that the bending stress caused by gravity is proportional to R^2/t, where R is the wafer radius and t is the thickness. Table 1 shows a comparison of R^2/t for 200mm and 300mm wafers, where R^2/t for the 200mm wafer is normalized to 1. Since the diameter of the 300mm wafer has increased significantly but the thickness has changed very little, the bending stress is approximately twice that of the 200mm wafer. This means that larger wafer sizes are more prone to slippage under the same conditions, and therefore, necessary measures should be taken to suppress slippage in large wafers.

Effects of Crystal Boat Material and Structure

Vertical furnaces are capable of processing large quantities of wafers, which are horizontally placed on crystal boats and then transported into the reaction chamber for thermal treatment. Crystal boats are made of different materials, including Si, SiC, and SiO2. Although SiO2 is more cost-effective, its thermal stability at higher temperatures (>1000℃) is poor. Additionally, the thermal expansion coefficients of silicon dioxide and silicon are different, which can cause relative motion between the wafer and the boat during the thermal treatment process, leading to scratches on the back of the wafer. In contrast, the material of the silicon boat is the same as that of the silicon wafer, so relative motion is less likely to occur during thermal treatment. Fukada [23] reported the impact of crystal boat materials on slip line length for 200mm wafers in the following order: Si~SiC＜SiO2. Therefore, compared to quartz, SiC and Si have more similar thermal expansion coefficients and higher thermal stability, making them widely used in high-temperature process equipment.

Wafers are horizontally placed on the boat, with only a small portion of each wafer supported by the boat, usually in a point or ring shape. The remaining portion of the wafer bends due to its weight, and bending stress is thus generated. On the other hand, slip dislocations usually occur in the wafer support area and grow at high temperatures. At these locations, the superposition of supporting force, bending force, and contact force can result in extremely high shear stress. Therefore, the magnitude of elastic stress in the wafer material is strongly influenced by the geometric structure of the wafer support, which is important for reducing gravity-related stress [17].

Typical support for 300mm wafers includes conventional boats with three or four supporting rods. The main issue for supporting large-diameter wafers on these boats is to prevent dislocations from occurring in the wafers. Fukada et al. [23] used thin plate line elasticity theory to calculate the shear stress caused by wafer bending in traditional three-point and four-point supports for 300mm wafers. As shown in Figure 6, the stress is highest for the conventional support method, at 2.3MPa, followed by the three-point support at 1.6MPa and the four-point support at 1.1MPa. They believe that traditional and three-point supports are not suitable for 300mm wafers because slippage is inevitable.

Wafer Diameter 2R(mm)	Thickness t(μm)	Volume (cm³)	R^2/t
200	725	22.78	1.0
300	775	54.78	2.1

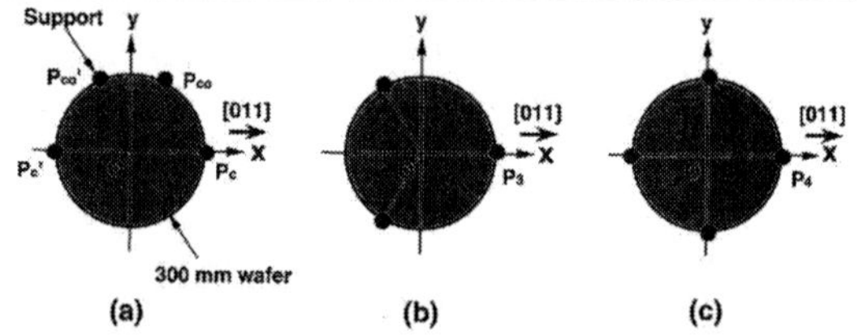

Figure 6: Three methods to support silicon wafers: conventional support (a), three-point support (b), and four-point support(c)

Improving the support method of 300 mm wafers is important because the gravitational stress increases greatly in comparison with that of 200 mm wafers. The ring-like support is an effective method for slip-free processes in 300 mm wafers. Akatsuka et al. [24] showed that a slip took place and was extended at wafer periphery, provided that 300 mm wafers were annealed on a conventional four-point support even at the lowest ramping up rate (1℃/min) from 900 to 1200℃. On the other hand, when 300 mm wafers were supported on a ring-like boat, in which bending stress was reduced considerably more than that of the conventional support, almost no slip was observed at 1200℃ (Fig 7).

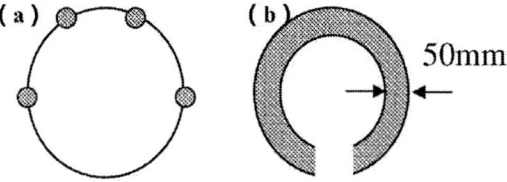

Figure 7: Support methods for 300mm wafers: (a) is Four-point; (b) is ring-like

On the other hand, the position of the support points is also important for reducing slip. Takeda et al. [20] found that moving the support points of the wafer's periphery inward by a certain distance significantly reduces the stress caused by gravity. The self-weight stress on a horizontally supported wafer can be expressed as:

$$\sigma_g = \frac{R \times R}{t} F_\sigma(\frac{x}{R}, \frac{b}{R}, \ v, \ \rho)$$

Where R is the radius of the silicon wafer, t is the thickness of the wafer, x is the radial coordinate of the wafer, b is the radial coordinate of the support position, v and ρ are the Poisson's ratio and density of silicon, respectively. As shown in Figure 8, the research indicates that the gravity-related stress is minimized when the support point position is at b/R=0.7, and is maximized when supported at the edge or center.

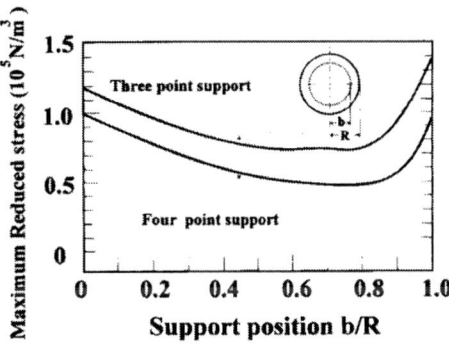

Figure 8: Maximum value of radial stress component when four supporting points are positioned within the wafer plane.

In summary, reducing gravity-induced stress is an important approach to controlling slip. At the support configuration with several points more than three, slip control is difficult, because there are some cases in which actual contact points are not more than three points and the support points may not be in symmetric positions.

However, by moving the support points from the periphery to the inside (b/R=0.7), significant reduction in gravity-induced stress can be achieved. [20] Therefore, we believe that inner or annular support would be the ideal choice for 300 mm wafer heat treatment.

Thermal Stress

Thermal stress is the stress caused by temperature changes. Thermal stress from T0 to Tf can be calculated by:

$$\sigma_t = E\alpha1(T_0 - T_f) = E\alpha1\Delta T$$

Where E is the elastic modulus and α1 is the linear thermal expansion coefficient. The thermal stress comes from the temperature gradient inside the wafer. The temperature gradient is usually generated by rapid heating and cooling. In this case, the external temperature of the wafer changes faster than the internal temperature, and the resulting size difference constrains the free expansion or contraction of adjacent areas. For example, during rapid heating, the wafer's outer region is hotter, causing it to expand more than the inner region and generating compressive stress. Conversely, during rapid cooling, the surface produces tensile stress.

As shown in Figure 9, thermal stress in silicon wafer batches during furnace treatment was studied. The wafers are first transferred to the boat at room temperature, and then the boat is moved into a reaction chamber at a certain temperature. It can be observed that the thermal stress of the wafer during the vertical furnace process is related to the heating power, insertion speed, wafer spacing, and withdrawal speed. [26~27]

Therefore, to reduce slip during the wafer processing, the thermal stress should be reduced, i.e., the temperature gradient inside the wafer should be reduced. Specifically, this can be achieved by reducing the temperature and velocity of the boat during the process to reduce the temperature difference and thermal plastic deformation inside the wafer. On the other hand, increasing the vertical spacing of wafers on the boat can improve the heat conduction rate and also reduce the temperature difference inside the wafer to a certain extent. In addition, reducing the temperature rise and fall rate can prevent an increase in thermal stress caused by a large temperature difference between the edge and center of the wafer.

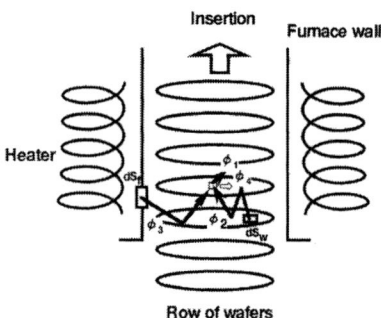

Figure 9: Heat transfer between a row of wafers and the heater, schematically shown in a vertical-type furnace.

Conclusion

Using larger wafers are more economical option to reduce the cost and increase the productivity of integrated circuits (ICs). However, larger wafer diameter also means more strict gravitational stress and thermal stress control are necessary. To reduce the risk of slip defects occur during high-temperature processing, which means device yield lost, below factors should be noticed for control slip line defects effectively: wafer strength, gravity-dependent stress, and thermal stress. Based on the formation mechanism of slip lines, we also proposes an optimization method by combining above factors to control slip defects produced in high-temperature process, especially for the vertical furnaces widely used to control the production cost finally.

REFERENCES

[1] Fujise J, Ko B, Ono T, et al., The Critical Shear Stress for Slip Generation due to Scratches in Silicon Wafers. ECS Journal of Solid State Science and Technology 2020, 9 (5), 055012.

[2] Fischer A, Kissinger G, Ritter G, et al., Plastic deformation in 200mm silicon wafers arising from mechanical loads in vertical-type and horizontal-type furnaces. Materials Science and Engineering: B 2009, 159-160, 103-106.

[3] Jurkschat K, Senkader S, Wilshaw P R, et al., Onset of slip in silicon containing oxide precipitates. Journal of Applied Physics 2001, 90 (7), 3219-3225.

[4] Aghabekyan A V, Ayvazyan G EVardanyan A H In Distribution of slip dislocations in thermally deformed silicon wafers, 2002 23rd International Conference on Microelectronics. Proceedings (Cat. No.02TH8595), 12-15 May 2002; 2002; pp 555-557 vol.2.

[5] Hu S M, A method for finding critical stresses of dislocation movement. Applied Physics Letters 1977,

31 (3), 139-141.

[6] Fukuda TOhsawa A, Mechanical strength of silicon crystals with oxygen and/or germanium impurities. Applied Physics Letters 1992, 60 (10), 1184-1186.

[7] Hu S M, Temperature dependence of critical stress in oxygen‐free silicon. Journal of Applied Physics 1978, 49 (11), 5678-5679.

[8] Senkader S, Jurkschat K, Gambaro D, et al., On the locking of dislocations by oxygen in silicon. Philosophical Magazine A 2001, 81 (3), 759-775.

[9] Fischer A, Richter H, Shalynin A, et al., Upper yield point of large diameter silicon. Microelectronic Engineering 2001, 56 (1), 117-122.

[10] Hu S M, Dislocation pinning effect of oxygen atoms in silicon. Applied Physics Letters 1977, 31 (2), 53-55.

[11] Harada HSumino K, Indentation rosettes and dislocation locking by oxygen in silicon. Journal of Applied Physics 1982, 53 (7), 4838-4842.

[12] Leroy B, Warpage of Silicon Wafers. Journal of the Electrochemical Society 1980, 127 (4), 961.

[13] Nilson R HGriffiths S K, Scaling wafer stresses and thermal processes to large wafers1Presented at the TCMCTF '97.1. Thin Solid Films 1998, 315 (1), 286-293.

[14] Shimizu H, Watanabe TKakui Y, Warpage of Czochralski-Grown Silicon Wafers as Affected by Oxygen Precipitation. Japanese Journal of Applied Physics 1985, 24 (Part 1, No. 7), 815-821.

[15] Sueoka K, Akatsuka M, Katahama H, et al., Dependence of Mechanical Strength of Czochralski Silicon Wafers on the Temperature of Oxygen Precipitation Annealing. Journal of the Electrochemical Society 2019, 144 (3), 1111-1120.

[16] Fujise J, Ko B, Ono T, et al., Measurement and empirical equation of critical stresses for slip generation from oxide precipitates in silicon wafers. Japanese Journal of Applied Physics 2018, 57 (3), 035501.

[17] Ono T, Sugimura W, Kihara T, et al., Wafer Strength and Slip Generation Behavior in 300 mm Wafers. ECS Transactions 2006, 2 (2), 109-122.

[18] Chiou H D, Improving Axial Oxygen Precipitation Uniformity in CZ Silicon Crystals Using the S‐Curve Concept. Journal of the Electrochemical Society 1994, 141 (1), 173-178.

[19] Fukuda THikazutani K-i, The Analysis of Slip Extension and Induced Stress in 300 mm Diameter Wafers on Three-point Symmetrical Support. Japanese Journal of Applied Physics 2000, 39 (Part 1, No. 3A), 999-1005.

[20] Takeda R, 300 mm Diameter Hydrogen Annealed Silicon Wafers. Journal of the Electrochemical Society 1997, 144 (10), L280.

[21] Yoo W S, Fukada T, Yokoyama I, et al., Thermal

Behavior of Large-Diameter Silicon Wafers during High-Temperature Rapid Thermal Processing in Single Wafer Furnace. Japanese Journal of Applied Physics 2002, 41 (Part 1, No. 7A), 4442-4449.

[22] Fukuda T, the Relationship between the Bending Stress in Silicon Wafers and the Mechanical Strength of Silicon Crystals. Japanese Journal of Applied Physics 1995, 34 (Part 1, No. 6A), 3209-3215.

[23] Fukuda T, the Analysis of Bending Stress and Mechanical Property of Ultralarge Diameter Silicon Wafers at High Temperatures. Japanese Journal of Applied Physics 1996, 35 (Part 1, No. 7), 3799-3806.

[24] Akatsuka M, Sueoka K, Adachi N, et al., Mechanical properties of 300 mm wafers. Microelectronic Engineering 2001, 56 (1), 99-107.

[25] Hu S M, Temperature Distribution and Stresses in Circular Wafers in a Row during Radiative Cooling. Journal of Applied Physics 1969, 40 (11), 4413-4423.

[26] Hirasawa S, Kieda S, Watanabe T, et al., Temperature distribution in semiconductor wafers heated in a vertical diffusion furnace. IEEE Transactions on Semiconductor Manufacturing 1993, 6 (3), 226-232.

[27] Akatsuka M, Sueoka K, Katahama H, et al., Calculation of Slip Length in 300 mm Silicon Wafers during Thermal Processes. Journal of the Electrochemical Society 2019, 146 (7), 2683-2688.

THE STUDY OF SILICON NITRIDE FILMS DEPOSITED IN BATCH ALD SYSTEM

Shiyao Cheng, Wei Kuai, Wenxu Duan, Xinyang Wang, Xiaomeng Liu, Shuo Cheng, Yuanxiang Xie, Xiaoping Shi
Beijing NAURA Microelectronics Equipment Co., Ltd, Beijing, China
chengshiyao@naura.com

ABSTRACT

Silicon nitride (Si_3N_4) is commonly applied as insulators and barriers in IC manufacture. It can be deposited by atomic layer deposition to achieve an excellent step coverage especially in high aspect ratio structure.[1] ALD process usually has a longer process time as lower throughput. To compensate this, batch ALD system such as vertical furnace has been designed for ALD process with a batch up to more than 100 wafers. However, the complex dynamic and chemisorption mechanism of the reactant gases alternatively blowing on the wafer surface makes it challenging to achieve better Si_3N_4 film uniformity and higher quality. A Couple of flow mode comparison experiments on the Si source gas flush flow and soaking was conducted. The results show that flush flow improves the wafer to wafer thickness uniformity within a batch significantly. However, the contrast experiments indicate a better within wafer thickness uniformity with APC open. Based on these results, comparison experiments with different exposure time of Si source and NH_3 as precursors were performed. The results demonstrate that as the exposure time of Si source on the wafer increases, the wet etch rate (WER) of Si_3N_4 decreases which represents improved film densities, along with a steady increase in film uniformity. This provides a significant guidance for semiconductor device fabrication.

INTRODUCTION

Silicon nitride is a critical material applied in many IC components such as dielectric layer, charge storage layer, stress liner, hard mask layer, barrier layer and passivation layer.[1] As the transistor integration density of a chip is increasing, the Si_3N_4 deposition process like chemical vapor deposition (CVD), faces great challenges to keep the robust device operation. In order to meet the design challenges, ALD has been applied for Si_3N_4 deposition in last few years due to its unique atomic level control in film thickness.[2] Comparing to conventional CVD, ALD can create high quality film with an excellent uniformity at low temperature particularly in complex 3D structure and high aspect ratio structure. As the ALD Si_3N_4 deposition process technology becomes mature, throughput and other economic considerations are taken into account. Subsequently, batch ALD system has been introduced for its high throughput and economy process compared with single wafer system. DRAM and flash memory industries have been the main driver for batch ALD as their requirements for the high uniformity of Si_3N_4 film.

Though ALD process has tremendous promising vantages, ALD process does not always behave ideally in practical production. When batch ALD system put on production, small variations on temperature could influence the growth rate significantly because it may induce thermal decomposition of the precursors which can be detrimental to the film. Furthermore, precursor source types and its dosage in each cycle, temperature homogeneity cross wafer and cross boat, pressure control of the tube during a deposition cycle, remain vital factors for obtaining high quality and uniformity film. It is getting more difficult to achieve high uniformity film among all the wafers in one batch and from batches to batches comparing to single wafer chamber system. Because in the batch process, a mass of precursor gas with moderate evaporation temperature is required to spread evenly to each wafer surface along the boat. This needs an advanced gas pumping system and reasonable gas transmitting system especially designed for the tube process. Consequently, it's necessary to tune ALD process parameters to achieve an optimum resulting film to meet various product demands, In this article, we aim to investigate the performance of the deposited film in terms of its thickness and uniformity at various deposition temperatures, specifically 600°C, 620°C, 630°C, and 650°C. Furthermore, we examine the impact of different Si source dosages on the thickness of the Si_3N_4 film. Lastly, we compare the influence of two pressure control systems, namely the Si source gas storage system (flush flow) and the auto pressure controller (APC) system (soaking), on film uniformity. The Si source gas storage system, that can store Si source gas before each cycle, utilizes a pneumatic valve controlled by the program to release the Si source gas to the tube when the gas pressure set to a pre-defined value. During an ALD cycle, the pressure in the tube is maintained at around 0.2 Torr before each cycle, and the Si source gas is dispersed instantly by the corresponding nozzles on each wafer. This is intended to increase the gas flow of the upper nozzles, which are comparatively further away from the storage system. The APC system serves as a pressure control system during the deposition process. , There are two tube pressure control methods during ALD process: Si source gas can be introduced into the tube and the flow velocity on the wafer surface can be adjusted when the APC valve is on. When the valve is off, the Si source gas is trapped in the tube (soaking).

EXPERIMENTAL

A batch reactor system, was used to deposit silicon nitride films on brand new single crystal (100) silicon wafers (300mm). N_2 served as a purge gas to purge the remaining precursor gas in the tube after each half deposition cycle. The gases were regulated using a mass flow controller (MFC) to ensure a precise flux.

Three monitor wafers are placed in the top, middle, and bottom of the boat, to evaluate the batch process performance. During Si_3N_4 film deposition process, wafers in the tube are heated by an outer barrel-shaped furnace embedded with multiple heating coils that radiated heat through the transparent quartz tube to wafers. The tube pressure was set to 0.1 Torr before each deposition cycle.

The thickness, within-wafer uniformity, and wafer-to-wafer uniformity of Si_3N_4 films were measured using an ellipsometer mapping of 49 points with 3 mm edge exclusion. The wet etch rate was evaluated through immersion and submersion in a wet bath filled with a diluted 100:1 49% hydrofluoric acid (HF) solution.

RESULTS AND DISCUSSION

Temperature study

Experiments are designed to verify temperature impacts, 4 different temperatures are selected. As shown in Figure 1, it is demonstrated that the temperature increase from 600°C to 650°C leads to an increase in the thickness of the Si_3N_4 film. These results suggest that the film growth is dependent on temperature. However, the uniformity does not exhibit a clear tendency with temperature and the best performance among these tests is observed at 620°C deposition compared to the other three deposition temperatures.

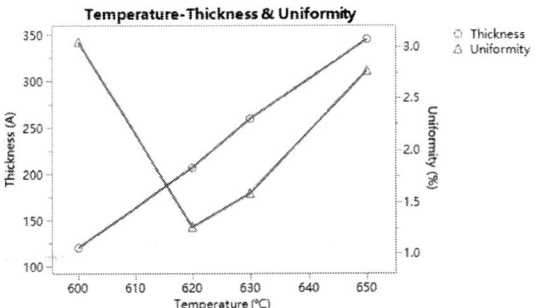

Figure 1: Thickness and uniformity with different temperature

Si Mass flow study

Considering an excellent uniformity and reasonable film thickness observed in 620°C deposition process, the subsequent tests were conducted under the same temperature condition. The quantity of Si source also

influences the film's performance. By tuning the Si source mass flow rate, the inflow of Si source into the tube could be changed. It can result in different film thicknesses and uniformities. As the Figure 2 illustrates, increasing the Si source mass flow rate leads to an increase in film thickness. Uniformity also increases with Si source mass flow rate generally.

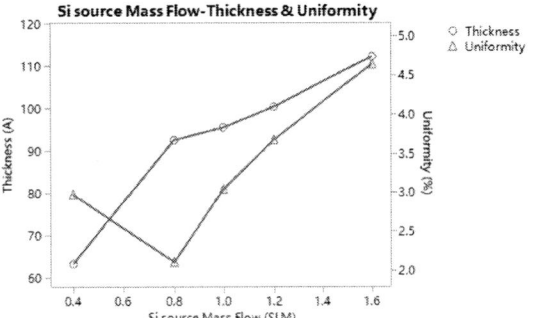

Figure 2: Thickness and uniformity with different Si source mass flow

Combination of soaking and flush flow

Figure 3 illustrates the results of four different mode tests conducted with varying combinations of APC and storage system conditions by using the same amount of NH_3 and Si source. In the first test, the APC was closed without the storage system and with soaking, Si source continually flowed into the tube.

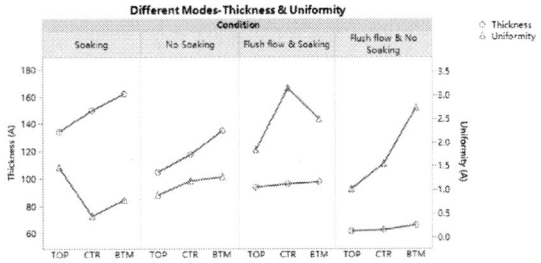

Figure 3: Thickness & uniformity with different modes

The results show a thicker film comparing to those in test 2. The thickness decreases from bottom to top in these two tests. It is due to the decreasing pressure gradient from bottom to top which leads to faster Si source spraying speeds for the nozzles on the bottom of the cylinder, consequently resulting in more Si source being deposited on the bottom wafer. In test 3 and test 4 Si source storage system is applied, APC valve is closed during Si source introduction in Test 3, while it is open in Test 4. Test 3 yields thinner film thicknesses because they were only exposed to Si source for 2 seconds even though more Si source was introduced during this period compared to tests 1 and 2. Test 4, with the APC open, resulted in a thinner film thickness compared to test 3. The film thickness of tests 3 and 4 had lower variation from bottom to top compared to tests 1 and 2. This demonstrates that the storage system can improve wafer to wafer uniformity within a batch. With regards to within wafer uniformity, the results do not clearly indicate a correlation with wafer placement or combinations of soaking and storage system. Future tests will focus on improving within wafer uniformity based on the conditions of test 3, which exhibited excellent film thickness variation and growth rate compared with other tests overall.

Flush flow time

Several experiments are conducted on the batch ALD system to investigate the impact of Si source exposure time on film uniformity. Various Si source gas release times were tested under the condition of same Si source mass flow and injection time to the storage system to ensure that the exposure of Si source in the tube did not change significantly. The quantity of Si source gas was estimated by monitoring pressure variation in the gas storage system before and after each cycle of Si source release, using the ideal gas law ($PV = nRT$). The difference of the storage system were all approximately 435 Torr for release times of 2 s, 3 s, 4 s, and 6 s, confirming a consistent quantity of Si source gas. Shorter release time corresponded to higher steady pressure in the storage system at full load, resulting in a greater pressure drop to drive the precursor flow into the tube with a higher gas flow rate consequently. This high flow rate facilitates the precursors' distribution to the top wafers in the tube through the injector, tending to improve the wafer-to-wafer film thickness uniformity.

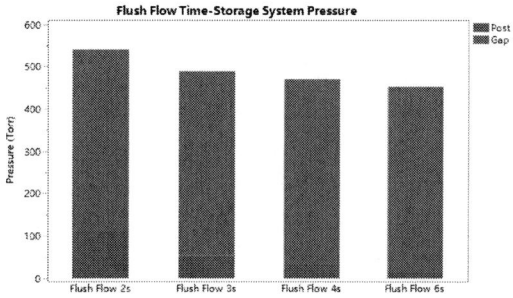

Figure 4: Pressure gap of storage system with different release time

The study investigated the effect of Si source release time on the thickness and uniformity of deposited Si_3N_4 films. Results indicate a nearly linear increase in film thickness when the Si source release time was extended from 2 seconds to 6 seconds in Figure 5. This is attributed to the adequate exposure time for Si source absorption on the wafer surface. Moreover, the uniformity of the film improved with an increasing release time, and when the Si

source release time was at 4 seconds, all the monitor wafers' uniformity were below 2% exhibiting excellent results.

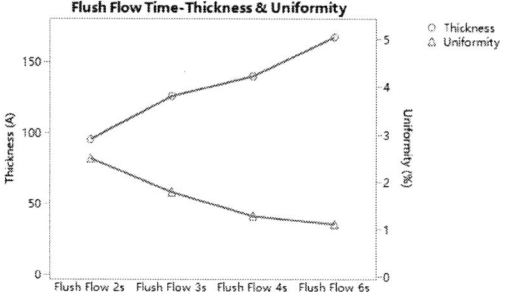

Figure 5: Thickness and uniformity with different Si source release time

As shown in Figure 6 the release time increased from 2 seconds to 4 seconds, the wet etching rate of the Si_3N_4 film decreased in dilute HF solution, indicating an increase in film density. However, when the Si source released to 6 seconds, the wet etching rate increases.

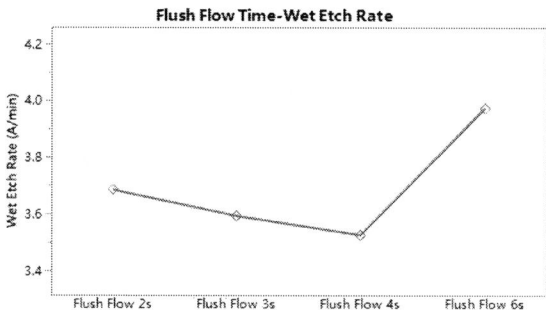

Figure 6: Wet etching rate of Si3N4 film with different flush flow time

TEM for Step Coverage

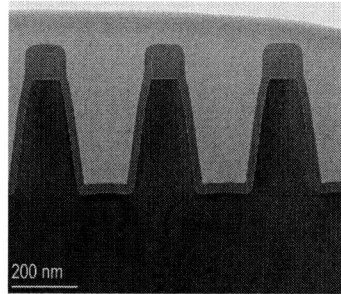

Figure 7: TEM image post ALD SIN film deposition

Figure 7 shows the TEM image of the Si_3N_4 film deposited on the structure wafer. It can be seen that the deposited film is relatively uniform in thickness both at the bottom and side walls, and the film surface is smooth without obvious pin holes, thus characterizing the better film quality and step coverage of the batch ALD process.

CONCLUSION

In this article, a couple of flow mode comparison experiments were conducted on the Si source gas storage system and tube pumping system. The experiments showed that the flush flow improves the wafer-to-wafer thickness uniformity in the whole boat significantly. Additionally, the results of comparison experiments with different exposure times of Si source showed that an extended exposure time of Si source on the wafer results in a lower wet etching rate of Si_3N_4 films. It represents improved film densities along with a steady increase in film uniformity. These results provide significant guidance for semiconductor device fabrication, particularly for obtaining high uniformity film in batch system. The investigation of the impact of Si source dosage and the influence of two pressure control systems on film uniformity can serve as a reference parameter for various product demands. However, it should be noted that variations in temperature could significantly affect the growth rate, indicating the need for precise temperature control during the deposition process.

In conclusion, Si_3N_4 deposition using ALD has tremendous advantages, but several factors must be considered to achieve high uniformity film, particularly in batch ALD system. These experiments provide valuable insights into improving film uniformity and density, which can be utilized in the semiconductor industry. Precise temperature control, precursor source types, and pressure control during the deposition cycle must be considered to obtain high-quality and uniformity films.

REFERENCES

[1] Meng, X., Byun, Y. C., Kim, H. S., Lee, J. S., Lucero, A. T., Cheng, L., & Kim, J. (2016). Atomic Layer Deposition of Silicon Nitride Thin Films: A Review of Recent Progress, Challenges, and Outlooks. Materials (Basel, Switzerland), 9(12), 1007.

[2] Amirzada, M.R., Tatzel, A., Viereck, V. et al. Surface roughness analysis of SiO2 for PECVD, PVD and IBD on different substrates. Appl Nanosci 6, 215–222 (2016).

INFLUENCE OF ION IMPLANTATION ON VOID DEFECT FORMATION IN EPITAXIALLY GROWN SILICON

Zeqi Zha[1]*, Zhenhui Wang[1]*, Ya Wang[1]

[1] Semiconductor manufacturing International (Beijing) Corporation, Beijing 100176, China

*Corresponding Author's Email: zhazeqi@163.com, zhenhui_wang@smics.com

ABSTRACT

Nowadays, silicon epitaxy process is widely used in the manufacturing of high voltage BCD chips. As a result, void defects in epitaxial layer have attracted growing interests of researchers in the semiconductor fabrication. In this paper, we reported one kind of void defect along a straight line in silicon epitaxial layer and proposed a possible formation mechanism. It is found that there is a strong correlation between the lattice damage in antimony implant process and the occurrence of the void defect. Furthermore, a solution to prevent void defects formation is also discussed and provided.

INTRODUCTION

Because of the fact that the epitaxial silicon layer does not contain mirror-polished Czochralski (CZ) grown-in defects, the silicon epitaxy process has been widely used for the high voltage BCD chips in the past decade [1]. The highly n-doped wafers are used as substrates for epitaxial layers which are mainly needed for high power devices. This is because that the NBL (n+ buried layer) located below the source region of the LDMOS has the effect of preventing breakdown between the source region and the p-substrate on the BCD chips [2].

Previous study show that silicon epitaxial layer defect is formed by oxygen contamination, silicon interstitial clusters and vacancy aggregates. Our recent manufacturing data revealed that antimony ion implantation process for n+ buried layer is a contributing factor to silicon epitaxial layer void defects. In this study, a series of experiments were carried out to demonstrate the effect of implantation on the formation of void defect along a straight line in silicon epitaxial layer. In addition, based on the location of silicon epitaxial layer void defect, a possible mechanism for the effect of implantation on void defect formation is proposed and a preventing solution is also provided.

EXPERIMENTS AND RESULTS

Figure 1 shows a standard flow from n+ ion implantation to silicon epitaxy process to form void defects. A silicon wafer was first masked with photoresist (PR) and then implanted with antimony (Figure 1a). After removal of the PR, the antimony-implanted silicon was annealed to active the implanted ions in dry-oxide ambient (Figure 1b). The SiO2 was also formed in the antimony drive-in process. The oxide layer in wafer needs to be removed by wet etching (HF acid) before silicon epitaxy process (Figure 1c). Silicon epitaxial layer was then grown on the n+ doped substrate (Figure 1d). DFI (dark field optical inspection) and SEM were carried out to scan and review the defect respectively.

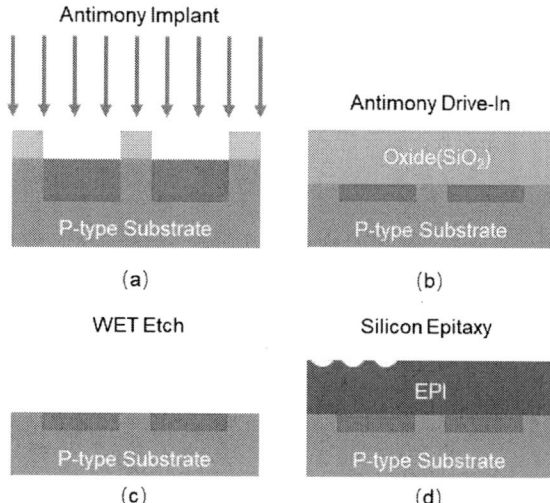

Figure 1: Standard void defect formation process flow diagram from ion implantation to silicon epitaxy process.

As shown in Figure 2a, unlike traditional epitaxial layer defect, the void defect is distributed at the wafer edge rather than randomly on the wafer surface. Another special feature is that the distribution of void defect usually exhibits the line-shape characteristic. Obviously, the SEM image in Figure 2b shows that the void defect has an inverted pyramid-like quadrilateral symmetric structure. The size of the defect is about 5μm. To further characterize the depth information of this void defect, we performed TEM cross-section analysis. As shown in Figure 2c, TEM image show that the depth of void defect is about 130 nm. And no dislocations of the silicon epitaxial layer are observed in the longitudinal cross-section of void defect. As can be seen from the SEM and TEM data, the region where void defect located is not associated with a specific crystal orientation. This is because if it was related to a specific crystal orientation, there would not be just one straight line (void defect located) in defect map.

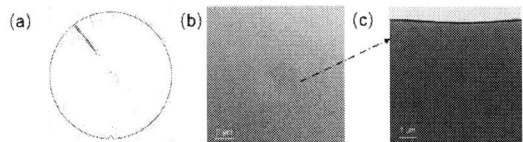

Figure 2: (a) wafer void defect map of wafer; (b) typical defect morphology of silicon epitaxy void defect using SEM; (c) TEM cross-section image of the void defect in (b).

We reviewed the characteristics of the machine passing through the manufacturing process and found that the position of this line was related to the flow direction of HF acid from the wet etching machine as show insert image of Figure 3e. In order to further investigate the formation mechanism, TEM was applied to obtain the cross-section image of wafer after wet etching process. As shown in Figure 3a and Figure 3b, the uneven surface was observed at the region of the wafer edge. In contrast, the surface at the wafer center area is very flat. High-resolution TEM analysis of the uneven areas in wafer edge revealed the presence of obvious grain boundaries (the arrow in Figure 3c) after implant antimony. Figure 3d shows that there were no obvious grain boundaries by means of high-resolution TEM, although there were also amorphous regions. It indicates that the implantation lattice damage of the wafer edge is more difficult to repair than the center area.

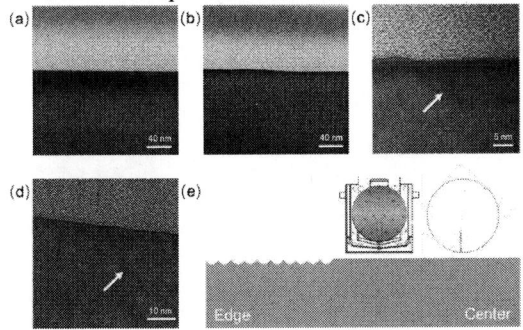

Figure 3: TEM image taken at the area of wafer center (a) and wafer edge (b) after wet etching process; High-resolution TEM image taken at the area of wafer edge (c) and wafer center (d) after wet etching process; (e) Schematic diagram of wafer surface flatness after wet etching process. The inset images are the flow direction of HF acid in wet etching process and the corresponding void defect map in epitaxial layer.

Base on previous discussion, a possible formation mechanism of void defect is proposed. First of all, due to the residual lattice damage by implant process, the surface at wafer edge becomes uneven after oxide removal in HF solutions. As previous paper reported, Si (100) surfaces becomes atomically rough upon HF etching, with coexistence of mono-, di-, and even trihydride species, and are found to be less stable towards contamination [3,4]. In the flow direction of HF acid, the contamination in solution was easily adsorbed in uneven surface. The growth of silicon epitaxial layer was affected by the contamination, and was then detected by DFI as void defects.

Table I is the split table to verify the theories about residual lattice damage by implant process for the fail model. The design split 2 experiment skipped the antimony ion implantation process and the result show that the void defect on silicon epitaxial layer almost disappeared. It confirms that the lattice damage caused by antimony ion implantation is a key factor for the formation of void defect. To solve epitaxial layer void defect at the wafer edge, the uniformity of lattice damage in implant process needs to be improved. Based on the design of implant tool, it can be known that the antimony ion beam is uniformly implanted into the wafer. Due to the presence of self-annealing effects during ion implantation, the uniformity of lattice damage in implant process is strong correlation to the thermal conductivity of the ESC (electrostatic chuck) [5]. As shown in Figure 4a, the backside N_2 gas setting of ESC is an implant factor to uniformity of ESC temperature [5]. Increasing the backside gas flow may be a solution for eliminating void defect.

TABLE I IMPLANTATION DAMAGE SPLIT TABLE

Condition	Design Split 1	Design Split 2	Design Split 3
Baseline	X		
Skip Antimony Implant		X	
Change ESC Backside Gas Cooling Setting			X
Result	Void Defect along a straight line	Void Defect Free	Void Defect Free

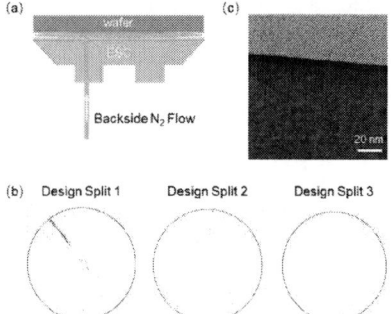

Figure 4: (a) The schematic of implant ESC backside gas cooling. (b) Implantation damage mode split result: defect map after epitaxy process. (c) TEM image taken at the area of wafer edge on condition 3 after wet etching process.

Figure 4b is epitaxial layer void defect map result by the split. The design split 3 experiment increased the backside gas flow of ESC in antimony ion implantation process. It is found that the void defect along a straight line disappeared, which means the uniformity of lattice damage in ion-implanted silicon have been improved. This is further confirmed by the TEM (Figure 4c) taken at the area of wafer edge on condition 3 after wet etching process. As shown in Figure 4c, the surface of the wafer edge is very flat.

CONCLUSIONS

In this paper, the formation mechanism of void defect along a straight line in epitaxial layer is proposed. A strong correlation was found between the residual lattice damage and the occurrence of the void defect. Based on the TEM analysis of wafer before epitaxy process, we proposed that this kind of void defect result from ununiformed lattice damage in antimony implant process. A method of increasing the backside gas flow was applied to solve the problem of ununiformed lattice damage of wafer. The test production is taken on this condition and void defect along a straight line disappear.

REFERENCES

[1] Disney D, Shen Z J. *IEEE Transactions on Power Electronics*, vol. 28, 2013, pp. 4168-4181.

[2] Ko C J, Lee S Y, Park I Y, Park C E, Jun B K, Lee Y J, Kang C H, Lee J O, Kim N J, Yoo K D. *2008 20th International Symposium on Power Semiconductor Devices and IC's. IEEE*, 2008, pp. 103-106.

[3] Chabal Y J, Higashi G S, Raghavachari K, Burrows V A. *Journal of Vacuum Science & Technology A: Vacuum, Surfaces, and Films*, vol. 7, 1989, pp. 2104-2109.

[4] Zhang X, Garfunkel E, Chabal Y J, Christman S B, Chaban E E. *Applied Physics Letters*, vol. 79, 2001, pp. 4051-4053.

[5] Lee K W, Ameen M S, Rubin L M, Dwight D R, Hong R, Reece R N, Yoon D. *Materials Science in Semiconductor Processing*, vol.117, 2020, pp. 105164.

THE EFFECT OF SIGE SICONI PRE-CLEAN TIME ON PLANNER LOGIC DEVICE PERFORMANCE STUDY

Xuechun Zhang[1]*, Weichi Cheng[2], Li Ning[1], and Jingang Wang[1]

[1]Semiconductor Manufacturing North China (Beijing) Corporation, Beijing 100176, China

[2] Semiconductor Technology Innovation Center (Beijing) Corporation, Beijing 100176, China

*Corresponding Author's Email: xuechun_zhang@smics.com

ABSTRACT

In this paper, different SiCoNi pre-cleaning time before the SiGe epitaxial growth process has been tried, and the effect of the Idsat and performance of the integrated SiGe (full-SiGe) and partial SiGe (par-SiGe) device was studied on a 28 nm planar logic device. The results showed that the Idsat of the core device nominal full-SiGe increases with SiGe pre-cleaning time, while that par-SiGe is just the opposite.

The performance of the Core device nominal PMOS full-SiGe device got a significant improvement of more than 10% during the SiGe SiCoNi pre-cleaning time increased by around 20 seconds from upper and lower limits in our experiments on the 28 nm HKMG platform. Nevertheless, the performance of the PMOS par-SiGe was kept the same for increased pre-cleaning time.

INTRODUCTION

Due to the semiconductor chip's continuous demand for size shrink and performance improvement. The MOSFET performance gain has been achieved by aggressive scaling of the device feature sizes at and above the 90nm technology node. However, below the 90nm technology node, performance gains have significantly increased by strained-Si engineering such as embedded SiGe source/drain (eSiGe S/D), dual stress liner (DSL), and the stress memorization technique (SMT) [1]–[3].

Among those methods, the eSiGe S/D is the most commonly used technique for PMOS extreme performance improvement [4]–[6], especially on technology nodes with a small pitch like 28nm.

In the eSiGe technique, there have been many studies on how to improve p-type channel mobility via eSiGe process control, such as the shape of eSiGe, Ge concentrations [7]–[8], and the SiGe pre-cleaning are rarely mentioned. Nevertheless, most studies only focus on the stress provided by the full-SiGe and ignore the performance of par-SiGe. As shown in Fig.1, the growth environment of full-SiGe and par-SiGe is different, and the impact of pre-cleaning on them is also different.

Therefore, the difficulty of balancing the PMOS Idsat between the full-SiGe and par-SiGe to improve LOD (Length of Diffusion) on the embedded SiGe process becomes the key topic on planar logic devices.

The focus of this paper will study the adjustment effect of SiGe SiCoNi pre-cleaning time on the Idsat ratio of par-SiGe to full-SiGe (LOD) and the performance of them on planar 28 nm HKMG logic devices.

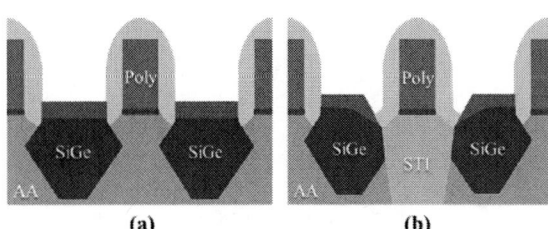

Figure 1: A schematic diagram of SiGe: (a) full-SiGe; (b) par-SiGe

EXPERIMENT

To study the effect of the Idsat and performance of the full-SiGe and par-SiGe, five split conditions were designed as shown in TABLE I. All wafers analyzed in this study were manufactured based on a 28 nm HKMG technology.

TABLE I.   SiGe SiCoNi PRE-CLEANING TIME SPLIT CONDITIONS

Split	Condition
T--	T-10 seconds
T-	T-8 seconds
T	T
T+	T+8 seconds
T++	T+13 seconds

RESULTS AND DISCUSSION

LOD by SiGe SiCoNi pre-cleaning time

For planar logic devices, the balance of PMOS Idsat between the full-SiGe and par-SiGe is crucial. Accordingly, LOD becomes a key parameter. The calculation method is the Idsat ratio of par-SiGe to full-SiGe. Fig.2 shows the LOD of different SiGe SiCoNi pre-cleaning time on a 28 nm planar logic device. Obviously, as time increases, LOD shows a decreasing trend. LOD was reduced by 15% during SiCoNi time increased by around 20 seconds.

979-8-3503-1101-3/23 $31.00 © 2023 IEEE

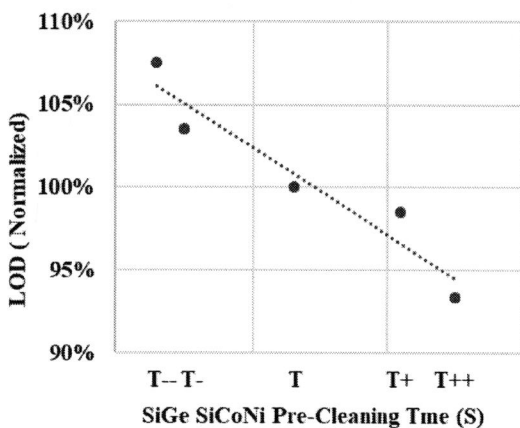

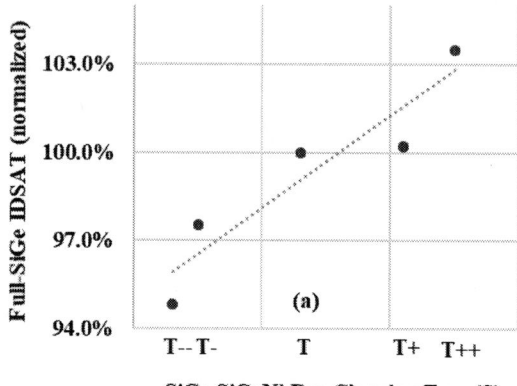

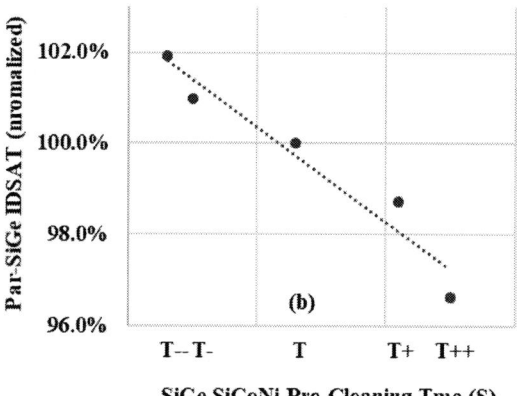

Figure 2: The LOD of different SiGe SiCoNi pre-cleaning time on a 28 nm planar logic device (Set the LOD to 100% as the SiGe SiCoNi pre-cleaning time is T)

Core device nominal P Idsat

Why does SiGe SiCoNi pre-cleaning time have a significant impact on LOD? As seen from Fig.3, For full-SiGe and par-SiGe, the Idsat of both has different performances as time increases. Fig.4a shows the effect of SiCoNi pre-cleaning time on core device nominal P Idsat. The full-SiGe Idsat increased along with time, while par-SiGe has the opposite result (Fig.4b).

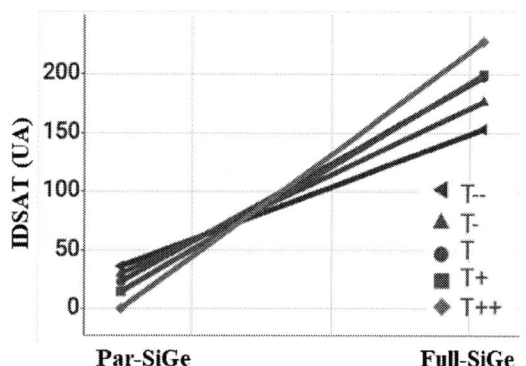

Figure 3: The core device nominal P Idsat of different SiGe SiCoNi pre-cleaning time

Figure 4: Core device nominal P Idsat of different SiGe SiCoNi pre-cleaning time: (a) full-SiGe; (b) par-SiGe. Set the IDSAT to 100% as the SiGe SiCoNi pre-cleaning time is T

Core device performance

Fig.5 shows the nominal core device (poly length 0.03um) performance of different SiGe SiCoNi pre-cleaning time split on full-SiGe. When pre-cleaning time increased 23s, then PMOS performance increased 8~10%. Besides, the performance tends to be stable as time goes on. Nevertheless, the performance of the PMOS par-SiGe was kept the same for increased pre-cleaning time.

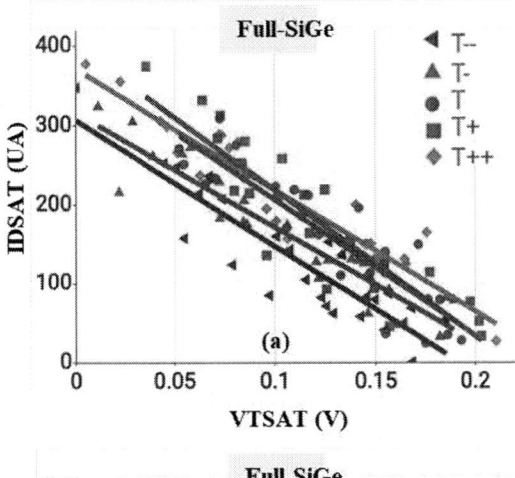

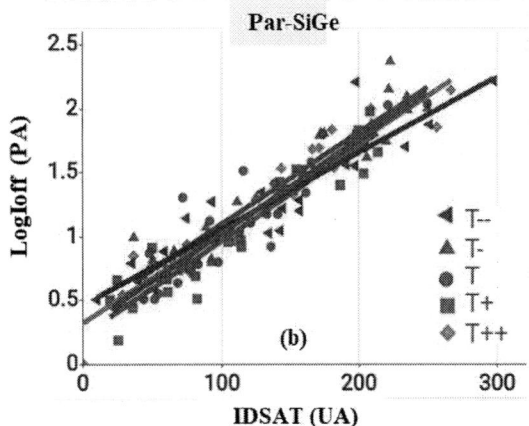

Figure 6: The par-SiGe performance on a planar 28 nm HKMG logic devices of different SiGe SiCoNi pre-cleaning time: (a) nominal P Vt vs. Id; (b) nominal P Id vs. Ioff

CONCLUSIONS

In this paper, different SiGe SiCoNi pre-cleaning time has been tried, and the effect of the Idsat and performance of the full-SiGe and par-SiGe device was studied on a 28 nm planar logic device. The results showed that the Idsat of the core device nominal full-SiGe increases with SiGe pre-cleaning time, while that par-SiGe is just the opposite. The performance of the Core device nominal PMOS full-SiGe device significantly improved more than 10% during the SiCoNi pre-cleaning time increased by around 20 seconds. Whereas the pre-cleaning time has no obvious impact on the Idsat of the PMOS par-SiGe. This paper opens a new idea to improve LOD on planar logic devices.

REFERENCES

[1] S. E. Thompson, G. Y. Sun, Y. S. Choi, and T. Nishida, "Uniaxial-process-induced strained-Si: Extending the CMOS roadmap," IEEE Trans. Electron Devices, vol. 53, no. 5, pp. 1010–1020, 2006.

[2] S. E. Thompson et al., "A logic nanotechnology featuring strained-silicon," IEEE Electron Device Lett., vol. 25, no. 4, pp. 191–193, 2004.

[3] C.-H. Chen, et al., "Stress memorization technique (SMT) by selectively strained-nitride capping for sub-65 nm high-performance strained-Si device application," in VLSI Symp. Tech. Dig., pp. 56–57, 2004.

[4] Satoru Mayuzumi, et al., " Mobility and Velocity Enhancement Effects of High Uniaxial Stress on Si (100) and (110) Substrates for Short-Channel pFETs", IEEE Trans. Electron Devices, vol. 57, no. 6, pp. 1295-1300, 2010

[5] H. Okamota, et al, "In situ doped embedded SiGe

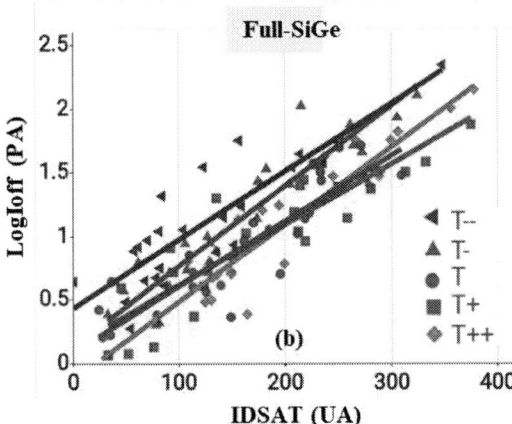

Figure 5: The full-SiGe performance on a planar 28 nm HKMG logic devices of different SiGe SiCoNi pre-cleaning time: (a) nominal P Vt vs. Id; (b) nominal P Id vs. Ioff

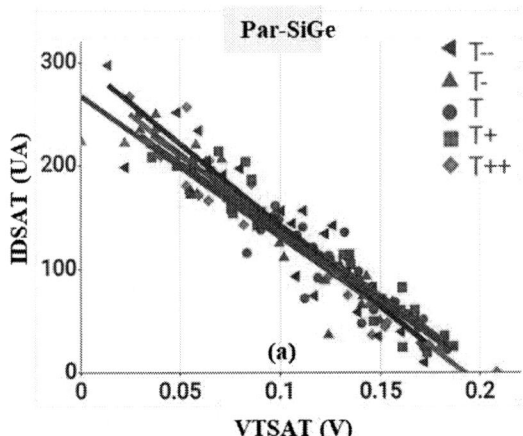

source/drain technique for 32 nm node p-channel metal–oxide–semiconductor field-effect-transistor," Jpn. J. Appl. Phys., vol. 47, no. 4, pp. 2564–2568, 2008.

[6] H. Okamoto, et al, "A study on aggressive proximity of embedded SiGe with comprehensive source and drain extension engineering for 32 nm-node high-performance pMOSFET technology," Solid State Electron., vol. 53, no. 7, pp. 712–716, 2009.

[7] J. Wang, et al., " Novel Channel-Stress Enhancement Technology with eSiGe S/D and Recessed Channel on Damascene Gate Process ", VLSI Tech. Dig., pp. 46-47, 2007.

[8] M. H. Lee, S. T. Chang, S. Maikap, C.-Y. Peng, and C.-H. Lee., "High Ge Content of SiGe Channel pMOSFETs on Si (110) Surfaces", IEEE Electron Device Lett., vol. 31, no. 2, pp. 141-143, 2010.

MECHANICAL PROPERTIES OF FLIP-CHIP BONDING STRUCTURES FOR MICRO-LED DEVICES: CU-CU BONDING WITH PASSIVATION LAYER AND INDIUM BUMPS BONDING

Kefeng Wang[1], Zehua Chen[1], Xiaoxiao Ji[2,3], Luqiao Yin[1,2], Xiuzhen Lu[1,], Jianhua Zhang[1,2]*

[1]School of microelectronics, Shanghai University, Shanghai, 201800, China

[2]Key Laboratory of Advanced Display and System Applications, Shanghai University, Ministry of Education, Shanghai, 200072, China

[3]School of Mechatronic Engineering and Automation, Shanghai University, Shanghai, 200072, China

*Corresponding Author's Email:xzlu@shu.edu.cn

ABSTRACT

A traditional bonding structure with indium bumps and a new bonding structure with passivation layer using Cu bumps were simulated to bond Micro-LED and substrate. Anand model and bilinear isotropic hardening model were used to characterize the viscoplastic behavior of In and the plastic behavior of composite structure simplified from the combination of Cu and Au, respectively. Mechanical properties of two kinds of micro-display devices with different bonding structures were analyzed by the ANSYS finite element method and the thermo-mechanical coupling field. The deformation and stress of the micro-display device using Cu-Cu bonding structure with passivation layer are reduced by 19.5% and 59.1% compared with the device bonded by In bumps.

INTRODUCTION

In recent years, Micro-LED devices had become a research hotspot in the field of display and visible light communication [1-3]. As the key technology to realize heterointegration, indium bump was widely used to bond Micro-LED chips and Si substrates[4]. However, as the pixel pitch decreased, the extrusion and overflow of indium produced during bonding process may lead to interconnection of the adjacent solder joints and the failure of the devices[5].

Cu-Cu direct bonding was regarded as the more promising integration technology for three dimensional integrated circuit (3DIC). The interconnection between solder joints could be avoided because of the high melting point of Cu[6]. However, the required high vacuum environment to prevent oxidation of Cu and high temperature for Cu-Cu direct bonding limit the application of Cu-Cu bonding in heterointegration[7,8].

A new bonding structure with passivation layer on Cu-Cu bonding interface was proposed to prevent oxidation and promote the diffusion of Cu[9]. Inert metals such as gold, silver or palladium were considered as passivation layer[10-12]. The technological process of inserting passivation layer on Cu-Cu bonding interface is shown in Figure 1[12]. Ti, Cu and passivation layer were deposited on silicon wafer by physical vapor deposition. During the thermal compression bonding process, Cu atoms diffuse to the bonding interface through the passivation layer to complete the Cu-Cu bonding.

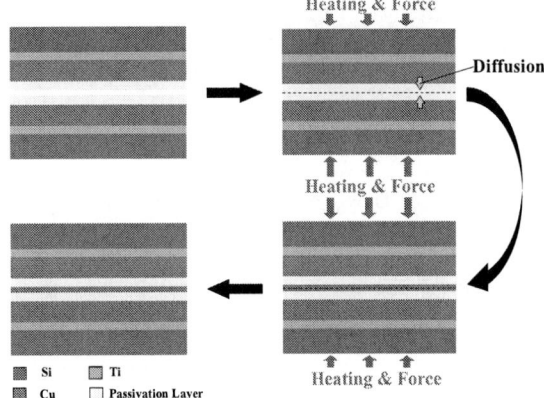

Figure 1: Schematic of Cu-Cu bonding with passivation layer.

Cu-Cu bonding structure with passivation layer was introduced to Micro-LED device to improve mechanical properties and prevent bridging failure of solder joints.

Mechanical properties of the micro-display devices with Cu-Cu bonding structure with passivation layer and micro-display device with indium bonding structure were analyzed by the ANSYS finite element method.

3D FINITE ELEMENT MODEL
Model & Parameters

Structure of the Micro-LED array bonded on the Si substrate through different bonding structures were used for ANSYS analysis. 1/4 model shown in Figure 2 was selected as the object of simulation because of the excellent symmetry of the overall structure. The micro-display device using In bonding structure was named Device A and the micro-display device using

979-8-3503-1101-3/23 $31.00 © 2023 IEEE

Cu-Cu bonding structure with passivation layer was named Device B.

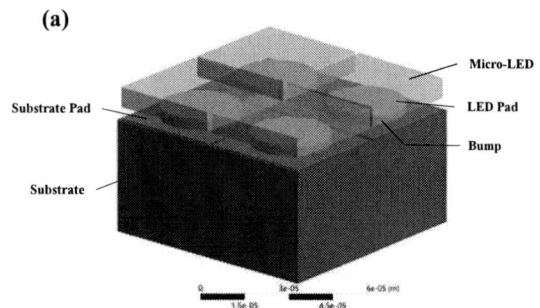

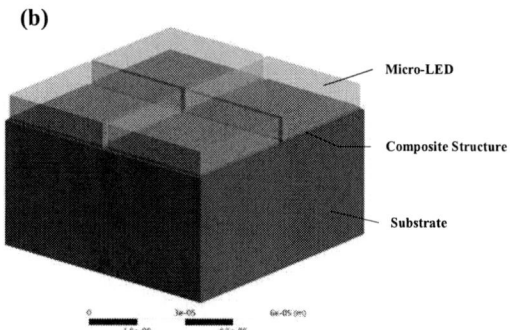

Figure 2: The 1/4 3D model of micro-display devices. (a) Device A (b) Device B

Considering the accuracy of the results and the simulation time as short as possible, the meshing size of 2μm was selected as the simulation parameter.

The details of materials and dimensions for the two models are shown in Table Ⅰ. The composite structure is simplified from the combination of Cu and Au. The thermodynamic parameters of composite structure are related to the thicknesses and the parameters of Cu and Au. The fitting formula of composite parameters is as follows:

$$X_{co} = \frac{H_A}{H_P}X_A + \frac{H_C}{H_P}X_C \qquad (1)$$

X_{co}, X_A and X_C are the thermodynamic parameters of composite structure, Au and Cu. H_P, H_A and H_C are the thicknesses of composite structure, Au layer and Cu layer, respectively.

TABLE I   DIMENSION OF MODEL PARTS

Parts	Material	Dimension/μm
Micro-LED	GaN	38×38×7.2
LED Pad	Au	Φ20×1
Substrate Pad	Au	38×38×0.4
Bump	In	Φ20×3
Substrate	Si	160×160×40
composite structure	Cu&Au	38×38×0.86

Material Properties

The melting points of In, Au and Cu are 156 ℃, 1064 ℃ and 1083 ℃. The viscoplastic behavior of In and the plastic behavior of Au and Cu were considered in the analysis of the mechanical properties of micro-display devices. Anand constitutive model was used to characterize the viscoplastic behavior of In. Table Ⅱ shows the Anand model parameters of In[13]. The bilinear isotropic hardening model was used to describe the plastic behaviors of Au, Cu and the composite structure. The required parameters of this model are shown in Table Ⅲ, and other linear parameters are shown in Table Ⅳ[2,13-15].

TABLE II ANAND MODEL CONSTANTS OF INDIUM

Description	Indium
So(MPa)	28.3
Q/R(K)	9369.7
A(SEC^-1)	2.33×108
ξ	49.97
m	0.3
ho(MPa)	500
Ŝ(MPa)	28.3
n	0.005
a	1

TABLE III   PARAMETERS OF BILINEAR MATERIALS

Material	Yield Strength (MPa)	Tangent Modulus (MPa)
Cu	172.3	517.1
Au	100	5500
composite structure	167.8	943.8

TABLE IV   LINEAR PARAMETERS OF MATERIALS

Material	E (GPa)	Poisson's ratio	CTE (10^{-6}/K)	ρ (kg/m³)
GaN	181	0.35	4.7	6070
In	10.6	0.45	33	7310
Si	190	0.28	2.6	2330
Cu	171	0.35	14.3	8900
Au	79.5	0.34	14	19320
composite structure	165	0.351	14.35	9728

TABLE V THERMODYNAMIC PARAMETERS OF MATERIALS

Material	Thermal conductivity (W/m·K)	Specific heat capacity (J/kg·K)
GaN	175	54.7
In	81.8	233
Si	150	729
Au	317	129
Cu	398	385
composite structure	394	367.8

Figure 3 shows the stress-strain curves of the three bilinear materials drawn according to the mechanical parameters shown in Table III. Most of material in the composite structure is Cu, so the curve of the fitted composite structure is similar to that of Cu. Table V shows the thermodynamic parameters of these materials[9].

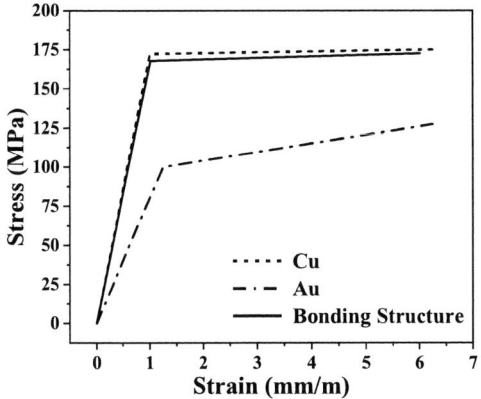

Figure 3: Stress-strain curves of bilinear materials

Loading and boundary conditions

The thermo-mechanical coupling field was used to analyze the temperature profile, deformation and stress distribution of the model. The external ambient temperature was 20 ℃. The heat generation rate of Micro-LED was set to 3.2×10^8 W/m³ to simulate the heat generation of Micro-LED. The convection coefficient of the lower surface of the Si substrate was 10 W/(m²·K) and the fixed support was added to the center of upper surface of the Si substrate[9].

RESULTS AND DISCUSSION

With the increase of temperature for Micro-LED, the heat emitted from the surface of the Si substrate to the environment increases gradually. The temperature of the whole structure reaches a fixed value according to a state of dynamic equilibrium of the device finally.

Figure 4 shows the deformation distribution of the two kinds of devices. During the heating process, the mismatch of the thermal expansion coefficient of the materials leads to the warping deformation of the devices. The deformation increases gradually from the center to the periphery, and it is proportional to the distance from the center of device.

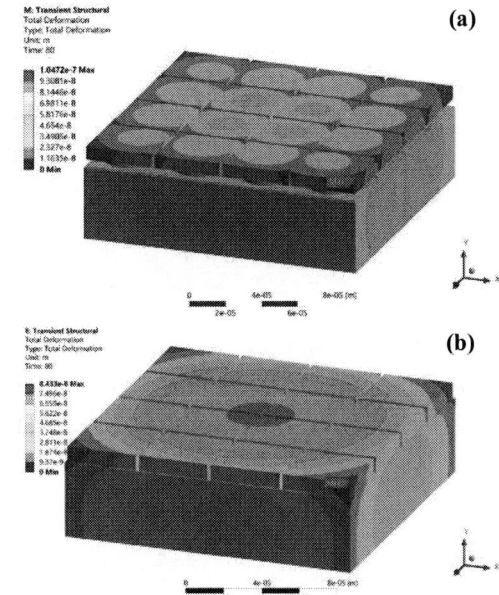

Figure 4: The deformation of micro-display devices.
(a) Device A (b) Device B.

Figure 5 shows the stress distribution of two kinds of devices. Figure 6 is the magnification of the local details of Figure 5. The maximum stress concentration point is located at the contact interface between Si Substrate and bonding structure.

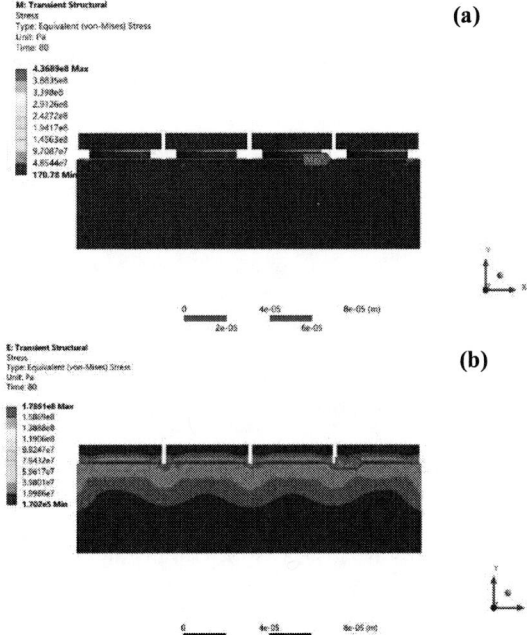

Figure 5: The stress of micro-display devices.
(a) Device A (b) Device B.

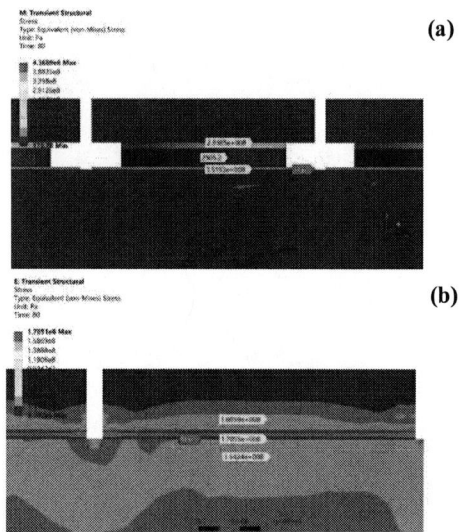

(a)

(b)

Figure 6: Schematic diagram of maximum stress area. (a) Device A (b) Device B

The steady-state temperature, strain and stress of micro-display devices with different bonding structures are shown in Figure 7. The steady-state temperatures of devices are both 227.93 ℃, which indicates that their heat dissipation performance is similar.

The maximum deformation of Device A is 104.72 nm, and that of Device B is 84.33 nm. In addition, stress of micro-display device reduces from 436.89 MPa to 178.51 MPa by using bonding structure with passivation layer to take place of indium bumps bonding structure.

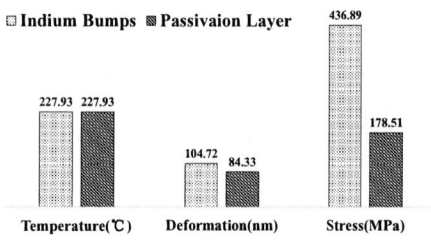

Figure 7: Properties of micro-display devices with different bonding structures.

CONCLUSIONS

The mechanical properties of two kinds of Micro-LED devices bonded by indium bumps and Cu-Cu bumps with passivation layer were investigated by ANSYS method.

The mechanical properties of Micro-LED device is improved effectively by using of Cu-Cu bonding structure with passivation layer compared with the device bonded by Indium bumps. The deformation of the device bonded by Cu-Cu bumps with passivation layer is reduced by 19.5% to the device bonded by Indium bumps, and the decrease of the stress reached 59.1%. The great potential of Cu-Cu bonding structure with the passivation layer in the field of micro-display and micro-size packaging is demonstrated by the simulation results.

ACKNOWLEDGEMENTS

This work was supported by the Science and Technology Commission of Shanghai Municipality Program (21511101302, 20010500100, 22511101000).

REFERENCES

[1] Mo, X.F., Xu et al. *Chinese Physics Letters*, Vol.34, No.11, 2017, pp. 102-105.

[2] X. Ji, F. Wang, L. Yin and J. Zhang. *China International Forum on Solid State Lighting*, Shenzhen, December.6-8, 2021, pp. 135-138.

[3] Pan, Z., Liu, J., Liu, X. et al. *International Conference on Display Technology (ICDT)*, Guangzhou, April.9-12, 2018, Vol. 51, No. 49, pp. 90-94.

[4] C.-J. Zhan, C.C. Chuang, et al. *Electronic Components and Technology Conference (ECTC)*, Las Vegas, June.1-4, 2010, pp. 1043-1049.

[5] Z. -J. Hong, D. Liu, H. -W. Hu. *Electronic Components and Technology Conference (ECTC)*, San Diego, June.1-4, 2021, pp. 347-352.

[6] P.R.Morrow et al. *IEEE Electron Device Letters*, vol. 27, no. 5, pp. 335–337.

[7] Y. -C. Huang et al. *Electronic Components and Technology Conference (ECTC)*, San Diego, June.1-4, 2021, pp. 377-382.

[8] Y. I. Kim, K. H. Yang and W. S. Lee. *International Reliability Physics Symposium*, 2004, pp. 667-668.

[9] Y.P Huang et al. *IEEE Electron Device Letters*, vol. 34, no. 12, pp. 1551-1553.

[10] Y. -P. Huang, Y. -S. Chien et al. *IEEE Transactions on Electron Devices*, vol. 62, no. 8, pp. 2587-2592.

[11] D.Liu, P.-C. Chen and K.-N.Chen, *International 3D Systems Integration Conference (3DIC)*, Sendai, October.8-10, 2019, pp. 1-4.

[12] T. -C. Chou et al. *IEEE Transactions on Components, Packaging and Manufacturing Technology*, vol. 11, no. 1, pp. 36-42.

[13] X. Xiao et al. International Conference on Electronic Packaging Technology (ICEPT), Xiamen, August.11-14, 2021, pp. 1-6.

[14] J. -S. Lan and M. -L. Wu. *International Microsystems, Packaging, Assembly and Circuits Technology Conference (IMPACT)*, Vienna, January.10, 2014, pp. 234-237.

[15] J. Liu, Q. Yao, P. Lin, Y. Cao and B. Lian. *International Conference on Electronic Packaging Technology (ICEPT)*, Wuhan, August.16-19, 2016, pp. 1330-1333.

A NOVEL METHOD TO OPTIMIZE SIGE PROFILE USING CO-IMPLANTATION

Zhiqiang Xiao[1,2]*, Long Feng[1]*, Cunzhe He[1], Jiaxing Xiao[1], Dongliang Gao[1], Mingying Liu[1]

[1]Semiconductor Manufacturing North China (Beijing) Corp. Beijing, China

[2]School of Integrated Circuits, Tsinghua University, Beijing, China

*Corresponding Author's Email: ZhiQiang_Xiao@smics.com;Frank_Feng@smics.com

ABSTRACT

The silicon oxide removal process prior to epitaxial growth of SiGe is critical to establish a defect-free surface for the epitaxy process. The introduction of *in-situ* Siconi process has excellent selectivity and queue time control over wet etch process on the native oxide etching, however, the undesirable removal of shallow trench isolation (STI) oxide also pose potential risks for the active area bridge and leakage. In this paper, an effective method to protect the STI morphology during the SiGe Siconi process was proposed with a novel modulation of oxide etch rate by ion implantation. A high tilt ion implantation was firstly performed on the STI side wall. Subsequently, during the Siconi process. the dopant ion can dramatically decrease the etch rate of STI dioxide and resultantly protect the STI profile from the oxide etching with minor change. The TEM result of pattern wafer confirmed that the STI dioxide with pre-implantation condition has better morphology than the baseline condition.

Keywords—ion implantation; Siconi; etch rate; oxide remove

INTRODUCTION

The introduction of embedding SiGe in adjacent source-drain(S/D) regions has become the key driver for improving transistor performance since its adoption of compressive uniaxial strain in the Si channel in 90 nm CMOS technology node [1]. To obtain high structural quality SiGe layers during epi process, a surface cleaning and oxide removing process before epitaxial growth is pre-requisite to reduced crystallographic defects, such as stacking fault, micro-twin, dislocation, or their combination. Since conventional *ex-situ* surface cleans of Si wafers with 100:1 DI: HF (DHF) cannot eliminate queue time constraints, the advent of *in-situ* cleans based on the SiCoNi process can boost the native oxide removal with high selectivity and grow defect-free epitaxial Si or SiGe layers.

Despite the high selectivity of SiCoNi process, there still exists undesirable STI oxide removal at the boundary of S/D and STI (Figure 1), which easily induce leakage even active area bridge. It is due to that the plasma NF3 and NH3 mixed gas can remove not only the surface oxide in silicon but also the STI oxide, which easily leads to subsequent SiGe growth extend to STI. In this paper, we develop an effective method to protect the STI profile during the SiGe SiCoNi process by a novel application of ion implantation. Typically, the etch selectivity ratio of the surface oxide to the STI oxide is around 0.9. By the pre ion doping to the STI exposed area in sigma

silicon, the etch rate of STI oxide in the SiCoNi process decreasing about 50% and accordingly the etch selectivity ratio of the surface oxide to the STI oxide increased to 1.76. With about 100% improvement to the etch selectivity ratio, the STI profile after epi process was effectively protected with minor change. This etch rate modification approach by ion implantation provide an alternative road for the fabrication of buried mask, sacrificial layers and etch stop layers.

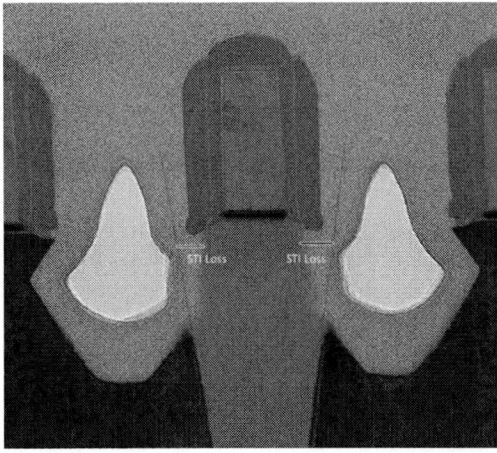

Figure 1. Typical STI profile after SiGe SiCoNi process

EXPERIMENT AND RESULT DISCUSSION

Ion implantation is not only the mostly used method for semiconductor doping but also of high interest to modify the film property. Many studies have been reported that ion implantation can evidently increase the etch rate of silicon oxide in the environment of diluted HF solution. However, for the environment of plasma NF3 and NH3 mixed gas by SiCoNi[TM] chamber, doping effect on the etch rate modification was complicated [2-4]. To explore the effect of ion implantation on silicon oxide, two set of blanket wafers were firstly deposited with 1000A silicon oxide with HARP and furnace respectively. The furnace oxide wafers were chosen as baseline sets (control group) for its etch rate is comparable with the surface oxide in the sigma area. The HARP oxide wafers were chosen as experiment groups as the STI oxide in this case was deposited with HARP oxide. The experiment group was split to two groups. The first group were splits without annealing. One wafer was chosen as baseline and the other four wafers were implanted with varied species and dosage. The second group were splits with annealing. The implant conditions were same as the first group and then all the wafer were annealed. The detail split condition are shown in Table 1.

979-8-3503-1101-3/23 $31.00 © 2023 IEEE

The film etch amount was collected after the plasma F radical etchant gas processing in SiCoNi chamber.

TABLE1. CONDITIONS FOR BARE WAFER ETCH RATE EXPERIMENT

Group	Oxide Type	IMP Conditon	Anneal
Baseline	Furnace	with/o	No
Group 1	HARP	with/o	No
		P,10KeV, 2E15,Tilt15	No
		N,10KeV, 2E15,Tilt15	No
		BF2,10KeV, 2E15,Tilt15	No
		BF2,10KeV, 3E15,Tilt15	No
Group 2	HARP	with/o	Yes
		P,10KeV, 2E15,Tilt15	Yes
		N,10KeV, 2E15,Tilt15	Yes
		BF2,10KeV, 2E15,Tilt15	Yes
		BF2,10KeV, 3E15,Tilt15	Yes

The etch rate result of silicon oxide prepared by HARP and furnace in SiCoNi chamber was shown in Figure 2. The etch amount of the non-implantation oxide wafers prepared by furnace were 30A after SiCoNi process. For the group 1 HARP oxide wafer split without annealing, the etch amount was much higher than the baseline due to the loose film quality of HARP oxide. It is interesting that the etch amount of the oxide wafers with BF2 implantation was much lower than the baseline HARP wafer while it was comparable for the wafers with P or N implantation. Furthermore, with the implant BF2 dose +, the etch amount was even lower. For the group 2 split with annealing, the etch amount of the non-implantation oxide wafer was 34A after SiCoNi process, with the etch selectivity ratio of the furnace oxide to the HARP oxide around 0.9. In consistent with the result of group 1, the etch amount for the conditions with P or N implantations was comparable with HARP oxide baseline in group 2 while it was much lower for the condition with BF2 implantation. Especially, with the BF2 implant dose + condition, the etch amount was as low as 17A and the etch selectivity ratio of the furnace oxide to the HARP oxide improved up to 1.76. It is noteworthy that the etch amount decreased with the implant dose increasing for both conditions. The etch rate modification with the BF2 implantation may be due to that the B or F ion suppressed the etch process of oxide with NH_4F or $NH_4F.HF$ plasma. However, it is still unclear and need further study for the mechanism of the implantation effect on the etch rate modification of HARP oxide.

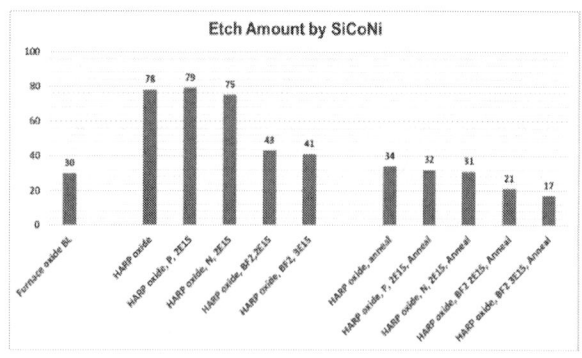

Figure 2. Silicon oxide film etch rate by SiCoNi process

In order to verify the effect of ion implantation modification on the etch rate of HARP oxide in pattern wafer, two sets of experiment were designed with the typical process flow as described in Figure 3. With the standard condition, the STI oxide was deposited following a normal HARP process without interrupt. For the STI pre-doping condition, silicon oxide with a thickness about 200A was firstly deposited to the STI trench, then a BF2+ dopant with energy 10KeV, dosage 3E15, tilt 15° was implanted, and finally the rest STI oxide was deposited. After STI anneal, gate oxide, poly, LDD and spacer loop, the two sets wafer came into the SiGe loop, which contains silicon recess, SiCoNi silicon oxide pre-clean and epitaxy SiGe growth. The SiGe cross-section profile after SiCoNi pre-clean was characterized by TEM.

- STI HARP deposition (w/i or w/o implantation)
- Gate OX & Poly
- LDD
- Spacer deposition
- SiGe loop

Figure 3. A typical flow before the SiGe epitaxy with or without ion implantation

Figure 4 shows the cross-section TEM results of the pattern wafers split. Compared with the baseline condition, the split with BF2 implantation during the STI HARP deposition show much lower STI over etch, indicating that the severe STI loss in baseline wafer is well suppressed by BF2 implantation in STI oxide. It is also noteworthy that the etch rate decrease of the STI oxide in the pattern wafer is higher than that in the bare HARP oxide wafer. As a result, the STI oxide profile was effectively protected with the etch rate modification effect of ion implantation.

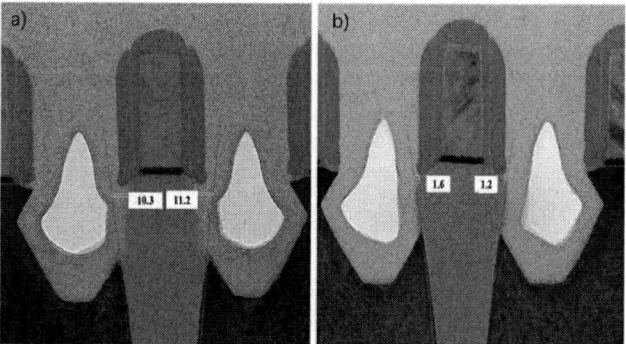

Figure 4. a) Condition without co-implantation show obvious over etch of STI profile; b) Condition with BF2 pre-implantation show minor over etch of STI profile

CONCLUSIONS

In this paper, the effect of ion implantation on etch rate modification of silicon oxide was investigated. The implantation of BF2 species greatly suppressed the etch rate of HARP oxide, while the implantation of P or N species almost had no effect on the oxide etch rate modification. Especially, the condition with BF2 dosage + implantation improved the etch selectivity ratio of the furnace oxide to the HARP oxide from 0.9 to 1.76. The positive effect of pre-implantation on the STI loss in SiCoNi process was also verified in the pattern wafer. With even higher etch selectivity ratio of the surface oxide to the STI oxide than bare wafer, the pattern wafer with STI pre-implantation had about 80~90% lower STI loss than the normal baseline condition. Though the mechanism is under further study, this etch rate modification approach by ion implantation provides an alternative road for the modulation of films characteristics and can be a good candidate for etch selectivity ratio improvement.

REFERENCES

[1] G. Tsutsui, S. Mochizuki, N. Loubet, S. W. Bedell, and D. K. Sadana, "Strain engineering in functional materials," AIP Adv. 9, 030701 (2019). https://doi.org/10.1063/1.5075637

[2] R. Charavel, J. Raskin. AIP Conference Proceedings 866, 325 (2006); https://doi.org/10.1063/1.2401523

[3] A. Losavio, B. Crivelli, and F. Cazzaniga, Appl. Phys. Lett. 74, 2453-2455 (1999) https://doi.org/10.1063/1.123878

[4] Y. He et al., 2014 International Workshop on Junction Technology (IWJT), 2014, pp. 1-4, doi: 10.1109/IWJT.2014.6842042.

A STUDY OF PARASITIC CAPACITANCE USING DIFFERENT BIT LINE SPACER INTEGRATION SCHEMES IN ADVANCED DRAM

Dempsey Deng[*], Qingpeng Wang, Yujia Zhong, Yu De Chen, Jacky Huang
Coventor Inc., A Lam Research Company, Shanghai, 201203 China
*Corresponding Author's Email: Dempsey.Deng@lamresearch.com

ABSTRACT

In this study, we evaluate parasitic capacitance between the node contact (NC) and the bit-line (BL) in advanced dynamic random-access memory (DRAM) devices. Process modeling was used to analyze capacitance performance under different integration approaches. The results indicated that bit-line contact (BLC) along with bit-line pitch walking (BL PW) have strong impact on DRAM parasitic capacitance. The use of both low-k spacers and airgap spacers in the integration scheme helped to reduce parasitic capacitance. Our study showed that the average improvement was 16.5% with low k spacer and 31.3% with airgap spacer, respectively.

INTRODUCTION

DRAM technology development is becoming increasingly difficult due to reduced DRAM cell size. A major problem in advanced DRAM is that the BL sensing margin and refresh times degrade with an increase in BL parasitic capacitance (C_b). The parasitic capacitance between the BL and the NC (C_{BL-NC}) is the main factor impacting C_b [1]. To reduce C_{BL-NC}, low-k spacers and airgap spacers have been proposed as potential solutions.

Traditionally, Si wafer based experimentation is used to evaluate a new process integration scheme, but this can be time-consuming and costly in practice. In this study, we use virtual process modeling with SEMulator3D®, instead of wafer-based experimentation, to analyze and propose new process integration schemes [2]. We review the C_{BL-NC} of a Nitride-Oxide-Nitride (NON) spacer and evaluate the potential improvements in C_{BL-NC} performance in using a low-k spacer or airgap spacer. The purpose of our study was to obtain a quantified comparison with different BL spacer integration schemes, to provide clear guidance to process developers.

DEVICE STRUCTURE AND SIMULATION METHODS

In this study, virtual 3D structures were constructed in the process modeling platform using a combination of layout data and process step data. Fig. 1 displays the final 3D structures simulated using a NON spacer (Fig 1a), a low-k spacer (Fig 1b) and an airgap spacer (Fig 1c). The NON spacer consists of a NON formation. For the low-k spacer, Nitride (SN) is replaced by a low k material, to create a low k-Oxide-low k spacer structure. To create the airgap spacer, we etch the dummy oxide (OX) spacer to form a Nitride-air-Nitride spacer structure.

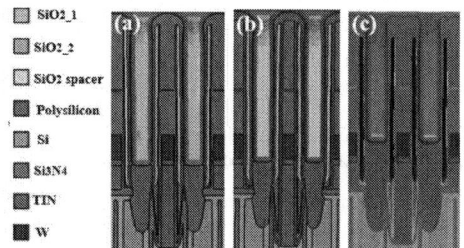

Figure1: (a) NON, (b) Low k and (c) Airgap spacer

Capacitance extraction was then executed to evaluate the C_{BL-NC} performance within a single bit cell for different process integration schemes. The extracted capacitance is proportional to the overlap area (A) and the effective dielectric constant (k), while it is inversely proportional to the distance (d) between the two components. The value of the capacitance (C) is reflected in the following formula:

$$C = \kappa A/d \qquad (1)$$

NON-SPACER SPLIT AND RESULTS

Figure 2(a) shows the relationship between the BLC X directional shift and single side C_{BL-NC}. The left NC to BL capacitance is identified as C_L while the right NC to BL capacitance is identified as C_R. Fig.2(b) shows the relationship between C_{BL-NC} and BLC X direction shift. Figures 2(c)-(g) display the simulation structure cross-sectional and top views of implementing a BLC shift from -4.5 to 4.5 nm. As shown in Fig. 2, when the BLC shifts from -4.5 to -2.5 nm, the distance(d) decreases due to the NC profile change and the C_L trends upward. In the BLC split range from -2.5 to 4.5 nm, the nitride proportion decreases, causing a decrease in the effective k and a downward trend in C_L. C_R behaves symmetrically with C_L. Fig.2 shows that the best condition can be found at a 0 nm BLC shift.

Since BL was generated by SADP process, BL pitch walking (PW, the difference between left space and right space) will be another impact factor for the parasitic capacitance. Figure 3(a) illustrates the relationship between C_L/C_R and BL PW. Figure 3(b) highlights the relationship between C_{BL-NC} and BL PW. Figures 3(c)-(e) display cross-sectional and top views of BL PW in the simulation structure using a range from -16 to 14 nm.

When BL PW is varied from -16 to 14 nm, the NC size increases, overlap area (A) will increase and the SN proportional increase makes the effective k trend up and the C_L will also trend up. The value of C_R across the pitch-walking range is symmetrical with C_L. As Figures 3 show, the best condition can be found at a value of 0 nm BL PW.

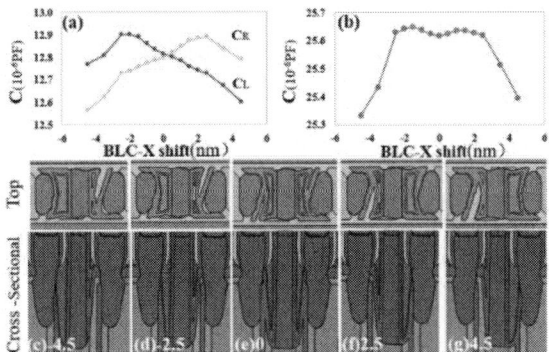

Figure 2: Relationships between BLC shift and (a) C_L / C_R, (b)C_{BL-NC}. (d)-(g) Different BLC shift simulation structures.

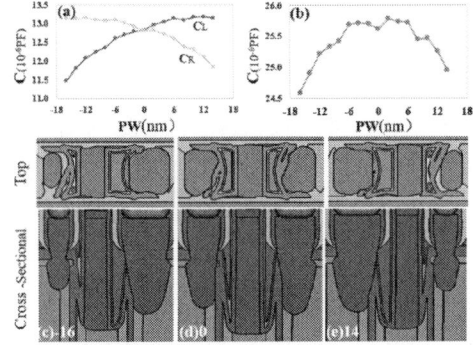

Figure 3: Relationships between BL PW and (a) C_L / C_R, (b) C_{BL-NC}. (d)~(f) Different PW shift simulation structure

Table 1: Relationship summary table between C_L and NON spacer structure

Varia ble	star t	end	k	A	d	C_L	structure impact	main factor	sensit ivity
BLCX	-4.5	4.5	↘	→	↪	⌢	lower part spacer	d then k	weak
BL PW	-16	14	↗	↗	↗	↗	Lower and upper part spacer	main A	strong

Table 1 displays a relationship summary between C_L and NON spacer structure. When BLC shifts from -4.5 to 4.5 nm, first the distance (d) decreases and C_L trends up, then the SN proportion decreases, the effective k decreases, and C_L trends down. The sensitivity of C_L to these variables is weak since the main impact is only the lower part of the NON spacer structure. When BL PW changes from -16 to 14 nm, C_L trends up sharply due to the impact

of pitch walking on both the upper and lower part of the NON spacer structure.

DIFFERENT SPACER STRUCTURE SPLITS AND RESULTS

Figure 4(a)-(d) illustrates the relationship between the C_{BL-NC} of NON, low-k, and airgap spacer and spacer thickness. The "golden" spacer thickness in each of these cases is respectively 2.5, 3.5, and 2.5 nm. When the spacer thickness trends up, C_{BL-NC} trends down, and the low-k and airgap spacers each show an improvement in the reduction of C_{BL-NC} (although the airgap spacer improves more).

Table 2 displays a table with the sensitivity of the C_{BL-NC} to spacer thickness. The sensitivity in NON and low-k spacers indicated that the SN on both sides is strongest since the SN unit thickness is doubled. The lower k value of OX compared to SN causes it display stronger sensitivity to spacer thickness. The outside SN shows the weakest capacitance sensitivity due to the impact being limited to the upper part of the spacer structure (as shown in Figure 1). The sensitivity of the airgap spacer indicated that the sandwiched OX is the strongest (of the airgap spacer test cases) since the dielectric constant of air is much smaller than OX. The double thickness change on both sides of the SN structure makes it a strong second airgap alternative, while the inside SN shows the weakest sensitivity among airgap options since increasing the airgap will introduce more dummy OX spacer etch residue than an outside SN scheme.

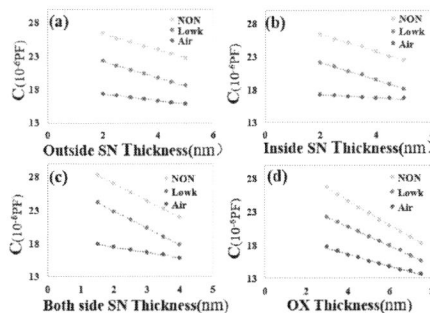

Figure 4:(a)~(d) Relationship between C_{BL-NC} of NON/low-k/airgap spacer and spacer thickness.

Table 2: Sensitivity (10^{-6} pF/nm) of spacer thickness

Variable	Outside SN/Low k	Inside SN/Low k	Both side SN/Low k	OX/ Air
NON	1.21	1.29	2.54	1.84
Low k	1.22	1.33	2.56	1.44
Airgap	0.52	0.20	0.81	0.90

Figures 5(a)-(d) highlights the relationship between an improvement of C_{BL-NC} and spacer thickness using a low-k/airgap spacer approach. When SN spacer thickness trends up, C_{BL-NC} improvements trend up in the low k

979-8-3503-1101-3/23 $31.00 © 2023 IEEE

spacer case and trend down in the airgap spacer due to the SN increase in the BL spacer proportion. When OX spacer thickness trends up, the improvement of C_{BL-NC} in the low k spacer approach trends down due to an SN decrease in BL spacer proportion and trends down in the airgap spacer due to an increase in the dummy OX spacer etch residue.

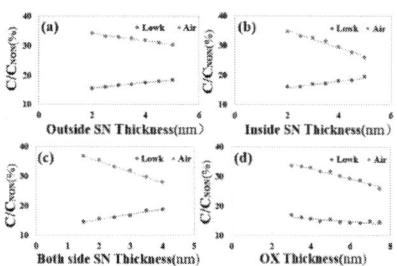

Figure 5:(a) ~(d) Relationship between the improvement of C_{BL-NC} and spacer thickness in low-k/airgap spacer.

Figure 6(a) displays the C_{BL-NC} of different spacer thicknesses under various integration schemes. Figure 6(b) shows the improvement of C_{BL-NC} with different spacer thicknesses using various integration schemes. As Figure 6(a) shows, C_{BL-NC} is in a range from 18.3 to 28.3×10^{-6} pF, with an average value of 23.9×10^{-6} pF in a NON spacer scheme. C_{BL-NC} is in a range from 15.7 to 24.1×10^{-6} pF, with an average value of 20.0×10^{-6} pF using a low-k spacer. C_{BL-NC} is in a range from 13.6 to 17.9×10^{-6} pF, with an average value of 16.3×10^{-6} pF in an airgap spacer configuration. As Figure 6(b) illustrates, the improvement of C_{BL-NC} using a low-k spacer goes from 14% to 20%, with an average of 16.5%, while the improvement of C_{BL-NC} using an airgap spacer ranges from 26% to 37%, with an average of 31.3%.

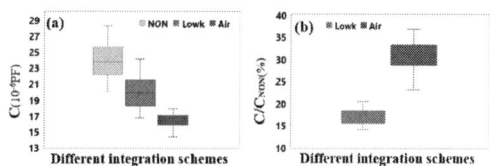

Figure 6:(a) C_{BL-NC}, and (b) The improvement of C_{BL-NC} using different spacer thicknesses under various integration schemes.

CONCLUSION

Virtual process modeling was used to study capacitance performance in DRAM processing. Our results indicated that BLC and BL PW will impact the C_{BL-NC} in NON BL spacer DRAM configurations. Capacitance is most sensitive to changes in BL PW, while the weakest sensitivity is found with BLC shifts, indicating that BL PW need be tightly controlled to give stable capacitance performance. Under varying integration schemes, low-k and airgap spacers can help reduce C_{BL-NC},

while the airgap spacer shows greater effect in reducing the capacitance than a low-k spacer. When the SN spacer thickness increases, reductions in C_{BL-NC} trend up in a low-k spacer structure and trend down in airgap spacer due to the SN proportional increase in the BL spacer structure. When the OX spacer thickness increases, improvements in reducing C_{BL-NC} trend down both in the low-k and airgap spacer structures. The average capacitance improvement of a low k spacer is 16.5% while that of the airgap spacer is 31.3%. Overall, air gap spacer is a good choice since it provides not only the lowest parasitic capacitance but also the lowest sensitivities between spacer thickness and the capacitance.

REFERENCES

[1] Q. Han, M. Cai, B. Wu and K. Cao. 2020 IEEE 15th International Conference on Solid-State & Integrated Circuit Technology (ICSICT)2020, pp. 1-3.

[2] Q. Wang, Y. De Chen, J. Huang, B. Vincent and J. Ervin. 2022 China Semiconductor Technology International Conference (CSTIC)2022, pp. 1-4.

IMPROVEMENT OF SIGE RELAXATION BY A NEW CLAMPING FILM DEPOSITION PROCESS METHOD

Zhiqiang Xiao[1,2*], Cunzhe He[1*], Dongliang Gao[1], Haitao Yan[1], Zhenchao Sui[1], Xin Zhang[1].
[1]Semiconductor Manufacturing North China (Beijing) Corp.
Beijing, China
[2]School of Integrated Circuits, Tsinghua University, Beijing, China
*Corresponding Author's Email: ZhiQiang_Xiao@smics.com; Kevin_HCZ@smics.com

ABSTRACT

SiGe has been widely used as stressors in source/drain (S/D) regions of Metal–Oxide-Semiconductor Field Effect Transistor (MOSFET) to enhance the carrier mobility of channel. However, relaxation of SiGe films led to less strain and the increase of hole mobility can not achieve expectations. In this paper, a method to prevent SiGe relaxation is provided. The method, which we call clamping film deposition process (CFD), consists of two parts: the common germanium concentration gradient down step is used to eliminate the lattice constant mismatch between films, while the low temperature silicon film deposition is used to eliminate the thermal mismatch between SiGe and silicon cap. The CFD can completely solve the problem of SiGe relaxation. Compared to common SiGe film deposition, the device performance is greatly improved (~8%) by this process, pattern wafer yield is baseline level. And this method is compatible with more advanced GAA process.

INTRODUCTION

SiGe is mostly used to enhance hole mobility of PMOS FET. Strain engineering through embedded SiGe (eSiGe) plays a critical role in improving the performance of pFET devices at advanced nodes [1], [2]. SiGe relaxation is bad for device performance because relaxation will discount the Idsat increase.

SiGe dislocation is the common relaxation source as a result of the lattice mismatch and thermal mismatch [3]. Low Ge% buffer layer or Ge% gradient layer is always used to prevent SiGe dislocation. However, the buffer layer can decrease the lattice mismatch, but the post high temp Si Cap film (for NiSi formation) will induce the thermal mismatch between SiGe and Si layer. Low temperature Si Cap might be a solution [3], but the growth rate is much lower and not friendly for fab mass production. In this study, a new SIGE clamping film deposition process is provided and this method can obviously prevent the SiGe relaxation and even enhance the device performance of PMOS.

EXPERIMENTS

Figure 1. showed the SiGe deposition step flow difference between normal process and the CFD method. For normal baseline (BL) process, Seed and Bulk layer, with low and high Ge% independently, are firstly deposited, then high temperature Si Cap layer is followed after Bulk layer. For CFD method, Ge% gradient down film and Si Cap layer with low process temperature are introduced before high temperature Si Cap deposition. In the experiments, film deposition on bare wafer and pattern wafer are both studied. SIMS test is used to calculate Ge/B concentration, while TEM is used to measure the film thickness and check relaxation condition. Bright field inspection (BFI) method is used to check the tiny particle defects after SiGe deposition, which characterizes the selectivity of the epitaxy process. Wafer Acceptance Test (WAT) test is used to verify the contribution of CFD film on device performance.

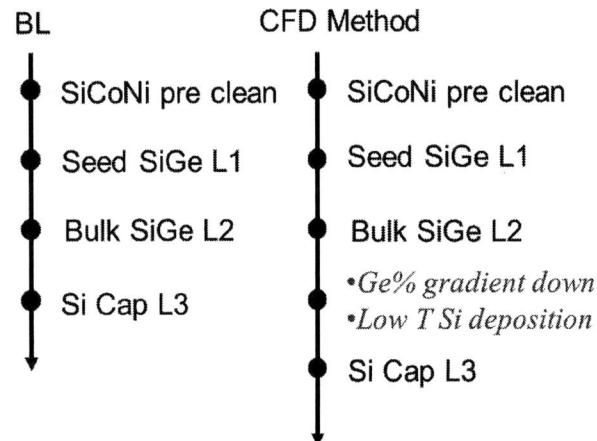

Figure 1: SiGe deposition step flow comparison between BL and CFD method (The added step is blue-marked)

RESULTS AND DISCUSSION

Figure 2 shows the BL film stack of SiGe/Si. The surface of L1 is smooth, but that of L2 and L3 is groove-like, indicating L2 and L3 has suffered strain relaxation. Jin et al. found the relaxation of SiGe film can be induced by lattice mismatch and thermal mismatch [3]. From Figure 3, L1 only and L1+L2 combined film are not relaxed because the buffer layer L1, with Ge% much lower than L2, reduces the lattice mismatch between Si substrate and high Ge% L2.

979-8-3503-1101-3/23 $31.00 © 2023 IEEE

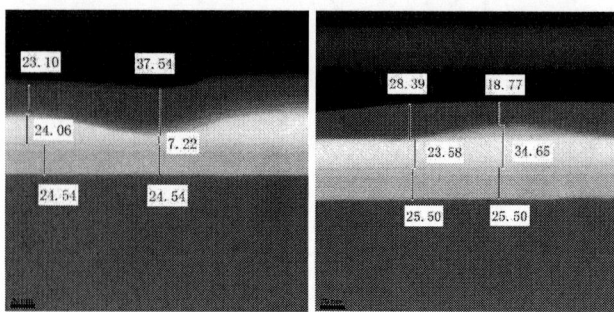

Figure 2: SiGe/Si film relaxation for BL method

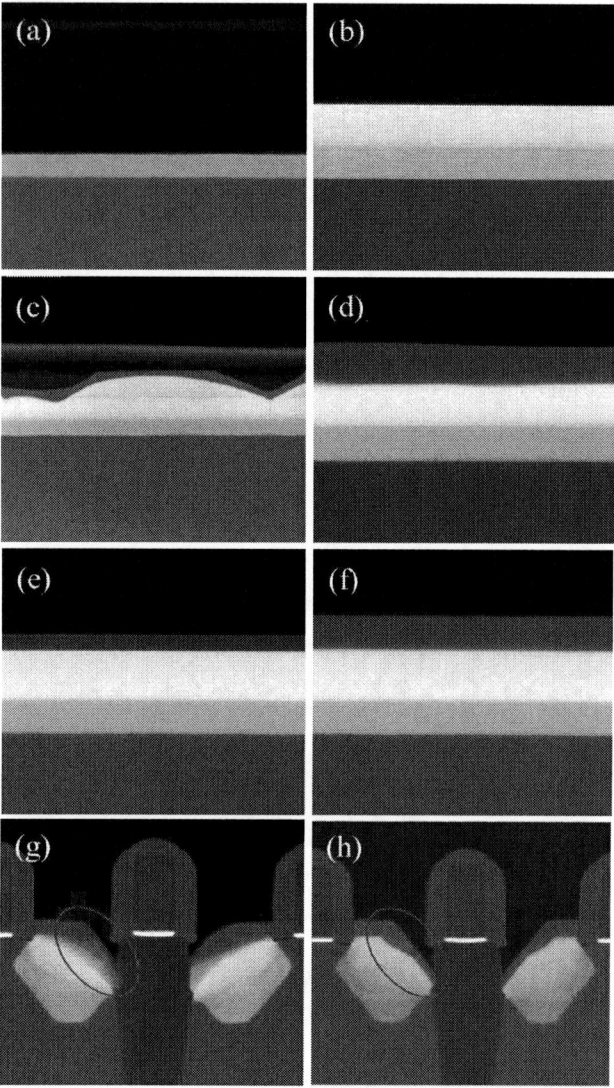

Figure 3: SiGe film relaxation root cause check a) L1; b) L1+ L2; c) L1+L2+L3 with BL method; d) L1+L2+ Ge% gradient layer +L3; e) L1+L2+Ge% gradient down layer +L3 (low T); f) L1+L2+L3 with CFD method; g) pattern wafer SA85 with BL method; h) pattern wafer SA85 with CFD method.

However, when L3 is deposited on L2 surface, both L2 and L3 relaxed (Figure 3(c)). This might be induced by the lattice mismatch between L2 and L3. To solve this problem, a GeH4 gas flow rate gradient down step is introduced between them, and we find the film get more smooth but still suffer some relaxation (Figure 3(d)). As a result, lattice mismatch between L2 and L3 is not the dominant factor for relaxation, while

thermal mismatch between L2 and L3 might be the key index, and a method to reduce the thermal budget between the films is needed. Firstly, L3 process temperature is decreased equally to that of L2. Figure 3(e) shows no relaxation happened, but much more process time would be used and the Wafer Per Hour (WPH) becomes smaller when L2 and L3 use the same lower temperature, which is not good for IC mass production. Secondly, a thin film, with several nanometers and with proper process temperature, is deposited before L3. TEM image of blank wafer, Figure 3(f), shows this method successfully inhibits the SiGe film from relaxation and the WPH keeps unchanged. Figure 3(g) and 3(h), compared with BL method, TEM image of SA85 test key on pattern wafer shows the SiGe film at trench location does not suffer relaxation.

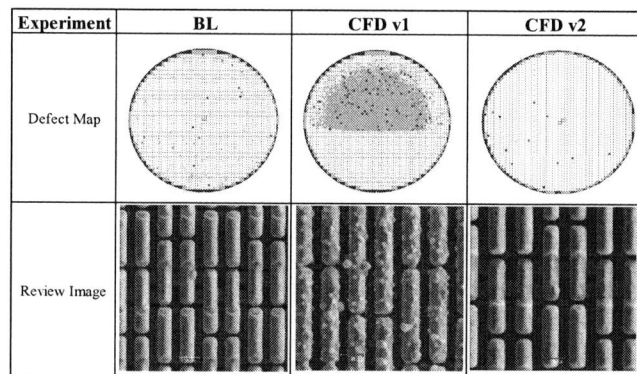

Figure 4: SiGe tiny particle defects comparison between BL and CFD method

Figure 4 shows the tiny particle defects of pattern wafer after SiGe film deposition. Tiny particles are the typical defect for SiGe epitaxy process, which comes from film deposition low selectivity on polyline end or SiN loss during trench opening by etching and wet cleaning [4]. BL defect map is clean, while that of CFD v1 locates in all SRAM area within wafer. When there are tiny particles, it will affect the yield and reliability of the chip though contact open or NiSi missing. So we need do recipe optimization by tuning process parameters during CFD deposition, and CFD v2 shows the comparable defect count with BL condition.

Figure 5 shows the normalized WAT and yield data of the CFD method. We can see that, the yield of the optimized CFD method is BL level. Device performance improvement mainly happens at SA85 location, especially at PMOS area. Compare with BL condition, Idsat is increased by 8.68%, while Ioff decreased by ~10%. Lan Jin found the junction leakage was obviously decreased when it processed under smooth surface condition comparing with the groove-like rough surface [3]. NMOS Idsat is almost not impacted by CFD method, but the Ioff was also improved. NMOS has no SiGe film deposition, as a result, a further study about the improvement is needed. CFD method improves the lattice mismatch and the thermal mismatch between L2 and L3, reduces the film relaxation, the compress strain can be retained in SiGe film and transferred to channel, and the device performance is finally improved. On the other hand, (111) plane is well formed (Figure 3 (h) red circle marked), and it decreases the contact interface area between L2 and STI HARP, which prohibits the doped boron out-gassing. As a result, both SiGe film relaxation and boron out-gassing reduction, bring the device performance improvement.

Experiment	BL		CFD v2	
Test location	SA215	SA85	SA215	SA85
P-Idsat	1	1	1.010	1.087
Log(P-Ioff)	1	1	0.997	0.907
N-Idsat	1	1	0.999	0.995
Log(N-Ioff)	1	1	0.969	0.967
Yield%				

Figure 5: device performance and yield comparison between BL and CFD v2

For 5nm process and above, Gate all around (GAA) technique will be used, one of the most key process is alternating Si/SiGe film deposition, which is usually called super-lattice structure. The alternating deposition of Si/SiGe is easy to suffer film relaxation. When CFD method is introduced to this process, as Figure 6 shows, no film relaxation is found in 3 or even in 6 alternating Si/SiGe stacking films. So this is a very promising technique for more advanced process.

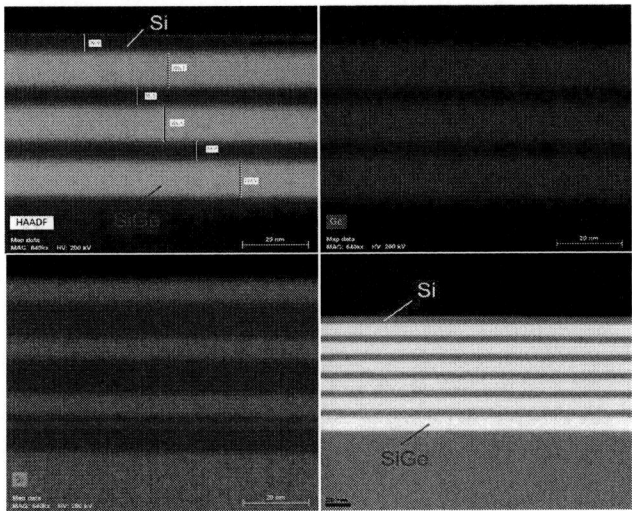

Figure 6: GAA SiGe/Si film stacking test use CFD method

CONCLUSION

In this paper a new method for prevent SiGe film relaxation was provided. SiGe relaxation can release the strain and was bad for the device performance. In the Clamping film deposition process, the common germanium concentration gradient down step was used to eliminate the lattice constant mismatch between films, while the low temperature silicon layer deposition was used to eliminate the thermal mismatch between SiGe and Silicon Cap. When the CFD method was used, the (111) plane of SiGe and Si Cap was well formed, and the out-gassing of boron was inhibited, finally the device performance was improved by 8.68%. CFD method is a promising technique and is compatible with more advanced GAA process.

REFERENCES

[1] D. Zhang, T. White and B.Y. Nguyen. Embedded source/drain SiGe stressor devices on SOI: Integrations, performance, and analyses. IEEE transactions on electron devices, 2006, pp. 3020-3024.

[2] C.Y. Cheng, Y.K. Fangs, J.C. Hsieh, H. Hsia, Y.M. Sheu, W.T. Lu and S.S. Lin. Investigation and localization of the SiGe source/drain (S/D) strain-induced defects in PMOSFET with 45-nm CMOS technology. IEEE electron device letters, 2007, pp. 408-411.

[3] L. Jin, H. Tu, Y. He, J. Lin, Y. He, W. Lu and J. Wu. Investigation of groove surface induced by strain relaxation in selective epitaxy SiGe process. In 2012 12th International Workshop on Junction Technology. IEEE, 2012, pp. 234-237.

[4] Q. Yan, C. Kun, H. Chen, L. Yin and W. Kai. Defect Principle and Improvement of 28nm Germanium Silicon Epitaxial Growth Process. CSTIC, 2021, pp. 1-2.

SOME METHODS TO REDUCE MICRO SCRATCH DEFECT FOR VIA CONTACT TUNGSTEN CHEMICAL MECHANICAL PLANARIZATION PROCESS

ZhiJie Zhang, Le Ning*, HongDi Wang, ZhiYang Liang

Semiconductor Manufacturing North China(Beijing) Corporation

No.18 WenChang Rd., Beijing Economic-Technological Development Area, 100176, China

*Corresponding Author's Email: Lance_Ning@smnchina.com

ABSTRACT

Chemical Mechanical Planarization (CMP) is a very important and essential part of modern integrated circuit fabrication process, and along with CMP process side-effects such as scratch defect appears inevitably. Contact Tungsten CMP (CT W CMP) process emphasizes both microcosmic topography and surface planarity. Scratch defects of via CT W CMP would induce yield loss by resulting in contact bridge failure. Several factors contributing to the formation of CMP scratch defect under mass production condition is generally introduced in this work, then particularly this paper shares a relationship between slurry particulates' size and count distribution by different layer of potted slurry and moreover indicates how slurry settling time during its storage affects this distribution, which explains the fact that the longer the slurry is aged and the higher layer of a slurry drum is used, the less scratch defects pattern wafers may suffer. Finally, an effective method to break such fact by alternating a new CMP polishing pad combined with a certain tungsten CMP slurry is also discussed here, supporting the opinion that polishing pad design has a great influence on the wafer microcosmic topography thanks to a better performance of the polishing byproduct's removal, with these methods by wafer micro scratch defect count was reduced by 89% on 28nm logic product wafers.

Key Words: CMP, micro scratches, slurry particulate, slurry settling, polishing pads

BACKGROUND

CMP scratch defect reduction is a historic and durable topic of semiconductor manufacturing. During the practice of fab process maintain, engineers usually implement a classification of macro scratch and random scratch. As to a wafer defect map, macro scratch shapes line patterns while random scratch has a random distribution. And in terms of defect SEM (Scanning Electron Microscope) images, random scratch can be divided into deep scratch and micro scratch. In this study, a deep scratch defect is defined to have a size of over 1um on its SEM image while a micro scratch defect's size is under 0.5um. Table I roughly illuminates the concept of such scratch classification. We observed a boom of micro scratch along with 28nm logic production, making micro scratch reduction a must.

TABLE I: CLASSIFICATION OF CMP SCRATCH DEFECT

Scratch types		Characteristic of scratch		
		patterns on wafer	SEM images	Size /um
Macro scratch				>0.5
Random scratch	Deep scratch			>0.5
	Micro scratch			<0.5

CMP equipment has a long history. Figure 1 shows a typical CMP polishing system [1]. There are so many factors that contribute to the formation of CMP scratch in mass production situation, such as solid content of slurry, slurry abrasive size, slurry agglomerating, the cleanness of slurry delivery system, the cleanness of polish module (head cover, slurry arm, disk arm and head parts themselves etc.), pad conditioning disk status, pad condition recipe, wafer in-coming backside cleanness and in-film buried particle and so on.

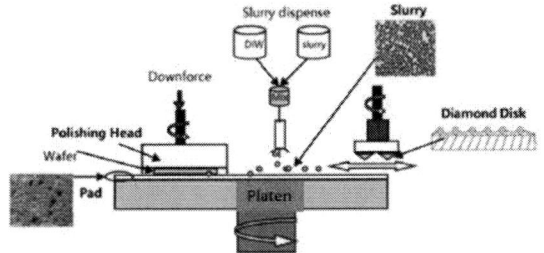

Figure 1 The schematic diagram of typical CMP polishing system

Lots of studies have been conducted to figure out solution to keep scratch defect at a low level to achieve a better yield performance of product wafer. For example, adding an additional slurry filter to the slurry pipeline right before the slurry is dispensed onto the polishing pad [2], or enhancing the post scrubber clean process of the in-coming film deposition step to heighten the cleanness of wafer backside [3], or increasing the removal amount of oxide buffing step with a soft pad to eliminate the scratches from polishing process with hard pads [4].

CT W CMP process relies on chemical reaction as much as mechanical interactions [5]. Nevertheless, the ever diversifying product types and technical platform and tightening defect criteria of SMIC fabs bring new requirement of CT W CMP scratch defect performance for 28nm technology.

In this article, in order to improve micro scratch defect performance at CT W CMP for 28nm production wafer, two methods are developed. Firstly, based on process maintain experience and the micro scratch classification, the micro scratch of 28nm product was supposed to be related with slurry particles. Secondly, since consumables setup optimization is another way for process tuning, alternating the hard pads to a new type CIP pad is also studied in this paper.

EXPERIMENT

In order to verify the slurry particle contribution of slurry and new type pad's effect on scratch reduction, a standard 5 zones tool was used for 300mm CT W CMP process by using a three-step polishing approach, while a bright field inspection equipment (BFI) was used for in-line defect detection. And a standard Scanning Electron Microscope equipment (SEM) was used for defect recognition.

2.1 Slurry particle test of different depths of slurry DRUM (200L)

SMIC lab can provide test service for all fab modules, where special instrument is used to detect particle size and count, thus the particle size distribution can be measured. In order to clarify how the settling time or aging time (from the time when a drum of slurry was manufactured to the time when it was utilized in POU slurry supply system) influences the particle distribution in a slurry DRUM, as figure 2 shows, upper layer and lower layer of drummed slurry with different aging time was sampled and tested in the lab, and table II is the experiment design. The slurry is a typical colloidal tungsten slurry.

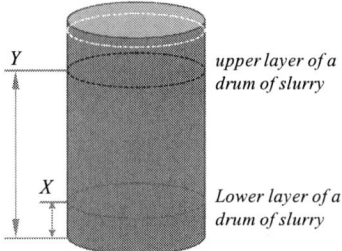

Figure 2 The sampling method for test of a drum of slurry

TABLE II: SAMPLING PLAN FOR SLURRY PARTICLE TEST

Test No.	Condition	
	Slurry aging time	Slurry layer
A	6.8X	upper
B	6.8X	lower
C	4X	upper
D	4X	lower

2.2 Defect-reducing CIP pad test

Known is that polishing pad design can make a big difference to the output of planarization including CMP scratch defect performance, so we specially chose a new type of industry's standard pad (CIP Pad) to do test to check if it could reduce micro scratch defect for 28nm product. Table III is the major pad characteristic.

TABLE III: PAD PROPERTIES COMPARISON

Pad Type	Characteristic			
	Pad Grooves	Pore Size	Compressibility	Lubricating addictive
Baseline Pad	Concentric	3X	2X	NO
CIP Pad	Concentric+ Radial	~1.7X	0.8X	Yes

RESULT AND DISCUSSION

3.1 Different sized slurry particle count distribution by depth of the slurry drum

This tungsten slurry is a colloidal silica based slurry and is used without agitation in slurry drums, so slurry particles do agglomerate and sedimentate [6].

As figure 3 shows, slurry sedimentation is both about time and space. As the settled time changes from 4X to 6.8X, 10um/5um/2um sized particle counts of lower layer slurry increased to 2.8 times/2.1 times/4.5 times.

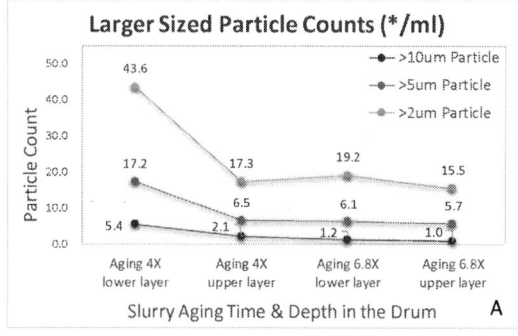

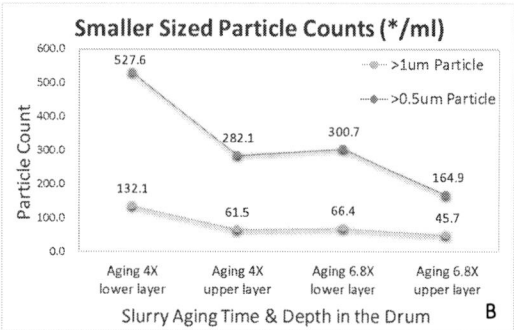

Figure 3 different size particle count of upper and lower layer of slurry with 6.8X and 4.8X aging time

In other words, it means the longer a drum of slurry

is settled and the higher layer of the drum is used, the less large particles there is in the slurry. So for mass production situation, shortening the outlet pipe by which slurry is absorbed from the slurry drum and then utilizing just the supernatant of the aged slurry will prevent micro scratch's forming during CMP material removal. 3.2 Taking advantage of slurry sedimentation to reduce product wafer's micro scratch defect

In order to use just the supernatant of a slurry drum, we come up with a new scheme of slurry supply as figure 4 (B) shows, the distance from the end of the pipe line to the top of the drum is 30*, while in the old scheme it is 80cm as figure 4 (A) shows. It will make sure that only the upper layer of the slurry drum is used.

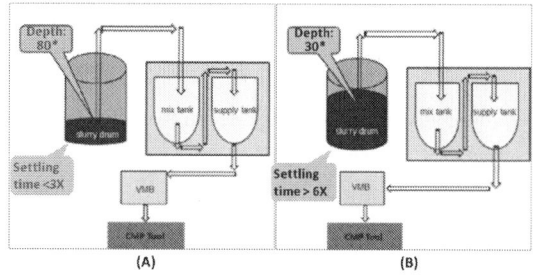

Figure 4 special slurry utilization at slurry local supply (A)old slurry utilization that the deeper layer of a slurry drum is used. (B)new slurry utilization that would just deliver upper layer of slurry drum to polish pads

Moreover, in the new scheme, the slurry used was settled for over 6X. To sum up, this method utilized the supernatant of longer aged slurry (settling time >6X) instead of fresh slurry (settling time <3X).

Thanks to BFI and SEM tools, the scratch defect performance of this new method is obtained. As figure 5 shows, wafer level micro scratch defect level is reduced by about 89%, it also shows good consistency.

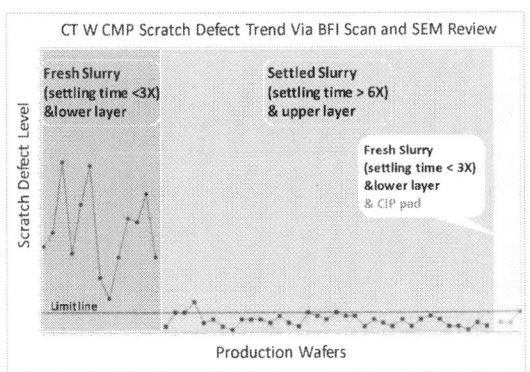

Figure 5 Scratch defect level comparison at same product wafers between the new and old methods of slurry utilization

The special slurry drum utilization sometimes may bring problems, such as increased loading of slurry drums' transport and the lack of storage room in the fab.

3.3 Scratch improvement with CIP hard Pad

To avoid the side-effects mentioned above, we aimed to pursue the CIP pad as another method to reduce micro scratch defect of 28nm technology.

Again as figure 5 shows, 3 pieces of product wafers' BFI inline micro scratch defect data is shown in comparison with the data of the previous new method and the baseline condition. The CIP pad data is represented by the green line (marked with yellow). It is obvious that adopting a superior type of hard pad can also remarkably reduce the scratch defect level to below the limit line. The selection of the CIP pad is successful mainly due to the following two facts.

Firstly, the CIP pad has both concentric grooves and radial grooves, as figure 6 shows, which greatly benefits the flow and distributing of diluted slurry solution and the move of byproduct.

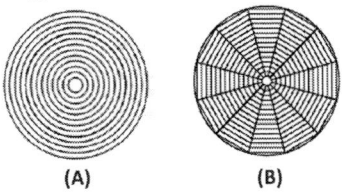

Figure 6 a scheme of pad design
(A) Cut concentric grooves of Baseline pad. (B) extra radial grooves of the CIP pad

Secondly, particular additives added to the top pad lubricates the contact surface area, therefore the polishing friction is reduced and also they help the polish byproduct to move away from the active area so the scratch source is reduced, as figure 7 shows.

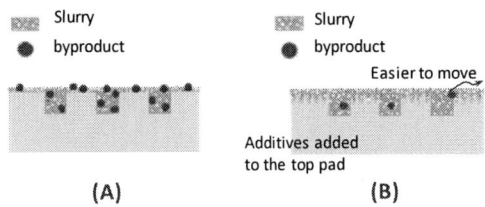

Figure 7 a scheme of pad surface
(A) The Baseline pad surface. (B) The CIP pad surface

CONCLUSION

This work illuminates that without agitation the mentioned colloidal silica based tungsten CMP slurry had an obvious effect of agglomerating and sedimentation. And then a new scheme of slurry utilization taking advantage of this effect was set up. And finally a new CIP pad was chosen and used for test to check in-line defect. it was proved both the new scheme and the new pad are successful in the reduction of 28nm product's micro scratch defect by 89%.

REFERENCES

[1] W.R. Sun, Y. Y., H. Li, W. Zhao, R. Xiao, J. Zhang, H.F. Zhou, J.X. Fang, Y. Zhang. CMP Scratch Improve in Advanced Technology Nodes. 2022 CSTIC

[2] Y. M. Sub; B. Y. T. Hian; T. K. Wui; L. W. Chye; A. B. Minhar; L. H. Jin; T. K. Yong. Reduction the Micro-sized Scratches using Optimal Design of POU Dual Filtration at STI CMP. 2018 13th IMPACT

[3] Y. Meng; L. Zhang; Y. Li; W. Zhang; H. Zhou; J. Fang. Impact of Bevel Condition on STI CMP Scratch. 2020 CSTIC

[4] T. L. Neo; S. Y. Shang; C. M. Chong; M. Huang; C. M. Chen; F. J. Hsu. CMP defect reduction by micro-scratch control using new monitoring technique. 2001 ISSM

[5] Jianfeng Luo; D.A. Dornfeld. Effects of abrasive size distribution in chemical mechanical planarization: modeling and verification. IEEE Transactions on

Semiconductor Manufacturing, vol. 16, no. 3, pp. 145-558, Aug. 2003

[6] W. Che, Y. Guo, A. Chandra and A. Bastawros, "A scratch intersection model of material removal during chemical mechanical planarization (CMP)", J. Manuf. Sci. Eng., vol. 127, no. 3, pp. 645-754, Oct. 2004

STUDY ON THE MECHANISM OF CMP INDUCED W SEAM AT ADVANCED TECHNOLOGY NODE

Shaojia Zhu*, Yurong Que, Feng Shi, Mingfei Yu, Jian Zhang, Jingxun Fang, Yu Zhang

Advanced Module Technology Development, Shanghai Huali Integrated Circuit Corporation, Shanghai 201314, China

*Corresponding Author's Email: zhushaojia@hlmc.cn

ABSTRACT

W seam is a common defect post WCMP at advanced technology node, which seriously affects the electrical performance of the device and the yield of the chip. The mainly suspected directions are the poor ability of the W gap fill with the CD shrink, or slurry corrosion during the WCMP process. In this paper, we focused the study on influences of WCMP process included oxide loss amount or corrosion caused by slurry soak time increase. First, the different oxide loss amount influence was studied, five wafers with the same incoming condition (gap fill) were processed with the same WCMP recipe body, and given different buffing polish time 20s, 30s, 40s, 50s and 65s for different oxide loss amount. Meanwhile, the five wafer were soaked different time with WCMP slurry. Then the defect was scanned by KLA. The results showed that the number of seam increased significantly when the polish time above 50s. The ILD film thickness of the 5 wafers decreased linearly with the increase of polish time, which proved that seam was buried in the middle of the W plug and gradually exposed on the film surface with the increase of oxide loss amount. By comparison, replace the polish step with slurry soaking step, the seam defect count shows no significant change with the increasing soaking time. The results show that seam is mainly strongly related to the WCMP oxide loss amount instead of WCMP slurry corrosion.

Keywords—WCMP; Seam defect; ILD thickness; Slurry corrosion

INTRODUCTION

Since its emergence in the 1980s, CMP technology has become indispensable in ultra-large-scale integrated circuits, and its applications have also increased along with the progress of scaling. In the CMP process, a wafer is pressed down and rotated against a rotating polyurethane pad that is saturated with abrasive slurry particles and a chemical solution to achieve both local and global planarization [1].

Generally, W CMP process mainly includes two parts: the first part is a bulk polishing process, to remove most of the surface of W and barriers layer; the second part is the oxide buffing polish to control the local plug protrusion [2]. W CMP mainly through the combination of electrochemical corrosion and mechanical polishing

surface effect, achieve the goal of removing W and oxide on the surface of the polishing. The surface recess on W plug is a common problem in polishing, this kind of defect will lead to higher contact resistance, caused by devices and metal wire connection fails, will seriously affect the product yield. Oxide buffing polishing process with the reversed selectivity to W CMP process, which has lower W RR than oxide film, helps to form W plug, reduce the risk of W recess [3].

However, with the development of semiconductor process node, the key of the contact hole size has been reduced to below 28 nm, contact hole depth-to-width ratio more than 4:1. Seams buried in tungsten plug is also becoming more and more serious. CVD tungsten gap fill face huge challenges. Furthermore, the seam will be grinding out slowly by W CMP. CMP corrosion from slurry and chemical may cause the seam become serious. In this paper, the correlation of W CMP and seams defect have been investigated [4-5].

EXPERIMENTAL

300mm wafers were used in this study, 28nm pattern wafers for inline thickness, defect, TEM evaluation. As shown in figure 1 CMP processes were carried out by EBARA F-REX 300 polisher using two platens process. First platen process was used for W removal and W clearing using eddy current end-point detection (R-ECM) on hard pad. Second platen was dedicated to barrier removal using a soft pad.

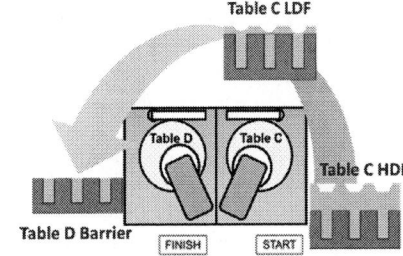

Figure 1: CMP equipment polishing process flow diagram

Oxide thickness was measured in patterned test box using a KLA-Tencor Aleris8510. Inline defect was measured using KLA PUMA9980. W resistivity measurements for thickness calculation were done using NAPSON WS3000. W seam on pattern wafer characterization was conducted using cross-sectional transmission microscopy (TEM).

RESULT AND DISCUSSION

The seam defect scanned post WCMP become more and more worse at advanced technology node. As shown in figure 1a, obvious seam defect can be seen at the middle of metal line. Further, the seam position in the metal line was confirmed by TEM, the image shows a carefully seam defect appear at almost every tungsten plug. Because the TEM position is randomly choosing. So, the real seam defect performance should be more serious than TEM.

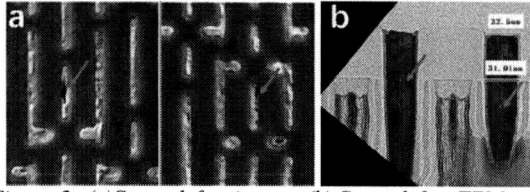

Figure 2: (a)Seam defect image; (b) Seam defect TEM;

Generally, W CMP process mainly include two parts: the first part is high pressure polishing process(HDF), to remove most of the surface of W, then change to low pressure polishing(LDF) with over polish(OP) to remove the remaining tungsten and barriers layers; the second part is the oxide buffing polish to control the local plug protrusion. Firstly, we study the correlation of seams with over polish time(OP) and buffing time. As shown in figure 3, we found the buffing polish time changes have the most significant influence. When the buffing polish time increase, the seams defect increase significantly.

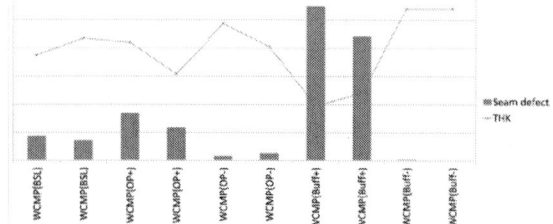

Figure 3: The correlation of Seam with WCMP split

Based on above conclusion, we are mainly concentrated in the experiment at buffing step. The different oxide loss amount influence was studied, five wafers with the same WCMP recipe given different buffing time 20s~65s for different oxide loss. As shown as figure 4a and 4b , with the increasing buffing polish time, the profile was stable, the relationship between film thickness and the grinding time was shown a linear relationship. Then defect was scanned and reviewed by KLA, the 20s, 30s, 40s split wafer shows no obvious seam defect, but when the buffing polish time increasing above 50s the seam began to appear. 65s split shows even more than ten thousand (figure 4c). So the buffing 50s' oxide loss amount is a critical value which WCMP can't exceed. As a control experiment, the five wafer were soaked different time with WCMP buffing slurry to check if the slurry corrosion increase seam defect. As shown in figure 4d , compared with figure 4c we can see that the seam defect count no newly increase. This may be the slurry we used have added W protective agent. The results confirm that seam is mainly strongly related to the WCMP oxide loss amount instead of WCMP slurry corrosion.

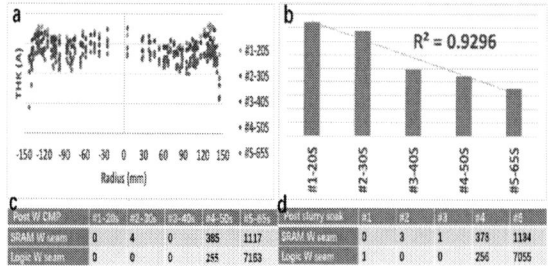

Figure 4: (a)Post WCMP profile; (b) Post WCMP THK; (c)Post WCMP Seam defect; (d) Slurry soak seam defect;

CONCLUSION

In this paper, we focused the study on influences of WCMP process include oxide loss amount or corrosion caused with slurry soak time increase. First, the different oxide loss amount influence was studied, five wafers with the same incoming condition (gap fill) were processed with the same WCMP recipe body, and given different polish time 20s, 30s, 40s, 50s and 65s for different oxide loss amount. Meanwhile, the five wafer were soaked different time with WCMP slurry. Then the defect was scanned by KLA. The results showed that the number of seam increased significantly when the polish time above 50s. The ILD film thickness of the 5 wafers decreased linearly with the increase of polish time, which proved that seam was buried in the middle of the W plug and gradually exposed on the film surface with the increase of oxide loss amount. By comparison, replace the polish step with slurry soaking step, the seam defect count shows no significant change with the increasing soaking time. The results show that seam is mainly strongly related to the WCMP oxide loss amount instead of WCMP slurry corrosion.

REFERENCES

[1] S. Guha et al., Annu. Rev. Mater. Res., Vol. 39, 2009, p181-202.

[2] K. Mistry et al., IEDM Tech. Dig., 2007, pp. 247-25.

[3] C. Auth et al., Symp. on VLSI Tech., 2008, pp. 128-129.

[4] Proceedings of the 18ih CSSDERC, J Phys Coltoq C4, Suppl 9, Tome 49 p171-201 and p489-525 plus references cited therein.

[5] R de Werdt et al., 1987 IEDM Technical Digest, p532.

STUDY ON THE MECHAMISM OF SIN RESIDUE FOR ILD0CMP

Yurong Que[1*], Xing Ma, Jian Zhang, Hu Li, Jingxu Fang, Yu Zhang

Advanced Module Technology Dept., Shanghai Huali Integrated Circuit Corp., Shanghai 201314, China

*Corresponding Author's Email: queyurong@hlmc.cn

ABSTRACT

In the gate-last scheme, inter-layer dielectric level zero (ILD0) chemical mechanical polishing (CMP) for replacing metal gate (RMG) application has been developed to meet the criteria of high-k metal gate (HKMG) devices at 28nm technology node. The gate height uniformity, loading and dishing performance control are important process challenges. However, the opening of poly-Si gate is the principal factor in the construction of HKMG. Moreover, it's a single-plank bridge that dummy poly-Si can be removed to deposit work function and aluminum metal. Long-term data shows the window of ILD0 CMP silicon nitride (SIN) residue is very narrow due to it's too hard for CMP to cover pre-layers loading, such as STI dishing, step height, photo resist etch back (PREB) SIN horn, NMOS and PMOS (N/P) loading, oxide deposition thickness (THK) loading. In this paper, the inducement of SIN residue was analyzed and discussed in detail, which will supply inspiration for resolve this system issue.

INTRODUCTION

High-k metal gate (HKMG) integration using a replacement metal gate (RMG) approach was recently developed by Intel, who had the first 45nm HKMG processor in volume production in 2007[1-2]. The chemical mechanical polishing (CMP) is a pivotal role of the RMG approach for defining metal gate structures. Two RMG CMP processes, called inter-layer dielectric level zero (ILD0) CMP processing before dummy poly-Si removed and aluminum metal (AL) CMP implementing after work function metal deposition[3-4], have been developed to fabricate the HKMG devices. In our scheme, ILD0 CMP need to ensure the opening of poly-Si gate that directly influence whether the dummy poly-Si can be removed completely to deposit metal gate or not. The product yield of the HKMG devices is particular sensitive to the silicon nitride (SIN) residue defect, which is a system defect. Shallow trench isolation (STI) dishing, step height, photo resistor etch back (PREB) SIN horn, NMOS and PMOS (N/P) loading, oxide deposition thickness (THK) loading and even pattern design can be a key factor. This article will discuss the root cause of SIN residue formation and how to decrease the level of this major defect.

RESULT AND DISCUSSION

In RMG loop, gate height is controlled through ILD0 CMP/ dummy poly-Si remove (DPR) and Al CMP. Thus, each process must control the remained thickness and uniformity accurately to attain the cross-loop well cooperation. For ILD0 CMP process, some pre-layer processes have directly or indirectly effects to the result of CMP. For example, dishing and step height originated from STI loop will lead to the uneven foundation of poly-Si. Secondly, the complex PREB process may cause SIN horn in some situations. On the other hand, it may cause poly-Si gate rounding and device N/P loading. Moreover, Contact Etching Stop Layer (CESL) SIN quality will also influence CMP end-point (EP) catching efficiency. Last but not least, oxide deposition may affect loading and dishing level. In order to intensify the ILD0 CMP process, it should not only choose the suitable configuration for smooth process, but also build up the health pre-layer steps.

In the worldwide market share, ILD0 CMP is almost 100% using AMAT LK tool which consists of three polish platens. The allocation is shown as Figure 1. The 1st platen is using hard pad conjugate to high selectivity oxide slurry, whose task is to remove the bulk oxide above CESL SIN through EP system. The 2nd platen is also utilizing hard pad combine with non-selectivity slurry to remove the oxide/SIN film above poly-Si and control poly-Si gate height by time. A chemical clean agent is introduced in 3rd platen to modify the hydrophobic to hydrophilic surface for defect improvement.

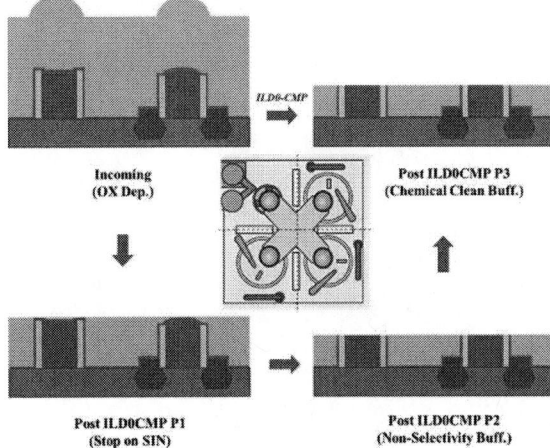

Figure 1: The configuration of ILD0CMP polish module

As an integral part of the RMG process flow, ILD0 CMP is critical for successfully forming high-k metal gate structures, this step need to expose dummy poly-Si gates, so the following etch step can remove the dummy gates for work function materials and AL deposition (Figure 2a). If SIN film is remaining above poly-Si post ILD0 CMP, there will be residue defect of poly-Si after DPR etch, which will affect the filling of work function and AL, which will lead to serious deviations in the yield of manufactured goods (Figure 2b). Of course, if the remove amount of ILD0 CMP is insufficient, it will induce SIN residue. For example, the remove rate profile of ILD0 CMP exist center/edge special map, wafer edge with slower remove rate will easily suffer SIN residue. But in most cases, the defect still occur even ILD0 CMP process and inline shows no abnormal. Multiple pre-layer processes may narrow ILD0 CMP process window and have significant influence for SIN residue, following content will introduce the various root cause of SIN residue.

a)

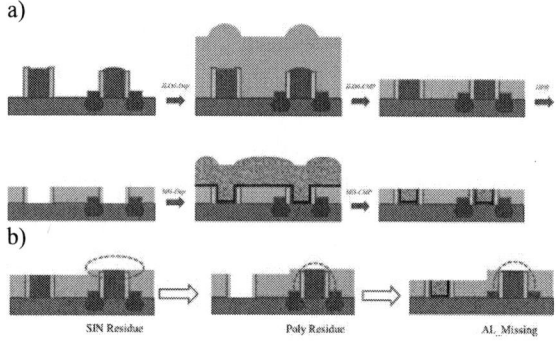

b)

Figure 2: Schematic representation of a) ILD0 & MG loop, b) SIN residue risk

PREB process need to etch gate oxideide and leave photo resist (PR) partially on S/D, which is used to protect S/D surface from etch damage. However, the short channel (SC) and long channel (LC) PR deposition loading and gate oxide height loading originated from pre layer SiGe loop leads to the margin process window of PREB. So SIN horn on NMOS, poly-Si rounding on PMOS and N/P gate height loading is unavoidable, and this phenomenon is the most common causation of SIN residue post ILD0 CMP, the schematic is shown as Figure 3. oxide residue post ILD0 CMP 1st platen polish is easy to occur on poly-Si gate with spacer SIN higher than poly-Si, while the 2nd platen is supposed to polish SIN, open the poly-Si gate and control poly-Si height to target with fix time. Therefore, the poly-Si gate with oxide residue post 1st platen polish will easily suffer SIN residue. Although poly-Si rounding is the opposite issue of SIN horn post PREB, which is over etch, poly-Si gate suffer rounding is still easily to cause SIN residue post ILD0 CMP. This situation is attributed to the high possibility of oxide residue on poly-Si gate with rounding post 1st platen polish. Except PREB loading issue, oxide deposition loading also is a key factor to induce oxide residue post 1st platen polish. Theoretically, 1st platen can extend over polish time to cover oxide residue issue, but it will enhance oxide dishing and induce metal residue post AL CMP. Notably, metal residue will result in short circuit of metal connection and the transistor may not work. Thus, in order to enlarge SIN residue window for PREB and oxide deposition loading issue, the gate height loss of ILD0 CMP need to be sufficient, which is of great importance for 2nd platen polish to cover some degree of oxide residue and decrease the ratio of SIN residue post ILD0 CMP.

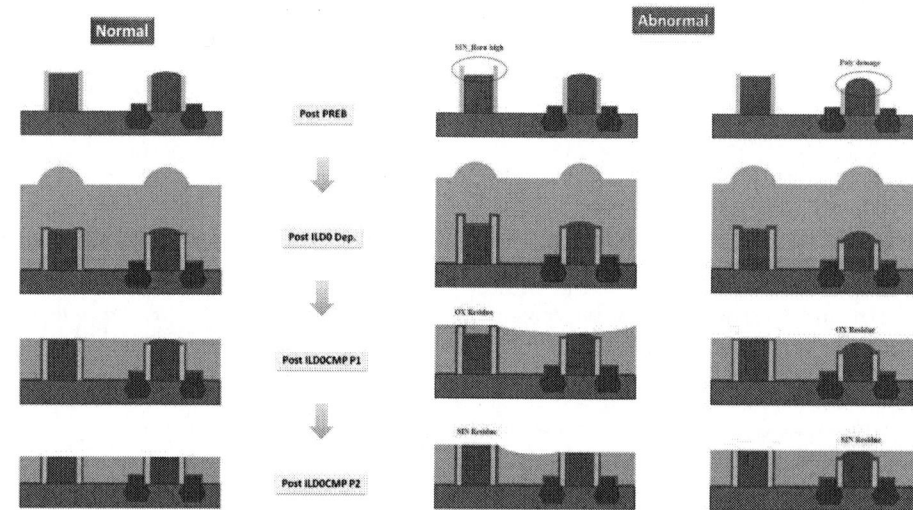

Figure 3: Mechanism schematic of SIN Residue induced by PREB Loop

STI loop, used to achieve device isolation, is like laying a foundation for transistor fabrication and RMG loop. An excellent flat surface is required underneath the gate at STI level to minimize the non-alignment of the poly-Si line. However, dishing and step height is unavoidable post STI loop. Dishing, grow out of STI CMP, is defined as the oxide loss relative to the level of the neighboring nitride space, wide trench or open structure. Poly-Si gate located in width STI area with worse dishing is easily suffered SIN residue (Figure 4) for the poly-Si is trapped underneath and fail to touch SIN post ILD0 CMP 1st platen polish, which result in oxide residue and the following SIN residue. Step height refers to height difference between active areas (AA) and STI on dense trenches possess the same mechanism of dishing. As a result, poly-Si gate located in lower area is easier to suffer SIN residue post ILD0 CMP.

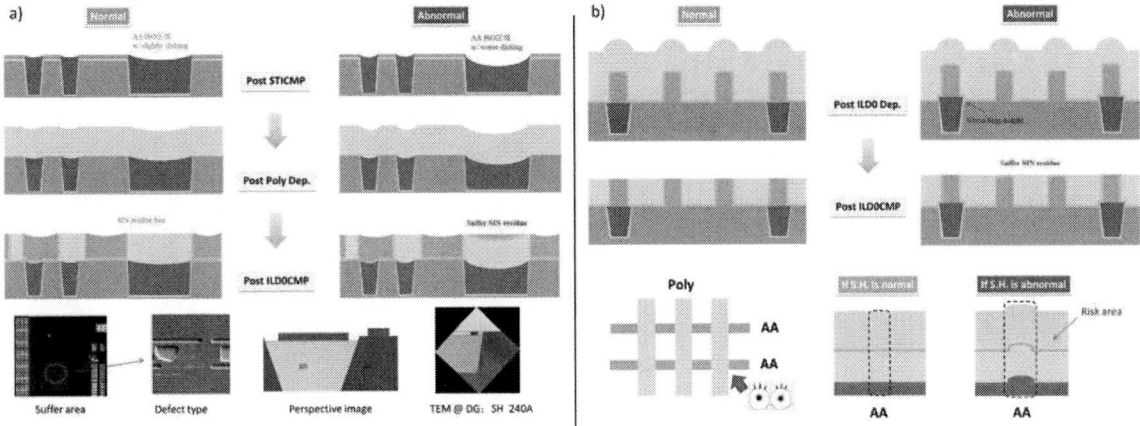

Figure 4: Mechanism schematic of SIN Residue induced by STI Loop a) dishing, b) step-height

CONCLUSION

For 28nm HKMG development, RMG is the most important loop to control gate height for device performance tuning. As an integral part of the RMG process flow, ILD0 CMP is critical for successfully forming high-k metal gate and it is important that all nitride covering the poly gate is fully removed as any residue can have an adverse impact on the subsequent poly etch/metal deposition step. As discussed, SIN residue post ILD0 CMP is an integrated defect, not only PREB loop, but also STI loop will have significant influence on this defect window. More stable, accurate process and cross-loop cooperation is requested to enlarge ILD0 CMP process window. In the ILD0 CMP, Platen 1 is remove bulk oxide and stop on SIN with high selectivity slurry. Over-polish time need to control for dishing concern, since which may induce metal residues in the next metal CMP. Platen 2 is a buffer CMP to remove oxide/SIN, this step need to open poly-Si and control poly-Si gate height to target base on integration flow. To cover pre-layer's loading (STI loop step height, dishing, PREB loop N/P loading, etc.), poly-Si gate height loss for ILD0 CMP need calculated based on cross-loop process window. Except process window, we also need to take pattern design into consideration, for example, poly-Si should be avoided locating on width shallow trench. In general, SIN residue post ILD0 CMP is an integration process window issue, it's hard to solve only count on CMP itself, we need to make a concrete analysis of a concrete problem.

ACKNOWLEDGEMENTS

All the experiment data was collected by Y. Que in HLMC. Y. Que, J. Zhang, H. Zhou, J. Fang and Y. Zhang have well discussed and analyzed these data to come out this article.

REFERENCES

[1] T. Y. Hoffman, Integrating High-k /Metal Gates: Gate-First or Gate-Last? Solid State Technology, 2010, 53, 20.

[2] C.Auth, 45nm High-k + Metal Gate Strain-Enhanced Transistors, Symp. VLSI Dig., 2008, 128.

[3] D. Lammers, Gate First or Gate Last: Technologist Debate High-k, Semiconductor International, 2010, 33.

[4] J. Steigerwald, Chemical Mechanical Polish: The Enabling Technology, Proceedings of IEDM, 2008.

SLURRY SYSTEM ESTABLISHMENT AND OPTIMIZATION FOR ADVANCED COBALT INTERCONNECT METALLIZATION

Lifei Zhang[1,2], Tongqing Wang[1,2], and Xinchun Lu[1,2]*

[1]Hwatsing Technology Co., Ltd, Tianjin, China

[2]State Key Laboratory of Tribology in Advanced Equipment, Tsinghua University, Beijing, China

*Corresponding Author's Email: xclu@tsinghua.edu.cn

ABSTRACT

Performing a chemical mechanical polishing (CMP) process for copper (Cu) interconnect metallization with cobalt (Co) diffusion barrier layer is well developed. However, the research on CMP process and associated polishing slurry system for Co as a novel interconnect wiring metal for sub-7 nm semiconductor device nodes, is still in its infancy. In this study, the establishment and optimization process of polishing slurry system for advanced Co interconnects have been presented on the foundation of a variety of mechanism analysis, including electrochemical survey, X-ray photoelectron spectroscopy measurement, surface wettability characterization, and adsorption isotherm calculation. As application verification, the final proposed CMP slurries for Co interconnects demonstrate the satisfactory material removal rates (MRR), excellent polished surface qualities, minimized particle residues, and flawless microstructures without galvanic corrosion.

INTRODUCTION

Metallization scheme of interconnects in middle of the line (MOL) and back end of the line (BEOL) has become one of the primary limiting factors of performance and yield for the advanced semiconductor manufacturing where technology nodes are lower than 7 nm. With narrower feature sizes, the major challenges for tungsten (W) with a titanium/titanium nitride (Ti/TiN) bilayer in MOL include the growing impact of parasitic contact resistance and the increasing difficulty for conformal deposition of bulk W gap-fill. Furthermore, the resistance of Cu in BEOL is rapidly approaching its limit due to the growing diffusive surface and grain boundary scattering of conduction electrons at such critical dimensions [1]. As a result, multiple material systems are being considered by the industry as promising candidates to replace the conventional W/Ti/TiN stack and Cu wiring metal, where Co presents the most important advantages on scaling contact resistance, thinner barrier layer, and higher electro-migration reliability [2].

Generally, the CMP process for Co interconnects can be mainly divided into two steps, consisting of rapid removal of bulk Co at a MRR of greater than 2000 Å/min and smooth polishing of heterogeneous materials (Co/Ti/TiN/dielectric) at a low Co MRR with a high removal selectivity between barrier/dielectric and Co.

Furthermore, several aspects during Co CMP process demand to be paid attention, involving galvanic corrosion, number of particle residues, and surface qualities. Here, attention is given to studying the CMP process for Co interconnects with Ti/TiN as the barrier layer and plasma-enhanced tetraethylorthosilicate (TEOS) as the oxide dielectric layer. Multiple analytical techniques were used to establish and optimize the polishing slurry system for two-steps CMP of Co interconnects. Furthermore, Co/Ti/TiN/TEOS patterned wafers and 12-inches production-level wafers were employed to validate the performance of our proposed slurries.

EXPERIMENTAL

The CMP experiments were performed on a T11 Polisher (Hwatsing Technology) with a 7-zone polishing head, applying a slurry flow rate of 300 mL/min and platen/head rotational speed of 93/87 rpm. Except for 12-inches blanket wafers, a specific kind of patterned wafers which has different line width and space was used in this study, whose two minimum arrays are presented in Figure 1. Potentiodynamic polarization curves were acquired using an electrochemical station (Metrohm) equipped with a three-electrode cell. The element types on Co surfaces were characterized by X-ray photoelectron spectroscopy (XPS, Ulvac-Phi).

0.50*0.50 μm array

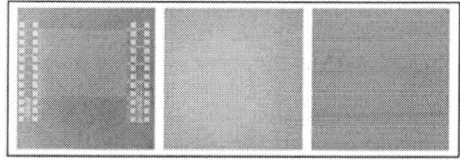

0.18*0.18 μm array

Figure 1: Two minimum arrays on used patterned wafers.

An atomic force microscopy (AFM, Bruker) was applied to map the surface morphology after polishing where a scan rate of 1 Hz was used on a 5×5 μm wafer area to obtain the AFM images. Surfscan SP5 wafer defect detection system (KLA-Tencor) was employed to display the number of particles presented on wafer surfaces, where the whole detected particle size ranges from 90 to 338 nm. The defects of interconnect structures were investigated by focused ion beam (FIB) and transmission electron microscopy (TEM). In particular, the samples for TEM were prepared by FIB. Figure 2 shows the top-view and cross-sectional profiles of the prepared patterned arrays.

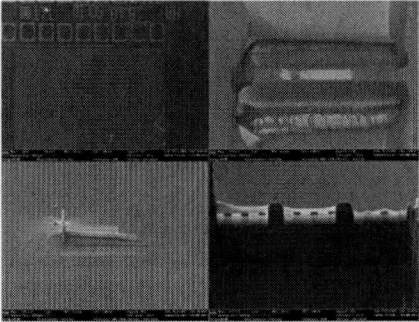

Figure 2: Profiles of the prepared patterned arrays.

RESULTS AND DISCUSSION

Establishment of Slurry System by Complexation and Inhibition Analysis

To design a new slurry system, finding an effective combination of complexing agent and corrosion inhibitor is the only way which must be passed. Considering the particularity and vulnerability of Co, satisfactory MRRs, removal selectivity, and galvanic corrosion potential were treated as the prior decisive factors. Glycine (GLY), dipotassium ethylenediaminetetraacetic acid (EDTA-2K), ammonium sulphate (AMS), nitrilotriacetic acid (NTA), citric acid (CA), and nitrilotriacetic acid trisodium salt (NTA-Na) were employed in this initial polishing tests and electrochemical characterization.

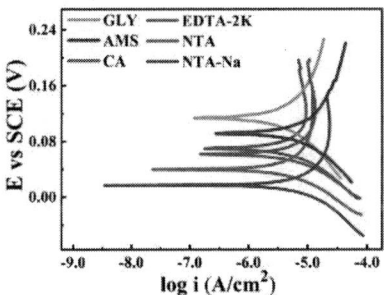

Figure 3: Potentiodynamic polarization curves of Co with different complexing agents.

Figure 3 shows a set of potentiodynamic polarization curves of Co with different complexing agents, where the calculated corrosion current densities present the complexation ability of each agent. Combined with the polarization curves of barrier layer, the minimal corrosion potential gap between Co and Ti could be obtained, indicting the lowest risk of galvanic corrosion. The determination of inhibitor could be realized by the same

approach, accompanied by the satisfaction of MRRs and selectivity. Through a lot of experimentation and research, a mixture of hydrogen peroxide, EDTA-2K, and potassium oleate (PO) in alkaline condition was suggested to conduct as the preselected scheme for Co slurries.

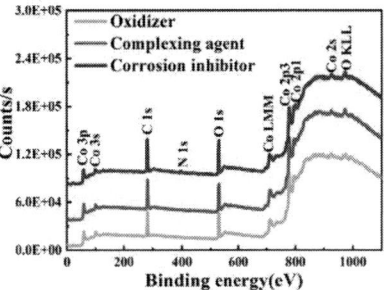

Figure 4: XPS spectra of Co films with different solutions.

The interaction mechanisms between Co surfaces and each component were explored by XPS survey and adsorption isotherm. Figure 4 displays the XPS spectra of Co surfaces after treated by different solutions, where the elements of Co, C, N, and O are recorded and marked. Typically, EDTA-2K could easily form soluble complexes with the ratio of 1:1 between metal ions and its ligand. Because of the soluble production, there is no reflection on the formation of Co surfaces. However, since PO possesses a hydrophobic chain of 18 carbon and a hydrophilic carboxylic acid end group, the carbon and oxygen signal become stronger than that in other conditions, probably originating from the adsorption of PO and the production of Co oxides.

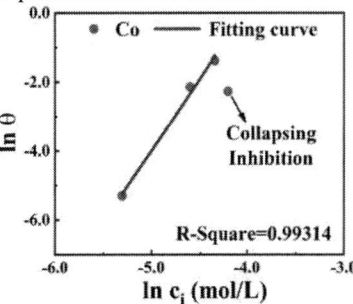

Figure 5: The Freundlich adsorption isotherm for Co.

Figure 5 presents the relationship between the logarithm of Co surface coverage (θ) and the logarithm of PO concentration (c_i). The intercept of the fitting curve provides the logarithm of adsorption equilibrium constant K, which is usually used to calculate the Gibbs energy of adsorption (ΔG^0_{ads}) to express the adsorption behavior. With the effect of low concentration PO, the calculated value of ΔG^0_{ads} is equal to -52.6 kJ/mol, confirming the spontaneity of the adsorption process for the PO inhibitor via chemisorption. However, with the excessive PO added, the stability of the inhibition effect has collapsed.

Improvement of Slurry System by Adding Surfactants

To improve and optimize our basic slurry system,

979-8-3503-1101-3/23 $31.00 © 2023 IEEE   323

various surfactants were introduced, aiming to control the number of particle residues and surface qualities after CMP process. The initial criteria for each surfactant selection adding in our basic formula is to check its influence on polishing results, including MRRs and removal selectivity. After investigation, three surfactants meet the material removal requirements, involving primary alcobol ethoxylate (AEO), polyvinyl pyrrolidone (PVP), and Tween-80 (T-80).

Figure 6: Images of measured contact angles.

To assess the effect of added surfactants on the wettability of Co films, the contact angles on fresh Co wafers were investigated, as shown in Figure 6. The maximum contact angle on clean Co wafer is ~40 with the addition of AEO, while the least contact angle of ~36 can be observed when PVP was introduced. Contact angle which presents surface energy of wafers is very vital in post-CMP cleaning process, where higher surface energy represents smaller contact angle, leading to better cleaning performance.

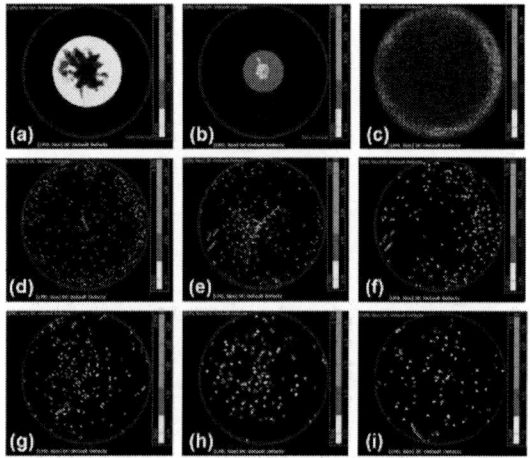

Figure 7: Particle residues on polished wafers.

Figure 7 (a)~(c) depicts the topographic images of 12-inches wafers polished by slurries only containing oxidizer, complexing agent, and inhibitor, where either the particle residues are overloaded or the number of particles are more than 10000. Figure 7 (d)~(e), (f)~(g), (h)~(i) present the particle residues after polishing by the three surfactants in sequence, where the least number of particles is around 160 with the addition of PVP, demonstrating that higher surface energy leads to better surface conditions after polishing. Furthermore, the surface topography and roughness of wafers polished with PVP were presented in Figure 8, showing excellent surface qualities.

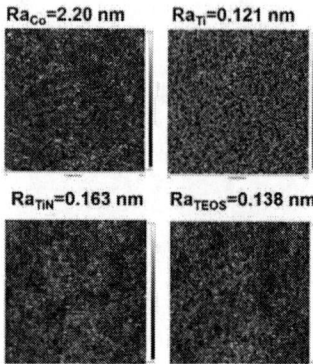

Figure 8: AFM images of polished wafers.

Polishing Performance of Slurry on Co/Ti/TiN/TEOS Stack

Co/Ti/TiN/TEOS patterned wafers were employed to validate the performance of our proposed slurry. Figure 9 presents the TEM images of 0.50*0.50 μm and 0.18*0.18 μm arrays after polishing. Although the erosion defect of ~20 nm both occurred in two arrays, indicating the TEOS MRR is too rapid, there is no Co loss, dishing, fang and galvanic corrosion in the stack, validating the excellent polishing performance of our proposed slurry system.

Figure 9: TEM images of patterned arrays.

CONCLUSIONS

In this work, the establishment and optimization of slurry system for Co interconnects have been clearly demonstrated by a variety of mechanism analysis. The proposed slurry system was validated by 12-inches production-level wafers and Co patterned wafers, showing the satisfactory MRRs, excellent polished surface qualities, minimized particle residues, and flawless microstructures.

REFERENCES

[1] Josell D, Brongersma S H, Tőkei Z. *Annual Review of Materials Research*, 2009, 39: 231-254.

[2] Kelly J, Chen J H C, Huang H, et al. *IEEE*, 2016: 40-42.

PATTERN LOADING IMPROVEMENT FOR CU CMP PROCESS

Lei Zhang[*], Yu Yang, Jian Zhang, Jingxun Fang, and Yu Zhang
Advanced Module Technology Development, Shanghai Huali Integrated Circuit Corporation,
Shanghai 201314, China
*Corresponding Author's Email: zhanglei_td2@hlmc.cn

ABSTRACT

Chemical mechanical polishing (CMP) is becoming a widely used technology to meet the precise machining in various applications. And also, the topography of pattern wafer surface shows very serious challenges. In semiconductor device, optimize post copper CMP uniformity performance, dishing and pattern loading control have been evaluated during copper CMP process. However, it is difficult to improve the pattern loading of copper and oxide area during copper CMP process, because of the different selectivity of CMP slurry. It is well known that polishing process may cause pattern loading of a layer to be planarization due to uneven distribution of device structures and thus reducing the effectiveness. This paper resent how to improve pattern loading during different pattern density of copper and oxide area, control with optimized barrier slurry and film stack schemes. To achieve the global, as well as local, planarization of wafer surface many innovative technologies have been improvement. A robust copper CMP method with better post CMP profile, tighten metal line sheet resistance (RS), smooth copper surface, and pattern loading control has been evaluated during copper CMP processing. Experiment results shown that there was no pattern loading between dense line area and ISO line area, and better dishing performance with optimized CMP polish methodology.

Keywords—Cu CMP; Pattern loading improvement; Selection ratio; Planarization

INTRODUCTION

In multilevel integrated circuit manufacturing, a planar surface preceding the next layer is very important to avoid topographical margin issues. Especially in copper layer planarization, the bulk copper layer is removed in the process of high polishing rate. High copper selectivity slurry can remove most surface copper and repair the pattern loading of different areas. The topography may cause the increasing or variation of wiring resistance and sort between the wiring upper layer, such as erosion or dishing post copper CMP. Poor topography will cause challenges in lithography and wiring for next layer. [1-2]

For pattern loading improvement, one of the solutions is add dummy pattern, which can improve the pattern fill methodology to improve the planarity of a given layer. However, dummy pattern adds capacitive load, parasitic effects on both digital circuits and analog. [3-4] Other solution is optimizing barrier metal CMP process, with a soft pad to get lower and stability remove rate. Barrier process decide the final defect and topography performance. Selectivity of different films has been identified as key influencing factors in CMP loading improvement. [5]

In this paper an optimized methodology is presented to reduce the pattern loading issue, that balance different films' polish remove rate during barrier metal step. This paper analyses correlation between selectivity and films' type, slurry flow rate, head/platen rotate speed, or down force. Improve removal rate stability and adjustable selectivity for process demanded.

EXPERIMENTAL

CMP polishing mainly consists of two steps during processing. First, preliminary polishing step to remove bulk copper with a high remove rate, the main role for preliminary polishing has a high selectivity to stop on barrier layer, which used as endpoint stop layer, as shown in figure 1(a & b). Real time process control system can well control surface uniformity for metal clear on barrier layer. The second step is barrier metal polish, for different films show different remove rate, post barrier metal CMP shows pattern loading at different locations, as shown in figure 1(c).

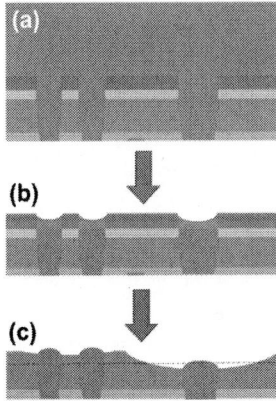

Figure 1: Cu CMP polishing process diagram. (a) Pre Cu CMP; (b) Post bulk Cu CMP, with cu dishing; (c) Post barrier metal CMP, with pattern loading

Figure 2 shows the TEM sectional view of different locations of pattern wafer. Figure 2(a) is bulk Cu area, and the thickness is normal. 2(b) is overlay mark area, which is

---

979-8-3503-1101-3/23 $31.00 © 2023 IEEE

design around bulk oxide area. TEM result shows this kind of structure is thinner than target, wafer edge has even worse performance. 2(c) is ISO Cu line area, which shows normal thickness.

Wafer edge overlay performance worse due to mark image damage during Cu CMP polishing, which suffer pattern loading issue because different remove rate on several kinds of films. TEM shows wafer overlay mark area thinner ~200A than anchor points, the reason is overlay mark area too ISO, need keep action to improve ISO-dense loading.

In attempting to solve the problems caused by pattern loading, CMP and CVD analyses correlation between selectivity and film types, slurry flow rate, head/platen rotate speed, or down force to improve removal rate stability and adjustable selectivity for process demanded.

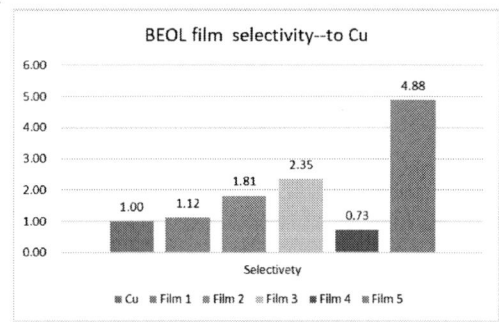

Figure 2: Different structure's TEM images after Cu CMP polishing process. (a) Bulk Cu pad; (b) OVL mark pad; (c) ISO pad

RESULT AND DISCUSSION

Barrier metal usual consist of TiN/NFDARC/Dense Low-K or other oxide films and Cu. In baseline CMP process condition, dense Low-K remove rate is much higher than copper, cause bulk oxide area great dishing maybe form weak point (high residue risk on overlap layer). Different films' selectivity to Cu as shown in figure 3.

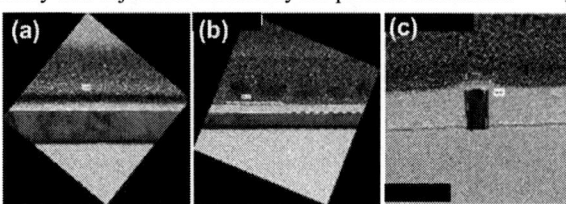

Figure 3: Different films selectivity to Cu

CMP research different Low-K film remove rate trend for new process development. Current dense Low-K remove rate is much higher than Cu (selectivity is 2.4), low pattern density area shows worse dishing defect. Figure 4 shows the correlation between Cu thickness and Cu density, which shows direct correlation.

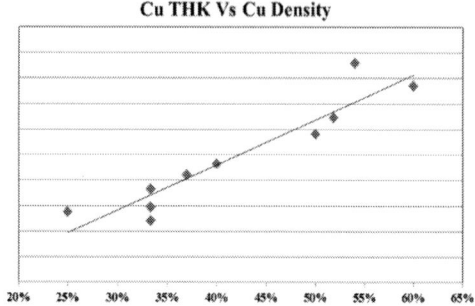

Figure 4: Correlation between Cu thickness and Cu density

Compare to Low-K 575A, Low-K 515A shows better pattern loading performance post CMP, due to less CMP polish amount, as shown in figure 5.

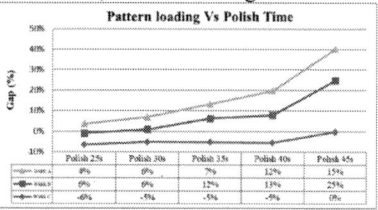

Figure 5: Correlation between pattern loading and Cu CMP polish time

On the other side, over polish split shows different thickness profile is no obvious change as shown in figure 6, which shows optimize over polish may improve pattern loading issue.

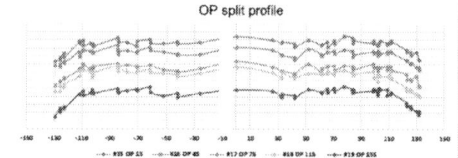

Figure 6: Post Cu CMP profile with different over polish split

Figure 7 is through pitch WAT results by pattern density of different oxide film loss analysis. Change film deposition method can obvious improve pattern loading performance, and Low-K remove amount shows correlation with pattern loading.

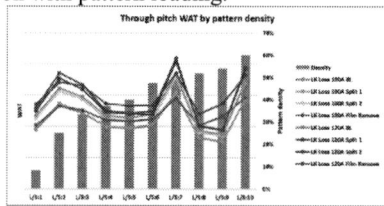

Figure 7: Through pitch WAT by pattern density of different oxide film loss analysis

As split 1 condition, dense Low-K film selectivity with copper improved from 2.4 to 1.3, K shift reduce from 3.15 to 3.0, hardness reduces from 2.8 to 2.2.

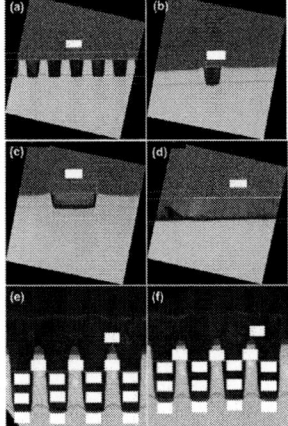

Figure 8: Different structure's TEM images after Cu CMP polishing process. (a) SRAM area; (b) ISO pad; (c) OVL mark pad; (d) Bulk Cu pad; (e) Thinner low-k CD performance; (f) Thicker low-k CD performance

Split 2 condition, new Low-K film improved Dense-ISO test key WAT gap from 32.8% to 22.9%, dry etch and wet etch rate all comparable to baseline film, pattern wafer etch profile comparable to baseline performance.

The gap of overlay mark area and anchor points drop from 84A to 48A due to scribe line area add dummy pattern. Bulk copper area Center-Edge gap drop from 88A to 66A due to scribe line area add dummy pattern, as shown in figure 8.

Figure 9 is the topography analysis simulation map of pattern wafer. Split result show both add dummy pattern, new low-k film and optimize polish condition can improve pattern loading performance.

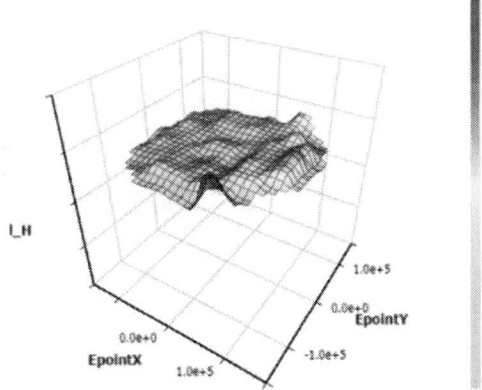

Figure 9: Topography analysis simulation map of pattern wafer

CONCLUSION

The planarization performance is influenced by topography characteristics, line/space width, pattern density, slurry chemistry, rotation speed, pad type, force/pressure, etc. CMP research different Low-K films remove rate trend for new process development. Compare to thicker low-k thinner thickness shows better pattern loading performance post CMP, due to less CMP polish amount. On the other side, over polish split shows different thickness profile is no obvious change, which shows optimize over polish may improve pattern loading issue. Change film deposition method can obvious improve pattern loading performance, and Low-K remove amount shows correlation with pattern loading. Dense low-k film selectivity with copper improved from 2.4 to 1.3, K shift reduce from 3.15 to 3.0, hardness reduces from 2.8 to 2.2. new Low-K film improved Dense-ISO test key WAT gap from 32.8% to 22.9%, dry etch and wet etch rate all comparable to baseline film, pattern wafer etch profile comparable to baseline performance. The gap of overlay mark area and anchor points drop from 84A to 48A due to scribe line area add dummy pattern. Bulk copper area Center-Edge gap drop from 88A to 66A due to scribe line area add dummy pattern. Add dummy pattern, new low-k film and optimize polish condition can improve pattern loading performance.

REFERENCES

[1] L. Zhang, et al. *Research and Solution of STI CMP Dishing and Uniformity Improvement for 28LP*, IEEE Conference Publications, 2017.

[2] L. Zhang, et al. *Study and Improvement on Tungsten Recess in CMP Process*, IEEE Conference Publications, 2019.

[3] L. Zhang, et al. *Pattern Loading Effect Optimization of BEOL Cu CMP in Technology Node*, IEEE Conference Publications, 2020.

[4] L. Zhang, et al. *Copper corrosion issue analysis and study on advanced CMP process*, IEEE Conference Publications, 2021.

[5] L. Zhang, et al. *An optimized method for Cu CMP dishing improvement*, IEEE Conference Publications, 2022.

A FEM MODEL OF MICRO-GALVANIC CORROSION EVOLUTION AT RU/CU INTERFACE IN H2O2 CMP SOLUTION

Shuo Gao, Qinhua Miao, Boyu Wen and Jie Cheng[*]

School of Mechanical Electronic and Information Engineering, China University of Mining & Technology-Beijing, Beijing 100083, China
*Corresponding Author's Email: jiecheng@cumtb.edu.cn

ABSTRACT

Our previous experiments demonstrate the galvanic corrosion occurred at the Ru (as the barrier) and Cu interface is one of the main causes of the interfacial defects during chemical mechanical polishing (CMP) process. In this work, a finite element method (FEM) model of the galvanic corrosion evolution at the Ru/Cu interface in H_2O_2 solutions was established to study the galvanic corrosion behavior at the Ru/Cu interface in H_2O_2-based slurry during CMP as a function of H_2O_2 concentrations. Results show the interfacial revolutions of Ru/Cu with the prolonged corrosion time. Compared with Ru, Cu acts as the anode and is more easily to be corroded, leading to a direct dissolution of Cu and accumulation of corrosion deposits ($Cu_2O/CuO/Cu(OH)_2$) on Cu. Considering both the dissolution of Cu and the corrosion product deposition, the resulted Cu surface is slighting above its initial position. When the concentration of H_2O_2 is 2 wt.%, the thickness of Cu deposits is 1.52 nm and the Cu dissolution depth reaches 1.49 nm, resulting in the subsequent Cu surface profile is 0.75 Å above the initial position of Cu. The results could verify our pervious experimental research. Without external mechanical force, the galvanic corrosion of Cu will reach equilibrium due to the formation of corrosion deposits. During CMP, the dynamic formation and removal of the surface film will accelerate the galvanic corrosion of Cu, leading to the severe interfacial defects like "fangs".

Keywords—FEM model; Ru/Cu couples; micro-galvanic corrosion; interface evolution; H_2O_2

INTRODUCTION

With the development of interconnect technology, the requirements for the technical nodes have become increasingly strict [1]. Ruthenium (Ru), which has been considered as one of the most promising barrier materials, can directly electroplating copper (Cu) without seed layer to achieve conformal and uniform precipitation of Cu [2][4]. Therefore, in the multi-layer structure of integrated circuits (IC), Ru and Cu come into direct contact and undergo chemical mechanical polishing (CMP) process under the infiltration of slurry. Galvanic corrosion is inevitable between Ru and Cu because of corrosive chemical composition and abrasive medium in slurry, leading to the interfacial defects on the wafer [3].

As a common mode of corrosion failure, galvanic corrosion occurs when two metals are electrically connected, implying an accelerated corrosion of the anode, i.e., metal with lower potential. Related studies have shown that compared with Ru, Cu has a lower corrosion potential and therefore acts as the anode when they are coupled [3],[4]. Hydrogen peroxide (H_2O_2) is the most used oxidant in conventional Cu planarization processes in CMP slurry [5]. Certain corrosion potential difference (ΔE_{corr}) could exist between Cu and Ru in H_2O_2-based slurries. The experimental results [4] show that the difference in the corrosion potentials of the Ru/Cu couple is about 0.1 V in the solution containing 5 mM benzotriazole (BTA), 3.0 wt.% H_2O_2 at pH 10. The potential difference is non-negligible for much small microstructures during CMP, and therefore the galvanic

corrosion phenomenon of Ru/Cu couple immersing in H_2O_2-based slurry deserves our attention.

The finite element method (FEM) is an effective method for the research of galvanic corrosion. The micro-galvanic corrosion at the interface of the interconnect structure can be quickly obtained by using the numerical simulation method. Researchers have conducted detailed research on bimetallic galvanic corrosion utilizing FEM [6][7] during the past decades. These studies proved that FEM could investigate the galvanic corrosion phenomenon from the microscopic/nanoscopic aspects and reveal the underlying mechanism to support the relevant experimental results.

Our work constructed a finite element model to describe the galvanic corrosion phenomenon of Ru/Cu couple immersed in H_2O_2-based slurry, considering the dissolution of electrode interfaces and precipitation of Cu corrosion deposits. COMSOL Multiphysics® is used to obtain the numerical solution of the model. Results show the evolution of corrosion depth and the corrosion deposits of the Ru/Cu coupling as a function of H_2O_2 concentrations.

EXPERIMENTAL

The potentiodynamic polarization test of Ru and Cu were first conducted. This can provide electrochemical reaction kinetic parameters for the subsequent FEM simulations. To study the effect of different concentrations of H_2O_2 on the galvanic corrosion, various concentrations of H_2O_2 (0.15 wt.%, 1.0 wt.%, 2.0 wt.% and 4.0 wt.%) were selected for slurry preparation. Abrasive particles were removed from the slurry to avoid the adsorption on the metal surfaces [4]. The pH value was adjusted using KOH and HNO_3 to 9. The conductivities of the slurries were measured. Potentiodynamic polarization tests were performed using an electrochemical workstation (Princeton, VersaSTAT 3F, USA) with a three-electrode system. The Ag/AgCl electrode (3.5 M KCl) was used as the reference electrode, the platinum electrode was used as the counter electrode, and the area of the working electrode was with a diameter of 5 mm. Before the test, the samples were polished to obtain a smooth surface, then ultrasonicated in absolute ethanol for 10 min, and dried with compressed air. During the measurement, the open circuit potential (OCP) was first measured, and the dynamical polarization test was performed after OCP stabilization.

MODEL DEVELOPMENT

A FEM model was developed to quantitatively describe and simulate the evolution of galvanic coupling effect of Ru/Cu. Below are the physicochemical processes considered in this model: a) the electrochemical reactions on the electrode; b) the mass transfer and electro-migration occurring in the electrolyte; c) the precipitation of corrosion deposits on the solid surface.

Definitions and Assumptions of Micro-model

The accelerated corrosion of Cu can be affected by area ratio of cathode to anode (A_c/A_a). Our work defined the width of Cu and Ru wiring on a 2D model based on the data from the 2015 International Technology Roadmap for Semiconductors (ITRS

979-8-3503-1101-3/23 $31.00 © 2023 IEEE

2015) [9]. Furthermore, certain definitions and assumptions were made as follows. Our work defined the width of the cathode as 1.9 nm and the anode 26 nm in the initial simulation, which theoretically undergoes the most severe galvanic corrosion condition. Considering the influence of mesh deformation of two direct-contacted metals and the dishing of Cu generated during CMP process on galvanic corrosion, the initial Cu surface was 0.1 nm lower than Ru, which is also set as the reference position in the analysis of profile changes of the Cu surface. Figure 1 shows the schematic diagram of the model construction. Furthermore, certain definitions and assumptions were made as follows.

a) The model was axisymmetric and only one side of the symmetry axis was built to economize computing resources.

b) The model focused only on the galvanic coupling effect between one cathodic Ru barrier and the corresponding anodic Cu interconnect it covers.

c) The electrolyte was set according to our experiments, and the initial pH value was 9.

d) The dissolution of O_2 in atmosphere and its effect on galvanic corrosion were ignored.

e) The dissolution reaction of the cathode was not contemplated.

f) Dilute solution theory was applicable, in which the convection in the static electrolyte could be ignored.

g) The porosity of porous deposits was assumed to be constant.

h) The formed Cu_2O could be dissolved again, and CuO and $Cu(OH)_2$ were stable once deposited on the electrode surface.

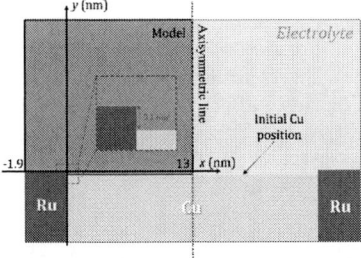

Figure 1: Schematic diagram of the model construction. The contact point of cathode and anode is set at the origin, and Cu acted as the anode is on the positive half axis of the x-axis. The initial Cu surface is 0.1 nm lower than Ru.

Local Electrochemical Reactions on Electrode Surface

Cathodic reduction of H_2O_2, as a typical of oxidizer, took place on the Ru surface. The local anodic dissolution of Cu occurring at the anode was taken into accounted in the kinetic simulation, while the dissolution of the corrosion deposits Cu_2O was also considered. The COMSOL software allows us to use the measured potentiodynamic polarization curves as direct input. The local electrochemical reactions were coupled according to the condition the real-time total current on the electrode surface is zero, which was significant due to the ever-changing cathodic and anodic area.

The anodic electrode reactions are described in Eqs. (1-3).

$$Cu \Leftrightarrow Cu^+ + e^- \qquad (1)$$
$$Cu \Leftrightarrow Cu^{2+} + e^- \qquad (2)$$
$$Cu_2O + H_2O \Leftrightarrow 2Cu^{2+} + 2OH^- + 2e^- \qquad (3)$$

The cathode reaction considered in the modeling is as follows.

$$H_2O_2 + 2e^- \Leftrightarrow 2OH^- \qquad (4)$$

Mass Transfer and Electro-migration

The active reactions on the surface of the electrode resulted in a concentration difference between various parts of the solution, which provided a driving force for the transport of substances in the liquid phase. Nernst-Planck equations for conservation of all species, i.e., Cu^+, Cu^{2+}, OH^- in the electrolyte were employed as governing equation, which includes diffusion, convection (not considered due to static corrosion), and migration

terms (Eq. (5))

$$\partial c_i / \partial t = -\nabla \cdot \left(-D_i \nabla c_i - z_i m_i F c_i \nabla \phi\right) + R_i \qquad (5)$$

where t is time, F is the Faraday constant, while c_i, D_i, z_i, m_i are the concentrations, diffusion coefficient, change number and electrical mobility for species i, respectively. The electrical mobility is given by Nernst-Einstein relationship shown in Eq. (6).

$$m_i = D_i / RT \qquad (6)$$

Furthermore, the potential conditions governing equation provided the additional equation to solve the potential ϕ in Eq. (7)

$$\frac{\partial}{\partial t} \nabla^2 \phi + \frac{\sigma}{\upsilon} \nabla^2 \phi = 0 \qquad (7)$$

where σ is the conductivity of electrolyte, and υ is the relative permittivity. The control equation in the calculation domain and the boundary conditions of mass transport and electromigration are shown in Figure 2.

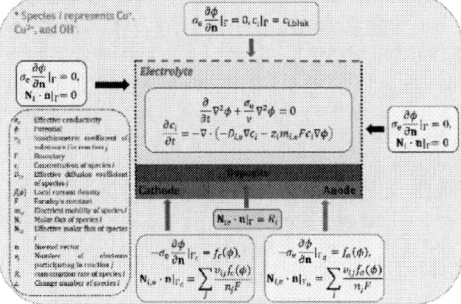

Figure 2: Schematic diagram of control equation and boundary conditions of mass transfer and electromigration in calculation domain.

Some terms needed to be corrected in the presence of deposition layer. Effective conductivity σ_e was obtained according to Arcbie's empirical law [10],[11], as shown in Eq. (8)

$$\sigma_e = s_L^m \varepsilon^n \sigma \qquad (8)$$

where s_L is fluid saturation, m is the cementation exponent, n is the saturation coefficient, and ε is the porosity of the deposits. The MacMullin number N_M was used to define the average properties of transfer phenomenon in the deposits, which was defined as follows [10][12]

$$N_M = \tau / \varepsilon \qquad (9)$$

where τ is the tortuosity of the deposits, which was set at 1 in this model. Correspondingly, diffusion coefficient of deposition layer $D_{i,e}$ could be corrected (Eq. (10)). Therefore, other relevant terms could be corrected accordingly.

$$D_{i,e} = (1-\varepsilon)D_i + \varepsilon D_i / N_M \qquad (10)$$

Furthermore, if the precipitation of corrosion deposits only occurs on the deposit interface, the consumption of species i was calculated according to the reaction rate of deposits (Eqs. (11-13)).

$$R_{Cu^+} = -R_{CuOH} \qquad (11)$$
$$R_{Cu^{2+}} = -R_{Cu(OH)_2} \qquad (12)$$
$$R_{OH^-} = -R_{CuOH} - 2R_{Cu(OH)_2} \qquad (13)$$

Precipitation of Corrosion Deposits on Solid Surface

Cu^+, Cu^{2+}, and OH^- were generated from anodic dissolution of Cu and the cathodic reaction of H_2O_2, and the species diffuse in the electrolyte through diffusion and migration, which leaded to the formation of Cu_2O, CuO, and $Cu(OH)_2$. This model only considered the impact of corrosion deposits formed on the electrode surface on its activity.

Cu^+ and Cu^{2+} would convert to their corresponding oxides when they deposit. Each transformation could be seen as a two-step reaction, including the formation of hydroxide and its subsequent dehydration and transformation into oxide. The precipitation conditions were given based on the equilibrium conditions of ion precipitation dissolution, i.e., the ions began to deposit and formed hydroxides when the dependent variable $\xi_n = \prod c_{ij}^{v_{ij}} / k_{sp} - 1 > 0$ (n = CuOH, Cu(OH)$_2$). The reaction rate could be calculated by Eq. (14) as follows

$$R_n|_\Gamma = k_s \left(\prod c_{ij}^{v_{ij}} - k_{sp} \right) \cdot \delta(\xi_n) \qquad (14)$$

where k_n is the reaction rate constant of deposits n, and $\delta(\xi_n)$ is a step function which can be shown as Eq. (15).

$$\delta(\xi_n) = \begin{cases} 0, \xi_n \le 0 \\ 1, \xi_n > 0 \end{cases} \qquad (15)$$

CuOH was rapidly dehydrated and transform into Cu$_2$O. The transformation was reckoned as fast enough in our model, i.e., the consumption rate of CuOH equaled to its formation rate. The experimental results confirmed that Cu(OH)$_2$ occupied a proportion in the deposits, so Cu^{2+} involved in the deposition reaction did not completely convert to CuO. Our model assumed that the corrosion deposits X of Cu^{2+} consisted of CuO and Cu(OH)$_2$ in a mass ratio of 1:1. The formation rate of X was equal to that of $R_{Cu(OH)2}$ calculated by Eq. (14). Furthermore, we considered that the precipitation rate of Cu$_2$O and X ($R_{dep,s}$, s = CuO, X) was fast enough, i.e., $R_{dep,s} = R_s$.

Our work divided the deposition process of corrosion deposits into two stages. Step 1, during the anodic dissolution of Cu, Cu$^+$, and Cu^{2+} ions were produced. Precipitation of ions occurred once the precipitation conditions were satisfied. Step 2, as Cu^{2+} ions began to deposit in Step 1, a portion of Cu$_2$O dissolved. Figure 3 shows the schematic diagram of the formation of corrosion deposits layer.

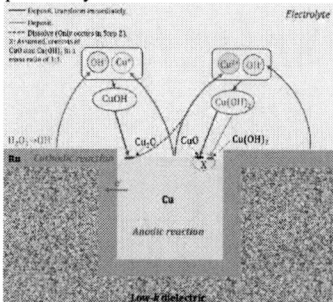

Figure 3: Schematic diagram of the formation mechanism of corrosion deposits layer on the Ru/Cu structure.

The arbitrary Langrangian-Eulerian (ALE) method was introduced into the model to accomplish the moving boundary trace. The normal vector of the active dissolution interface movement was obtained by solving the Faraday equation. The growth rate of deposits could be calculated by Eq. (16)

$$u_{dep} = \sum \frac{R_s M_s}{\rho_s} \qquad (16)$$

where M_s and ρ_s are the molar mass and density of corrosion deposits, respectively.

RESULTS AND DISCUSSION

Figure 4 shows the potentiodynamic polarization curve of Cu and Ru in different concentrations of H$_2$O$_2$ solutions at pH 9. Results show that ΔE_{corr} between Cu and Ru is about 0.1 V, which can indicate that Cu acts as anode compared with Ru, and is more susceptible to be corroded, which can lead to a loss of Cu, especially at the interface. ΔE_{corr} also increases as the concentration of H$_2$O$_2$ increases, indicating that the corrosion of Cu interface will be more serious with the increased concentration of H$_2$O$_2$. Data from potentiodynamic polarization tests provided the model with kinetic parameters for the electrochemical reaction.

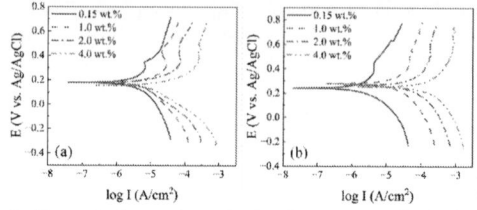

Figure 4: Potentiodynamic polarization curve of (a) Cu and (b) Ru with different concentrations of H$_2$O$_2$ at pH 9.

Concentration of Hydroxyl

The concentration and distribution of OH$^-$ can reflect the deposition of Cu$^+$ and Cu^{2+} ions. Figure 5 shows the OH$^-$ concentration distribution in the electrolyte after soaking the couple in slurry with different concentrations of H$_2$O$_2$ for 2 min. At different H$_2$O$_2$ concentrations, the concentration distribution of OH$^-$ ions is that near the cathode is higher, while the anode is lower. The reason for the distribution is the reduction reaction of H$_2$O$_2$ at the cathode. Furthermore, the concentration of OH$^-$ ion varies greatly at different H$_2$O$_2$ concentrations. When the concentration of H$_2$O$_2$ is 0.15 wt.%, there is little difference in the concentration of OH at various positions in the electrolyte, indicating that the amount of OH$^-$ generated on the cathode is not significant. As the concentration of H$_2$O$_2$ increases, the concentration of OH$^-$ near the cathode also increases slowly, reaching at ca. 1.003×10^{-5} M at a H$_2$O$_2$ concentration of 4.0 wt.%. This illustrates that within the range of H$_2$O$_2$ concentration studied, as it increases, the reduction reaction intensity of H$_2$O$_2$ on the cathode gets stronger. It should be noted that the concentration of OH$^-$ near Cu surface increases obviously with the increased concentration of H$_2$O$_2$, compared with that on Ru surfaces. Therefore, the concentration distribution of OH$^-$ near Cu is more significantly affected compared with Ru. The concentration of OH- will affect the deposition of Cu$^+$ and Cu^{2+}, and higher concentrations of OH$^-$ will facilitate the process of CuO/Cu$_2$O/Cu(OH)$_2$ formation.

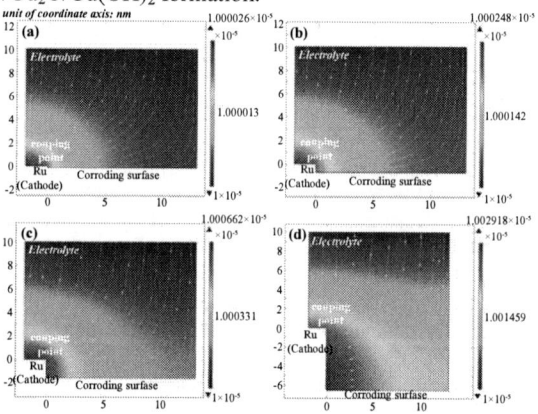

Figure 5: Concentration of OH (unit: M) after 2 min in (a) 0.15 wt.%, (b) 1.0 wt.%, (c) 2.0 wt.%, and (d) 4.0 wt.% H$_2$O$_2$.

Dissolution of Anode Surface and Precipitation of Corrosion Deposits

The corrosion of Cu will be accelerated when it is connected to Ru, and the accelerated corrosion of Cu will initiate from the Cu/Ru interface to our knowledge. The corrosion depth, and the accumulation of deposits can intuitively reflect the severity of galvanic corrosion of Ru/Cu couples. The corrosion depth indicates the profiles of Cu surface in consideration of the accelerated dissolution of Cu. The Cu surface profile change refers to the Cu surface position with reference to the initial position of Cu surface indicated in Fig. 1, considering both the Cu dissolution and the formation of corrosion deposits. We define the corrosion depth and height of Cu surface at time t, $x = x_0$ as the difference between the y-coordinate of the anode interface and deposits surface at that time and the y-coordinate of the anode interface at time 0, respectively. Figure 6 shows the corrosion depth and Cu surface profile changes after 2 min in different concentrations of H_2O_2 solutions. The surface of Cu descends compared with its initial position due to the direct dissolution. When the concentration of H_2O_2 solution is 0.15 wt.%, an extremely low corrosion depth (ca. 0.5 Å) can be formed. As the concentration of H_2O_2 solution increases to 2 wt.%, the corrosion of Cu becomes significant (ca. 6.0 Å). At a concentration of 4.0 wt.% H_2O_2, the corrosion depth of Cu exceeds the barrier thickness, reaching 64.3 Å. Such corrosion depth is non-negligible because the scale of the interconnect is only several nanometers. The Cu direct dissolution will lead to the sharp trenches along the sidewall of each Cu line, which is so-call "fangs" [13] interfacial defect after CMP.

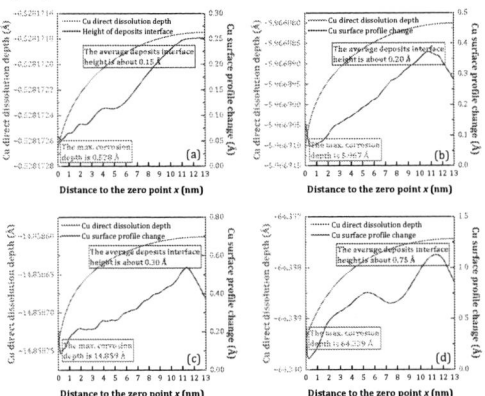

Figure 6: The direct dissolution depth and surface profile change of Cu in (a) 0.15 wt.%, (b) 1.0 wt.%, (c) 2.0 wt.%, and (d) 4.0 wt.% H_2O_2 solutions after 2 min.

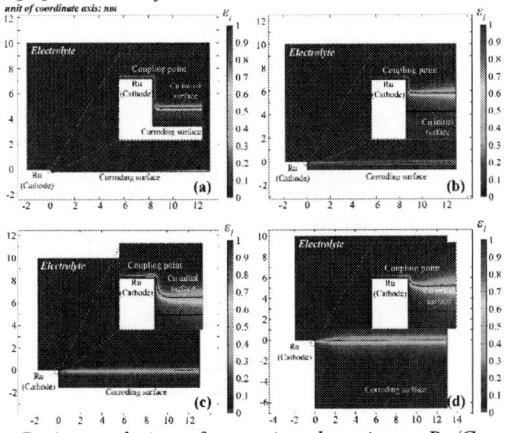

Figure 7: Accumulation of corrosion deposits on Ru/Cu surface in (a) 0.15 wt.%, (b) 1.0 wt.%, (c) 2.0 wt.%, and (d) 4.0 wt.% H_2O_2 solutions after 2 min. $\varepsilon_l = 0$, $\varepsilon_l = 1$, $\varepsilon_l = 0.5$ represents the electrolyte, the deposits, the interface between the deposits and the electrolyte, respectively.

TABLE I.   ESTIMATED DATA OF CU BASE ON FIGURE 6

H_2O_2 (wt.%)	Cu direct dissolution depth (Å)	Cu surface profile change (Å)	Thickness of Cu deposits (Å)
0.15	0.53	0.15	0.68
1.0	5.97	0.20	6.17
2.0	14.86	0.30	15.16
4.0	64.34	0.75	65.09

We summarized the Cu dissolution depth, Cu surface profile change and the calculation of Cu deposits thickness based on the simulation results from Fig. 6, as are shown in Table I. It is noteworthy that both the Cu dissolution depth and the thickness of the deposits layer on Cu increase when H_2O_2 concentration becomes higher. The thickness of Cu deposits in 2 wt.% H_2O_2 is 1.52 nm and the Cu dissolution depth reaches 1.49 nm, resulting in the subsequent Cu surface profile is 0.75 Å above the initial

position of Cu. The results are beyond our expectation because previously we believe the galvanic corrosion effect will definitely lead to loss of Cu along the Ru/Cu interface, but the simultaneous formation of Cu corrosion deposits will accumulate on the Cu surface, leading to the slight bulge of Cu over its initial position. The Cu corrosion deposits will in turn slow down the Cu dissolution, which is also call the passivation effect. It should be noted that there happens not only corrosion of metals, but also the mechanical removal of the surface layers during the real CMP process. Therefore, we should expect the formed surface deposits on Cu will be accelerated, resulting in serious interfacial defects afterwards.

Figure 7 shows the deposits accumulate on the Cu electrode surfaces after 2 min in H_2O_2 solutions. The deposits do not show a more significant accumulation near the coupling point ($x = 0$) compared to the far end ($x = 13$ nm) on Cu. Two reasons could explain this. The first reason is the re-dissolution of Cu_2O closing to the coupling point is much faster than that of the far end on Cu. The second is the scales we use in the modeling dimension is quite small (1.9 nm for Ru and 13 nm for Cu), the whole Cu surface is not sensitive to its distance of the coupling point. Furthermore, a small amount of corrosion deposits could also be found on the Ru surface, which might be due to the ion diffusion in the electrolytes. The results predicted in the model are consistent with our previous results [3].

CONCLUSIONS

To investigate the micro-galvanic corrosion between Cu and Ru in H_2O_2 solutions, a FEM model was constructed to study the galvanic corrosion evolution at different concentrations of H_2O_2. The production of OH^- raises slightly as the concentrations of H_2O_2 increases. Higher H_2O_2 concentration leads to more severe Cu direct dissolution and a thicker corrosion deposits layer. Without the external mechanical abrasion, the resulted Cu surface profile is slightly above its initial position, taking into account both the deposition of the corrosion products and the direct Cu dissolution. The simulation results are not sensitive to its geometric factors due to the nano-scale modeling dimension. The simulation results could verify our previous experimental study.

ACKNOWLEDGEMENTS

The work was supported by the National Natural Science Foundation of China (No. 52075037), Natural Science Foundation of Beijing Municipality (No. 3222017), State Key Laboratory of Tribology in Advanced Equipment (No. SKLTKF21A01), and Fundamental Research Funds for the Central Universities (No. 2021XJJD01).

REFERENCES

[1] Y. Tian, J. Zhou, C. Wang, H. Li, C. Xu, Y. Li, and Q. Liu. *ECS J. Solid State Sci. Technol.*, vol. 11, 2022, 034006.

[2] T. N. Arunagiri, Y. Zhang, O. Chyan, M. El-Bouanani, M. J. Kim, K. H. Chen, C. T. Wu, and L. C. Chen. *Appl. Phys. Lett.*, vol. 86, 2005, 083104.

[3] J. Cheng, J. Pan, T. Wang, and X. Lu. *Corros. Sci.*, vol. 137, 2018, pp. 184-193.

[4] L. Jiang, Y. He, Y. Li, and J. Luo. *Appl. Surf. Sci.*, vol. 317, 2014, pp. 332-337.

[5] S. Pandija, D. Roy, and S. V. Babu. *Microelectron. Eng.*, vol. 86, 2009, pp. 367-373.

[6] K. B. Deshpande. *Corros. Sci.*, vol. 52, 2010, pp. 3514-3522.

[7] L. Yin, Y. Jin, C. Leygraf, and J. Pan. *Electrochim. Acta.*, vol. 192, 2016, pp. 310-318.

[8] W. Li, D. Li, Z. Yu, Y. Xie, F. Liu, and Y. Jin. *J. Electroanal. Chem.*, vol. 882, 2021, 114977.

[9] ITRS International Technology Working Groups. *2015*

International Technology Roadmap for Semiconductors (ITRS), Semiconductor Industry Association, 2015.

[10] W. Sun, G. Liu, L. Wang, T. Wu, and Y. Liu. *J. Solid. State. Electrochem.*, vol. 17, 2013, pp. 829-840.

[11] G. E. Archie, *Trans. Am. Inst. Min. Metall. Pet. Eng.*, vol. 146, 1942, pp. 54-62.

[12] R. B. MacMullin. *AIChE. J.*, vol. 2, 1956, pp. 393-403.

[13] J. Cheng, B. Wang, T. Wang, C. Li, and X. Lu. *ECS J. Solid. State. Sci. Technol.*, vol. 7(11), 2018, P634-P639

EFFECTS OF PROCESS TO MATERIAL REMOVAL IN CMP: MODELLING AND EXPERIMENTS

Yanming Ren[1,2,3], Yiran Liu[2]; Zijun Guan[2], Lei Zhu[2,3], Yuanda Gao[2], Wenjie Yu[2,3], Weimin Li[2,3]*

[1]School of Materials and Chemistry, University of Shanghai for Science and Technology, Shanghai 200093, China

[2]Shanghai Institute of IC Materials Co., Ltd, Shanghai 201899, China

[3]Shanghai Institute of Microsystem and Information Technology, Chinese Academy of Sciences, Shanghai 200050, China

*Corresponding Author's Email: weimin.li@mail.sim.ac.cn

ABSTRACT

Chemical mechanical polishing (CMP) performance can be affected by many factors. In this work, computation modeling, wafer characterization and process experiments are combined to systematically study the influence of process parameters on material removal. Pressure and rotational speed are selected as representative process parameters, whose influence on material removal is studied and verified by both computation and experiments.

INTRODUCTION

With the increasing demand of local and global planarization of the wafer, CMP process has been immensely improved. One of the methods to improve within-wafer nonuniformity (WIWNU) is to divide the wafer area in polishing head into multiple annular regions along the wafer radial direction, providing precise control by applying different pressure in respective regions. The pressure can be set up as zone1-zone5 from the edge to the center, as shown in Figure 1.

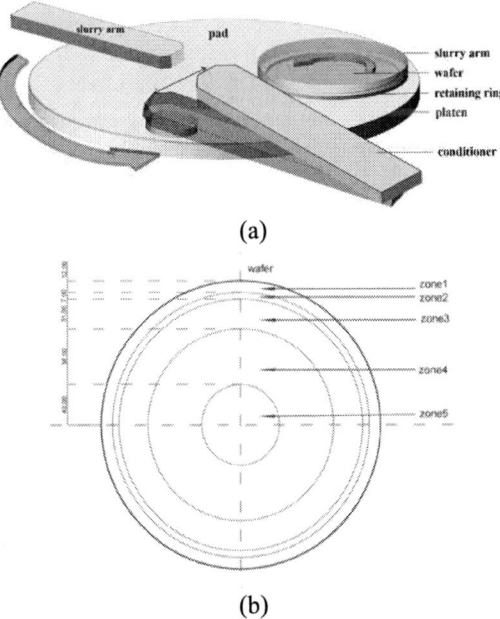

(a)

(b)

Figure 1: Schematic diagram of CMP process

Material removal rate (MRR) is one of the key factors to

evaluate wafer quality in CMP. It is necessary to study the relationship between CMP process factors and MRR. The most frequently referenced equation for MRR is Preston equation.

$$MRR = K_p \cdot P \cdot v \qquad (1)$$

Where P is the pressure applied to the wafer surface, v is the relative velocity of the wafer surface to pad, and K_p is the Preston coefficient.

The relationship between process setup and MRR has been established based on theoretical models and experiments[1]–[4]. However, most research only studied the relationship between the process factors and the global average MRR. Technical know-how on local flatness requires a more prevailing paradigm to study.

Further research on local flatness requires simultaneous consideration of theoretical model, wafer characterization, and process experiments. Based on this method, the effects of retaining ring on MRR has been systematically studied , proving effectiveness of this research method [5].

In this paper, the theoretical derivation and experiments are combined, aiming at build more accurate structure-activity relationship. Pressure and rotational speed are selected as representative process parameters. The experiments are carried out based on high-volume manufacturing machine, and the advanced detection equipment is used to characterized local material removal condition.

BASIC THEORY

The model and experimental analysis developed in this study are based on CMP motion path calculation, tribology, and elastic mechanics, focusing on studying the influence of rotational speed and pressure.

The instantaneous MRR can rarely be measured. This study takes the removed thickness during CMP process, ΔH, as the total removal amount (MR). The relationship between MRR and MR can be expressed as follows:

$$MRR = \frac{MR}{t} = \frac{\Delta H}{t} = \frac{n \cdot V_{single-particle}}{t} \qquad (2)$$

Where $V_{single-particle}$ is the material removed volume of single abrasive particle. Usually, the shape of the abrasive particles in slurry is spherical [6]. Based on the abrasive wear, the wafer is subjected by the normal load applied from the spherical abrasive particles, which leads to the surface wear

[7].

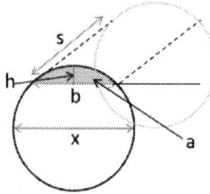

Figure 2: Model of material removed by a single abrasive particle

Figure2 shows the amount of material removed by a single abrasive particle, where h is the compaction depth of the particles, b is the indentation width, a is the indentation section area, x is the abrasive particle size, s is the slip distance of the abrasive particle on wafer surface. The volume of the removed material of $V_{single-particle}$ is:

$$V_{single-particle} = a \cdot s \qquad (3)$$

CALCULATION MODEL

The scratch depth of the abrasive particle on wafer is much smaller than the abrasive particle radius[8], [9]. The calculation model uses the relative displacement S of a specific position A on the macroscale wafer to pad, representing the slip distance of the abrasive particle to the wafer surface.

$$S = \int v(t) \cdot dt \qquad (4)$$

Where v is the relative instantaneous velocity of a specific position on the macroscale wafer with pad, t is the CMP process time. The material removal amount under the corresponding process conditions is:

$$MR = k \cdot P \cdot S = k \cdot P \cdot \int v(t) \cdot dt \qquad (5)$$

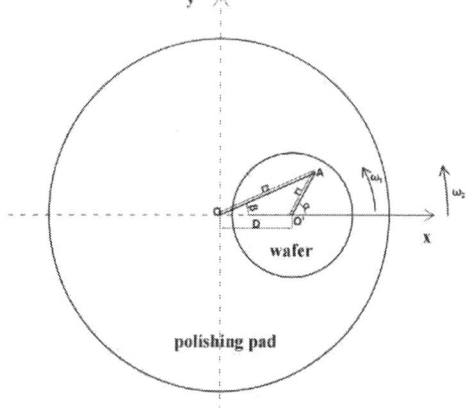

Figure 3: Schematic diagram of relative position of wafer and pad

In order to calculate the relative speed, the speed of wafer and pad and the relative position should be determined. Based on the geometric structure shown in Fig.3, the coordinate system can be established. The radius of the wafer area is 150mm, and the distance from wafer center O' to pad center O is D=180mm. O-O' direction is set as the x-axis direction, and the direction perpendicular to O-O' is set as the y-axis. O is set as coordinate origin. Suppose position A is a certain position in the wafer region. The angle between A-O and O-O is α, the coordinates of A in the model is:

$$A: (D + r_1\cos(\alpha), r_1\cos(\alpha)) \qquad (6)$$

The relative speed of point A on the wafer is determined by the wafer speed ω_1 and pad speed ω_2. To calculate the relative displacement S of point A in the wafer per second, the angular velocity needs to be converted into linear speed. The relative linear velocity of point A can be calculated based on kinematics:

$$v_A = (v_X^2 + v_Y^2)^{0.5} = (\omega_1^2 \cdot r_1^2 + \omega_2^2 \cdot r_2^2 - 2\omega_1 \cdot \omega_2 \cdot r_1 \cdot r_2)^{0.5} \quad (7)$$

The relative displacement of point A versus the polishing pad is:

$$S = \int v_A(t) \cdot dt \qquad (8)$$

Another key process parameter affecting the amount of material removal is pressure. Base on G-W contact model and elastic mechanics theory, the abrasive particles are assumed to be embedded in the wafer and pad in this research [10].The contact model is demonstrated in Figure4.

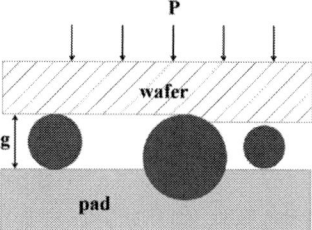

Figure 4: Schematic of contact model in CMP process

In the contact process, the abrasive can affect MRR by changing the gap g between pad and wafer. In the gap between pad and wafer, there are many abrasive particles with a certain size distribution. The pressure affects the size of the gap, whose change has an effect on the number of abrasive particles in contact with the pad and wafer[1], [3]. The abrasive particles, in contact with pad and wafer, are involved in material removal

The gap g between pad and the wafer, is determined by the abrasive particle embedded pad depth and wafer depth. From the Hertz contact theory and the sphere elastic semi-space contact model, the indentation depth of wafer d_w and that of pad d_p can be calculated respectively.

$$d_w = F / \pi \cdot x \cdot H_w \qquad (9)$$

$$d_p = F / \pi \cdot x \cdot H_p \qquad (10)$$

Where H_w and H_p are the wafer hardness and pad hardness, respectively. x is the particle size. Normal load of spherical abrasive F and overall indentation depth are given by:

$$F = 0.25 \cdot \pi \cdot P \cdot x^2 \qquad (11)$$

$$d = d_p + d_w = (F/(\pi \cdot x)) \cdot (1 / H_w + 1 / H_w) \qquad (12)$$

The wafer hardness is much greater than that of the pad. In this case, the depth of the abrasive particle into polished wafer is much less than that of pad. Hence the small indentation depth of the abrasive particles on the wafer can be ignored.

The particle size distribution in experiments usually presents as normal distribution[8], [11]. Therefore, the initial particle size distribution is assumed to match a Gaussian distribution when establishing the model of effective particles.

The gap g is affected by the hardness of the pad, applied load, and the particle size distribution, and can be calculated as follows[12]:

$$g = ((H_p - 0.25P) / H_p) \cdot (x_{avg} + 3\sigma) \tag{13}$$

Where x_{avg} is the mean size and σ is the standard deviation. The effective particles involved in the polishing process are the particles with size x larger than the gap g, and the probability density of these particles is:

$$p(x \geq g) = 1 - p(g) = f(g; x_{avg}, \sigma) = \frac{1}{\sigma\sqrt{2\pi}} \int_{-\infty}^{x} e^{-0.5 \left(\frac{g - x_{avg}}{\sigma}\right)^2} \tag{14}$$

Combining relative displacement and particle probability density calculation, the amount of material removal (MR) can be expressed as follows:

$$MR = k \cdot p\,(x \geq g) \cdot S$$

$$= k \cdot \frac{1}{\sigma\sqrt{2\pi}} \cdot \int_{-\infty}^{x} e^{-0.5\left(\frac{\frac{H_p - 0.25P}{H_p}(x_{avg}+3\sigma) - x_{avg}}{\sigma}\right)} \cdot \int v(t) \cdot dt \tag{15}$$

MATERIALS AND METHODS

In this study, high-volume manufacturing polishing machine (Universal-300 X, Hwatsing Technology Co.Ltd., Tianjin, China), is used to perform CMP process experiments. Two pads with different hardness are selected in the CMP process experiments. The two kinds of pads are hard pad A with 90D shore hardness and soft pad B with 30D shore hardness. The slurry with SiO2 abrasive (NP7050, NITTA DuPont, Osaka, Japan) is used in the experiments. The MR of wafers are measured by Bare Wafer Geometry Metrology Systems (WaferSight2+, KLA Tencor, Milpitas, CA, USA). The effects of pressure and rotation speed on the MRR are studied. Detailed process parameters are shown in the Table 1.

RESULTS AND DISCUSSION

Effect of rotational speed on MRR

From the motion equation (5), the global MR is positively correlated with the relative displacement S between wafer and pad. To study the effect of the rotational speed on MRR at different positions of the wafer, relative velocity changes in different r_1 position are calculated. The results are shown in Fig5(a), and the results of relative displacement S are shown in Fig5(b).

Combining the results of Figure 6 and equation (5), it can be concluded that the rotation speed has no effect on the MRR at different positions of the wafer. This presumption is also validated by experiments in Fig6.

TABLE I

N O.	PROCESS PARAMETER of EXPERIMENTS						
	ω_1	ω_2	P_{zone1}	P_{zone2}	P_{zone3}	P_{zone4}	P_{zone5}
#1	123	119	5.8	2.98	2.92	2.82	2.83
#2	123	119	6.96	3.576	3.504	3.384	3.396
#3	53	52	5.8	2.98	2.92	2.82	2.83

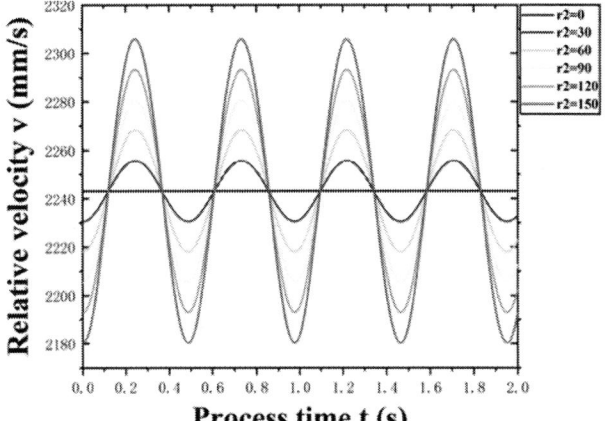

(a)

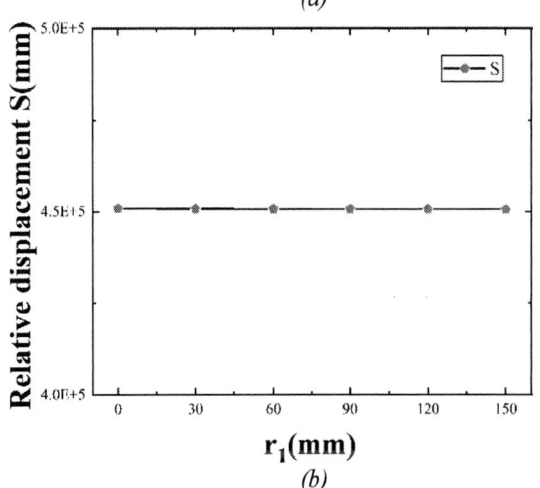

(b)

Figure 5: Relative velocity and displacement diagram for different radial positions on wafer

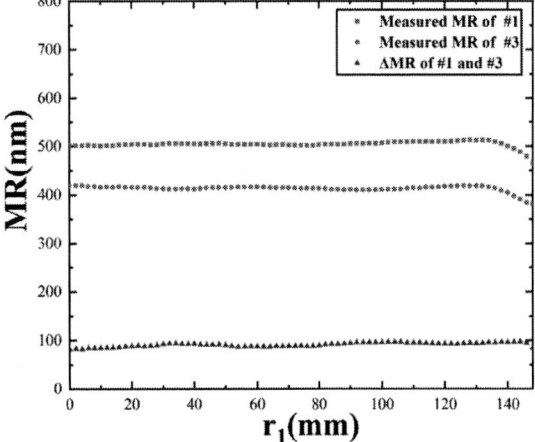

Figure 6: The relationship between relative velocity and MR

979-8-3503-1101-3/23 $31.00 © 2023 IEEE

According to the theoretical and experimental results of Fig. 5 and Fig. 6, in CMP process, changing the relative velocity can influence the MRR of the global wafer, but could not affect the material removal condition in a certain area of the wafer.

Effect of the pressure on the MR

This section studies the effect of changing the global pressure on MRR. The CMP experiments use two pads with different hardness, pad A with higher hardness and pad B with lower hardness. To study the effect of the global pressure change on MRR, experiments are conducted under process conditions #1 and #2, and ΔMR is defined as MR(#2) - MR(#1). The results are shown in Figure 7.

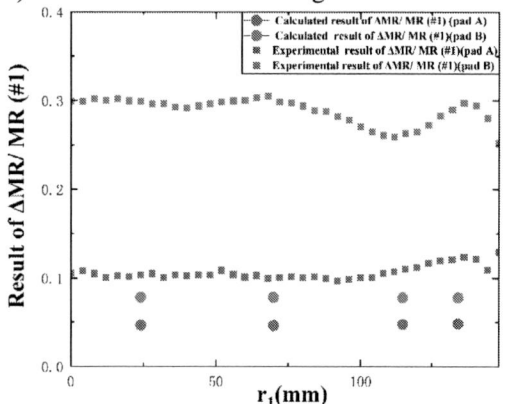

Figure 7: Effect of the global pressure change on material removal

Both the experimental results of pad A and B show that the MR of the global wafer is improved by changing the global pressure. The results are consistent in trend with the results of the relative MR calculation model that established above.

In addition, both experiment and calculation show that pad B with low hardness is more sensitive to the change of the global pressure. As the global pressure increases, the change of ΔMR (pad B) in global wafer is more significant. As the ΔP is constant, ΔMR/ MR (#1) of pad B is about 30% while that of pad A is only about 10%. This phenomenon is consistent with the elastic contact mechanics conclusions.

CONCLUSION

This paper focuses on the effect of process parameters (rotation speed and pressure) on CMP material removal amount. Combining material removal calculation and process experiment results, the conclusion can be drawn as follows:
1. Changing the relative velocity can influence the MR of the global wafer, but will not affect the local MR condition.
2. The global-wafer material removal rate is positively correlated with the change of the global pressure on the polishing head. The hardness of pad plays an important role in MR. It can be deduced from calculation model that the soft pad has a larger change in contact probability under the same pressure change, leading to a larger MRR change. This means soft pad could provide more flexible process control.

ACKNOWLEDGEMENTS

The research was funded by Science and Technology Innovation Plan of Shanghai Science and Technology Commission, Grant number 21DZ1101200. The authors also wish to acknowledge the Shanghai Institute of IC Materials Shanghai for financial support.

REFERENCES

[1] J. Luo and D. A. Dornfeld, "Material removal mechanism in chemical mechanical polishing: theory and modeling," *IEEE Trans. Semicond. Manuf.*, vol. 14, no. 2, pp. 112–133, 2001.

[2] J. Seok, C. P. Sukam, A. T. Kim, J. A. Tichy, and T. S. Cale, "Multiscale material removal modeling of chemical mechanical polishing," *Wear*, vol. 254, no. 3–4, pp. 307–320, 2003.

[3] W. Fan and D. Boning, "Multiscale modeling of chemical mechanical planarization (CMP)," in *Advances in Chemical Mechanical Planarization (CMP)*, Elsevier, pp. 137–167, 2016.

[4] D. Tamboli, G. Banerjee, and M. Waddell, "Novel Interpretations of CMP Removal Rate Dependencies on Slurry Particle Size and Concentration," *Electrochem. Solid-State Lett.*, vol. 7, no. 10, p. F62, 2004.

[5] S. Zhang *et al.*, "Numerical Analysis of the Effect of Retaining Ring Structure on the Chemical Mechanical Polishing Abrasive Motion State," *Materials*, vol. 16, no. 1, p. 62, 2023.

[6] J. C. Yang, D. W. Oh, G. W. Lee, C. L. Song, and T. Kim, "Step height removal mechanism of chemical mechanical planarization (CMP) for sub-nano-surface finish," *Wear*, vol. 268, no. 3–4, pp. 505–510, 2010.

[7] T. Yu, A. F. Bastawros, and A. Chandra, "Experimental and modeling characterization of wear and life expectancy of electroplated CBN grinding wheels," *Int. J. Mach. Tools Manuf.*, vol. 121, pp. 70–80, 2017.

[8] G. Ahmadi and X. Xia, "A model for mechanical wear and abrasive particle adhesion during the chemical mechanical polishing process," *J. Electrochem. Soc.*, vol. 148, no. 3, p. G99, 2001.

[9] G. B. Basim, J. J. Adler, U. Mahajan, R. K. Singh, and B. M. Moudgil, "Effect of particle size of chemical mechanical polishing slurries for enhanced polishing with minimal defects," *J. Electrochem. Soc.*, vol. 147, no. 9, p. 3523, 2000.

[10] J. A. Greenwood and J. P. Williamson, "Contact of nominally flat surfaces," *Proc. R. Soc. Lond. Ser. Math. Phys. Sci.*, vol. 295, no. 1442, pp. 300–319, 1966.

[11] M. Bastaninejad and G. Ahmadi, "Modeling the effects of abrasive size distribution, adhesion, and surface plastic deformation on chemical-mechanical polishing," *J. Electrochem. Soc.*, vol. 152, no. 9, p. G720, 2005.

[12] J. Luo and D. A. Dornfeld, "Effects of abrasive size distribution in chemical mechanical planarization: modeling and verification," *IEEE Trans. Semicond. Manuf.*, vol. 16, no. 3, pp. 469–476, 2003.

IMPROVING 300mm Si WAFER PLANARIZATION PROCESS WITH A WHOLISTIC APPROACH

Zijun Guan[1]; Yiran Liu[1]; Jiaming Fan[1]; Yuanda Gao[1]; Lei Zhu[1,2]; Wenjie Yu[1,2]; Weimin Li[1, 2]*

[1]Shanghai Institute of IC Materials Co., Ltd., Shanghai 201899, China

[2]Shanghai Institute of Microsystem and Information Technology, Chinese Academy of Sciences, Shanghai 200050, China

*Corresponding Author's Email weimin.li@sicm.com.cn

ABSTRACT

Despite the widespread usage of chemical mechanical polishing (CMP) on wafer planarization, the understanding of mechanism behind CMP remains incomplete, particularly with regards to affecting the global flatness of silicon substrate wafers. To address the knowledge gap, this study systematically investigates the final CMP process of 300 mm Si wafers focusing on the impact of various process parameters on overall planarity. With the assistance from a numerical model for the material removal, it is observed that several factors, including retaining ring design, pad characteristics, rotation speed, slurry flow rate, and polishing pressure all have an impact on the material removal rate (MRR). However, these factors do not necessarily change the relative MRR profile. Overall, our findings show that with a wholistic approach, the final polish (FP) process' effect on the global planarization of 300mm Si wafers changes from worsening by 16nm to improving by 35nm in average.

Keywords—CMP; Wafer flatness; Material Removal Rate, Slurry; Pad; Retaining Ring.

INTRODUCTION

Global flatness back ideal range (GBIR), the thickness range of a wafer as defined by the SEMI standard, is a measure of the global flatness of 300mm Si wafers. The lower the GBIR, the less variation across the wafer for subsequent chip manufacturing processes. Especially for the advanced photolithography, uneven topography can result in pattern distortion or poor overlay.

CMP is a surface treatment technology that is widely used to improve wafer planarization. In a Si wafer FP process, the wafer is attached to the polishing head (Head) and pressed face-down to the polishing pad (Pad). The Pad is a porous polymer material that can absorb slurry. The Pad and the wafer are both rotating around their respective centers, and the abrasive particles and the chemicals in the slurry act together on the wafer surface to remove the surface material. Abrasive particle size reduction is essential in each stage for achieving a more detailed surface treatment effect. The process materials (slurry and Pad) and process parameters, such as the pressure between abrasive particles and the wafer surface, play a crucial role in the MRR. MRR can be affected by various factors, including Pad compression, abrasive particle size and shape, and slurry composition and flow rate.

It has been studied that the relationship of material removal rate and process materials (slurry and Pad) and process parameters can be described as follows[1]:

$$MRR = \frac{4}{3}kq(\frac{f_s}{C})(\frac{R_p}{\sigma_p})^{\frac{1}{2}}\frac{PV_{re,avg}}{E_{pw}}(1-\frac{\zeta}{2})^{\frac{3}{2}}\int\limits_{D_{cr}}^{\infty}\emptyset(D)D^2dD \quad (1)$$

Where, P is the pressure the Head exserting onto the Pad, $V_{re,avg}$ is the relative speed of the wafer to the Pad, both are process parameters. f_s is related to the surface porosity of the Pad, R_p is related to the size of the Pad surface asperities, σ_p reflects the distribution of the Pad asperity heights, E_{pw} is the composite Young's modulus of the Pad and the wafer. The integral term is related to the distribution of the abrasive particles, where D is the diameter of the particle, $\Phi(D)$ is probability density of the particle sizes, D_{cr} is the critical diameter of particles that contribute to the material removal. q is the area density of the particles. Z is a factor related to the abrasive particle indentation depth, which is proportional to the Pad hardness. These are the factors related to the materials.

When the Si wafer global flatness is concerned, it is important to tune the relative MRR profile across the whole wafer to compensate the incoming wafer's thickness profile. Typical incoming wafer thickness profile is shown in Fig. 1. While there are studies on some of the factors affecting MRR profile, there is a lack of systematic study of both material and process related factors and the combinations thereof on the surface flatness due to the complexity of the CMP process. As a result, FP typically results in a slight increase in GBIR, instead of improving it.

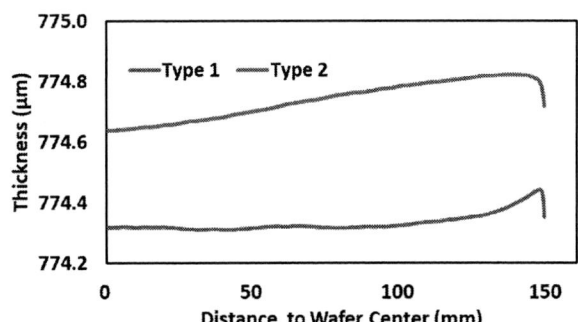

Figure 1: Typical 300mm Si wafer thickness profile before FP

In this work, slurry abrasive distribution, Pad properties,

Pad and wafer rotation speeds, slurry flow rate, Head pressure, and the interactions of these factors are studied for the impacts on MRR. An optimized FP improves GBIR by an average of 30nm, while the reference FP worsens the GBIR by an average of 16nm or worse. This study provides valuable insights into the optimization of CMP processes to improve the quality of silicon wafers used in advance IC manufacturing.

EXPERIMENTAL

In this work, Si wafer FP were performed on a 300mm CMP tool (Universal-300 X by Hwatsing, Tianjin, China). The materials including Pads, slurries and retaining rings were sourced commercially. Two types of the Pads were used for the experiments, and the detailed physical parameters are shown in the Table I. For the retaining rings with groove designs by our modeling, only the grooving was done by customization. The pressure applied on the head can be set up as zone1-zone5 from the edge to the center, as shown in Table II.

The materials removal amount (MR) of the wafers were measured by Bare Wafer Geometry Metrology Systems (WaferSight2+, KLA Tencor, Milpitas, CA, USA).

TABLE I.

Experimental Pad Characteristics			
Pad properties	Unit	Pad A	Pad B
Surface porosity	%	42.3	28.4
Surface roughness	μm	32.2	16.3
Mean peak height	μm	27.7	16.6
Mean pore depth	μm	87.4	39.8
Median pore area	mm^2	2208.7	1894.6
Mean pore area	μm^2	6862.4	5074.2
Hardness	Shore D	80.5	89.0

TABLE II.

Distribution of Head Pressure Zoning	
Pressure Zoning	Distance to Wafer Center
ZONE 1	150~138mm
ZONE 2	138~130mm
ZONE 3	130~100mm
ZONE 4	130~40mm
ZONE 5	40~0mm

RESULTS AND DISCUSSIONS

Abrasive distribution effect

The Ring is attached to the Head and used to confine the wafer during the spinning. It has grooves to allow slurry passing through. Based on our computational fluid mechanics (CFD) modeling[2], the amounts of abrasives both entering and exiting the Ring increase with the openings on the Ring (either by increasing the groove size or the number), but the net result is more abrasive entering. The added abrasives can distribute through the space inside the Ring, even between the wafer and the Pad. This results in higher MRR across the wafer without significant change of the MRR profile, as shown in Fig. 2. As the Ring opening keeps increasing, the abrasives accumulate more near the wafer edge, causing an acceleration of the edge MRR.

When the incoming wafers exhibit a thickness profile with a low center and high edge, using more openings on the Ring can achieve higher edge MRR without changing process parameters, and does not affect the MRR in other locations, providing greater process window.

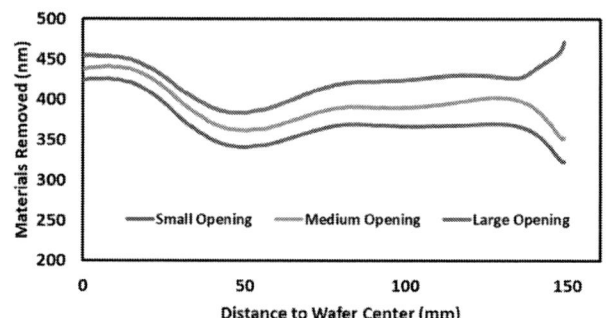

Figure 2: Materials removed by FP with different Ring grooves.

Pad effect

Two types of Pads are measured and compared. Pad A has comparable hardness to Pad B, with Shore D hardness of 80.5 and 89, respectively. Pad A has more surface pore areas than Pad B, 42.3% vs. 28.4%; larger pore size, 2200μm median pore area vs. 1900μm; and deeper pores, 87μm average pore depth vs. 40μm. MRR of Pad A with 460nm total MR is slower than that of Pad B with 490nm MR. This can be explained by more effective contacts with the abrasives in Pad B's shallower surface pores than in Pad A. The surface topology of a Pad should be considered along with its hardness when considering the impact to MRR and other polishing results.

Rotation speed and Slurry Flow effect

As shown in Fig. 3, changing the Pad/Wafer rotation speeds from 93/87rpm to 53/52rpm, MRR is reduced significantly, but the relative MRR profiles remains the same. It can be deducted that a higher relative velocity can result in a higher Material Removal Rate (MRR). Meanwhile, changing the slurry flow rate between 300ml/min and 400ml/min has no significant change in either MRR or MRR profile.

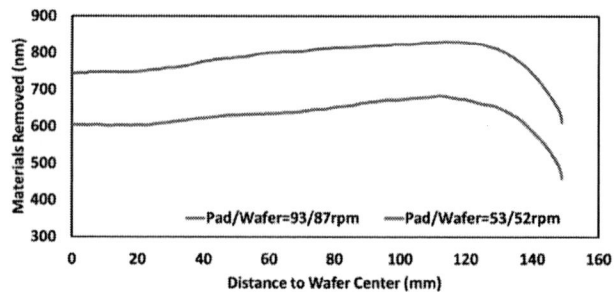

(a)

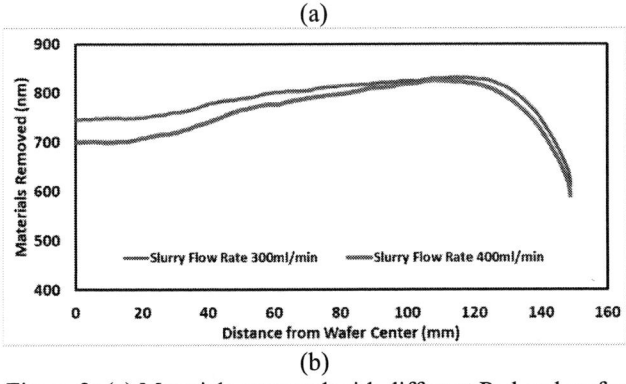

(b)

Figure 3: (a) Materials removed with different Pad and wafer rotation speed, and (b) Materials removed with different slurry flow rate.

Head Pressure effect

The Head of the CMP tool used in this work has five concentrical pressure zones. As shown in Fig. 4, even though the Zone 1 region is 138-150mm from the wafer center, increasing only the Zone 1 pressure from 4.0psi to 4.8psi increases the MRR across the whole wafer. In contrast, reducing Zone 1 pressure to 3.2psi only reduces the Zone 1 MRR significantly. If the incoming wafers have Type 2 thickness profile as shown in Fig. 1, MRR profile with higher Zone 1 pressure may match better. But if the incoming wafers have Type 1 thickness profile, Head pressure adjustment may not improve the GBIR. But a combination of the Head pressure with large Ring opening may achieve better results.

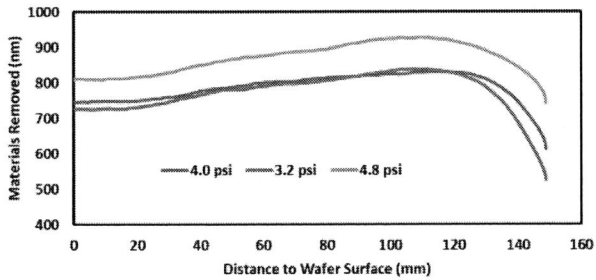

Figure 4: Materials removed with different Head Zone 1 Pressure

As shown in Fig. 5, a typical reference FP process yields GBIR worsening by average of 70nm and 16nm for different incoming wafer thickness profiles. But with experiment process 1, GBIR is worsened by only 1nm in average. With experiment process 2, GBIR is improved by an average of 35nm.

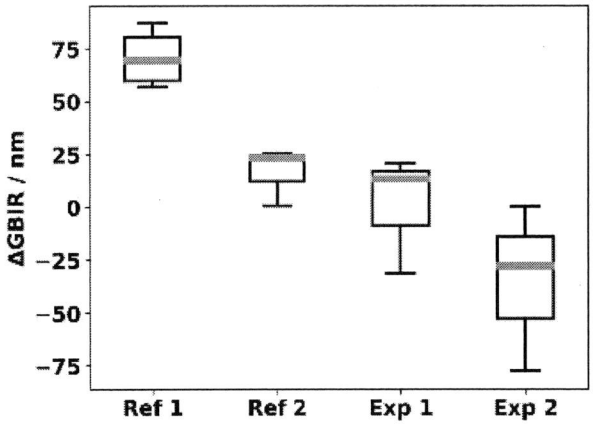

Figure 5: Change of GBIR by FP with different processes

CONCLUSION

In conclusion, the effects of the Ring on the abrasive distribution and the MRR profile are modeled and confirmed by experiments; the hardness and the surface structure of the Pad are important to the MRR and the relative MRR profile; Process parameters such as rotating speed and slurry flow rate have little impact on the MRR profile. High Head Zone 1 pressure increases the MRR of the whole wafer, but doesn't change the MRR profile, while low Zone 1 pressure reduces mainly the edge MRR. Taking a wholistic approach to tune the FP process, GBIR of 300mm Si wafer can be improved for over 30nm by average compared to worsened 16nm by average for the reference process.

REFERENCES

[1] W. Fan and D. Boning, "Multiscale modeling of chemical mechanical planarization (CMP)," in Advances in Chemical Mechanical Planarization (CMP), Elsevier, 2016, pp. 137–167.

[2] S. Zhang, Y. Liu, W. Li, et al. "Numerical Analysis of the Effect of Retaining Ring Structure on the Chemical Mechanical Polishing Abrasive Motion State", [J]. Materials, 2023, 16(1): 62

RESEARCH PROGRESS AND CHALLENGES OF CHEMICAL MECHANICAL POLISHING FOR SILICON CARBIDE WAFER

Lijuan Zhang, The chairman of Shanghai XinQian Semiconductor Co.,Ltd.

*Corresponding Author's Email: leahzhang@xinqiansemi.com

ABSTRACT

Silicon carbide (SiC), as the representative of 3rd generation semiconductor materials was widely used in many sophisticated technology fields. As the essential substrate materials, SiC wafer requires an atomic-level ultra-smooth surface. However, chemical mechanical polishing (CMP) efficacy was frequently limited by its high hardness, strength and chemical stability. Exploiting the CMP process can benefit for the enhancement of polishing efficiency. Herein, we summarized the recent progress for SiC wafer CMP, with a special emphasis on the characteristics and mechanisms. Future challenges and opportunities are also discussed in detail. We believe these SiC CMP technologies offer a good chance for the production translation.

INTRODUCTION

Silicon materials were widely used in photovoltaic power generation, memory, discrete devise, integrated circuits and other fields due to their unique physicochemical and fabrication technology since 1950s [1]. Most of chips used were made of silicon. However, silicon devices will fail when operated at the temperature above 250 °C by it's narrow bandgap (1.12 eV). Thus, the 2nd semiconductor materials, including gallium (GaAs) and indium antimonide (InSb), attracted more attentions because of their high mobility. Although the 2nd generation semiconductor has outstanding performance, the high cost and limited wafer size constrained their industrial applications. Silicon carbide (SiC), as the representative of 3rd generation semiconductor materials stood out via the wider bandgap [2].

SiC is an inorganic semiconductor material composed of carbon (C) and silicon (Si) atoms, which is the only stable compound in Si and C systems. SiC is a typical covalent bond compound with the bond energy as high as 104 kcal mol^{-1} [3]. Thus, SiC is a chemical property stable crystalline material that is nearly insoluble in any solvent at room temperature. On the other hand, the wide bandgap (3 ~ 3.33 eV), high breakdown electric filed (2.5 ~5 mV cm^{-1}), electron drift velocity (2 x 10^7 cm s^{-1}) and thermal conductivity (4.5 ~4.9 W cm^{-1} °C^{-1}) endowed single crystal SiC is the preferred material for functional devices in 5th generation (5G) communications, aeronautics, high voltage, new energy vehicles and other fields (Figure 1).

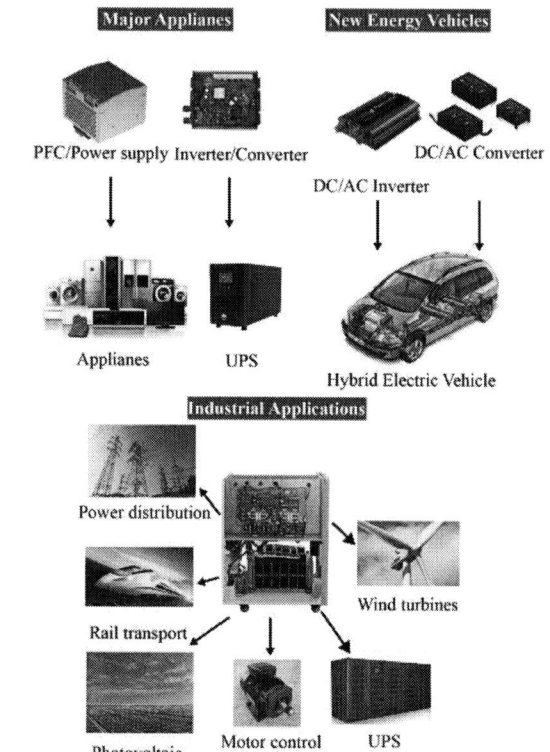

Figure 1: Applications of SiC power devices

Compared with the 1st generation (silicon, germanium) and the 2nd generation (GaAs, InSb) materials, the characterizations are below:

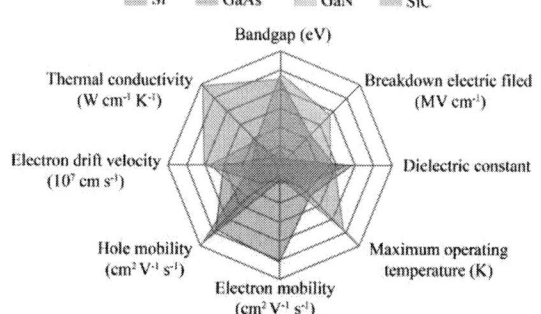

Figure 2: Characterizations of different semiconductor materials.

It can be included from Figure 2 that, SiC is superior to conventional silicon materials, in addition to the mobility.

The single crystal SiC substrate needs to go through cutting, grinding, polishing and other processes (Figure 3). The processing technology determines the surface quality of the wafer. In the fabrication of semiconductor devices, it is necessary to grow epitaxial layers on the surface of wafer by epitaxial materials which put forward higher requirements to the atom arrangement and heterogeneous structure. Any defect could affect the quality of the epitaxial layer which would decrease the performance of the final devices. It is necessary to control the total thickness variation (TTV) ≤ 15 µm, Si surface roughness (Ra) ≤ 0.5nm and C surface Ra ≤ 0.3 nm (10 µm × 10 µm). Moreover, the surface required good geometric and performance integrity. As it was illustrated above, SiC has high hardness and stable chemical property. It's a big challenge to grind and polish SiC wafers. Chemical mechanical polishing (CMP) is an effective global planarization technology, responsible for the key process to realize ultra-smooth and defect-free of the SiC wafer [4].

Figure 3: SiC wafer manufacturing process.

CMP TECHNOLOGY

Numerous CMP strategies, such as traditional CMP [5], plasma assisted polishing (PAP) [6], catalyst-referred etching (CARE) [7], ultraviolet assisted chemical mechanical polishing (UV-CMP) [8], chemical mechanical polishing based on Fenton reaction (Fenton-CMP) and electro-chemical mechanical polishing (ECMP) were widely investigated. They can be divided into intrinsic type (traditional CMP, CARE and Fenton-CMP) and equipment auxiliary type (PAP, UV-CMP and ECMP).

The traditional CMP usually polished the Si surface with alkaline silica solutions (pH 9 -13) on the soft pad (Figure 4). The mechanism is that the pre-treated Si atoms on the surface covalent bonded with three C atoms, one of which is exposed to form a single bond that can be reconstituted or removed by reactions. Especially in alkaline polishing solutions, the bonding force between Si and C can be weaken and the hydroxyl hydroxide (OH-)

can oxidize the silicon atoms. Then, the oxidized layer was removed by the abrasives to achieve super-smooth surface. However, the other surface (C surface) can not be polished by the chemical reactions. Meanwhile, the orientation of single crystal SiC is anisotropic, and the mechanism of different oxidants is not clear, it is very difficult for the crystal SiC CMP.

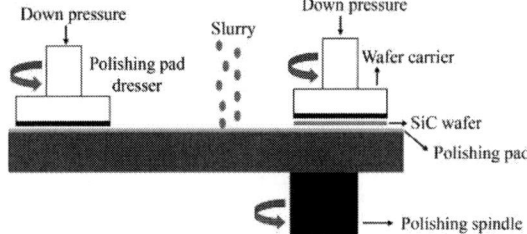

Figure 4: Principle of traditional chemical mechanical polishing.

CARE was conducted by adding the catalyst (platinum) and etching agent (hydrofluoric acid or H_2O) to the polishing pad (Figure 5). The hydrofluoric acid (HF) can ionize into the fluorine ion (F^-), hydrogen ion (H^+) and active substance (h^+). H_2O can decompose into hydroxyl radicals (•OH), H^+ and h^+ with the catalyst. Subsequently, the h^+ oxide the crystal SiC into SiO_2 layer which can be transformed to H_2SiF_6 with HF. Lastly, the oxide layer can be removed to realize the ultra-precision SiC CMP. Due to the extremely short life of the active substance, etched layer can only form on the surface in contact with the catalyst, resulting in the relative low remove rate (RR).

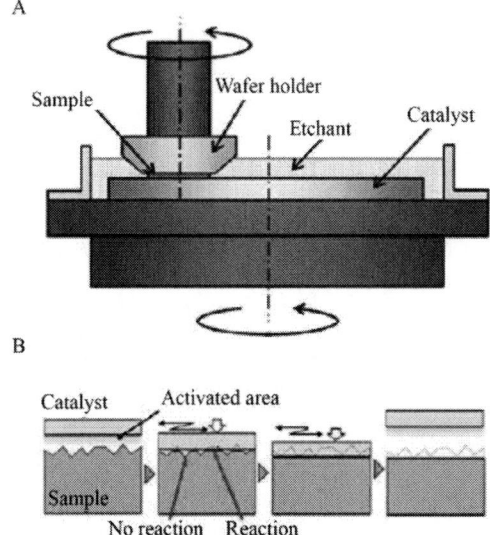

Figure 5: The schematic of catalyst-referred etching: schematic diagram of (A) processing equipment and (B) material removal mechanism [9].

Fenton-CMP is a chemical enhanced CMP technology by Fenton reactions. With the catalyst of Fe^{2+}, H_2O_2 can decompose into large amount of •OH which can oxide the surface SiC to form a soft and weak SiO_2 layer (Figure 6). Thus, the mechanical polishing can remove the oxidize SiO_2 with abrasives. As a pity, the Fenton reaction has the maximum reaction rate at the acid pH. The alkaline CMP solutions may have the limited effect. Moreover, the produced •OH would decrease with the Fenton regents. It's not easy to keep stable in practical applications.

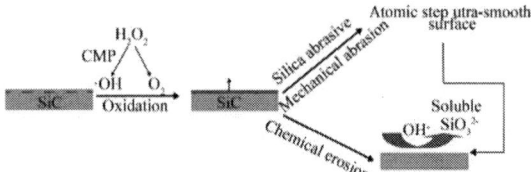

Figure 6: The schematic of Fenton-CMP [10].

PAP introduced reactive gas (CF_4, He or O_2) into the plasma produced by the high frequency power source (RF) to form active free radicals (•F, •OH) (Figure 7). In view of the strong oxidation capacity, the active radicals could oxide the SiC to the soft layer (SiO_2, $Si_4C_{4-x}O_2$ or SiF_4), which can be removed by abrasives to obtain high surface quality. The RR was controlled by the active radicals and mechanical process. However, the formation rate of the active radicals is too slow to support the efficient polishing process. Moreover, the high cost and expensive equipment limit the application of the PAP technology.

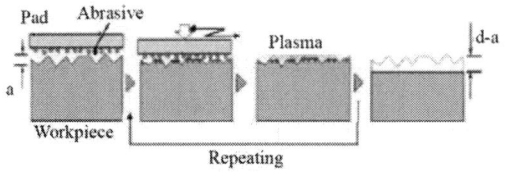

Figure 7: The removal mechanism of plasma assisted polishing [11].

UV-CMP was another equipment assisted technology combined with UV and oxidants. TiO_2 was usually used as catalyst to produce electron (e^-) – hole (h^+) couples by energy level transition. Meanwhile, the oxidants (H_2O_2, $KMnO_4$, O_3 or F_2) can produce free radicles (•OH). The single crystal SiC wafer is oxidized by the combination of the couples and the free radicles. Then, the oxide layer could be removed under mechanical polishing to achieve ultra-precision process. However, the formation rate of the active substances is still too low.

Aimed to improve the formation rate of the oxidize layer, the researchers applied direct electric field to oxide the SiC on the anode, which was called as ECMP. Compared which other technology, the chemical process could be controlled and enhanced by the electrochemical ways. It was reported that, ECMP could provide less subsurface defects.

CONCLUSION AND PERSPECTIVE

In summary, we introduced some ultra-precision technology for SiC CMP. The auxiliary assisted processes can improve the remove efficiency of the SiC wafer compared with the traditional CMP. However, the CMP effect varied with different planarization process, resulting in the lack of predictability and stability. And, there is no consensus on the mechanism of material removal in conventional CMP technology. It's a challenge to transform the technologies into industrial applications. In future, we must pay more attention to study the effects of plasma, catalyst, UV and electric field on the CMP process. The mechanism must be revealed by quantitative and qualitative characterizations. Moreover, the subsurface defect layers are usually inevitable during the CMP process, which was often ignored. The accurate detection and evaluation of subsurface defect layers are very important for optimizing the CMP process and controlling the cost. The fully study of the subsurface defects may promote the development of the SiC CMP in the future.

ACKNOWLEDGEMENTS

I appreciate the help from Dr Li given to me, ,and without my Xin Qian teams great work on this study , I couldn't done this study by myself

REFERENCES

[1] H. Bencherif, F. Pezzimenti, L. Dehimi, C. Della. *Appl. Phys. A*, vol. 126, 2020, pp: 854-861.

[2] H. Matsunami. *Proceeding of the Japan Academy, Series B*, vol. 96, 2020, pp: 235-254.

[3] F. Roccaforte, P. Fiorenza, G. Greco. *Microelectronic Engineering*, vol. 187, 2018, pp: 66-77.

[4] J. PAN, Q. YAN, W. LI. *Micromachines*, vol. 10, 2019, pp: 332-338.

[5] X. CHEN, X. XU, X. HU. *Materials Science and Engineering: B*, vol. 142, 2007, pp:28-30.

[6] N. LIU, H. YAMADA, N. YOSHITAKA. *Diamond and Related Materials*, vol. 119, 2021, pp: 108-113.

[7] T. DAISETSU T, P. VAN, Y. KAZUTO. *International Journal of Automation Technology*, vol. 15, 2021, pp: 74-79.

[8] Z. YUAN, Y. HE, X. SUN. *Materials and Manufacturing Processes*, vol. 33, 2018, pp: 1214-1222.

[9] H. Hara, Y. Sano, H. Mimura. *Materials Science Forum*, vol. 556, 2007, pp: 749-751.

[10] A. Kubota, K. Yagi, J. Murata. *Journal of Electronic Materials*, vol. 38, 2009, pp: 159-163.

[11] K. YAMAMURA, T. TAKIGUCHI, M. UEDA. *CIRP Annals-Manufacturing Technology*, vol. 60, 2011, pp: 571-574.

Research on the Dispersion Stability and Polishing Performance of Ceria Slurry

Min Liu[1,2], Baoguo Zhang[1,2*], Shitong Liu[1,2], Dexing Cui[1,2], Wenhao Xian[1,2], Pengfei Wu[1,2,] Ye Wang[1,2]

[1]School of Electronics and Information Engineering, Hebei University of Technology
Tianjin 300130, People's Republic of China
[2]Tianjin Key Laboratory of Electronic Materials and Devices, Tianjin 300130,
People's Republic of China
*Corresponding Author's Email: bgzhang2000@yahoo.com

ABSTRACT

The slurry is based on nano cerium oxide as the abrasive. The effect of adding different organic acids (acetic acid, lactic acid and oxalic acid) during wet milling on the dispersion performance of CeO_2 slurry is investigated. It is found that when acetic acid as pH adjuster and it can also as the dispersant. The weight ratio of nano cerium oxide to milling media is 1:4 and the pH is 3.0. After 6 hours of wet milling, the average particle size of CeO_2 slurry is 229.1 nm with a Zeta potential of 41 mV. The slurry is formulated with 1 wt% CeO_2 at pH 4, the material removal rate of quartz wafer is 413.96 nm/min.

INTRODUCTION

In Chemical Mechanical Polishing, the slurry is one of the key elements. The composition, pH, particle size and dispersion stability of the slurry have a significant impact on the CMP process. Cerium oxide nano-abrasive plays a very important role in the field of chemical mechanical polishing because of its high polishing efficiency, excellent surface quality, clean operating environment. K9 glass is one of the most common materials used for optical component [1, 2].

Z. J. Wang et al. investigated the effect of mass fraction of cerium oxide abrasive and additives on the polishing rate of quartz wafer. It is found that the removal rate of quartz wafer increases significantly with the increase of mass fraction of abrasive. The removal rate of quartz wafer reaches 97.9 nm/min when abrasive mass fraction is 20 %. When adding FA/O Ⅰ chelating agent and FA/O active agent, the removal rate is 93.4 nm/min [3]. Z. X. Yang et al. investigated the effect of ceria slurry on the polishing performance of quartz wafer. It is found that when slurry with 1 wt% CeO_2 at pH 5.0, the Zeta potential of the abrasive is 35.6 mV and removal rate of quartz wafer is 248.9 nm [4].

The polishing rate of quartz wafer of slurry consisting cerium oxide as an abrasive still needs to be improved. Moreover, because of the high surface energy of cerium oxide, it is highly prone to agglomeration. Therefore, the powders based on rare earth element oxides made currently are unstable [5]. Therefore, in this paper, good dispersion stability of CeO_2 slurry was prepared by adding organic acids and its polishing performance on quartz wafer was investigated.

EXPERIMENTAL

Slurries preparation. - The CeO_2 abrasive is provided by Yiyang Hongyuan Rare Earth Investment Co., Ltd. The acetic acid (analytical purity) used in this paper is provided by Tianjin Jiuding Chemical Co., Ltd. Oxalic acid (analytical purity) and lactic acid (analytical purity) are provided by Tianjin Kermel Chemical Reagent Co., Ltd. The alkaline pH adjuster is KOH (analytical purity) which is provided by Shanghai Yien Chemical Technology Co., Ltd.

Polishing experiment. - The chemical mechanical polishing experiments are carried out on RuiXuan SSP-500 milling and polishing machine. The diameter of quartz wafer (JGS2) used is 100 mm and the thickness of it is 0.7 mm, and the SiO_2 purity is greater than 99%. After polishing, the wafer is cleaned with deionized water and blown drying with nitrogen gas. The polishing time of wafer is 5 min at 4 psi, with a slurry flow rate of 100 ml/min. The speed of polishing pad is 50 r/min and the polishing head is 50 r/min. The mass of the quartz wafer is weighed before and after polishing using Mettle Toledo AB204-N electronic analytical balance and calculating the average value of three times. The material removal rate (unit: nm/min) is calculated according to equation (1).

$$MRR = \frac{\Delta m}{\rho s t} \qquad (1)$$

where Δm is the reduction in the weight of the quartz wafer after polishing (Unit: nm/min), ρ is the density of SiO_2 (2.2 g/cm^3), s is the area of SiO_2 layer (Unit: cm^2), and t is the polishing time.

RESULTS AND DISCUSSION

Firstly, the CeO_2 abrasives were subjected to wet ball milling to study the effect of adding different organic acids (acetic acid, lactic acid and oxalic acid) on the dispersion performance of the CeO_2 slurry. The weight ratio of nano cerium oxide to milling media was 1:4 and the pH was 3.0. And the ball milling time was 6h. The average particle size of CeO_2 was measured when concentration of CeO_2 is 1 wt% and the pH value is 4.0. As shown in Figure 1, using acetic acid as the pH adjuster and dispersant, the average particle sizing of CeO_2 abrasive is 229.1 nm, with a Zeta potential of 41 mV. The slurry has good dispersion stability. The average particle sizing of CeO_2 is 325.9 nm and 593.4 nm when lactic acid and oxalic acid is used as pH adjuster and dispersant, respectively. They are a little bigger than that of acetic acid. The Zeta potential is 27.3 mV and 23.6 mV of them and the slurries above show obvious significant delamination.

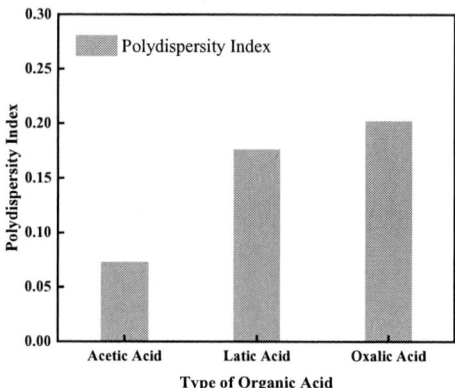

Figure 2: Effect of different organic acids on polydispersity index of CeO_2 abrasives

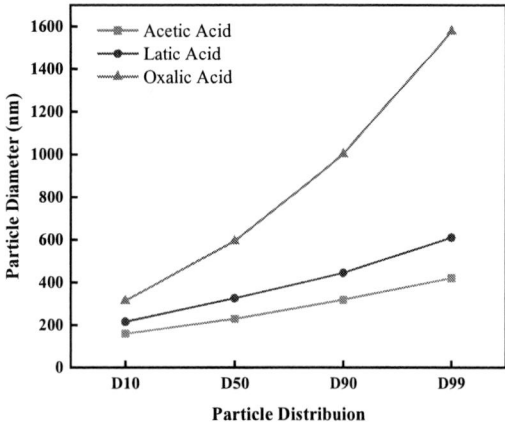

Figure 1: Effect of different organic acids on particle size distribution of CeO_2 abrasives

The polydispersion index is a dimensionless value which reflects the width of the particle size distribution of the slurry [6]. If PI < 0.1, the sample is monodisperse, if 0.1 < PI < 0.4, the sample is polydisperse, and if PI > 0.4, it indicates that the particle size distribution measured is not very accurate. It can be concluded from Figure 2 that when acetic acid is used as the dispersant, the PI value of the abrasive is 0.073. The value is less than 0.1, indicating that the sample is monodisperse and has good dispersion stability. When lactic acid and oxalic acid are used as the dispersant, the PI value of the abrasive is 0.176 and 0.202 respectively, which shows poor dispersion effect. Therefore, the experiments were all carried out on the basis of acetic acid as dispersant and pH adjuster.

The dispersion behavior of acetic acid, lactic acid and oxalic acid on cerium dioxide abrasive is mainly through electrostatic forces and the double electric layer effect. CeO_2 has a positive surface charge at pH 4.0 due to its PZC (isoelectric point) is 7.9. Through electrostatic forces, the CeO_2 particles will repel each other and keep slurry stable. All three additives are carboxylic acids, and their carboxylate ions can adsorb onto the surface of CeO_2 particles, further improving the dispersion of the abrasive through the double electric layer effect.

Acetic acid was chosen as the pH adjuster and dispersant and the effect of different pH on the removal rate of quartz wafer was investigated at a CeO_2 abrasive concentration of 1 wt%, as shown in Figure 3. It is found that as the pH of the slurry increases from 3.0 to 7.0, the removal rate of SiO_2 is on a trend of increasing and then decreasing. In particular, the highest removal rate of 413.96 nm/min is achieved at pH 4.0.

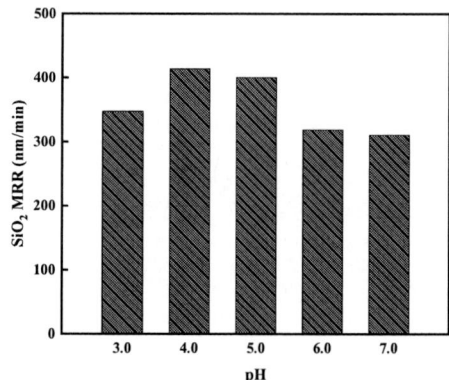

Figure.3: The removal rate of quartz glass at different pH

From this it can be concluded that the optimum components for the slurry is that acetic acid as pH adjuster and dispersant, a CeO_2 abrasive concentration of 1 wt% and pH 4.0.

CONCLUSIONS

In this paper, CeO_2 slurry was prepared by wet ball milling method. Acetic acid was chosen as pH adjuster and dispersant and the slurry had good dispersion effect. Acetic acid, as an organic carboxylic acid, improves the dispersion performance of the slurry through the adsorption of carboxylate ions on the CeO_2 surface and forming a double electric layer effect. The average particle sizing of CeO_2 after ball milling is 229.1 nm and the Zeta potential is 41 mV. The maximum removal rate for quartz wafer is 413.96 nm/min when abrasive concentration is 1 wt% at pH 4.0.

ACKNOWLEDGEMENTS

This work was supported by the Major National Science and Technology Special Projects (No. 2016ZX02301003-004-007), Natural Science Foundation of Tianjin, China (16JCYBJC16100, 18JCTPJC57000). The authors also thank the teachers and classmates for their helpful suggestions.

REFERENCES

[1] S. Qu. F. W. Meng, L. Y. Chen, Z. L. Ma, Z. X. Wang, M. Li, T. B. Yu, J. Zhao. *CERAMICS INTERNATIONAL,* vol. 48, 2022, pp. 19945-19953.

[2] X. G. Guo, Y. J. Wei, Z. J. Jin, D. M. Guo, W. Maosen. *The International Journal of Advanced Manufacturing Technology, 2013,* pp. 1277–1283.

[3] Z. J. Wang, S. L. Wang, C. W. Wang, W. Q. Zhang, H. Zheng. *Micronanoelectronic Technology,* vol. 54, 2017, pp. 48-52.

[4] Z. X. Yang, B. G. Zhang, X. F. Yang, Y. Li, H. R. Li. *ECS Journal of Solid State Science and Technology,* vol. 10, 2021, pp. 093005.

[5] M. I. Lebedeva, L. A. Arzhatkina, E. L. Dzidziguri, E. N. Sidorova. *Nanotechnologies in Russia,* 2014.

[6] T. Mudalige, H. Qu, D. Van Haute, S. M. Ansar, A. Paredes, and T. Ingle. *Nanomaterials for food applications, 2019,* pp. 313-353.

Study on the Slurry for Chemical Mechanical Polishing of Sapphire Wafer

Wenhao Xian[1,2], Baoguo Zhang[1,2], Liu Min[1,2], Dexing Cui[1,2], Pengfei Wu[1,2], Ye Wang[1,2]*

[1]School of Electronics and Information Engineering, Hebei University of Technology
Tianjin 300130, People's Republic of China
[2]Tianjin Key Laboratory of Electronic Materials and Devices, Tianjin 300130,
People's Republic of China

*Corresponding Author's Email: bgzhang2000@yahoo.com

ABSTRACT

The effect of Al_2O_3 polishing slurry on sapphire wafer CMP has been studied in this paper. It is found that a high material removal rate (MRR) can be achieved with this low weight concentration Al_2O_3-based slurry. By changing the particle size of the alumina abrasive, the MRR of sapphire was improved. The experimental results show that when the mean diameter of Al_2O_3-based slurry is about 500 nm, the MRR can reach 5.1 μm/h.

INTRODUCTION

With the development of the semiconductor industry, sapphire, an important substrate material for LED manufacturing[1], increasingly requires high-precision surface processing technology to achieve a flattened surface that meets the industry's needs of the industry. Chemical mechanical polishing (CMP) is a key technology widely used in sapphire surface processing today[2]. Highly efficient surface smoothing of materials can be achieved in CMP through the synergistic effect of mechanical friction and chemical etching.

Polishing slurry is the most critical factor in CMP technology, and the abrasive is extremely important in polishing slurry. It is worth mentioning that the shape, hardness and particle size of the abrasive will greatly affect the CMP results[3]. In order to analyze the patterns more comprehensively, the effects of several different types of alumina abrasives on the material removal rate (MRR) of sapphire CMP have been studied in this paper.

The main material component of sapphire is α-Al_2O_3, a colorless, high hardness single crystal material with a Mohs hardness of 9. The most commonly used abrasives in sapphire CMP polishing slurry are colloidal silica and alumina powder. However, the Mohs hardness of silica is about 6~7, so the polishing efficiency of colloidal silica is lower compared with alumina powder[4]–[6].

EXPERIMENTAL

Materials Preparation

A 4-inch diameter sapphire wafer with a thickness of 670 ~ 680 μm was used in this study. Three different types of self-made alumina powders (TA-1, TA-2, TA-3) were used for comparison tests to investigate the effects of the shape and structure of alumina on the MRR of sapphire CMP. KOH and HNO_3 (Tianjin Damao Chemical Reagent Technology Co., Ltd, analytical grade) were used as pH adjusters for the polishing slurry.

Polishing Experiment

CMP experiments on sapphire were performed with an RX-SSP500 polishing machine (produced by Wuxi Ruixuan Lapping Technology Co., Ltd., China). After polishing, the sapphire wafers were cleaned with deionized water and quickly dried with high-pressure nitrogen gas. In the polishing experiments, the pH of the polishing slurry was adjusted to 12.0, the polishing pressure was set to 6.1 psi, the flow rate of the polishing solution was 80 ml/min, and the polishing time was 40 min. Then, the platen and carrier speeds were 50 rpm and 45 rpm, respectively.

The mass of the sapphire before and after polishing was measured using a high-precision electronic analytical balance (Mettler Toledo AB204-N), and the MRR was calculated using Equation 1.

$$MRR = \frac{m_1 - m_2}{\rho \pi r^2 t} \qquad (1)$$

where m_1 is the mass of the sapphire wafer before polishing and m_2 is the mass after polishing. ρ and t are the density of sapphire ($\rho = 3.96\ g/cm^3$) and polishing time, respectively. r is the radius of the wafer.

Characteristics Analysis

The mean diameter of the polishing slurry was tested with a laser nanoparticle size analyzer NIcomp-380 (produced by PSS Inc., U.S.A.). The microscopic morphology of the alumina particles in the slurry was observed and analyzed using a Sigma-500 scanning electron microscope.

RESULTS AND DISCUSSION

Based on the TA-1 polishing slurry, three types of slurries with different particle sizes were prepared, named TA-1-a, TA-1-b and TA-1-c, respectively, as shown in Figure 1. The highest MRR (5104.3 nm/h) was obtained for TA-1-a and the lowest MRR (2351.3 nm/h) was obtained for TA-1-c without any additional additives.

Meanwhile, the mean diameter of the three slurries is also shown in Figure 1, and the MRR decreases as the mean diameter decreases.

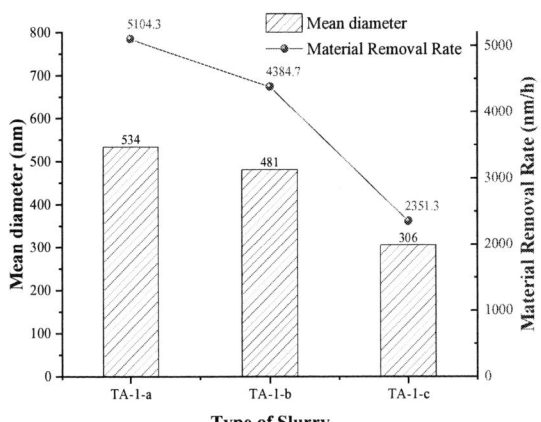

Figure 1: MRR corresponding to three mean diameters of alumina abrasives for TA-1

However, this trend is only suitable for mean diameter in the range of 300 ~ 600 nm, because too large mean diameter usually leads to a reduction in the dispersion stability of the polishing slurry, which in turn leads to a reduction in MRR. For example, if the mean diameter of the polishing slurry is larger than 1 μm, the alumina abrasives will settle down quickly during the CMP process, as shown in Figure 2.

Figure 2: Settling phenomenon of alumina abrasive in polishing solution

Based on the above study, comparative CMP tests were carried out on TA-1, TA-2 and TA-3 and the MRR results after polishing are shown in Figure 3.1, where TA-3 had the highest MRR (5776.5 nm/h) and TA-2 had the lowest MRR (4147.8 nm/h).

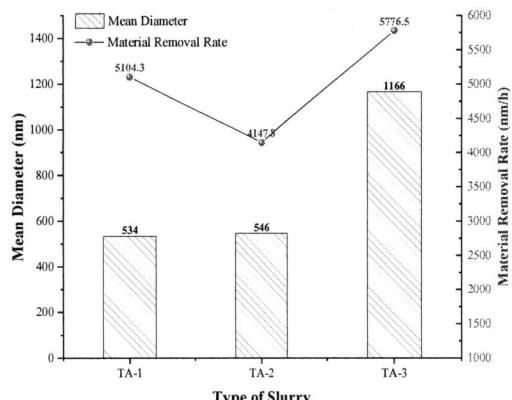

Figure 3: MRR and D50 of three types of alumina(TA-1, TA-2, TA-3) polishing slurry

From Figures 3, it can be found that the variation pattern of the mean diameter does not match the pattern of MRR when the type of alumina abrasive is different. This phenomenon is due to the fact that in addition to the difference in the mean diameter of the three alumina abrasives, the microstructure of the abrasives themselves is also very different.

SEM test images corresponding to the three types of alumina abrasives are shown in Figure 4. The shape and structure of the alumina abrasive can be clearly observed under the magnification of 10000 times.

Structural characteristics of TA-1: smaller and sharper particle size, agglomeration phenomenon between small particles and better dispersion. The TA-2 particles are uniform in size and have more rounded ends. In addition, the dispersion between small particles is not very good. The dispersion between TA-3 particles is the worst among the three types, but the small particles in the agglomerated particles have a special shape, mostly in the shape of "T" and "H".

Compared with TA-2, some large-size agglomerated alumina particles are present in both TA-1 and TA-3, and the results of both TA-1 and TA-3 reach the level of 5 μm /h from the MRR point of view.

It is worth mentioning that the large mean diameter of TA-3 is due to the agglomeration of Al_2O_3 particles at the microscopic level. It can be found that the agglomerates have many small, long, and striped particles, as shown in (c) of Figure 4. And such small particles can also be observed in the SEM test images of TA-1.

Therefore, these small, long, and striped particles can effectively improve the MRR of sapphire CMP.

Figure 4: SEM test images of three types of alumina slurries; (a) is the image of TA-1, (b) is the image of TA-2, (c) is the image of TA-3

CONCLUSIONS

By changing the mean diameter of TA-1 alumina abrasive, it was found that the MRR decreases with decreasing mean diameter without adding additional additives. The effects of three types of alumina abrasives on the CMP of sapphire were also compared, and it was found that the particle shape in the alumina abrasives also significantly affects the MRR. The slightly sharp crystal shape allows for a better mechanical friction process on the sapphire surface, which in turn accelerates the removal of the abrasive from the sapphire material.

ACKNOWLEDGEMENTS

This work was supported by the Major National Science and Technology Special Projects (No. 2016ZX02301003-004-007), Natural Science Foundation of Tianjin, China (16JCYBJC16100, 18JCTPJC57000). The authors also thank the teachers and classmates for their helpful suggestions.

REFERENCES

[1] Z. Chen, L. Cao, J. Yuan, B. Lyu, W. Hang, and J. Wang, *Micromachines*, vol. 11, 2020, p. 759.

[2] M. Krishnan, J. W. Nalaskowski, and L. M. Cook, *Chem. Rev.*, vol. 110, 2010, pp. 178–204.

[3] X. Jiang, X. Long, and Z. Tan, *Chin. J. Lasers*, vol. 48, 2021, p. 0401014.

[4] C. Zhou, Q. Zhang, C. He, and Y. Li, *Optik*, vol. 125, 2014, pp. 4064–4068.

[5] S. Hong, D. Han, and K.-S. Jang, *Wear*, vol. 466–467, 2021, p. 203590.

[6] E. Kim, J. Lee, Y. Park, C. Shin, J. Yang, and T. Kim, *Powder Technol.*, vol. 381, 2021, pp. 451–458.

PAD SURFACE VARIATION AND ITS EFFECT ON SIO$_2$ REMOVAL RATE IN CERIA-BASED CMP SLURRY

Chenchen Yang, Yu Yao and EngHoe Tan

Semiconductor Manufacturing Beijing Corporation, Beijing, China.

Chenchen_Yang@smbcs.com

ABSTRACT

Ceria based slurry have been extensively applied to chemical mechanical planarization (CMP) such as STI (shallow trench isolation) and inter-level dielectric (ILD) process. Taken bulk SiO$_2$ into account, high removal rate is one of the important factors. The material removal rate depends upon the surface properties of the polishing pad and the conditioning process. Without regeneration of the pad surface during polishing, the removal rate of oxide rapidly increases. By evaluating the regeneration of the pad surface during polishing, the relationship between polishing pad surface feature regeneration and the effect on polish rates is analyzed. In addition, pad is one of the key factors to affect removal rate. Removal rate has a stronger correlation with condition process, when pad with larger groove density and smaller pore size.

Keywords—Ceria, chemical mechanical planarization (CMP), SiO$_2$ removal rate, pad

INTRODUCTION

Chemical mechanical planarization (CMP) has been broadly used in semiconductor manufacturing to achieve planarity to meet photolithography requirements.[1] Target materials mainly include oxide (SiO$_2$), tungsten (W) and copper (Cu).[2] The material removal rate depends on applied pressure, relative velocity between pad and wafer, slurry characteristics and pad properties in CMP process.[3] Ceria based slurries show excellent planarization capability with admirable removal rate in oxide STI and ILD CMP processes, owing to the chemical affinity of ceria for oxide.[4] Several studies have reported the mechanism of material removal in CMP processes.[5, 6] For mechanically driven processes, the materials removal rate correlates to the roughness of the pad surface. Rough pad surface is essential to maintain high material removal rate.[7] However, ceria-based SiO$_2$ oxide CMP processes show an interesting contrast phenomenon. Taken into consideration of pad surface roughness, conditioning process can remove the debris from the pad surface and regenerate pad surface quality. The polishing pad plays a key role in the mechanical aspects of polishing. In this paper, for ceria based SiO$_2$ oxide CMP process, the relationship between polishing pad surface feature regeneration and the effect on polish rates will be analyzed.

EXPERIMENTAL

All CMP processes were carried out on CMP polishing tool in clean-room environment. Ceria-based slurry was used in polishing SiO$_2$ oxide wafers, which was diluted with DI water. Polishing pad A and B were prepared with circular grooves. Both in-situ and ex-situ conditioning at a lbs were evaluated for a diamond conditioning disk.

RESULT AND DISCUSSION

Pad and conditioning process are two key factors affecting the material removal rate. Firstly, pad A was used to investigate the removal rate of SiO$_2$ oxide film under different conditioning process. Ceria-based slurry dilution with DI water was used in polishing SiO$_2$ oxide wafers. As we all known that pad surface roughness can be effectively adjusted by pad conditioning. And the pad conditioning process can refresh pad surface during CMP. Thus, different conditioning process processes were set to study the effect on removal rate of SiO$_2$ oxide film. As shown in Figure 1, the removal rate of SiO$_2$ oxide film under different conditioning process was obtained. In the case of 100% insitu conditioning, the removal rate of SiO$_2$ oxide was 1.0. In the case of 75% insitu conditioning, the removal rate of SiO$_2$ oxide was increased up to 1.2. The removal rate of SiO2 oxide film increased gradually as the ratio of insitu conditioning ratio decreased. And the removal rate of SiO$_2$ oxide film under 75% insitu conditioning process was closed to the removal rate under 50% insitu conditioning. Similarly, the removal rates under 25% insitu and exsitu conditioning were close. As shown in Figure 2, good linear fitting curves can be obtained by linear fitting of removal rate under different conditioning process, in which the slope was 0.086. The results suggest that rough surface is bad for the removal rate of SiO$_2$ oxide film, when the ceria slurry was used, vice versa.

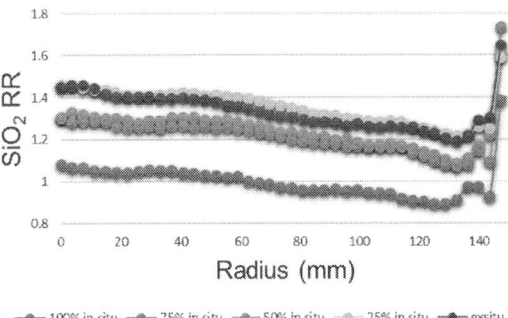

Figure 1 SiO$_2$ oxide removal rate profile of pad A under different conditioning process.

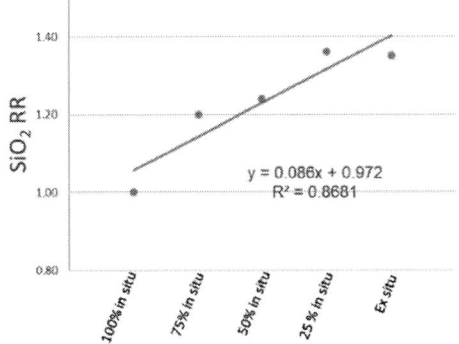

Figure 2 Linear fitting of SiO$_2$ oxide removal rate of pad A under different conditioning process.

In addition, Pad B was also used to investigate the removal rate of oxide film under different conditioning process. As shown in Figure 3, the removal rate of oxide film under different conditioning process was obtained. Similarly, the removal rate of oxide film increased gradually as the ratio of insitu conditioning ratio decreased. In the case of exsitu conditioning process, the fastest removal rate of oxide film was obtained. As shown in Figure 4, excellent linear fitting curves can be obtained by linear fitting of removal rate under different conditioning process, in which the slope was 0.1152, which was larger compared to that obtained from pad A. The result manifest that the change of conditioning process has a more distinct effect on the removal rate of pad B. The results suggest that rough surface is bad for the removal rate of oxide film, when the ceria slurry was used, vice versa.

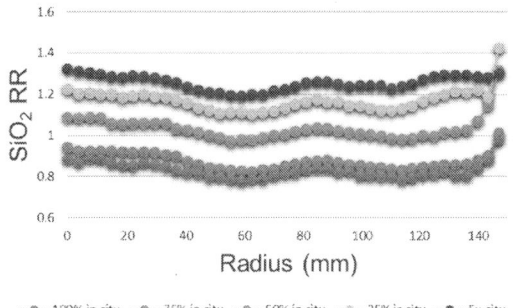

Figure 3 SiO_2 oxide removal rate profile of pad B under different conditioning process.

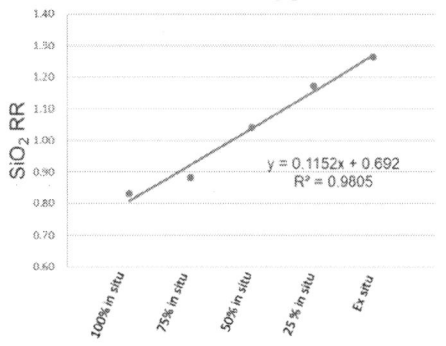

Figure 4 Linear fitting of SiO_2 oxide removal rate of pad B under different conditioning process.

It is known that the pad surface consists of asperities and pores. And the conditioning process is used to maintain the feature of pad surface. It is supposed that there is excellent affinity between CeO_2 and SiO_2 oxides, which is benefit to break the chemical bonds on the SiO_2 surface. It is known that Si-OH groups exit on SiO2 surface, while the Ce-OH groups cover CeO_2 surface. On the interface, CeO_2 react with SiO_2 to form the Ce-O-Si bond which is stronger than the Si-O-Si bond. Finally, the SiO_2 can be removed under the mechanical abrasion force[8], as shown in Figure 5A. In the case of 100% insitu conditioning, the required roughness of pad surface can be restored. On the one hand, the rough surface can remain more slurry. In this case, more inhibitor in slurry which remain pad surface adsorbs on the surface of ceria abrasives forming passive layer to limit removal rate. On the other hand, the rough surface of pad is bad for the transport of ceria abrasives (Figure 5B). The ceria abrasive cannot be transported enough. Thus the reaction between CeO_2 and SiO_2 is suppressed as well as that mechanical force is reduce, which lead to low removal rate of

oxide film. On the contrary, without conditioning process, the surface of pad is "glazed" which reduce the surface roughness and the surface of pad became smooth. The smooth surface can remain less slurry. In this case, less inhibitor adsorbed on the ceria abrasives cannot form effective passive layer to limit removal rate. The glazed pad surface with low roughness is benefit to the transport of ceria abrasive (Figure 5B). In this case, the ceria abrasive can be transported enough which increase the react between CeO_2 and SiO_2 as well as mechanical force. Thus removal rate of oxide film is high. In 100% insitu conditioner, the surface of pad is more rough and less abrasives can be transport on the pad surface which cause lower removal rate. In exsitu conditioner, the pad surface is glazed and more abrasives can be transport on the pad surface which cause higher removal rate.

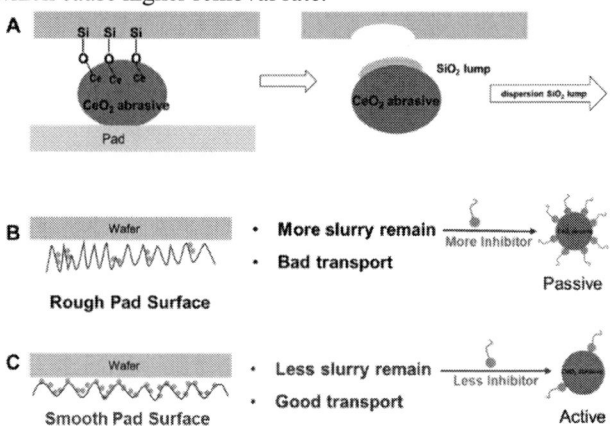

Figure 5 Schematic illustration of polishing mechanism of CeO_2 abrasive (A), abrasives transport on rough pad surface (B), and abrasives transport on glazed pad surface.

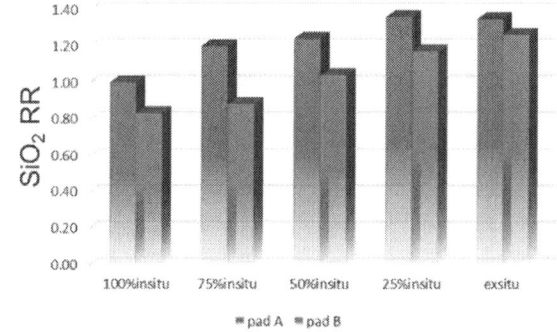

Figure 6 The removal rate as a function of different pads.

Under same condition, the removal rate of pad A is faster than that of pad B (Figure 6). This might be owing to the different feature on pad surface. As shown in Figure 7, pore on surface pad A is larger than that on surface pad B. The pitch of circular grooves on the surface pad A is larger than that on the surface pad B. When the areas of pad A and B keep identical, the results suggest that the groove density on the surface pad B is larger. Therefore, removal rate of pad B has a stronger correlation with conditioning process.

979-8-3503-1101-3/23 $31.00 © 2023 IEEE   350

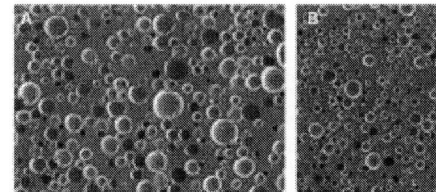

Figure 7 SEM image of the polishing pad surface (A) pad A, (B) pad B.

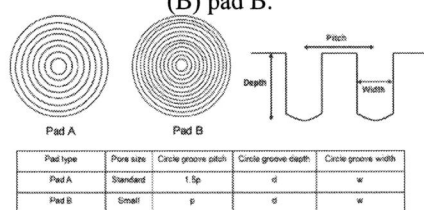

Pad type	Pore size	Circle groove pitch	Circle groove depth	Circle groove width
Pad A	Standard	1.5p	d	w
Pad B	Small	p	d	w

Figure 8 Schematic illustration of circle groove on surface of pad A and pad B and the table of surface parameter of pad.

CONCLUSION

In ceria slurry system, in the case of exsitu conditioning process, high SiO_2 oxide removal rate can be obtained. However, in the case of 100% insitu conditioning process, low SiO_2 oxide removal rate can be obtained with. Pad is one of the important factor to affect removal rate. In this paper, removal rate has a stronger correlation with condition process, when pad with larger groove density and smaller pore size.

REFERENCES

[1]K. Qin, B. Moudgil, C.W. Park, "A chemical mechanical polishing model incorporating both the chemical and mechanical effects". Thin Solid Films vol. 446, pp. 277–286, 2004.

[2]X. Jin, L. M. Keer, Q. Wang, "A 3D EHL simulation of CMP". J. Electrochem. Soc. vol. 152, pp. G7–G15, 2005.

[3]M. R. Oliver, Chemical-mechanical planarization of semiconductor materials". Springer, Berlin Heidelberg. 2004.

[4]L. M. Cook, "Chemical processes in glass polishing," J. Non-Cryst. Solids, 1990, 120, 152-171.

[5]J. Luo and D. A. Dornfeld "Material removal regions in Chemical Mechanical Planarization for Submicron Integrated Circuit Fabrication: Coupling Effects of Slurry Chemicals, Abrasive Size Distribution and Wafer-pad Contact Area," IEEE Trans. Semicond. Manuf. Vol. 16, pp. 45-56, 2003.

[6]A. Philipossian and S. Olsen, "Fundamental Tribological and Removal Rate Studies of Inter-Layer Dielectric Chemical Mechanical Planarization," Jpn. J. Appl. Phys. vol. 42, pp. 6371–6379, 2003.

[7]J. McGrath and C. Davis, "Polishing pad surface characterization in chemical mechanical planarization," J. Mater. Processing Technol. vol. 153, pp. 666–673, 2004.

[8]J. Cheng, Y. Li, X. Lu, "Characterization of Lanthanide Elements Doped Ceria Nanoparticles and Its Performance in Chemical Mechanical Polishing as Novel Abrasive Particles", IEEE, 2018.

IMPACT OF SLURRY FOR DISHING REDUCTION DURING CU CMP

Yu Yao, Chenchen Yang, and EngHoe Tan
Semiconductor Manufacturing Beijing Corporation, Beijing, China.
Giggs_Yao@smbcs.com

ABSTRACT

The chemical mechanical planarization (CMP) is one of the key process to achieve global and local surface planarization in integrated circuits (IC) manufacture. Defects such as dishing pits are easily formed on the Cu surface. Reducing Cu dishing of Cu CMP for removing thick Cu has been studied in this paper. A neutral pH slurry with good corrosion resistance and small particle size abrasives was investigated. The Atomic Force Microscope (AFM) and Transmission Electron Microscope (TEM) results indicate that the slurry can reduce the depth of dishing pit effectively.

Keywords—copper, dishing, chemical mechanical planarization (CMP).

INTRODUCTION

Copper chemical mechanical planarization (CMP) has been widely used as the major process for local as well as global planarization of integrated circuits (ICs).[1] As shown in Figure 1, in Cu CMP process, firstly, most of the bulk copper film can be removed. Secondly, copper film which remained on barrier layer can be removed completely which stop on the barrier layer via the end point technique. Lastly, the barrier layer can be removed, after that, the low K dielectric material is exposed.[2] In despite of that Cu CMP processes have been broadly applied in ICs manufacture, there are still many issues to be overcome such as organic residues, surface particles, scratches, erosion, and dishing. Thereinto, dishing pit are easily formed on the Cu surface.[3,4] In this work, a neutral pH slurry with good corrosion resistance and small particle size abrasives was investigated.

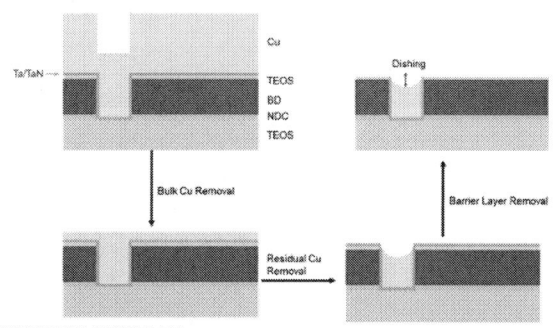

EXPERIMENTAL

All CMP processes were carried out on CMP polishing tool in clean-room environment. The slurries used in Cu polishing process were investigated. The slurries, pads and disks were commericial available. Polishing slurry flow rate was X ml/min. Atomic Fore Mircroscope (AFM) and Transmission Electron Mircroscope (TEM) were employed to characterize the dishing of cu surface.

RESULT AND DISCUSSION

For bulk Cu, high Cu removal rate and minimum metal dishing are necessary. To meet the requirement, slurry is the key enabling material for CMP. In this work, in order to minimize Cu dishing, two different slurries were used to study. Figure 2 show the dishing result after CMP when slurry 1 was used as slurry. The center, middle and edge metal dishing is 2.73θ Å, 2.77θ Å and 2.69θ Å, respectively. Figure 2 show the AFM images of dishing result after CMP when slurry 2 was selected as slurry. The center, middle and edge metal dishing is 0.74θ Å, 0.42θ Å and 0.19θ Å, respectively. The dishing result can also be observed from its TEM. As shown in Figure 3, after slurry 1 CMP processes, the thickness of Cu film on dishing pad was 3.2θ Å. However, when slurry 2 was selected as slurry after Cu CMP processes, the thickness of Cu film on dishing pad was 4.5θ Å. Compared to slurry 1, the results suggests that the slurry 2 can reduce dishing.

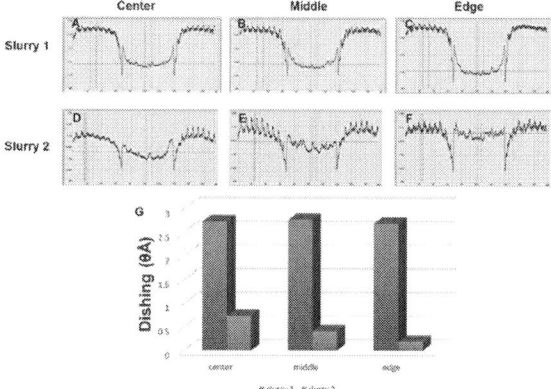

Figure 2 (A-C) Cemter, Middle and Edge AFM images of dishing pad, respectively, after slurry 1 Cu CMP processes; (D-F) Cemter, Middle and Edge AFM images of dishing pad, respectively, after slurry 2 Cu CMP processes.(G) The dishing as a function of different slurries.

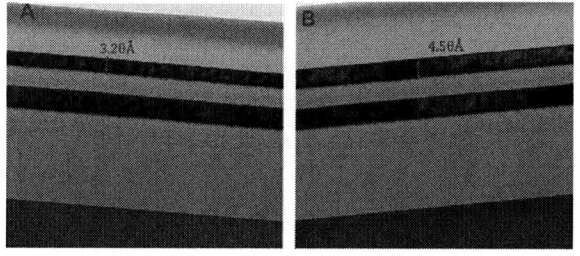

Figure 3 TEM images of dishing pad.(A) Slurry 1, (B) Slurry 2.

For etching at acidic pH, Cu will undergo direct dissolution as shown in below chemical reaction which leads to high removal rate of Cu layer:

$$Cu + 2H^+ = Cu^{2+} + H_2 \qquad [1]$$

Besides, the oxidizing reactions are as follow:

$$Cu + H_2O_2 = CuO + H_2O \qquad [2]$$
$$2Cu + H_2O_2 = Cu_2O + H_2O \qquad [3]$$

$$Cu_2O + H_2O_2 = 2CuO + H_2O \qquad [4]$$
$$CuO + H_2O = Cu(OH)_2 \qquad [5]$$

The oxide will react with complexing agents which forms thin soft film of insoluble Cu complexes covering the Cu surface as a passive layer. For Cu CMP, inhibitors play a key role to realize planarization. And corrosion inhibitors will compete against the oxidants to inhibit the degree of oxidation. Compared to slurry 1, slurry 2 can reduce dishing 73% in center, 85% in middle and 96% in edge, respectively. This result indicates that the concave regions is not effectively protected by slurry 1, which lead to high chemical etch rates. As for slurry 2, based on the synergetic passivation mechanism, the concave areas are protected and have lower reaction active energy and mass transfer energy. Hence, the material removal rate in convex area is faster than the chemical etch rate of the concave area, which is benefit for planarization.[5] In addition, as shown in Table 1, the particle size of abrasives in slurry 1 is large, which is larger than that in slurry 2. It is known that the smaller particle size can reduce dishing.

Table 1 The parameters of slurries.

CONCLUSION

Reducing Cu dishing of Cu CMP for removing thick Cu has been studied in this paper. A new slurry with good corrosion resistance due to neutral pH, small particle size abrasives and proprietary corrosion inhibitor technology was investigated. Based on the Atomic Force Microscope (AFM) and Transmission Electron Microscope (TEM) measurements, the results indicate that the slurry 2 can reduce the depth of dishing pit effectively.

REFERENCE

[1] J. Cheng et al., "Material removal mechanism of copper chemical mechanical polishing in a periodate-based slurry," Appl. Surf. Sci., vol. 337, pp. 130, 2015.

[2] X. Luan, Y. Liu, B. Zhang, S.Wang, X. Niu, C.Wang, and J.Wang, "Investigation of the barrier slurry with better defect performance and facilitating post-CMP cleaning," Microelectron. Eng., vol. 170, pp. 21, 2017.

[3] C. Zhang, P. Feng, and J. Zhang, "Ultrasonic vibration-assisted scratch-induced characteristics of C-plane sapphire with a spherical indenter," Int. J. Mach. Tools Manuf., vol. 64, pp. 38, 2013.

[4] T. Sun, Y. Zhuang,W. Li, and A. Philipossian, "Investigation of eccentric PVA brush behaviors in post-Cu CMP cleaning," Microelectron. Eng., vol. 100, pp. 20. 2012. 2013.

[5] C. Yang, X. Niu, J.i Zhou, J. Wang, Z. Huo,and Y. Lu, "Synergistic Action Mechanism and Effect of Ammonium Dodecyl Sulfate and 1,2,4-triazole in Alkaline Slurry on Step Height Reduction for Cu CMP," ECS Journal of Solid State Science and Technology, vol. 9, pp. 034010, 2020,.

EFFECT OF ABRASIVE ON THE CMP PERFORMANCE OF C-PLANE (0001) GAN FLIM

Jianghao Liu[1,2], Xinhuan Niu[1,2*], Ni Zhan[1,2], Yida Zou[1,2], Yebo Zhu[1,2]*

[1]School of Electronics and Information Engineering, Hebei University of Technology, Tianjin 300130, China

[2]Tianjin Key Laboratory of Electronic Materials and Devices, Tianjin 300130, China

*1296819173@qq.com, xhniu@hebut.edu.cn

ABSTRACT

In gallium nitride (GaN) chemical mechanical planarization (CMP) process, abrasive directly determines the material removal rate (MRR) and the surface quality. In the present research, the effects of SiO_2, Al_2O_3 and their mixture on MRR and surface quality was analyzed. The experimental results verified the addition of abrasive significantly increased the MRR, and SiO_2 obtained slower MRR and better surface quality than Al_2O_3 after polishing. Moreover, SiO_2 mixed with Al_2O_3 could obtain higher MRR and lower surface roughness.

Keywords—Abrasive; Gallium nitride (GaN); Material removal rate; Surface morphology; Chemical Mechanical Planarization (CMP)

INTRODUCTION

As the representative third-generation semiconductor material, gallium nitride (GaN) has excellent properties such as wide band and high breakdown electric field strength. The performance of GaN-based devices is directly influenced by the quality of the GaN surface, so it is essential to achieve efficient flattening of GaN surfaces [1]. At present, chemical mechanical planarization (CMP) is the only way to achieve global planarization of GaN and the slurry is the most important consumable in the CMP process [2]. Slurry typically is composed of additives such as abrasive, complexing agents and oxidizing agents, in particular, abrasive enhances the mechanical action of the CMP process, and the performance of other chemical agents is based on abrasive [3].

Gong et al [4]. found that polishing GaN with SiO_2 abrasive resulted in atomically stepped topography on the Ga surface with line roughness Ra as low as 0.07 nm, and this atomic step shape was shown to be determined by the crystal structure of GaN itself. Zou et al [5]. investigated a stepwise polishing method using Al_2O_3 and SiO_2 in the first step and SiO_2 in the second step. A smooth surface with a linear roughness Ra below 0.07 nm was obtained, indicating that the two-step polishing with hard and then soft abrasive could balance the material removal rate (MRR) and roughness. K. Asghar et al [6]. compared the performance of SiO_2 and Al_2O_3, and the results showed that the MRR of GaN was 39 nm/h and 85 nm/h with a Ra of 0.7 nm and 1.3 nm, respectively, demonstrating that although Al_2O_3 could provide higher MRR, SiO_2 could be used to prepare defect-free and atomically smooth GaN surfaces.

From the above studies, it is easy to conclude that high MRR and low surface roughness cannot be achieved at the same time and it is necessary to further optimize the properties of the abrasive. Therefore, in this paper, higher MRR and lower surface roughness were obtained by adjusting the abrasive concentration and compounding on the basis of SiO_2 and Al_2O_3.

EXPERIMENTAL

Polishing experiments—N-type non-doped c-plane (0001) GaN templates, deposited on 2-inch-diameter sapphire was polished using X62 S82×305-D-S polisher and Suba600 polishing pad. The optimized process parameters were shown in Table 1. The used slurry with pH 10 was composed of colloidal silica (SiO_2, particle size between 70 nm-80 nm), aluminium oxide abrasives (Al_2O_3, particle size between 60 nm-70 nm or 220-240 nm), potassium hydroxide (KOH), phosphoric acid (H_3PO_4). After CMP, GaN wafer was dried off by N_2 air stream. Removal rate (RR) was calculated by equation (1):

$$MRR = \frac{\Delta m}{\rho \times s \times t} \quad (1)$$

Here, *MRR* represents GaN MRR, Δm/g is the weight difference before and after polishing. ρ/(g/cm$_3$) is density of GaN, s/cm^2 is the wafer size and t/h is polishing time.

Table 1 Process parameters of CMP.

Parameter	Condition
Polishing time	40 min
Polishing head speed	50 r/min
Platen rotation speeds	45 r/min
Down force	5.8 psi
Slurry flow rate	80 ml/min
Polishing temperature	22-28 ℃

CMP performance measurement—Agilent 5600LS atom force microscope (AFM) was used for measuring the surface morphology and roughness of GaN after polishing.

RESULTS AND DISCUSSION

Effect of abrasive on GaN MRR—Abrasives enhance the mechanical effects of polishing process, which are extremely important in GaN CMP, the effects of other chemicals are based on abrasives. The variation of the GaN MRR with different abrasive concentrations was shown in Fig 1. The MRR increased with the increased of abrasive concentration. The slurry containing 1 wt% Al_2O_3 and 10 wt% SiO_2 had the similar MRR of 44.63 nm/h and 48.78 nm/h respectively. However, the MRR of 2 wt% Al_2O_3 was much higher than 20 wt% SiO_2, reaching 121.95 nm/h. It is worth noting that the MRR of 0.5 wt% Al_2O_3 was only 24.39 nm/h, which caused by the low concentration of Al_2O_3 abrasive. When the slurry was spread on polishing pad, the actual effective contact surface with the GaN wafer was small, resulting in extremely weak mechanical effects during the polishing process, leading to very low MRR.

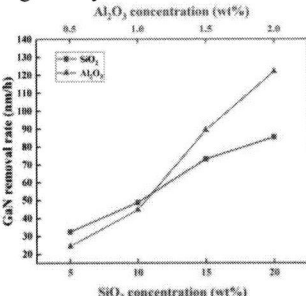

Fig. 1. The variation of the MRR of GaN with different abrasive concentration.

Following above experiments, the effect of different particle sizes of Al_2O_3 on GaN MRR was further investigated. As can be seen from Fig. 2, the mechanical removal of the large particle size Al_2O_3 during CMP process was greater than small particle size Al_2O_3, and the MRR all increased with increasing Al_2O_3 abrasive concentration, using 2 wt% larger particle size Al_2O_3 obtained 162.56 nm/h MRR.

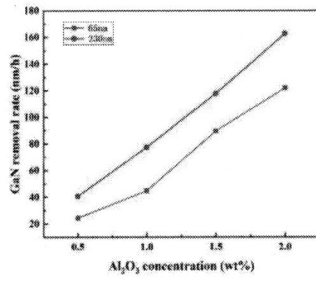

Fig. 2. The comparison of the MRR of GaN when using Al_2O_3 of different particle size in different concentrations.

Effect of abrasive on GaN surface quality—Post-polishing surface roughness by three abrasives (concentration of maximum MRR) were tested for AFM, the results were shown in Fig. 3. The surface roughness (Sq) after polishing by 20 wt% SiO_2, small and large particle size Al_2O_3 were 1.27 nm 1.74 nm and 2.36 nm respectively, which indicated that although the

concentration of SiO_2 was much higher than Al_2O_3, the damage to GaN surface by Al_2O_3 was greater than SiO_2 due to the higher hardness.

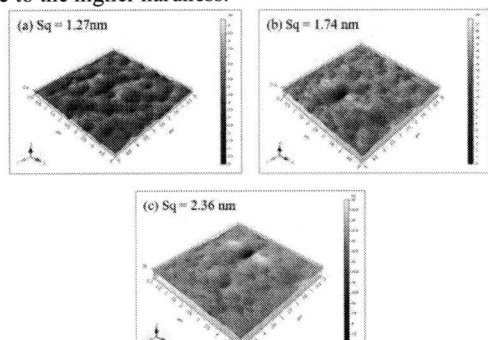

Fig. 3. Surface morphology of GaN after polishing with different abrasive (a) 20 wt% SiO_2, (b) 2 wt% small particle size Al_2O_3, (c) 2 wt% big particle size Al_2O_3.

GaN MRR compounding with different ratios of SiO_2 and Al_2O_3—As shown in Fig. 4, adding Al_2O_3 to SiO_2 could effectively increase the MRR, which was in line with expectations. In addition, 10 wt% SiO_2 + 1 wt% Al_2O_3 and 5 wt% SiO_2 + 1.5 wt% Al_2O_3 as abrasives obtained higher MRR than 2 wt% Al_2O_3.

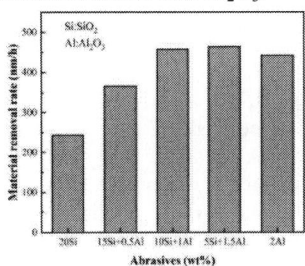

Fig. 4. MRR of GaN using mixed abrasive of SiO_2 and Al_2O_3 in different proportions.

The principle is shown in Fig. 5, SiO_2 effectively fills the gaps between the large Al_2O_3 particles during CMP and high concentration of SiO_2 and the large number of small spherical SiO_2 increase the effective contact area between the abrasive and the wafer surface, thus increasing the MRR of GaN.

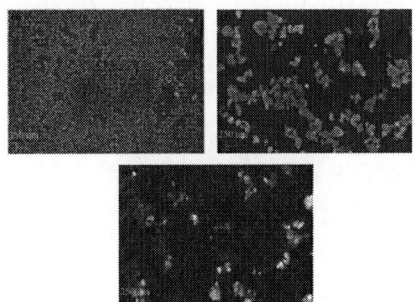

Fig. 5 SEM images of abrasives (a) 20 wt% SiO_2, (b) 2 wt% Al_2O_3, (c) SiO_2 + Al_2O_3.

Sq of GaN compounding with different ratios of SiO₂ and Al₂O₃—As can be seen from Fig. 6, the Sq increased to 1.1 nm after polishing by 15 wt% SiO_2 + 0.5 wt% Al_2O_3 group. The Sq of the 10 wt% SiO_2 + 1 wt% Al_2O_3 group was 1.27 nm and 5 wt% SiO_2 + 1.5 wt% Al_2O_3 group was 1.32 nm. It could be concluded that increasing the proportion of Al_2O_3 in slurry would increases the MRR but also raise the Sq, and the use of large particles of Al_2O_3 will tend to leave deep pits on post-polishing surface.

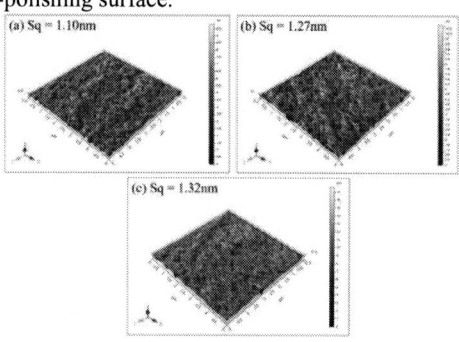

Fig. 6. Surface morphology of GaN after polishing using mixed abrasives of SiO_2 and Al_2O_3 in different proportions (a) 15 wt% SiO_2 + 0.5 wt% Al_2O_3, (b) 10 wt% SiO_2 + 1 wt% Al_2O_3, (c) 5 wt% SiO_2 + 1.5 wt% Al_2O_3.

CONCLUSIONS

The effects of three abrasives on MRR and surface quality of GaN were studied. For single abrasives, the addition of SiO_2 or Al_2O_3 significantly increased the MRR, and the SiO_2 abrasive had slower MRR than Al_2O_3 but better post-polishing surface quality. For compound abrasives, higher SiO_2 concentrations combined with lower Al_2O_3 concentrations could effectively increase the MRR up to 450 nm/h with 1.32 nm of Sq, utilizes the advantages of both abrasives.

ACKNOWLEDGEMENTS

This work was supported by the Major National Science and Technology Special Projects (No. 2016ZX02301003-004-007), National Natural Science Foundation of China (No. 62074049), Natural Science Foundation of Hebei Province (No. F2021202009), Scientific Research Program of Tianjin Education Commition (No. 2019KJ094).

REFERENCES

[1] L.L. Cheng, Z.H. Liu, G.Z. Xu, H.J. Zhong, C.Y. Zhang, K.B. Chen, W.T. Song, K. Xu. *Semiconductor Materials*, vol. 44, 2019, pp. 790-794.

[2] S. Hayashi, T. Koga., M.S. Goorsky. *Journal of The Electrochemical Society*, vol. 155, 2008, pp. 113-116.

[3] H. Aida, H. Takeda, K. Koyama, S. Haruji, K. Kazuhiko, D. Toshiro. *Journal of the electrochemical Society*, vol. 158, 2011, pp. 1206-1212.

[4] H. Gong, G.S. Pan, C.L. Zou, Z. Yan, X. Li. *Surfaces and Interfaces*, vol. 6, 2017, pp. 197-201.

[5] C.L. Zou, G.S. Pan, X.L. Shi, G. Hua, Z. Yan. *Proceedings of the Institution of Mechanical Engineers part J-Journal of Engineering Tribology*, vol. 228, 2014, pp. 1144-1150.

[6] K. Asghar, M. Qasim, D. Das. *ECS Journal of Solid State Science and Technology*, vol. 3, 2014, pp. P277.

ANALYSIS OF THE ADSORPTION AND PASSIVATION MECHANISM OF JFCE ON COPPER SURFACE IN ALKALINE CMP SLURRY

Ni Zhan[1,2]*, Xinhuan Niu[1,2]*, Yinchan Zhang[1,2], Fu Luo[1,2], Han Yan[1,2]

[1]School of Electronics and Information Engineering, Hebei University of Technology, Tianjin 300130, China

[2]Tianjin Key Laboratory of Electronic Materials and Devices, Tianjin 300130, China

*2460518061@qq.com, xhniu@hebut.edu.com

ABSTRACT

The mechanism of adsorption and passivation of the non-ionic surfactant fatty alcohol polyoxyethylene ether (JFCE) on copper (Cu) surfaces was investigated by contact angle and Cu surface measurements. The adsorption inhibition model of JFCE on Cu surface was established by scanning electron microscopy (SEM), potential polarization curve and X-ray photoelectron spectroscopy (XPS) analysis, and the adsorption and passivation mechanism of JFCE as a surfactant in glycine-based alkaline Cu polishing slurry was reasonably revealed.

Keywords—Chemical Mechanical Planarization (CMP); JFCE surfactant; adsorption mechanism; copper

INTRODUCTION

In integrated circuit (IC) fabrication, Cu as a representative interconnecting material has attractive properties to improve the power consumption, signal transmission speed and reliability of electronic devices due to its excellent electrical properties compared with original aluminum [1]. In the present, chemical mechanical planarization (CMP) is the essential method to achieve the global planarization of the Cu film, and the slurry is an influential factor in the performance of CMP [2]. Slurry generally consists of additives such as abrasives, surfactants, oxidizers, complexing agents, and corrosion inhibitors, among which surfactants can be effectively adsorbed on the wafer and particle surfaces, forming an adsorption layer that protects the surface from chemical erosion and significantly improves the wafer surface uniformity and planarization [3]. Y.S. Cheng et al [4] found that surfactant JFCE could greatly improve the wetting and diffusion properties of the cleaning solution and increase the electrostatic repulsion between the particles and the cobalt (Co) surface to accelerate the removal of particles when studying the particle removal problem during the cleaning process after Co CMP. In addition, J.K. Zhou et al [5] also investigated the synergistic effect between JFCE and chitosan (CTS), and the experimental results showed that JFCE can increase the hydrophilicity of polishing slurry and improve the complexation or inhibition of CTS. However, there is a lack of systematic research on the mechanism of JFCE action on Cu surface.

Therefore, it is necessary to conduct a comprehensive understanding on the mechanism of JFCE adsorption and passivation on metal surfaces to provide a theoretical basis for future surfactant research. Accordingly, a series of methods was presented to comprehensively analyze the mechanism of JFCE adsorption and passivation on Cu layer surfaces and verified it by XPS measurements.

EXPERIMENT

Materials, slurry and electrochemical equipment. —Each working electrode of Cu was prepared from the material of commercial purity(4N), delivered in the form of a rod with a 5-mm diameter. The sample was mechanically pretreated by grinding (with SiC paper up to 2000 grit). Before the experiment, Cu samples were rinsed in ultrasonic cleaning with citric acid (CA) concentration of 1%, DI water, degreased in alcohol, and dried. Without considering the effect of colloidal silica, slurry at pH 9 were prepared with 20 ml/L oxidizing agent H_2O_2 (the mass fraction of 30 wt %) and different concentrations of JFCE (0, 1.5, 3, 5, 7 ml/L) in KOH diluted solution. In order to acquire dynamic potential curves (Tafel plot), it carried out on the CHI660E electrochemical workstation with a three-electrode glass cell (purchased from Shanghai Chenghua, Co Ltd).

Contact angle and surface tension measurements. —Static angle analyzer (JC2000D, Shanghai Zhongchen Inc.) was used to measure the contact angle and surface tension of different solutions dropping on the surface of the Cu coupons.

Scanning electron microscopy (SEM). —Scanning electron microscopy (SEM, Zeiss Sigma 500/VP) was applied to observe the surface morphology of Cu film.

X-ray photoelectron spectroscopy (XPS). —X-ray photoelectron spectroscopy (XPS, PHI 250Xi) was used to obtain the composition information of Cu surface after treatment in different solutions.

RESULTS AND DISCUSSION

In order to analyze the action mechanism of the JFCE on the surface of Cu wafers, the wetting effect of JFCE on Cu surface was investigated. The variation of the contact angle between the Cu surface and slurry and the surface tension of the solution with different concentration of JFCE was measured as shown in Fig. 1.

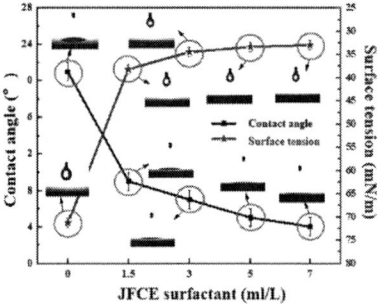

Fig.1. Effect of different concentration of JFCE on the contact angle between wafer and slurry and the surface tension of the solution.

It can be seen that as the concentration of JFCE increases from 0 to 3 ml/L, the contact angle between the solution and Cu surface drops sharply from 21.0° to 7.0°, while the surface tension of the solution plummets from 71.3 mN/m to 34.5 mN/m. However, the change was not apparent when the concentration of JFCE was increased from 3 ml/L to 7 ml/L. It can be concluded that by adding JFCE the hydrophilicity of the Cu film surface can be effectively improved, resulting in a more uniform contact between the polishing slurry and the Cu surface.

In order to further understand the effect of JFCE on the Cu surface, the static effect of JFCE on the Cu film surface was observed by SEM test after treatment with different solutions as shown in Fig. 2. It can be seen from Fig. 2(a) that after immersion in pH 10 20 ml/L, H_2O_2, and 2.3 wt% glycine solutions for 5 min, the Cu surface was porous and discontinuous, and was severely oxidized and corroded. However, as shown in Fig. 2(b), it can be observed that the corrosion holes on the Cu surface were filled and the surface become uniform and continuous after adding 3 ml/L JFCE. Therefore, it can be concluded that JFCE can produce adsorption in the recessed part of the Cu surface, effectively protecting the surface from excessive chemical corrosion, which is conducive to achieving global flatness. But the mechanism of action of JFCE needed to be further explored.

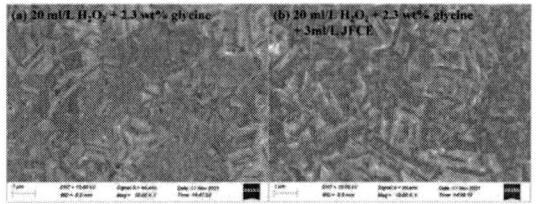

Fig.2. SEM images of Cu surface after immersion in solution with or without JFCE for 5 min.

To reveal the mechanism of JFCE inhibition on Cu surface, the electrochemical behavior of Cu was tested. The potential polarization of Cu specimens in solution containing 20 ml/L H_2O_2, 2.3 wt% glycine, and different concentrations of JFCE was shown in Fig. 3. The data of the corrosion current density (I_{corr}), corrosion potential (E_{corr}) and corrosion inhibition efficiency (θ) (calculated by Equation (1)) were listed in Table 1.

$$\theta = 1 - \frac{i_{corr,\ with\ JFCE}}{i_{corr,\ without\ JFCE}} \qquad (1)$$

Table 1. E_{corr}, I_{corr} and θ of Cu at different JFCE concentrations.

JFCE (ml/L)	E_{corr} (V)	I_{corr} (A)	θ
0	-0.004	4.674×10^{-4}	-
1.5	0.013	4.369×10^{-4}	0.065
3	0.013	4.108×10^{-4}	0.121
5	0.015	3.982×10^{-4}	0.148
7	0.020	2.632×10^{-4}	0.437

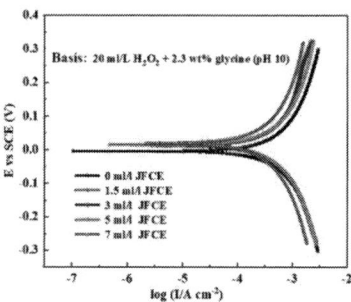

Fig.3. The potentiodynamic polarization curves of Cu in solutions with 20 ml/L H_2O_2, 2.3 wt% glycine and different concentrations of JFCE.

As can be seen from the table, as the JFCE concentration increases from 0 ml/L to 7 ml/l, the E_{corr} increases from -4 mV to 20 mV, the I_{corr} decreases from 467.4 $\mu A/cm^2$ to 263.2 $\mu A/cm^2$, and the θ increases continuously. Both the increase in E_{corr} and the decrease in I_{corr} indicates the formation of a passivation film on the Cu electrode surface, which inhibited corrosion of the Cu surface. The presence of a stable current over a range of potentials also confirmed that the inhibition of Cu by JFCE was related to the formation of a passivation film on the Cu surface.

The surface composition of the Cu film was examined by XPS technique in order to further study the interface reaction between JFCE and Cu surface. After being treated with a solution at pH 10, 20 ml/L H_2O_2, 2.3 wt% glycine (without or with 3 ml/L JFCE), the Cu 2p3/2 plots were shown in Fig. 4(a) and (b), respectively. As

979-8-3503-1101-3/23 $31.00 © 2023 IEEE

shown in Fig. 4(a), the surface of Cu is occupied by its oxides, and the binding energy of 935.9 eV, 932.4 eV, and 934.2 eV represents Cu-glycine, CuO and Cu/Cu$_2$O, respectively. In addition, as shown in Fig. 4(b), with the addition of JFCE, a new peak appears at 932.1 eV (the corresponding substance of the binding energy was not identified in the database). Besides, the complete XPS spectrum and the content of each element were compared and the results are shown in Figs. 4(c) and (d).

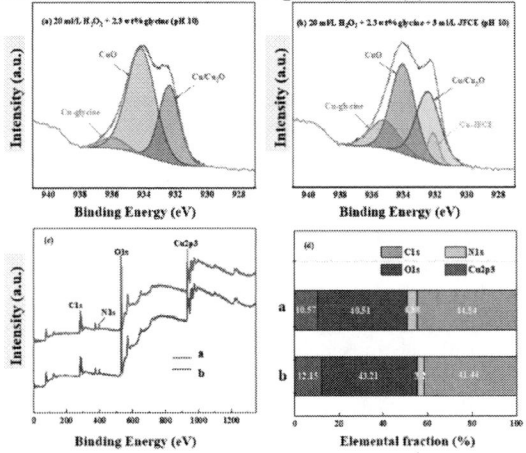

Fig.4. (a) and (b) are the peak fits of XPS Cu 2p3/2 spectra, (c) and (d) are the complete XPS spectra of (a) (b) and the corresponding percentages of each element, respectively.

It can be seen that compared with sample (a), the peak intensity and elemental content of Cu and oxygen (O) of specimen (b) increase significantly, while the peak intensity and elemental content of nitrogen (N) and carbon (C) decrease significantly with the addition of JFCE. It is reasonable to infer that the new peak in Fig. 4(b) is the product of Cu-JFCE absorption on Cu surface. In addition, according to the chemical formula of JFCE and glycine, it can be inferred that the significant decrease in N peak intensity is due to the spatial occupancy of JFCE hindered the complexation and dissolution of glycine on the Cu surface, which was caused by the physical adsorption and chemical adsorption of JFCE on the Cu surface.

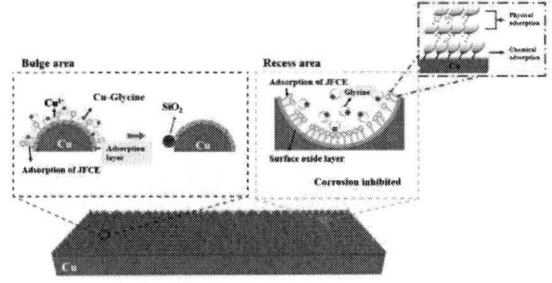

Fig.5. The adsorption inhibition model of JFCE on Cu surface.

Combined with the improvement of JFCE surface quality after polishing and its analysis of the inhibition mechanism on Cu surface, the adsorption inhibition model of JFCE on Cu surface was established, as depicted in Fig. 5. As can be seen from the Fig. 5, the JFCE had a dual effect on Cu removal. The concave regions of the Cu surface were protected by the JFCE chemistry and physisorption film due to surfactant dispersion and coalescence, and the reaction and mass transfer energy was low, resulting in insufficient contact between the abrasive, the complexing agent, and the Cu surface. Therefore, the mechanical action of the concave regions was small, and the chemical corrosion reaction was also inhibited. However, for the convex parts, the JFCE adsorption was dispersed and the primary physical adsorption was relatively weak, easily removed by the mechanical action of the abrasive. As a result, there was a higher Cu removal rate in the convex areas and a lower Cu removal rate in the concave areas during dynamic CMP, ultimately improving the planarization of the Cu film CMP and achieving a lower surface roughness with a higher surface uniformity.

REFERENCES

[1] Q.Z. Xu, L. Chen, F. Yang, H. Cao. *Microelectronic Engineering*, vol. 183-184, 2017, pp. 1–11.

[2] J. Li, Y.H Liu, Y. Pan, X.C. Lu. *Applied Surface Science*, vol. 293, 2014, pp. 287–292.

[3] G. Yang, H.X. Wang, N. Wang, R. Sun, C-P. Wong. *Journal of Alloys and Compounds*, vol. 770, 2019, pp. 175–182.

[4] Y.S. Cheng, S.L. Wang, H.L. Li, C.W. Wang, Y.D. Yang, S.S. Lei, S. Li. *Colloids and Surfaces A: Physicochemical and Engineering Aspects*, vol. 627, 2021, pp. 127189-127198.

[5] J.K. Zhou, X.H. Niu, Z. Wang, Y.Q. Cui, J.C. Wang, C.H. Yang, Z.Q. Huo, R. Wang. *Colloids and Surfaces A: Physicochemical and Engineering Aspects*, vol. 586, 2020, pp. 124293–124305.

EFFECT OF SURFACTANTS ON CMP PROPERTIES OF M-PLANE SAPPHIRE

Yida Zou[1,2], Xinhuan Niu[1,2*], Ziyang Hou[1,2], Minghui Qu[1,2], Ni Zhan[1,2], Jianghao Liu[1,2]*

[1]School of Electronics and Information Engineering, Hebei University of Technology, Tianjin 300130, China

[2]Tianjin Key Laboratory of Electronic Materials and Devices, Tianjin 300130, China

*383401447@qq.com, xhniu@hebut.edu.com

ABSTRACT

M-plane sapphire is commonly used as a substrate material for gallium nitride and zinc oxide. Chemical mechanical polishing (CMP) is one of the most effective methods to achieve atomic-scale smooth surface. Each component of the slurry plays an important role in the CMP process. So, the effect of three surfactants, JFCE, SDBS and CTAB on the CMP removal rate and surface roughness of M-plane sapphire was investigated. From the result, it is found surfactant can improve the removal rate and surface quality of M-plane sapphire. Meanwhile, JFCE is slightly better than the other two in the same condition.

Keywords—JFCE surfactant; SDBS surfactant; CTAB surfactant; M-plane Sapphire; Chemical Mechanical polishing (CMP)

INTRODUCTION

Sapphire (α-Al_2O_3 single crystal) has been widely applied in semiconductor device, light emitting diodes (LED), solid lasers, infrared window and precision optics due to its excellent optical, chemical and mechanical properties such as high hardness, thermal stability, chemical inertness, great electrical and dielectric properties[1]. M-plane sapphire is commonly used as a substrate material for gallium nitride and zinc oxide. However, because of the requirements of high efficiency and surface quality for sapphire, surface ultra-precision machining has brought greater challenges[2]. Chemical mechanical polishing (CMP) is becoming one of the most important technologies because of its ability to obtain super-smooth and non-damaged wafer surface, and slurry is an essential factor influencing the performance of CMP[3]. Conventional slurry contains abrasives, complexing agents and surfactants, among which surfactants, due to their wetting, emulsifying and dispersing properties, can significantly improve the uniformity and planarization effect of the wafer surface[4].

Xin Zhao et al[5] investigated that effect of anionic surfactant concentration on r-plane sapphire removal rates, and the experimental results showed that anionic surfactant could improve removal rate and surface quality due to its good dispersion and stability to prevent nano-SiO_2 sol agglomeration. GuoMei Chen et al[6]

studied the effect of amphoteric surfactant NL on the CMP of A-plane sapphire wafer in the pH 6 to 12. The results showed that NL can inhibit the removal of sapphire at pH 6 to 10. At pH 11 to 12, NL can improve MRR of sapphire. When pH was 12, the removal rate of sapphire was the highest, and the surface roughness after polishing was 0.807 nm. YaWen Bai et al[7] researched the effect of cationic surfactant CTAB on the CMP of A-plane sapphire wafer in the pH 6 to 11. The results showed that CTAB can inhibit the removal of sapphire at pH 6 to 8. At pH 9 to 11, CTAB can improve MRR of sapphire. When pH was 9, the removal rate of sapphire was the highest, and the surface roughness after polishing was 0.810 nm. YaQi Cui et al[8] studied that the effect of non-ionic surfactants JFCE, anionic surfactants LAS and cationic surfactants DTAC on A-plane sapphire. It is found that the three surfactants can improve the removal rate of sapphire and reduce the surface roughness when the volume fraction of the three surfactants was in the range of 0% to 0.6%, and the polishing effect was the best when the volume fraction of 0.4% DTAC was used. Therefore, combined with previous studies, it was reasonable to explore the influence of surfactant on the removal rate of sapphire.

In this paper, the effects of different surfactants on CMP of C-plane sapphire and M-plane sapphire were investigated. At the same time, the comparison of different surfactants on surface quality of M-plane sapphire was studied.

EXPERIMENTAL

Two-inch single crystal commercial C-plane sapphire wafer and M-plane sapphire wafer were used for experiments. Polishing experiments were performed on X62 S82×305-D-S single-side CMP polisher produced by Suzhou Herriot with a Suba600 polishing pad, and sapphire wafers were put in inlaid layer holes on the wax-free polishing template adsorption film.

Alkaline sapphire slurry prepared in the lab was used and nano-SiO_2 sol was selected as abrasive, whose concentration was 40 wt%. The average particle size of the abrasive was 80-90 nm. KOH was used as pH regulator to adjust the pH value of the slurry to 10.5. Surfactants were added to the sapphire slurry to enhance chemical action. Each group of experiment was repeated at least three times to take the average.

979-8-3503-1101-3/23 $31.00 © 2023 IEEE

Sapphire CMP process parameters were shown in Table I. Professional electronic balance was used to measure the weight of sapphire substrate before and after polishing, with a precision of 0.1 mg (AUY120 ASSY). The surface topography and roughness Sq were measured by Agilent 5600LS atomic force microscopy (AFM).

Table I. Process parameters of Sapphire CMP

Parameters	Conditions
Polishing time	30 min
Polishing head speed	50 rpm
Platen rotation speed	45 rpm
Slurry flow rate	160 mL/min
Downward pressure	0.1 Mpa
Upward pressure	0.06 Mpa

Material removal rate (MRR) is determined by the Equation (1):

$$MRR = \frac{\Delta m \times 10^4}{\rho \pi r^2 t} \qquad (1)$$

where Δm (g) is the mass loss, t is the polishing time (t = 1/2 h), ρ is the sapphire density (ρ = 3.98 g/cm^3), r is the radius of sapphire substrate (r = 2.54 cm), and MRR(μm/h) is the corresponding removal rate.

RESULT AND DISCUSSION

Effect of surfactants concentration on sapphire removal rate.

The ratio of deionized water to nano-SiO$_2$ sol was 1:1. The surfactants with different concentration were added to the slurry. According to Figure 1, when the concentration of SDBS increased from 0 wt% to 0.2 wt%, the MRR of sapphire on C-plane and M-plane showed an upward trend, with MRR$_C$ = 4.288 μm/h and MRR$_M$ = 3.929 μm/h. This is because after the addition of SDBS, the repulsive force between particles becomes larger, the dispersion of abrasive in slurry is enhanced, and thus the mechanical force is enhanced, and the MRR is improved.

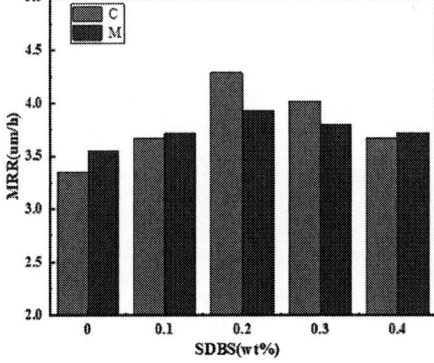

Figure 1: Effect of SDBS surfactant concentration on sapphire MRR.

As shown in Figure 2, the MRR of C-plane sapphire increases from 3.919 μm/h to 4.887 μm/h, and the maximum removal rate of M-plane sapphire is 4.513 μm/h when the concentration is 0.004 wt%. This is because the addition of cationic surfactants makes the surface of SiO$_2$ abrasive adsorb positive charge, particle size increases, and mechanical action increases the MRR. When the concentration of cationic surfactant continues to increase, the agglomeration of abrasive in polishing fluid reduces the number of effective particles and the MRR begins to decline. In addition, the large amount of foam and agglomerative abrasive produced in the polishing process is very unfavorable to the polishing effect.

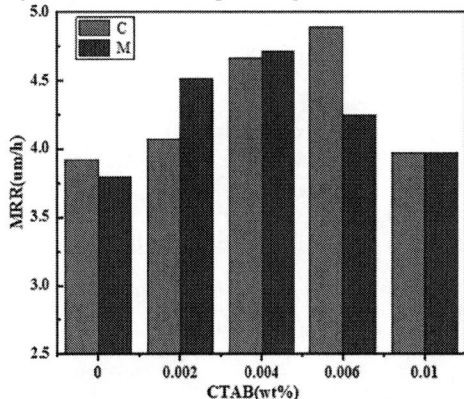

Figure 2: Effect of CTAB surfactant concentration on sapphire MRR.

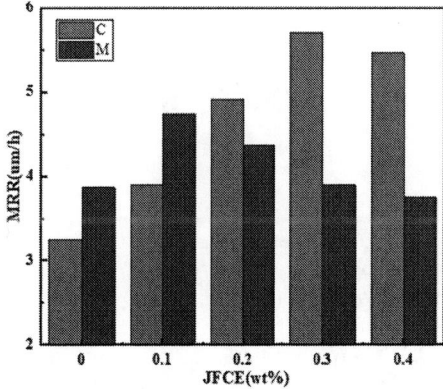

Figure 3: Effect of JFCE surfactant concentration on sapphire MRR.

As shown in Figure 3, the MRR of C-plane sapphire is significantly improved when the concentration of non-ionic surfactant is 0.3 wt%, and the highest MRR of M-plane sapphire is 4.738 μm/h when the concentration is 0.1 wt%. This is due to the enhancement of dispersion of non-ionic surfactants to abrasives, as well as the stronger contact between wafers and abrasives and the enhancement of mass transfer effect caused by the hydrophilic groups on the macromolecular chains. The effect of surfactants on different crystal surface is different, because the anisotropy of sapphire wafer material itself, the wafer surface atomic arrangement period is different.

It is obvious that 0.1 wt% JFCE has the best removal effect on M-plane sapphire. Although the removal rate of CTAB is comparable to that of JFCE, the large amount of foam and agglomeration of abrasives during CMP is not good for polishing quality. Based on the above results, the non-ionic surfactant JFCE has the best effect on the CMP MRR of M-plane sapphire.

Effect of surfactants on sapphire roughness.

The effect of CMP on the MRR and surface quality of sapphire wafer should be considered. Therefore, the surface roughness and morphology of M-plane sapphire were measured under the optimal surfactant concentration condition, and the results were shown in Figure 4. As can be seen from Figure 4, the surface roughness of M-plane sapphire after polishing with anionic and non-ionic surfactant is less than 0.3 nm, while the surface roughness of M-plane sapphire after adding cationic surfactant is higher than the other two kinds.

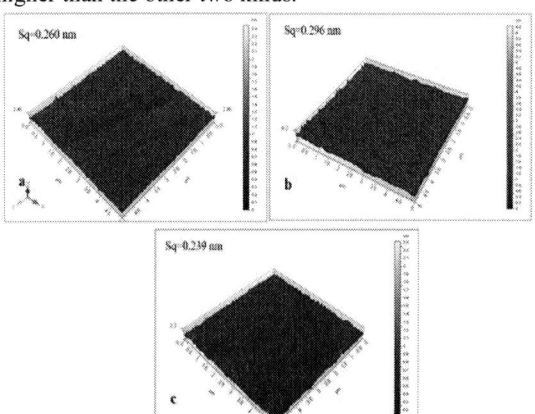

Figure 4: Surface quality of M-plane sapphire after polishing with optimal surfactant concentration: (a) SDBS; (b) CTAB; (c) JFCE

CONCLUSION

The effects of SDBS, CTAB and JFCE on the properties of sapphire CMP were studied. The results show that the MRR of M-plane sapphire can be improved by proper concentration of SDBS and JFCE, but the maximum MRR of M-plane sapphire is 4.738 μm/h when the concentration of non-ionic surfactant JFCE is 0.1 wt%. Three surfactants can improve the surface quality of M-plane sapphire CMP, and the non-ionic surfactant JFCE has the best effect. According to the two indexes of

MRR and surface quality, the non-ionic surfactant JFCE has the best effect on the CMP of M-plane sapphire.

ACKNOWLEDGMENTS

This work was supported by the Major National Science and Technology Special Projects (No. 2016ZX02301003-004-007), National Natural Science Foundation of China (No. 62074049), Natural Science Foundation of Hebei Province (No. F2021202009), Scientific Research Program of Tianjin Education Commition (No. 2019KJ094). The authors also thank the teachers and classmates for their helpful suggestions.

REFERENCES

[1] Y.Q. Cui, X.H. Niu, J.K. Zhou, Z. Wang, R. Wang, J. Zhang. *ECS Journal of Solid State Science and Technology*, vol. 8, 2019, pp.488–495.

[2] Y.N. Lu, X.H. Niu, C.H. Yang, Z.Q. Huo, Y.Q. Cui, J.K. Zhou, Z. Wang. *ECS Journal of Solid State Science and Technology*, vol. 9, 2020, pp.064006.

[3] Z.Y. Hou, X.H. Niu, Y.N. Lu, Y.C. Zhang, Y.B. Zhu. *ECS Journal of Solid State Science and Technology*, vol. 10, 2021, pp.104001.

[4] G. Yang, H.X. Wang, N. Wang, R. Sun, C-P. Wong. *Journal of Alloys and Compounds*, vol. 770, 2019, pp. 175–182.

[5] X. Zhao, X.H. Niu, D. Yin, J.C. Wang, K. Zhang. *ECS Journal of Solid State Science and Technology*, vol. 7, 2018, pp.135–141.

[6] G.M. Chen, C.K. Du, Z.F. Ni, Y.W. Bai, Y.X. Liu, Y.W. Zhao. *Modern Manufacturing Engineering*, vol. 12, 2020, pp.83–87.

[7] Y.W. Bai, G.M. Chen, K. Teng, Z.F. Ni. *JOURNAL OF SYNTHETIC CRYSTALS*, vol. 47, 2018, pp.470–475.

[8] Y.Q. Cui, X.H. Niu, Z. Wang, J.K. Zhou. *Semiconductor Technology*, vol. 44, 2019, pp.883–887,898.

Methods for Fin Etching profile maintaining and measurement

Yun Xu, Tong Wu, Fairy Chen

Semiconductor Manufacturing International (Shanghai) Corporation Failure Analysis Laboratory

Zhangjiang Road #18, Pudong New Area, Shanghai, P. R. China, 201203

Email: Yun_Xu@SMICS.COM

ABSTRACT

Characterization of nano-scale device meets increasingly challenging with the decreasing dimension of transistors. The measurement and monitoring of silicon-Fin CD (critical dimension) play an important roles during FinFET process researching and developing. Fin CD measurement is based on TEM images, in which, the sub-angstrom layers have too limited contrast to distinguish the interface between Silicon Fin and isolation SiO2 or Fin profile maintain film. Therefore, the definition of those interface cannot be standardized cause of TEM sample thickness variation, Fin deformation and TEM imaging error just like sphere and chromatic aberration. In this job, firstly, TEM sample preparation was taken with profiling maintain film which has the optimized internal stress and good gap filling for high aspect ratio Fin Etching step. Secondly, Sobel model based interface identification system was setup for auto-measurement which has good data variation performance which is within 2% of the target size to meet the needs of process development.

Keywords—FinFET, critical dimension, auto-measurement

INTRODUCTION

Within FinFET (Fin field-effect transistor), the gate electrode is present in both sides of the Fin which brings better control of short channel effects and better carrier mobility, along with significant reduction of current leakage. As the channel for carrier movement, Fin plays a critical role and its characterization is necessary reference for process during the chip fabrication. For example, visualization of Fin bending provides new thought for FCVD/CVD recipe optimizing. In addition, the changes CD (Critical dimension) of Fins etched by different recipes were monitored can point the way of its tuning. Currently, major methods of CD measurement for nano-scale profile of IC includes CD-SEM (Scanning Electron Microscope), AFM (Atomic Force Microscope) and TEM (Transmission Electron Microscope). There are two major challenges of Fin characterization with TEM imaging. First one, in the process of sample preparation in FIB (Focused Ion Beam), Fin bending is caused by the mismatch between Fin filling material and Fin. Second one, the Fin–STI (Shallow Trench Isolation) boundary is difficult to distinguish because of their overlapping with each other along the channel or thickness direction.

A set of methods, include sustaining Fin profile undamaged during the sample preparation and system assisted CD measuring, were developed to compromise those issues.

Fin-Etch Sample Characterization Optimization

A. Improved filler coating method for TEM sample preparation

The structure of Fin-Etch process site is characterized by high aspect ratio and small width as shown in Fig.1. For monitoring the Fin-Etch process, appropriate filler is required to strengthen the surface profile to prevent Fin deformation caused by scanning electron beam. Filler also has internal stress, which can also damage the silicon Fin so that Fin cannot maintain its original form. Four different filler materials are selected and detailed trials listed below.

1) The trench filling ability of all 4 candidates
2) The thermal expansion coefficient material that best matches silicon-Fin is developed for filling
3) The effectiveness of desolvation condition of different organic filler
4) Stress analysis of filling materials
5) Optimization of filling material coating method.

TABLE I
Result of filling experiment

Sample Number	SN1	SN2	SN3	SN4	SN5
Filler	Glue 1	Glue 2	Organic material 1	Organic material 2	Organic material 2
Coating Method	Manual	Manual	Spin	Spin	Spin
Curing Condition	Bake	Bake (High temperature)	Bake	Bake	Bake (increase time)
Result					
Comment	No	No	No	No	Yes

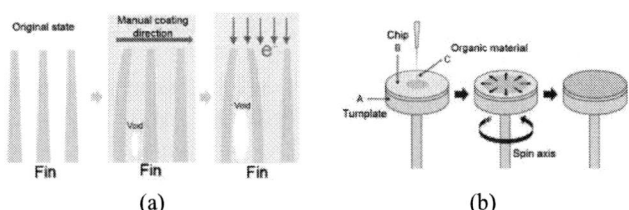

(a)   (b)

Fig.1. Coating method: (a) Manual coating; (b) Spinning coating

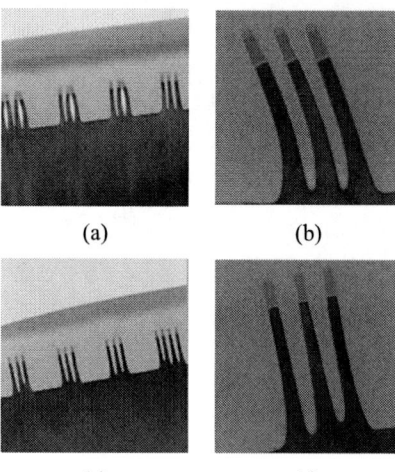

Fig.2. Optimized sample preparation by FIB, pre-improvement: (a) (b); post-improvement: (c) (d)

Experimental parameters and results are shown in Table I and Fig.2. According to the result, an organic filler with two major advantages is selected. First, it is capable of filling trench with high aspect ratio with no void present. Second, the silicon Fin was not deformed during its desolvation process. It can be found out from the desolvation trail that, lowering the desolvation temperature and adjust its duration will greatly reduce the deformation of silicon Fin. The coating method trial indicates that, porous and sparse can be caused by manual coating the filler, which would be aggravated by the electron beam action during manual coating and then intensify the porosity and finally lead the result the Fin is bend. With assistance of spin coating, the damage to silicon Fin during the process of sample preparation in FIB is significantly reduced by this set of filler coating method.

B. Damage-free FIB milling method and improved TEM imaging

The Fin-high dielectric constant material boundary becomes difficult to distinguish due to their overlapping along the thickness direction. What's more, the recognition on diffraction pattern of silicon crystal lattice are affected by Fin thickness and the amorphous layer on the front surface. In the progress of sample thinning, the Ga+ ions collide with the atoms on the sample and sputter them away from the sample surface. Then deposited ions onto the sample squeeze the atoms indigenous to the sample away from their original positions. Interstitial and vacancy defects are caused by those incident ions and thus increase the disorder within the solid, which is equivalent to the amorphous layer in this case [1]. Both sides of the TEM sample are subjected to FIB thinning, and amorphous layers are produced on both sides walls of the sample, forming a sandwich-like structure. As shown in Fig3, the middle of the TEM sample is a crystalline layer and both sides are amorphous layers formed by ion beam damage. The thickness of amorphous layer is proportional to the thinning ion beam voltage during the sample preparation. According to a study, 30keV thinning voltage can generate 20-30nm amorphous layer, 5keV can generate 4-5nm amorphous and 1-2nm amorphous silicon results from 2keV voltage [2-5]. As shown

in Fig.3 and Fig.4, the interference of amorphous layer can be effectively improved by reducing the cleaning voltage, and the atomic details can be shown more clearly.

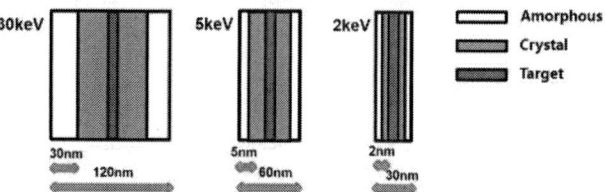

Fig.3. Thickness of amorphous layer under different thinning voltage

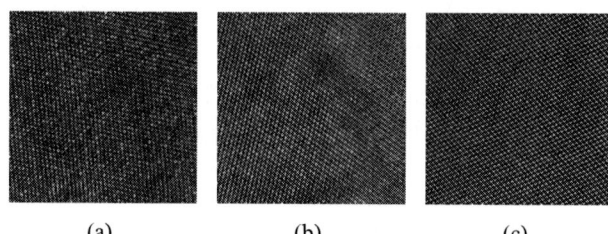

Fig.4. TEM imaging of sample with different thinning voltage: (a) 30keV; (b) 5keV; (c) 2keV

Sample should be made within 30nm thickness in order to avoid overlap of Fin boundary. Low thinning voltage such as 2keV can slice the sample thin enough without leaving thick amorphous layer and curl the entire sample. Fig.5 demonstrates the entire thinning sequence, alone with where to switch to lower ion beam voltage. The samples were first milled using 30keV till 100nm and further milled using 5keV and 2keV to reduce the total lamella thickness to 30nm [6]. This method is timesaving and can reduce the amorphous thickness to an ideal level.

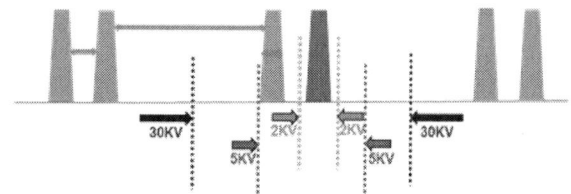

Fig.5. Relationship between sample preparation process and voltage used

In the TEM imaging, the stage must be tilted to an angle that the silicon {110} plane perpendicular to the incident beam. The inappropriate defocus would distort the Fin boundary, and it can be adjusted by observing the integrity of silicon lattice as shown in Fig.6.

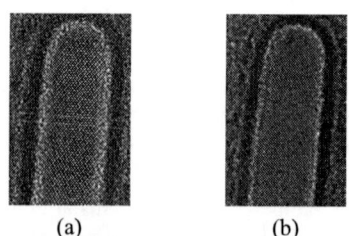

Fig.6.TEM imaging silicon lattice at different defocus (a): On focus ; (b): Under focus

The optimized FIB and TEM technology mainly highlights the difference of Fin boundary contrast to the maximum extent by reducing the proportion of amorphous layer, preparing ultra-thin samples, and optimizing the defocus amount, so as to improve the interface identification during measurement.

Development of auto-measurement system based on DM script

Fin CD measurement should be based on the most optimized TEM image. Unlike traditional grey-scale based image processing, part of this system rotates the image with the base on the (100) spot of FFT of silicon substrate. It can rotate the image faster and more accurate than the conventional way as shown in the Fig.7. It links the (100) spot and the center then calculates the angle between the central–vertical line. As shown in Fig.8, human error caused by manual way will be eliminated by this new method in a highly repeatable way.

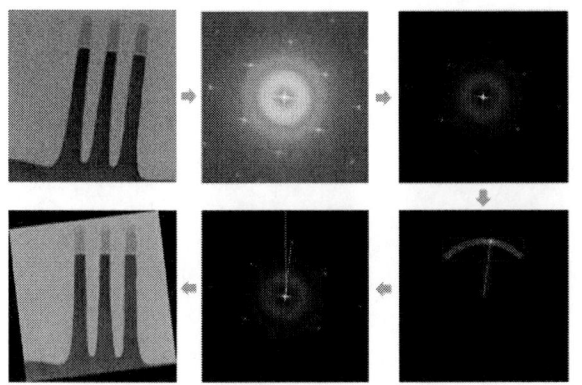

Fig.7. Demonstration for image rotating procedure carry out by the system

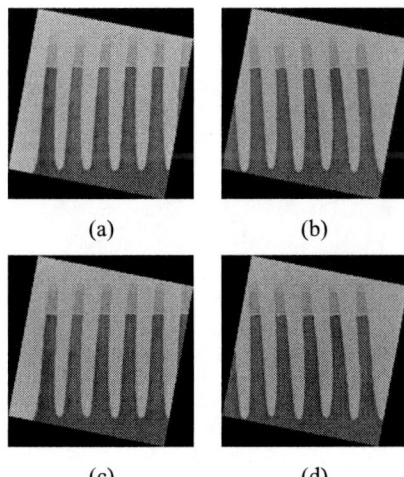

Fig.8.The rotation angle of the same sample by different methods, Conventional way: (a) 10.64° (b) 10.80°; system way: (c) 10.79° (d) 10.79°

The Gauss blur and Sobel model are combined to erase noise and highlight the Fin boundary. The relationship between Gauss blur parameter and noise erasing is shown in Fig.9. The Sobel model gets the intensity gradient of one pixel and its vicinity to highlight the intensity difference and its direction.

With this, the human error, which contains 6 to 8 pixels, is lowered to about 3pixles.

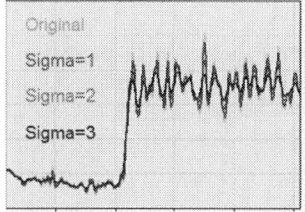

Fig.9. Relationship between Gauss blur parameter and image noise

Using the debugged DM (Digital Micrograph) Script, after the system assisted Fin boundary recognition, a few measurement parameters, includes the space between CD labels, the starting point from the tip of Fin and the number of labels required are set for the incoming measurement command. The command is carried out for every selected Fin. The results are exported to any chart drawing software such as Microsoft Office Excel to reconstruct the Fin profile as shown in Fig.10.

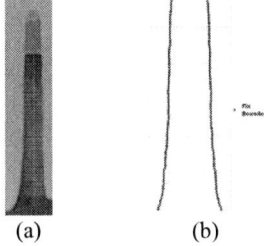

(a)   (b)

Fig.10. Final result of Fin measurement and profile reconstruction: (a) TEM picture; (b) Profile reconstruction model

Unlike any other manual measurement method, massive data collection and analysis for characterization of Fin profile can be carried out by this tool. What's more, those reconstructed profile can be stacked for the comparison between recipe parameters as shown in Fig.11.

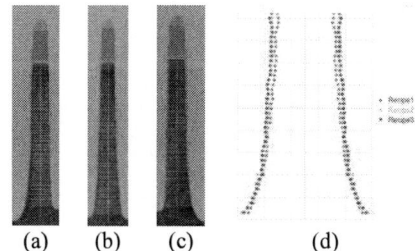

(a)   (b)   (c)   (d)

Fig.11.The reconstruction of Fin profile made by different Etching recipe: (a) Recipe 1; (b) Recipe 2; (c) Recipe 3; (d) Comparison result

This system can always be improved to add new functionality such as monitoring the channel width of Fin.

CONCLUSIONS

TEM sample preparation of Fin Etching step was optimized by profile maintaining material and coating process. Based on the damage-free pre-sample preparation and optimized FIB milling parameters, effective data for process tuning can be provided by the qualified TEM imaging practically at Fin Etching step. And then auto-measurement make sure data variation under controlled less than 2% of target dimension.

REFERENCES

[1] Song Yun, and Lu Hai wei, "Analysis of damage induced by FIB irradiation in microprocess," Journal of Fudan University (Natural Science), 2007, vol. 46, No.1, pp. 119.

[2] S. RUBANOV, and P.R. MUNROE, "FIB-induced damage in silicon," Journal of Microscopy, 2008, vol.214, No.3, pp. 213-221.

[3] S. Rajsiri, "FIB damage in silicon: Amorphization or Redeposition," Microscopy and Microanalysis, 2002, Vol. 8, pp.50-51.

[4] J.F. Ziegler, J.P. Biersack, and U. Littmark, The Stopping and Range of Ions in Solids. 2009, New York: Pergamon Press.

[5] Q. Gao, "FIB Milling During TEM Sample Preparation," in IEEE 42nd Annual International Reliability. 2004: Phoenix. p. Physics Symposium.

[6] Tee Irene, Li Kun, and Liu Pan, "Study of Low-kv Cleaning Method to Improve TEM Sample Prepared by FIB," IEEE Proceeding of 16th IPFA, 2009.

INVESTIGATION OF DIFFERENT GATE BIAS ON PMOS HCI PERFORMANCE

Lei Li *, Canny Chen, and Atman Zhao

Semiconductor Manufacturing International (Shanghai) Corp. No.18 Zhangjiang Road
Pudong New Area Shanghai, China
*Corresponding Author's Email: Lei_Li4@smics.com

ABSTRACT

In this work, hot-carrier injection (HCI) effect of HV p-channel CMOS was studied. We found the change of both saturation drain current (I_{dsat}) and constant current threshold voltage (V_{tlin}) shew opposite trend under the same drain bias (V_d) and different gate bias (V_g), which indicated that there were different dominant degradation mechanisms. The HCI dependence on V_g was also investigated through harge pumping (CP) technique and Technology Computer Aided Design (TCAD) simulation. It was proved that interface traps and positive holes trapping were together responsible for the degradation at 1.1 times V_g operation voltage ($1.1xV_{gop}$), while negative electrons trapping induced by relatively larger vertical electric field were predominant for the degradation at maximum gate current (I_{gmax}) near the drain side.

INTRODUCTION

Since HCI effect can impose restriction on device scaling, HCI reliability has progressively been a major concern in early CMOS processes [1-3]. The purpose of this paper is to discover the generated defect charges of either interface traps or oxide traps under different V_g bias. Fundamental understanding of HCI damage locations and degradation mechanisms can be achieved through a variety of data comparison, so as to get relatively general principles on HCI effect of our HV p-channel CMOS device.

The CP technique for MOS devices has been recently applied to separate the effects of interface state and oxide charge in MOS transistors. In this paper, the variable base level CP method is adequate for the characterization of PMOS HCI degradation phenomena [4-8]. The basic experimental setup for charge pumping measurements is represented in Fig. 1. The gate of p-channel CMOS is connected to a 100kHz frequency and 6V amplitude pulse generator with increasing base level (V_{base}). While the drain and source are grounded and the measured substrate current is regarded as charge-pumping current (I_{cp}). For a given PMOS, the minority holes coming from the source and the drain when device is in the inversion state and the captured carriers will be repetitive recombined with majority electrons when the device is in the accumulation state. Therefore, the I_{cp} is proportional to the interface trap density. An increase in the saturation level of I_{cp} after stress indicates the existence of interface state density (D_{it}),

while a left or right shift of I_{cp} curve indicates positive or negative oxide charge. The defect properties can be identified by CP characteristics after and before degradation of HCI stress.

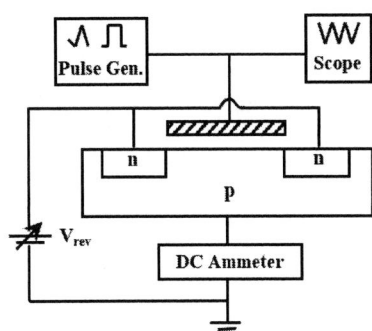

Figure 1: Basic experimental setup for charge pumping measurements.

EXPERIMENT

The schematic cross-section of the investigated HV p-channel CMOS device in this paper is shown in Fig. 2. This device implements the standard symmetrical source and drain regions. The deep P type drift (DDP) regions are designed lightly doped to sustain high voltage drop, which allow equal 18V operation voltage on both gate and drain sides. A gate oxide thickness of 380Å is defined and whereas the channel width and length are defined as 20 and 1.9μm.

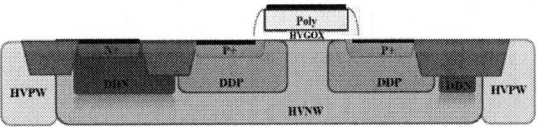

Figure 2: The schematic cross-section of the HV p-channel CMOS.

The stress bias of gate is performed at I_{gmax} and $1.1xV_{gop}$ with a given drain bias ($1.1xV_{dop}$) at room temperature for HV PMOS. The stress time is set as 10000 seconds. The saturation drain current (I_{dsat}) and constant current threshold voltage (V_{tlin}) shifts are used as electrical parameters to represent the HCI degradation. I_{dsat} is monitored at $V_d=V_g=18V$. V_{tlin} is the gate voltage applied

---

979-8-3503-1101-3/23 $31.00 © 2023 IEEE   367

to the device at which the drain current is equal to $0.1\mu A$ times the ratio of gate width (W) to gate length (L).

RESULTS AND DISCUSSION

Fig. 3 indicated I_{dsat} and V_{tci} degradation of HV PMOS with two different gate bias combinations and both electrical degradations show opposite trends along with stress time. When HV PMOS was stressed with gate bias condition of maximum I_g, I_{dsat} increased and V_{tlin} decreased with stress time. This electrical characteristic was a signature of negative electrons trapping inducing higher I_{dsat}. While under $V_g=1.1xV_{gop}$, I_{dsat} decreased and V_{tlin} increased with stress time. It suggested that interface traps or positive holes trapping were dominant under this bias condition. Clearly from the measured results, the worst I_{dsat} degradation occurred at $V_g=I_{gmax}$ because I_{dsat} indicated carrier quantities and migration characteristics. The negative electrons were trapped from channel into gate oxide. While the worst V_{tlin} degradation occurred when V_g reached high voltage of 19.8V ($1.1xV_{gop}$) because V_{tlin} reflected much more the quality of the oxide performance and interface traps and positive holes trapping were together predominant for the degradation

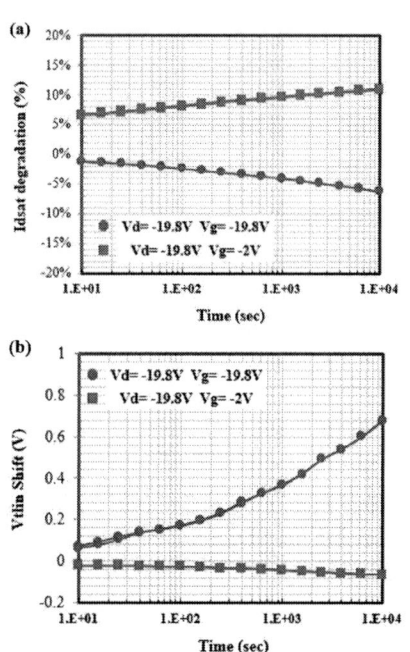

Figure 3: (a) I_{dsat} and (b) V_{tlin} versus HCI stress time.

CP technique is usually adapted to verify the types of defects such as number of interface traps (N_{it}) and number of oxide charge (N_{ot}). We also use normalized CP data to amplify oxide defects from right or left shift of the edges. The relationship between CP current (I_{cp}) and pulse base

voltage at various stress times was shown in Fig. 4. When under $V_g=1.1xV_{gop}$, we can see a significant increase in the saturation level from the CP curve, it was revealed that interface defects were dominant. From the normalization CP method, it can be seen that the CP curve shifted slightly to the right and then to the left, indicated that there was an influence of electrons injection at the initial stage of stress, and the effect of holes injection at the later under this stress condition. In general, the interface defects played a dominant role during the entire stress period. While $V_g=I_{gmax}$, no obvious change was showed in CP curve. It's known that the CP performance is the switching characteristic of the interface between inversion and accumulation. There was a recombination under $V_g=I_{gmax}$ because electrons injection attracted holes, so the CP current did not change significantly. The normalization CP curves can be amplified the changes of oxide layer defects because of slightly right shift of the edges of I_{cp}, which confirmed the effect of electrons injection.

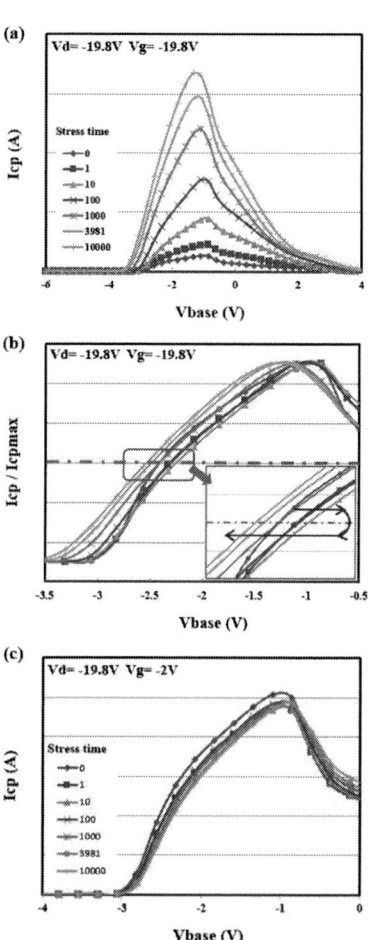

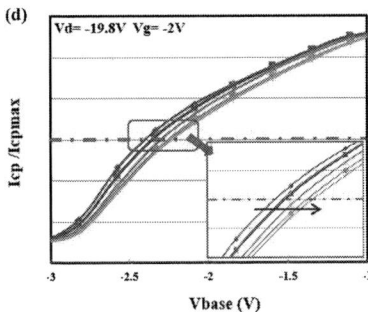

(d) Vd= -19.8V Vg= -2V

Icp /Icpmax

Vbase (V)

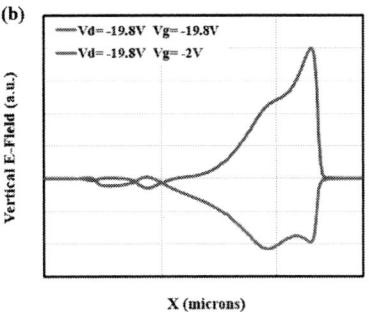

(b)
Vd= -19.8V Vg= -19.8V
Vd= -19.8V Vg= -2V

Vertical E-Field (a.u.)

X (microns)

Figure 4: CP currents as a function of pulse base voltage stressed at (a) V_g=-19.8V and (c) V_g=-2V at various stress time. The normalized CP currents were reflected at (b) V_g=-19.8V and (d) V_g=-2V

Figure 5: (a) TCAD simulation of impact ionization and (b) vertical electric field along the Si/SiO$_2$ interface.

We used TCAD simulation to explore the impact ionization and electric field under different V_g bias. From the simulation diagram shown in Fig. 5, it was clearly that the position of the maximum impact ionization gradually moved towards the drain side when the gate bias decreased from 19.8V to 2V. It was suspected that the voltage drop between gate and drain (ΔV_{gd}) became larger with V_g decreased; the drain region gradually took the dominant position, resulting in a stronger impact ionization region near the drain. We can also obtain relevant information about the vertical electric field through the TCAD simulation. It was implied that when V_g=I_{gmax}, the direction of the vertical electric field was from gate to the substrate along the Si/SiO$_2$ interface and it was conducive to the injection of electrons. While V_g=$1.1xV_{gop}$, the direction of the vertical electric field was from the substrate to gate, which was beneficial to holes injection. The simulation results of this TCAD were also consistent with our HCI and CP results.

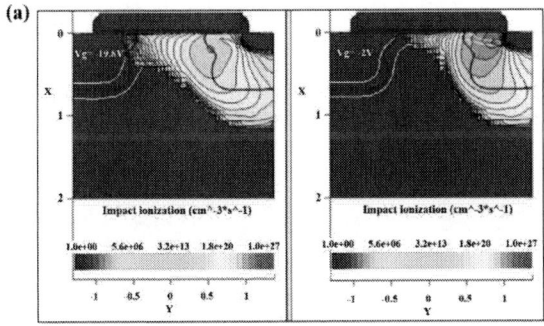

CONCLUSIONS

In this paper, from HCI degradation behavior associated with CP and TCAD simulation results, we find that both interface traps and positive holes trapping are observed under V_g=$1.1xV_{gop}$ during HCI stress. While the electrons trapping is more dominant at I_{gmax}. The vertical electric field is opposite when the gate bias decreased from 19.8V to 2V, which leads to different trap mechanism. At the same time, different impact ionization will also produce different numbers of interface states.

ACKNOWLEDGEMENTS

At the end of the paper, I would like to express my sincere thanks to reliability, special technology and TCAD groups for their generous support in this paper.

REFERENCES

[1] E. Riedlberger, R. Keller, H. Reisinger, W. Gustin, A. Spitzer, and M. Stecher. *IEEE IRPS*, p.175, 2010.

[2] J. Y. Jia et al. *Proceedings of the 20th IEEE International Symposium on the Physical and Failure Analysis of Integrated Circuits (IPFA)*, Suzhou, China, 2013, pp. 701-704.

[3] K. N. Quader, P. K. Ko, and C. Hu. *IEDM*, pp.5II-S14, December 1993.

[4] G. Groeseneken, H. E. Maes, N. Beltran, and R. F. De Keersmaecker. *IEEE Trans. Electron Devices*, vol ED-31, NO. 1, pp. 42-53, Jan. 1984.

[5] X. Lu, M. Wang, K. Sun and L. Lu. *IEEE International Reliability Physics Symposium*, Anaheim, CA, 2010, pp. 1040-1043.

[6] J. S. Brugler and P. G. Jespers. *IEEE Trans. Electron Devices*, vol. ED-16, p. 297, 1969.

[7] A. B. M. Elliot. *Solid-Stare EIecrron*, vol. 19, p. 241, 1976.

[8] L. Li, S. Zhou and K. Yang. *2021 China Semiconductor Technology International Conference (CSTIC)*, Shanghai, China, 2021, pp. 1-3.

AN EFFICIENT TOOL FOR GENERATING TEST PROGRAM TO SAVE MARGINAL FAIL CHIPS

Hanyan Chen[1]*

[1]Advantest, Shanghai 201210, China

Corresponding Author's Email: hanyan.chen@advantest.com

ABSTRACT

As for engineering verification, such as three-temperature test, corner test and wafer factory transfer, some experimental processing has been done on the chip, resulting in some parameters of the chip being slightly shifted compared with normal bin1 chip. As a result, some marginal fail would occur if tested by normal FT (Final Test) program. But these chips are good at EVB (Evaluation Board) verification. In this case, if need a test program to cover these chips, test engineers need to increase power voltage or slow frequency for the corresponding test item. In SMT7 (Smartest7) program, it may need to modify hundreds of level or timing spec (specification) values and generate a test program with a new, loosen spec value, which would take at least one day. This paper introduces a solution that can provide a simple excel format tool for input modification, a flexible operator configuration to deal with spec value and develop a new test program with the click of a few buttons. It's highly integrated, comprehensively improves program development efficiency and is easy to use.

Keywords—test program; Smartest7; spec value; V93000; tool.

INTRODUCTION

Background

During the engineering verification stage, such as three-temperature test, corner test and wafer factory process adjustment, some experimental processing has been done on the chip, resulting in some parameters of the chip being slightly shifted compared with normal bin1 chip. As a result, some marginal fail would occur if tested by normal FT program. But these chips performed well at EVB verification. In this case, if a SMT7 test program is required to cover these chips, test engineers need to incorporate a slight change by increasing power voltage or slowing frequency for the test item. The traditional method requires test engineers to locate each failed test suite, select the corresponding level spec and timing spec value, edit them in the level/timing setup editor, download to the V93000 hardware and then retest to pass all test items to ensure no quality issue, as shown in figure 1.

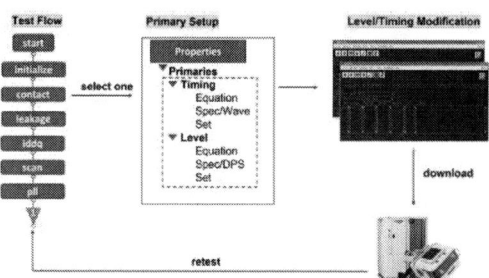

Figure 1: Spec Value Modification

However, as a complete test program contains hundreds of test suites, test engineers repeat this process to modify hundreds of levels or timing specs and make a new test program with newly loosened specs for these chips. In general, this process takes at least one day. If more marginal items fail, more time is needed.

Legacy Solution Disadvantages

Before the current tool was developed, engineers used to insert a spec change code at the beginning of each failed test method and restore it at end. This method uses PSLV (Power Supply Level) firmware commands to get the existing level settings and increase the voltage slightly with a DC drift at each run, such as 200mV. While complex interface operations can be replaced by code control in this solution, it still takes time to insert the code into each failed test method and rerun each failed test suite. An obvious disadvantage is that it can only modify the level, not the timing, and it does not support mathematical operations, only simply increases the voltage. If the result of one run does not pass, it needs to be repeated several times as shown in figure 2.

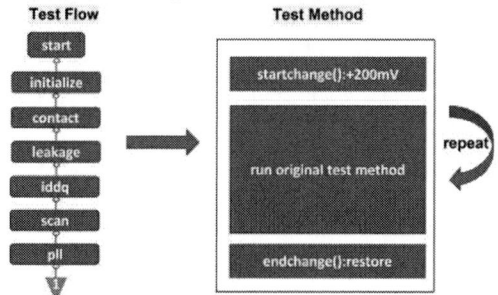

Figure 2: Insert Spec Change Code

979-8-3503-1101-3/23 $31.00 © 2023 IEEE

Current Solution Overview

For this situation, we designed a solution to improve the efficiency of the program update. First, use a tool written in python to export all level and timing spec as a reference. Then, enter all the spec values that need to be edited in a CSV (Comma-Separated Values) file, including the operations that need to be performed to get the desired values. Then, when test engineers run the whole test flow, the first test suite "initialize" of the test flow calls the "SPECSTRUCT.cpp" method to update all spec values which are applied to the corresponding test suite. The detailed process is shown in figure 3. As a result, the desired spec value will take effect and the marginal failure item will pass. This solution has some advantages:

- This tool designs an input interface based on the CSV file. Test engineers don't need to be familiar with the SMT7 program structure and can easily start with the program at-hand.
- This tool is developed in C++ which is the programming language used by the V93000 SMT7 eclipse platform. Therefore, it can be well integrated into the original SMT7 code without the need to install other programming languages.
- Less manual work while the test program updates, which reduces the risk of a change error occurring.

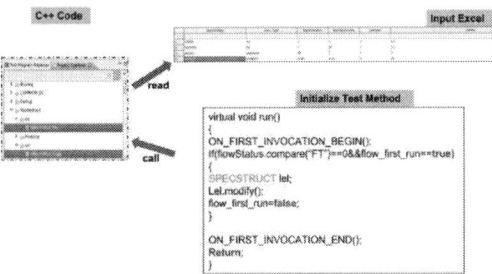

Figure 3: Overview Process

Input File Structure Design

First, it can provide a simple GUI (Graphical User Interface) that allows the user to enter all marginal fail items at once. Then, test engineers can edit expressions consisting of mathematical operators and spec values, such as adding a certain offset or multiplying by N. Meanwhile, the input file could be read and processed by C++ code. Considering all the above requirements, we chose CSV as the input file format.

In the V93000 SMT7 test program, each test suite must have two primary setups: level and timing. For the level and single-port timing setup, we can retrieve the spec values by using the function member "getSpecValue()" of "LEVEL_SPEC/TIMING SPEC" API (Application Programming Interface)[1]. In this case, API only needs the equation set number, spec set number and spec

variable to obtain a specific value. However, the "SPECIFICATION()" API for multi-port timing requires the spec variable, port name and spec name as input parameters. Therefore, the input CSV is designed to include the columns for spec variable, spec type, equation/port-name, spec/spec-name, operator and offset.

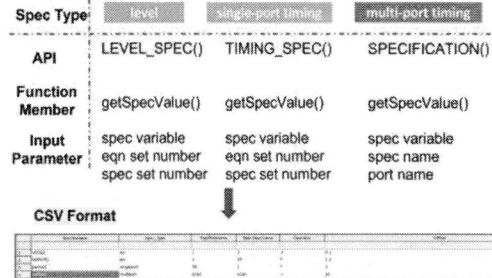

Figure 4: Similarities and Differences between APIs

Firmware Task Selection

The key part of this tool is the command that is selected to change the hardware when given a modified value. For the V93000 systems, there are two types of commands: test method API and firmware commands.

Test method API is a variety of customized functions that the smartest software supplies, which is an extension of the standard C++ programming language to control the V93000 tester hardware and calculate data in test method programs based on UTM (Universal Test Method)[2]. "LEVEL_SPEC" or "TIMING_SPEC" API both offer a "change()" member function to change the value of the specified spec.

Firmware commands are used for low level control of the test system with the "FW_TASK" API. When you use the V93000 from the user interface, the interaction between the software and the hardware operates transparently. The setup data and the test data that you enter in the software windows is downloaded to the hardware and you see the test results either as result values or as an error message in the report window. What actually happens is that a series of firmware commands are converted by a Unix process into specific actions in the hardware[3].

Therefore, test method API is combined and wrapped by multiple firmware commands, and we can use firmware commands directly to perform the functions. We use SVLR (Spec Values at Runtime) to set a base variable to the supplied value and recalculates the timing or levels[4].

For this solution, we chose firmware commands instead of test method API because API needs to act on single-port spec or multi-port spec through "LEVEL_SPEC/TIMING_SPEC" or "SPECIFICATION()" APIs respectively. SVLR can act on both types of specifications directly through the "spec_type" input parameter. The most important reason is that we may change the primary set up and restore it due to test conditions in other test suites. It is important to note that specifications changed at the beginning of the test program cannot be restored throughout the program run. "LEVEL_SPEC" and "TIMING_SPEC" test method APIs cache values in order to enable the "restore()" functionality. This caching is not available on the firmware level, so the "restore()" function in other test suites will not affect the change of beginning. Example is shown in figure 5.

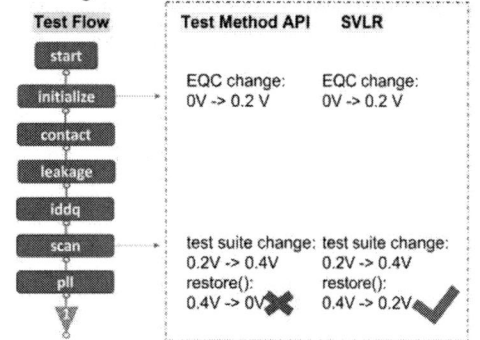

Figure 5: "Restore()" Difference between Test Method API and SVLR

Test Program Structure

"SPECSTRUCT.CPP" and "SPECSTRUCT.H" were created to read the contents of the input excel and use SVLR to modify spec values.

In "SPECSTRUCT.H", we have defined three map variables: level, single-port timing and multi-port timing. The key of these map variables is string, and the value is the map object "map<string,specProperties>specVar" which is used to manage the variables in each line of the CSV. The key of "specVar" is the spec variable that needs to be modified, and the value "specProperties" is a structure that defines the variables from CSV, which includes operator, equation number, spec number, port name, spec name, offset value and "process()" function. "process()" is to deal with operator to get modified value.

In "SPECSTRUCT.CPP", there are two functions: "SPECSTRUCT()" and "modify()". "SPECSTRUCT()" is used to get the content of each column in the CSV, assign it to the corresponding variable, and then put the variable into a different map variable according to the spec type, as shown at the top of figure 6. In "modify()", first, it retrieves the original spec value and uses "process()" function to get new spec value. Then the map variables of level, single-port timing and multi-port timing are traversed respectively. In the case of level or single-port timing, only the equation number, spec number and new value need to be extracted from the map variables and then applied with SVLR. In the case of multi-port timing, SVLR needs to be applied with spec name and the combination of spec variable and port name, as shown at

the bottom of figure 6. After traversal, all the changes are downloaded to the hardware at once with "FLUSH()" at the end of "modify()".

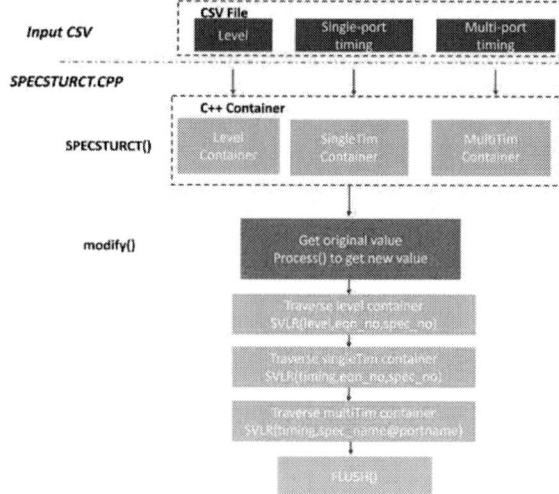

Figure 6: Test Program Flow Chart

Application Scenarios

"SPECSTRUCT.CPP" is called at the first test suite. During the test program run, spec value modified or not is controlled by global variables "flow_first_run", as figure 7 shows. After modification, it can be seen in the hardware view that the levels or timings applied to the DUT have been changed and the new test program was created automatically.

In summary, with this tool, test engineers do not need to be familiar with the SMT7 program structure and can easily start with the program at-hand. It has been widely used, not only for one kind of chips but also unlimited to the V93000 SMT version. Hundreds or even thousands of test suites' new specs can be developed at back by simply clicking a few buttons which saves engineers a lot of time.

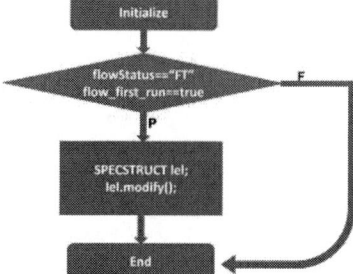

Figure 7: Initialize Flow Chart

ACKNOWLEDGEMENTS

I would like to express my gratitude to all those who helped me in the development of this tool. My deepest gratitude goes first to Beryl Xu, my manager, for her constant encouragement and guidance. Second, I gratefully acknowledge the help of Nick Song and Steve Xie, who has offered me valuable suggestions during the design and implementation phases of this tool.

REFERENCES

[1] ADVANTEST. *Test Documentation Center LEVEL_SPEC*, Rev.7.6.4, topic. 29495.

[2] ADVANTEST. *Test Documentation Center Test Method Program Introduction*, Rev.7.6.5, topic. 71342.

[3] ADVANTEST. *Test Documentation Center Command Reference*, Rev.7.6.5, topic. 103977.

[4] ADVANTEST. *Test Documentation Center SVLR*, Rev.7.6.5, topic. 98764.

RESEARCH ON TDDB PHYSICAL MECHANISM OF 28HKMG MOSFET

Ting Wan [1]*, Hao Jiang [1], Yueqin Zhu [1], and Ke Zhou [1]

[1] Shanghai Huali Integrated Circuit Corporation, Shanghai, China

Corresponding Author's Email: wanting@hlmc.cn

ABSTRACT

TDDB (Time dependent dielectric breakdown) issue has been one of the most important reliability issues as the gate dielectrics are changed to HK/IL (HfO_2/ Inter layer SiO_2) stack for solving intolerable tunnel leakage problems. In this paper, the physical explanation of TDDB current jump was interpreted through the analysis of trap generation in 28nm HKMG technology. Obviously SILC (Stress induced leakage current) increase on the I-T curve of NMOSFET was found and there existed two current jump phenomenon on NMOS, while SILC effect on PMOSFET was not obvious and only one current jump was observed. The first current jump had a strong correction to shallow traps generated in HK layer during TDDB test and the generation of deep traps result in the second current jump. The deep trap generation was responsible for the permanent damage and the formation of percolation path during the gate oxide degradation. This study provides guidance for TDDB lifetime prediction.

INTRODUCTION

Metal gate technology was proposed to solve the compatibility issue between high-k dielectric and conventional polysilicon gate. The use of metal gate instead of polysilicon gate can eliminate the remote Coulomb's scattering effect, effectively suppress the decline of channel carrier mobility caused by surface soft phonon scattering in HK dielectric, and possibly solve the non-modulation problem of threshold voltage caused by Fermi level pinning effect. Compared with polysilicon gate, HKMG has higher electron and hole mobility, appropriate threshold voltage, and higher driving current performance. In order to meet people's needs of high performance, low power consumption and low cost electronic products; the integration of integrated circuits is constantly improved, which makes the feature size of transistor get smaller and smaller. To overcome the problem of gate leakage current, the HK dielectric of HfO_2 becomes the most suitable alternative materials. However, the low crystallization temperature of HfO_2 is against to the high temperature treatment in the subsequent process. The direct contact interface between HfO_2 and the Si substrate or electrode is poor. In addition, it is easy to produce defects such as high concentration of oxygen vacancies during the HfO_2 film deposition process. These material characteristics cause serious reliability problems on MOSFET [1-2].

With the improvement of the manufacturing process, the external defects of gate oxide introduced by the process are gradually reduced. The intrinsic breakdown of gate oxide has become the focus of the oxide reliability research. TDDB has been widely concerned by the industry and academia. SILC caused during the oxide degradation process has become the main reliability problem because the addition of HK layer has introduced a large number of oxygen vacancy defects in the gate stack [3-5]. In this paper, the TDDB of MOSFET with IL/HK (HfO_2) as gate oxide dielectric is tested, the relationship between the current jump of MOSFET gate and the shallow and deep traps generated in the HK dielectric layer is studied, and the judgment criteria for the TDDB failure of IL/HK gate oxide is proposed.

EXPERIMENT

The N- and P- MOSFET with Hf-based gate dielectric were fabricated using HKMG technology with N28 generation. For TDDB testing, the measure time to failure is extracted, based on devices stressed on the gate terminal while source/drain/bulk terminals were grounded in the range from 85℃ to 150℃.

RESULTS AND DISCUSSION

N/P MOSFET TDDB breakdown curve analysis

Gate oxide breakdown mode can be divided into HBD (Hard Breakdown) and SBD (Soft Breakdown) according to the magnitude of a sudden gate current increase. Gate oxide breakdown is a very important aspect of device reliability. How to define and distinguish SBD and HBD is very important. It is generally believed that SBD is defined as the magnitude of a sudden current increase during breakdown is small ($Ig(t) \geq 2\sim5 Ig(t-1)$), which usually occurs when the first conductive path is formed; When the current increases by more than 1 order of magnitude, that is, $Ig(t) \geq 10 Ig(t-1)$, it is defined as HBD, which usually occurs when multiple conductive paths are aggregated in parallel or a single conductive path with strong conductivity formed; SILC phenomenon usually occurs in the device where the gate oxide is formed by HK dielectric. The gate current gradually increases for a period of time before HBD or SBD occurs and the SILC phenomenon is considered as a precursor of SBD and HBD [6]. The device function will not completely lose after SBD. However, the device and even the entire circuit will fail after HBD. SBD and HBD are randomly generated in the entire gate oxide area. The difference lies in the size of the breakdown point and the conductivity of the conductive path. The aggregation of multiple SBDs will lead to the occurrence of HBD. After each breakdown mode occurs, it will maintain this state for hundreds or even thousands of seconds until the next breakdown mode occurs.

The accuracy of device life prediction is related to the selection of breakdown criteria. Dielectric breakdown mainly caused by HBD when the gate oxide is thick. With the dielectric becomes thinner, especially after the introduction of HK dielectrics, multiple breakdowns will occur simultaneously. Figure 1 is the breakdown curves of the NMOSFET (a) and PMOSFET (b) studied in this paper. Devices were stressed at the same stress voltage and measured at different SILC voltages. It can be seen from Figure 1 (a) that the NMOSFET exists obviously SILC phenomenon: the gate leakage current under low electric field increases significantly after the MOS structure was subjected to a high electric field stress. The breakdown curve of NMOSFET consists of SILC, SBD and HBD. Interestingly, the larger the SILC voltage, the more obvious the first current jump (SBD) occurs. However, the SILC phenomenon of the PMOSFET in Figure 1(b) was not obvious and the breakdown curve was only composed of HBD. There was no phenomenon of two current jumps on PMOSFETT measured under different SILC voltages.

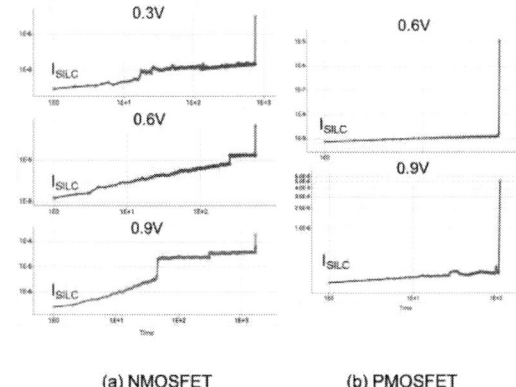

Figure 1: TDDB I-T curves of N/P MOSFET under various SILC voltages

Theoretical analysis of IL/HK dielectric breakdown

For traditional SiO_2 dielectric, SBD/HBD is caused by the formation of a conductive path between the cathode and anode when the trap density in the gate oxide layer reaches a critical value. A large current will flow through the oxide after the conductive path is formed, resulting in a thermal effect. The local temperature in the oxide will rise, prompting more traps and conductive paths to form and gather with each other. Finally, the huge current will cause the entire device to be completely destroyed [7]. For HK/IL dielectric, shallow traps and deep traps will be formed. The oxygen vacancy defects formed near the conduction band of the HK layer are shallow traps, and the defects near the center of the forbidden band of the HK layer and IL layer are Deep traps [8]. In the SILC region of

Figure 1(a), the gradual increase in NMOSFET current was caused by the accumulation of electronic defects in the HK layer. During the TDDB Stress process, the HK layer simultaneously generated deep traps and shallow traps [9]. Figure 2(a) is the energy band diagram of NMOSFET at different SILC voltages. When $V_{SILC}=0.9V$, substrate injected electrons were tunneling through IL layer into HK layer, easily interacted with HK shallow traps, which form a percolation path and lead to the first current jump (SBD). The accumulation of HK deep traps and IL traps form a complete conductive pathway, resulting in the second current jump (HBD). However, when $V_{SILC}=0.3V$, substrate injection electrons were close to the HK deep traps but far from shallow traps. Hence two current jump phenomenon was not obviously when V_{SILC} was low. For PMOSFET, as shown in Figure 2(b), substrate injected holes were hard to tunnel through IL into HK layer, resulting in no SILC effect. Since substrate injection holes were close to the HK deep traps, only one current jump was observed.

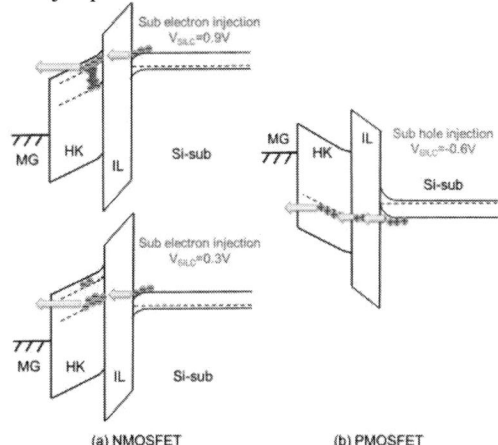

Figure 2: shallow trap and deep trap under various SILC voltages on the HK/IL band diagrams

Analysis of MOSFET TDDB breakdown curves at different temperatures

NMOSFET TDDB measurements were performed at the same stress and SILC voltage but different temperatures (85°C and 150°C). The typical I-T curves are shown in Figure 3. It can be seen from the figure that when the temperature was low (85°C), SILC phenomenon was inconspicuous, indicating that the trap density formed by the HK layer was low. With the increase of temperature (150°C), the SILC effect was obvious and there were two current jumps (SBD and HBD) at high temperature. However, only one current jump was observed at low temperature. Since SBD is mainly caused by shallow traps and HBD is mainly triggered by deep traps. We therefore conclude that the concentration of shallow traps generated at high temperature was higher while the concentration of deep traps generated at low temperature was higher.

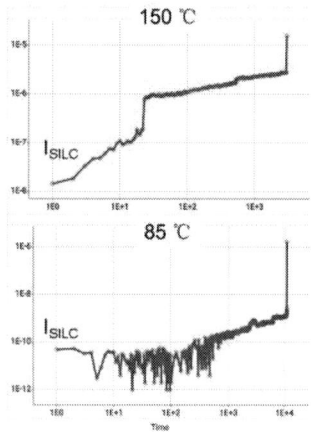

Figure 3: The temperature effect on TDDB I-T curves

Influence of breakdown criteria on the TDDB lifetime prediction

The accuracy of device TDDB lifetime prediction is related to the selection of the breakdown criteria. It is easy to cause over-shoot or over-optimistic if the breakdown criteria are not selected properly. Figure 4 is the Weibull distribution diagram taking the time of the first gate current jump and the second current jump as the TDDB TTF (Time to Failure) when the SILC voltage of the NMOSFET was 0.9V. The TTF corresponding to the first current jump was shorter than the second current jump. Hence the derived TDDB lifetime will be shorter than that of the second current jump. Therefore, the breakdown judgment criteria are particularly important. The shallow traps generated in HK form an incomplete current penetration path, which was a non-permanent damage. Since devices works in alternating current (AC) mode, some trapped charges will get de-trapped out of the gate oxide. The formation of current penetration path is delayed because damage formed by the shallow trap is recovered [10]. Therefore, it is more reasonable to select the time corresponding to the second jump current as the TTF of NMOSFET TDDB test.

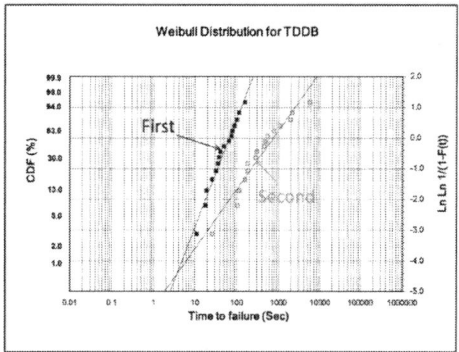

Figure 4: Weibull distribution of different breakdown determination criteria (first current jump and second current jump)

CONCLUSIONS

This paper studied the TDDB characteristics of MOSFETs with IL/HK (HfO2) as the gate oxide. Our measurements indicate that the I-T curve of the NMOSFET will jump twice at a lower SILC voltage. The first current jump was due to the leakage paths formed by the shallow traps. The second current jump was caused by the deep trap density in the gate dielectric layer reaching a critical value, which made a conductive path between the cathode and the anode. By comparing the I-T curves at different TDDB test temperatures, we found that more shallow traps formed at higher temperature; meanwhile the SILC effect was obvious. Based on the actual situation, this paper proposes that it is more reasonable to take the time corresponding to the second current jump as TTF for an accurate lifetime prediction of HK/IL oxides.

ACKNOWLEDGEMENTS

I would like to appreciate senior engineer Hao Jiang, Yueqin Zhu, Ke Zhou for their support and advice on this experiment.

REFERENCES

[1] A. Kerber, E. Cartier. *Device and Materials Reliability*, IEEE Transactions on, vol. 9, no. 2, pp. 147-162, June 2009.

[2] E. Cartier et al. *in Electron Devices Meeting (IEDM)*, 2011 IEEE International, Dec. 2011, pp. 18.4.1-18.4.4.

[3] S. Mukhopadhyay et al. *International Reliability Physics Symposium (IRPS)*, 2014 IEEE International, June. 2014, pp. GD.3.1-GD.3.11.

[4] J. Yang et al. *International Reliability Physics Symposium (IRPS)*, 2012 IEEE International, April 2012, pp. 5D.4.1-5D.4.7.

[5] K. Joshi et al. *International Reliability Physics Symposium (IRPS)*, 2013 IEEE International, 2013, pp. 4C.2.1-4C.2.10.

[6] Ernest Y. Wu. *IEEE transactions on electron devices*, vol. 66, 2019, pp. 4535-4545.

[7] Maserjian J and Zamani N. *JAP*, 1982, pp. 53:559.

[8] C. H. Yang, S. C. Chen, Y. S. Tsai et al. *International Reliability Physics Symposium*, 2016 IEEE International, April. 2016, pp. 7A.1.1-7A.1.6.

[9] C. H. Yang, P. S. Chien, Y. S. Choet al. *International Reliability Physics Symposium*, 2022 IEEE International, 2022, pp. P28-1- P28-5.

[10] C. H. Yang, S. C. Chen, Y. S. Tsai et al. International Reliability Physics Symposium, 2017 IEEE International April. 2017, pp. 3C.4.1-3C.4.6.

RESERVOIR EFFECT STUDY ON ELECTRO-MIGRATION BEHAVIOR OF ALCU INTERCONNECTS

Jizhou Li[*]*, Kitty Wang and Weihai Fan*

Semiconductor Manufacturing International (Shanghai) Corp. No.18 Zhangjiang Road

Pudong New Area Shanghai, China

*Corresponding Author's Email: JiZhou_Li@smics.com

ABSTRACT

Reliability risk is considered as the bottleneck in high performance ICs, especially for automotive application. One of the major concerns for reliability comes from the lack of enough EM (electro-migration) resistance. DFR (Design for reliability) is an effective way to approach better reliability performance. In this paper, structures with different reservoir layouts are used in EM test to demonstrate the mechanism of reservoir. Longer EM TTF (time-to-failure) can be achieved by adding reservoir part to provide metal ions to postpone the time that resistance increase reaches failure criterion, and TTF improvement is related to the area of reservoir.

INTRODUCTION

Since the increasing demands from commercial, industry and automotive fields for greater intelligence, ICs are needed with higher spec of both performance and reliability. From the view of reliability, Electro-migration of metallization interconnects is one of the major concerns. For analog ICs, mature nodes such as 0.18um and 0.15um are more beneficial options based on the cost and performance. AlCu metallization interconnects is widely used in these nodes and has been developed for many years. Therefore, there is little space for EM resistance further improvement through process refining. Design for reliability (DFR) is another way to achieve better reliability performance, which mainly relies on two fundamental theories: Blech effect and reservoir effect. For the reservoir effect, metal ions migrate from reservoir area to fill the void formed by EM so as to prolong the time when the resistance of metallization interconnects increase to failure criterion [1-5].

In this paper, we focus on the reservoir effect for AlCu metallization interconnects. Several kinds of test structures are designed with different reservoir layouts. EM tests are carried out for these test structures to find the relationship between reservoir design and EM performance. Significant improvements can be achieved with reservoir layout and the TTF increases as the area of reservoir becomes bigger and bigger. Obviously different distributions of TTF (time-to-fail) and R-t (resistance shift vs. time) curves divides test samples into two groups. The results of FA (failure analysis) help us reveal the mechanism of reservoir effect. We will discuss this in detail in following chapters.

EXPERIMENT

The EM test structures are fabricated by AlCu interconnects. We build the test structures with two metal layers (M1 and M2) and one via layer(Via1), and choose downstream for this experiment. Five kinds of layouts are used in our design as shown in Figure 1. The test structures have different ways to obtain reservoir areas: a) normal structure as baseline; b) extension along line end (length direction); c) extension along enclosure (width direction); d) extension for both line end and enclosure; e) rectangle reservoir with trapezoid connection. The experiments are carried out using package level equipments with stress temperature of 250 °C and current density of $0.80\,\mathrm{MA/cm^2}$ in Qualitau Mira. 20% increase of resistance shift is defined as EM failure criterion. A minimum of 15 valid time-to-fail data points are required for each test.

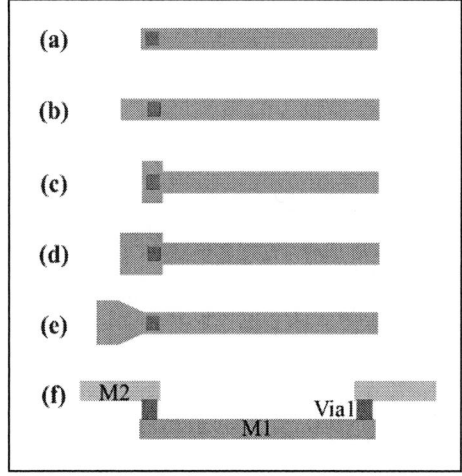

Figure 1: EM test structure layouts: (a)-(e) different reservoir design, (f) cross section.

RESULTS AND DISCUSSION

A. EM test data analysis

The TTF distributions of EM test for all test structures are shown in Figure 2. For the test structure (a), the TTF distribution shows very good linearity as expected of typical EM test result, and the small deviation implies the uniformity of samples. For the test structure (b) and (c), obvious gap between the short TTF group (mode A) and long TTF group (mode B) can be found, which means EM failure for both structure (b) and (c) can be attributed to two distinct failure mechanisms simultaneously. The mode A groups of structure (b) and (c) have similar TTFs compared with that of the test structure (a) under the same stress condition. Based on this phenomenon, the location of EM induced void for the short TTF groups of structure (b) and (c) are expected to be consistent with that of the test structure (a). Furthermore, the mode B groups of structure (b) and (c) have longer TTFs than those of mode

A groups, which may result from the design of reservoir. For test structure (d) and (e), the TTF distributions also show distinct bi-modal profile. The mode A groups of structure (b)-(e) have comparable TTFs, while mode B groups of structure (d) and (e) show significantly better EM resistance compared with mode A groups, and the gaps between these two groups is even larger than those of structure (b) and (c). All structures with reservoir consist of two different parts of TTF distribution, one part is comparable with structure (a) without reservoir and the other part shows obvious improvement for EM resistance. Hence the reservoir effect does not occurs in all the samples in our experiments.

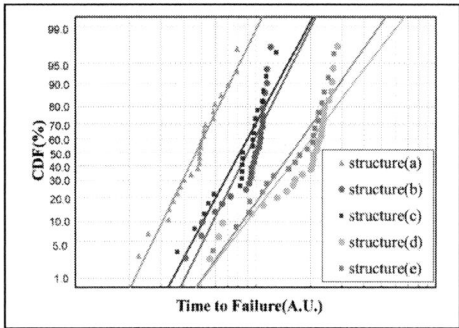

Figure 2: TTF distributions of EM test for structure(a)-(e).

Figure 3. shows the R-t curves of all tests. For structures with reservoir, mode A groups and mode B groups have different trend during EM testing. The resistance increases gradually after EM stress in mode A group, while the resistance of mode B group keeps unchanged at first, then followed by suddenly increase. Considering that all the samples are from the same wafer, there is no difference in process, and the metal ion migration velocity is identical under the same stress condition. The different R-t curves of mode A and mode B may be induced by different locations of void formation. From previous study on EM behavior, void nucleating and growing in or under via usually cause resistance increase rapidly, meanwhile, resistance increases gradually when void formation in metal line. These conclusions can fully agree with our data. We will confirm it through failure analysis technique in next chapter. It should be noted that there are also two kinds of R-t curves in structure (a), one shows gradual increase and another increases sharply.

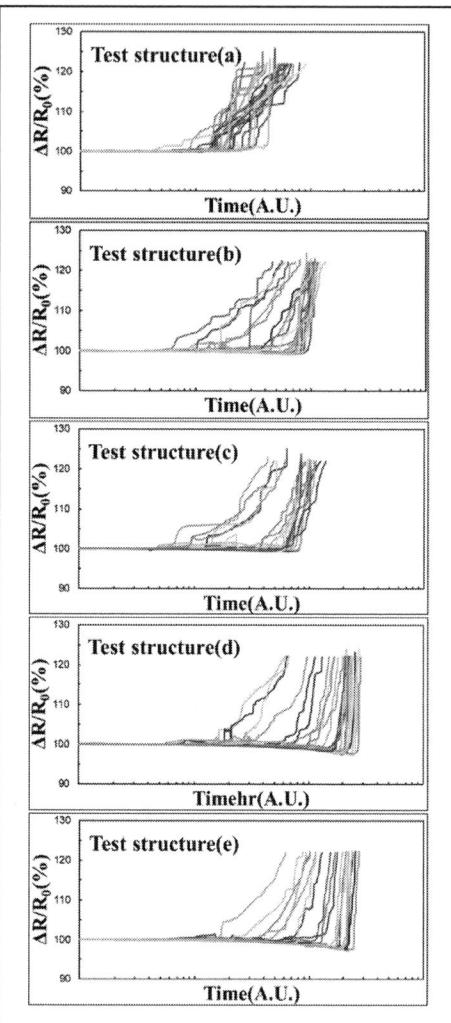

Figure 3: R-t curves of EM test for structure(a)-(e).

B. Failure analysis and discussion

By dividing samples into mode A group and mode B group after EM experiments, failure analysis is performed to reveal the root cause for the different performance of all test structures in our experiments. Through FIB and SEM, the voids causing EM failure are shown in Figure 4. For mode A groups of structure (b)-(e), all voids are located in metal line without extending to the area under via1. Meanwhile, for mode B groups of structure (b)-(e), all voids are located under via1, and metal ions in each reservoir region have been nearly depleted. Based on these failure analysis results, the differences of TTF and R-t curves between mode A groups and mode B groups can be explained. When void is nucleated and grows away from via, metal ions in reservoir can not migrate to refill the void timely. When the void is big enough to cause 20% resistance increase, which can be called as critical size, EM failure occurs. TTF of this mode is just the time for metal ion depletion in metal line to form void as large as critical size. As void grows, resistance increases gradually. When void nucleates and grows under via, metal ions in reservoir can persistently migrate to refill the void until metal ion of reservoir area is totally depleted. TTF of this mode is the total time needed to deplete metal ion in reservoir region and to form void with critical size. Thus, TTFs of mode B groups are much more longer than TTFs of mode A groups. Before emptying reservoir area, resistance is stable and R-t curve keeps fixed. Since conductive path under via is significantly smaller than that in metal line, critical size under via is therefore smaller than in metal line. After depletion of reservoir region, resistance increases dramatically as void grows under via, which leading to steep rise of R-t curve.

As mentioned above, there are also two kinds of R-t curves in structure (a), one shows gradual increase and another sharp increases. In Figure 4 for structure (a), void away from via (related to gradual increase R-t curve) and void under via (related to increases sharply R-t curve) can be found at the same time. Because structure (a) has no reservoir area, there is no obvious gap between TTFs of two groups.

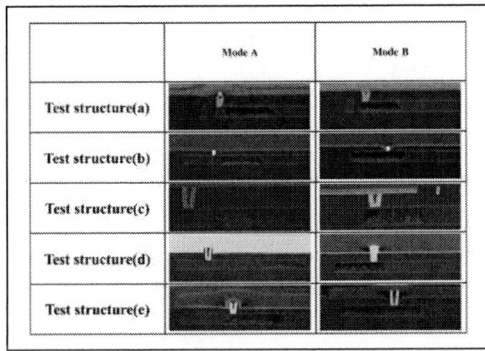

Figure 4: Failure analysis for mode A and mode B groups (for mode A of structure (e), void initially nucleates away from via since reservoir keeps intact).

Based on the discussion above, layout with reservoir can improve EM TTF through providing additional metal ions to be depleted to postpone the time for resistance increase to failure criterion. When different test structures are stressed in same condition, the velocities of metal ion migrating are similar. Therefore, the time to deplete metal ion in reservoir region is related to the area of reservoir. The mode B groups data of all test structures are extracted

to calculate the MTTFs (median-time-to-failure) without mode A groups data. MTTFs are linear with reservoir areas as shown in Figure 5.

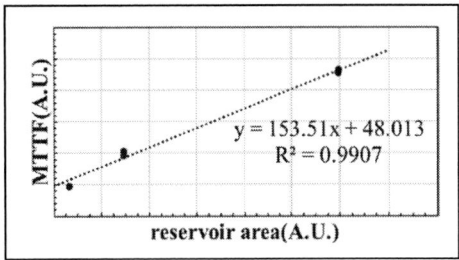

Figure 5: Relationship between reservoir area and MTTF of mode B group samples.

CONCLUSION

Test structures with different reservoir layouts are used in EM test to demonstrate the mechanism of reservoir effect. For the TTF distributions of structures with reservoir, there are two distinct groups with obvious gap between them, which means EM failure can be attributed to different failure mechanisms simultaneously. Based on R-t curves and failure analysis, voids located in metal line without extending to the area under via1 are responsible for short TTF group, meanwhile, voids are located under via1 and metal ion in every reservoir have been nearly depleted for long TTF group. Reservoir can greatly improve EM TTF by providing metal ions to postpone the time that resistance increase reaches failure criterion only when void nucleates near reservoir. Moreover, TTF improvement is related to the area of reservoir.

REFERENCES

[1] Jens Lienig. *Proceedings of the 2013 ACM International Symposium on International Symposium on Physical Design*, Stateline, March 2013, pp. 33-40.

[2] H. V. Nguyen, C. Salm, R. Wenzel, A. J. Mouthaan, and F. G. Kuper. *Microelectronics Reliability,* vol. 42, 2002, pp. 1421-1425.

[3] Wei Shao, A.V. Vairagar, Chih-Hang Tung, Ze-Liang Xie, Ahila Krishnamoorthy, and S.G. Mhaisalkar. *Surface and Coatings Technology*, Vol. 198, 2005, pp. 257-261.

[4] Michael J. Dion. *Microelectronics Reliability*, Vol.41, 2001, pp.805-814.

[5] V. Girault, F. Terrier, and D. Ney. *Microelectronics Reliability*, Vol.48, 2008, pp.219-224.

A UNIVERSAL AUTO TEST PROGRAM GENERATION ON ADVANTEST V93000 ATE PLATFORM

Xin Song[1*], Yefang Wang[1], Hanyan Chen[1]

[1]Advantest (China) Co. Ltd, Shanghai 201203, China

*Corresponding Author's Email: Nick.Song@advantest.com

ABSTRACT

The semiconductor industry is in a period of rapid growth. 5G, IOT and HPC/AI need to keep up with the changing request of the consumer market. As a core component, IC accelerates the speed of upgrading and iteration.

ATE test is an important part of production. The test program development is the core of the whole process. How to improve the efficiency of test program development is the concern of many Fables at present. Therefore, ATPG (Auto Test Program Generation) has been urgent request of many febless.

Based on the principle of ATPG, this paper introduces the relevant requirements analysis, framework design and result. Based on the V93000 platform of Advantest, the standard specifications of ATPG are shown.

Keywords—ATPG, ATE, V93000

INTRODUCTION

ATPG Overall

ATPG (Auto Test Program Generation) is a way to improve test program development efficiency. Test program is divided into "Test Flow", "Pin Config", "Level Set" and "Timing Set" and "Test Method" on V93000 platform.

TABLE I.   TEST PROGRAM STRUCTURE

Program Part	Description
Test Flow	Test suite execution sequence
Pin Config	Test resource assignment on ATE
Level Set	DC set of Power and I/O
Timing Set	AC set of I/O
Test Method	Test resource control

Actually, "Test Method" is code base and too complex and flexible. That means it is difficult to auto generate. The other four part could implement with ATPG solution.

The ATPG solution is a set of python scripts that correspond to four V93000 program part. Each script has input file and output file interface. User obtains the generated V93000 program module by configuring the input file and executing the script.

ATPG Flow

The ATPG test flow solution could generate ATE test flow on Smartest 7 environment on V93000 platform. User needs a test flow configuration file which could describe test flow information. The information could be divided 3 parts.

- Flow Information: Define flow name, global variables, and various files (pin, level, timing and pattern) to be loaded.
- Testsuite Information: Define information about testsuite in flow. Such as suite name, suite sequence, level and timing used by suite.
- Testmethod Information: Define relevant parameters of test suites.

Based on this information, we can complete the information collection of ATE test flow.

ATPG Flow -- Framework

This ATPG solution design a framework to achieve input file reading and storage, flow information process and flow file generation. Input file is flow configuration file and could be red by ATPG solution. Then the storage information will be used to sperate some items as V93000 flow file structure (Table II). After reassembly, the generated test flow can become recognized by V93000.

TABLE II.   V93000 FLOW STRUCTURE

Flow Part	Description
Flags	Global variable definition
Testmethodparameters	Parameter for test
Testmethods	Test method file to be called
Testsuites	Level/Timing and local flag
Testflow	Flow sequence
Binning	Hardware Bin setting
Context	Other files associated with Flow

The specific calling relationship is shown in figure 1.

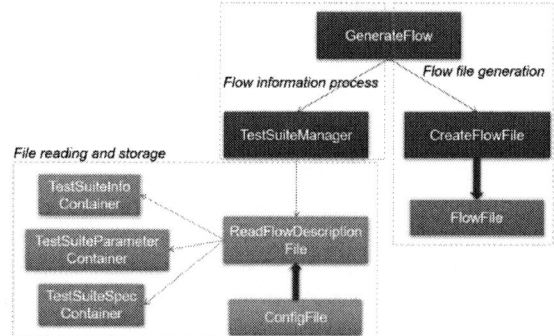

Figure 1: ATE Flow Framework

ATPG Flow – Interface

The ATPG solution's interface is also very simple. User just specify the config file path and click "run" button then get the generated flow file.

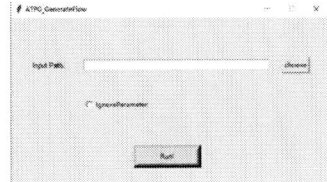

Figure 2: ATPG Flow solution interface

ATPG Pin Config

ATE Pin Config generation costs at least one day manually, especially for AI/HPC device, the process will cost more time due to massive channels. ATPG solution can improve setup efficiency and has some advantages as below.

- ATPG Pin Config solution can help generate pin setup files which is suitable for Smartest 7 of Advantest [1].
- The solution processes digital channels, analog channels, DPS channels, utility channels, groups definitions.
- This solution processes multisite pin files generation and site deleting.

ATE Pin Config -- Framework

Framework works like building blocks for users to get V93000 pin config files.

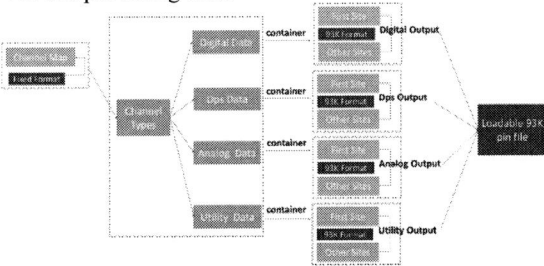

Figure 3: ATE Pattern Iteration Process

As expected, we get LB vendor channel map firstly. ATPG solution analyzes channel map and process key information and stores it in different data structures. Combined with the V93000 unique string, preserved channel information in data structures will be translated into V93000 pin file by ATPG solution.

ATE Pin Config -- Channel Map Description Table

LB vendors always provide channel map file for resource definitions of tester channel. But the format is not uniform. So, we design a common simple template which could be accepted by different PCB vendors. (Figure 4)

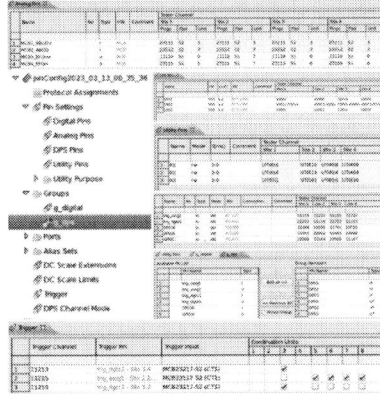

Figure 4: Channel Map File Template

User must follow some filling rules which could be realized by ATPG solution. Such as Digital pin should start with "D", Power pin should include "P" and so on.

ATE Pin Config – Data Storage & Generation

The specific V93000 format of each channel in detail is shown in Figure 5. There are specified channel information like Dps, analog, digital, trigger pins in Smartest 7.

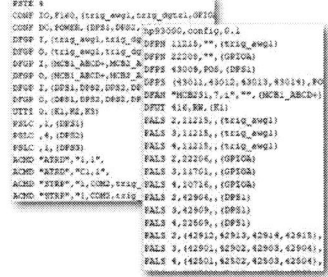

Figure 5: Output Loadable Example

The above information in Figure 5 is actually stored in Smartest as text information shown below in Figure 6, which can be generated by the ATPG solution to achieve rapid offline generation and modified by the ATPG solution to match the ATE resources.

Figure 6: Output Text View Example

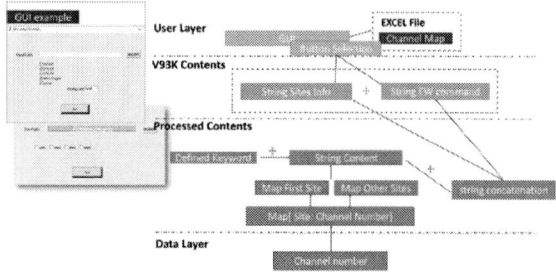

Figure 7: Data Container Design

As shown above in figure 7, a single common data container includes information like channel number and keywords. For different channel types, independent data containers will be added to be more compatible.

When the test program pin config does not match that on OSAT, ATPG solution can automatically modify the pin config file to achieve matching.

ATPG Level – Input File

For V93000 program, the level setup assigns values for digital I/O pins and power supply voltages to DUT. It can be organized as one single file, or its individual level equation sets and test data blocks can be stored in separate files [2]. In the latter case, "master file" is a file to manage all level files. For this solution, we use "master file" to organize different level files as figure 8 shown. Therefore, the input file is in excel format and it's designed that one sheet corresponds to one equation set, which requires three modules: DPS module, level sets module and specs module.

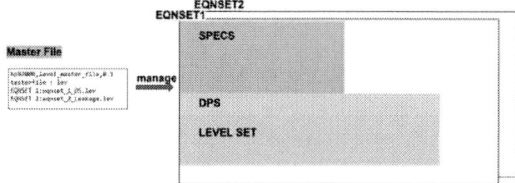

Figure 8: Master File Manages Level Files

DPS module uses expressions(variables) to specific the DPS pins levels and parameters of the DPS pins resources to define the behavior of DPS pins. Normal resource parameters can be grouped by functions as voltage, current, connections and timings, voltage settling time and voltage slew rate.

Level sets module assigns values to the level resources of I/O pins using expressions. It includes two parts: pin assignment and its feature such as vih/l, voh/l and ioh/l.

Specs module links the spec variables and its units used in the DPS module and the level sets module. We also merge the values of the spec variables which are defined in spec set in this module. The relationship between these three modules is shown in figure 9.

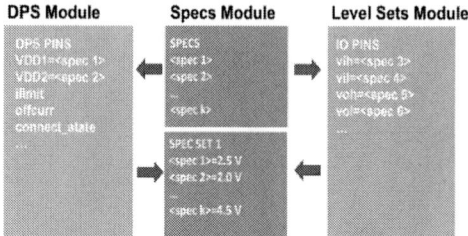

Figure 9: Three Modules Relationship

ATPG Level -- Data and Storage

After all the level related information is filled in excel, this solution programmed in python will process the information and generate the corresponding level files.

First, the entire excel will be analyzed by "ReadLevelDescriptionFileTotal" class, and "ReadLevelDescriptionFileBysheet" class will process the content of each sheet page separately and create a corresponding "sheet container". And one "sheet container" includes three types of containers, which are "specsetcontainer", "DPSsettingcontainer" and "levelsetcontainerDic". "levelsetcontainerDic" manages multiple "levelsetcontainer", because in V93000 program, one equation set may contain multiple level sets. After processing all the sheets, we will get a "sheetContainerDic".

Then "ReadLevelDescriptionFileBysheet" class splits and stores the settings of the spec set, DPS set and level set into the corresponding containers through various fill functions, as shown in figure 10.

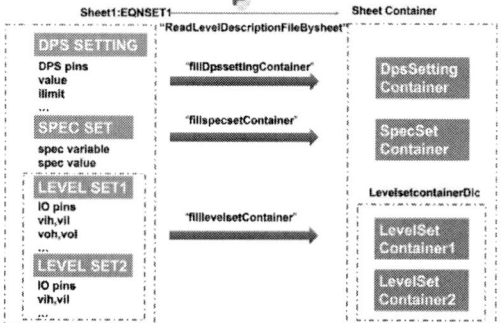

Figure 10: Data Classification and Storage

AFPG Level --File Generation

After storing all parameters to corresponding containers, we need to retrieve and print them based on the format requirements of V93000 level file. The structure of an equation-based level setup includes three key parts: Specs module, DPS module and level sets module. Figure 11 below illustrates the structure of an equation-based level setup.

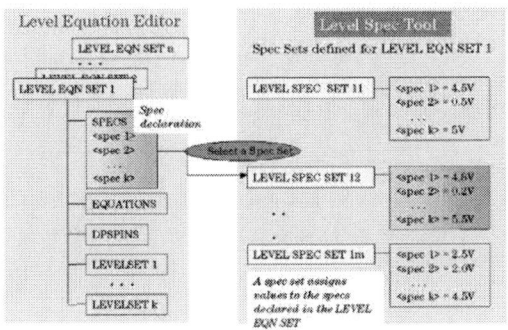

Figure 11: Structure of Level Setup

As mentioned above, each container provides a filling function to store the corresponding level settings. It also provides access function as an interface to get the settings of each container. "GenerateLevelBysheet" class iterates through the contents of each container in sequence, first setting DPS set, then level set and then spec set. Next the contents are printed to level file in turn, as shown in figure 12. After all level files are generated, "GenerateLevelTotal" will generate a master file to manage all of them.

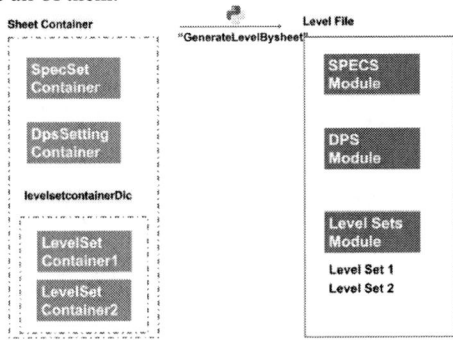

Figure 12: Level File Generation

APTG Level -- TMS

For SoC (System on Chip), different power supply domains have different power up and down sequence requirements. ATPG should consider how to define parameters to represent different power up/down sequences.

In V93000, to specify setup times and delays for DPS pins, it provides two options: DPS connect setup time or voltage ramp. But it can't set within the same set in one DPS set of one channel [3]. If only power up/down sequence is considered, rather than the power up/down slope, we can only use the "t_ms "parameter to generate a graph showing the DPS pin directly power up/down after "t_ms" time, as shown at the top of figure 13. Alternatively, the voltage ramp parameter is used to specify setup times and slew rates separately for rising and falling DPS voltage transitions. As the bottom of figure 13 shown, to facilitate user understanding, we convert original voltage ramp parameters

(vout_rise_settling_t_ms, vout_rise_t_ms_per_volt) of V93000 into "up_starttime" and "up_waittime". "up_starttime" represents total time of pin voltage rising and waiting. And the same applies to power off. With this function, the user does not have to understand V93000 up and down settings but can double check by simple graphical comparison.

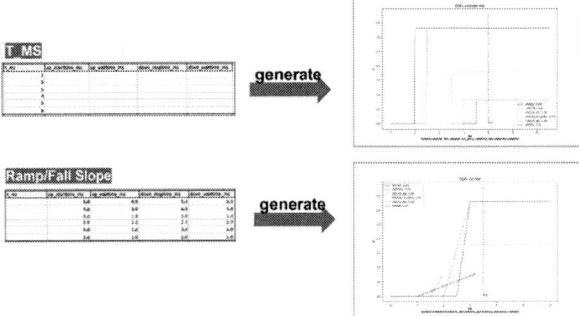

Figure 13: Power Up/Down Diagram

APTG Level – Interface

The final interface of ATPG level is shown in figure 14. User only need to fill in the input excel and load it; the level file can be automatically generated.

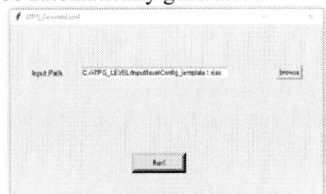

Figure 14: ATPG Level GUI

APTG Timing

The automatic generation of timing is relatively less complex. Because timing is bound to pattern. Using various pattern tools, when generating patterns, the corresponding timing will also be generated.

However, the number of timings generated can be very large, ranging from tens to hundreds, so it is necessary to contact the "master file" feature of the V93000 to organize them [4]. The master file is shown in figure 15.

```
hp93000,timing_master_file,0.1
testerfile: demo_project.tim

EQNSET 1 :DC/test500.tim
WAVETABLE "test500" :DC/test500.tim

EQNSET 1 [Remap 2]:DC/test501.tim
WAVETABLE "test501" :DC/test501.tim

EQNSET 1 [Remap 3]:DC/test502.tim
WAVETABLE "test502" :DC/test502.tim
```

Figure 15: V93000 Timing Master File

In this figure, there are three timing files, each file has

"equation set" and "wavetable set". A set of "equation set" and "wavetable set" correspond at least one pattern.

Each "equation set" has unique number so some "equation set" number will be remapped to other number. Usually, the timing file is classified and placed under different folders.

Based on these features, ATPG Timing solution achieve three core features.

- Auto generate timing master file
- Auto remap equation set number
- Auto find available timing file and record path

Besides this, ATPG solution also consider user will continue to add new timing file that means timing master file could be auto refreshed.

APTG Timing -- Framework

ATPG Timing solution has two key features: create new master file and refresh exist master file. "GenerateMasterFile" is main class. It controls "CreateMasterFile" class and "RefreshExistFile" class.

"TimingFileParser" class is used to parse all timing files and record file informations. "TimingFileManager" record file name to file list.

"RefreshMasterFile" class is used to refresh exist timing master file. All information in master file will be reserverd in "MasterFileInfo" class. The unused timing files will be checked and move to backup folder by "MoveTimingFile2Backup" class.

The details framework of ATPG timing solution is shown in figure 16.

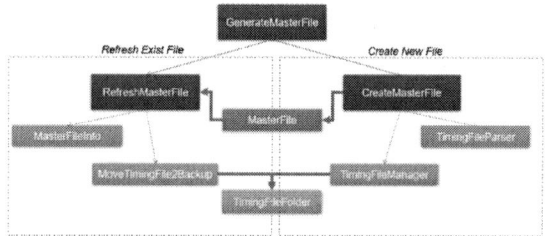

Figure 16: V93000 Timing Master File

ATPG Timing – Interface

The ATPG Timing solution's interface is as below. It could achieve "create new file" and "refresh exist file" automatically.

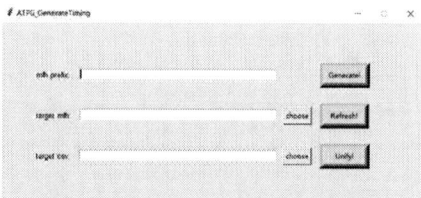

Figure 17: ATPG Timing solution interface

CONCLUSION

This ATPG solution summarize a universal input information and shows high efficiency. It is based on V93000 platform and could be used for major ICs.

ACKNOWLEDGEMENTS

We would like to express our gratitude to Beryl Xu, whose brilliant ideas and perceptive observations have proved immensely constructive.

REFERENCES

[1] ADVANTEST. *Test Documentation Center: Pin configuration file*, Rev7.5.4, topic. 94263.

[2] ADVANTEST. *Test Documentation Center: Level setup master file*, Rev7.6.4, topic. 91598.

[3] ADVANTEST. *Test Documentation Center: DPS connect setup time and voltage ramps*, Rev7.6.4, topic. 125205.

[4] ADVANTEST. *Test Documentation Center: Master file format for timing setup*, Rev7.5.5, topic. 91498.

Innovation Test Technology for Ultra-High-Speed ADC on ATE

Yanyan.Chang[1], Tianyu.Chen[2], Jiaying.Xiang[3], Yichen.Xiao[4]*
[1]SA, Advantest (China) Co., Ltd, Shanghai 201203, China
Corresponding Author's Email: yanyan.chang@advantest.com

ABSTRACT

Typically, traditional mix signal test theory on ATE cannot fully meet the accuracy requirement of ultra-high-speed ADC's dynamic performance. Extra complex factors must be taken into consideration, such as chip intrinsic spur and test system introduced spur, insertion loss of ADC high-speed input signal caused by ATE load board hardware bandwidth limitation, etc. Hence, this paper presents two innovative test technologies on the ATE platform to support ultra-high-speed ADC's accuracy test and post-calibration scenery.

First, this paper presents accurate and reliable algorithms for the identification and separation of the spectrum spur of ultra-high-speed ADC, which has been verified on the ATE test platform.

Secondly, this paper proposes a new de-embedding technique based on S parameter to compensate for high-speed input stimulus signal loss quickly.

Keywords—Spur separation; S parameter; ultra-high-speed ADC; ATE

INTRODUCTION

With the development of the 5G and optical communication market, the sampling rate of ADC in chips is increasing, immensely promoting the research and innovation of new test methodologies. ATE (Automatic Test Equipment) plays an important role in meeting the complicated parameters of test in the semiconductor industry. Therefore, accurately testing the dynamic performance of ultra-high-speed ADC on ATE is very essential.

The ATE tester is a combination of workstations and test instruments, which contain different types of test cards. The host computer of the workstation controls the test instrument by executing a test command in smartest software. Then, the workstation uploads the test results quickly and reliably after the test is complete. The connection between the chip and the ATE tester requires load board, and there is a socket on the load board to place and fix the chip. Figure 1 shows the ATE test system.

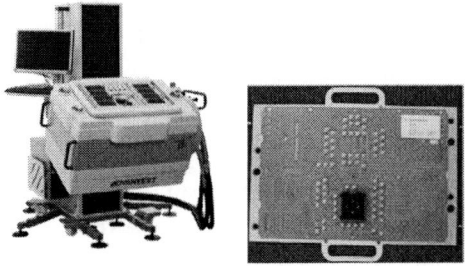

Figure 1. ATE Tester(left) and load board hardware(right)

When testing chips in large quantities with ATE, it is usually necessary to test multiple sites in parallel, and each chip may have multiple channels of high-speed ADCs. Figure 2 shows that testing a 2-site, 4-channel chip is equivalent to testing 8 high-speed ADCs in parallel.

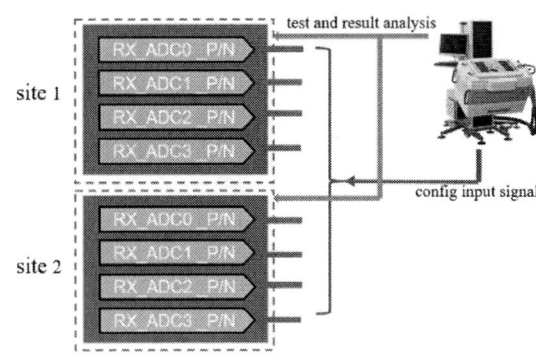

Figure 2. ATE test multi-site, multi-channel ADCs in parallel

To test the performance of ultra-high-speed ADC with decent accuracy, spurious caused by environments (such as load board hardware) and ultra-high-speed ADC architectures should be identified and separated. The spurious should not be treated as an error introduced by the high-speed ADC design or manufacturing process. The following questions should receive attention before obtaining ADC performance in ATE test:

1) Intrinsic spur caused by the mismatch of time-interleaved analog to digital converter (TI-ADC); the mismatch mainly includes offset mismatch error, gain mismatch error and sampling clock phase mismatch error [1].

2) Determined spur caused by hardware or other known clock source, such as system clock and reference clock on load board.

3) The known large burr caused by ADC ultra-high-speed input signal should be detected by a low-cost solution.

4) The high-speed input signal of ADC is transferred among ATE load board hardware, while the hardware environment will bring a non-negligible loss.

TEST CHALLENGE AND SOLUTION

Spurious due to interleaving structure

To achieve higher sampling rates, the ADC is internally composed of multiple sub-ADCs, which operate in an interleaved manner. Since the sub-ADCs cannot be identical, the spur caused by sub-ADCs mismatch will occurs in high-speed ADC [1][2]. Figure 3 shows the TI-ADC converting the input sine wave. Exp. three sub-ADC as an example.

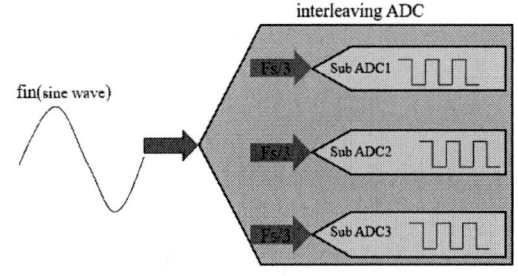

Figure 3. TI-ADC converts the input sine wave

979-8-3503-1101-3/23 $31.00 © 2023 IEEE

The input signal passes through the multi-way sub-ADC and each sub-ADC convert the input signal with similar but not identical dc offset individually. There is an ADC offset error between sub-ADCs, Figure 4 shows the signals sampled by the three sub-ADCs; the output signal has fluctuations. Figure 5 shows the input signal passes through the multi-way sub-ADC with clock phase mismatch.

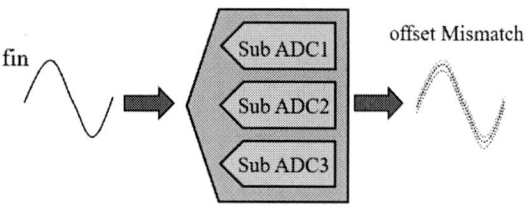

Figure 4. the input signal passes through an ADC with dc offset

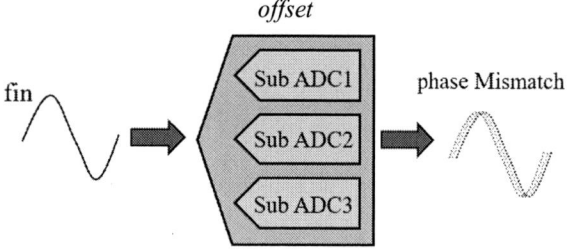

Figure 5. the input signal passes through an ADC with clock phase mismatch

When the ATE obtains the wave of high-speed ADC through the digital capture pattern, it first need to separate the spurious caused by sub-ADC mismatch. Smartest API can accurately obtain the output signal spectrum. Knowing the number N of sub ADCs, the position of spurious is mainly distributed in formula (1) [1], K=1,2,3…

$$Fspur = +/-Fin + K * Fs / N \quad (1)$$

The smartest use DSP_FFT() API obtain the spectrum of output signal, as shown in figure 6, smartest finds the main tone fin in the spectrum automatically. ADC sample rate Fs is known by configuration of smartest. Then, the position of spurious can be calculated according to formula (1).

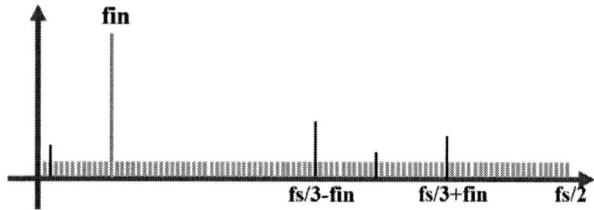

Figure 6. ADC spectrum in ATE

After finding these spurious in the spectrum, smartest will ignore these spur power when calculating the ADC performance. A common approach is to mark and clear these spurious in the spectrum. Depending on the RBW of the spectrum, it may need clear spur sidelobes.

Spurious caused by the system clock crosstalk

When testing high-speed ADC, ATE needs to provide the system clock and reference clock to the chip. The reference clock is the source for ADC to obtain its own sampling clock. The system clock drives the chip to read and write ram and so on. The system clock and reference clock provided to the chip by ATE must pass through the load board hardware and socket. When these clocks pass through the load board, they will leak into the ADC input signal circuit on the load board. So, smartest need to separate the reference clock and the system clock in the output signal spectrum.

The reference clock is typically multiplied by N (N=1,2,3…) inside the chip to obtain the sampling clock required by the ADC, and the sampling clock will often be coupled with the output signal, smartest need separate N times clock of reference clock.

The spurious brought by the reference clock and the system clock has a fixed position and is easily identified. Figure 7 shows the ATE ADC spectrum with the spurious caused by the reference clock.

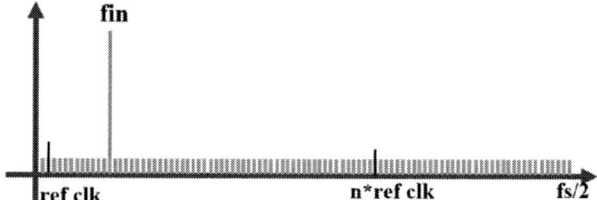

Figure 7. ADC spectrum in ATE with reference clock

Known large burr caused by ultra-high-speed input signal

In this test, the input signal of the ADC has a known large burr. The burr introduced by the input signal can be clearly observed by smartest in the time domain of the output signal.

The ideal signal corresponding to ADC actual signal is fitted by a trigonometric function theorem in smartest. When the actual signal burr is larger than a user setting threshold, smartest will detected the burr using the ideal signal.

The output signal is known to be a sine wave, it is assumed that the ideal signal fitted according to the actual output signal of ADC conforms to formula (2):

$$Y = A * \sin(2 * \pi * F * X + \varphi) + C \quad (2)$$

In formula (2), A represents the peak amplitude of signal, F represents the signal frequency, φ represents the initial phase of the signal, C represents the dc component of the signal. When the number of known burrs in the time domain is less than 15 (the number of specimens is 32,768), and the error is <=3.5%. The coefficients A, F, C can be obtained through the actual output signal.

The spectrum0 is obtained by smartest API. Then, find the main tone, calculate frequency f0 and power P0 of the main tone, and find the position index I0 of the main tone. According to the known ADC sampling rate fs0, we obtain the following:

$$A = P0 / \sqrt{2}$$
$$F = f0$$
$$X = n / fs0, (n = 0,1,2,3...)$$
$$C = spectrum0[0]$$

After the DSP_FFT() API calculation, the imaginary part Img0 and the real part Real0 of the main tone are obtained, φ can be calculated.

$$\varphi = a\tan 2(\text{Im} g0 / \text{Re} al0) + \pi / 2$$

The fitted curve of ideal signal is:

$$Y = P0 * \sqrt{2} * \sin[2 * \pi * f0 * n / fs0 + \varphi] + spectrum0[0]$$

The error threshold of the actual signal and the ideal signal is set through the smartest flow parameter. If the error between the actual signal and ideal signal is greater than the threshold, ATE will consider that the signal at this time is affected by the input signal burr, and replace the actual signal at this time with the ideal signal.

Quick compensation of input signal loss

The ultra-high-speed input signal of ADC must pass through the load board and socket to the input terminal of the chip. Due to the high input signal frequency, the chip will receive the input signal with a large loss (up to 1db loss at 5G). This loss needs to be compensated. The loss varies greatly at different frequencies. When the frequency of the input signal is changed. The amplitude of the input signal needs to be recompensated.

During the ATE verification test, it is necessary to test multiple sites and multiple channels of ADC. It is also necessary to observe the conversion ability of ADC with many different frequency signals. In figure 2, the amplitude of eight parallel input signals needs to be accurately compensated.

The traditional compensation is a process of multiple trims. First, Config amplitude of ADC input signal and received the signal with a golden chip, then upload and analyzes the measured result. Next, ATE determines whether the input signal received by the chip meets the demand. If it does not meet the ADC input demand, the amplitude of the input signal must be adjusted. Figure 8 shows the traditional compensation input signal flow.

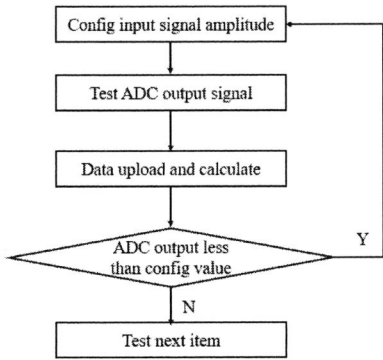

Figure 8. Traditional trim flow chart

The ADC of each channel usually needs to be adjusted 3 to 5 times to obtain the compensation value, which increases the test cost.

When the chip is tested in ATE, the load board and sockets are fixed. A new de-embedding technique based on S parameter is proposed to quickly compensate for high-speed input stimulus signal loss. By measuring the S parameter of the input signal path on the load board, the loss of the input signal on the load board can be calculated.

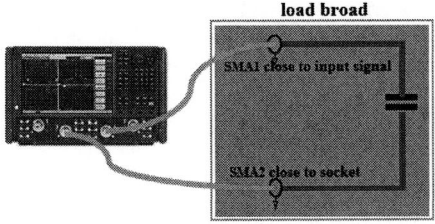

Figure 9. Testing the S-parameter

The path loss of one channel in different frequencies shows in table 1, same as the other site and the other channels.

TABLE I. PATH LOSS

Freq	SMA1-SMA2
0 HZ	0.0000
500 MHz	-0.3800
1.0 GHz	-0.5050
1.5 GHz	-0.5820
2.0 GHz	-0.63500
2.5 GHz	-0.74000
3 GHz	-0.80500
3.5 GHz	-0.87670
4.0 GHz	-0.94700
...	...

The compensation value at a finite frequency can be fitted to the compensation value at each frequency by an interpolation algorithm. As shown in figure 10.

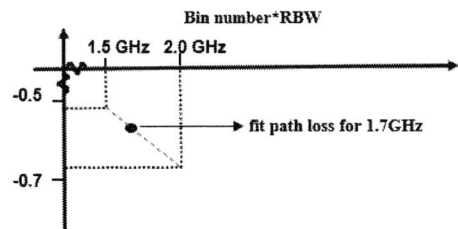

Figure 10. compensation curve after fitting

The generated de-embedded file is based on the compensation curve in ATE. When the input signal amplitude is configured, ATE automatically compensates the config amplitude value of multi-channel and different frequency input signals.

CONCLUSION

The paper described several handily solutions to test the performance of ultra-high-speed ADC on ATE with decent accuracy. The spurious introduced by the ultra-high-speed ADC structure and the spurious introduced by the load board hardware could be separated with specific treatment. Finally, A new de-embedding technique based on S parameter is proposed to quickly compensate for high-speed input stimulus signal loss is also introduced in this essay.

ACKNOWLEDGEMENTS

Authors wish to thank Advantest for its support on this project.

REFERENCES

[1] "Explicit Analysis of Channel Mismatch Effects in Time-Interleaved ADC Systems", Naoki Kurosawa, etc. 2001, IEEE.

[2] 3. Ki, S. C., Ko, S. H., & Yoo, H. J. (2013). A 20 GS/s 8-bit interleaved time-interleaved SAR ADC with background calibration. IEEE Journal of Solid-State Circuits, 48(10), 2512-2522.

Computer Vision Technology Supported Rapid DRAM Capacitor Analyzing System Based On TEM Image

Zhi-Yuan Gui[1], Chang Xu [*], Han Yan [2], and Zhi-Yu Li[3]

Fujian Jinhua Integrated Circuit Co., Ltd., Jinjiang, Quanzhou, China

*Corresponding Author's Email: xander.xu@jhicc.com

ABSTRAC

Each downsizing of the DRAM capacitor circle will bring great benefits along with many production related process problems at the same time. In the R&D stage, tremendous accurate metrology results, such as critical dimensions of capacitor circles, are necessary to be obtained. However, DRAM capacitors are fabricated by bunches of different process steps, the final internal structure of capacitor array cannot be obtained by surface scan metrology tools. In DRAM capacitor structure design stage, cross-section TEM image analysis will be used to perform new design verification and improvement plan tracking. To manually measure details such as critical dimensions in TEM images is very time consuming. In order to improve the efficiency of analysis, this paper developed a rapid DRAM capacitor analyzing system, which can obtain the critical dimensions of capacitor circles quickly, in batches and with visibility.

Keywords—DRAM capacitor; metrology; TEM; critical dimensions; visibility

INTRODUCTION

Since Gordon Moore's original description of "Moore's Law" in 1965 [1], transistor dimensions have been scaled down roughly every twenty-four months. Though it goes slowly in Dynamic Random Access Memory (DRAM) circuit, each version of product upgrade also brings a lot of benefits such as lower power consumption and stronger performance. However smaller size of storage node (SN) in DRAM are prone to leakage from capacitor storage units, in order to optimize the capacitor structure design, metrology schemes must be established to support the analysis.

Traditional in-line critical dimension(CD) metrology tools, such as optical critical dimension (OCD) and critical dimension scanning electron microscope, while still indispensable, cannot match the sub-angstrom spatial resolution and sub-surface 3-dimention (3D) analytical capabilities provided by transmission electron microscopy (TEM). But widespread use of TEM analysis has generally been limited by long cycle time and requires high levels of operator interaction [2]. In the process of research and design (R&D), a large volume of engineering wafer capacitor CD is needed to make next decision. However, existing personnel can take a long time to perform manual CD measurement, which are not efficient. So, this article introduces a rapid DRAM capacitor analyzing system based on TEM image, with the support of computer vision technology, large number of results tally with those obtained from manual measurement can be quickly accomplished. Due to corporate confidentiality requirement, all data and images presented in this paper have been artificially processed.

INTELLIGENT IMAGE PROCESSING METHOD

TEM images are obtained by slicing the DRAM wafer, as it shown in Fig 1-1. There is an obvious border between dielectric layer and poly-silicon, and image grayscale is different between each independent material. Taking advantage of this, the capacitance circle can be split and critical dimensions can be obtained.

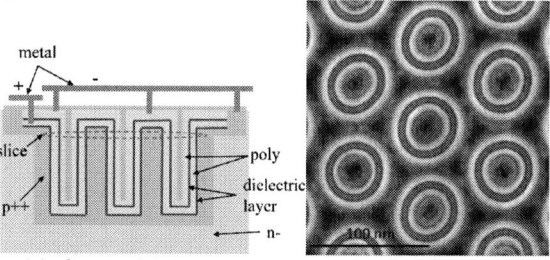

(a) slice position   (b) TEM image

Fig: 1-1 SN structure and orignal TEM image

Locating Capacitance Circle

In the Fig 1-1(a), outer dielectric layer is surrounded with p++ filling, and in TEM image, p++ area is darker then the dielectric layer. In order to separate each independent parts, firstly, de-noising the TEM image to make the boundary clearer [3], and then images processed by adaptive threshold method with GAUSSIAN algorithm, each pixel can be calculated using the following formula:

$$f(x,y) = \frac{1}{2\pi\sigma_1\sigma_2}exp\left(-\frac{1}{2}[\frac{(x-u_1)^2}{\sigma_1^2} + \frac{(y-u_2)^2}{\sigma_2^2}]\right)$$

x: horizontal distance between the neighborhood pixels and the center point

y: vertical spacing

u1: horizontal average pixels

u2: vertical average pixels

σ_1: horizontal direction pixel standard deviation

σ_2: vertical direction pixel standard deviation

The final binary image is shown in Fig 1-2(a) [4], and after applying morphological edge detection method, contours can be obtained as shown in Fig 1-2(b) [5].

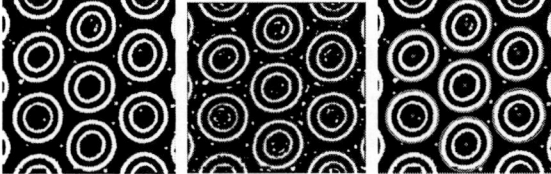

(a) binary   (b)contour (c)minimum circumscribed
Fig: 1-2 capacitance circle picture processing

It is obvious that the outer dielectric layer contour has the biggest surrounding area, by getting each contour's minimum circumscribed circle in Fig 1-2(c), the outer dielectric layer can be selected. And then define the circle center as the center point of the capacitance.

X/Y Direction Calibration

As it shows in Fig 1-3(a), based on layout file all capacitor circles align to vertical direction.

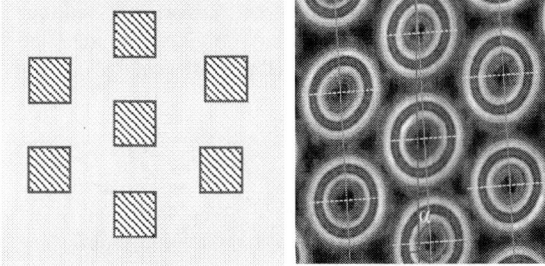

(a) layout file   (b) TEM measure
Fig: 1-3 measure alignment

However, the TEM images obtained after slicing will have some artificial angle deviation due to manual operation. Error angle α as shown in Fig 1-3(b), angular bias calibration is required to maintain good consistency with the test results from the manual measurements.

After obtaining the coordinates of the center of the circle, calculate the slope of the straight line that aligns the center of the circle vertically using the line detection method based on HOUGH transform. First, convert Cartesian coordinates(x, y) into polar coordinates(ρ, θ), a polar point corresponds to a line in Hough space, and a point in Hough space corresponds to a line in Cartesian coordinates, Select point(ρ, θ) which is converged by as many lines as possible in Hough space, finally, the fitted linear equation is obtained.

$$\begin{cases} x = \rho \cos\theta \\ y = \rho \sin\theta \end{cases} \qquad \rho = x * \cos\theta + y * \sin\theta$$

Then calculate the average slope value based on all fitted lines. Adjust all straight lines slope value to average value follow by using Linear Least Squares Regression to obtain the intercept value so that all lines can fit corresponding circle centers well.

The equation of the straight line can be expressed as:

$$y = ax + b$$

a is equal to the average slope and b can be calculated by function:

$$\begin{cases} \emptyset(a,b) = \sum_{i=1}^{n}(yi - a - bx_i)^2 \\ \dfrac{\partial\emptyset}{\partial b} = 0 \end{cases}$$

After that Center of Y-direction reference lines with uniform slope are shown in Fig 1-3(b). X-direction reference line should be perpendicular to the Y, and with the help of circle center coordinate, calculate and draw X reference line as shown in yellow color dotted line.

Get Outer Poly Measure Border

With the center of the circle are determined, TEM image is divided into several parts each contains one SN as shown in Fig 1-4. As poly rings is darker then surrounded dielectric layer, in Fig 1-4 (a), two inner and outer contour circles are selected after contours area filtered and combined with Fig 1-2 (b) result, the outer contour is the measuring border needed for CD calculation and area comparison.

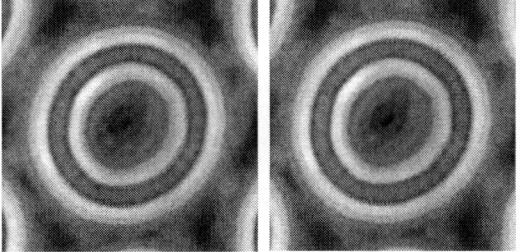

(a) ring point contour   (b) key circle point contour
Fig: 1-4 find key circle border

Calculate CD In X/Y Direction

In the previous calculation, the measurement border and reference line in X and Y directions are determined. When calculating Y_CD, the border is divided into two parts by X reference line. All distances between border point and X reference line are calculated for both parts. As Fig 1-5(a) shows, the two longest distances are selected and marked as Y1 & Y2. The critical dimension CD_Y=Y1+Y2. And for CD_X= X1+X2 is obtained similarly as illustrated in Fig 1-5(b) below.

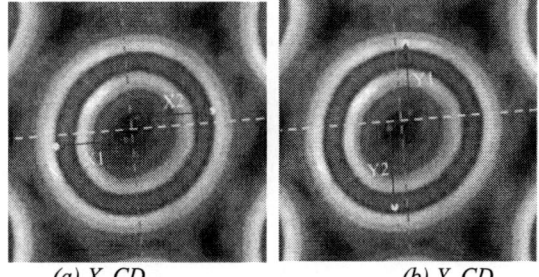

(a) X_CD   (b) Y_CD
Fig: 1-5 CD measure method

979-8-3503-1101-3/23 $31.00 © 2023 IEEE

RESULT AND COMPARISON

In this section, the paper will begin to present the measurements and demonstrate the reliability of measurement result.

In order to facilitate engineers to verify the results, all measured capacitance are highlighted by yellow border line, and measurement results with measuring points are marked on the capacitance as shown in Fig 2-1. After the measurement is completed, individual CD value, average CD value, and other calculated values of each capacitor circles will be exported to excel, so that engineers can find the overall measurement of all capacitor circles in the first time and find the specific deviation of the data.

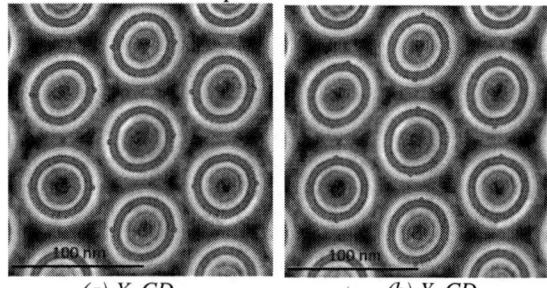

(a) X_CD   *(b) Y_CD*

Fig: 2-1 measure display

To verify the reliability of the system measurement data, several wafers in different conditions are selected. After slicing the wafer and choosing the same area to get TEM images, results are measured by both human and rapid measurement system separately. Detail comparison of the average value of the results is shown in table I.

TABLE I

CD measurement result comparison

SYS X_CD	HUMAN X_CD	X_CD Bias	SYS Y_CD	HUMAN Y_CD	Y_CD Bias
48.34	48.81	-0.47	47.20	47.57	-0.37
48.41	48.58	-0.17	47.06	47.31	-0.25
49.58	49.76	-0.18	48.47	49.00	-0.53
49.50	49.56	-0.06	48.35	48.63	-0.28
49.37	49.54	-0.17	48.17	48.00	0.17
48.49	48.84	-0.35	47.95	47.73	0.22
48.51	48.11	0.40	47.91	47.65	0.26
48.88	49.17	-0.29	47.64	48.37	-0.73
48.22	48.20	0.02	47.78	47.44	0.36

CD bias summary data is shown in Fig 2-2. All bias are Controlled within 1 nm, mean square error is 0.25nm and 0.39nm in X/Y direction, mean average percentage error is 0.39% and 0.27% in X/Y direction. The error is within the referential tolerance and the system measurement result has proved to be accurate and reliable.

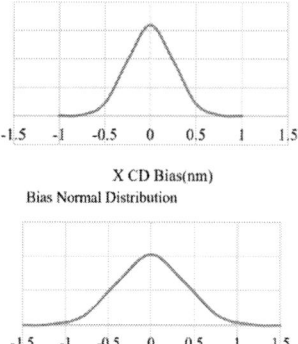

Fig: 2-2 bias distribution

ACKNOWLEDFEMENTS

We would like to show our gratitude to our manager Haw-Jyue Luo for sharing his wisdom and expertise with us during the course of this research. We also thank our colleagues from JHICC who provided insight and expertise that strongly supported the research.

REFERENCES

[1] Kenslea A, Hakala C, Zhong Z, et al. CD-TEM: Characterizing impact of TEM sample preparation on CD metrology[C]// 2018 29th Annual SEMI Advanced Semiconductor Manufacturing Conference (ASMC). 2018.J. Clerk Maxwell, A Treatise on Electricity and Magnetism, 3rd ed., vol. 2. Oxford: Clarendon, 1892, pp.68–73.

[2] Zhang N, Pu S, Akin B. An Automated Multi-Device Characterization System for Reliability Assessment of Power Semiconductors[C]// 2021 IEEE 13th International Symposium on Diagnostics for Electrical Machines, Power Electronics and Drives (SDEMPED). 0.K. Elissa, "Title of paper if known," unpublished.

[3] C. Xu, Y. -F. Zhang and C. -R. Luo, "An Adaptive Denoising System for Sub-nm Scale Failure Analysis Based on TEM Image," 2020 China Semiconductor Technology International Conference (CSTIC), 2020, pp. 1-3, doi: 10.1109/CSTIC49141.2020.9282463.

[4] Bradley, D., & Roth, G. (2007). Adaptive Thresholding using the Integral Image. Journal of Graphics Tools, 12(2), 13–21. doi:10.1080/2151237x.2007.10129236.

[5] Satoshi Suzuki, Keiichi Abe, Topological structural analysis of digitized binary images by border following, Computer Vision, Graphics, and Image Processing,Volume 29, Issue 3,1985,Page 396, ISSN 0734-189X.

A NOVEL MODEL-MATCHING BASED SCRATCH TOOL TRACING SYSTEM

Shi-Qiang He[1], Yan-Qiu Zhang[*], Xiao-Lei Zhang[2], Chic-Kuo Fang[3]

Fujian Jinhua Integrated Circuit Co., Ltd, Jinjiang, Quzhou, Fujian, China

Email: tc.zhang@jhicc.com

ABSTRACT

In semiconductor manufacturing process, wafer will inevitably contact some parts of the process tool and cause mechanical scratches [1], which will affect the yield. In order to identify the process tool that may cause wafer scratches, it is necessary to stop all the process steps that wafer through and conduct scratch tests, thus affecting production line capacity. The paper proposes a novel model-matching based scratch tool tracing system. This system comprehensively consider the scratch pattern of defect map, the common process tool, the matching degree between the contact point of process tool and the scratch position of the wafer, and the wafer through count of the process tool to build an analysis system for automatically checking the process tool that may cause scratches for wafer. Through simulation experiment, this system can effectively screen out the process tool that may cause wafer scratches, improve the efficiency of engineers to check out the process tool may cause scratch, and reduce the number of process tool stopped for scratch test.

Keywords—Wafer Scratch; Defect Map; Process Tool; Common Tool; Matching Degree;

INTRODUCTION

Semiconductor manufacturing industry is the manufacturing industry with the highest degree of automation at present. The entire manufacturing process is fully automated to avoid bringing particles or other impurities into the cleanroom environment and affecting the stability of the process. Wafer entering and exiting process tool or wafer transferring between different steps are completed by different automation equipment. If there have some problem with process tool, when wafer through this process tool, the point where the process tool contacts with wafer may scratch the wafer surface and affects the subsequent manufacturing process.

At present, the tracing root cause of wafer scratch is mainly to recognize the pattern of wafer scratch through defect map, and then analyze the root cause of the scratch manually from the scratch pattern. In this paper, based on the recognition of the scratch type by the defect map, a mathematical model is constructed to speculate the wafer scratch by calculating the common process tool probability based on following factors: number of scratch wafers run through same process tool, the matching degree between the contact point of process tool with wafer and the scratch position on the defect map, the number of times that the scratched wafer through different process tool. The scratch search system mainly includes the following procedures as shown in the figure 1:

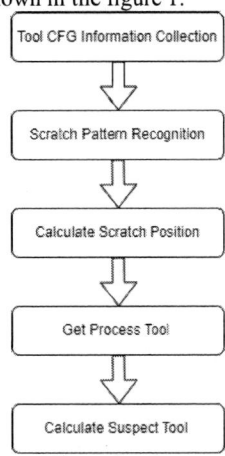

Figure 1: process flow of wafer scratch tool tracing

Using this system to analyze wafer scratch information, we can quickly identify the process tool and components that cause wafer scratch, thereby improving the target process tool to prevent recurrence. Continuously promote the improvement of process tool and process defects, reduce the probability of scratch and improve the yield of product.

WAFER SCRATCH TOOL SEARCH ROOT CAUSE ANALYSIS SYSTERM

Tool Config (CFG) Information Collection

The most important step of the system is to collect the detail of the contact point between the process tool and wafer to build up tool CFG information. Without this information, it is impossible to calculate the size matching degree between the wafer scratch position and contact point of the process tool. As illustrated in Figure 2, it is a schematic diagram of the contact point between the process tool and wafer.

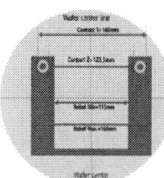

Figure 2: contact point information between the process tool and wafer

The process tool CFG information mainly includes:

1. The process tool size and position of the contact point with wafer. By considering size and position information the unmatched process tool can be filtered out.
2. The pattern of the contact point. By recognition the pattern of the wafer scratch from defect map and then matching it with the pattern of the contact points, filter the process tool that unmatched.
3. Whether the contact point position rotates during contact with wafer. For example if the pattern of the scratch is arc, it can rise the suspicious coefficient of the target process tool.

Scratch Pattern Recognition

The categories of scratch pattern are: straight line, circle, arc and other pattern. For different pattern of scratches, there are some differences in calculation of process tool size matching.

The scratch defect type is the most challenging work because the position, shape, size and curvature vary widely from one scratch to another [2]. In the defect map, there are usually many random defects besides scratch defects. Therefore, these random defect points need to be filtered out before scratch pattern recognition. This paper uses K-means [3] algorithm to remove isolated points, and then fitting defect points with least square method [4] to recognition the defect pattern.

Calculate Scratch Position Data

For different scratch pattern of defect, there will be different analysis methods to calculate the wafer scratch position.

As shown in the figure 3, if the scratch pattern is linear, we will calculate distance of wafer center point to the scratch line as the scratch position data.

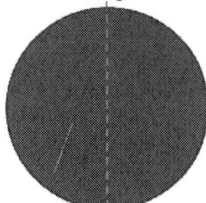

Figure 3: line scratch pattern

As shown in the figure 4, if the scratch pattern is circle or other irregular pattern. We will calculate the maximum and minimum distance from the defect points to the center

perpendicular of the wafer and the point of wafer center as the scratch position data.

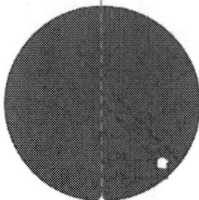

Figure 4: circle scratch pattern

As shown in the figure 5, if the pattern of the scratch is an arc, the radius of the circle formed by the scattered points of the arc is taken as the scratch position data.

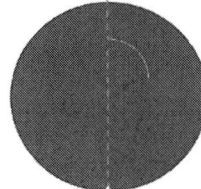

Figure 5:arc scratch pattern

The Model Of Calculating Suspect Tool

The suspicion index of the scratched process tool is mainly include of the common process tool, the matching degree between the contact point of process tool with wafers and the wafer scratched position, and the wafer through count of the process tool.

The parameter of common process tool (T):

$$T_{tool} = \frac{C_{though\,wafer}}{C_{total\,wafer}}$$

$C_{though\,wafer}$: wafer though count of the process tool

$C_{total\,wafer}$: total wafer count

The parameter of matching degree (M):

$$S_{min,\,component} - \tau < S_{calc} < S_{min,\,component} + \tau$$

OR

$$S_{max,\,component} - \tau < S_{calc} < S_{max,\,component} + \tau$$

$$M_{tool,wafer} = \varepsilon$$

$$S_{min,\,component} + \tau \leq S_{calc} \leq S_{max,\,component} - \tau$$

$$M_{tool,wafer} = \lambda * \varepsilon$$

$S_{min,\,component}$, $S_{max,\,component}$ represents the maximum and minimum values of the contact point between the process tool and wafer, as shown in the figure 6.

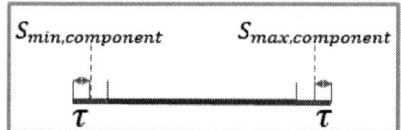

Figure 6: maximum and minimum values of the contact point between the process tool and wafer

τ: calculate tolerance

S_{calc}: calculate scratch position

ε: custom matching parameters

λ: attenuation factor of matching degree

The ratio of the wafer through count of the target process tool to the number of all process tool which wafer through (φ):

$$\varphi_{tool} = \frac{C_{through,tool}}{\sum_1^n C_{through,tool_i}}$$

$C_{through,tool}$: through tool count of suffer wafer

n: total through process tool count

h: wafer through count of the process tool

The suspect tool of wafer scratch model(P):

$$P = T_{tool} * (\alpha * \frac{\sum_1^h M_{tool,wafer_j}}{h} + \beta * \varphi_{tool})$$
$$\alpha + \beta = 1$$

α: the weights of common process tool and matching degree of process tool and wafer contact point.

β: the weights of φ.

SIMULATION AND RESULTS

In this part, the paper is using a real scratch case to verify the effectiveness of the system. This scratches case has 4 affected wafers. The defect maps of the 4 wafers are shown as figure 7, and the experimental results show in table 1.

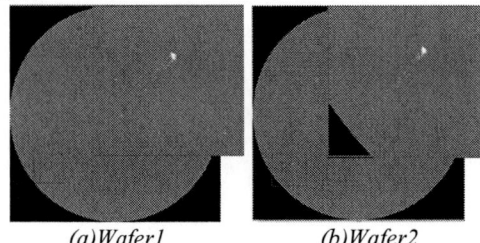

(a)Wafer1   *(b)Wafer2*

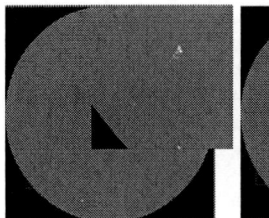 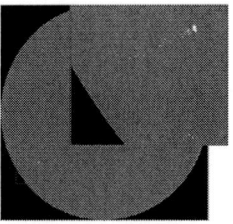

(c)Wafer3   *(d)Wafer4*

Figure 7: scratch of defect map

It can be seen from table 1, the actual scratch process tool suspect coefficient rank first in the calculation results. Each parameter constituting of the calculation scratch model (P) for the ET-TOOLA are relatively high compare with other process tool. Matching degree (M) of The ET-TOOLA is 0.95 means the process tool strongly matches scratch position. If matching degree (M) equal to 1 means the process tool perfect match scratch position. Tool through ratio (T) is 1.0 it means that the all scratch wafers have run through this process tool. The count of four wafers through machine ET-TOOLA six times is due to this tool has been using for muti-process steps.

REFERENCES

[1]. Danna M , Design O . Meeting the Challenge of Rising Costs of Process Tool Development and Implementation with a New Design and Manufacturing Paradigm.

[2]. Sun C , Sun M , Chen H T . Direct Voxel Grid Optimization: Super-fast Convergence for Radiance Fields Reconstruction[C]// 2021.

[3]. Wei J , Chen H P . A Hybrid Hierarchical k-means Clustering Algorithm[J]. Journal of Hohai University Changzhou, 2007.

[4]. Liu A . A NEW RELATIVE EFFICIENCY OF LEAST SQUARE ESTIMATE IN LINEAR MODELS[J]. Chinese Journal of Applied Probability and Statistics, 1989.

Table 1: the result of the scratch case simulation

Wafer List	Suspect Tool	Suspect Component	P	$M_{tool,wafer}$	T_{tool}	Tool Through Count	Actual Tool
Wafer 1 Wafer 2 Wafer 3 Wafer 4	ET-TOOLA	TM Robot	88.65	0.95	1.0	6.0	ET-TOOLA
	WET-TOOLA	FI Robot	79.51	0.86	1.0	4.0	
	ET-TOOTB	HCRA	77.71	0.84	1.0	4.0	
	PH-TOOLC	CRA	61.93	0.90	0.75	3.0	
	CMP-TOOLA	MPRA	35.87	0.78	0.5	2.0	

FASTER AU-AL IMC GROWTH UNDER CHLORINE ENVIRONMENT

Liao Jinzhi Lois[1]*, Wang Bisheng[2], Zhang Xi[3], Hua Younan[4], and Li Xiaomin[5]

[1,3,4] WinTech Nano-Technology Services Pte. Ltd.,10 Science Park Road, #03-26, The Alpha Science Park II, Singapore 117684

[2]Huawei Technologies Co Ltd, Bantian Huawei Base, Longgang District, Shenzhen, China 518129

[5]Wintech-Nano (Suzhou) Co., Ltd., Room 507, Building 09, Northwest District, Nano City, No. 99, Jinjihu Ave, Suzhou Industrial Park, Suzhou, China 215123.

*Corresponding Author's Email: lois@wintech-nano.com

ABSTRACT

Interestingly, this study found that Gold-Aluminum (Au-Al) IMCS grew much faster at the presence of higher content of chlorin (Cl). To the best of the author's knowledge, there is no study reported on the faster Au-Al IMC growth under Cl environment. It is found that the Cl is a major factor contributing to the IMCs growth. It is suggested that Cl lowered down the activation energy of atom solid-diffusion, and accelerated the Au-Al atom inter-diffusion. As a result, the IMC growth is faster under environment higher content of Cl.

INTRODUCTION

Wire bonding is the predominant mode of interconnection in microelectronic packaging due to its technology maturity and cost-effectiveness. Gold (Au) bonding wire is well known for its corrosion resistance and have been used in microelectronic packaging for more than 30 years. Au wire normally bonded to the aluminum (Al) pad. The intermetallic compound (IMC) growth of Au-Al is comparatively faster than that of other types of IMCs, like copper-aluminum (Cu-Al), silver-aluminum (Ag-Al). It is well-known that Au-Al IMCs growth is affected by factors such as temperature and voltage, and the Au–Al IMC growth is widely characterized and reported [1-3]. However, so far there is no report on the chlorine (Cl) effect on Au-Al IMCs growth rate. This paper discussed the influence of Cl effect on the Au-Al IMC growth rate. Different contents of Cl were purposely added in to the epoxy mold compounds (EMCs), which were applied to encapsulate the Au-Al units after wire bond. Two types of reliability tests were conducted, i.e. biased highly accelerated stress test (bHAST), and high temperature storage test (HTS). In these reliability tests, Au-Al encountered more failure after bHAST obviously, compared to HTS. However, it is interesting to find that the growth of Au-Al IMCS was faster in EMC with high Cl content compared to that of lower Cl content. To the authors' best knowledge, this phenomenon has not been reported. The mechanism of facilitation of IMC growth by Cl was discussed in this paper. It is suggested that Cl lowered down the activation energy of atom solid-diffusion, therefor accelerated the Au-Al atom inter-diffusion. As a result, the IMC grow was faster under higher content of Cl.

EXPERIMENTAL

Materials

0.8mil Au (purity > 99.99%) wire was used in this study. The wires were ball bonded to Small Out-line Package (SOP) lead frame, with 0.6um Al-0.5Cu using K&S ProCu bonder. The wire-bonded samples were then epoxy encapsulated. Two different contents of Cl, i.e. <20ppm and 500ppm, were purposely added into the EMC, respectively. Table 1 lists the experimental legs. The purpose is to study the Cl effect on Au-Al IMC growth speed.

TABLE I. EXPERIMENTAL LEGS

Leg#	Wire bond	EMC contamination
1	Au-Al	Clean (<20ppm Cl)
2	Au-Al	500ppm Cl

Reliability tests

To accelerate the IMC growth, two types of reliability tests, that is biased highly accelerated stress test (bHAST) and high temperature storage test (HTS) were conducted. The corresponding stress test conditions are tabulated in Table 2. Electrical test was performed on the unit at intervals to check whether the unit failed or passed. If the electrical resistance of the sample was larger than 20% of the time 0 sample, it was considered failed. After the electrical test, the failed samples were submitted for failure analysis and the good samples were subjected to continual reliability test.

TABLE II. RELIABILITY TESTS

#	Reliability tests	Condition	Test duration	Stress
1	bHAST	130ºC/85%RH/20V	192h	Temperature, humidity, electrical
2	HTS	150ºC	2000h	Temperature

Failure analysis

To further study the failure mechanism, the tested samples were prepared by epoxy mounting, ion milling and focus ion beam (FIB). Scanning electron microscope (SEM) and transmission electron microscopy (TEM) equipped with Energy-dispersive X-ray spectroscopy (EDX) were utilized to characterized the samples. SEM model is FEI Helios NanoLab 600i, and TEM model is FEI Tecnai G2 F20, respectively.

RESULTS AND DISCUSSION
bHAST

Table 1 presents the Au-Al bHAST results. It can be seen that leg2 (500ppm Cl) encountered failure, which was expected, as the Cl content is high in this leg. Fig. 1 shows the optical and SEM images of leg1 and leg2 at time 0, before reliability tests. No abnormality was observed at the ball-pad interface. The average IMC thickness of both samples is about 1.5µm. Fig. 2 and Fig. 3 shows the optical and SEM images of leg1 and leg2 after bHAST 96h and 192h, respectively. Here it is noticed that the average IMC thickness of leg2 is 2.2 µm after bHAST 96h, and 2.8um after bHAST 192h. Whilst, the average IMC thickness of leg1 is 1.7µm after bHAST 96h, and 1.8um after bHAST 192h. IMC thickness of leg2 is much thicker than that of leg1. Interfacial crack was observed of leg2 after bHAST 96h and 192h. The crack was at the top of IMCs. There are many studies on the Au-Al ball corrosion due to Cl and moisture [4-6]. Cracking usually starts at the Au ball bond periphery, and it will propagate toward the center of the Au ball bond. The proposed corrosion mechanism is described as below [4].

$$Au_yAl_x + Cl- + H_2O \rightarrow Au + AlClO + H_2 \qquad (1)$$

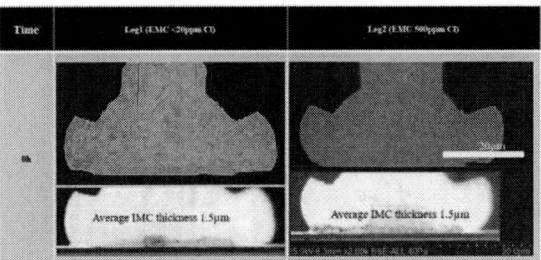

Figure 1: SEM and optical images of leg1 and leg2 at time 0.

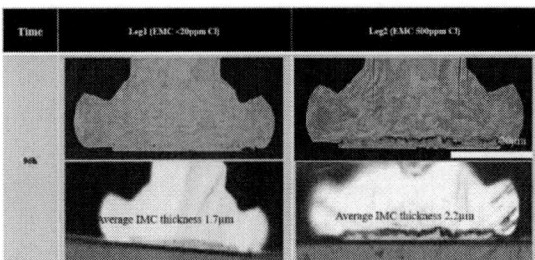

Figure 2: SEM and optical images of leg1 and leg2 after bHAST 96h.

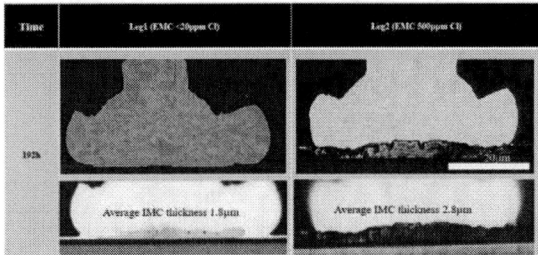

Figure 3: SEM and optical images of leg1 and leg2 after bHAST 192h.

Fig. 4 compared the IMC thickness of leg1 and leg 2 during bHAST test. It is interesting and noticeable that there was a significant IMCs thickness difference between leg1 and leg2. The difference between leg1 and leg2 wass that there were 500ppm Cl in the EMC of leg2. From the data, it showed Cl accelerated the IMC growth during bHAST.

TABLE III. BHAST RESULTS

DOE legs		bHAST (130°C, 85% RH, 20V)			
Leg#	EMC contamination	0h	96h	192h	Remark
1	Control (<20ppm Cl, S)	0/25	0/25	0/23	Pass
2	500ppm Cl	0/25	6/25	20/23	Fail

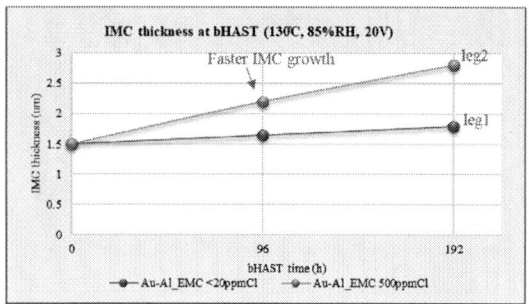

Figure 4: IMC thickness Au-Al during bHAST. Faster IMC growth was observed for the leg2 with high Cl content.

HTS

Table 2 presents the Au-Al HTS results. It can be seen that leg2 (500ppm Cl) encountered failure, while there is no failure of leg1. Fig. 5 and Fig. 6 shows the optical and SEM images of leg1 and leg2 after HTS 24h and 48h, respectively. No abnormality was observed from the images. The IMC thickness of both samples after HTS 24h is about 1.5-1.6um. However, the IMC thickness of leg2 is much thicker than that of leg 1 after HTS 48h. IMC thickness of leg 2 is about 2.2um, while leg 1 is about 1.6um.

Fig. 7, Fig. 8 and Fig. 9 shows the optical and SEM images of leg1 and leg2 after HTS 72h, 192h and 1000h, respectively. There is an obvious crack at the interface of leg2 after HTS 1000h. Here it is noticed that the average IMC thickness of leg2 was much higher than that of leg1.

TABLE IV.   HTS RESULTS

DOE legs		HTS (150°C)							
Leg#	EMC contamination	0h	96h	192h	504h	1000h	1500h	2000h	Remark
1	Control (<20ppm Cl, S)	0/25	0/19	0/17	0/15	0/13	0/12	0/10	Pass
2	500ppm Cl	0/25	0/19	0/17	1/15	13/13	12/12	10/10	Fail

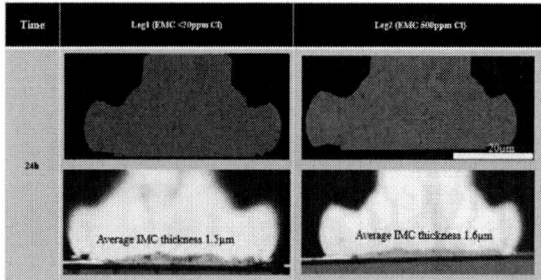

Figure 5: SEM and optical images of leg1 and leg2 after HTS 24h.

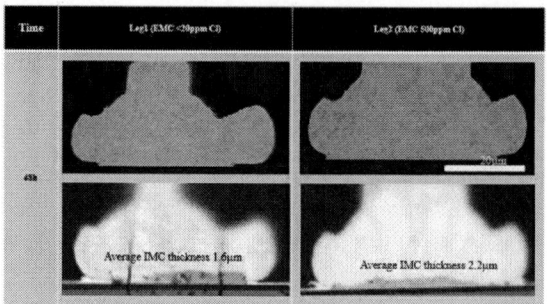

Figure 6: SEM and optical images of leg1 and leg2 after HTS 48h.

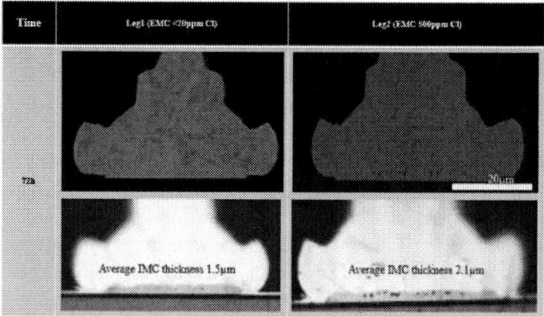

Figure 7: SEM and optical images of leg1 and leg2 after HTS 72h.

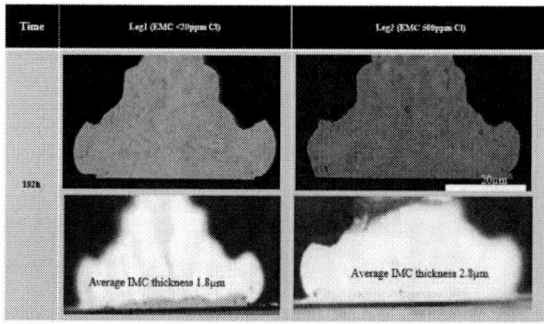

Figure 8: SEM and optical images of leg1 and leg2 after HTS 192h.

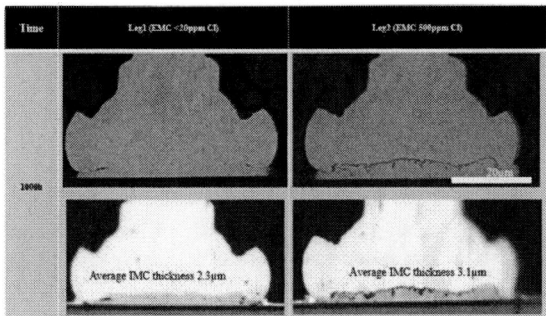

Figure 9: SEM and optical images of leg1 and leg2 after HTS 1000h.

Fig. 10 presents the average IMC thickness of leg1 and leg2 during HTS. It is obvious that the IMC thickness of leg2 grew much faster than leg1 during HTS test. It is suggested that Cl helped lower the activation energy barrier, increasing the atom diffusion and IMC growth. With higher Cl content, the atoms diffusion is much easier, therefore higher IMC thickness.

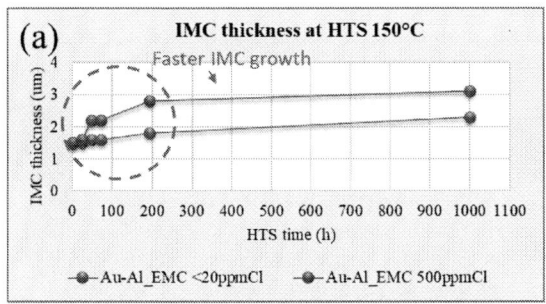

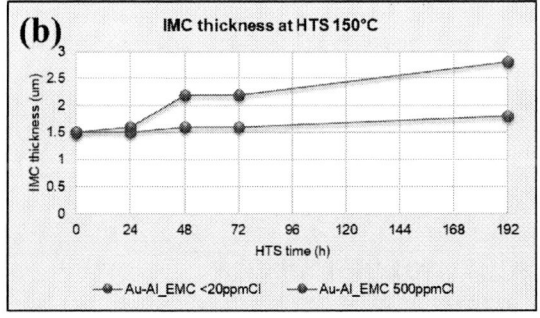

Figure 10: (a) IMC thickness comparison of Au-Al leg1 and leg2 during HTS from 0 to1000h, and (b) enlarged IMC thickness comparison from 0 to 192h.

CONCLUSIONS

In this work, the effect of Cl on Au-Al IMC growth

was studied during reliability tests of bHAST and HTS. It is found that with higher Cl content EMC, the Au-Al IMC thickness is much higher than that of lower Cl content EMC. It is suggested that Cl helped lower the activation energy barrier, increasing the atom diffusion and IMC growth. With higher Cl content, the atoms diffusion is much easier, therefore higher IMC thickness.

ACKNOWLEDGEMENTS

The authors would like to express their sincere gratitude to Wintech-nano for their support for this study.

REFERENCES

[1] C.L. Gan, Ng EK, Chan BL, Classe FC, Kwuanjai T, Hashim U, Nanomaterials, 2013, pp.1–9

[2] B.K. Appelt, A. Tseng, C.H. Chen, Y.S. Lai, Microelectronics Reliability, 2011, pp. 13-20.

[3] C.D. Breach, Gold Bulletin, 2010, 43:150–168

[4] J. Z. Liao, X. Zhang, B. S. Wang, X. M. Li, Y. N. Hua, C. Fu, W. K. Tee, B. H. Yee, S. L. Mao, The 17th IEEE International Conference on IC Design and Technology, June 17th–19th, 2019, Suzhou, China

[5] L. G. Chong, F. Classe, B. L. Chan, Uda Hashim, Gold Ball, 2013, Vol. 46, pp. 103-115.

[6] M. H. Lue, C. T. Huang, S. T. Huang & K. C. Hsieh, Journal of Electronic Materials, vol. 33, 2004, pp.1111-1117.

THE VERIFICATION OF TDDB ACCELERATION MODEL IN ULTRATHIN GATE DIELECTRIC

Wen Ying*, Canny Chen, Atman Zhao

Q&R, Semiconductor Manufacturing International Corporation (SMIC) Shanghai 201203, China

*Corresponding Author's Email: Wen_Ying@smics.com

ABSTRACT

The usage of Time-dependent dielectric breakdown (TDDB) acceleration model has intensively debated for many decades. According to JEP122H, the thickness of gate dielectric (GD) between 2 ~ 4 nm is the controversial area, as power law model is used for thickness under 2 nm while E model for above 4 nm. Choosing appropriate acceleration model is significant in lifetime projection, thus we took three experiments to verify which model was the appropriate one between 2 ~ 4 nm. Moreover, as each model has its own physical mechanism, it would be helpful to know more about the detail of dielectric breakdown.

INTRODUCTION

Gate dielectric (GD) is a crucial layer in a Metal Oxide Semiconductor Field Effect Transistor (MOSFET). With the aggressive scaling in microelectronics, GD is thinner and thinner to meet the demand of higher performance. In order to evaluate the lifetime of GD, accelerating test is an essential method, which applies a larger stress than actual to accelerate the breakdown of GD. The stress could be voltage, current or temperature, and voltage is used more frequently compared with current, as it is in accordance with practical use.

It is crucial to choose correct statistic distribution and acceleration model to project the lifetime under actual stress from accelerated stress. So far, Weibull distribution[1] is widely proven as the suitable distribution model, but the acceleration model is extremely controversial. There are four main models: E model[2], sqrt E model[3], 1/E model[4] and power law model[5], with their own physical mechanism: thermochemical model, Poole-Frenkel emission, anode hole injection and anode hydrogen release model, respectively. Each model has its own limitation and according to JEP122H, E model is suitable for GD thicker than 4 nm, while power law model is suitable for GD thinner than 2 nm, but there is controversy on which model is the right one between 2~4 nm.

The lifetime projected by E model and power law model has huge difference, so we must figure out which acceleration model is the correct one between 2 ~ 4 nm GD. In this paper, we took three experiments to verify it. Combining these results, we found that NMOS and PMOS had different acceleration model, even though they had almost the same GD thickness. It means there are other factors on acceleration model except thickness. As each acceleration model corresponds to a unique physical mechanism, it could help us get insight into the detail and influence factor of the breakdown of GD.

EXPERIMENTAL

The GD of NMOS and PMOS we used was around 3.5 nm.

Experiment I: low voltage test. GD Breakdown mechanism under high stress may be different from low stress, but acceleration test usually applies much higher stress than actual to shorten test time, so the higher stress, the more it deviates from true lifetime. In this experiment, we first fit acceleration test data by different models, then applied low stress to verify whose simulated $T_{63\%}$ meet the experimental data. GD area in this experiment keeps the same.

Experiment II: Self-Consistent Acceleration Poisson Statistic (SCAPS)[6]. Although long-term test under low stress is precise, it is very time-consuming, thus, SCAPS method that based on Poisson area scaling raised. Concluding from Poisson area scaling, the voltage or electric field acceleration factor is independent of GD area. We chose a series of GD areas and evaluated their voltage or electric field acceleration factor. The model, who had minimum difference on acceleration factor under different areas, was the most suitable acceleration model.

Experiment III: translation from breakdown voltage (Vbd) to breakdown time (Tbd)[7]. Vbd and Tbd are measured by voltage ramp (Vramp) and TDDB test, respectively. They can convert to each other through unique formula of each model. If the converted Tbd is consistent with actual Tbd, the formula and its corresponding model are correct.

All the above tests were carried out under accumulation mode.

RESULT AND DISCUSSION

Figure 1 showed the result of Experiment I. We used three high voltages to fit the parameters of E model, 1/E model and power law model, as sqrt E model was not suitable to describe GD breakdown, and the corresponding curves were shown in Figure 1. The gap among three models would become larger with voltage being lower, so the $T_{63\%}$ under low voltage could be used to verify the correct model, and we applied four equidistant voltages below 3.1 V to test. As a result, for NMOS, most data fell on the trend line of E model, which mean E model was the right one. However, it was undecidable for PMOS as all data were between E model and power law model.

$T_{63\%}$ versus voltage or electric field with several different areas in the forms of three models were plotted in Figure 2, and the slope of each straight line was the corresponding voltage or electric field acceleration factor. Theoretically, the fitting lines should be parallel as the acceleration factor was independent of areas[6], but actually, there was no exactly parallel lines due to test error. In order to minimize the effect of test error, we used variance to confirm which group had the closest acceleration factor, and the result was list in Table I. The value of acceleration factor was normalized through dividing by the acceleration factor of 1 um^2, because the acceleration factors of three models had different orders of magnitude.

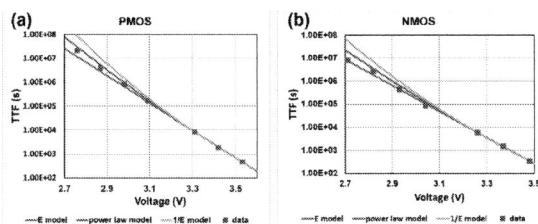

Figure 1: Experiment I: The trend lines of three models and $T_{63\%}$ under different voltages for (a) PMOS and (b) NMOS

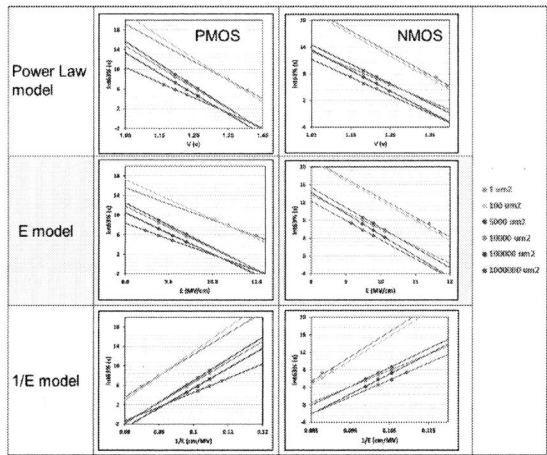

Figure 2: Experiment II: Self-Consistent Acceleration Poisson Statistics. The voltage/electric field acceleration factor under different areas

As Table I showed, E model had the smallest variance for NMOS, which was consistent with the result of Experiment I, and for PMOS, power law model had the smallest variance. Combining with Experiment I, we could conclude that power law model was more suitable

for PMOS.

TABLE I.   NORMALIZED ACCELERATION FACTORS AND THEIR VARIANCE OF THREE MODELS IN EXPERIMENT II

	PMOS			NMOS		
	power law	E model	1/E model	power law	E model	1/E model
1 um2	1.000	1.000	1.000	1.000	1.000	1.000
100 um2	1.179	1.214	1.145	0.999	1.027	0.973
5000 um2	1.189	1.382	1.023	0.886	1.029	0.762
10000 um2	1.122	1.306	0.965	0.759	0.882	0.653
100000 um2	1.094	1.272	0.942	0.937	1.089	0.807
1000000 um2	0.814	0.977	0.677	0.823	0.987	0.686
Variance	0.100	0.139	0.120	0.047	0.024	0.105

To confirm it further, we carried out Experiment III. The Vbd and Tbd were tested by Vramp and TDDB, respectively, and Vbd was converted to Tbd by Equation (1) for power law model, and by Equation (2) for E model[7]. The distribution of actual Tbd and converted Tbd was shown in Figure 3. As expected, power law model had more coincident distribution than E model. That is to say, power law model was more suitable than E model for PMOS.

$$T_{BD}(V_{REF}) = \frac{V_{REF}}{R(n+1)} * \left(\frac{V_{BD}}{V_{REF}}\right)^{n+1} \qquad (1)$$

$$T_{BD}(V_{REF}) = \frac{1}{\gamma R} * e^{-\gamma V_{REF}} * (e^{\gamma V_{BD}} - 1) \qquad (2)$$

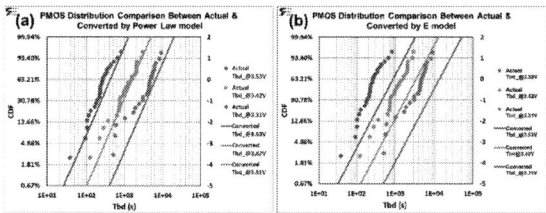

Figure 3: Experiment III: Vbd to Tbd. Comparison between actual Tbd and converted Tbd. The result showed (a) power law model had more coincident distribution than (b) E model

Concluded from the above three experiments, it was interesting to see that NMOS and PMOS had different acceleration models, which mean they had different breakdown mechanisms, even though with almost the same GD thickness. The difference might come from the polarity of applied voltage, and it deserved further study.

CONCLUSION

We took three experiments to verify the correct acceleration model of GD thickness around 3.5 nm, both in NMOS and PMOS. Experiment I should be the most accurate but time-consuming among the three experiments. If there was a need for quick verification, Experiment II and III were better. The results showed that E model was right for NMOS, while power law model for PMOS. It was a little confuse but interesting, because material, thickness and voltage value were generally thought to be the critical factors of GD breakdown mechanism, but the above result introduced a new factor, which might be related to the polarity of gate voltage. It might be a novel topic for further research, and develop a new perspective for GD breakdown.

REFERENCES

[1] E. Wu, R. Vollertsen, and J. Suñé, *Reliability Wearout Mechanisms in Advanced CMOS Technologies*, S. Tewksbury and J. Brewer, Eds. Hoboken, NJ, USA: Wiley, 2009, chs. 2–3, pp. 71–329.

[2] J. W. McPherson and R. B. Khamankar, *Semicond. Sci. Technol.*,vol. 15, no. 5, 2000, pp. 462–470.

[3] S. M. Sze, *J. Appl. Phys.*, 38: 1967, pp 2951–2956.

[4] I.-C. Chen, S. E. Holland, and C. Hu, *IEEE Trans. Electron Devices,* vol. 32, no. 2, Feb. 1985, pp. 413–422.

[5] E. Y. Wu, J. Aitken, E. Nowak, A. Vayshenker, P. Varekamp, G. Hueckel, J. McKenna, D. Harmon, L. K. Han, C. Mostrose, R. Dufresne and R.-P. Vollertsen. *Digest of the 2000 Int. Electron Device Meeting*, 2000, pp 54–57.

[6] E. Wu, A. Vayshenker, E. Nowak, J. Suñé, R.-P. Vollertsen, W. Lai and D. Harmon, *IEEE Trans. Electron Devices*, vol. 49, no. 12, Dec. 2002, pp. 2244–2253.

[7] A. Kim, E. Wu, B. Li and B. Linder, *2019 IEEE International Reliability Physics Symposium (IRPS)*, Monterey, CA, USA, 2019, pp. 1-5.

THE DESIGN-BASED INSPECTION STRATEGY FOR CU VOID DEFECTS REDUCTION

Xingdi Zhang[1], Hunglin Chen[1], Yin Long[1], and Kai Wang[1]

Shanghai Huali Integrated Circuit Corporation, Shanghai 201314, China

ABSTRACT

With the decrease of line width, it is a big challenge for Cu gap-filling in Back-end-of-line (BEOL) which can induce Cu void defects. Poor Cu gap-filling can cause yield loss and problems of reliability, so Cu void defects should be reduced as soon as possible in early stage of research and development. The influencing factor is various for Cu gap-filling, so many experiments are necessary for reduction of Cu void defects. The accuracy and timeliness of results after Cu void defects inspection are important. In this paper, the design is introduced into Cu void defects inspection to develop a design-based strategy. Using the new strategy not only greatly improves accuracy, but also greatly saves time, which can speed up the research and development progress.

INTRODUCTION

The tolerance of tiny physical defects in the integrated circuit (IC) manufacturing becomes smaller and smaller, since advanced semiconductor process continually drive the line width shrinking. In 55nm technology, tiny Cu void defects are gentle, but they would become fatal killers in advanced node. There are various root causes for formation of Cu void in different locations. The various solutions could be used to reduce Cu void defects, so the accurate classifications of Cu void defects are necessary in the procedure of defect reduction.[1]-[3] Images of Cu void defects in different locations are showed in Fig1.

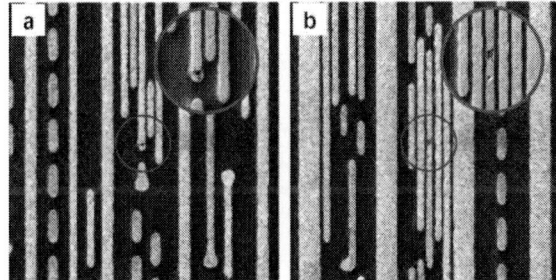

Fig1 Images of Cu void defects (a)in the middle of metal line;(b)in the end of metal line

The signal of defect is stronger than background noise, so defect can be inspected after filtering the noise. But signal of background noise is also strong in some layers, so the noise is difficultly filtered. When defects are inspected, some noise is also inspected which leads to the higher total count. The inspected noise is called as nuisance. The high nuisance ratio can affect the accuracy of results. [4]-[8] Metal void defects are inspected post Cu CMP. Because of the transparent inter metal dielectric (IMD) and rough copper grains, the nuisance count is high post Cu CMP inspection which can influence the accuracy of counted defects counts.

In this paper, an design-based strategy is developed to inspect metal void defects. Using the new strategy not only greatly improves accuracy, but also greatly saves time, which can speed up the research and development progress. Metal void defects were reduced rapidly by using the design-based strategy.

DESIGN-BASED INSPECTION STRATEGY

Traditional inspection strategy is the combination of inspection and review, but design-based strategy is only inspection. The procedure comparison of two different strategy in fig2.

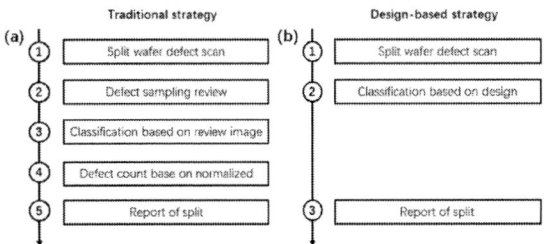

Fig2 The procedure comparison of two different strategy (a) traditional strategy and (b) design-based strategy

In order to obtain accurate defect quantity, It is optimal to review all defects which are from inspection, but the idea is unavailable in any foundry because of capacity constraint. The common practice is to use sampling review firstly, and then normalize to obtain defect count after defect classification. The normalized defect count is usually regarded as the actual defect count. The calculation method is as follows:

$$Normalized\ count = \frac{Classified\ count}{reviewed\ count} \times total\ count$$

For example, the total count is 500, and 100 of them are sampled to review. After defect classification, DOI (defect of interesting) is 25, so the normalized defect count is 125. In general, the normalized defect count is comparable with actual defect count.

As shown above, the nuisance count is high post Cu CMP inspection which can influence the accuracy of results. Two post Cu CMP wafers were selected to confirm the influence of nuisance. There were about 500 defects on wafer A, and about 5000 defects on wafer B. The defect maps are showed in Fig3.

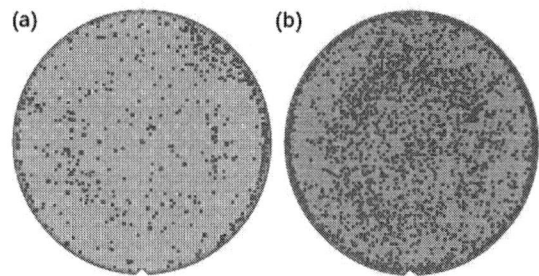

Fig3 The defect maps of two different wafers (a) wafer A and (b) wafer B

Firstly, traditional sampling review method was chosen to obtain normalized defects count of metal void. Then, all defects were reviewed on two wafers to confirm the accuracy of normalized defect count. As the result showed that the normalized count were comparable with actual count on wafer A, but the normalized count were higher than actual count on wafer B. (Fig4) The influence of nuisance is obvious for normalized defect count.

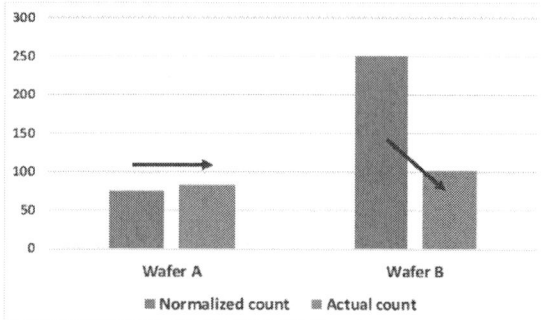

Fig4 The comparison of normalized count and actual count on different wafers

For eliminating the influence of nuisance, design-based strategy was developed. In design-based strategy, every defect have a design image which can be used to classify defects. The strategy is similar to the method of totally reviewing because every defect have a design image. Differently, design images can be obtained during inspection process, and additional operation such as reviewing in SEM can't be required. Furthermore, defect location can be marked in design image during inspection process. As mentioned above, there are various root causes for formation of Cu void in different locations, so it is instructive that classifying metal void in different location to develop solutions. For wafers with high

nuisance, the accurate count of single defect can't be obtained in traditional strategy, let alone the multiple defects. As shown in Fig5, marked location (in red dashed box) in design images matched with location of defects in SEM images.

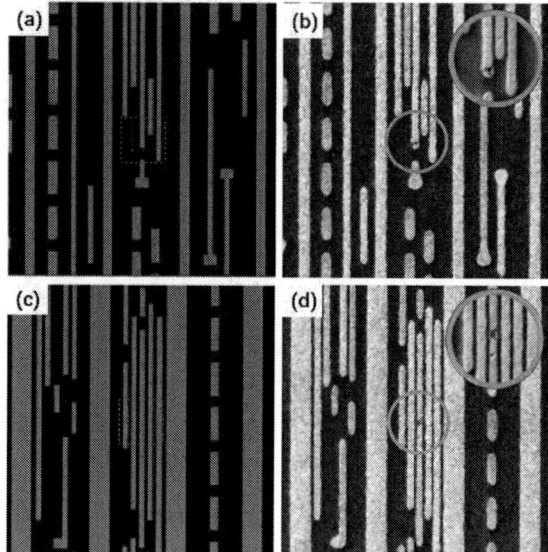

Fig5 Design images of Cu void defects (a) in the end of line and (c) in the middle of line; SEM images of Cu void defects (b) in the end of line and (d) in the middle of line

The comparison of defects quantity counted by design-based strategy and actual count is shown in Fig6. (The wafer B was selected to collect data) The total metal void count, especially metal void count in different location, are comparable with actual count. The effectiveness and practicability of design-based strategy are confirmed well.

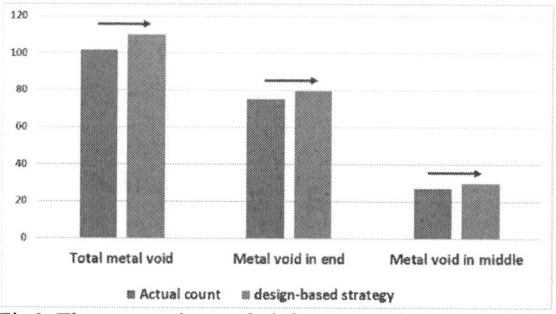

Fig6 The comparison of defects quantity counted by design-based strategy and actual count

In addition to accuracy, timeliness is another advantage of design-based strategy. The review step can be skipped because of classification by design image. It takes about 40 minutes to review one wafer, so 90 minutes can be saved on a wafer at least if the waiting time and inspecting time are considered. For an experiment with 8pcs wafer, the results can be showed 720 min in advance which can speed up the research and development progress.

CONCLUSION

In this paper, an design-based strategy is developed to inspect metal void defects. Using the new strategy not only greatly improves accuracy, but also greatly saves time, which can speed up the research and development progress. Metal void defects were reduced rapidly by using the design-based strategy.

ACKNOWLEDGEMENTS

Thank you very much for the leadership and colleagues for their support of my work

REFERENCES

[1] Weiye He, Beichao Zhang, Jian Kang et al., "The contributions of barrier resputter for BEOL integration", *ECS Transactions*, vol. 44, no. 1, pp. 487-492, 2012.

[2] Yu Bao, Xuezhen Jing, Jingjing Tan et al., "Optimization of Metallization Processes for 28-nm-node Low-k /Cu Multilevel Interconnects", ECS Transactions, vol. 44, no. 1, pp. 477-480, 2012.

[3] Xuezhen Jing, Jingjing Tan, Jiquan Liu, 32/28NM BEOL CD GAP-FILL CHALLENGES FOR METAL FILM, CSTIC, 2015.

[4] I. Bezel et al, "High power laser-sustained plasma lightsources for KLA-Tencor broadband inspection tools" 2015Conference on Lasers and Electro-Optics (CLEO), 2015

[5] A. Srivastava, H. Nguyen, T. Herrmann, R. Kirsch, and R. M. Kini, "Inline inspection of DRC generated hotspots," 2015 26th Annual SEMI Advanced Semiconductor Manufacturing Conference (ASMC), Saratoga Springs, NY, pp. 336-339, 2015.

[6] R. Thaiphan and T. Phetkaew, "Comparative Analysis of Discretization Algorithms on Decision Tree," 2018 IEEE/ACIS 17th International Conference on Computer and Information Science (ICIS), Singapore, pp. 63-67, 2018

[7] A.S. More, Dipti P. Rana, "Review of random forest classification techniques to resolve data imbalance," 1st International Conference on Intelligent Systems and Information Management (ICISIM) 2017

[8] SangHyun Lee et al. "Machine Learning Approaches for Nuisance filtering in Inline Defect Inspection" 2019 30th Annual SEMI Advanced Semiconductor Manufacturing Conference (ASMC)

ANOMALY DETECTION OF NON-NORMAL DISTRIBUTION WAFER ACCEPTANCE TEST DATA VIA GMM-BASED METHOD

Junjun Zhuang[1], Yong Wang[1], Guiyun Mao[1], Xu Chen[1], Yansheng Wang[1], Zhengying Wei[1]*
[1] Shanghai Huali Microelectronics Corporation, Shanghai, China
*Corresponding Author's Email: chenxu@hlmc.cn
† The authors contributed to the paper equally

ABSTRACT

This paper focuses on techniques for automatically anomaly detection of wafer acceptance test (WAT) data from a production database based on Gaussian Mixture Model (GMM). An integrated analysis scheme is developed to facilitate interpretation of the results. It consists of four phases: Kernel Density Estimation, parameter initialization ,GMM iteration until convergence, and anomaly detection. A field data case study shows that this analysis scheme is able to diagnose the non-normal distribution data of WAT which is difficult to detected by system process control (SPC) rule, which is helpful to engineers take sight into the problems of specific manufacturing process, and is also very useful in process development and debugging.

Keywords—Anomaly Detection, GMM, Wafer Acceptance Test, Non-normal Distribution

INTRODUCTION

Semiconductor manufacturing is one of the most important industries in the world [1]. Among the processes of semiconductor manufacturing, quality control is significant for its cost saving and in-time delivery [2]. Among these quality control steps, the CP process determines the wafer yield, but this process needs tremendous time spent on expensive and specialized equipment. The result of wafer yield based on WAT parameters is therefore used by engineers to reduce manufacturing time and production costs spent on CP process [3][4].

As the size of integrated circuits continues decreasing and the processing technology becomes more complicated, the number of parameters that needs to be tested is gradually increased during the WAT process, the corresponding time consumption and test cost are increased at the same times [5]. In view of the large amount of WAT parameters, the distribution of parameters is more complex with the variation of different machines and recipes (Figure 1). Sometimes the WAT data has multiple distributions or it has multiple peaks. It does not always have one peak or Gaussian distribution, and one can notice that by looking at the data set. It will look like there are multiple peaks happening here and there. There are two peak points and the data seems to be going up and down twice or maybe three times or four

times. In addition, the WAT results, especially the fail parameters provide aided information for diagnosing root causes of low yield in semiconductor manufacturing and help engineers to fine-tune the manufacturing process [6]. Therefore, it is important to pick out the anomaly WAT data after test and feedback the results to engineers as soon as possible. In practice, traditional methods based on statistical process control (SPC) are limited in gaussian distribution parameters, so it is not suitable for non-normally distributed data. Gaussian mixture models (GMM) are probabilistic concept used to model real-world data sets. GMMs are a generalization of Gaussian distributions and can be used to represent any data set that can be clustered into multiple Gaussian distributions, in other words, the distributions in Figure 1 can be cluster into several gaussian distributions theoretically.

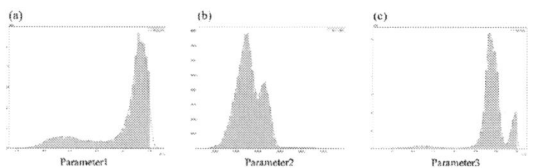

Figure 1. The different distributions of several parameters of WAT.

In this paper, we cluster the input WAT data into several Gaussian distributions using a Gaussian mixture model (GMM) based method, and then applied SPC on different gaussian distributions to detect the anomaly WAT data. Firstly, the Kernel Density Estimation (KDE) is employed to calculate the initial cluster number of GMM model. Secondly, the parameters, such as mean, sigma, weight, are initialized by K-means. Then, iterations of GMM are displayed until the convergence of all parameter. For evaluating the multiple Gaussian distributions of WAT data is whether similar to the original distribution, the Kullback-Leibler (KL) divergence is utilized to calculate the difference between the original distribution and the multiple Gaussian distribution. Until the distribution satisfied the criteria we had set, the parameters would be record for the anomaly detection. The results on our method is discussed on Section **EXPERIMENT**

METHODPLOGY
Kullback-Leibler Divergence

The divergence between the probability density functions (PDFs) of healthy and test data can be achieved by the KLD computation between the two distributions. For discrimination between two continuous probability density functions (PDFs) f(r) and g(r) of a random variable r, the Kullback-Leibler Information (KLI) is defined as:

$$I(f\|g) = \int f(r)\log f(r)g(r)dr \qquad (1)$$

The KL Divergence (KLD) is then defined as the symmetric version of the KL Information [11]:

$$KLD(f,g) = I(f\|g) + I(g\|f) \qquad (2)$$

Kernel Density Estimation(KDE)

Kernel density estimation is the process of estimating an unknown probability density function using a kernel function $K(u)$. While a histogram counts the number of data points in somewhat arbitrary regions, a kernel density estimate is a function defined as the sum of a kernel function on every data point. The kernel function typically exhibits the following properties:

1. Symmetry such that $K(u)=K(-u)$
2. Normalization such that $\int_{-\infty}^{\infty} K(u)\,du = 1$
3. Monotonically decreasing such that $K'(u)<0$ when $u>0$
4. Expected value equal to zero such that E[K]=0

Here we have used kernel="gaussian", as seen above. Mathematically, a kernel is a positive function K (x; h), which is controlled by the bandwidth parameter h. Given this kernel form, the density estimate at a point y within a group of points x_i; i=1,2,3,...N is given by:

$$\rho_K(y) = \sum_{i=1}^{N} K(y - x_i; h) \qquad (3)$$

The bandwidth here acts as a smoothing parameter, controlling the tradeoff between bias and variance in the result. A large bandwidth leads to a very smooth (i.e. high-bias) density distribution. A small bandwidth leads to an unsmooth (i.e. high-variance) density distribution.

When the initial cluster number is confirmed by KDE, another operation should be token to filter the redundant peaks. As shown in Figure 2(a), if the difference between the maximum point and nearest minimum point is less than 1% of the maximum value of the entire distribution, this peak will be redundant. The other case in Figure2(b), if the cumulative probability density on one side of the maximum point is less than 1%, this peak will be redundant also. Finally, the peak number will be confirmed.

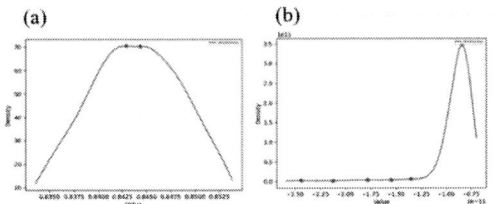

Figure 2. The different distributions with redundant peaks.

K-Means

The K-means algorithm identifies a certain number of centroids within a data set, a centroid being the arithmetic mean of all the data points belonging to a particular cluster. The algorithm then allocates every data point to the nearest cluster as it attempts to keep the clusters as small as possible (the 'means' in K-means refers to the task of averaging the data or finding the centroid). At the same time, K-means attempts to keep the other clusters as different as possible.

It works as follows:

1. The K-means algorithm begins by initializing all the coordinates to "K" cluster centers. (The K number is an input variable and the locations can also be given as input.)
2. With every pass of the algorithm, each point is assigned to its nearest cluster center.
3. The cluster centers are then updated to be the "centers" of all the points assigned to it in that pass. This is done by re-calculating the cluster centers as the average of the points in each respective cluster.
4. The algorithm repeats until there's a minimum change of the cluster centers from the last iteration.

Gaussian Mixture Model

The Gaussian mixture model is a probabilistic model that assumes all the data points are generated from a mix of Gaussian distributions with unknown parameters. A Gaussian mixture model can be used for clustering, which is the task of grouping a set of data points into clusters. GMMs can be used to find clusters in data sets where the clusters may not be clearly defined. Additionally, GMMs can be used to estimate the probability that a new data point belongs to each cluster.

If we obtained initial cluster numbers of WAT data by the method KDE mentioned above, for example three. So we have three initial Gaussian Distribution as GD1, GD2, GD3 having initial mean as μ1, μ2, μ3 and initial variance 1,2,3. Then the final mean and variance of each distribution will be defined after several iterations until model convergence. Then, for a given set of data points GMM will identify the probability of each data point belonging to each of these distributions follow the formula above:

$$N(\mu,\Sigma) = \frac{1}{(2\pi)^{d/2}\sqrt{|\Sigma|}}\exp(-\frac{1}{2}(x-\mu)^T\Sigma^{-1}(x-\mu)) \qquad (4)$$

Where μ = Mean

Σ = Covariance Matrix of the Gaussian

d = The number of features in our data

x = The number of data points

and the probability given in a mixture of K Gaussian where K is a number of distributions:

$$p(x) = \sum_{j=1}^{K} w_j \cdot N(x|\mu_j, \Sigma j) \qquad (5)$$

where w_j is the prior probability of the jth Gaussian

Once we multiply the probability distribution function of d-dimension by W, the prior probability of each of our gaussians, it will give us the probability value X for a given X data point. If we were to plot multiple Gaussian distributions, it would be multiple bell curves. What we really want is a single continuous curve that consists of multiple bell curves. Once we have that huge continuous curve then for the given data points, it can tell us the probability that it is going to belong to a specific class.

EXPERIMENT

As described above, the WAT data from the previous 6 months were used as baseline after removing the outliers. Firstly, the KDE method was applied to initialize the dimensions of multiply Gaussian distributions, which was equal to the peak number of KDE. Meanwhile, the redundant peaks that were shown in Figure 2 were filtered to obtain the final dimension n that would be passed to GMM. The brief pipeline was shown in Figure 3. The purpose of initialization by KDE was to reduce the iteration times of GMM, because we could start these parameters as 1-d dimension and increase the dimension after every convergence. Considering the time consuming, KDE can give a rough number of initial dimension rather than taking it from 1.

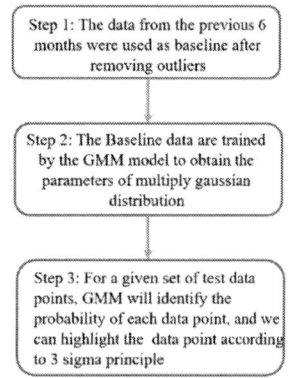

Figure 3. The brief pipeline of this work

After obtaining the dimension n, we used K-means to initialize the mean, sigma, weight of each Gaussian dimension with the cluster number n. GMM iterated with initial mean, sigma, weight of each dimension until the parameters converging. Then the KL divergence was applied to evaluate the deviation between multiple gaussian distributions and the original distribution from raw data. If the value of KL divergence is less than 0.5, the two distributions are considered to be similar, otherwise the cluster number will add 1 to the next iteration, until the KL divergence satisfied the given criteria (Figure 4).

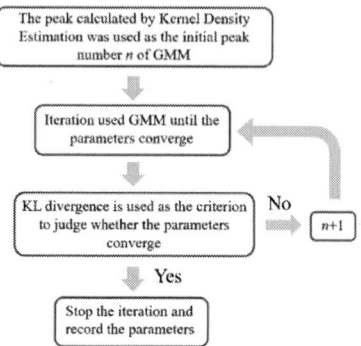

Figure 4. The iteration flow of GMM.

Figure 5 showed the results of multiple Gaussian distributions with the original distribution, the distribution in green curve was raw data, and the areas in other colors were multiple Gaussian distributions. As we can see, the distribution in Figure 5(a) was divided into two Gaussian distributions, and the other was clustered into by three Gaussian distributions (Figure 5(b)), which illustrated that the GMM was an efficient method to cluster the data with any distribution to multiple Gaussian distributions.

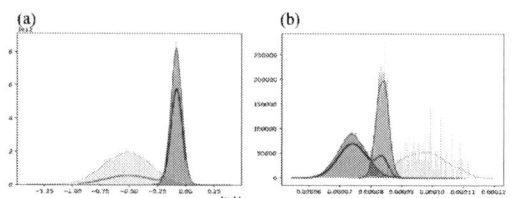

Figure 5. The result of GMM. The original distributions (green curve) and the Gaussian distributions (areas in color) that are fitted with GMM.

After getting the multiple Gaussian distributions, for a given data point, we could calculate its probability value in every Gaussian distribution according to formula (5). In this study, we followed the 3 sigma principle to pick out the outliers, in other words, if the probability value of data point was less than 0.03% in any Gaussian distribution, it would be identified as an outlier. The highlight result in Figure 6 on real world WAT data showed that the method was appropriate to highlighting the outliers of non-normal distributions.

Figure 6. The outliers are highlighted by our method.

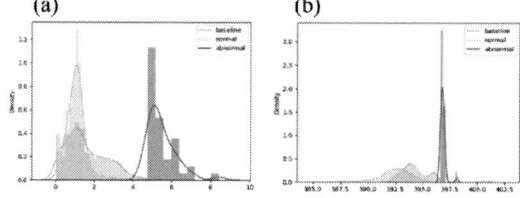

CONCLUSIONS

An innovative method which is based on GMM is proposed to detect the anomaly data after WAT test. It is composed with kernel density estimation, K-means and Gaussian mixture models. An empirical study shows that the proposed method can effectively improve the efficiency and accuracy of anomaly detection of WAT data with non-normal distribution because we raise an idea that converting the non-normal distribution to multiple Gaussian distributions. It can be considered as powerful method for implementing anomaly defection 1-d dimension test data in the semiconductor manufacturing.

REFERENCES

[1] Ying, Z. , et al. "Model Quality Evaluation in Semiconductor Manufacturing Process With EWMA Run-to-Run Control." IEEE Transactions on Semiconductor Manufacturing 30.1(2017):8-16.

[2] Ruth, J. , and R. Berndt . "Quality control for ultrafiltration of ultrapure water production for high end semiconductor manufacturing." 2016 27th Annual SEMI Advanced Semiconductor Manufacturing Conference (ASMC) IEEE, 2016.

[3] Tan, F. , et al. "Survey on Run-to-Run Control Algorithms in High-Mix Semiconductor Manufacturing Processes." IEEE Transactions on Industrial Informatics 11.6(2015):1-1.

[4] Moyne, J. , J. Samantaray , and M. Armacost . "Big Data Capabilities Applied to Semiconductor Manufacturing Advanced Process Control." IEEE Transactions on Semiconductor Manufacturing 29.4(2016):283-291.

[5] Zhu, Q. , et al. "Optimal Scheduling of Complex Multi-Cluster Tools Based on Timed Resource-Oriented Petri Nets." IEEE Access 4(2017):2096-2109.

[6] Yang, F. J. , et al. "Petri Net-Based Efficient Determination of Optimal Schedule for Transport-dominant Single-Arm Multi-Cluster Tools." IEEE Access PP.99(2017):1-1.

IMPACT OF INTERFACE TRAP DENSITY ON THE ENDURANCE OF HFO₂/SI FEFETS

Jiaqi Zheng[1], Yue Peng[2], Yanbin Yang[3], Dawei Gao[1], Rui Zhang[1,*], and Genquan Han[2]

[1]School of Micro- and Nano-Electronics, Zhejiang University, Hangzhou 311200, China
[2] School of Microelectronics, Xidian University, Xi'an 710071, China
[3]Institute of Zhejiang Intelligence Lab, Chengdu 610213, China
*Corresponding Author's Email: ruizhang@zju.edu.cn

ABSTRACT

The endurance of HfO_2/Si FeFETs are analyzed from the view point of MOS interface quality. It is found that endurance degradation in the FeFETs is strongly correlated with interface trap density (D_{it}) at MOS interface, suggesting that the MOS interface passivation is mandatory for improving the performance of FeFETs.

INTRODUCTION

HfO_2-based ferroelectric field-effect transistors (FeFETs) are considered as a promising candidate for next-generation nonvolatile memory (NVM), because of a number of advantages such as low operation voltage, CMOS-compatible and faster operating speed [1][4]. However, limited endurance significantly restricts the application of FeFETs. Typically, the memory window of the HfO_2-based FeFETs closes after about 10^4-10^5 program/erase cycles [5]. Although many papers have mentioned the correlation between D_{it} in the ferroelectric gate stack and endurance in theory [6], there is still no article that directly confirms the determinant role of D_{it} on endurance fatigue with experimental data. In this paper, the impact of D_{it} on the endurance of HfO_2/Si FeFETs are investigated experimentally. It is found that the D_{it} of 10^{12} cm⁻²eV⁻¹ may induce a memory window close within ~10^8 operation cycles, suggesting that the MOS interface passivation is mandatory for FeFET devices.

EXPERIMENT

The HfO_2/Si FeFETs were fabricated using a gate first process (Fig. 1). The (100) p-Si wafer (1-10 Ω·cm) was cleaned using a standard RCA process, followed by ALD deposition of HfO_2 ferroelectric film. Tungsten was deposited by sputtering and patterned as the metal gate.

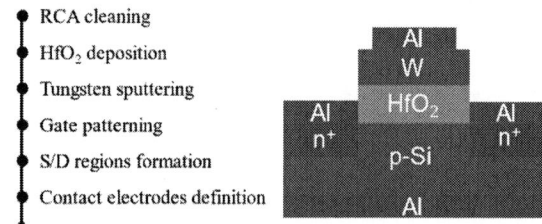

Figure 1: Structure and fabrication process for the HfO₂-based FeFETs.

After that, the S/D regions were formed by phosphorous ion implantation followed by rapid thermal annealing. Finally, the Al contact pads were deposited for S/D, gate and back contact.

INTERFACE TRAP DENSITY AND ENDURANCE MEASUREMENTS

Fig. 2(a) shows the I_d-V_g curves measured from a FeFET device. A normal nMOSFET operation has been confirmed with an ON/OFF ratio of ~10^4. Fig. 2(b) shows the sweep scan of the I_d-V_g curve. A clear memory window of ~1 V is observed, representing the excellent storage characteristics of the FeFETs.

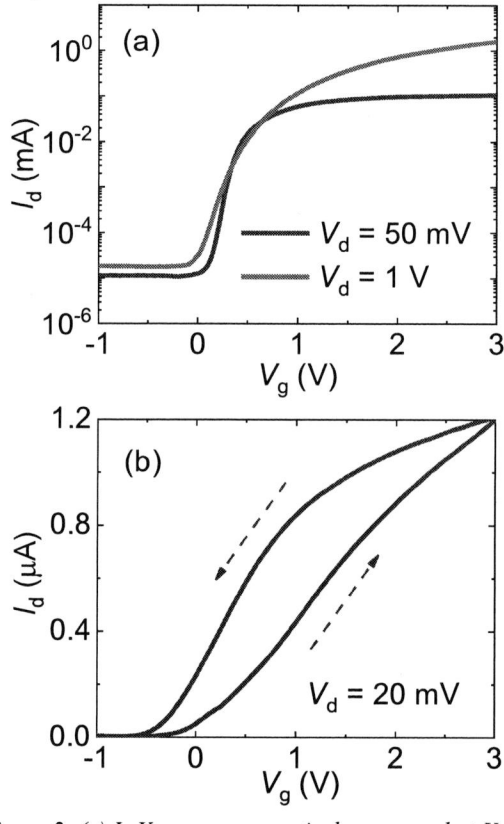

Figure 2: (a) I_d-V_g curves respectively measured at V_d = 50 mv and 1 V. (b) I_d-V_g curve with high /low V_{th} state obtained by forward/backward V_g scanning.

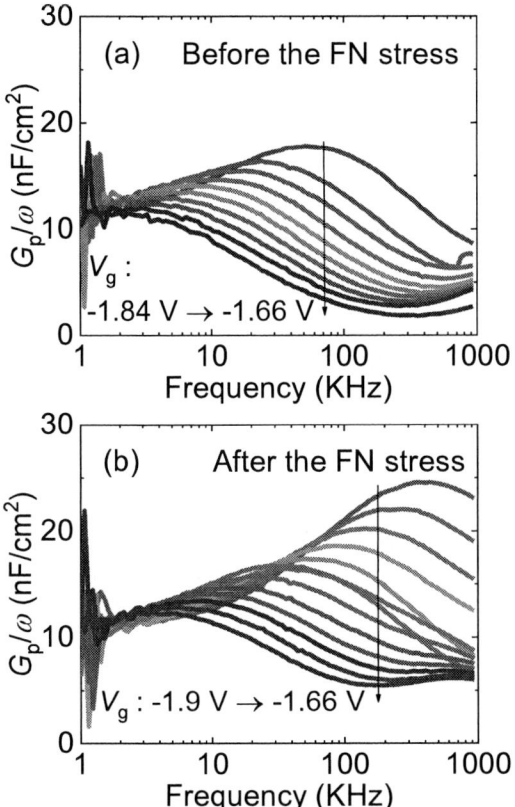

Figure 3: Frequency dependence of G_p/ω (a) before the FN stress while V_g ranging from -1.84 V to -1.66 V and (b) after the FN stress while V_g ranging from -1.9 V to -1.66 V.

In order to characterize the impact of D_{it} on the endurance of FeFETs, the different positive Flower-Nordheim (FN) stress was applied to generate different amounts of D_{it} at the HfO_2/Si MOS interface. The low temperature conductance method was employed to quantitatively evaluate the D_{it}, with considering the surface potential fluctuation [7][9]. Fig. 3(a) and (b) show the full conductance curves taken from the gate stack of a FeFET before and after the FN stress. It is confirmed that the conductance peak increase after the FN stress, attributing to the D_{it} increase at the HfO_2/Si MOS interface. Fig. 4(a) and (b) show the D_{it} distribution of energy and the time constant τ_p of the HfO_2/Si gate stack after different FN stress conditions. The D_{it} changes from 10^{11} to 10^{12} $cm^{-2}eV^{-1}$ after different FN stress conditions. On the other hand, these D_{it} exhibit similar τ_p, meaning that the FN stress-induced interface traps represent similar properties with the original ones in the gate stack.

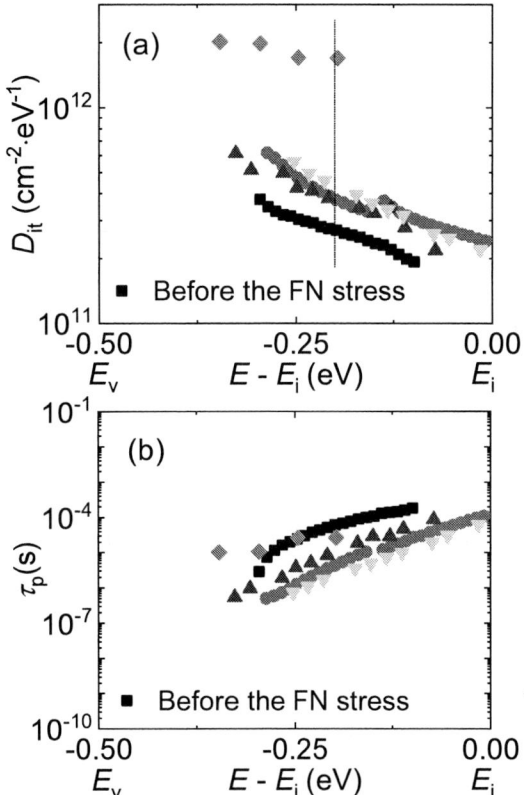

Figure 4: Distribution of (a) D_{it} and (b) τ_p in the energy band at the HfO_2/Si gate stack after different FN stress conditions.

Fig. 5(a) and (b) show the P-V curves of the HfO_2/Si FeFETs with different D_{it} in the gate stack. It is found that the FeFETs with a larger D_{it} exhibit a clear P_r decrease after 3.4×10^5 program-erase cycles. In contrast, the device with lower D_{it} does not show clear P_r degradation

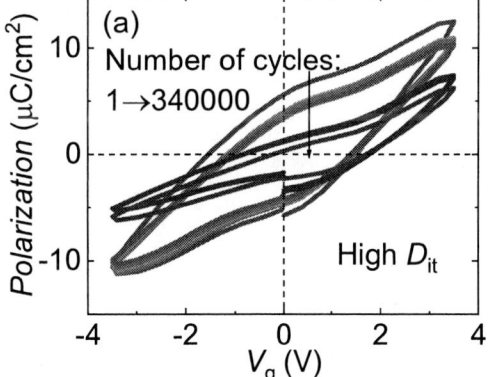

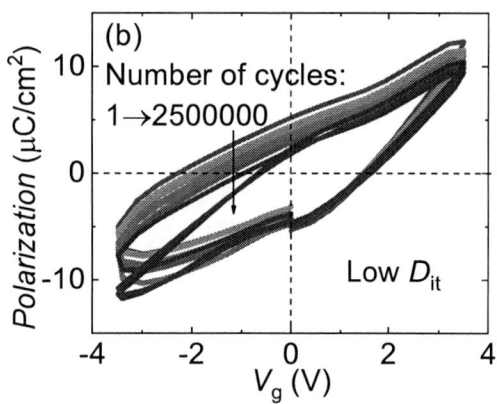

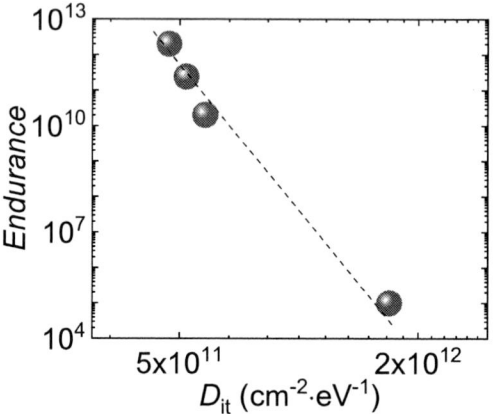

Figure 5: Polarization-V_g curves after a different number of cycles of triangular waves (a) under high D_{it} condition and (b) under low D_{it} condition.

Figure 7: The relationship between endurance and D_{it} after different FN stress conditions. Note that the D_{it} takes the value at E_i-0.2 eV from Figure 4(a).

even after a 2.5×10^6 program-erase cycles. The fitting process is carried out with an exponential relationship between P_r and the cycle number to determine the endurance of FeFETs at a condition of $2P_r = 1 \ \mu C/cm^2$ (Fig. 6) [10]. As shown in Fig. 7, a clear relationship is obtained between endurance and D_{it}, and a D_{it} (at E_i-0.2 eV) of $10^{12} \ cm^{-2}eV^{-1}$ is mandatory to guarantee an endurance of $\sim 10^8$.

CONCLUSION

In this study, the impact of MOS interface quality on the endurance of HfO$_2$/Si FeFETs is investigated. It is confirmed that the FeFETs endurance is significantly degraded with increasing the D_{it} at MOS interface. These results suggest that the MOS interface passivation is a sufficient approach to enhance the endurance of FeFETs.

ACKNOWLEDGEMENT

The authors would greatly appreciate the kind guidance from Prof. Hanming Wu. This work was supported by Natural Science Foundation of Sichuan Province (No. 2021YJ0093).

REFERENCE

[1] Y. -S. Liu, *IEEE TED*, vol. 68, pp. 1639, 2021.
[2] T. Hatanaka, *VLSI Symp.*, 2009, pp. 78.
[3] E. Yurchuk, *IEEE TED*, vol. 61, pp. 3699, 2014.
[4] J. Müller, *VLSI Symp.*, 2012, pp. 25.
[5] S. Mueller, in IEEE TED, vol. 60, pp. 4199, 2013.
[6] H. Zhou, 2020 *International Memory Workshop (IMW)*, 2020, pp. 1.
[7] J. Li, *IEEE JEDS*, vol. 8, pp. 350, 2020.
[8] H. Bae, *IEEE EDL*, vol. 33, pp. 1138, 2012.
[9] Y-C. Chien, *IEEE International Symposium on the Physical and Failure Analysis of Integrated Circuits (IPFA)*, 2020, pp. 1.
[10] K-Y. Hsiang, *IEEE IRPS*, 2022, pp. P9-1.

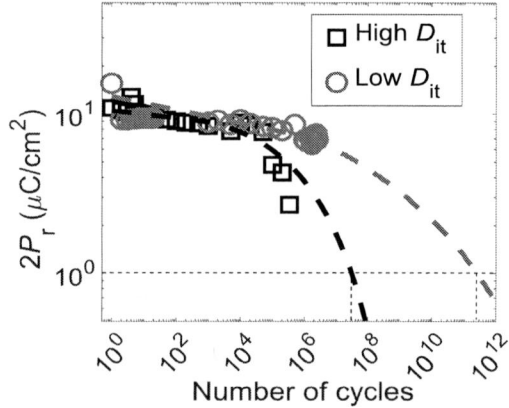

Figure 6: The fatigue characteristics of FeFETs for high D_{it} and low D_{it} conditions. Note that the dashed lines are the fitting curves, and the endurance is taken when $2P_r = 1 \ \mu C/cm^2$.

Reliability Research on micro bump and c4bump in Large-Size 2.5D FCBGA

Xiang Li[1,2]*, Zhuqiu Wang[2], Xiao He[2], Dan Yang[2], Na Mei[1]

[1] State Key Laboratory of Mobile Network and Mobile Multimedia Technology, Shenzhen 518055, China
[2] Department of Reliability Engineering, Sanechips, Shenzhen 518055, China
*Corresponding Author's Email: li.xiang24@sanechips.com.cn

ABSTRACT

The electronic components using flip chip bonding technology have high pin density, good thermal conductivity and excellent electrical performance, and are widely used in the telecommunications field. In this paper, ~5000mm² area 2.5D FCBGA is taken as the research object, and the failure phenomenon of its key interconnect structure under temperature cycle is tested and simulated. In the test, it was found that ubump failed earlier than c4bump, and there were underfill delamination on both sides of the copper pillar with different bump structures. Then, the impact of underfill delamination on the stress of the bump is studied through finite element analysis, and the failure risks of the two kinds of bump structures are compared. Finally, it is proposed that proper underfill materials can reduce the failure risk of bump.

INTRODUCTION

With the advent of the high-density packaging era, 2.5D FCBGA has become an important foundation to support the development of future advanced packaging. For a large-size 2.5D FCBGA, the chip is interconnected to the interposer through ubump, and the interposer is interconnected to the substrate through c4bump. The reliability of both bump structures is critical[1,2]. Due to the CTE of interposer does not match the chip and substrate, the bump failure problems under different temperature stress can be divided into two types: over-stress and fatigue. In this temperature cycle test, it is found that the bump failure is caused by underfill delamination, which belong to thermal stress failure.

Real time monitoring of resistance value was conducted in the temperature cycle test. The temperature condition was -40-125 °C. At ~1050cycles, it was found that the ubump daisy chain failed for the first time, while the c4bump daisy chain failed for the first time at ~1250cycles, and the failures were concentrated in the corner chain. Before cutting the ubump chain, it is found that the corner of the interface between die and ubmp are abnormal through SAM (Scanning Acoustic Microscope), which is suspected to be cracks, as shown in Figure 1. Therefore, when the chip is sliced, cracks are found at the corners of underfill and the ubump and c4bump of outermost corners, as shown in Figure 2 and Figure 3. As can be seen from Figure 2(c) and Figure 2(d), the alpad of ubump on the die side has obvious cracks caused by over-stress, and the delamination between underfill and PI, as well as between underfill and pillar sidewall, can also be seen on both sides of the copper pillar. After preliminary analysis, the delamination of underfill on both sides of the pillar reduces the constraint on the pillar, resulting in excessive pillar stress, which is the main reason for cracks of alpad. And the delamination between underfill and PI on the left side of pillar is usually caused by the crack at the corner of underfill extending inward, as shown in Figure 2(a) and Figure 2(b). Similarly, it can also be seen in Figure 3(b) and Figure 3(c) that c4bump pillar has obvious cracks on the interposer side, and the delamination of underfill

can also be seen on both sides of the pillar. Therefore, it is also considered that the crack of c4bump pillar is caused by the delamination of underfill on both sides of pillar, while the corner of underfill tends to crack and extend internally due to stress concentration, as shown in Figure 3(a).

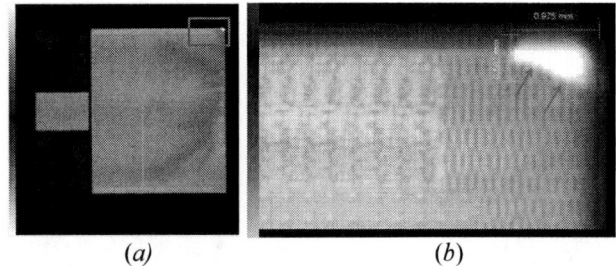

Figure 1: SAT graphs of interface between die and ubump(a) top view of ubump(b) details of corner ubump

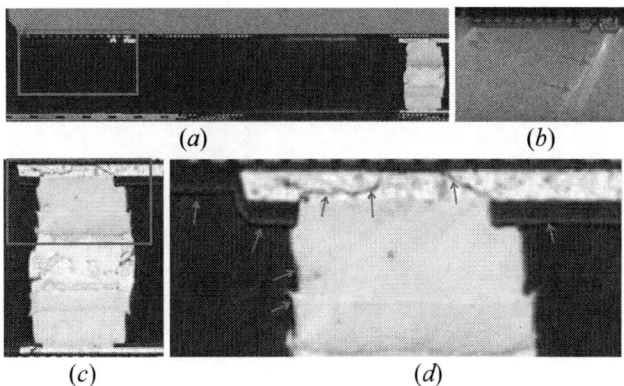

Figure 2: OM graphs of underfill and ubump(a) cross-section of underfill delamination(b) delamination details of underfill(c) cross-section of failure ubump(d) crack details of alpad

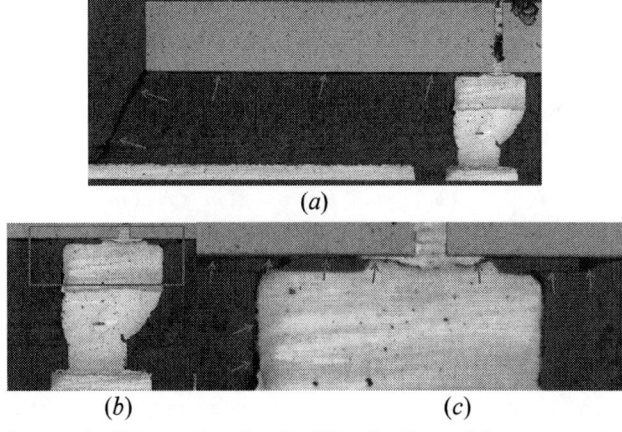

Figure 3: OM graphs of underfill and c4bump(a) cross-section of underfill delamination(b) cross-section of failure c4bump(c) crack details of copper pillar

SIMULATION

To analyze the effect of underfill *delamination*, different bump structures and underfill materials on the stress of ubump and c4bump, we can establish a model of large-size 2.5D FCBGA through finite elements analysis, as shown in Figure 3. In the large-size 2.5D FCBGA, the minimum ubump diameter is only ~25 um, which is three orders of magnitude different from the global model size. In addition, there are too many ubumps and c4bumps. When the finite element method is used to analyze the packaging structure, grid division and calculation are difficult. Therefore, the ubump and c4bump arrays of 15*15 are established in 1/4 global model at the same time[3], as shown in Figure 4(a) and Figure 4(b). the submodel method is used to perform detailed modeling analysis on the local ubump and c4bump of the highest risk, as shown in Figure 4(c) and Figure 4(d).

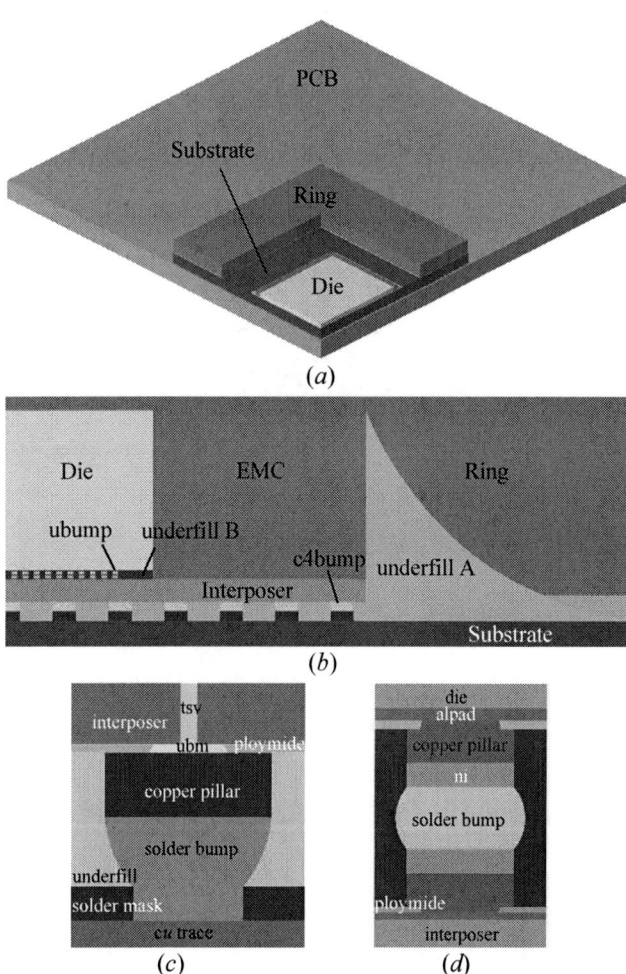

Figure 4: The model and cross-section structure(a)global model(b) cross-section of global model(c) cross-section of c4bump submodel(d) cross-section of ubump submodel

As can be seen from Figure 5(a) and Figure 5(b), the overall maximum stress of the underfill at the ubump and c4bump locations is at the corner, and the underfill stress at the ubump location is greater than that at c4bump, indicating that the underfill at the ubump location is easier to crack at the corner, which is consistent with the previous analysis results. From Figure 5(c) and Figure 5(d), it can be seen that the tensile stress at the underfill and interposer interfaces between c4bumps is relatively large, as is the tensile stress at the

underfill and die interfaces between ubumps, indicating that the underfill around ubumps and c4bumps has a delamination risk dominated by tensile stress. In addition, from Figure 5(e), Figure 5(f), Figure 5(g) and Figure 5(h), it can be seen that the tensile stress in the horizontal direction of the UF around ubump and c4bump is also large, indicating that there is also a risk of delamination dominated by tensile stress on the sides and underfill of the two types of bump. It can be seen that the stress risk location of underfill in the simulation is consistent with the delamination results of previous tests. Moreover, it can be seen that the mises stress and tensile stress of underfill around ubump are greater than those of underfill around c4bump, indicating that the delamination risk of underfill around ubump is higher, which is more likely to lead to the failure of ubump.

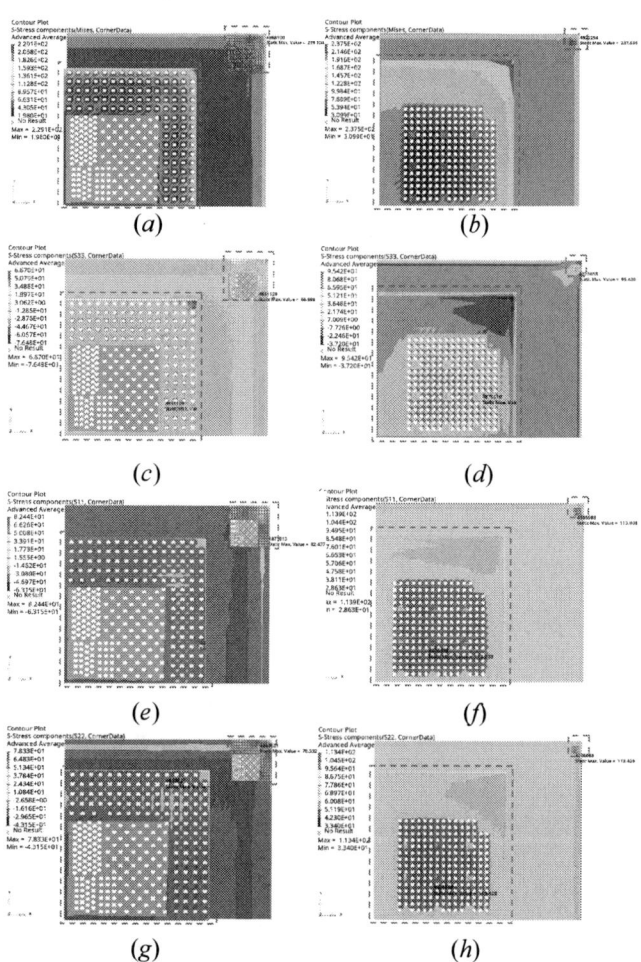

Figure 5: stress in simulation(a) von Mises of underfill around c4bump(b) von Mises of underfill around ubump(c) S33 of underfill around c4bump(d) S33 of underfill around ubump(e) S13 of underfill around c4bump(f) S13 of underfill around ubump(g) S23 of underfill around c4bump(h) S23 of underfill around ubump

As shown in Figure 6(a) and Figure 6(b), in the simulation, the maximum mises stress location of ubump is on the alpad, while the maximum stress location of c4bump is on the ubm. The stress risk location is also consistent with the crack location of the bump in the previous test. Moreover, the mises stress of alpad is greater than ubm, which further indicates that ubump has a higher failure risk compared to c4bump.

979-8-3503-1101-3/23 $31.00 © 2023 IEEE

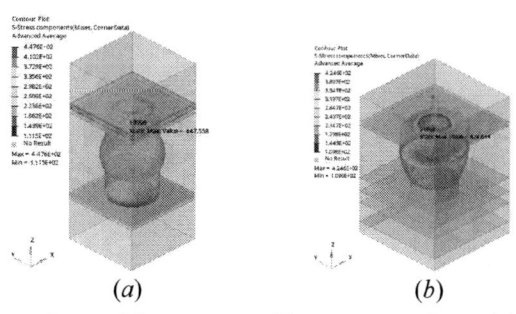

Figure 6: von Mises stress of bump in simulation(a) ubump submodel(b) c4bump submodel

SIMULATION RESULTS AND DISCUSSION
Effect of underfill delamination at different locations

In order to further study the impact of underfill delamination on bump failure, two different delamination models are established for ubump and c4bump. Figure 7(a) and Figure 7(b) show one side delamination model and complete delamination model of the c4bump, respectively. Figure 8(a) and Figure 8(b) show one side delamination model and complete delamination model of the ubump, respectively. Delamination simulation is simplified by not establishing common nodes between finite element mesh. To avoid mesh penetration, frictional contacts are established at the delamination interface.

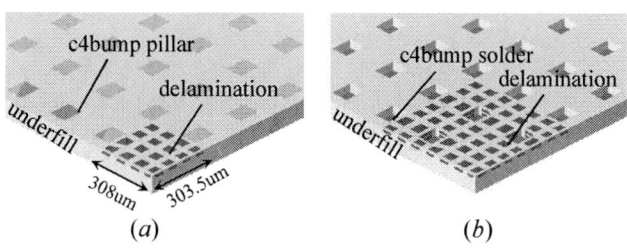

Figure 7: underfill delamination in simulation(a) delamination between underfill and interposer on one side of c4bump(b) delamination between underfill and interposer around c4bump and between underfill and c4bump sidewall

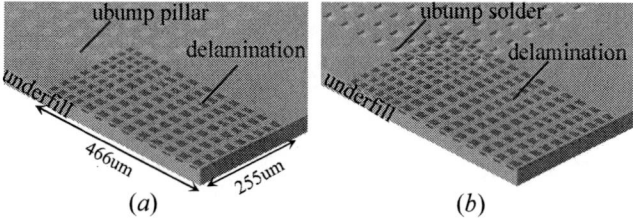

Figure 8: underfill delamination in simulation(a) delamination between underfill and interposer on one side of ubump (b) delamination between underfill and interposer around ubump and between underfill and ubump sidewall

As shown in Figure 9, the maximum mises stress locations for three different models of c4bump are on the pillar, which is consistent with the crack location of the test. Compare the results of Figure 9(a) and Figure 9(b), when underfill and interposer delaminate at the corners, the mises stress on the pillar increases by 2.4%, while the maximum stress position on the pillar moves from the chip side to the corner side. Compare the results of Figure 9(b) and Figure 9(c), it can be seen that the mises stress of complete delamination model increases by

7.9%, which means that the risk of failure is higher, and the maximum stress position on the pillar returns to the chip side from the corner side.

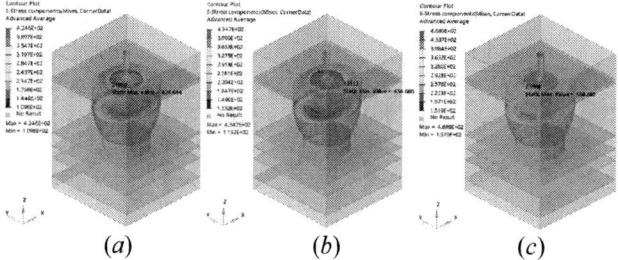

Figure 9: von Mises stress of different delamination models for c4bump in simulation(a) no delamination model(b) one side delamination model(c) complete delamination model

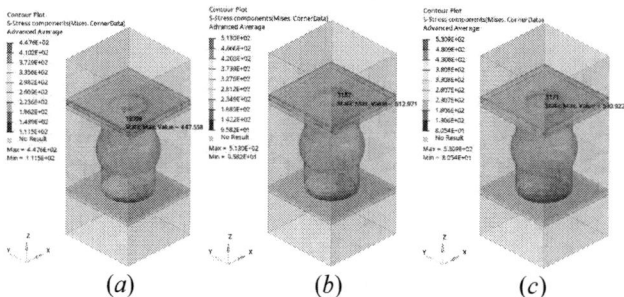

Figure 10: von Mises stress of different delamination models for ubump in simulation(a) no delamination model(b) one side delamination model(c) complete delamination model

Similarly, as can be seen from Figure 10, the maximum mises stress locations for three different models of ubump are on the alpad, which is also consistent with the crack location of the test. Compare the results of Figure 10(a) and Figure 10(b), when underfill and die delaminate at the corners, the mises stress on the alpad increases by 14.6%, while the maximum stress position on the alpad moves from the chip side to the corner side. Compare the results of Figure 10(b) and Figure 10(c), it can be seen that the mises stress of complete delamination model increases by 3.5%, which means that the risk of failure is higher.

Comparing the results of underfill delamination at different locations, it can be seen that both ubump and c4bump have the greatest stress risk after underfill is fully delaminated. Compared to c4bump, the stress in one side delamination model of ubump increases more after delamination, and its maximum stress position is always at the corner side of alpad. From Figure 7 and Figure 8, it can be calculated that the delamination area of ubump in one side delamination model is 27.1% larger than that of c4bump, indicating that the maximum stress of the bump is greatly affected by the area of the underfill delamination, which means that the bump is more likely to crack first on the side where the underfill delamination is more severe.

Effects of different bump structures and underfill materials

In order to fully study the effects of different bump structures and properties of underfill materials on the stress of underfill and bump, as shown in Tables I and II, materials A, B, C, and D are commonly used in engineering. The underfill materials for ubump and c4bump in group 1 are consistent with the test. By comparing the results of groups 2, 3, and 4, it can be seen that the stress of the alpad on ubump is always higher than that of the pillar on c4bump. This indicates that structurally, the failure risk of alpad on ubump is higher, which is also consistent with the previous test results. In addition, comparing group 1, Grou2, and group 5, as well as group 1, Grou3, and group 6, it can be seen that using different underfill materials on the same layer of ubump or c4bump has little impact on the stress of the bump and underfill on the other layer. Therefore, the underfill of ubump and c4 bump can be kept consistent, and the effects of the material properties of underfill on the stress of the same layer of underfill and bump can be studied separately for ubump and c4bump.

TABLE I
MATERIAL PROPERTIES OF DIFFERENT UNDERFILL

Underfill material	Tg (°C)	CTE (ppm/°C)		Young's modulus (Gpa)	
		$\alpha 1$	$\alpha 2$	$E1$	$E2$
Real_A	t	$a1$	$a2$	$e1$	$e2$
Real_B	t	$a1+3$	$a2+15$	$e1$	$e2-0.05$
Real_C	$t-5$	$a1+7$	$a2+45$	$e1-4$	$e2-0.13$
Test_D	$t+15$	$a1+7$	$a2+45$	$e1-4$	$e2-0.13$

TABLE II
STRESS OF UNDERFILL AND BUMP UNDER DIFFERENT UNDERFILL MATERIAL

Group	underfill	Mises stress (MPa)			
		underfill		alpad	pillar
	ubump, c4bump	ubump	c4bump	ubump	c4bump
1	B, A	237.5	229.1	447.6	424.6
2	A, A	230.3	229.3	437.8	424.7
3	B, B	239.8	248.5	448.5	427.3
4	C, C	175.6	238.3	491	435.3
5	C, A	176.6	229.4	488.9	423.2
6	B, C	236.3	237.9	448.9	435.2
7	D, D	164.7	209.9	468.6	431.7

For the c4bump structure, comparing group 2 with group 3, it can be found that due to the decrease in CTE of underfill, the thermal stress generated during the cooling process is smaller, and the stress at the corners of the underfill around c4 bump is significantly reduced by 7.7%, while the stress of pillar is only reduced by 0.6%. Comparing group 4 with group 3, it can be seen that after the modulus of underfill decreases, the stress at its corners decreases by 4.1%. Moreover, due to the reduction of the stress of the underfill itself, the constraints imposed on c4bump are reduced, resulting in an increase in the stress shared by c4bump, while the stress of pillar increases by 1.9%. In addition, comparing group 4 with group 2, the stress at the corners of the underfill increased by 3.8%, and the stress of pillar also increased by 2.5%. The main reason is that the CTE difference between material A and material C is greater than that between material C and B. From the above three groups of comparisons, it can be seen that the CTE and modulus of underfill have a significant impact on the stress of underfill and bump in the same layer, but the trend of CTE and modulus's impact on the stress of bump is opposite. Similarly,

the same rule can be seen for ubump structures. In order to study the effect of the Tg of underfill on the stress of bump and underfill, material D was added as a comparison to material C, as shown in Table I. Through the results of group 7 and group 4, it can be seen that after the Tg of underfill increases, the stress of ubump, c4bump, and the same layer of underfill decreases significantly. Therefore, in practical engineering applications, it is recommended to increase the Tg of underfill as much as possible, reduce its CTE value, and appropriately reduce its modulus, thereby reducing the stress of the bump and underfill.

Conclusion

In large-size 2.5D FCBGA, both ubump and c4bump have a greater failure risk of over-stress. Therefore, the influence of thermal stress on the failure risk of ubump and c4bump was studied by finite element method, and their failure behavior was analyzed. When the underfill at the corner position delaminates on one side of the bump, the stress of the bump increases significantly compared to that without delamination, and the delamination area has a significant impact on the stress of the bump. In addition, when the underfill around the bump completely delaminates, the stress of the bump increases compared to that of incomplete delamination. At this time, the bump is most likely to fail due to over-stress. From the perspective of packaging structure, ubump is more likely to fail than c4 bump, and different underfill materials have little impact on the bump and underfill of the other layer. However, the material properties of different underfills have a significant impact on the failure risk of bumps on the same layer. Therefore, reducing the CTE of underfill, increasing its Tg point, and appropriately reducing its modulus can reduce the failure risk of the bump. However, in actual production, the modulus of underfill generally increases with the decrease of CTE, which can lead to excessive stress in underfill and increase the delamination risk of underfill. Therefore, in practical engineering, it is necessary to comprehensively consider the impact of different underfill materials and select appropriate underfill materials to ensure the interconnection reliability of ubump and c4bump structures in 2.5D packaging.

References

[1] R. Katkar, M. Huynh, and L. Mirkarimi, "Reliability of Cu Pillar on Substrate Interconnects in High Performance Flip Chip Pacakges", Electronic Components and Technology Conference, pp. 965-970, 2011.

[2] H.Y. Gwee and N.K. Kay, "A Sample Preparation on Decapsulation Methodology for Effective Failure Analysis on Thin Small Leadless (TSLP) Flip Chip Package with Copper Pillar (CuP) Bump Interconnect Technology", 40th Int'l Symp. for Testing and Failure Analysis (ISTFA) pp.100-104, 2014.

[3] J WEN REN, "The Geometrical Effects of Bumps on the Fatigue Life of Flip-chip Packages by Taguchi Method", International Conference on Network-based Information Systems. IEEE Computer Society, 2006.

DESIGN AND OPTIMIZATION OF RC TRIGGERED MV-NMOS FOR 28NM CMOS TECHNOLOGY ESD PROTECTION

Jia Zhu[1]*, Lanying Wei[1], Yang Li[1], Jun Wu[1], Kun Wang[1], Wei Chen[1]
[1] Semiconductor Manufacturing International Corporation, Beijing, China
*Corresponding Author's Email: zhujia2100@163.com

ABSTRACT

An effective design of RC triggered medium voltage ESD nMOS power clamp in 28nm high voltage CMOS technology is presented in this work. Through transmission line plus test, it is found that there are two modes in the conduction of the power clamp NMOS: MOS channel conduction and parasitic NPN turn on conduction. Different conduction modes given the device very different robustness. Herein, by optimizing the layout and process condition, the ESD power clamp performance (It2) significantly improved from 0.2A to 1.17A with same area. This is because the parasitic NPN can be turned on in time to release most of the current.

INTRODUCTION

Electrostatic discharge (ESD) device design has been a challenging aspect for their straitly ESD design window in advanced CMOS technologies. For an example, gate grounded NMOS (GGNMOS) is one of the classical ESD protection schemes for their advantages of area saving and process compatible. As shown in Fig.1, The GGNMOS design window must meet Vdd<Vhold<Vtrigger<GOX BV. However, the condition is hardly to meet, such as medium voltage (MV, 8V in this paper) MOS in 28nm high voltage (HV) CMOS technology, the holding voltage of GGNMOS goes lower than operation voltage, because of strong snapback and latch-up immunity issues [1,2]. So we need to searching another way for MV ESD protection strategy. RC-triggered power clamps (PC) have been widely used to provide a higher turn-on speed and a low-impedance path from VDD to GND when under ESD current [3,4]. A typical RC triggered nMOS power clamp circuit is shon in Fig.2, when a ESD event occurs, the gate voltage is logic high to turn on the big nMOS to discharge the ESD current. This kind of power clamp circuit has no snapback characteristic, and thus no risk of latch-up. However, it needs a large area to discharge the current around the channel. Because the channel conduction is less robust than parasitic NPN in GGNMOS [5]. In addition, the resistor and capacitance are also area consuming. Between complex design and area efficiency, many scholars have made optimization attempts [6,7]. In this work, we present 3 types of the big nMOS, which the ESD current conduction modes can be divided into MOS channel and parasitic NPN turn on. Transmission line plus

(TLP) test and technology computer-aided design (TCAD) simulation were used to analyze the physical mechanism and the relative principle of the result.

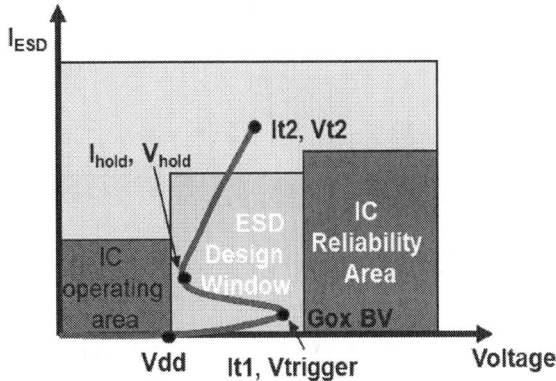

Figure 1: ESD design window and safe operation area.

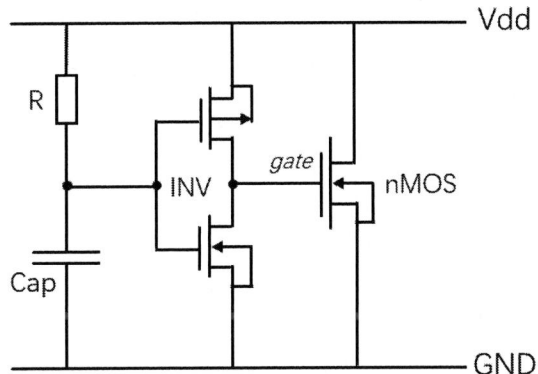

Figure 2: Typical RC triggered nMOS power clamp circuit.

EXPERIMENTS AND THE RESULTS

The test chip was made by 28nm HV CMOS technology consisting with 28V/8V/3.3V/0.9V devices. The 8V nMOS used in this paper is referred as MV device with total width 316 µm and length 0.9 µm. Fig.3 shows the schematic cross-sections of the 3 types MV nMOS with design parameters defined.

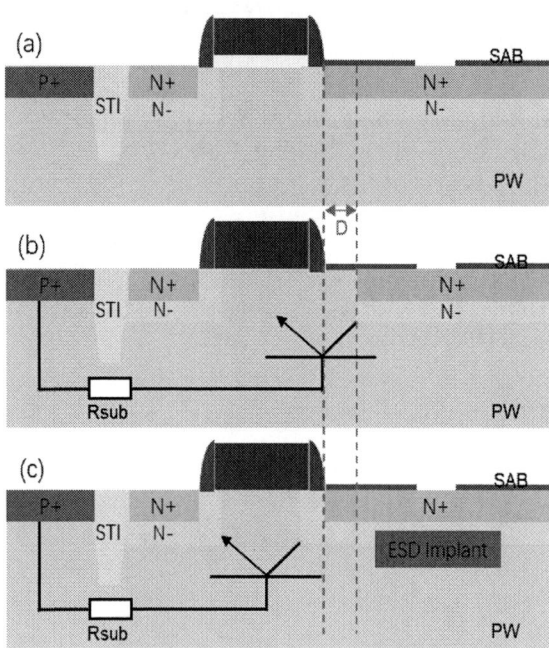

Figure 3: Schematic cross-sections of the 3 types MV nMOS.

Salicide block (SAB) layer at the drain is for ESD protection purpose. Fig.3(a) is the normal MV ESD nMOS. During the Vdd to GND TLP stress, the voltage at node GATE can be rapidly raised to high level, and the channel of the nMOS is formed. Most of the ESD current flows through the channel. The parasitic NPN may turn on weakly, due to the less bulk current. As shown in Fig.4, this kind of nMOS robustness is poor and the It2 is only 0.2A. In order to improve the It2 by turning on the parasitic NPN, we modified the layout of the device, as shown in Fig.3(b) and Fig.3(c). For details, Fig.2(b), the difference to (a) is that there is a gap between N+ and ploy near the drain region. The width of the gap is defined by ESDMK layer which can be controlled by layout design. During the TLP stress, although the inversion layer will be formed under the gate, there is a gap between channel and drain. So that the current can't go from drain to channel directly. The current must be conducted by turning on the parasitic NPN. So that the current conduction changes from surface to bulk. Fig. 4 shows the It2 can reach 1.17A (red line), which is ~6X higher than type (a). For Fig.3(c), a ESD implant was added to reduce the drain to bulk breakdown voltage. When the TLP voltage increasing from 0 to 9V, the current flows through the channel which is same to type (a). As shown in Fig.4, the TLP curves of the type (a) and (c) is coincident before 9V. With the TLP voltage further increasing, the drain to bulk will breakdown, resulting in large bulk current.

Simultaneously, the parasitic NPN turning on, because I*Rsub>Vbe satisfied. In other words, this kind of nMOS will also turn on the parasitic NPN conduction mode before it thermal breakdown near the channel. So the It2 can also reach 1.17A. However, the disadvantage of type (c) is that the drain to bulk breakdown voltage must lower than thermal breakdown, to ensure parasitic NPN turn on before device burnout.

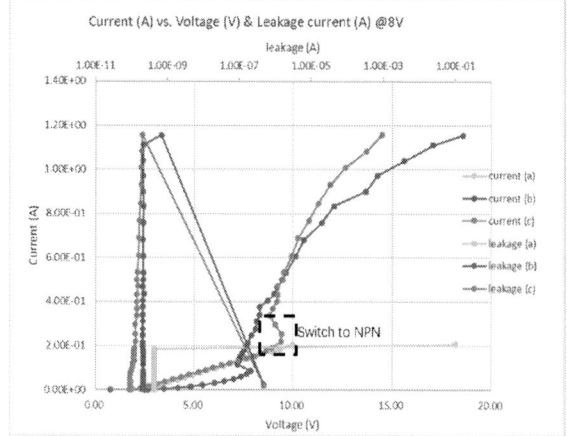

Figure 4: TLP curve of the PC circuits with the 3 types MV nMOS.

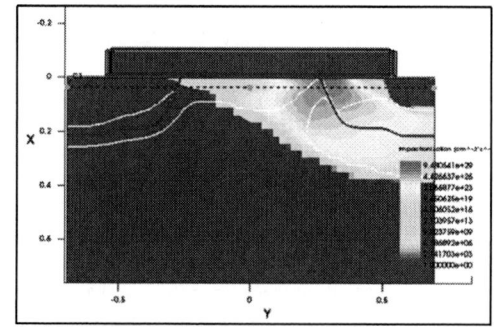

Figure 5: The TCAD simulation.

The TCAD simulation is performed to better understand the physical mechanism of thermal breakdown. On the one hand, for a normal MV nMOS working in saturation region, the channel current is near the silicon and gate oxide surface, and along with channel hot carrier injected [8, 9]. Some electrons and holes in the region can gain enough energy from the high electric field to surmount the interface barrier and enter the SiO2 layer. As for NMOS, the injection from Si to SiO2 are hot electrons and the concomitant hot holes by impact ionization will conduct to bulk. Destructive gate oxide breakdown will occur near the weak point where the most impact ionization region (red highlight region in Fig.5). on the other hand, the current conduction of MOS is under the

979-8-3503-1101-3/23 $31.00 © 2023 IEEE

gate oxide, which the thermal conductance is far less than silicon. So that the MOS conduction model is easier to burnout.

To verify the model, we also collected the TLP curve of the 3 kinds of nMOS as GGNMOS. As shown in Fig.6, the current (a), (b) and (c) is corresponding to nMOS type (a), (b) and (c) in the Fig.3. the 3 types of GGNMOS are parasitic NPN model. But the different is that: (I) the trigger voltage of type (c) nMOS is about 8.5V which is match with Fig.4 green curve. The lower trigger voltage is because of additional ESD implant, which decreased the drain to bulk breakdown voltage. When Vtrigger occurring, the current conduction mode will switch to parasitic NPN. (II) the Vtrigger and Vhold of type (b) is higher than (a). Comparing the layout of (a) and (b), the width of base area is difference. The longer base width will induce higher Vhold.

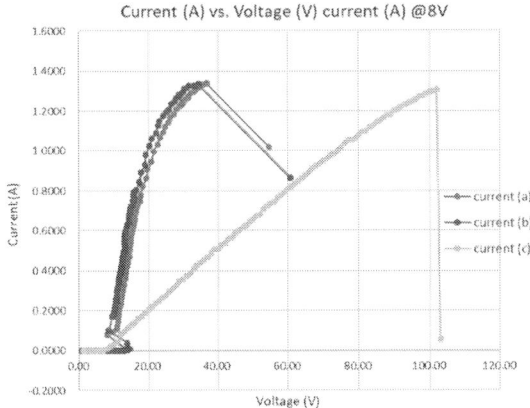

Figure 6: TLP curve of the 3 types MV nMOS as GGNMOS.

CONCLUSION

In this paper, we demonstrated 3 types of MV big nMOS in RC triggered ESD PC circuit. TLP characteristic was used to analyzed their current conduction modes, which the ESD current conduction modes can be divided into MOS channel and parasitic NPN turn on. The later has better robustness because of bulk current conduction. The ESD performance of parasitic NPN conduction mode is about 6X higher than channel conduction. By modifying the layout design, we can control the ESD current conduction. This work can give a well direction for improving ESD devices robustness designing.

ACKNOWLEDGEMENTS

We thank Department of WAT for test assistant. .

REFERENCES

[1] J. Li, R. Gauthier, S. Mitra, C. Putnam, K. Chatty, R. Halbach, and C. Seguin. *IEEE Electrical Overstress/Electrostatic Discharge Symposium*, September, 2006, pp. 179-185.

[2] F. Ma, Y. Han, B. Song, S. Dong, M. Miao, J. Zheng, and K. Zhu. *Microelectronics Reliability*, 2011, 51(12), 2124-2128

[3] J. Liu, H. Fan, J. Li, L. Jiang, and B. Zhang. *IEEE 8th International Conference on ASIC*, October, 2009, pp. 1047-1050.

[4] G. Lu, Y. Wang, L. Zhang, Y. Wang, R. Huang, and X. Zhang, *IEEE International Symposium on Circuits and Systems*, 2018, pp. 1-5.

[5] W. Wei, Y. Wang, X. Chen, Y. Zheng, J. Li, Cao. P, and W. Cao. *Journal of Electrical Engineering & Technology*, 2021, vol. 16(3), 1583-1589.

[6] S. P. Karalkar, V. Ganesan, M. Paul, K. Hwang, and R. Gauthier. *IEEE International Reliability Physics Symposium*, 2021, pp. 1-5.

[7] T. H. Lai, L. A. Chen, T. H. Tang, and K. C. Su, *IEEE International Reliability Physics Symposium*, 2011, pp. 4C-1.

[8] C. Ih-Chin, C. J. Yeol, and H. Chenming, *IEEE transactions on electron devices*, 1988, 35(12), 2253-2258.

[9] K. H. Oh, C. Duvvury, K. Banerjee, and R. W. Dutton, *IEEE International Electron Devices Meeting. Technical Digest*, December, 2001, pp. 14-2.

Reliability analysis of metal thermal interface materials for ultra-large size FCLGA package

Keqing Ouyang[2], Zhuolun Wu[1,2]*, Zhuqiu Wang[2], Weilun Wang[3], Dan Yang[2], Na Mei[1]

[1]State Key Laboratory of Mobile Network and Mobile Multimedia Technology, Shenzhen 518055, China

[2]Department of Reliability Engineering, Sanechips Technology Co., Ltd., Shenzhen 518055, China

[3]Department of Packaging and Testing Engineering, Sanechips Technology Co., Ltd., Shenzhen 518055, China

Corresponding Author's Email:wu.zhuolun@sanechips.com.cn

ABSTRACT

During a ultra-size flip chip land grid array(FCLGA) package level temperature cycling(TC), we found that metal thermal interface material(TIM) had obviously degraded by ultrasonic scanning. TIM coverage decreased from the 90% initially to 20% at 1000 cycles. scanning electron microscope(SEM) image showed that the crack starts from the to TIM interface and propagates inside the TIM or near the intermetallic compound(IMC) of lid to TIM interface.The results of finite element analysis(FEA) shows a severe plastic strain at the corners and center of the die during the temperature change, which will lead to the crack initiation. Finally, five process parameters are analyzed by finite element method and response surface method, who have significant impact on TIM reliability.

INTRODUCTION

Nowadays, there are many challenges in the application of large size flip-chip(FC) packages, such as solder ball thermal fatigue crack caused by coefficient of thermal expansion(CTE) mismatch between printed circuit board(PCB) and package, and Head in Pillow(HIP) caused by package warping. FCLGA packaging can bring a new thinking to solve these problems. With the increase of chip power density and package size, the thermal conductivity of TIM is a very important parameter. Compared with traditional organic TIM, the thermal conductivity of indium TIM has been greatly improved[1]. However, due to the characteristics of metal, TIM itself has many reliability problems[2]. Especially in the TC test[3], the fatigue cracking of metal itself lead to more obvious chip overtemperature in use. In this article, We discussed the failure mechanism of TIM in the TCT combining with the means of finite element analysis. Based on the classical metal fatigue failure theory, the problem of TIM coverage degeneration caused by thermal fatigue was analyzed, which provide guidance for the reliability improvement of indium TIM.

RESEARCH METHOD

The 2D Lid package used in TCT is about 100mm*100mm in size, and its TIM is made of pure indium. The conditions of -40-125°C were selected and 1000 cycles were executed to fully observe the coverage evolution of TIM. We established a quarter-symmetric model, in which the TIM material is elastic-plastic, and the other components are elastic. The simulation of several cycles at - 40 °C - 125 °C is carried out, and the TIM equivalent plastic strain increment of the last cycle is taken for analysis. Finally, for the controllable variables in the design process, we carried out a response surface design to studied the factors that have a greater impact on the thermo-mechanical reliability of TIM, and obtained the optimized parameters, which provided the reference for the improvement of TIM coverage in the future.

RESULTS AND DISCUSSION

Failure analysis

Representative sample's ultrasonic scanning result and its binary result were selected from 25 samples to show the delamination shape as Figure 1. At TC0, the TIM coverage is more than 90%, but it can be found that with the increase of the cycle number, the TIM coverage gradually decreases. When the number of cycles reaches 1000, the TIM coverage is only about 20%, and a very serious delamination can be found, which is symmetrically distributed.

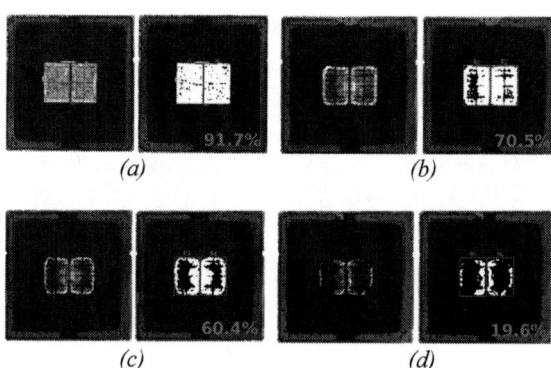

Figure 1: Ultrasonic scanning result(left) and its binary result(right): (a)TC0; (b)TC250; (c)TC500; (d)TC1000

In order to explain the reason of TIM delamination, the lid of package after TCT was removed and fracture morphology was observed by optical microscope(OM) and SEM. Figure 2(a) shows TIM fracture morphology on die side，there is a severe plastic deformation as show in Figure 2(b) ,(c) and (d) ，which shows that most cracks propagate in the interior of indium. Which is also proved that it is also a important way to improve the fatigue performance of indium by reducing indium stress. The cross-section morphology of TIM crack initiation was observed as shown in Figure 3, and we found that delamination tend to initiate at the interface of die to TIM, but fracture mainly occurs in the interior of indium body during propagation.

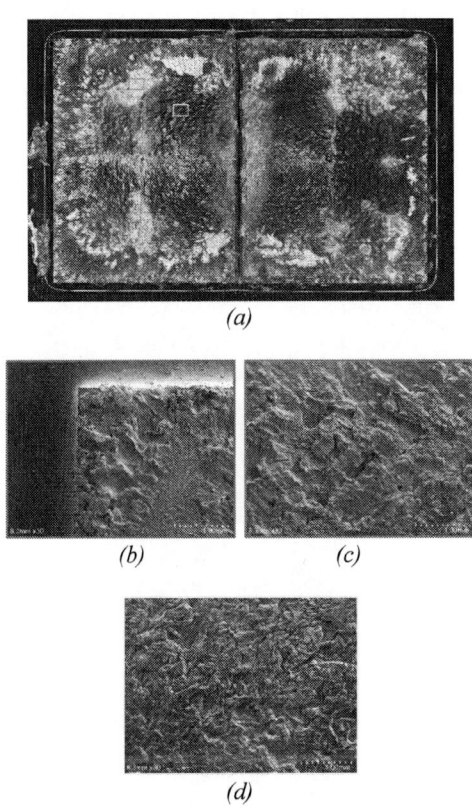

Figure 2: Fracture morphology of TIM on die side: (a)Overall morphology; (b)detail in red line area; (c)detail in blue line area; (d)detail in yellow line area;

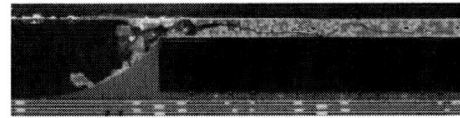

Figure 3: Cross-section morphology of TIM crack initiation

Simulation and response surface method

The ABAQUS software was used to analyze the stress of TIM during the TCT, In order to intuitively reflect the position where TIM is prone to failure in the TCT process, equivalent plastic strain increment distribution nephogram in last cycle in the simulation is made for analysis as shown in Figure 4. It can be found that the corner are the positions where the plastic strain increment is the largest, and the cracks are prone to initiate and propagate at these positions. In addition, in the internal area of die, the position with relatively serious plastic strain will also be prone to occur as shown in Figure 4(a).

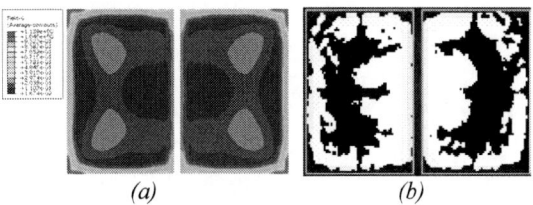

(a)   (b)

Figure 4: Comparison of simulation and ultrasonic scanning binary result: (a)simulation result; (b) ultrasonic scanning binary result

On the other hand, it is also found that the maximum plastic strain increment is located at the interface between TIM and die, which indicates that cracks are easier to initiate at the interface of TIM to die, as shown in Figure 5. This is also consistent with the micrographs and cross-section morphology results above.

Figure 5: Distribution of plastic strain increment in the thickness direction of TIM corner

Finally, we use the response surface method to optimize five parameters, which include indium thickness(200-400um), lid thickness(1-3mm), lid foot width(6-12mm), substrate CTE(12.5-15.5ppm/K), adhesive young's modulus(20MPa-100MPa). The structure description is shown in Figure 6.

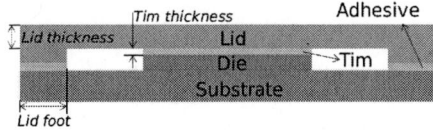

Figure 6 FCLGA structure diagram

We selected effects with a p-value of less than 0.05, than compares the significance of the main effect and the interaction on the TIM plastic strain by scaled estimate, a larger absolute value represents a more significant effectas, as shown in *Figure 7*. Lid thickness have the most significant effects on TIM plastic strain, followed by TIM thickness, substrate CTE, lid foot width, adhesive modulus. Whether they are positively or negatively correlated with plastic strain can be judged by the positive and negative of scaled estimates, and there are also some interactions, which is diffcult to explain its mechanism.

Scaled Estimates

Term	Scaled Estimate		Std Error	t Ratio	Prob>|t|
Intercept	1.0197329		0.007385	138.08	<.0001*
lid thickness(1,3)	0.2904111		0.005205	55.79	<.0001*
substrate cte(12.5,15.5)	0.0812444		0.005205	15.61	<.0001*
Tim thickness*Tim thickness	0.0443355		0.012015	3.69	0.0020*
lid thickness*substrate cte	0.032975		0.005521	5.97	<.0001*
adhesive modulus(20,100)	-0.003589		0.005205	-0.69	0.5004
substrate cte*adhesive modulus	-0.01815		0.005521	-3.29	0.0046*
lid thickness*lid foot width	-0.020475		0.005521	-3.71	0.0019*
lid foot width(6,12)	-0.055689		0.005205	-10.70	<.0001*
Tim thickness*lid thickness	-0.058475		0.005521	-10.59	<.0001*
lid thickness*lid thickness	-0.105664		0.012015	-8.79	<.0001*
Tim thickness(200,400)	-0.158411		0.005205	-30.43	<.0001*

Figure 7 Scaled estimates of effect

For interactions, we can understand them in response surfaces as shown in Figure 8. We found that when the substrate CTE is small, increasing the adhesive young's modulus will increase plastic strain and reduce TIM reliability. However, when the substrate CTE is large, increasing the adhesive modulus will reduce plastic strain and improve TIM reliability. This is the case that interaction is negative. In addition, when the lid thickness is large, increasing the thickness of TIM has a more significant effect on improving the reliability of TIM than that when the thickness of lid thickness is small.

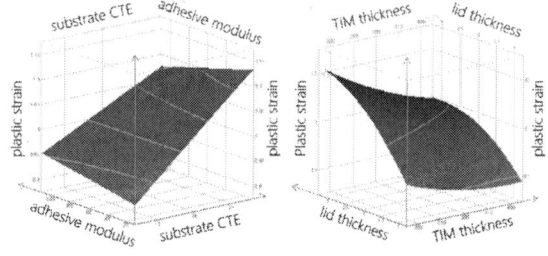

Figure 8 Actual by predicted plot of plastic strain

CONCLUSION

The coverage of indium TIM in the TC test of the oversize FCLGA package was seriously reduced, which was caused by the thermal fatigue fracture of indium. We analyzed the influence factors of the plastic strain of the TIM by response surface method, and found that lid thickness, indium thickness, substrate CTE, lid foot width, and adhesive young's modulus were all significant influence factors, There are also some non-negligible second-order interactions. it is necessary to adopt RSM method In the reliability optimization process of TIM.

REFERENCES

[1] Y. Kim et al., "Metal Thermal Interface Material for the Next Generation FCBGA," 2021 IEEE 71st Electronic Components and Technology Conference (ECTC), 2021.

[2] Y. Luo, F. Wang, W. Wang, N. Mei and J. Fang, "Study on the Coverage of Metal Thermal Interface Material for Ultra-Large FCBGA Packaging," 2022 23rd International Conference on Electronic Packaging Technology (ICEPT), 2022.

[3] C. A. Yang, C. R. Kao, H. Nishikawa and C. C. Lee, "High Reliability Sintered Silver-Indium Bonding with Anti-Oxidation Property for High Temperature Applications," 2018 IEEE 68th Electronic Components and Technology Conference (ECTC), 2018

IMPACT OF INTERFACE TRAPS GENERATION ON FLICKER NOISE DEGRADATION IN SI PMOSFETS

Yi Jiang[1], Luping Wang[1], Yanbin Yang[2], Dawei Gao[1], and Rui Zhang[1]*

[1] School of Micro- and Nano-Electronics, Zhejiang University, Hangzhou 310000, China
[2] Institute of Zhejiang Intelligence Lab, Chengdu 610213, China
*Corresponding Author's Email: ruizhang@zju.edu.cn

ABSTRACT

In this study, we examined the impact of interface states to the low frequency in Si pMOSFETs by applying a negative bias temperature instability (NBTI) stress. The subthreshold swing (S factor) degradation, the effective carrier mobility (μ_{eff}) and the normalized drain current noise power spectral density (S_{Id}/I_d^2) under different NBTI stress voltage (V_S) were characterized in devices. It is found that the S_{Id}/I_d^2 increase with the stress voltage increases, due to the variation of interface trap density (D_{it}) increase.

INTRODUCTION

The flicker noise (1/f noise) in metal -oxide-semiconductor -field-effect transistors (MOSFETs) is one of the most important key factors affecting the electrical performance of modern analog mixed-signal (AMS) and radio frequency (RF) circuits [1]. It is found that the 1/f noise exhits a strong correlation with the NBTI behaviors in Si pMOSFETs, since the NBTI is the key aging mechanism in the devices resulting in an increase of interface traps and oxide trpas, and the 1/f noise is strongly related with the interface quality (Fig. 1).

In this work, the relationship between the 1/f noise and the NBTI stress is examined, from a view point of interface trap density generation. The 1/f noise degradation is confirmed for the Si pMOSFETs with increasing of the interface traps at SiO₂/Si MOS interface.

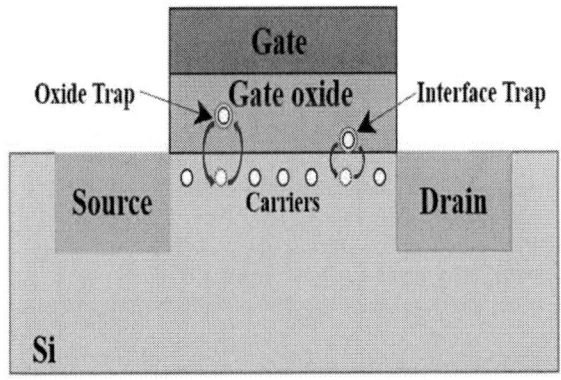

Figure 1: The schematic illustration of the mobile carriers in channel trapped/detrapped by the traps near the SiO₂/Si interface randomly.

MOS INTERFACE DEGRADATION DURING STRESS

The Si pMOSFETs was fabricated with a 10-nm-thick thermal oxidation SiO₂/Si gate stack and a gate length of 3 μm. Fig.2(a) and (b) show the I_d-V_d and I_d-V_g characteristics curve measured form a Si pMOSFET. The Si pMOSFETs show S factor of 129.50 mV/dec, suggesting a superior interface quality for the original SiO₂/Si interface.

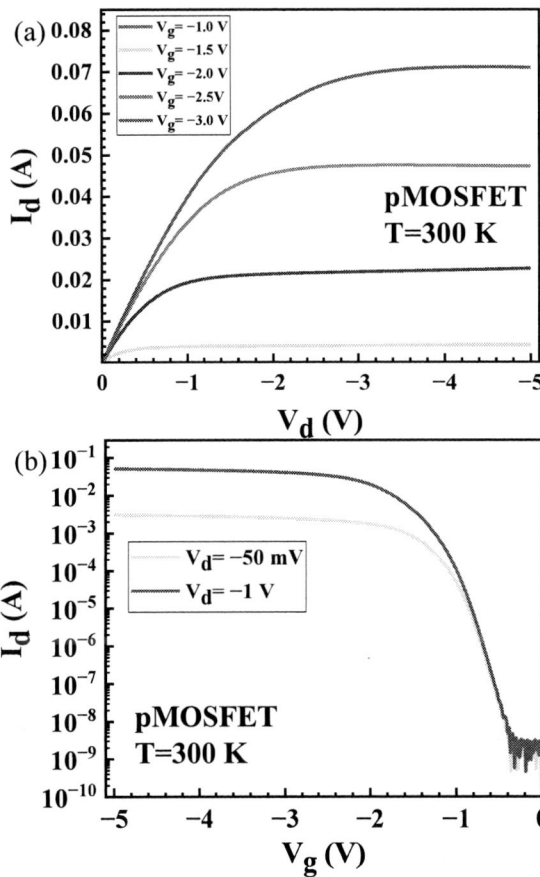

Figure 2:(a) The initial I_d-V_d curves of Si pMOSFETs before NBTI stress. (b) The initial I_d-V_g curves of Si pMOSFETs before NBTI stress.

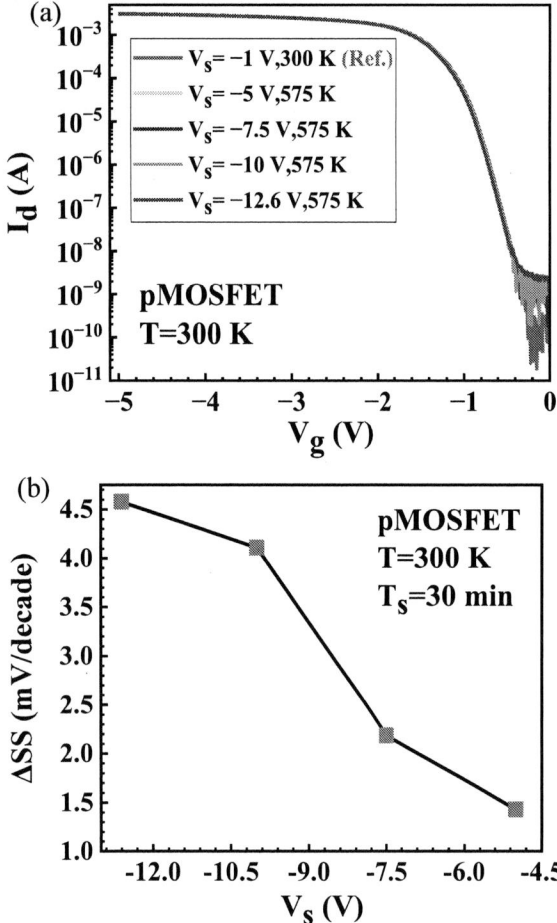

Figure 3: (a) The I_d-V_g curve of Si pMOSFETs with different stress condition during NBTI stress. (b) The ΔS factor of the Si pMOSFETs with different stress voltages during NBTI stress.

The measure-stress-measure (MSM) scheme was applied to the Si pMOSFETs with a stress time (T_S) of 30 min under a temperature of 575 K. Fig. 3(a) shows the I_d-V_g curves after different stress voltages, from which the ΔS factor was evaluated (Fig. 3(b)). The ΔD_{it} was evaluated from the ΔS factor using the following equation [2]:

$$\Delta D_{it} = \frac{\Delta S \cdot C_{ox}}{ln(10)kTq} \qquad (1)$$

As shown in Fig. 4(a), the ΔD_{it} increases with increasing the stress voltage, and a ΔD_{it} of ~2×10^{11} cm^{-2}eV^{-1} was confirmed for the device stressed at −12.6 V. The μ_{eff} was measured for the Si pMOSFETs after stress (Fig. 4(b)). It is also confirmed from the mobility

degradation at low N_S region that the NBTI stress induces extra generation of MOS interface traps [3].

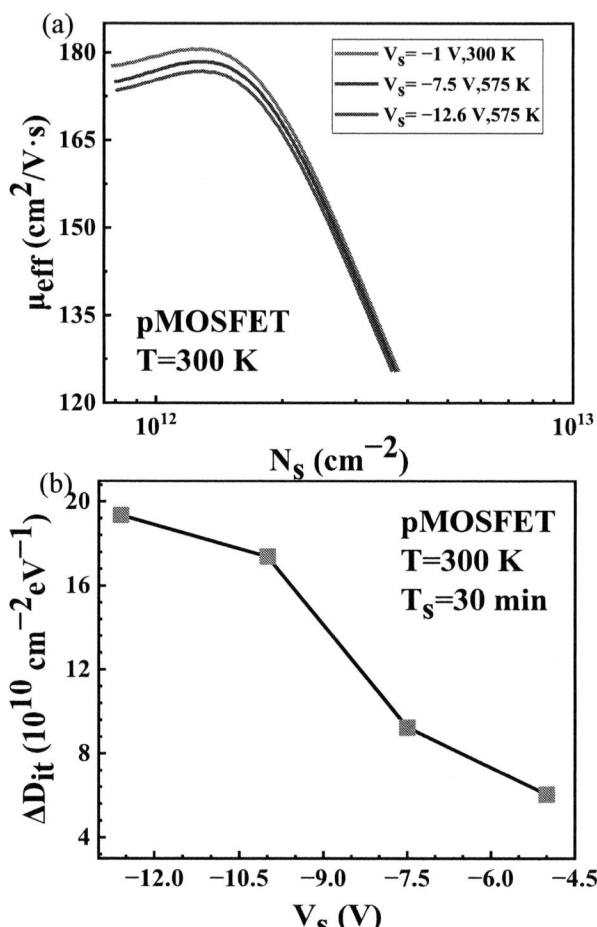

Figure 4: (a) The ΔD_{it} of the Si pMOSFETs with different stress voltages during NBTI stress. (b) The μ_{eff} versus N_S of Si pMOSFETs with different stress condition during NBTI stress.

RELATIONSHIP BETWEEN INTERFACE TRAPS DENSITY AND FLICKER NOISE

The S_{Id}/I_d^2 was evaluated at different V_g and V_d bias conditions of: (1) subthreshold region (V_g−V_{th}=0.2 V, V_d=−0.1 V); (2) linear region (V_g−V_{th}=−0.2 V, V_d=−0.1 V) and (3) saturation region (V_g−V_{th}=−0.2 V, V_d=−0.5 V), by means of Wiener-Khintchine theorem (Fig. 5) [4]. It is observed that the S_{Id}/I_d^2 represent a $f^{-\beta}$ dependence for all bias conditions, with fitted coefficiency β of 0.97~1.02 and the classical value is 0.8~1.2. There results suggest that the Si pMOSFETs after NBTI stress still follows a 1/f noise behavior, and it can be interpreted that the S_{Id}/I_d^2 increase is attributable to the carriers trapping and de-trapping procedures by the D_{it} at SiO$_2$/Si interface.

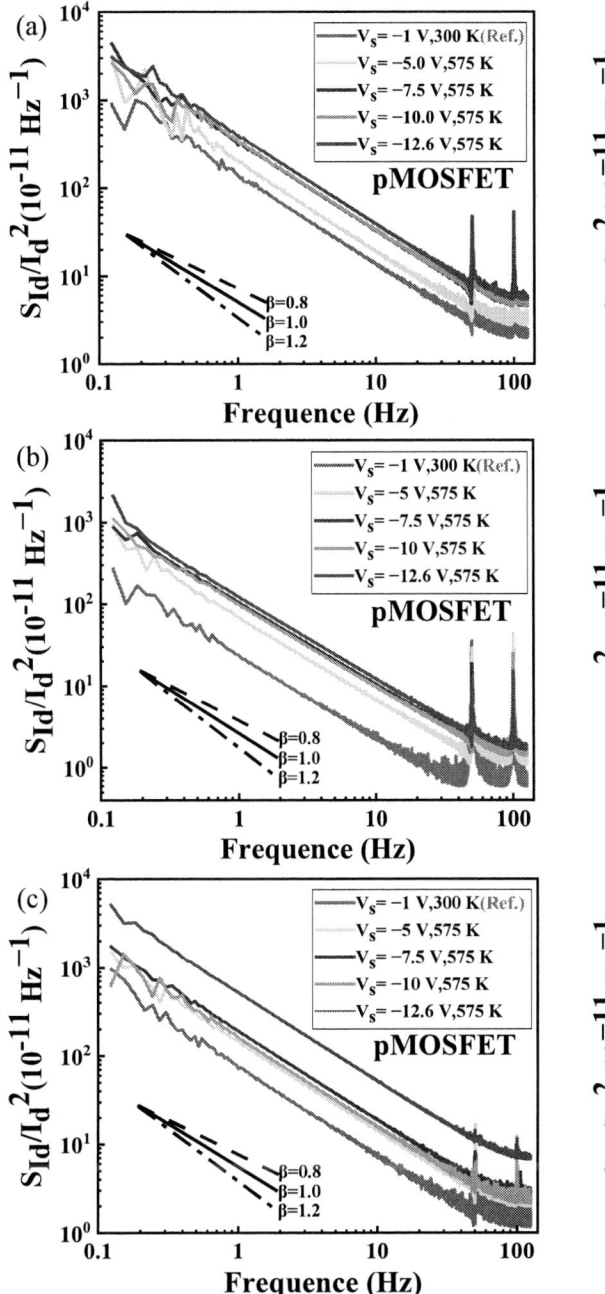

Figure 5: The S_{Id}/I_d^2 of the Si pMOSFETs in (a) subthreshold region, (b) linear region and (c) saturation region.

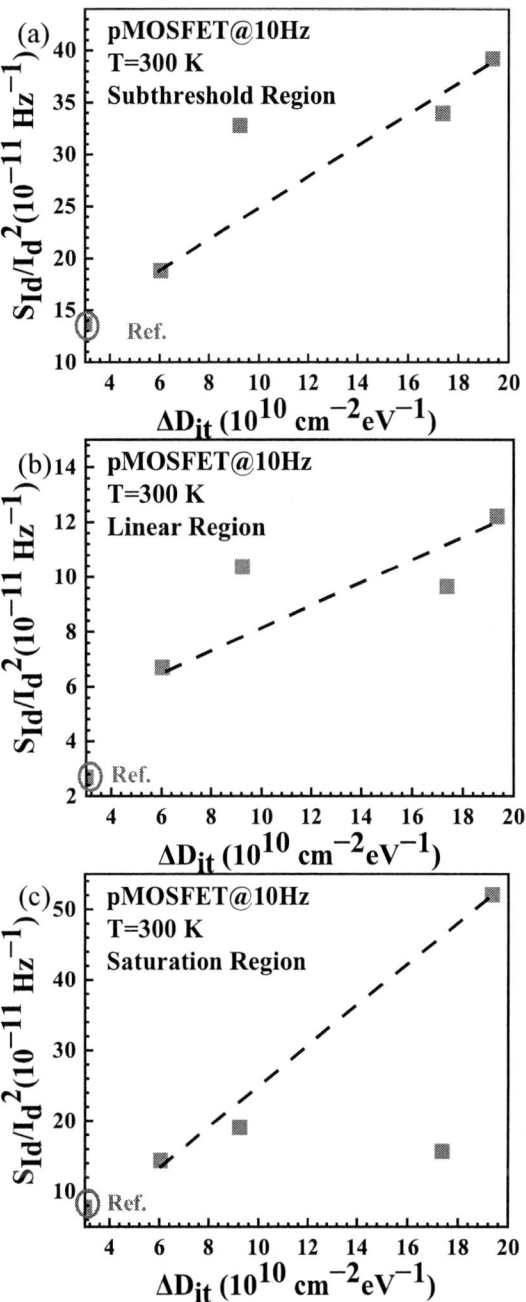

Figure 6: The S_{Id}/I_d^2 at 10 Hz versus ΔD_{it} of the Si pMOSFETs in (a) subthreshold region, (b) linear region and (c) saturation region.

Fig. 6 summarizes the S_{Id}/I_d^2 at 10 Hz for the Si pMOSFETs with different ΔD_{it} at the subthreshold, linear and saturation regions, respectively. A strongly correlation between the S_{Id}/I_d^2 and the ΔD_{it} was observed directly for all regions. The S_{Id}/I_d^2 at different regions can be interpreted to represent a linear law to ΔD_{it} using the number fluctuation model [5] and unified model [6, 7], as shown in equation (2) ~ (4).

979-8-3503-1101-3/23 $31.00 © 2023 IEEE   423

Subthreshold region:

$$S_{Id}(f) = q^4 \frac{N_t(E_{fn})}{kT\gamma f WL} \frac{I_d^2}{(C_{ox}+C_{it}+C_d)^2} \qquad (2)$$

Linear and saturation region:

$$S_{Id}(f) = \frac{I_d^2}{f WLN^2} \frac{N_t(E_{fn})_{eff}}{\gamma} \qquad (3)$$

$$S_{Id}(f) = \frac{kTI_d^2}{\gamma f WL}(\frac{1}{N}+\alpha\mu)^2 N_t(E_{fn}) \qquad (4)$$

Since the oxide traps and the interface traps may generated during the NBTI simultaneously, the fine agreement between the S_{Id}/I_d^2-ΔD_{it} relationships by measurement and by theoretical calculation suggest that the D_{it} generation is the key factor resulting the $1/f$ noise degradation in Si pMOSFETs.

CONCLUSION

In this study, the $1/f$ noise degradation is analyzed with applying the NBTI stress for Si pMOSFETs. It is found that the $1/f$ noise is strongly affected by MOS interface traps, suggesting that the SiO_2 interfacial layer degradation during electrical stress could be a critical issue for $1/f$ noise degradation even in advanced high-k/Si gate stacks.

ACKNOWLEDGEMENT

The authors would great appretiate the kind guidance from Prof. Hanming Wu. This work was supported by Natural Science Foundation of Sichuan Province (No. 2021YJ0093).

REFERENCE

[1] D. K. Schroder and J. A. Babcock, "Negative bias temperature instability: Road to cross in deep submicron silicon semiconductor manufacturing," J. Appl. Phys., vol. 94, no. 1, pp. 1–18, Jul. 2003, doi: 10.1063/1.1567461.

[2] R. J. Van Overstraeten, G. J. Declerck, and P. A. Muls, "Theory of the MOS transistor in weak inversion-new method to determine the number of surface states," IEEE Trans. Electron Devices, vol. 22, no. 5, pp. 282–288, May 1975, doi: 10.1109/T-ED.1975.18119.

[3] Z. Ji, J. F. Zhang, W. D. Zhang, B. Kaczer, S. De Gendt, and G. Groeseneken, "Interface States Beyond Band Gap and Their Impact on Charge Carrier Mobility in MOSFETs," IEEE Trans. Electron Devices, vol. 59, no. 3, pp. 783–790, Mar. 2012, doi: 10.1109/TED.2011.2177839.

[4] M. J. Kirton and M. J. Uren, "Noise in solid-state microstructures: A new perspective on individual defects, interface states and low-frequency ($1/f$) noise," Adv. Phys., vol. 38, no. 4, pp. 367–468, Jan. 1989, doi: 10.1080/00018738900101122.

[5] A. L. McWhorter, "$1/f$ noise and germanium surface properties," in Semiconductor Surface Physics. Philadelphia: University of Pennsylvania Press, 1957, pp. 207-228.

[6] K. K. Hung, P. K. Ko, C. Hu, and Y. C. Cheng, "A unified model for the flicker noise in metal-oxide-semiconductor field-effect transistors," IEEE Trans. Electron Devices, vol. 37, no. 3, pp. 654–665, Mar. 1990, doi: 10.1109/16. 47770.

[7] M. Valenza, A. Hoffmann, D. Sodini, A. Laigle, F. Martinez, and D. Rigaud, "Overview of the impact of downscaling technology on $1/f$ noise in p-MOSFETs to 90 nm," IEE Proc. - Circuits Devices Syst., vol. 151, no. 2, p. 102, 2004, doi: 10.1049/ip-cds: 20040459.

RESEARCH ON HOT CARRIER INJECTION OPTIMIZATION OF 28HKMG TECHNOLOGY

Weiwei Ma[1]*, Yang Li[1], Ran Huang[1], Yamin Cao[1], Wei Zhou[1]

[1]Shanghai Huali Integrated Circuit Corporation, Shanghai, China
Corresponding Author's Email: maweiwei@hlmc.cn

ABSTRACT

With the aggressive scaling down of the gate dielectric and introducing of metal gate, reliability issues especially NBTI (Negative Bias Temperature Instability) and HCI (Hot Carrier Injection) become serious challenges. In this study, the critical role of metal gate work function metal (WFM) removing process is investigated in 28 nm HKMG technology. It's found that HCI performance of IO NMOS is dominated by the integrity of metal gate film stacks, which is much more influential than implant profile modulation as commonly used to solve HCI issue in 28 nm Poly SiON technology. And then, we propose an aluminum diffusion of metal gate and dipole formation model to explain this phenomenon.

INTRODUCTION

With the continuously scaling down of transistors, high-k dielectrics and metal gates are introduced as new gate stacks for solving intolerable tunnel leakage problems [1-4]. Meanwhile, reliability problems such as NBTI and HCI have become major concerns when realizing highly reliable integrated CMOS devices [5-7]. Lots of works have been done to reveal the mechanism behand HCI, like implant profile, voltage of gate and drain and traps in gate oxide [8-10].

In this study, the critical role of work function metal removing process was investigated in 28 nm HKMG technology. And an aluminum diffusion of metal gate and dipole formation model is proposed to explain this phenomenon.

EXPERIMENT

Wafers of different gate process integrations were prepared. All wafers analyzed in this study were manufactured based on HLIC 28 nm HKMG technology. And the HCI lifetimes of samples were tested by common practice in the industry.

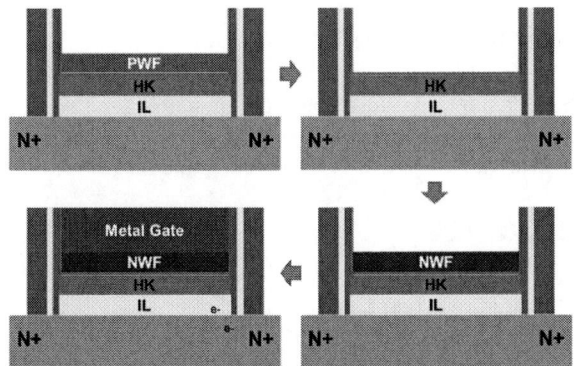

Figure 1. Simplified NMOS Metal Gate Process Flow

A simplified NMOS metal gate process flow was shown in figure 1, dummy poly removing process of NMOS and PMOS was carried out in the same step. P-type work function metal (PWF) was deposited in the whole poly trench including NMOS area. Therefore, a work function metal removing step was necessary for NMOS area. And these removing processes should be critical to NMOS reliability performance. Thus the healthiness of this removing step need to be verified firstly. For this purpose, a pure NMOS wafer, without PWF deposition and also skip PWF removing, was prepared.

Detailed experiment splits were carried out as table 1.

Condition	Related Process Step	
	NWF Remove	**Extra Thermal Post NIO LDD**
Split1	Pure NMOS	N/A
Split2	BSL	N/A
Split3	BSL	Yes
Split4	CIP	N/A
Split5	CIP	Yes

Table 1. Experiment Split Table

RESULTS AND DISCUSSION

1. Effect of Metal Gate Integrity on HCI

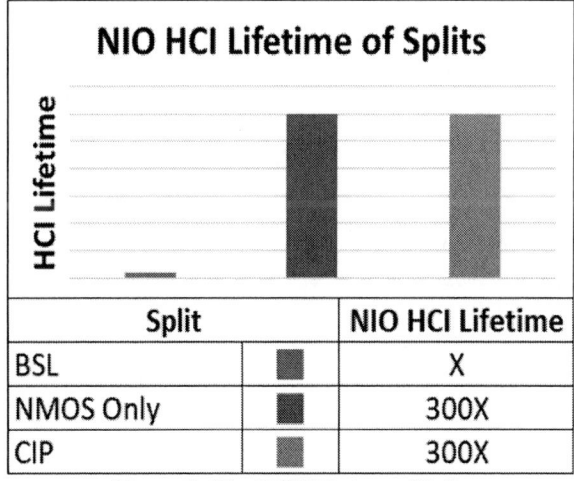

Split		NIO HCI Lifetime
BSL		X
NMOS Only		300X
CIP		300X

Figure 2. The HCI Lifetime of Splits

As shown in figure 2, different splits of metal gate process integrations had significant different HCI performance. The pure NMOS sample had 300 times longer HCI lifetime than the baseline (BSL) sample, which revealed the p-type work function metal removing process of NMOS had fundamental effect on HCI performance. And it was also proved the unhealthiness of BSL process.

979-8-3503-1101-3/23 $31.00 © 2023 IEEE

To figure out the root cause of such significant gap of HCI lifetimes between BSL and pure NMOS samples by step TEM check was carried out. As shown in figure 3, damage on bottom barrier metal (BBM) layer of IO NMOS was detected, which was on the top corner of metal gate. It gave us an index of the possibility of BBM damage on long channel device bottom. And actually BBM at these places would be polished before metal gate was finally formed so that TEM of final NMOS could not see such damage.

Extraordinary efforts had been done to optimize WFM removing processes. Further, sample of optimized WFM removing process also demonstrated same order of improvement of HCI.

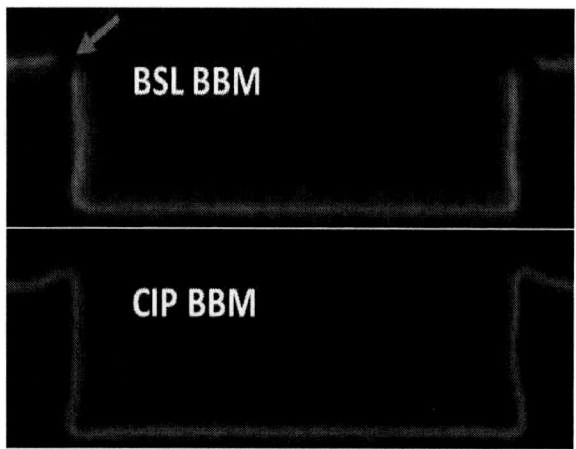

Figure 3. BBM Metal of IO NMOS post PWF Remove

2. The Isub Results of Splits

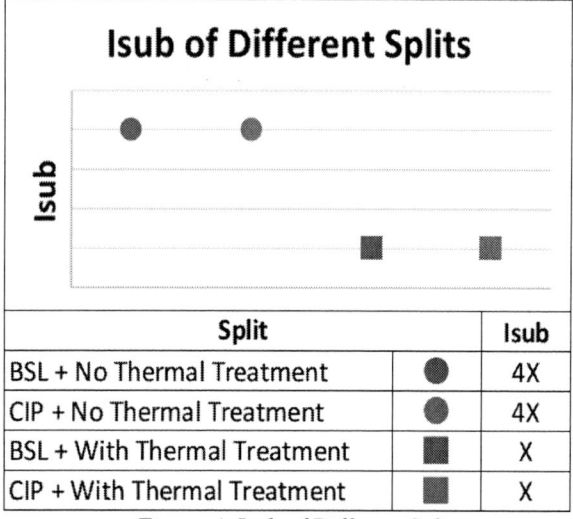

Split		Isub
BSL + No Thermal Treatment	●	4X
CIP + No Thermal Treatment	●	4X
BSL + With Thermal Treatment	■	X
CIP + With Thermal Treatment	■	X

Figure 4. Isub of Different Splits

Before HKMG technology Isub was a key indicator parameter of HCI performance. Thus, the Isub results were also tested and checked as shown in figure 4. And no difference was found between BSL and CIP samples, which meant the HCI improvement was not benefited from silicon channel. Thermal splits could see Isub improvement as commonly expected, which due to implant profile modulation. However, such level of improvement was not enough to make a difference to HCI performance of HKMG NMOS.

3. Proposed Model

Based on these analyses, we proposed an aluminum diffusion of metal gate and dipole formation model to explain this phenomenon as shown in figure5. To be specific, the dipoles of Al-Hf [6, 11-12] continuously formed during reliability stress introduced extra charge in HK layer, which not only damage gate oxide integrity but also provided extra attraction to hot carriers. Therefore, in 28 nm HKMG technology the healthiness of metal gate integration become essential to HCI performance.

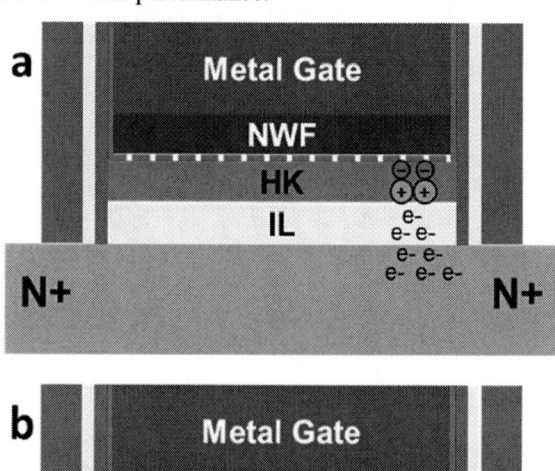

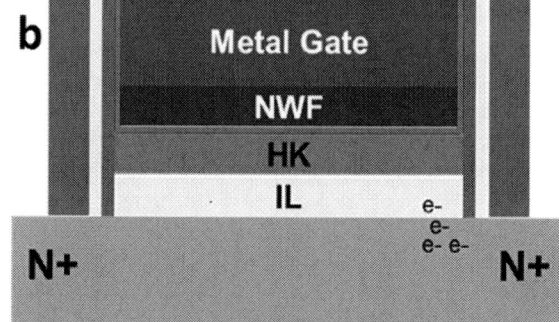

Figure 5. Model Schematic Diagram: a. Dipole Induced HCI Worse; b. Better Performance of CIP

CONCLUSION

We proposed an aluminum diffusion of metal gate and dipole formation model to explain the internal mechanism of HCI improvement in 28 nm HKMG technology. In 28 nm HKMG technology the healthiness of metal gate integration become essential to HCI performance.

ACKNOWLEDGEMENTS

I would like to appreciate senior engineer Yang Li, Ran Huang, Yamin Cao, Wei Zhou for their support and advice on this experiment. At the same time, I am grateful to the reliability team of HLIC for their testing and discussion.

REFERENCES

[1] "Work Function Setting in High-k Metal Gate Devices," Complementary Metal Oxide Semiconductor Chaper 3, 2018.

[2] "Challenges for The Integration of Metal Gate Electrodes," International Electron Devices Meeting, 2004.

[3] "Characteristics and Mechanism of Tunable Work Function Gate Electrodes Using a Bilayer Metal Structure on SiO2 and HfO2," Electron Device Letters, 2005.

[4] "Dipole-induced Modulation of Effective Work Function of Metal Gate in Junctionless FETs," AIP Advances, 2020.

[5] "NBTI enhancement by nitrogen incorporation into ultrathin gate oxide for 0.10-pm gate CMOS generation," IEEE VLSI Symp, 2000.

[6] "Research on Reliability Optimization Mechanism of 28HKMG Technology," CSTIC, 2022.

[7] "Negative bias temperature instability: Road to cross in deep submicron silicon semiconductor manufacturing," Journal of Applied Physics, 2003.

[8] "Understanding Hot Carrier Reliability in FinFET Technology from Trap-based Approach." IEDM, 2021.

[9] "Improvement of 28HKMG NIO device HCI by Implant Scheme and Sequence Optimize Doping Profile and E-field," CSTIC, 2018.

[10] "Hot Carrier Degradation in Cryo-CMOS," IRPS, 2020.

[11] "Dipole-induced Modulation of Effective Work Function of Metal Gate in Junctionless FETs," AIP Advances, 2020.

[12] "Interface Dipole Engineering in Metal Gate/High-k Stacks," Chin Sci Bull, 2012.

APPLICATION OF PICOSECOND ULTRASONIC TECHNOLOGY FOR CMOS IMAGE SENSORS

Johnny Mu[1], Kaixing Song[3], Johnny Jin[1], Cheolkyu Kim[2], Yaodong Huang[3], Hong Hong[1]

[1]Onto Innovation, Floor 3, Building 3, 690 Bibo Road, Pudong New District, Shanghai, 201203 China
[2]Onto Innovation, 16-6, Sunae-dong, Bundang-gu, Sungnam-si, Gyunggi-do, 3965 Korea
[3]GalaxyCore, No. 198 Xinyuan South Road, Pudong New District, Shanghai, 201203 China
*Corresponding Author's Email: johnny.mu@ontoinnovation.com

ABSTRACT

Picosecond Ultrasonic Technology (PULSE™ Technology) has been widely used in metal, thin-film metrology because of its unique advantages, such as being a rapid, non-contact and non-destructive technology with the capability for simultaneous multiple-layer measurement. In this paper, we describe the advantages of PULSE Technology in the manufacturing of backside illuminated (BSI) CMOS image sensors (CIS). In a front-side illuminated (FSI) sensor, the light reaches the photo diode active region through the passivation, metallization and inter-dielectric layers. The coupling of light from the front side of the sensor results in a loss of light, which in turn results in a reduction in quantum efficiency (QE). Light scattering at the metal layers also contributes to a significant amount of crosstalk, resulting in reduction of the sensor's signal-to-noise ratio (SNR). A BSI sensor contains the same components as a FSI sensor, but the sensor's metals are located behind the photodiode. Deposition and dry etching to the thin tungsten (W) film grid at the physical boundaries of each pixel is a common practice for traditional manufacturers. Chemical mechanical polishing (CMP) prior to dry etching provides better top-profile control of the W grid film, resulting in a better SNR and performance. For BSI application, the adoption of PULSE Technology for inline W thickness measurements has proven to be key for device-level process control and yield improvement. Pulse technology use the small spot size which enable measurements in a 15μm site size and an ability to map within wafer uniformity profiles to the wafer edge. Besides this, Picosecond ultrasonic measurements are rapid and provide excellent repeatability and long-term stability, making it possible to achieve the high-sampling rate required in a high-volume manufacturing environment.

INTRODUCTION

CMOS image sensors consist of an array of light-sensitive pixels. Each pixel consists of a photo diode (PD), which is the light-sensitive element, and several control transistors. The PD collects and stores the photo-carriers, while the control transistors are used for setting the exposure time, transforming charge to voltage and readout control sequence. The pixel array is connected to the control circuit by several metallization layers. Each metallization layer is separated by inter-dielectric material. The passivation layer, which is placed above all the metallization layers and isolates the chip from environmental hazards, serves as the interface between the chip and the outside world.

In a front-side illuminated (FSI) sensor [Figure 1(a)], the light reaches the PD active region through the passivation, metallization and inter-dielectric layers. There are several loss mechanisms associated with the coupling of light from the front side of the sensor: first, the reflection of light from the passivation layer; second, the reflection of light from the metal control lines surrounding the PD; and third, the shift and reflectance of light coming from large angles to the sensor, especially at the edges of the sensor. This angled light travels through the thick inter-dielectric layer, and it may be collected by a neighboring pixel. These three mechanisms cause a reduction in the maximum available photons or, in other words, a quantum efficiency (QE) reduction, and contribute to crosstalk and, as a result, a reduction of the sensor's SNR [1].

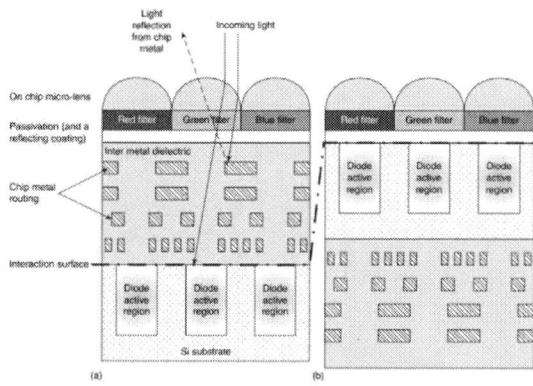

Figure1: Illustration of (a) frontside vs (b) backside illuminated sensor.

A BSI sensor contains the same components as a FSI sensor, but the BSI sensor's metals are located behind the photodiode [Figure 1(b)]. During manufacturing, the wafer is flipped upside down so the metallization and passivation layers are located beyond the PD with respect to the light. The manufacturing of such a sensor is highly complicated since it requires processing steps for both sides of the wafer. The main technological challenges associated with BSI involve the bonding of the wafer to a holding wafer and the thinning of the original wafer to a thickness of a few microns. In addition, there are also some issues related to the optimization of the sensor's optical and electrical performance, such as significant amounts of inherent color cross talk, which degrades the image quality, and excess dark current coming from the additional interface. In the past, these problems prevented the mass market from manufacturing of BSI sensors. Consequently, until recent years, BSI was used in niche high-end markets where low-light performance could not be compromised.

979-8-3503-1101-3/23 $31.00 © 2023 IEEE

PULSE Technology, implemented in the Echo™ system, is a non-contact, non-destructive pump-probe laser acoustic technique for the measurement of metal film thickness. Interested readers may find additional details in the literature [2].

PUSLE Technology is a proven workhorse in semiconductor fabs around the world. A 0.2 picosecond (ps) laser pulse (pump) is focused to a small (12μm x16μm) spot onto a wafer surface to create a sharp acoustic wave. The acoustic wave travels away from the surface through the film at the speed of sound. Upon interfacing with another material, a portion of the acoustic wave is reflected and returns to the surface, while the rest of the acoustic wave is transmitted. The probe pulse detects this reflected acoustic wave as it reaches the wafer surface. The detector can detect the change in optical reflectivity (REF) caused by the strain of an acoustic wave, or the deflection of the reflected probe beam caused by the deformation of the surface due to the acoustic wave using a position sensitive detector (PSD). Both modes, REF and PSD, are used in characterizing metal films. By knowing the speed of the sound in the material and the arrival time of the echoes, the thickness of the film is readily determined using the first principles technique. Depending on the application, information on film density can also be obtained by examining the damping rate of the echoes, while details regarding surface roughness can be obtained by analyzing the width of the echoes. The latest Echo system improvement includes some additional modifications to the experimental setup to enhance SNR for applications with excessive surface roughness.

Using the PULSE technique, we have performed high-resolution line scans (0.5 mm edge exclusion) on different types of metal films commonly used in BSI CIS and demonstrated 3 sigma (3σ) repeatability performance for thickness < 0.3 ~ 0.6%, as shown in Table 1. The accuracy of the technique has been correlated to cross-section scanning electron microscopy (SEM) with $R^2 > 0.95$.

TABLE 1: Typical dynamic repeatability performance for PULSE measurement for commonly used metal films in BSI CIS.

Film	Thickness [Å]	Repeatability [3 sigma%]
TiN	50~500	<0.6%
W	500~5000	<0.3%
AlCu	1000~32000	<0.3%
Ti	50~500	<0.6%
NiV	1000~3000	<0.3%
Ag	5000~15000	<0.3%

APPLICATIONS IN BACKSIDE

ILLUMI-NATED CMOS IMAGE SENSORS

W metal film thickness measurement using PULSE Technology

In this paper, we demonstrate how PULSE Technology can provide a unique in-line metrology solution in CIS applications because of the advantages described previously. Additionally, in high-volume manufacturing one of the critical requirements is the ability to have the robust capability to cover process variations in addition to excellent repeatability, long-term stability and tool-to-tool matching. The ability to measure multi-layer stacks eliminates the need to conduct measurements on monitor wafers and provides direct feedback for process monitoring and control.

Traditional W grid forming process is shown in Figure 2. After the deep trench formed by etching (A1), thin DARC layer and liner oxide are deposited (A2) followed by thin TiN barrier and W (A3). Excess W metal film is removed by CMP (A4).

According to recent research [3], using dry etch to remove excess W film (B4) results in better final optics and SNR on the image sensor. However, because W thickness is not uniform across the wafer, the W recess in the deep trench will make it difficult to control the final W grid critical dimension.

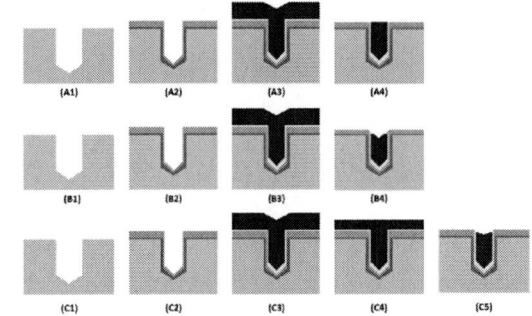

Figure 2: W grid forming process.

Hence, a CMP process is used to form a uniform W surface prior to the dry etch step and for controlling the W recess in deep trench (C4). This makes the post-CMP inline W thickness measurements extremely critical to the final W grid critical dimension and W grid profile.

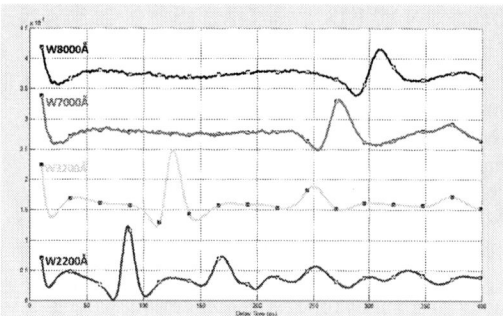

Figure 3(a): Raw signal and modeling examples of measurement on W film using PULSE Technology.

PULSE technology has been adopted as the process tool-of-record for W thickness measurement prior to W etch back. Figure 3(a) shows W measurement raw signal (REF) data from 2,200Å to 8,000Å films. The acoustic echoes located at ~90ps-320ps on the various traces correspond to the sound wave reflected from the bottom of W film. Figure 3(b) shows the modeled fit (black curve) to the measured signal (green curve). The peak at ~120ps is used to calculate the thickness of the W film and is modeled at ~3,000Å. Figure 3(c) shows the W thickness profiles of 13 points across the wafer.

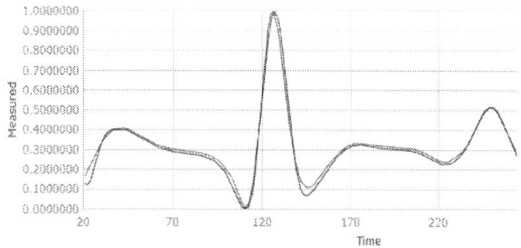

Figure 3(b): Modeling example of measurement on W film using PULSE Technology.

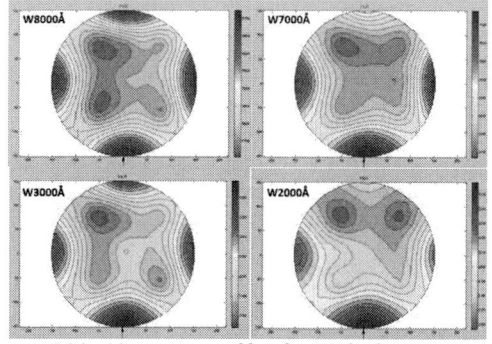

Figure 3(c): 13-points profiles for W thickness 2,000Å ~ 8,000Å.

Repeatability of thickness measurement by PULSE Technology

PULSE Technology has been widely used for the measurement of Al, W, NiV, Ag, Ti, TiN and tungsten silicide (WSi). Table 2(a), 2(b), 2(c) ,2(d)and 2(e) show the typical dynamic thickness repeatability and precision [3 sigma (3σ)] performance for W, NiV, Ag, and Ti. The results meet the process' repeatability requirements.

TABLE 2(a): Dynamic repeatability for W 3,300Å thickness measurement based on nine (9) points.

Repeatability: W3300 Å										
	PT1	PT2	PT3	PT4	PT5	PT6	PT7	PT8	PT9	
X [mm]	0	0	50	0	-50	0	90	0	-90	
Y [mm]	0	50	0	-50	0	90	0	-90	0	AVG
Repeat 1	3289.3	3169.4	3195.2	3166.1	3254.3	3172.7	3198.9	3189.3	3292.8	3214.2
Repeat 2	3288.4	3172.3	3197.8	3167.8	3256.8	3172.8	3199.1	3189.8	3289.0	3214.9
Repeat 3	3288.5	3173.3	3198.2	3167.5	3257.3	3173.1	3199.7	3190.5	3288.6	3215.2
Repeat 4	3289.4	3171.4	3194.9	3166.3	3253.6	3174.4	3198.8	3191.8	3288.7	3214.4
Repeat 5	3289.4	3171.4	3197.5	3166.5	3255.2	3174.4	3199.5	3190.6	3289.0	3214.8
Repeat 6	3289.5	3170.0	3197.4	3166.7	3253.6	3175.3	3198.3	3192.0	3289.1	3214.8
Repeat 7	3288.5	3171.6	3195.6	3166.5	3257.5	3173.5	3199.1	3191.9	3287.9	3214.7
Repeat 8	3289.3	3171.4	3194.5	3165.5	3257.0	3174.7	3199.0	3191.7	3288.8	3214.7
Repeat 9	3290.2	3171.8	3196.1	3166.6	3255.9	3174.0	3198.8	3189.8	3288.7	3214.6
Repeat 10	3282.8	3172.8	3195.5	3165.9	3256.5	3173.7	3199.0	3191.5	3289.3	3214.6
Repeat 11	3289.5	3172.3	3197.4	3166.8	3256.0	3172.0	3198.6	3189.8	3288.9	3214.6
Repeat 12	3290.5	3170.0	3195.1	3165.3	3257.6	3173.2	3199.1	3189.6	3289.5	3214.4
Repeat 13	3289.3	3171.1	3194.6	3166.7	3258.8	3170.7	3198.6	3189.6	3288.7	3214.2
Repeat 14	3289.5	3171.0	3195.4	3166.7	3258.6	3170.9	3198.2	3188.9	3288.5	3214.1
Repeat 15	3290.2	3171.7	3195.3	3167.0	3253.0	3172.9	3198.1	3190.1	3288.9	3214.1
AVG	3289.2	3171.4	3196.0	3166.5	3256.2	3173.2	3198.9	3190.5	3289.1	3214.6
STDEV [Å]	0.9	1.1	1.3	0.7	1.7	1.3	0.5	1.0	1.1	0.3
3*STDEV [%]	0.08%	0.10%	0.12%	0.06%	0.16%	0.12%	0.04%	0.10%	0.10%	0.03%

TABLE 2(b): Dynamic repeatability for NiV 2,700Å thickness measurement based on nine (9) points.

Repeatability : NiV2700 Å										
	PT1	PT2	PT3	PT4	PT5	PT6	PT7	PT8	PT9	
X [mm]	0	0	50	0	-50	0	90	0	-90	
Y [mm]	0	50	0	-50	0	90	0	-90	0	AVG
Repeat 1	2809.8	2800.9	2747.5	2743.2	2795.6	2733.5	2647.5	2641.8	2726.5	2738.5
Repeat 2	2810.3	2800.4	2747.1	2743.4	2795.4	2733.2	2646.4	2641.6	2724.7	2738.1
Repeat 3	2811.3	2801.3	2747.1	2743.7	2796.0	2733.8	2646.9	2641.8	2724.9	2738.5
Repeat 4	2810.2	2800.2	2746.9	2743.0	2795.2	2732.6	2646.6	2641.8	2723.9	2737.9
Repeat 5	2810.4	2800.7	2746.8	2743.0	2795.5	2733.8	2647.7	2642.3	2724.5	2738.3
Repeat 6	2811.1	2800.1	2746.6	2744.1	2796.1	2732.8	2646.3	2641.7	2723.9	2738.1
Repeat 7	2810.0	2800.1	2746.1	2743.7	2796.3	2733.4	2646.4	2641.4	2724.9	2738.3
Repeat 8	2810.8	2800.6	2747.0	2743.6	2796.4	2733.1	2647.2	2641.9	2724.9	2738.4
Repeat 9	2810.7	2801.5	2746.9	2743.6	2796.7	2733.7	2646.9	2642.3	2725.2	2738.3
Repeat 10	2810.3	2800.6	2747.2	2742.9	2795.9	2733.1	2645.4	2641.7	2724.8	2738.0
Repeat 11	2811.1	2801.1	2747.2	2743.4	2795.3	2733.6	2646.5	2641.6	2724.7	2738.3
Repeat 12	2811.0	2801.0	2747.8	2743.7	2795.7	2733.2	2646.5	2642.2	2725.3	2738.5
Repeat 13	2811.1	2801.4	2747.0	2743.7	2796.0	2733.7	2646.8	2642.3	2725.4	2738.6
Repeat 14	2811.6	2801.6	2747.7	2743.5	2795.9	2733.5	2646.8	2642.1	2725.3	2738.7
Repeat 15	2811.7	2801.4	2748.0	2743.9	2796.4	2734.0	2647.0	2642.5	2725.2	2738.9
AVG	2810.8	2800.8	2747.1	2743.5	2795.9	2733.4	2646.7	2642.0	2724.9	2738.3
STDEV [Å]	0.6	0.5	0.5	0.3	0.4	0.4	0.5	0.3	0.6	0.3
3*STDEV [%]	0.06%	0.05%	0.05%	0.04%	0.04%	0.04%	0.06%	0.03%	0.07%	0.03%

TABLE 2(c): Dynamic repeatability for Ag 12,000Å thickness measurement based on nine (9) points.

Repeatability : Ag12000 Å										
	PT1	PT2	PT3	PT4	PT5	PT6	PT7	PT8	PT9	
X [mm]	0	0	50	0	-50	0	90	0	-90	
Y [mm]	0	50	0	-50	0	90	0	-90	0	AVG
Repeat 1	12950.8	12661.5	12722.9	12894.7	12858.6	12260.3	12302.5	12583.9	12530.5	12640.6
Repeat 2	12957.2	12705.8	12700.8	12894.2	12845.4	12263.0	12299.8	12585.6	12536.1	12643.1
Repeat 3	12947.5	12703.4	12714.1	12890.4	12871.5	12286.7	12302.0	12574.9	12534.2	12647.2
Repeat 4	12955.1	12700.8	12724.9	12900.9	12871.3	12283.1	12309.9	12576.4	12544.3	12651.9
Repeat 5	12971.3	12677.0	12729.1	12904.5	12856.7	12274.4	12297.6	12597.5	12547.0	12650.6
Repeat 6	12974.6	12673.6	12719.6	12900.7	12855.8	12273.5	12293.5	12582.6	12544.3	12646.4
Repeat 7	12961.5	12673.6	12728.0	12889.9	12861.1	12275.9	12305.5	12597.4	12553.3	12649.4
Repeat 8	12986.8	12710.7	12709.4	12887.8	12890.5	12279.9	12302.4	12586.0	12538.4	12654.7
Repeat 9	12972.2	12692.3	12737.7	12885.2	12848.7	12284.7	12289.9	12594.2	12535.1	12648.9
Repeat 10	12966.8	12699.8	12714.5	12910.0	12876.0	12270.6	12301.7	12574.0	12550.5	12651.5
Repeat 11	12973.1	12682.2	12728.4	12900.4	12872.2	12275.1	12313.3	12588.4	12540.8	12652.7
Repeat 12	12965.7	12691.0	12713.9	12864.2	12860.4	12266.7	12311.5	12589.0	12526.1	12643.2
Repeat 13	12967.9	12687.4	12738.8	12890.4	12875.0	12273.6	12295.3	12586.8	12554.5	12652.2
Repeat 14	12959.9	12699.4	12719.8	12888.7	12870.2	12271.6	12296.9	12594.7	12540.0	12649.0
Repeat 15	12976.7	12700.8	12713.4	12904.7	12876.9	12282.1	12307.2	12557.8	12537.8	12650.8
AVG	12965.8	12690.6	12721.1	12893.9	12866.0	12274.7	12301.8	12584.6	12540.9	12648.8
STDEV [Å]	10.6	14.3	10.4	11.1	12.1	7.7	6.7	10.6	8.2	4.0
3*STDEV [%]	0.24%	0.34%	0.25%	0.26%	0.28%	0.19%	0.16%	0.25%	0.20%	0.09%

TABLE 2(d): Dynamic repeatability for Ti 1200Å thickness measurement based on nine (9) points.

	PT1	PT2	PT3	PT4	PT5	PT6	PT7	PT8	PT9	
				Repeatability : Ti1200 Å						
X [mm]	0	0	50	0	-50	0	90	0	-90	
Y [mm]	0	50	0	-50	0	90	0	-90	0	AVG
Repeat 1	1230.8	1208.5	1219.6	1229.8	1217.9	1172.9	1187.3	1207.8	1189.2	1207.1
Repeat 2	1231.1	1210.9	1219.2	1230.8	1219.5	1171.7	1187.4	1207.6	1189.8	1207.6
Repeat 3	1233.6	1209.3	1219.5	1230.9	1219.8	1172.2	1187.2	1208.1	1189.2	1207.8
Repeat 4	1232.5	1210.6	1219.4	1231.1	1219.9	1172.7	1187.2	1208.4	1189.9	1208.0
Repeat 5	1232.9	1210.4	1219.7	1231.4	1219.6	1171.9	1187.2	1208.2	1189.8	1207.9
Repeat 6	1233.0	1209.6	1219.9	1231.0	1218.9	1171.3	1187.3	1207.6	1189.2	1207.5
Repeat 7	1232.3	1207.3	1217.9	1229.7	1217.8	1172.2	1185.3	1208.0	1187.7	1206.5
Repeat 8	1231.8	1209.5	1220.5	1231.7	1219.9	1172.7	1186.8	1207.7	1190.4	1207.9
Repeat 9	1232.4	1209.6	1220.2	1230.5	1218.9	1172.2	1186.2	1207.6	1189.2	1207.4
Repeat 10	1233.5	1210.0	1219.3	1231.5	1219.6	1172.9	1188.0	1208.4	1190.7	1208.2
Repeat 11	1234.2	1209.5	1220.9	1230.7	1218.2	1172.0	1187.2	1208.1	1188.7	1207.7
Repeat 12	1233.8	1208.8	1219.3	1230.9	1218.6	1172.9	1187.7	1208.4	1189.9	1207.8
Repeat 13	1233.9	1209.4	1219.3	1230.3	1219.0	1172.6	1187.0	1208.1	1190.3	1207.8
Repeat 14	1233.9	1209.3	1220.4	1231.0	1219.5	1173.2	1186.7	1207.7	1189.4	1207.9
Repeat 15	1233.6	1210.3	1219.1	1230.2	1219.1	1171.4	1186.0	1207.9	1189.7	1207.5
AVG	1232.9	1209.5	1219.6	1230.8	1219.1	1172.3	1187.0	1208.0	1189.5	1207.6
STDEV [Å]	1.1	0.9	0.7	0.6	0.7	0.6	0.7	0.3	0.7	0.4
3*STDEV [%]	0.26%	0.22%	0.18%	0.14%	0.17%	0.15%	0.17%	0.07%	0.19%	0.10%

TABLE 2(e): Precision for Ti 1,300Å, NiV 2,800Å, Ag 13,000Å and W 3,300Å thickness measurement based on one (1) point.

	Ti	NiV	Ag	W
	1312.1	2809.7	12916.2	3199.1
	1307.9	2809.7	12913.2	3199.5
	1310.0	2809.4	12916.5	3199.5
	1309.5	2809.2	12913.0	3199.6
	1310.2	2809.1	12914.9	3199.6
	1307.5	2808.8	12918.4	3199.9
	1309.8	2808.5	12915.7	3199.4
	1306.3	2809.0	12915.8	3199.8
	1309.3	2808.4	12910.3	3199.8
	1308.8	2807.9	12916.6	3199.7
	1308.6	2808.5	12917.5	3199.6
	1307.8	2808.3	12918.2	3199.6
	1309.4	2807.5	12914.7	3200.0
	1306.7	2808.4	12917.5	3199.6
	1309.0	2808.5	12911.7	3200.0
AVG	1308.9	2808.7	12915.3	3199.7
STDEV [Å]	1.5	0.6	2.4	0.2
3*STDEV [%]	0.34%	0.07%	0.06%	0.02%

PULSE Technology thickness measurement correlation with SEM

Figure 4 shows the SEM correlation of PULSE Technology thickness measurement of W films. We can see a correlation with $R^2=0.99$ that indicates very strong correlation of the PULSE Technology measurement with the SEM measurement. To protect the confidentiality of the data, we have not shown the actual thickness values, but the correlation was validated across the process window.

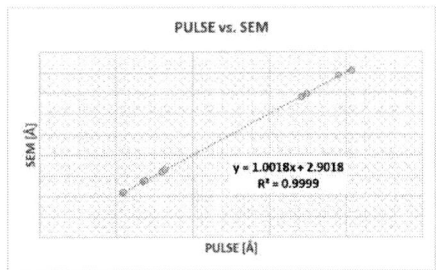

Figure 4: Correlation of PULSE Technology thickness measurement with SEM thickness measurement.

CONCLUSION

In summary, PULSE Technology has been successfully used for thickness measurement of W grid metal film in the CIS BSI process. The repeatability of this technique can meet the stringent demands for inline process control. Beside W inline measurement, we have demonstrated thickness measurement capability for TiN, W, TaN, Al, Ti, NiV and Ag.

REFERENCE

[1] A. Lahav, A. Fenigstein, A. Strum. *High Performance Silicon Imaging*, Woodhead Publishing, 2014, pp98-123.

[2] J. Dai, R. Mair, K. Park, X. Zeng, P. Mukundhan, C. Kim and T. Kryman. *2018 China Semiconductor Technology International Conference (CSTIC)*, Shanghai, China, 2018, pp. 1-3.

[3] Qian. Zhang, Kaiqu. Ang. *Application of IC*, 2021, 038(007):43-47.

NEUTRON IRRADIATION INDUCED CARRIER REMOVAL AND DEEP-LEVEL TRAPS IN N-GAN SCHOTTKY BARRIER DIODES

Jin Sui[1,2,3], Jiaxiang Chen[1,2,3], Haolan Qu[1,2,3], Ruohan Zhang[1,5], Min Zhu[1,2,3], Xing Lu[4]
and Xinbo Zou[1,5]*

[1]SIST, ShanghaiTech University, Shanghai 201210, China
[2] Shanghai Institute of Microsystem and Information Technology, CAS, Shanghai 200050, China
[3] School of Microelectronics, University of Chinese Academy of Sciences, Beijing 100049, China
[4] School of Electronics and Information Technology, Sun Yat-sen Univ., Guangzhou 510275, China
[5] Shanghai Engineering Center of Energy Efficient and Custom AI IC, Shanghai 200031, China
*Corresponding Author's Email: zouxb@shanghaitech.edu.cn
(Jin Sui and Jiaxiang Chen contributed equally to this work.)

ABSTRACT

Effects of 14.9 MeV neutron irradiation on the carrier concentration (N_S) and deep-level traps were analyzed for n-GaN Schottky barrier diodes (SBDs). Neutron irradiation caused a minor positive shift of threshold voltage and typically unchanged reverse leakage current. As the irradiation fluence was increased up to 8×10^{14} n/cm^2, the net carrier concentration was significantly decreased, showing carrier removal effect. Concentration of two shallow traps (E1 and E2) in the GaN epi-layer was enhanced upon neutron irradiation, as revealed by deep-level transient spectroscopy (DLTS). A new deep-level trap E4 (E_C-0.64 eV) was spotted for neutron-irradiated samples. Analysis of DLTS amplitude suggested that E4 was associated with extended defects rather than point defects. The results indicate that the GaN SBDs are promising for operations in high-dose neutron radiation environments.

INTRODUCTION

High energy particles have been extensively used in medical diagnostics/treatment, defense, and space applications. Harsh environment of high-fluence radiation sets new reliability requirements on associated electronic systems. Irradiation of various energetic particles, such as neutrons, may induce deep-level traps and has a significant influence on the degradation of semiconductor devices and related electronics. Gallium nitride (GaN), due to its wide energy bandgap, strong atomic bonds, and thermal stability, has been regarded as a feasible material for high-radiation applications [1, 2]. Therefore, it is of scientific and practical significance to understand the nature of defects and failure induced by neutron irradiation in GaN devices. There have been some studies about neutron irradiation effects on III-V semiconductor devices, but the beam energy of neutron irradiation is mostly lower than 10 MeV [3, 4]. Device degradation and deep-level traps in GaN diodes induced by high energy neutron irradiation are still limited in the literature.

In this paper, we investigated the carrier removal effects and deep-level traps induced by high energy (14.9 MeV) neutron irradiation on the n-GaN Schottky barrier diodes (SBDs). With increasing neutron irradiation fluence, the electrical properties and carrier removal effect were studied. Deep-level transient spectroscopy (DLTS) was employed to determine the trap properties. The possible origins of the deep-level trap induced by neutron irradiation were also investigated.

DEVICE AND EXPERIMENT

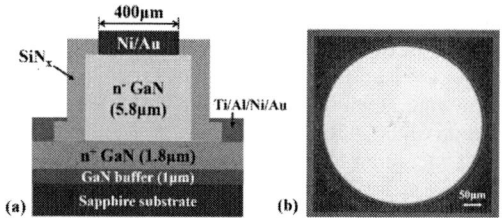

Figure 1: (a) Schematic cross-section of GaN quasi-vertical Schottky barrier diodes. (b) An optical image of a pristine device under test with a mesa diameter of 400 μm.

Figure 1 (a) is the cross-sectional schematic image of GaN SBD, and Figure 1 (b) shows the optical image of a pristine device. The GaN SBDs in this work were grown on 2-inch sapphire substrate by metal organic chemical vapor deposition (MOCVD). The epi-layer structure consists of a 1-μm-thick GaN buffer layer, a 1.8-μm-thick n^+-GaN layer with electron concentration of 5×10^{18} cm^{-3}, and a 5.8-μm-thick unintentionally-doped n^--GaN layer with nominal carrier concentration of 5.3×10^{15} cm^{-3}. After etching and sidewall passivation with SiN_x, Ti/Al/Ni/Au and 400-μm diameter Ni/Au layer were deposited on the exposed n^+-GaN layer and the mesa serving as cathode and anode respectively. The Ti/Al/Ni/Au Ohmic contact was annealed at 850 ˚C for 30 s, whereas the Ni/Au Schottky contact was left unannealed. To study influence of neutron irradiation, two GaN SBDs were irradiated by fast neutrons at 14.9 MeV with a total fluence of $5×10^{14}$ n/cm^2 and $8×10^{14}$ n/cm^2, respectively.

Electrical characteristics of GaN SBDs w/wo neutron irradiation were measured using a Keysight B1500A parameter analyzer. The devices were further sent to DLTS analyzer measurement to study trap properties.

RESULT AND DISCUSSION

Figure 2 compares the current-voltage (*I-V*) and *1/C²-V* characteristics of the GaN SBDs with and without fast neutron irradiation. In forward *I-V* characteristics, with increasing neutron irradiation fluence, threshold voltages of samples were extracted as 0.67 V, 0.78 V, and 0.8 V respectively, given 1 A/cm^2 as the threshold current density. Meanwhile, the neutron irradiation exerted negligible influence on the leakage current of the GaN SBDs, as shown in Figure 2(b).

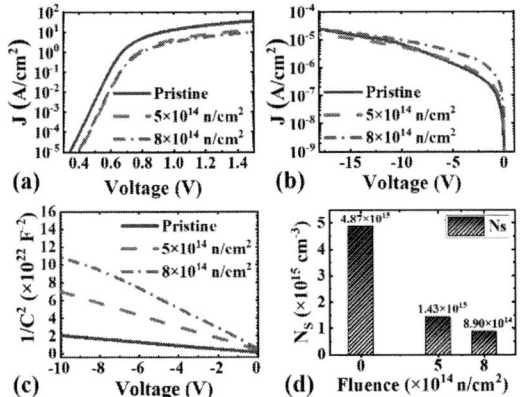

Figure 2: (a) Forward I-V, (b) reverse I-V, (c) 1/C²-V characteristics, and (d) carrier concentration of three GaN SBDs with different neutron irradiation fluence at 300K.

The *1/C²-V* curves are plotted in Figure 2(c), which show good linearity for devices w/wo irradiation, indicating a uniform distribution of carrier concentration in all three samples. The carrier concentration N_S can be obtained by [5]:

$$\frac{d(1/C^2)}{dV} = \frac{2}{\varepsilon_r \varepsilon_0 q A^2 N_S} \tag{1}$$

where ε_r and ε_0 are relative and vacuum permittivity, respectively. q is the elementary charge, and A is the anode area. Carrier concentration was decreased with increasing fluence of irradiation, showing a substantial carrier removal effect, as displayed in Figure 2(d). The carrier removal rate R_C could be determined by correlating the radiation fluence (Φ), initial carrier concentration (n_0) and final carrier concentration after irradiation (n), through the equation $R_C = (n - n_0)/\Phi$ [3]. R_C was extracted to be 5.17±1.35 cm^{-1} for the irradiation condition in this study.

To analyze trap properties of the three samples, DLTS signal was recorded with a reverse bias (V_R) of -6 V, a filling pulse height (V_P) of -1 V and a filling pulse width (t_p) of 100 ms, from 77 to 350 K as shown in Figure 3(a-c). Trap properties were extracted by the Arrhenius plot, as shown in Figure 3(d) and summarized in TABLE I.

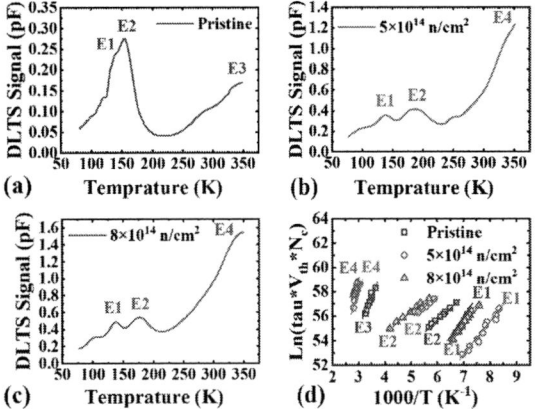

Figure 3: DLTS spectra of (a) pristine and irradiated samples with different irradiation fluence of (b) $5×10^{14}$ n/cm^2 and (c) $8×10^{14}$ n/cm^2. (d) Arrhenius plot obtained from DLTS data of three samples.

Both E1 and E2 could be found in all three samples. It was revealed that trap concentration (N_T) of E1 and E2 in samples upon $5×10^{14}$ and $8×10^{14}$ n/cm^2 neutron irradiation are both higher than the pristine sample. The physical origins of traps E1 and E2 have been presumed to be associated with oxygen impurities or nitrogen-vacancy in GaN material [6-8]. Upon neutron irradiation, the trap concentration and capture cross-section of E1 have been

TABLE I.   DETAILED DLTS RESULTS OF SAMPLES

Irradiation fluence	Trap Properties		
	Activation energy (eV)	Capture cross-section (cm²)	N_T (cm^{-3})
Pristine	0.21 (E1)	$2.74×10^{-16}$	$5.82×10^{13}$
	0.14 (E2)	$1.60×10^{-21}$	$6.10×10^{13}$
	0.46 (E3)	$1.18×10^{-16}$	$5.53×10^{13}$
$5×10^{14}$ n/cm^2	0.23 (E1)	$1.49×10^{-15}$	$8.37×10^{13}$
	0.15 (E2)	$3.32×10^{-21}$	$1.90×10^{14}$
	0.63 (E4)	$1.65×10^{-16}$	$2.90×10^{14}$
$8×10^{14}$ n/cm^2	0.25 (E1)	$9.29×10^{-16}$	$1.72×10^{14}$
	0.15 (E2)	$7.21×10^{-22}$	$2.33×10^{14}$
	0.64 (E4)	$9.05×10^{-17}$	$7.00×10^{14}$

greatly increased, compared with the pristine one. Higher capture cross-section of E1 after neutron irradiation indicates larger probability of capturing electrons, which could lead to a downshift of carrier concentration. For E2, the trap concentration was also increased upon neutron irradiation, however, the capture cross-section was insignificantly affected. It is noted that the amplitude of DLTS spectra which is proportional to N_T increases with elevating irradiation fluence.

After irradiation, a newly-detected distinct peak in DLTS spectra labeled as E4 was spotted as a dominant trap. The DLTS signal indicated that the activation energy of E4 was 0.64 eV, featuring the highest N_T of 7.00×10^{14} cm^{-3} among all the detected traps, however, E3 in the pristine sample was not observed anymore. This result is consistent with trap E_C-0.65 eV identified for a neutron-irradiated GaN SBD with a fluence of 1×10^{14} n/cm^2 fast neutrons in the literature [9]. E4 is also similar to trap E_C-0.66 eV with N_T of 3.8×10^{14} cm^{-3}, which was observed in hydride vapor-phase epitaxy (HVPE) GaN, and was considered as recombination center [10].

The kinetics of carriers captured into E4 was studied by means of recording the dependence of the capacitance transient amplitude $\Delta C(t_p)$ on the filling pulse width t_p, as shown in Figure 4 (a) and (b). ΔC_{max} represents the capacitance transient amplitude when traps are completely filled with a long t_p. A nonlinear relation between $ln(1-\Delta C(t_p)/\Delta C_{max})$ and t_p was observed. Furthermore, the amplitude of capacitance transient with respect to logarithmic t_p was plotted in Figure 4 (b), exhibiting a linear relationship for over five orders of t_p. The above results indicate that trap E4 shows capture behavior of an extended defect, such as a grain boundary-related defect, or a dislocation rather than an isolated point defect [11].

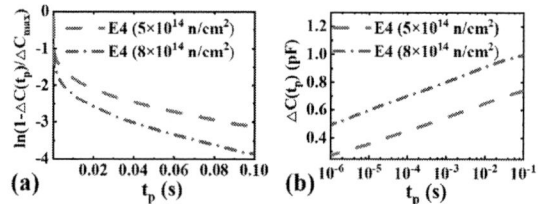

Figure 4: The curves of (a) $ln(1-\Delta C(t_p)/\Delta C_{max})$ versus t_p and (b) amplitude of DLTS transient $\Delta C(t_p)$ with respect to t_p in logarithmic terms of E4 at 350K.

CONCLUSION

To summarize, impact of 14.9 MeV neutron irradiation on n-GaN Schottky diodes on sapphire substrates has been investigated. Positive shift of threshold voltage and carrier removal effect was observed after neutron irradiation with fluence of 5×10^{14} n/cm^2 to 8×10^{14} n/cm^2. Using DLTS, it was found that the trap concentration and capture cross-section of deep-level trap E1 had been greatly increased, compared with the pristine sample. A newly-generated deep-level trap E4 (E_C-0.64 eV) by neutron irradiation was also identified as holding the highest trap concentration. The logarithmic

dependence of DLTS transient amplitude on pulse width suggested that trap E4 is associated with extended defects. The results indicate that the GaN-based SBDs are quite promising for operations in extreme radiation conditions.

ACKNOWLEDGEMENTS

This work was supported by ShanghaiTech University Startup Fund 2017F0203-000-14, the National Natural Science Foundation of China (Grant No. 52131303), Natural Science Foundation of Shanghai (Grant No. 22ZR1442300), and in part by CAS Strategic Science and Technology Program under Grant No. XDA18000000.

REFERENCES

[1] X. Fu, B. Wei, J. Kang, W. Wang, G. Tang, Q. Li, F. Chen and M. Li, *Results Phys.*, **38**, 2022, 105574.

[2] J. Chen, W. Huang, H. Qu, Y. Zhang, J. Zhou, B. Chen and X. Zou, *Appl. Phys. Lett.*, **120**, 2022, 212105.

[3] S. J. Pearton, F. Ren, E. Patrick, M. E. Law and A. Y. Polyakov, *ECS J. Solid State Sci. Technol.*, **5**, 2016, Q35.

[4] Y. Ren, L. Zhou, K. Zhang, L. Chen, X. Ouyang, Z. Chen, B. Zhang and X. Lu, *physica status solidi (a)*, **217**, 2020, 1900701.

[5] J. Chen, M. Zhu, X. Lu and X. Zou, *Appl. Phys. Lett.*, **116**, 2020, 062102.

[6] H. K. Cho, C. S. Kim and C. H. Hong, *J. Appl. Phys.*, **94**, 2003, 1485-1489.

[7] S. Li, J. D. Zhang, C. D. Beling, K. Wang, R. X. Wang, M. Gong and C. K. Sarkar, *J. Appl. Phys.*, **98**, 2005, 093517.

[8] M. Zhu, Y. Ren, L. Zhou, J. Chen, H. Guo, L. Zhu, B. Chen, L. Chen, X. Lu and X. Zou, *Microelectron. Reliab.*, **125**, 2021, 114345.

[9] C.-H. Lin, E. J. Katz, J. Qiu, Z. Zhang, U. K. Mishra, L. Cao and L. J. Brillson, *Appl. Phys. Lett.*, **103**, 2013, 162106.

[10] P. Hacke, T. Detchprohm, K. Hiramatsu, N. Sawaki, K. Tadatomo and K. Miyake, *J. Appl. Phys.*, **76**, 1994, 304-309.

[11] S. Heo, J. Chung, H.-I. Lee, J. Lee, J.-B. Park, E. Cho, K. Kim, S. H. Kim, G. S. Park, D. Lee, J. Lee, J. Nam, J. Yang, D. Lee, H. Y. Cho, H. J. Kang, P.-H. Choi and B.-D. Choi, *Sci. Rep.*, **6**, 2016, 30554.

METAVIT-TRANS: A FRAMEWORK FOR MIXED-TYPE DEFECT DETECTION OF WAFERS WITH VISION TRANSFORMER COMBINED WITH META-LEARNING AND TRANSFER LEARNING

Junfeng Zhao[1,2], Lixin Tang[1]*

[1] National Frontiers Science Center for Industrial Intelligence and Systems Optimization,
Northeastern University, Shenyang, 110819, China

[2] Key Laboratory of Data Analytics and Optimization for Smart Industry (Northeastern University),
Ministry of Education, Shenyang, 110819, China

*Corresponding Author's Email: zhaojunfenglv@outlook.com

ABSTRACT

The defect detection of semiconductor wafer patterns is essential in the process of chip manufacturing. With the improvement of the IC process and design level, the types of wafer surface defects become more complex. In addition, the sample size of wafer map defect data is usually small, and the class is unbalanced, which poses higher challenges to the existing methods. To effectively identify mixed-type defects, a mixed-type defect detection framework with Vision Transformer (ViT) combined with meta-learning and transfer learning (MetaViT-Trans) is proposed. The results show that the MetaViT-Trans framework has a good defect feature extraction ability for imbalanced small-scale wafer map defect data, and has a good multi-scale information learning ability, which can effectively identify a variety of mixed-type defects.

INTRODUCTION

Wafer surface defect detection is a very important step in semiconductor manufacturing. In the process of wafer processing, scratch damage, particle contamination, rotation defects, and edge defects may occur on the wafer surface. If not detected and treated in time, it will not only greatly affect the final performance of the product, but also cause a huge waste of production costs. A lot of previous studies have focused on single defects on the wafer surface. For example, Tsai et al. [1] proposed an automatic visual detection scheme for polycrystalline solar wafers based on mean displacement technology, which is used for the detection of fingerprint and contamination defects on the wafer surface. Li et al. [2] proposed a wavelet-based image defect detection method for polycrystalline solar wafers, which can effectively detect fingerprint, pollutant, and saw mark defects. However, with the improvement of integrated circuit technology and design level, the frequency and types of wafer surface defects increase and become more complex in the semiconductor manufacturing process, and various defect patterns are often mixed. In recent years, with the progress of hardware and algorithm, many excellent research results have been presented for the detection of mixed defects. For example, Kong et al. [3] introduced a convolutional neural network and template-matching method to locate and classify mixed defects. Wang et al. [4] proposed a deforming convolution network (DC-Net), which can effectively selectively sample and decompose mixed defects. Chiu et al. [5] researched and developed an integrated segmentation model based on the Mask R-CNN instance, which could accurately classify and locate defect patterns on wafer maps with limited training data. However, the defect data of wafer map in the actual production process is usually small in sample size and unbalanced in the category, the existing methods have some shortcomings. To effectively identify mixed-type defects with small sample data sets, a mixed-type defects detection framework with Vision Transformer (ViT) combined with meta-learning and transfer learning (MetaViT-Trans) is proposed.

PRINCIPLE AND ANALYSIS

A. MetaViT-Trans framework

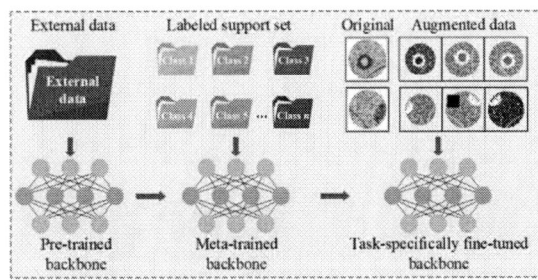

Fig. 1: MetaViT-Trans framework

As can be seen in Fig. 1, the execution process of the proposed MetaViT-Trans framework mainly consists of three stages, which are pre-training, meta-learning, and fine-tuning (FT). Specifically, to build a mixed-type defects detection model for wafer surface, it is necessary to use the self-supervised loss to pre-train features on un-labeled external data to extract backbone network ViT. Then, the ProtoNet (PN) loss in meta-learning is used to conduct meta-training for the feature extraction backbone ViT on the labeled simulated small sample task. By constructing the feature mapping function f, ProtoNet maps the data point *x* to the m-dimensional feature space, thus realizing the classification. The expression of *x* belonging to class *k* is:

$$P\big(y=k\,|\,x\big)=\frac{\exp\big(-d\big(f(x),c_k\big)\big)}{\sum_{k'}\exp\big(-d\big(f(x),c_{k'}\big)\big)} \qquad (1)$$

where, *d* is the cosine distance, c_k is the prototype of class *k*, and c_k is defined as:

$$c_k=\frac{1}{N_k}\sum_{i:y_i=k}f\big(x_i\big),\qquad N_k=\sum_{i:y_i=k}1 \qquad (2)$$

Finally, a mixed-type defects feature extraction backbone network is deployed on a new small sample task, and the network structure and parameters are fine-tuned based on the enhanced data set of each defect category, so the final detection model can achieve our expected results.

B. Experimental setting

The experimental dataset is the Mixed WM-38[4]. The dataset contains approximately 38,000 wafer maps of 52×52 dimensions with a total of 38 defect types, including 1 normal defect type, 8 single-type defects, and 29 mixed-type defects. Here, all mixed-type defects consist of 8 single-type defects, each of which is described in detail in Table I.

TABLE I. DEFECT PATTERN DESCRIPTION

No.	Type	Name	ID	Label
1		Normal	C1	[0 0 0 0 0 0 0 0]
2		Center(C)	C2	[1 0 0 0 0 0 0 0]
3		Donut(D)	C3	[0 1 0 0 0 0 0 0]
4	Single-type	Edge_Loc(EL)	C4	[0 0 1 0 0 0 0 0]
5		Edge_Ring(ER)	C5	[0 0 0 1 0 0 0 0]
6		Loc(L)	C6	[0 0 0 0 1 0 0 0]
7		Near_Full(NF)	C7	[0 0 0 0 0 1 0 0]
8		Scratch(S)	C8	[0 0 0 0 0 0 1 0]
9		Random(R)	C9	[0 0 0 0 0 0 0 1]

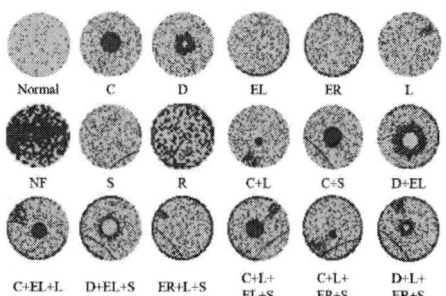

Fig. 2: Wafer maps of different defect patterns

Each position in the wafer map has three values of 0, 1, and 2, which respectively represent the blank spot, the normal mode that passes the electrical test, and the bad mode that fails the electrical test. The wafer maps of the normal type, 8 single defect types, and some mixed-type defects are shown in Fig. 2.

To verify the performance of the proposed deep learning framework on small-sample learning, we first use the ImageNet1K (IN1K) [6] dataset based on DINO [7] to generate the pre-training model of the backbone network. Then we randomly divided 24 types of samples in the Mixed WM-38 dataset into 24 training, 6 validation, and 8 test, and two experiments were designed: 1) Cross-domain sample experiment and 2) in-domain sample experiment, the difference between the two lies in that we selected different data sets CIFAR-FS [8] (64 training, 16 validation, and 20 test) and Mixed WM-38 respectively in the meta-training stage. The number of training and verification tasks was set as 500 and 120 respectively, and the test data in the meta-test stage were the same as in the Mixed WM-38 dataset. In the experiment, the K-way-N-shot problem was taken as the training and evaluation benchmark (K represents the number of categories in the task and N represents the number of samples contained in each category), and the training rounds were set to 100. To evaluate the small-sample classification performance of the model, 500 evaluation tasks were assigned from the divided test set. The evaluation index is the average classification accuracy on the task. Finally, to further optimize the performance of the model, deploying the feature backbone on novel few-shot tasks with optional fine-tuning on the augmented support set of each task.

C. Results and analysis

In the cross-domain sample experiment, the change of the loss curve of the model based on the CIFAR-FS dataset in the process of meta-training is shown in Fig. 3. In the figure, Avg-Loss represents the average loss of 500 training tasks in the training process. The overall training loss shows a downward trend, indicating that the classification performance of the model on the data set is gradually improved.

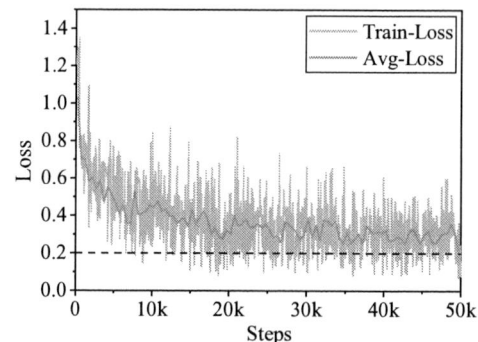

Fig. 3: Change curve of loss in the meta-training

Next, we first applied the test set divided in the Mixed WM-38 dataset to verify the performance of the training model on cross-domain samples, and then fine-tuned the operation to further improve the performance of the model. The specific experimental results are shown in Table II.

TABLE II. PERFORMANCE OF THE MODEL ON CROSS-DOMAIN SAMPLES

No.	Backbone	Pre Train	Meta Train (CIFAR-FS)	Mixed WM-38 5/5
1	ViT	DINO (IN1K)	PN	82.92
2			PN+FT(lr=0.01)	83.60
3			PN+FT(lr=0.001)	82.79
4			PN+FT(Auto)	85.68

It can be seen from Table II that using ProtoNet loss for meta-training of the backbone network ViT can effectively improve the classification performance of the pre-training model. By changing the super parameters and operation modes in the fine-tuning operation, the classification performance of the training model on the cross-domain small sample data set can be further improved, which proves the effectiveness of the proposed method.

Through the observation of the in-domain sample experiment, the change of loss curve of the model based on the Mixed WM-38 dataset in the process of meta-training is shown in Fig. 4. We can still conclude from the figure that with the increase of training epochs, the fitting ability of the model to the category of the dataset is gradually improved, and the performance improvement is better than that of the cross-domain sample experiment.

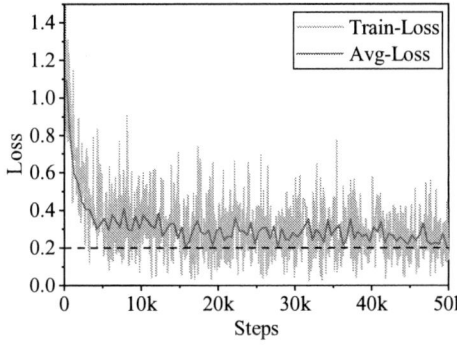

Fig. 4: Change curve of loss in the meta-training

Similarly, we once again applied the test set divided in the Mixed WM-38 dataset to verify the performance of the training model. After verification, to further improve the performance of the model, the same fine-tuning operations as in the above tests were carried out. The specific experimental results are shown in Table III.

TABLE III. PERFORMANCE OF THE MODEL ON IN-DOMAIN SAMPLE

No.	Backbone	Pre Train	Meta Train (Mixed WM-38)	Mixed WM-38 5/5
1	ViT	DINO (IN1K)	PN	99.07
2			PN+FT(lr=0.01)	95.35
3			PN+FT(lr=0.001)	99.37
4			PN+FT(Auto)	99.41

As can be seen from Table III, compared with the cross-domain sample experiment, the model obtained from the in-domain sample experiment has better performance, indicating that the proposed framework can learn the prior knowledge of the training set more effectively for the data in the same domain, to guide the training director of the model and obtain a classification model with fewer samples with better performance.

CONCLUSION

Based on the ViT backbone network, combined with meta-learning and transfer learning, this paper established a deep learning framework, MetaViT-Trans, which can be used for mixed-type defects detection of wafer maps under the condition of limited samples (few sample datasets). The experimental results show that the framework is effective in discriminating and detecting mixed defects of wafer maps. It is further demonstrated that for the limited wafer map samples of other defect types obtained in the actual semiconductor manufacturing process, the corresponding detection model can be quickly constructed based on the MetaViT-Trans framework and applied to the actual production, thus improving the overall yield and cost efficiency.

FUNDING

This research was supported by the Major Program of National Natural Science Foundation of China (72192830, 72192835), and the 111 Project (B16009).

REFERENCES

[1] D. M. Tsai, J. Y. Luo. *IEEE T. Ind. Inform.*, vol. 7, no. 1, 2010, pp. 125-135.
[2] W. C. Li, D. M. Tsai. *Pattern Recogn.*, vol. 45, no. 2, 2012, pp. 742-756.
[3] Y Kong, D Ni. *Proceedings of SMILE2019*, Hang zhou, April 19-21, 2019, pp. 4-8.
[4] J. Wang, C. Xu, Z. Yang, J. Zhang, X. Li. *IEEE T. Semiconduct. M.*, vol. 33, no. 4, 2020, pp. 587-596.
[5] M. C. Chiu, T. M. Chen. *IEEE T. Semiconduct. M.*, vol. 34, no. 4, 2021, pp. 455-463.
[6] Y. Chen, Z. Liu, H. Xu, T. Darrell, X. Wang. *Proceedings of ICCV2021*, Montreal, Oct. 10-17, 2021, pp. 9062-9071.
[7] M. Caron, H. Touvron, I. Misra, H. Jégou, J. Mairal, P. Bojanowski, et al. *Proceedings of ICCV2021*, Montreal, Oct. 10-17, 2021, pp. 9650- 9660.
[8] L. Bertinetto, J. Henriques, P. H. Torr, A. Vedaldi. *arXiv preprint arXiv:1805.08136*, 2018.

LITHOGRAPHY HOTSPOT DETECTION BASED ON TRANSFER LEARNING WITH HIGH RESOLUTION NETWORKS

Hongzhe Wang[1,2,], Lixin Tang[1]*

[1] National Frontiers Science Center for Industrial Intelligence and Systems Optimization, Northeastern University, Shenyang, 110819, China.
[2] The Key Laboratory of Data Analytics and Optimization for Smart Industry (Northeastern University), Ministry of Education, Shenyang, 110819, China.
*Corresponding Author's Email: wamghongzhe@163.com

ABSTRACT

As integrated circuit (IC) technology continues to advance, lithography hotspot detection is of importance in physical verification flow and can affect the turn-around time and the yield of IC manufacturing. In this paper, a deep learning-based lithography hotspot detection method is proposed to overcome the problem of unbalanced positive and negative samples in detection. The proposed method uses the High-Resolution Network (HR-Net18) and pre-trained model from ImageNet to improve transferability. ICCAD 2012 benchmark suits is used for model train and test. The experimental results show that the proposed method performs well in terms of the values of AUC and precision.

INTRODUCTION

The lithography machine is an important equipment used in the manufacture of integrated circuits. Its function is to engrave the mask pattern on the photoresist on the wafer. With the development of integrated circuits, the size of transistors has reached below 10 nanometers, and the number of transistors has grown to tens of billions [1]. Due to the increasing integration of integrated circuits, the traditional optical lithography inspection technology greatly limits the design and manufacture of integrated circuits. Affected by the layout design and lithography process, the lithography results of some patterns in the layout are quite different from the target layout, which may easily lead to short circuit or open circuit problems, resulting in lithography hot spots. Hot spot detection and layout correction in the layout design stage can effectively affect the turnaround time and yield of integrated circuit manufacturing.

In the research of lithography hotspot detection, it is mainly divided into lithography simulation-based, pattern matching-based and learning-based methods. The lithography simulation-based method mainly uses physical simulation techniques to predict lithography results on wafers[2], and this method has high detection accuracy, but is computationally complex and time-consuming. The pattern matching-based method detects lithography hotspots by comparing the similarity between lithography layout and lithography hotspots[3], which has some advantages but performs poorly in

complex scale circuit layouts and is ineffective for unknown hotspot patterns. The learning-based method mainly uses machine learning techniques[4], using existing layout data for training, and the trained model is used for detection, which has a certain generalization ability, good detection performance, and has the advantage of fast detection speed.

Deep neural networks have been shown to have good feature extraction capabilities, and in general, deep learning methods have more network parameters and require more training data. This problem can be effectively overcome using a transfer learning approach, where fine-tuning of pre-trained models using existing data can achieve accuracy comparable to that of training from scratch.

In this paper, we propose a transfer learning-based method for lithography hotspot detection using a pre-trained HR-Net [5] network model based on ImageNet, using the ICCAD 2012 benchmark suite as training data [6]. It performs well in terms of accuracy, precision, recall, F1 score and AUC value

METHOLOGY

The proposed method for lithography hotspot detection is shown in Fig.1. For the complete layout data, the sliding window method is used to cut it into some small blocks of the layout, and they are sent to the pre-trained network model for transfer learning.

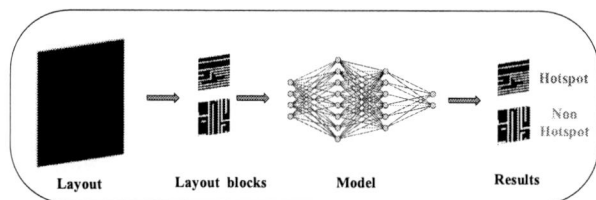

Fig.1. The proposed mothed for hotspot detection

Dataset: The ICCAD 2012 benchmark suite contains a total of five sub-benchmark problems[6], sub-benchmark 1 is based on a 32nm layout design, and sub-benchmarks 2-4 are based on a 28nm layout design. The specific parameters in the modified test suite are shown in Table I. The sub-benchmark contains a total of

1204 hotspot images and 17096 images that don't contain hotspots. The specific form is shown in Fig.2. From Table I, it can be seen that the sample distribution in this test suite and its imbalance, which poses a challenge to the learning algorithm.

TABLE I The ICCAD Benchmark Suit

	Train Set			Test Set		
	HS	NHS	Total	HS	NHS	Total
Sub 1	99	340	439	226	319	545
Sub 2	174	5285	5459	498	4146	4644
Sub 3	909	4643	5552	1808	3541	3649
Sub 4	95	4452	4547	177	3386	3563
Sub 5	26	2716	2202	41	2111	2152
Total	1303	16896	18199	2750	13503	16203

To alleviate the data imbalance, the proposed method augments the lithographic hotspot data using mirror flip and rotation to expand the dataset without changing the presentation of the original data.

The structure of the HR-NET[5] is shown in Fig.3. At

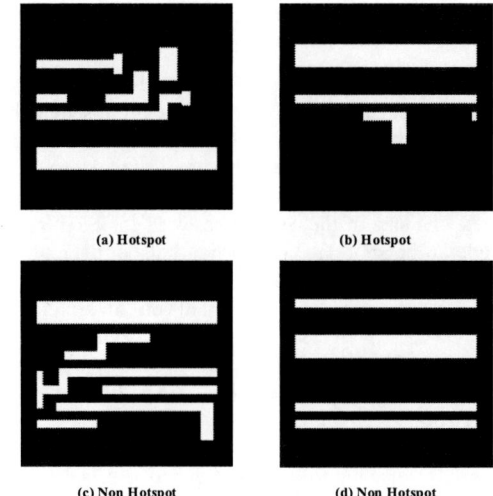

(a) Hotspot

(b) Hotspot

(c) Non Hotspot

(d) Non Hotspot

Fig.2. The example of the ICCAD

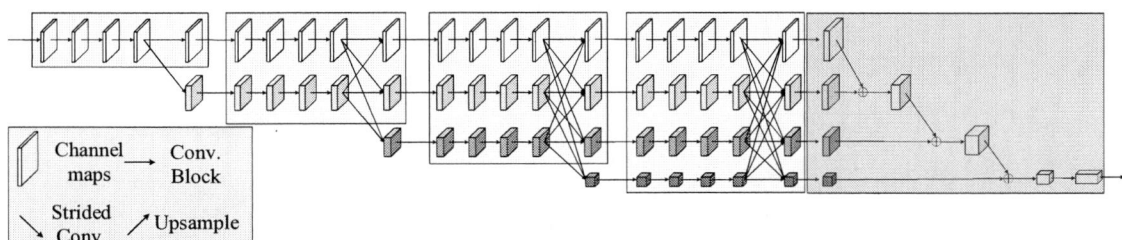

Fig.3. The structure of the high-resolution network

the beginning of the network, the input layer of the HR-NET contains two convolutional layers with stride 2, so the resolution of the high-resolution branch is reduced to 1/4 of the original. The HR-NET with down-sampling at the end of stage 1 to reduce the resolution to 1/8, down-sampling at the end of stage 2 to reduce the resolution to 1/16, and down-sampling at the end of stage 3 to reduce the resolution to 1/32. The entire network structure consists of a total of four parallel branches with different resolutions.

As shown in Fig.4, multi-resolution parallel branch fusion is the fusion of feature maps of multi-resolution parallel branches, which mainly includes three components: up-sampling, down-sampling and flat-level direct connection. This module is placed between stage 2 and stage 4, when the network is down-sampled and low-resolution branches are added before the end of the current stage, and the network structure at this time is roughly as shown in Fig.4, for the high-resolution paths, mainly the fusion and low-resolution feature maps after up-sampling and the current feature maps after the convolution operation of that path. For the low-resolution network, it can be divided into two cases, when the path is in the middle (i.e., it has a lower resolution than the current

path and a higher resolution than the current path), it is a fusion of the feature map of the high-resolution down-sampling, the feature map of the current path after the convolution operation and the feature map of the lower resolution after the up-sampling; When the path is at the bottom (i.e., the current path is the lowest resolution), it is the fusion of feature maps of all high-resolution paths by down-sampling of different sizes.

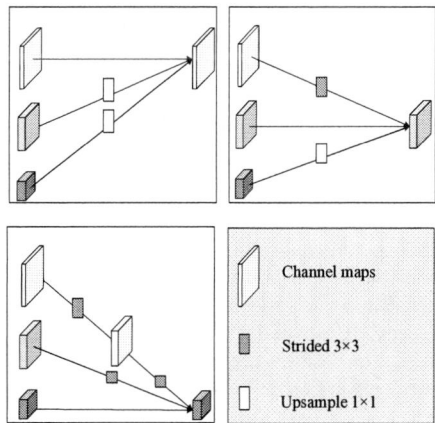

Fig.4. Multi-resolution feature fusion module

EXPERIMENT AND DISCUSSION

To validate the effectiveness of the proposed method, the HR-Net w18 network pre-trained with the ImageNet dataset was used as the pre-trained model, and the ICCAD 2012 test suite was used for training and validation of the model. The size of the training set used for each training include all hotspot data and all non-hotspot data, and the test set was the original test set. The model training and validation process was performed on an Intel Core I7-11700 CPU with wsl2-ubuntu 22.04, Python 3.7, and PyTorch 1.10 software environment. Cross entropy with weights and SGD optimizer were used for the training of the network.

To better detect and weigh the performance of the network, we use the following accuracy, precision, recall, F1 score, and AUC values to evaluate the final results. Since the number of non-hotspots in the test sample is more than ten times the number of hotspots, we mainly focus on recall, precision, F1 score.

$$Acc = \frac{TP + TN}{TP + TN + FP + FN} \#(1)$$

$$Precision = \frac{TP}{TP + FN} \#(2)$$

$$Recall = \frac{TP}{TP + FP} \#(3)$$

$$F1 = \frac{2Precision * Recall}{Precision + Recall} \#(4)$$

The definitions in formulas (1)-(4) are as follows:
TP: the lithography hotspot is judged as a lithography hotspot.
TN: the non-lithography hotspot is judged as a lithography hot- spot.
FP: the non-lithography hot spot is judged as a lithography hot- spot.
FN: the lithographic hot spot is judged as a non-lithographic hotspot.

TABLE II   Results of HR-Net

Data	Acc	Precision	Recall	F1	AUC
Sub 1	92.1%	99.5%	63.1%	77.3%	86.3%
Sub 2	99.1%	92.1%	90.8%	91.4%	95.8%
Sub 3	97.8%	98.0%	89.7%	93.7%	97.9%
Sub 4	99.4%	88.6%	85.9%	87.3%	94.1%
Sub 5	99.5%	94.8%	65.1%	77.1%	97.1%

The experimental results are shown in Table II.

From the results in Table II, it can be seen that the HR-Net used in this paper has good performance in terms of accuracy, recall and F1 value, while retaining the high-resolution features of the plates. Due to the presence of multiple resolution branches, the model retains most of the information of the input features, effectively improving the classification performance of the network.

It can also be seen from Table II that the proposed algorithm performs differently under different sub-datasets due to the imbalance of the dataset distribution. Especially in sub1 and sub5, the performance is poor, because the amount of data in sub1 and the hotspot data in sub5 is too small. The data in Table I can support our conclusion.

CONCLUSION

In summary, a high-resolution network-based approach was used for the lithography hotspot detection problem, alleviating the need for deep neural networks for data by using a transfer learning approach, and the method used was tested based on the ICCAD 2012 test suite. The deep network with retained high resolution proved to have better results in terms of accuracy and recall of detection.

ACKNOWLEDGEMENTS

This research was supported by the Major Program of National Natural Science Foundation of China (72192830, 72192835), and the 111 Project (B16009)

REFERENCES

[1] M. Neisser. *J. Micro/ Nanolithogr. MEMS MOEMS*, vol. 20, 2021, p. 044601.
[2] S. Nakamura, et al. *SPIE Advanced Lithography 2012*, San Jose, February 11-16 ,2012, pp. 233-238.
[3] W.Y. Wen, et al. *IEEE Trans. Comput.-Aided Des. Integr. Circuits Syst.*, vol. 33, 2014, pp. 1671-1680.
[4] M. Shin, J. H. Lee. *J. Micro/ Nanolithogr. MEMS MOEMS*, vol. 15, 2016, p. 043507.
[5] J. Wang, et al. *IEEE Trans. Pattern Anal. Mach. Intell.*, vol. 43, 2020, pp.3349-3364.
[6] J.A. Torres. *2012 IEEE/ACM International Conference on Computer-Aided Design (ICCAD)*, San Jose, November 5-8, 2012, pp. 349-350.

A Methdology for Testing Scan Chain with Diagnostic Enhanced Structure

Keqing Ouyang[1,2]*, Minqiang Peng[1,2], Shuai Wang[1,2], Guohua Zhou[1,2], and Kai Wang[1,2]

[1]State Key Laboratory of Mobile Network and Mobile Multimedia Technology, Shenzhen, Guangdong, P. R. China

[2] Dept.of back-end design, Sanechips Technology Co., Ltd., Shenzhen, Guangdong, P. R. China

*Corresponding Author's Email: ouyangkeqing@sanechips.com.cn

ABSTRACT

Scan chains have emerged as an indispensable component of very-large-scale-integration (VLSI) circuits, commonly utilized for the purposes of testing and diagnosis. However, despite their crucial role as internal test circuits, scan chains are not immune to faults, similar to other logic. In fact, it has been reported that scan chain faults are responsible for as much as 50% of chip failures [1], particularly in advanced process nodes. Therefore, it is of utmost importance to accurately locate and diagnose faults within the scan chains to improve manufacturing yield. In this work, a novel structure-based scan chain diagnosis technique aiming to identify and isolate scan chains' faults with high accuracy and efficiency is proposed. During the implementation of this innovative approach, the performance and reliability of VLSI circuits are significantly enhanced, ensuring optimal yield and customer satisfaction.

INTRODUCTION

Scan chains are commonly used in modern electronic design for testing the fabrication of digital circuits. They provide a convenient and efficient method for validating the operation of a design, ensuring that it meets the required specifications. Despite the benefits of scan chains, they are also prone to faults and can present significant challenges when it comes to diagnosing and troubleshooting issues.

Diagnosing faults in scan chains can be a complex and time-consuming process, as it requires a detailed understanding of the underlying circuitry and how it operates. Moreover, the size and complexity of VLSI circuits can make it difficult to identify specific faults within a scan chain. These factors can lead to extended debug times and increased cost.

One of the most common faults in scan chains is a stuck-at fault, where a particular bit in the chain is stuck at either a logic 0 or 1. This type of fault can be caused by a variety of issues, such as transistor defects, manufacturing process variations, or design errors [2]. Other types of faults may exist in scan chains include delay faults, bridging faults, and transition faults [3], each with their own unique set of causes and diagnostic challenges.

To diagnose faults within scan chains, a variety of techniques and tools are employed. One such approach involves software-based algorithms that utilize chain testing to identify faulty scan flip-flops. Hyeonchan Lim et al. proposed a method using two-stage artificial neural networks (ANNs) to precisely locate the location of such faults [4]. However, it is well known that ANN-based algorithms are susceptible to overfitting, and the result of information gain may be biased towards characteristics with more data. Another diagnostic technique involves hardware-based circuits. Yu Huang et al. introduced the concept of a two-dimensional scan architecture that enables more efficient testing and diagnosis of defective scan cells [5]. Unfortunately, this method requires test data to be shifted in four different directions, leading to increased test time and cost.

To address these limitations, in this work, an enhanced diagnostic structure is proposed and a methodology that accurately identifies faults within scan chains is demonstrated. This approach leverages advanced testing techniques to improve accuracy and efficiency while minimizing design complexity. By utilizing this methodology, manufacturers can improve the quality of their products while reducing the costs and delays associated with traditional scan chain testing methods.

PROPOSED METHOD

The proposed method consists of two steps: configuring the diagnostic-enhanced scan flip flop (DESFF), and subsequently, analyzing any faulty DESFFs. This method is not limited to detecting just stuck-at faults but can also identify transition faults.

A. Configuration of DESFF

In order to achieve diagnosis of scan chains, the structure of a DESFF is proposed as shown in Figure 1a. It consists of a normal scan register, several muxes and an exclusive OR(XOR) logic.

Figure 1b illustrates the scan chain diagnostic network consisting of DESFFs. The Q output of one DESFF is connected to the D input of the next DESFF in a chain, with the chain direction assumed to be from top to bottom. Within each vertically aligned scan chain, any fault-free scan chain can be considered as a dedicated debug chain when its Dedicated_sel is active. Moreover, DESFFs at the same horizontal position but in different chains are designated as the same level DESFFs. In this diagnostic approach, an XOR operation on the outputs of adjacent DESFFs in sequence is conducted, and the results are subsequently transfered to the next scan chain of the same level. This process will be repeated until it reaches the dedicated debug scan cell.

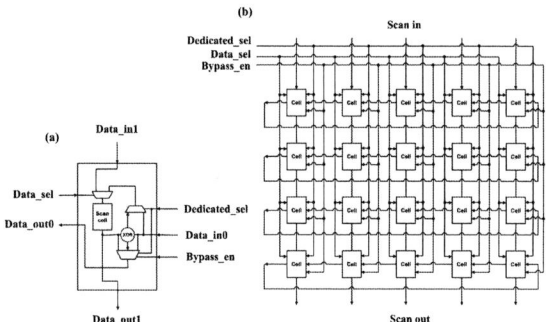

Figure 1. (a) Proposed DESFF structure (b) The configuration of scan chain diagnostic network

Each DESFF is accompanied by several muxes with their selector pins that are controlled by three primary input signals: Data_sel, Dedicated_sel and Bypass_en. Normally, Bypass_en is low during the diagnostic scan chain network operation. In certain complex debug scenarios, the corresponding scan chain can be bypassed by enabling the bypass signal. The control pin Data_sel is utilized to set the target DESFF to either 0 or 1. This is achieved through an XOR with the right adjacent DESFF result, using a specific shift pattern. When Data_sel is high, the test pattern will be launched into the DESFF from the output of the upper stage DESFF. Conversely, when Data_sel is low, the corresponding scan chain will be set to a fixed value.

B. Faulty DESFFs analysis

A common sequence used for chain flush patterns is '00110011...' which can detect stuck-at faults, as well as setup-time and hold-time faults [6-7]. However, research has demonstrated that this simple pattern may not be sufficient for identifying all internal defects within the scan cells and their interconnections [8]. To achieve higher defect coverage, more complex patterns can be employed. However, for the purposes of this paper, only the standard '00110011...' shift patterns are utilized.

As related to ensuring generalization, the stuck fault arises great attention as illustrated in Figure 2. The design includes five scan chains, numbered 0 through 4, each comprising DESFFs, with the fifth being a dedicated debug scan chain. The second scan chain is identified with two DESFFs exhibiting a stuck fault, indicated by red markers in the figure. The test signals and scan in/out patterns of each scan chain are shown in Table I. It is worth noting that the Bypass_en of all scan chains are 0 and only the Dedicated_sel of Chain 4 is 1. For each scan chain with 4 DESFFs that requires testing, the test sequence '00110011' is applied. Following several clock cycles, the test pattern is shifted out from the opposite end

of the scan chain. At this stage, the output pattern will be verified.

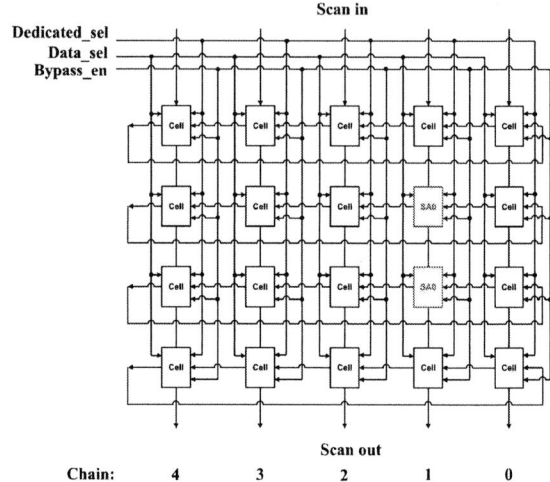

Figure 2. An example of diagnosis of two stuck-at 0 fault DESFFs

Evidently, the output sequence of the chain No.1 consists entirely of 0s, while the output sequence of the remaining chains matches the input. Accordingly, it can be concluded that the second chain contains a stuck-at-0 fault, while the other scan chains are fault-free.

TABLE I: TEST SIGNALS AND DIAGNOSIS RESPONSE OF SCAN CHAINS

Chain	Data_sel	Scan_in	Scan_out
0	1	00110011	00110011
1	1	00110011	00000000
2	1	00110011	00110011
3	1	00110011	00110011
4	0	none	00110011

Given that a stuck-at-0 faulty scan chain has been identified, the next step is to pinpoint the precise location of the fault within the scan chain. It is widely recognized that when a test sequence detects its initial stuck at 0 fault, its output will be 0 and subsequently propagate downstream, leading to the corruption of other data. Consequently, locating the cell responsible for the fault becomes increasingly problematic.

To resolve this issue, the Data_sel signal of the scan chain No.1 where the faulty cell is situated to 0 need to be set. All the DESFFs are reset first. Then, a test pattern of all 1s is applied to the scan chain on the right(chain No.0 shown in Table II), and all cells will be expected taking on an input data value that equals the XOR output of the adjacent cell on the right, which is 1.

As shown in Table II, when the remaining scan chains

are also supplied with test patterns of all 0s, any faulty scan chain will exhibit dedicated debug cell values of 1 for all fault scan cells at the same level. By using this approach, we can accurately determine the position of each stuck-at fault cell.

TABLE II: TEST SIGNALS AND DIAGNOSIS RESPONSE OF SCAN CHAINS

Chain	Data_sel	Scan_in	Scan_out
0	1	1111	1111
1	0	xxxx	0000
2	1	0000	0000
3	1	0000	0000
4	0	none	0110

Employing the same diagnostic methodology, one can accurately pinpoint cells with other fault types as well. The correlation between the output reactions of diverse fault types and the input sequence of '00110011...' is presented in Table 3. By analyzing the input sequence of 00110011, the fault type can be determined and the value of DESFFs can be configured on the corresponding chain to obtain precise location information.

TABLE III: DIAGNOSIS RESPONSE OF SCAN CHAIN WITH DIFFERENT FAULTY TYPES

Fault types	Scan_in	Scan_out
Stuck-at 1	001100110011	111111111111
Stuck-at 0	001100110011	000000000000
Fast-to-rise	001100110011	X01110111011
Fast-to-fall	001100110011	X00100010001
Slow-to-rise	001100110011	00100010001X
Slow-to-fall	001100110011	01110111011X

CONCLUSION

In summary, the scan chain diagnostic method proposed in this paper has shown promising results in accurately locating and identifying various fault types in VLSI circuits. The implementation of DESFFs and dedicated debug chain has proved to be both effective and efficient, as it can be easily integrated into existing design flows without significant overhead in terms of area or power consumption. Moreover, the method is flexible and can be adapted to diagnose different types of faults by modifying the input sequence or the diagnostic algorithm. Overall, the results of this study demonstrate the potential of the scan chain diagnostic method as a reliable and practical solution for fault diagnosis in digital circuits.

ACKNOWLEDGEMENTS

The author wishes to thank State Key Laboratory of Mobile Network and Mobile Multimedia Technology and Sanechips Technology Co., Ltd. for the support on this paper. We also thank all members of DFT group for their great support on this work.

REFERENCES

[1] H. Lim, S. Jang, S. Kim and S. Kang. *International SoC Design Conference*, 2019, pp. 295-296.

[2] K. S. Kim, S. Mitra and P. G. Ryan. *IEEE Design & Test of Computers*, 2003, vol. 20, pp. 8-16.

[3] H. Lim, T. H. Kim, S. Kim and S. Kang. *International SoC Design Conference*, 2020, pp. 57-58.

[4] H. T. Vierhaus, W. Meyer and U. Glaser. *Proceedings of IEEE International Test Conference*, 1993, pp. 83-91.

[5] Y. Huang and W. T. Cheng. *International Symposium on VLSI Design, Automation and Test*, 2017, pp. 1-4

[6] J. S. Yang and S. Y.Huang. *International Conference on Computer Design*, 2005, pp. 157-160.

[7] C. W. Tzeng and S. Y. Huang. *IEEE Transactions on Circuits and Systems II*, vol. 54, 2007, pp. 690-694.

[8] R. Guo, L. Lai, H. Yu and W. T. Cheng. *IEEE International Test Conference*, 2008, pp. 1-10.

AN END-TO-END DETECTION APPROACH FOR MICROPIPE DEFECT OF SIC WAFERS VIA FUSING MULTIPLE HIERARCHICAL FEATURES

W. X. Shi[1], T. G. Zhao[1], and J. W. Zhang[1]*

[1]State Key Laboratory of Precision Measurement, Department of Precision Instruments, Tsinghua University, Beijing 100084, P. R. China
*Corresponding Author's Email: zhangjw@tsinghua.edu.cn

ABSTRACT

In this paper, we present an object detection model for detecting micropipe defect (MP) on silicon carbide wafers. The model is based on the Faster R-CNN architecture, with ResNet34/50 as the backbone. We propose an Improved Feature Pyramid Network (IFPN), which can perceive more location details of defects and improve the detection performance of small objects. Our model achieved a Mean Average Precision (mAP) of 80.8/87.0% with ResNet34/50 as the backbone on our self-made MP dataset (THU-MP-DET). On test data set, we obtained 92% recall and 95% precision, and were able to detect micropipes at a speed of 10 fps using a single GPU.

INTRODUCTION

Micropipe Defect

Silicon carbide (SiC) wafers are crucial components in power electronic devices due to their excellent properties, such as high thermal conductivity, high temperature resistance, and high voltage breakdown. However, the fabrication of SiC wafers is complex and challenging, and defects can easily be introduced during the manufacturing process. One of the most critical defects in SiC wafers is micropipe defects, which are small tubular voids that occur during the crystal growth process.

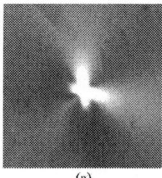 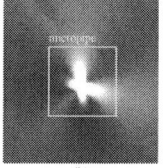

Figure 1: Micropipe Defect

In Figure 1(a), a micropipe defect is shown, as photographed by crossed polarizing microscopy. Figure 1(b) displays the result after defect detection, including classification and location.

Defect Detection

The detection of micropipe defects is important because they can significantly impact the performance and reliability of SiC devices.

Traditional detection methods for MP defects usually involve corrosion technology. After soaking in a corrosive solution for a certain period, micropipes on the surface can be exposed, observed and analyzed using tools such as scanning electron microscope. For example, a study conducted by Mahajan S. et al. in 2013 investigated the use of molten KOH etching of 6H n-SiC for revealing defects as a function of temperature and time.

In SiC-related production processes, non-destructive, fast, and accurate MP detection is crucial. Methods commonly used in factories today include techniques such as optical microscopy, scanning electron microscopy, and X-ray diffraction. Additionally, traditional defect detection methods rely on visual inspection by experts, which is time-consuming, subjective, and prone to human errors. Therefore, there is a need to develop new, rapid, accurate, and non-destructive MP detection methods.

Currently, some scholars have begun to use machine learning methods or computer vision technology to improve the ability of automatic detection. For example, literature [1] developed a cross-polarized nondestructive optical stress technique and used a combination of image processing and machine learning to detect and classify micropipes in SiC crystals. Literature [2] proposed an automatic detection algorithm for dislocation contrast of n-type SiC wafers in birefringence images in 2021, which can detect large changes in contrast levels near the dislocation contrast. These methods have the advantages of fast detection speed, high accuracy and are expected to be an important direction for MP detection in the future.

In recent years, deep learning-based object detection models have shown great potential in automated defect detection. Among them, Faster R-CNN is a widely used architecture due to its excellent performance and fast inference speed.

In this paper, we propose an end-to-end model for detecting MP defects in SiC wafers. This is an exploration of the application of deep learning-based computer vision technology in MP detection and verifies the feasibility of deep learning object detection in high-precision non-destructive online MP detection.

MATERIALS AND METHODS

Dataset

We have prepared a dataset named THU-MP-DET, which includes 4254 images with 2026 annotated MP defects. The data is obtained from 20 4-inch SiC wafers and 10 6-inch SiC wafers, covering MP defects of various sizes and shapes.

Our team has established the transmission microscope platform, and the images captured under the 2.5x objective lens are 4096×3000 pixels, 8-bit single-channel images.

We have collected two sets of images under both crossed polarized and non-polarized light simultaneously, as shown in Figure 2. By comparing the defect positions of the two sets of images, we can distinguish MP from particles, pits, and other stress lines. The image is then automatically annotated using the pre-trained YOLOv5 model. Finally, we manually adjust the labeling to ensure the dataset's quick and accurate completion.The dataset is divided into a training set of 3404 images, a validation set of 425 images, and a testing set of 425 images.

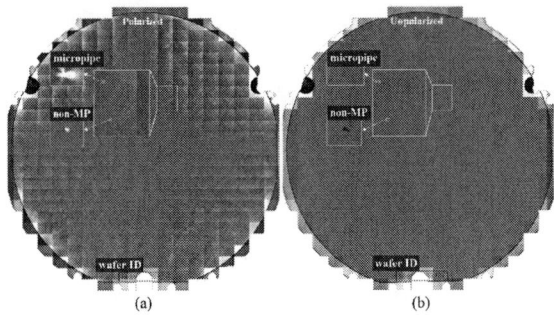

Figure 2: SiC Wafer Map

Model Architecture

Our model is based on the Faster R-CNN architecture, which is a two-stage model that provides excellent accuracy and speed. In the first stage, the region proposal, we use the ResNet34/50 backbone to extract feature maps from the input images[3]. Then we apply an improved feature pyramid network (IFPN) to the feature maps extracted from layer1.2, layer2.3, layer3.5, and layer4.2 to obtain multi-scale feature maps.

Unlike FPN, IFPN adds additional fusion of lower-level information, in addition to top-down feature fusion. Down-sampling uses bilinear interpolation, which better preserves image information and focuses more on small target information compared to nearest neighbor interpolation used by FPN. The resulting multi-scale feature maps are used to generate region proposals.

In the second stage, we use a region of interest (RoI) pooling layer to extract fixed-length features from each region proposal. Then we use these features to classify the proposal as a micropipe defect or background and to predict the bounding box coordinates of the defect. We use an 8-anchor design with anchor sizes of 32, 48, 64, 96, 128, 256, 384, and 512 to improve the model's accuracy.

Finally, the class probability and bounding box regression parameters are obtained. The MP Defect Detection Network is shown in Figure 3.

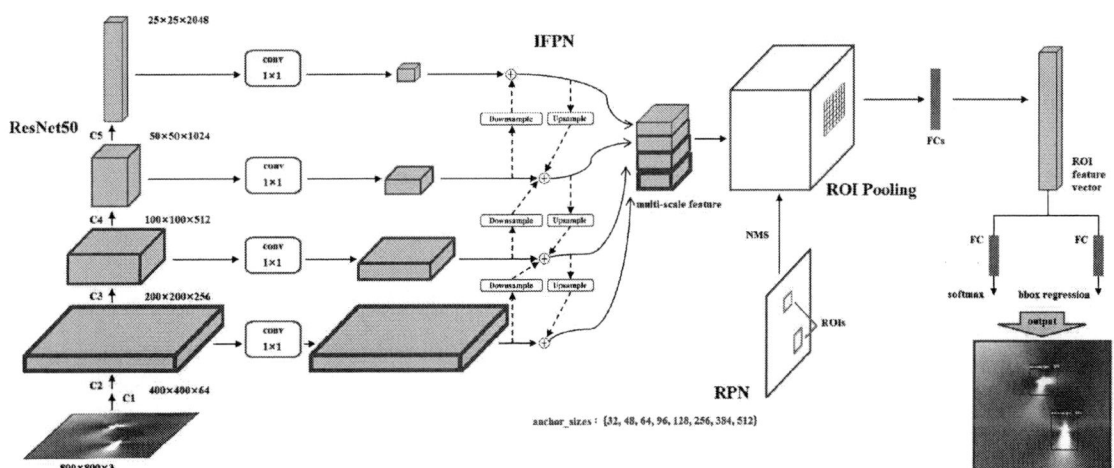

Figure 3: MP Defect Detection Network

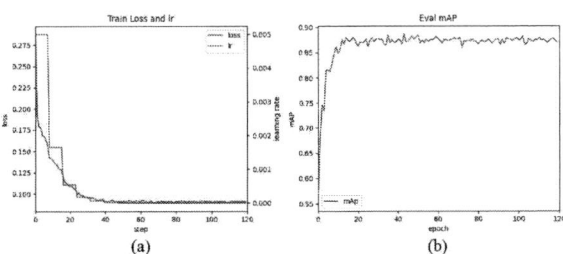

Figure 4: The curves of train loss, learning rate and mAP

Training

We trained our model using the PyTorch framework on a single Tesla V100-SXM2-32GB GPU. We used stochastic gradient descent (SGD) as the optimizer with a learning rate of 0.005, momentum of 0.9, and weight decay of 0.0001. We also used a StepLR learning rate schedule, where the learning rate was multiplied by 0.33 every 8 epochs. The model was trained for a total of 120 epochs with a batch size of 4.

As shown in Figure 4(a), the loss keeps decreasing to about 0.09. We also compared the loss reduction under different learning rate step-sizes, specifically 3, 5, 8, 10, and 20, the results were the best when the step-size was 8.

Evaluation

We evaluated our model's performance on the validation set using Mean Average Precision (mAP) as the metric. mAP is a widely-used metric in object detection that calculates the average precision (AP) for each class and then takes the mean over all classes.

$$AP = \int P(r)dr \qquad (1)$$

AP is the integral under the PR curve that is drawn by the recall and precision under different prediction probabilities. The recall is TP/(TP+FN), and the precision is TP/(TP+FP), where TP stands for true positives, FN for false negatives, and FP for false positives.

Results

IFPN combines features from various levels into a multilevel feature, which is effective for improving detection. In Table I, we used an IoU threshold of 0.5 and 0.5:0.95 for the evaluation. The model combining IFPN outperforms the other models, indicating that the multi-scale feature is effective for improving the accuracy of detection. In addition, low-level features should be paid more attention to for MP detection because they provide richer location information than high-level features.

The mAP of the model with ResNet50-IFPN as the backbone is 87%, as shown in Figure 4 (b), which is 6.2% higher than the model with ResNet34-IFPN.

TABLE I. DETECTION RESULTS ON THU-MP-DET

Backbone	mAP(0.5)	mAP(0.5:0.95)	FPS
Resnet34	54.6%	15.0%	18
Resnet50	62.8%	26.1%	15
Resnet34+IFPN	80.8%	40.5%	11
Resnet50+IFPN	87.0%	47.3%	10

On the testing set, we predicted the images using our model with ResNet50-IFPN as the backbone. As shown in Figure 5, (a), (b), (c) shows some correctly detected micropipes with a probability close to 100%. Figure 5 (d), (e), (f) shows some failure cases, where other defects such as particles, pits, and stress lines have been caught, and overlapping bounding boxes appear in some pictures.

However, we noticed that the probability of misidentifying defects was mostly below 80%, implying that finer predictions could be achieved by tuning the prediction parameters. Finally, we achieved a recall rate of 92%, precision rate of 95%, and speed of 10 fps using a single GPU. The results have reached the detection accuracy required by the market, and the speed is 20 times faster than the result of the literature [1].

This is an attempt to apply deep learning object

detection to the MP defects detection. The model's data distribution covers dozens of processes, and the detection range is much larger than existing research methods. It has high precision and fast speed, and does not require frequent manual parameter adjustment.

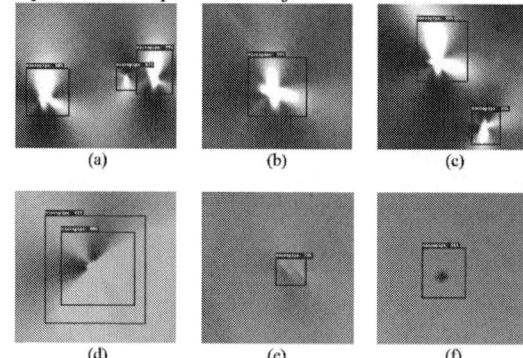

Figure 5: Examples of detection results on THU-MP-DET. For each defect, the box is the bounding box indicating its location and the label is the class score.

CONCLUSION

In this paper, we propose a deep learning-based object detection model for detecting MP defects on SiC wafers. The proposed model is based on the Faster R-CNN architecture with a ResNet backbone network and IFPN integration. The model was trained using an anchor-based approach with eight different anchor sizes. The proposed model achieved an mAP score of 87% on the THU-MP-DET dataset, as well as a recall rate of 92%, precision rate of 95%, and speed of 10 fps on the testing set, demonstrating its effectiveness in detecting MP.

Future work will focus on expanding our dataset, improving the model's performance on other types of defects, and exploring the possibility of applying the model in real-time defect detection systems.

ACKNOWLEDGEMENTS

We are deeply grateful to CSTIC and all those who have contributed to this research project. We acknowledge their invaluable contributions.

REFERENCES

[1] Kubota, T., et al. A nondestructive automated defect detection system for silicon carbide wafers. Machine Vision and Applications 16, 170-176 (2005).

[2] Kawata A, Murayama K, Sumitani S, Harada S (2021) Design of automatic detection algorithm for dislocation contrasts in birefringence images of SiC wafers. Jpn J Appl Phys 60(SB):SBBD06.

[3] Y. He, et al. "An end-to-end steel sur face defect detection approach via fusing multiple hierarchical fea tures," IEEE Trans. Instrum. Meas., vol. 69, no. 4, pp. 1493–1504, Apr. 2020.

A NOVEL METHOD TO ACHIEVE HIGH EFFICIENT ITERATION OF MBIST PATTERN

Minqiang Peng[1,2], Keqing Ouyang[1,2]*, Feilong Pan[1,2], Guohua Zhou[1,2], Lei Chen[1,2]

[1]State Key Laboratory of Mobile Network and Mobile Multimedia Technology, Shenzhen, Guangdong, P. R. China

[2]Sanechips Technology Co., Ltd, Shenzhen, Guangdong, P. R. China

*Corresponding Author's Email: ouyangkeqing@sanechips.com.cn

ABSTRACT

This is a common method to insert BIST logic in the chip for detecting memory faults during production. Power consumption and dynamic voltage drop (IR_drop) are the two obstacles limiting testing of all memories in a chip simultaneously with increasing area of memories. Current alternative method is to divide memories into several parts and test serially. However, it will take considerable time on group division and pattern iteration. In this paper, a novel idea is proposed to achieve high-efficiency iteration of MBIST pattern. With modified construction of mbist circuits, MBIST pattern was splitted into configuration part and main part. MBIST pattern could be updated easily by adjusting configuration, thus reducing the time of pattern iteration.

INTRODUCTION

Chip test is a very important step in silicon production. Ideal chip test should take as little time as possible in testing and debugging. The chip mainly consists of combination logics, sequential logics and memories. In most cases, the production faults of logic are tested by SCAN technology and memory faults are detected by Memory Built-in Self Test (MBIST).

MBIST logic is a self-test circuit inserted into the chip which can generate pseudo-random sequence to the embedded memories and analyze the output to determine whether there is a fault in the memories[1]. There are no other additional IO ports needed except JTAG interface. Instructions are delivered to MBIST circuits by TDI and the MBIST circuits will generate internal stimuli to test memories.

In order to reduce test time of MBIST, the memories of the whole chip could be tested parallelly. However, this method may only be suitable for a chip with a few memories or a chip with a small area. It is difficult to test all memories simultaneously for a chip with a large amount of memories due to the limitation of power consumption and dynamic voltage drop (IR_drop). Current alternative method is to divide memories into several groups and test serially[2][3][4]. However, this method will lead to a trade-off between time cost and chip power, in order to maximize benifits, the best practice is to achieve a balance between test time and test power. It is a general flow of MBIST pattern debugging in chip test as shown in Figure 1. Firstly, DFT engineers need to divide all MBIST controllers of full chip into several groups. The initial groups are usually determined according to the result of power analyze supported by backend engineers. Secondly, the MBIST patterns are generated based on the initial groups. This step usually needs to spend a lot of time especially for a chip with a large die area or mem area, as the MBIST database of the full chip is definitely massive and it is necessary to generate mbist patterns. These first three steps are mainly completed by DFT engineers. After generating all MBIST patterns, DFT engineers delivery patterns to ATE engineers. ATE engineers also need to convert patterns into the corresponding format which the tester could identify. Then DFT engineers and ATE engineers verify all test patterns together. If all these patterns are brought up in different conditions of process, voltage, and temperature

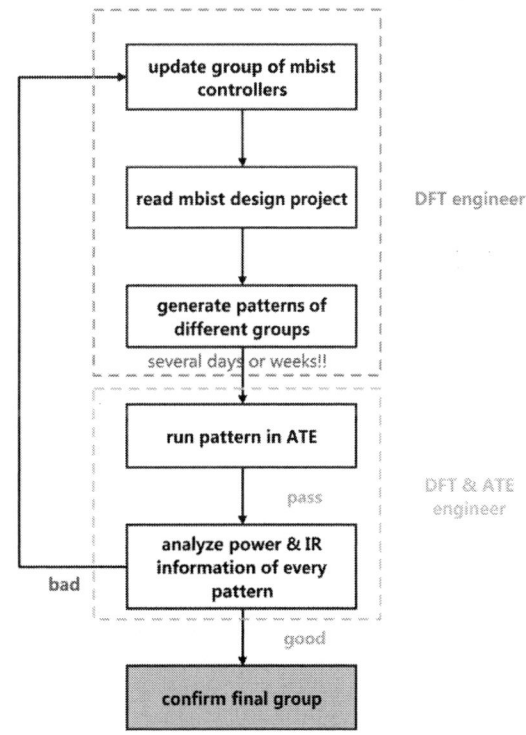

Figure1 the flow of MBIST pattern debugging

(PVT). DFT engineers need to analyze the reality power information of every groups and confirm the final groups. The iteration of patterns is necessary if the patterns are not passed or the power information are not statisfied. Each iteration will cost a lot of time, thus leading to an increasment of chip time-to-market.

According to the Figure 1, the flow of MBIST pattern verification is not a simple process, which needs co-working of DFT teams and ATE teams. And the flow also takes a considerable time. The majority time cost is on the process of pattern generation, thus the key to time saving could be solved from two aspects. The first one is to reduce iteration of patterns and the other one is to speed up the process. In this paper, we propose a novel idea to solve the above problem. Firstly, MBIST circuit is modified by adding extra control logic. The control logic could mask the signals of MBIST test_enable from MBIST Controller and test results of every MBIST sub-controllers to determine whether the sub-controller will be tested. Correspondingly, mbist pattern concludes config part and mainbody. Config part is a description of the blocks involved in a group and mainbody describes the instructions of BIST test. Therefore, the mbist pattern could be iterated easliy by updating config part only or changed in ATE tester. This method could reduce time spent on pattern regeneration and improve test efficiency.

ARCHITECTURE OF MODIFIED MBIST CIRCUIT

Architecture of General MBIST Circuit

A general MBIST circuit mainly consists of one MBIST controller and several sub-controllers, MBIST controller is the central processor which control sub-controllers parallelly. MBIST controller will transfer

test_enable signals to sub-controllers and receive test results of every sub-controllers. Sub-controller is mainly responsible for memory test. Controling the flow of test sequences, generating self-test pattern, and comparing test data are all responsible by sub-controller and other underlying logic. Every sub-controller is independent and just influenced by MBIST controller, therefore testing all memory simultaneously or partly is possible. When MBIST controller distributes test_enable signal to one or several sub-controllers with high speed clock required by memory, all memories under the sub-controllers will be tested with one algorithm and compare output with expected response, the results will be transfered to MBIST controller and then shifted to output port for observing.

Architecture of Modified MBIST Circuit

The New MBIST circuit is modified with extra user-defined control logic based on general MBIST circuit. As displayed in Figure2, user-defined control logic is a flexible circuit which could mask not only test_enable signals to all sub-controllers but also test results to MBIST controller. The high speed clock signals of memories below each sub-controllers are also masked. Every mask signal to sub-controllers is independent and just be controlled by user-defined control logic. User-defined control logic could be configured easily through JTAG interface or other test interface.

With the new MBIST architecture, MBIST controller will not have the only access to each sub-controllers. In this condition, testing the memories under a sub-controller will need other configurations. For example, if the memories to be tested are under control of sub-controller_0, MBIST controller needs to send test_enable signal to sub-controller_0 and func clock

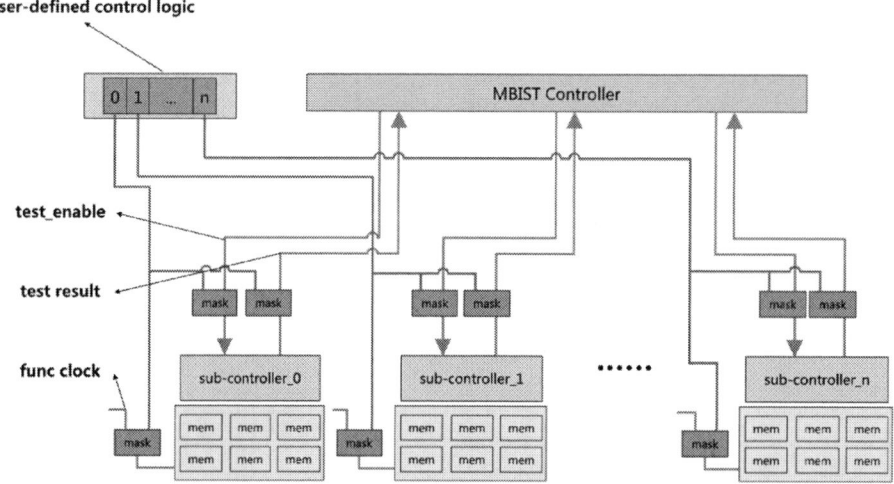

Figure2 The architecture of modified MBIST circuit

should also be provided firstly. The mask signal of sub-controller_0 from user-defined control logic should be inactive. After the above conditions are met, the memories under sub-controller_0 will be tested and the sub-controller_0 will have a feedback to MBIST controller. If the mask signal of sub-controller_0 is asserted, the memories under sub-controller_0 will not perform BIST and MBIST controller also will not receive test result from sub-controller_0.

Although memories under sub-controllers will not perform BIST if the MBIST controller does not delivery test_enble signals to the sub-controllers with the primary MBIST architecture, the modified MBIST circuit is more flexible, convenient, and adjustable.

NEW FLOW OF PATTERN GENERATION AND ITERATION

The flow of MBIST pattern generation and iteration will be changeg based on the modified architecture of MBIST circuit. The new flow is exhibited in Figure3. The complete MBIST patterns consist of config_part and main_part. Config_part mainly includes configuration of user-defined control logic and more specifically the mask signal of every sub-controller is active or inactive just as its name implies. The MBIST design database of full_chip is not necessary for generation of config_part, therefore the time spent for config_part is quite a few. Main_part is similar to previous condition, it needs reading MBIST design database and taking considerable time. The difference is that MBIST controller will send test_enable signal to all sub-controllers. The memories under a sub-controller will perform BIST or not is determined by

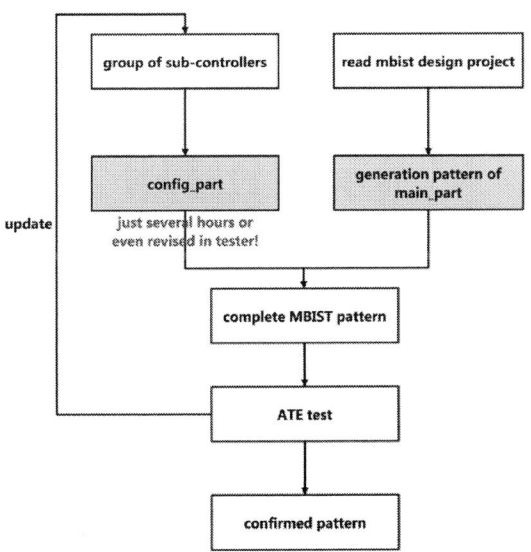

Figure3 new flow of MBIST pattern generation and iteration

mask signal from user-defined control logic other than MBIST controller, thus the main_part only need to generate once.

The different config_parts are generated according to different groups of sub-controllers at the beginning. When running MBIST patterns in tester, config_part needs to be run first and then main_part. For a pattern needs to be iterated due to problems of power or IR_drop, only changing the mask signal configuration of config_part could achieve an alteration of sub-controllers in a group and main_part do not need to be iterated. It will take just several hours or even revised in the tester if there are detailed comments in config_part. The time of the pattern iteration reduces a lot and the efficiency of MSBIT pattern bring-up improves greatly.

SUMMARY

In this article, we proposed a novel method to archive a high efficient iteration of MBIST patterns. MBIST circuit is modified with extra user-defined control logic to achieve more flexible controlment. MBIST patterns are divided into config_part and main_part. Iteration of MBIST patterns are converted to iteration of config_parts based on the modified MBIST circuit. This method could reduce time spent on pattern regeneration and accelarate chip time-to-market.

ACKNOWLEDGEMENTS

This work is supported by the department of back-end design of Sanechips Technology Co., Ltd. Thank all the team members.

REFERENCES

[1] H. Wu. *2022 China Semiconductor Technology International Conference (CSTIC)*, Shanghai, China, 2022, pp. 1-3.

[2] K. S. Das and P. Prakash. *2019 IEEE International Conference on Electronics, Computing and Communication Technologies (CONECCT)*, Bangalore, India, 2019, pp. 1-6.

[3] Kahng A B, Kang I. *2014 Design, Automation & Test in Europe Conference & Exhibition (DATE)*, IEEE, 2014, pp. 1-6.

[4] Zaourar L, Kieffer Y, Wenzel A. *2011 3rd Asia Symposium on Quality Electronic Design (ASQED)*, IEEE, 2011, pp. 46-53.

[5] Yeh C H, Cheng C H, Huang S H. *2016 5th International Symposium on Next-Generation Electronics (ISNE)*, IEEE, 2016, pp. 1-2.

CALIBRATION OF PITCH STANDARDS OF SEM FOR SEMICONDUCTOR DIMENSION METROLOGY APPLICATION

Wei Li[1], Yang Qu[2], and Yushu Shi[1]*

[1] Division of Center for Advanced Measurement Science，
National Institute of Metrology, Beijing 100029, China
[2] Institute of Microelectronics, Chinese Academy of Sciences (CAS), Beijing, China
*Corresponding Author's Email: liwei@nim.ac.cn

ABSTRACT

Dimension metrology is important for all stages of semiconductor technology. SEM is an effective tool for defect inspection and 3D dimension metrology for semiconductor. However, the SEM's magnification varies significantly and is influenced by different factors including acceleration voltage and magnification. In this study, calibration standards are used to evaluated the SEM measurement deviation for MEMS metrology from micro to nano scale. A traceable method for the micro to nano scale standards calibration for SEM by a metrological AFM is proposed. With this standards system, the grating pitches are calibrated and can be used as standard for dimension metrology based on SEM.

INTRODUCTION

MEMS device, such as accelerometers, gyroscopes are increasing used in the smartphones and other electronics. With the development of MEMS fabrication technique, the aspect ratio of feature on wafer is increasing. Also, the shrinkage of FinFET structure is a challenge for measurement. These complex 3-dimensional (3D) structure in semiconductor needs high accurate instrument to determine the dimension parameters. Optical and electron beam inspection tool are widely utilized for the inspection and metrology in semiconductor. Optical inspection methods are usually based on optical wave model and need to be calibrated by standards to evaluate the performance. SEM as an image based tool, with high resolution and large variation range of magnification, is used as a powerful in-line or off line measurement tool in semiconductor industry，to inspect the complex 3D structure. However, in SEM the electron beam scanning is controlled by electro-magnetic field and the image range is very sensitive to the parameter of measurement. A higher level of calibration is to be performed on the SEM image, and then SEM can be used for accurate measurement for high aspect ratio features [1].

In this work, the dimension deviation of SEM image is studied at different magnification, and to correct this deviation a method of calibration the SEM images is introduced. The standard of micron scale is calibrated with a standard measurement system called metrological Atomic Force Microscope (mAFM) [2,3]. The traceability of the measurement persists the comparability of SEM images and as standard method for the measurement of the micro and nano scale devices at various magnification.

EVALUATION OF THE SEM IMAGE DEVIATION

The cross-section of MEMS device is usually imaged and measured by SEM for the dimension determination. For high aspect-ratio structure such as comb fingers, several tens of magnification variation is usually need to determine the height and width of the feature size with adequate image pixels. To investigate the image deviation of SEM image, a serial of images of grating pitches are acquired at different magnification. The image magnification is varied from 1 k to 40 k while the voltages of the acceleration are varied from 10 kV and 20 kV, corresponding to the horizontal field of view from about 100 μm to 2.8 μm. Figure 1 demonstrates a 2 dimensional (2D) grating pitch standards image acquired by SEM. The pitch is 10 μm and the width of the feature is 5 μm. In this work the edge position of the feature at the same side is analyzed manually with the software of the SEM, for the bright edge on the image derived from the interaction between electrons and the sharp edge on the sample. However, the pitch can also be determined by calculation of the centroid of the square feature with image processing algorithm. The pitches measured at different horizontal rows of the images show the uniformity of the sample with a relative standard deviation less than 0.2%.

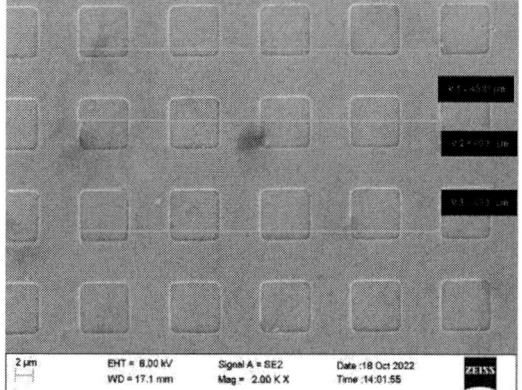

Fig.1. SEM image of a 2D grating with nominally 10 μm pitch. The dimensions of the grating are measured at different rows.

CALIBRATION OF DEVIATION OF SEM IMAGES

Standards calibrated by standard measurement system

The accurate of SEM can be determined by compare the measured pitch with the calibrated pitch. The traceable dimension measurement of standard is realized by a metrological AFM developed in National Institute of Metrology [4]. The measurement system utilizing optical interferometer as position sensor to monitor the 3 movement degrees of freedom of the sample stage. The standard measurement systems have been developed in some national metrological institutes in the past decades to facilitate the micro-nano dimensional metrology and some international comparisons have been carried out in order to support the semiconductor with accurate metrology technology [5].

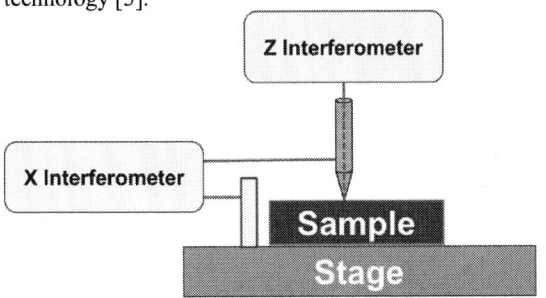

Figure 2: principle of the metrological AFM with sample scanning mode, tracing the measurand to SI unit.

The system allows for both tip scanning and sample scanning modes, facilitating evaluating samples various dimension. For large scanning range of tens of micron, sample scanning motion is utilized. The measuring mirror of the interferometer fixed on the sample stage reflects light beam from interferometer. The displacement of the sample stage is detected by the differential interferometer in 3 axes. The coordinates of the 3D topography traced by the tip is synchronously recorded by touching the surface of the sample during the lateral scanning. The feature size of the sample is measured, and the value can be traced directly to the international unit of length using the frequency-stabilized laser by the interferometer.

A nano-size silicon tip probes the surface topography. Compared with piezo tube scanning, utilizing this sample scanning mode in three orthogonal axes, the standard measurement system eliminates the scanning curvature and the interferometers can trace the dimension to SI unit. Figure 3 is an example of the scanned topography of a 10 μm 2D grating pitch standard measured by the metrological AFM. The scanned range of the image is 100 μm×10 μm showing only one row of the structure. The grating pitch can be determined by evaluating the average pitch with centroid method.

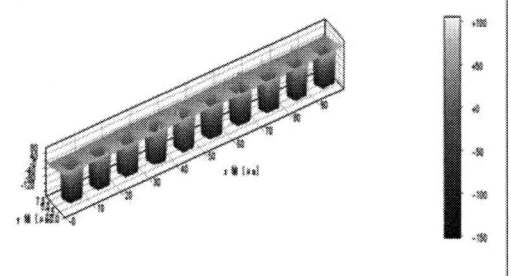

Fig.3. Topography of a 2D grating with 10 μm pitch calibrated by metrological AFM, using laser interferometers to measure the displacement of the sample stage and tracing the measurand to SI unit.

TABLE I. SEM DEVIATION EVALUATED WITH STANDARD SAMPLE AT 10KV.

Magnification	Pixel size	Deviation
1k	0.112μm	1.0%
2k	0.075μm	0.4%
5k	0.022μm	0.4%
16k	0.0070μm	0.4%
32k	0.0035μm	0.5%

With this standard the deviation of the SEM can be evaluated at different magnification and voltage (Tab.1). The standard is a ruler to calibrate the dimensions measured by the SEM.

SEM correction by standards

We analyzed different images with magnification from 1 k to 40 k, with a serial of standards with nominal pitch from 10 μm to 250 nm. The measured deviation is shown in Table 1. The deviation is about 1.0 % at magnification as low as 1k but reduced to 0.4% when the magnification increases. It was also found that the scanning fields of the image varies at different electron energy due to the landing energy of the accelerated electrons. For 10 keV electron landing energy, the measured pitch is larger than at 20 keV, resulting an expanded measured dimension.

The uniformity of the standards should be less than the variation. Otherwise, with a standard of large local pitch variation, the calibration should be performed on the same region of interest to minimize the influence from sample inhomogeneity of the artefacts.

With the standard calibrated by metrological AFM, the image scale of the SEM can be corrected. Then the SEM provides traceable dimension measurement. Figure 3 shows an SEM image of pitch structure on silicon wafer,

979-8-3503-1101-3/23 $31.00 © 2023 IEEE

manufactured by deep reactive ion etching. The cross-section will the vertical sidewall is scanned with SEM. The high aspect ratio feature is common in MEMs and is designed for the application of optical inspection tool evaluation for MEMS device. The dimension parameters of the feature can be measured with the calibrated SEM. For the height measurement, the magnification is about 1.4 k, while the trench width is not adequate resolution and about 9k magnification is suitable. Thus, after the magnification dependent deviation of magnification is corrected, the dimension parameters of the trenches are determined with 4.0 μm pitch and 42.6 μm depth, resulting an aspect ratio larger than 10.

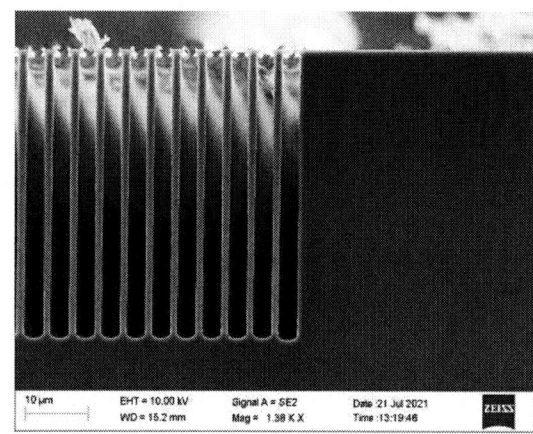

Fig. 4. SEM image of cross section of an etched high aspect ratio grating structure.

CONCLUSION

In this paper a calibration method for SEM images based on a metrological AFM is introduced, which provides traceable measurand to the SI Unit. The calibrated grating standards can provide accurate SEM scale for dimensional measurement application particularly for those with significant dimension variation. A serial of standards with various dimension corresponding to different magnification of SEM image can be used to investigate the deviation of SEM respect the magnification. The deviation of image is found to vary with magnification, so the SEM need to be calibrated at different magnification for quantitative measurement.

ACKNOWLEDGEMENTS

We are grateful to Sitian. Gao from National Institute of Metrology for providing kind suggestions and supports in the standard system development. This study was supported by the National Key Research and Development Program of China, Grant No. 2017YFB1104702 and Development Program of the Ministry of Science and Technology of China (2011BAK15B02).

REFERENCES

[1] W. Häßler-Grohne et al. Proc. SPIE Microlithography, 2004, pp. 426-436.

[2] S. Gao, Q. Li, W. Li, et al. 2013, Proc. of SPIE Vol. 8729, pp.872905.

[3] K. Sugawara. et al. Journal of the Chinese Society of Mechanical Engineers, Series C: Transactions of the Chinese Society of Mechanical Engineers 2006, 27-5.

[4] M. Lu, S. Gao, Q. Li, et al, Proc. SPIE 8759, 2012, pp. 87594Y.

[5] J. Garnaes. Metrologia, 2008, vol. 45, Tech. Suppl., 04003.

ULTRA-WIDEBAND (UWB) TEST SOLUTION ON V93000

Kevin Yan, Daniel Sun

Business Development & Center of Expertise, Advantest (China) Co., Ltd,
Shanghai 201203, China
Corresponding Author's Email: Kevin.Yan@advantest.com, Daniel.Sun@advantest.com

ABSTRACT

Ultra-wideband (UWB), as defined by IEEE802.15.4/4z standard, is a short-range RF technology for wireless communication that can be applied to detect the location of people, devices, and assets with unrivaled precision. The location-based services such as access control and real-time indoor track & tracing are very attractive for the smartphone and automotive market. The increasing popularity of UWB has resulted in the availability of multiple UWB radio devices.

This paper investigates the fundamentals of UWB by providing an overview of the technology, including its RF performance and specifications as defined in the standard. Next, this paper examines the UWB industry and market trends before focusing on ATE requests for UWB chips. Finally, this paper concludes by discussing how V93000 ATE test solutions for wideband RF applications can meet the demands of this growing market.

UWB TECHOLOGY OVERVIEW
What is UWB (Ultra-Wideband)?

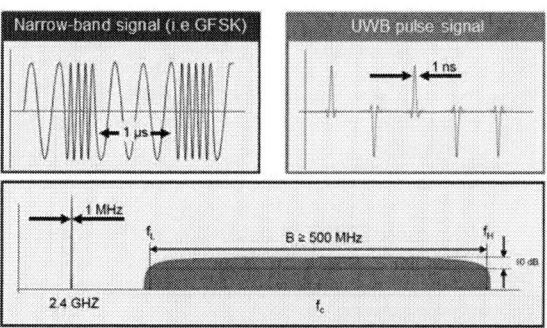

Figure 1: UWB Definition

Ultra-wideband (UWB) is a radio technology that uses a very low energy level for short-range, high-bandwidth communications over a large portion of the radio spectrum. While UWB has traditional applications in non-cooperative radar imaging, recent applications target sensor data collection, precision locating and tracking applications. UWB support has started to appear in high-end smartphones.

UWB transmits information across a wide bandwidth (>500 MHz), allowing for the transmission of a large amount of signal energy without interfering with conventional narrowband and carrier wave transmission in the same frequency band. Regulatory limits in many countries allow for this efficient use of radio bandwidth at low power levels, thus enabling high-data-rate personal area network (PAN) wireless connectivity, longer-range low-data-rate applications, and radar and imaging systems, coexisting transparently with existing communications systems.

UWB is known as pulse radio. The FCC and the International Telecommunication Union Radiocommunication Sector (ITU-R) currently define UWB as an antenna transmission for which emitted signal bandwidth exceeds the lesser of 500 MHz or 20% of the arithmetic center frequency.

UWB applications can cross many service categories. These include secure hands-free access, indoor navigation, hands-free Payments, credential sharing, item tracking, etc. Among these categories, indoor navigation/positional is the most popular application.

UWB high level specifications

Band group* (decimal)	Channel number (decimal)	Center frequency, f_c (MHz)	Band width (MHz)	Mandatory/Optional
0	0	499.2	499.2	Mandatory below 1 GHz
1	1	3494.4	499.2	Optional
	2	3993.6	499.2	Optional
	3	4492.8	499.2	Mandatory in low band
	4	3993.6	1331.2	Optional
2	5	6489.6	499.2	Optional
	6	6988.8	499.2	Optional
	7	6489.6	1081.6	Optional
	8	7488.0	499.2	Optional
	9	7987.2	499.2	Mandatory in high band
	10	8486.4	499.2	Optional
	11	7987.2	1331.2	Optional
	12	8985.6	499.2	Optional
	13	9484.8	499.2	Optional
	14	9984.0	499.2	Optional
	15	9484.8	1354.97	Optional

Table 1: UWB PHY band allocation

Parameter	Value
Center Frequency Range (HRP UWB)	6489.6 MHz ~ 9484.8 MHz (Most UWB products currently available on the market focus on the high band group)
Channel Bandwidth	500MHz (typical) up to > 1GHz
Transmit Output Power	< -41.3 dBm/MHz
Modulation Type	BPM and BPSK (Burst Position Modulation- Binary Phase Shift Keying)
Data Rates	110kbps, 425kbps, 850kbps, 1.7Mbps, 6.81Mbps, 27.24Mbps
Range	10m ~ 100m
Positional Accuracy	< 30cm

Table 2: UWB high level specifications (802.15.4z)

The physical layer of UWB is available in two flavors: the IEEE 802.15.4z standard low-rate pulse (LRP) repetition frequency and high-rate pulse (HRP) repetition frequency. The HRP UWB was becoming successful in various industrial applications for ranging and low-power device-to-device communication.

The HRP UWB physical layer uses an impulse radio signaling scheme with band-limited pulses. It defines operating frequencies in three different bands and 16 channel numbers. Frequencies are categorized into sub-GHz band (channel 0), Low band (channels 1 to 4) and high band (UWB channels 5 to 15). UWB uses an unlicensed spectrum, meaning anyone can implement communication links using a UWB transmitter and UWB receiver, provided the system operates within frequency and power limits.

The HRP UWB PHY uses a combination of burst position modulation (BPM) and binary phase-shift keying (BPSK) to modulate symbols. Each symbol comprises an active burst of UWB pulses, and the variable-length bursts support various data rates.

In China, the Radio Administration of the Ministry of Industry and Information Technology of the People's Republic of China issued the Regulations on Radio Administration of UWB Equipment (Draft for Comments) in January 2023. The new regulations set the frequency band of domestic UWB use at 7235~8750MHz, and the restrictions on transmission power. These regulations will also affect the domestic UWB industry.

UWB market & application trend

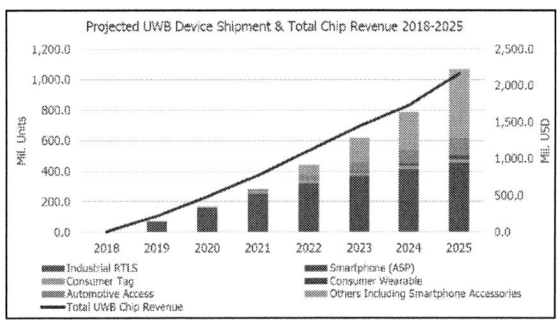

Figure 2: UWB Device Shipment& Total Chip Revenue

UWB technology is poised for significant growth in the coming years, with an adoption curve projected to surpass 1 billion devices annually by 2025. This growth is expected to generate chipset revenues of over $2 billion per year, with a projected 40% CAGR for UWB unit shipments. While UWB technology has potential applications in both the business market as well as the consumer markets, its key focus is expected to be on serving the consumer market. This is due to the much higher volume of consumer devices that could benefit from UWB technology, such as smartphones, wearables, and other personal electronic devices. These devices will benefit from the precision locating and tracking capabilities that UWB offers.

ATE TEST REQUIREMENTS FOR UWB CHIPS

Block diagram of a typical UWB chip

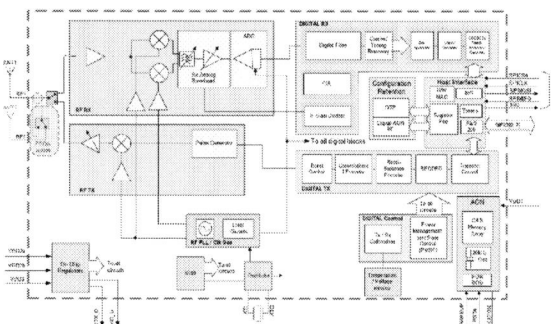

Figure 3: IC block diagram of a UWB transceiver (source: DW3000 datasheet by QORVO)

A typical UWB transceiver consists of an analog front end containing a receiver, a transmitter and a digital back end that interfaces to an off-chip host processor. Also, a TX/RX switch is used to connect the receiver or transmitter to the antenna port.

The receiver includes an RF front end which amplifies the received signal in a low-noise amplifier before down-converting it directly to the baseband. The transmit pulse train is generated by applying digitally encoded transmit data to the analog pulse generator. PLL provides the RF local oscillator signals for the RX Mixer and the TX RF frequency carrier to the Tx mixer.

More advanced UWB modules use two RF antenna ports and are used for Phase Difference of Arrival (PDoA) applications. Some modules even have three or four RF antennae to improve positioning accuracy.

RF test requirements for UWB chips

UWB RF test requirements generally differ from traditional IoT devices, like Bluetooth, Wi-Fi, and 4G/5G cellular. However, UWB test specifications need to follow the standard IEEE802.15.4/4z. From a high-volume mass production point of view, most UWB chips focus on the below test categories.

979-8-3503-1101-3/23 $31.00 © 2023 IEEE   454

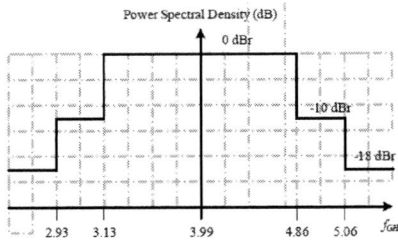

Figure 4: transmit spectrum mask

Transmit measurements: These tests are to ensure that the device meets all the emissions rules that are released by the FCC. Typical test items consist of transmit power spectral density (PSD) mask, which measures maximum power density, maximum power density mask, and transmit center frequency tolerance which checks the carrier frequency offset.

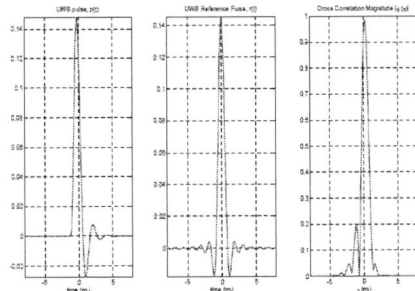

Figure 5: compliant pulse example

Pulse-related measurements are essential to ensure the interoperability and performance of UWB devices. These tests are time domain-based analyses, which means the pulse shape and different pulse parameters are important metrics to ensure that other devices communicate. Typical test items consist of baseband impulse response, including pulse main lobe width, pulse side lobe power, and NMSE. Other tests are performed on chip rate clock and chip carrier alignment tests for chip clock error and chip frequency offset.

Direct receiver measurements are not specified in the 802.15.4z standard. A typical receiver test measures a minimum sensitive power level that the device can operate with minimum error. One common way to test this is to send a minimum power stimulus and measures the device's packet error rate (PER).

Another common test is time of flight (ToF), which is a positioning method based on two ways of ranging. Such tests are used to characterize the positioning performance. In mass production, they are often replaced by phase shifts between RX antennas.

Test challenges for UWB chips.

Obviously, there are three test challenges for UWB chips.

Challenge No.1: high RF frequency. According to the standard IEEE802.15.4/4z, the frequency of UWB devices has three different bands: Band 0 at 500MHz, Band 1 at 3.5GHz to 4.5GHz, and Band 2 at 6.5GHz to 10GHz, the highest frequency being up to 10.6GHz. With traditional ATE RF instruments having a <6Ghz range, this can be a challenge.

Challenge No.2: ultra-wide bandwidth. With all 16 channels using a variety of bandwidths ranging from 500MHz to 1.35GHz. This requires that the test instruments also have ultra-wide analysis capabilities. Most ATE RF instruments can only cover a <200MHz bandwidth, so such wide bandwidth poses a strong challenge.

Challenge No.3: time domain and frequency domain demodulation. UWB devices need to be tested to "transmit PSD mask" by frequency domain and tested baseband impulse response by time domain. This requires that the test software has complex algorithms to handle such demodulations, as well as efficient architecture to cope with huge data processing.

V93000 TEST SOLUTION FOR UWB CHIPSETS

In general, facing these test challenges for UWB chips, the V93000 has a comprehensive solution from hardware to software.

Frequency and bandwidth coverage by V93000 hardware

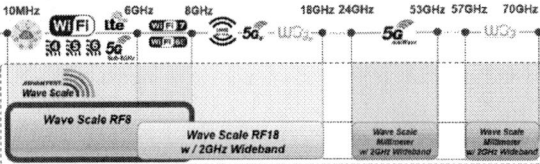

Figure 6: V93000 Wave Scale wireless solutions

From a hardware point of view, V93000 wave scale RF instruments can cover the frequency range from 10MHz to 70GHz, as well as cover bandwidth up to 2GHz.

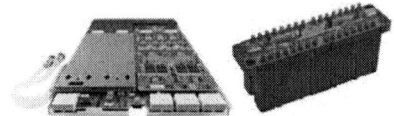

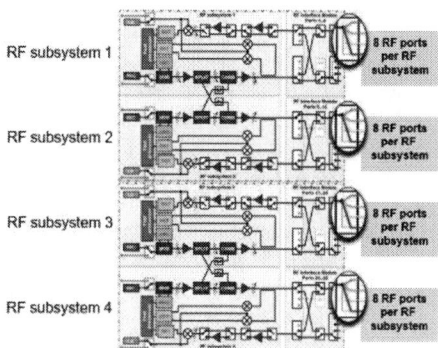

Figure 7: V93000 Wave Scale RF8 Card

The Wave Scale RF8 card supports 10MHz to 8GHz frequency stimulus and measurements with 200MHz bandwidth. A Wave Scale RF8 consists of a Wave Scale RF base card and an RF interface module, it consists of four independent RF subsystems, and each subsystem has 8 bi-directional RF ports, which include one vector signal analyzer, one vector signal generator, and one vector signal network analyzer. The whole instrument has 32 bi-directional RF ports in total.

Figure 8: V93000 Wave Scale RF18 Card

The Wave Scale RF18 card supports 5.85GHz to 18GHz frequency stimulus and measurements with a 200MHz bandwidth. A Wave Scale RF18 consists of a Wave Scale RF base card which is the same as Wace Scale RF8 and an 18GHz RF interface module. A card has 16 RF ports.

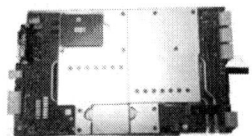

Figure 9: V93000 Wave Scale Wideband Card

The Wave Scale Wideband card is optional, it can co-work with the Wave Scale RF18 and extend the bandwidth to 2GHz. The card also has built-in event triggering for asynchronous UWB chips' TX packets.

Therefore, V93000 Wave Scale RF8, RF18, and Wideband cards can cover the RF frequency and ultra-wide bandwidth requirements of HRP UWB. The V93000 Wave scale RF's 128 max RF ports can achieve high multi-site parallel testing.

Time domain and frequency domain analysis for UWB signals by V93000 Software

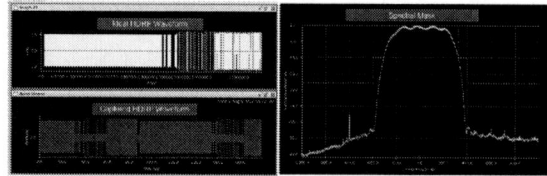

Figure 10: UWB demodulation library

The V93000 Software, Smartest8, contains the UWB demodulation library. It can analyze the UWB signal by time domain and frequency domain and come out with the results of UWB key test items (e.g. transmit PSD mask, transmit center frequency tolerance, baseband impulse response, chip rate clock, and chip carrier alignment…). The spectrum mask also can be printed for intuitive debugging.

So, for "Challenge No.3: time domain and frequency domain demodulation," the V93000 Software, Smartest8, has the capability to handle these demodulations. Also, Smartest8 naturally supports hidden upload of captured waveforms and multi-threaded background processing. This will significantly improve test efficiency and throughput by uploading and processing the previous captured data while capturing the next measurement. In addition, besides such TX test items, for receiver and ToF-related coverage, they can be calculated by Smartest8's existing test methods.

SUMMARY

HRP UWB technology focuses on ultra-high precision positioning and security applications. The frequency range is up to 10.6GHz with a bandwidth up to 1.35GHz. Most currently available UWB products focus on the high band group at 7.2-8.25GHz with 500 MHz bandwidth. RF test requirements of UWB should follow 802.15.4/4z, which includes TX measurements, Pulsed-related measurements, RX measurements, and ToF measurements.

V93000 can cover all the UWB test requirements. The hardware and software have capabilities to handle TX measurements and pulsed-related measurements, and existing test methods can easily cover RX and ToF measurements. Meanwhile, abundant tester resources can achieve high multi-site parallel approaches.

ACKNOWLEDGEMENTS

We would like to acknowledge and give our warmest

979-8-3503-1101-3/23 $31.00 © 2023 IEEE   456

thanks to Frank Goh, who supports the UWB V93000 solution and provided professional guidance to this paper, Frank Goh is a principal consultant at the Center of Expertise Asia from Advantest Singapore.

REFERENCES

[1] IEEE, "IEEE Standard for Low-Rate Wireless Networks" IEEE Std 802.15.4 -2020, pp. 1–800, 2020.

[2] IEEE, "IEEE Standard for Low-Rate Wireless Networks – Amendment 1: Enhanced Ultra Wideband (UWB) Physical Layers (PHYs) and Associated Ranging Techniques," IEEE Std 802.15.4z-2020 (Amendment to IEEE Std 802.15.4-2020), pp. 1–174, 2020.

[3] ADVANTEST, "SmartTest 8.6.0 Documentation", 2023.

[4] ADVANTEST – Edwin Lowery, Joe Kelly, Max Seminario, Frank Goh, and Daniel Sun, "ID_84_Overview of the UWB-HRP Test Solution using WSWB.pdf', pp. 1–27, 2022.

[5] Keysight – Eric Hus, "IEEE 802.15.4 HRP UWB Ranging Process and Measurements", Accessed March,2023, [Online]. Available: https://blogs.keysight.com/blogs/tech/rfmw.entry.html/2021/07/25/ieee_802_15_4_hrpuw-veMI.html.

[6] Keysight – Eric Hus, "An Overview of the IEEE 802.15.4 HRP UWB Standard", Accessed March,2023, [Online]. Available: https://blogs.keysight.com/blogs/tech/rfmw.entry.html/2021/07/28/an_overview_of_ieee-J7ac.html.

[7] WIKIPEDIA, "Ultra-wideband", Accessed March,2023, [Online]. Available: https://en.wikipedia.org/wiki/Ultra-wideband.

[8] FCC, "AMENDED Comments of The Ultra Wide Band (UWB) Alliance Before The Federal Communications Commission FURTHER NOTICE OF PROPOSED RULEMAKAING Mid-Band Spectrum Between 3.7 and 24 GHz.", Accessed March,2023, [Online]. Available: https://www.fcc.gov/ecfs/filing/107142666226784.

[9] QORVO, "DW3000 datasheet.", Accessed March,2023, [Online].Available: https://www.qorvo.com/products/d/da008142.

[10] NXP, "Ultra-Wideband(UWB)", Accessed March,2023, [Online]. Available: https://www.nxp.com/applications/enabling-technologies/connectivity/ultra-wideband-uwb:UWB.

[11] Rohde&Schwarz, "浅谈超宽带技术与测试方法" Accessed March,2023, [Online]. Available: https://www.eettaiwan.com/20210118ta31-ultra-wideband-technology-and-testing.

Applications of Picosecond Laser Acoustics to Power Semiconductor Device: IGBT and MOSFET

Johnny Dai[1], Cheolkyu Kim[2], Priya Mukundhan[1]
[1]Onto Innovation, 550 Clark Drive, Budd Lake, NJ 07828, USA
[2]Onto Innovation, 16-6, Sunae-dong, Bundang-gu, Sungnam-si,Gyunggi-do, 3965 Korea
*Corresponding Author's Email: johnny.dai@ontoinnovation.com

ABSTRACT

Picosecond Ultrasonics (PULSE™) Technology has been the unique metrology tool for single-layer and multilayer metal films in semiconductor process control. [1] PULSE Technology is an industry benchmark for metal metrology and is a tool-of-record in multiple device segments, including logic, radio frequency (RF), memory, microelectromechanical systems (MEMS) and flash. In addition to thickness, the non-destructive technique has been adopted to provide elastic modulus, which is a critical parameter for process control. For example, low-k materials in logic and DRAM, amorphous carbon (a-C) in 3D NAND and AlN in RF filter devices are all exclusively characterized in-line using Picosecond Ultrasonics Technology.

In recent years, power semiconductor applications have expanded from industrial control and consumer electronics to energy, rail, smart grid, inverter home appliances, electrical vehicles and other markets. The market size for global power components is estimated to be $97.2 billion (U.S.) by 2030, with a compound annual growth rate of 6.30% by Market Research Future (MRFR).[2] As the automobile industry transitions from traditional cars to electric vehicles, the increase in automotive electronics has been beneficial to the power semiconductor market. The value of semiconductor components for a pure electric vehicle can be $750 (U.S.); the value of power semiconductor alone is about $413 (U.S.), which is about six times the price of a power semiconductor for a traditional car.

In this paper, we demonstrate how PULSE Technology can be applied to the production of power semiconductors, specifically insulated gate bipolar transistors (IGBT) and metal oxide semiconductor field effect transistors (MOSFET), targeting both the frontside (gate metallization) and backside (backside metallization). We show how PULSE Technology is capable of being used for standard power semiconductor applications involving single-layer metal films and offers excellent repeatability, long-term stability, and fast and reliable thickness profile measurement. For a gate metallization application with a three-to-five layer stack, we demonstrate PULSE Technology's unique advantage when it comes to measuring multiple layers with a single measurement, while offering excellent repeatability and long-term stability. We also show how PULSE Technology offers improved repeatability and throughput by using dual modulation and/or crossed polarization for relatively rough films, such as aluminum, and backside metallization (BSM) stacks.

INTRODUCTION

Looking to the future, the most promising power semiconductor devices will be IGBT and MOSFET modules. IGBT is a three-terminal semiconductor device used in various electronic circuits for the switching and amplification of signals. It has three terminals: an emitter (E), collector (C) and gate (G). MOSFET is a four-terminal semiconductor switching device used for switching and amplifying the signals in electronic circuits. The four terminals of the MOSFET are: a source (S), drain (D), gate (G) and body (or substrate). IGBT and MOSFET are applied differently depending on power, voltage, current, frequency, switching speed, application environment and cost [3].

The metallization layer deposited on the die of power devices (IGBT and MOSFET) has two main functions. First, it connects elementary cells constituting the power dies to the source (MOSFET) or emitter (IGBT). It also allows for the welding of bond wires on the chip. The metallization also allows thermal conduction. Power devices get very hot, so the adhesive layer needs to withstand high temperatures over a long lifetime of use. The thickness of these metal films have been controlled tightly for both electrical and thermal purposes. For advanced power devices to achieve best-in-class conduction loss, it is important for the collector-emitter saturation voltage to be as low as possible. Metallization film thickness and even stress are key to process control. [4]

The materials used for the BSM stacks vary depending on the device type and the solder used in the subsequent assembly stage, but the most common materials are Ti-NiV-Ag, and for some devices, a thin layer of Al or Al alloy is deposited prior to the Ti-NiV-Ag stack. Among them, Ti is used as the adhesion layer to the silicon chip. Nickel vanadium, NiV, offers a barrier that prevents the diffusion of silver, Ag, into silicon. The silver makes up the bonding layer and will be bonded to tin-silver, Sn-Ag. Because power devices run high currents at high-operating temperatures, these layers have been properly controlled for property and thickness, such as void-free attachment, to enhance thermal conductivity, excellent adhesion of the back metal to the silicon, high thermal conductivity adhesive, and a high melting temperature to withstand the operating temperatures of the device and minimize diffusion.

PULSE Technology, as used by the Echo™ metrology system, is a non-contact, non-destructive, first principles technique. It is capable of measuring metal films from 50Å to 20μm, with the option to extend to 35μm. The laser beam is focused to a tight spot (8μm x 10μm) on the wafer, enabling direct measurements on devices (15μm). Measurements take a few seconds per site, and the high throughput allows the entire wafer to be mapped in minutes.

979-8-3503-1101-3/23 $31.00 © 2023 IEEE

In the standard configuration, also known as single modulation, the pump beam is modulated at 5.5MHz, the probe beam is unmodulated and the signal is demodulated at the same frequency as the pump beam. However, when the films are rough, the surface tends to scatter the measurement beams excessively, reducing the overall signal-to-noise ratio (SNR). To improve SNR, mini maps are introduced to average out the surface roughness, or scan times are increased to acquire better quality signals at the expense of measurement throughput. In the new and improved configuration, depending on the application, one can use both the standard single-modulation configuration and a dual-modulation configuration at the recipe level. In dual modulation, the pump beam is modulated at 5MHz, the probe beam is modulated at 0.5MHz and the signal is demodulated at 5.5MHz; this has been demonstrated to be an effective approach. In the modified configuration, any scattering from both pump and probe beams are filtered out, and the SNR is significantly improved and contributes directly to performance and throughput gains. Switching between standard single modulation and dual modulation is achieved at the film-stack level using the appropriate settings. This ensures that thinner film performance is still not impacted. Other improvements include the ability to change the polarization of the pump and probe beam from a standard configuration (pump-s, probe-p) to a cross-polarization (pump-p, probe-s) configuration at the film-stack level. Films that show stronger absorption of the p-polarized light and a stronger reflectivity of the s-polarized light benefit from SNR improvements, which helps with both repeatability and throughput.

We have previously discussed [5,6] a wide range of applications of PULSE Technology for the RF and MEMS markets because of the technology's unique capability to simultaneously measure multilayer metal thickness and the acoustic velocity of dielectric and piezoelectric films. In this paper, we demonstrate how PULSE Technology is uniquely suited to monitor the metallization process in power semiconductor devices.

APPLICATION OF PULSE TECHNOLOGY TO POWER SEMICONDUCTORS

1. Metal film thickness measurement using PULSE Technology

PULSE Technology has been adopted as the process tool-of-record for measuring the metal thickness of power devices. It has been widely used to measure metal films such as AlCu, Ti, NiV, Ag, TiW, TiN, W, WSi, Ni and others. For commonly used metal films in power semiconductor devices, including both frontside and backside applications, PULSE Technology offers a repeatability of 3 sigma at ~0.1-0.3%. Figure 1 shows selected repeatability performances of some commonly used metal films in power devices. The measurements are stable and do not show any trends over the three-day period. Long-term stability performance data seen on customer statistical process control (SPC) charts are comparable to the data shown here.

Typical measurement times for single films is ~2s per site, which enables high-resolution wafer mapping for process tool qualification. Figure 2 shows the thickness profiles of commonly used film, Al, Ti, NiV and Ag, using 49-point wafer uniformity maps.

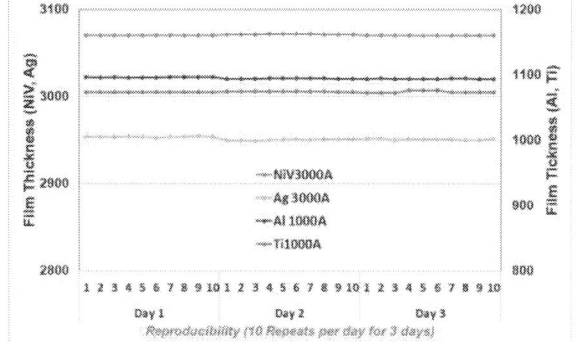

Figure 1. Reproducibility data for common metal films. Data was collected at 10 repeats per day for three (3) days. The long-term stability performance for all above films is comparable to reproducibility and is 3 sigma < 0.30%.

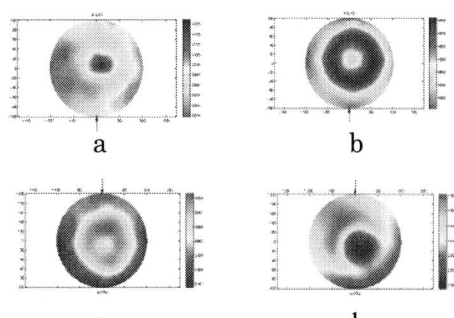

Figure 2. 49-point maps of single metal layer (a). Al 1,000Å, (b). Ti 1,000Å, (c). NiV 3,000Å, and (d). Ag 1,500Å.

2. Performance improvement for repeatability and throughput

As explained in the previous section, with the introduction of film-stack level control of dual modulation and cross-polarization, we are able to improve the performance of PULSE Technology on rough films. In this section, we discuss the specific applications that benefit from the use of these enhanced features. Table 1 summarizes the significant improvement to both repeatability and throughput using the new dual-modulation and cross-polarization feature compared to a standard configuration. The 3 sigma of static and dynamic repeatability improved by 10 times and five (5) times, respectively, for measurements using dual modulation and cross-polarization compared to a standard configuration. Throughput improved by more than five (5) times.

	Stadnard Configuration	Dual Modulation Cross Polarization
Static (3σ)	0.90%	0.08%
Dynamic (3σ)	1.50%	0.27%
MAM Time (s)	45	5.3
13 Point TPT (WPH)	6	34.6

Table 1. Static and dynamic repeatability and throughput comparison of standard measurement configuration and new feature using dual modulation and cross polarization.

We have applied similar measurement controls using the new feature for a wide range of thickness measurements and show similar and robust results. The measurement results also match very well to the reference thickness of $R^2=1.0$, with a slope of 1.0 and offset to 0.0.

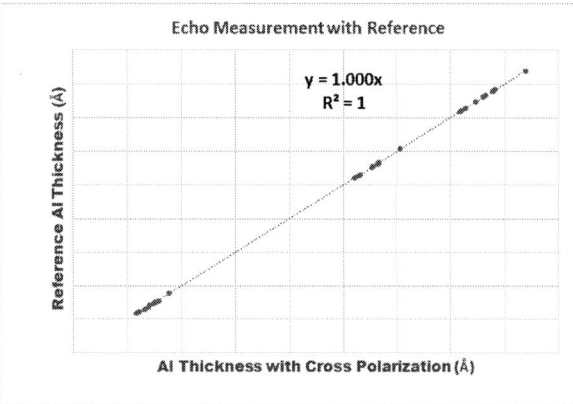

Figure 3. Measurement correlation of the new dual-modulation and cross-polarization feature to reference thickness.

3. Multilayer measurement for gate metallization

For power semiconductors, gate metallization often requires multiple metal layers. PULSE Technology has been widely used to monitor gate/source metallization. Table 2 shows the excellent repeatability of PULSE Technology's static and dynamic measurements for a multilayer stack of Ti/TiN/AlCu/TiN. We can see that 3 sigma of thickness for all top three layers is less than 0.5%. The 3 sigma for the bottom Ti layer is less than 1%. One of the biggest advantages of PULSE Technology is the ability to measure the multilayer metal thickness of repeating materials. For example, in the current stack we were able to individually discriminate the top TiN from the bottom TiN. Alternative metrology tools cannot discern such a difference and can only report a total TiN thickness and, as a result, are not useful for actual device monitoring.

Layer	BTM Ti 300Å	BTM TiN 500Å	AlCu 3500Å	Top TiN 350Å
Static Repeatability(3σ)	0.95%	0.27%	0.04%	0.38%
Dynamic Repeatability(3σ)	0.52%	0.13%	0.03%	0.47%

Table 2. Typical static and dynamic repeatability performance of PULSE measurements for a commonly used multilayer metal film stack of Ti/TiN/AlCu/TiN in power devices.

4. PULSE Technology measurement on BSM tri-layer stack

Several metal layers are used for backside metallization (BSM) and offer good adhesion, electrical properties and long-term reliability. The most commonly used stacks for BSM are Ti/NiV/Ag or Al/Ti/NiV/Ag. Quite often, the backside metal surfaces feature rough grain structures compared to the front side. In a standard configuration, these measurements can be prohibitively slow for high-volume manufacturing and, in some extreme cases, are not able to be measured. With PULSE Technology's recently commercialized dual-modulation and cross-polarization

feature, we can significantly improve SNR related to the measurement of BSM. Table 3 summarizes both static and dynamic repeatability when measuring Ti/NiV/Ag films. The corresponding 49-point thickness profiles are shown in Figure 4.

Layer	Ti 1000Å	NiV 3000Å	Ag 1500Å
Static Repeatability(3σ)	0.73%	0.42%	0.46%
Dynamic Repeatability(3σ)	0.57%	0.09%	0.21%

Table 3. Static and dynamic repeatability performance for PULSE measurement for commonly used multilayer metal film stack Ti/NiV/Ag in power devices.

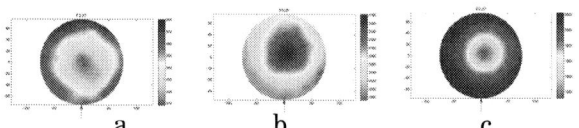

Figure 4. 49 - point maps of (a). Ti 1,000Å, (b). NiV 3,000Å, and (c). Ag 1,500Å.

PULSE Technology has been a metal film metrology workhorse and has carved out a unique space because of its capabilities, such as non-contact, non-destructive, simultaneous multilayer measurement, thickness and sound velocity measurement, fast throughput, and excellent gage capable repeatability and reproducibility. PULSE Technology has been widely adopted for the front-end metallization and backside metallization of power semiconductors. The development of PULSE Technology's dual modulation and cross-polarization capability enhances repeatability and throughput and enables additional capability for more challenging BSM stacks.

REFERENCES

[1] C. Thomsen, H. T. Grahn, H. J. Maris, J. Tauc, Phys. Rev. B, vol. 34, 1986, pp. 4129-4138

[2].https://www.globenewswire.com/en/search/organization/Market%2520Research%2520Future

[3].S. Pietranico, S. Lefebvre, S. Pommier, M. Bouaroudj, S. Bontemps, Microelectronics Reliability, vol. 51, issues 9–11, 2011, pp. 1824-1829

[4]. S. P. Lee, David Goh, T. F. Lim, Published 12 March 2019, 2019 Electron Devices Technology and Manufacturing Conference (EDTM)

[5] J. Dai, Johnny Mu, Cheolkyu Kim, Priya Mukundhan, March 14-15, 2021 CSTIC, Shanghai, China

[6] J. Dai, P. Mukundhan, R. Mair, M. Mehendale, C. Wang, E. Wang, C. Kim, March 14-15, 2020 CSTIC, Shanghai, China

RC-Triggered Silicon Controlled Rectifier-based ESD Clamp with Fast Transient Reaction

Lingran Pan[1], Wenwen Zhang[2], Da-Wei Lai[2], Yidan Liang[1], Feijun Zheng[1]*

[1]Zhejiang University, Hangzhou, 310012, China
[2]Pride Silicon Technology Co., Ltd, Hangzhou, 310058, China
*Corresponding author's E-mail: zhengfj@zju.edu.cn

ABSTRACT

We demonstrate an RC-triggered Silicon Controlled Rectifier-based electrostatic-discharge (ESD) protection device with a fast transient reaction in a 0.18-um bulk CMOS technology. The proposed ESD device is simply formed by adding a gate-grounded NMOS (ggNMOS) into the silicon controlled rectifier (SCR) device. By connecting both the drain of the ggNMOS and the anode (P$^+$) of the SCR to the ESD zapping pin, the ESD conduction mechanism is determined by the SCR in parallel with the ggNMOS during a 100-ns normal transmission line pulse (TLP) event. By contrast, a ggNMOS (parasitic NPN) dominates the triggering and reacts quickly with respect to the traditional RC-triggered SCR device during a 2.5-ns very fast transmission line pulse (VF-TLP) event.

INTRODUCTION

ESD rail-clamp (BIG-FET) [1] is a well-known supply protection with a lower trigger voltage (V_{TI}), but those high leakage currents and large areas are incapable for some area and/or leakage-limited applications. SCRs are well-known ESD devices with low leakage currents, good area efficiency, and ESD robustness. However, the V_{TI} was too high to protect those devices to be protected (victims). RC-triggered SCR protection devices [2]-[3] have been demonstrated for V_{TI} reduction in a normal-TLP (Human Body Model (HBM)-like) event, but these traditional SCR-based ESD devices cannot react quickly in a VF-TLP (Charge Device Model (CDM)-like) fast transient event [4]. In this work, we propose an alternative RC-triggered-based ESD device by adding a ggNMOS into the SCR. By doing so, the proposed ESD device can react quickly in a VF-TLP (CDM-like) fast transient event when compared to the traditional RC-triggered SCR device.

DEVICE DESIGN AND OPERATION

The schematic view of a traditional RC-triggered SCR device is shown in Figure 1 (a). The Tri.N and Tri.P node is pulled high potential (VDD) via a pMOS transistor (Mp1) at on-state and pulled low potential (VSS) through the Rg, respectively. Therefore, all the ESD supply clamps are disabled during normal operation.

During a positive ESD stress from the VDD pin to the VSS pin, the Tri.N and Tri.P node is pulled low potential (VSS) via the nMOS transistor (Mn1) at on-state and pulled high potential (VDD) through the turn-on pMOS transistor (Mp2) simultaneously, respectively. By doing so, V_{TI} of the RC-triggered SCR device can be reduced much lower with respect to the standalone SCR device triggered by avalanching breakdown. The traditional RC-SCR device ("Tra. ESD"), as shown in Figure 1 (a/b) can be expected to have a robust ESD current capability (I_{t2}) in a TLP (HBM-like) event, but not in a VF-TLP (CDM-like) event due to poor fast reaction. In order to enhance turn-on speed against the fast transient event, the proposed ESD device ("Pro. ESD"), as shown in Figure 1 (c) has been implemented by adding a ggNMOS into the SCR device. In fact, the ggNMOS has a superior fast transient behavior [5] and a lower clamping voltage with respect to the traditional SCR-based ESD devices. As a result, a lower voltage overshoot and a fast turn-on behavior can be expected for the "Pro. ESD" in a VF-TLP (CDM-like) event.

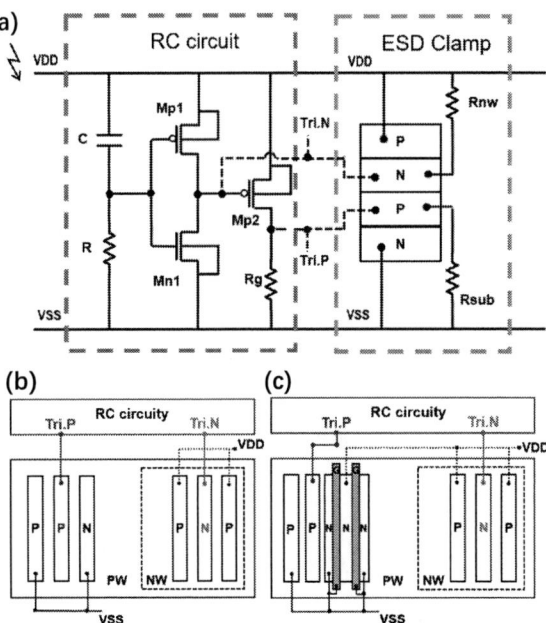

Figure 1: Schematic view of (a) traditional RC-triggered SCR device. The corresponding layout view and together with the RC-circuitry (b) traditional SCR "Tra. ESD" and (c) proposed ESD device "Pro. ESD".

EXPERIMENT RESULTS

In order to understand the ESD conduction mechanism of the "Pro. ESD", a 100-ns normal TLP and 2.5-ns VF-TLP IV characteristics is shown in Figure 2 and Figure 3(a), respectively. Both sizes of the SCR and RC circuits in the "Tra. ESD" and "Pro. ESD" are identical.

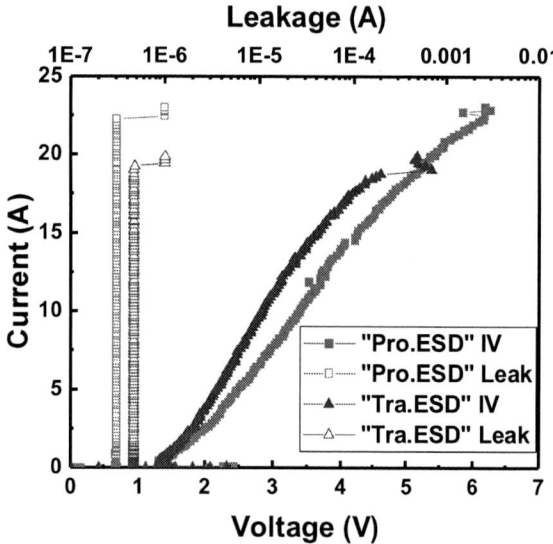

Figure 2: Normal TLP IV ($t_{rise} = 10$-ns; $t_{pulse} = 100$-ns) characteristics of the traditional and proposed ESD device at 25 °C.

It is well known that the traditional RC-SCRs can fully react in a normal TLP event with a small on-resistance (R_{on}) [6]. The behavior of the "Pro. ESD" in a 100-ns TLP event is identical to the "Tra. ESD", which demonstrates that the triggering for both devices is determined by RC-SCR. Both devices have same trigger voltage (V_{T1}) and hold voltage (V_h). In addition, I_{t2} of the "Pro. ESD" is at around 22 A which is higher than the "Tra. ESD". The extra I_{t2} for the "Pro. ESD" is due to ggNMOS added in parallel between the VDD pin and the VSS pin.

The IV characteristics for both "Tra. ESD" and "Pro. ESD" devices during a 2.5-ns VF-TLP is shown in Figure 3(a). It is clear that there are two conduction mechanisms for the "Pro. ESD". Initially, SCR action dominates the conduction until 6 V, then the ggNMOS (parasitic NPN) takes over the rest of the ESD current since 6 V, which is the V_h of the ggNMOS.

It can be seen that both devices show the same slope and intersect the x-axis at 3.7 V (as shown in "Tra. ESD" tangent), which is the V_h of the SCR device. In addition, the R_{on} for both devices are 0.3 Ω during a VF-TLP event when compared to the normal TLP, 0.15 Ω as shown in Figure 2. It further demonstrates that the traditional RC-triggered SCR device is fully incapable to respond during a VF-TLP event. The ggNMOS (parasitic NPN) of the "Pro. ESD" triggers and takes over the rest of the ESD current since 6 V, which is the V_h of the ggNMOS. In addition, R_{on} of the "Pro. ESD" is decreased from 0.3 Ω to 0.1 Ω. As a result, the I_{t2} of the "Pro. ESD" is around 27 A which is much higher with respect to the "Tra. ESD" (at 10 A), as shown in Figure 3 (a).

(a)

(b)

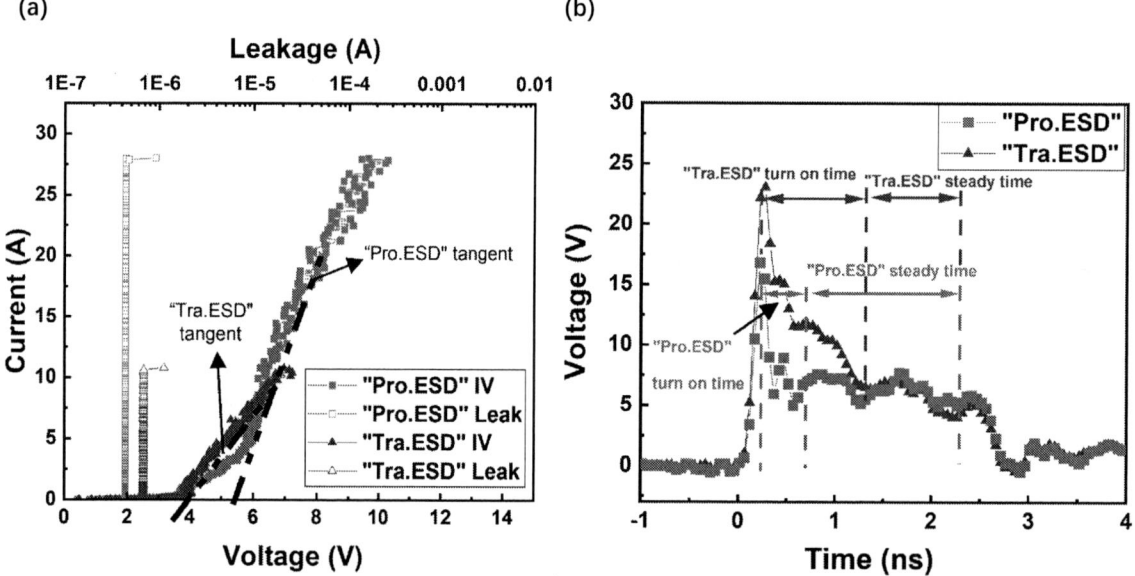

Figure 3: (a) VF-TLP IV ($t_{rise} = 100$-ps; $t_{pulse} = 2.5$-ns) and (b) voltage-waveform (captured at 5 A) characteristics of the traditional and proposed ESD device at 25 °C.

A voltage-waveform characteristics of the "Tra. ESD" and "Pro. ESD" captured at around 5 A (as shown in Figure 3(a)) is shown in Figure 3(b). It can be seen that the overshoot of the "Pro. ESD" is at 17 V which is much smaller when compared to the "Tra. ESD" at 23 V. The turn-on speed is defined in between 90% and 10% of the falling edge until the steady state. It is clear that the "Tra. ESD" triggers at around 23 V, and then dissipates the energy and reaches a steady state at roughly 6 V at around 1.3 ns (turn-on speed of the "Tra. ESD"). A high voltage overshoot is associated with SCR action. By contrast, the ESD conduction for the "Pro. ESD" is dominated by the parasitic NPN, which triggers at around 0.2 ns with a voltage overshoot at 17 V, then dissipates the energy and clamps back to a steady state around 0.6 ns, 5.5 V (V_h of the ggNMOS). In addition, the turn-on speed of the "Pro. ESD" is improved from 1.3 ns to 0.4 ns.

CONCLUSION

In this work, an alternative RC-triggered SCR-based ESD devices have been successfully verified in 0.18-um bulk CMOS technology. The proposed ESD device is designed and fabricated by adding a ggNMOS into the SCR device. The ggNMOS (parasitic NPN) of the proposed ESD device dominates conduction with respect to the traditional RC-SCR device in a 2.5-ns VF-TLP (CDM-like) event. As a result, a smaller R_{on}, a lower voltage overshoot, a faster turn-on speed and greater ESD robustness can be achieved for the proposed ESD device.

REFERENCES

[1] E. R. Worley, R. Gupta, B. Jones, R. Kjar, C. Nguyen, and M. Tennyson, "Sub-micron chip ESD protection schemes which avoid avalanching junctions," *Electrical Overstress/Electrostatic Discharge Symposium (EOS/ESD)*, pp. 13–20, Sep. 1995. DOI: 10.1109/EOSESD.1995.478263

[2] Salcedo et al., "APPARATUS AND METHOD FOR ELECTRONIC CIRCUIT PROTECTION," US 8,320,091 B2, Nov. 27, 2012.

[3] Pee-Ya Tan, M. Indrajit, Pian-Hong Li, and S.H. Voldman, "RC-triggered PNP and NPN simultaneously switched silicon controlled rectifier ESD networks for sub-0.18/spl mu/m technology," *Proceedings of the 12th International Symposium on the Physical and Failure Analysis of Integrated Circuits (IPFA)*, pp. 71–75, June 2005. DOI: 10.1109/IPFA.2005.1469134

[4] Ming-Dou Ker and Kuo-Chun Hsu, "Substrate-triggered SCR device for on-chip ESD protection in fully silicided sub-0.25-/spl mu/m CMOS process," *IEEE Transactions on Electron Devices*, vol. 50, no. 2, pp. 397-405, Feb. 2003, doi: 10.1109/TED.2003.809028.

[5] D. -W. Lai, G. de Raad, S. Sque, W. Peters and T. Smedes, "High-Voltage ESD Protection Device With Fast Transient Reaction and High Holding Voltage," *IEEE Transactions on Electron Devices*, vol. 66, no. 7, pp. 2884-2891, July 2019, doi: 10.1109/TED.2019.2917264.

[6] H. Feng, R. Zhan, Q. Wu, G. Chen, X. Guan and A. Z. Wang, "RC-SCR: a novel low-voltage ESD protection circuit with new triggering mechanism," *Asia-Pacific Conference on Circuits and Systems, Denpasar, Indonesia*, 2002, pp. 97-100 vol.2, doi: 10.1109/APCCAS.2002.1115133.

VIRTUAL METROLOGY MODELING FOR CVD FILM THICKNESS WITH LASSO-GAUSSIAN PROCESS REGRESSION

Shijia Yan, Cong Luo*, Sen Wang, Shenglan Ding, Lei Li, Juan Ai, Qiang Sheng, Qing Xia, Zhi Li,
Qilin Chen, Shilin Li, Hongwei Dai, Yuting Zhong
Wuhan Xinxin Semiconductor Manufacturing Co., Ltd.(XMC), Wuhan, China
*Corresponding Author's Email: Sunny_Luo@xmcwh.com

ABSTRACT

In order to achieve statistical process control in high volume semiconductor manufacturing, it is of great significance to detect the film thickness of wafers and further achieve early warning of abnormal conditions. Virtual metrology (VM), based on the production parameters, has been utilized for process monitoring and control for the last decades. It can realize the real-time online prediction of measurement value, with the existing sampling system, to reducing the sampling rate and ultimately compress cycle time. However, most of existing VM strategies for CVD usually use global nonlinear models, which are difficult to catch sudden changes of the fluctuating production process. Moreover, the machine learning methods used in existing VM can only obtain the predicted values, but cannot judge the confidence of the results. This also results in a low accuracy of traditional approaches when dealt with the high-dimensional data with small sizes. Based on the just-in-time learning framework, this paper proposes the Lasso-Gaussian Process Regression(LGPR) and develops an intelligent VM system, which can realize the real-time prediction of film thickness in CVD process. The experiment results show that the LGPR is more stable and higher accurate.

INTRODUCTION

Due to the limitation of Moore's Law and the strain of production capacity, semiconductor manufacturing are facing fierce competition. The market demand has increased significantly, while the product life cycle and manufacturing cycle have shortened significantly has been shortened significantly, which brings new challenges to the semiconductor manufacturing. In addition, customized products force manufacturers to reduce batches, but improve the defect rate of products.

In order to improve the yield, the fab must measure the wafer to ensure the quality of the product at each stage of the production process. There are two main quality control strategies in the process: (1) system control: to measure each wafer; (2) batch control: to select some wafer to measure between batches, and to use the result of measurement of a small number of wafer to present the whole batch. Measurement includes physical measurement and virtual measurement. Although physical measurement can provide accurate information, it is faced with the problem that the measurement tool is limited by

the space in the clean room and the cost of production. Therefore, the measurement delay and sampling ration are low. They also make it impossible to detect the variation and shift of the process in real time.

In order to overcome the limitation of physical measurement, virtual measurement (VM) is used as real-time fully measurement. It predicts the results of measurement by analyzing the data of status variable identifications (SVID) and fault detection and classification (FDC). Haoshu[1] et al took advantage of Just-in-time (JIT) model-based strategy to predict the material Removal Rate (MRR) of Chemical Mechanical Planarization (CMP). They first queried similar samples from the historical datasets. Chinghsien[2] et al proposed a based on combination of tree-based ensemble mo Sumika et[3] al used kernel SVM (kSVM) to build CVD (Chemical vapor deposition) process VM model. dels virtual physical vapor deposition metrology method.

However, there are the following problems in the current use of VM technology to predict film thickness: 1) The existing VM technology usually adopts global non-linear prediction model. However, it's difficult to adapt to sudden changes in the process. 2) The machine learning method used in the existing VM technology can obtain the predictive value, but it could not judge the confidence value of it. 3) In practical application, the accuracy of the model is lower when the small sample was used for modeling with high-dimensional features. Therefore, we proposed a Lasso-GPR algorithm based on JIT to predict the thickness of CVD process wafers.

The composition of this paper is as follows: the second part introduces data preprocessing. The third part introduces the Lasso-GPR algorithm based on JIT, the fourth part introduces the experimental results, and finally, the fifth part is the summary.

DATA PREPARATION

Due to the confidentiality of information, the details of the process involved in this study are private. The process studied belongs to chemical vapor deposition (CVD). This process is usually used to deposit thin films on wafers. It involves many complex chemical reactions. XMC (Wuhan Xinxin Integrated Circuit Manufacturing Co., Ltd.) provides the data of the deposition process, and uses four descriptive statistical terms for feature extraction, including average value, variance, minimum value and maximum value. There are 49 variables in total.

It should be noted that the input characteristic data are the original data provided without conversion. The data directly provided by XMC is in the form of four statistical terms widely used in semiconductor manufacturing. The output is the average film thickness. For proprietary reasons, the names of input and output variables are not directly showed in this document.

In similar sample query, it is necessary to normalize the data. Because the scale of features is not completely consistent. The data points are mapped to the [0,1]. The formula is shown below. It can improve accuracy of the model by eliminating the dimensional influence between features.

$$X_{std} = \frac{X - X.\min(axis = 0)}{X.\max(axis = 0) - X.\min(axis = 0)} \quad (1)$$

$$X_{scaled} = X_{std} * (max - min) + min \quad (2)$$

When modeling, we need to standardize the data. We make the processed data conform to the standard normal distribution. The formula is as follows. It can improve model performance.

$$x^* = \frac{x - \mu}{\sigma} \quad (3)$$

JUST-IN TIME LASSO-GAUSSIAN PROCESS REGRESSION

Just-in Time Modeling

When the process changes, it is difficult to update the model in time using the traditional global dataset modeling method. At the same time, the training model is time-consuming and laborious. To solve these problems, JIT modeling is used to automatically build local models.

In JIT modeling, we selected 50 historical sample points as the initial historical data set of the first query point. Note that each sampling point represents a single wafer. We sort the sampling points in order according to the process start time to reflect the actual situation. When the sample is predicted, the constructed local model is discarded. At the same time, the sample will be dynamically added to the historical data set. So as to continuously update the historical dataset. It could avoid the decline of model accuracy due to system variation.

Since the direct application of global data modeling in Gaussian process regression would increase the calculation of loading, we adopt the local modeling method. Before building the local model, we first searched the similar sample set from the historical dataset. The search was divided into two steps. First, we searched the k1 samples closest to the current query point time.

Then we found the most similar k2 samples in the historical dataset according to the input characteristics of the query points. We used Euclidean distance to measure similarity. The distance between samples was calculated as follows. We used BallTree algorithm to search similar samples. BallTree would first divide the sample space into

several sub-nodes according to the distance. It then would find the sub-node containing the Target from top to bottom, and would found the nearest observation point from this node. According to the distance, it would query the most similar k2 samples in turn.

$$d = \sqrt{\sum_{i=1}^{n} (x_i - y_i)^2} \quad (4)$$

Gaussian Process Regression

Gaussian process regression is a nonparametric model that uses Gaussian process prior to regress data. For dataset D: (X, Y), let $f = [f(x_1), f(x_2), \cdots, f(x_n)]$, the set of x_i, which to be predicted, was defined X^*, the predict value was f^*. According to Bayesian formula (5), there are:

$$p(f^*|f) = \frac{p(f|f^*)p(f^*)}{p(f)} = \frac{p(f, f^*)}{p(f)} \quad (5)$$

Assume that the prior of function f was Gaussian process $f(\cdot)$, $f \sim N(\mu, K)$, μ was the mean vector, and K was the covariance matrix. Then according to the prior probability distribution of $f^* \sim N(\mu^*, K^*)$ and $f \sim N(\mu, K)$ to calculate the posterior probability distribution of f^*. The joint probability distribution of the two is as follows, σ^2 was the standard deviation vector

$$\begin{bmatrix} y \\ f^* \end{bmatrix} \sim N\left(0, \begin{bmatrix} K(X,X) + \sigma^2 I & K(X, X^*) \\ K(X^*, X) & k(X^*, X^*) \end{bmatrix}\right) \quad (6)$$

We calculated the edge distribution of f^* for the above joint distribution. From the edge distribution property of the joint normal distribution, we could get:

$$p(f^*|X, y, X^*, \sigma^2) = N[f^*|\overline{f^*}, cov(f^*)] \quad (7)$$

$$\overline{f^*} = k(X^*, X)(K + \sigma^2 I)^{-1} y \quad (8)$$

$$cov(f^*) = k(X^*, X^*) - k(X^*, X)(K + \sigma^2 I)^{-1} k(X, X^*) \quad (9)$$

The above formula was the prediction form of GPR. It was also the finite dimensional distribution of the function space posterior to the test sample.

Lasso

In this paper, Lasso algorithm is used for feature dimensionality reduction, which is a compression estimation. It obtains a more refined model by constructing a penalty function, which makes it compress some regression coefficients. It forces the sum of absolute values of coefficients to be less than a fixed value. It also sets some regression coefficients to zero. Therefore, the advantages of subset contraction are retained. This is a biased estimate for processing data with complex collinearity.

Lasso's objective function is as follows. It is based on the coordinate axis descent method to solve the solution that reaches the minimum value. It compresses the coefficients of unimportant variables to 0. So as to screen out important variables. It solves the sparsity problem of high-dimensional data.

$$Q(\beta) = \|y - X\beta\|^2 + \lambda\|\beta\|_1 \qquad (10)$$

Where y is Observation vector of n×1, X is Design matrix of n×p, β is Weight parameter vector of p×1 to be calculated, λ is a regularization parameter that controls the sparsity of the estimated model coefficients.

Lasso-GPR VM Model Based on JIT

For CVD process data, this paper adopted Just-in-TIME learning modeling strategy. Instead of the traditional global model, we built an efficient local VM model. Because the sample distribution of semiconductor process data is uneven. The traditional global modeling strategy will reduce the accuracy of the model. But the modeling strategy of Just-in-TIME learning is applicable to small dataset modeling. Gaussian process regression uses the joint probability distribution of similar sample sets and samples to obtain the predicted value. At the same time, the output of Gaussian process regression is a posterior probability distribution. It can not only obtain the predicted value, but also obtain the confidence of the predicted value.

This paper proposed a Lasso-GPR virtual measurement model framework based on the timely learning strategy, as shown in the figure 1. Under the timely learning strategy, we first selected the approximate data set from the historical data set and the sample to be predicted. After data standardization, we built real-time modeling. We built a two-level regression prediction model. It first used lasso model to reduce the dimension of input features. It then applies the filtered features to the prediction of Gaussian process regression model. The real-time prediction results, including prediction value and confidence, are obtained. We compared the error between the actual measured value and the predicted value. In order to prevent systematic abnormal errors from interfering with the accuracy of the model, we did not update the samples with abnormal errors to the historical dataset. After completing the prediction, the model would be discarded and moved to the next sample for the above online modeling and prediction.

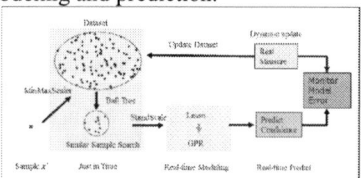

Figure 1: Lasso-GPR virtual measurement model framework

EXPERIMENT

We conducted experiments in the semiconductor CVD process to verify the Lasso-GPR based on JIT algorithm proposed in this article. Due to the company's confidentiality regulations, detailed information about this process was not disclosed herein. In order to predict the thickness of the thin film deposited by wafer after completing the CVD process. We analyzed the deposition process. And we collected characteristic data for 49 sensor signals related to thickness. We collected the amount of data collected by the sensor over an 18-month period. There are 1103 wafers in total. Each wafer has its true thickness measurement.

Comparison of results between JIT Lasso-GPR and global modeling algorithms

We split the dataset applied to the global modeling algorithm, using 70% as a training set and 30% as a test set. The training set was used to train different regression models, including linear models such as LinearRegression, Lasso, Ridge, and nonlinear integration models such as GradientBoost, RandomForest, AdaBoost, ExtraTree, DecisionTree, DecisionTree, and KNNRegressor. Then we evaluated the effectiveness of the model on a test set. For JIT Lasso-GPR model testing, we first ranked 30% of the test sets by CVD process start time. Then we selected the previous 50 sample data as the initial dataset for the first wafer as the training set of the model. Then we dynamically updated the dataset, model and predicted the thickness of the wafer.

Model	R2	MAE
JIT Lasso-GPR	0.97	6.27
GradientBoost	0.97	10.54
RandomForest	0.97	10.62
Ridge	0.97	10.65
LinearRegression	0.97	10.88
Bagging	0.97	11.44
Lasso	0.96	12.20
AdaBoost	0.96	12.20
ExtraTree	0.95	14.18
DecisionTree	0.95	14.74
KNNRegressor	0.89	18.57
SVR	-0.21	63.90
MLPRegressor	0.00	75.21

In addition, this chapter adopted two commonly used prediction error indicators to compare the estimated performance of the three virtual measurement models. The corresponding indicator comparison results were shown in Table 3.1. The two indicators were the correlation coefficient R^2 and the Mean Absolute Error (MAPE):

$$R^2 = \frac{SSR}{SST} = \frac{\sum(\hat{y}_i - \bar{y})^2}{\sum(y_i - \bar{y})^2} \qquad (11)$$

$$MAE = \frac{1}{n}\sum|\hat{y}_i - y_i| \qquad (12)$$

Where SSR was the sum of regression squares and SST was the total sum of squares, $\hat{y}_i$ was the predicted value, $\bar{y}$ is the mean of the true value, $\bar{y}$ is the true value. The evaluation results of the model were shown in the table. It could be seen that the MAE of JIT Lasso-GPR is the smallest, only 6.27, which meant that the prediction accuracy of the model is the highest.

JIT Lasso-GPR Model Prediction Effect

The JIT Lasso-GPR proposed in this paper could dynamically reduce the dimension of feature space. The figure 2 showed the characteristic number of each test sample after dimensionality reduction during prediction. Compared to all 49 features used by other models, this model used only 4-9 features. It reduced features by 82% to 92%. It preserved the most important features for modeling. Therefore, it avoided reducing model accuracy due to feature redundancy.

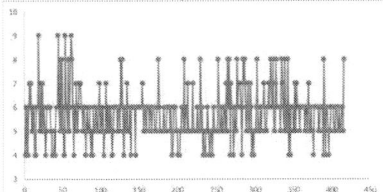

Figure 2: The characteristic number

The frequency histogram of the selected feature for each test sample after feature dimensionality reduction was shown in the figure 3. Among them, features 1 and 3 were selected by most test samples. Features 3, 5, 6, 8, 9, and 11 were also selected by one-third of the test samples. On the other hand, there were 18 features that have no contribution at all and have no correlation with the thickness of the film. And they were eliminated by the model. Referring to the actual physical meaning of the feature, there was a significant correlation between the film thickness and the relevant data in the CVD process, such as reflection capacitance, power, chamber pressure, temperature, and gas flow.

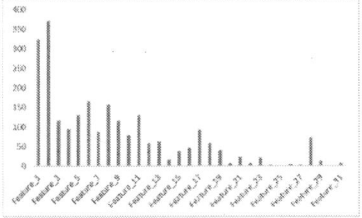

Figure 3: The frequency histogram of the features

Compare the prediction effect of the model with the actual measured value, as shown in the figure 4. We could see that the trend of the predicted value of the model was very close to the actual value. This indicates that the model could solve the challenge of forecasting on small samples. At the same time, the model also outputted the confidence interval of each prediction sample, as shown in the figure 5. Therefore, we could judge the reliability of the predicted value

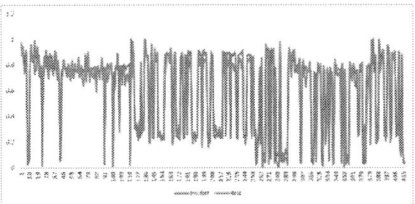

Figure 4: Comparison the prediction value with the actual measured value

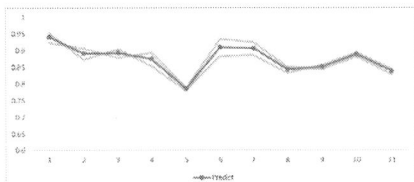

Figure 5: The confidence of the predict

CONCLUSION

In this paper, virtual measurement of CVD process thin film deposition thickness in semiconductor manufacturing faces problems such as small sample size, high feature dimensions, and process fluctuations that affect model accuracy. We propose a Lasso-GPR virtual measurement algorithm based on a timely learning modeling strategy. This algorithm can dynamically model complex semiconductor process fluctuations by selecting an approximate sample set. At the same time, it performs feature dimensionality reduction through Lasso, eliminating redundant features, and improving model performance. It then uses Gaussian process regression to perform joint probability distribution statistics on the selected features, outputting predictive values and confidence levels. After experimental verification, the JIT-Lasso GPR algorithm has better results than commonly used regression algorithms based on global modeling.

REFERENCES

[1] Cai, H., et al. "Adaptive Virtual Metrology Method Based on Just-in-time Reference and Particle Filter for Semiconductor Manufacturing." *Measurement* 168(2020).

[2] Chen, Yi Zheng. "Virtual metrology of semiconductor PVD process based on combination of tree-based ensemble model." *ISA Transactions* 103.1(2020).

[3] Arima, S., et al. "Feature Extractions from a High-dimension Low-samples Data for Multi-dimension Virtual Metrology." *2019 Joint International Symposium on e-Manufacturing & Design Collaboration(eMDC) & Semiconductor Manufacturing (ISSM) 2019.*

A REAL-TIME DETECTION METHOD FOR WAFER PROBE REFERENCE DIE SHIFT

Deguang Zheng, Kuan Lu, Bo Zhong, Shuxin Liu and Xiaofeng Liang*

Wafer Test Department, NXP Semiconductors, Tianjin 300385, China
*Corresponding Author's Email: xiaofeng.liang@nxp.com

ABSTRACT

Wafer-level Probe is a fundamental process in the semiconductor backend manufacturing that intends to sort out good and bad dies by electrical testing then write die locations with bin codes in a wafer map. With the rapid miniaturization of integrated circuits (IC), Probe Reference Die Shift (RDS) event has become a big talking point recently in semiconductor backend factories. Known methods of Probe-RDS detection are based on analysis of a fully tested wafer map, which happens post-stage. This paper proposes a method of real-time detection on Probe-RDS by introducing an optimized test path and zonal algorithms. The experimental results proved good accuracy of Probe-RDS detection in the extremely early stage as anticipated.

INTRODUCTION

In the Semiconductor industry, wafer-level Probe testing is a critical process to check the functionality of Integrated Circuits (IC) in each die on a wafer and output an electronic wafer map [1]. A typical wafer manufacturing process flow is shown in Figure 1.

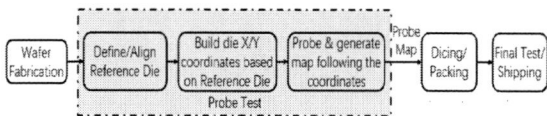

Figure 1: Wafer Backend flow

In wafer-level Probe, a reference die is being used to align map coordinates with IC die locations on a physical wafer which is crucial to the following Assembly Dicing and Final Test process. The correct coordinates need to be carried out between Probe and Assembly so that only good dies will be packaged in Assembly process. It is called reference die shift (RDS) if the Probe reference die is misaligned with the physical location, the failure mode of RDS is illustrated in Figure 2. The quality risk is very high since bad dies can be shipped to customer if RDS happens during Probe process.

There are all sorts of errors that can induce Probe-RDS, so both prevention and detection aspects are being studied in hope of minimizing the risk. The majority of preventive direction raises alignment scores by calibrating the prober vision system, selecting a unique pattern on the wafer and calculating the distance between the wafer center and the designated pattern [2], etc.

Another aspect of detection focuses on wafer result map analysis after the entire lot is fully probed.

As illustrated in Figure 2, an obvious crescent-shaped pattern consisting of continuous failure bins can be observed in the corresponding wafer edge when a Probe-RDS happens. Consequently, SBL(statistical bins limit) can be defined to catch Probe-RDS which is widely used in the wafer Probe industry today [3]. It is a key indicator to reflect fabrication process shift and reliability excursion [4]. SBL is an overall screening algorithm post-probe to alarm any anomaly of exceeded bin quantity, but it cannot judge whether the bin outlier is Probe-RDS related or not. Therefore, we introduce specific bins only for Probe-RDS event, which are called Maverick Bins. From the Probe-RDS sample in Figure 2, we observe that Probe-RDS would be accompanied with multiple Maverick Bin8 failure dies. Then the electric bad dies can be picked up as the good ones after Assembly Dicing process, it will cause serious quality incident.

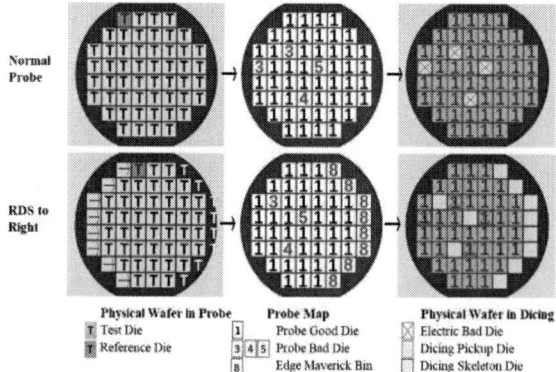

Figure 2 : Normal Probe vs. RDS Probe

Morn etc. provided a "Wafer Edge Yield Comparison" method to improve the efficiency on WLCSP wafers [2], by calculating the yield difference between one half-edge and the opposite half-edge side to predict the Probe-RDS event. This method exhibited better accuracy of detection as edge yield does have a certain trend drop as long as Probe-RDS happens. Objectively, yield is not a rigorous variable to measure RDS as the yield loss may come from a variety of failures rather than RDS itself to confuse the judgment. Consequently, it is still a Post-Probe detection methodology leading to inevitable losses.

Earlier Probe-RDS detection is extremely necessary to Probe areas where corrective actions can be taken in place to minimize the loss. Similar to the way of working on real-time fallout bin monitoring to identify wafer or tester issues [5], and inline probe mark inspection and controlling [6], a new scenario is raised in this paper to detect Probe-RDS with an advanced algorithm based on Maverick Bins in each one-quarter edge along with the most optimized test path. More importantly, this method achieves in-process detection instead of post-map analysis.

METHODOLOGY

Typically, the Probe-RDS could occur in any direction due to various failure causes. There are four cardinal directions (Top, Bottom, Left, Right) with a shift in a single axis and another four intermediate directions (Top-Left, Top-Right, Bottom-Left, Bottom-Right) which are shifted in dual-axes. As shown in Figure 3, a demo was involved to produce RDS intentionally by one die offset in all eight directions.

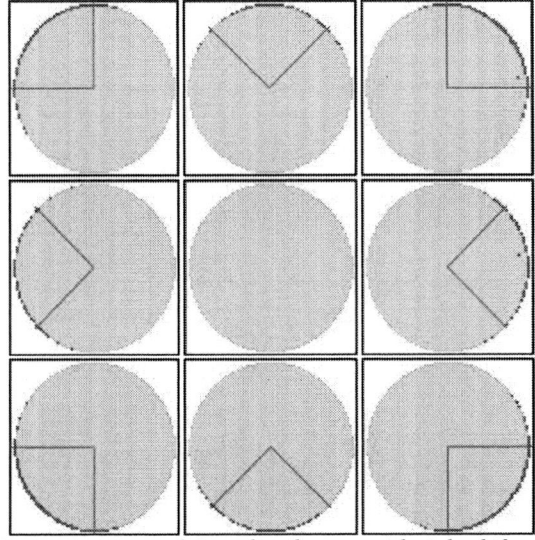

Figure 3: Maverick Bins distributions with eight shift directions.

As illustrated in Figure 3, those Maverick Bins from the edge area have a strong correlation to Probe-RDS events which can give us a precise signal for Probe-RDS detection. To be more specific, the prober indexes out of testable area and touchdown on partial or ugly dies which are located in wafer edge when a real RDS happens. These Maverick Bins form a clear boundary pattern along with the same shift direction and location of each Maverick Bin can be categorized into the quadrant it belongs to.

The Maverick Bins located in the wafer central area are not valuable in prediction of Probe-RDS. On the contrary, they become noises to impact the accuracy of detection. A distance range is to be defined to screen out non-edge dies. For instance, the distance between the target die and center die shall be close to the wafer radius of 100mm and 150mm respectively in the case of 200mm and 300mm slices which are mainstream wafer sizes in the semiconductor industry today. Otherwise, it is not an edge die to be excluded.

Design of the Proposed Detection Model

A Cartesian coordinate system is used to locate the fail bins on the wafer map as it is shown in Figure 4.

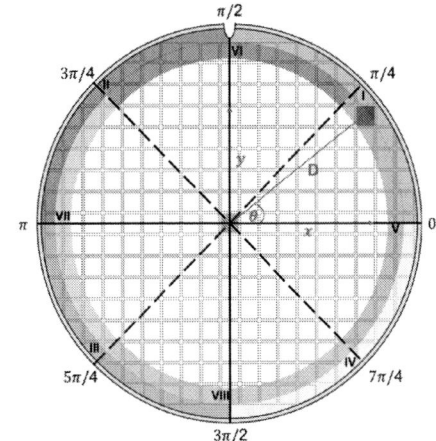

Figure 4: Eight zones for eight directions

For any die on the wafer map, it can be expressed as $Die(D, \theta)$.

$$D = f(x, y) \quad (1)$$
$$\theta = g(x, y) \quad (2)$$

Where,

x — x coordinate on the wafer map

y — y coordinate on the wafer map

D — Distance between testing die to wafer center

θ — Angle between Die(x, y) to x positive direction

$$D = \sqrt{x^2 + y^2} \quad (3)$$

The function range of the arccos is from 0 to π, To express the $(0, 2\pi)$ range for all the areas of the wafer, the below expression is used.

$$\theta = \begin{cases} \arccos \frac{x}{\sqrt{x^2+y^2}}, y \geq 0 \\ 2\pi - \arccos \frac{x}{\sqrt{x^2+y^2}}, y < 0 \end{cases} \quad (4)$$

TABLE I. LIST OF WAFER ZONE

Zone(i)	Range of θ	Region of Interest
I	$0 \leq \theta < \pi/2$	Top Right
II	$\pi/2 \leq \theta < \pi$	Top Left
III	$\pi \leq \theta < 3\pi/2$	Bottom Left
IV	$3\pi/2 \leq \theta < 2\pi$	Bottom Right
V	$7\pi/4 \leq \theta < 2\pi$ or $0 \leq \theta < \pi/4$	Right
VI	$\pi/4 \leq \theta < 3\pi/4$	Top
VII	$3\pi/4 \leq \theta < 5\pi/4$	Left
VIII	$5\pi/4 \leq \theta < 7\pi/4$	Bottom
70% of R < D < R (wafer radius) for all of Zones		

As shown in Figure 4, eight zones are defined to locate any given edge $Die(D, \theta)$. Table I lists all θ value for each defined $Zone(i)$. $Die(D, \theta)$ and $Zone(i)$ will be used in the following detection model.

Process of the Proposed Detection Mode

According to the above analysis of Probe-RDS, we understand that the disparity of Maverick Bins along the wafer edge is a very significant feature in Probe-RDS induced area-dependent failure. We anticipate diagnosing the reality of Probe-RDS on condition that the Maverick Bins can be quantified and managed by control limits. With respect to Maverick Bins in each zone, a threshold can be defined to alarm the Probe-RDS. Thus, a Probe-RDS detection flow is proposed to catch such quality cases in Figure 5. The monitor system is enabled as long as the probe starts on a wafer and resets while the wafer testing is ended. Each shift direction has an independent counter to record the sum of edge Maverick Bins. Once any of the counters exceeds its respective threshold, an alarm will be triggered to cease Probe testing on the current wafer immediately.

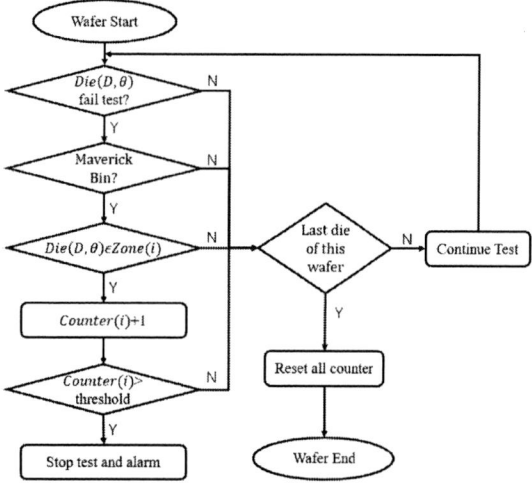

Figure 5: Reference die shift detection flow

Theoretically, a consistent threshold control among Probe-RDS directions might be working based on one specific device type. However, it is quite distinctive in actual conditions that the edge die quantity can vary a lot among the Probe-RDS directions caused by center die offset, die size, wafer mask layout, etc. Therefore, the threshold of the edge Maverick Bin in each direction can be figured out by individual experiments.

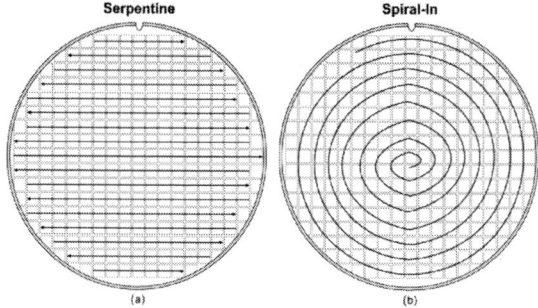

Figure 6: Serpentine test path(a) vs. Spiral-in test path(b)

Additionally, the test path is also a critical factor to be considered. The traditional test path as shown in Figure 6(a) follows a serpentine stepping sequence to complete the test across the whole wafer. Generally, the wafer probe starts from the leftmost die in the first row and goes along the right direction then jumps to the next row, starts from the rightmost die and goes along the left direction, and so on. With this serpentine test path, Probe-RDS in top edges can be detected at early stages, but detection for other edges is lagging. The response time of detection in bottom edges is almost equivalent to the test time of an entire wafer.

To achieve early-stage detection of Probe-RDS, a spiral-in test path is introduced to optimize the stepping sequence as shown in Figure 6(b). When a wafer probe starts, edge dies are prioritized to probe along the clockwise direction, and then index to the inner area of the wafer. Using this new test path, a significant boost in response time of Probe-RDS detection is promising.

RESULTS & DISCUSSION

The proposed method is validated using real wafer test data of NXP products. To explore the influence of different test paths, we conduct a Probe-RDS experiment using a selected NXP product. The trigger points of the Probe-RDS event in all eight directions are indicated in Figure 7.

As explained in the previous paragraph, the detection response time varies a lot among the shift directions decided by the test path and alarm threshold. With the traditional serpentine path, the Probe-RDS cases were detected in a range between 9% and 99.2% completion on a single wafer. Such tremendous variation in detection response time cannot be adopted in semiconductor industry applications.

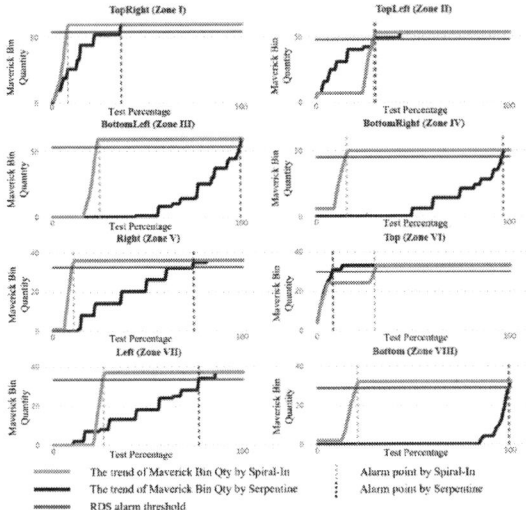

Figure 7: Probe-RDS alarm trigger points

Given the new spiral-in test path, the RDS detections were brought forward to a range between 9% to 30.3% completion. Detections in six shift directions are dramatically improved with earlier-stages and only detection in Straight-Top direction exhibits longer response time which is negligible compared–with the overall promotion.

On the other hand, 5 NXP products were selected for trial runs in a period with the new Probe-RDS algorithm enabled to validate the reliability of the proposed detection model. The experimental setups and results are summarized in Table II. The accuracy of detection and confusion matrix are primary considerations being used to measure the effectiveness.

As is revealed in Table II, the actual accuracy of RDS detection reached over 99% on all these NXP products. In the meanwhile, the false detection rate calculated by the confusion matrix remains at an extremely low level is even more valuable which means the alarm induced by real Probe-RDS can be isolated from other conditions without engineering disposition to the utmost. Given all the above, it has been proven that the new Probe-RDS detection method is effective and efficient.

TABLE II. ACCURACY OF RDS DETECTION BY THE PROPOSED METHOD ON 5NXP PRODUCTS

Product	Wafer	Alarms	RDS	Accuracy	Confusion Matrix
P-1	1617	3	1	99.88%	$\begin{bmatrix} 1 & 0 \\ 2 & 1614 \end{bmatrix}$
P-2	1472	1	0	99.93%	$\begin{bmatrix} 0 & 0 \\ 1 & 1471 \end{bmatrix}$
P-3	1620	1	1	100.00%	$\begin{bmatrix} 1 & 0 \\ 0 & 1619 \end{bmatrix}$
P-4	7440	5	2	99.96%	$\begin{bmatrix} 2 & 0 \\ 3 & 7435 \end{bmatrix}$
P-5	3667	3	1	99.95%	$\begin{bmatrix} 1 & 0 \\ 2 & 3664 \end{bmatrix}$

CONCLUSION

Improving the detection rate of Probe Reference Die Shift is a persistent topic in the semiconductor wafer test area. Probe-RDS always happens and may cause fatal results in Customer Quality Incidents (CQI), financial loss, and reputation damage. So far, almost all the traditional methods to detect Probe-RDS are post-screening based on wafer map analysis, pattern recognition, SBL alarm, etc. In another word, a full wafer even the whole lot has to be 100% tested with the wrong setup when a Probe-RDS occurs. In this paper, a breakthrough idea is raised to achieve early detection instead of post-Probe disposition. Considering the benefit of real-time stop and fix for the loss minimization, it is highly recommended to fan out this method to wafer test operation.

The proposal is focus on detection method only, auto-correction of Probe-RDS was not extended in this paper, it is a new direction worthy to explore in future research.

REFERENCE

[1] Sudarshan Bahukudumbi, *Wafer-Level Testing and Test During Burn-In for Integrated Circuits*, Artech, 2010.

[2] M. Jin, W. He, J. Qiao, W. K. Chien and S. Zhao, *"Wafer level package wafer probing shift error-proof quality control,"* 2015 IEEE International Conference on Industrial Engineering and Engineering Management (IEEM), 2015, pp. 939-942

[3] Muriel, P. Garcia, O. Maire-Richard, M. Monleon and M. Recio, *"Statistical bin analysis on wafer Probe,"* 2001 IEEE/SEMI Advanced Semiconductor Manufacturing Conference (IEEE Cat. No.01CH37160), 2001, pp. 187-192.

[4] S. Illyes and D. A. G. Baglee, *"Statistical bin limits-an approach to wafer disposition in IC fabrication,"* in IEEE Transactions on Semiconductor Manufacturing, vol. 5, no. 1, pp. 59-61, Feb. 1992

[5] Q. Khasawneh, J. Dworak, P. Gui, B. Williams, A. C. Elliott and A. Muthaiah, *"Real-time monitoring of test fallout data to quickly identify tester and yield issues in a multi-site environment,"* 2018 IEEE 36th VLSI Test Symposium (VTS), 2018, pp. 1-6

[6] B. Zhong, D. Zheng, X. Dai and Y. Hu, *"A Novel Automatic Probe-to-Pad Alignment Error Correction Approach,"* in IEEE Transactions on Semiconductor Manufacturing, vol. 35, no. 1, pp. 146-148, Feb. 2022

979-8-3503-1101-3/23 $31.00 © 2023 IEEE

NOVEL LOCALIZATION APPROACHES IN METAL –INSULATOR-METAL STRUCTURE FAILURE ANALYSIS

Lvye Fang*, Hongtao Qian and Qinqin Yu

Semiconductor Manufacturing International (Shanghai) Corp. No.18 Zhangjiang Road
Pudong New Area Shanghai, China
*Corresponding Author's Email: Lvye_Fang@smics.com

ABSTRACT

To effectively improve the quality and performance of products, improving the capability and efficiency of failure analysis (FA) is impending. Routine analysis method may not suitable for every case and is short of efficiency. In this paper, we mainly introduce hot spot localization methods to solve different MIM cases. For abnormal capacitance cases, PVC is a convenient localization way. For small-scale cases, EBAC with differential amplifiers could be applied. For cases hard to localize fail site, deprocessing is suggested. Finally, in-situ TEM is introduced as a potential equipment to study MIM breakdown mechanism.

INTRODUCTION

With the rapid development of integrated circuits, superior metal-insulator-metal (MIM) which equips higher-capacitance density has been extensively used in field of mixed-signal IC to match smaller scale integration [1]. As a key of semiconductor chip fabrication, there is huge demands in MIM research and development. Whereas, high fail ratio always results in low fabrication efficiency. To strengthen industrial research and production capability to meet the standards of superb quality and diversification, it is of extremely necessity to analyzing failure mode. In other words, improving the capability and efficiency of failure analysis (FA) plays a hugely important role.

Routine FA analysis always contains background obtaining, analysis and report composing (Figure 2a). There are four basic technological parts with analysis that empower us to figure out the cases: electrical test, physical localization, sample preparation and defect character -ization.

Commonly, background information such as test items (involving spec, fail value, fail type), fail position, layout and thickness of every layer should be first acquired. Those information comes from electrical test includes wafer acceptance test (WAT), reliability (RE) test and electrical failure analysis (EFA) confirmation. Figure 1a shows the schematic diagram of MIM circuit connection that top plate and bottom plate are connected separately to different pad. Figure 1b and c demonstrates the equivalent circuit diagram related to WAT test items. MIM capacitance is measured by adding sweep voltage of AC 100 kHz on the top plate while the bottom plate connects

to ground (Figure 1c). The capacitance density can be calculated. According to the principle of break-down voltage (BV) test that leakage current density could be acquired when a sweep scan voltage (0-40 V) is added to top plate. The voltage between two plates while the leakage current reach to 100 pA/μm2 is defined as BVMIM (Figure 1c). As for RE test, it usually includes deterioration test, Vramp and time dependence dielectric breakdown (TDDB) test by which we could figure out the break-down voltage and lifetime of device.

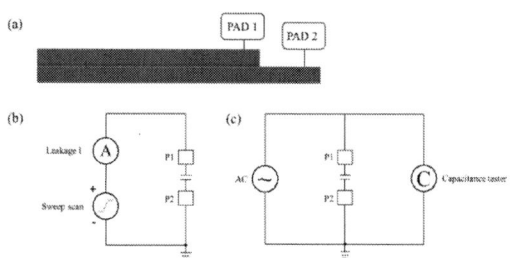

Figure 1: (a) Schematic diagram of MIM structure; (b) Equivalent circuit diagram of BV_{MIM} test; (c) Equivalent circuit diagram of capacitance test.

As the integration degree is daily larger and critical dimension is reduced, traditional method by microscopes is like searching for a needle in a haystack to further analyze. To improve analysis efficiency, here comes a pivotal step - failure localization. Optical beam induced resistance change (OBIRCH) is always applied in MIM FA of large-scale process. The principle of OBIRCH is that fault area could produce a different resistivity change induced by confocal laser. The optical signal can be converted into electrical signal and displayed in screen. By stacking this electrical signal generated mapping overlay with optically scanned images, fault area is obvious to be localized [2]. As showed in Figure 2b, we could see that hot spot was detected in MIM array. After delayer process of top metal layers, abnormality was inspected by OM/SEM. From SEM image (Figure 2c), metal defect, which has same contrast with W, was speculated. To confirm our surmise and find out which step was wrong in process flow, cross-section analysis followed. Samples usually prepared by focus ion beam (FIB) equipment. During cross-section milling process, we could approximately speculate defect-appearing loop.

979-8-3503-1101-3/23 $31.00 © 2023 IEEE

Additionally, the combination of high-resolution TEM image and energy dispersive spectroscopy (EDS) helps affirming failure mode to improve industrial inline process.

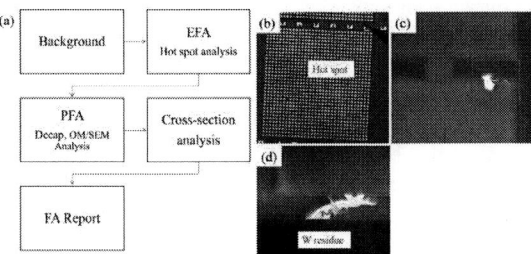

Figure 2: (a) Flow of routine FA analysis; (b) Hot spot image; (c) SEM image; (d) Cross-section image prepared by FIB.

Given the variety of failure modes, routine FA analysis cannot meet all demands. To improve the efficiency of analysis and enrich the means of FA, other methods are introduced as followed.

OTHER FAILURE ANALYSIS METHODS
PVC on capacitance fail case study

Voltage Contrast is an effective method in the field of failure analysis. Passive VC (PVC) is widely used because it could be simply operated by SEM (or FIB) and don't need to connect to any other supply [5]. PVC uses the electron beam to either charge a conductor positively or negatively at low accelerating voltages from 500 volts to 2 kilovolts. Structures with different electrical properties will have different brightness. For example, a floating conductor always acquires a positive voltage potential under the influence of primary beam in SEM under low acceleration voltage, thus it shows a dark contrast because the conductor emit fewer secondary electrons.

In this case, schematic diagram of six parallel capacitor is introduced as followed Figure 3. Pad 1 and 3 were connected to capacitance measuring tester. As we all known, total capacitance is equal to the sum of capacitance in parallel.

Final WAT results showed that capacitance of failed sites seemed proportionally decreased related to capacitor quantities. As 1 fF/um^2 is for 6 capacitors connected in parallel, the total capacitance reduces by 16.7% with each removal of an individual capacitor. Accordingly, we supposed that via which connect MIM array with metal lines may have something wrong. The die with 16.7% reduced capacitance was chosen to analysis. Based on the principle of PVC, different via status should have different brightness. From Figure 3b, we could see that abnormal dark VC was found by SEM after delayering samples to via-layer. This was consistence with our speculation before. The capacitor with abnormal dark VC seems that

it's via were of the floating status. Compared with random cut, this set of methods is significantly faster and more accurate in fail site localization.

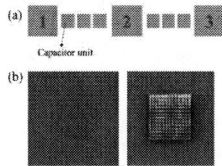

Figure 3: (a) Schematic diagram of six parallel capacitors; (b) SEM images.

PVC method has advantages of simple operation, is well suited for via open cases. Sometimes it is applied to double confirming fail units as well.

Nano-probing technology on leakage case study

Nano-probing technique, which combine nano-probing with SEM, also includes electron beam induced current (EBAC) and DI-EBAC, has become a potential method for accurate localization. It works when abnormal sites absorb electron from electron beam penetrated sample surface and then EBAC image formed via probe to amplifier [3-7]. EBAC measurements are mainly divided into two modes according to the types of amplifier. Figure 4a shows the mode that one probe connected to current amplifier. It can amplify current path for visualization. Figure 4b shows the mode that has two probes connect to amplifiers so that it has stronger capability to display abnormal position as for high resistance points. This structure was also designed to detect leakage failure.

Previous paper has studied EBAC images of breakdown transistors that provides some reference because the structure of metal oxide semiconductor is associated with MIM [7]. MIM structure, which is polished to via layer above top plate, can have its failure site localized by both modes in principle. However, one probe method always has no obvious signals of leakage point and just qualitatively represented leakage issue or detected open sites as showed in Figure 4c. Finally, hot spot was found in the position EBAC signal revealed by differential amplifier mode (Figure 4d). Compared with one-probe EBAC signal, two-probe EBAC with differential amplifier has a sharper contrast and more precise localization in the nature of things. Defect finally found (Figure 4e), which double confirmed EBAC results.

OBIRCH is not suggested to be applied in small-scale cases, because pulse current produced by machine may seriously damage fail sites. EBAC method has advantages of high accuracy and relatively low enhanced damage, is well suited for small-scale cases that cannot localize failure site by SEM.

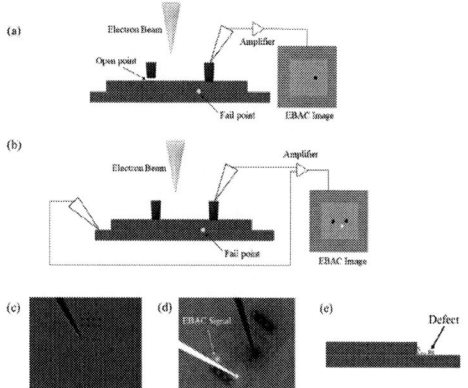

Figure 4: The schematic circuit diagram of each EBAC amplifier including current amplifier (a), differential amplifier (b). (c) Current-amplifier EBAC image of breakdown MIM structure. (d) Differential-amplifier EBAC image of breakdown MIM structure. (e) Schematic diagram of MIM Cross-section.

Delayer method on single-unit fail case study

In this case, MIM test key was designed as one single unit. Multiple layers of dummy metal layers were mandatorily deposited below and above due to design requirements. As showed in Figure 5a and b, OBIRCH failed to localize fail site precisely. It may because dummy metal resist the emission of signal to some extent. Moreover, the current change in lead may higher than fail site, which resulted in wrong hot spot.

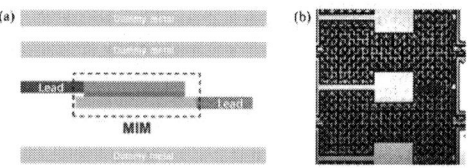

Figure 5: (a) Schematic diagram of MIM structure; (b) OBIRCH image.

Fail to narrow the analysis range, we have tried to polish samples to nearly top-plate layer, but nothing worthful found. In this situation, delayering was chosen because it is easier to control film flatness in single unit. After slightly wiping off the top plate by acid in a few seconds, fail site was observed in dielectric layer. It should be noted that long acid soaking time may destroy bottom layers.

If there is no hot spot in single- or few-units structures, this method could be haven a try.

In-suit TEM on fail mechanism study

All cases above are related to manufacture process such as defect, bad profile caused by etch or film deposition etc. instead of the intrinsic break-down of materials. In-suit TEM may be a potential characterization technology to study MIM mechanism, like the impact of in-homogenous microstructure, in-homogenous element distribution, porosity, defect concentration and electron trap. Up to now, there is few indagation in industrial field for its high research cost or immature study system. While observation of dynamical breakdown process under different conditions like various grain size, dielectric material, hillock. etc., may helps industrial process improvement.

CONCLUSION

FA routine method could cover large parts of cases, but a few other cases cannot be solved. To improve the efficiency of failure analysis and better support industry to fine tune process, we introduce several novel methods for MIM analysis in this paper. For abnormal capacitance cases, PVC is a convenient localization way if capacitance of fail sites decreased regularly. For small-scale cases that cannot localize failure site by SEM, EBAC with differential amplifiers could be used. In consideration with several studies where hot spot is hard to localization, it can be suggested to delayering. Finally, to better acknowledge the breakdown mechanism of MIM, we suppose in-situ TEM will be a potential way to study.

ACKNOWLEDGEMENTS

Extremely appreciate joint efforts of colleagues Hongmei Bai, Ningning Song and others.

REFERENCES

[1] J. A. Babcock, et al. *IEEE*, vol. 22, 2001, pp. 230-232.

[2] C. T. Yeh, et al. *IEEE E-manufacturing & Design Collaboration Symposium*, 2016.

[3] P. K. Tan, et al. *IEEE International Integrated Reliability Workshop Final Report (IIRW)*, 2014.

[4] L. Tian, et al. *IEEE International Symposium on the Physical and Failure Analysis of Integrated Circuits (IPFA)*, 2018, pp. 1-5.

[5] L. Chang, et al. *IEEE 22nd International Symposium on the Physical and Failure Analysis of Integrated Circuits*, 2015.

[6] R. Rosenkranz. *Journal of Materials Science: Materials in Electronics*, vol. 22, 2011, pp.1523-1535.

[7] J. Fuse., et al. *IEEE 24th International Symposium on the Physical and Failure Analysis of Integrated Circuits (IPFA)*, 2017.

THE IMPROVEMENT STUDY OF UTS CIS BEVEL PEELING DEFECT BASED ON THE APPLICATION OF SEM API

Xianghua Hu[*], Guangzhi He, Jingfeng Wang, Qiliang Ni

Shanghai Huali microelectronics Corporation, Shanghai 201210, China

*Corresponding Author's Email: huxianghua@hlmc.cn

ABSTRACT

A bevel peeling defect formation during backside film remove processes on UTS CIS Logic products was introduced based on the using of SEM API function. The reason of the formation and its affect factors of peeling defect were studied in this paper. Chemical 1 with high concentration was used to clear up backside film, LTO/Poly/SIN, in backside remove processes, these acids seeped into the metal film body, then eroded metal film into bump, through the interfaces of films bevel remained. The metal bump would break down to peeling source after 4days' waiting time. Bevel etch before wafer package, were studied to induce bevel peeling. The experiment results showed that, peeling defect would be free when bevel etch process was added.

Keywords-Bevel; Peeling; UTS CIS; SEM API; Backside;

INTRODUCTION

Along with the development of integrated circuit technology in VSLI and chip manufacturing cost control requirement, 12 inches size has gradually become the mainstream around the world, and the size may develop to 18 inches, even bigger in the near future. Bigger size have bigger wafer edge area and the manufacturing process becomes more difficult. The edge and bevel location defects, such as peeling, crater, non-uniformity [1] etc., often appeared during the processing and reduced the chip yield. At the same time, the consumer goods, such as phone, digital camera, and other mobile devices, which would using image sensor chips requested higher and higher resolution of COMS image sensor (CIS), which drive the CIS type evolution to BSI, UTS from FSI[2,3] during past decades. A bevel peeling defect would appeared and dropped on surface during wafer bonding has been reported [4], which related to non-clear problems on the wafer edge, included lithography photoresist residue, oxide and metal film remain. These falling defects would cause bonding bubble and yield loss, even scrapped at serious situation. A lot of developers used bevel trimming processing to solve these problems. For UTS products, before logic and pixel wafer combined, thinning process is necessary used to reduce the thickness, which involving LTO/Poly/SIN backside film remove. But no report show that peeling defect appeared and effect yield caused by these special steps.

In this paper, we used SEM API function to collect and analysis defect data, find where is the source, when it getting worse, what is the influence factors. At the end, a suspect mode of the mechanism of defects was put forward and a improve action was implanted based on the study.

12 inch wafer edge introduction

The area about 3 mm（12inch wafer is 5mm）on the edge of the wafer is wafer edge , which contains 5 parts, ① area is top near edge,②area is top bevel, ③area is bevel, ④area is bottom bevel, ⑤area is bottom near edge. As far from wafer center, wafer edge area film stack always disorder, and no active die on these areas, as well as more active dies close to the boundary with the chip size became smaller and smaller.

During the manufactory, the wafer edge will become weak after diffusion deposition, photolithography, dry etch, wet etch and CMP processes, especially lithography and electrochemical deposition EBR process to clean edge with different criteria, it will be seriously influence the defect status on the edge of the wafer, many typical defects such as flake, void, crake etc., appears after STI etch. Especially to the BEOL, complex condition with repeat processes of Metal1, 2, 3…, induced defect source and peeling or bubble. Wafer edge profile OM and SEM image, have been taken at TM, as figure 1 cartoon image shown, film stack messy and un-uniform.

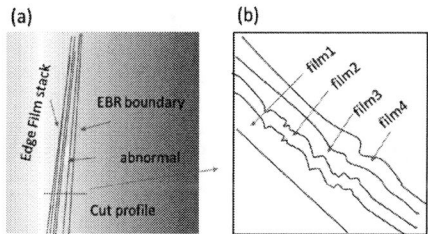

Figure 1: wafer edge check at BEOL TM (a) top view cartoon image, (b) Profile view cartoon image

Scanning electron microscope (SEM) and auto process inspection (API) introduction

SEM used focused electron beam bombard sample surface, collected secondary electron, scattering electrons, which were produced by the interaction between electron and sample surface atom, to observant and analyze sample surface or section morphology. At the same time, SEM could do component analysis combine with EDS. so SEM is one of the main instruments for microstructure analysis

979-8-3503-1101-3/23 $31.00 © 2023 IEEE

which has been widely used in semiconductor, materials, metallurgy, mineral, biology and other fields.

SEM API review is an auto address review function without inspection defect map, which run through API recipe at machine local position when we determined the certain locations and then collected SEM and OM images. SEM API include 2 types, surface API and edge API, which are showed in figure 2. Surface API could get images of the targets on wafer surface, while those images were got on wafer edge when edge API was used.

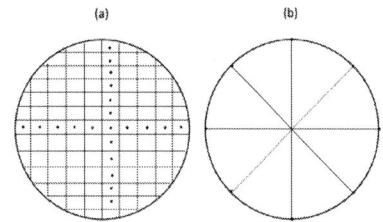

Figure 2: SEM API review (a) surface API (b) edge API

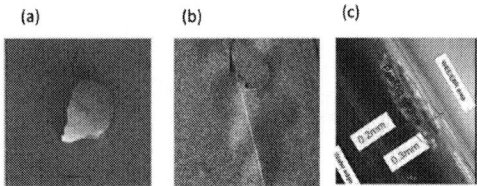

Figure 3: UTSL peeling defect (a) SEM image of peeling on wafer surface (b) SEM and OM image of peeling source on wafer edge

The peeling defect of UTS CIS

Before the bonding process of Pixel and logic wafer, we found peeling defect on wafer surface, and suspect peeling sources are on the edge areas. EDX analysis showed that peeing is Si, N. As figure 3 shown, these peeling source locations are random around the far edge of wafer, while CIS produces without the same type, which tell us that the special processes of UTS CIS may be the murderer.

A remained metal film, separated from its internal, was found through TEM analysis method, as figure 4 shown, which was from metal DEP process and failure removed at metal etch and CMP. The smooth characteristics of the RV film profile explain that crack appeared after RV DEP which reveal the maybe process step.

The formation mechanism research of peeling

We observed the defect performance on wafer edge step by step of backside film remove processes (LTO/poly/SIN) used SEM edge API, and bump sources

were found after the first backside remove process and chemical 1(C1) was used. These bump keep stable when the second and third process completed, and the observations showed that C1 play a crucial role on the generation of edge peeling defect. We suspected C1 intruded the edge remained metal film, caused an abnormal stress change on the wafer edge, then film deformed and teared from the weak points after a few days. As showed in table 1, physical Characteristics of bumps have not changed during the three process steps, while 4days later, it started to burst. This data show that, peeling would getting worse after some waiting time.

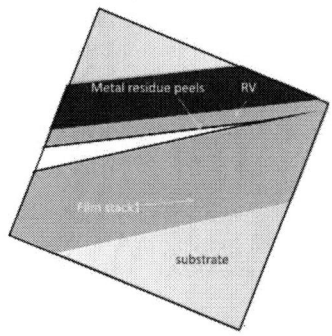

Figure 4: TEM analysis (cartoon image) of peeling source

Table 1 the observation of peeling source step by step

Step	Process step	Chemical	performance
1	1st API		normal
2	1st film remove	C1	
3	2nd API		Bump appears
4	2nd film remove	C2	
5	3rd API		Bump keep stable
6	3rd film remove	C3	
7	4th API		Bump keep stable
8	API	After 4days	Peeling defects were observed

The experiment of improvement and the result discussion

As it showed in the front of this paper, peeling was caused by the edge remain film corroded by C1, the improvement of the wafer edge residue is the critical goal.

We used bevel etch to eliminate residue before backside film remove steps that C1 would be used. As figure 5 shown, 200 edge images on etch wafers was collected by SEM API, some peeling sources were observed base on Baseline (BSL) condition, while 0ea peeling source appeared on the wafers edge which with bevel etch condition.

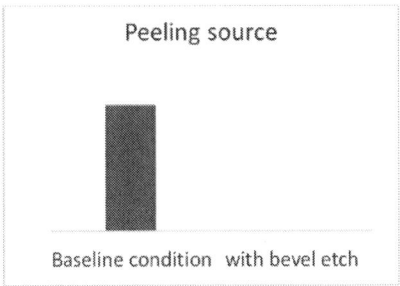

Figure 5: The peeling defect improvement after adding bevel etch process

CONCLUSION

In this paper, a bevel peeling defect formation during backside film remove processes on UTS CIS Logic products was introduced based on the using of SEM API function. Metal remain film on the wafer edge was eroded by chemical 1 with high concentration which used in backside film (LTO/POLY/SIN) remove processes, and an abnormal stress change was caused on the wafer edge and then deformed, teared from the weak points after 4 days. Bevel etch process before package was used to induce bevel peeling. Experiment results showed that peeling defect would be free when bevel etch process before backside remove take place.

REFERENCES

[1] M.F. Hsu and J.H. Yang, Taiwan Semiconductor Manufacturing Co. Ltd. (TSMC), Hsinchu, Taiwan; E. Yang, H. Chen, M. Ng, M. Li and C. Perry-Sullivan, KLA-Tencor Corp., Milpitas, Calif. 2010.

[2] Xiaofeng Yuan, Qiang Zhang, Jing 'an Hao, 2017 China Semiconductor Technology International Conference (CSTIC).

[3] Jami, Kalyan, et al., Solid State Technology, October 2009.

[4] Yiling Sun, Jihong Zhang, Yu Jiang, Fulong Qiao, Keqiang He, Zhigang Zhang, Kang Huang, Yushan, 2020 China Semiconductor Technology International Conference (CSTIC).

GENERAL CHIP DIGITAL DATA OBTAINING SOLUTION ON ATE

Steve Xie
ADVANTEST, Shanghai, China
*Corresponding Author's Email: Steve.Xie@advantest.com

ABSTRACT

Digital Data Obtaining is a necessary process in Chip Test. Many modules test in Chip Test need this data for spec analyzing, like ADC, PLL and Protocol Interface. A convenient Digital Data Obtaining developing tool will greatly improve the usability and efficiency of the ATE test program.

In traditional ATE test programs like Advantest's V93000 SmarTest7 already has provided 2 efficient solutions for implementation. Each of them has its own benefit on some occasion.

For greatly improving user experience of Digital Data Obtaining implementation and more efficiently using existed digital data obtaining solution on ATE, we are developing a general library that makes the configuration of Digital Data Obtaining independent from ATE's native structure on the SmarTest7 environment.

INTRODUCTION

In the ATE operation environment, obtaining a serial or parallel logical data from DUT output waveform often requires a set of necessary operation. To better understand new solution, it is important to understand the working mode of ATE while obtaining digital data

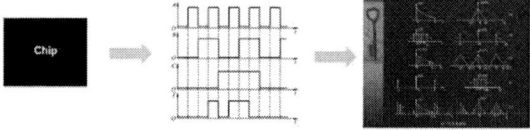

Figure 1: obtaining digital data

The SmarTest7 environment of the V93000 is a very popular ATE system. In the SmarTest7 environment we have 2 normal development kits for implementing Digital Data Obtaining actions: API solution and sRDI solution. In API solution, the test engineer manually places a capture point at the specific time point of a vector. Depending on the format of output data, capture data need set as serial or parallel mode. After that, in each capture data section, the developer also needs to open reference vector view and choose insert capture variable for creating data container of capture result. In this step, some property of data package format can be directly configured. For test method developing, only need write fixed code to retrieve result from ATE.

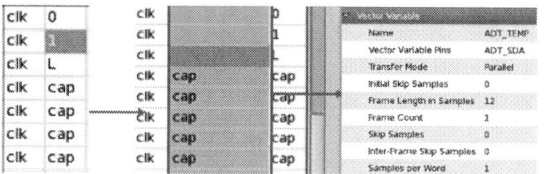

Figure 2: Manual Work of API Solution

If user prefer code base solution, sRDI solutions are very popular within the circle of testing engineers. In the sRDI solution, the entire operation explained above will use test methods for implementation. sRDI has a simple code to configure digital data capture and various interfaces to optimize the capture process. If some capture action needs to be shared from different tests, reference codes are easy to move and modify.

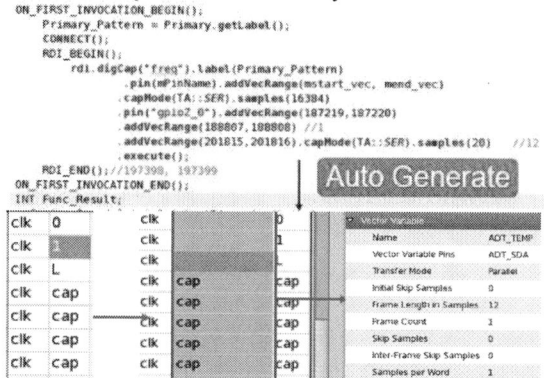

Figure 3: Advantatge of sRDI solution

Due to sRDI is a code-based solution, meaning a command needing execution must be according to its reference code. If the test condition has changed, the test code will also require an immediate update. Updating test methods need strict code verifying. Sometimes, this update only needs to modify an individual parameter,

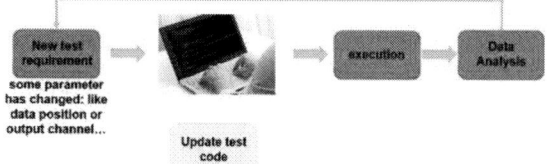

Figure 4: Update Flow of Code Base Solution

especially when the capture action is performed in multiple clock domain of the DUT. Due to the editing rule of sRDI, each test action in different clock domains require that the reference code be inputted into a different sentence because sRDI uses a unique ID to distinguish different clock domains. In different clock domains, test

---

979-8-3503-1101-3/23 $31.00 © 2023 IEEE   478

channels must be associated in different IDs (so-called multiport). That means we need to distribute different pins into different parts of the code in sRDI format which represents different clock domains.

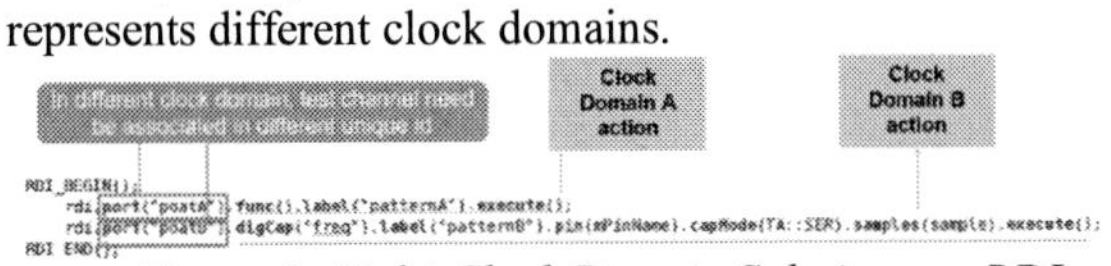

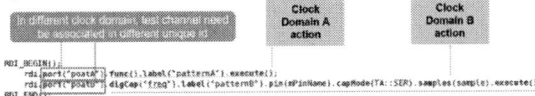

Figure 5: Multi-Clock Domain Solution on sRDI

According to the previous discussion, both solutions have their own characteristic. The API solution requires manual work and simplified. sRDI does not need extra manual work consumption. But test condition has changed, they need to update the test method because some parameters have changed.

For improving the user experience of Digital Data Obtaining implementation, we have developed a general library that makes Digital Data Obtaining independent from ATE native structures and more efficient for the user. In this library, we are designing 2 sets of parameter input interface.

First is dynamic suite property in SmarTest7 test flow perspective, in which users can easily set parameters (like observe channel, data structure or data position) by inputting data in a blank box with instructions or select on the pull-down menu using this suite property interface. All these blank boxes and pull-down menus are dynamically displayed, meaning it will adjust content when you input different parameters (this setup can guide users to finish all parameter filling).

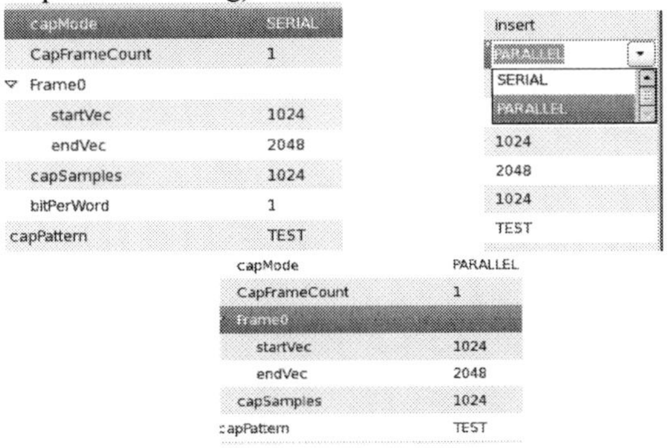

Figure 6: Dynamic Suite Property

Second is the CSV file input format, in which users can set parameters in a specified CSV file. In this file, users only need to write several intuitional parameters that reference test, like test suite name, capture channel, data position…. All these parameters are known on the ATE level. It will greatly improve the work efficiency of the testing engineer as the copy/paste action becomes easy in the CSV file). The CSV file has an advantage on the quality checking process, as there are many tools for CSV file checking.

After finishing these parameter setups with the above interface, we only need write a simple constant sentence in

the test-method file to start the test of digital data obtaining. In this sentence, we can switch 2 sets of parameters, only by modifying an input parameter. If this test has multiple time domains, the library will make a mapping relation between each time domain and different port reference code. This process implements in the background, so users only need to input test channel name, and the library will analyze the channel name for subtract port information. This is library support serial, parallel data output format. It also supports both output data formats at the same time, meaning some test channels will capture data in serial mode while other test channels capture data in parallel mode. It also supports input in all serial channel names as 1 parameter even if these channels belong to a different clock domain. This library supports multiple vectors doing data capture in 1 test item, which simplifies the test method structure as users do not have to increase code content for extra vectors. To retrieve output data, it provides several convenient result retrieval interfaces. These interfaces support different levels of result output (test suite, pattern, pin group, single pin).

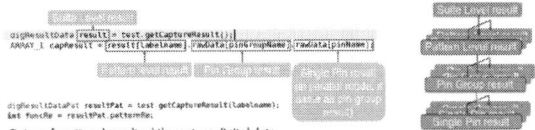

Figure 7: Output Data Retrieve Structure

And it integrates some useful macros for improving user experience while retrieving raw data (it can replace complex multiple-loop code, needed by multiple level data retrieving).

Figure 8: useful macros

This library also supports frequency calculate based on capture data. In this function, it does not need input parameter like period or x- mode that reference with the ATE environment. Instead, it will perform the analysis in the background without calling. This solution can return functional results while capturing digital data at same time, even capturing channel and functional result channel in different clock domains.

This solution consists of 2 classes: One is created for the CSV format input file, and the other one supports both CSV format input file and dynamic suite property in SmarTest7 test flow perspective. The first class is a basic class that takes charge of the Implementation of the underlying level. We create a multilevel data container for

979-8-3503-1101-3/23 $31.00 © 2023 IEEE   479

storing all raw data that is stored in the CSV file. This data is defined as static variable and only initialize 1 time. This data classifies 3 levels – suite, pattern, and pin groups. The bottom level pin group contains pin group reference information that have the same digital data obtaining rules (like the same capture position and data package rule). We also have a defined variety data structure for storing detailed information about each data capture test item extracted from the multilevel data container. If these data structures are defined as static variables too, it will save execution time. And this class will create individual objects for each test item as each object has an independent variable space for storing final parameters that belong to each test item, so it can prevent data intervention between different test items. These final parameters are transferred from the CSV file or input from the test suite parameter. During the execution, this class severed 3 different API functions for fitting different test requirement - Fully test item execution, single vector digital data obtaining execution, and single vector digital data obtaining execution with functional test result. All the API functions can select the data source CSV or dynamic suite property. This class integrated 4 resulting retrieve functions, and each function can get data from different levels: fully test item result, single vector result, result of single channel group in single vector result, and result of single channel in single vector.

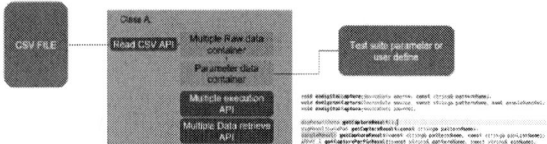

Figure 9: CSV File Reference Structure

Another class is a the single-function class, created for realizing functions for collecting capture condition data from a dynamic suite property. This class inherited from test method class that belonged to the SMT7 library (this relation help class can use test suite property controlling API). After that, this class uses multiple test suite property controlling API to get capture condition data from the suite property. And for the convenient input of this data, we set dynamic showing functions on some properties. Properties needed for input will be dynamically modifying based on the content written by users. In suite properties interface, we add selections of data sources: CSV or suite properties. That means users can using CSV file as a data source even running a second class. This class creates lots of data containers for storing information from suite property, and these data containers will store independently between different test items because this class will generate individual objects for the test items calling it. The first class is involved in second the class and is used for implementing digital data capture actions at an underlying level. So, the previously defined data container will transmit to first class. Second also has multiple execution functions and data retrieve functions. Each of these functions is mapping with the same function in the first class and it is going to call reference function in the background.

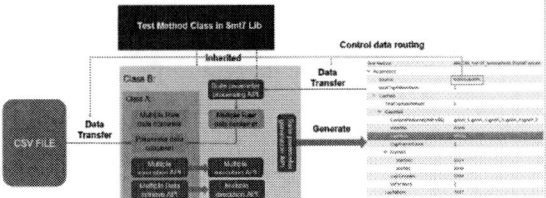

Figure 10: Suite Property Reference Structure

Based on this library, developers can create a digital capture test rapidly even if he is a beginner.

In summary, there are several features as follows:

- Greatly reducing development time of Digital Data Obtaining test (about 80%)
- Friendly to beginning ATE test program developers, especially on parameter setup process
- Flexible input data source for different case (CSV/property view), and easy to check.
- Multiple execution and result retrieve interface for different usage
- Stronger universality on different implementation occasions (no test method level modifies in different test program), improving the safety of the test program

Now, this library has implemented in a lot of real cases of chip test programs and is receiving very good feedback from testing engineers.

ACKNOWLEDGEMENTS

I would like to extend my sincere gratitude to Fang Yanfen and Song Xin for providing instructive advice and useful suggestions on this solution. I am deeply grateful to their help in the completion of this solution.

REFERENCES

[1] SmarTest Documentation Center, Topic 98751.
[2] SmarTest Documentation Center, Topic 98748.
[3] SmarTest Documentation Center, Topic 98615.
[4] SmarTest Documentation Center, Topic 98552.
[5] SmarTest Documentation Center, Topic 98815.
[6] SmarTest Documentation Center, Topic 98663.
[7] NICHOLAS A. SOLTER. C++Advanced Programming [M] Machinery Industry Press, 2006
[8] Bruce Eckel, Chuck Allison, Eckel, etc C++ programming idea: practical programming technology [J] Machinery Industry Press, 2006

AN EFFICIENT PROTOCOL FRAMEWORK SOLUTION ON V93000

Jun Chen[1]*, Xin Song[2], and Yanfen Fang[2]

[2]Advantest, Shanghai 201210, China

[3]Advantest, Shanghai 201210, China

*Corresponding Author's Email: jun.chen@advantest.com

ABSTRACT

With the development of the electronics industry, a system on a chip (SoC) is often composed of several different hardware components to achieve various functions, which require data communication between different parts. Given this background, a variety of recognized standard protocol interfaces have been promoted to realize hardware communication, such as JTAG/I2C/SPI/UART, etc. This paper provides a CBB (common build block) framework solution for all protocol-related test suites on the V93000. This solution can greatly improve programming development efficiency and reduce engineers' human resources.

Keywords –Soc, protocol, CBB framework, V93000, protocol solution

INTRODUCTION

SoC test brief introduction

SoC is an integrated circuit that integrates most or all components of a computer or other electronic systems. An SoC is often composed of a central processing unit (CPU), graphic processing unit (GPU), I/O interfaces, memory interfaces, Wi-Fi, Bluetooth, and sometimes includes a digital or analog signal processing unit [1]. Figure 1 shows the internal structure of a standard SoC.

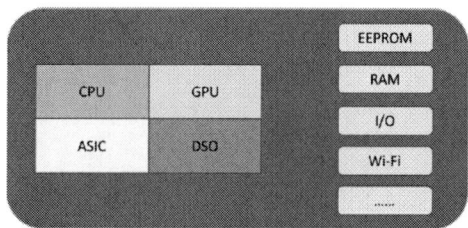

Figure1: SoC structure

As Moore's law states, as semiconductor density increases, so does the functionality of an SoC device, which creates a challenge for Automatic Test Equipment (ATE) testing. As a part of semiconductor ecosystem, spanning the design, test, and manufacturing environment, the ATE system ensures that chip devices perform and function as intended once they're circulated in the market, making it a significant part to the overall ecosystem.

ATE pattern iteration

To realize communication between different SoC modules, a variety of recognized protocols have been developed, such as JTAG/I2C/SPI/UART. A protocol is a set of formal rules to describe how to transmit data between sender(s) and receiver(s), which defines a standardized syntax, semantics, and synchronization of communication between two devices or modules [2]. In the process of ATE testing, SoCs are always required to config its internal register to set a certain mode or verify a device's function, which is essential for DC, RF, analog or other types of test suites.

The traditional way to config registers is as follows: Firstly, the design for testability (DFT) engineer provides simulated vectors (such as .wgl\.stil\.vcd format file) using a pattern generator tool, which calls pattern in ATE testing. Secondly, ATE engineers convert it to the specified format patterns that could work on ATE to verify the chip's function with the pattern conversion tool. After converting patterns from DFT to ATE, the generated patterns are loaded into the tester and are subjected to debug and testing. Finally, ATE engineers feedback the result of the pattern (such as pass or fail, fail cycles, and pin margin) to DFT engineers by using the ATE debug tool, such as timing diagram, pattern editor, error map.

After analyzing the pattern's result, DFT engineers update the new vector file when it comes to some error information. Then the updated pattern goes through another cycle of this process. This process is called "ATE pattern iteration" [3]. Figure 2 shows the overall process of ATE pattern iteration.

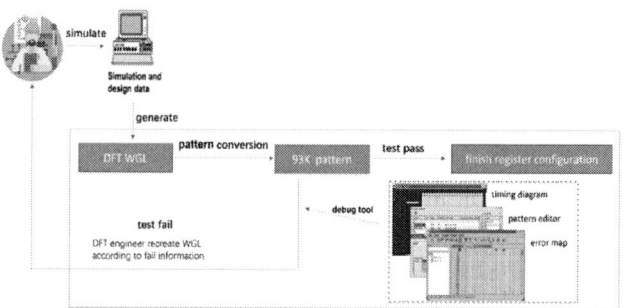

Figure2 ATE pattern iteration process

Using "ATE pattern iteration" to verify register config related test suites, setup and simulation may require many hours for each setup on ATE. Besides that, multiple steps within the long process could result in any error type which will cause incorrect results that are challenging to debug. So, it usually costs the same amount of time to debug patterns, which deeply affects the production testing period of SoC device.

979-8-3503-1101-3/23 $31.00 © 2023 IEEE

Protocol-aware test solution based on V93000

Besides debug patterns, DFT engineers also provide protocol template waveforms, then ATE engineers design different protocol interfaces to realize register programming for the specified complex protocols like JTAG, SPI, I²C on ATE's software environment.

To simulate the hardware's communication via different protocols on the ATE platform, Advantest's V93000 has promoted solutions for protocol communication [4]. Using C++ programming language, SmarTest7(SMT7) is the software package accompanying the V93000 series tester. After understanding the messaging component of protocol, i.e., write, read, idle, the ATE engineer provides protocol interfaces by predefining setups which incorporate Timing, Level, and protocol definition in the SMT7 software environment. Therefore, engineers only provide the transaction information to setup the protocol-based test suites, which includes the sequences of protocol transactions, address, and data. Figure 3 shows the general protocol-based data communication between ATE and chip. This process facilitates exchanging test setups between the tester and bench environment. At the same time, DFT engineers only need to provide a configuration list instead of a pattern, and the returned value can be captured directly.

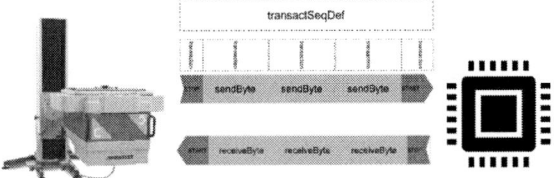

Figure3 protocol-based communication on ATE

With SMT7's protocol-aware test solution, register programming could be much easier to realize and faster to debug on ATE [5]. It also makes the generation of test setups for protocols more efficient and less error prone, greatly reducing the debug time compared to the traditional way of ATE pattern iteration.

It would be more efficient if ATE engineers coordinate with DFT engineers to debug together, so that DFT engineer could update register configuration according to the returned register's value. It would also be efficient if DFT engineers are familiar with SMT7's software environment so that they could debug on the tester's environment themselves. However, using this way to debug register related tests would, without any doubt, directly increase the human resources or learning cost.

PROTOCOL CBB SOLUTION

The protocol CBB introduction

As mentioned last paragraph, SMT7's protocol-aware

solutions [5] have great benefits to registering config-related test and don't need any specific hardware, but it would create a human resources burden. Another weakness of this mode is that different protocols always need several specified API interfaces, and they are always developed by different ATE engineers, which would make it difficult to ensure a code's quality and to maintain the code.

To solve these problems, we have promoted an efficient protocol CBB software solution to realize register programming on V93000 easily. CBB refers to components, modules, technologies, and other related design results that can be shared between different products and systems [6].

With this framework, a base class called "ProtocolInterfaceBase" has been promoted to manage all protocols, such as JTAG, SPI, I2C, UART. "ProtocolInterfaceBase" class have provided several virtual functions, such as write (unsigned long long Addr, unsigned long long Data). The first parameter represents the value of address, while the later parameter represents the value of data. Different protocol interfaces, as a derived class, are reconstructed from the base class to implement virtual functions according to the protocol's waveform template. Figure 4 shows the Unified Modeling Language (UML) of our protocol CBB.

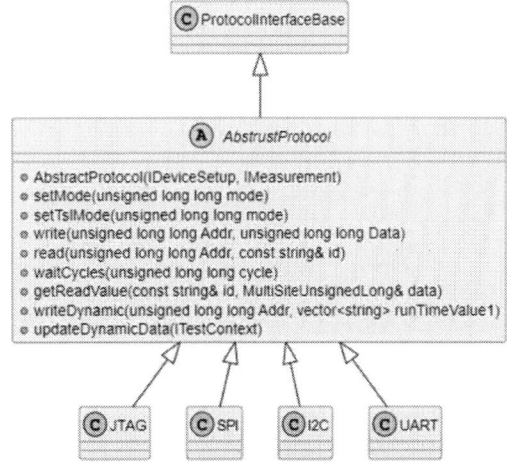

Figure4 Protocol CBB UML

Based on the CBB structure, we promoted a solution for all protocol test on ATE. With the new protocol CBB solution, patterns which consist of register related information would be automatically generated based on the Comma Separated Values (csv) file.

New protocol solution

To create or update the register configuration using our new protocol solution, instead of the SMT7's C++ code, the DFT engineers only need to fill a CSV file, named "*ProtocolConfig Table*". A CSV file is a plain text file that stores data by delimiting data entries with commas which could be opened in text editors, spreadsheet programs like Excel, or other specialized applications, so it is very easy understood and flexible for them to config register. The CSV's column includes *Test suite name, Type, Register_name, Register_address, Register_data, Mask and Wait_time*, every value filed represents the information to be configued.

Figure 5 shows the process of pattern conversion from CSV file into pattern sequence. The red circled line in figure 5 (a) means write 0x130 into the register (*XIF_CLKCTRL_TS_CLK_CTRL*) which address is 0xff646190, then wait 5us. After that, write another certain value to the specified register which is stored in the next row. Once the CSV file is finished, a sequence of patterns will be automatically generated by "*ProtocolConfig Table*" using the protocol CBB solution as a testmethod to run in the SMT7 environment. Finally, the register's (to be read) value would be collected into a CSV or console window and fed back to DFT engineers. After that, the engineer chooses to update (fix) the new register configuration or succussed to config register configuration according to the returned result.

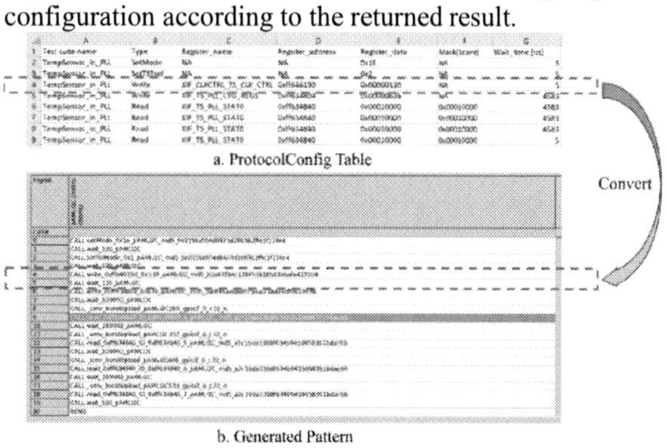

Figure5 Pattern conversion

The advantages of this solution are as follows:

1. This solution provides a friendly user CSV interface for DFT engineers, which would support DFT engineers debugging on an ATE tester while they are unfamiliar with the tester's software environment

2. This framework has defined a protocol CBB bottom frame, which unifies the access interfaces of all protocols, and formulates corresponding specifications, making it much easier for program specification and maintenance.

3. Compared with the traditional pattern iteration way and protocol-aware test solution, this solution could greatly reduce the pattern iteration period and human resource costs, which could significantly improve project development efficiency.

SUMMARY

This paper has proposed an efficient protocol solution based on a unified bottom frame to ease the register programming on the V93000's Smartest7 environment. With the new solution, project development efficiency can be improved greatly.

By now, this solution has been widely used for registering config-related test suites on the ATE tester. With this solution, DFT engineers could debug themselves by simply filling out the CSV file and updating the pattern if needed. This is very user-friendly for all engineers, regardless of one's experience or technical background. This solution would significantly reduce engineers' learning cost and greatly reduce the "ATE pattern iteration" period. In a real case, using the protocol CBB framework solution could save 80% of ATE protocol-related pattern debug and iteration time.

ACKNOWLEDGEMENTS

I would like to thank my colleague, Fang Yanfen, for his great contribution to the whole CBB framework. I also deeply appreciate Nick Song's guidance on my thesis.

REFERENCES

[1] A. Nezar and M. Creighton, "System on chip: challenges and design for manufacturing," *Fifth International Workshop on System-on-Chip for Real-Time Applications (IWSOC'05)*, Banff, AB, Canada, 2005, pp. 54-59, doi: 10.1109/IWSOC.2005.101.

[2] F. Leens, "An introduction to I2C and SPI protocols," in *IEEE Instrumentation & Measurement Magazine*, vol. 12, no. 1, pp. 8-13, February 2009, doi: 10.1109/MIM.2009.4762946.

[3] X. Song and M. Cao, "Complex Protocol Construct System on ATE Platform," *2021 China Semiconductor Technology International Conference (CSTIC)*, Shanghai, China, 2021, pp. 1-4, doi: 10.1109/CSTIC52283.2021.9461465.

[4] J.-C. Fernandez, C. Jard, T. Jéron, and C. Viho. An experiment in automatic generation of test suites for protocols with verification technology. *Journal of Science of Computer Programming-Special Issue on Industrial Relevant Applications of Formal Analysis Techniques*, 29, p. 123–146, 1997.

[5] J. Xiang, Y. Xiao, J. Sun, Q. Xia and Y. Fang, "High Inheritability and Flexibility Smarttest8 Testmethod Library on Ultra-High-Speed Serdes Measurement," *2022 China Semiconductor Technology International Conference (CSTIC)*, Shanghai, China, 2022, pp. 1-4, doi: 10.1109/CSTIC55103.2022.9856891.

[6] H. Garavel. OPEN/CÆSAR: An Open Software Architecture for Verification, Simulation, and Testing. In B. Steffen, editor, *Proceedings of the First International Conference on Tools and Algorithms for the Construction and Analysis of Systems (TACAS'98)*, LNCS vol. 1384, p. 68–84, March 1998.

Ultra-high-throughput inline probe metrology and inspection on EUV resist

Andrew Humphries[1], John Cossins[1], Lei Feng[1]*
[1] Infinitesima Limited, Abingdon, Oxfordshire, United Kingdom
*Corresponding Author's Email: lei.feng@infinitesima.com

ABSTRACT

The accelerating pace of adoption of EUV lithography and in particular the imminent introduction of High NA EUV lithography have resulted in the need for new metrology capability applied to EUV resist images post exposure and develop, prior to pattern transfer. Long established metrology and inspection techniques based on optical or electron imaging are reaching the limits of their resolution both laterally and vertically. They are not easily capable of producing detailed images of local variations in linewidth, resist height and roughness caused by EUV photon stochastics, materials inhomogeneities and mask variations. Understanding and controlling these local variations is key in achieving the required precision in CD, EPE and other dimensional or positional parameters that must be attained in semiconductor device fabrication.

In this work, we will also present a novel, contactless probe metrology technique that simultaneously achieves very high spatial resolution and does so at data rates far superior to existing probe metrology solutions. This new technique uses ultra – fine probes, achieves lateral and vertical resolution of below 0.1 nm and has been able to successfully image L/S structures to a pitch of 24 nm. Also presented are examples of microbridges and microbreaks in EUV resist images resolved over the full depth of the photoresist. Dimensional analysis of EUV resist profiles such as LWR, LER, Line Top and Line Bottom (scumming) roughness are also presented. Complete resist profiles are presented, and it will be demonstrated that such profiles allow the extraction of hitherto unavailable data such as LWR as a function of vertical position on the line.

The emerging metrology gap is not limited to photolithography. Within an increasing requirement for surface flatness prior to resist coating, CMP performance requirements are approximately three times as stringent for

High NA EUV than for its predecessor. In this work we will demonstrate results for several applications, achieving high throughput without sacrificing metrology accuracy and image fidelity. Using an innovative thermally actuated cantilever and displacement interferometer system in three axes, high speed probe operation and long probe lifetime can be achieved, the latter by reducing the tip/sample contact time. The typical image time on 50 µm images is below 2 seconds. By taking 9 sites per wafer and 50 µm

FoV images, a high throughput probe microscopy system measuring more than 20 wafers per hour is achieved.

Keywords—AFM; Probe Microscopy; EUV; High throughput; CMP; Lithograph; Metrology; Inspection.

INTRODUCTION

The heart of the invention is a new cantilever and probe architecture It has several innovations that enable high data rate operation [1][2]. First, the cantilever consists of a bimorph, for which the upper and lower layers have mismatched coefficients of thermal expansion. When an Infra-Red laser beam is played upon this bimorph, it flexes in response. The IR beam can be modulated, and the thermal dissipation from the cantilever is rapid enough that very high modulation frequencies can be attained. For this study, drive frequencies of up to 600 kHz were used, though in principle, even higher frequencies are possible. Further, the modulation of the cantilever is not restricted to sinusoidal profiles; more complex modulation such as sawtooth or triangular profiles are possible, and these can be combined in arbitrary ways.

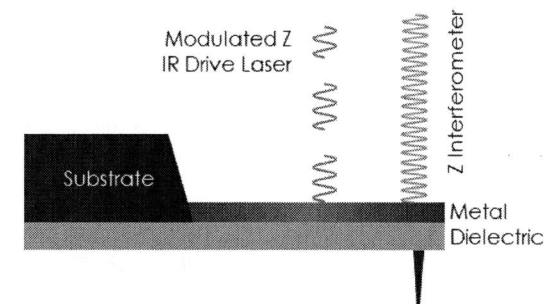

Figure 1 Schematic graph of modulated IR driven cantilever and interferometer Z detection.

High bandwidth probe actuation was demonstrated to permit data rates of up to 260 kPs^{-1}, meaning that the pixel rate of such a measurement device is up to two orders of magnitude higher than that for conventional AFMs (depending on application).

To be able to maintain high accuracy while scanning and oscillating at such high frequencies, a Z interferometer with a bandwidth of several MHz is used to directly measure the vertical position of the tip by reflection from the top of the cantilever directly above the probe. Interferometry in the Z axis, when combined with interferometry in the X and Y axes ensures extremely high (sub 1 nm) measurement accuracy in all axes, as well as eliminating many sources of image drift that can plague conventional AFM architectures.

Finally, the high response time of the cantilever, combined with the ability to sense the approach of a high topography feature such as a rising sidewall, permits the rapid withdrawal of the probe from the measurement sample, while simultaneously detecting and measuring the edge. This gives greatly reduced probe to topography interaction forces, which in turn can improve probe lifetime by at least an order of magnitude, allowing useful tasks to be performed with acceptable probe costs per unit measurement area.

EXPERIMENTAL SETUP

The new cantilever and probe assembly was installed in a low noise, automated metrology system that combines a high-speed wafer stage, optical pattern recognition for the location of the wafer image and a capacitance sensor - based rapid wafer surface detection system (all outside the scope of this paper). After ensuring that all modules and data collection channels were functioning to specification, the system was used to image a variety of high interest samples from semiconductor manufacturing processes.[3] These included EUV photoresist images, with L/S structures to 24 nm pitch and Contact Holes to a nominal dimension of 20 nm. Also imaged were various post CMP structures, including for advanced packaging applications. An assessment of imaging data rate as a function of topography was performed. Probe dimensions were selected for best access to the features in question. Where possible, results were compared with SEM or other data.

RESULTS

Imaging data rate was evaluated as a function of measurement type and the results are summarized below. Numerous parameters were used to determine the quality of the images obtained, including trench access, trench bottom figure, depth accuracy (against an industry standard) and image symmetry.

Measurement Type	Data Rate
Nominally flat, surface roughness	260 kPs⁻¹
Low topography, low spat. freq. (e.g. post CMP)	260 kPs⁻¹
Medium topo, high SF (e.g. EUV resist)	to 25 kPs⁻¹
High topography high SF (e.g. FEOL transistor architecture)	to 10 kPs⁻¹

These results confirm the high data rate capability of the invention, and from this point the system was used to evaluate various short – loop process wafers.

EUV Lithography

Line / space to P24 and contact hole images to 20 nm were imaged after probe selection and optimization of system setup. Imaging quality was assessed and compared to CD SEM images for the same samples, where possible. Using a probe deconvolution algorithm, we were able to show for two different probe architectures, very good agreement in lateral CD when compared to a CD SEM

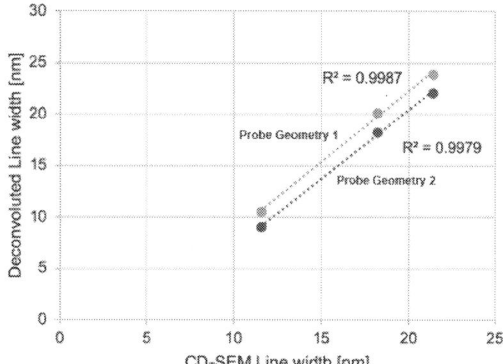

Figure 2 Rapid Probe Microscope (RPM) correlation to CD-SEM measurement results

The system was then assessed for 3D imaging data and showed excellent (below 1 nm) Z resolution. It was immediately clear that in evaluating EUV resist images, details not visible at all to SEM systems became easily visible.

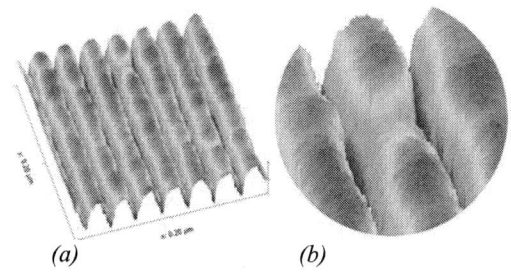

Figure 3 (a) Pitch 28nm line/space Image Showing Near Microbreak, (b) detail and Analysis of Imaging Defect

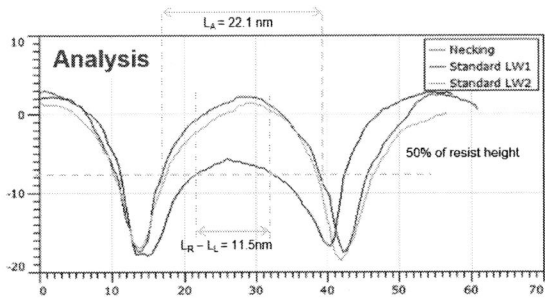

Local linewidth loss of 10.6 nm

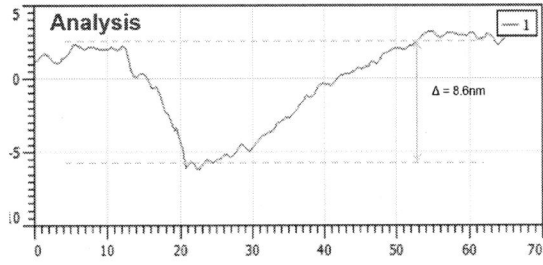

Local resist thickness loss of 8.6 nm

Figure 4 Top: Linewidth loss on the microbridge. Bottom: Resist loss on the microbridge in figure 3

Even in the small areas of EUV resist initially imaged, local, presumably stochastics related imaging defects were seen. For the P28 L/S image, several near line breaks were detected an analyzed. In the example shown above, the local resist loss was 8.6 nm. When imaging still smaller structures in thinner resist, several microbridges were detected and imaged.

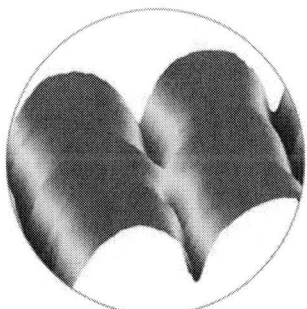

Figure 5 Pitch 24nm line/space Image Showing Microbridge

As part of the evaluation of the capability of the system, Contact Holes (CHs) of various dimensions and pitches were also imaged. CHs are more challenging to image because of probe / sidewall interactions, but with some experimentation with the probe length, stiffness and

profile, good results of full CH access down to 23 nm were obtained. Immediately apparent were stochastics induced depth, profile and diameter variations that are not easily visible with CD SEMs. Below are some example images for CHs.

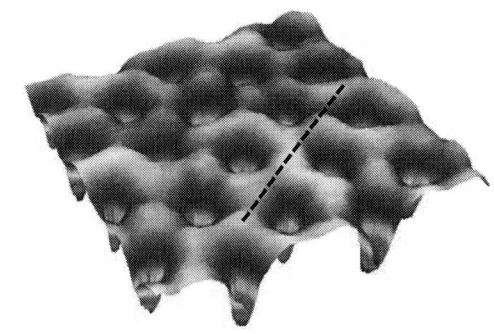

Top Resist Thickness Variation of 6.4 nm P-V

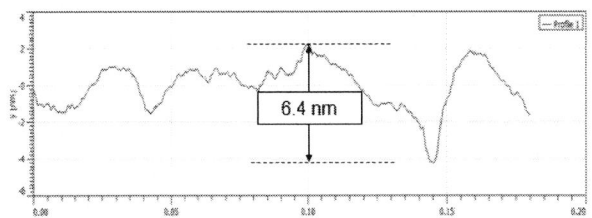

Figure 6 Top: CD26 Hex Array CHs Image, Bottom: Resist thickness variation analysis on top surface.

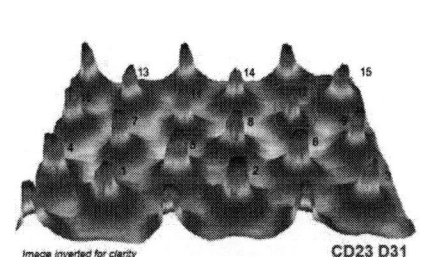

CH	Depth	CD
1	31.02	25.92
2	30.86	25.92
3	30.35	24.34
4	31.03	23.95
5	31.30	29.06
6	31.59	27.88
7	30.01	19.63
8	33.37	26.70
9	30.93	23.56
10	26.72	22.38
11	31.30	20.03
12	29.90	15.31
13	31.06	24.34
14	31.06	26.31
15	30.96	29.44
Average	30.90	24.32
Std. Dev	0.99	3.80

Figure 7 Inverted CH Image Showing Stochastics Effects

Other Imaging Capabilities

The ability to measure line roughness at arbitrary heights in a profile permits the calculation of LWR / LER as a function of Z. In the data below, such a set of measurements was taken for an EUV L/S structure. This information is useful in the patterning process.

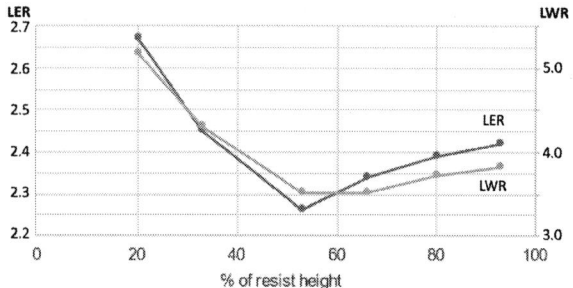

Figure 8 LER / LWR vs. Profile Height

Top line roughness data were extracted for numerous line types. In this example, a severe line roughness can be seen in organic EUV CAR.

Figure 9 An example of Top Line Roughness

The superior Z plane imaging capability permits the viewing of detailed CH profiles, providing valuable information to the patterning engineer.

In this example, a CD26 nm contact is imaged to the substrate.

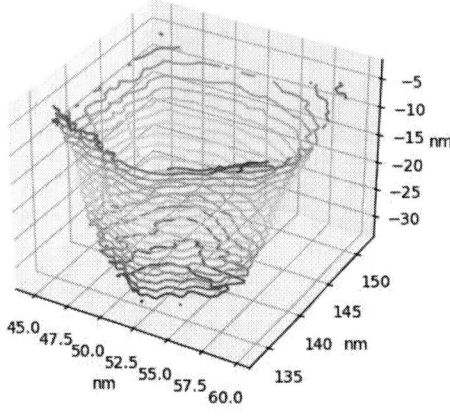

Figure 10 Detailed Contact Hole Profiling

THROUGHPUT CAPABILITY

Throughput of the system is strongly dependent on the scan lateral resolution and area to be scanned on the wafer. In all cases, the system was found to be at least 10 times faster than conventional AFMs, and up to 100 times faster on low topography structures. In the example below, a measurement protocol compatible with litho in – line metrology is modelled, and a throughput of 46 WPH is attained. This was confirmed experimentally.

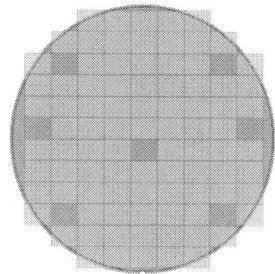

7 Sites per Wafer
0.5 micron FOV
128 x 512 pixels
3 Site Global
Alignment
46 WPH

Figure 11 Fully automated wafer map of an example of inline process control application

CONCLUSION

A novel probe metrology technique has been developed and has been shown to enable high speed, high accuracy probe based imaging of structures of interest in semiconductor manufacturing. High-accuracy 3D images have been collected at throughputs compatible with in-line metrology and inspection requirements, but without sacrificing accuracy in Z, Y and Z resolution. The technique enables the extraction of measurement parameters not typically possible with prior techniques.
If you have any questions regarding your manuscript, please do not hesitate to contact Dr Lei Feng at lei.feng@infinitesima.com

ACKNOWLEDGEMENTS

We would like to express our gratitude to our colleagues and EUV manufacturers who provided invaluable support and advice throughout the project. We are also grateful to the reviewers for their insightful comments and suggestions, which helped to improve the quality of this paper.

REFERENCES

[1] A. Humphris, B. Zhao, D. Catto, J. Howard-Knight, P. Kohli, and J. Hobbs, *Review of Scientific Instruments,* vol. 82, no. 4, p. 043710, Apr. 2011,

[2] A. Humphris, L. Feng, M. Tedaldi, L. Mudarikwa, D. Ockwell and J. Goulden. *Proc. SPIE 11325 Metrology, Inspection, and Process Control for Microlithography XXXIV*, 2020, 113252H.

[3] A. Humphris et al., Proc SPIE 11325 Metrology, Inspection, and Process Control for Microlithography XXXIV, Mar. 2020, 113251M.

STUDY ON E-BEAM INDUCED DEPOSITION WITH GAS INJECTION SYSTEM

Fan Zhang*, Yun Xu, Hongtao Qian

Semiconductor Manufacturing International (Shanghai) Corporation, Shanghai 201203, China
*Corresponding Author's Email: Fan_Zhang2@SMICS.com

ABSTRACT

The Transmission electron microscopy (TEM) plays the most important role in observing and analyzing nano-scale samples of FinFET technology, which are mainly prepared by dual-beam system consisting both e-beam and I-beam columns. However, the lamella may include unexpected tier or elements caused by e-beam induced deposition with GIS during sample preparation. So, this paper intends to reveal the impurity from the deposition by utilizing standard TEM imaging with EDS method on a series experiment. Not only the components of the unexpected layer are uncovered, but the mechanism behind the phenomenon is illustrated. This study also proposed a brief workflow to overcome the side effects of beam induced deposition to achieve damage-free TEM samples.

INTRODUCTION

Traditional planar Complementary Metal Oxide Semiconductor (CMOS) technology has achieved ceiling by past decades due to its short channel effects and leakage resulted from the squeezing of gate length. The growing of short-channel and current leakage problems of CMOS based transistors make it impossible to continue further scaling down.Therefore, the planar CMOS technology is hardly to keep up with the Moor's law, to meet tomorrow's requirements including higher speed, and lower supply voltage [1]–[3]. Fortunately, the FinFET technology, three-dimensional (3D) structure with multi-gate, has achieved huge progress, which is an innovative successor of CMOS to achieve the gate length under 20nm. In other words, the FinFET technology is existing for future's system-on-chip applications[3], [4].

However, increased the aspect ratio of Fin size and reduced pitch size increase the complexity of fabrication process which require large amounts of observations and analysis on each process, such as, the thickness of the high-K material, the profile of the Fin shape, the composition of the metal gate, the uniformity of certain layers, and the size of contacts. As a result, TEM is required in failure analysis (FA) department to analyze the FinFET samples in nano-scale, in addition to dual-beam system (SEM and FIB) which is widely used to prepare extremely thin samples for above TEM applications.

Right now, one of the biggest challenges for FA department is to prepare a good nano- scale TEM sample without damage or induced impurity that will definitely introduce error and uncertainty in TEM observations. Commonly, an electron beam (e-beam) or ion beam (I-beam) induced protective layer is firstly deposited on surface of the sample. Reasonably, how these layers effect the sample is an urgent question needs be answered, which is not shown in existing literature. Consequently, this article is trying to reveal the side effects using current deposition techniques.

Recently, a poor film uniformity case was reported by hard mask etching module. The metrology errors were not acceptable in terms of both long channel and short channel device measurement, such as shown in Fig.1, there was an unexpected layer observed through TEM, which deviated the measuring results. Both samples in Fig.1 were covered by a e-beam induced tungsten (W) deposition layer as protection tier. Accordingly, a sequence of experiments were introduced to investigate the root causes for this mysterious event.

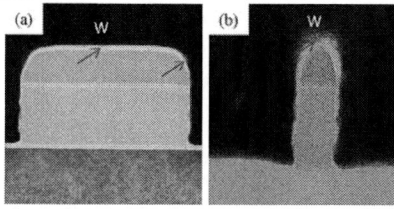

Figure 1. Poly Etching hard mask with GIS tungsten deposition layer, an un-expected layer between protection layer GIS tungsten and hard mark layer (a) Long channel, (b) Short channel.

EXPERIMENTS AND ANALYSIS

The elemental distribution from E-beam deposition layer on sample top is shown in Fig.2, which contains C (70%), O (10%) and N (3%) in addition to tungsten. This EDX result indicates that physical-chemical reactions are taking place during the E-beam coating, which will introduce some unexpected elements or compounds during deposition.

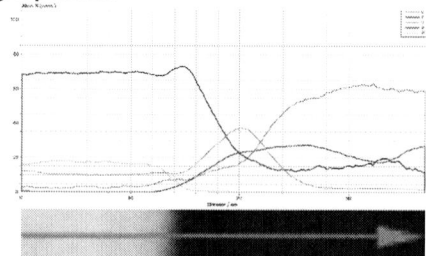

Figure 2. EDX Line-scan for E-beam deposition layer

Experiment shows that sample with SiO2 deposited on its top has great uniformity as shown in Fig. 3 (a), (b) and (d), which support the fact that abnormal layer (bright contrast) may be precipitated from GIS. Also, compare with Fig. 1(a), Fig.3 (c) shows the layer shrank when the E-beam deposition material is Platinum.

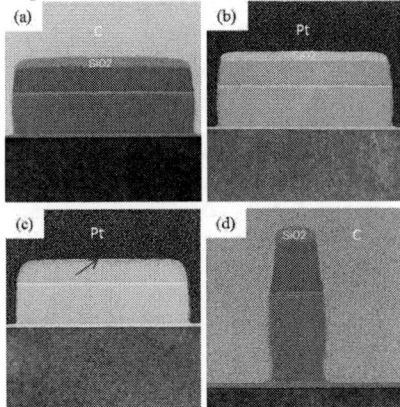

Figure 3. Post hard mask etching, protection layer split (a) (d) deposition with Carbon, (b) (c) deposition with Platinum.

Unexpected layer may relate to the E-beam deposition layer, another sample from EPI module can verify the conjecture, TEM image shows as Fig.4 (a), (b), (c) . Compared Fig.1(a) with Fig. 4(b), a conclusion shows a thicker non-uniform film is formed where E-Beam deposition layer is W rather than Pt, though there is the same E-Beam deposition voltage and current. Also, non-uniform film may be thinner when E-Beam protected layer deposited on the SiO substrate. A explanation that the K value of SiN is 8.7 which bigger than SiO's 3.9, the former may trap more electron and absorb enough contaminant which lead to a thicker non-uniform film.

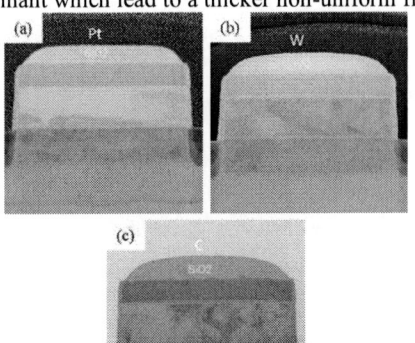

Figure 4. Dummy poly pattern (a), (b), (c), E-beam coated with Platinum, Tungsten and Carbon.

EDX line-scan profile of dummy poly pattern shown in Fig.5 (a) indicate E-beam deposition of Platinum layer consist of 70% C, 13% O, 7% N and 10% Pt. Also, abnormal film has the elements of C, O, and Si. And the line-scan in Fig.5 (b) shows that E-Beam deposition of Tungsten consists of C, O, N and W, but the atomic percentage of C is 95% while W only has 2%. In addition, abnormal film has C, O, Si and less W. It's sure that GIS atoms interact with sample top surface atoms [5]. Fig.4(c) and Fig.5(c) shows E-beam deposition with Carbon is the most moderate way to protect the sample from damage caused by atomic interactions. For the sample coated with Tungsten, there still exists a poor uniformity layer with contrast at the interface. It can be confirmed that the root cause of the layer precipitation is the side effect of GIS deposition.

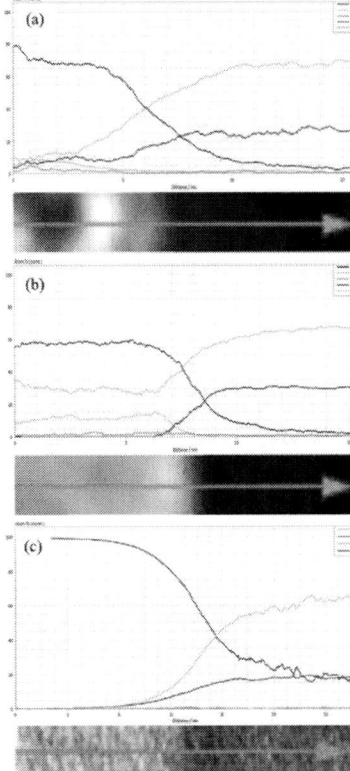

Figure 5. EDX Line-scan profile for E-beam deposition layer, (a), (b) and (c) coated with Platinum, Tungsten and Carbon respectively.

GIS DEPOSITION MECHANISM

GIS source characteristics are shown in Table 1 [6]–[8]. The GIS deposition is accompanied with decomposing the adsorbed gas and sputtering the substrate. A brief illustration of deposition model is shown as Fig. 6. After the GIS is heated by current, the organic gas molecules inside the nozzle tube are injected down to the sample surface, then result to physical chemical adsorption and desorption. That means, the gas molecules would accept the energy from E-Beam so that multiple collisions will occur. When the energy is greater than the molecular bond energy, the molecules are decomposed. Theoretically, organic gasses are decomposed into both volatile and nonvolatile type, the former is mainly pumped away, while the latter forms a protective layer deposited on the sample surface. The nonvolatile Pt atoms gather together during the process of nucleation. But an action that the volatile part of GIS can't spread out in time thus hinders the process where the crystal nucleus is congregated to an "island". The Pt organic gas molecules have a high ratio of carbon, so as to the gap between Pt particles and the interface between Pt particles and samples filled with bright contrast non-uniform layer. In addition, higher deposition rate would cause that the size of Pt particles is larger than W particles. However, the smaller W particle with narrower gap and more interfaces make it next to impossible for volatile parts to pump away in time. Those volatile gases will eventually diffuse into the sample top surface. E-beam scanning with over-abundance of energy leads to acute chemical reaction, atom collision and Coulomb force adsorption for contamination comes from chamber interior, as shown Fig. 7. Deposited atoms interact with sample surface atoms to generate interactive diffusion; a thick bright contrast layer is then formed.

Table 1 : GIS source characteristics

GIS	W(CO)6	Platinum(C9H16Pt)
Molecule structure	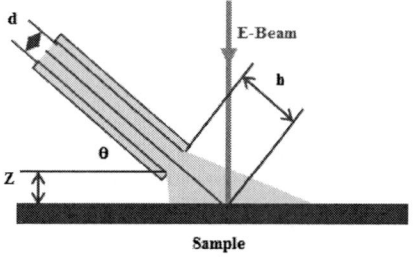	
Physical properties	Lower resistivity than Platinum, Solid	High hardness, Chemically resistant Solid
Operates current	70-100 pA/μm^2	2-6 pA/μm^2
Operates temperature	50℃	38-42℃

Figure 6. The deposition model of the GIS: Z - Needle height; h - needle distance; d - needle diameter.

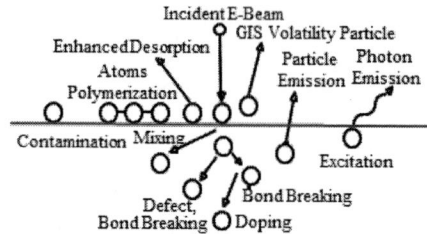

Figure 7. Gas molecules, nonvolatile and substrate materials atoms interactive reaction at E-beam irradiation.

Based on the GIS deposition mechanism, the phenomenon of Fig.1 that the precipitation layer is thicker on the pattern top than sidewall can be explained. Normally, the E-beam radiates perpendicularly towards the pattern, while the angle between E-Beam and the sidewall is significantly lower than 90 degrees, shown as Fig.8. This will cause massive electron scattering and great dose dropping. Before the first formation of the tungsten layer, the pollutant, which is from the chamber interior. And the larger beam dose will lead the pattern top to absorb more pollutants than the sidewall by static electricity. Right after the W layer is formed, the decomposition rate of the GIS becomes higher and volatile compounds from GIS can be chemically absorbed. However, due to the lower decomposition rate caused by low dose on the sidewall, the chemical absorption rate of GIS volatile substances is slow on the sidewall during the W layer growth,which ultimately reduce the thickness of precipitation layer on the sidewall.

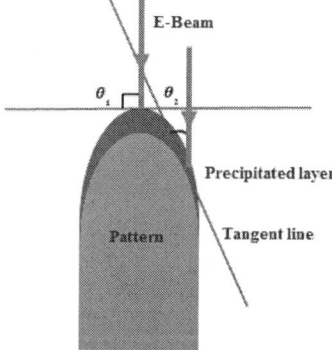

Figure 8. The angle between E-beam and pattern tangent line.

In addition, E-beam voltage and current are researched, shown as Fig. 9. The bright contrast of precipitation layer is the thickest when the E-beam current is 6.4nA. The layer becomes thinner as the E-Beam current is reduced to 3.2nA and finally the layer is at its thinnest when the current is 1.6nA. The contamination chiefly comes from chamber interior and volatile function group of GIS. During the deposition process, smaller current means lower efficiency of GIS gas molecules decomposition and slower rate of deposition. This will grant sample surface longer time to absorb those pollutants from GIS and a relatively thicker precipitation is formed underneath the Tungsten layer. Experiments shows smaller voltage of E-beam with insufficient energy could not break the chemical bonds of organic gases effectively. Higher voltage of E-beam maybe leads to a lower deposition rate and sample surface damage [9]. So, 5kv and 3.2nA is comprehensively the most appropriate voltage and current combination for E-beam deposition.

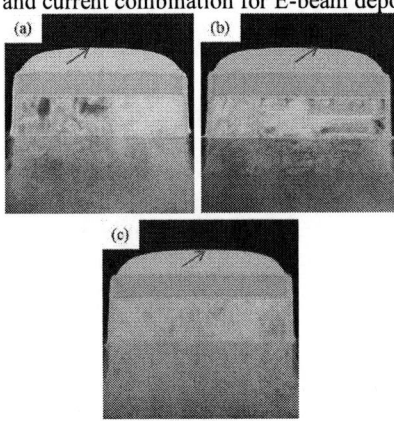

Figure 9. Dummy poly pattern (a), (b), (c), E-beam coated with Tungsten of 1.6 nA , 3.2 nA and 6.4 nA.

CONCLUSION

In this paper, the root causes of FIB GIS's side effects are clarified. Poor uniformity layer is formed by precipitation from the volatile types from GIS and contamination absorption comes from chamber interior. E-beam scanning would intensify the adsorption.

According to the experiments' results, the unexpected layer precipitation caused by GIS gas are strongly related to the deposition elements of GIS molecule, E-beam current and sample surface topography. Massive electron scattering and great dose dropping lead to a precipitation layer which is thicker on the pattern top than sidewall. 5kv and 3.2nA are the most appropriate voltage and current combination for E-beam deposition.

The sample can be coated by atomic layer deposition (ALD) using HfO followed by e-beam induced carbon deposition layer. Afterwards, tungsten or platinum deposition can be applied on the sample that suffers insufficient signals to improve both composition and topography contrast. Importantly, E-beam induced GIS gas deposition could be replaced by customized bond, such as SOC, PR, BARK and TiN growing in Fab, for specific cases including low K material.

ACKNOWLEDGEMENTS

The authors would like to reserve acknowledgement and thank the support of FA team from SMIC. Also, we are appreciated to the review of Dr Wu from Thermofisher

REFERENCES

[1] C. H. Chu, P. S. Kuo, T. F. Chang, J. X. Yang, J. Chang, and R. Yang, "Failure analysis of IC contains FinFET," in 2017 IEEE 24th International Symposium on the Physical and Failure Analysis of Integrated Circuits (IPFA), 2017, pp. 1–4, doi: 10.1109/IPFA.2017.8060056.

[2] G. Tshagharyan, G. Harutyunyan, S. Shoukourian, and Y. Zorian, "Overview study on fault modeling and test methodology development for FinFET-based memories," in 2015 IEEE East-West Design Test Symposium (EWDTS), 2015, pp. 1–4, doi: 10.1109/EWDTS.2015.7493149.

[3] A. K. Kuna, K. Kandpal, and K. B. R. Teja, "An investigation of FinFET based digital circuits for low power applications," in 2017 International Conference on Circuit, Power and Computing Technologies (ICCPCT), 2017, pp. 1–6, doi: 10.1109/ICCPCT.2017.8074280.

[4] D. Anandani, A. Kumar, and V. S. K. Bhaaskaran, "Gating techniques for 6T SRAM cell using different modes of FinFET," in 2015 International Conference on Advances in Computing, Communications and Informatics (ICACCI), 2015, pp. 483–487, doi: 10.1109/ICACCI.2015.7275655.

[5] H. F. Winters and J. W. Coburn., "Impulsive excitation of FeCp+2 and SiMe+3 during surface induced dissociation at organic multilayers." Surface Sci. Rep., 14 (1992), 161–270.

[6] D. K. Stewart, L. A. Stern and J. C. Morgan., "Electron Beam X-ray and Ion Beam Technologies," Sub Micrometer Lithogr. VIII Proc SPIE, vol. 1089, pp. 18–25, 1989.

[7] T. Tao, J. S. Ro, J. Melngailis, Z. Xue and H. Kaesz. J., "Classical and quantum transport in focused-ion-beam- deposited Pt Nano interconnects," Vac Sci Technol, vol. B, 8, no. 1990, pp. 1826–9.

[8] J. Puretz and L. W. Swanson. J. ., "Focused ion beam deposition of Pt containing films," Vac Sci Technol B, vol. 10 (1992), pp. 2695–8.

[9] "Platinum Deposition Technical Note." (Hillsboro, OR: FEI Company, 2003).

A SIMULATION STUDY ON THE THERMAL EFFECTIVENESS OF GRAPHENE-BASED FILMS IN INTELLIGENT POWER MODULES

*Jie Bao[1] *, Juan Hu[1], Yunyan Zhou[1], Yuan Xu[1]*

[1]Engineering Technology Research Center of Intelligent Microsystem of Anhui Province, Huangshan University, Huangshan 245041, China
*Corresponding Author's Email: 105034@hsu.edu.cn

ABSTRACT

Intelligent Power Modules (IPMs) integrate gate drive circuits and power transistors into one package. A large amount of heat generated from the power devices can affect the performance of the entire module. Graphene, an outstanding lateral heat spreader, can be employed in power modules to improve the thermal dissipation. However, due to its lower vertical thermal conductivity, graphene would bring different cooling effects depending on its geometry and placement. This paper studies the heat dissipation in three different IPMs with graphene-based film (GBF) placed on the power chips, on the different layers of direct bonded copper (DBC) substrate or on the copper substrate, and on the bottom of the IPMs. The aim is to further improve the packaging structures. The best heat dissipation can be achieved when placing GBFs on the materials with poor thermal conductivity. This is also verified by the equivalent spreading thermal resistance calculation.

INTRODUCTION

The volume and weight of a power system using insulated gate bipolar transistors (IGBTs) or metal oxide semiconductor field effect transistors (MOSFETs) can be further reduced with the use of intelligent power modules (IPMs). The level of integration and reliability can also be improved. IPMs are widely used in motor drives and various types of switched-mode power supplies [1-2]. IPMs usually incorporate the gate drive circuits and multiple power devices in the same package. The smaller form factor leads to high power density and possible severe local heating. If the heat generated inside these modules cannot be removed effectively, it will affect the electrical performance, eventually leading to module failure [3-4]. Therefore, thermal management of IPMs has become an important issue.

This paper is organized as follows. The models for three different IPM packages are first established. The model parameters and thermal simulation conditions are described in Section II. The placement of high lateral thermal conductivity graphene-based film (GBF) to enhance the cooling of IPMs is investigated in Section III. The influence of different GBF placements and parameters on heat dissipation is simulated. The thermal resistance is also calculated. Finally, conclusions are given in Section IV.

MODELING THE STRUCTURES

Three commercially available IPMs are used as references in this study. The x-ray images of the chip layout inside the modules are shown in Fig. 1. Module I contains one gate driver integrated circuit (IC), six IGBTs and six fly-wheel diodes (FWDs). Due to the electrical connection, three groups of power devices are mounted on the same upper copper layer of the direct bonded copper (DBC) substrate. The collector electrodes are connected to the P terminal. The collectors of the other three groups are separately connected to the three phase output terminals U, V and W.

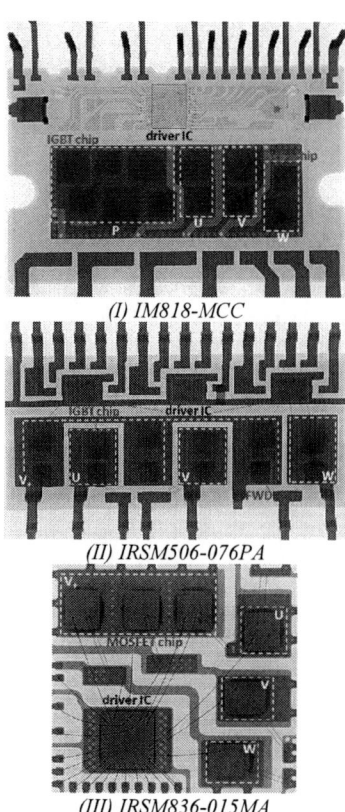

(I) IM818-MCC

(II) IRSM506-076PA

(III) IRSM836-015MA

Figure 1: X-ray images of the die layout inside the modules

Module II contains three half-bridge gate driver ICs, six IGBTs and six FWDs. Module III contains one three-

979-8-3503-1101-3/23 $31.00 © 2023 IEEE

phase gate driver IC and six MOSFETs. In these two packages the power devices are mounted on the copper substrates.

COMSOL is employed to model the IPM packages. The parameters used in the models are listed in Table I. Referring to the absolute maximum ratings on the product datasheet, Module I was simulated with a power dissipation of 67.5 W for each IGBT chip, and 10 W for each FWD chip. From Fig. 2 (a), the COMSOL thermal simulation shows that the maximum temperature of the IPM is approximately 110 ˚C, concentrated on the IGBT chips. Module II was simulated with a power dissipation of 16 W for each IGBT chip, and 3 W for each FWD chip. As shown in Fig. 2 (b), the maximum temperature of the IPM is approximately 134 ˚C. In Module III, the power dissipation for each MOSFET chip is 11 W. The maximum temperature of this IPM is about 85 ˚C as shown in Fig. 2 (c). The top of Module I and the PCBs in Module II and III are attached to a finned heat sink as described in Table I and is forced cool by a fan. The air flow speed through the heat sink is set to 5 m/s with an ambient temperature of 25 ˚C and an atmospheric pressure of 1 Bar [5].

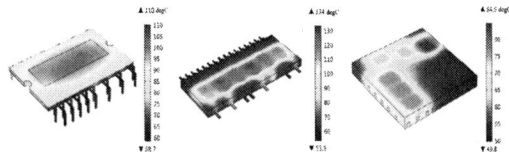

(a) IM818-MCC (b) IRSM506-076PA (c) IRSM836-015MA

Figure 2: Temperature distribution in the IPMs

TABLE I.   POWER MODULE GEOMETRIES

Model	Model parameters	Size ($l \times w \times h$)
I	IGBT chip	7mm×7mm×100μm
	FWD chip	5mm×5mm×100μm
	Gate Driver IC	6.5mm×9mm×200μm
	DBC Cu thickness	300 μm
	DBC Al_2O_3 thickness	680 μm
	Solder thickness	100 μm
	TIM thickness	30 μm
	Heat sink (cold plate)	100mm×80mm×2mm
	Heat sink (fins)	100mm×1mm×8mm
II	IGBT chip	2mm×2mm×100μm
	FWD chip	1.5mm×1.5mm×350μm
	Gate Driver IC	1.2mm×1mm×350μm
	Cu substrate thickness	200 μm
	Solder thickness	100 μm
	TIM thickness	30 μm
	Heat sink (cold plate)	50mm×50mm×2mm
	Heat sink (fins)	50mm×1mm×8mm

Model	Model parameters	Size ($l \times w \times h$)
III	MOSFET chip	1.65mm×2mm×150μm
	Gate Driver IC	2.6mm×3mm×230μm
	Cu substrate thickness	200 μm
	Solder thickness	100 μm
	TIM thickness	30 μm
	Heat sink (cold plate)	40mm×40mm×2mm
	Heat sink (fins)	40mm×1mm×8mm

INFLUENCE OF THE GBFS

The thermal conductivity of graphene is anisotropic. It is high in-plane, but low in the cross-plane. Therefore, its main role is to extract the heat from local hot spots laterally and further dissipate them through other materials, i.e. it changes the heat conduction path of the packaging structure. The different geometries of the GBFs are designed to alter the in-plane heat conduction and the cross-plane phonon scattering. They will impact on the thermal resistance of the heat dissipation path [6-7]. The thinner the GBF, the higher the in-plane thermal conductivity but the lower the heat capacity. A series of GBF products from Smart High Tech AB, Sweden are used as the reference material in the simulation. In this paper, the GBF is chosen to have a thickness of 50 μm, with an in-plane thermal conductivity of 1700 W/ m·K and a cross-plane thermal conductivity of 10 W/ m·K.

In the study on Module I, GBFs are placed (a) on the surface of the IGBT chips, (b) on the upper copper surface, (c) on the Al_2O_3 surface, and (d) on the bottom of the DBC substrate as shown in Fig. 3. In the studies on Module II and III, GBFs are placed on the surface of the IGBT chips, on the surface of copper substrates, and on the bottom of IPMs.

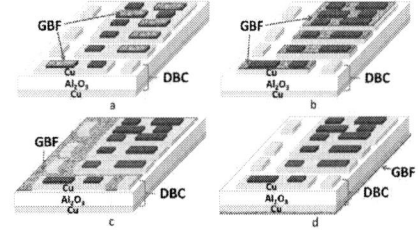

(a) On the IGBT chips, (b) on the upper copper surface of the DBC, (c) on the Al_2O_3 surface of the DBC, (d) on the bottom of the DBC

Figure 3: Diagram of the GBF placements

It is found that placing GBF on the ceramic surface of the DBC substrate resulted in the largest temperature drop in Module I. The peak temperature drop in Module II with placing GBF on the bottom of the IPM is the largest. However, in Module III, the placements of GBFs have little effect on the peak temperature of the IPM.

When heat is transferred from a small hot spot to a much

larger area, the one-dimensional thermal resistance analysis is inaccurate. However, in most situations, it will be useful if an equivalent one-dimensional thermal resistance can be defined [8]. Taking Module I as an example, and from the thermal resistance calculation point of view, the total thermal resistance from the IGBT chips to the ambient can be expressed as in (1). R_{sp1} represents the equivalent spreading thermal resistance (ESTR) from the point heat source to the entire chip. R_{sp2} represents the ESTR from the chip to the upper copper layer of the DBC substrate. R_{sp3} represents the ESTR from the copper layer to the ceramic layer of the DBC substrate. R_{sp4} represents the ESTR from the bottom of the DBC substrate to the TIM. R_{sp5} represents the ESTR from the TIM to the heat sink.

$$R_{total}=R_{sp1}+R_{chip}+R_{solder}+R_{Cu1}+R_{sp2}+R_{Al_2O_3}+R_{sp3}+$$

$$R_{Cu2}+R_{sp4}+R_{TIM}+R_{sp5}+R_{hs} \qquad (1)$$

The placements of GBF have a significant improvement on the ESTRs. In general, the ESTR is larger than the thermal resistance of the materials. The reduction of the ESTRs with GBF in the IPM can lead to a decrease in the total thermal resistance. When GBF is placed at different positions in the IPM, the degree of the ESTR reduction is different. Analyzing the influence of thermal resistance brought by GBF, it can be found that the total decrease in thermal resistance when the GBF is placed on the ceramic surface of the DBC substrate is the most significant.

The area of the ceramic layer is also an important factor affecting the heat dissipation of the GBF. As the area increases, the thermal resistance of the DBC decreases and the peak temperature decreases as shown in Fig. 4. However, the larger the ceramic surface, the more the influence the GBF will have on the ESTR. Comparing to the DBC without GBF, the total thermal resistance is smaller with the placement of the GBF on the ceramic surface. Therefore, the temperature drop brought by the GBF increases significantly with the increase in the ceramic layer area.

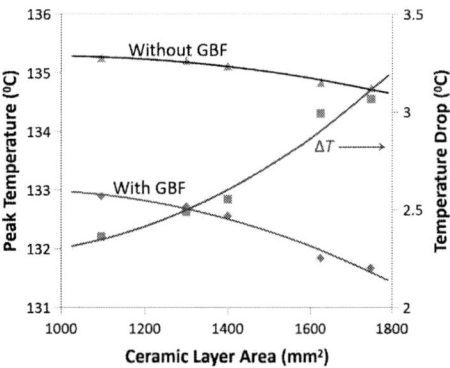

Figure 4: The peak temperature in Module I as a function of the ceramic layer area.

CONCLUSIONS

Based on the commercial IPMs, thermal simulations were carried out. The graphene-based films with high lateral thermal conductivity were placed on the surface of power devices, on the different surfaces of the DBC or the copper substrates and on the bottom of the IPMs. It was found that placing the GBFs on the materials with poor thermal conductivity provides the largest reduction in peak temperature.

With the increased power dissipation in IPMs, thermal extraction becomes more and more important. The use of GBFs provides a more uniform heat flow around the power devices. This work provides useful guidelines for the applications of graphene-based films to improve the thermal performance of next generation IPMs.

ACKNOWLEDGEMENTS

This research is funded by the Major Natural Science Research Program of Anhui Province (2022AH040270), Anhui Excellent Young Talents Project (gxyq2022087), and the Open Fund Projects of Engineering Technology Research Center of Intelligent Microsystem of Anhui Province (MSZXXM2002).

REFERENCES

[1] M. Jiao, Y. Li, J. Yu, J. Xie, P. Zeng and Z. Zhao. *7th Electronic System-Integration Technology Conference*, Dresden, September 18-21, 2018, pp. 1-5.

[2] M. Tsukizawa, H. Sasaki, J. Lee, J. Jeong and H. Bae. *5th International Conference on Electric Power and Energy Conversion Systems*, Kitakyushu, April 23-25, 2018, pp. 1-4.

[3] S. Young, Z. Chen, T. Lee, B. Choo and J. Son. *International Exhibition and Conference for Power Electronics, Intelligent Motion, Renewable Energy and Energy Management*, Shanghai, June 26-28, 2019, pp. 1-4.

[4] B. Bidouche, Y. Avenas, M. Essakili and L. Dupont. *IEEE Applied Power Electronics Conference and Expositon*, Tampa, March 26-30, 2017, pp. 2317-2322.

[5] S. Kang. *7th International Conference on Integrated Power Electronics Systems*, Nuremberg, March 6-8, 2012, pp. 1-8.

[6] Y. Zhang, H. Liu, L. Tan, Y. Zhang, K. Jeppson, B. Wei and J. Liu. *Materials*, vol. 13, 2020, pp. 104.

[7] Y. Xu, J. Bao, R. Ning, L. Hou, Z. Chen and B. Zhou. *International Conference on Electronic Packaging Technology*, Hongkong, August 11-15, 2019, pp. 1-4.

[8] Y. Shabany. Heat transfer: thermal management of electronics, CRC Press, December 17, 2009, pp. 88-91.

Effects of different catalysts on epoxy molding compound

Yangyang duan[1]*, Wei Tan[1], Xingming cheng[1,] Lanxia Li[1], Hongjie Liu[1], Dandan Fan[1], Lingling Liu[1], Xiaojuan Jiang[1], Liang Cui[1], Xingzhi Cui[1]

Jiangsu Hua Hai Cheng Ke Advanced Material Co. Ltd., Lianyungang 222047, China

*Corresponding Author's Email: yangyang.duan@hhck-em.com

ABSTRACT

There are many factors that affect the glass transition temperature (Tg) and adhesion (EMC) of epoxy molding compound (EMC). The type of epoxy resin is critical, but the effects of catalysts are often overlooked. In this paper, the effects of different types of catalysts on Tg and adhesion were studied. The studies have shown that imidazole has the highest Tg, while amines and organic phosphorus have relatively low Tg. In terms of bonding performance, both imidazole and amines have good adhesion to copper, while organic phosphorus has good adhesion to silver. In addition, the modulus and CTE were also affected.

Keywords: epoxy molding compound; catalyst; Tg; adhesion

INTRODUCTION

With the rapid development of semiconductor devices and integrated circuits, EMC has become the mainstream packaging material today. For any packaging form, Tg and adhesion are two important characteristics of EMC. There are many factors that have effects on the Glass transition temperature and adhesion of EMC. The type of epoxy resin is critical, but the role of catalysts is often overlooked.

In this paper, different types of catalysts were selected to study its effects on the Glass transition temperature and adhesion such as imidazole, amines and organophosphorus. Except for different types of catalysts, other raw materials are the same. It should be clarified that the catalyst used in this article is actually a curing accelerator.

EXPERIMENTAL

The catalysts selected in this paper are 2E4MZ (imidazole), DBU (amines) and TPP (organophosphorus). Epoxy resin has the structure of Epoxy Cresol Novolac Resins (ECN). Hardener has the structure of Phenol Novolac Resins (PN). The filler content was amorphous silica. The coupling agent is KH560. The release agent is Carnauba Wax.

The ratio of Epoxy value / hydroxyl value was 1.2. The filler content is about 80%, the coupling agent content is 0.3%, and the release agent content is 0.2%.

The preparation of epoxy molding compound was prepared as described in reference [1]. For a typical process, all the ingredients were weighted up according to the formulation and mixed in the high-speed mixing machine. Afterwards, the mixture is fed into twin screws extruder under heating conditions for extrusion, and the discharged material is then subjected to sheet forming, cooling, pre braking, granulation, and post mixing.

Standard transfer molding techniques were used to fabricate the test specimens. All test specimens were prepared with in mold cure time of 140sec at 175℃ followed by a six-hour post mold cure under the same temperature condition [5].

For thermodynamic properties, Tg were measured by by DMA. Coefficient of thermal expansion (CTE) were measured by TMA. TA instrument model rises from 40℃ to 260℃ at a rate of 5℃ per minute. α1 takes the CTE below Tg and α2 takes the CTE above Tg.

Adhesion was measured by Tab Pull [1-2]. A Tab pull sample can be taken as two pieces of lead frames connected by EMC as shown in Figure 1. The enclosed area is encapsulated by EMC and the shadowed area on the lead frame surface is attached by EMC. The tensile test was done with a Universal testing machine (AGS-5kNA, Shimadzu Corporation). The tensile force was reported as the indication of adhesion force between epoxy molding compound and lead frame. The lead frame surface are plated by copper or silver.

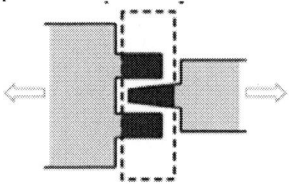

Figure 1: The schematic graph of a tap pull sample for the tensile test. Two pieces of lead frames are connected by EMC. The enclosed area is encapsulated by EMC and the shadowed area on the lead frame surface is attached by EMC.

RESULT AND DISCUSSION

Different type of catalysts on Tg

The effects of different catalysts on Tg were shown on Figure 2. As shown in Figure 2, imidazole (2E4MZ) has the highest Tg, while amines (DBU) and organic phosphorus (TPP) have relatively lower Tg. Table 1 shows the value of Tg and other performance of different catalysts. In the 2E4MZ catalyst system, the EMC has the highest Tg, the highest modulus and the lowest CTE. In the DBU catalyst system, the modulus was the lowest. In the TPP catalyst, EMC have the highest CTE (α2 above Tg) and middle modulus.

When phenolic resin cures epoxy resin, Tg and Coefficient of Thermal Expansion (CTE) of the cured product vary greatly due to different catalyst types. In the 2E4MZ catalyst system, Tg and network density increase with the increase of epoxy resin. This is because when 2E4MZ is used as a catalyst, there is an excess of epoxy resin, which undergoes reactions between epoxy groups and phenol, epoxy groups and hydroxyl groups, and polymerization reactions between epoxy groups. TPP as catalyst, epoxy group does not react with phenol, when the epoxy group is excessive, the unreacted epoxy group and phenol group increase, the crosslinking density of the cured product decreases, and the concentration of Tg and network decreases. Another reason is that the insufficient phenolic group causes the loss of TPP activity. DBU is a specific tertiary amine with the lowest Tg.

TABLE I. THE VALUE OF Tg FOR DIFFERENT CATALYSTS

Catalysts	2E4MZ	DBU	TPP
Tg(℃)	189	163	165
α1(×10-6/℃)	13	13	13
α2(×10-6/℃)	48	51	56
Storage Modulus (25℃)	20100	21700	22600
Storage Modulus (175℃)	9980	2590	3050
Storage Modulus (260℃)	1990	1380	1590

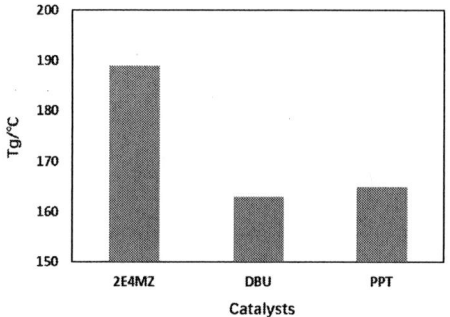

Figure 2: The effects of different catalysts on Tg

Different type of catalysts on adhesion

The impact of the different type of catalysts on adhesion were studied. Figure 3 shows the different type of catalysts on adhesion of copper (Cu) and silver (Ag). Table 2 is the value of adhesion for different catalysts. In terms of bonding performance, both 2E4MZ and DBU have excellent adhesion performance to copper, while TPP has excellent adhesion performance to silver. In general, TPP is relatively high for both copper and silver.

TABLE II. THE VALUE OF ADHESION ON CU/AG FOR DIFFERENT CATALYSTS.

Adhesion	Cu/N	Ag/N
2E4MZ	500	300
DBU	550	300
TPP	400	500

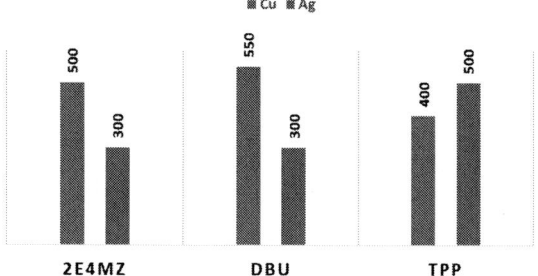

Figure 3: The effects of different catalysts on adhesion.

CONCLUSION

The Tg and adhesion to copper and silver varied with the EMC catalyst system. In addition, the modulus and CTE were also affected. This is because different catalysts have different catalytic mechanisms.

In the 2E4MZ catalyst system, EMC has the highest Tg, highest modulus, and the lowest CTE (α2 above Tg), with excellent adhesion to copper and relatively poor bonding to silver. In the DBU catalyst system, Tg and modulus are the lowest, with excellent adhesion to copper and poor adhesion to silver. In the TPP catalyst system, EMC has the lower Tg, middle modulus and the maximal CTE (α2 above Tg). For adhesion properties, the adhesion to silver is the best. Different catalyst systems can be selected according to the needs of the product.

REFERENCES

[1] W. Tan, H. Liu, Y. Duan. *The study on the mold ability and reliability of epoxy molding compound.* CSTIC, 2017.

[2] X. Cheng, Z. Wang, W. Tan, H. Liu, Y. Duan. *Study on the high reliability performance and high conductivity epoxy molding compound.* CSTIC, 2018.

[3] M. Hu. *Epoxy curing agent and curing agent.* Chemical Industry Press, 2011.

ROUGH NICKEL PPF FOR MOLD ADHESION IMPROVMENT

Wei-Gang Wu[1]*, Tsz-Chun Lo[1], Ka-Kiu So[1], Fai-Lung Ting[1] and Maria Rzeznik[2]

[1] 15 On Lok Mun Street, On Lok Tsuen, Fanling, Hong Kong SAR, China

[2] 455 Forest Street, Marlborough, MA 01752 USA

* david.w.wu@dupont.com; DuPont de Nemours, Inc.

ABSTRACT

This paper describes the development of a rough nickel plating process for palladium-preplated leadframe (PPF) that provides a consistent surface roughness at the nano- or submicron scale. This process enables outstanding adhesion with epoxy molding compound (EMC) even after Moisture Sensitivity Level-1 treatment, as well as good bonding and solderability performance. Fracture surface analysis indicated that the fracture mode of rough nickel PPF was cohesive failure inside EMC, while the interfacial failure occurred at traditional PPF and EMC interface. Here we demonstrate rough nickel PPF is an effective technique to enhance the mold adhesion for use in leadframe-based integrated circuit (IC) packages where high reliability is required.

INTRODUCTION

Since the early 1990s, palladium-preplated leadframe (PPF) with Ni/Pd/Au finishes [1-3] was introduced to the market as an alternative method to selective silver plating. The major advantages included a) a lead-free / environmental-friendly alternative; b) no tin whisker risk; c) quick turn-around time and cost reduction by omitting solder plating on outer-leads; and d) improved solderability performance, especially after heat treatment or long-term storage. The adoption of this technique, however, is hindered by the fluctuating price of noble metals and the poor adhesion force between epoxy molding compound (EMC) and the leadframe substrate, which may result in delamination, cracking or the "popcorn" phenomena, ultimately causing device failure.

Estimates [4, 5] state that there are roughly around 1000+ chips in a non-electric vehicle and at least twice as many in an electric one, assembled in various electronic control units in a car for safety, powertrain, electrical, comfort, infotainment and connectivity systems. With the rise of electric vehicles and the miniaturization of semiconductor devices, there is a strong driver for high reliability leadframe-based Integrated Circuits (IC) packages where the devices must withstand harsh environments and function correctly at least 10 years longer than consumer chips.

To meet that need, researchers had studied various methods known in the art. Such methods include improving EMC formulations [6, 7], or utilizing special leadframe designs [8, 9] and chemical bonding routes [10]. Nevertheless, it may not enable the high reliability or meet the MSL-1 compliance (moisture sensitivity level-1, 85 ℃

and 85% relative humidity for 168 hours, IPC/JEDEC J-STD-20) to achieve industry standards of adhesion between EMC and the PPF leadframe. The gold and palladium layers of PPF leadframes are substantially inert metals, so it is hard to enhance mold adhesion by modifying chemical bonding. On the other hand, no copper surface in PPF leadframes is exposed for the brown oxide or copper roughening treatment. The feasible means is to utilize a mechanical interlocking effect [11-13] by roughening the copper or nickel substrate surface. Ma [13] has reported a rough PPF leadframe by plating Ni/Pd/Au layers on the rough substrate treated by micro-etching or sandblasting processes, but the drawback is that roughened substrate would be gradually leveled with the increased nickel deposit thickness.

This paper presents the development and characterization of a rough nickel plating process for lead frames. The performance evaluation on mold adhesion improvement, wire bonding and solderability are also discussed.

EXPERIMENTAL

Sample Preparation & Characterization

The C194 Cu alloy substrates were treated with standard DuPont PPF process including electrocleaning, activation, nickel plating, palladium plating and gold plating, to obtain a multilayer structure of Ni 0.75µm/Pd 10nm/Au 3nm. All chemicals used for this development were DuPont Electronic Materials products, unless otherwise noted. Specifically, the rough nickel bath was a newly developed formulation, while Nikal™ SC Nickel Electroplating Process was used as nickel control.

The plated thickness was measured by Hitachi FT9500X XRF. The surface roughness was analyzed by an Olympus 3D Laser Microscope-LEXT OLS5000-LAF, 1D-line profile and 3D surface topology were mapped with 50x objective magnification. Surface morphology and composition were characterized by Scanning Electron Microscope (SEM) and Energy Dispersive X-ray (EDX) using a Zeiss Sigma 300, respectively.

Button Shear Test

Button shear tests were carried out according to the SEMI G69-0996 standard where the button height was 3 mm with a diameter of 3-3.5 mm, shear height used was 20% of the button = 600 µm, and shear speed was 85 µm/second. The plated substrates were cut into a 6 mm x 27 mm strip, acknowledging the mold box requirement.

Then EMC-G700LA molding compound (Sumikon bakelite Co. Ltd) was molded into a button shape, followed by the post-mold curing at 175 ºC for 6 hours. The half with button shaped molding was taken out as the as-is sample. The other half underwent exposure to MSL-1 conditions (85 ºC and 85% relative humidity for 168 hours). The button shear was done using a Nordson Dage 4000 multi-purpose bond tester.

Wire Bonding & Pull Test

A gold bonding wire with a diameter of 25 μm (Heraeus AW-14) was solder bonded onto the sample surface using a K&S manual wire bonder Model 4524. The pull test was then carried out using a Nordson Dage 4000 multi-purpose bond tester.

Contact Angle Test

Water contact angle testing was carried out using a JY-PHb contact angle analyzer. An ultrapure water droplet of 10 μL, with a resistivity of 18.2 MΩ·cm, was dropped onto the sample. Photos were taken using a built-in camera at 2x magnification.

Solderability Test

Dip-and-look solderability tests were carried out using a 6 mm x 27 mm strip. Samples were first dipped into Alpha 100 flux for 5 seconds. About 10 mm of the sample was dipped into SAC305 soldering bath for another 5 seconds.

RESULTS

Deposit Appearance & Roughness

 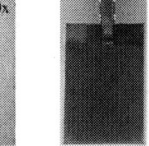 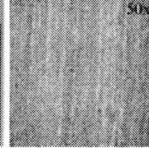

(a) Traditional Nickel Deposit   (b) Rough Nickel Deposit

Figure 1: Appearance of traditional nickel (left) and rough nickel deposits (right), and optical microscope images(50x)

Traditional nickel and rough nickel deposits were prepared under the same conditions of 10 ASD, 60 ºC and 0.75 μm in thickness. It was observed from Figure 1 that traditional nickel was semi-bright and smooth; while rough nickel was matte and blackish grey under visual inspection, appearing brownish yellow with enhanced linear pattern under optical microscope.

The line profile in Figure 2 indicates that the relative height of bare copper or traditional nickel does not fluctuate much; while rough nickel demonstrates many taller large peaks with smaller peaks developed on them.

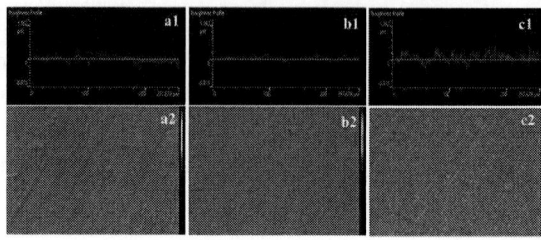

(a) Bare Copper   (b) Traditional Nickel Deposit   (c) Rough Nickel Deposit

Figure 2: 1-Dimension line profile and 3-Dimension surface topography of bare copper (left), traditional nickel (middle) and rough nickel (right) deposits

TABLE I.   PROFILE AND AREA ROUGHNESS

Sample (μm)	Profile Roughness			Area Roughness		
	Rz	*Ra*	*Rzjis*	*Sz*	*Sa*	*Sdr %*
Bare Cu	0.377	0.044	0.267	1.628	0.050	0.450
Trad. Ni	0.352	0.045	0.237	1.214	0.052	0.580
Rough Ni	0.169	0.009	0.072	1.627	0.095	7.071

Note: Rz–maximum height, Ra–arithmetic mean deviation in profile roughness, Rzjis–ten points mean roughness; Sz–maximum height, Sa–arithmetical mean height in area roughness, Sdr–developed interfacial area ratio.

The 3-D surface topography and SEM image (Figure 4) provide a more intuitive illustration on the surface morphology. As shown in Table I, bare copper and traditional nickel have similar values in most roughness parameters, suggesting that they are nearly the same in roughness. Meanwhile, rough nickel was observed to have a significant increase of Sa from 0.052 μm to 0.095 μm and Sdr from 0.58% to 7.07%, which was attributed to the cone shaped grains generated on the substrate. Moreover, it was found that rough nickel deposit roughness was controllable via the adjustment of working parameters or bath compositions.

Button Shear Test

It is generally believed that the adhesion force of bare copper to EMC is better than that of traditional PPF. The result given in Figure 3 verifies it for both as-is and MSL-1 samples. As expected, rough nickel PPF shows an excellent, unchanged adhesion force even after harsh MSL-1 treatment. As a comparison, bare copper and traditional PPF demonstrate a shear force reduction of 22% and 36%, respectively.

Figure 4 illustrates the surface morphology and cross-sectional view of EMC buttons on bare copper, traditional PPF and rough nickel PPF. The red-dotted line indicates the possible boundary between the Ni/Pd/Au deposits and the Cu substrate. It could be observed that both bare copper and traditional PPF was relatively smooth, while rough nickel PPF was covered by dense cone shaped grains with a mean peak height of 755 nm and a mean peak width of 352 nm. From the cross-sectional images, it was observed that bare Cu and rough nickel PPF were both firmly attached to EMC without any delamination. By comparison, there was a significant delamination at the interface of traditional PPF and EMC, which was probably caused by the stress from sample cutting or polishing. It suggests traditional PPF has a relatively poor adhesion to EMC.

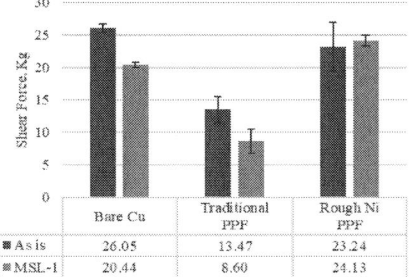

Figure 3: Button shear test results with and without MSL-1 treatment

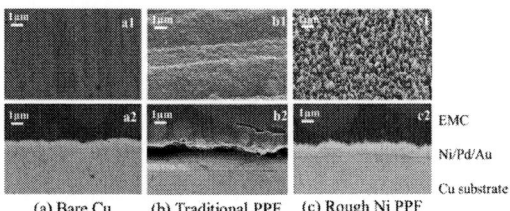

Figure 4: Surface morphology and sectional view of EMC buttons on the substrates of bare copper, traditional PPF and rough nickel PPF (5000x)

Fracture Surface Analysis

To further investigate the mechanism of mold adhesion improvement on rough nickel PPF, fracture surface analysis was conducted using SEM and EDX. After button shear the fracture surface on the substrates were divided into two regions, namely R1 and R2, which represented the blank surface and white area/black circle edge respectively. Then EDX was carried out on both R1 and R2 by full frame mode or point detection mode shown in a red crosshair. The SEM images and elemental analysis are presented in Figure 5 and Table II, respectively.

It was found that bare copper R1 was relatively clean, even under SEM inspection, the majority element was Cu. It means that EMC was completely detached from the copper substrate, an interfacial failure occurring at the interface. In R2, it looks in white color and not a few EMC residue can be observed, with increased element signals in Si, C and O. It implies that EMC was firmly attached to the substrate, hence a cohesive failure occurred.

In traditional PPF, R1 was relatively clean, where the majority element was Ni, with only trace amounts of Si, C and O found, indicating that an interfacial failure has occurred. R2 shows a thin layer toward the edge of

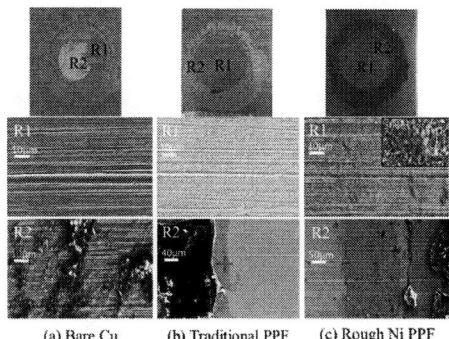

Figure 5: Optical microscope and SEM images of after button shear on the substrates of bare copper, traditional PPF and rough Ni PPF; shear direction from right to left

the circle, with the small increment of C, O and Si. Together with the extremely small area of R2, it was determined that the fracture mode was partially cohesive failure, where the adhesion of traditional PPF to EMC was poor.

In rough nickel PPF, R1 shows similar EDX readings to traditional PPF. However, black dots and white silicon balls can be seen under 5000x magnification, which is further shown by the significant increment of C, O and Si. The broad area of R2 around the circle edge exhibits high amount of C, O and Si content, indicating that a cohesive failure occurred. In conclusion, the fracture mode of rough nickel PPF was cohesive failure, thus the adhesion force to EMC was better than traditional PPF.

Functional Tests

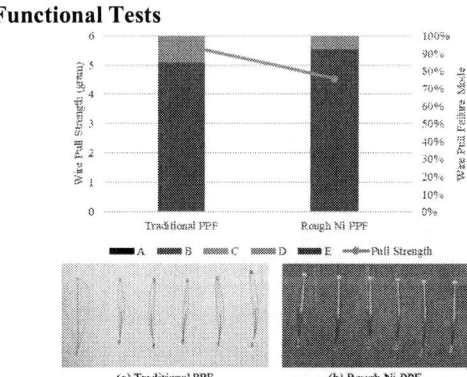

Figure 6: Au wire bonding and wire pull test results (wire pull failure mode, A: bond metal lift, B: neck break, C: span break, D: stitch break, E: wedge metal lift)

One concern of the roughening process was that it might deteriorate the bonding ability. Figure 6 shows that the gold wire was successfully bonded on rough nickel PPF, the wire pull strength did not diminish significantly, with acceptable failure modes of neck or stitch break.

TABLE II.   ELEMENTAL ANALYSIS ON FRACTURE SURFACES ON THE SUBSTRATES OF BARE COPPER, TRADITIONAL PPF AND ROUGH NI PPF

Element, wt%	Region		CK	OK	SiK	AuK	PdL	NiK	CuK	Fracture Surface
Bare Cu	blank	R1	1.86	0.72	/	/	/	/	97.43	Cu-EMC, Interfacial
	white	R2	17.4	17.3	45.08	/	/	/	20.14	EMC, Cohesive
Traditional PPF	blank	R1	2.39	0.84	0.13	4.71	1.93	82.99	7.01	PPF-EMC, Interfacial
	edge	R2	6.75	1.91	1.06	4.33	1.77	78.33	5.85	EMC, Partially Cohesive
Rough Ni PPF	blank	R1	3.03	0.91	0.17	4.92	2.46	83.72	4.78	EMC, Partially Cohesive
	edge	R2	12.55	5.18	3.68	3.23	2.01	69.23	4.13	EMC, Cohesive
	Si	R1	9.66	25.64	19.09	3.29	1.99	37.85	2.49	EMC, Cohesive

Another concern was that a roughened surface might lead to serious epoxy bleed out because of high surface energy. From contact angle tests shown in Figure 7, it was observed that the water droplet contact angle of bare copper was relatively low, representing a hydrophilic surface. Rough nickel PPF had a higher contact angle compared to traditional PPF. This implies that rough nickel PPF surface was hydrophobic in nature, which may be beneficial to reduce epoxy bleed out. There is a similar observation reported by Vivet [14] where the apparent contact angle increases when the surface roughness increases. The deep valleys formed by the cone shaped structure would trap the air between the substrate surface and spreading liquid, ultimately increasing the surface hydrophobicity.

The third concern was whether a roughened surface would reduce the solderability. From the dip-and-look test shown in Figure 7, rough nickel PPF had a comparable solderability performance to traditional PPF and slightly better than bare copper.

Figure 7: Contact angle (upper) and solderability (lower) tests on bare copper, traditional PPF and rough Ni PPF

CONCLUSION

A unique DuPont rough nickel plating process for PPF was developed to improve mold adhesion between EMC and the leadframe. It offers a matte, blackish grey and roughened deposit with a dense cone shaped structure. Button shear test verified that rough nickel PPF exhibits excellent adhesion with EMC even after MSL-1 treatment. The fracture surface analysis revealed that cohesive failure occurs at the rough nickel PPF interface, while interfacial failure occurs at the traditional PPF interface. The bondability and solderability of rough nickel PPF were not significantly affected by the roughened surface. Instead, it increased the hydrophobicity, which may be beneficial to reduce epoxy bleed out. In summary, rough nickel PPF is an effective technique to enhance the mold adhesion, for use in leadframe-based IC packages where high reliability is required.

REFERENCE

[1] D. C. Abbott, R. M. Brook, N. McLelland, and J. S. Wiley, *IEEE Trans. CHMT*, vol. 14, 1991, pp. 567.

[2] A. Murata, D. C. Abbott. *Technical Proceedings, Semicon Japan*, 1990, pp. 415.

[3] Douglas Romm, Bernhard Lange, Donald Abbott. *Texas Instruments, Application Report*. SZZA026 – July 2001.

[4] How Many Chips Are in Our Cars. *Electronics Sourcing*, May 4, 2022.

[5] Jack Ewing, Neal E. Boudette. *The New Yorker Times*, April 23, 2021.

[6] Kim, G., Hurley, J., Dhoble, A. *56th Electronic Components and Technology Conference*, 2006, pp. 1436-1441.

[7] H.J. Liu, W. Tan, L.X. Li, et al. *ECS Transactions*, vol. 60, No 1, 2014, pp. 781-786.

[8] Huang, Y.-L., Ajayan, M., Chien, B.-H. C., & Lin, W.-C. *2017 IEEE 67th Electronic Components and Technology Conference (ECTC)*, 2017, pp. 2199-2204.

[9] J. Fauty, J. Knapp, J. Yoder. *The International journal of microcircuits and electronic packaging*, vol. 25, No 1, 2022, pp. 51-79.

[10] Peng He, Matthew M. F. Yuen. *2011 IEEE 61st Electronic Components and Technology Conference (ECTC)*, 2011, pp. 651-655.

[11] Senthil Kangavel, Dan Hart, *Chip Scale Review*, 2017, pp. 22-25.

[12] Lewis Chan, Kwan Yiu Fai and Yau Chun Ho. *IEEE/ CPMT International Electronics Manufacturing Technology Conference*, 2008, pp. 1-6.

[13] Rui Ma, Mingyu Li, Lin Fang, et al. *Journal of Adhesion Science and Technology*, vol. 30, No 4, 2016, pp. 422–433.

[14] L. Vivet et al. *Applied Surface Science*, vol. 287, 2013, pp. 13-21.

ROUGH SILVER FOR IMPROVED LEAD-FRAME RELIABILITY

Fai-Lung Ting[1]*, Ka-Kiu So[1], Tsz-Chun Lo[1], Wei-Gang Wu, [1] and Maria Rzeznik[2]

[1] 15 On Lok Mun Street, On Lok Tsuen, Fanling, Hong Kong SAR, China

[2] 455 Forest Street, Marlborough, MA 01752 USA

failung.ting@dupont.com, Dupont de Nemours, Inc.

ABSTRACT

Delamination or cracks within miniaturized semiconductor devices may occur due to the weaker adhesion of the Ag-Epoxy molding compound (EMC) interface compared to the Cu-EMC interface. To enhance the reliability of Integrated Circuit (IC) packages, adhesion promotion via surface organic/inorganic modifications to the silver surface have been the subject of many studies. An alternate approach is to improve the mold adhesion by roughening the silver surface. This paper described a new electrolytic rough silver-plating process. Roughened silver deposits, controlled by adjusting plating parameters, were obtained directly without loose particles. The resultant roughness demonstrated a positive correlation to the adhesion force of Ag-EMC.

INTRODUCTION

Lead-frames are used to mount and process semiconductor die or chips in the production of semiconductor devices. They connect the chip electrically to external devices via leads. Certain types of metal deposit were introduced to the lead-frame industry, to produce spot silver/solder-coated lead-frames and palladium pre-plated lead-fames (PPF). Conventionally, silver plating is applied to the entirety or part of the surface of a copper or copper alloy lead-frame to diminish the influence of copper diffusion, thereby assuring a strong interconnection between the chip and lead-fames through wire bonds (such as Au wire or Cu wire). Once the semiconductor die or chips are mounted onto the lead-frame base, the semiconductor devices are encapsulated with a plastic Epoxy Molding Compound (EMC) to form a package. To meet the high reliability requirement, good adhesion between lead-frame base and the EMC is key to securing proper functioning of integrated circuit (IC) devices. During the lifetime of the package, ambient moisture may be absorbed at the interface between the EMC and the lead-frame base. Moisture absorption and retention inside the device results in trapped moisture which is then vaporized at elevated temperatures. This exerts tremendous internal package stress, which may lead to delamination or cracking of the package, a so-called "popcorn" effect. IPC/JEDEC defined a standard classification of moisture sensitivity levels (MSLs) of leaded IC packages for recognizing the tendency of a given package to delaminate. According to the standard (J-STD-020D), moisture sensitivity of the package was categorized into 8 levels, with the MSL increasing with the vulnerability of the package to delaminate. MSL 1 represents the highest standard where packages are to be free of delamination [1].

To improve package reliability, various processes and methods have been studied. Such methods include improving EMC formulations [2,3], utilizing special lead-frame designs [4,5] chemical bonding approaches [6], and mechanical interlocking [7-9]. However, no evidence indicates that these approaches can meet MSL-1 compliance (85⁰C/ 85% Relative Humidity/ 168 hours aging). One method to achieve a positive effect on interfacial adhesion [10-12] is by roughening the copper surface chemically or electrochemically. However, the relative area of the copper surface to silver surface exposed to the EMC has shrunk as package device miniaturization has become a trend. Therefore, Ag-EMC adhesion is more critical to achieve package reliability. Due to the weaker adhesion on Ag-EMC interfaces compared with Cu-EMC, there is a need for a method to enhance adhesion on the silver surface to EMC in semiconductor packaging.

This paper focuses on the development and characterization of a rough silver deposit obtained by directly plating rough silver on a lead-frame substrate. Performance evaluation on mold adhesion improvement and wire bonding results will be described.

EXPERIMENTAL

Sample Preparation & Characterization

The C194 Cu alloy substrates were treated with standard DuPont spot silver plating process including electro-cleaning, activation, silver strike plating, and silver plating, to obtain 3µm Ag. All chemicals used were DuPont Electronic Materials products unless otherwise noted. Specifically, the rough silver bath was a newly developed formulation, while SIlVERJET™ 220 SE Silver Electroplating Process was used as the traditional silver control.

Adhesion of plated silver deposits to Cu alloy substrates was assessed by a tape test with Scotch® 550 Transparent Tape. After removal of the tape from the plated silver deposit, the amount of residual loose particles on the tape was evaluated by visual inspection. The plated thickness was measured using a Hitachi FT9500X XRF. The surface roughness was analyzed by Olympus 3D Laser Microscope-LEXT OLS5000-LAF, 3D surface topology was mapped with 50x objective magnification.

Surface morphology was characterized using a Zeiss Sigma 300 Scanning Electron Microscope (SEM).

Button Shear Test

Button shear test was carried out according to the SEMI G69-0996 standard. The button height was 3 mm with a diameter of 3-3.5 mm, shear height was 20% of the button (equal to 600 µm), and shear speed was 85 µm/second. The plated substrates were cut into 6 mm x 27 mm strips, taking into account the mold box requirement. Then EMC-G700LA molding compound (Sumikon bakelite Co. Ltd) was molded into a button shape, followed by post-mold curing at 175°C for 6 hours. The half with button shaped molding were taken out as the As-is sample. The other half underwent the full manufacturing process. The full process samples experienced a thermal load (175°C for 2 hours and 230°C for 30minutes) after the silver electrolytic plating, and then proceeded to the molding process with post-mold curing. The molded samples were then exposed to MSL-1 conditions (85 °C and 85% relative humidity for 168 hours) followed by a reflow (3 times) process (peak temperature is 260°C). The button shear test was conducted using a Nordson Dage 4000 multi-purpose bond tester.

Wire Bonding & Pull Test

A gold bonding wire with a diameter of 25 µm (Heraeus AW-14) was solder bonded to the sample surface using a K&S Model 4524 manual wire bonder. The pull test was then carried out using a Nordson Dage 4000 multipurpose bond tester.

Contact Angle Test

Water contact angle tests were carried out using a JY-PHb contact angle analyzer. An ultrapure water droplet of 10 µL, with a resistivity of 18.2 MΩ·cm, was dropped onto the sample. Photos were taken using a built-in camera at 2x magnification.

RESULTS

Deposit Appearance & Tape Test

Using the new developed rough silver-plating process samples were prepared at 60°C with varied current densities (80ASD, 100ASD and 120ASD). Three (3.0) µm deposits were confirmed by XRF measurement. Figure 1 shows that the rough silver was a uniform white and matte deposit. No adverse impact on silver deposit adhesion, no voids, and no residuals particles were found on adhesive tape using the new developed rough silver-plating process at current densities of 80 ~ 120 ASD. Superior adhesion of the electrolytic silver deposit on copper substrate was also confirmed.

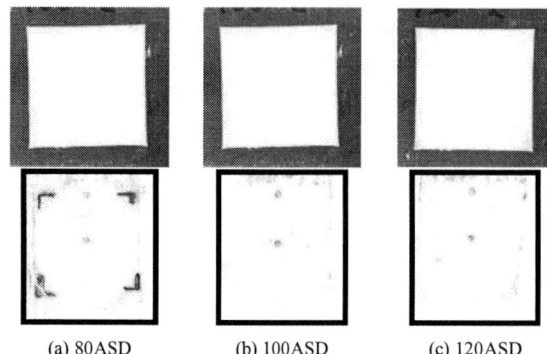

| (a) 80ASD | (b) 100ASD | (c) 120ASD |

Figure 1: Appearance and Tape Test results of direct electroplated rough Ag deposit at (a) 80ASD, (b) 100ASD, and (c) 120ASD

TABLE I. AREAL ROUGHNESS OF VARIES METAL DEPOSITS

Sample	Current Density	Areal Roughness		
		Sa (µm)	Sdr(%)	Sz (µm)
Bare Cu	---	0.071	0.95	1.37
Trad. Ag	100ASD	0.115	1.10	1.71
Rough Ag	80ASD	0.168	16.42	3.92
Rough Ag	100ASD	0.199	20.77	4.15
Rough Ag	120ASD	0.264	28.58	5.01

Note: Sa: Arithmetical mean height in area roughness; Sdr: Developed interfacial area ratio; Sz: Maximum height

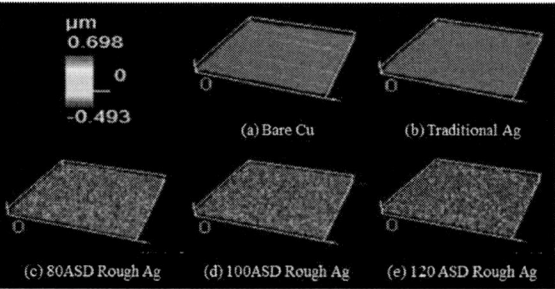

Figure 2: 3-Dimension surface topography of (a) bare copper; (b) traditional Ag and- Rough Ag deposits at (c) 80ASD; (d) 100ASD; (e) 120ASD

3-D surface topography images (Figure 2) and the roughness results in Table I show the bare copper and traditional silver deposit have similar surface profiles and values under most roughness parameters, suggesting that traditional silver itself resembles the bare copper substrate. Rough silver deposits, however, show a significant increase of Sa from 0.115 µm to 0.199 µm, and Sdr from 1.10% to 20.77%, based on 100ASD electrolytic plating. Such high roughness is contributed by the acicular shaped grains confirmed by the surface morphology inspection (Figure 3). Meanwhile, it is observed that the higher the current density, the higher the roughness, which was shown in table I, that Sa of plated rough silver at 80ASD, 100ASD, and 120ASD are 0.168 μm, 0.199 μm and 0.264 μm, while Sdr are 16.42%, 20.77% and 28.58% respectively. It illustrated that the roughness of the rough silver deposit can be controllable via the adjustment of current density.

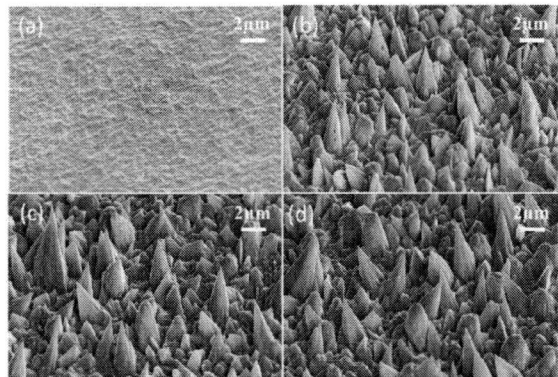

Figure 3: Surface morphology images (Tilt 52⁰, 5000x) of (a) Traditional Ag deposit, Rough Ag deposit at (b) 80ASD; (c) 100ASD; and (d) 120ASD

Button Shear Test

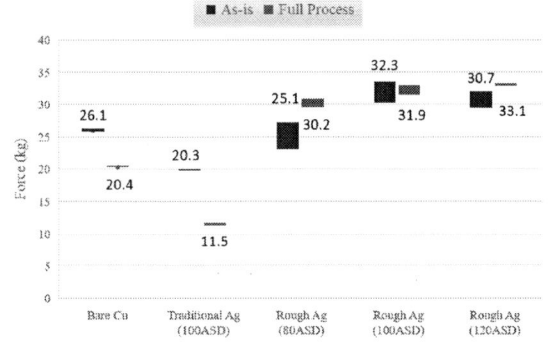

Figure 4: Button Shear Test results among surface of bare Cu, traditional Ag, and rough Ag at 80ASD, 100ASD and 120ASD

Molded samples were subjected to a reflow process (3 times) heating at 260⁰C peak temperature after absorbing dampness from a humid environment during MSL-1 moisture aging to simulate the moisture that may trigger cracking between silver deposits and the EMC, known as the "popcorn effect", posing a risk on interfacial adhesion. To investigate the influence of MSL-1 moisture aging process on the adhesion of silver deposits and EMC, button shear testing was utilized to quantify the amount of shear force required to separate the molded EMC on plated silver deposits, as well as the bare copper substrate. As shown in Figure 4, for as-is samples, traditional Ag shows weaker adhesion force than the copper surface, while rough silver deposits provide an improved adhesion force to EMC and compared to the copper surface, especially for the rough silver deposits plated by 100ASD and 120ASD. With the full process treatment, which includes MSL-1 moisture aging process, and reflow process, the adhesion force between bare Cu and EMC has been reduced from 26.1kg to 20.4kg (~22% reduction), and between traditional Ag and EMC has been reduced from 20.3kg to 11.5kg (~44% reduction). However, rough Ag at current densities between 80ASD to 120ASD showed excellent adhesion sustainability with the adhesion force maintained at or above 30kg. Hence, the plated rough silver deposit enhances high reliability performance. Interestingly, it is also observed that there is a positive correlation between the adhesion force to molding compound and lead-frame surface roughness as displayed in Figure 5.

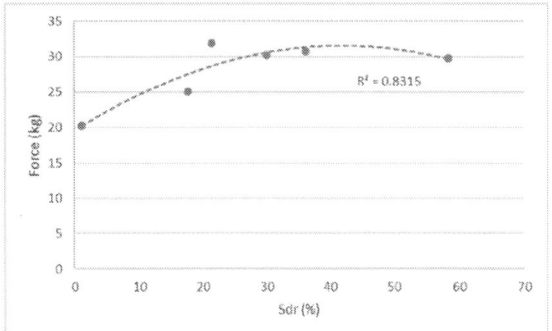

Figure 5: Correlation between Adhesion Force to EMC and Lead-frame Surface Roughness

It can be verified that rough silver deposits generated from the new silver electrolyte are promising to achieve high package reliability and reduce the risk of delamination.

Functional Tests

One concern of the roughening process is that it might deteriorate the bonding ability. Figure 6 shows that gold wires can be successfully bonded to rough Ag. At current densities between 80 to 120ASD, the wire pull strength does not decrease appreciably, with acceptable failure modes of neck or stitch break.

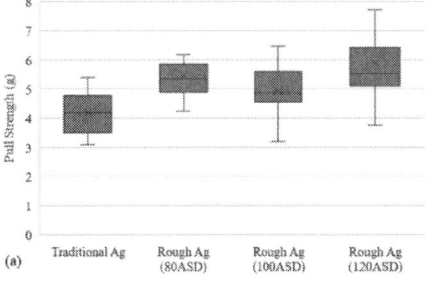

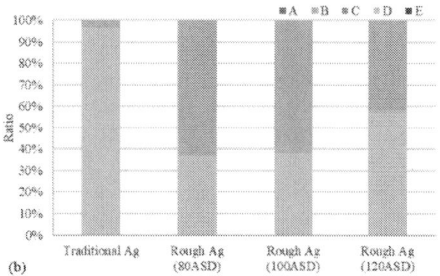

Figure 6: Au wire bonding performance of plated Ag deposit on (a) Pull Strength Results, and (b) Failure Mode Results (A: bond metal lift; B: neck break; C: span break; D: stitch break; E: wedge metal lift)

Another concern is that a roughened surface might lead to a serious epoxy bleed out (EBO) due to the high surface energy. Contact angle testing shown in Figure 7 illustrates the water droplet contact angle on rough silver deposits have a significant increase compared to traditional Ag or bare copper. Higher contact angles indicate that the rough Ag surface exhibits hydrophobic behavior, which is potentially beneficial on reducing EBO. There is a similar observation reported by Vivet [13] that the apparent contact angle value increases when the surface roughness increases. The deep valleys formed by the aciculate structure traps the air between the substrate surface and spreading liquid, ultimately to increase the hydrophobicity.

Figure 7: Contact angle test on (a) bare Cu, (b) traditional Ag and (c) rough Ag deposit

CONCLUSION

A recently developed high-speed electrolytic rough silver-plating process for lead-frames is a one-step silver electrodeposition that provides highly adhesive, controllable rough silver deposits. It offers a white matte and roughened deposit without loose particles on the deposit surface. Mechanical interlocking effects are provided by the aciculate structure of the silver grain morphology and strengthens the adhesion between the silver coated lead-frame and molding materials. Such roughened silver surface showed an insignificant effect on the wire bonding ability. The higher hydrophobicity may be beneficial to reduce epoxy bleed out. MSL-1 moisture aging and reflow processes showed less significant effect on the adhesion ability. In summary, rough silver deposits are an effective technique to enhance the mold adhesion,

for use in lead-frame-based IC packages where high reliability is required.

ACKNOWLEDGEMENTS

The authors would like to thank Dr. Jeffery Lo and the team in EPACK LAB of Hong Kong University of Science and Technology for their assistance in the molding process and button shear test.

REFERENCES

[1] IPC, "JOINT IPC/JEDEC STANDARD FOR MOISTURE/ REFLOW SENSITIVITY CLASSIFICATION FOR NONHERMETIC SURFACE-MOUNT DEVICES," 2007.

[2] Kim, G., Hurley, J., Dhoble, A. *56th Electronic Components and Technology Conference*, 2006, pp. 1436-1441.

[3] H.J. Liu, W. Tan, L.X. Li, et al. *ECS Transactions*, vol. 60, No 1, 2014, pp. 781-786.

[4] Huang, Y.-L., Ajayan, M., Chien, B.-H. C., & Lin, W.-C. *2017 IEEE 67th Electronic Components and Technology Conference (ECTC)*, 2017, pp. 2199-2204.

[5] J. Fauty, J. Knapp, J. Yoder. *The International journal of microcircuits and electronic packaging*, vol. 25, No 1, 2022, pp. 51-79.

[6] Peng He, Matthew M. F. Yuen. *2011 IEEE 61st Electronic Components and Technology Conference (ECTC)*, 2011, pp. 651-655.

[7] Senthil Kangavel, Dan Hart, *Chip Scale Review*, 2017, pp. 22-25.

[8] Lewis Chan, Kwan Yiu Fai and Yau Chun Ho. *IEEE/ CPMT International Electronics Manufacturing Technology Conference*, 2008, pp. 1-6.

[9] Rui Ma, Mingyu Li, Lin Fang, et al. *Journal of Adhesion Science and Technology*, vol. 30, No 4, 2016, pp. 422–433.

[10] C. W. B. W.-G. N. Lam, Europe Patent EP1820884A1, 22 August 2007.

[11] Y. F. K. C. C. H. Y. C. Lee, United States Patent US7691679B2, 6 April 2010.

[12] Y. F. K. C. C. C. C. Lee, United States Patent US8012886B2, 6 September 2011.

[13] L. Vivet et al. *Applied Surface Science*, vol. 287, 2013, pp. 13-21.

PRINTABLE COPPER SINTERING PASTE FOR HIGH-POWER DIE-ATTACH APPLICATION

Li Ma[1]*, Hongyun Li[1], Min Yao[1]*, Fen Chen[1], Xuelian Han[1], Yan Liu[1]
[1]Indium Corporation (Suzhou), Jiangsu, China
*Corresponding Author's Email: mma@indium.com; dyao@indium.com

ABSTRACT

Low-temperature copper (Cu) sintering paste with high thermal and electrical conductivity is promising for die-attach applications in power electronics and third-generation semiconductor devices. Large-size die-attach applications for IGBT modules request high-efficiency printing transfer processes. This paper details the influence of printing process parameters to obtain high-volume and flat Cu paste deposition. Ideal paste shape and flatness were achieved with an 8 kg squeegee pressure and squeegee speeds less than 30 mm/s. Flux vehicle type and squeegee blade angle also affected the printing process, the optimal blade angle is 45°. The smaller blade angle induced a much higher vertical force on paste, and lower squeegee speed gave sufficient time to deposit in the aperture, all which caused the paste to be evenly distributed in the stencil apertures. The pressure-assisted Cu sintering paste achieved a high shear strength above 50 MPa for 3 x 3 mm joints and above 30 MPa for 5 x 5 mm joints, each of which being suitable for multiple metallized substrates such as gold (Au), silver (Ag) and Cu. The printable Cu paste succeed in the challenge of oxidation susceptibility.

Keywords—Copper sintering; Print; Die attach; shear strength; bonding material

INTRODUCTION

As high power devices are being widely demanded in numerous fields such as electric trains, hybrid and electric vehicles, elevators, and renewable energy systems, as well as in many consumer electronics. Silicon carbide (SiC) and gallium nitride (GaN) were developed to replace silicon (Si) as their advantages in term of higher voltage blocking, higher switching frequency for new high power modules which could be leveraged at high temperatures [1]. The conventional Tin-based solders could no longer meet the requirement as die-attach joint which attributes to their limited efficiency and operation temperature.

Based on the technological requirement, silver (Ag) sintering materials were developed to replace Sn-based solders as die-attach materials for new high-power modules [2]. Sintered Ag bonding often demonstrated higher shear strength and an improved resistance to thermal fatigue and creep than Tin-based solders. However, the material has two significant limitations: high material costs and electrochemical migration. Furthermore, Ag sinter paste showed lower shear strength on copper (Cu) and nickel (Ni) surfaces. Cu sinter material was then researched to improve upon the limitations of an Ag-based paste [3, 4, 5].

Responding to these new application demands, we have developed both pressure-less and pressure-assisted Cu sintering pastes with high thermal and electrical conductivity. Papers detailing the pressure-less and pressure-assisted Cu sintering pastes were reported at ECTC 2018 and ECTC 2022, respectively. Both printing and dispensing are common paste deposit methods for solder materials in Surface Mount Technology (SMT) which also leveraged to sinter materials. Printing exhibits higher process efficiency compared to dispensing due to the particle size of Ag or Cu in the sinter material. These particles are nanometers to several microns in diameter, which is much smaller than solder particles. Therefore, the parameters and processes designed for solder paste would not be appropriate for sinter material. In this paper, we show the appropriate printing parameters and processes necessary to obtain an ideal deposit shape and high shear strength in Cu sintering paste, including squeegee pressure, squeegee speed, squeegee fixture design and flux vehicle type.

EXPERIMENT DESIGN

Materials

The pressure Cu sintering paste was prepared by mixing Cu particles with different flux vehicles. The Cu particles are commercially available in the range of 1–20 μm. A specially designed flux vehicle, which resisted the oxidation of the Cu paste during the storing, printing, and sintering processes, was mixed with the Cu particle to obtain Cu sinter paste. Figure 1 demonstrates the paste packed in jar. The names of the tested materials are noted as Cu-FA and Cu-FB, which represent Cu particles mixed with flux A and flux B, respectively. Figure 2 displays the test board used for printing, which had 9 pads of size 5 x 5 mm.

Figure 1: Pressure Cu sintering paste in jar

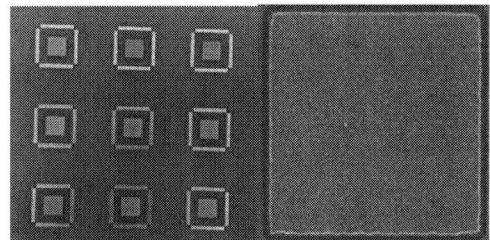

Figure 2: Test board for printing (left: the whole view of 9 pads, right: the pad with printed Cu paste)

Printing Equipment and Parameters

A DEK Horizon printer was used to investigate the influence of print parameters on the Cu sinter paste. Tested parameters included squeegee pressure, print speed, squeegee blade angle, separation distance, and separation speed. Each test board was examined by a KOH Young SPI. The photograph of deposited Cu sinter paste was taken by a Keyence Microscope.

Printing Test Procedure

The printing procedure is described as below:
1) Hand-mix Cu sinter paste for two minutes
2) Establish and input printer parameters
3) Print Cu sinter paste with the stencil (100 um thickness)
4) Examine paste deposition with KOH Young SPI
5) Photograph each pad of deposited paste with the Keyence Microscope

Design of Experiment (DOE) for Printing Parameters

TABLE I.   DESIGN OF EXPERIMENT FOR PRINTING

Squeegee Pressure	Print Speed	Squeegee blade angle	Separation distance	Separation Speed
6 kg	20 mm/s	60°	1 mm	2 mm/s
8 kg	30 mm/s	45°	2 mm	5 mm/s
9 kg	50 mm/s	—	5 mm	7 mm/s

RESULTS
Squeegee Pressure

TABLE II.   EFFECT OF SQUEEGEE PRESSURE WITH CU-FA PASTE

Squeegee pressure	6 kg	8 kg	9 kg
Squeegee speed	30 mm/s	30 mm/s	30 mm/s
Blade angle	60°	60°	60°
Separation distance	2 mm	2 mm	2 mm
Separation speed	2 mm/s	2 mm/s	2 mm/s
Printing results			

Table 2 demonstrates the effect of squeegee pressure on deposited Cu sinter paste shape with Cu-FA paste. A squeegee pressure of 8 kg obtained high quality deposition. Increasing squeegee pressure to 9 kg resulted in solder paste being scooped away, yielding a lower solder deposition. The 6 kg squeegee pressure exhibits "dog ears" of Cu paste deposition, with an unevenly deposited surface potentially caused by the paste not adhering to the substrate.

Squeegee Speed

TABLE III.   EFFECT OF SQUEEGEE SPEEDS WITH CU-FA PASTE

Squeegee pressure	8 kg	8 kg	8 kg
Squeegee speed	20 mm/s	30 mm/s	50 mm/s
Blade angle	60°	60°	60°
Separation distance	2 mm	2 mm	2 mm
Separation speed	2 mm/s	2 mm/s	2 mm/s
Results			

Table 3 shows the effects of squeegee speed on the deposition shape with Cu-FA paste. Flat paste deposition was obtained at squeegee speeds between 20-30 mm/s. At higher squeegee speeds (e. g. 50 mm/s), the Cu paste did not have sufficient time to deposit in the aperture and form a flat surface, again resulting in a dog-ear appearance and an uneven deposition surface. Therefore, the die could not connect to the Cu paste and thereby negatively contributed to sinter performance.

Blade angle

TABLE IV.   EFFECT OF BLADE ANGLE WITH CU-FA PASTE

Squeegee pressure	8 kg	8kg
Squeegee speed	30 mm/s	30 mm/s
Blade angle	45°	60°
Separation distance	2 mm	2 mm
Separation speed	2 mm/s	2 mm/s
Printing results		

The squeegee blade angle affects the vertical force applied on the paste. Smaller angles result in greater forces. Table 4 illustrates the influence of the vertical force, which was tested by applying blades with SMT industry standard angles of 45° and 60°. The deposition shape of Cu paste exhibited better flatness with the 45° blade angle. The smaller blade angle induced a much higher vertical force, which caused the paste to be evenly distributed in the stencil apertures.

Separation distance

TABLE V.   EFFECT OF SEPARATION DISTANCE WITH CU-FA PASTE

Squeegee pressure	8 kg	8 kg	8 kg
Squeegee speed	30 mm/s	30 mm/s	30 mm/s
Blade angle	45°	45°	45°
Separation distance	5 mm	2 mm	1 mm
Separation speed	2 mm/s	2 mm/s	2 mm/s
Printing results			

Separation speed is defined as the lift-down speed of the PCB from the stencil immediately following the completion of the printing process. Table 5 shows that the separation speed affects the deposition of the Cu paste. A separation distance of 2 mm produced a flatter deposition shape. The lower separation distance of 1mm resulted in an uneven deposition layer with a toothed edge, while the higher separation distance of 5 mm led to a thick deposition layer with the paste stream on the surface.

Separation speed

TABLE VI.   EFFECT OF SEPARATION SPEED WITH CU-FA PASTE

Squeegee pressure	8 kg	8 kg	8 kg
Squeegee speed	30 mm/s	30 mm/s	30 mm/s
Blade angle	45°	45°	45°
Separation distance	2 mm	2 mm	2 mm
Separation speed	1 mm/s	2 mm/s	5 mm/s
Printing results			

The speed of separation between stencil and test board after printing is also critical to achieve an ideal deposition shape. Much higher separation speeds result in paste clogging in or near the stencil apertures. Higher speeds also result in tailing and high edges around the paste deposits [6]. Table 6 demonstrates the effect of separation speed on the Cu-FA paste. A separation speed of 2 mm/s produced an improved surface flatness for the Cu sinter paste, whereas 5 mm/s caused high edges around the paste deposit. These edges induced overflow after the die attached for wet process sintering and die cracking post-attachment for dry process sintering.

Summary of the printing parameters

Using the DOE employed in this paper, the printing parameters for this novel Cu sintering paste were optimized to maximize deposition shape flatness. The optimal parameters are a squeegee pressure of 8 kg, squeegee speed of 30 mm/s, squeegee blade angle of 45°, separation distance of 2 mm, and a separation speed of less than or equal to 2 mm/s.

Effect of different flux vehicles of Cu sinter paste

Flux vehicles developed for the Cu powder also impacted printing performance. Table 7 demonstrates the printing performance of Cu-FA and Cu-FB. Significantly, Cu-FA can be rolled well with the optimized print parameters, while Cu-FB paste cannot be rolled during print and lacks sufficient deposit quantity on the substrate. Furthermore, the paste Cu-FB on stencil cannot be easily cleaned.

TABLE VII. COMPARISON OF CU-FA AND CU-FB PASTE

Paste type	Paste on stencil	Deposit shape
Cu-FA		
Cu-FB		

The Cu paste deposit height is inspected by a KOH young Meister-S to evaluate Cu-FA and Cu-FB print performance. Cu-FA paste consistently achieved a higher deposit height than Cu-FB (Figure 3). Figure 4 shows the coplanarity of deposited Cu paste, which was also examined by the KOH young Meister. The color in the 3D view presents the paste deposit height, similar colors means similar heights. The resulting of Cu-FA paste performed better overall and was flatter than Cu-FB, with a height concentrated at about 90 um across the pad. Cu-FB paste deposit height ranged from 60 um in the center to 90 um at the edge, producing an unfavorable heterogeneity.

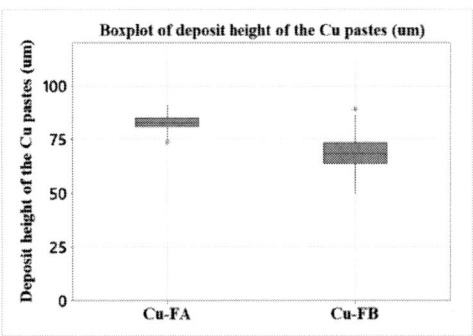

Figure 3: Deposit height from KOH young inspection

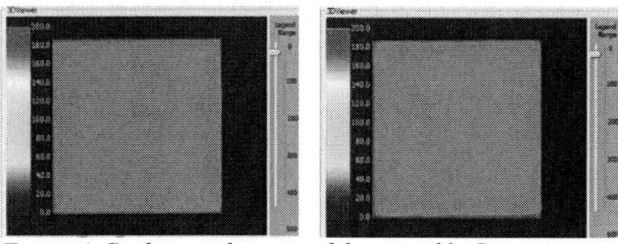

Figure 4: Surface coplanarity of the printable Cu paste
(Left: 3D overview of printed Cu-FA; Right: 3D overview of printed Cu-FB)

Shear strength of the printable Cu paste

For shear strength testing, sintered Cu joint samples were prepared using the printable Cu-FA paste. Details of the die, substrate, and sintering profile are presented in Table 8. The sintered Cu joint shear strength was evaluated by implementing XYZ Condor Sigma after the sintering process concluded.

TABLE VIII.   SHEAR STRENGTH OF THE PRINTABLE CU-FA PASTE

Material	Printable Cu-FA sintering paste	
Die	Cu test die	3 x 3 mm 5 x 5 mm
Substrate	Bare Cu	23.5 x 23.5 mm
	ENIG	23.5 x 23.5 mm
Pressure Sintering	Pre-dry	100°C / 10 min / N2
	Sintering	260°C / 5 min / 15MPa / N2
Shear Test	Cu die 3 x 3 mm / bare Cu substrate	60 MPa
	Cu die 5 x 5 mm / bare Cu substrate	30 MPa
	Cu die 3 x 3 mm / ENIG substrate	40 MPa

The shear strength test results (Table 8) demonstrate that the Cu-FA paste can be sintered well with Cu dies on both bare Cu substrates and Au substrates. The shear strength was measured up to 60 MPa with a 3 x 3 mm Cu die and 30 MPa with a 5 x 5 mm Cu die. Each was subjected to a 15 MPa sintering pressure. Sintering pressure reinforces the contact between adjacent Cu particles and between Cu particles to both die and substrates. The SEM-captured image of the cross-sectioned sintered joint microstructure (Figure 5) shows there is no large peeling or voids observed in the bonding area. There was also no deterioration of the sintered joint with the 3 x 3 mm Cu die. A relatively lower shear strength of 40 MPa was measured with a 3 x 3 mm Cu die on the Au surface substrate. This is a result of the micromechanical structure differences between the two substrates. The similar lattice matching between the Cu paste and the Cu (lattice constant: 0.3615 nm) substrate helps to reinforce the sintering joint [7, 8]. Conversely, the Au substrate has a different lattice geometry (lattice constant: 0.4078 nm) which is less compatible with Cu sintering paste.

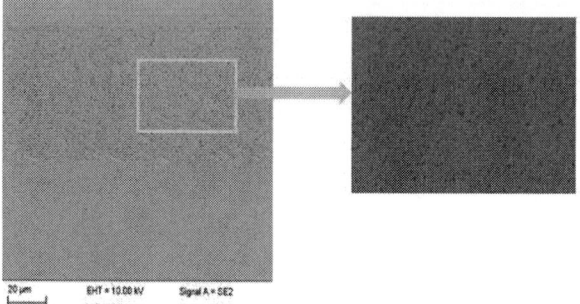

Figure 5: The microstructure of cross-sectioned sintered joint
(Left: original SEM capture; right: enlarged view)

CONCLUSION

To improve transfer efficiency, a printable Cu sintering paste was developed for large-size die-attach application for IGBT modules. By using optimal printing parameters, this printable Cu sintering paste (Cu-FA) can be printed with a flat surface, good deposit shape and proper deposit height corresponding to the stencil thickness. Moreover, a high shear stress of 50 MPa was obtained for paste with a Cu substrate and 40 MPa for paste with an ENIG substrate. The pressure Cu sintering paste is appropriate for different metallized substrates such as gold (Au), Ag and Cu. The printable Cu paste has succeed in the challenge of oxidation susceptibility, it can be applied in printing technology as a sinter material.

REFERENCES

[1] H. S. Chin, Y. Cheong, A. B. Ismail, Metall. Mater. Trans. 41B, "Treatise on Electricity and Magnetism", 3rd ed., vol. 2. Oxford: Clarendon, 1892, pp.68–73,824-832,2010J. Clerk Maxwell.

[2] S. J. Krumbein, "Metallic electromigration phenomena," 33rd Meeting of the IEEE Holm Conference on Electrical Contacts, September 1987K. Elissa.

[3] Junjie Li, Tielin Shi, Xing Yu, Chaoliang Cheng, Jinhu Fan, Guanglan Liao,Zirong Tang* "Low-temperature and low-pressure Cu-Cu bonding by pure Cu nanosolder paste for wafer-level packaging". 2377-5726/17 © 2017 IEEE DOI 10.1109/ECTC. 2017. 35.

[4] Dai Ishikawa, Hideo Nakako, "Copper Die-Bonding Sinter Paste: Sintering and Bonding Properties", 978-1-5386-6814-6/18 2018 IEEE.

[5] Yang Zuo, Jun Shen. "Effect of different sizes of Cu nanoparticles on the shear strength of Cu-Cu joints", Volume 199, 15 July 2017, Pages 13-16 Material Letters.

[6] Sekharan Gopal, Jafri Mohd Rohani, Sha'Ri Mohd Yusof, Zailis Abu Bakar, "Optimization of Solder Paste Printing Parameters Using Design Of Experiments (DOE)", Jurnal Teknologi, 43(A) Dis. 2006: 11–20.

[7] T. Furukawa, M. Shiraishi, Y. Yasuda, A. Konno, M. Mori, T. Morita, S. Watanabe, T. Arai, M. Nakamura, D. Kawase, "High power density side-gate HIGT modules with sintered Cu having superior high-temperature reliability to sintered Ag", IEEE (2017) 263–266, ttps://doi.org/10.23919/ISPSD.2017.7988910.

[8] Y. Kobayashi, T. Shirochi, T. Maeda, Y. Yasuda, T. Morita, "Microstructure of metallic copper nanoparticles/metallic disc interface in metal–metal bonding using them", Surf. Interface Anal. 45 (2013) 1424–1428, https://doi.org/10.1002/ sia.5299.

INTEGRATING HIGH FREQUENCY RADAR CHIP USING LAMINATED SUBSTRATE TRANSITIONS FOR SYSTEM-IN-PACKAGE DESIGN

Zhiqiang Fang and Boping Wu
JCET Group Co., Ltd., Shanghai 201206, China
Email: johnny.fang@jcetglobal.com, boping.wu@jcetglobal.com

ABSTRACT

This paper presents a novel transition method for interconnecting high-frequency chips within the 90-110 GHz frequency range using planar laminate substrate in advanced packaging. The transitions were designed to connect 50-Ohm grounded coplanar waveguide outputs, and were assembled and characterized using a millimeter-wave chip-scaled package (CSP) and laminated stack-up. Compared to traditional front-end RF pads, the transitions showed minimal performance degradation and energy loss. Each transition was optimized to support an operational bandwidth, with an average insertion loss of 1.5dB and return loss of 15dB. A 3D electromagnetic simulation software was used to optimize the design for the 100 GHz F-band. An air cavity lid could be added for additional protection. The generic and reciprocal transition design could also be used for packaging other electronic or optical components, and coupling electromagnetic waves from the WR to a GCPW.

INTRODUCTION

Ultra-broadband interconnects for chip-packages operating up to 350 GHz are feasible today [1]. This is achieved by utilizing advanced packaging technologies that minimize interconnect dimensions and electrical length. A prerequisite is to use a comprehensive solution that allows for accurate structure features and sizes. The flip-chip concept offers the best bandwidth and heterogeneous integration approaches, combining several semiconductor technologies in the same package. This can achieve similar and better performance for inter-chip communications [2]. For bandwidths above the 100 GHz, various packaging prototypes using chip scaling and integrated waveguide are possible. A broadband on-chip matching network is a key element in maintaining the intrinsic performance of signaling interconnects for high frequency radar applications. With these optimized RF transitions, compact waveguide-in-package can be mounted directly onto evaluation boards for ultra-high frequency circuit integrations. Multi-chip modules (MCM) or system-in-package (SiP) with numerous interconnect circuits could be produced at millimeter-wave frequencies over large bandwidths without the need for specialized PA chip and wire bond technology.

GCPW-TO-WR SLOT TRANSITION

This substrate design for a transition utilizes a double-slot antenna structure and is based on ultra-thin laminated package technology, as depicted in Figure 1. The initial step involves etching the double-slots at the back-side ground plane of the substrate (HL972LF, 150um of dielectric thickness), illustrated in Figure 2(b). The RF signal is then coupled into the rectangular waveguide directly through these slots. The transition also includes two matching-stubs placed on top of the slots, as shown in Figure 2(a). The precise placement of these matching-stubs over the coupling-slots reduces radiation loss and improves the coupling efficiency from the grounded coplanar to the rectangular waveguide.

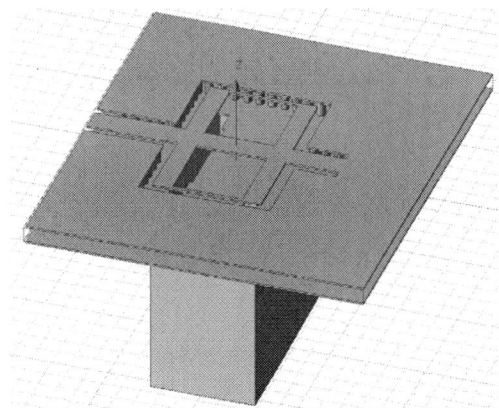

Figure 1: A planar grounded coplanar waveguide to rectangular waveguide (GCPW-WR8) transition in SiP.

The Theoretical Mechanism of the Structure

The double-slot structure's symmetry ensures that the power is split equally between both stubs, resulting in an even field distribution in both directions: the GCPW input port and the short-circuited stub port. Waves moving towards the stubs' center are out of phase and cancel each other out, resulting in minimal E-field at the center point. Short-circuited stubs are advantageous over open-circuited ones because they are easier to implement and less prone to the electromagnetic radiation leakage. The transition was further optimized by adding a shorter notch to the coupling slots towards the center of the rectangular waveguide and a capacitive load of ~45 Ω GCPW line before the antenna structure for improved matching. The transitions were numerically analyzed and optimized using a commercial full-wave 3D electromagnetic field simulation tool based on the finite element method (ANSYS HFSS).

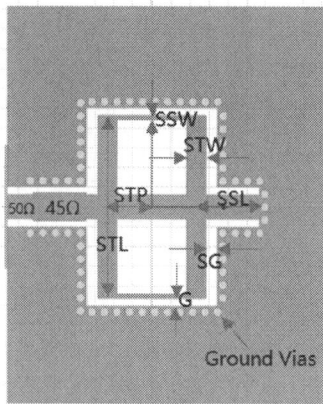

(a) Top view

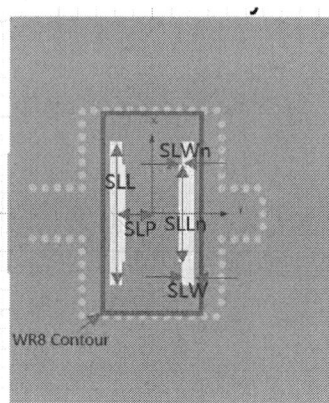

(b) Bottom view

Figure 2: Package design of a 100GHz (GCPW-WR8) waveguide transition.

The Optimizing Method and Analytical Discussion

To design high-frequency and wideband millimeter-wave circuitry, it is important to use ultra-thin substrate dielectric to maintain low dispersion for GCPW-WR8 transitions. The simulation of driven modal considers the low dielectric loss tangent of the substrate for both the GCPW port and the WG port. An insertion loss (IL) in the range of about 1.5 dB and a return loss (RL) of about 15 dB have been achieved from 90 GHz to 110 GHz, as shown in Fig. 5. The translated effective wavelength ranges from 2.7mm to 3.3mm, with the center operating wavelength at 3mm. This SiP design surpasses the coupling bandwidth of the single probe-feeding transition built on a PCB presented in [3].

Dispersion can occur in non-homogeneous interfaces, such as GCPW or micro-strip transmission-line (TL),

resulting in a non-linear propagation constant. As the frequency changes, energy will propagate at different group velocities, causing different arrival times for the energies at different frequencies. It is important to minimize the deviation of the phase constant as a function of the radiating frequency (group delay). The thicker the substrate, the more inhomogeneous the TL will be, thus resulting in a more dispersive signaling performance [4].

The use of a double-slot structure on the ground plane of the substrate can increase the coupling bandwidth and overcome the poor return loss problem that is caused by using a single-slot structure, as shown in Fig. 3(a). By accurately placing the matching stubs over the coupling-slots, radiation loss can be decreased and the coupling efficiency from the coplanar to the rectangular waveguide can be improved. Additionally, metalized vias are used to surround the transition structure, as shown in Fig. 3(b), in order to prevent surface mode propagation and energy leakage [5]-[7].

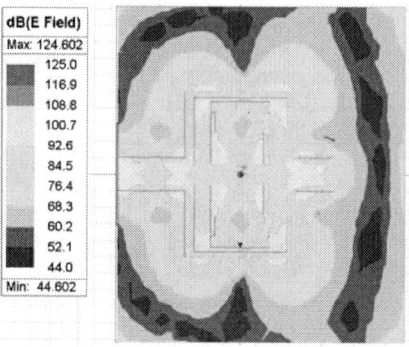

(a) Top view

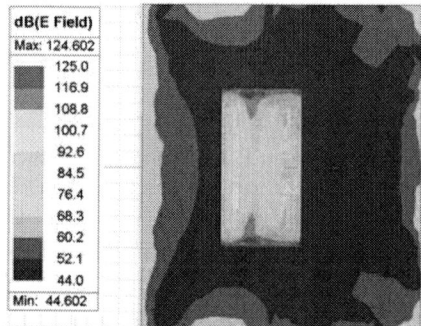

(b) Bottom view

Figure 3: Electric field distribution at 100GHz, Via diameter is 90μm, pitch distance is 130μm

The mode transition takes place from quasi-TEM mode on planar micro-strip line to TE10 mode in waveguide as depicted in Fig. 4. However, due to the reversed phase of wave propagation of E-field in the two slots, the rectangular TE10 mode is excited instead of the other modes, while the mixed TM mode only occurs at the transition area. By properly sizing the slots and patches to the optimal values shown in Table I, the phase shift is adjusted, resulting in the desired wave propagation of the fundamental TE10 mode into the rectangular waveguide.

TABLE I.   DESIGN PARAMETERS OF THE GCPW-WR8 TRANSITION.

Parameter	Value (um)	Description	
SLL	1500	slot	length
SLW	144		width
SLP	460		position
SLLn	1019	slot notch	length
SLWn	33		width
STL	1856	stub	length
STW	202		width
STP	466		position
SSL	672	stub short	length
SSW	40		width
SG	120	stub gap	
G	80	GCPW	

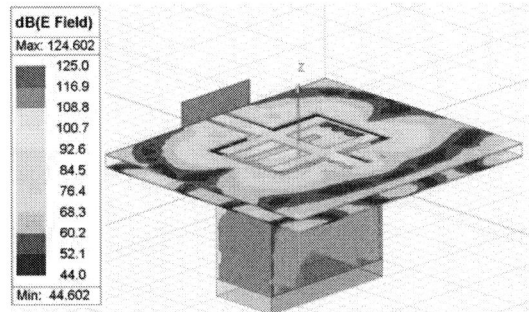

Figure 4: 3D Electric field distribution at 100GHz.

Metallized vias are positioned in the transition region to ensure a solid connection between the top and bottom grounds, creating effective shielding. These surrounding vias are placed to allow for a broad transition from grounded coplanar waveguide to micro-strip line, while also facilitating heat dissipation to the bottom metal ground.

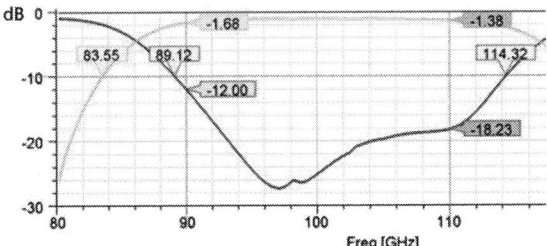

Figure 5: S-parameters of the GCPW-WR8 transition design at 100 GHz. (substrate DK =3.38, DF =0.004)

CONCLUSIONS

The GCPW to WR8 package substrate transition for F-band applications has been developed using the planar Mitsubishi HL972LF-(LD) BT laminate. The transition offers an average insertion loss of -1.5 dB and return loss of -15 dB, which have been precisely optimized through 3D electromagnetic modeling. The chip-to-waveguide transition structure provides ultra wideband performance between 90~110GHz and is less susceptible to assembly and fabrication issues. This generic and reciprocal design demonstrates the use of high-frequency radar chips at 100 GHz range using planar laminated substrate in advanced packaging.

ACKNOWLEDGEMENTS

We would like to express our gratitude to JCET colleagues who provided technical discussions and support throughout our research.

REFERENCES

[1] T. Tajima, H. -J. Song and M. Yaita, "Design and Analysis of LTCC-Integrated Planar Microstrip-to-Waveguide Transition at 300 GHz," *IEEE Trans. Microw. Theory Techniques,* vol. 64, no. 1, pp. 106-114, Jan. 2016.

[2] Y. Zhang, D. Zhao and P. Reynaert, "A Flip-Chip Packaging Design with Waveguide Output on Single-Layer Alumina Board for E-Band Applications," *IEEE Trans. Microw. Theory Techniques*, vol.64, no.4, pp.1255-1264, April 2016.

[3] B. Khani, S. Makhlouf, A. G. Steffan, J. Honecker and A. Stöhr, "Planar 0.05–1.1 THz Laminate-Based Transition Designs for Integrating High-Frequency Photodiodes with Rectangular Waveguides," *Journal of Lightwave Technology*, vol.37, no.3, pp.1037-1044, 1 Feb, 2019.

[4] A. Hassona *et al.*, "Nongalvanic Generic Packaging Solution Demonstrated in a Fully Integrated D-Band Receiver," *IEEE Trans. Terahertz Science and Technology*, vol.10, no.3, pp.321-330, May 2020.

[5] W. Heinrich *et al.*, "Connecting Chips with More Than 100 GHz Bandwidth," *IEEE Journal of Microwaves*, vol.1, no.1, pp.364-373, Jan. 2021.

[6] C. Wang, X. Yi, M. Kim, Q. B. Yang and R. Han, "A Terahertz Molecular Clock on CMOS Using High-Harmonic-Order Interrogation of Rotational Transition for Medium-/Long-Term Stability Enhancement," *IEEE Journal of Solid-State Circuits*, vol.56, no.2, pp. 566-580, Feb. 2021.

[7] X. Yi *et al.*, "Emerging Terahertz Integrated Systems in Silicon," *IEEE Trans. Circuits and Systems I: Regular Papers*, vol.68, no.9, pp.3537-3550, Sept. 2021.

ELECTROMAGNETIC INTERFERENCE SHIELDING SOLUTION FOR SYSTEM-IN-PACKAGE

Lihong Liu [1]*, Jiongjiong Gu [1], and Boping Wu[2]
[1]JCET Group Co., Ltd., Jiangyin 214437, China
[2] JCET Group Co., Ltd., Shanghai 201206, China
* Email: Lisa.liu@jcetglobal.com Patrick.gu@jcetglobal.com

ABSTRACT

In the era of 5th Generation Mobile Communication Technology, communication equipment needs to support multiple frequency bands and integrate hundreds of components. Electromagnetic interference (EMI) between devices has become challenging in seriously-tight space. Many SiP devices began to adopt conformal shielding, compartment shielding or wire cage to achieve the purpose of EMI shielding and miniaturization. However, for a good shielding solution, there are many factors to consider, such as shielding size, material, ground connection, and distance from device. Otherwise, there will be poor shielding and interference issues, even affecting the modular performance.

This paper presented measurement and simulation methods for conformal and compartment shielding in SiP package. Regarding conformal shielding, which is applied in sub6G band, analyze the effects of grounding the shielding coating, the thickness of the molding, and the sputter material on shielding effectiveness (SE). In the case of compartment shielding, filling materials of shielding wall and vertical are analyzed at high frequencies. In future 6G and mm-Wave applications, SiPs will become complex to keep high performance with innovations.

INTRODUCTION

In the era of 5G communication, SiP technology meets the requirements for higher-density module integration while also accounting for the characteristics of being lightweight, thin, and small. However, the compact space shortens the distance between noise sources and sensitive circuits, which can easily lead to EMI problems, resulting in system failure or failure to pass EMC standards. Therefore, reasonable electromagnetic shielding design is necessary in package design.

This paper studied the factors that affect SE in conformal and compartment shielding based on SiP package structure. Firstly, the paper demonstrated measurement and simulation methods for conformal SE, and analyzed the effects of grounding the shielding coating, the thickness of the molding, and the sputter material on shielding effectiveness. Additionally, this paper presented a measurement and validation solution for compartment SE, discussed the use of conductive adhesive to fill the shielding wall and densely packed vertical wires for compartment shielding.

CONFORMAL SHIELDING

Recently for advanced SiP , sputter coating is widely used. This coating sputters shielding metal onto the top and four sides of the package, achieving an ideal shielding effect without increasing the package thickness , illustrated in Figure 1.

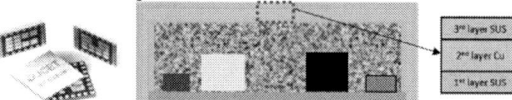

Figure 1: SiP Package with sputter shielding coating

Conformal SE measurement solution and definition

Near-field scanning is a common method in the industry to measure SE. This technology directly measures the electromagnetic radiation strength in space through an electric/magnetic probe. However, commercially available probes only provide a single frequency point and a single component (horizontal/vertical) electric/magnetic field distribution for each scan. The probe needs to be replaced multiple times and the scanning system needs to be restarted to measure electromagnetic field distribution comprehensively, which easily introduces measurement errors. Therefore, this paper simulated the horizontal magnetic field component of a simple transmission line to correlate with measurement results and verified the simulation method's rationality. Then, analyze different shielding structures and materials through simulation to obtain the electric/magnetic field SE results.

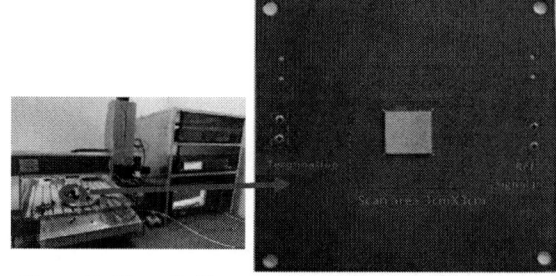

Figure 2: Near-field scanning system for SiP package

The test system for package near-field scanning included an N9041B Signal Analyzer, an N5182B RF Vector Signal Generator, an FLS106 IC EMI Scanner, and an ICR HH 100-27 horizontal loop magnetic field probe. The packaged transmission line is located at the center of the test board and had a size of 1.5cm * 1.5cm. The near-field magnetic field probe scanned a plane located 2mm above the PCB board with a size of 3cm * 3cm, as shown in Figure 2.

The conformal shielding simulation used HFSS three-dimensional full-wave simulation method. Initially, an equivalent model was constructed, and the material's electrical parameters were defined in HFSS. Additionally, The shielding layer thickness was in the um-level and was set as the boundary's Finite Conductivity in HFSS, which allowed for more precise and efficient calculations in HFSS. Table 1 showed the correlation between simulation and measurement, indicating that the magnetic field distribution at 1GHz was consistent between simulation and test, and the calculated SE values were also relatively close. Therefore, the simulation method can be used to analyze the SE.

TABLE I.   CORRELATION BETWEEN SIMULATION AND MEASUREMENT RESULTS

	Measurement Result	*Simulation Result*
Normal Package		
Conformal Shield Package		
SE(dB)	20.07	19.5

There are various definitions of SE commonly used in the industry for near-field scanning. For this paper, SE is calculated by comparing the maximum field strength on the scanning surface, which considers the effect of the maximum radiation field on the device. The equation is as follows,

$$SE_E = 20log_{10}(E_{0\,max}/E_{i\,max})$$
$$SE_H = 20log_{10}(H_{0\,max}/H_{i\,max})$$

$E_{0\,max}/H_{0\,max}$, $E_{i\,max}/H_{i\,max}$ represent the maximum electric/magnetic field value on the scanning plane before and after shielding respectively.

Comparison of different package shielding solutions

Both electric and magnetic fields need to be considered in near-field shielding. In this study, we designed a SiP package and observed the changes in electric/magnetic SE by varying the grounding method of the shielding coating, molding height, and coating material. The SiP package consisted of two high-resistance silicon dies and spiral inductance coils that were fabricated on the dies. The high-resistance silicon dies were mounted on the substrate using a wire bonding process. The SiP size was 4mm*5mm, test board size was 5cm*5cm, as shown in Figure 3. Furthermore, all SE values in this section were calculated using the above equation, the observation plane was 4mm*4mm located 2mm above the PCB.

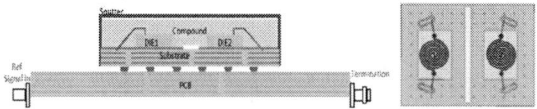

Figure 3: SiP package design

1. Near-field electric shielding

One method for electric shielding is to place a well-grounded metal plate between the radiation source and susceptor to suppress the coupling of parasitic capacitance and achieve the purpose of electric field shielding[1]. In SiP design, the grounding of the shielding coating is mainly connected through the exposed copper on the side of the substrate. We varied the grounding method between the shielding coat and the ground plane layer of the package, as shown in Table 2. LEG#1 had a floating shielding coating, LEG#2 had the shielding coating connected with the ground plane via 4 plating lines, LEG#3 had the shielding coating fully connected with the ground plane, and LEG#4 had the shielding coat fully connected with the ground plane and also connected with ground via at package edge. The results showed that LEG #1 had the worst near-field SE since the shielding coating was not connected with ground. As the contact area between the shielding coating and the substrate ground plane increased, the near-field SE improved to some extent.

TABLE II.   THE IMPACT OF GROUNDING METHOD

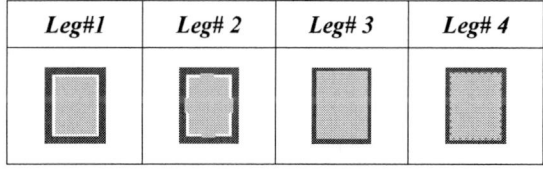

Leg#1	*Leg# 2*	*Leg# 3*	*Leg# 4*

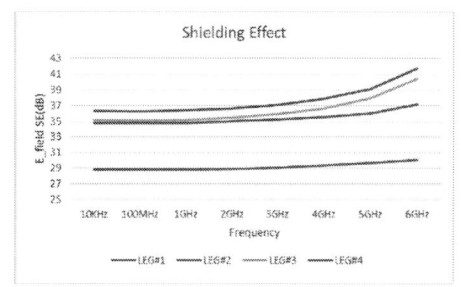

Figure 4: E_field SE for different grounding methods

By varying the molding height, which is the distance between the shielding coating and the radiation source, while keeping the LEG#4 setting constant, it was observed that a lower coating height leads to better electric field SE.

TABLE III.   THE IMPACT OF MOLDING HEIGHT

Molding Height	Leg# 4	Leg #5	Leg #6
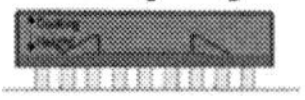 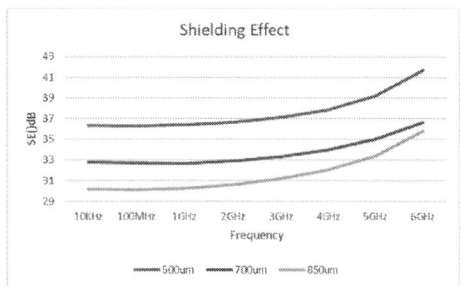	500um	700um	850um

Figure 5: E_field SE for different molding height

2. Near-field magnetic shielding

Magnetic field shielding can be categorized into low frequency and high frequency shielding. For high frequency magnetic field shielding, good conductors are necessary to reduce eddy current impedance. When comparing the SE of copper and nickel in the range of 1 GHz to 6 GHz, we found that copper had better magnetic SE than nickel, with a difference of about 5~10 dB.

TABLE IV.   THE IMPACT OF COATING MATERIAL

Coating material	Copper	Nickel
Conductivity(S/m)	5.8e7	1.45e7

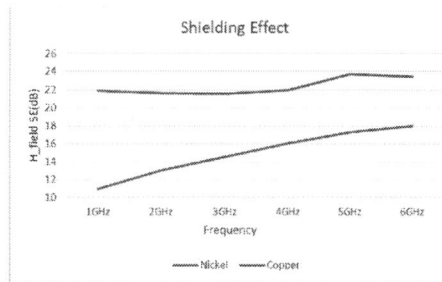

Figure 6: H_field SE for different coating material

In addition, high-frequency currents tend to flow near the surface of the conductor due to skin effect, which means that the shielding layer doesn't need to be too thick for high-frequency magnetic field shielding. Using the LEG#4 design as a base, we varied the thickness of the copper coating, and the results showed in Figure 7. For frequencies below 1 GHz, the impact of coating thickness on SE increased as the frequency decreased. At 1 MHz and 10 MHz, SE was below 5 dB with a coating thickness of 2 μm or less, while it achieved more than 10 dB with a thickness of 6.6 μm or greater. At 100 MHz, the impact of shielding thickness on SE decrease. At 1 GHz and above, the SE values were the same for all coating thicknesses. As according to the skin depth at 1 MHz/10 MHz/100 MHz/1 GHz respectively estimated according to skin effect are 66 μm/20 μm/6.6 μm/2 μm. Therefore, when selecting the thickness of the shielding coating, we should consider the frequency of the radiation source and the accuracy of the sputtering process comprehensively.

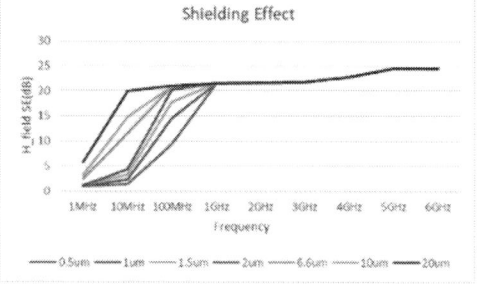

Figure 7: H_field SE for coating thickness

COMPARTMENT SHIELDING

To achieve more flexible shielding effectiveness between different functional blocks, compartment shielding technology is used. This technology divides the shielding cavity into small cavities to reduce its size, and the resonant frequency will be higher than the system frequency, avoiding electromagnetic resonance and making the system more stable.

Conformal SE measurement solution and definition

The equation for compartment SE is defined as follows,

$$SE(dB) = S21_{(TV1)} - S21_{(TV2)}$$

S21 insertion loss is a measure of the power loss from the transmission (Tx) source to the receiver (Rx). TV1 and TV2 respectively represent S21 without and with EMI shielding walls[2].

As shown in Figure 8, we conducted SE measurements on a SiP package with compartment shielding, which utilized conductive adhesive filling technology, and had a material conductivity of 111.11S/m. The DUT was placed on a probe station, and two high-frequency probe tips respectively pointed at the TX/RX source pin. The coaxial interfaces of the probe were connected to the Keysight N5247B VNA via cables. The measurement range was set at 20~40 GHz. At the same time, we built an equivalent model in HFSS and found that the measurement S21 results were close to the simulation results. Therefore, we analyzed the SE of different conductive materials and different spacing vertical wire processes through simulation.

Figure 8: Test system for compartment SE

Comparison of Compartment Shielding structures

This experiment compared four different compartment shielding structures, as shown in Table 4. LEG# A and B were both shielding walls, with LEG#A filled with conductive adhesive in the laser groove, having a conductivity of 111.11 S/m. LEG#B was filled with silver ink material, which had high conductivity of 6.3e7 S/m, but due to material characteristics, only 3-5um of silver ink adhered to the side wall of the laser groove after high temperature baking. LEG#C and D both used vertical wires for shielding, with 250um bonding wire space for LEG#C, and 70um for LEG#D.

TABLE V.   Different compartment shielding structures

Legs#	A	B	C	D
structures				
Conductivity S/m	111.11	6.1e7	5.8e7	5.8e7

The SE simulation results for different shielding structures were presented in Figure 9.

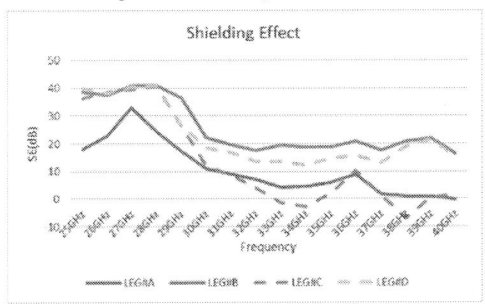

Figure 9: SE for different compartment shielding structures

In the 25GHz to 28GHz frequency range, the SE of LEG#B, C, and D were similar, at around 40dB, which were 10-20dB higher than LEG#A. As from a power perspective, the SE equivalent formula can be expressed as follows,

$$SE(dB) = 10log_{10}(P_0/P_I)$$
$$=^{(1)} SE_{skin\,effect} + E_{H\,effect}$$

$(1) SE_{skin\,effect} = K/\delta$, K is a complex function and δ is expressed as $\delta = \frac{1}{\sqrt{\pi f u \sigma}}$, δ =Skin depth, σ =Electric conductivity, μ= Permeability of the conductor. Here SE_H can be ignored, therefore SE∝ σ[3].Conductive adhesive of LEG#A has lower conductivity compared to silver and copper, resulting in a smaller shielding SE.

Beyond 28GHz, the SE values of all structures decreased with increasing frequency. However, LEG#B had the best SE, maintaining around 20dB, with LEG#D coming in second at about 15dB. Once the frequency exceeded 30GHz, the SE of LEG#C dropped to below 10dB. This decrease in shielding performance is attributed to the wavelength of radiation electromagnetic waves, which decreases as the frequency increases. When the wavelength becomes comparable to the wire bonding space, the shielding performance may decline.

CONCLUSIONS

This paper presented measurement and simulation methods for conformal and compartment shielding in SiP. By correlating simulation and measurement results, the factors affecting shielding effectiveness were analyzed through simulation. Regarding conformal shielding, connecting the shielding coating to the substrate ground and reducing the thickness of the molding are beneficial for improving electrical shielding effectiveness. While for magnetic shielding, high-conductivity materials are beneficial for high-frequency magnetic shielding, and the thickness of the shielding material should be considered in conjunction with the signal frequency. In case of compartment shielding, good conductive filling materials and densely packed vertical wires can achieve good shielding effectiveness at high frequencies.

ACKNOWLEDGEMENTS

I'd like to express my gratitude to colleagues who provided necessary data and materials during the research, as well as collaborated with me to overcome any challenges we faced, offering their invaluable help and expertise.

REFERENCES

[1] Guanglu Li. *Key points of electric field, magnetic field and electromagnetic field are described briefly.* 1994-2008 China Academic Journal Electronic Publishing House, Jan 2007, pp.46-49.

[2] Yi He. *Study on a Conformal Shielding Structure With Conductive Adhesive Coated on Molding Compound in 3-D Packages.* IEEE Transactions on electromagnetic compatibility, Vol. 58, No. 2, Apr 2016, pp. 442-446.

[3] Jay Li. *EMI Shielding Technology in 5G RF System in Package Module.* IEEE 70th ECTC,2020, pp.931-937.

A Composite Photodector with Wide Dynamic Range and Small Area for Dynamic Vision Sensor Application

Yaping Chen,Xiaona Zhu,Shaofeng Yu**

School of Microelectronics, Fudan University, Shanghai 201203, China
*Corresponding Author's Email:{xiaona_zhu,shaofeng_yu}@fudan.edu.cn

ABSTRACT

In this paper, we design and simulate a photodetector with a photodiode-body-biased MOSFET (PD-MOSFET) and apply it to the dynamic vision camera (DVS). The parameters of the process and simulation are adjusted for the DC characteristics, spectral characteristics and transient response of PD-MOSFET. Compared with traditional photodiode (PD), the photocurrent of PD-MOSFET can be increased from $1.4e^{-12}A \sim 6.8e^{-7}A$. The spectral peak is around 550 nm and the latency range from 6us (10^5lx) to 0.2s (0.1lx). DVS improves the dynamic range of PD-MOSFET to 120 dB compared to traditional DVS.

INTRODUCTION

Dynamic vision sensor (DVS) event camera is a new type of image sensor different from the traditional camera. It detects the change of light intensity through the photodiode (PD) of fixed area size. When the change of light intensity compared with the last event output exceeds the threshold, the polarity and position information generated by the event will be output. Because DVS uses logarithmic photoreceptors and the output reading of events are based on asynchronous adress-event representation (AER) architecture, DVS can realize lower time latency and wider dynamic range, which means that the DVS can operate under high-speed scenes and challenging lighting conditions.

The development trend of DVS is to reduce the pixel area and improve the resolution [1]. As the pixel area is smaller and smaller, the photosensitive area of PD is smaller and smaller. The photocurrent is proportional to the photosensitive area of PD and decreases accordingly. Even the photocurrent is smaller than the leakage current of MOSFET in logarithmic photoreceptors, which leads to background noise increase and dynamic range decay. The dynamic range decayed to 90dB when the pixel size was reduced to 4.95um [2].

This paper reports a photodiode-body-biased MOSFET（PD-MOSFET）that takes the p region of PD as the floating potential region adjusted by light intensity regulation, and then adjusts the substrate voltage of MOSFET next to PD, so that the light intensity can affect the drain current of MOSFET, and finally the drain current is input as the light current into the pixel processing circuit. In this paper, we adjust the process parameters by analyzing the DC, spectral and transient characteristics to verify that this device has great advantages in improving the dynamic range of DVS with small pixel area.

DEVICE STRUCTURE

The detailed structure of PD-MOSFET is shown in Figure 1. The whole structure is realized in the N well of P substrate. Both the P substrate and the N well are grounded outside, which can isolate and control the substrate of MOSFET at a low potential. The optical window is above the photodiode with a size setting of 3um × 3um, the polarity value of light is 0.5 and the wavelength is 550nm. The N$^+$ doping of the photodiode is $2e^{17}$cm^3 and the cathode external voltage is 1v. The source of the MOSFET is grounded, the drain terminal is connected to 0.3V, and the gate external voltage is -2.5V, so that the MOSFET works in the off-current region. Upon light exposure, electron-hole pairs are generated. Under the effect of the potential, part of the hole flows to the N well, part flows to the source terminal of MOSFET, while others accumulate to increase the potential in the P region and finally reach equilibrium. The change of the potential in the P region causes the output current of the MOSFET to change, which means the light intensity can change the output current of MOSFET.

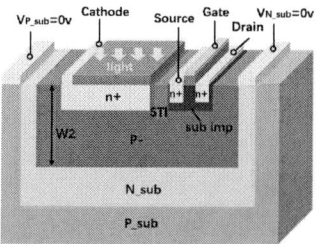

Fig. 1. The structure of PD-MOSFET photodector.

DC Characteristic

Figure 2 shows the relationship between the potential of the P region with light intensity under different doping concentrations in the P region. The potential in the P region is mainly determined by the longitudinal phototransistor structure in the photosensitive area, so the change of potential reflects the ratio of photogenerated current and triode leakage current. As the doping concentration in the P region changed from $1e^{12}$cm^3 to $1e^{16}$cm^3, the ability of PD-MOSFET to detect the lowest light intensity becomes higher and then lower, and finally the doping concentration in the P region is set to $1e^{15}$cm^3.

Figure 3 shows the relationship between the drain current of MOSFET and light intensity at different channel doping concentrations. The MOSFET in the off-current state is equivalent to a transistor, and adjusting the channel concentration is equivalent to adjusting the base region concentration. As the channel concentration changes from $1e^{16}cm^3$ to $1e^{17}cm^3$, the overall current of Ids becomes lower, the range span becomes higher. Since the leakage current of the logarithmic photoreceptor is on the order of pA, the minimum current of Ids should be greater than $1e^{12}A$, and the final channel doping concentration should be set to $7e^{16}cm^3$.

Figure 4 shows the relationship between the drain current of the MOSFET and the light intensity at different doping concentrations of source-drain terminal. The adjustment of the source-drain doping concentration is equivalent to the adjustment of the emission-collector region concentration in transistor. As the source-drain doping concentration increases from $1e^{18}cm^3$ to $1e^{19}cm^3$, the overall current of Ids becomes higher and the range span becomes smaller. The photocurrent that the logarithmic photoreceptor can respond to normally is not greater than the order of uA, so the maximum current of Ids should be less than $1e^{-6}A$. Since the source-drain doping concentrations should not be too low, and the adjustment of the source-drain doping concentration has little effect on the DC range of Ids, this paper does not adjust the source-drain doping, but only changes the channel doping. The source-drain doping concentration is set to $1e^{19}cm^3$.

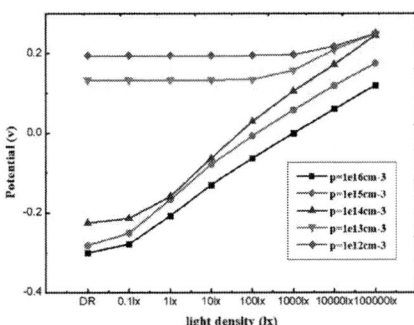

Fig. 2. *The potential with light intensity at different doping concentrations of P region.*

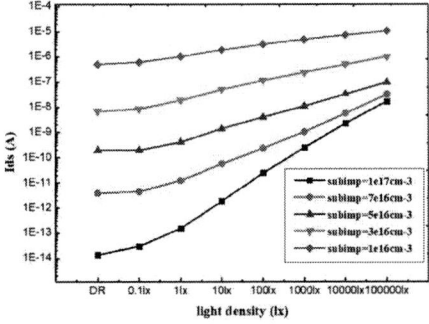

Fig. 3. *Ids with light intensity at different channel doping*

concentrations.

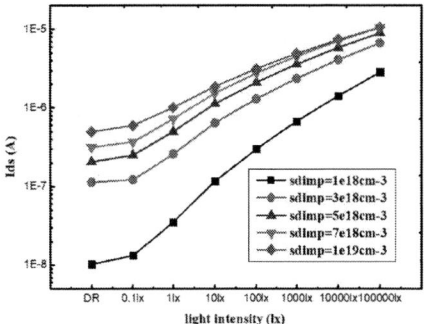

Fig. 4. *Ids with light intensity at different source-drain doping concentrations.*

Spectrum Response

There are two main depletion regions of photosensitivity: one is the depletion region of the photodiode, and the other is the depletion region between the P region and the N well. Among them, the depletion region of the photodiode is inverse bias, and its position depends on the depth of N^+ doping; The depletion region between the P region and the N well is positively biased, and the position depends on the P region depth W2. The peak spectral response of the human eye is 550nm, and the photodetector should adjust the spectrum to match the spectral response of the human eye as much as possible.

Figure 5 shows the spectral response under different W2 an illumination intensity of $0.0146mW/cm^2$. As the depth of W2 ranges from 4um to 2um, Ids becomes slightly larger and the peak shifts left from 620nm to 550nm, and finally sets W2 to 2.5um.

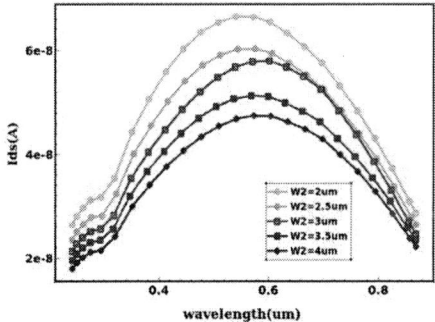

Fig. 5. *Ids with wavelegth at different depth of W2*

AC Characteristic and Circuit Implementation

DVS is mostly used for high-speed image capture, so the internal photodetector delay should be low to ensure high bandwidth [3]. The working process of this photodetector is divided into two steps: first, the photogenerated carrier charges the P region to raise the potential; Second, the MOSFET changes the output current based on the change of substrate potential. Even though the PD-MOSFET can maintain a high gain in low light, it requires a low photocurrent to charge the capacitor. Therefore, the higher the light intensity, the smaller the delay of Ids.

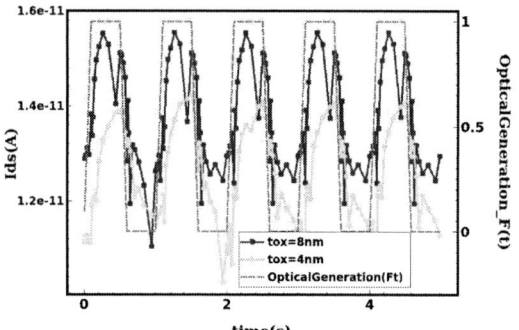

Fig. 6. Transient response at different tox.

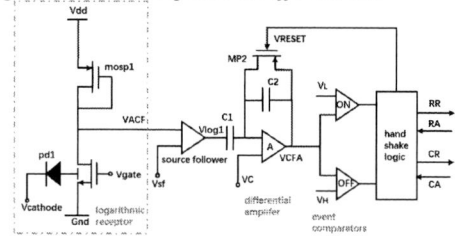

Fig. 7. Abstracted pixel schematic of DVS based on PD-MOSFET.

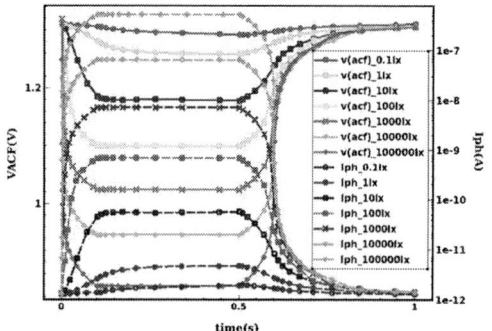

Fig. 8. Transient response of VACF and Iph.

Cox is one of the important factors affecting the response latency. Figure 6 shows the instantaneous current response at a light intensity of $1.46e^{-8}W/cm^2$ and an oxide layer thickness of 4nm and 8nm. Upgrading from 4nm to 8nm makes Ids larger, but the latency is shortened from 0.3s to 0.1s.

Figure 7 shows the internal circuit diagram of the DVS pixel [4]. The drain terminal of the PD-MOSFET is connected to the drain and gate terminals of mosp1. Ids serves as the current source of mosp1, and VACF outputs the subthreshold voltage of mosp1. The Vdd was set to 1.8v.

Figure 8 shows the transient response of VACF and Iph (Ids) at different light intensity simulated by Mixmode, with the light intensity range increasing from 0.1lx to 10^5lx. As the light intensity increases, the VACF decreases and the delay of Ids decreases.

Figure 9 shows that the VACF based on PD and PD-MOSFET changes with light intensity, which can be observed in the range of 0.1lx~10^5lx. The VACF after using PD-MOSFETs remains linear overall, and the dynamic range is extended to 120dB.

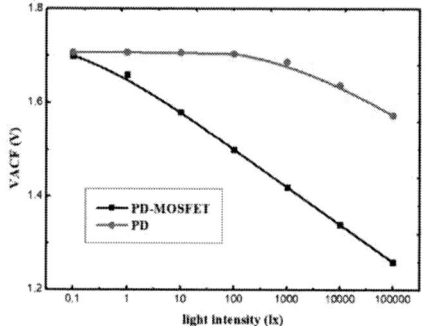

Fig. 9. VACF with light intensity of PD and PD-MOSFET.

CONCLUSION

This paper introduces a new type of PD-MOSFET hybrid structure photodetector. By adjusting the doping concentration in P region, channel doping concentration, P region depth W2, and gate oxide thickness, so that under the light intensity of 0.1lx~10^5lx, we make its photocurrent in the range of $1.4e^{-12}$A~$6.8e^{-7}$A, the spectral peak is 550nm, and the response delay range is 0.1s~5us. DVS with PD-MOSFET extend the dynamic range to 120dB.

REFERENCES

[1] Chi-Hang Chan,Lin Cheng,Wei Deng, et al. (2022). *Journal of Semiconductors,* 2022, vol.7.

[2] Y. Suh et al., *2020 IEEE International Symposium on Circuits and Systems(ISCAS)* , 2020, pp.1-5.

[3] M. Akrarai, N. Margotat, G. Sicard and L. Fesquet, *New Circuits and Systems Conference (NEWCAS),* 2020, pp. 238-241.

[4] P. Lichtsteiner, C. Posch and T. Delbruck, *Journal of Solid-State Circuits,* vol.43, Feb. 2008, pp.566-576.

IMPROVE THE BREAKDOWN VOLTAGE OF SILICON PIXEL SENSOR WITH OPTIMIZED MULTI-GUARD RINGS

Peng Sun[1,2], Gaobo Xu[1,2], Jianyu Fu[2], Mingzheng Ding[1], Yinan Yan[1,2],*
Luoyun zhang[1,2], and Huaxiang Yin[1,2]*

[1]Integrated Circuit Advanced Process R&D Center, Institute of Microelectronics,
Chinese Academy of Sciences, Beijing 100029, China
[2]University of Chinese Academy of Sciences, Beijing 100049, China
*Corresponding Author's Email: yinhuaxiang@ime.ac.cn, xugaobo@ime.ac.cn;

ABSTRACT

The silicon pixel sensor (SPS) for X-ray free electron laser detection requires ultra-high operating voltage. In this paper, for an optimized multi-guard rings, a SPS with ultra-high breakdown voltage and low leakage is obtained. The effect of the arrangement of the multi-guard rings on the SPS breakdown is investigated by TCAD simulation. It is found that SPS with a gradually increasing gap from inner ring to outer ring has a higher breakdown voltage. Guided by this, the multi-guard ring structure of SPS is optimized and applied into the manufacture of devices. Finally, the optimized SPS exhibits an ultra-high breakdown voltage of exceeding 1,800 V; and the dark current of per pixel is approximately 2.0 pA at a bias voltage of 1,000 V.

Keywords—silicon pixel sensor (SPS); Multi-guard rings; High voltage; Breakdown; X-ray free electron laser (XFEL);

INTRODUCTION

The X-ray free electron lasers (XFEL) could emit ultra-high brightness and ultra-short duration photons at one pulse duration. The excellent performance of XFEL provides support for ultra-fast dynamic structure imaging analysis of material science [1, 2]. Silicon pixel sensor (SPS) is the core device in the XFEL detection system. However, the XFEL with high intensity photons would induce a "plasma effect" in silicon, delaying the charge collection time of readout electrode of pixel, and further declining the imaging frame rate of the detection system. For the silicon P-I-N sensor of 280 μm thickness, applying a bias voltage of 500 V on the sensor can effectively suppress the plasma effect [3]. For the SPS with a thickness of 500 μm, it may require a higher operating voltage of up to 1,000 V to suppress the plasma effect [4]. To ensure that the SPS operates at such high voltages, an enhanced multi-guard ring design is required to improve its hard breakdown voltage.

In this paper, at first, the effect of the arrangement of the multi-guard rings on the SPS breakdown is analyzed via Sentaurus TCAD tools. In second, guided by the analysis, the multi-guard ring structure of the SPS is optimized and a fabricated SPS achieve an ultra-high breakdown voltage

over 1,800 V.

SIMULATIONS AND ANALYSIS

The Sentaurus TCAD tools is used to simulate and analyze the effect of the arrangement of the multi-guard ring structure on the silicon pixel sensor (SPS) breakdown. As illustrated in Fig. 1, a SPS usually consists of pixel region, current collection ring (CCR) and multi-guard ring (GR) structure; the pixel region is the sensitive area of the sensor, the CCR and are the non-sensitive area of the sensor; a reversed bias voltage is applied on cathode, the anode of pixel is connected with readout electronics, CCR is grounded and the GRs are floating. The sensor is made on a 500 μm thick, n-type <100> silicon substrate with a high bulk resistivity of 13000 Ω·cm and a relatively low concentration of oxide layer fixed charge (10^{11} cm^{-2}) is applied for simulations. The size of the pixel is 200 μm, the width of p+ area of the CCR and GR are 120 μm, 25 μm respectively.

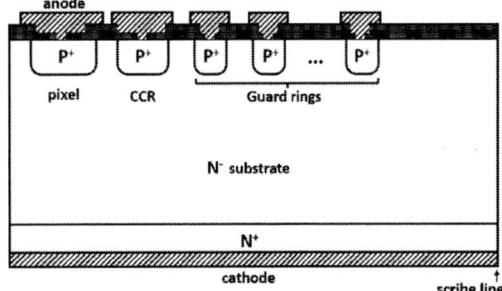

Fig. 1. *Schematic cross-section of the SPS used in simulations.*

Two different types of arrangements of devices with 10 guard rings are simulated and investigated. One type of arrangement of device #1 with the equal ring gap, and the ring gap is 35 μm; the other type of arrangement of device #2 with the gradually increasing ring gap from the inner ring to the outer ring, the gap from the first ring to the second ring is 21.5 μm, and thereafter the ring gap is gradually increased on the basis of 3.0 μm; the total length of both types of arrangements from CCR to the last GR is 600 μm. Fig. 2 shows that the simulated I-V plots of the device

---

979-8-3503-1101-3/23 $31.00 © 2023 IEEE

#1 and device #2. As shown in Fig. 2 (a)，the dark current of the pixels both device #1 and device # 2 maintain a low level approximately 26 pA at a bias voltage of 1,000 V, which attribute to the protection function of CCR and GRs. As shown in Fig. 2 (b)，a hard breakdown is observed in CCR, the CCRs of device #1 and device #2 suffer a hard breakdown at 1,450 and 2,000 V, respectively. The simulated results reveal that using gradually increasing ring gap from the inner ring to the outer ring can further improve the hard breakdown performance of the SPS; which attributed to the fact that applying this type arrangement allows the high electrostatic potential at the edge of the sensor to fall more smoothly to the grounded electrostatic potential at the CCR, and thereby reduces the surface electric field at the p-n junction location on the CCR side near the GR, delaying the avalanche multiplication effect triggered by high electric field inside the bulk silicon.

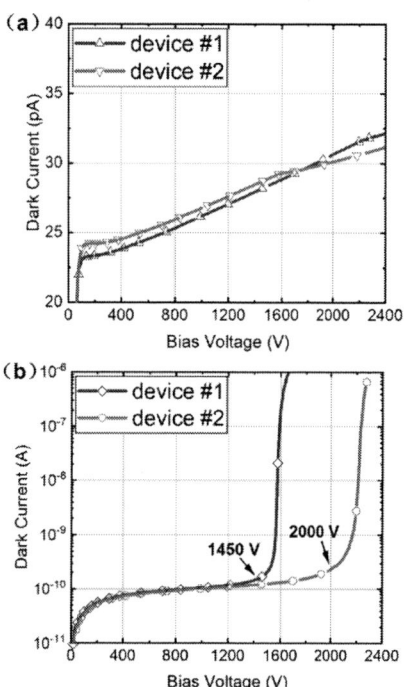

Fig. 2. *Simulated I-V plots of **(a)** pixel and **(b)** CCR.*

The Fig. 3 provides the surface electric field distribution below 50 nm of Si/SiO$_2$ interface of device #1 and device #2 at a bias voltage of 2,000 V. It can be seen that the maximum value of the electric field strength occurs on the side of the CCR adjacent to the GR, with a maximum electric field strength of 407,331 V·cm^{-1} for device #1 and 387,195 V·cm^{-1} for device #2, so that device #1 is more susceptible to avalanche breakdown.

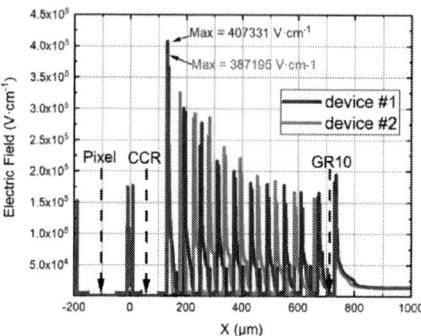

Fig. 3. *The surface electric field distribution below 50 nm of Si/SiO$_2$ interface of device #1 and device #2 at a bias voltage of 2,000 V.*

Fig. 4 presents the avalanche breakdown voltage of the CCR with 6 GRs，8 GRs，10 GRs，12 GRs，the arrangement type of the GRs is that gradually increasing ring gap from the inner ring to the outer ring. The breakdown voltage of CCR progressively increases with the number of GRs, and the bias voltage when the breakdown just occurred is approximately 1,925 V, 1,970 V, 2,000 V, 2,010 V.

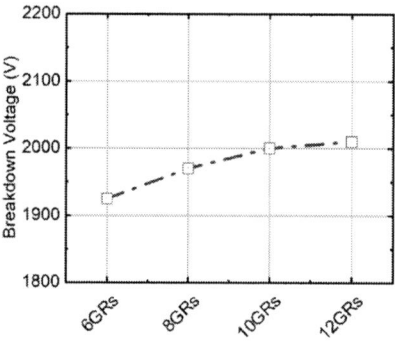

Fig. 4. *Simulated breakdown voltage of CCR with 6, 8, 10, 12 GRs, and the arrangement type of GRs is that gradually increasing ring gap from the inner ring to the outer ring.*

EXPERIMENTAL

According to the analysis, applying the arrangement of the multi-guard ring structure with gradually increasing ring gap from the inner ring to the outer ring could greatly improve the breakdown voltage of SPS; although increasing the number of GRs can increase the breakdown voltage, the improvement effect is not obvious when the number of GRs is increased to 10. Therefore, a 10 GRs structure with gradually increasing ring gap from the inner ring to the outer ring can be used as an optimized multi-guard ring structure on SPS.

Two types of interconnected 8×8 array SPSs (device #1

and device #2) with different multi-guard rings arrangements are fabricated, the parameters of devices are the same as those used in the simulation; device #1 represents SPS with equal gap multi-guard rings and device #2 represents SPS with optimized multi-guard rings. Fig. 5 presents the image of an interconnected 8×8 array SPS with optimized multi-guard ring structure.

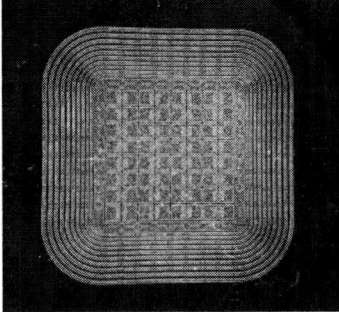

Fig. 5. *The 8×8 array SPS with optimized multi-guard rings.*

Device #1 and device #2 are dicing from wafer and tested, the distance of the dicing edge to the pixel region are both about 1500 μm，which ensures that the depletion zone of the SPS does not reach the dicing edge at ultra-high bias voltage. Aglient Technologies B1505A Power Device Analyzer are used to test the I-V characteristics of the SPSs.

Fig. 6 (a) and (b) present the measured I-V characteristics of the pixel region and CCR, the test results reveal that breakdown voltages of device #1 and device #2 are 1412 V and 1818 V, respectively. It confirmed that the SPS through optimizing could achieve an ultra-high operating voltage, the breakdown voltage of the CCR and pixel region are both up to 1,800 V; as well as the dark current of per pixel is approximately 2.0 pA at a bias voltage of 1000 V. The test results indicate that the optimized SPS successfully meet the design objectives.

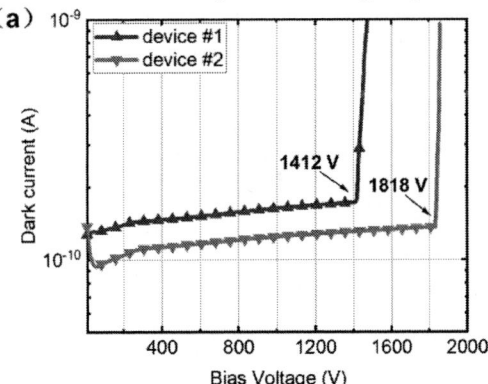

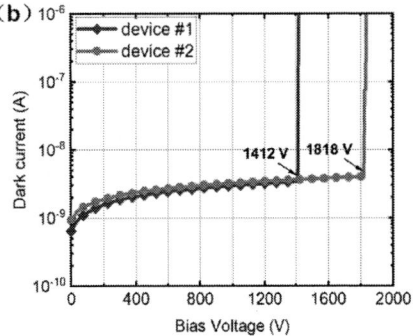

Fig. 6. *Measured I-V characteristics of (a) pixel region and (b) CCR. Device #1 represents SPS with equal gap multi-guard rings and device #2 represents SPS with optimized multi-guard rings.*

CONCLUSION

In this work, a silicon pixel sensor (SPS) with optimized multi-guard ring structure for ultra-high voltage and low leakage current operation is designed and successfully fabricated. The final test results indicate that the optimized SPS has a breakdown voltage of more than 1,800 V; and a dark current of 2.0 pA per pixel at a bias voltage of 1,000 V. It provides a reliable support for improving the imaging frame rate of XFEL detection system.

ACKNOWLEDGEMENTS

This work is supported in by National Natural Science Foundation of China under Grant 22127901. The silicon pixel sensors in this paper were fabricated in the Integrated Circuit Advanced Process R&D Center, Institute of Microelectronics, Chinese Academy of Sciences.

REFERENCES

[1] Feldhaus J, et al. AMO science at the FLASH and European XFEL free-electron laser facilities. Journal Of Physics B-Atomic Molecular And Optical Physics. 2013;46.

[2] Young L, et al. Roadmap of ultrafast x-ray atomic and molecular physics. Journal Of Physics B-Atomic Molecular And Optical Physics. 2018;51.

[3] Becker J, Eckstein D, Klanner R, Steinbruck G. Impact of plasma effects on the performance of silicon sensors at an X-ray FEL. Nuclear Instruments & Methods In Physics Research Section a-Accelerators Spectrometers Detectors And Associated Equipment. 2010;615:230-6.

[4] Klanner R, Becker J, Fretwurst E, Pintilie I, Pohlsen T, Schwandt J, et al. Challenges for silicon pixel sensors at the European XFEL. Nuclear Instruments & Methods In Physics Research Section a-Accelerators Spectrometers Detectors And Associated Equipment. 2013;730:2-7.

979-8-3503-1101-3/23 $31.00 © 2023 IEEE

STUDY ON IMPROVEMENT OF DARK COUNT RATE FOR SILICON PHOTOMULTIPLIER

Xing Chen*, Zhigao Wang

Semiconductor Manufacturing International Corp., Beijing, China
*Corresponding Author's Email: Xing_CX@smics.com

ABSTRACT

The dark count rate (DCR) is a major factor limiting the application of silicon photomultiplier(SiPM). It is widely believed that one of the major origins of DCR is generation effects from Shockley-Read-Hall (SRH) defects near the depletion region, on which plasma induced damage (PID) of the interfacial regions including STI, DTI and photodiode surface plays a substantial role. Normally for backside illuminated (BSI) SiPM, high-k films induced hole accumulation at the interface of DTI-Si is employed to reduce DCR, however, this option is not available for front-side illuminated (FSI) SiPM due to contamination concerns. Therefore, systematic studies are conducted in this paper for improving DCR by process optimization, finally the DCR is improved significantly.

INTRODUCTION

Silicon Photomultipliers(SiPMs) are arrays of hundreds to thousands of Single Photon Avalanche Diodes (SPADs) connected in parallel, each SPAD with an integrated resistor for passive quenching, referred to as microcells. All microcells are connected with a common anode and cathode, giving a signal proportional to the number of fired cells. SiPM has acquired increasing research interest in last years due to its advantages such as compactness, low operational voltage (tens of volts) and insensitivity to magnetic fields, thus making it a good alternative for the photomultiplier tube (PMT) in medical, high-energy physics, spectroscopy, biology and light detection and ranging (LiDAR) applications. In such applications, SiPM has the advantages of single photon sensitivity, high photon detection efficiency (PDE), high gain, good time resolution and high dynamic range.

Each SPAD in SiPM is a P-N junction reversely biased above the breakdown voltage (BV), in other words, works in Geiger mode. The junction is properly designed thus allowing the breakdown only takes place in the central active area of the SPAD. Before the detection of photons, the electric field is very high (in the order of 10^5 V/cm) that when a single carrier is injected or generated into the depletion layer, an electron-hole pair is created, both carriers will drift in the depletion layer and induce an impact ionization which generates a second electron-hole pair and so on, thus building up an avalanche and a self-sustaining avalanche current can be triggered. Afterwards, when the increasing avalanche current flows through the quenching resistor in series with the SPAD,

the bias voltage on the junction drops near or below the BV, thus quenching the avalanche.

There are several noise sources in SiPMs, among which the dark count rate (DCR) is a major factor limiting the application of SiPMs and is in requirement of research efforts for its improvement. The DCR or dark current (DC) in SiPM normally originates from thermal generation from bulk of the cell, generation effects from Shockley-Read-Hall (SRH) defects in the depletion region, and diffusion of carriers from the quasi-neutral boundaries of the junction [1]. In order to improve DCR for SiPMs, interface and surface passivation and plasma optimization during etch process is normally employed.

In this paper, the sources of DCR in SiPM are explored from its physical structure and related process. Moreover, several process optimization approaches correspondent with the sources are employed to study their improvement on DCR, and finally the DCR is reduced significantly.

ANALYSIS OF DCR IN SIPM

A. Structure of SiPM

The SiPM studied in this paper is a front side illuminated (FSI) structure, the structure of a SiPM cell is shown in Fig.1. The SiPM cell consists of a SPAD as photodiode, a high resistance poly as quenching resistor, a shallow trench isolation (STI) and a front side deep trench isolation (DTI) as pixel optical isolation and electrical isolation.

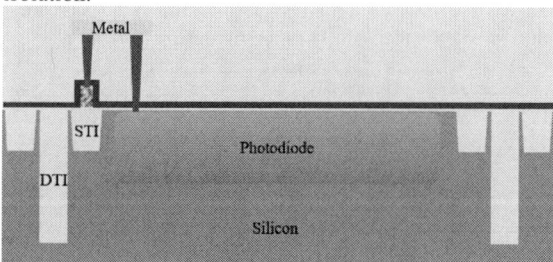

Figure 1: Structural schematic of a Front-Side-Illuminated (FSI) SiPM, consisting of photodiode, Poly Resistor, STI and front-side DTI

The process flow of SiPM is shown in Fig.2, where STI and DTI formation are followed by Poly-Si patterning and photodiode implantation, finally the BEOL process completes the FSI SiPM.

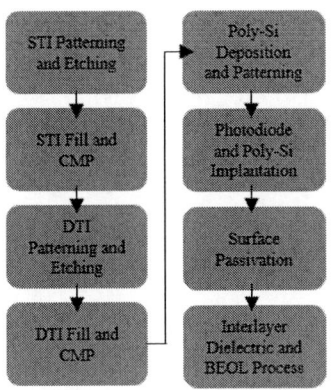

Figure 2: A simplified process flow of FSI SiPM

B. Sources of DCR in SiPM

The behavior and mechanism of DCR in SiPM/SPAD is similar with the dark current (DC) in CMOS image sensor (CIS) in many ways. It is believed that one of the major origins of dark current is traps in band gap of semiconductor, which is created by Si-SiO2 interfacial defects and dangling bonds, where thermal generation and recombination of free carriers often locate [2-3]. With regard to CMOS process of SiPM, the interfacial defects correlate well with the plasma induced damage during the plasma etching process such as poly etching, high density plasma oxide process, and high energy implantation used for formation of photodiode. In the FSI SiPM structure studied here, 3 major regions which are damaged by plasma process are shown in Fig.3, which are the STI sidewall interface, DTI side wall interface, and the surface above photodiode.

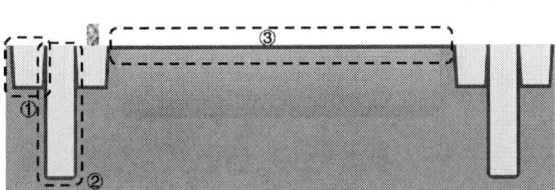

Figure 3: Regions of plasma induced damage as sources of DCR in SiPM formation process

IMPROVEMENT OF DCR FOR SIPM

Plasma induced damage(PID) is one of the major origins of interfacial defects and traps, therefore, the reduction or repair of PID is crucial for improvement of DCR. PID events during the CMOS process can generate neutral or charged traps in the surface oxide of SiPM, which can subsequently serve as generation and recombination center of dark current [4]. Therefore, it is reasonable to conduct some experimental studies on the reduction of PID and the passivation of traps induced by PID.

From the analysis above, 3 major PID damaged regions are taken into consideration. Moreover, 3 types of process optimization are employed including passivation of silicon interfacial dangling bonds by thermal oxide, add N_2 anneal at certain steps for inducing hydrogen related interfacial trap passivation, and surface P type implantation for generating positive charged layer around STI or DTI to screen the defective sidewalls and edges of STI/DTI from the depletion region of photodiode [5].

TABLE I. DCR IMPROVEMENT SPLITS DIVEDED BY OPTIMIZED REGION

Region	Improvement Condition
STI Sidewall Interface	High temperature N_2 anneal after STI-CMP
	STI fill with low density plasma SiO2
	STI sidewall high dose P type implant
DTI Sidewall Interface	DTI liner SiO_2 with thermal dry oxide
	DTI liner SiO_2 with thermal wet oxide
	DTI sidewall high dose P type implant
	High temperature N_2 anneal after DTI liner oxide
Photodiode Surface	Silicon surface passivation with thermal dry oxide
	Silicon surface passivation with thermal wet oxide
	High temperature N_2 anneal after ILD-SIN

As shown in Table.1, process splits or combination of splits are employed to study the improvement of DCR. The DCR results of the splits are shown in Fig.4, the DCR is reduced from 43.7 Hz/um^2(Baseline process with no optimization) to 1.3 Hz/um^2.

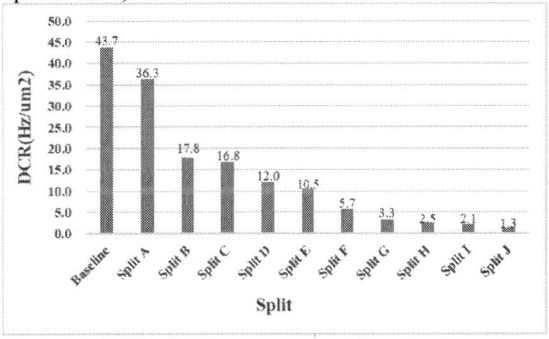

Figure 4: DCR of different process splits

CONCLUSION

In this paper, based on a sub-90nm node CMOS process of SiPM manufacture, the process optimization methods are explored. By employing several plasma reduction and trap passivation methods, the DCR of the FSI SiPM studied here is improved significantly, which can be an inspiration for further studies regarding to DCR reduction of SiPM.

REFERENCES

[1] E. Sciacca, A. C. Giudice, D. Sanfilippo, F. Zappa, S. Lombardo, R. Cosentino, C. Di Franco, M. Ghioni, G. Fallica, G. Bonanno, S. Cova, and E. Rimini, "Silicon planar technology for single-photon optical detectors" IEEE Trans. Electron Devices, vol. 50, no. 4, pp. 918–925, 2003.

[2] V. Goiffon, C. Virmontois, P. Magnan, Member, S.Girard, and P. Paillet, Senior, "Analysis of Total Dose-Induced Dark Current in CMOS Image Sensors From Interface defects and Trapped Charge Density Measurements, " IEEE Trans. Nuclear Science, vol. 57, no. 6, Dec. 2010.

[3] S. S. Park, and J. D. Lee, "The Effects of Deuterium Annealing on the Reduction of Dark Currents in the CMOS APS Hyuck In Kwon, O. Jun Kwon, Hyungcheol Shin, Byung-Gook Park, " IEEE Trans. Electron Devices, vol. 51, no. 8, Aug. 2004.

[4] A.Martin, "Plasma induced charging damage: From appropriate MOS test structures to antenna design rules, a comprehensive process qualification procedure" in Proc. 2020 International Integrated Reliability Workshop, pp. 1-8.

[5] C. R. Moon, J. W. Jung, D. W. Kwon, 1. Yoo, D. H. Lee, and K. Kim, "Application of Plasma-Doping (PLAD) Technique to Reduce Dark Current of CMOS Image Sensors, " IEEE Trans. Electron Device Letters, vol. 28, no. 2, Feb. 2007.

PROCESS OPTIMIZATION AND PERFORMANCE IMPROVEMENT OF CMOS MICROBOLOMETER WITH A SALICIDED POLYSILICON THERMISTOR

Jiang Lan[1], Haolan Ma[1], Yaozu Guo[1], Ke Wang[1], Feng Yan[1], Yiming Liao[2]* and Xiaoli Ji[1]

[1]School of the Electronic Science and Engineering, Nanjing University
[2]School of Electronic and Optical Engineering, Nanjing University of Science and Technology
*Corresponding Author's Email: liaoyiming@njust.edu.cn

ABSTRACT

In this study, the absorptivity of a polysilicon microbolometer based on standard CMOS technology at a wavelength of 7-13 μm is improved by the deposition of silicon nitride (Si_3N_4). The influence of process temperature and thickness on absorptivity of microbolometer are investigated. Simulation results show that the absorptivity of the microbolometer increases significantly with the increased thickness of Si_3N_4 in the wavelength range of 8-10 μm and 10.5-13 μm. The experimental results demonstrate that the average detectivity of the microbolometer in the range of 10-13 μm can be further improved by 87.10% when the deposition temperature of Si_3N_4 film is increased to 300℃. When the thickness of the Si_3N_4 film is increased to 300 nm, the average detectivity of the microbolometer in the range of 10-13 μm can be improved by 152.55%.

INTRODUCTION

The microbolometer based on the standard CMOS process has attracted extensive attention because of its advantages such as easy implementation, no refrigeration, low cost, low power consumption, lightweight, and small size [1,2]. However, the infrared absorptivity of the microbolometer is too low, which seriously hinders its large-scale applications. To solve this problem, a method of depositing silicon nitride (Si_3N_4) film on the microbolometer surface is proposed. Because Si_3N_4 material has good infrared absorptivity and is compatible with standard CMOS process [3,4].

This paper mainly studies the effect of deposition temperature and thickness of Si_3N_4 on microbolometers' infrared (IR) absorptivity. Simulation results show that the infrared absorptivity of the microbolometer increases with the deposition temperature and thickness of Si_3N_4 film.

DESIGN AND FABRICATION

In this paper, polysilicon is used as the thermal material, SiO_2 medium is selected as the IR absorber and support arm, and the first layer of metallic aluminum is used as the hard mask. Figure 1 shows a 3D schematic diagram of a polysilicon microbolometer.

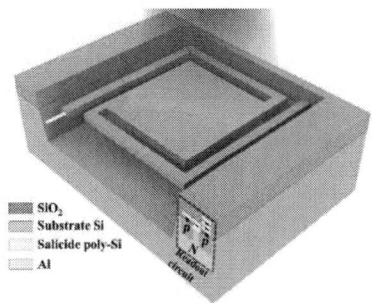

Figure 1: 3D schematic of polysilicon microbolometer

In this paper, the effect of the thickness of the Si_3N_4 film on the IR absorptivity of the microbolometer is simulated using FDTD software to determine the optimal film thickness. The absorber size is set to 40 μm×40 μm with a thickness of 1.75 μm, and the thermal isolation cavity is set to 70 μm. The material of absorber is SiO_2, the substrate material is set to Si, and the dielectric parameters of the material use the default settings provided by FDTD. The light source is plane-wave. Set the boundary conditions to perfect boundary matching layer. In the simulation, the thickness of Si_3N_4 film varies from 100 nm to 600 nm. As can be seen from Figure 2, the absorptivity of the microbolometer increases with the thickness of the Si_3N_4 film. When the thickness of the Si_3N_4 film is 600 nm, the absorptivity is close to saturation, reaching 95% at 9 μm.

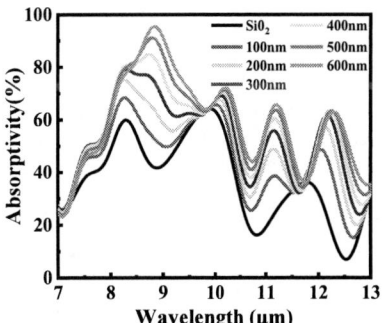

Figure 2: Simulated IR absorptivity for Si_3N_4 thickness from 0 to 600 nm

Figure 3 shows a basic flowchart for the fabrication of a polysilicon microbolometer. First, N and P wells are prepared by ion implantation and annealing processes, and then gate oxygen layer is generated by dry oxidation, and then a polysilicon layer is deposited on top of the gate oxygen layer. The source and drain regions are ion-implanted, and the polysilicon is metalized, followed by the deposition of a SiO_2 dielectric layer on top, and the preparation of through holes and a metallic Al linkage layer is completed.

The SiO_2 dielectric layer and Si_3N_4 passivation layer above the metal Al mask are removed by ICP etching process until the Si substrate is completely exposed. Then the microbolometer silicon substrate is etched with TMAH solution to form a thermally isolated cavity. Finally, a Si_3N_4 film is deposited on the surface of the microbolometer by PECVD. Figure 4 shows the microbolometer with Si_3N_4 film deposited and its bottom thermally isolated cavity.

Figure 3: Process flow for the fabrication of the microbolometer

Well implantation and annealing
Gate Silicon oxide growth
Poly Si gate formation and salicide
Dielectric SiO_2 deposition and polish
Al Interconnect layer deposition
Si_3N_4 passivation layer deposition
ICP dry etch
TMAH wet etch
Remove the mask

Figure 4: Microbolometer with Si_3N_4 film deposited and its bottom thermally isolated cavity

RESULTS AND DISCUSSIONS

The optoelectronic performance of the microbolometer is characterized by testing its voltage responsivity and detectivity. The chip is encapsulated in a double-row inline ceramic tube housing, and placed in a suitable vacuum cavity. A vacuum pump is used with an internal air pressure below 3.75×10^{-2} Torr. Infrared radiation of 7-13 μm is generated by quantum cascade laser and modulated by chopper. The signal voltages at different wavelengths are measured with a lock-in amplifier. The voltage responsivity at different bias currents is calculated according to the following equation:

$$R_v = \frac{\Delta V}{P_{in}} \quad (1)$$

where R_v is the voltage responsivity, ΔV is the output voltage, and P_{in} is the input power of the infrared laser.

we can obtain the detectivity D^* of the microbolometer according to the following equation[5]:

$$D^* = \frac{R_v \sqrt{A_d \Delta f}}{V_n} \quad (2)$$

Where D^* is the detectivity, A_d is the area of the absorber, Δf is the electrical bandwidth, and V_n is the noise voltage. Figure 5 shows the detectivity of the microbolometer without Si_3N_4 films, with 100℃/300nm Si_3N_4 films, and with 300℃/300nm Si_3N_4 films. When the temperature is below 100°C, the intrinsic stress of the Si_3N_4 film is larger, resulting in tensile stress, which makes it difficult to deposit the film. When the deposition temperature is higher than 400°C, the growth of the Si_3N_4 film is uneven, which is prone to cracking. Moreover, the processing temperature for integrated circuit chips should not exceed 400°C. Therefore, we chose a substrate temperature below 300°C for comparison. The results show that there is a very significant increase in the detectivity at 10-13 μm of the microbolometer relative to that of the without Si_3N_4 film. When the deposition temperature is 100°C, the detectivity of the microbolometer in the range of 10-13 μm increased by 32.70% compared to the microbolometer without Si_3N_4 film. When the deposition temperature is 300°C, the detectivity of the microbolometer in the range of 10-13 μm increased by 148.78% compared to the microbolometer without Si_3N_4 film. Si_3N_4 films deposited at a substrate temperature of 300□ showed a more significant improvement in the detectivity of the microbolometer.

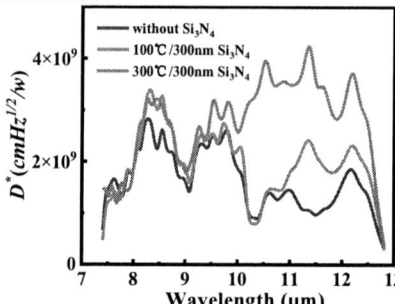

Figure 5: Detectivity of microbolometer as a function of wavelength

Figure 6 shows the detectivity of microbolometers with different thicknesses of Si_3N_4 films in the wavelength range of 7-13 μm at the deposition temperature of 300□. The results show that the deposition thickness has a limited effect on the performance enhancement of the microbolometer in the 8-9 μm wavelength band, and the enhancement is 15%. As the thickness of the deposited Si₃N₄ film increases, the detectivity of the microbolometer increases at 10-13 μm. The detectivity at 10-13 μm of microbolometer with 100 nm Si₃N₄ film improves by 88.02% compared to the microbolometer without Si₃N₄. When 300 nm Si₃N₄ film is deposited, the detectivity at 10-13 μm improves by 152.55%, the detectivity of the microbolometer at a wavelength of 10.5 μm improves by 208.82%, and the D^* reaches 4.2×10^9 cmHz$^{1/2}$/W.

However, when the thickness of the Si₃N₄ film is further increased to 600 nm, the average detectivity in the range of 7-13 μm only increases by 11% compared with the microbolometer with 300 nm Si₃N₄ film. This result reveals that the improved detectivity is limited when the thickness of Si₃N₄ exceeds 300 nm. Moreover, an excessively thick Si₃N₄ film will affect the supporting effect of the supporting arm of the microbolometer. During the preparation of the microbolometer with 600 nm Si₃N₄ film, the phenomenon of supporting arm bending and collapse occurred. Therefore, 300 nm is the optimal thickness for depositing Si₃N₄ film.

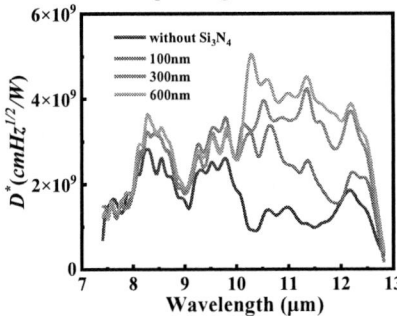

Figure 6: Detectivity of microbolometer as a function of wavelength

Figure 7 shows the frequency responsivity curves of microbolometers with 300nm Si₃N₄ films and without Si₃N₄ films. We fit the data according to the following formula:

$$R_v(f) = \frac{R_v}{(1+4\pi^2 f^2 \tau^2)^{1/2}} \quad (3)$$

Where $R_v(f)$ represents the voltage responsivity at different frequencies, f is the frequency, and τ is the thermal time constant, the points in the figure are the experimental data, and the curves are the results obtained by fitting. According to the formula fitting calculation, the thermal time constant of the microbolometer before and after the deposition of Si₃N₄ film did not change basically, which is still 33ms, indicating that the microbolometer performance after the deposition of Si₃N₄ film is stable.

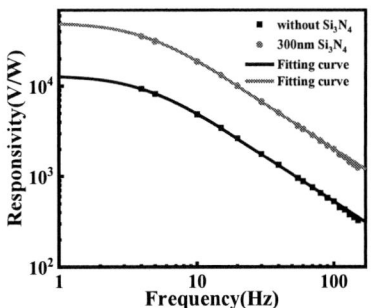

Figure 7: Frequency responsivity of the microbolometer

CONCLUSION

In this study, the detectivity of a polysilicon microbolometer based on the standard CMOS process at 7-13 μm is improved by depositing Si₃N₄ film. The experimental results show that the optimum temperature and thickness of Si₃N₄ films are 300°C and 300 nm, respectively. The responsivity of the microbolometer at the wavelength of 10-13 μm is significantly improved by 152.55%. The detectivity of the microbolometer with 300 nm Si₃N₄ at 10.5 μm improves by 208.82% and the D^* reaches 4.2×10^9 cmHz$^{1/2}$/W. Besides, the performance of microbolometer is stable before and after the deposition of Si₃N₄ film.

REFERENCES

[1] A Baobre. Silicon Based Uncooled Microbolometer[D]. New Jersey Institute of Technology. 2018.

[2] Y. Z. Guo, M. C. Luo, H. L. Ma, et. al, "Microbolometer with a salicided polysilicon thermistor in CMOS technology," Optical Express, vol. 29, November 2021, pp. 37787-37796.

[3] Abdel-Rahman M, Al-Khalli N, Zia M F, et al. Fabrication and design of vanadium oxide microbolometer[C]//AIP Conference Proceedings. AIP Publishing LLC, 2017, 1809(1): 020001.

[4] Y. Z. Guo, H. L. Ma, J. Lan, et al. Impact of Various Thermistors on the Properties of Resistive Microbolometers Fabricated by CMOS Process. Micromachines. 2022; 13(11):1869.

[5] Y. Z. Guo, H. L. Ma, K. Wang, et al., "Performance-enhanced polysilicon microbolometer in CMOS technology with a grating structure." in IEEE Photonics Journal, doi: 10.1109/JPHOT.2023.3244634.

INVESTIGATION OF VERTICALLY STACKED HORIZONTAL GATE-ALL-AROUND SI NANOSHEET ION SENSITIVE FIELD EFFECT TRANSISTOR FOR DETECTION OF C-REACTIVE PROTEIN

Yang Liu[1,2], Qingzhu Zhang[1], Junjie Li[1], Cinan Wu[2*], Lei Cao[1], Yanna Luo[1], Zhaohao Zhang[1], Shuhua Wei[3], Qianhui Wei[4], Jiaxin Yao[1], Jiawei Hu[1,3], Meiyan Qin[1], Enxu Liu[1], Yanchu Han[1,2], LianLian Li[1], YingLu Li[1,3], Tao Yang[1], Na Zhou[1], Jianfeng Gao[1], Junfeng Li[1]*

[1] Key Laboratory of Microelectronic Devices and Integration Technology, Institute of Microelectronics, Chinese Academy of Sciences, Beijing 100029, China;

[2] College of Big Data and Information Engineering, Guizhou University Guiyang 550025, China;

[3] School of Information Science and Technology, North China University of Technology, Beijing 100144, China;

[4] State Key Laboratory of Advanced Materials for Smart Sensing GRINM Group Co. Ltd., Beijing 100088, China.

*Corresponding Author's Email: lijunjie@ime.ac.cn; cnwu@gzu.edu.cn

ABSTRACT

In this work, based on advanced gate-all-around (GAA) technology, the extended sensing gate (ESG) GAA Si nanosheet (SiNS) ion sensitive field effect transistor (ISFET) sensor was fabricated and reported for the first time. Due to the GAA structures and the vertically-stacked SiNS channels, the average sensitivity of the ESG GAA SiNS ISFET sensor can be reached 58.8 mV/pH, which provides good gate control and ultra-sensitive detection capabilities. In addition, the actual minimum detection concentration of C-reactive protein (CRP) by ESG GAA SiNS ISFET sensor in 1 × PBS environment is as low as 100pg/mL, which is much lower than the normal concentration of human body.

Keywords—extended sensing gate (ESG); gate-all-around Si nanosheet (GAA SiNS); ion sensitive field effect transistor (ISFET); sensitivity; stability;

INTRODUCTION

Ion sensitive field effect transistors (ISFETs) have great application prospects for biomolecular detection, due to their high sensitivity, small size, and quick response, which are suitable for diagnose disease of biology [1]. a mass of advanced nanomaterials, e.g. one-dimensional nanowires [2], two-dimensional materials [3], have been adopted to fabricate ultra-sensitive biosensor for the detection of various biological markers such as pH, antigen protein and DNA, which aroused the great interest of scientists.

Conventionally, opening gate based ISFETs with receptors face many issues, which caused by exposing the delicate gate oxide directly to environments for sensing [4], leading to low stability, high noise, and poor stability for detection of biological markers. However, extended sensing gate (ESG) ISFETs configured in an extended-gate architecture with the transistor gate extended through metal/via stacks to a remote metal sensing layer are competitive advantage in noise reduction, sensing layer stability. GAA is considered to be the most potential structure beyond 3 nm node. Because of its excellent gate control ability, it can significantly suppress the short channel effect and has the advantages of high integration and low working voltage. The ESG GAA SiNS ISFET can not only reduce the impact of environments for sensing on the device, working voltage and the sensitive layer has good stability, but also improve the sensing sensitivity and achieve directly integration with logic circuits.

In this paper, we fabricated a ESG GAA SiNS ISFET sensor based on the advanced GAA technology, which is fabricated by CMOS process. A 58.8 mV/pH of the average sensitivity is achieved. The results demonstrate that the ESG GAA SiNS ISFET sensor has obviously detected C-reactive protein (CRP) in 1 × PBS and 200 s, with the actual lowest limit of detection (LOD) as low as 100 pg/mL, thus showing excellent performance in detection.

EXPERIMENTAL

ESG GAA SiNS ISFET SENSOR PREPARATION

The ESG GAA SiNS ISFET sensors with replace metal gate (RMG) were fabricated on 200 mm Si (100) wafers, and the fabrication flow of the sensors was based on a GAA flow [5], except for a few modified fabrication steps after the metal contact process, as shown in Fig. 1(a). Multi-layer SiGe/Si stacks with 18 nm SiGe and 12 nm Si were grown. Afterwards, the patterns were transferred into beneath multilayered structure of SiO_2, SiGe/Si stacks and Si substrate by multi-steps reactive ion etching (RIE) processes. Later, an amorphous silicon dummy gate was formed, and then removes the dummy gate in the gate-last process. The SiNSs channels were formed by wet etching of SiGe materials in the fins with multi-layer SiGe/Si stacks. And then, high-k/metal gates (HKMG) stacks were deposited by atomic layer deposition (ALD) method, which forms stacked GAA SiNSs devices. The designed ESGs with 100μm × 100μm in size were formed by depositing metal (Al) and dry etching. After the deposition of SiO_2 and etch procedures, sensitive materials (HfO_2) were deposited by ALD with 10 nm(The geometric parameters of ESG GAA SiNS ISFET are shown in Table 1). The schematic diagram of the ESG GAA SiNS ISFET sensor are shown in Fig. 1(b).

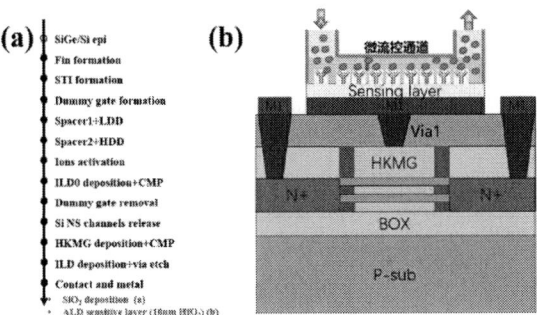

Figure 1: Process flow for the fabrication of the GAA SiNS ISFET sensor(a); The schematic diagram of the ESG GAA SiNS ISFET (b).

TABLE I
GEOMETRICAL PARAMETERS FOR ESG GAA SINS ISFET

Symbol	Geometrical Parameters	Values
L_G	Gate length	60 nm
W_{NS}	Nanosheet width	20 nm
T_{NS}	Nanosheet thickness	12 nm
N	Nanosheet numbers	2
S_{ESG}	Size of extended sensing gate	$1\times10^4\,\mu m^2$
T_{SL}	Thickness of sensing layer	10 nm

CHARSCTERIZATION OF ELECTRICAL PROPERTIES

In experiment, a Agilent 4156B semiconductor parameter analyzer was used to characterize the electrical characteristics of the fabricated ISFET sensors. The electrical properties of the prepared ESG GAA SiNS ISFET devices are explored. The typical transfer curves (I_D-V_G) and output curves (I_D-V_D) of a L_g is 60 nm n-type sensor are shown in Fig. 2(a) and (b), respectively. Compared with some common commercial ISFETs, the working voltage and threshold voltage of the device is smaller, which conducive to consume low power. The value of threshold voltage(V_t) is 0.139 V; The subthreshold swing (*SS*) is 95.2 mV/dec; The ratio of the on-state and off-state current (I_{on}/I_{off}) ratio is over 3.87×10^5 and Drain-Induced Barrier Lowering (*DIBL*) is

12.7 mV/dec. Meanwhile, with low leakage current, demonstrating a good controllability of the gate.

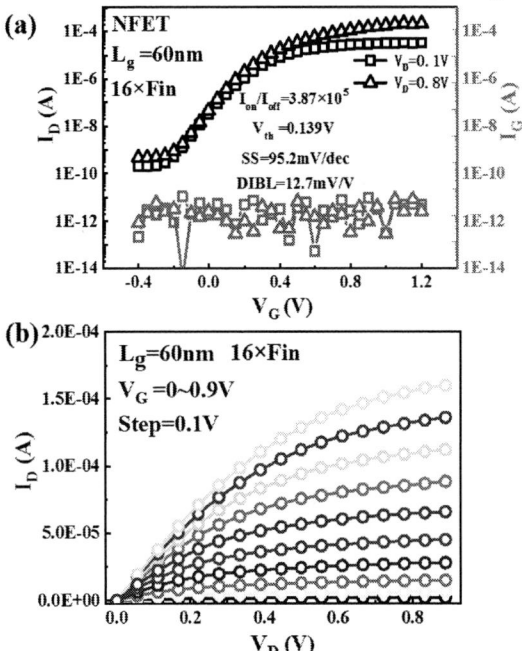

Figure 2: The typical transfer curves (I_D-V_G) (a);and output curves (I_D-V_D) (b).

SENSITIVITY CHARACTERIZATION

The pH sensing characteristics of the sensor were explored by using the prepared ESG GAA SiNS ISFET sensors to detect different pH=7-5-3-7-9-11 solutions. The transfer curves of the sensors measured by sweeping the V_G were carried out in various pH buffer solutions (see Fig.4). During the detection, the V_D was set to 0.1V, and obtain the transfer curves under different pH solutions. The average sensitivity is as high as 58.8 mV/pH, which near the Nernst limit (~59.2 mV/pH).

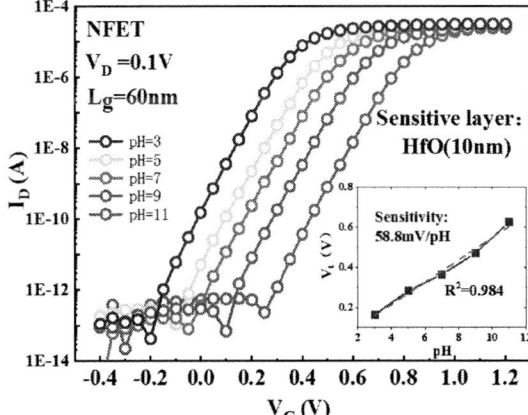

Figure 3. Transfer curves of the ESG GAA SiNS ISFET sensors in various pH buffer solutions.

DETECTION OF C-REACTIVE PROTEIN

In the bioassay of CRP, CRP-MAB (20 μg/mL) antibody was immobilized at $V_D=0.2V$ (working voltage of 0.2V) and $V_G=0V$. After adding 100 pg/mL, 1 ng/mL, 10 ng/mL, 100 ng/mL, 1 μg/mL, 10 μg/mL, 20 μg/mL, ESG GAA SiNS ISFET sensor outputs an obvious current gradient response in 200 s (see Fig.4).

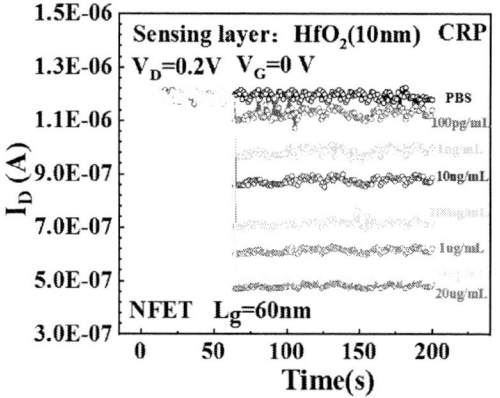

Figure 4. Current response corresponding to different CRP concentrations

$\Delta I_D/I_{D0}$ is extracted from the current gradient response of different concentrations of CRP, where ΔI_D is the current change in the testing process, and I_{D0} is the mean value of the initial current. With the increase of CRP concentration, the $\Delta I_D/I_{D0}$ response decreases (see Fig.5). There is a good linear relationship in the concentration range of 100 pg/ml~20 μg/mL, and the linear fitting degree is $R^2=0.994$. When the concentration is more than 10 μg/mL, $\Delta I_D/I_{D0}$ decreases slowly, which is limited by the fixed amount of antibodies in the sensitive layer.

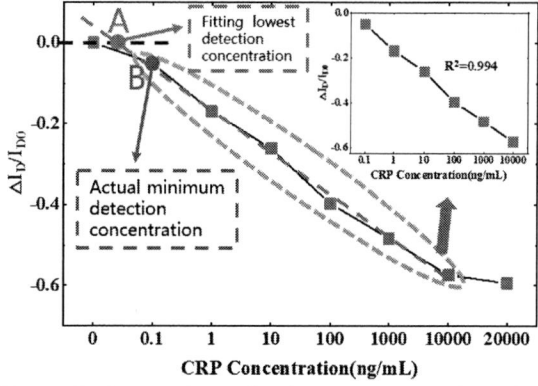

Figure 5. Curve relationship between extracted $\Delta I_D/I_{D0}$ and CRP concentration

It is concluded that the lowest detection concentration of CRP is (point A), and the actual minimum detection concentration of this question is 100 pg/mL (point B), which is much lower than the normal human CRP protein concentration (< 8μg/mL). When the concentration is greater than 10 μg/mL, although the decrease of $\Delta I_D/I_{D0}$ slows down, the maximum detection limit is greater than 20 μg/mL. In theory, the maximum detection concentration can be controlled by changing the area of the extended gate, so as to predict a variety of diseases in the actual detection process. This means that the ESG GAA SiNS ISFET sensor has excellent controllability to the detection concentration range of CRP, and preliminarily verifies the feasibility of ESG GAA SiNS ISFET sensor in practical application.

CONCLUSION

The ESG GAA SiNS ISFET sensor based on the advanced GAA technology is fabricated by CMOS process. The threshold voltage of the device is smaller ($V_t=0.139\ V$) and the ratio of the on-state and off-state current (I_{on}/I_{off}) ratio is over 3.87×10^5. This helps to obtain higher sensitivity at low operating voltage. The sensitivity is as high as 58.8 mV/pH. The C-reactive protein (CRP) can be detected in real time by ESG GAA SiNS ISFET sensor, which is as low as the actual minimum of 100 pg/mL. It shows excellent sensing characteristics.

ACKNOWLEDGEMENTS

This work was supported in part by the Strategic Priority Research Program of the Chinese Academy of Sciences under Grant (XDA0330300), in part by the Supported Technical Talents Project of the Chinese Academy of Sciences (E2YR01X001), in part by the development Project of Plasma PR Strip Equipment Gate All Around(GAA) Nanowire Release Function (E2SH01X).

REFERENCES

[1] O. Synhaivska, et al., "M. Detection of Cu2+ Ions with GGH Peptide Realized with Si-Nanoribbon ISFET" Sensors 2019, 19, 4022.

[2] A. Moumen, et al., "Catalyst–Assisted vapor liquid solid growth of α-Bi2O3 nanowires for acetone and ethanol detection" Sensor Actuat B-Chem. 346 (2021) 130432.

[3] Tu H L, Zhao H B, Fan Y Y, et al. Recent developments in nonferrous metals and related materials for biomedical applications in China: a review[J]. Rare Metals, 2022, 41(5): 1410-1433.

[4] N. Gao, et al., "General Strategy for Biodetection in High Ionic Strength Solutions Using Transistor-Based Nanoelectronic Sensors" Nano Lett. 15(3) (2015) 2143-2148.

[5] Q.Z. Zhang, et al., "Structural Optimization and Electrical Characteristics Analysis of Multilayer Stacked GAA NS Devices" Nanomaterials. 2021; 11(3):646.

MONOLITHIC 3D INTEGRATION OF DENDRITIC NEURAL NETWORK WITH MEMRISTIVE SYNAPSE, DENDRITE AND SOMA ON SI CMOS

Tingyu Li, Jianshi Tang*, Junhao Chen, Xinyi Li, Han Zhao, Yue Xi, Wen Sun, Yijun Li,
Qingtian Zhang, Bin Gao, He Qian and Huaqiang Wu

School of Integrated Circuits, Beijing Innovation Center for Future Chips (ICFC), BNRist, Tsinghua
University, Beijing, China
*E-mail: jtang@tsinghua.edu.cn

ABSTRACT

We report a monolithic three-dimensional integration of dendritic neural network (M3D-DNN) with memristors-based artificial synapse, dendrite and soma on top of Si-based CMOS logic. The Si CMOS layer served as control logic fabricated in foundry. A 1k-bit artificial synaptic array was built with HfO_2-based nonvolatile memristors to implement computing-in-memory (CIM). In addition, TiO_x-based memristive artificial dendrite and NbO_xN_y-based memristive artificial soma were adopted to implement the dendritic neuron (DN) layer to process postsynaptic signals. Both the CIM and DN layers were fabricated using a BEOL-compatible process. The structural integrity and proper function of each layer in the M3D-DNN were verified. Our work demonstrates a promising architecture to efficiently implement bio-plausible artificial neural networks (ANNs).

INTRODUCTION

The rise of artificial intelligence (AI) with deep learning demands for ever increasing computing power and energy efficiency. This imposes critical challenges for conventional computing hardware based on von Neumann architecture. Inspired by human brain, neuromorphic computing with bio-mimicking devices, such as memristors, emerges as a promising paradigm to break the von Neumann bottleneck and build energy-efficient AI chips [1]. Tremendous progress has been made in the past decade to use various memristors to implement ANNs with orders of magnitudes higher energy efficiency than CPU and GPU [2-4]. Most prior works have been focused on memristor-based artificial synapses with the advantage of CIM. It should be noted that, besides synapse, dendrite and soma also play vital roles in the signal processing in biological neural networks. Their functions, such as the signal filtering and nonlinear integration of dendrite, are indispensable for the extremely low power of human brain. Recently, a novel dendritic neural network (DNN) with memristors-based artificial synapse, dendrite and soma was proposed as a more bio-plausible ANN. A board-level DNN system was built to demonstrate the classifications of both static images and dynamic human motions [5-6], exhibiting significant advantages in accuracy and energy efficiency by incorporating dendrites.

In this work, inspired by the 3D nature and complex topography of brain, we demonstrate an M3D-DNN with memristors-based artificial synapse, dendrite and soma on top of Si-based CMOS logic. The ultra-dense inter-layer vias (ILVs) in M3D could facilitate the high bandwidth data transfer across different layers with significantly reduced latency and power consumption. The memristor devices were carefully optimized for these three critical computing units to fulfill their functions.

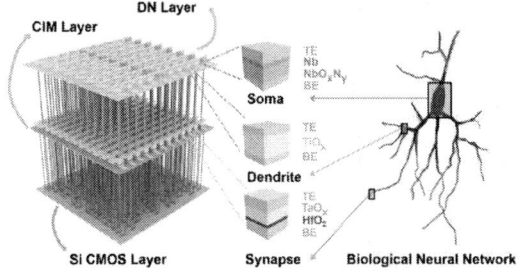

Figure 1: The architecture of M3D-DNN and its correspondence with biological neural network.

FABRICATION OF M3D-DNN

To start, the Si CMOS layer was fabricated using a standard 130nm CMOS foundry process for logic and control. The process was stopped at M4 with W vias exposed after chemical mechanical polishing (CMP). The CIM layer was then fabricated with $TiN/TaO_x/HfO_2/TiN$ memristors. The 8nm HfO_2 serving as the resistive switching layer was deposited by atomic layer deposition (ALD) at 300 °C, followed by sputtering 45nm TaO_x serving as the thermal enhanced layer. The TiN top and bottom electrodes (TE and BE) were deposited by sputtering. The memristors were then patterned by reactive ion etching (RIE) and passivated.

After that, the DN layer consisting of $Ti/TiO_x/Pd$ dendrite devices and $Pt/Nb/NbO_xN_y/Pd$ soma devices was fabricated. First, 50nm Pd was evaporated as the BE of dendrites. 30nm TiO_x was sputtered using Ti target in Ar/O_2 followed by deposition of 30nm Ti as the TE of dendrite. Next, Pd was evaporated as the BE of soma. Then, 50nm NbO_xN_y was sputtered using Nb target in $Ar/O_2/N_2$ followed by the deposition of 10nm Nb. The proportion of Ar, O_2 and N_2 was strictly controlled and the thin Nb interface layer improved the yield of soma devices

979-8-3503-1101-3/23 $31.00 © 2023 IEEE

[6-7]. Finally, Pt was deposited as the TE of soma. The process flow and chip images are presented in Figure 2.

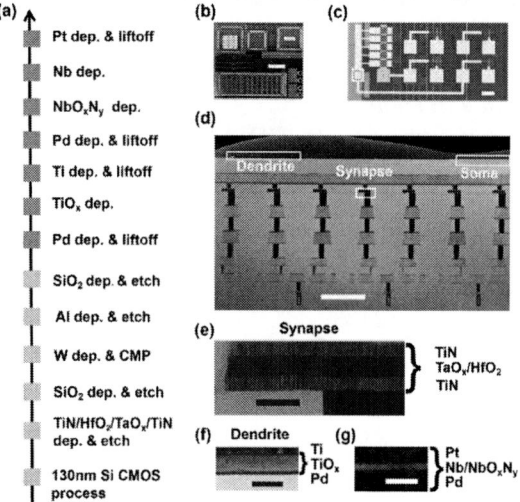

Figure 2: (a) The fabrication process flow of M3D-DNN. (b) Photo of the fabricated chip. Scale bar is 2 mm. (c) Optical image of memristors. Scale bar is 100 um. (d) Cross-sectional image of each layer. Scale bar is 2 um. (e) TEM image of HfO₂-based artificial synapse in the CIM layer (f) TEM image of TiOₓ-based artificial dendrite and (g) NbOₓNᵧ-based artificial soma in the DN layer. Scale bars in (e), (f), (g) are 50 nm.

CHARACTERIZATIONS OF M3D-DNN

Electrical properties of each layer in the chip were measured. Figure 3a shows the analog switching characteristics of the memristive synapse in the CIM layer. Figure 3b presents the programmability of 8 representative conductance states (~3bits) read for 100 cycles, where 64 devices were measured for each state. Figure 3c-d show the mapping result and corresponding error of a 32×32 matrix using the 1k-bit one-transistor-one-resistor (1T1R) synaptic array. These results confirm the excellent analog switching characteristics of artificial synapses.

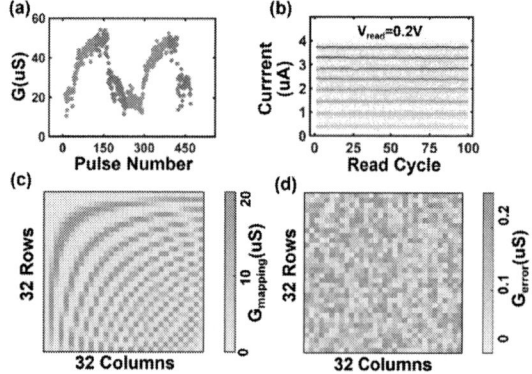

Figure 3: (a) Analog switching characteristics of a typical

cell in the 1T1R synaptic array. (b) Read noise of 8 states. (c) Mapping result $G_{mapping}$ and (d) corresponding error G_{error} when mapping a 32×32 matrix in the 1k-bit array.

For the DN layer, we first characterized the dendrite device as shown in Figure 4. The device remained off when applying a voltage below the threshold (~3V), and turned on when the bias went above the threshold (e.g. 4V) [5]. It also exhibited a nonlinear current integration behavior resembling biological dendrite (Figure 4b). Figure 5a-b illustrate the relatively small cycle-to-cycle and device-to-device variability. Figure 5c depicts the dendrite device size dependence of the current response, confirming the interfacial switching mechanism. It also provided a knob to tune the device resistance to match with the soma device in the DN unit.

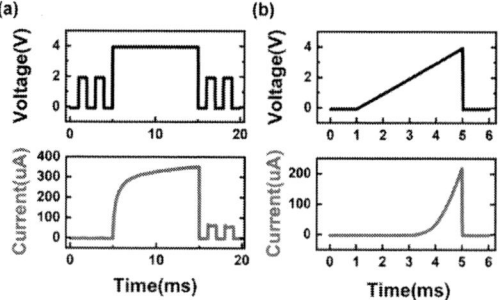

Figure 4: (a) Current response of the artificial dendrite device in the off and on states, exhibiting a filtering property. (b) Current response of the dendrite device, showing a nonlinear integration behavior.

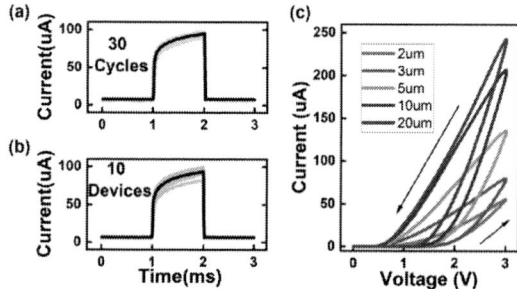

Figure 5: (a) Measured 30 cycles of a typical dendrite device with the size of 3um×3um under a 1ms pulse. (b) Measured 10 devices with the same size of 3um×3um. (c) I-V characteristics of dendrite devices with different sizes.

Furthermore, Figure 6a shows the threshold firing property of the soma device with a large window (~1V) between the threshold voltage (V_{th}) and hold voltage (V_{hold}). Besides, oscillation neuron characteristics were also demonstrated in Figure 6b-c. As the bias voltage increased, the oscillation frequency also increased linearly. The stable oscillations indicate low variability in V_{th} and V_{hold}, thanks to the N dopants in the NbOₓNᵧ layer that help confine the migration of oxygen vacancies [7].

979-8-3503-1101-3/23 $31.00 © 2023 IEEE

NEURAL NETWORK SIMULATION

Using the above characterized artificial synapse, dendrite and soma units, a bio-plausible DNN can be implemented as illustrated in Figure 7a-b. Figure 7c shows the neuron-firing rate using the street-view house numbers (SVHN) dataset to benchmark the performance of our M3D-DNN. A high accuracy of ~89.7% was achieved by incorporating artificial dendrite, which helps improve the accuracy and reduce power consumption [5]. Figure 8 reveals that M3D-DNN could achieve 2923× lower power consumption than GPU and 4.3× faster speed than 2D baseline owing to ultra-dense ILVs and ultra-high on-chip bandwidth in M3D architecture [5,8-10].

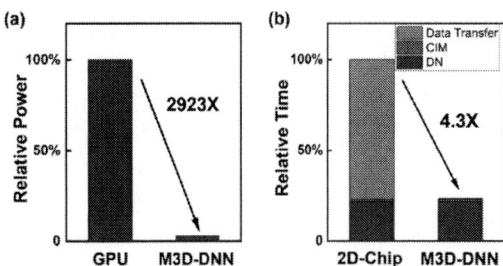

Figure 8: (a) Power consumption benchmark of M3D-DNN and GPU executing the same network. (b) Execution time of M3D-DNN and 2D baseline.

CONCLUSION

To sum up, we have designed and fabricated an M3D-DNN chip with HfO_2-based memristive synapse in the CIM layer, TiO_x-based dendrite and NbO_xN_y-based soma in the DN layer on top of Si CMOS logic layer. These three different types of memristors were carefully engineered to be compatible with BEOL process. Structural integrity and electrical properties were characterized to verify the performance of M3D-DNN. The presented M3D-DNN architecture could endow neuromorphic computing hardware with enhanced performance as well as significantly reduced energy consumption and latency.

ACKNOWLEDGMENT

This work was in part supported by the National Key Research and Development Program of China (2021YFF1200803), National Natural Science Foundation of China (92264201).

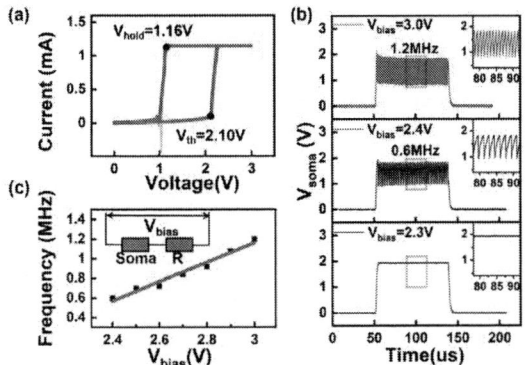

Figure 6: (a) I-V characteristic of a soma device with the size of 10um×10um. (b) Output waveforms of the oscillation neuron circuit under 100us-wide pulses with amplitudes of 3.0 V, 2.4V and 2.3V. (c) Oscillation frequency of the neuron circuit (R=3kΩ) under different V_{bias}.

REFERENCES

[1] J. Tang et al., *Adv. Mater.*, vol. 31, pp. 1902761, 2019.

[2] M. Prezioso et al., *Nature*, vol. 521, pp. 61-4, 2015.

[3] P. Yao et al., *Nature*, vol. 577, pp. 641-646, 2020.

[4] W. Wan et al., *Nature*, vol. 608, pp. 504-512, 2022.

[5] X. Li, J. Tang, Q. Zhang et al., *Nat. Nanotechnol.*, vol. 15, pp. 776-782, 2020.

[6] X. Li et al., *Adv. Mater.*, pp. 2203684, 2022.

[7] J. Chen et al., *in IEEE Trans. Electron Devices*, vol. 69, pp. 6686-6692, 2022.

[8] Y. -J. Lee, P. Morrow and S. K. Lim, *2012 IEEE/ACM ICCAD*, pp. 539-546, 2012.

[9] M. Shulaker, G. Hills, R. Park et al., *Nature*, vol. 547, pp. 74-78, 2017.

[10] Y. Li et al., *2021 IEEE IEDM*, pp. 21.5.1-21.5.4, 2021.

[11] R. An et al., *2022 IEEE IEDM*, pp. 18.1.1-18.1.4, 2022.

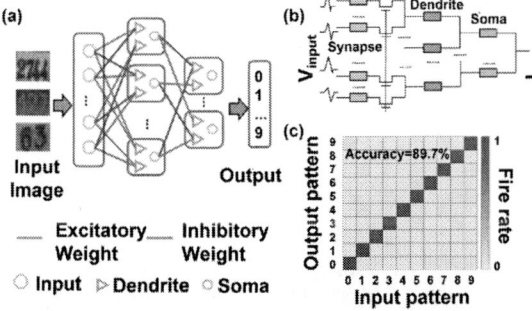

Figure 7: (a) Schematic of the implemented DNN where synapses represent tunable weights, dendrites process hierarchical post-synaptic information and somas provide the integration and firing function to yield the final output. (b) The equivalent circuit model. (c) Firing rate and recognition accuracy of SVHN dataset for M3D-DNN.

SIMULATION INVESTIGATION ON THE CHARACTERISTICS OF GAN-BASED MULTI-QUANTUM WELLS MICRO-LEDS

Pengfei Ye[1], Youshan Gui[2], Yue Li[1], Ding Chen[1], Jinghao Yu[2], Yi Tong[2], and Haixia Da[1]*

[1] College of Electronic and Optical Engineering & College of Flexible Electronics (Future Technology), Nanjing University of Posts and Telecommunications, Nanjing 210023, China.
[2] College of Integrated Circuit Science and Engineering, Nanjing University of Posts and Telecommunications, Nanjing 210023, China.
*Corresponding Author's Email: eledah@njupt.edu.cn

ABSTRACT

Micro-LED has been recognized as a promising candidate for next-generation display and emerging communication technology owing to its excellent properties, such as high modulation bandwidth, high saturation current density and outstanding luminous power efficiency. In this study, the characteristics of the GaN-based multi-quantum wells micro-LEDs with various epitaxial structures were investigated based on the theoretical simulations using Silvaco-TCAD software. The structural parameters including Al content in $Al_xGa_{1-x}N$ of the electron-blocking layer, In content in $In_yGa_{1-y}N$ of the quantum well layer, and quantum-well periods are regulated to reveal their influences on the optical and electrical properties of the device. These are for the purpose of exploring the optimal performance of micro-LEDs to implement multifunction applications, for instance, micro-display and visible light communication.

INTRODUCTION

Micro-LED display is the most marketable technology compared with traditional large-size light-emitting diode (LED), liquid crystal display (LCD) and organic light-emitting diode (OLED), which has excellent characteristics of self-luminescence without backlight, and lower power consumption, longer life, higher saturation current density, and faster response time [1-3]. As a light source for communication, its high light-output power and higher modulation bandwidth have been confirmed to achieve high-speed data transmission in free-space visible light communication and underwater optical wireless communication, making it more competitive in the industry of 6G communication [4-7].

For GaN-based micro-LEDs, careful consideration should be given to the heterojunction structure with InGaN/GaN multi-quantum wells (MQWs), as it plays a significant role in determining the energy band structure, which ultimately impacts the performance of the device [8]. Moreover, the implementation of an electron blocking layer (EBL) is crucial to effectively prevent carrier leakage, which may result in undesirable leakage currents that degrade the optical and electrical characteristics of the micro-LEDs [9].

In this study, to investigate the impacts of the EBL and MQWs on the performance of GaN-based micro-LEDs in terms of design parameters, ATLAS build-in Silvaco-TCAD was used to simulate the devices by adjusting the Al content in $Al_xGa_{1-x}N$ of EBL ($0<x<1$), the In content in $In_yGa_{1-y}N$ for $In_yGa_{1-y}N$/GaN MQWs ($0<y<1$), and the number of periods of MQWs, respectively. The obtained results provide theoretical guidance for fabricating high-performance micro-LED devices.

SIMULATION MODELING AND MATERIAL PROPERTIES

The ATLAS module of the commercial Silvaco-TCAD is used for the simulation of micro-LEDs to study the electrical and optical properties. A typical GaN-based micro-LED device model is designed, shown in Figure 1a. The structure is consisted of, a sapphire substrate, an unintentionally doped GaN buffer layer (u-GaN), an n-type doped GaN layer (n-GaN), an In content-tunable $In_yGa_{1-y}N$/GaN MQWs layer ($0<y<1$), a p-type doped $Al_xGa_{1-x}N$ for EBL (p-$Al_xGa_{1-x}N$, $0<x<1$), a p-type doped GaN layer (p-GaN), a current spreading layer of indium tin oxide (ITO), as well as metal (Au) contact layers. Figure 1b illustrates a partial enlargement of the simulation model of micro-LEDs in Silvaco TCAD.

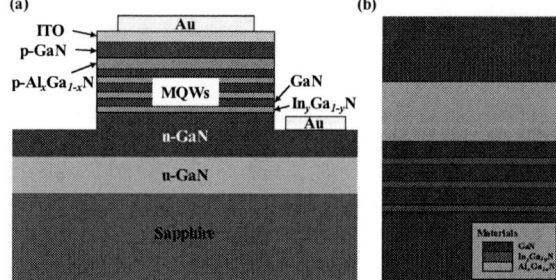

Figure 1: (a) Schematic diagram of the GaN-based MQWs micro-LED; (b) partial zoomed-in view of a two-dimensional device model for Silvaco simulation

RESULTS AND DISCUSSION

The simulation process is carried out by adjusting the

structural parameters so that the Al component of the EBL and the In component of the MQWs are respectively varied in steps from 0.10 to 0.35. In Figure 2a, as the Al content of the EBL increases, the threshold voltage of the device becomes significantly larger, gradually increasing from 2.68 V ($x = 0.10$) to 4.11 V ($x = 0.35$), and the forward series resistance remains nearly constant. Figure 2b demonstrates a red-shift trend in the peak of the electroluminescence (EL) spectra, and the corresponding intensity of normalized power spectrum decreases when the Al content increases from 0.10 to 0.35. It can be explained that the gradual increase of the Al content in $Al_xGa_{1-x}N$ of EBL raises the potential barrier of the valence band for $Al_xGa_{1-x}N$, which further impedes hole injection into the active region. Related to this, the decrease in hole concentration then affects the efficiency of the device [9].

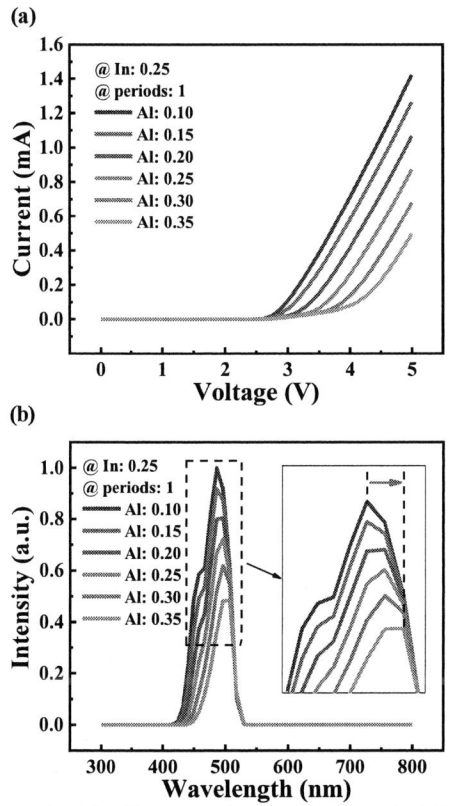

Figure 2: (a) I-V curves of micro-LEDs with different Al contents; (b) Normalized EL spectra of micro-LEDs with different Al contents

As shown in Figure 3a, the threshold voltage exhibits minimal variation, from 3.19 V ($y = 0.10$) to 3.47 V ($y = 0.35$), with the In content in the $In_yGa_{1-y}N/GaN$ of MQWs gradually increasing, while the forward series resistance gradually increases as a result. The significant red-shift of the peak of the EL spectra is shown in Figure 3b, and the intensity of normalized power spectrum gradually decreases. It is generally considered that, on one hand, the increase in the In content leads to the formation of mismatched dislocations and the prominent role of nonradiative recombination centers, severely reducing the carrier lifetime and resulting in an increase in the threshold voltage of micro-LEDs [10, 11]. On the other hand, the increase in the In content enhances the strength of the polarization field and intensifies the quantum confined Stark effect (QCSE), causing a red-shift in the peak of emission spectrum [12].

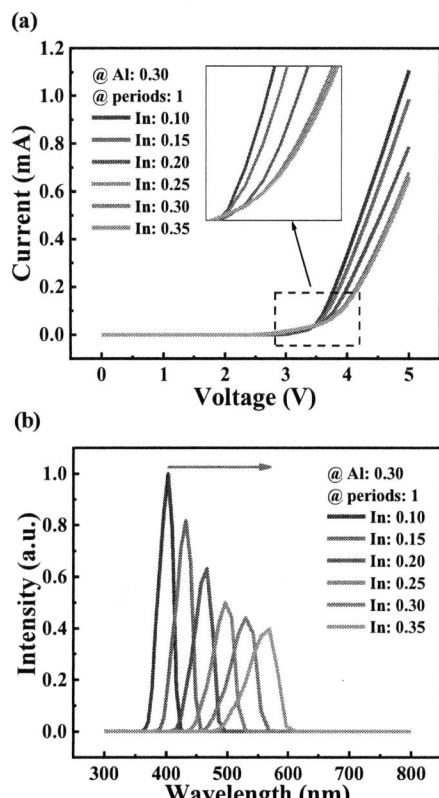

Figure 3: (a) I-V curves of micro-LEDs with different In contents; (b) Normalized EL spectra of micro-LEDs with different In contents

In addition, Emission spectra of micro-LEDs with different quantum-well periods ranging from 1 to 4 were simulated separately, as shown in Figure 4. As the number of periods decreases, a blue-shift trend is observed in the peak of the spectrum, and the full width at half maximum (FWHM) of the spectrum gradually widens. The decrease in the period number results in an increased concentration of carriers in the active region, where the carrier relaxation time is significantly shorter than their lifetime. This further weakens the QCSE, and the intensity of the polarization effect is correspondingly weakened, which results in a blue-shift of the spectral peak [13].

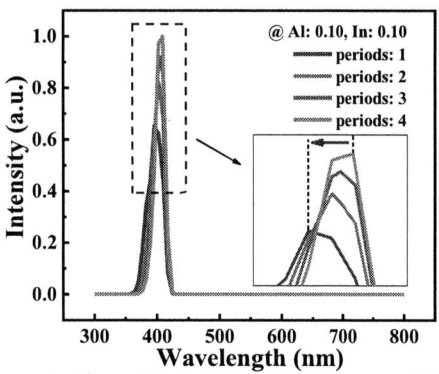

Figure 4: Normalized EL spectra of micro-LEDs with different quantum-well periods

CONCLUSION

In this paper, simulations of GaN-based MQWs micro-LEDs with different structural parameters were conducted using Silvaco-TCAD. The research findings demonstrate that with an increase in the Al content, the threshold voltage of the micro-LEDs increases, the peak wavelength of the emission spectra exhibits a red-shift, and the intensity gradually decreases. As the In content increases, the spectral peak also exhibits a significant red-shift. Moreover, a decrease in the number of quantum-well periods leads to a blue-shift in the spectral peak and a gradual broadening of the FWHM. The above research provides theoretical guidance for the preparation and application of high-performance micro-LEDs, particularly in the realms of micro-display and visible light communication.

REFERENCES

[1] P. Tian, D. McKendry, Z. Gong, et al. *Journal of Applied Physics*, vol. 115, 2014, 033112.

[2] K. Ding, V. Avrutin, N. Izyumskaya, et al. *Applied Sciences*, vol. 9, 2019, 1206.

[3] Z. Wang, X. Shan, X. Cui, and P. Tian. *Journal of Semiconductors*, vol. 41, 2020, 041606.

[4] X. Liu, R. Lin, H. Chen, et al. *ACS Photonics*, vol. 6, 2019, pp. 3186-3195.

[5] S. K. James, Y. M. Huang, T. Ahmed, et al. *Applied Sciences*, vol. 10, 2020, 7384.

[6] P. Tian, X. Liu, S. Yi, et al. *Optics Express*, vol. 25, 2017, pp. 1193-1201.

[7] L. Wang, Z. Wei, C. Chen, et al. *Photonics Research*, vol. 9, 2021, pp. 792-802.

[8] A. Vaitkevičius, J. Mickevičius, D. Dobrovolskas, et al. *Journal of Applied Physics*, vol. 115, 2014, 213512.

[9] S. Han, D. Lee, S. Lee, et al. *Applied Physics Letters*, vol. 94, 2009, 231123.

[10] D. Cherns, S. Henley, F. Ponce. *Applied Physics Letters*, vol. 78, 2001, pp. 2691-2693.

[11] J. Wierer, A. Fischer, and D. Koleske. *Applied physics letters*, vol. 96, 2010, 051107.

[12] S. Nakamura. *Science*, vol. 281, 1998, pp. 956-961.

[13] M. Leroux, N. Grandjean, M. Laügt, et al. *Physical Review B*, vol. 58, 1998, R13371.

NEAR-INFRARED SENSITIVITY ENHANCEMENT OF CMOS IMAGE SENSOR WITH GERMANIUM ON SILICON STRUCTURE

Hui Chen*, Chenchen Qiu, Zhengying Wei, Chang Sun, Jun Qian, **and** *Yufei Peng*

Shanghai Huali Microelectronics Corporation, Shanghai 200433, China

*Corresponding Author's Email: chenhui@hlmc.cn

ABSTRACT

Silicon sensor has gained attention due to an increased demand for near-infrared CMOS image sensors for mobile imaging, digital cameras, industrial surveillance, and intelligent driving et.al. However, due to the low absorption rate of silicon for wavelength near 700 nm ~ 1100 nm, the silicon sensor has low detection efficiency which limits its application in near-infrared region. The detection efficiency represents the ability of an image sensor to convert photons into electrons and directly determines the image quality of image sensor. It is not only dependent on the wavelength, but also dependent on material. To increase the detection efficiency of the silicon sensor, germanium was widespread used because of the same lattice structure as silicon and higher optically responsive than silicon. In this paper, a new type of the image sensor with germanium integrated on silicon (Ge on Si) image sensor structure is presented and the detection efficiency of Ge on Si structure sensor was studied. It was demonstrated that the detection efficiency of Ge on Si structure sensor is 2.3 ~ 600 times higher than that of the silicon sensor for near-infrared wavelength of 700 nm ~ 1100 nm.

INTRODUCTION

With the increasing demand for ultra-high resolution image sensors (CIS), the pixel size of image sensors is getting smaller and smaller, and the minimum pixel size has reached about 0.56 μm at present [1-3]. The reduction of pixel size continuously reduces the detection efficiency of short wave near infrared light (SW-NIR, wavelength range from 700 nm to 1100 nm). This will seriously reduce the quality of the image, resulting in image color distortion. Therefore, how to ensure the detection efficiency of small pixel CIS to near infrared light becomes the key to restrict the further development of small pixel CIS. To increase the detection efficiency of the silicon sensor, germanium was widespread used because of the same lattice structure as silicon and higher optically responsive than silicon [4-7]. In this paper, a new type of image sensor with germanium integrated on silicon (Ge on Si) image sensor structure is presented and the detection efficiency of Ge on Si structure sensor was studied. Compared with conventional silicon-based CIS, the detection efficiency of the novel Ge on Si structure sensor for SW-NIR light is improved by 2.3 ~ 600 times, which

can greatly improve the imaging performance of small pixel CIS products. This method is also suitable for CIS field of large pixel security monitoring which has high demand for short wave near infrared light detection.

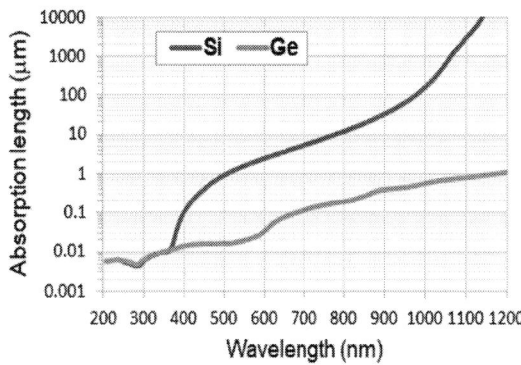

Figure 1: Comparison of the Absorption length of Si and Ge for different wavelengths.

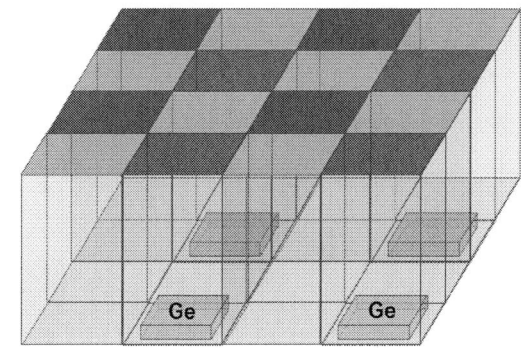

Figure 2: The schematics diagram of the novel Ge on Si image sensor.

As shown in Figure 1, the absorption length of silicon for short-wave near infrared light ranged from 5 μm to 2800 μm, lead to the detection efficiency of conventional silicon based CIS (~ 3 μm) for short-wave near infrared light greatly reduced. The near-infrared light with wavelengths greater than 700nm cannot be effectively detected by conventional silicon-based CIS, which seriously restricts the key to the further development of small-pixel CIS. Because of the absorption length germanium for SW-NIR light is just 0.1 μm ~ 0.8 μm, and the band gap of germanium is 0.66 eV, which is suitable for the absorption of SW-NIR to generate photo electrons. Therefore, the detection efficiency of SW-NIR light can be greatly improved by introducing Ge process into conventional silicon-based CIS. The schematic diagram of the novel germanium silicon-based CMOS image sensor (Ge on Si-based CIS) structure proposed in this paper is shown in Figure 2. The novel Ge on Si-based CIS is composed by adding etched filled germanium (0.8 μm) process under the corresponding pixel in the red light detection region. The Si thickness is 2.2 μm, and the total thickness of the Ge/Si stack is 3 μm.

STRUCTURE AND METHOD

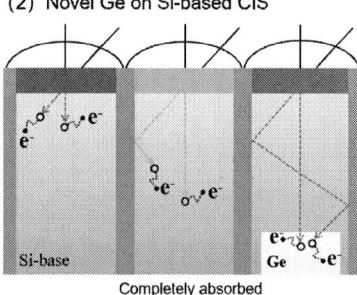

Figure 3: Comparison diagram of the detection for RGB wavelengths between novel Ge on Si-based CIS and conventional Si-based CIS.

To increase the detection efficiency of the image sensor for SW-NIR light, Ge process was introduced into the conventional Si-based CIS. Figure 3 shows a schematic comparison of the detection for RGB wavelength light between the novel Ge on Si-based CIS and the conventional Si-based CIS. It can be seen from Figure 3 (1) that in conventional Si-based CIS, part of the red light cannot be effectively absorbed and transmitted. This is due to the long absorption length of red light in Si (One absorption length of 700 nm wavelength in Si is 5 μm). In order to improve the absorption rate of CIS for red

light, Ge is introduced at the pixel corresponding to the red color filter. Since the absorption length of red light by Ge is only 0.1 μm at a wavelength of 700 nm, the introduction of 0.8 μm Ge can effectively promotes the red light absorption in CIS. The schematic of red light absorption in the novel Ge on Si-based CIS is shown in Figure 3 (2). In this study, the TCAD software package was used to simulate the process of photoelectron photoelectrons generated in the image sensor [8].

RESULTS AND DISCUSSION

Figure 4 shows the comparison of detection efficiency between conventional Si-based CIS with 3 m thickness and the novel Ge on Si-based CIS with the same thickness (0.8 μm Ge and 2.2 μm silicon stack, total thickness is also 3 μm) in the range of 300 ~ 1100 nm wavelength. It can be seen that the detection efficiency of the novel Ge on Si-based CIS is 2.3 ~ 600 times higher than that of the conventional Si-based CIS in the SW-NIR wavelength. Compared with the conventional Si-based CIS, the detection efficiency of novel Ge on Si-based CIS for 700 nm light was increased by about 2.3 times from 43% to 100%, and the detection efficiency of higher wavelength 1100 nm light has been increased from about 0.105% to 63%, which is 600 times.

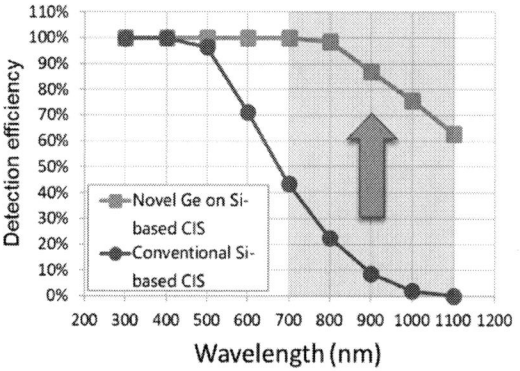

Figure 4: Comparison of detection efficiency for wavelength in range of 300 nm to 1100 nm between novel Ge on Si-based CIS and conventional Si-based CIS.

CONCLUSIONS

The low detection efficiency of conventional small pixel Si-based CIS for short-wave near-infrared light leads to imaging color distortion and seriously reduces imaging performance. In this paper, a novel Ge on Si-based CIS is presented to improve the detection efficiency of by adding Ge process. Compared with conventional Si-based CIS, the detection efficiency of the novel Ge on Si-based CIS for SW-NIR light is improved by 2.3 ~ 600 times, and the imaging performance can be greatly improved. This method is also suitable for large-pixel security monitoring CIS field which has high demand for short-wave near-infrared light detection.

ACKNOWLEDGEMENTS

The authors would like to expresses their deep gratitude to the Huali Microelectronics Corporation for proving an excellent research environment to complete this work.

REFERENCES

[1] J. Xu, Q. Chen and Z.Y. Gao, IEEE Electron Devices Society, vol. 123, 2021, pp. 27-35.

[2] S.Y. Chai and S.H. Cho, Crystals, 2021, pp. 1106.

[3] E.Ponizovskaya Devine, A. S.Mayet, A. Rawat, A. Ahamed, S. Wang, A. F. Elrefaie, T. Yamada and M.S. Islam, IEEE Sensors Journal, 2022, pp. 1-6.

[4] A. Köllner, Z.Yu, M.Oehme, J. Anders, M. Kaschel, J. Schulze and J.N. Burghartz, IEEE Sensors Journal, 2019, pp. 1-4.

[5] C. Jan, P. Bai, J. Choi, G. Curello, S. Jacobs, J. Jeong, K. Johnson and D. Jones, et al, Proceedings of the IEEE International Electron Device Meeting, 2005, pp. 65.

[6] L. Colace, G. Masini, V. Cencelli, F. DeNotaristefani and G. Assanto, IEEE Journal of Quantum Electronics, vol. 43(4), 2007, p. 311.

[7] R. Kaufmann, G. Isella, A. Sánchez-Amores, S. Neukom, A. Neels, L. Neumann and A. Brenzikofer, Journal of Applied Physics, vol. 110, 2011, pp. 023107.

[8] SILVACQ International - ATHENA user's guide.

DIFFERENTIAL EVOLUTION WITH MULTIVARIATE GAUSSIAN SAMPLING FOR SENSOR ARRANGEMENT

Kuiling Du[1,2,], Gang Tang[3]*

1. National Frontiers Science Center for Industrial Intelligence and Systems Optimization, Northeastern University, Shenyang, 110819, China.
2. Key Laboratory of Data Analytics and Optimization for Smart Industry (Northeastern University), Ministry of Education, Shenyang, 110819, China.
3. China Construction Fifth Engineering Bureau Ltd, Changsha, Hunan, China 410004
*Corresponding Author's E-mail: Kuiling0649@163.com

ABSTRACT

Sensor arrangement problems with the unknown number of sensors and obstacles in an area to be covered by sensors, which are complicated for common evolutionary algorithms to search for the global optimum. This paper presented a new evolutionary algorithm to solve the problem, which utilized hybrid mutation and a specific crossover strategy. Sampling the best individual of the population using the multivariate Gaussian distribution during the mutation process, and the individuals generated by multivariate Gaussian sampling would skip the crossover to selection. It has been proved that the new algorithm has better solution quality and faster convergent speed than classic differential evolution (DE).

Keywords—differential evolution; sensor arrangement; local search; multivariate Gaussian distribution; global optimization.

INTRODUCTION

Sensor arrangement, a combinational problem aims to find a set of sensor coordinates so that the area can be covered as much as possible with fewer sensors. The research on this problem can be applied in different fields, such as base-station localization problems in wireless communication, deployment of tags/antennas in radio-frequency identification (RFID) systems, camera arrangement in vision systems, and so on [1].

Different ways have been used to solve sensor arrangement problems, and some of the most promising methods are evolutionary algorithms, because of their stochastic search and parallel computation. The evolutionary algorithms mentioned above include genetic algorithms (GAs) [2], particle swarm optimization (PSO) 错误!未找到引用源。, simulated annealing (SA) [4], differential evolution (DE) [5], etc. These algorithms can be applied in multi-objective optimization and single-objective optimization.

Developing from univariate Gaussian distribution (or Gaussian distribution), multivariate Gaussian distribution plays an important role in statistics. Moreover, the researchers use it as local research to sample data near targets, with the aim of accelerating the convergence of evolutionary algorithms [6].

In this paper, we introduce a differential evolution algorithm that the multivariate Gaussian sampling and specific crossover are adopted, this algorithm is named differential evolution with multivariate Gaussian sampling (DEMGS). Experiments were conducted to inspect the performance of DEMGS. The results show that multivariate Gaussian sampling accelerates the optimization process, and improves the quality of the solutions.

METHODOLOGY

1. Optimization Model of Sensor Arrangement

In evolution algorithms, there are a lot of candidate solutions, also called individuals or chromosomes. A floating vector X_{gen}^i is created to represent one individual, X_{gen}^i means the i_{th} individual in gen_{th} generation, details about the vector are as follows:

$$X_{gen}^i = \{P_{gen}^1, P_{gen}^2, P_{gen}^3, ..., P_{gen}^N, n\} \tag{1}$$

where n is the number of sensors within a predefined range, N is the maximum of n. P_{gen}^j denotes a point which is the location of the j_{th} sensor in a rectangular area. P_{gen}^j is shown by

$$P_{gen}^j = (x_{gen}^j, y_{gen}^j) \tag{2a}$$

$$-15 < x_{gen}^j < 15 \tag{2b}$$

$$-10 < y_{gen}^j < 10 \tag{2c}$$

$$(x_{gen}^j, y_{gen}^j) \notin A_o \tag{2d}$$

where the search space of P_{gen}^j is continuous, n is discrete by contrast. A_o denotes the area of obstacles inside a rectangular map, in other words, sensors will not be placed in A_o, and the sensing area of the sensor cannot cover A_o.

To evaluate individual fitness comprehensively, uncovered area ratio and invalid coverage ratio are used.

The uncovered area ratio U is calculated as follows:

$$U = \frac{A_r - A_o - A_c}{A_r - A_o} \times 100 \qquad (3)$$

where A_r is the area of the whole rectangular map, and A_c denotes the area covered by sensors, in the form of circles.

Invalid coverage ratio I is calculated as follows:

$$I = \frac{n\pi r^2 - A_c}{n\pi r^2} \times 100 \qquad (4)$$

where r is the sensing radius of the sensor. I indicates the area blocked by obstacles and the area of sensors interference.

A single objective function $F(X)$ for evaluating individuals is created, which is a function to be minimized. It should be noted that only the first n points $\{P_{gen}^j \mid j=1, 2,\ldots, n\}$ participate in the fitness evaluation for every individual. Details about $F(X)$ are as follows:

$$F(X) = 0.5U + 0.5I \qquad (5)$$

2. Proposed DEMGS

The DEMGS follows the rule of classic DE, except that DEMGS uses hybrid mutation and a specific crossover that does not affect the individuals generated by multivariate Gaussian sampling (MGS). The hybrid mutation and specific crossover is shown in Fig 1.

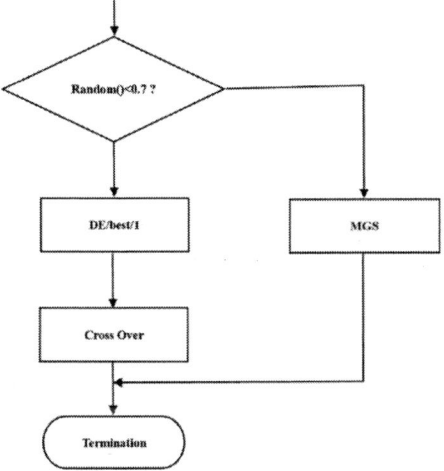

Figure. 1: The hybrid mutation and specific crossover

EXPERIMENT AND DISCUSSION

In this section, a simple but typical sensor arrangement experiment was conducted first to compare DEMGS with DE. Then we verified the performance of DEMGS in a complicated and practical sensor arrangement problem.

1. Experiment Conditions

Matplotlib in Python 3.8 was used to draw sensor layouts. The parameters of both DE and DEMGS for all experiments were as follows:

    population size $NP=160$;
    termination condition *generations*=40;
    mutation strategy DE/best/1;
    scale factor $F=0.5$;
    crossover rate $CR=0.9$;
    number of sensors $n \in [10,18]$;
    sensor sensing radius $r=4$ meters;
    map size 30 meters×20 meters.

2. Comparative Experiment

A map with one black box obstacle was loaded to examine the advantages of DEMGS over DE. Then the experiment was conducted. Results are shown in Fig. 2.

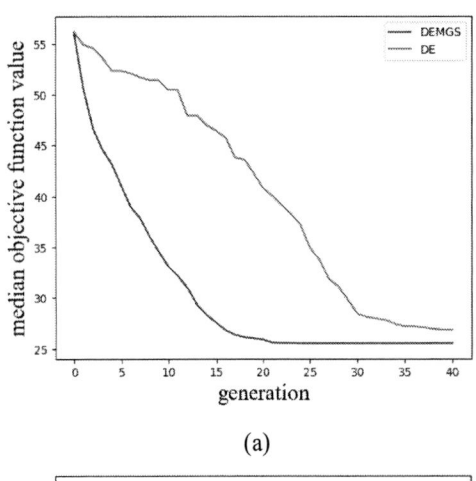

(a)

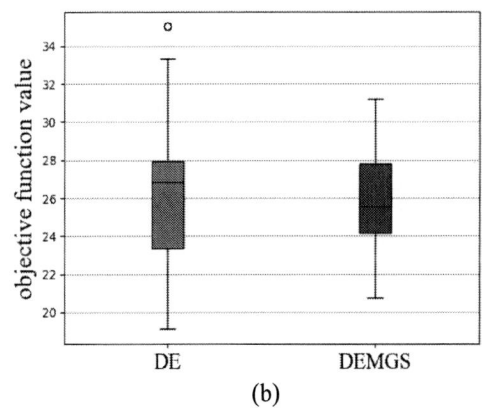

(b)

Figure. 2: (a) median objective function value of every generation over 30 runs. (b) boxplot of optimization results collected from 30 runs.

In Fig. 2(a), DEMGS have searched the optimal solution after 23 generations, DE converges slowly after 40 generations. From Fig. 2(b), we can see that DEMGS have results with smaller variance and median values than DE. In addition, an outlier exists in DE.

TABLE I. MEAN, MEDIAN, AND STANDARD DEVIATION
OF OPTIMIZATION RESULTS OVER 30 RUNS

	Mean	*Median*	*Standard Deviation*
DE	26.40	26.84	3.94
DEMGS	25.76	25.54	2.94

The optimization results from 30 runs were analyzed statistically, as shown in Table I.

Fig.2 and Table I have proved that DEMGS has faster convergence, higher fitness solutions, and more robust solution distribution than DE.

3. Application of DEMGS

We used a sensor arrangement problem that is close to the real scenario to test the performance of DEMGS. There were some rectangular and circular obstacles on this map, making the sensor arrangement task more complicated. The map and the optimization result using DEMGS are shown in Fig. 3.

(a)

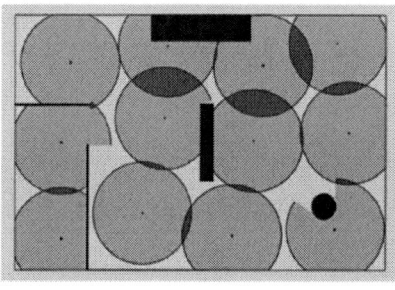

(b)

Figure. 3: (a) the map to be covered, A_o=26.72. (b) the optimization results of DEMGS, n=12, U=13.88, I=18.15, F=16.05.

From Fig. 3, we can see that sensors are placed evenly, *U* and *I* are reduced as much as possible. The above experiment shows the practical value of DEMGS.

FUNDING

This research was supported by the Major Program of National Natural Science Foundation of China (72192830, 72192835), and the 111 Project (B16009).

CONCLUSION

In general, a hybrid evolution algorithm was presented to optimize the sensor arrangement problem. Multivariate Gaussian sampling was applied in the mutation of DE, which sampled the best individual in the current population. To examine the proposed DEMGS, experiments were performed. It has been proved that DEMGS converges faster and had better solutions compared with DE. Furthermore, DEMGS works well in complicated sensor arrangement scenario.

REFERENCES

[1] Lee, Joon-Yong, Joon-Hong Seok, and Ju-Jang Lee. "Multiobjective optimization approach for sensor arrangement in a complex indoor environment." *IEEE Transactions on Systems, Man, and Cybernetics*, Part C (Applications and Reviews) 42.2 (2011): 174-186.

[2] Hanh, N. T., Binh, H. T. T., Hoai, N. X. & Palaniswami, M. S. "An efficient genetic algorithm for maximizing area coverage in wireless sensor networks," *Inf. Sci.* 488, 58–75 (2019).

[3] Philippe Blanloeuil, Nur A. E. Nurhazli, Martin Veidt, "Particle swarm optimization for optimal sensor placement in ultrasonic SHM systems," Proc. SPIE 9804, *Nondestructive Characterization and Monitoring of Advanced Materials, Aerospace, and Civil Infrastructure 2016*, 98040E (8 April 2016).

[4] Tong, K. H., Bakhary, N., Kueh, A. B. H., & Yassin, A. Y. "Optimal sensor placement for mode shapes using improved simulated annealing," *Smart Struct.* Syst 13.3 (2014): 389-406.

[5] Qiao, Dapeng, and Grantham KH Pang. "A modified differential evolution with heuristic algorithm for nonconvex optimization on sensor network localization," *IEEE Transactions on Vehicular Technology* 65.3 (2015): 1676-1689.

[6] R. A. Krohling, "Gaussian swarm: a novel particle swarm optimization algorithm," *IEEE Conference on Cybernetics and Intelligent Systems*, 2004., 2004, pp. 372-376 vol.1, doi: 10.1109/ICCIS.2004.1460443.

A 2A 4MHZ DUAL-PHASE ZDS HYSTERETIC DC-DC BUCK CONVERTER WITH PEAK EFFICIENCY ABOVE 90%

Yanye Chen[1], Quan Sun[2], Changyou Men[2] and Lenian He[1, 2]*

[1]School of Micro-Nano Electronics, Zhejiang University, Hangzhou 310058, China
[2] Hangzhou Vango Technology Inc., Hangzhou 310052, China
*Corresponding Author's Email: helenian@zju.edu.cn

ABSTRACT

This paper presents a 2A 4MHz dual-phase zero-delay synchronized (ZDS) hysteretic DC-DC buck converter for large current application such as outdoor units of air conditioners. With an on-chip synchronizer for phase synchronization, the proposed buck converter employs a ZDS hysteretic feedback control loop for accurate voltage regulation and a phase switcher cooperated with an operating mode switcher for high efficiency in full load range. The 2A 4MHz dual-phase buck converter was designed in TSMC 55nm process. The post-simulation results show that the peak efficiency reaches 90.1% when the load is 450mA, and the efficiency is almost higher than 80% in the full load range. Accurate current balance is achieved between two phases, which solves the problem of heating concentration effectively.

INTRODUCTION

With the application of more complex control algorithms and AI learning, the operating speed of the control chip of the air conditioner will be greatly improved and the power consumption will also exceed 1A. At the same time, the outdoor unit's chip still needs to ensure high reliability under external lightning strikes and surges. Improving system design backup and redundancy is an important means to improve reliability. To handle these requirements, high switching frequency and multiphase buck converter [1]-[7] is widely used. Although the multiphase converter can reduce the size of passive components, improve the speed of transient response and reliability of design, it also brings a series of problems in current balance, multiphase control, load efficiency, etc. In a multiphase converter, any mismatch between phases may make the phase currents differ largely, causing thermal imbalance and safety issue. More phases also mean more complex control. Only by accurately stagger each phase can the output voltage ripple be minimized, and the advantage of small output voltage ripple of multiphase converter can be maximized. What's more, how to improve full load efficiency is also a hot research direction.

Many current equalization strategies have been proposed to achieve equal average current of each phase. The R_{on} of power transistor is estimated indirectly by feedback input and output voltage, and then the current of each phase is adjusted by digital algorithm in [5]. It depends on multiple high-precision ADCs and complex algorithms, which performs poorly in terms of cost and

integration. Reference [6] senses the current of the upper transistor and the lower transistor at the same time and take the average value by superposition. But it is extremely challenging to achieve accurate full cycle sensing.

As for phase synchronization, there are also many methods have been proposed. Reference [2] consists of a synchronizer to set the phase and frequency of each single-phase converter. While reference [7] uses a delay-locked loop (DLL) for phase synchronization.

In this paper, a 2A 4MHz dual-phase buck converter is proposed, in which the number of switching phases and operating mode is determined by the sensed average load current. The dc resistance (DCR) of switching inductor is used to detect the average current, which is also utilized to balance the load current among the two switching phases.

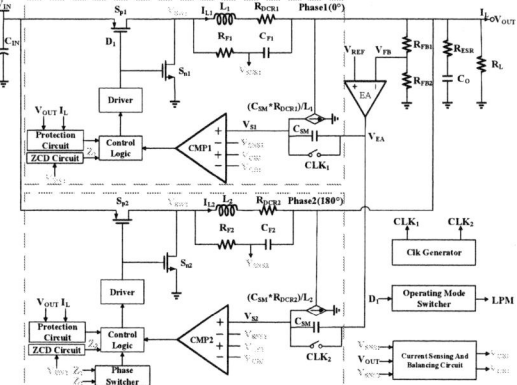

Figure 1: System architecture diagram of the dual-phase DC–DC buck converter.

DUAL-PHASE ZDS HYSTERETIC DC-DC BUCK CONVERTER

System Architecture

The proposed 4MHz dual-phase buck converter is shown in Fig. 1, which consists of 2 ZDS single-phase sub-converters [1] with 2MHz 180-interleaved clock signals (CLK_1, CLK_2).

ZDS control ensures clock synchronization and achieves stability in the whole duty cycle range compared with peak/valley current mode PWM control [1]. Assuming there is no mismatch between two phase currents, the working principle of the proposed buck converter is shown in Fig. 2. V_{SNS} is a ripple compensation signal whose DC value is equal to the output voltage V_{OUT}. By setting $R_F*C_F=L/R_{DCR}$, the AC value of V_{SNS} contains inductor current ripple. Take the first phase as an example, V_{SNS1} increases during DT and decreases during (1-D)T in the steady state[2]. At t_1, the rising edge of CLK_1 arrives and V_{S1} is initialized as the output voltage V_{EA} of the error amplifier (EA). The V_{SNS1} at this moment determines the lower limit of the hysteresis window (V_{L1}). Then V_{S1} drops with a fixed slope due to capacitor discharge until the next rising edge comes. The falling slope k_{VS1} of V_{S1} is

$$k_{VS1} = \frac{V_{out}R_{DCR1}}{L_1} = \frac{V_{out}}{R_{F1}C_{F1}} \quad (1)$$

So V_{S1} down tracks with the same slope of V_{SNS1}, creating a near-zero hysteretic window for (1-D)T. At t_2, V_{S1} intersects with V_{SNS2} during the falling process and V_{CMP1} is turned to low level. The V_{SNS1} at this moment determines the higher limit of the hysteresis window (V_{L2}).

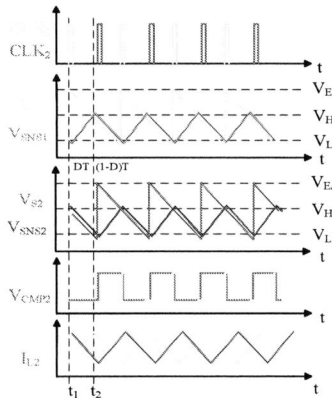

Figure 2: Waveform diagrams of control mode based on ZDS hysteresis.

It can be seen that CLK_1 fixes the switching frequency so that the output voltage noise can be easily eliminated. EA can improve the accuracy of the output voltage by adjusting the loop slightly. In addition, V_{SNS} is able to exit the hysteretic boundary without delay, that results in quicker system response.

Average Inductor Current Sensing Circuit

It is necessary to design an accurate inductor current sensing circuit because both dual-phase current balancing and phase switching need to detect the inductor current first. Considering the instantaneous current sensing method is more difficult to design and has higher power consumption, the average current sensing method is adopted in this design.

Fig. 3 shows the proposed inductor current sensing circuit based on average current sensing, in which resistance R_F and capacitor C_F are connected in parallel at both ends of the inductor. By setting $L/R_{DCR}=R_F*C_F$, the voltage of the capacitor C_F is equal to $I_{L,AVG} \times R_{DCR}$. The amplifier, RC filter, PMOS transistors M_1 and M_3, resistor R_1 and capacitor C_F form a negative feedback loop to make the voltage of the capacitor C_F equal to the voltage of the resistor R_1. If $R_1:R_2=1:1$, V_{CS} is equal to $I_{L,AVG} \times R_{DCR}$.

Though V_{CS} contains inductor current and DCR

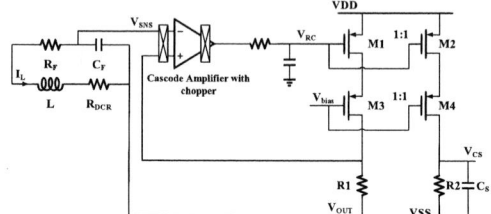

Figure 3: Structure of accurate average current sensing using chopper technique and RC filter.

information, the input offset voltage of the amplifier will have a significant impact on the sensing accuracy because they belong to the same order of magnitude. So the chopper technique is adopted to reduce offset voltage and a RC filter is added to eliminate the ripple caused by the chopper technique, as shown in Fig. 3 [4].

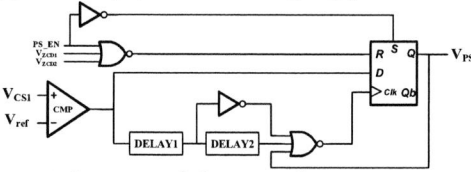

Figure 4: Structure of phase switcher.

Phase Switcher

Dual-phase or multiphase buck converter needs phase switcher, which can automatically adjust the opened phase number according to the load and improve the system efficiency in full load range. The proposed phase switcher is shown in Fig. 4, which consists of a comparator and some digital devices. When the zero-current-detection (ZCD) circuit detects the inductor currents of both phases are lower than zero, which indicates that the load is light, the phase switcher will turn off the second phase to increase the inductor current of the first phase to improve efficiency. When the inductor current increases to a certain extent, the efficiency will decrease significantly with the rapid increase of conduction loss. A suitable switching point is selected to switch single-phase to dual-phase, so that the load current carried by each converter can be in a better efficiency range.

Operating Mode Switcher

The proposed dual-phase buck converter has two modes: normal mode and low power mode, which are determined by the operating mode switcher. As shown in Fig. 5, The operating mode switcher detects the off time D_1 of the power transistor S_{p1} within a cycle, which is related to the load. When the load is light enough to make D_1 longer than the delay and maintain 8 consecutive cycles, the carry signal CA of the 3-bit counter turns to high level so that LPM is also set to high level, which means the system will switch to low power mode to maintain high efficiency in very light load. When D_1 is shorter than the delay and maintain 8 consecutive cycles, system will exit low power mode.

979-8-3503-1101-3/23 $31.00 © 2023 IEEE

546

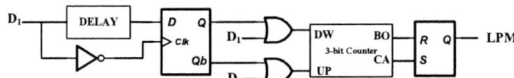

Figure 5: Structure of operating mode switcher.

PERFORMANCE VERIFICATIONS

The proposed 4MHz dual-phase buck converter is designed in TSMC 55nm process. With an input voltage V_{IN} of 3.3V, the output voltage V_{OUT} is 1.2V typically and the maximum load current is 2A.

The simulation results of the current sensing and balancing circuit are shown in Fig. 6. When 30mV offset is added to the inputs of the comparator CMP1, the difference of currents between two phases is 0.385A. When the current balancing circuit starts working, it adjusts each phase's duty cycle to make the difference decrease gradually until the current balance is achieved.

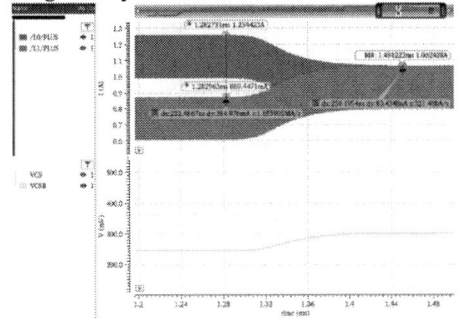

Figure 6: Simulation results of the current sensing and balancing circuit.

Fig. 7 shows how the phase switcher adjusts the opened phase number according to the load. When the load current changes from 1A to 150mA, both phases enter the DCM operation, so the output of phase switcher (VPS) turns to low level to turn off the second phase. When the load current changes from 150mA to 1A, the sensing voltage of the first phase inductor current gradually rises to exceed the reference voltage and the comparator's output (VCMP) turns to high level, making VPS turn to high level to open second phase.

The efficiency curve of the proposed dual-phase buck converter varying with load is shown in Fig. 8, demonstrating 90.1% peak efficiency when the load is 450mA and increased efficiency in full load range due to the phase switcher and the operating mode switcher. Table I provides the performance comparison of the proposed dual-phase converter with other works, from which we can see that the proposed dual-phase DC-DC buck converter achieves higher efficiency in full load range.

CONCLUSION

A 2A 4MHz ZDS hysteretic dual-phase buck converter is presented and verified in this paper. The converter can automatically adjust the opened phase number according to the load and improve the system efficiency in full load range. In addition, the current in

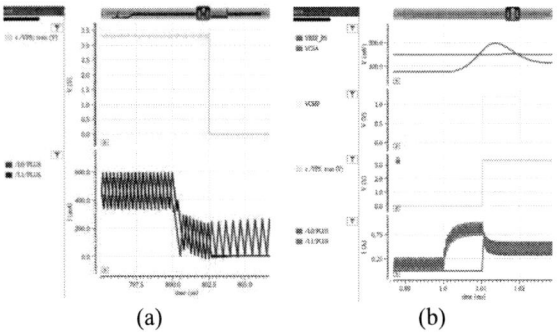

| (a) | (b) |

Figure 7: Simulation results of the phase switcher (a)1A to 150mA (b)150mA to 1A.

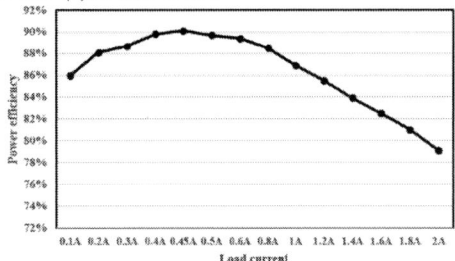

Figure 8: The efficiency curve of proposed dual-phase buck converter varying with load.

TABLE I. PERFORMANCE COMPARISONS

Parameter	[1]	[4]	This work
Process	0.18μm	0.13μm	**55nm**
Control	ZDS	PWM	**ZDS**
Phase Number	4	4	**2**
V_{IN}(V)	3.3	1.2	**3-3.6**
V_{OUT}(V)	0.7-2.5	0.6-1.05	**0.9-1.5**
f_{SW}(Hz/Phase)	40M	2.25M	**2M**
Inductance	78nH	0.47uH	**1.5uH**
C_{OUT}(uF)	0.94	88	**10**
Peak Efficiency(%)	86.1	91.1	**90.1**
Light-Load Efficiency(%)	67.2	57	**86.0**
Heavy-Load Efficiency(%)	64.3	80.0	**79.1**

each phase is well balanced to avoid the problem of heat concentration.

REFERENCES

[1] M. K. Song et al., *ISSCC*, pp. 80-81, 2014.
[2] G. Schrom et al., *APESC*, pp. 4702-4707, 2004.
[3] G. Mao et al., *IPEC*, pp. 2456-2460, 2022.
[4] Y. -S. Roh et al., *Proceedings of the IEEE*, pp. 5159-5169, 2015.
[5] J. Gordillo et al., *Proceedings of the IEEE*, pp. 3480-3489, 2017.
[6] C. Huang et al., *ISSCC*, pp. 362-363, 2013.
[7] M. K. Song et al., *ASP-DAC*, pp. 9-10, 2016.

IMPROVE SPARSE IMPLICIT PROJECTION VIA INCOMPLETE CHOLESKY FACTORIZATION

Yang Yang[1], Fan Yang[1]* and Xuan Zeng[1]

[1]School of Microelectronics, Fudan University, Shanghai 200433, China

*Corresponding Author's Email: yangfan@fudan.edu.cn

ABSTRACT

Model order reduction is widely used to speed up the simulation of complex RC network with large number of terminals. Eliminating the internal nodes is a class of important reduction methods for RC network with large number of terminals. However, the reduced system matrices would be dense ones due to the introduction of fill-ins during elimination of internal nodes. In this paper, we propose to introduce incomplete Cholesky factorization into the elimination-based approach Sparse Implicit Projection (SIP) to produce a sparser reduced RC network with fewer passive elements. Compared with traditional approaches, our proposed method can achieve sparser reduction results with acceptable accuracy, and experiments show the benefits in the subsequent simulation.

INTRODUCTION

As the complexity of circuits continually grows, it is inevitable to take the effects of interconnection into account, which makes the simulation of large-scale RC network very expensive. Lots of works have been proposed to speed up such simulation process. Among them, Model Order Reduction (MOR) is a quite popular strategy. Within acceptable accuracy loss, MOR turns a large RC network into a reduced one and accelerate the subsequent simulation process [1]. Recent years has seen the development of MOR [2-9].

There are mainly two classes of MOR methods. One is based on Krylov subspace projection [2, 7] and the other is based on elimination [3, 5, 6, 8, 9]. With the carefully selected projector, the large system matrix can be reduced to a much smaller one with main features preserved. Although a reduced RC network at any accuracy can be generated via this strategy, there exists a challenge when there exist large number of ports in the RC system. For Krylov subspace projection method, the orders of the reduced systems are proportional to the number of ports. Since the reduced-order matrices are dense ones for Krylov subspace-based approaches, the simulation time of the reduced-order systems can even be longer than the original ones. Compared with projection-based method, methods based on elimination are much more friendly for RC systems with large number of ports. However, reduced RC network generated by such way is not accurate enough in many cases.

In this paper, we introduce Incomplete Cholesky Factorization (ICF) into the SIP process to substitute the Exact Cholesky Factorization (ECF). Reduced RC network generated this way can further reduce considerable number of passive elements further, which can further accelerate the subsequent simulations. The rest of the paper is organized as follows, MOR and Sparse Cholesky Factorization (SCF) are briefly reviewed in Background, which is followed by the review of Sparse Implicit Projection. Then Modified ICF (MICF) and MICF based MOR approach is detailed. Experiments follow and Conclusion comes the last.

BACKGROUND

Model Order Reduction

Considering a passive RC network with n internal nodes and p ports, we can describe it in Laplace domain by the system of equations

$$\begin{cases} GX(s) + sCX(s) = BU(s) \\ Y(s) = B^T X(s) \end{cases}, \quad (1)$$

where $X(s) \in R^{m \times 1}$, $m = n + p$ contains the voltage information of all the nodes. $U(s) \in R^{p \times 1}$ and $Y(s) \in R^{p \times 1}$ collect all ports currents and port voltages, respectively. $G \in R^{m \times m}$ represents conductance matrix and $C \in R^{m \times m}$ represents capacitor matrix. Besides, $B \in R^{m \times p}$ connects the ports with the full nodes. The transfer function for (1) is

$$H(s) = B^T(G + sC)^{-1}B. \quad (2)$$

MOR aims to approximate the transfer function $H(s)$ with smaller matrix G_{red} and C_{red}.

Generally, both G and C can be divided into submatrices such as internal matrix, port matrix and connection matrix. Their block form can be described as

$$G = \begin{bmatrix} G_I & G_C \\ G_C^T & G_P \end{bmatrix}, C = \begin{bmatrix} C_I & C_C \\ C_C^T & C_P \end{bmatrix}, \quad (3)$$

Sparse Implicit Projection

Projection-based MOR methods can efficiently deal with the RC network with small number of ports. SIP is an elimination-based MOR method, which can also be expressed as a projection-based approach. It demonstrates that with carefully selected projector, the projection process is equivalent to seek the Schur complement for the conductance matrix and then the same projector works for the capacitor matrix [8]. Substitute matrix G and matrix C with block matrix form in (3), we arrive at

$$G + sC = \begin{bmatrix} G_I & G_C \\ G_C{}^T & G_P \end{bmatrix} + s\begin{bmatrix} C_I & C_C \\ C_C{}^T & C_P \end{bmatrix}. \quad (9)$$

As shown in (9), the block matrix representing internal nodes locates in the top left and the bottom right one represents ports. After constructing G and C via stamping, node reordering is carried out to assure that internal nodes are in the front part and ports are in the back part. SIP is used to compress the information in the internal nodes and the information between ports and the internal nodes into the information between ports. It is proved that with the carefully selected projector

$$V_P = \begin{bmatrix} -G_I{}^{-1}G_C \\ I \end{bmatrix}, \quad (10)$$

the reduced matrix for matrix G can be described as

$$G_{red} = G_P - G_C{}^T G_I{}^{-1} G_C. \quad (11)$$

With the same projector, reduced capacitor matrix can be expressed as

$$C_{red} = C_P - G_C{}^T G_I{}^{-1} C_C + G_C{}^T G_I{}^{-1} C_I G_I{}^{-1} G_C - C_C{}^T G_I{}^{-1} G_C \quad (12)$$

Assign $B = G_C{}^T G_I{}^{-1}$, then (12) can be simplified as

$$C_{red} = C_P - BC_C + BC_I B^T - C_C{}^T B^T \quad (13)$$

As shown in (11) and (12), the inversion of internal block matrix G_I occurs frequently. Solving the inversion of G_I directly is always substituted by decomposing G_I via ECF followed by seeking the reversion of a triangular matrix, which is more computation-friendly.

IMPROVE SPARSE IMPLICT PROJECTION

Incomplete Cholesky Factorization

Iteration methods are widely used in solving the sparse symmetric linear system $Ax=b$ [10]. Incomplete matrix factorization is widely used in iteration methods to generate the target preconditioner. For a sparse symmetric positive definite matrix, Incomplete Cholesky Factorization (ICF) is often the right choice to get the final preconditioner [11].

Different from ECF, the incomplete one executes the similar decomposition process with much less fill-ins. Looking back the factorization process listed in (6)-(8), it is not difficult to draw a conclusion that the kth column in the factor L is dependent on the k-1th column. When the first k columns in L are denser than those in A, the remaining columns from $k+1$ to n will become much denser and even full, which increases the burden of the solver.

To alleviate the pressure brought by ECF, ICF is widely adopted instead via dropping some fill-ins. Based on the dropping strategy, ICF can be divided into static incomplete one and dynamic one. For the static strategy, the nonzero pattern of A is decided by the predetermined nonzero pattern, that is, nonzero elements which are not in such pattern are dropped directly. The final nonzero pattern for the factor L is usually determined by the level of fill in. When the level of fill in is set zero, the final pattern for $L+L^T$ is the same as that in A. For the dynamic strategy, nonzero elements in the factor L are dropped by predetermined threshold. Once the threshold is decided, elements that are larger than the threshold will be preserved and elements that are less than the threshold will be dropped. Compared with the static one, the dynamic one focuses more on the numerical information and is adopted by both MATLAB and OCTAVE. Instead of all the columns in L share the same dropping threshold, all the elements in the same column share the same dropping threshold. For the kth column, its dropping threshold is

$$d_{thresh}(k) = d_{scale} * |A(k:n, k)|_1, \quad (14)$$

where d_{scale} is a user-predetermined numerical value shared by each column.

However, either static dropping strategy or dynamic strategy depends on predetermined pattern or numerical value to carry out, which may drop some information of the system matrix at the same time. To preserve more information of the original system matrix A in the factor L, Modified Incomplete Cholesky Factorization (MICF) is proposed [12]. The core idea of such technique is to assure the preconditioner $M=LL^T$ with the same row sum as A. When the row sum of M does not meet the requirement, the diagonal element of L is modified to make the row sum of M equal that of A, that is,

$$LL^T e = Ae, \quad (15)$$

where e is a column vector all the elements of which are 1.

In this paper, such modified technique is combined with traditional dynamic dropping strategy. The implementation is based in the built-in function *ichol* in MATLAB.

Improve Sparse Implicit Projection via MICF

Now it comes back to the solution of the inversion of G_I, that is, $G_I{}^{-1}$. First, MICF is imposed on G_I and we get

$$L_I = micf(G_I). \quad (16)$$

Construct the inverse of G_I with L_I, we can get

$$\left(G_I^I\right)' = \left(L_I^T\right)^{-1}(L_I)^{-1} \quad (17)$$

Further, (17) can be introduced in (10), which leads to

$$V_P{}' = \begin{bmatrix} -\left(L_I^T\right)^{-1}(L_I)^{-1}G_C \\ I \end{bmatrix} \quad (18)$$

Once the projector $V_P{}'$ is generated, congruence transformation can be carried out on the original system matrices as follows.

$$G_{red} = \left(V_P{}'\right)^T G V_P{}' \quad (19)$$

$$C_{red} = \left(V_P{}'\right)^T C V_P{}' \quad (20)$$

$$B_{red} = \left(V_P{}'\right)^T B \quad (21)$$

Table 1 Comparisons with SIP.

Test Case	Type	nodes	ports	nnz(G)	nnz(C)	Simulation Time(s)	Speedup with Orig.	Average Relative Error
dl_bfpn_761833	Orig.	2445	213	6925	3028	18.8	/	/
	SIP	213	213	45369	45369	11.2	1.68x	1.00e-4
	iSIP	213	213	387	3049	3.02	6.23x	5.91e-5
dpd_best_5527905	Orig.	3197	137	9731	3433	17.6	/	/
	SIP	137	137	18769	18769	4.45	3.96x	1.00e-4
	iSIP	137	137	159	363	0.651	27.0x	2.65e-5
dpd_best_5534514	Orig.	3087	125	9323	3327	15.8	/	/
	SIP	125	125	15625	15625	4.14	3.82x	1.00e-4
	iSIP	125	125	181	439	0.801	19.7x	2.61e-5

EXPERIMENTS

To verify our proposed strategy, three test cases with large number of ports are adopted. Experiments are carried out at 11th Gen Intel(R) Core Intel(R) Xeon(R) Gold CPU @ 2.90GHz. All the algorithms are implemented in MATLAB 2021b. All the test cases are expanded at 1e4 Hz and simulated from 1e3 Hz to 1e5 Hz with 1000 points in each simulation process. Results are shown in Table 1. Compared with the reduced systems generated by SIP, our proposed strategy leads to much less capacitors and resistors in reduced systems for all the test cases. What's more, our proposed approach can achieve 3.71x-6.82x speedup over SIP method without burden on accuracy.

CONCLUSION

In this paper, we propose to introduce MICF to the computation process of SIP. On the one hand, subsequent MOR process becomes sparser. On the other hand, reduced matrices are sparser, which is more friendly to subsequent simulation. Experiments demonstrate that our proposed approach can achieve significant speedup over SIP without burden on accuracy.

ACKNOWLEDGEMENT

This research is supported partly by National Key R&D Program of China 2020YFA0711900, 2020YFA0711901, partly by National Natural Science Foundation of China (NSFC) research projects 62141407, 61974032, 61929102, 61822402 and 62090025.

REFERENCES

[1] Schilders, W. H., Van der Vorst, H. A., & Rommes. "Model order reduction: theory, research aspects and applications", vol. 13, p. 13., 2008, Berlin: springer.

[2] P. Feldmann and R. W. Freund, "Efficient linear circuit analysis by Padé approximation via the Lanczos process, "*IEEE Trans. Computer-Aided Design Integr. Circuits Syst.*, vol. 14, no. 5, pp. 639–649, May 1995.

[3] L. Hao and G. Shi, "High-Dimensional Extension of the TICER Algorithm," in *IEEE Transactions on Circuits and Systems I: Regular Papers*, vol. 68, no.

11, pp. 4722-4734, Nov. 2021, doi: 10.1109/TCSI.2021.3106390.

[4] C. S. Amin, M. H. Chowdhury, and Y. I. Ismail, "Realizable reduction of interconnect circuits including self and mutual inductances, " *IEEE Trans. Computer-Aided Design Integr. Circuits Syst.*, vol. 24, no. 2, pp. 271–277, Feb. 2005.

[5] N. Sheehan, "Realizable reduction of RC networks, "*IEEE Trans. Computer-Aided Design Integr. Circuits Syst.*, vol. 26, no. 8, pp. 1393–1407, Aug. 2007.

[6] K. J. Kerns and A. T. Yang, "Stable and efficient reduction of large, multiport RC networks by pole analysis via congruence transformations," in *IEEE Transactions on Computer-Aided Design of Integrated Circuits and Systems*, vol. 16, no. 7, pp. 734-744, July 1997, doi: 10.1109/43.644034.

[7] Altan Odabasioglu, Mustafa Celik and Lawrence T. Pileggi, "PRIMA: Passive reduced-order interconnect macromodeling algorithm", *IEEE Trans. Computer-Aided Design*, vol. 17, no. 8, pp. 645-654, Aug. 1998.

[8] Z. Ye, D. Vasilyev, Z. Zhu and J. R. Phillips, "Sparse Implicit Projection (SIP) for reduction of general many-terminal networks," 2008 *IEEE/ACM International Conference on Computer-Aided Design*, 2008,pp.736-743,doi:10.1109/ICCAD.2008.4681658.

[9] B. N. Sheehan, "TICER: Realizable reduction of extracted RC circuits," 1999 *IEEE/ACM International Conference on Computer-Aided Design. Digest of Technical Papers* (Cat. No.99CH37051), 1999, pp. 200-203, doi: 10.1109/ICCAD.1999.810649.

[10] S. Yousef. "Iterative Methods for Sparse Linear Systems". Society for Industrial and Applied Mathematics, second edition, 2003.

[11] M. Benzi, "Preconditioning Techniques for Large Linear Systems: A Survey". *Journal of Computational Physics* 182, no.2 (2002): 418-477.

[12] J. Chen, F. Schafer, J. Huang, and M. Desbrun, "Multiscale Cholesky preconditioning for ill-conditioned problems". 2021, *ACM Trans. Graph.*,40, 4, Article 81, 13 pages

HIGH EFFICIENT AUTOMATIC POWER/GROUND LAYOUT ROUTING ALGORITHM FOR ANALOG ICS

Jiaxin. Zuo[1], Fei. Li[2] and Jing. Wan[1]*

[1]State key lab of ASIC and System, School of Information Science and Engineering, Fudan University, Shanghai, China

[2]Suzhou Foohu Technology Co., Ltd.

*Corresponding Author's Email: jingwan@fudan.edu.cn

ABSTRACT

In this work, we explored an efficient automatic layout routing algorithm for connecting the power and ground pins in analog integrated circuits. A rectilinear minimal spanning tree (RMST) algorithm for two sets of pins is developed, in which minimal spanning tree is used to form the initial connections between pins. The obstacle-avoiding maze routing algorithm is used to break and reconnect the power and ground nets to avoid any short circuit. The genetic algorithm (GA) is further introduced to optimize the total connection wirelength. We also expanding the wire width to avoid electromigration and IR-drop.

INTRODUCTION

Automation design of digital circuit has been developed for many years, in which power and ground routing can be generated and analyzed by power grids [1]. However, due to the complexity of analog circuits, studies on automation design of analog integrated circuits are rare. Routing for power and ground pins in analog circuit is usually accomplished manually. Due to the increasing scale and complexity of analog integrated circuits, manual design becomes less efficient. So, the power and ground pins routing automation for analog circuit design is very important.

In recent years, there has been little research on the automatic power and ground layout routing for Analog ICs. Previous routing methods can be classified into two major categories, one is based on maze-routing algorithm, and the other is based on the channel connectivity graph [2]. The first method abstracts the routing problem into a maze routing problem, such as A* search algorithm. This method is more suitable for net with only two pins [3]. The second method on connectivity graphs are based on the channel model, such as Dijkstra's algorithm. This method is more suitable for nets with more than two pins [4]. But this method requires a fully connected graph as input.

More and more researcher has begun to apply artificial intelligence algorithms to solve the problem of analog integrated circuit layout design automation [5]. Genetic algorithm transforms the process of solving the problem into the process of crossing and mutation of chromosome genes in biological evolution, and it can achieve optimized results efficiently [6].

METHODOLOGY

algorithm overview

In this paper, we provide an automatic power/ground layout routing algorithm for analog circuits combining tree search algorithm and genetic algorithm. This algorithm can be used to automatically connect power and ground pins without any short circuit between these two nets. The total wirelength is further used as the optimization target in genetic algorithm, exploring the best connection route between pins. Figure 1 shows the overview of this algorithm.

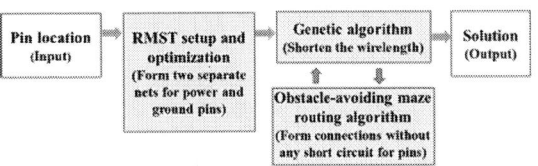

Figure 1: Automatic power and ground layout routing algorithm overview

Rectilinear spanning tree setup and optimization

The power and ground pins need to be connected respectively to form power and ground net without any short. Meanwhile, the total connection wire-length should be minimized. First, two initial minimum spanning trees are generated by Kruskal's algorithm [7]. We sort edges between two pins of the same type by Manhattan distance, then add them one at a time to the tree while avoiding circle connection. Then, take the dictionary containing pins and iterate through each level to find neighbors and distance, by pin indexing pairs [8]. After going through all the pins, we can get a tree connects all pins of the same type using pin-to-pin connections that are composed of vertical and horizontal segments.

The overlap between power and ground nets can cause short circuit and thus need to be avoided. There are two methods of rectilinear connection between two points that are not horizontal or vertical. So, some of the unnecessary overlap violations can be reduced. For other problematic nets with overlap, we can iteratively remove these nets.

979-8-3503-1101-3/23 $31.00 © 2023 IEEE

Rerouting algorithm

The connected two nets are broken at the overlap points. Thus, the original nets are decomposed into sub-trees. Multi-source multi-sink maze routing has been proposed by Min Pan and Chris Chu and widely used in maze routing problems [9]. Based on this algorithm, we added obstacle avoiding function, and developed a novel obstacle-avoiding maze routing algorithm. The broken sub-trees are reconnected by the obstacle-avoiding multi-source multi-sink maze routing algorithm to form the power and ground nets without any short circuit.

The input of rerouting algorithm is the connection and location information between pins. We first get sub-tree groups based on breakpoints and connection information. Then, the algorithm starts from one sub-tree, avoiding obstacles and another type of sub-trees, and find the nearest sub-tree with the same type. The relationship between each node and its neighbor node is put in the dictionary in order to find the path between sub-trees, until all breakpoints are covered by the tree.

Wirelength optimization

In each of the overlap point between power and ground, the power net is constantly broken and reconnected by the maze routing algorithm to avoid any short circuit. However, we can use genetic algorithm to choose which net should be broken and further optimize the connection wirelength. For each overlap point between power and ground nets, the choice of net for breaking is encoded in binary form in the population of GA. In each generation, the broking net is determined by the binary code at every overlap point, and the power/ground pins are reconnected by the maze routing algorithm. The wirelength after rerouting is used as the fitness to determine the next generation in the GA algorithm.

Wire widths optimization

Electromigration and voltage drop are two major issues in analog integrated circuits. Electromigration may causes permanently open or short circuits due to excessive current densities. IR-drop causes insufficient power supply. We expanding the wire width to avoid electromigration and IR-drop according to the requirements of current density.

RESULTS AND DISCUSSIONS

Figure 2(a) shows the location of power pins and ground pins, which is typically available in the metal layer of analog ICs. Two shapes in the figure represent two types of pins, such as power and ground. Figure 2(b) shows two minimum spanning trees without any circle connection for power and ground pins.

Figure 3 shows the relationship between generation and fitness. The population size is 20 and 25 generations are used during the optimization. As generation increases, the total wirelength gradually decreases, and tends to be stable around the 18th generation.

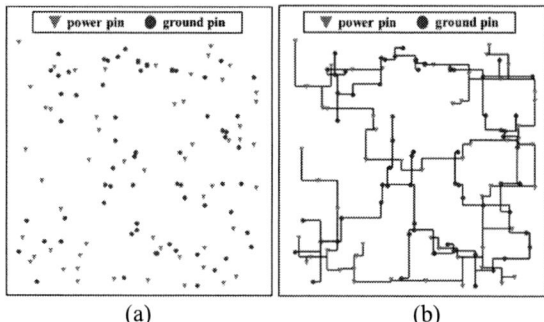

(a)   (b)

Figure 2: (a) Positions of power and ground pins;(b) Two minimum spanning trees for connecting two sets of pins;

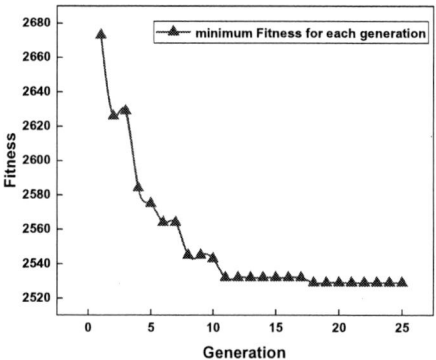

Figure 3: Minimum fitness for each generation

Figure 4 shows the sub-trees after the power and ground nets are broken at each overlap point. After the optimization by GA algorithm, the net is reconnected and the total connection wirelength is minimized. Figure 5 shows that final connected nets of power and ground without any overlap and shortest connection length.

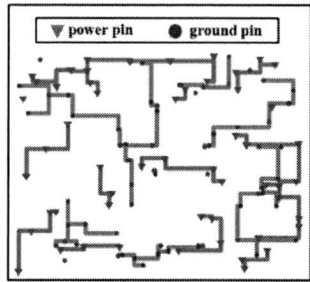

Figure 4: The result after breaking the overlap point;

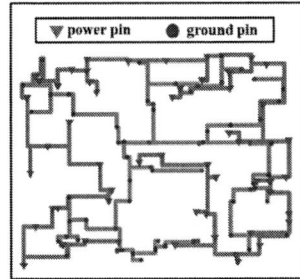

Figure 5: The result after reconnection assisted by GA;

SUMMARY

In this paper, we have developed an efficient method to automatically connect power and ground pins in the ICs. Three algorithms are used in this work. The rectilinear minimal spanning tree algorithm is used to connect all power/ground pins to form two separate nets. Then, the obstacle-avoiding maze routing algorithm is developed to form connections without any short circuit. The genetic algorithm is further used to determine the breaking point in order to shorten the wirelength. The method developed in our work is not only efficient for power/ground connections, but also a very useful reference for other layout auto-routing regimes.

ACKNOWLEDGEMENTS

This work was supported by the Natural Science Foundation of Shanghai (23ZR1405900), Shanghai Science and Technology Commission "explorer project" (21TS1401300). Guangdong Province Research and Development in Key Fields from Guangdong Greater Bay Area Institute of Integrated Circuit and System(No.2021B0101280002) and Guangzhou City Research and Development Program in Key Field (No.20210302001).

REFERENCES

[1] R. Dutta, M and Marek-Sadowska, *26th ACM/IEEE Design Automation Conference*, 1989, pp.783–786.

[2] Lin C W, Huang S L and Hsu K C, *IEEE Transactions on Computer-Aided Design of Integrated Circuits and Systems,* 27(11),2008, pp.643-653.

[3] Nawaf Hazim Barnouti, Sinan Sameer Mahmood Al-Dabbagh and Mustafa Abdul Sahib Naser, *Journal of Computer and Communications*, 4(11), 2016, pp.15-25.

[4] Elizabeth Nurmiyati Tamatjita and Aditya Wikan Mahastama, *ComTech*, 7(3), 2016, pp.161-171.

[5] Li T and Ge Z, IEEE Computer Society,2009, pp.89-92.

[6] Oltean G, Hintea S and Sipos E, *Knowledge-based & Intelligent Information & Engineering Systems, International Conference*, Kes, Santiago, Chile, September,2009, Part II.

[7] Paryati and Salahddine Krit, *MATEC Web of Conferences*,2021, pp.348.

[8] Xiong Xiao-hua, Ning Ai-bing and Ma Liang, *2009 4th International Conference on Computer Science & Education*,2009, pp.469-473.

[9] Min Pan and Chris Chu, *2007 Asia and South Pacific Design Automation Conference*, 2007, pp.250-255.

Implementing Boolean Function by Ternary Content Addressable Memory with Approximate Match

Jian Shi[1], Weikang Qian[1,2,*]

[1]University of Michigan-SJTU Joint Institute and [2]MoE Key Lab of AI, Shanghai Jiao Tong University, China
Email: {timeshi, qianwk}@sjtu.edu.cn; [*]corresponding author

Abstract—**Ternary content addressable memory (TCAM) is a widely used component for high-speed lookup operation. In this work, we advocate a novel use of TCAM, *i.e.*, for implementing a Boolean function. We further leverage approximate match to reduce the resource usage. To achieve this, two extra columns are added to the TCAM-based architecture. The experimental results show that to support the implementation of any 4-input Boolean functions, the proposed architecture can reduce 37.5% rows and 12.5% bit cells over the conventional architecture.**

Index Terms—**Boolean function, TCAM, approximate match**

I. INTRODUCTION

Ternary content addressable memory (TCAM) is a special memory that can be searched by content instead of location. It is widely used in high-speed search operation such as data compression, network router, and image processing [1].

The design of efficient TCAM architecture has attracted much attention recently. For example, Chang *et al.* introduce a TCAM structure with lower power consumption in peripheral circuits [2]. Ghofrani *et al.* propose an approximate match technique for TCAM built with nonvolatile devices, which enables the match of more inputs [3]. However, all existing architectures are only used as a lookup table for storing frequently-used patterns. In this work, we advocate the use of TCAM to implement a Boolean function. The basic idea is to represent a function as an optimized sum-of-product (SOP) expression and then use TCAM to store the product terms in the expression. Furthermore, we explore the approximate match technique to reduce the total number of bit cells in a TCAM-based architecture. To achieve this, two extra columns are added to the TCAM-based architecture.

II. BACKGROUND

A TCAM compares an input pattern with the stored patterns and activates the *matchline (ML)* of the matching pattern. The left part of Fig. 1 shows a TCAM design. It consists of k rows, each storing an n-bit word and associated with a ML. The input pattern is fed through the *searchlines (SLs)*. If it matches a word, the corresponding ML is activated.

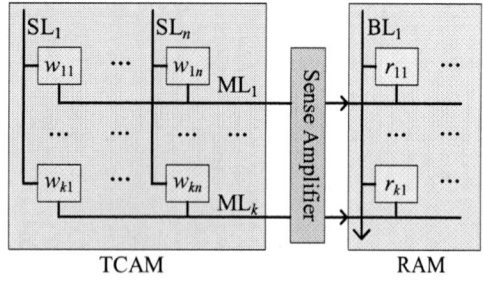

Fig. 1: An n-input TCAM associated with a RAM.

The TCAM can be associated with a RAM shown in the right part of Fig. 1. Once the i-th ML is activated, the RAM outputs the value stored in its i-th row through the *bitlines (BLs)*. In fact, the i-th BL performs an OR operation on its bit cells enabled by the corresponding MLs, *i.e.*,

$$\text{BL}_i = r_{1i} \cdot \text{ML}_1 + \cdots + r_{ki} \cdot \text{ML}_k, \qquad (1)$$

In this work, we consider a nonvolatile TCAM implementation using 2 transistors and 2 memristors (2T-2R) in each bit cell [4]. For a 2T-2R TCAM, the MLs are first pre-charged. If the input pattern has bit differences over the stored word, the corresponding ML is discharged by the mismatch bit cells in the row. The number of mismatch bit cells equals the *Hamming distance (HD)* between the input pattern and the stored word. The larger the HD, the faster the ML is discharged. Fig. 2 shows how the voltage of ML drops with time for different HDs between the input pattern and the stored word. Clearly, the voltage decrease of the 1-HD case is much slower than the other HD cases. Ghofrani *et al.* proposed to shorten the sampling period for approximate computing [3]. In this situation, the voltage of the ML for the 1-HD case is still high when sampled, and consequently, an input pattern that is 1 HD from the stored word can be treated as matching with the stored word. This technique is called *1-HD match* in TCAM. In contrast, an input pattern that is at least 2 HDs from the stored word cannot activate the ML of the word.

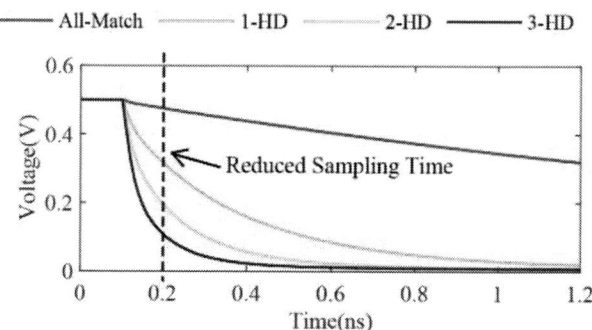

Fig. 2: The ML voltages under different HD cases.

III. METHODOLOGY

In this work, we advocate the use of TCAM to implement a Boolean function. Consider the Boolean function represented by a Karnaugh map shown in Table I(a). Its simplest SOP expression is

$$f([a,b,c,d]) = \bar{a} \cdot b \cdot \bar{c} + \bar{a} \cdot b \cdot d + \bar{a} \cdot \bar{c} \cdot d + b \cdot \bar{c} \cdot d \quad (2)$$

979-8-3503-1101-3/23 $31.00 © 2023 IEEE
554

TABLE I: The Karnaugh maps of two 4-input Boolean functions.

(a)

$cd\backslash ab$	00	01	11	10
00	0	1	0	0
01	1	1	1	0
11	0	1	0	0
10	0	0	0	0

(b)

$cd\backslash ab$	00	01	11	10
00	0	0	1	0
01	0	1	1	1
11	0	0	1	0
10	1	0	0	0

We can implement the SOP by a TCAM-based architecture shown in Fig. 3. The TCAM part has 4 rows storing the product terms in the SOP, i.e., $[0, 1, 0, X]$, $[0, 1, X, 1]$, $[0, X, 0, 1]$, and $[X, 1, 0, 1]$, where X represents an input don't care bit. For the RAM part, each bit cell stores 1. With such a setup, by Eq. (1), the final output equals the given Boolean function.

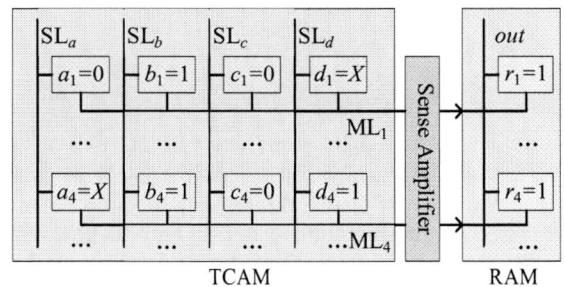

Fig. 3: A TCAM-based architecture to implement the Boolean function in Eq. (2).

However, by exploiting the 1-HD match, we can reduce the number of needed rows to 1. Specifically, we only need to store the word $[0, 1, 0, 1]$ in the TCAM and store 1 in the bit cell of the associated RAM. For any input pattern in the ON set of the Boolean function f shown in Table I(a), since it either equals $[0, 1, 0, 1]$ or is 1 HD from $[0, 1, 0, 1]$, by 1-HD match, it activates the ML of the stored word $[0, 1, 0, 1]$. Consequently, the final output is 1. For any input pattern in the OFF set, since it is at least 2 HDs from $[0, 1, 0, 1]$, it cannot activate the ML, leading to a final output of 0.

The above example shows that if 5 on-set input patterns form a cross-like pattern in the Karnaugh map, as shown in Table I(a), then by exploiting the 1-HD match, we can reduce the number of rows stored in the TCAM. We call this technique *cross pattern-enabled row reduction*. Inspired by the above example, we try to exploit 1-HD match and cross pattern-enabled row reduction to minimize the hardware cost.

However, if the 1-HD match is enabled in the TCAM, some Boolean functions cannot be correctly implemented. Consider the one shown in Table I(b). To implement it, the pattern $[0, 0, 1, 0]$ or its 1-HD neighbour should be stored in the TCAM. Then, some OFF set input patterns can also activate the ML. For example, if the pattern $[0, 0, 1, 0]$ is stored, then the input pattern $[0, 0, 1, 1]$ can also activate the ML of the row, causing a wrong output value of 1. To implement an arbitrary Boolean function, we propose a solution in Section III-A by introducing an extra column in the TCAM. Furthermore, we notice that for some cases, it is impossible to exploit cross pattern-enabled row reduction. To address this, we propose

an improved design with an extra column in the RAM in Section III-B.

A. Extra Column in TCAM for Arbitrary Boolean Function Implementation

To implement an arbitrary Boolean function, we introduce an extra column in the TCAM. Then, each row in the TCAM has an extra bit cell ε. When we want to get the output of an input pattern, the input pattern is appended with an extra bit $\delta = 0$ and fed into the TCAM. We apply the following rules to configure the TCAM.

1) When a pattern v is in the ON set of the Boolean function and some of its 1-HD neighbours are in the OFF set, the pattern is stored in a row r as usual with the extra bit cell ε of the row storing 1. By this configuration, when we look for the output of the input pattern v, the input to the TCAM is v appended with a 0. In this case, the input and the stored word only have 1 bit difference, which occurs at the extra column. Therefore, the input activates the ML of row r, and the TCAM outputs a 1, which is the correct output for the input pattern v. When we look for the output of a 1-HD neighbour u of v in the OFF set, the input to the TCAM is u appended with a 0. In this case, the input is 2 HDs from the word stored at row r. Thus, it does not activate the ML of row r, and the TCAM outputs a 0, which is the correct output for the input pattern u.

2) When a pattern v and all of its 1-HD neighbours belong to the ON set, the pattern is stored in a row r as usual with the extra bit cell ε of the row storing 0. By this configuration, when we look for any input pattern u within 1 HD from v, the input to the TCAM is u appended with a 0, which is within 1 HD from the word stored at row r. Thus, the ML of row r is activated, and the TCAM outputs a 1, which is the correct output for the input pattern u.

With the proposed technique, only two rows are needed in the TCAM to implement the Boolean function shown in Table I(b). The corresponding stored patterns are

$$[a, b, c, d, \varepsilon] = [0, 0, 1, 0, 1] \text{ and } [a, b, c, d, \varepsilon] = [1, 1, 0, 1, 0],$$

where the first row can only be activated by the input pattern $[0, 0, 1, 0]$, and the second row can be activated by the remaining ON set input patterns.

B. Extra Column in RAM for Row Reduction

Although by introducing an extra column in the TCAM, we can implement an arbitrary Boolean function, a target function sometimes has no cross pattern that enables row reduction, e.g., the Boolean function in Table II(a).

To maximally exploit the cross pattern-enabled row reduction, we propose an improved architecture shown in Fig. 4, where we add an extra column to the RAM part. Furthermore, the final output out_c is the XOR of the output out' of the original column in the RAM and the output σ of the extra column in the RAM, i.e., $out_c = out' \oplus \sigma$. We call the architecture *approximate match-based TCAM (AM-TCAM)*.

With the new architecture, we can exploit the cross pattern-enabled row reduction. Consider the Boolean function in

979-8-3503-1101-3/23 $31.00 © 2023 IEEE

TABLE II: A 4-input Boolean function: (a) its Karnaugh map; (b) the configuration of AM-TCAM to implement it.

(a)

$cd\backslash ab$	00	01	11	10
00	0	**1**	0	0
01	**1**	**1**	**0**	0
11	0	**1**	0	0
10	0	0	0	0

(b)

	TCAM					RAM	
a	b	c	d	ε		out'	σ
0	1	0	1	0		1	0
1	1	0	1	1		1	1

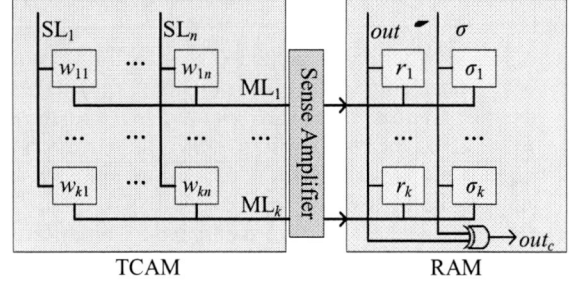

Fig. 4: The proposed AM-TCAM architecture.

Table II(a) again. The basic idea is to first treat the output of the input pattern $[1, 1, 0, 1]$ as 1, which leads to a cross pattern that enables row reduction, and then correct the wrong output value of the pattern $[1, 1, 0, 1]$. To achieve this, we configure the TCAM and the RAM as shown in Table II(b). Specifically, by treating the output of the input pattern $[1, 1, 0, 1]$ as 1, we identify a cross pattern and hence, can reduce the number of rows by storing the center of the cross pattern, *i.e.*, $[0, 1, 0, 1]$, in a row of the TCAM with the extra bit cell ε set as 0. For this row, the original and extra RAM bit cells store 1 and 0, respectively. To correct the wrong output value for the pattern $[1, 1, 0, 1]$, the pattern is stored in another row of the TCAM with the extra bit cell ε set as 1. The corresponding original and extra RAM bit cells both store 1. Under such a configuration, we have:

1) For the ON set input patterns $[0, 1, 0, 0]$, $[0, 0, 0, 1]$, $[0, 1, 0, 1]$, and $[0, 1, 1, 1]$, the first row in Table II(b) is activated due to at most 1 HD from the input, while the second row is not due to at least 2 HDs from the input. By Eq. (1), the RAM outputs are $out' = 1$ and $\sigma = 0$. Therefore, the final output is $out_c = out' \oplus \sigma = 1$.

2) For the input pattern $[1, 1, 0, 1]$, both rows in Table II(b) are activated due to 1-HD match. By Eq. (1), the RAM outputs are $out' = 1$ and $\sigma = 1$. Therefore, the final output is $out_c = out' \oplus \sigma = 0$.

Thus, AM-TCAM realizes the target Boolean function. Furthermore, it needs 10 bit cells in the TCAM and 4 bit cells in the RAM. In contrast, the *conventional architecture* without exploiting 1-HD match needs to store 3 words, *i.e.*, $[0, 1, 0, X]$, $[0, 1, X, 1]$, and $[0, X, 0, 1]$, leading to 12 bit cells in the TCAM and 3 bit cells in the RAM. This example shows the benefit of AM-TCAM in resource usage reduction.

IV. EXPERIMENTAL RESULTS

In this section, we compared the resource usage between AM-TCAM and the conventional architecture, measured by the total number of bit cells in the TCAM and the RAM, for 4-input Boolean functions. It can be shown that all Boolean functions in the same negation-permutation-negation (NPN) equivalence class need the same total number of bit cells. Thus, for each NPN equivalence class, we choose one function in it as the test case, which is the one with the minimum ON set size in the class. All the test cases are simplified by ESPRESSO [5]. We find that the following function needs the maximum number of rows for the conventional architecture:

$$f([a, b, c, d]) = a \oplus b \oplus c \oplus d. \qquad (3)$$

Indeed, it needs 8 rows. However, using AM-TCAM, only 4 rows are required, and its configuration is shown in Table III.

TABLE III: The configuration of AM-TCAM to implement the Boolean function in Eq. (3).

	TCAM					RAM	
a	b	c	d	ε		out'	σ
0	0	1	1	0		1	0
0	0	1	1	1		1	1
1	1	0	0	0		1	0
1	1	0	0	1		1	1

Moreover, we manually configure AM-TCAM for the selected representative functions of all the NPN equivalence classes. We want to obtain the maximum number of rows needed over all the functions, since an architecture with the number of rows equal to that maximum value can implement any 4-input Boolean function. We find that the maximum number of rows needed is 5. Thus, to support the implementation of any 4-input Boolean function, the total number of cells in AM-TCAM is $5 \times 7 = 35$. The conventional architecture needs 8 rows to support the implementation of any 4-input function, and the total number of cells it contains is $8 \times 5 = 40$. Therefore, AM-TCAM reduces 37.5% rows and 12.5% bit cells over the conventional architecture.

V. CONCLUSION

This work proposes to implement Boolean functions by a TCAM-based architecture. To reduce the resource usage, it exploits 1-HD match and contains an extra column in the TCAM part and an extra column in the RAM part. The proposed architecture can realize single-output Boolean functions with a limited number of inputs. Thus, it has the potential to replace the conventional lookup tables used in FPGA. Currently, the configuration of the proposed architecture for a given Boolean function is done manually. Our future work will develop an algorithm for automatic configuration.

REFERENCES

[1] K. Pagiamtzis and A. Sheikholeslami, "Content-addressable memory (CAM) circuits and architectures: A tutorial and survey," *JSSC*, vol. 41, no. 3, pp. 712–727, 2006.

[2] M.-F. Chang *et al.*, "A 3T1R nonvolatile TCAM using MLC ReRAM for frequent-off instant-on filters in IoT and big-data processing," *JSSC*, vol. 52, no. 6, pp. 1664–1679, 2017.

[3] A. Ghofrani *et al.*, "Associative memristive memory for approximate computing in GPUs," *JETCAS*, vol. 6, no. 2, pp. 222–234, 2016.

[4] J. Li *et al.*, "1 Mb 0.41 μm² 2T-2R cell nonvolatile TCAM with two-bit encoding and clocked self-referenced sensing," *JSSC*, vol. 49, no. 4, pp. 896–907, 2014.

[5] R. K. Brayton *et al.*, *Logic Minimization Algorithms for VLSI Synthesis*. Kluwer Academic Publishers, 1984.

979-8-3503-1101-3/23 $31.00 © 2023 IEEE

Verification of 100Gb/s Data-Rate Transceiving through Silicon-Photonic Module in an FPGA Platform

Xuhui Liu [1], Chun-Zhang Chen [1,2,*], Xiaoli Fang [1], Liang Wang [1], Quan Pan [1,3] and Hanming Wu [1,2]

[1] *Peng Cheng Laboratory, Shenzhen 518000, China*
[2] *School of Micro-Nano Electronics, Zhejiang University, Hangzhou, China*
[3] *School of Microelectronics, Southern University of Science and Technology, Shenzhen, China*
*Corresponding Author: chenchzh@pcl.ac.cn, czchen126@126.com

ABSTRACT

This study conducts research on the performance of NRZ 100Gb/s modulation, as well as its scalability to PAM4 400Gb/s and 800Gb/s data-rate. The FPGA verification platform consists of a CMOS SerDes, an InfiniBand Extended Data Rate (IB EDR) module, a 4-lane QSFP28 compliance board, and a loop-back on-board optics (OBO) 100G Evaluation Board, with electrical/optical I/O assembly or silicon-photonic (Si-Ph) chip module. The FPGA platform follows the Common Electrical I/O (CEI) standard. For high-speed data-rate of 100 Gbps through 4×25G SerDes TX/RX, the eye diagrams and the bit error rate (BER), of Si-Ph loop-back and fiber-Si-Ph loop-back, are measured respectively. The results lead to discussions on the applicability the newest Compute Express Link (CXL) 3.0 in FPGA verification platform, which supports NRZ and PAM4 modulations. It is also suitable to evaluate essential characteristics and applications, such as the delay and interconnect concerns in co-packaged optics (CPO) technology, of IP integration and chiplet packaging.

Keywords: 100G Ethernet, NRZ, FPGA, SerDes, TX/RX, Si-Ph, OBO, PCIe, CXL, CPO

INTRODUCTION

With increased use in big data generation scenarios, and video/audio applications such as constant social media usage and internet applications, the Ethernet link speed has been progressed at 25, 50, 100, 200, 400 to 800 Gigabit Ethernet (GbE) and above, defined by Ethernet Technology Consortium (ETC) [1], at the endpoint of data centers (DCs) and high performance computing (HPC). In DCs, the performance and the revenues of 100GbE and 400GbE are seen overtaking 200GbE (Fig. 1), as 200GbE was approved later than 400GbE by IEEE 802.3 Working Group [2].

The link speed of 800G Ethernet (800GbE) is coming to be a leading communication technology in the next 3 to 5 years between silicon and photonic (or simply Si-Ph) interconnect. Gigabit Ethernet continues upgraded and progressed for a required high-speed data-rate (8, 16, 32, 64 GT/s; GT, gigatransfer). The latest PCIe 6.0

technology, released in 2021, offers a data rate at 64GT/s for various applications typically in DCs and HPC, artificial intelligence/ Machine Learning (AI/ML), Internet of Things (IoT), automotive and military/ aerospace industry.

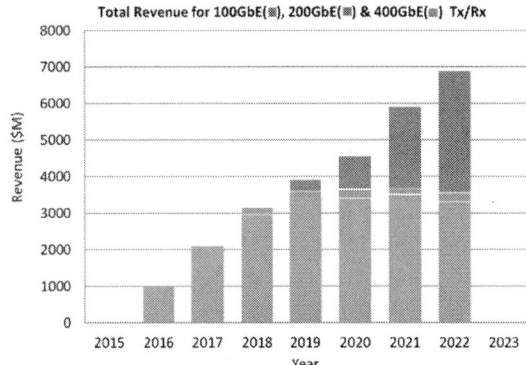

Figure 1: Revenues of Ethernet Transceivers (TX/RX) in 2016-2022 (Redrawn from Ovum data)

For 800GbE requiring 100 GB/s of bandwidth, as the PCIe 6.0 supports 64 GT/s, the 128 GB/s envelope of a ×16 PCIe 6.0 link withstands, through the high-speed interface (I/F) I/O, such as SerDes transceiver (TX/RX). SerDes in PCIe interconnection is a prominent cost-effective and scalable solution for data-intensive markets. The key role SerDes is to be tested and verified for transceiving characteristics, including its performance and functionality such as signal integrity, signal attenuation, reflection, impedance matching, and jitter etc. This helps designers overall aim for lower costs and power, improve performance and utilization (of endpoints), and simplify the design.

In addition, for interconnect and high data-rate transfers in/between processors and/or memories, various open standards are established, such as OpenCAPI (2016), CCIX (2016), Gen-Z (2016), Nvlink (NVL, 2015), and a converged new standard Compute Express Link (CXL, 2019) originally by Intel [3] (Table 1). CXL 3.0 (in a fixed packet (FLIT) interface), is an industry-supported Cache-Coherent Interconnect (CCI) protocol for Processors, Memory Expansion and Accelerators. Based

---

979-8-3503-1101-3/23 $31.00 © 2023 IEEE

on PCIe 5.0, CXL is a high-bandwidth, low-latency serial bus interconnect between host processors and devices such as accelerators, memory controllers/buffers, and I/O devices; thus it contains three components (I/O, memory, and cache).

Table 1: Interconnect Speed of CXL Standard

Yr, Ver	Speed	x1	x16	FLIT
2019, 1.x	32 GT/s	3.938 GB/s	63.015 GB/s	68 Byte
2020, 2.0	32 GT/s	3.938 GB/s	63.015 GB/s	68 Byte
2022, 3.0	64 GT/s	7.563 GB/s	121.0 GB/s	256 Byte

For high-speed I/F IP in PCIe 4.0 and CXL 2.0 based architecture, it is essential to set up the verification of 100Gb/s data-rate transceiving through a silicon-photonic (Si-Ph) module, which is scalable and extensible for 400Gb/s or 800 Gb/s in future experiments. This work reports the verification tests of FPGA platform [4], following the Common Electrical I/O (CEI) standard, i.e. the CEI interconnection of die-to-die (D2D), or multi-chip module (MCM <25mm, NRZ), as well as chip-to-optical element (C2OE) for below extra short reach (XSR <50mm, PAM4).

The measurement result confirms that the proposed verification platform supports up to 4×25 Gb/s in NRZ modulation and is suitable to evaluate essential characteristics of high-speed data-rate of 100 Gbps through 4×25G SerDes TX/RX. Besides, further discussion is made on the characteristic and features of high speed IPs based on the measurement.

METHODOLOGY

The diagram of measurement flow is shown below in Fig. 2. The FPGA platform is composed of four parts, i) the Software–Hardware (S/W–H/W) environment; ii) electronic-optic (*e2o*) and optic-electronic (*o2e*) interfaces with 100G InfiniBand Extended Data Rate (IB EDR); iii) HiLink network solution for TX/RX 4 × 25Gbps module with QSFP28 ports and iv) Lightpass on-board optics (OBO) module from I-PEX (Fig. 2).

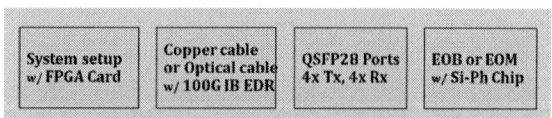

Figure 2: Diagram of the Verification Flow

As described previously [4], the FPGA board Alveo U50 (Xilinx) is used for the verification study (Fig. 3). It supports PCIe Gen4, and embeds with 8 GB of HBM2 for a bandwidth of 460GB/s, The FPGA board is installed in FunsionServer Pro (model 2288H V5) machine with a Linux OS (Fig. 3, panel A).

The high-speed data is transmitted either via Scheme A, or Scheme B. In Scheme A, a Mellanox copper cable (up to 5m Reach distance), is attached to a pair of

connectors that meets IB EDR modulation (NRZ) per industry TX/RX standard (Nvidia Mellanox). In Scheme B, a fiber cable is using a pair of connectors, pluggable optical modules (POMs). The POM for *e2p* signal conversion is plugged at the input of the fiber. A second POM is for *p2e* signal conversion and receiving that plugged at the output of the fiber cable (Fig. 3B).

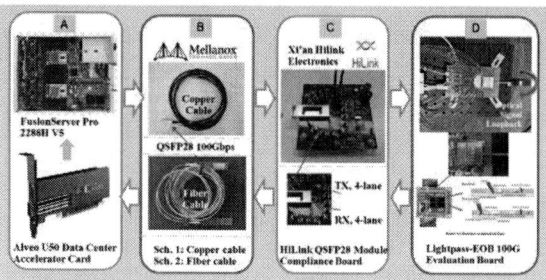

Figure 3: Components of the FPGA-Based Verification Platform

The second POM is then connected to HiLink QSFP28 module compliance board for 4×25Gbps transceiving (Fig. 3C).

In the final setup, the Lightpass evaluation board from I-PEX [5] is used to encompass 4×25Gbps *e2p* and *p2e* on a Si-Ph chip. Either embedded optical blade (EOB, with horizontal connector, Fig. 3D), or alternatively embedded optical module (EOM, using socket connector, not shown) is used and tested. The optical I/O core (fingertip-size optical module or Si-Ph chip) is provided with optical signal feedback loops. The Si-Ph was made on 300mm SOI wafer, lithographic with ArF immersion technique [6].

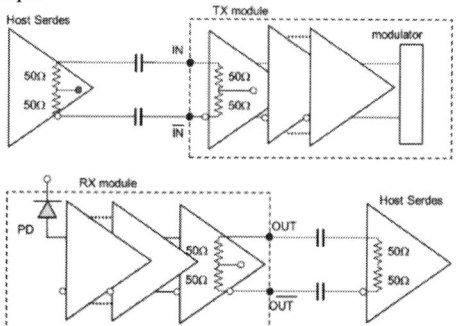

Figure 4: Evaluation Board of EOM 100G from I-PEX

There are total 16 optical cables connected to the Si-Ph chip (Fig. 3D), 4 pairs for 4× TX and 4 pairs for 4× RX. Each of the 4 pairs (a positive P and a negative N) of the optical cables is used in 1-lane TX module for electronic signal through a differential circuit (Fig. 4). Similar to the signal loop-back, each pair of the optical cables is used in 1-lane RX module. A photo-diode (PD) is also used in each RX lane (Fig. 4).

RESULTS AND DISCUSSION

In practice, the verification flow depicted in Fig. 3 can be done with two measurement schemes. In Scheme A, with an IBERT (Integrated Bit Error Ratio Tester, Xilinx) generating pseudo random binary sequence (PRBS31), the eye diagram of 100Gbps is observed via a PC or an oscilloscope (Fig. 5).

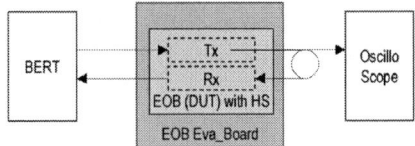

Figure 5: Setup of 100G Verification (Scheme A), Data format PRBS31.

Following Scheme A, the IBERT generated PRBS31 code are sent by 1×TX module through EOB to an optical cable. The eye diagram is measured at its loop-back of 1×RX module, on an oscilloscope (Keysight Infiniium). The measured eye diagrams are shown in Fig. 6 (a) at speed of 16 Gbps, and in Fig. 6 (b) at speed of 25 Gbps, respectively.

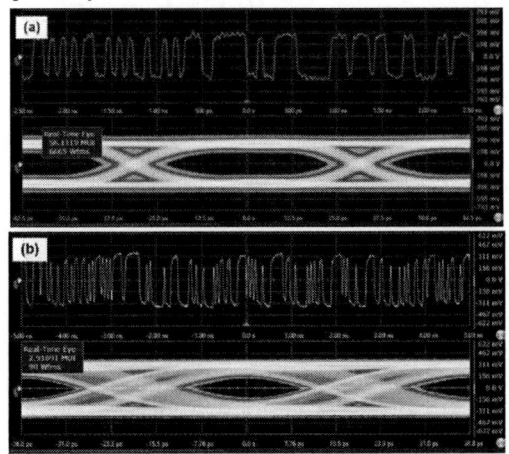

Figure 6: (a) Eye diagram of IBERT data-rate at 16 Gb/s (for testing the environment); (b) Eye diagram of IBERT at 25 Gb/s

At 16Gbps data-rate, the mean value of the eye height is 368.5mV, the eye width 0.75 UI (46.9ps), and the V_{p-p} 967.5mV (Fig. 6a). At 25 Gbps data-rate (Fig. 6b), the mean value of the eye height is 213mV, the eye width 0.416 UI (16.64ps).

In Scheme B, as shown in Fig. 3 and Fig. 4, the IBERT code PRBS31 generated by Alveo U50 inside the FusionServer (Fig. 3A), transmitted via the fiber cables (FAST Photonics, Fig. 3B), through HiLink 4×TX/4×RX module (Fig. 3C), through loopback of EOM (Fig. 3D), and coming back to U50. The eye diagram viewed on a PC at FPGA is shown in Fig. 7. After 30 minutes transmitting, for lane 3, the measured UI is 0.737, and BER is 9.56E-10.

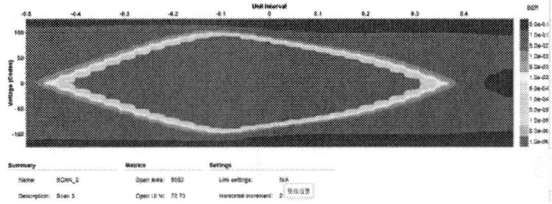

Figure 7: Eye diagram of the loop-back via fiber cable, 4×TX/RX and Si-Ph Chip (Scheme B)

The results of the present study show that our home-made FPGA-based verification platform is suitable and feasible used to evaluate essential characteristics of high-speed data rate of 100 Gbps through 4×25G SerDes TX/RX. The verification system in this work fits the purpose to evaluate characteristics of eye diagrams (height, width/UI), as well as jitters, and BERs etc.

The setup and test are using an OBO system that is in line with co-packaged optics (CPO) technology[7], which focuses the advanced packaging technique to shorten the interconnects and delays at I/F IP, to fulfill the optimization of electronics and photonics.

As high-speed I/F IP plays a key role in DCs, HPC, and Automotive etc., the verification and integration of IP in an SoC, and system co-design optimization are all critical steps. Though the results of Scheme A and Scheme B are preliminary, the study is encouraging and is continued for 400GbE (8×56G, PAM4) Ethernet interconnects, to analyze with PCIe analyzer and to meet with CXL protocol features.

Acknowledgments: Usage of *I-PEX* products is based on a mutual NDA. Part of measurements was done at *Pangomicro* and *Xuan Wu*.

REFERENCES

[1] ETC, https://ethernettechnologyconsortium.org/
[2] IEEE 802.3 Ethernet Working Group, https://grouper.ieee.org/groups/802/3/
[3] Compute Express Link (CXL) 3.0, 2022. https://www.computeexpresslink.org/
[4] Chen, C.-Z., Liu Xuhui and Wu, Hanming, 2022, "An FPGA-Based Verification Platform for High-Speed Interface IPs," 2022 China Semicon. Tech. Int. Conf. (CSTIC, June 14-15, 2022, Shanghai)
[5] I-PEX confidential. LIGHTPASS™-EOM 100G I-PEX Embedded Optical Module. 2021.
[6] Nakamura, Takahiro, et al, Fingertip-Size Optical Module, "Optical I/O Core", and Its Application in FPGA (invited), IEICE Trans. Electron., Vol.E102-C, No. 4 April 2019.
[7] Tan, M., Xu, J., Liu, S. et al. Co-packaged optics (CPO): status, challenges, and solutions. Front. Optoelectron. 16, 1 (2023).

ARTIFICIAL NEURAL NETWORK COMPACT MODELING METHODOLOGY FOR COMPLEMENTARY FIELD EFFECT TRANSISTOR

Ouwen Tao[1], Xiaona Zhu[2]*, Yage Zhao[2], Rongzheng Ding[2], Shaofeng Yu[2], Ye Lu[1]*

[1] School of Information Science and Technology, Fudan University, Shanghai 200433, China
[2] School of Microelectronics, Fudan University, Shanghai 200433, China
*Corresponding Authors' Email: xiaona_zhu@fudan.edu.cn; lu_ye@fudan.edu.cn

ABSTRACT

Device compact model (DCM) is the key bridge between process technology and circuit design. Conventional DCM requires detailed understanding of device structure and its corresponding physics for underline devices. And This is particularly difficult for complementary field effect transistor (CFET) due to entangled pair device in a stacked structure. We develop a novel artificial neural network (ANN)-based modeling methodology for CFET cells instead of individual transistor to overcome this obstacle. The created model captures the CFET electrical characteristics accurately with fast turnaround time. The standard cells and various logic circuits are simulated and studied based on the new model.

INTRODUCTION

CFET is one of the most promising options for extending Moore's law beyond 3nm technology node [1]. Various investigations have been focused on its process challenges and device level performance capabilities [2-3]. Fundamentally different from previous advanced nodes, a basic CFET cell block has a three-dimensional transistor structure, one transistor stacks on top of another with a common gate. It is difficult to model individual transistor in this unique structure with conventional approaches due to entangled device parasitic. In this work, we propose a novel ANN based modeling methodology to directly model the entire CFET cell. New model captures the CFET characteristics in high precision. Further, various logic circuits are built with the cell and simulations are successfully accomplished using created models.

CFET CELL MODELING

The CFET cell structures in circuits' applications can be divided into four basic cases. These cases are shown in Table I. In order to build digital logic cells with CFETs, the CFET structure of the common gate and separate drain (type3), as shown in Fig. 1, needs to be modeled. We propose a data driven scheme for the model creation. I(V) or Q(V) data between each CFET cell terminals are first generated using Sentaurus Device TCAD tool, and they are further used as the training data to generate an ANN based compact model.

TABLE I.   DIFFERENT CELL STRUCTURES OF CFET

Structure type	Structural features
Type1 structure	Separate gate and separate drain
Type2 structure	Separate gate and common drain
Type3 structure	Common gate and separate drain
Type4 structure	Common gate and common drain

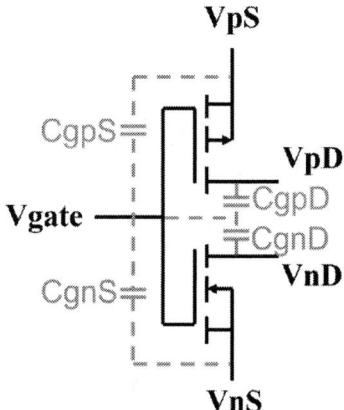

Figure 1: Schematic of the type3 *structure in CFET, CgpS, CgnS, CgpD, CgnD are the main parasitic capacitances*

The I(V) electrical characteristics of the CFET ranges several orders of magnitude, and ANN tends to overweight the large numbers, and this causes the large model error when device is operated at close or below threshold voltage. Thus, data processing is required, such as using a logarithmic function or a power function to compress the data range.

In this work, a power function is chosen as the compression function. Using power functions with exponents such as 1/3, 1/5, and 1/7, has two advantages over using logarithmic functions: (1) Such power functions can handle data throughout the entire real number domain, which is a larger range than the logarithmic function's domain of non-negative numbers. This is particularly important for training the electrical characteristics model of CFETs, which exhibit changes in port current direction under different voltage input combinations. (2) Power functions can directly output a zero point, while logarithmic functions require shifting and other methods to output a zero point. This means that using power functions as compression functions can better reflect data characteristics near the zero point after compression, resulting in more accurate fitting of data near the zero point by ANN models. However, power functions have one disadvantage compared to logarithmic functions. The function graph of $f(x) = x^{(1/3)}$, is shown in Fig. 2, which demonstrates that a power function with a 1/3 exponent can handle data within a range of 5 orders of magnitude. Compared to logarithmic functions, the compression capability of power functions is limited. However, this drawback can be addressed by selecting a suitable power exponent. For example, a power function with a 1/7 exponent can handle data within a range of 13 orders of magnitude.

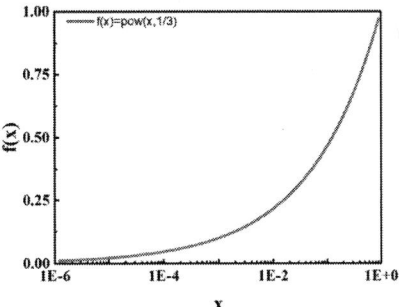

Figure 2: The function graph of f(x) = x^(1/3)

On the other hand, the relationship between charge and corresponding node voltage (Q(V)) is used to model the AC characteristic of the device. Referring to the multi-gradient training method in Ref. [4] and Ref. [5][4], a neural network modeling scheme using only capacitance and corresponding node voltage relationship (C(V)) is proposed and applied in this work. The loss function is shown in the following equation:

$$loss = \frac{1}{n}\sum_{i=1}^{n}(grad_i[Q] - grad_i'[Q])^2 \quad (1)$$

That means the neural network will automatically obtain a Q-V that conforms to C-V during the training process, which will reduce the amount of data required for modeling and reduce the difficulty of data acquisition without affecting the accuracy of the neural network model. Fig. 3 shows the structure of Q-V neural network model proposed in this work which only requires C-V data.

Fig. 4 shows the model output and TCAD data comparisons of the DC part and AC part of the TCAD simulation data of type3 structure in CFET. The model well captures the data, and the error is < 4%.

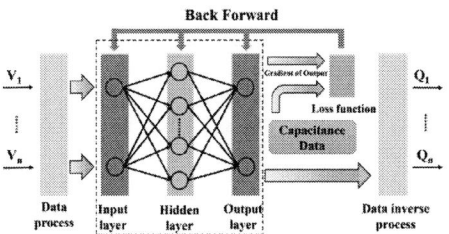

Figure 3: Illustration of Q-V neural network proposed in this work

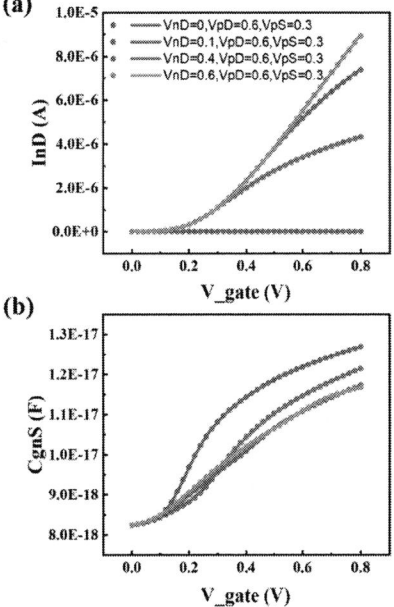

Figure 4: Example neural network model fitting results of CFET type3 structure TCAD data. The dots represent the TCAD data, and the solid lines represent the neural network fitting results. (a)InD-Vgate; (b)CgnS-Vgate

DIGITAL CIRCUIT SIMULATION APPLICATION

In order to simulate the neural network model established in this work in digital circuits based on CFET units, the CFET neural network model needs to be converted into a compact CFET model in Verilog-a language, which is a format that Spice simulators such as the commercial HSPICE can accept. BY converting the CFET I-V and C-V part neural networks established in the previous section into Verilog-A code in an automated manner, the external voltage and port current relationship of the CFET compact model is defined by the input-output relationship of the I-V part neural network of the CFET. Similarly, the internal charge model of the CFET compact model can be defined by the Q-V part neural network of the CFET, thereby realizing the definition of the internal parasitic capacitance.

Fig. 5 is the schematic of several basic logic circuits formed with CFET cells, and their simulation is accomplished by using commercial HSpice simulator. Additionally, in the circuits simulated in this work, all type4 structures were replaced with type3 structures by shorted the VpD and VpS ports in it. This was done to effectively avoid any matching issues between the type4 and type3 structures. Fig. 6 shows the transient simulation waveform of the NAND cell, which demonstrates that the developed CFET compact model works well in practical circuits. This enables the simulation of more complex logic circuits formed with CFET cells.

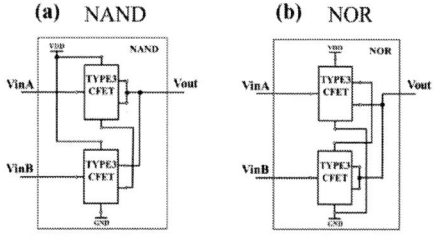

Figure 5: Basic logic gate composed of CFET cells.; (a)Nand; (b)Nor

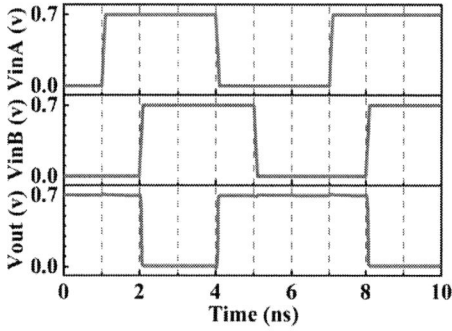

Figure 6: Transient simulation waveform of CFET NAND Circuit. VinA and VinB represent the waveforms of the two input signals of the NAND circuit, Vout is the waveform of the output signal

To verify the application of the developed CFET compact model in larger scale sequence circuits, a 2-bit asynchronous counter circuit based on D latch circuit was selected. The D latch circuit can be constructed by combining several CFET logic gates shown in Fig. 5. By using two such D latch circuits, a D flip-flop circuit can be constructed, which in turn can be used to construct a 2-bit CFET asynchronous counter circuit. Fig. 7 shows the transient simulation waveform of the 2-bit CFET asynchronous counter. These simulation results demonstrate that the developed CFET compact model works well for both combinational and sequential digital circuit simulations.

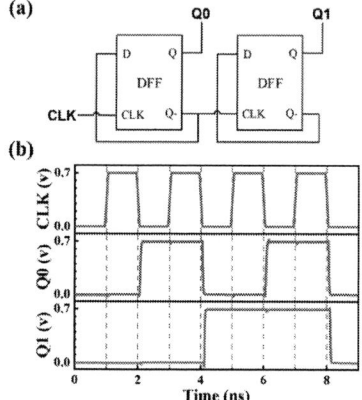

Figure 7: (a)Schematic of 2-bit counter. (b)Transient simulation waveform of the CFET 2-bit counter circuit. CLK represents the waveform of the clock signal of the counter circuit, Q0 and Q1 are the waveforms of the output signals, representing the first and second bits of the count value in binary, respectively

CONCLUSIONS

By treating CFET basic cell as a unit, detailed parasitic effects within cell can be omitted for the DCM creation. We propose ANN based modeling methodology for CFET DCM. The created model exhibits high accuracy. Furthermore, the entire model creation process is highly automated, resulting in a rapid creation process. This allows for early-stage Design Technology Co-Optimization (DTCO) capability for CFET devices.

REFERENCES

[1] Ryckaert J, Schuddinck P, Weckx P, et al. The Complementary FET (CFET) for CMOS scaling beyond N3[C].2018 IEEE Symposium on Vlsi Technology,2018: 141-142.

[2] Schuddinck P, Zografos O, Weckx P, et al. Device-, circuit-& block-level evaluation of CFET in a 4 track library[C].2019 Symposium on VLSI Technology,2019: T204-T205.

[3] Salahuddin S M, Shaik K A, Gupta A, et al.SRAM with buried power distribution to improve write margin and performance in advanced technology nodes[J].IEEE Electron Device Letters,2019, 40 (8): 1261-1264.

[4] Yang Q, Qi G, Gan W, et al.Transistor compact model based on multigradient neural network and its application in SPICE circuit simulations for gate-all-around Si cold source FETs[J].IEEE Transactions on Electron Devices,2021, 68 (9): 4181-4188.

[5] Wang J, Kim Y-H, Ryu J, et al.Artificial neural network-based compact modeling methodology for advanced transistors[J].IEEE Transactions on Electron Devices,2021, 68 (3): 1318-1325.

A 14.7mW 4Gb/s/lane WIRELESS THROUGH SILICON INTERFACE FOR MEMORY CUBE EXPLOITING 16-QAM AND MAGNETIC RESONANCE

Chonghui Sun [1], Rushuo Tao [1], Kun Yang [1], Xuhui Liu[2], C.-Z. Chen[2], Xiaolei. Zhu[1]*
[1]School of Micro-Nano Electronics, Zhejiang University, Hangzhou 311200, China
[2] Peng Cheng Laboratory, Shenzhen 518000, China
*Corresponding Author's Email: xl_zhu@zju.edu.cn

ABSTRACT

A 14.7mW 4Gb/s/lane wireless TSI (through silicon interface) employing magnetic resonance and 16-QAM (quadrature amplitude modulation) is presented for 3D (three dimensional) stacked memory cube. In this scenario, taking advantage of broadband low noise signal approach and high order modulation, 4x symbol data rate and long-distance transmission could be obtained simultaneously. The transmitter based on source follower common-mode amplitude modulator is developed to increase the voltage headroom and the saturated output power. Besides, a double Gm-boost LNA is proposed to achieve high power efficiency and ultrahigh sensitivity, respectively. The proposed interface prototype is designed with 55nm CMOS process and achieves a maximum data rate of 4Gb/s/lane at transfer distance of 300um while dissipating 14.72mW from a 1.2V supply.

INTRODUCTION

With the rapid growth of scaling memory capacity and low delay cache demand, high density integration of memory and processor is getting more popular. 3D stacked memory cube, such as HBM (high bandwidth memory), is so far the most successful commercial application based on TSV (though silicon via) technology, in spite of that TSV craft is costly and easily meets open-contact failure. TSV is suitable for massive parallel connection of power and ground grid while it has the possibility to degrade the reliability of data connection.

The magnetic resonance based wireless through silicon interface has been demonstrated as a potential solution for data communication in multi-chip stacking[1]. However, insufficient spectrum utilization of NRZ (non-return-zero) based design[1] may cause inter-symbol interference. Besides, there is less consideration on the low-noise scheme in reported works, which leads to the tradeoff between the transfer distance and data rate. Hence, the thickness of each stacked chip in such NRZ system is required to be ultra-thin, e.g., within 100 um, which results in a low yield in manufacturing.

To address those problem, A 16-QAM TSI is proposed to support 3D-stack for HPC (high-performance computing system) depicted in Fig.1. Conventionally, 16-QAM signal is generated by RF-DAC[2]. To accustom with the signal channel among stacked chips, the transmitter utilizes source follower based CAM (common-mode

amplitude modulator) which consists with only active circuits and capacitors instead of area hungry components such as transformer, power combiner. On the other hand, a low noise receiver along with wideband impedance matching is designed to ensure the capability of demodulating a tiny signal.

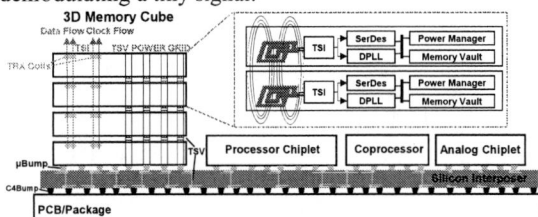

Figure 1: HPC System integrated with Memory Cube

TSI ARCHITECTURE

TSI Overview

Fig.2 shows the integral architecture of proposed TSI transceiver. A full transceiver is composed mainly with T/RX local oscillator, frequency divider, 16-QAM transmitter, LNA, passive mixer and LPF. Driven by a 2GHz clock, a PRBS-7 generator is integrated on the base die to produce four parallel 1Gbps test data streams. The data streams are then transferred to sub die via the coupling inductor. A 9-elements model is utilized to describe magnetic coupling inductors precisely.

Transmitter

The proposed transmitter has several main functional parts. The carrier wave generator consists of a LC-VCO and CML Frequency divider. Output frequency of VCO is adjusted roughly from 10GHz to 16GHz by digital control bits, which decides the num of capacitors involved in LC tank. Accurate frequency could be achieved by tuning the voltage of varactor to cover the variable PVT. Desirable wave has a peak-to-peak amplitude of 630mV at 6GHz.

Passing through two amplifier paths with calibrated buffer, original orthogonal wave duplicated to half gain wave and unit gain wave. Phase shift caused by feedthrough is dramatically reduced by using parallel amplifier.

Figure.3a indicates SF (source follower) based CAM. In this diagram, duplicated waves are loaded on common-mode voltages by switch-controlled SF. SF based common-mode voltage modulator achieves a better area efficiency compared with reported RF modulator in

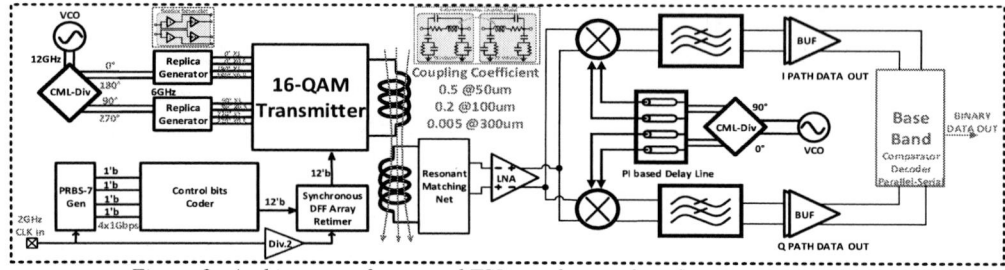

Figure 2: Architecture of proposed TSI interface and its data transmission path

Fig.3b[3] and Fig.3c. Besides, input replicas with different amplitudes negate the concern of non-linear and parasite effect in practice.

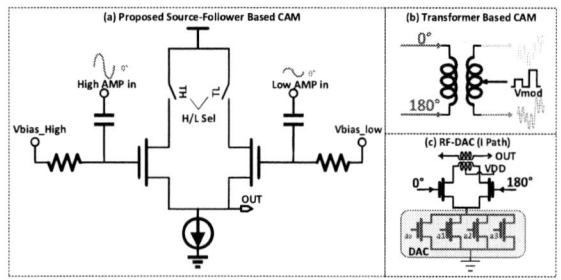

Figure 3: (a)Proposed Source-follower based CAM (b)Transformer Based CAM (c)Conventional RF-DAC

Followed the encode combinations shown in Fig.4, the codes translated from PRBS-7 stream control 8 CAMs inside I/Q path PAM4 modulators. Pre-modulated waves generated by CAMs would be finally delivered to power transistors complex which consists power transistors and compensation capacitors. As shown in Fig.4, a single path PAM4 modulator is composed with quadruple CAMs and a power transistor complex. There are two PAM-4 modulators in the proposed QAM transmitter. The signal is transmitted by a symmetric differential inductor. Tx inductor is implemented with the stander RF inductor from library, whose self-resonant frequency and inductance are 18GHz and 2.5nH, respectively. From the perspective of simulation by 3D full wave simulator, the slope of inductance is stable enough in band (Δ0.1nH).

Unlikely RF-DAC, which is common in millimeter-wave frequency, performance degradation caused by cascade switch transistors is avoided. In addition, the voltage headroom of the TX transistor and saturated output power is increased significantly. SF based CAM makes it possible to get rid of the silicon area consuming component like spiral transformer, in spite of an inherent 3dB insertion loss introduced by the directly combining process of the I/Q path signals. The maximum differential peak-to-peak output voltage between inductor's two terminals is 1.1V.

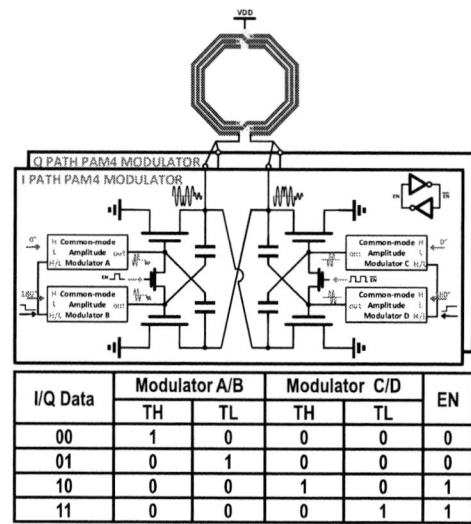

I/Q Data	Modulator A/B		Modulator C/D		EN
	TH	TL	TH	TL	
00	1	0	0	0	0
01	0	1	0	0	0
10	0	0	1	0	1
11	0	0	0	1	1

Figure 4: TSI 16-QAM transmitter with CAM and its encoding scheme of control signal

Receiver

As shown in Fig.2, this system is compatible with most of the general RF frontends, consisting of PI-based delay line, LNA, Mixer and LPF. However, consideration on low power and silicon area saving has been taken into this design. Therefore, a CG-topology LNA shown in Fig.5, is employed to get broadband RF characters. The Gm-boost technique is used in the LNA to save both area and energy. The equivalent Gm of composite transistor reaches to 20mS dissipating only 1.1mA with 1.2V supply, while maintaining an excellent impedance matching at 50Ω from 100MHz to 20GHz. Besides, the capacitor cross coupling technique is also utilized to achieve extra gain without additional power consumption. Furthermore, no other auxiliary feedback circuits are needed except for a voltage bias. Based on simulation, the Gain/S_{11} versus frequency for the proposed LNA is plotted in Fig.5, which shows a desired performance.

Except for the LNA, a double-balanced passive mixer is required to down convert the received RF signal to baseband signal, which consists of Gm stage, main mixer and TIA (transimpedance amplifier). In order to enhance the driving strength of proposed LNA, a DHVB[4] buffer is used to serve as the Gm stage that converts the output signal of LNA to current signal. Flowing though the main mixer, the current signal is then delivered to TIA, which is applied to obtain baseband voltage signal. In the meantime, the baseband voltage signal is shifted to a calibrated common mode voltage by feedback circuits inside TIA. Afterward, the baseband signal with desired input common-mode voltage level is delivered to the super source follower based LPF [5]. According to Nyquist' law, the corner frequency of LPF is set at 2.5GHz.

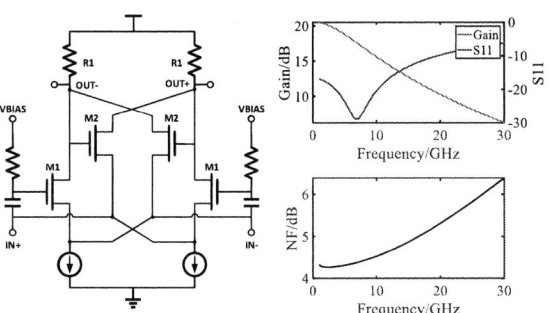

Figure 5: Ultralow power LNA and corresponding merits

POST SIMULATION RESULT

The proposed TSI is designed in TSMC 55nm CMOS process. The I path output signal is shown in figure.6. The BER (bit error rate) is estimated at 10^{-8}. Particular power consuming and layout including core circuit, auxiliary bias circuit and IO are depicted in Fig.7. According to the Momentum simulation result, coefficient of coupling is 0.005, which indicates that distance existing between T/RX coils is nearly 300um. Finally, performance of this TSI and previously published inner-tier inductive coupling data transceiver are summarized in Table I.

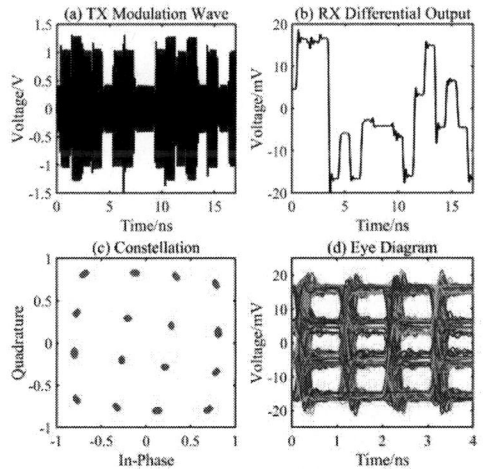

Figure 6: a. transmit waveform b. output baseband signal c. constellation d. eye diagram of output signal

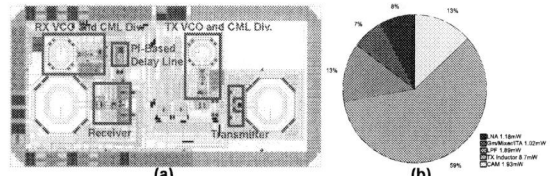

Figure 7: a. Layout of TSI b. Detailed power consumption

TABLE I.   PERFORMANCE AND COMPARISON WITH STATE-OF-THE-ART

Metric	JSSC 2019[1]	VLSI 2020[6]	This Work
Modulation	NRZ	BPSK	QAM
Process	40nm	65nm	55nm
Data Rate	3.6Gbps	1.2Gbps	4 Gbps
Distance	10~80um	80um	300um
BER	10^{-12}	10^{-6}	10^{-8}
Area	$0.04mm^2$	$0.06mm^2$	$0.11mm^2$
Efficient	2pJ/b	——	3.6pJ/b

CONCLUSION

An inductive coupling inner-tier TSI for memory cube is proposed. The proposed wireless TSI achieves a maximum data rate of 4Gb/s/lane at transfer distance of 300 um, while maintaining a favorable efficiency at 3.6pJ/bits. Compared with reported works, this study shows the possibility of increasing the number of layers up to 10 for the wireless TSI based 3D-stacked memory cube without suffering the low yield from ultra-thin polish process required by stacked chips.

ACKNOWLEDGMENTS

This work is supported by the National Natural Science Foundation of China (No.U20A20220), the Major Scientific Research Project of Zhejiang Lab (No.2019KC0AD02) and the Major Scientific Research Project of Zhejiang Province (No.2022C01048)

REFERENCES

[1] K. Ueyoshi et al., *IEEE Journal of Solid-State Circuits*, vol. 54, no. 1, pp. 186-196, Jan. 2019.

[2] Thakkar C et al., *IEEE Journal of Solid-State Circuits*, vol. 54, no. 12, pp. 3565-3576, Dec. 2019.

[3] X. Meng et al., *IEEE Transactions on Circuits and Systems I: Regular Papers*, vol. 67, no. 6, pp. 1835-1845, June 2020.

[4] D. Im et al., *IEEE Transactions on Microwave Theory & Techniques.*, vol. 57, no. 11, pp.2633–2642, Nov. 2009.

[5] M. De Matteis et al., *IEEE Journal of Solid-State Circuits*, vol. 50, no. 7, pp. 1516-1524, July 2015.

[6] B. J. Fletcher et al., *2020 IEEE Symposium on VLSI Circuits*, 2020, pp. 1-2.

A HARDWARE ACCELERATOR FOR STANDARD CONVOLUTION AND DEPTHWISE CONVOLUTION

Fubang An[1], Wei Cao[2], Xuegong Zhou[2]* and Lingli Wang[1]*

[1]School of Microelectronics, Fudan University, Shanghai 200433, China
[2]Institute of Big data, Fudan University, Shanghai 200433, China
*Corresponding Author's Email: llwang@fudan.edu.cn; zhouxg@fudan.edu.cn

ABSTRACT

In this paper, a CNN hardware accelerator for standard convolution and depthwise convolution is proposed. The accelerator can support two different data flow modes. A computation array composed of DSP is designed to support the parallel strategy of input channel and output channel for standard convolution efficiently. A *Membank* architecture is designed to make the computation array more efficient for the parallel strategy of input feature map and kernel for depthwise convolution. The accelerator can accelerate EfficientNet on Xilinx Alveo U280 at the system clock of 300MHz and the DSP clock of 600MHz. The results show that the accelerator can achieve 1.14× throughput and 1.12× throughput/DSP compared with the latest FPGA-based accelerator on the same data center accelerator card.

INTRODUCTION

With lower computation complexity and fewer memory accesses than previous CNNs, lightweight convolutional neural networks have become hot research topics, such as EfficientNet[1].

Lightweight convolutional neural networks mainly contain standard convolution(STC) and depthwise convolution(DWC). The principle of DWC is that spatial and channel correlations can be decoupled and separately realized. Compared to STC, DWC can save the numbers of parameters and Multiply-Accumulate Operations (MACs) but can guarantee high precisions.

Although the performances of lightweight CNNs are excellent, the existing hardware accelerator architectures are not efficient to support both standard and depthwise convolutions.

The contributions of this work are as follows:

1. The reconfigurable CNN accelerator proposed in this paper supports two parallel modes: parallel on the input channel and output channel dimensions and parallel on the input feature map(IFM) and kernel dimensions. As a result, the accelerator is more flexible and efficient for the standard convolution and depthwise convolution.

2. This paper proposes a *Membank* architecture composed of multiple register chains, which can improve the performance of DWC on the basis of the basic architecture.

The rest of the paper is organized as follows. Section II introduces the analysis of design methods of different convolution modes. In section III, the new hardware architecture compatible with two modes is presented. Section IV analyzes the performance of the proposed architecture compared with the previous work. Section V concludes the paper.

DESIGN SPACE ANALYSIS

As shown in Figure 1, each input channel has to perform a convolution with one specific kernel firstly in STC, and then the result is the sum of the convolution results from all input channels. Because the kernel sizes of STC in different networks are often different and there are lots of pointwise convolutions in lightweight CNNs, parallelism on the feature map dimension is not universal. Based on these computational features, it is natural to adopt the method of parallelism on the channel dimension. It is easy to ensure that data from adjacent input channels are simultaneously broadcast to the PE array.

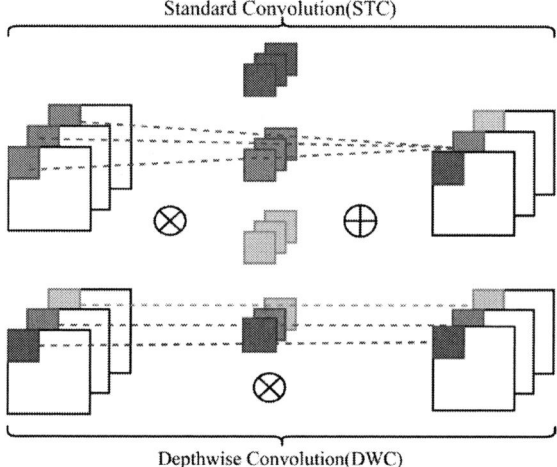

Figure 1: Standard and Depthwise Convolutions

However, in DWC, performing the convolution for each input channel individually is the main step. Each output channel corresponds to an input channel independently, so we have to parallel on the input feature map and kernel dimensions. To parallel on the input feature map, it is necessary to ensure that the adjacent input data on the IFM can be simultaneously sent to the PE

979-8-3503-1101-3/23 $31.00 © 2023 IEEE   566

array. Therefore, we need to design a new data flow mode to broadcast the input data.

HARDWARE ARCHITECTURE

Basic Hardware Architecture

The proposed system architecture of the hardware accelerator and the PE architecture are presented in Figure 2 and Figure 3. The system architecture references the method of LETA[2]. The computational kernel is composed of 16 processing elements(PEs). Each PE contains 4 *Micro_PE* cells. The INT8 quantization method is used in this paper. The DSP48E2 slice in Xilinx Alveo U280 has a 27×18 multiplier, where two INT8 multiply operations with two separate factors and a common factor can be packed. Under the condition that the width of the multiplication result is 16 bits, the maximum width of the sum result is 18 bits[3]. In addition, the frequency of the DSP is twice that of the main system of FPGA. As a result, each DSP can be equivalent to at most four INT8 multipliers in a system clock cycle while computing STC.

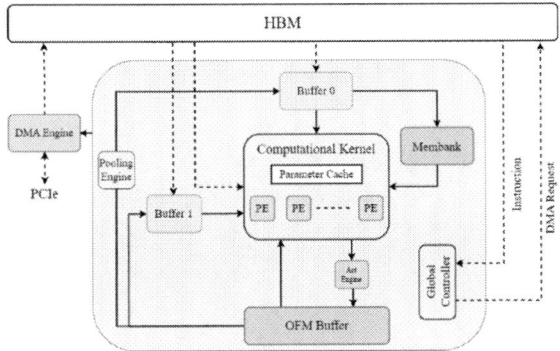

Figure 2: The hardware architecture

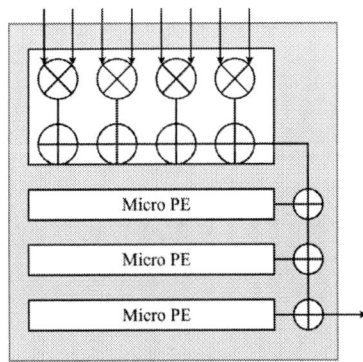

Figure 3: The PE architecture

During the standard convolution, the input data are broadcast to different PEs, so as to support the parallel computation on the dimension of the output channel. In a single PE, the input data are multiplied with the corresponding weights which are stored in the parameter caches to realize the parallel computation on the input channel dimension. Each DSP is connected with 4 weight caches from 4 different output channels to increase parallelisms and make full use of the computing ability. Because the numbers of input and output channels in neural networks are often integer multiples of 16, setting 16 DSPs in a PE and 16 PE units in an array can ensure the maximum theoretical utilization of computing resources along the direction of input and output channels. However, if this architecture is used to calculate the depthwise convolution, the utilization of the PE unit will be very low.

Membank Architecture

Because the depthwise convolution does not need to add data along the direction of input channels, it is more suitable for the parallel computation on the 2-D feature map in a PE. Based on the original hardware architecture, we propose the *Membank* architecture to support the parallel computation on the dimensions of the input feature map and convolutional kernel of DWC.

A three-line register array, which contains nine registers, is used to form the *Membank*. The *Membank* is presented in Figure 4. When the data transfer starts, each row of data is transferred to the register array in turn by every clock cycle. When the three columns of registers are filled, the data are sent to the PE for calculation. The parameters ICP and OCP indicate the parallelisms of input channel and output channel. Each PE unit corresponds to an input feature map. Because there are 9 DSPs actually used in each PE, the utilization is 9/16.

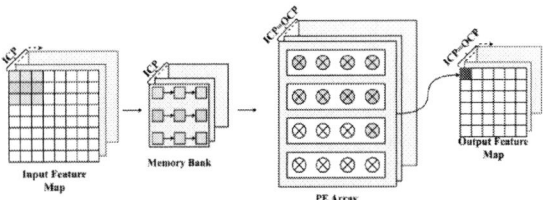

Figure 4: The architecture of Membank

Considering that each new row needs to waste cycles to fill the *Membank*, we add the fourth register chain and design the bidirectional register chain to optimize the architecture, which is called *wrap-reverse*. In addition, because the data in *Membank* should remain the same for two cycles when a row is moved to the next, the register chain is also set to *self-stop*. Each row of data on the output feature map is corresponding to three rows of the *Membank*. As a result, switching between different rows does not waste cycles. While computing a row of data on the output feature map, three rows of data in the *Membank* are selected and transferred into the PE. The bidirectional register chain is presented in Figure 5 and the process of moving rows is shown in Figure 6.

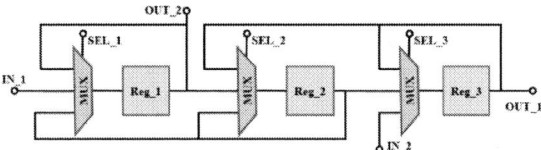

Figure 5: The bidirectional register chain

Another benefit of this architecture is that it saves the register resources. In the method of line-buffer in paper[4], it needs $((W + PAD) \times 2 + 3)$ registers in the 3×3 DWC. W is the width of IFM. PAD is the width of Zero Padding of IFM. It also needs more cycles to fill the register array. The number of registers of *Membank* is not affected by the size of input feature map, and it can also save the numbers of registers and latency.

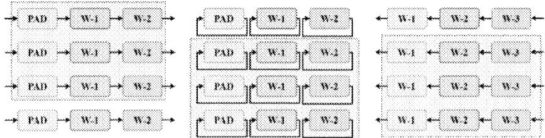

Figure 6: The process of moving rows in Membank

In addition, we notice that the above architecture can only support 3×3 DWC efficiently. Another common kernel size of DWC in lightweight CNNs is 5×5. We can extend *Membank* to 25 registers and use two adjacent PEs to compute one pixel on the output feature map.

RESULTS

The complete design of the hardware accelerator is implemented on Xilinx Alveo U280 data center accelerator card by Verilog HDL. The main system of the proposed hardware runs at 300 MHz, while the frequency of the DSP kernel is 600 MHz. We choose the EfficientNet-B3 for testing, and the performances compared with other typical lightweight CNN hardware accelerators are shown in Table I, where paper[5] can accelerate EfficientNet-B0 and LETA[2] can accelerate EfficientNet-B3. The accelerator proposed in this paper can achieve the performance of 62.01 IPS and the throughput of 227.73 GOPS which is 1.14 times compared with the latest EfficientNet accelerator LETA[2] on the same FPGA data center accelerator card. The throughput/DSP of our work is 0.419 GOPS/DSP which is 1.12 times compared with LETA[2].

CONCLUSION

In this paper, we propose a new hardware architecture for standard convolution and depthwise convolution. Firstly, a PE array is proposed to support the parallel strategy of input channel and output channel for standard convolution efficiently. Next, the *Membank* architecture is designed to improve the performance of depthwise convolution. Finally, we choose the EfficientNet-B3 for testing and the results show that our work can achieve 1.14× GOPS and 1.12× GOPS/DSP compared with the latest FPGA-based EfficientNet accelerator LETA[2].

TABLE I
PERFORMANCE COMPARISONS

	Paper[5]	LETA[2]	This Work
Platform	Xilinx XCVU440	Xilinx XCVU37P	Xilinx XCVU37P
Network	ENet-B0	ENet-B3	ENet-B3
Precision	INT16	INT8	INT8
DSP	1008*	534	544
Frequency	180MHz	300MHz	300MHz
MAC(B)	0.78	3.67	3.67
Latency(ms)	N/A	18.4	16.13
GOPS	180.3	199.6	227.73
GOPS/DSP	0.179	0.374	0.419
IPS	231.2	54.34	62.01

* DSP numbers of the accelerator using INT16 quantization methods are unified to INT8 for comparisons.

ACKNOWLEDGEMENTS

This work is supported by the National Key R&D Program of China under grants 2022YFB4500903 and the National Natural Science Foundation of China under grants 61971143 and 62174035.

REFERENCES

[1] M. Tan and Q. V. Le, "EfficientNet: Rethinking model scaling for convolutional neural networks," in International Conference on Machine Learning (ICML), 2019, pp. 6105–6114.

[2] J. Gao et al., "LETA: A lightweight exchangeable-track accelerator for efficientnet based on FPGA," 2021 International Conference on Field-Programmable Technology (ICFPT), 2021, pp. 1-9, doi: 10.1109/ICFPT52863.2021.9609919.

[3] Y. Fu et al., "Deep learning with INT8 optimization on Xilinx devices," Tech. Rep., 2017. [Online]. Available:https://www.xilinx.com/support/documentation/white_papers/wp486-deep-learning-int8.pdf.

[4] L. Bai, Y. Zhao and X. Huang, "A CNN accelerator on FPGA using depthwise separable convolution," in IEEE Transactions on Circuits and Systems II: Express Briefs, vol. 65, no. 10, pp. 1415-1419, 2018.

[5] S. Shivapakash et al., "A power efficiency enhancements of a multi-bit accelerator for memory prohibitive deep neural networks," in IEEE Open Journal of Circuits and Systems, vol. 2, pp. 156-169, 2021, doi: 10.1109/OJCAS.2020.3047225.

A MULTI-LAYER STACKED 3-D SRAM SYSTEM BASED ON WIRELESS TRANSCEIVER USING INDUCTIVELY COUPLED INTERFACE IN 22-NM CMOS

Kun Yang[1], Chonghui Sun[1], Rushuo Tao[1], Jiannan Guo[4], Cheng Yang[4], D.Ma[2,3], Xiaolei Zhu[1,3]

[1] School of Micro-Nano Electronics, Zhejiang University, Hangzhou 311200, China
[2] School of Computer Science And Technology, Zhejiang University, Hangzhou 311200, China
[3] Zhejiang Lab, Hangzhou 311121, China
[4] JCET Group Co., Ltd, Shanghai, China
xl_zhu@zju.edu.cn

ABSTRACT

This paper proposes a three-dimensional SRAM system based on inductively coupled technology. The data, clock, and energy signals are delivered simultaneously through the multi-layer of the stacked chip via inductive coupling interface. By using the 22nm CMOS process, it supports the stacking of up to five layers of a chip with 25μm in thickness for each. Simulation results show the proposed 3-D SRAM system achieves a maximum data rate of 1Gb/s while dissipating around 400μW for memory read/write from a 0.9V power supply. The Bit Error Rate (BER) is less than 10^{-14} and the energy delivery efficiency is larger than 10% within the distance of 100um.

Keywords—Inductor, wireless communication, SRAM, 3-D package, power system stability

INTRODUCTION

The existing 2-D system-on-chip (SoC) and system-in-package (SiP) design methodologies, through silicon via (TSV) integration and μbumps currently in production, may have large parasitics and larger pitch than what is needed for high-performance applications. [1]. Multiple CMOS ICs for die-2-die (D2D, or chip-2-chip, C2C) integration, typically using 65nm to 7nm processes, are assembled in today's SoC designs. The emergence of three-dimensional stacked packaging technologies has increased for the application of 3-D SRAM system in recent years.

This work focuses on the inductively coupled interface for 3-D stacked SRAM integration, which supports highly efficient delivery of the clock, data and power through the multi-layer of stacked chips. As shown in Fig. 1, the system consists of one master chip and several slave chips stacked in a 3-D structure. The upper layers of the chip are slave chips without contacts between each other and only for wireless data/energy delivery, while the master chip is externally linked and powered from a PCB. The data, clock and power are delivered from the master chip to the slave chip via inductively linked interfaces. In this arrangement, both logic and memory chips can be used in multiple systems, and the stacking of

chips with distinct functionalities allows for high-bandwidth external memory [2].

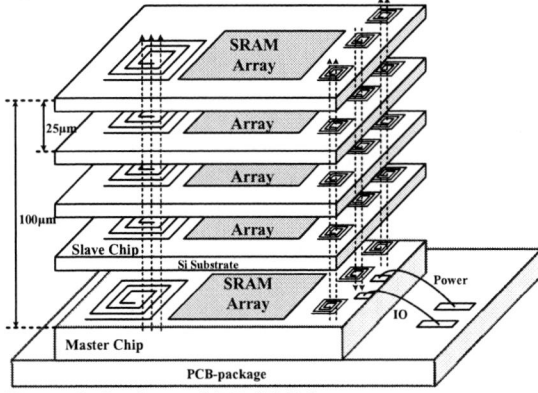

Figure 1: Package of 3-D SRAM

SYSTEM ARCHITECTURE

Overview

Fig. 2 depicts the circuit structure of the entire system, which consists of the transceiver, SRAM and the peripheral circuits. On the master chip, the data signal, chip-select signal, address signal, and read/write control signal is packaged and delivered to the slave chip by the data receia the serializer.

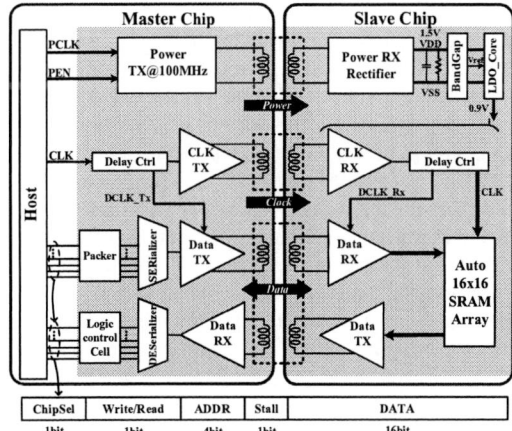

Figure 2:Circuit Structure of The Entire System

Fig. 3 depicts a 16x16-bit SRAM system with an 8-bit counter driven by the clock, with the lower 4 bits connected to the 4-to-16 decoder of the bit line for selecting the sequential bit and the upper 4 bits connected to the 4-to-16 decoder of the word line for selecting the 16-bit read/write data port. The data transceiver with a clock link employs a Bi-phase modulation (BPM) circuit to provide a noise-immune current pulse signal for each bit of data, resulting in a lower BER [3]. The host side provides the 100M energy clock signal and the drive signal PEN to transmit the energy signal for the energy link. The receiving inductor then outputs a stable supply voltage for the SRAM system via the rectifier and the corresponding low dropout regulator (LDO).

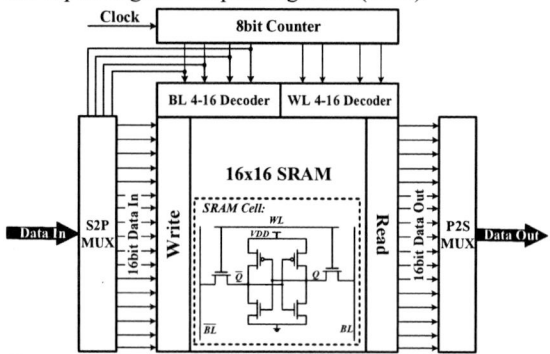

Figure 3: 16x16-bit SRAM System

On-chip Inductors for Wireless Communication

Most of the on-chip inductors utilized in this design are planar spiral inductors created by spirally wrapping metal wires using both the top and second top layers provided from the process library. During the simulation, the CMOS process substrate model is established in the electromagnetic field simulation software HFSS and construct various inductor models to meet the requirements from the specified parameters, such as the series resistance R, on-chip inductor inductance L, quality factor Q, and self-resonance frequency f_{SR}. Table I lists the simulation parameters of the inductor for each channel. The resistance value must be low enough for an inductor on the transmitter side to generate a sufficient amount of current. Consequently, the wiring of the transmitter inductor should be wide enough while has fewer number of turns. Increasing the number of turns while narrowing the metal width for receiver inductors will increase self and mutual inductance until their self-resonant frequency exceeds twice the signal bandwidth [4]. Unlike the data link, the clock link only transmits a narrow band signal at 1GHz. Therefore, the inductor of the clock link can be constructed in light with the maximum possible Q value, which maximizes the cross-conductance and effectively suppresses the ambient noise.

TABLE I   SIMULATION PARAMETERS OF INDUCTORS

Channel	L(nH)	R(Ω)	f_{SR}(GHz)	Q
Data TX	5.8	4	7.7	9.1
Data RX	7.85	5	6.4	9.9
CLK	33.5	33	1.7	6.35
Power TX	12.3	1.6	1.4	7.7
Power RX	380	145.2	0.22	2.6

Wireless Power Transceiver

Fig. 4 shows the schematic diagram of energy transmitter circuits. The energy transmitter is an H-bridge circuit, whereas the receiver is an NMOS cross-gate rectifier [5]. The four transistors on the transmitter side should be sized pretty large to drive sufficient currents for the inductor. The large-sized transistor increases the likelihood that the circuit generates an extensive, thorough current during the transition, which will damage the transistors and cause the system to crash. In order to prevent a large current, the dead zone control circuit is implemented to feed the equivalent transistor gate signal back to the opposite end via the level shifter circuit. The signal is then applied to control a logic circuit. These two driving signals are utilized to generate a sufficient delay to avoid dead zone for the circuits.

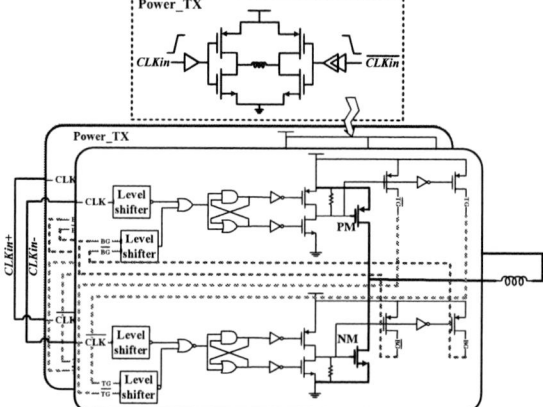

Figure 4:Schematic Diagram of Energy Transmitter

SIMULATION RESULTS

The transient simulation results for the overall energy, clock, data transceiver, and SRAM system are depicted in Fig. 5. The energy link operates under 1.5V while the internal core circuit operates at 0.9V power supply. The energy transmitter starts working in 1μs. Line 2 shows the received voltage from the rectifier of the energy receiver with a peak value of approximately 1.5V, and line 3 shows the internal power supply voltage of 0.9V at the output of the LDO.

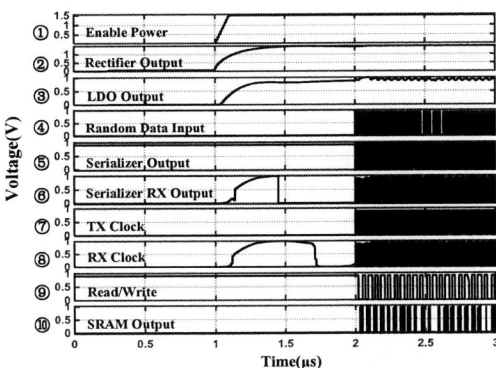

Figure 5: Overall Transient Simulation Results

The data and clock signals are transferred in 2μs, and Fig. 6 depicts the specific data waveform. First, on the master chip side, 16-bit random data is generated by the transmitter, based on the combination of the chip-select signal and other control signals. The random data is then received/unpacked by the slave chip and fed sequentially into the SRAM for regular reading/writing. By simulating and calculating, the read/write operations in the SRAM dissipate average power of 400uW from a 0.9V power supply. If two layers of chips are allocated by the distance of 100um, the equivalent coupling coefficient K is around 0.2. In such a case, the recovery power on the receiver side is larger than 10 mW which is sufficient for the power requirement from a 16x16-bit SRAM. The prototype is designed in 22nm CMOS and as illustrated in Fig. 7, it occupies around 4.2mm^2 of silicon area. The functionality and performance are also verified by performing a post-layout simulation.

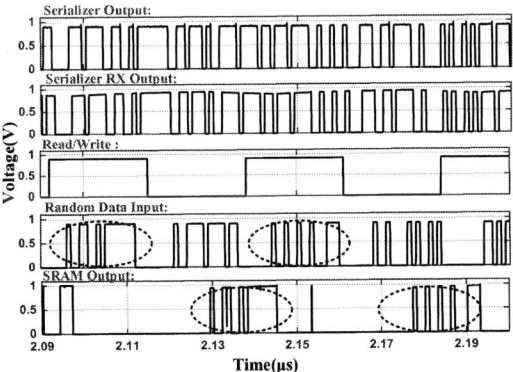

Figure 6: Transient Simulation of Data Transmission

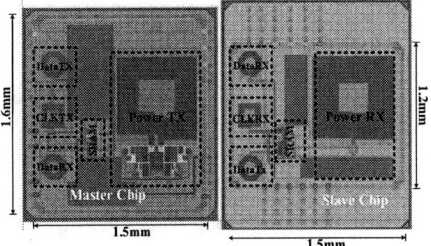

Figure 7: Layout of 3-D SRAM

CONCLUSION

This work proposes an inductive coupling-based wireless interconnect interface for 3-D SRAM integration. The data, clock, and energy signals are delivered simultaneously through the multi-layer of stacked chip. The simulation results show the design not only has the correct functionality but also achieves a high data rate with low BER. Not limited to the 3-D SRAM integration, the proposed wireless interconnect interface could be extended to various applications for both homogeneous and heterogeneous integration.

ACKNOWLEDGEMENTS

This work is supported by the National Natural Science Foundation of China (No.U20A20220), the Major Scientific Research Project of Zhejiang Lab (No.2019KC0AD02) and the Major Scientific Research Project of Zhejiang Province (No.2022C01048)

REFERENCES

[1] R. Agarwal et al., *3D Packaging for Heterogeneous Integration*, 2022 IEEE 72nd Electronic Components and Technology Conference (ECTC), 2022, pp. 1103-1107.

[2] K. Ueyoshi et al., *QUEST: Multi-Purpose Log-Quantized DNN Inference Engine Stacked on 96-MB 3-D SRAM Using Inductive Coupling Technology in 40-nm CMOS*, IEEE Journal of Solid-State Circuits, 2019, vol. 54, no. 1, pp. 186-196.

[3] N. Miura et al., *A 1 Tb/s 3 W Inductive-Coupling Transceiver for 3D-Stacked Inter-Chip Clock and Data Link*, IEEE Journal of Solid-State Circuits, 2007, vol. 42, no. 1, pp. 111-122.

[4] B. J. Fletcher, S. Das and T. Mak, *Design and Optimization of Inductive-Coupling Links for 3-D-ICs*, IEEE Transactions on Very Large Scale Integration (VLSI) Systems, 2019, vol. 27, no. 3, pp. 711-723.

[5] X. Li, C. -Y. Tsui and W. -H. Ki, *A 13.56 MHz Wireless Power Transfer System With Reconfigurable Resonant Regulating Rectifier and Wireless Power Control for Implantable Medical Devices*, IEEE Journal of Solid-State Circuits, 2015, vol. 50, no. 4, pp. 978-989.

AN ADAPTIVE CONTROLLED CHIP-LEVEL WIRELESS POWER TRANSFER SYSTEM WITH DPID CONTROLLER FOR WIRELESS 3-D STACKED CHIPS

Rushuo Tao[1], Chonghui Sun[1], Kun Yang[1], Cheng Yang[2], Jiannan Guo[2] and Xiaolei Zhu[*]
[1] School of Micro-Nano Electronics, Zhejiang University, Hangzhou 311200, China
[2] JCET Group Co., Ltd, Shanghai, China
*Corresponding Author's Email: xl_zhu@zju.edu.cn

ABSTRACT

In this paper, we presented an adaptive controlled chip-level wireless power transfer (WPT) system with a digital proportional-Integral-derivative (DPID) controller for next generation of high-density wireless three-dimensional (3-D) stacked semiconductor technology. The WPT is designed in UMC 55 nm CMOS process with an area of 0.3 mm^2. Simulation result shows the received power is controllable varying from 0 to the peak value of 43mW with a settling time less than 2 us from a 2.5 V power supply. It achieves 15% power efficiency and over 90% of power utilization ratio.

Keywords—wireless power transfer; 3-D stacked chips; adaptive control; WPT system; DPID.

INTRODUCTION

With the continuous progress of CMOS technology, the size of IC chips is getting smaller and smaller [1]. Therefore, 3-D stacked packaging technology has developed to realize continuous growth on performance, integration, and low power of IC chips [2].

In recent years, inter-chip wireless interconnection technology has been greatly developed, and the data communication link of 3-D stacked chips based on inductor coupling interconnection has been widely studied [3]. In previous chip-level WPT system studies, an open-loop system without feedback circuit is proposed in [4] and [5]. For further study of inter-chip WPT system, the received voltage is fed back to the transmitter in the form of digital signal after sampling, coding and other processing to make up a feedback loop to enhance the stability of the transfer system [6]. However, a system with simple feedback circuit is not enough to improve the performance of power utilization ratio.

This paper proposed a power frequency controllable chip-level wireless power transfer (WPT) scheme using digital proportional-Integral-derivative (DPID) control algorithm for wireless interconnected 3-D stacked IC. The prototype achieves the received power controllable which varies from 0 to the peak value of 43mW with a settling time less than 2 us from a 2.5 V power supply. It achieves 15% power efficiency and over 90% of power utilization ratio.

DESIGN AND ANALYSIS OF THE WPT CONTROL SYSTEM

System Architecture

The block diagram of the DPID controlled WPT system is shown in Figure 1. In daughter die, a comparator and level shift array are designed to convert the received voltage into a digital signal, and then the data transmitter coil sends it to mom die. In mother die, a data receiver sends the feedback signal to DPID controller which calculates the control code of VCO and frequency divider. Then the VCO and frequency divider generate a clock of power transmitter which is utilized to control the transmitting power.

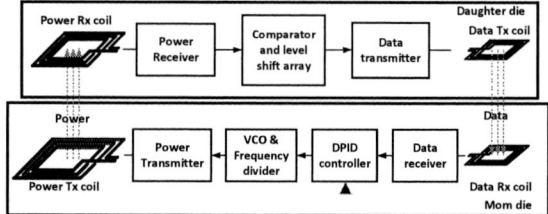

Figure 1: Block diagram of the DPID controlled WPT system.

Design of Power Transceiver

The Power transceiver circuit is shown in Figure 2. An H-bridge circuit is designed in transmitter to change DC voltage into AC current, and the transmitter coil L_T produces a magnetic field. The receiver coil L_R senses the changing magnetic field and two body dynamic control diodes and two NMOS transistors are designed in receiver to improve the delivery efficiency.

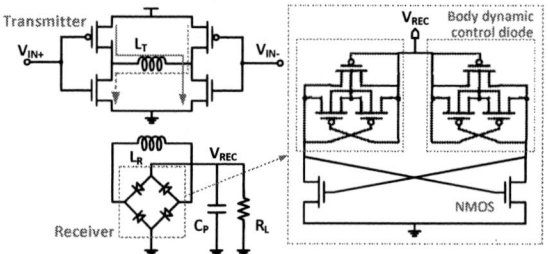

Figure 2: Power transceiver circuit design.

The body dynamic control technique is utilized to avoid latch-up and substantially reduce the leakage going through the substrate, even if the voltage swing is very large on the terminal. The rectifier includes a pair of NMOS transistors working under high over-drive voltage which achieves a better power conversion capability.

Design of Power Transceiver Coils

A transmitter (Tx) coil and a receiver (Rx) coil are implemented using the thickest top metal layer as shown in Figure 3, a nine-element equivalent circuit is utilized by ADS simulator for modeling the inductance parameters of the coils. The parameters of the coils are shown in Table I when the distance between two coils is 50 μm. The frequency of magnetic-field resonance coupling is lower than 100MHz for this design.

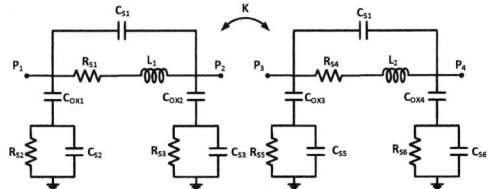

Figure 3: Nine-element equivalent circuit provided by ADS simulator.

TABLE I. PARAMETER OF POWER TRANSCEIVER COILNINE-ELEMENT EQUIVALENT CIRCUIT

L_1	L_2	R_{S1}	R_{S2}	R_{S3}
10.0nH	189nH	2.76ohm	307ohm	31.3ohm
R_{S4}	R_{S5}	R_{S6}	C_{OX1}	C_{OX2}
31.7ohm	158ohm	194ohm	2.93pF	2.93pF
C_{OX3}	C_{OX4}	C_{S1}	C_{S2}	C_{S3}
163fF	162fF	245fF	78.3fF	156fF
C_{S4}	C_{S5}	C_{S6}	K	
14.1fF	5.93pF	5.94pF	0.8	

Design of Digital Proportional-Integral-Derivative Controller

A stable received voltage at the supply voltage is required to ensure the regular operation of load circuit. However, an amount of received power is wasted if the working state of the load circuit changed. It is clear that the received power is directly proportional to the clock frequency as shown in Figure 4.

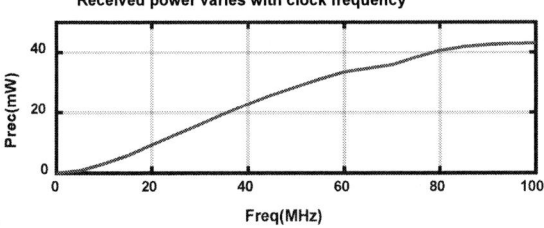

Figure 4: Received power varies with clock frequency.

The DPID control signal changes clock frequency of the WPT system to regulate the received power and improve the power utilization ratio. The power utilization ratio η is given by Eq.1.

$$\eta = \frac{P_{used}}{P_{rec}} \qquad (1)$$

Where P_{used} is supply power of the load circuit, and P_{rec} is the received power. We can calculate the DPID control signal according to the received voltage and the reference voltage. Digital PID controller is usually obtained by discretization of analog PID controller. The output signal $u(t)$ of analog PID controller is given by Eq.2.

$$u(t) = K_p e(t) + K_i \int_0^t e(t)dt + K_d \frac{de(t)}{dt} \qquad (2)$$

Where K_p, K_i and K_d are the proportional, integral and derivative factor, respectively. $e(t)$ is the error signal between reference signal and feedback signal. According to Eq.2, The output of DPID controller is derived as Eq.3.

$$u[k] = u[k-1] + K_p\big[e[k] - e[k-1]\big] + K_i e[k] + K_d\big[e[k] - 2e[k-1] - e[k-2]\big] \qquad (3)$$

Where $u[k-1]$ is the delay signal of $u[k]$, likewise, $e[k-1]$ and $e[k-2]$ are delay signals of $e[k]$. Eq.3 can be further simplified to Eq.4.

$$u[k] = u[k-1] + K_0 e[k] + K_1 e[k-1] + K_2 e[k-2] \qquad (4)$$

Where K_0, K_1 and K_2 are expressed as follows:

$$K_0 = K_p + K_d \qquad (5)$$
$$K_1 = -K_p + K_i - 2K_d \qquad (6)$$
$$K_2 = K_d \qquad (7)$$

The block diagram of DPID controller and the Multiplier-Accumulator state transition diagram are shown in Figure 5 and Figure 6, respectively.

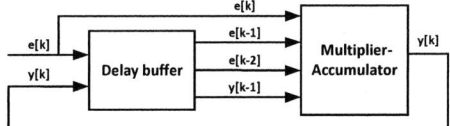

Figure 5: Block diagram of DPID controller. Delay signals are inputted in Multiplier-Accumulator to obtain y[k].

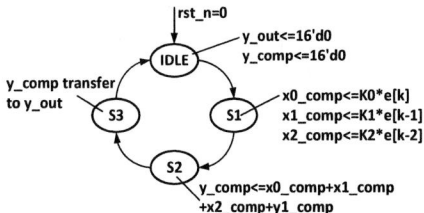

Figure 6: State transition diagram of Multiplier-Accumulator. IDLE: rst_n=0, output signal y_out=0. S1: calculate x0_comp, x1_comp and x2_comp. S2: calculate y_comp; S3: transfer y_comp to y_out. Attention, all signal with suffix 'comp' are complement code.

The layout of the WPT system with DPID controller is shown in Figure 7. It occupies an area of 0.48 μm² while the total area of the prototype is around 0.3 mm².

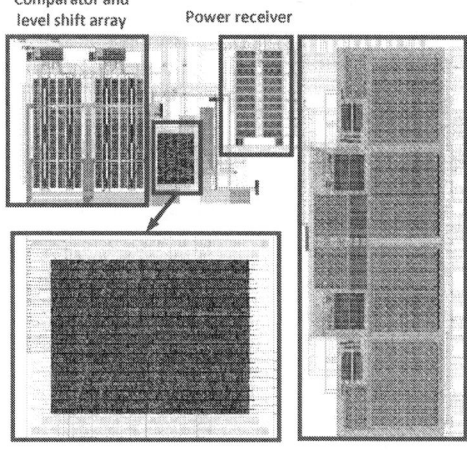

Comparator and level shift array | Power receiver

DPID controller | Power transmitter

Figure 7: layout of the WPT system with DPID controller.

SIMULATION RESULT

The received voltage and power transmitter clock frequency varies with time are shown in Figure 8 when the equivalent resistance of load circuit changes from 150 ohm to 300 ohm.

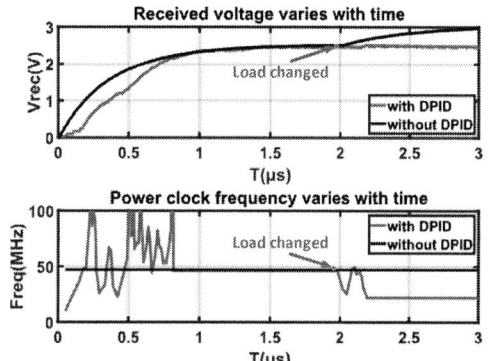

Figure 8: Received voltage and power frequency varies with time.

The received voltage is stable at 2.5V when the load changes by frequency controlled of DPID controller. Typically, in Figure 8, there is 14mW power reduced with the stability of received voltage of 2.5V. Table II is the performance summary of this design compared with the reported work in recent years.

CONCLUSION

A chip-level WPT system scheme with DPID controller is proposed for 3-D stacked Chips. The system achieves 43mW maximum power transmission with a settling time less than 2 μs from a 2.5 V power supply. It also achieves 15% power efficiency and over 90% of power utilization ratio.

TABLE II.
PERFORMANCE COMPARED WITH PREVIOUS WORKS

	This Work	**[4]**	**[5]**
Control scheme	DPID	Null	Null
Distance	50μm	20μm	15μm
Settling time	<2μs	~	~
Frequency	0~100MHz	1GHz	140MHz
$P_{rec,max}$	43mW	12mW	36mW
Effeciency	15%	14%	10%
CMOS process	55nm	65nm	180nm
Power utilization ratio	90%	~	~

ACKNOWLEDGMENT

This work is supported by the National Natural Science Foundation of China (No.U20A20220), the Major Scientific Research Project of Zhejiang Lab (No.2019KC0AD02) and the Major Scientific Research Project of Zhejiang Province (No.2022C01048).

REFERENCES

[1] K. Rupp and S. Selberherr, "The economic limit to moore's law," in IEEE Transactions on Semiconductor Manufacturing, vol. 24, no. 1, pp. 1-4, Feb. 2011.

[2] L. Li, P. Ton, M. Nagar and P. Chia, "Reliability challenges in 2.5D and 3D IC integration," 2017 IEEE 67th Electronic Components and Technology Conference (ECTC), 2017, pp. 1504-1509.

[3] N. Miura et al., "A 1 Tb/s 3 W Inductive-Coupling transceiver for 3D-stacked inter-chip clock and data link," in IEEE Journal of Solid-State Circuits, vol. 42, no. 1, pp. 111-122, Jan. 2007.

[4] L. Sun, J. Xu, W. Mao, S. Zou, P. Lv and X. Zhu, "1GHz wireless power delivery using 0.2×0.2mm2 on-chip inductor for 3-D stacked chips," 2015 IEEE International Conference on Electron Devices and Solid-State Circuits (EDSSC), 2015, pp. 475-478.

[5] Y. Yuan et al., "Simultaneous 6Gb/s data and 10mW power transmission using nested clover coils for non-contact memory card," 2010 Symposium on VLSI Circuits, 2010, pp. 199-200.

[6] X. Li, C. -Y. Tsui and W. -H. Ki, "A 13.56 MHz wireless power transfer system with reconfigurable resonant regulating rectifier and wireless power control for implantable medical devices," in IEEE Journal of Solid-State Circuits, vol. 50, no. 4, pp. 978-989, April 2015.

AN IMPROVED NOISE CANCELING STURDY 2-1 MASH SIGMA-DELTA MODULATOR WITH MULTI-BIT SAR QUANTIZER

Tengteng Mu[1]*, Lianxi Liu[2]

School of Microelectronics, Xidian University, Xi 'an 710071, China

*Corresponding Author's Email: 20111223144@stu.xidian.edu.cn

ABSTRACT

An improved delay-based noise canceling (DNC) sturdy 2-1 MASH Sigma-Delta modulator is designed in this paper, which eliminates the quantization noise of the first noise shaping loop compared with the traditional structure. SAR quantizer and data weighted average (DWA) are used to reduce the nonlinear problem of power consumption and feedback DAC. The circuit design uses 0.18μm CMOS standard technology, the working power supply voltage is 3.3V, the sampling rate is 2.56MHz, the signal bandwidth is 20kHz, the peak SNDR is 102.2dB, the significant bit is 16.68bit and the overall power consumption is 2.87mW.

INTRODUCTION

With the rapid development of the IoTs, the demand for data processing such as sensors and audio is constantly increasing, so is the demand for high resolution and low power analog-to-digital converter (ADC). Sigma-Delta modulator is widely used because of its high precision and signal-to-noise ratio. The stability condition of the high-order single-loop Sigma-Delta modulator is limited by the accuracy of the loop coefficient, and the MASH structure composed of multiple low-order single-loop modulators has good stability[1]. Multi-bit Flash quantizer can reduce the inherent noise leakage in the modulator but has a high power consumption. SAR quantizer does not require multiple comparators, which is conducive to the design of low power consumption[2].

In this paper, a new delayed multi-bit quantized 2-1 MASH Sigma-Delta modulator for noise cancellation is designed, which provides a noise shaping capability greater than 90dB and reduces the noise cancellation module [3]. The 5-bit SAR quantizer in the first stage reduces the inherent noise leakage in the first stage modulator and improves the noise shaping effect. The data weight averaging algorithm (DWA) was adopted to reduce the nonlinearity caused by the mismatch of feedback multi-bit DAC and effectively improve SNDR[1].

DESIGN OF 2-1 MASH SIGMA-DELTA MODULATOR

MASH Sigma-Delta modulator

The traditional MASH Sigma-Delta modulator and Sturdy-MASH modulator are shown in Figure 1[1].

The dotted line in Figure 1 shows the traditional Sigma-Delta MASH architecture. By combining the analog output from the cascaded two loop filters, the high-order noise shaping property is realized. Ideally, the output is:

$$Y = STF_1STF_2X - NTF_1NTF_2E_2 \qquad (1)$$

However, if the analog filter does not match the digital filter, the output end will appear $(NTF_1STF_{2D}-STF_2NTF_{1D})E_1$. In order to reduce this error, it is necessary to design more than 80 dB high gain operational amplifier and accurate transmission function.

Sturdy MASH Sigma-Delta modulator, shown in the solid line section in Figure 1, removes the digital logic structure at the back end of the modulator.

$$Y_{MASH} = STF_1X + NTF_1NTF_2 \left(E_1 - E_2\right) \qquad (2)$$

The advantage of this structure is that the digital cancellation logic is removed to remove the noise leakage caused by the device matching problem between analog and digital, but there will still be level 1 quantization noise in the final output.

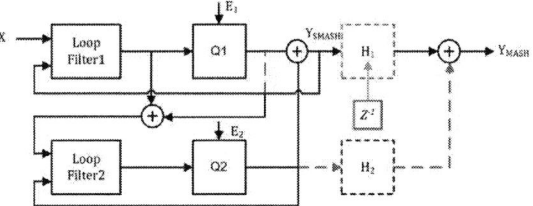

Figure 1: Traditional MASH and Sturdy MASH

Delay based noise canceling sturdy MASH

In order to completely eliminate the first-level quantization noise,[4] proposed the delay based noise canceling sturdy MASH(DNC SMASH), which retained H1 in Figure 1 Sturdy-MASH and set it as z^{-1}. The output of loop 1 is delayed by one clock. The output expression of DNC-SMASH is:

$$Y_{NDC} = Z^{-1}STF_1X + NTF_1(Z^{-1} - STF_2)E_1$$
$$- NTF_1NTF_2E_2 \qquad (3)$$

E_1 can be completely eliminated by selecting appropriate parameters. In this way, only the second quantization noise will appear in the output, which can

979-8-3503-1101-3/23 $31.00 © 2023 IEEE

well increase the noise shaping ability of the modulator.

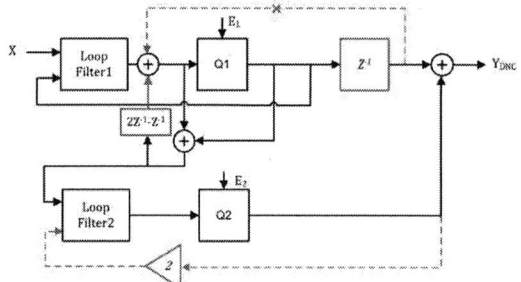

Figure 2: Proposed *Improved Delay DNC SMASH*

It is not easy to implement STF$_2$=1 in the discrete time domain without delay, so add delay (z^{-1}) to NTF1 to implement STF$_2$=z^{-1}. However, the second level output feedback to the second level input without delay leads to timing problems in the discrete time domain implementation. Paper [3] proposes improvements as shown in the dotted line in Figure 2. NTF$_1$ remains the same as before, but the transfer function from the second level to the final output (NTF$_1$') becomes ($1+z^{-1}$)NTF$_1$', and STF$_2$ and NTF$_2$ acquire additional terms of ($1+z^{-1}$) in their denominators. The additional terms of ($1+z^{-1}$) cancel out completely at the final output. Only the noise-shaped E$_2$ appears at the final output. This paper proposes a new structure based on the above structure, as shown in the solid line in Figure 2. The delay feedback of the first level is eliminated, and the error feedback structure is added, while the double feedback of the second level is maintained.

Figure 3 shows the improved delayed DNC SMASH modelThe transfer function of the modulator output is:

$$Y = X + \left(1 - Z^{-1}\right)^3 E_2 / \left(1 - Z^{-1} + 1/2\right) \quad (4)$$

Due to the local compensation path, ($1+z^{-1}$)NTF1 shows the second-order noise shaping property with an additional term ($1+z^{-1}$). Compared with the traditional cascade structure, it can reduce the error leakage caused by the mismatch of analog and digital circuits, reduce the voltage swing in the loop, reduce the requirement of amplifier gain, and improve the power efficiency on the premise of ensuring the noise shaping ability.

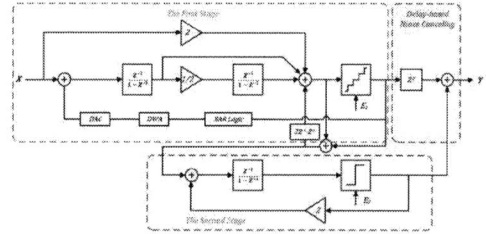

Figure 3: Proposed Improved Delay DNC SMASH model

CIRCUIT DESIGN
Modulator circuit design

The simplified schematic diagram of the modulator proposed in this paper is shown in Figure 4. The modulator uses the DNC SMASH two-stage structure to improve stability, reduce noise leakage drop, and lower voltage swing. In the first stage, a 5-bit SAR quantizer is used to solve the mismatch problem of multiple comparator components, simplify the structure and reduce power consumption. Φ1 and Φ2 are two phase non - overlapping clocks. Φ1d and Φ2d are delay clocks, which can reduce the influence of switching charge injection. The input signal is sampled in phase Φ1 and integrated in phase Φ2. During phase Φ1, half of the Ci in the feedback DAC is charged to the Vcm voltage, while the other half is charged to the zero voltage in preparation for the voltage feedback. In the first phase, the DWA controls the feedback DAC to improve linearity. At the falling edge of phase Φ2, the quantization operation is triggered and the quantizer outputs the thermometer code [5].

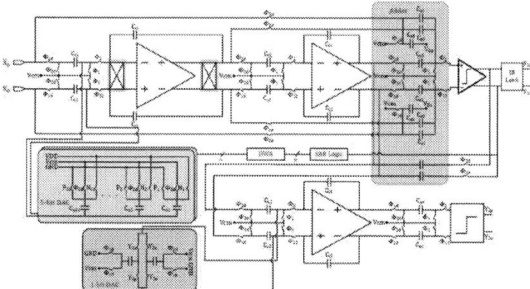

Figure 4: Proposed Improved Delay DNC SMASH circuit

Second order data weighted average algorithm

The SAR quantizer output 5-bit binary code stream is converted to 31 bit thermometer code. The data register is controlled by the pointer, which in cyclic shift always points to the next bit of the last elector capacity [6]. The schematic diagram of DWA shift is shown in Figure 5. The weighted average code output by the DWA generates a feedback DAC switch control signal in the middle phase through the control logic in Figure 4.

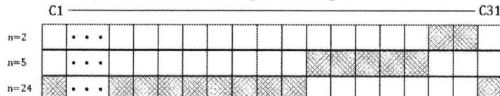

Figure 5: Displacement diagram of DWA algorithm

SIMULATION RESULTS

The modulator adopts 0.18 μm CMOS standard technology and is simulated by Spectre emulator. The supply voltage is 3.3V. When the input is 20kHz, the spectrum diagram of the integrator's single-terminal output is depicted in Figure 6. As can be seen from

Figure 7, when the operational amplifier gain reaches 40 dB, the signal-to-noise ratio basically does not change, while the traditional 2-1 cascade structure modulator gain needs to reach more than 60dB, indicating that the junction does not need high operational amplifier to compensate for the mismatch between analog and digital. As can be seen from Figure 8, at the sampling rate of 2.56MHz and the signal bandwidth of 20kHz, the peak SNDR reaches 102.2dB and the significant bits reach 16.68bit.Table I provides a performance comparison with other works.

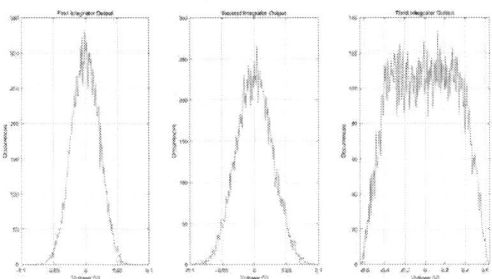

Figure 6:The integrator output spectrum

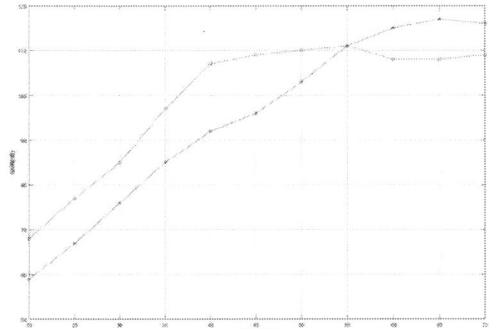

Figure 7: Effect of operational amplifier gain on signal to noise ratio of modulator

TABLE I.   TABLE STYLES PERFORMANCE COMPARISON

	[2]	[3]	[5]	This work
Tech(nm)	180	350	180	180
Supply(V)	1	3.3	3.3	3.3
Power(mW)	0.35	0.528	5.85	2.87
Sampling Freq (MHz)	3.07	1	4.41	2.56
Bandwidth (kHz)	24	22.05	22.05	20
SNDR (dB)	92	110	108.8	102.2
FoM[a](dB)	170.36	171	173.78	170.6

a.FOM = SNDR + 10 ∗ log(Bw/power)

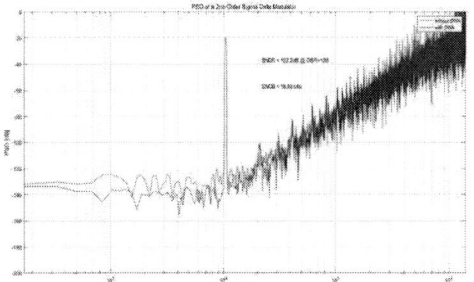

Figure 8: Simulated FFT result at 2.56MS/s for the proposed ADC

CONCLUSION

In this paper, a novel delay noise cancellation multi-bit quantized 2-1 MASH Sigma-Delta modulator is designed for sensing measurement. Compared with traditional structures, it provides sufficient noise shaping capability, reduces noise cancellation modules and reduces the effect of amplifier gain. In the first stage, a 5-bit SAR quantizer is used to reduce the inherent noise leakage in the modulator and improve the noise shaping effect. DWA algorithm is used to reduce the nonlinearity caused by the mismatch of feedback multi-bit DAC, CHS reduces the 1/f noise in bandwidth, and effectively improves SNDR. In the sampling rate of 2.56MHz and signal bandwidth of 20kHz, the peak SNDR reaches 102.2dB, the significant bits reach 16.68bit, and the overall power consumption is 2.87mW.

REFERENCES

[1] Pavan, Schreier, C. Temes, *Undersanding Delta-Sigma Data Converters,IEEE Press and Wiley Inter science*, 2020, pp. 96-107.

[2] Liyuan Liu , Dongmei Li, A 1V 350μW 92dB SNDR 24 kHz Δ Σ Modulator in 0.18μm CMOS, *IEEE Asian Solid-State Circuits Conference*, Beijing, China, Nov, 2010, pp. 14.

[3] Ki-Hoon Seo, Il-Hoon Jang, An Incremental Zoom Sturdy MASH ADC, *IEEE 60th International Midwest Symposium on Circuits and Systems (MWSCAS)*, Aug, 2017, pp. 1013-1016.

[4] C. Han and N. Maghari, Delay based noise cancelling sturdy MASH delta-sigma modulator, *IEEE Electronic Lett*, vol.50, Feb.2014, pp 351-353.

[5] Yuxiang Tang,Xiaofei Chen, A 108-dB SNDR 2-1 MASH ΔΣ Modulator with First-stage Multibit for Audio Application, *2018 3rd International Conference on Integrated Circuits and Microsystems*, Nov, 2018, pp. 336-340.

[6] Liu Mingyang, Wang Xiaosong, u Yu. A high precision multi-bit sigma-delta modulator with mismatch-shaping DACs, Application of Electronic Technique, vol.47(9), 2021, pp. 25-29.

Post-training Quantization or Quantization-aware Training? That is the Question

Xiaotian Zhao, Ruge Xu and Xinfei Guo*
University of Michigan – Shanghai Jiao Tong University Joint Institute
Shanghai Jiao Tong University, Shanghai, China
{xiaotian.zhao, schrodinger, xinfei.guo}@sjtu.edu.cn, *Corresponding author

Abstract—Quantization has been demonstrated to be one of the most effective model compression solutions that can potentially be adapted to support large models on a resource-constrained edge device while maintaining a minimal power budget. There are two forms of quantization: post-training quantization (PTQ) and quantization-aware training (QAT). The former starts from a trained model with floating-point computation and then gets quantized afterward, while the latter compensates for the quantization-related errors by training the neural network using the quantized version in the forward pass during training. Though QAT is able to produce accuracy benefits, it suffers from a long training process and less flexibility during deployment. Traditionally, researchers usually make the one-time bold decision between QAT and PTQ depending on the quantized bit-width and hardware requirement. In this work, we observed that even though the hardware cost is approximately the same for various quantization schemes, the sensitivity to training for each quantized layer is different. This leads to that certain scheme requires QAT more than others. We argue that it is necessary to look into this dimension by measuring the accuracy difference for each layer under QAT and PTQ conditions. In this paper, we introduce a methodology to provide a systematic and explainable way to quantify the tradeoffs between the quantization forms. This is especially beneficial for evaluating a layer-wise mixed-precision quantization (MPQ) scheme, where different bit-widths across are allowed and the search space is enormous.

I. INTRODUCTION

As intensive computing tasks such as AI and machine learning computations started to shift toward the edge, a big challenge for hardware designers is how to fit these large models into resource-constrained tiny devices. These constraints pushed the development of various model compression techniques [1]. Among these, quantization appears to be an effective solution that reduces the bit-widths of weights and activations by allowing an accuracy drop. It leads to a significant reduction in memory, lower network latency, and better power efficiency [2]. Despite the benefits, the tradeoff with quantization is always between accuracy loss and improvements in hardware resources in terms of latency, memory usage, and power. Uniform quantization realization is based on uniformly spaced quantization levels across the model, a well-accepted scheme is the INT8 model [3]. However, if we push this further by allowing lower quantization levels such as INT4 or below, the accuracy drop will be significant and unacceptable, layer-wise mixed-precision quantization (MPQ) appeared to be an alternative solution where each layer is quantized with different bit precision, thus it is able to deliver

very fine-granular tradeoff exploration between accuracy and hardware resources. The quantization can be performed either by retraining the model, a process called Quantization-Aware Training (QAT) [4], or done without re-training, a process called Post-Training Quantization (PTQ). It is arguably true that QAT is preferred over PTQ method for good accuracy. At the same time, QAT incurs huge computational costs of re-training the models. To recover the accuracy, the re-training process requires several hundred epochs especially for lower-bit quantization levels such as INT4 and INT2 [2]. In addition, re-training is costly from economic and environmental perspectives [5]. The decision for QAT vs. PTQ is usually based on quantization levels and hardware requirements, and it also depends on the lifetime of a model. While for mixed-precision quantization schemes that support all different quantization levels from lower to high bit widths, it appears that the decision is more binary. While this can lead to unnecessary training processes that can be totally avoided for certain model schemes. In this paper, we postulate that different MPQ schemes with similar hardware cost or similar accuracy will perform differently with respect to the PTQ or QAT necessity. To qualify this study, we propose a metric denoted as $\mathcal{P}$ to measure the influence of QAT on accuracy over PTQ. By performing experiments, we indeed proved our hypothesis and observed strong layer-wise sensitivity to $\mathcal{P}$ for various quantization schemes. With this observation, we develop a new methodology for searching the optimal layer-wise MPQ scheme by introducing a new dimension called QAT necessity. This methodology also provides a qualitative way of analyzing the impact of training on various quantized models.

II. APPROACH AND RESULTS

The goal of an optimal MPQ scheme is to obtain higher accuracy while still consuming fewer hardware resources such as computation power. BOPs (bit operations) have been widely used as a proxy of hardware resources [6], [7]. It provides an approximation of the sum of the bit operations of the convolution and activation layers throughout the training or inference processes. Previous work has studied intensively how to make tradeoffs between BOPs and accuracy [6], [8]. While there is still a missing dimension for evaluating the quantization scheme based on its sensitivity to training. We argue that this analysis is necessary for avoiding non-required

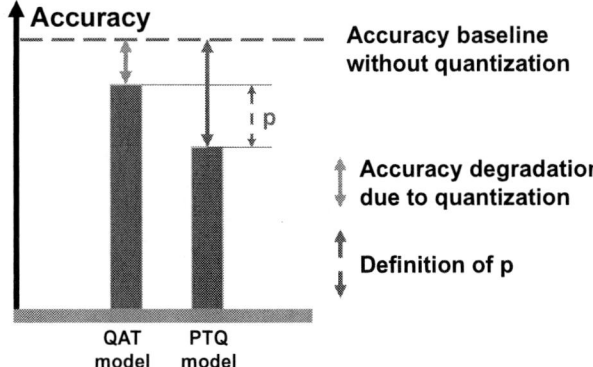

Fig. 1: Illustration of newly defined recoverable accuracy $\mathcal{P}$

training since PTQ is easier to deploy and avoids the in-situ training for many edge devices. The following sections will present our study along with conclusions.

A. Recoverable Accuracy $\mathcal{P}$

We introduce a new metric called recoverable accuracy (denoted as $\mathcal{P}$) to characterize the sensitivity of the quantized models to training. It is defined in the following:

$$\mathcal{P} = QAT_{Accuracy} - PTQ_{Accuracy} \qquad (1)$$

This metric reflects the accuracy difference for a model with and without quantization-aware training. It is thus can be treated as an indicator of sensitivity to quantization with training. As shown in Fig. 1, both QAT and PTQ models suffer accuracy drops, while QAT is able to recover the errors during the training, thus it produces higher accuracy than PTQ. This difference does depend on the quantization scheme and should be examined when deciding whether to run QAT.

B. Experiments and Results

To perform the study, a state-of-the-art MPQ framework HAWQ-V3 is chosen for examining various layer-wise quantization schemes [6]. The framework supports INT4 and INT8 quantization levels. ResNet18 [9] has been widely used as a lightweight model that runs at the edge and is selected for running our experiments. The ImageNet is used as the dataset [10]. The training is conducted with the Nvidia 3090 graphic card, and the inference runs with the AMD Ryzen 9 5900X CPU.

Various quantization schemes are applied to the model with PAT and PTQ strategies respectively. Fig. 2 shows two cases that various models (denoted as model_number) that achieve similar BOPs (computation costs) since the same number of layers are quantized with INT4 (the rest of the layers are all quantized to 8 bits). Depending on the combinations of quantized layers, the inference accuracy with PTQ only (left axis) shows differences. The right axis shows the recoverable accuracy $\mathcal{P}$ for each quantized model. While results show that in most cases the higher the PTQ inference accuracy, the least

possible it requires QAT (the smaller $\mathcal{P}$ is), there are few outliers in both cases (marked as a red dot) which indicate that the QAT is more beneficial even though the PTQ inference accuracy is already high, such as model_11. This validates the need to use recoverable accuracy $\mathcal{P}$ to measure the sensitivity of models

To further examine the aforementioned behaviors, three representative outlier models are selected and the quantized layers with lower bit levels (4 bit) are listed in Fig. 3. To make comparisons, their neighboring models from Fig. 2 are also included in the table. It is worth mentioning that all three outlier models quantized two common layers (highlighted) with lower bits, and the rest models only share one common quantized layer. This implies that there is a layer-wise sensitivity to the introduced recoverable accuracy. The sensitivity of models can be inferred from the layers they quantify, and therefore this sensitivity can be used as a new metric to evaluate MPQ schemes. The study also demonstrates that the quantization scheme that leads to similar BOPs exhibits very diverse behaviors with respect to recoverable errors.

C. MPQ Search Space with $\mathcal{P}$

In this paper, we make the assumption that PTQ is always preferred if it is able to provide similar or the same accuracy when compared to QAT. Thus the newly introduced indicator $\mathcal{P}$ adds another dimension to the MPQ search space. To determine $\mathcal{P}$ for the whole model, we propose to sample the layer-wise sensitivity for each available quantization bit level to $\mathcal{P}$ first, which is a one-time process for each neural network architecture. Once layer-wise sensitivity is obtained for a given network model, this can be combined with any other searching algorithm for deciding the layer-wise quantization schemes. Fig. 4 is an example where we integrate this new metric in the HAWQ-V3 framework [6]. On top of the inference accuracy (denoted as accuracy in the figure) and BOPs, $\mathcal{P}$ provides a global view of training necessary for each quantized model. It clearly indicates that there exist models that achieve higher accuracy and lower BOPs with smaller recoverable accuracy. Through such methodology, optimal quantization schemes can be obtained by taking into consideration of the training necessity.

III. Conclusions

In this paper, we introduced a new metric called recoverable accuracy to qualify the influence of training on a quantized model. We demonstrated that this metric can be used as a strong indicator for evaluating the MPQ scheme. We also observed that different combinations of quantized layers exhibit diverse behaviors with respect to this metric. We proposed to employ recoverable accuracy as a new dimension for evaluating quantization models in the vast search space. As for future work, we are planning to examine this metric further by looking into other edge AI models, and we are also working on developing a full MPQ search methodology that is hardware-aware and integrates multiple dimensions including layer-wise sensitivity to accuracy and recoverable accuracy

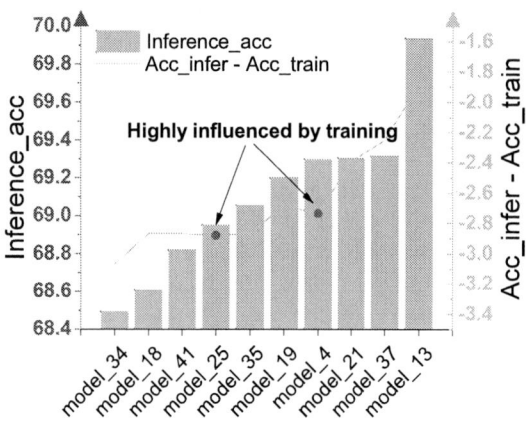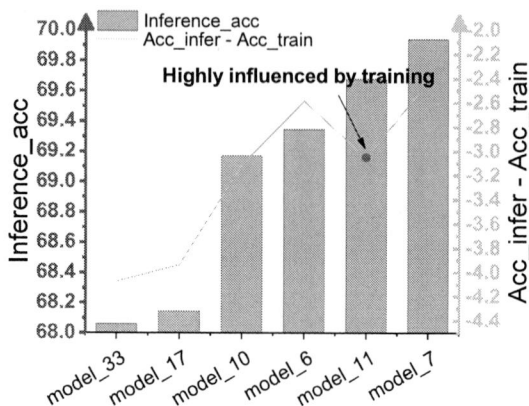

Fig. 2: PTQ Accuracy (marked as Inference Accuracy) vs. Recoverable Accuracy $\mathcal{P}$ for models with the same BOPs (Left: BOPs=107.67; Right: BOPs=106.75). Please note that it shows $-\mathcal{P}$ for presentation purposes on the right axis.

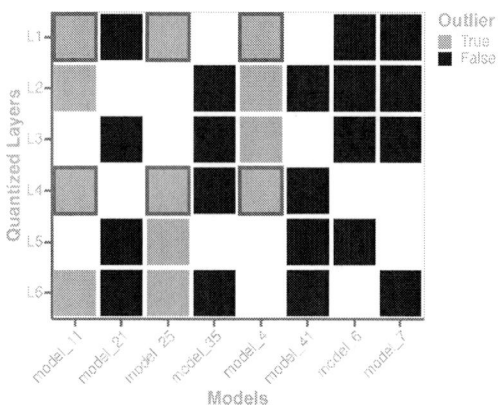

Fig. 3: Comparison between outlier models and neighboring models. The colored block represents the layer (y-axis) being quantized in the model (x-axis).

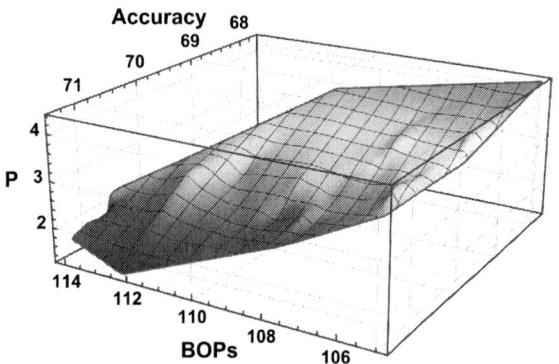

Fig. 4: Search space after adding $\mathcal{P}$. It shows that tere are no linear relationship between $\mathcal{P}$, BOPs, and Accuracy.

within the search. The ultimate goal of this research is to support large models to run at the edge via hardware/software co-compression methodology.

ACKNOWLEDGEMENT

This work is funded in part by an UM-SJTU Startup Fund, in part by an SJTU Explore-X Research Grant D6040004/038, and in part by a CCF-Tencent Open Fund No. RAGR20220114. Special thanks to Mr. Yimin Gao and Prof. Mircea Stan from the University of Virginia and Dr. Vaibhav Verma from Qualcomm for their help and feedback on this work.

REFERENCES

[1] S. Han, H. Mao, and W. Dally, "Compressing deep neural networks with pruning, trained quantization and huffman coding. arxiv 2015," *arXiv preprint arXiv:1510.00149*.

[2] A. Gholami, S. Kim, Z. Dong, Z. Yao, M. W. Mahoney, and K. Keutzer, "A survey of quantization methods for efficient neural network inference," *arXiv preprint arXiv:2103.13630*, 2021.

[3] B. Noune, P. Jones, D. Justus, D. Masters, and C. Luschi, "8-bit numerical formats for deep neural networks," *arXiv preprint arXiv:2206.02915*, 2022.

[4] B. Jacob, S. Kligys, B. Chen, M. Zhu, M. Tang, A. Howard, H. Adam, and D. Kalenichenko, "Quantization and training of neural networks for efficient integer-arithmetic-only inference," in *Proceedings of the IEEE conference on computer vision and pattern recognition*, 2018, pp. 2704–2713.

[5] D. Patterson, J. Gonzalez, Q. Le, C. Liang, L.-M. Munguia, D. Rothchild, D. So, M. Texier, and J. Dean, "Carbon emissions and large neural network training," *arXiv preprint arXiv:2104.10350*, 2021.

[6] Z. Yao, Z. Dong, Z. Zheng, A. Gholami, J. Yu, E. Tan, L. Wang, Q. Huang, Y. Wang, M. Mahoney *et al.*, "Hawq-v3: Dyadic neural network quantization," in *International Conference on Machine Learning*. PMLR, 2021, pp. 11 875–11 886.

[7] M. Van Baalen, C. Louizos, M. Nagel, R. A. Amjad, Y. Wang, T. Blankevoort, and M. Welling, "Bayesian bits: Unifying quantization and pruning," *Advances in neural information processing systems*, vol. 33, pp. 5741–5752, 2020.

[8] C. Baskin, N. Liss, E. Schwartz, E. Zheltonozhskii, R. Giryes, A. M. Bronstein, and A. Mendelson, "UNIQ: Uniform Noise Injection for Non-Uniform Qantization of Neural Networks," *ACM Transactions on Computer Systems*, vol. 37, no. 1-4, jun 2021.

[9] K. He, X. Zhang, S. Ren, and J. Sun, "Deep residual learning for image recognition," in *Proceedings of the IEEE conference on computer vision and pattern recognition*, 2016, pp. 770–778.

[10] O. Russakovsky, J. Deng, H. Su, J. Krause, S. Satheesh, S. Ma, Z. Huang, A. Karpathy, A. Khosla, M. Bernstein, A. C. Berg, and L. Fei-Fei, "ImageNet Large Scale Visual Recognition Challenge," *International Journal of Computer Vision (IJCV)*, vol. 115, no. 3, pp. 211–252, 2015.

A FRONT-END FOR 1.5GSPS 12BIT PIPELINED ADC

Xiuheng Wu[12], Xuan Guo[1*], Fangyuan Xu[12], Zeyu Li[12], Hanbo Jia[1], Xinyu Liu[1]

[1] Institute of Microelectronics of The Chinese Academy of Sciences, Beijing 100029, China

[2] School of Integrated Circuits, University of Chinese Academy of Sciences, Beijing 100049, China

Corresponding Author's Email: guoxuan@ime.ac.cn

ABSTRACT

In this paper, the design of a front-end for 1.5GSPS 12bit Pipelined ADC is presented. The front-end circuit consists of an input buffer circuit and a Track and Hold Amplifier (THA) circuit. A common mode voltage (V_{CM})stablizing technique is applied to improve stability of input buffer over process, voltage and temperature (PVT) variations. To boost high frequency performance, a high frequency small signal elimination technique using dummy transistors is applied to the THA circuit. To maximize high speed performance, best performance of 40nm devices is taken advantage of in the front-end circuit by a new set of supply power and ground suiting for thin oxide device. This work is implemented in 40nm CMOS process with a 1.8V power supply. The THA circuit exhibits 71dB SFDR and 10.25bit ENOB at 1.5GHz.

INTRODUCTION

In the era of information, wireless communication technology has developed rapidly. As the communication systems is modernizing, a great demand for high-speed and high-resolution analog-to-digital converter (ADC), which is the key module in a communication system, is being created. Compared with ADCs adopting other architecture such as flash or SAR, the pipelined ADCs can provide optimal trade-off between high-speed and high-resolution.

A pipelined ADC consists of a front-end circuit, several stages including Multiple Digital to Analog Converters (MDAC), and a flash stage. As a key part of Pipelined ADC, the front-end circuit has deep influences on conversion performance. The front-end is composed of input buffer circuit and THA circuit. In a front-end circuit working properly, V_{CM} of input buffer circuit is equal to V_{CM} of THA circuit. However, PVT variations changes V_{CM} of input buffer circuit, which has negative effects on circuit performance. In this paper, a V_{CM}-stable technique is adopted to keep this output voltage at a fixed value. High speed performance of this front-end circuit brings high frequency signal problem to THA circuit. To eliminate high frequency small signal in THA circuit, new structure with dummy transistors is employed in switched capacitor circuit. Besides different architectures within front-end circuit, device performance also has a great influence on high-speed sampling. In order to maximize sampling speed, best performance of 40nm devices is taken in this front-end circuit. A new set of supply voltage and ground are generated to realize this aim.

FRONT-END ARCHITECTURE

The 1.5GSPS 12bit pipelined ADC presented in figure 1 consists of a front-end circuit, 3 stages including a Sub-ADC and a MDAC, and a 4bit flash stage. To increase conversion performance of the proposed ADC, a front-end circuit with higher speed is designed in this paper.

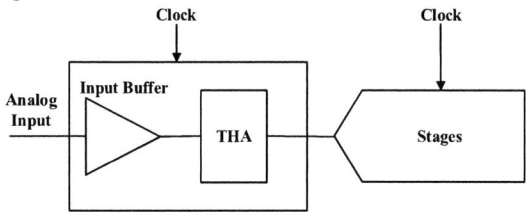

Figure 1: 1.5GSPS 12bit pipelined ADC architecture

As the key part in the pipelined ADC, a high-speed front-end circuit, is composed of an input buffer circuit, and a THA circuit. The input buffer circuit is employed to supply large amplitude signal and to isolate kick-back noise. The THA circuit is employed to sample analog signals.

Within the input buffer circuit, the threshold voltage (V_{TH}) of NMOS transistor will be influenced by PVT variations, which will make the V_{CM} of input buffer non-constant. To solve this potential problem, a V_{CM}-stable structure is employed in this input buffer circuit. The negative feedback architecture consists of a source follower, an operational amplifier, a PMOS transistor and two resistances.

The THA circuit is composed of a high gain operational amplifier and switched capacitor circuit. In the high-speed working environment, some high frequency small signals are passed into the switched capacitor circuit and influence sample results of THA circuit. To remove this interference, a high frequency small signal elimination structure is adopted within the switched capacitor circuit. As shown in figure 3, two dummy transistors are employed to connect the source and the drain ends of two different switches in different ends.

To improve the high-speed performance of the front-end circuit, the gain bandwidth product (GBW) of the THA circuit needs to be increased significantly. The GBW of THA is influenced by the RC constant of devices adopted in THA. Thus, besides optimizing circuit structure, improving device performance is considered as

979-8-3503-1101-3/23 $31.00 © 2023 IEEE

another idea in this paper. To get the minimum RC constant of 40nm CMOS device, thin oxide device is employed in THA. However, the thin oxide device can only work properly at a voltage difference of 0.9V. To ensure the THA circuit adopting new devices can sample and hold at a voltage difference of 1.8V, a new set of supply voltage and ground are generated by a low dropout regulator (LDO) circuit. The new supply and ground voltages have the same common-mode voltage equal to 0.9V as the whole front-end circuit.

CIRCUITS IMPLEMENTATION

Input Buffer V_{CM}-Stable technique

As the first part of the front-end circuit, the input buffer is employed to supply large amplitude signal to increase ENOB. In addition, input buffer can also isolate the kick-back signal to reduce its impact on the signal source.

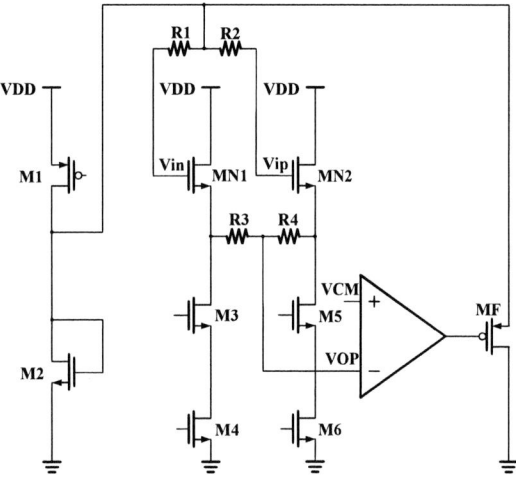

Figure 2: Input buffer with V_{CM}-stable structure

The threshold voltage (Vth) of the NMOS transistor MN1 and MN2 changes with the PVT variation, which forces the output common-mode voltage offset the constant 0.9V. To keep V_{CM} stable, as shown in figure 2, a close-loop architecture is employed to design this V_{CM}-stable input buffer circuit. The V_{CM}-stable loop is composed of an operational amplifier, a PMOS transistor MF and two resistances R1 and R2. The operational amplifier is adopted to subtract VOP from VCM (equal to 0.9V). If VOP is higher than 0.9V, MF will decrease Vin. If VOP is lower than 0.9V, MF will increase Vin and Vip. After this negative feedback process, VOP will keep constant at 0.9V over PVT variations. Thus, the robustness of the input buffer is optimized.

High Frequency Small Signal Elimination Technique

Two dummy transistors DUM1 and DUM2 are adopted to offset the high frequency small signal. During the sampling and holding process, high frequency small signal passes through parasitic capacitances Cds of the bootstrap switches and influenced the sampling results. When MO1 and MO2 are added in THA, high frequency small signal also passes through Cds of DUM1 and DUM2 and offset the small signal with reverse phase. Thus, THA's performance on high frequency is improved significantly. This high-frequency small signal offsetting technique is described by schematic diagram in figure 3.

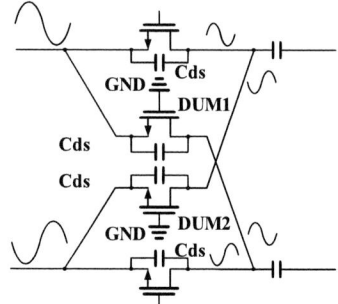

Figure 3: High-frequency signal offsetting technique

Application of 40nm Device Best Performance

Besides optimizing architecture of THA, improving performance of devices is another effective way to increase sampling speed of the front-end circuit. As shown in [2], because of the lower resistance, the thin-oxide device works faster and consumes less power compared to thick-oxide device. Due to this performance, the capacitor of thin-oxide device charges and discharges faster, which makes a quicker switching device. Thus, the GBW of THA can be increased. However, the thin-oxide device can operate reliably with lower voltage difference than the voltage difference amplifier is suitable for. In this paper, the voltage difference of high gain amplifier and whole THA circuit is 1.8V. To adapt to the voltage difference (equal to 0.9V) of 40nm thin-oxide devices, a new set of supply voltage and ground is set.

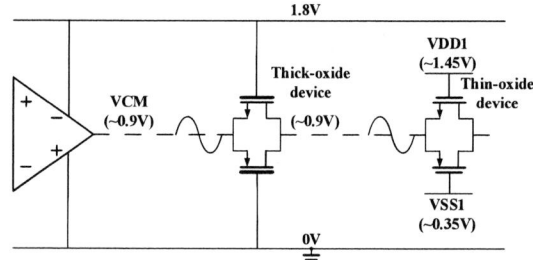

Figure 4: A new set of voltages adapt to thin-oxide device

As shown in figure 4, to accommodate the large signal swing of the amplifier, the midpoint voltage of this new set of supply voltage and ground (VDD1 and VSS1) is designed to be the same as the amplifier output common-mode voltage equal to 900mV.

The VDD1 and VSS1 are generated by two LDOs. The reference voltages for the LDOs are level-shifted up and down with a constant 0.45V from the common-mode voltage. This fixed voltage difference is generated by a bandgap circuit to keep the voltage difference itself and two new-generating voltages (VDD1 and VSS1) constant over PVT variations.

SIMULATION RESULT

As shown in figure 5, the simulation results show SFDR and ENOB variations of the THA circuit as the input signal frequency changes from 0GHz to 1.5GHz. When the frequency is 1.498GHz, the THA achieves 71dB SFDR and 10.25bit ENOB.

The simulation results in figure 6 shows the output spectrum of the THA circuit when the input signal frequency is 152.3MHz. Except the triple frequency harmonic at 450MHz, there's little interference generated at other frequency.

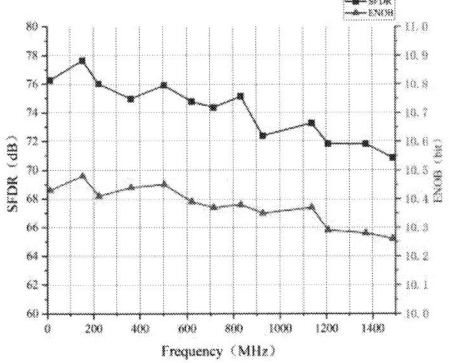

Figure 5: Simulation result of THA circuit

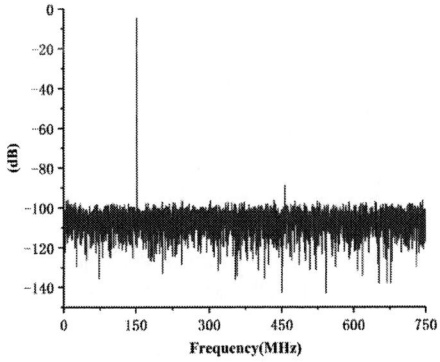

Figure 6: Output spectrum of THA circuit at 152.3MHz

CONCLUSION

This paper describes a design of the front-end circuit of a 1.5GSPS 12bit pipelined ADC. Within the front-end circuit, output common-mode voltage of the input buffer circuit is stabilized by a VCM-stable structure over PVT variations. The THA circuit is composed of a high gain operational amplifier and switched capacitor circuit. To remove interference from high speed working environment, a high frequency small signal elimination structure is employed in this THA. To maximum high speed performance, the best performance of 40nm device is taken advantage of in this THA. A new set of supply voltage and ground is generated to suit for new thin-oxide devices. The simulation results of the THA circuit of this pipelined ADC show that the SFDR is 71dB and the ENOB is 10.25bit at the input signal frequency of 1.5GHz. The simulation results also show that there's little interference for the output spectrum of THA circuit when the input signal frequency is 152.3MHz.

ACKNOWLEDGEMENTS

This work was supported by the Key Research and Development Plan of Shandong Province, China, grant number 2022CXGC010107, and Youth Innovation Promotion Association, Chinese Academy of Sciences, grant number 2021113.

REFERENCES

[1] A. M. A. Ali et al. "A 14 Bit 1 GS/s RF Sampling Pipelined ADC With Background Calibration," Solid-State Circuits, IEEE Journal of 49.12(2014): 2857-2867.

[2] C. -Y. Chen, J. Wu, J. -J. Hung, T. Li, W. Liu and W. -T. Shih. "A 12-Bit 3 GS/s pipeline ADC with 0.4 mm² and 500 mW in 40 nm digital CMOS," Solid-State Circuits, IEEE Journal of 47.4(2012): 1013-1021.

[3] X. Zheng et al. "A 14-bit 250 MS/s IF Sampling Pipelined ADC in 180 nm CMOS Process," IEEE Transactions on Circuits and Systems I: Regular Papers of 63.9(2016): 1381-1392.

[4] B. D. Sahoo and B. Razavi. "A 10-b 1-GHz 33-mW CMOS ADC," Solid-State Circuits, IEEE Journal of 48.6(2013): 1442-1452.

[5] A. Verma and B. Razavi. "A 10b 500MHz 55mW CMOS ADC," 2009 IEEE International Solid-State Circuits Conference - Digest of Technical Papers, (2009): 84-85.

Logic Circuit Simulation based on Semi-Tensor Product

Ruibing Zhang, Hongyang Pan, and Zhufei Chu
EECS, Ningbo University, Ningbo 315211, China
Email: chuzhufei@nbu.edu.cn

ABSTRACT

Simulation of logic circuits is a critical part of logic synthesis and verification, as it reduces the amount of time needed for various computations. However, it has been found to be difficult to simulate k-input lookup table (k-LUT) networks. This paper proposes a k-LUT logic circuit simulation method based on semi-tensor products (STPs) of matrices, which supports simulation of all nodes or only primary outputs (POs). By improving the computational efficiency of STP of matrices, the matrix encoding algorithm can accelerate all-nodes simulation. For only-PO simulation, we can run faster by partitioning the circuit based on multiple fan-out nodes. Experimental results on EPFL benchmark suites indicate that we were able to improve CPU time by 4.0x (4.8x maximum) and 7.4x (16.6x maximum) when we simulate all nodes and only POs, respectively.

Index Terms—**logic circuit simulation; semi-tensor product of matrices; k-LUT network**

INTRODUCTION

Logic synthesis play an important role in electronic design automation (EDA), and extensive research has been done on optimizing logic networks since the emergence of this field. Typical methods of logic synthesis include both algebraic and Boolean method. The algebraic method treat Boolean functions as polynomials and optimize the logic network locally. The Boolean method uses Boolean logic to improve optimization quality, where simulation is used to reduce the time required for various Boolean computations. Therefore, logic circuit simulation plays an imperative role in logic synthesis, e.g., for SAT sweeping and Boolean resubstitution. Through a combination of structural hashing, simulation, and SAT queries, SAT sweeping simplifies logic networks by merging graph vertices from inputs to outputs [1]. In Boolean resubstitution, the function of a node is expressed using other nodes (called divisors) in the logic network. It is typical to use window simulation technology to compute the truth tables for each node in a Boolean network to determine which nodes have equivalent functions [2].

Modern arithmetic logic units support bitwise logic operations to enhance simulation efficiency today. In contrast,

This work was supported by the NSFC under Grant NOs 62274100 and 61871242.

simulating k-LUT networks is difficult since it does not take advantage of this feature. In this paper, we propose a k-LUT logic circuit simulation based on the semi-tensor product (STP) of matrices. The k-LUT network can be easily represented and simulated by STP. A logic matrix can be used to define the Boolean variables in order to prove the basic properties of logic using the STP method [3]. Moreover, we use the matrix encoding algorithm to improve the efficiency of STP-based simulation and use the cut algorithm to only simulate the value of POs.

BACKGROUND

Logic Circuit Simulation

A simulation pattern is a collection of Boolean values assigned to each primary input (PI) of a network. Logic circuit simulation is done by visiting nodes in a topological order and computing the output values using their input values. In practice, several simulation patterns can be bundled together by using machine words, instead of a single bit, to represent a sequence of Boolean values. This way, 32 or 64 patterns can be computed for a node within a single CPU instruction using bitwise logical operations supported by modern arithmetic logic units. The simulation signature of a node is an ordered set of values produced at the node under each simulation pattern. A set of simulation patterns is exhaustive if it covers all possible combinations of value assignment, which requires 2^k patterns for k PIs. The simulation signatures produced by simulating an exhaustive pattern set are also called truth tables and they completely specify the Boolean functions of the nodes [4].

Simulation can be done globally in the entire network or locally in a small window. In the former case, the simulation pattern set is possibly nonexhaustive because 2^{16} patterns are already impractical to handle, but the number of PIs is usually larger than 16. To use an exhaustive set of patterns, simulation must be restricted to a window of less than 16 (typically 8 to 10) leaf nodes.

k-LUT Network

In LUT-based FPGA (Field-Programmable Gate Array), the basic logic block is a k-LUT that can implement any Boolean function of up to k variables [5]. The technology mapping for LUT-based FPGAs is to generate a mapping of a set of Boolean functions onto k-LUTs, where their Boolean functions are described as a Boolean network $G = (V, E)$, defined as a Directed Acyclic Graph (DAG).

Semi-Tensor Product of Matrices

This section discusses the matrix form of logic formulas, also known as the STP form. The real matrix with $m \times n$ dimensions is represented by $M^{m \times n}$. The basic condition for the matrix product of $X^{m \times n}$ and $Y^{p \times q}$ is that $n = p$, however, the STP can produce matrices in any dimension. First, we denote the set of Boolean variables S_v.

$$S_V : \{True = \begin{bmatrix} 1 \\ 0 \end{bmatrix}, False = \begin{bmatrix} 0 \\ 1 \end{bmatrix}\}.$$

Definition 1: Let $X \in M^{m \times n}$ and $Y \in M^{p \times q}$, the semi-tensor product of X and Y, denoted by $X \ltimes Y$, defined as

$$X \ltimes Y = (Xc \otimes I_{t/n}) \cdot (Y \otimes I_{t/p}),$$

where I represents the identity matrix, t is the least common multiples of n and p, and $\otimes$ is the Kronecker product of two arbitrary dimensional matrices [6].

Definition 2: A 2×2^n matrix is called a logic matrix if all its columns are elements in S_V.

Definition 3: A logic matrix M_σ in which columns are consistent with the truth table of a logic operation σ is called the structural matrix of σ.

Theorem 0.1: Any logical expression $L(x_1, x_2, x_3, ..., x_n)$ that contains the free variable $x_1, x_2, x_3, ..., x_n$ can be uniquely represented as a canonical form as

$$L(x_1, x_2, x_3, ..., x_n) = M_L x_1 x_2 x_3 ... x_n,$$

where M_L is a 2×2^n structural matrix.

PROPOSED METHOD

Any Boolean function can be converted into its STP form by their structural matrices. Therefore, the k-LUT network can be represented by the STP of matrices. We also use the command `lut_mapping` in ALSO[1] to convert the logic circuit into a k-LUT network.

Method A: Matrix Coding

In the process of converting a Boolean function into a canonical matrix using STP method, all of the matrices that appear in between have only one "1" in each column.

Here, we can use a logical matrix to represent the truth table information of the nodes in the k-LUT network, so that we can use the method of matrix operation to handle this network. In light of the unique characteristics of the logical matrix, we propose an encoding method to optimize the representation of the matrix. We implement a complete set of matrix operation rules. The encoding rules are as follows:

Definition 4: The encoding result of $A^{m \times n}$ is a vector of rows $[x_0, x_1, ..., x_n]$. The value of x_0 is equal to m, which represents the number of rows in matrix A, and the value of $x_i, i \in [1, n]$ represent the number of rows minus 1 in which the '1' of column i of the matrix A(in order to facilitate computer processing).

[1] https://github.com/nbulsi/also

This encoding method not only preserves all the information of the logical matrix, but also greatly reduces the space complexity.

Example 1. Let $B = \begin{bmatrix} 1 & 1 & 1 & 0 \\ 0 & 0 & 0 & 1 \end{bmatrix}$. The encoding result of B is $\begin{bmatrix} 2 & 0 & 0 & 0 & 1 \end{bmatrix}$.

Now let's give the definition of the Kronecker product and the semi-tensor product in the case of matrix coding. Consider two matrices $A^{m \times n}$ and $B^{p \times q}$ that satisfy the coding conditions and the definition of the semi-tensor product, and their coding results are $A' = [m, x_1, x_2, ..., x_n]$ and $B' = [p, y_1, y_2, ..., y_q]$, respectively.

Definition 5: Using the above encoded matrix A' and B', $C = A' \otimes B'$ is calculated as follows:

$$A' \otimes B' = [mp, x_1 p + y_1, ..., x_1 p + y_n, ..., x_m p + y_1, ..., x_m p + y_n].$$

The result is a row vector of length $n \times q + 1$. The time and space complexity of this operation is $O(n \times q + 1)(O(m \times n \times p \times q)$ before coding).

Definition 6: Using the above encoded matrix A' and B', $C = A' \ltimes B'$ is calculated as follows:

(1)if $n\%p = 0$, let $t = n/p$, we divide A' into $r = n/t$ blcoks, express as $A' = [m, A^0, A^1, ..., A^{r-1}]$, where $A^i = [x_{it+1}, x_{it+2}, ..., x_{2it}]$. Then

$$A' \ltimes B' = [m, A^{y_1}, A^{y_2}, ..., A^{y_q}].$$

(2)if $p\%n = 0$, we calculate by definition 1 and matrix coding.

We can complete the transformation of theorem 1 by using only the above two operations. By means of matrix coding, we can greatly speed up operations and have the ability to deal with larger Boolean functions.

Method B: Cut based on Multi-Fanout

Assume that there is a node with a number of fanouts of n. This node will be accessed $n + 1$ times in all nodes simulation, including one time computation of its output and n times extraction of its value. In some applications, we only need the PO values from the circuit simulation, and simulating every node of the circuit would be a waste of resources. Therefore, we use STP to compute the truth tables of each cut of the circuit using multiple fanout nodes as the boundary. The simulator can only work on nodes with multiple fanouts, which efficiently obtain the value of POs at a low cost.

The algorithm of Logic Circuit Simulator based on Semi-Tensor Product is shown in Algorithm 1. The input of the algorithm is a k-LUT logic circuit C to be simulated, a simulation vector set P and simulation mode *mode*. The output is the circuit simulation information S obtained according to the choice of mode. If we need to get a complete simulation(S) information of the circuit, we visit all the nodes in the circuit in a topological order and use their input information to calculate the output information(line 2). Otherwise, when we only need to get the simulation information of the PO, we perform the calculation by following the steps below. First, we calculate the size *limit* of a cut based on the number of simulated vectors.

Because it also takes time to compute the truth table of cuts, this ensures that the STP method is more efficient than direct simulation(line 5). Second, we take multiple fan-out nodes and *limit* as the boundary to cut circuit C, so that each cut is a tree structure with leaf node no larger than *limit*, and the root node of each cut is stored in *root* set(line 6). Then, we use the STP method mentioned earlier to calculate the truth table of all cuts in the *root* set(line 7); Finally, each cut node in the root collection is accessed in topological order and its output is calculated based on the input information(line 8). At this point, the algorithm ends and the program returns the desired simulation information.

Algorithm 1: STP-based logic circuit simulator

Input: a logic circuit C, a set of simulation P,
    mode(all nodes a or only POs p)

Output: simulation information(S) corresponding to
    the selection mode

1 **if** mode=a **then**
2 | S=sim_all_nodes(C,P);
3 **else**
4 | n=P.size();
5 | $limit$=$log(n)$;
6 | $root$=circuit_cut($C,limit$);
7 | compute_tt($C,root$);
8 | S=sim_POs(C,P);
9 **end**
10 return S;

EXPERIMENTAL RESULT

The source code of our simulator[2] are publicly available. All experiments were performed on an Intel(R) Xeon(R) Silver 4210R CPU @ 2.40GHz with 64GB of main memory. The results are shown in Table I. The EPFL combinational benchmark suites are used for experimental evaluation [7]. The "PI/PO" and the "Nodes" columns lists the number of PI/PO and logic nodes of the benchmarks. We respectively give the running time of logic circuit simulation by mockturtle[3] and our proposed logic circuit simulation in the matrix coding method and the cut method. Finally, we use "T(s)" and "x" to indicate the amount of CPU time that has been accelerated. Each benchmark is simulated with randomly generated 10000 simulation patterns. On average, CPU time is reduced by 4.0x (4.8x maximum) with method A and 7.4x (16.6x maximum) with method B, respectively. The improvement of method B is more significant since it only simulates POs.

SUMMARY

In this paper, a method based on semi-tensor product is proposed to simulate k-LUT networks. We first put forward the matrix coding technique in the field of Boolean algebra by applying the theory of semi-tensor product. This technique will

[2]https://gitee.com/rui-bing-zhang/simulation-stp.git
[3]https://github.com/lsils/mockturtle.git

TABLE I
LOGIC SIMULATION RESULTS FOR EPFL BENCHMARKS.

Benchmarks	PI/PO	Nodes	Mockturtle T (s)	Method A T (s)	Method A x	Method B T (s)	Method B x
adder	256/129	641	0.268	0.061	4.39	0.041	6.54
bar	135/128	3366	1.391	0.304	4.58	0.084	16.56
div	128/128	44459	18.056	4.179	4.32	2.388	7.56
hyp	256/128	160806	66.761	18.149	3.68	11.625	5.47
log2	32/32	27677	10.989	2.799	3.93	1.688	6.51
max	512/130	2665	1.088	0.228	4.77	0.121	8.99
multiplier	128/128	22848	9.432	2.236	4.22	1.511	6.24
sin	24/25	4793	1.885	0.435	4.33	0.284	6.64
sqrt	128/64	19810	8.135	1.845	4.41	1.455	5.59
square	64/128	15798	6.337	1.507	4.21	0.979	6.47
Voter	1001/1	9341	4.161	1.039	4.01	0.675	6.16
router	60/30	277	0.106	0.029	3.66	0.016	6.63
priority	128/8	1090	0.442	0.124	3.56	0.063	7.02
mem_ctrl	1204/1231	48489	20.326	5.479	3.71	2.676	7.59
int2float	11/7	259	0.106	0.026	4.08	0.013	8.15
i2c	147/142	1484	0.565	0.148	3.82	0.074	7.64
dec	8/256	584	0.154	0.054	2.85	0.029	5.31
ctrl	7/26	166	0.061	0.017	3.59	0.007	8.71
cavlc	10/11	753	0.304	0.078	3.89	0.039	7.79
arbiter	156/129	12355	4.741	1.331	3.56	0.691	6.86
Total		377661	155.308	40.068		24.459	
Improvment					3.98		7.42

not lose any matrix information and greatly reduce the space and time complexity of computer implementation, and almost complete a series of operations of realizing matrix with linear time complexity. This technique can also pave the way for other researches in the field of Boolean algebra by using the theory of semi-tensor product. The proposed circuit cutting technology can achieve a great degree of information reuse and improve the simulation speed. Using EPFL benchmarks,on average,CPU time is reduced by 4.0x (4.8x maximum) with method A and 7.4x (16.6x maximum) with method B, respectively.

REFERENCES

[1] Q. Zhu, N. Kitchen, A. Kuehlmann, and A. Sangiovanni-Vincentelli, "Sat sweeping with local observability don't-cares," in *Proceedings of the 43rd Annual Design Automation Conference*, 2006, pp. 229–234.
[2] E. Testa, L. Amarú, M. Soeken, A. Mishchenko, P. Vuillod, P.-E. Gaillardon, and G. De Micheli, "Extending boolean methods for scalable logic synthesis," *IEEE Access*, vol. 8, pp. 226 828–226 844, 2020.
[3] Z. Liu and D. Cheng, "Canonical form of boolean networks," in *2019 Chinese Control Conference (CCC)*. IEEE, 2019, pp. 1801–1806.
[4] S.-Y. Lee, H. Riener, A. Mishchenko, R. K. Brayton, and G. De Micheli, "A simulation-guided paradigm for logic synthesis and verification," *IEEE Transactions on Computer-Aided Design of Integrated Circuits and Systems*, pp. 1–1, 2021.
[5] A. H. Farrahi and M. Sarrafzadeh, "Tdd: a technology dependent decomposition algorithm for lut-based fpgas," in *Proceedings. Tenth Annual IEEE International ASIC Conference and Exhibit (Cat. No. 97TH8334)*. IEEE, 1997, pp. 206–209.
[6] C. F. Van Loan, "The ubiquitous kronecker product," *Journal of computational and applied mathematics*, vol. 123, no. 1-2, pp. 85–100, 2000.
[7] L. Amarú, P.-E. Gaillardon, and G. De Micheli, "The epfl combinational benchmark suite," in *Proceedings of the 24th International Workshop on Logic & Synthesis (IWLS)*, no. CONF, 2015.

979-8-3503-1101-3/23 $31.00 © 2023 IEEE

CirSAT: An Efficient Circuit-based SAT Solver via Fanout-driven Decision Heuristic

Kunmei Hu, Zhufei Chu

EECS, Ningbo University, Ningbo 315211, China

Email: chuzhufei@nbu.edu.cn

ABSTRACT

Circuit-based Boolean satisfiability (SAT) solver is efficient for solving electronic design automation (EDA) problems. Compared to widely-used conjunctive normal form (CNF)-based SAT solvers, circuit-based SAT is rarely publicly available. In this paper, we propose an open-source circuit-based SAT solver, CirSAT, which implements several efficient SAT algorithms directly on circuits. In particular, we make use of the fanout of logic gates as a heuristic to guide conflict decision. Experimental results on ISCAS'85 benchmark suites indicate our method achieves 47.7x (upto 267.8x) CPU time acceleration and saves 18.9x time compared to MiniSAT.

Index Terms—**Boolean satisfiability; circuit-based SAT Solver; EDA**

INTRODUCTION

The satisfiability problem has been extensively studied in recent decades. Renowned SAT solvers are designed based upon the *conjunctive normal form* (CNF) [1] , which is the *product of sum* (PoS) logic expressions. For many *electronic design automation* (EDA) applications, applying SAT to a circuit-oriented problem often requires converting the gate-level netlist to its corresponding CNF format. Circuit-to-CNF transformations remove not only topological and structure information, but they also increase solving steps and unnecessary overhead, which adversely affects SAT solving.

As compared to CNF-based SAT solvers, circuit-based SAT solvers are rare. QuteSat [2] and NIMO [3] are examples of circuit-based SAT solvers proposed in the last decade, but their source codes are not publicly available. Recently, a deep integration of circuit simulator and SAT solver is proposed for logic synthesis problem [4], in which a tailor-made SAT solver with circuit information significantly accelerates the logic synthesis procedure. Consequently, circuit-based SAT solvers are of paramount importance for EDA applications, as well as for general SAT problems.

In this paper, we propose an efficient circuit-based SAT solver, **CirSAT**, which implements classical SAT strategies directly on the circuit data structure. Our solver relies on *conflict-driven circuit learning* (CDCL), a two-pointer watching scheme for efficient *Boolean constraint propagation*

This work was supported by the NSFC under Grant NOs 62274100 and 61871242.

(BCP), and a fanout-driven decision heuristic. We compare the proposed CirSAT with a CNF-based SAT solver, MiniSAT, over ISCAS' 85 benchmark suites. In terms of CPU time of solving, our circuit-based solver achieves 47.7x (267.8x maximum) CPU time acceleration and save 18.9x time on average, respectively.

BACKGROUND

Boolean Satisfiability

Given a Boolean function F over several Boolean variables, the SAT problem is to answer whether a set of variable assignments exists to make F evaluate to be True. A literal is the positive or negative form of Boolean variables, such as a or $\bar{b}$. A clause is a disjunction of literals. For example, $\phi_1 = a \vee \bar{b}$ is a clause. The Boolean formula $F = \phi_1 \wedge \phi_2 \ldots \wedge \phi_n$ disjunctively connected by n clauses is a CNF expression of a Boolean function.

A simple CNF formula $F = (a \vee b) \wedge (b \vee c) \wedge (\bar{a} \vee \bar{b} \vee \bar{c})$ over Boolean variables a, b, and c can be evaluated to be True if a and c are True, and b is False. Hence, F is satisfiable, otherwise F is proven unsatisfiable if no assignments exist to make F evaluated to be True.

CNF-based SAT solver

SAT solvers are programs that solve SAT problems. CNF-based SAT solvers (CNF-SAT) require a CNF formula as input and output a satisfiable assignment or an unsatisfiable conclusion. As shown in Fig. 1, given a specific problem, we first need to define Boolean variables and add constraints over these Boolean variables. Each clause can be seen as a constraint. After encoding the problem into a CNF formula, SAT solvers are used to solve the problem to find solutions. After receiving a SAT solution, we decode it to obtain the original problem's answer. Hence, the encoding and decoding steps cause considerable unnecessary overhead. The CNF encoding is low-level and does not preserve domain-specific information or the structure of the original problem, which will be detrimental to solving SAT problems.

Circuit-based SAT solver

The circuit-based approach requires us to determine whether one or more of the primary input (PI) assignments can result in all primary outputs (PO) being true. As shown in Fig. 1, given a Boolean circuit as input, the circuit SAT solver outputs the

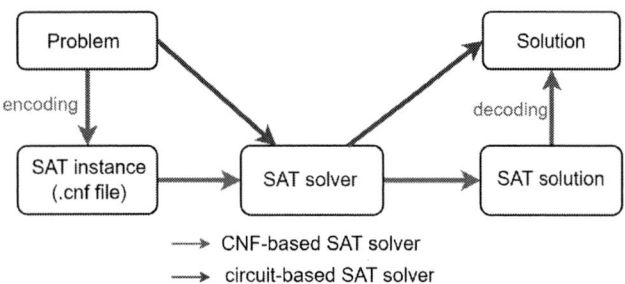

Fig. 1. The steps of CNF-based and circuit-based SAT solvers

original problem solution without generating CNF formulas. CNF-based SATs require additional steps of transformation, whereas circuit-based SATs can calculate and obtain the solution directly. On the other hand, we can utilize circuit structure information to develop more efficient SAT algorithms. As many traditional EDA problems are circuit structure-oriented, circuit SAT solvers seem to be more promising.

PROPOSED METHOD

Inspired by the previous work of QuteSAT [2], this paper implements and improves the *conflict driven clause learning* (CDCL) [5] algorithm of CNF-SAT on the circuit, making full use of the topological structure information. The structural information given be a circuit include connectivity of logic gates, the number of logic gates, fanin/fanout of logic gates, and the logic gate types. Our method differs from QuteSAT in the following aspects.

1) We consider a heterogeneous logic network consisting of single-input NOT, two-input XOR, and multi-input AND and OR gates.
2) We propose a heuristic based on fanout numbers of logic gates to enable fast conflict learning.
3) We make use of the learned OR gate instead of AND gate as an additional clause for further processing.
4) We propose a PO priority-based execution algorithm that achieves fast solving of both SAT and UNSAT instances.

Fanout-driven decision heuristic

A SAT solver assigns '0' or '1' to Boolean variables by making decisions. When making a decision in SAT solving, it might be more efficient to select a variable and assign it a value that is more likely to cause a conflict. So that the conflict could be detected and learned earlier. In the circuit structure, the fanouts of a logic gate serves as a critical parameter that reflects the complexity of a circuit's logic gate connections. By utilizing Boolean constraint propagation with more variables having a higher fanouts, more assignment information for signal lines can be deduced during decision reasoning. Therefore, by incorporating information about the fanouts of logic gates when evaluating decision signal lines, more effective decisions can be made. The decision heuristic used by CirSAT is a combination of fanout of logic gates and improved Variable State Independent Decaying Sum (VSIDS) [6] technique,

where literal activities are scored by fanout numbers before any conflict occurs, not just by increasing activity when a conflict occurs like the original VSIDS.

Conflict-driven circuit learning

A necessary assignment deduced by reasoning is an implication. Consecutive implications result in constraint propagation. During propagation, having a conflict means a variable is implied to be 0 and 1 simultaneously by another assignment. For example, variable a implies variable c should be 0, but another variable b indicates c must be 1. Thus, a conflict happens. Modern CNF-based SAT solvers often utilize CDCL to deal with conflicts. The conflict is analyzed and learned in such a way that an additional clause is added to the SAT instance CNF to avoid resolving the same conflict later. By contrast, we improve and implement it on circuit data structures. When a conflict occurs, conflict analysis examines implication dependencies on implication graph [7]. And it also characterizes the root cause of conflict as a learned OR gate with an output equal to 1. For instance, if $a = 0$ implies $c = 0$ and $b = 1$ implies $c = 1$, we add an OR gate with output equal to 1, i.e., $a \vee \bar{b} = 1$. The OR gate can naturally be represented in disjunction form, which is more processing-friendly than the AND gate.

Two-pointer watching schema

Solvers use a 'watch pointer scheme' to eliminate unnecessary monitoring of the gates for fast BCP [5]. Rather than monitoring all assignments in a gate, we only monitor the first two signals using two pointers, which correspond to the '2-literal watching' clauses in the CNF-based SAT solver. The gate should only be visited when the pointed signal is assigned the corresponding watch value. watch value means that when the signal is assigned a watching value, the assignment of other associated signal lines can be directly inferred. For example, a 3-input AND gate as shown in Fig. 2, assumed that $x1$ and $y1$ are selected as the monitoring signals for this gate with two pointers, the watch values for $x1$, $x2$, and $x3$ are 0, while the watch value for $y1$ is 1. The AND gate will be visited for checking only when $x1$ is assigned a watching-value of 0 or $y1$ is assigned a watching-value of 1. In other words, this gate will not be visited when any other assignment occurs.

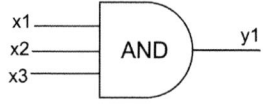

Fig. 2. A three-input AND gate.

PO priority-based execution

The goal of solving circuit-SAT is to find one or more assignments of PIs that result in all POs being assigned as 1. To achieve this, we propose a priority-based PO evaluation algorithm. During algorithm execution, we prioritize the inference of all POs assignments by traversing the POs one by

979-8-3503-1101-3/23 $31.00 © 2023 IEEE

TABLE I
Comparison Results Between MiniSAT and CirSAT(ms)

Benchmarks	#Gates	PI/PO	Solution	MiniSAT Time	Method A Time	Method A x	Method B Time	Method B x
c17	6	5/2	sat	0.274	0.094	3.0	0.037	7.4
c432	160	36/7	sat	2.945	0.852	3.5	0.198	14.9
c880	383	60/26	unsat	1.694	0.281	6.0	0.027	62.7
c1908	880	33/25	sat	7.249	5.362	1.3	3.363	2.2
c1355	546	41/32	sat	3.956	0.942	4.0	0.315	12.6
c2670	1193	233/140	unsat	4.045	1.325	3.0	0.0151	267.8
c3540	3553	207/5	sat	41.738	12.752	3.3	4.697	8.9
c5315	2307	178/123	unsat	6.481	1.547	4.2	0.061	106.2
c6288	2416	32/32	unsat	16.706	3.252	3.3	0.082	203.7
c7552	3512	207/108	unsat	33.654	12.329	2.7	1.061	5.6
rc2670	1198	233/5	sat	8.257	3.851	2.1	1.463	5.6
rc3540	1671	50/2	sat	22.132	11.384	2.0	4.643	4.8
rc6288	2420	32/4	sat	48.334	18.852	2.6	7.016	6.9
rc7552	3517	207/5	sat	43.071	15.634	2.8	6.657	6.5
Average						2.9x		47.7x

TABLE II
Comparison Results Between MiniSAT and CirSAT for Total Solving Time (ms)

Benchmarks	CNF encoding	MiniSAT solving	MiniSAT Total	Our's Total
c17	16.24	0.274	16.514	0.037
c432	17.06	2.945	20.005	0.198
c880	17.58	1.694	19.274	0.027
c1908	18.26	7.249	25.509	3.363
c1355	18.88	3.956	22.836	0.315
c2670	19.28	4.045	23.325	0.0151
c3540	28.97	41.738	70.708	4.697
c5315	24.12	6.481	30.601	0.061
c6288	31.72	16.706	48.426	0.082
c7552	28.68	33.654	59.334	1.061
rc2670	19.81	8.257	28.067	1.463
rc3540	21.16	22.132	43.292	4.643
rc6288	29.03	48.334	77.364	7.016
rc7552	32.01	43.071	75.081	6.657
Total			560.336	29.635
Ratio			1	18.9x

one and assigning them as 1, then execute the propagation of Boolean constraints on the circuit. If a conflict arises during this process, the CirSAT returns UNSAT immediately to achieve fast convergence of UNSAT instance problems. If there is no conflict and all PIs have been assigned, it returns SAT.

EXPERIMENTAL RESULT

Experimental evaluation is conducted using the combinational circuits of ISCAS' 85 benchmark suites and several randomly modified designs (e.g., rc7552), in which both SAT and UNSAT circuits are included. We use MiniSAT[1] as a baseline for comparison, and our circuit-based SAT solver, CirSAT[2], is publicly available. The results are presented in Tables I and II. All experiments are performed on a Intel(R) Xeon(R) Silver 4210R CPU @ 2.40GHz with 64GB of main memory.

For a fair comparison, we use `write_cnf` command in ABC[3] to convert a circuit in BENCH file format into a CNF. The experimental results are shown in Table I, where CPU time are listed in milliseconds and 'x' indicate the accelerate rate (times). The column 'Method A' is the basic version without circuit information, while 'Method B' is the full version of our circuit-based SAT solver with fanout-driven heuristic. On average, 'Method A' achieves 2.9x (up to 6x) CPU time speed-up, whereas 'Method B' gets 47.7x (upto 267.8x) CPU time acceleration.

Note that CNF-based SAT solving needs CNF encoding, solving, and decoding. However, our circuit-based CirSAT can be straightly applied to circuit netlist structure and obtain the solutions of primal problem without any additional transformation overhead. The results shown in Table II indicate that encoding of a circuit into CNF increases significant CPU time overhead. For example, for the c17 circuit with only 5 inputs

[1] https://github.com/niklasso/minisat, 2013
[2] https://github.com/77sup/circuitSolver, 2022
[3] https://github.com/berkeley-abc/abc

and 2 outputs, the CNF encoding time is 16.24ms. Although the size is small, its CNF conversion overhead is comparable to that of the c880 circuit with a problem size more than 10 times larger (17.58ms). Under the condition of taking into account the CNF encoding time, compared with MiniSAT, our CirSAT can save 18.9x time on average.

CONCLUSION

In this paper, we present a circuit-based SAT solver and demonstrate it to be high performance in circuit-oriented SAT problems. With the efficient and extensible circuit-based SAT solver, we can conduct more circuit-oriented research without additional overhead of problem transformation in future work. The possible directions include: (1) Attempt to solve the problem of Combinational logic Equivalence Checking(CEC) with our proposed SAT, (2) Add more circuit-based heuristics technologies and efficient CNF-SAT algorithms into solver for acceleration, (3) Combining circuit SAT with Automatic Test Pattern Generation (ATPG) tools to achieve the rapid generation of high-quality test patterns.

REFERENCES

[1] G. Audemard and L. Simon, "Sat solver glucose 3.0 (2013)."
[2] C.-A. Wu, T.-H. Lin, C.-C. Lee, and C.-Y. Huang, "Qutesat: a robust circuit-based sat solver for complex circuit structure," in *2007 Design, Automation & Test in Europe Conference & Exhibition*. IEEE, 2007, pp. 1–6.
[3] F. Lu, L.-C. Wang, K.-T. Cheng, and R.-Y. Huang, "A circuit sat solver with signal correlation guided learning," in *2003 Design, Automation and Test in Europe Conference and Exhibition*. IEEE, 2003, pp. 892–897.
[4] H.-T. Zhang, J.-H. R. Jiang, L. Amarú, A. Mischenko, and R. Brayton, "Deep integration of circuit simulator and sat solver," in *2021 58th ACM/IEEE Design Automation Conference (DAC)*. IEEE, 2021, pp. 877–882.
[5] J. P. Marques Silva and K. A. Sakallah, "Grasp—a new search algorithm for satisfiability," in *The Best of ICCAD*. Springer, 2003, pp. 73–89.
[6] M. W. Moskewicz, C. F. Madigan, Y. Zhao, L. Zhang, and S. Malik, "Chaff: Engineering an efficient sat solver," in *Proceedings of the 38th annual Design Automation Conference*, 2001, pp. 530–535.
[7] R. I. Brafman, "A simplifier for propositional formulas with many binary clauses," *IEEE Transactions on Systems, Man, and Cybernetics, Part B (Cybernetics)*, vol. 34, no. 1, pp. 52–59, 2004.

FAST NOC ROUTER LATENCY ESTIMATION USING MACHINE LEARNING

Yang Li, and Pingqiang Zhou[*]

School of Information Science and Technology, ShanghaiTech University, Shanghai, China
*Corresponding Author's Email: zhoupq@shanghaitech.edu.cn

ABSTRACT

The Network-on-Chip (NoC) is prevailing in the current communication system of multi-core processors. However, the traditional methods for estimating NoC performance, such as simulator-based and analytical methods, suffer from either great runtime cost or imprecise issues. To address these problems, we propose a neural network model to estimate the queue waiting time of a router's input channel by capturing both the traffic and architecture characteristics. Our method achieves high accuracy (with an average error of only 9%) and efficiency (with only 0.5ms inference time). Compared to the state-of-art SVR method, our model has an accuracy enhancement of 15% to 30% and can capture both the traffic and architecture characteristics.

INTRODUCTION

To achieve lower energy and latency consumption compared to the classical bus, Networks-on-Chip (NoC) is the primary choice as a communication system [1] in current multi-core and asynchronous architectures. NoC designers need to search a huge number of parameters until the designed NoC meets the specification. However, obtaining an accurate NoC latency efficiently during the design iterations presents significant challenges.

Traditionally, designers use simulators [2-3] to evaluate the latency with high accuracy, but simulations run hours to days, making the process extremely slow. Some researchers investigate analytical models [4-5] to quickly estimate NoC latency, but these models require certain traffic assumptions, or the error rate will increase. Machine learning is now commonly used in regression problems, with Neural Networks (NN) offering high accuracy and negligible inference time [6]. Some research groups [7-8] have tried SVR-based methods to predict the NoC latency, but these models only capture the traffic pattern feature without considering the architecture feature.

This paper presents a NN-based performance estimation model for NoC routers. The model allows designers to efficiently obtain the channel queuing time of the router by catching both the traffic flows and NoC architecture parameters. The NN model replaces the most time-consuming simulation part to obtain the path latency. With an average accuracy of around 91% and an inference time of only 0.5ms, the NN engine offers a fast router latency estimator model that can be used for design iterations and optimizations.

METHODOLOGY

Router Architecture

The router, which connects Network Interfaces (NI) and other routers, is the most important building block of NoC. As shown in Figure 1, a simplified router consists of control logic, such as route computing, switch allocation and arbitrator units, as well as a data path that handles the packet storage and movement. The router has five ports four for packet transition and one for injection. Each port has both input and output channels with corresponding buffers.

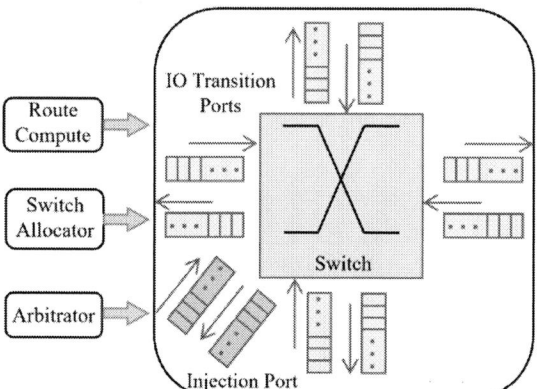

Figure 1: Simplified input queue router architecture with five ports

Problem Formulation

Given a specific traffic pattern and router design specifications, we can evaluate each path delay of the customized NoC to detect the congestion and iterate the design. For a specific traffic path from the source router r_s to the destination router r_d, the path latency L_{sd} is defined by Eq. (1) [3], where the corresponding parameter definitions are shown in Table I.

$$L_{sd} = Qtime_s + \sum_{(i,k) \in \prod sd}(Qtime_{ik} + T) \qquad (1)$$

$Qtime_s$ denotes the channel queuing time, which is the time interval of a packet arriving at the input buffer to finish the switch in a router. In conventional simulators, the measurement of $Qtime_{ik}$ involves a sophisticated traffic transition process including packet injection, route computation, and switch allocation to switch transition,

---

This work was supported by the NSFC under award 62074100.

TABLE I.   PARAMETERS AND DEFINITIONS

Parameters	Definitions
$Qtime_s$	Injection channel queuing time at source router
$Qtime_{ik}$	Channel queuing time of input port i of router k
λ_i	Traffic arrival rate from input port i of router
F_{ij}	Packet forwarding probability from input port i to output port j
T	Constant packet service time

leading to heavy time consumption. To obtain latency efficiently and accurately, we propose building the neural network model to estimate the $Qtime_{ik}$.

Neural Network Modeling of Router Latency

We are targeting mesh-based wormhole switching with an input queue router, using XY routing, which is a prevalent choice for NoC design.

The router's performance depends on the applied traffic patterns and its architecture parameters. To capture these two perspectives, we extract features from them. For traffic patterns, the traffic arrival rate λ_i at each router's port and the forward probability F_{ij} [4], as defined in Table I. These features capture the most significant transition characteristic of the traffic. Additionally, we extract the source and destination of every flow and packet arrival time distribution. As for the router architectures, adjusting the packet size and input buffer size always has a great impact on performance optimization.

Since we are aiming to build a router model that can work well with various traffic patterns, the random traffic flows are fed into the simulator to cover the traffic space as much as possible. Each router has 5 λ_is, and the F_{ij} matrix with 5x4 elements can be generated from these λ vectors, where 5 and 4 correspond to the number of injection input ports and transition output ports respectively. To avoid congestion, we collect random packet injection rates below a throughput of 0.02

TABLE II.   MODEL PARAMETERS AND VARIATIONS

Element	Parameter	Variations
[0:4]	λ_i	Random value of 0 ~ 0.02 (packet/cycle)
[5:24]	F_{ij}	Generated by related Injection Rate
[25]	Mesh Size	2x2, 4x4, 6x6
[26]	Packet Size	8, 16, 32
[27]	Buffer Size	16, 32, 48, 64

packets/cycle. The mesh size, packet size, and buffer size variations shown in Table II cover a large NoC design space.

The NN model is built with 28 input elements as shown in Table II, with output being the target prediction channel queuing time presented by 5 elements. The mean squared error (MSE) in equation (2) is chosen as the loss metric, which is the most popular performance metric in regression fitting problems. R2 score in equation (3) measures the effectiveness of the NN model, while the mean absolute percentage error (MAPE) in equation (4) indicates the error rate of estimation performance. A higher R2 score implies greater accuracy of the model.

$$MSE = \frac{1}{n}\sum(y - \hat{y}) \qquad (2)$$

$$R2\ Score = 1 - \frac{\sum(y-\hat{y})^2}{\sum(y-y^2)^2} \qquad (3)$$

$$MAPE = \frac{100\%}{n}\sum|\frac{y-\hat{y}}{y}| \qquad (4)$$

EVALUATION RESULTS
Experimental Setup

The experiments are conducted on the station with an i7-9700 core and running on a Linux UBUNTU 18.04 LTS OS at 3.4GHz. The data sets are generated from BookSim2 [2]. The router NN model is built on PyTorch.

By applying the parameter variations in Table II, 13,440 samples are collected and divided into 4:1 for training and validation respectively. The NN model used in this work consists of five full-connected layers with several sigmoid activation functions, where each layer contains 40 to 100 neurons. During the training stage, Kaiming weight initialization is applied, and the SGD optimizer with weight decay parameters is used.

The Evaluation of Model Performance

By training our proposed NN model at the minute level, we are able to achieve an MSE of 0.001 and an R2 score of 0.97. The model's inference time is 0.5ms on average. The error rate of $Qtime_{ik}$ ranges from 7% to 13%, with an average of 9%. To compare our model's performance, we also do experiments using the SVR method [7] with the same data set. The SVR model, which uses the RBF kernel to fit the non-linear regression, achieves a R2 score of 0.93. However, the average estimation error rate of $Qtime_{ik}$ is unacceptable, ranging from 25% to 44%. This indicates that the optimal SVR is unable to fit the prediction regression as accurately as our proposed model. Table III provides details of the comparison results.

Our presented NN model shows similar accurate estimation effects on both synthetic traffic, such as uniform and transpose, and application traffic, such as

979-8-3503-1101-3/23 $31.00 © 2023 IEEE

TABLE III. THE MODEL COMPARISON BETWEEN OUR PROPOSED MODEL AND SVR MODEL

Performance Metrics	SVR (RBF)	Our Proposed Model
MSE	0.032	0.001
R2 Score	0.93	0.97
MAPE (Error rate)	25%~44%	7%~13% (Avg. 9%)

PIP and MWD. Figure 2 visualize two traffic pattern scenarios with the prediction values and golden truths. In Figure 2(a), the uniform traffic pattern is applied on NoC with 3x3 mesh, 32-flit packet and 8-flit buffer capacity. By varying the injection rate uniformly from 0.001 to 0.019 with a 0.001 packet/cycle step, we compare the NN model estimation results with those of BookSim2. The error rate ranges from 0.06% to 12% and the average is 5.25%. In Figure 2(b), we use a 4x4 mesh and choose the longest path R0 to R15 for transpose traffic. The packet size and the buffer size are set as 64 flits and 8 flits, respectively. Before the throughput threshold of 0.005 packets/cycle, the error rate decreases. However, the network enters the congestion state rapidly after the threshold, leading to error rate fluctuations. The minimum difference between the simulation results and our proposed model is 2% while the maximum error rate is less than 21%, with an average error of 9.7%. For the application traffic patterns, the proposed model is tested on PIP with 3x3 mesh and MWD 4x4 mesh, showing the maximum error rate is less than 20% and the average error rate is 4.5%. Evaluating the NN model under various traffic patterns demonstrates the effectiveness and accuracy compared to the simulations.

In terms of runtime, our proposed method achieves a significant advantage, with an average inference time of only 0.5ms per process. In contrast, traditional simulations can cost hours to complete for large NoCs. This demonstrates the substantial speedup by our neural network model compared to simulators.

CONCLUSION AND FUTURE WORK

In this work, we propose a machine learning-based approach for building a general router channel queuing time prediction model. The model accurately predicts the channel queuing time and then obtain the NoC path latency under various traffic scenarios, including synthetic and application, with an error rate of less than 10%. Moreover, the inference time of our proposed model is only 0.5ms, significantly reducing prediction time costs. Since our current model only focuses the static traffic features, future work will consider dynamic latency prediction using machine learning techniques.

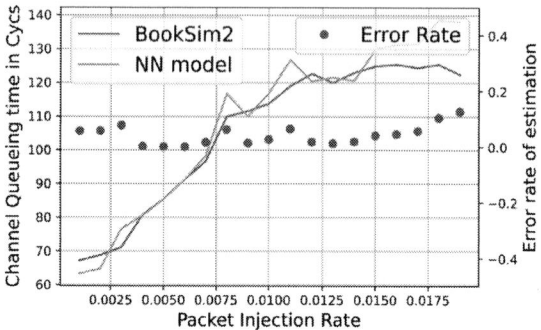

(a) Uniform, Path=R0->R8, 3x3mesh, P=32, B=8.

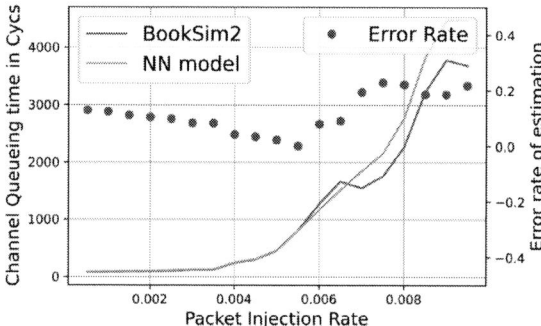

(b) Transpose, Path = R0->R15, 4x4 mesh, P = 64, B = 8.
Figure 2: The comparison between channel queuing time and BookSim2 with error rate per PIR. (a) Synthetic traffic: Uniform (b) Synthetic Traffic: Transpose

REFERENCES

[1] W. Dally and B. Towles, *Principles and Practices of Interconnection Networks.* Morgan Kaufmann, 2003.

[2] N. Jiang et al., "A detailed and flexible cycle-accurate network-on-chip simulator", *ISPASS, pp. 86-96,* 2013.

[3] K. -C. Chen et al., "NN-Noxim: High-Level Cycle-Accurate NoC-based Neural Networks Simulator", *11th International Workshop on NoCArc*, pp. 1-5, 2018.

[4] U. Y. Ogras et al., "Analytical Router Modeling for Networks-on-Chip Performance Analysis", *IEEE TCADICS*, pp. 1-6, 2010.

[5] Z. Qian et al., "A comprehensive and accurate latency model for Network-on-Chip performance analysis", *ASP-DAC,* pp. 323-328, 2014.

[6] LeCun, Y. et al., "Deep Learning", *Nature,* 521, 436–444, 2015.

[7] Z. Qian et al., "SVR-NoC: A performance analysis tool for network-on-chips using learning-based support vector regression model", *DATE,* pp. 354-357, 2013.

[8] Z. Qian et al., "Performance Evaluation of NoC-Based Multicore Systems: From Traffic Analysis to NoC Latency Modeling", *ACM Trans. Des. Autom. Electron. Syst.,* pp. 82-84, 2016.

LUTPLACE: AN IMPROVED LOOKUP TABLE-BASED PLACEMENT FOR ROUTABILITY

Yihang Qiu[1,2], Yan Xing[1]*, Shuting Cai[1], Xingquan Li[2], Xiaoming Xiong[1]

[1]School of Integrated Circuits, Guangdong University of Technology, Guangzhou 510006, China
[2] Peng Cheng Laboratory, Shenzhen 518000, China
*Corresponding Author's Email: yanxing@gdut.edu.cn

ABSTRACT

Congestion estimation is critical in routability-driven placement since it guides subsequent congestion optimization. This paper explores a Rectangular Uniform wire DensitY (RUDY) based algorithm for routability estimation that achieves a better balance between efficiency and accuracy compared to invoking a global router. We build a lookup table based on each net's pin count and aspect ratio to refine RUDY's wirelength computation and enhance RUDY's accuracy in congestion estimation. The experimental results on the DAC 2012 benchmarks show that the lookup table helps improve scaled HPWL by an average of 3.1% and routing congestion by an average of 2.3%.

INTRODUCTION

Standard cell placement is a significant challenge in modern very large-scale integration (VLSI) designs, especially in terms of routability. Routability-driven placement can be divided into two phases: congestion estimation and congestion optimization. Congestion optimization is based on the congestion map generated by congestion estimation, so the accuracy and efficiency of congestion estimation are essential. Modern placers[1][2] often call a global router to get the congestion map, but this method relies on the router's performance and is more time-consuming. In contrast, Rectangular Uniform wire DensitY (RUDY)[3] has a better compromise between accuracy and efficiency, so it is often used to estimate congestion.

However, RUDY's wirelength estimation is modeled as the half-perimeter wirelength (HPWL), while signal nets are routed as Steiner trees, which are ideally modeled as Rectilinear Steiner Minimum Trees (RSMT). There is a gap in accuracy between these two models when a net has more than three pins.

There are two methods for estimating RSMT. One is a construction method, such as FLUTE[4], which is accurate but slow. The other is a lookup table (LUT) method, by weighting HPWL according to a lookup table of scaling factors. The LUT parameters include pin count (PC) [5], and the aspect ratio (AR) of the net's bounding box[6]. Compared with the construction method, the LUT method can provide a better trade-off between accuracy and efficiency, reducing the design cycle required for multiple iterations in VLSI design.

Our work makes the following contributions.

1. We are the first to apply LUT, whose parameters are the pin count and aspect ratio of each net, to improve RUDY's wirelength computation. We then use the improved RUDY to guide congestion optimization.

2. We explore the effect of RSMT cost on routability optimization by LUTs. We find that different LUTs significantly influence routability, especially Total Overflow (TOF) and Maximal Overflow (MOF), highlighting the importance of accurate LUTs.

3. We conducted experiments on the DAC2012 benchmarks[7], and the results show that the LUT helped improve scaled HPWL (sHPWL) by 3.1% and routing congestion by 2.3% on average.

OVERVIEW OF OUR METHODOLOGY

Our Framework

The framework is shown in Figure 1. We implement our idea on DREAMPlace[8], using an improved RUDY (LUT-RUDY) model to generate a route utilization map and using cell inflation for routability optimization. After global placement and legalization, we call NTUplace4h[9] to finish detailed placement.

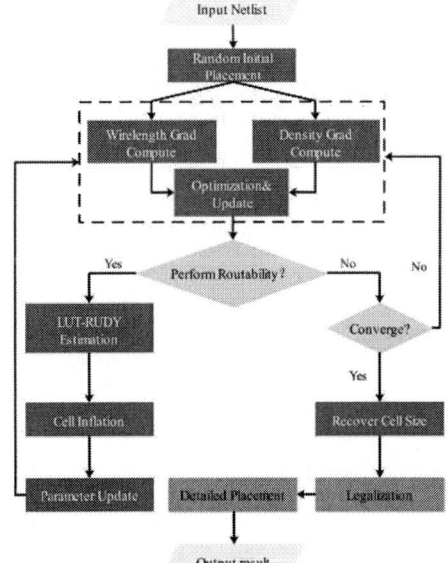

Figure 1: The framework of our proposed routability-driven placement algorithm.

Congestion Estimation Model

The RUDY of net n is shown in Equation (1). The $WireArea_n$ is calculated by HPWL times the wire width. The $BBoxArea_n$ is the area of the minimum enclosing rectangle of net n. Figure 2 depicts the elements for computing the RUDY of a net.

$$RUDY_n = \frac{WireArea_n}{BBoxArea_n} \qquad (1)$$

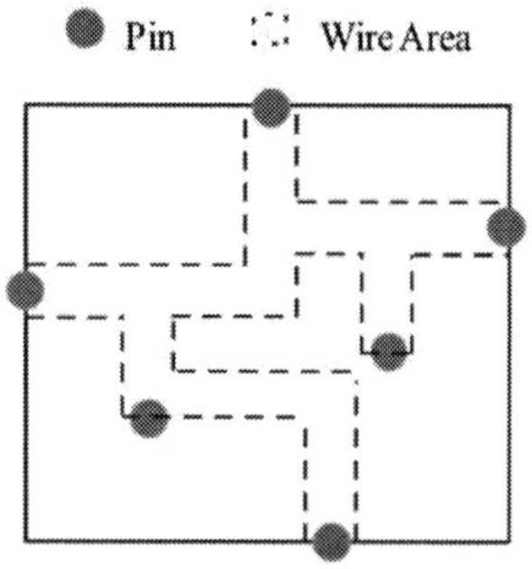

● Pin ⋮⋮ Wire Area

Figure 2: Calculation of Rectangular Uniform wire DensitY(RUDY) for a net.

Cheng[5] estimated the RSMT cost as a correction factor to the HPWL, and the factor is a function of pin count. Andrew further added each net's aspect ratio to improve RSMT cost accuracy and provided a new LUT in [6]. The new LUT is shown in Table I. Table I shows the average RSMT values for pointsets that have bounding box half-perimeter equal to one. By utilizing this new LUT, we can refine RUDY's wirelength computation and enhance RUDY's accuracy in congestion estimation. Finally, Equation (1) can be replaced with Equation (2).

$$LUTRUDY_n = LUT(PC, AR) * \frac{WireArea_n}{BBoxArea_n} \qquad (2)$$

Additionally, we need to calculate the ratio of the overlapping area between each net and the routing grid, and use Equation (3) to obtain the routing utilization map. The routing demand of an edge e on the routing utilization map is denoted as D(e) and can be obtained using LUTRUDY. The routing capacity of an edge e is denoted as C(e) and is typically defined in process technology files.

$$\frac{D(e)}{C(e)} = \sum_{i=1}^{m} LUTRUDY_i * \frac{Overlap(i)}{C(e)} \qquad (3)$$

TABLE I. WIRELENGTH LOOKUP TABLE USING PIN COUNT(PC) AND ASPECT RATIO(AR)

PC \ AR	1	2	4	10
4	1.06	1.05	1.03	1.01
5	1.13	1.11	1.07	1.03
6	1.19	1.16	1.11	1.05
8	1.32	1.27	1.18	1.08
10	1.42	1.36	1.25	1.12
15	1.66	1.59	1.41	1.21
20	1.87	1.78	1.57	1.29
30	2.22	2.10	1.84	1.45

Cell Inflation

The congestion map can be obtained according to Equation (3). Then all cells within congested grids are inflated using Equation (4). The cell inflation ratio is limited to a maximum of 2.5 and y_{super} is also set as 2.5. Besides, we prevent the total incremental area from being too large by dynamically adjusting the inflation rate. If the total incremental area exceeds the predefined maximum value, we reduce the inflation rate of each cell by a small value to decrease the size of inflation and recheck the total area. We repeat the above process until the total incremental area becomes smaller than the predefined maximum value. Figure 3 illustrates the process of cell inflation.

$$\frac{D'(e)}{C'(e)} = \max\left(\left(\frac{D(e)}{C(e)}\right)^{y_{super}}, 2.5\right) \qquad (4)$$

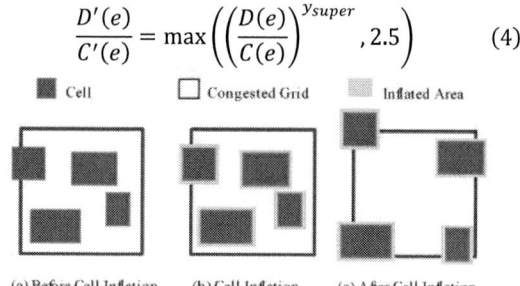

(a) Before Cell Inflation (b) Cell Inflation (c) After Cell Inflation

Figure 3: The process of cell inflation.

EXPERIMENTAL RESULTS

Our algorithm was implemented in C++ and Python. The experiments were performed on Intel(R) Xeon(R) Platinum 8380 CPU @ 2.30GHz, Nvidia A100 40GB PCIe GPU under Linux workstation. The benchmarks are from DAC2012 contest. Since superblue2 and superblue11 cannot get legal results with DREAMPlace, only the remaining eight cases will be considered.

We used four quality metrics and runtime to evaluate our placement results. These quality metrics are defined as shown in Equation (5-9). $ACE(x)$ is the average peak congestion defined in [7]. We used the evaluation script provided by DAC2012 contest to get all these values of quality metrics. And the metrics of runtime are recorded after finishing global placement.

$$RC = max(100, PWC) \qquad (5)$$

$$PWC = \frac{\sum(K_x * ACE(x))}{\sum K_x}, x = 0.5,1,2,5 \qquad (6)$$

$$sHPWL = HPWL * \left(1 + 0.03 * (RC - 100)\right) \qquad (7)$$

$$TOF = \sum Overflow(e) \qquad (8)$$

$$MOF = max\left(Overflow(e)\right) \qquad (9)$$

Our results are shown in Table II, and the results show that the LUT helps improve sHPWL by 3.1% and routing congestion by 2.3% on average. We also find that different LUTs significantly influence routability, especially TOF and MOF. Figure 4 shows the MOF map of superblue6. We can clearly find that our method is effective.

TABLE II. OUR EXPERIMENTAL RESULTS ON DAC2012 BENCHMARKS.

Case	Without LUT					With LUT				
	sHPWL	*RC*	*TOF*	*MOF*	*Time*	*sHPWL*	*RC*	*TOF*	*MOF*	*Time*
SB3	30.75	100.58	9922	10	463	32.14	100.41	6930	10	676
SB6	32.55	101.12	7852	28	486	32.48	100.55	4590	8	519
SB7	37.95	100.78	11318	20	497	38.55	100.76	12926	20	532
SB9	21.73	101.13	12672	12	322	21.77	100.57	6748	10	349
SB12	70.72	164.69	977304	92	523	64.95	156.78	855472	60	514
SB14	23.84	104.17	22482	16	295	22.62	101.90	13732	14	310
SB16	25.94	101.23	11958	20	309	26.37	100.75	9160	20	453
SB19	18.94	112.69	48304	34	200	15.47	103.82	17726	18	231
Normal	100%	100%	100%	100%	100%	96.9%	97.7%	84.2%	69.0%	115.8%

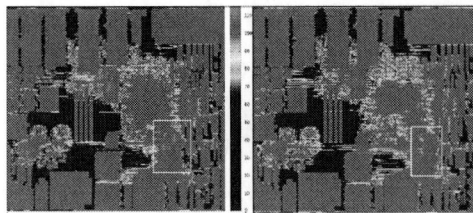

Figure 4: The maximal overflow map of superblue6. The left is the result without using lookup table, and the right is the result with using lookup table.

CONCLUSION

In this paper, we explore a RUDY-based algorithm for routability estimation, which better compromises efficiency and accuracy than invoking a global router. We build a lookup table based on each net's pin count and aspect ratio to improve RUDY's accuracy. The experimental results on the DAC 2012 benchmarks show that the lookup table helps improve scaled HPWL by 3.1% and routing congestion by 2.3% on average. In the future, we will continue to explore more accurate LUTs, and one of the methods is machine learning.

ACKNOWLEDGEMENTS

This project is supported in part by Key-Area Research and Development Program of Guangdong Province(NO. 2022B0701180001), the Major Key Project of PCL (PCL2021A08), the Ministry of Education's Cooperative Education Project (No. 202102172025), and the Guangdong-Hong Kong-Macao Joint Innovation Field Project (No. 2021A0505080006).

REFERENCES

[1] C.-K. Cheng, A. B. Kahng, I. Kang, and L. Wang, "RePlAce: Advancing Solution Quality and Routability Validation in Global Placement," *IEEE Trans. Comput.-Aided Des. Integr. Circuits Syst.*, vol. 38, no. 9, pp. 1717–1730, Sep. 2019, doi: 10.1109/TCAD.2018.2859220.

[2] X. He *et al.*, "Ripple 2.0: high quality routability-driven placement via global router integration," in *Proceedings of the 50th Annual Design Automation Conference on - DAC '13*, Austin, Texas: ACM Press, 2013, p. 1. doi: 10.1145/2463209.2488922.

[3] P. Spindler and F. M. Johannes, "Fast and Accurate Routing Demand Estimation for Efficient Routability-driven Placement," in *2007 Design, Automation & Test in Europe Conference & Exhibition*, Nice, France: IEEE, Apr. 2007, pp. 1–6. doi: 10.1109/DATE.2007.364463.

[4] C. Chu and Y.-C. Wong, "FLUTE: Fast Lookup Table Based Rectilinear Steiner Minimal Tree Algorithm for VLSI Design," *IEEE Transactions on Computer-Aided Design of Integrated Circuits and Systems*, vol. 27, no. 1, pp. 70–83, Jan. 2008, doi: 10.1109/TCAD.2007.907068.

[5] C. E. Cheng, "Risa: Accurate And Efficient Placement Routability Modeling," in *IEEE/ACM International Conference on Computer-Aided Design*, Nov. 1994, pp. 690–695. doi: 10.1109/ICCAD.1994.629897.

[6] A. E. Caldwell, A. B. Kahng, S. Mantik, I. L. Markov, and A. Zelikovsky, "On wirelength estimations for row-based placement," *IEEE Transactions on Computer-Aided Design of Integrated Circuits and Systems*, vol. 18, no. 9, pp. 1265–1278, Sep. 1999, doi: 10.1109/43.784119.

[7] N. Viswanathan, C. Alpert, C. Sze, Z. Li, and Y. Wei, "The DAC 2012 routability-driven placement contest and benchmark suite," in *Proceedings of the 49th Annual Design Automation Conference on - DAC '12*, San Francisco, California: ACM Press, 2012, p. 774. doi: 10.1145/2228360.2228500.

[8] Y. Lin *et al.*, "DREAMPlace: Deep Learning Toolkit-Enabled GPU Acceleration for Modern VLSI Placement," *IEEE Trans. Comput.-Aided Des. Integr. Circuits Syst.*, vol. 40, no. 4, pp. 748–761, Apr. 2021, doi: 10.1109/TCAD.2020.3003843.

[9] M.-K. Hsu *et al.*, "NTUplace4h: A Novel Routability-Driven Placement Algorithm for Hierarchical Mixed-Size Circuit Designs," *IEEE Transactions on Computer-Aided Design of Integrated Circuits and Systems*, vol. 33, no. 12, pp. 1914–1927, Dec. 2014, doi: 10.1109/TCAD.2014.2360453.

ACARM: A NOVEL SEMICONDUCTOR WAFER HANDLING ROBOT

Donglin Chen[1,2,], Lixin Tang[1], Dehong Cong[1,2] and Jingchao Qiao[1,3]*

[1] National Frontiers Science Center for Industrial Intelligence and Systems Optimization,
Northeastern University, Shenyang, 110819, China
[2] Key Laboratory of Data Analytics and Optimization for Smart Industry (Northeastern University),
Ministry of Education, Shenyang, 110819, China
[3] Liaoning Engineering Laboratory of Data Analytics and Optimization for Smart Industry,
Shenyang, 110819, China
*Corresponding Author's Email: 2210284@stu.neu.edu.cn

ABSTRACT

In the field of semiconductor manufacturing, highly automated production methods have become an inevitable trend of development thanks to their precise, stable and reliable characteristics. This paper focuses on the handling link in the semiconductor production process, designed a novel handling robot. Minimum Snap method is used to plan the end-effector trajectory of the robot, the control method of the robot is based on the position control, and the experiment environment is the co-simulation environment of Simulink and Adams. All the simulation testing of the planning and the control method has been done on the virtual prototype.

INTRODUCTION

The increasing semiconductor integration puts forward higher requirements in the accuracy of semiconductor processing. At the same time, after entering the information era, the industry has become increasingly dependent on integrated circuits, which also puts forward higher requirements in the efficiency of semiconductor processing. Under this background, this paper focuses on the handling link in the semiconductor production process, designed a novel wafer handling robot which meet the above requirements as Figure 1.

There have been some successful robot cases applied to wafer handling. In order to reduce the influence of errors due to mechanical transmission on the accuracy of the end-effector, Xiong Jing designed a wafer handling robot driven directly by motor [1], which shows excellent performance in terms of fast response and high flexibility. However, the configuration of this robot is series single arm, therefore, there are hidden dangers in the vertical stability of the end-effector. The double arm scheme proposed by Chao Tang can effectively alleviate this phenomenon [2], but the mechanism design of this robot is slightly complicated. In addition to continuous innovation and improvement in robot mechanism design, there are also many successful achievements in robot trajectory planning and motion control. Wen-Tao Ye applies polynomial interpolation method to robot trajectory planning [3], which can generate more complex trajectories. However, this method is relatively curing. On

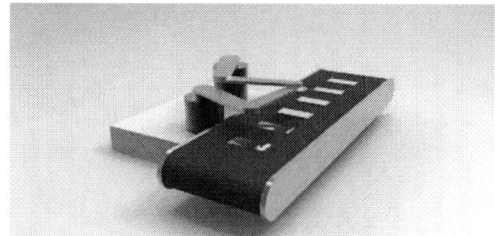

Figure 1: AcArm: a novel semiconductor wafer handling robot

the basis of using polynomial interpolation to plan the robot trajectory, Xiang Li uses optimization methods to make the trajectory better [4]. However, this two-stage approach is slightly redundant.

Based on the achievements of predecessors, this paper has made the following innovations and improvements:

1. The parallel configuration helps to maintain the stability of the end-effector in the vertical direction.

2. Use Minimum Snap method to plan the trajectory of the end-effector in Cartesian space.

MINIMUM SNAP

Minimum Snap trajectory planning method is an excellent algorithm proposed by Kuma [5]. The principle of this method is to divide a whole trajectory into n sub trajectories by $n+1$ key points, and set constraints at each key points to ensure the smoothness of the whole trajectory. Taking the x direction as an example, the mathematical expressions of the two sub trajectories are shown as follow:

$$\begin{cases} l_1(t) = a_{1,5}t^5 + a_{1,4}t^4 + a_{1,3}t^3 + a_{1,2}t^2 + a_{1,1}t^1 + a_{1,0}t^0 \\ l_2(t) = a_{2,5}t^5 + a_{2,4}t^4 + a_{2,3}t^3 + a_{2,2}t^2 + a_{2,1}t^1 + a_{2,0}t^0 \end{cases} \quad (1)$$

Then, a quadratic programming problem based on formula (1) can be constructed as follows:

$$\min\left\{ \int_0^{\frac{T_s}{2}} \left(l_1^{(4)}(t)\right)^2 dt + \int_{\frac{T_s}{2}}^{T_s} \left(l_2^{(4)}(t)\right)^2 dt \right\} \quad (2)$$

$$s.t. \quad l_1(0) = x^{cur} \quad (3)$$

$$l_1\left(T_s/2\right) = x^{cmd} \quad (4)$$

$$l_2\left(T_s\right) = x^{des} \quad (5)$$

$$l_1\left(T_s/2\right) = l_2\left(T_s/2\right) \quad (6)$$

$$l_1^{(1)}\left(T_s/2\right) = l_2^{(1)}\left(T_s/2\right) \quad (7)$$

$$l_1^{(2)}\left(T_s/2\right) = l_2^{(2)}\left(T_s/2\right) \quad (8)$$

$$l_1^{(1)}\left(0\right) = x'^{cur} \quad (9)$$

$$l_1^{(1)}\left(T_s/2\right) = x'^{cmd} \quad (10)$$

$$l_2^{(1)}\left(T_s\right) = x'^{des} \quad (11)$$

Formulas (3) to (9) constrain the key point from position, velocity and acceleration respectively to ensure high order continuity, thus ensuring the smoothness of the whole trajectory. For the end point of the whole trajectory, additional restrictions on acceleration are required:

$$l_n^{(2)}\left(T_{end}\right) = 0 \quad (12)$$

INVERSE KINEMATICS

After getting the trajectory of end-effector by Minimum Snap, desired angle of two actuators can be calculated by inverse kinematics. Figure 2 shows the process of inverse kinematics algorithm with an actuator an example:

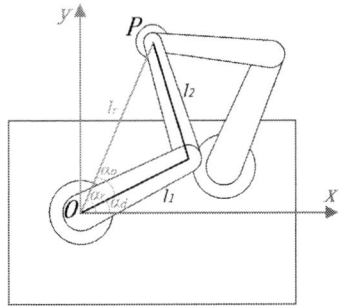

Figure 2: Process of inverse kinematics algorithm with an actuator an example

l_r can be calculated by the Pythagorean theorem based on the position of end-effector P, the result of Minimum Snap. Meanwhile, α_r can be calculated by the arctangent of the position of P. α_o and α_d are calculated as follows:

$$\alpha_o = \arccos\left[\left(l_1^2 + l_r^2 - l_2^2\right)/\left(2 \cdot l_1 \cdot l_r\right)\right] \quad (13)$$

$$\alpha_d = \alpha_r - \alpha_o \quad (14)$$

α_d is the control parameter of position control.

EXPERIMENTAL RESULTS

Robot information

The name of each component and the setting of the coordinate of AcArm are shown as Figure 3. The physical property of each component are listed in TABLE I.

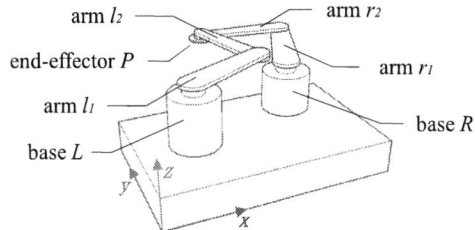

Figure 3: Each component and the setting of the coordinate of AcArm

TABLE I.   PHYSICAL PROPERTIES OF ACARM'S COMPONENTS

Component	Physical property	Value(m)
L	height	0.16
R	height	0.13
l_1	length	0.267
l_2	length	0.267
r_1	length	0.267
r_2	length	0.267

Trajectory

The trajectory of AcArm end-effector planned based on Minimum Snap is shown as Figure 4. End-effector of AcArm will pass through red dot, purple dot, cyan dot and blue dot in turn.

Figure 4: The trajectory of AcArm end-effector planned based on Minimum Snap

Virtual prototype and controller

The AcArm virtual prototype built in Adams is shown as Figure 5.

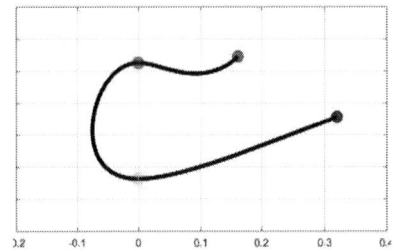

Figure 5: AcArm virtual prototype built in Adams

The controller built in Simulink that interacts with the virtual prototype is shown as Figure 6, and whose workflow is shown as Figure 7.

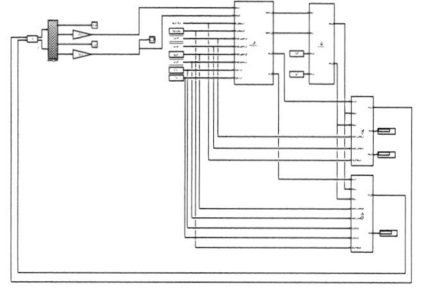

Figure 6: AcArm controller built in Simulink

In Figure 7, the modules in the red box only run once in a program working cycle, while the modules in the blue box will run once per iteration.

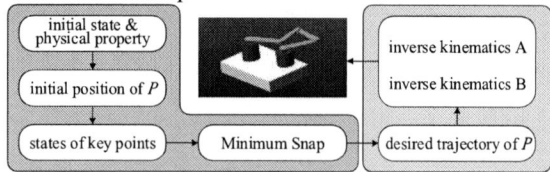

Figure 7: Workflow of AcArm controller

RESULT

The control values of base L and base R under the simulation environment of Adams and Simulink are shown as Figure 8 and Figure 9.

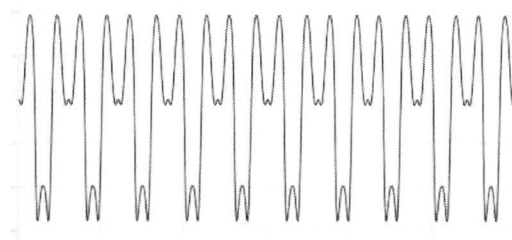

Figure 8: Change process of control value of base L

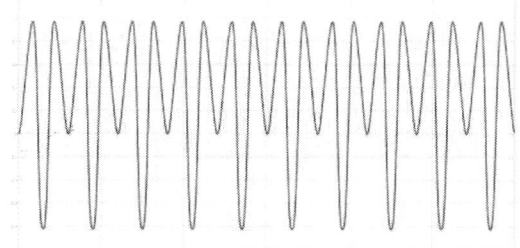

Figure 9: Change process of control value of base R

From Figure 8 and Figure 9, both actuators exhibit periodic trend in their motion, with continuous and smooth parameter changes that meet the conditions for physical implementation.

CONCLUSION

AcArm performs a complete planning task as shown in the Figure 10. The trajectory finally realized by the end-effector of AcArm is exactly the same as that trajectory planned in Figure 4, which proves that the method and robot designed in this paper is effective and feasible. However, there are still some deficiencies to be improved:

1. Design and deploy force control solution to achieve more compliant.

2. The simulation environment is ideal, after the real physical robot's development is completed, it needs to be further tested on the real physical robot.

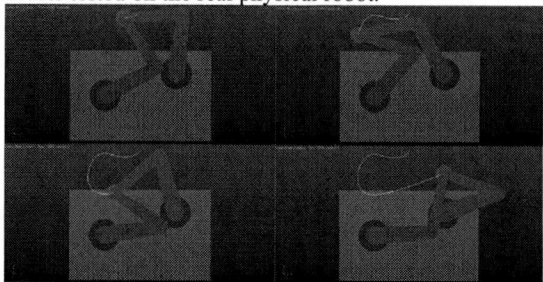

Figure 10: AcArm performs a complete planning task

ACKNOWLEDGEMENTS

This research was supported by the Major Program of National Natural Science Foundation of China (72192830, 72192831), and the 111 Project (B16009).

REFERENCES

[1] Jing X. Development and control system design of a wafer-handling robot[D]. Harbin Institute of Technology, 2010.

[2] Tang C. Kinematics simulation of a wafer handling manipulator[J]. *Machinery*, 2014, 52(5):4.

[3] Ye W.T. Obstacle avoidance trajectory planning and position pre-correction based on wafer & substrate Handle Robot[D]. Guangdong University of Technology, 2019.

[4] Li X. Research on key technologies of a new direct-drive wafer handling robot[D]. Donghua University, 2022.

[5] D. Mellinger and V. Kumar. *IEEE International Conference on Robotics & Automation*, Shanghai, May 9-13, 2011, pp. 2520-2525.

EFFICIENT PARTITIONING AND COMMUNICATION SCHEME-BASED DISTRIBUTED EDGE COMPUTING TO ACCELERATE DEEP NEURAL NETWORK

Xudong Lu[1], Cheng Zhuo[1]*

[1]Zhejiang University, Hangzhou, China

*Corresponding Author's Email: czhuo@zju.edu.cn

ABSTRACT

In the era of the Internet of Things (IoT), there are growing demands to deploying Deep Neural Network (DNN) on edge devices. However, edge devices typically have limited computing resources, which makes it challenging to perform the large computing workloads required for DNN inference. While several methods have been proposed to address this challenge, such as Server-Client pipeline, distributed computing, etc., the workload partitioning and transmission schemes for edge devices have not been well discussed. This paper proposes a distributed edge computing system that leverages efficient partitioning algorithms and a TCP-based communication mechanism to accelerate DNN inference. The experimental results demonstrate that the proposed system with 4 nodes reduces the computation delay by 74.39%. Furthermore, when the number of nodes increases from 2 to 4, the inference is accelerated by 1.98-3.90×, which is 12.98% better than the prior MoDNN system in [6].

INTRODUCTION

DNN is commonly employed in diverse applications such as image classification, object detection, and natural language processing (NLP), owing to their remarkable precision and portability. Nevertheless, with the increasing complexity of DNN models, their computational demands and parameter scale escalate rapidly. Furthermore, the expansion of industrial IoT has led to the deployment of DNN models in edge scenarios like autonomous driving, facial recognition payment, and industrial monitoring [1].

On the edge devices, a significant challenge is the lack of computing resources to efficiently handle the huge computing workloads of DNN models. Numerous studies have addressed the acceleration of DNN inference by implementing diverse techniques, including DNN model compression [2], designing lightweight models that balance precision and computation delay, such as ShuffleNet [3] and MobileNet [4], offloading intermediate data to a cloud server [5], and developing a distributed computing system with multiple sub-nodes sharing the workloads [6]. However, the workload partitioning and transmission schemes for edge devices have not been thoroughly discussed in these studies [7].

This paper introduces a distributed edge computing system that utilizes highly efficient model partitioning algorithms, a TCP-based communication mechanism, and parallel computing schemes. The experimental results indicate that as the number of nodes in the system grows from 2 to 4, DNN inference accelerates by a factor of 1.98 to 3.90. Additionally, compared to MoDNN, a prior distributed computing system, our proposed approach demonstrates a 12.98% performance boost [6].

BACKGROUNG AND MOTIVATION

The Client-Server architecture functions by transferring workloads to a cloud server with ample computing resources [5]. Although this architecture accelerates DNN inference computation effectively, the considerable amount of data transmission results in significant communication delays. In general, the Client-Server architecture compromises increased communication delay for decreased computation delay, resulting in an 11% acceleration in DNN inference.

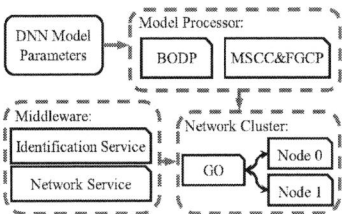

Figure 1: System overview of MoDNN [6]

Due to the constraints of the Client-Server architecture, some scholars have shifted their focus to developing distributed computing systems. MoDNN is a prominent distributed computing system designed to expedite DNN inference on mobile devices. The system partitions model parameters and allocates sub-tasks to worker nodes for parallel computation, as illustrated in Figure 1. Compared to the Client-Server architecture, MoDNN shows reduced latency owing to its narrower computing capacity gap and shorter communication distance between individual devices [6].

Despite the suitability of current distributed computing systems for GPUs or mobile devices, the hurdles of edge computing devices with limited computing capacity have not been adequately addressed. As a consequence, when implemented in edge scenarios, these systems may not offer optimal acceleration. Thus, a distributed edge computing system that enhances performance is presented in the next section.

PROPOSED DISTRIBUTED EDGE COMPUTING SYSTEM

The structure of the proposed distributed edge computing system is depicted in Figure 2, comprising a manager node *Node 0* and multiple worker nodes *Node x*. In addition to the regular computation, Node 0 is responsible for partitioning the convolutional layer (CL) and fully-connected layer (FL) to the manageable size at the beginning of the computing. The following operations are conducted in sequence once the input matrix is received in the system:

1) *Node 0* divides the matrix into multiple sub-tasks based on the model partitioning algorithms;

2) *Node 0* allocates sub-tasks to worker nodes, including itself;

3) *Node x* transmits the computation results to *Node 0*;

4) *Node 0* aggregates all the results and executes the remaining computations, such as MaxPooling, ReLU, Biased, and so on.

5) Repeat 1) - 4) for each subsequent layer until completion.

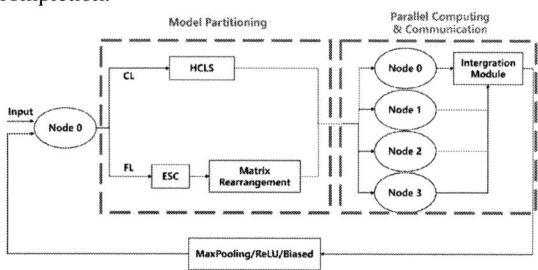

Figure 2: Structure of the proposed distributed edge computing system

For CL partitioning, the convolutional output matrix O is partitioned to match the computing ability CA_i (i=0,1,2...) of each worker node *Node i*. The partitioning schemes of 1-D, 2-D, and Depth Partitioning are depicted in Figure 3.

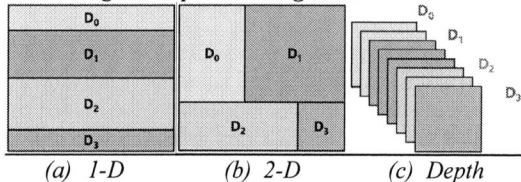

(a) 1-D   *(b) 2-D*   *(c) Depth*

Figure 3: Three CL partitioning schemes

Take a 4-node, zero-padding system as an example, for CL with input matrix of $r×c×d$ and convolution kernel of $k×k×d×d_o$ (d_o is the depth of output matrix), the size of output sub-matrix D_i (i=0,1,2,3) for 1-D and 2-D Partitioning is shown in (1) and (2), respectively.

$$D_i = (\frac{CA_i×r}{sum(CA_i)}, r, d_o) \quad (1)$$

$$D_i = (\frac{(CA_i+CA_{i\pm1})×r}{sum(CA_i)}, \frac{CA_i×r}{CA_i+CA_{i\pm1}}, d_o) \quad (2)$$

Furthermore, the amount of data transmission, denoted as T_i (i=0,1,2,3), is shown in (3) for 1-D partitioning and (4) for 2-D partitioning, while the redundant data for each scheme is calculated as $R_i = T_i - D_i$.

$$T_i = (\frac{CA_i×r}{sum(CA_i)} + k - 1, c, d_o) \quad (3)$$

$$T_i = (\frac{(CA_i+CA_{i+1})×r}{sum(CA_i)} + \frac{k-1}{2}, \frac{CA_i×r}{CA_i+CA_{i+1}} + \frac{k-1}{2}, d_o) \quad (4)$$

Combining Equations (1) - (4), we can derive the value of R_i for both 1-D and 2-D partitioning schemes, which are $3cd_o(k-1)$ and $d_o(r + c)(k - 1)$, respectively. Typically, for square matrix (i.e., $r = c$), R_i of 2-D equals $2cd_o(k-1) < 3cd_o(k-1)$, indicating that 2-D partitioning outperforms 1-D partitioning due to less redundant transmission. Furthermore, as the depth increases, depth partitioning performs more efficiently as channels become the main factor of computation delay [8]. Therefore, for VGG-16, we employ 2-D partitioning for the first 3 CLs and depth partitioning for the remain ones.

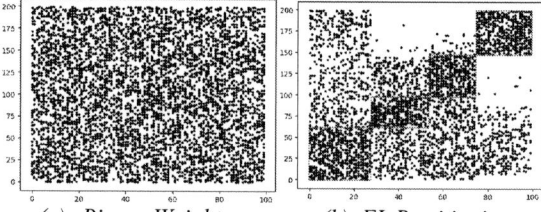

(a) Binary Weight   *(b) FL Partitioning*

Figure 4: Results of the proposed FL partitioning algorithm

For FL partitioning, we propose the Extended Spectral Clustering (ESC) algorithm. First, we convert the weight matrix $W_{r×c}$, as shown in Figure 4(a), into an undirected graph with $r+c$ neurons, and create an adjacency matrix A to remove the structural differences between rows and columns. Next, we normalize the Laplacian matrix L_{norm} and perform Eigen Value Decomposition (EVD) to obtain the eigenvector x_j, which contains information about the connections. We then combine the normalized x_j of the smallest eigenvalues into a $(r+c)×n$ matrix X_c. Subsequently, we apply K-Means clustering to X_c, and rearrange W based on the clustering results, resulting in partitioning W into n dense matrices W_i and an outlier sparse matrix O, as shown in Figure 4(b) [9].

A TCP-based communication mechanism has been developed to facilitate data transmission between nodes. Upon the joining of a new node, denoted by *Node x*, a connection request is sent to *Node 0*, and the communication process commences upon acceptance of the request.

EXPERIMENTAL RESULTS

The performance evaluation of the proposed distributed edge computing system is conducted using the VGG-16 model implemented in PyTorch on 1 to 4 ARM-based nodes. Additionally, the system's performance is compared to that of MoDNN to assess its efficiency.

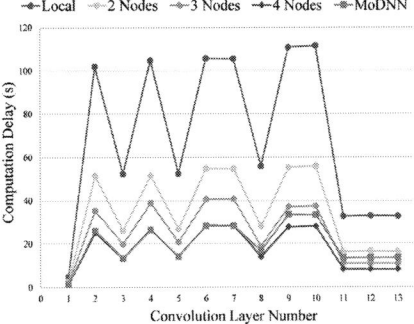

Figure 5: Computation delay T_c of CLs in VGG-16

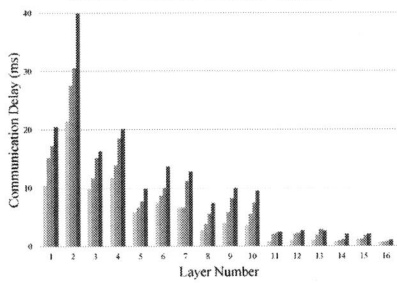

Figure 6: Communication delay T_s of layers in VGG-16

Figure 5 illustrates the variation in computation delay T_c of convolutional layers (CLs). With the increase in the number of nodes from 1 to 4, T_c demonstrates a linear decrease. Specifically, for the system with 2, 3, and 4 nodes, the computation time is accelerated on average by 1.98, 2.83, and 3.95 times, respectively. Notably, the proposed system outperforms MoDNN in terms of T_c for CLs beyond Conv8.

Figure 6 illustrates the communication delay T_s of the distributed edge computing system with 2-4 nodes and MoDNN. The value of T_s is mainly influenced by the overall data volume. As the number of nodes increases from 2 to 4, the size of the transmitted matrices also increases, resulting in a 29.7%-62% rise in T_s. Nevertheless, compared to MoDNN, the proposed system has achieved a 17.2% reduction in T_s.

The overall inference delay, $T_o = T_c + T_s$, and the acceleration ratio, R_a, are evaluated and presented in Table I.

TABLE I.   OVERALL DELAY OF THE PROPOSED SYSTEM

System	T_c	T_s	T_o	R_a
Local	904.75	0	904.75	1.00
2 Nodes	456.71	90.40	456.80	1.98
3 Nodes	324.37	116.99	324.49	2.79
4 Nodes	231.81	146.33	231.95	3.90
MoDNN	261.87	177.95	262.05	3.45

The results show that the VGG-16 inference is accelerated by 1.98-3.90 times when nodes are increased from 1 to 4, indicating a linear speedup with the number of nodes. Furthermore, compared to the existing 4-node MoDNN system, our design achieves a 12.98% improvement in the acceleration ratio.

CONCLUSION

In this paper, we design an efficient distributed edge computing system that utilizes an efficient partitioning and communication mechanism to accelerate deep neural network processing. Our approach employs model partitioning algorithms, a TCP-based communication mechanism, and parallel computing techniques. We evaluate the effectiveness of our system by deploying VGG-16 and conducting experiments that demonstrate a linear increase in DNN inference acceleration with the number of nodes. Furthermore, our design outperforms existing methods in terms of acceleration ratio.

ACKNOWLEDGEMENTS

We would like to thank the supports from NSFC (Grant No. 62034007, 61974133, and 62141404) and SGC Cooperation Project (Grant No. M-0612).

REFERENCES

[1] J. Chen and X. Ran, "Deep Learning With Edge Computing: A Review," *Proceedings of the IEEE*, 2019, pp. 1655-1674.

[2] Z. Qu, S. Cai, Q. Ji, et al. "Lightweight Urine Sediment Image Recognition Network Based on Deep Separable Residual Structure," *IEEE ICEMI*, 2021, pp. 152-157.

[3] X. Zhang, X. Zhou, M. Lin, et al. "ShuffleNet: An Extremely Efficient Convolutional Neural Network for Mobile Devices," *IEEE/CVF Conference on Computer Vision and Pattern Recognition*, 2018, pp. 6848-6856.

[4] D. Sinha and M. El-Sharkawy, "Thin MobileNet: An Enhanced MobileNet Architecture," *IEEE Annual UEMCON*, 2019, pp. 0280-0285.

[5] J. Hauswald, T. Manville, Q. Zheng, et al. "A hybrid approach to offloading mobile image classification," *IEEE International Conference on Acoustics, Speech and Signal Processing (ICASSP)*, 2014, pp. 8375-8379.

[6] J. Mao, X. Chen, K. W. Nixon, et al. "MoDNN: Local distributed mobile computing system for Deep Neural Network," *Design, Automation & Test in Europe Conference & Exhibition (DATE)*, 2017, pp. 1396-1401.

[7] E. Li, L. Zeng, Z. Zhou, et al. "Edge AI: On-Demand Accelerating Deep Neural Network Inference via Edge Computing," *IEEE Transactions on Wireless Communications*, 2020, pp. 447-457.

[8] Y. You, Z. Zhang, C. -J. Hsieh, et al. "Fast Deep Neural Network Training on Distributed Systems and Cloud TPUs," *IEEE Transactions on Parallel and Distributed Systems*, 2019, pp. 2449-2462.

[9] C. Chinrungrueng and C. H. Sequin, "Optimal adaptive k-means algorithm with dynamic adjustment of learning rate," *IEEE Transactions on Neural Networks*, 1995, pp. 157-169.

A HYBRID TRAINING FRAMEWORK FOR SPEEDING UP THE INFERENCE PROCESS OF SPIKING NEURAL NETWORKS

Ziwen Li, Yu Ma, and Pingqiang Zhou[*]

School of Information Science and Technology, ShanghaiTech University, Shanghai, China

*Corresponding Author's Email: zhoupq@shanghaitech.edu.cn

ABSTRACT

Spiking Neural Networks (SNNs) have recently attracted enormous research interest as their spike-based computing paradigm brings high energy efficiency. However, there is a trade-off between the accuracy and the inference speed of SNNs. The existing works on SNNs achieve high accuracy at the cost of low speed. In this paper, we first analyze the reason why high-accuracy SNNs require a large number of time steps. Then we propose a hybrid training framework to speed up the inference process and reduce the accuracy loss of SNNs. Experimental results show that the converted SNN with our proposed technique can improve the inference speed by 7X on CNET and 2.67X on VGG-16.

INTRODUCTION

The past two decades have witnessed the tremendous success of Artificial Neural Networks (ANNs) in various applications [1]. However, ANNs require substantial energy consumption in the inference process, which brings a challenge for deploying ANNs on resource-limited devices. Recently, Spiking Neural Networks (SNNs) have attracted significant research interest due to their high energy efficiency [2]. And researchers have considered SNNs as promising alternatives to ANNs for resource-limited devices.

There are three training methods for SNNs including Spike Timing Dependent Plasticity (STDP) based learning [3], spike-based error back-propagation algorithm [4], and ANN-to-SNN conversion method [5-8]. STDP and error back-propagation are not effective for large and deep SNNs, while the ANN-to-SNN conversion method has excellent scalability and yields high-accuracy SNNs. [5] first proposes the ANN-to-SNN conversion method by mapping the parameters from ANNs to SNNs. However, the converted SNNs require a large number of time steps and suffer accuracy loss compared to the source ANNs. To improve the accuracy, [6] develops the model-based and data-based parameter normalization methods. Building on this work, [7] discusses the influence of different normalization factors on the accuracy of SNNs and proposes a robust parameter normalization technique, which results in higher accuracy and inference speed of SNNs. Furthermore, [8] treats the normalization factors as trainable parameters of ANNs by adding trainable clipping

---

This work was supported by the NSFC under award 62074100.

layers (TCL). TCL limits the maximum activation value of each layer, thus improving the inference speed and accuracy of SNNs. Despite these efforts, the inference speed and accuracy of SNNs still require improvement.

In this paper, we first explain that the information transmitted by the converted SNN to the next layer is the quantized ANN activation value. Then we consider the quantization in the training process and propose a hybrid training framework. In our framework, ANN uses unquantized values and quantized values to perform forward propagation, respectively. The loss function used to perform backward propagation consists of the losses of the above two forward propagation processes. We evaluate the proposed technique on the CIFAR-10 dataset. Compared to [8], the converted SNN with our training framework can improve the inference speed by 7X on CNET and 2.67X on VGG-16 without accuracy loss.

PRELIMINARIES

Inference Process of ANNs and SNNs

The inference process of ANNs with ReLU function at layer l is described as follows:

$$a_i^l = max\,(0, \textstyle\sum_j w_{ij,A}^l a_j^{l-1} + b_{i,A}^l) \qquad (1)$$

a_i^l is the activation value of the i-th neuron. $w_{ij,A}^l$ and $b_{i,A}^l$ are the weight and bias of ANNs.

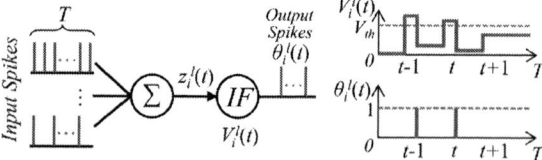

Figure 1: The inference process of SNNs.

Different from ANNs, SNNs simulate neurons with a series of spikes. In our work, we consider the Integrated-and-Fire (IF) model for spiking neurons. And the inference process of SNNs is illustrated in Fig. 1. The total number of inference time steps of SNNs is denoted as T. At a certain time step t, the i-th neuron in layer l integrates membrane potentials (MP) as follows.

$$z_i^l(t) = \textstyle\sum_j w_{ij,S}^l \theta_j^{l-1}(t) + b_{i,S}^l \qquad (2)$$

$\theta_j^{l-1}(t)$ is the spike input from the j-th neuron in layer l-1. $w_{ij,S}^l$ and $b_{i,S}^l$ are the synaptic weight and bias of the i-th spiking neuron in layer l. Once the MP exceeds the threshold potential V_{th}, the spiking neuron fires a spike to the following layer. The spike output of the i-th neuron in layer l can be formulated as

$$\theta_i^l(t) = U(V_i^l(t\text{-}1) + z_i^l(t) - V_{th}) \qquad (3)$$

where $U(x)$ is a unit step function and $V_i^l(t\text{-}1)$ is the residual MP at time step $t\text{-}1$. Then the MP of the spiking neuron is reset. We adopt the reset-by-subtraction [8] strategy and the residual MP $V_i^l(t)$ is computed as follows.

$$V_i^l(t) = V_i^l(t\text{-}1) + z_i^l(t) - V_{th}\theta_i^l(t) \qquad (4)$$

ANN-to-SNN Conversion Theory

During conversion, the threshold V_{th} is set to 1.0. We integrate the Eq. (4) over T, and divide it by T:

$$r_i^l(T) = \sum_j w_{ij,S}^l r_j^{l-1}(T) + b_{i,S}^l - \frac{V_i^l(T)}{T} \qquad (5)$$

where $r_i^l(T) = \frac{\sum_j^T \theta_i^l(t)}{T}$ is the fire rate of the i-th spiking neuron at layer l, and it falls into the interval of [0, 1]. When T is large, $\frac{V_i^l(T)}{T}$ is negligible. Comparing Eq. (1) to Eq. (5), the ANN can be converted to SNN by mapping the output of ReLU to the fire rate with the following parameter normalization rules [8]:

$$w_S^l = w_A^l \frac{\lambda^{l-1}}{\lambda^l} \; ; \; b_S^l = \frac{b_A^l}{\lambda^l} \qquad (6)$$

where λ^l is the maximum activation value of layer l in ANN. Following the practice in [8], we add the trainable clipping parameter after the ReLU function of each ANN layer to obtain the maximum activation value in Eq. (6).

METHODOLOGY

In this section, we first analyze the reason why the converted SNN requires a large T to achieve comparable accuracy as the source ANN. Based on the analysis, we propose a hybrid training framework to improve the inference speed and accuracy of SNN.

The conversion method applies Eq. (6) to convert the pre-trained ANN to SNN. The analog activation value a_i^l of ANN is mapped to the fire rate r_i^l of SNN. When T is small, the residual MP $V_i^l(T)$ in Eq. (5) cannot be neglected, and the converted SNN suffers significant accuracy loss. According to [8], the fire rate r_i^l can be approximated by the following equation:

$$r_i^l \approx \lfloor \frac{a_i^l}{\lambda^l} T \rfloor / T \qquad (7)$$

From Eq. (7), we can conclude that the analog activation value a_i^l of ANN is quantized to $\tilde{a}_i^l = r_i^l \lambda^l$ in SNN. For example, if $a_i^l = 1$ and $\lambda^l = 1.5$, the conversion loss rate $(a_i^l - \tilde{a}_i^l)/a_i^l$ of SNN is 1% with $T = 100$ while it is 25% with $T = 4$. Therefore, there is a trade-off relation between the accuracy and the total number of inference time steps.

A large T can reduce the conversion loss and ensure the high accuracy of the converted SNN. However, SNN with large T increases the computational overhead and inference time per classification. To overcome this challenge, we propose a hybrid training framework to improve the inference speed and improve the accuracy of

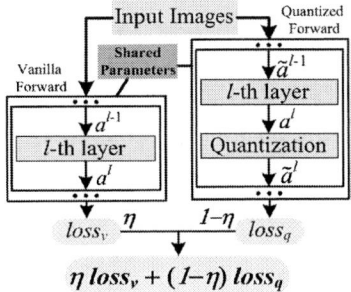

Figure 2: The proposed hybrid training framework.

SNN as well. The conversion loss in the inference process of the converted SNN with small T stems from the large difference between a_i^l in ANN and the quantized $\tilde{a}_i^l$ in SNN. Therefore, during the training process, we take the influence of the quantized $\tilde{a}_i^l$ into consideration.

The proposed training framework is illustrated in Fig. 2. Each forward propagation process during training contains two parts. One part is that ANN uses the unquantized activation value a_i^l to perform the forward propagation and obtains the vanilla training loss: $loss_v$. The other part is that ANN uses the quantized $\tilde{a}_i^l = \lfloor \frac{a_i^l}{\lambda^l} T_q \rfloor \lambda^l / T_q$ to perform the forward propagation and obtains the quantized training loss: $loss_q$. T_q is a hyperparameter and determines the quantization resolution. The ANNs in these two parts have shared weights and biases. Then we multiply $loss_v$ and $loss_q$ with coefficients η and $(1 - \eta)$, respectively. The total training loss becomes:

$$loss = \eta \, loss_v + (1 - \eta) \, loss_q \qquad (8)$$

A larger η indicates that the training process targets to optimize the network to be as accurate as the vanilla ANN while a smaller η represents the training process tends to optimize the network to learn the quantization process. The update rules of the parameters in the backward propagation of the training process are formulated as:

$$p^l = p^l - \alpha\eta \frac{\partial loss_v}{\partial p^l} - \alpha(1 - \eta) \frac{\partial loss_q}{\partial p^l} \qquad (9)$$

where p^l is the weight or bias and α is the learning rate.

EXPERIMENTAL RESULTS

Implementation Details

We evaluate our proposed training method on CNET [8] and VGG-16 with the CIFAR-10 dataset. We follow the training strategy in [8]. The analog input is fed to the first layer of SNN, which is known as direct encoding. We compute the integrated MP of neurons in the output layer and take the index of maximum value to classify the input image. The coefficient η in Eq. (8) is empirically set to 0.5. All the experiments are implemented on NVIDIA GeForce 2080Ti GPUs with the PyTorch framework.

TABLE I. PERFORMANCE COMPARISON BETWEEN THE PROPOSED METHOD AND PREVIOUS WORK ON CIFAR-10 DATASET.

Network	Method	ANN Acc.	Accuracy of SNN with different values of T									
			2	4	8	16	24	32	40	48	56	64
CNET	TCL [8]	91.52	84.92	89.07	90.46	91.21	91.33	91.39	91.42	91.49	**91.51**	91.51
	Ours	91.52	88.97	90.94	**91.52**	91.52	91.52	91.52	91.52	91.52	91.52	91.52
VGG-16	TCL [8]	93.29	83.59	87.33	90.67	92.37	92.85	93.01	93.15	93.15	93.20	**93.23**
	Ours	93.29	88.03	90.74	92.37	93.21	**93.29**	93.29	93.29	93.29	93.29	93.29

The Effect of the Quantized Parameter T_q

We train ANNs with different quantized parameters T_q on the CIFAR-10 and convert these ANNs to SNNs. The dash lines in Fig. 3 represent the accuracy of ANNs. When T_q is set to 2^0, the accuracy of ANNs tends to become lower: 87.38% on CNET and 85.17% on VGG-16. As T_q increases to 2^1, 2^2, and 2^3, the accuracy of ANNs can achieve the same value as the vanilla ANN without quantization: 91.52% on CNET and 93.29% on VGG-16. Fig. 3 also illustrates how the accuracy of converted SNNs changes regarding T. When $T_q = 2^0$, the converted SNN with T=4 can achieve comparable accuracy as ANN. However, the low accuracy of ANN limits the accuracy of the converted SNN. When T_q is equal to 2^1, 2^2, and 2^3, the converted SNNs can reach comparable accuracy to the vanilla ANN as the value of T increases. At the same time, the accuracy of SNN drops as T_q increases. For both CNET and VGG-16, the converted SNN with $T_q = 2^1$ can achieve comparable accuracy to vanilla ANN with the optimal inference speed.

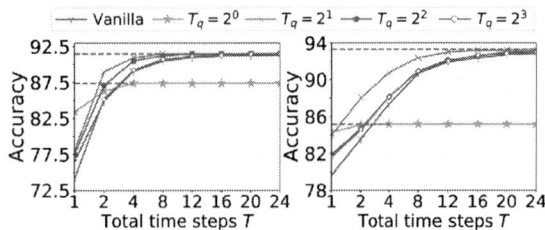

Figure 3: Comparison of the accuracy of SNNs with different T_q and T on CNET (left) and VGG-16 (right).

The Comparison with the Previous Method

We compare the results of our method with $T_q = 2^1$ to the method without quantization in [8]. Table I illustrates the accuracy of converted SNN with different values of T. Our hybrid training framework can improve the inference speed and reduce the accuracy loss of the converted SNN. For CNET, when T=56, the converted SNN in [8] achieves comparable accuracy as ANN with a conversion loss of 0.01%. However, with our hybrid training method, the accuracy of the converted SNN is the same as ANN with a smaller T=8. The proposed method can improve the inference speed of SNN by 7X on CNET. For VGG-16, the accuracy of the converted SNN in [8] only achieves 92.23% with a significant T=64 while the accuracy of the SNN converted by our training method reaches 93.29% with a smaller T=24. Our training framework can improve the inference speed of SNN by 2.67X on VGG-16.

CONCLUSION

In this paper, we focus on speeding up the inference process and reducing the accuracy loss of SNNs. We first analyze the conversion loss of SNNs. According to our analysis, we propose a hybrid training framework and apply it to the training process of ANNs. Experimental results show that our proposed technique can improve the inference speed of the converted SNNs by 7X on CNET and 2.67X on VGG-16 without accuracy loss.

REFERENCES

[1] LeCun, et al. "Deep learning." *Nature*, vol. 521, no. 7553, pp. 436–444, 2015.

[2] Akopyan, et al. "TrueNorth: Design and tool flow of a 65 mW 1 million neuron programmable neurosynaptic chip." *TCAD*, vol. 34, no. 10, pp. 1537–1557, 2015.

[3] Thiele, et al. "Event-based, timescale invariant unsupervised online deep learning with STDP." *Frontiers in computational neuroscience*, vol. 12, no. 46, pp. 1-13, 2018.

[4] Lee, et al. "Enabling spike-based backpropagation for training deep neural network architectures." *Frontiers in neuroscience*, vol. 14, no. 119, pp. 1-22, 2020.

[5] Cao, et al. "Spiking deep CNN for energy-efficient object recognition." *IJCV*, vol. 113, no. 1, pp. 54–66, 2015.

[6] Diehl, et al. "Fast-classifying, high-accuracy spiking deep networks through weight and threshold balancing." in *IJCNN*, pp. 1-8, 2015.

[7] Rueckauer, et al. "Conversion of continuous-valued deep networks to efficient event-driven networks for image classification." *Frontiers in neuroscience*, vol. 11, no. 682, pp. 1-12, 2017.

[8] Ho, et al, "TCL: an ANN-to-SNN conversion with trainable clipping layers." in *DAC*, pp. 793-798, 2021.

ATTENTION-BASED MECHANISM FOR TECHNOLOGY MAPPING OPTIMIZATION

Zhaohui Yang[1], Yinshui Xia[1]*, Mengke Wang[1], Chenghao Yang[1], and Xiaojing Zha[1]
[1]School of Ningbo University, Ningbo 315211, China
*Email: xiayinshui@nbu.edu.cn

ABSTRACT

Technology mapping is responsible for realizing logic functions by physics circuit units. In technology mapping, the search space of cuts is pruned mainly by heuristic algorithms. However, run-time efficiency can be estimated based on the number of cuts explored during technology mapping, and the number of cuts available per node is an exponential relationship between the graph size and the number of leaves, which leads to a trade-off between the number of cuts searched and the *Quality-of-Results* (QoRs). With this in mind, a method is proposed to transform the cut sorting problem into a classification problem using an attention mechanism. Enables better delay and area while considering fewer cuts in technology mapping. Adding attention mechanism to the classification problem can help us to learn feature representation better and pay attention to the needed information. Several tests have been conducted, and the experimental results show that compared with the state-of-the-art tool, the number of cuts used are reduced with the assistance of attention mechanism, and the average optimization of area and delay reaches 4% (up to 9%) and 5% (up to 24%) respectively.

INTRODUCTION

Logic synthesis in *Electronic Design Automation* (EDA) tools is to translate high-level descriptions into gate level netlists. It consists of three main phases: logic optimization, technology mapping, and gate-level optimization or post-mapping. Technology mapping is the process of converting technology-independent RTL circuit descriptions to those described using a specific technology library. The goal is to optimize delay and area.

Nowadays, the dominant approach for technology mapping is cut pruning heuristics [1]. These heuristics algorithms can effectively achieve good QoR with small runtime and memory usage. These algorithms need to be carefully developed by engineers with domain-specific knowledge based on extensive experimental analysis. And as the circuit size increases, the number of available cuts per node grows exponentially and it becomes very difficult to obtain good QoR. Neto W L [2] used CNN for the first time to optimize the technology mapping, eliminating the pruning search process. However, this approach cannot focus on the important parts of the cut.

With the emergence of attention mechanism, deep learning can discover connections between data. Deeping

learning model can extract more critical information and make more accurate judgments with the help of attention mechanism. In this work, the choice of cut problem is transformed into a multi-classification problem. Our goal is to train a deep learning network with an attention mechanism to output more accurate classification results. Experiments show that the method can optimize delay and area while exploring a small number of cuts.

BACKGROUND
Boolean Networks

Boolean network is a directed acyclic graph (DAG), denoted as $G = (N, E)$. Common types of DAGs for logic manipulation include *And-Inverter Graphs* (AIGs) and *Majority-Inverter Graphs* (MIGs) [3]. An AIG consists of two-input nodes representing logical conjunction, each node represents an AND gate. Nodes with no incoming edge are the *Primary inputs* (PIs). The solid line represents buffer, and the dotted line represents inverter. A simple AIG is shown in figure 1.

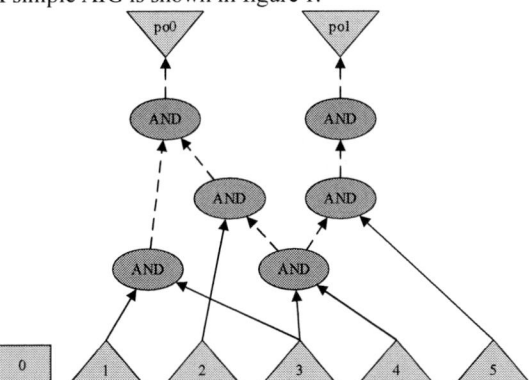

Figure 1: AIG of C17 circuit, 1,2,3,4,5 as input, po0 and po1 as output

Cut-based Technology

A cut c of a root node n is a pair (n, L), where L is the set of cut's leaves, such that any path from a PI to n passes through at least one leaf. A trivial cut of n is made up of itself. Therefore, PI has only trivial cuts, while each logical node has at least one trivial cut. A nontrivial cut, on the other hand, covers all nodes along the path from the root node to the leaf node, both root and non-leaf. A cut set is said to be k-feasible if it has no more than k leaf nodes. If all nodes of a cut set are contained within another cut set [4], we say the cut set is covered.

979-8-3503-1101-3/23 $31.00 © 2023 IEEE

Attention Mechanism

The attention mechanism is inspired by human biological systems, which tend to focus on different parts when large amounts of information is processed. People tend to selectively focus on one piece of information when and where they need it, while other perceived information is ignored. For example, when looking at a computer, people focus on the screen and ignore the keyboard and mouse. In deep learning, the function of attention mechanism is to focus attention on a part or a channel of a certain piece of input (image, text, etc.) to guide the reasoning process. Two kinds of attention mechanism have been tried in this work. One is channel attention mechanism [5], another is the combination of channel and spatial attention mechanism [6]. The latter, named CBAM, an effective feedforward convolutional neural network attention module, is used. Given an intermediate feature map, the CBAM module will infer the attention map along two independent dimensions (channel and space) in turn, and then the attention map is multiplied with the input feature map for adaptive feature optimization.

METHOD

Distribution of metrology data

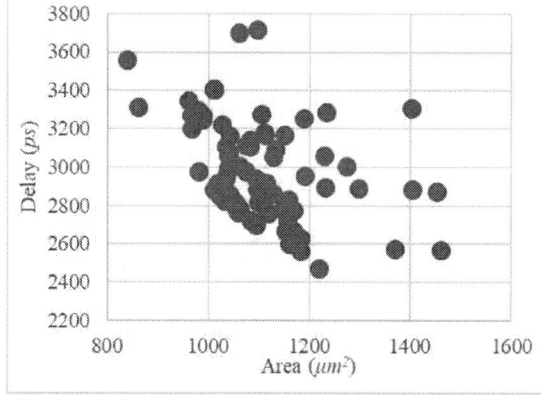

Figure 2:The solution space of the ADDER circuit is shown. Each data point represents a mapping result, the x axis represents the area size, and the y axis represents the delay size.

We explored the design space by adjusting the list of cuts on each node. ASAP 7nm PDK [7], an open-source standard cell library, is used in the technology mapping. The result of the mapping of the ADDER circuit is shown in Figure 2. The x axis represents the area, while the y axis represents the delay. Each data point represents the result of a mapping. Area and delay can be obtained by using ABC *stime* command [8]. As we can see, even on small circuits, the solution space is large.

Schematic of Algorithm

The framework of model training is based on reinforcement learning. We used ABC to generate different mappings, normalize them according to the product of delay and area, and label the cut by each mapping according to the normalized result. We modeled the problem as a classification problem. The information of each cut, such as level, number of nodes and number of leaf nodes, is input into the neural network. The convolutional layer is used to extract the features in the input. CBAM attention mechanism is added after the convolutional layer, and multiple dropout layer is used to prevent over-fitting and improve the generalization of the network. The neural network architecture used in this article is shown in Figure 3. The architecture of the attention mechanism used is described in Figure 4. We used the full connection layer to divide the results into 10 categories. Class 0 represents the cut that will improve the circuit performance, while Class 9 represents the cut that will reduce the circuit performance. We give the first 6 categories of classification results to ABC for mapping.

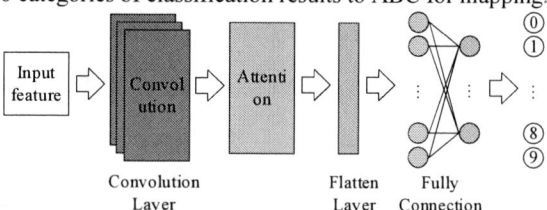

Figure 3: Shows the architecture of the deep learning network we use, and finally outputs 10 classes

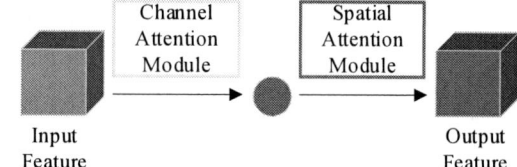

Figure 4: The overview of CBAM. The module has two sequential sub-modules: channel and spatial.

EXPERIMENTAL RESULT

Our approach can replace the cut needed to be considered in the ABC map command. Pytorch is used to train the deep learning model framework and 7nm standard cell library is used for technology mapping. We selected circuits from EPFL benchmark suite [9] and ISCAS '85 [10] benchmark suite and AES core. We tested a total of 13 circuits, we compared the proposed approach with the ABC and the SLAP: A supervised learning approach for priority cuts technology mapping [2], as shown in Table 1. Notice that we recured the SLAP to compare the circuits that did not appear in that paper. Compared with traditional heuristic implemented in ABC, on average we optimized 4% (up to 9%) in area and 5% (up to 24%) in delay. Compared to SLAP, we observed an average improvement in delay by 4% (up to

TABLE I. Shows our work in comparison with SLAP and ABC, with better results than both, marked in bold

Circuit	ABC [8]			SLAP [2]			This Work		
	Area (μm^2)	Delay (ps)	Cuts Used	Area (μm^2)	Delay (ps)	Cuts Used	Area (μm^2)	Delay (ps)	Cuts Used
adder	898.13	3770.65	10118	1031.33	3268.67	13522	980.71	**2879.71**	8935
AES	17321.27	638.09	264380	16489.63	594.64	145532	**16239.79**	610.96	183025
bar	2680.39	1114.90	42760	3083.23	923.82	10705	**2447.57**	**888.71**	9220
c3540	940.35	797.06	17161	956.47	777.73	14225	**903.88**	**744.14**	12220
c5315	1391.05	550.73	27586	1426.35	588.57	17416	**1354.89**	557.96	13544
c6288	3000.45	1265.88	95519	3023.54	1236.59	111609	**2781.40**	**1228.94**	96427
c7552	2045.00	817.46	52150	2002.01	800.09	30933	**1925.03**	**694.53**	20747
log2	26561.26	6797.77	1114075	27200.76	7238.41	798329	**25385.53**	6944.97	691055
max	2312.27	3809.03	39727	2292.44	3710.54	22521	**2175.51**	3739.75	21766
multiplier	25458.31	4649.10	833565	24021.07	4278.48	892650	**24591.10**	**4125.31**	524955
sin	5207.04	3955.57	191131	5087.60	3584.79	141788	**4969.18**	**3449.06**	106787
sqrt	20252.20	180518	591044	22353.10	180719	293746	20379.87	**172477**	160791
square	15744.07	3680.87	541321	15789.09	3023.01	380787	**14931.69**	**3019.02**	338903
Sum	123811	212365	3820537	124756	210744	2873763	**119066**	**201360**	2188375
Ratio	1	1	1	1.01	0.99	0.75	**0.96**	**0.95**	**0.57**

21%). Area is improved on average by 5% (up to 13%).

CONCLUSION

We apply deep learning with attention mechanism to optimize technology mapping. By testing 13 circuits and comparing them with ABC and SLAP, the proposed method optimizes area and delay, when cut used is reduced.

REFERENCES

[1] S. Chatterjee, A. Mishchenko, R. K. Brayton, X. Wang, and T. Kam, "Reducing structural bias in technology mapping," IEEE TCAD, vol. 25, no. 12, pp. 2894–2903, 2006.

[2] Neto W L, Moreira M T, Li Y, et al. SLAP: A Supervised Learning Approach for Priority Cuts Technology Mapping[C]//2021 58th ACM/IEEE Design Automation Conference (DAC). IEEE, 2021: 859-864.

[3] L. Amaru, P.-E. Gaillardon, and G. De Micheli, "Majority-inverter graph: A new paradigm for logic optimization," IEEE TCAD, vol. 35, no. 5, pp. 806–819, 2016.

[4] A. Mishchenko, S. Chatterjee, and R. K. Brayton, "Improvements to technology mapping for LUT-based FPGAs," IEEE Transactions on Computer-Aided Design of Integrated Circuits and Systems, vol.

26, no. 2, pp. 240–253, 2007.

[5] Hu J, Shen L, Sun G. Squeeze-and-excitation networks[C]//Proceedings of the IEEE conference on computer vision and pattern recognition. 2018: 7132-7141.

[6] Woo S, Park J, Lee J Y, et al. Cbam: Convolutional block attention module[C]//Proceedings of the European conference on computer vision (ECCV). 2018: 3-19.

[8] Brayton R, Mishchenko A. ABC: An academic industrial-strength verification tool[C]//International Conference on Computer Aided Verification. Springer, Berlin, Heidelberg, 2010: 24-40.

[7] L. Clark, V. Vashishtha, L. Shifren, A. Gujja, S. Sinha, B. Cline, C. Ramamurthy, and G. Yeric, "Asap7: A 7-nm finfet predictive process design kit," Microelectronics, vol. 53, pp. 105–115, 7 2016.

[9] Amarú L, Gaillardon P E, De Micheli G. The EPFL combinational benchmark suite[C]//Proceedings of the 24th International Workshop on Logic & Synthesis (IWLS). 2015 (CONF).

[10] Hansen M C, Yalcin H, Hayes J P. Unveiling the ISCAS-85 benchmarks: A case study in reverse engineering[J]. IEEE Design & Test of Computers, 1999, 16(3): 72-80.

AN EFFICIENT ATPG TECHNOLOGY BASED ON TIME DIVISION MULTIPLEXING METHOD

Minqiang Peng[1,2], Keqing Ouyang[1,2], Lunmao Zhou[1,2], Guohua Zhou[1,2]

[1]State Key Laboratory of Mobile Network and Mobile Multimedia Technology

[2]Sanechips Technology Co., Ltd

Shenzhen, Guangdong, P. R. China

*Corresponding Author's Email: ouyangkeqing@sanechips.com.cn

ABSTRACT

As IC complexity increases, test is emerging as large expenses in VLSI Circuits manufacturing, which affects the efficiency of chip shipments. In the traditional ATPG (Automatic Test Pattern Generation) process, the test efficiency of each pattern is related to the parallelism of the clocks at the capture stage while only the clocks without cross-talk that are considered to be the synchronous clock can be pulsed at the same time in the ATPG tool. It is obvious that the efficiency of the test pattern will reach the highest when all clocks work together in one pattern, which meas the parallelism reaches the maximum. In this paper, an efficient ATPG method named Capture Clock Time Division Mutiplexing Pseudosynchronization Control is presented, which can improve the working parallelism of test clocks in a single pattern and significantly reduce the test application time and test data volume.

INTRODUCTION

The test methods of digital circuits can be mainly divided into four categories: SCAN, MBIST, Boundary-SCAN and function testing, while SCAN test can account for 70% or more of the entire chip test time according to the circuit structure characteristics of the chip[1]. The DFT SCAN test technique is used to detect whether the logical functions of chip is correct. In the DFT design, the registers are serialized into the scan chains, and the patterns are automatically generated using ATPG tool. One effective way to reduce the number of SCAN pattern and test time is to improve ATPG efficiency.

In the traditional ATPG process,the test efficiency of each pattern is related to the parallelism of clocks in the Capture stage. A well-known conclusion is that the clock parallelism will reach its maximum and meanwhile the test efficiency of the pattern will reach the highest when all clocks can be toggled in one pattern. In traditional ATPG tools, only the clocks (at least two) without an interactive path or cross-talk are considered to be the synchronous clocks, so that parallel testing can be carried out. However, in complex circuits, the relationships and the interactions of each clock are complicated, resulting in a sharp increase in the number of patterns and a low test efficiency in SCAN testing.

During SCAN ATPG, the clocks are typically controlled by specifying the atpg tool to generate patterns based on the clock domain, that is, only one capture clock will be pulsed within a cycle[2]. Taking the module in Figure 1 as an example, it has three clocks: CLKA, CLKB and CLKC, and there are interaction paths between each two clocks. In the default ATPG behavior, at least three types of patterns are required to enable all the three clocks to toggle, because these clocks are incompatible to each other in ATPG and cannot be toggled in the same pattern, which is a serial pattern clock control behavior as shown in Figure 2. The main reason is that timing check is not usually done between two different clock domains. If any two clocks toggle at the same time (a parallel pattern clock control behavior), data transfer will occur between these two clocks. However, this data transfer is not guaranteed by timing and will result in a logic error.

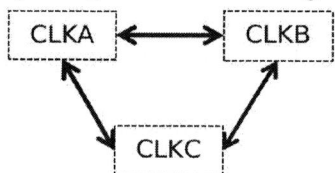

Figure 1. Diagram of clocks relationship in a module

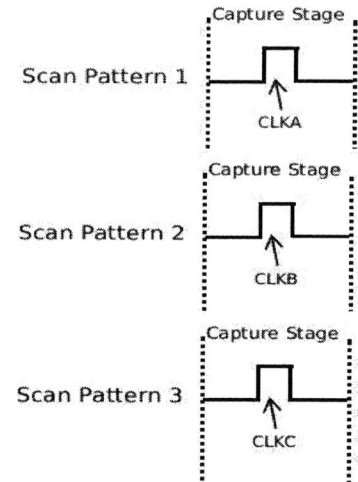

Figure 2. Diagram of different patterns and corresponding Capture clocks at Capture Stage

IMPLEMENTATION SCHEME

Generally, hierarchical DFT design solution is used for ultra-large-scale chips, and usually the number of clocks are large and clock interaction is complicated in the design[3]. This paper proposes an efficient ATPG method, which can improve the working parallelism of test clocks in a single pattern by time division multiplexing method. The implementation scheme includes an OCC (On-Chip Clock Controller) circuit with controllable delay and a matching working method.The OCC that can freely control the output delay of capture clock is proposed. As shown in Figure 3, the OCC circuit includes at least three inputs: 1)OCC Capture Enable, 2)Capture Clock,and 3)Shift Clock. The OCC Clock MUX unit is configured to switch between the OCC Shift Clock and the OCC Capture Clock. The OCC Clock Gating is used to control the Capture Clock pulse of the OCC output. The delay control logic is a component inserted into the input and output of the OCC to control the delay time. Therefore, when receiving the OCC Capture signal, the OCC output clock port can output the Capture Clock normally after a controllable time delay.

The working principle of this scheme to improve the parallelism of Caputre clocks is as follows: during the ATPG process, multiple Capture clocks which are controlled by OCC circuit with controllable delay can be toggled in the same pattern and avoid logical data sampling errors caused by timing problems in the meanwhile. During Capture stage, different clocks are controlled to be pulsed in diferent phases in turn and the delay between different phases is large enough to meet the maxdelay constraint between clocks. The data transfer between any two clocks can be regarded as normal because it can be considered that these OCC clocks work synchronously with each other in the ATPG process. This method is named as Capture Clock Time Division Mutiplexing Pseudosynchronization Control, as shown in Figure 4.

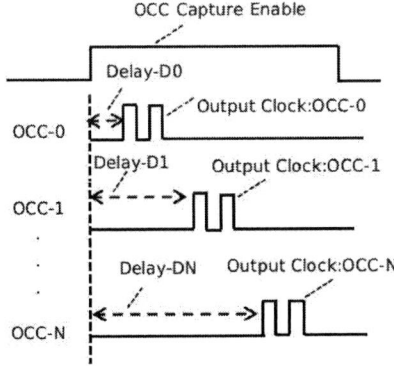

Figure 4. Operational timing for controllable delay of OCC

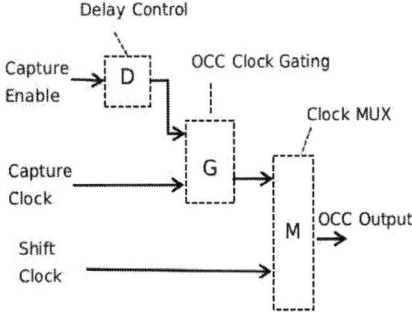

Figure 3. OCC Circuit with Controlled Delay

Furthermore, the implementation method steps are as follows:

a) In physical implementation, a unified max delay constraint Dmax is added between capture clocks of OCCs, that is, data transmission between capture clocks of any OCCs does not exceed Dmax.

b) Configure the capture clocks of different OCCs(OCC-0,OCC-1,...OCC-N), add different delays (D0, D1,...DN), and ensure that the delay control value between any two OCC clocks is greater than Dmax. The expression is as follows:

$|Dx-Dy| > Dmax$

In this case, data sampling between different OCCs is guaranteed by timing.

c) During ATPG process, different OCC clocks are defined to be synchronized.

d) During ATPG process, different OCC clocks are defined to correspond to different pulse cycle. For the normal two pulse transition ATPG, the pulse cycle is described as follows[4]:

OCC-0 pulse cycle: Cycle 0&cycle 1

OCC-1 pulse cycle: Cycle 2&cycle 3

....

OCC-N pulse cycle: Cycle 2N&cycle 2N +1

For a normal single pulse stuck at ATPG, the pulse cycle is described as follows:

OCC-0 pulse cycle: Cycle 0

OCC-1 pulse cycle: Cycle 1

....

OCC-N pulse cycle: Cycle N

This design above adopts CCD (Clock Control Definition) construction method to increase the definition of ATPG cycle of each Capture clocks, so as to achieve multi-capture clocks to be pulsed by stage in the same pattern,as shown in Figure 5.

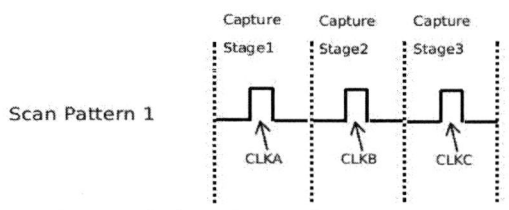

Figure 5.Capture clocks at Capture Stage in a pattern

In addition, the OCC circuit can implement toggle control in stages, and each OCC clock in which stage toggle can be configured by the jtag. Moreover, the wrapper chain design of the large-scale chip is required, so that under intest mode, only the Capture Clock Time Division Multiplexing Pseudosynchronization Control within each block needs to be implemented separately, and the Capture Clock Pseudosynchronization Control of the whole chip can be realized.

RESULTS AND DISCUSSION

Compared with the traditional atpg method, the verification result shows that the quantity of patterns can be reduced by 30% and the runtime of ATPG can be reduced by 26% under the same coverage. What's more, the more multiple clock interaction paths,the higher efficiency of the test pattern by using the above solution, as shown in TABLE I.

TABLE I. ATPG RESULT FOR 3 BLOCKS WITH 3 OCC CLOCK DOMAINS

Block	Instance	Traditional ATPG		Improved ATPG	
		Pattern Count	Run Time (s)	Pattern Count	Run Time (s)
A	1223201	1844	1525	1290	1128
B	2539899	6461	3341	4022	2270
C	3823254	9256	7230	5458	5044

CONCLUSION

In this paper, we propose a scheme that multiple capture clocks with cross-talk can be pulsed by stage in one pattern with the Time Division Multiplexing Pseudosynchronization Control method in the very large-scale chip intest atpg, so that most or even all clocks in the chip can be synchronized, which significantly improves the test efficiency of each pattern and reduces the total number of scan patterns.

ACKNOWLEDGEMENTS

This work is supported by the Backend Design Dept of Sanechips Technology Co., Ltd.

REFERENCES

[1] Yang, B. , et al., (2009). Research and design of soc dft. Electronics & Packaging.
[2] Waayers, et al,. (2010). Clock control architecture and ATPG for reducing pattern count in SoC designs with multiple clock domains. IEEE International Test Conference. IEEE.
[3] Remmers, et al., (2004). Hierarchical DFT Methodology - A Case Study. International Test Conference. IEEE.
[4] Mentor Graphics Corporation, Siemens Industry Software Inc., Tessent Scan and ATPG User's Manual,2020.3.

Learning-Based Performance and Power Model for Processor Microsecond DVFS

Yingtao Shen, An Zou

UM-SJTU JI, Shanghai Jiao Tong University, 200240, China

{doctorcoal,an.zou}@sjtu.edu.cn

Abstract

With the maturity of integrated voltage regulators, the processor dynamic voltage and frequency scaling (DVFS) rate can reach up to microsecond timescales. However, the state-of-the-art DVFS method, such as learning-based DVFS, lacks interpretability and cannot be directly deployed to microsecond timescales. Therefore, in this work, we propose a fine-grain per-core DVFS modeling approach that leverages information from the processor runtime states at the microsecond level to establish a practical optimum performance loss model. Experiments across different benchmarks demonstrate that the proposed model can achieve up to 21.8% energy saving with 5.8% performance loss.

Keywords—DVFS; energy-efficiency; fine-grain; processor; performance prediction

Introduction

Microprocessors have developed during the past two decades to reach better computing performance, and as a result, their power density has also improved at an accelerated rate. Therefore, energy efficiency and energy saving have become major concerns in modern microprocessors. The well-known per-core Dynamic Voltage and Frequency Scaling (DVFS) is widely used to optimize power and energy consumption in multi-core systems. Benefiting from integrated voltage regulators [1-2], microsecond timescales DVFS becomes practical in recent years. However, the existing state-of-art online fine-granularity (μs level) DVFS method based on reinforcement learning proposed in [3] and other complex modeling method lacks interpretability and is challenging to assess for optimality, which restricts its further optimization and generalization. Meanwhile the complexity in the DVFS strategies limits the practical performance of those methods especially at microsecond time scales.

In this paper, we propose a fine-grain per-core DVFS modeling approach that takes full advantage of information from microsecond level transient application and processor runtime states to establish a practical optimum performance loss prediction model at microsecond timescales.

Methodology

A. Modeling Platform

To establish the model for processor microsecond performance prediction and power management, we build a platform with the CPU architecture-level performance and power simulator *Sniper* v7.3 [4] (with Mcpat [5]) to construct a dataset with instruction count (or execution time) calibration point. The platform is designed to ensure that the obtained dataset can contain the most exact information about the possible performance loss of any DVFS operation at any running time point in a computationally feasible way under the simulator accuracy limitation. The first offline batch lets the benchmark run at each fixed frequency/voltage pair to generate the Fixed-VF dataset. The second offline batch will make the voltage and frequency of some specific cores change to the highest/lowest level at each certain run time point—thousands of offline running with different DVFS configurations form the Well-VFS dataset. Detailed processing of two datasets enables us to extract the performance loss and corresponding system runtime states in fine granularities. The proposed platform finds out and assesses the practical modeling optimal solution for an offline pre-trained model, allowing us to select and configure the modeling method so that we can close to this optimal solution as much as we want while considering model complexity.

B. Modeling Method

With framework and dataset, we perform learning-based regression to predict performance loss of specific V/F pair at some runtime middle point and be further classified for an optimal V/F pair at this runtime condition. The methodology can fit with various regression/classification methods to produce a model which takes the transient (and/or past) performance counter values in each core as input and output corresponding optimal per-core V/F pairs. In this work, we adopt two simple and typical methods, SVM (classification method) and Linear Regression (regression method), as the examination target of our methodology, considering a microsecond level overhead.

Evaluation

We evaluate the proposed modeling approach with experiments on an Intel Nehalem x86 processor with four core: [0, 1, 2, 3], four possible frequencies that can be adjusted per-core: [2000, 1800, 1600, 1400] (MHz), and the corresponding voltages: [1.20, 1.08, 0.96, 0.84] (V). The processor performance and power is simulated with Sniper v7.3 (with Mcpat).

A. Benchmark Characteristic

Our evaluated benchmarks are selected from Splash2, parsec, and NPB benchmark sets. However, in the current stage only *fft, radix, lu.cont,* and *ocean.cont* in Splash2 are heavily tested. Fig.1 demonstrates the performance property of *lu.cont* when run on 0 core and 2 core. The x-axis represents the shorten (say, 100000ns) performance penalty when applying DVFS to reduce the frequency and voltage to the lowest level. Y-axis represents the number of DVFS position in each performance penalty interval.

979-8-3503-1101-3/23 $31.00 © 2023 IEEE

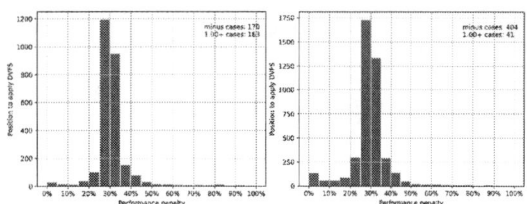

Fig. 1 core 0 (left) and core 2 (right) performance penalty of *lu.cont* at different DVFS point

For page limitation, we don't put the figure of core 1 and 3 here, but the situation of core 1 and 3 is similar to core 2. Evaluation of other benchmarks reveals the same fact: the chance to apply DVFS with a low penalty (within 10% or 20%) in core 0 is less than in other core.

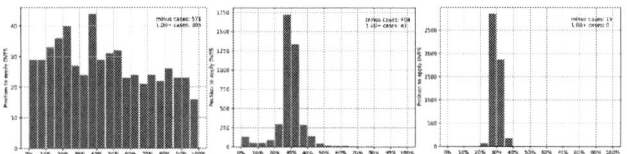

Fig. 2 2 core performance penalty of fft(left), lu.cont(middle) and radix(right) at different DVFS point

Fig. 2 here is to demonstrate the DVFS chance on core 2 of different benchmarks, which shows that the CPU/memory-bound condition of different benchmarks varies a lot and the DVFS strategy for three benchmarks should able to handle those variation.

B. Energy Efficiency of SVM and LR Trained on Our Platform

TABLE I
ENERGY EFFICIENCY OF SVM AND LR

method	model	benchmark	total dE(%)	total dT(%)
single startidx (lu.cont)	LR	Lu.cont	-26.08	10.38
multi startidx (lu.cont)	svm	Lu.cont	-21.78	5.82
	LR	Lu.cont	-24.59	8.25
multi startidx (radix)	svm	Radix	-19.61	10.94
	LR	Radix	-17.89	13.72
multi startidx (lu.cont + radix + fft)	svm	Lu.cont	-29.08	15.88
		Radix	-30.00	24.93
	LR	Lu.cont	-24.38	9.38
		Radix	-30.62	25.16
multi startidx (lu.cont+radix+fft+ocean.cont)	svm	Lu.cont	-29.76	16.01
		Radix	-30.41	24.98
	LR	Lu.cont	-37.92	20.36
		Radix	-30.95	25.18

The above table summarizes part of our evaluation results. "single startidx" means the double offline data batches fix their voltage and frequency at the highest level when not applying DVFS (default V/F pair); while "multi startidx" collects the dataset of each default V/F pair to do different prediction when application run at different V/F level. Because of the findings in Fig. 1 part, no single 0 core cases are evaluated. "Benchmark" means the online evaluation benchmark, which is different

with the training benchmark in the bracket of "method" part. "total dE" is the total static and dynamic energy change; "total dT" represent the total execution time change, which is to verify performance change.

According to TABLE I: experiments on different benchmarks from Splash2, parsec, and NPB benchmark sets demonstrate that our trained Linear Regression model can achieve up to 24.59% energy saving while the performance sacrifice is within 8.25% in a single benchmark evaluation; and the trained SVM model can achieve up to 21.78% energy saving in 5.82% performance loss. In mixture (three) benchmark evaluation, our Linear regression model also saves 24.38% energy within 9.38% performance loss.

Conclusion & Significance

This work develops a well-designed online power-performance model from microsecond DVFS which can automatically predict performance verification within different V/F pairs of each core, with a relatively low time and energy overhead. The contributions of this work are summarized as:

1. The classification is the performance loss and the model follows the assumptions of
   a) linear performance loss & frequency reduction relationship in specific running points.
   b) similar power reduction in the same level of frequency scaling.
   We take the power model into consideration and assess the power reduction together with performance loss. Also, a non-linear model of performance-frequency is constructed together with our performance model.

2. With the dataset, the modeling of SVM and linear regression follows the simplest approach without feature selection or PCA. More modeling methods are examined with a better configuration to enable the best modeling performance.

3. A multi-layer online per-core framework is further proposed based on the property of our optimal offline modeling solution.

4. The proposed modeling framework is synthesized to evaluate its performance in practice.

References

[1] Zou, An, Jingwen Leng, Yazhou Zu, Tao Tong, Vijay Janapa Reddi, David Brooks, Gu-Yeon Wei, and Xuan Zhang. "Ivory: Early-stage design space exploration tool for integrated voltage regulators." *In Proceedings of the 54th Annual Design Automation Conference* 2017, pp. 1-6. 2017.

[2] Zou, An, Jingwen Leng, Xin He, Yazhou Zu, Christopher D. Gill, Vijay Janapa Reddi, and Xuan Zhang. "Voltage-stacked power delivery systems: Reliability, efficiency, and power management." *IEEE Transactions on Computer-Aided Design of Integrated Circuits and Systems* 39, no. 12 (2020): 5142-5155.

[3] A. Zou, K. Garimella, B. Lee, C. Gill and X. Zhang, "F-LEMMA: Fast Learning-based Energy Management for Multi-/Many-core Processors," *2020 ACM/IEEE 2nd Workshop on Machine Learning for CAD*, doi: 10.1145/3380446.3430630.

[4] Trevor E Carlson, Wim Heirman, and Lieven Eeckhout. Sniper: Exploring the level of abstraction for scalable and accurate parallel multi-core simulation. *In SC, 2011.*

[5] Sheng Li, Jung Ho Ahn, Richard D Strong, Jay B Brockman, Dean M Tullsen, and Norman P Jouppi. Mcpat: an integrated power, area, and timing modeling framework for multicore and manycore architectures. *In MICRO, 2009.*

A HIGH-SENSITIVITY AND LARGE-DYNAMIC RANGE READOUT CIRCUIT FOR POLYSILICON-BASED MICROBOLOMETER

Wei Zhu[1], Ke Wang[1], Yaozu Guo[1], Sheng Xu[1], Feng Yan[1], Yiming Liao[2]* and Xiaoli Ji[1]

[1]School of the Electronic Science and Engineering, Nanjing University

[2]School of Electronic and Optical Engineering, Nanjing University of Science and Technology

*Corresponding Author's Email:liaoyiming@njust.edu.cn

ABSTRACT

In this paper, a high-sensitivity and large-dynamic range readout integrated circuit (ROIC) is designed for polysilicon microbolometer based on CMOS process. The ROIC employs a calibration-circuit which contains a correlated double sampling (CDS) circuit to suppress the pixel fixed pattern noise (FPN) and the offset voltage of the capacitive trans-impedance amplifier (CTIA) circuit. In addition, an automatic gain control (AGC) circuit is designed to further extend the dynamic range of the microbolometer. Results show that the ROIC can suppress 99.98% of the pixel FPN when the non-uniformity of the microbolometer is 0.2% and at least 99.45% of the offset voltage of the CTIA, while increasing the dynamic range by 20 dB due to the addition of the AGC.

INTRODUCTION

Compared with the microbolometer based on MEMS technology, the polysilicon microbolometer based on standard CMOS process has the advantages of low noise equivalent power, low cost, and high integration [1]. However, its infrared responsivity is lower than VOx and α-Si microbolometers so that the detected signal current is weak. In addition, by the deviation in the process, the resistances of different pixels are not exactly equal, resulting in pixel fixed pattern noise (FPN), which makes the microbolometer responses in the absence of infrared radiation. The pixel FPN not only limits the sensitivity, but also occupies part of the dynamic range. L. X. Zhang used an additional transistor to generate current to eliminate the current generated by pixel FPN [2], but the bias voltage of the transistor needs to be manually adjusted. S. Xu used a self-calibrating circuit to automatically suppress the pixel FPN [3], but the offset voltage of the CTIA was not considered, so that the voltage after calibration may actually be the result of including the offset voltage.

In this paper, a ROIC with a calibration-circuit is used to suppress both offset voltage of the CTIA and the pixel FPN. At the same time, an automatic gain control (AGC) circuit is added to the ROIC to extend the dynamic range.

DESIGN AND RESULTS

The calibration-circuit

Fig 1.a shows the traditional ROIC for microbolometer. One microbolometer pixel consists of a detection pixel R_s and a blind pixel R_b, and the bias voltages of the two pixels are equal.

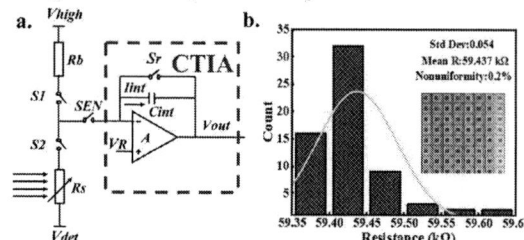

Fig 1.a: Traditional ROIC for one pixel of microbolometer
b: The frequency distribution histograms of the resistances

When R_s is exposed to infrared radiation, the microbolometer generates a current I_{int} and flows into C_{int} to integral. Assuming the integration time $t = t_2 - t_1$, V_R is the reference voltage, the output voltage can be shown as:

$$V_{out} = V_R - \frac{1}{C_{int}} \int_{t_1}^{t_2} I_{int}\, dt = V_R - \frac{I_{int}t}{C_{int}} \quad (1)$$

When the microbolometer is not exposed to infrared radiation, the R_s and R_b should be equal in resistance and the I_{int} is 0. But due to the process deviation of the pixel in the standard CMOS process, it is difficult to achieve the equal resistance, which leads to the pixel FPN. Fig 1.b shows the frequency distribution histograms of the resistances of the 8*8 polysilicon microbolometer. It can be seen that most of the resistances are distributed in the range of 59.40~59.65 kΩ, the average resistance is 59.437 kΩ and the non-uniformity is 0.2%. The pixel process deviation R_{FPN} will cause the pixel FPN current ΔI to:

$$\Delta I = \frac{V_{high} - V_R}{R_b} - \frac{V_R - V_{det}}{R_s + R_{FPN}} \quad (2)$$

Simulate the ROIC by setting R_b to 59.437 kΩ and R_s to 59.556 kΩ according to the measured pixel process deviation, V_R is set to 2.2 V, C_{int} is 5 pF and t is 100 μs. When the bias voltages of R_b and R_s are 1.5 V, the result of the output voltages is depicted in Fig 2.

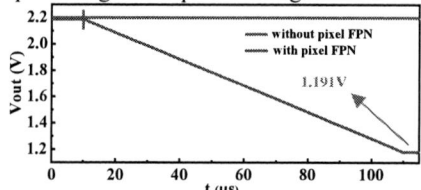

Fig 2: Output voltage curves of the traditional ROIC superimposed with pixel FPN and without pixel FPN

What we can see is that when the non-uniformity of the pixel is 0.2%, the pixel FPN voltage of the traditional ROIC for one pixel of polysilicon microbolometer is almost 1.009 V, which occupies 45.86% of the output dynamic range due to the reference voltage V_R of 2.2 V.

In fact, besides the pixel FPN, due to the uncertainty of the CMOS process, the actual size of the MOSFET in the CTIA differs from the designed size, which may generate the offset voltage V_{os} of the CTIA [4]. The V_{os} equivalents to superimposing on reference voltage V_R at the mV level, which must be suppressed to ensure the reset voltage of the ROIC is the reference voltage V_R we set.

In this paper, we design a calibration-circuit which contains a CDS circuit to suppress the pixel FPN and the offset voltage of the CTIA, as shown in Fig 3.

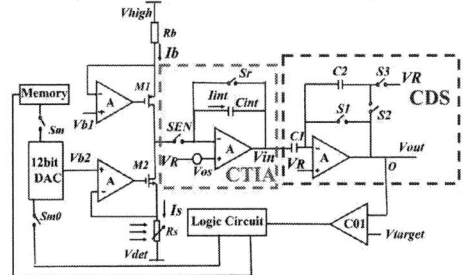

Fig 3: The calibration-circuit

The working process of CDS to suppress the offset voltage is as follows:

In the first stage of CDS, Sr and S2 turn off, S1 and S3 turn on. CTIA is integrating and the V_{in1} of CDS is:

$$V_{in1} = V_R + V_{os1} - \frac{I_{int}t}{C_{int}} \quad (3)$$

The charge on C_1 is $C_1(V_R + V_{os2} - V_{in1})$ and on C_2 is $C_2 V_{os2}$. V_{os2} is the offset voltage of the CDS.

In the second stage of CDS, Sr and S2 turn on, S1 and S3 turn off. The V_{in2} of CDS is the reset voltage of CTIA:

$$V_{in2} = V_R + V_{os1} \quad (4)$$

The charge on C_1 and C_2 turn to $C_1(V_R + V_{os2} - V_{in2})$ and $C_2(V_R + V_{os2} - V_{OUT})$ respectively.

Let the capacitor $C_1 = C_2$, according to the total charge constants in two stages stored in C_1 and C_2, the accurate output voltage is obtained:

$$V_{out} = V_R - \frac{C_1}{C_2}(V_{in1} - V_{in2}) = V_R - \frac{I_{int}}{C_{int}} \quad (5)$$

Assuming the signal current is 10 nA, set the working conditions of the CTIA to be the same as those in Fig 2 to test the CDS for its ability to suppress the offset voltage under the condition of 0 to 10 mV. Output voltage curves of the CTIA and the CDS are shown in Fig 4. It is easy to observe that the output voltage of the CTIA increases linearly as the offset voltage increases, but the output voltage of the CDS remains almost constant. When the offset voltage ranges from 0 to 10 mV, the CDS can suppress at most 99.9% and at least 99.45% of them.

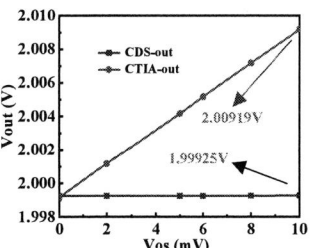

Fig 4: Output voltage curves of CTIA and CDS with Vos

After the reset voltage of the ROIC almost equals to the set reference voltage V_R due to the elimination of the offset voltage, we order a comparator C01, a 12-bit DAC, a Logic Circuit and a Memory to suppress the pixel FPN by limiting the FPN current ΔI to 0. The details of the working process are as followed:

Set V_{target} to the reference voltage V_R of the CTIA and fix the V_{b1}, the output of the logic circuit is first set to "100000000000" and input it to the DAC, the DAC converts the digital signal into a bias voltage ($V_{b2} - V_{det}$) of R_b, C01 compares the output voltage of CDS with V_{target}. Based on the comparison result, the logic circuit increases or decreases its digital output and controls the DAC to generate various voltage V_{b2} to produce different currents for counteracting ΔI, so that the output voltage of CDS can converge to V_{target}. A 12-bit DAC as shown in Fig 5 is used to make the calibration accuracy meet the requirements, which adopts the method of a 4-bit voltage scaling DAC and a 8-bit charge scaling DAC of combination. After calibration, the digital signal of the Logic Circuit is stored in Memory for direct use afterwards.

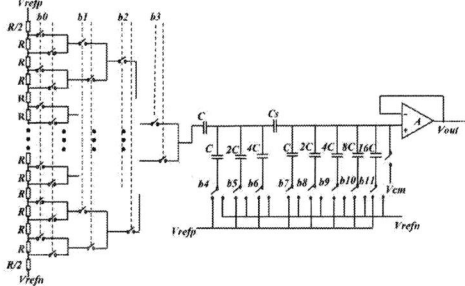

Fig 5: The 12-bit DAC

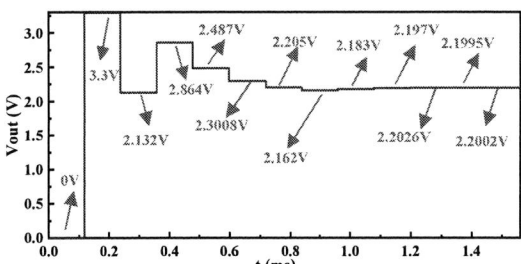

Fig 6: Output voltage curve of the calibration-circuit

Simulate the calibration-circuit under the same working conditions as in Fig 2, the result shown in Fig 6 indicates that the pixel FPN voltage can be limited to 0.0002 V from 1.009 V after calibrated, which means the calibration-circuit can suppress 99.98% of the pixel FPN when the non-uniformity of the microbolometer is 0.2%.

The automatic gain control circuit

To obtain high sensitivity of the microbolometer, the traditional ROIC increases the conversion gain by reducing the integration capacitor. However, small capacitor tends to saturate, which leads to the small dynamic range of ROIC.

The ROIC proposed in this paper adds an automatic gain control circuit (AGC) to increase dynamic range while keeping the sensitivity constant. The complete ROIC is shown in Fig 7.

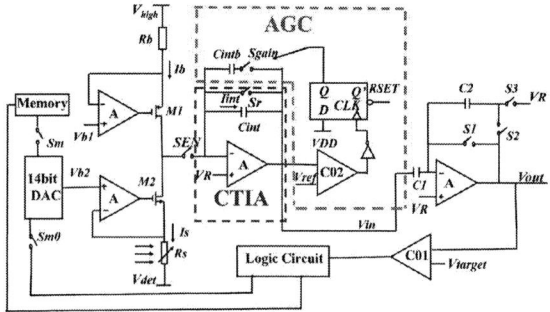

Fig 7: The ROIC designed in this paper

The AGC is mainly composed of a gain switch Sgain, a integration capacitor C_{intb}, a comparator C02 and an Edge-triggered D flip-flop. The C_{intb} is controlled by the Sgain and the C02 compares the output voltage of the CTIA with the gain conversion voltage V_{ref} If the output voltage is higher than V_{ref}, D flip-flop outputs low level and the Sgain keeps off. The ROIC works in high gain mode and employs the small integral capacitor C_{int} to obtain the high sensitivity. The D flip-flop resets in each integration stage of the CTIA. The output voltage is :

$$V_{out} = V_R - \frac{I_{int}t}{C_{int}} \qquad (6)$$

Otherwise, D flip-flop outputs high level and the Sgain turns on. The ROIC works in low gain mode and uses a big integrating capacitor to extend the dynamic range of ROIC by connecting C_{intb} and C_{int} in parallel. The output voltage is:

$$V_{out} = V_R - \frac{I_{int}t}{C_{int} + C_{intb}} \qquad (7)$$

Set C_{int} to 500 fF, C_{intb} to 4.5 pF, V_R to 2.2 V, V_{ref} to 250 mV and the integration time to 100 μs to simulate the ROIC proposed in this paper. Simulation results of the output voltages of the ROIC that calibrated before is shown in Fig 8. It can be seen that when the signal current

exceeds 9.675 nA, the ROIC can automatically switch from high-gain mode to low-gain mode. Since the sum of C_{int} and C_{intb} is 5 pF and ten times as much as C_{int} which is 500 fF, the theoretical maximum detectable current increases tenfold, while the minimum detectable current constants. That means the dynamic range can be improved by 20 dB due to the addition of the AGC.

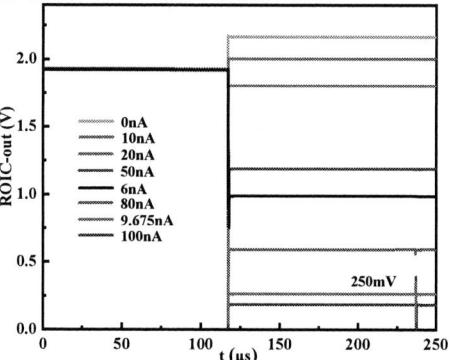

Fig 8: Output voltage curves of the ROIC with different currents

CONCLUSION

In this paper, a readout circuit based on the standard CMOS process is proposed. Simulation results show that the ROIC can effectively suppress the noise of the pixel FPN and offset voltage of the CTIA and increase the dynamic range while maintaining high sensitivity, which is suitable for polysilicon microbolometer.

REFERENCES

[1] Y. Z. Guo, M. C. Luo, H. L. Ma, et. al. Microbolometer with a salicided polysilicon thermistor in CMOS technology, *Optical Express*. vol. 29, November 2021, pp. 37787-37796.

[2] L. X. Zhang, Y. Yuan, J. Guo. Design of infrared detector readout circuit input stage with background dark current suppression (in Chinese), *Laser and Infrared*. 2021,51(01), pp. 69-73.

[3] S. Xu, Y. Z. Guo, X. S. Kong, H. Y. Zhu, et. al. Self-Calibration Readout Circuits for CMOS Microbolometers (in Chinese), *International Conference on Solid-State & Integrated Circuit Technology2022*, Nangjing, pp. 1-3.

[4] M. Zou, et al. Low-light-level CMOS imaging sensor with CTIA and digital correlated double sampling [J]. *Analog Integrated Circuits and Signal Processing*. 2019, 101(3), pp. 449-461.

A SCALABLE AND CONFIGURABLE LOW-POWER MIXED SIGNAL NEUROMORPHIC ACCELERATORS FOR SPIKING NEURAL NETWORK

Yekuan Chen[1], YiQi Meng[1], Yiling Chen[1], Xiaolei Zhu[1,2]

[1] School of Micro-Nano Electronics, Zhejiang University, Hangzhou 311200, China
[2] Zhejiang Lab, Hangzhou 311121, China
*Corresponding Author's Email: xl_zhu@zju.edu.cn

ABSTRACT

Spiking Neural Network (SNN) mimics the biologically inspired neural network that benefits for the computing with low power and high parallelism. In this paper, a scalable and configurable low-power mixed signal neuromorphic accelerator is proposed for the SNN application. It consists of 64 neurons, 64 synapses, 64×64 Static Random-Access Memory (SRAM) array, 64×64 Content Addressable Memory (CAM) array and a number of digital asynchronous routers. 15×64 neurons are fully connected by proposed scalable architecture. The proposed neuromorphic accelerator is designed in 65 nm CMOS process dissipating 4.8 mW at 100 MHz of clock frequency from a 1.2 V power supply.

INTRODUCTION

The Spiking Neural Network (SNN) becomes more and more popular due to the ultra-low power consumption and super high parallelism benefit from the biologically inspired characteristic. Compared to the Convolutional Neural Network (CNN), SNN performs the information communication and computation based on the spikes. [1] When a neuron receives multiple spikes in rapid succession, the membrane potential increases to a certain level, triggering a spike to its downstream neurons via interconnecting synapses[2]. In order to achieve fast speed and ultra-low power, it is necessary to implement a hardware accelerator instead of software simulation[3].

The mixed signal-based neuromorphic hardware architecture is proposed due to its better performance in light of power consumption and silicon area[4]. It basically consists of two critical blocks: the local processor and the digital router. The local processor aims to process the input data and generate spike package data[5]. The digital routers deliver these package data based on Address Event Representation (AER) protocol to determine the data flow for both inside chip and chip-chip throughout the whole scalable system. 15×64 neurons are fully connected by proposed scalable architecture. In practice, for the different tasks, the number of selected neurons, their connections and the data flow could be configured by varying the input package data. The prototype is designed in 65 nm standard CMOS process with the area size of 2x1 mm2. Simulation results shows the proposed neuromorphic accelerator dissipates 4.8 mW at 100 MHz of clock frequency from a 1.2 V power supply.

ARCHITECTURE AND CIRCUITS

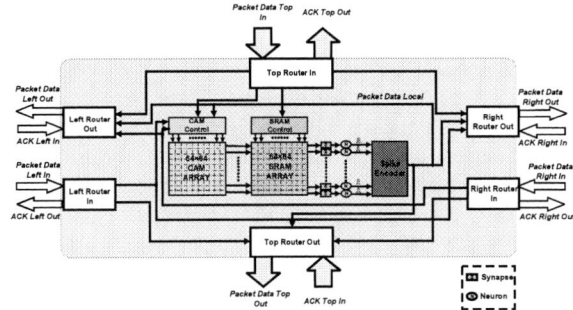

Fig.1. SNN chip architecture

Fig.1 shows the architecture of the SNN chip. The chip is composed of a 40Kb CAM array, a 20Kb SRAM array, a number of neurons and synapses, a spike encoder and some digital routers. The router decodes the input package data firstly and delivers the decoded data to controllers of the CAM and SRAM separately for further decoding in order to control the function of memory. Each CAM cell stores a 10-bit data for the address of the presynaptic neuron while SRAM cell stores a 5-bit weighted data The synapse circuit amplifies the spiking voltage signal from presynaptic neuron by a stored weight value and then converts this weighted spiking signal from voltage to an exponential current fed into its downstream neurons circuit. The neuron circuit generates a spiking voltage signal if its membrane potential increases to a certain configured level.

Router Block

The router block determines the selected neurons, their connections and data flow for both inside chip and chip-chip throughout the system. The topology for the chip-chip connection is based on a Binary Tree architecture shown in Fig.2, which supports arbitrary number of connected chips by changing the bit width of chip address in the packet data. The chips communicate with each other with packet data streams controlled by the top, left and right router blocks implemented on each chip. There are two steps for the whole system operation. Firstly, the SRAM and CAM configuration event packets are processed successively by the top router, which determines the selected neurons, their connections and

979-8-3503-1101-3/23 $31.00 © 2023 IEEE

weights. Once the configuration is completed, the system starts to normally operating by delivering the spiking event packet data through top, left and right router block.

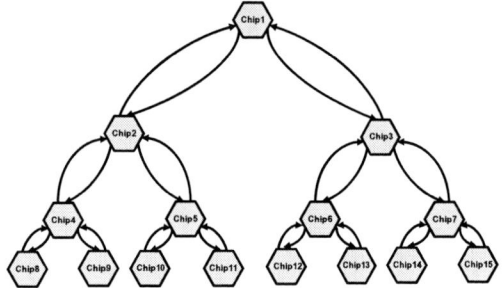

Fig.2. Chip interconnection

Communication between routers is based on Address Event Representation (AER) protocol. As shown in Fig. 3, the data sent to receiver is valid when the rising edge of the clock arrives. Then the receiver processes the data for a while and replies to the sender with a high ACK signal. Once the sender receives this answer signal, the data from sender will be reset to "0". Finally, the receiver will pull down the ACK signal, which means an information exchange completed.

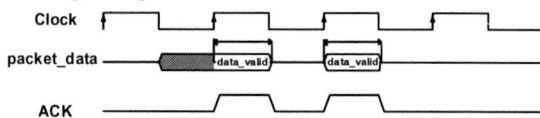

Fig.3. AER timing diagram

Synapse Circuit

The Synapse circuit aims to amplifies the spiking voltage signal from presynaptic neuron by a stored weight value and then converts this weighted spiking signal from voltage to an exponential current.

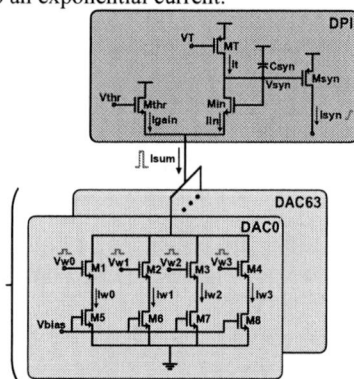

Fig.4. Synapse circuit

As shown in Fig.4, each synapse has 64 DACs with a shared DPI. For DAC circuit, it is a current mirror structure. The transistor M1-M4 are sized to a ratio of

1:2:4:8 resulting in four ratioed currents I_w0-I_w3. Four-bit weight signals (V_w3-V_w0) are applied on the gate of M1-M4 to determine the total amount of current I_W, which converting the weight information from digital voltage to an analog current. However, generated I_W is a pulse-type current which is not consistent with the characteristics of bio-currents. a Diff-Pair Integrator (DPI) circuit is utilized to generate a biological-like current from pulse-type current. All transistors in DPI operate in sub-threshold region and the output is expressed as:

$$I_{syn} = \begin{cases} I_{gain}\left(\dfrac{I_w}{I_a}-1\right)\left(1-e^{-\frac{t-t_i^-}{\mu}}\right) & \text{discharge stage} \\[2ex] I_{syn}^+ e^{-\frac{t-t_i^+}{\mu}} & \text{charge stage} \end{cases} \quad (1)$$

Where, I_w is the DAC output current, I_{gain} is the bias current and t_i is the time constant. The current magnitude can be regulated by configuring above parameters.

Neuron Circuit

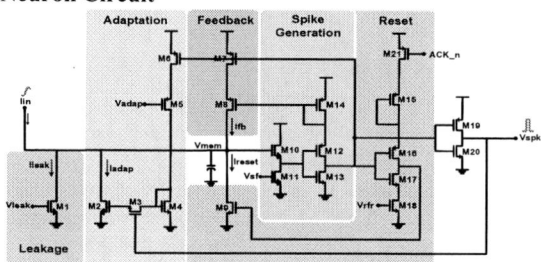

Fig.5. Neuron circuit

The neuron circuit shown in Fig.5 is based on I&F model and the equation of membrane potential can be expressed as:

$$C_{mem}\frac{dV_{mem}}{dt} = I_{in} - I_{leak} + I_{fb} - I_{adap} \quad (2)$$

Where C_{mem} simulates the neural membrane capacitance and I_{in} is the input synapse current. I_{leak} represents the leakage current and it is produced by M1. I_{fb} is the positive feedback current which can only be generated once the spike signal is high. I_{adap} is produced at the end of one spike cycle in order to control the spike response time in the next cycle. Only if the V_{mem} reaches the threshold voltage of the M10, the spike signal is generated. M10-M11 is a source follower and the value of V_{sf} determines the spiking threshold voltage. When the neuron receives the valid ACK_n signal, it will pull down the V_{spk} representing the end of one spiking cycle.

SIMULATION RESULTS

The SNN design is designed in TSMC 65nm CMOS process. The specifications of chip are given in Table I. We have configured the stored information in order to

979-8-3503-1101-3/23 $31.00 © 2023 IEEE

verify the system functionality by observing the spiking behavior for any assigned neuron throughout the whole system. Fig.6 shows the layout of SNN design and the total area is around 2mm×1mm. Fig.7 illustrates the decoding process for the several critical routers which generate correct control signals for the data flow. The simulation of neuron circuit is shown in Fig.8, where indicates the desired continuously spike signal is generated according to the variation of membrane potential. The whole chip for a typical operation dissipates 4.8 mW at 100 MHz of clock frequency from a 1.2 V power supply.

TABLE I   Chip specifications

Process	TSMC 65nm
Clock Frequency (MHZ)	100MHZ
Total number of neurons	64 (each chip)
SRAM memory size	20Kb
CAM memory size	40Kb
Spike generation time(ns)	175ns
Power consumption(mw)	4.8432mw
Number of chips supported for interconnection	15
Area	$2 \times 1 mm^2$

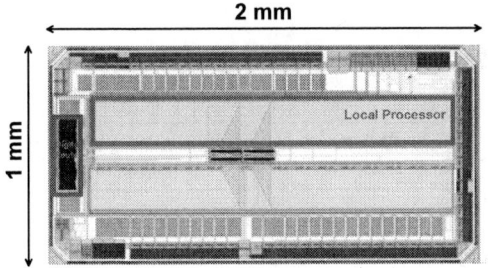

Fig.6. Layout of SNN chip

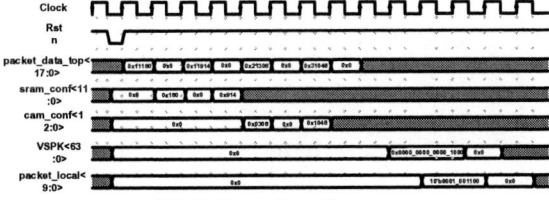

Fig.7. Decoding diagram

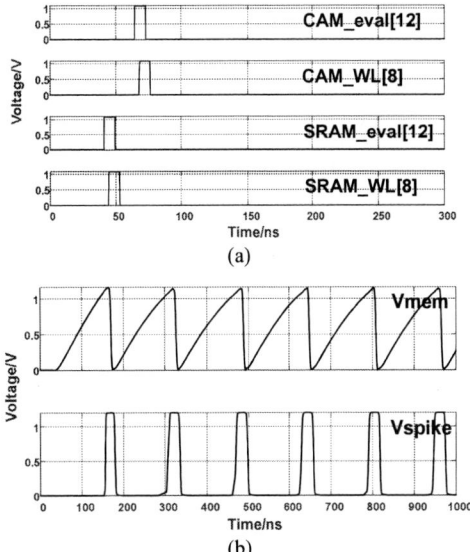

Fig.8. (a) Control waveform of memory cells (b) Neuromorphic signal

CONCLUSION

In this paper, a scalable and configurable low-power mixed signal neuromorphic accelerator is proposed for a good balance between the network flexibility and scalability. However, the proposed scalable architecture with the standard communication interface defined by the AER format also enables future extensions to a chiplets integration with AER bus connection, which could scale up size of the SNN on the system level to potentially millions of neurons instead of mere hundreds.

ACKNOWLEDGMENT

This work is supported by the National Natural Science Foundation of China (No.U20A20220)，the Major Scientific Research Project of Zhejiang Lab (No.2019KC0AD02) and the Major Scientific Research Project of Zhejiang Province (No.2022C01048).

REFERENCES

[1] Moradi S, Qiao N, et al. IEEE Transactions on Biomedical Circuits & Systems, 2017, pp. 99.

[2] N. Qiao, et al. 2017 IEEE SOI-3D-Subthreshold Microelectronics Technology Unified Conference (S3S), Burlingame, CA, USA, 2017, pp. 1-4.

[3] S. Moradi, G. Indiveri. IEEE Transactions on Biomedical Circuits and Systems, vol. 8, no. 1, pp. 98-107, Feb. 2014.

[4] M. V. Nair, et al. 2019 IEEE International Symposium on Circuits and Systems (ISCAS), 2019, pp.1-5.

[5] N. Qiao, G. Indiveri. 2016 IEEE Biomedical Circuits and Systems Conference (BioCAS), Shanghai, China, 2016, pp. 552-555.

RLCkt: An Analog Circuit Automatic Sizing Sage Based on Reinforcement Learning

Wangge Zuo[1], Lingge Liu[1], Fei Li[2], Yifei Huang[2], Liqian Zhang[2], Jing Wan[1]*

[1]School of Information Science and Engineering, State key lab of ASIC and System,
Fudan University, Shanghai 200433, China

[2] Suzhou Foohu Technology Co., Ltd. Shanghai 200433, China

*Corresponding Author's Email: jingwan@fudan.edu.cn

ABSTRACT

Automated solutions for sizing analog circuits have gained significant interest due to the labor-intensive nature of the task in a typical design cycle, especially with technology development and circuit scaling. This study introduces the RLCkt, an Analog Circuit Automatic Sizing Sage, which utilizes reinforcement learning (RL) and the SPICE simulator to automatically adjusts design parameters to meet performance metrics. After several hours of training, RLCkt runs 94 times faster than a traditional generic algorithm with comparable performance. Additionally, its generalization capability surpasses that of state-of-the-art methods.

INTRODUCTION

The design of analog circuits involves a complex process with multiple steps and long cycles. Circuit sizing is a crucial aspect of the design, which consumes a significant amount of time for designers. To address this issue, analog circuit sizing automation has gained attention in recent times [1-2]. However, constructing accurate expressions for multiple circuit performances [3] through programming is a challenging task and is becoming less popular. Genetic algorithms (GA) have been explored as an optimization method, but their stochastic nature makes convergence and recurrence uncertain [4]. Bayesian optimization is also a time-consuming approach to seeking parameter sets [5] due to its computational complexity. Reinforcement learning (RL) has been successful in introducing new techniques, such as the transferable architecture presented in Ref. 6, which leverages Graph Neural Networks (GNN) but suffers from challenging structure design and long convergence periods. Ref. [7] uses the proximal policy optimization (PPO) algorithm with discrete action and appears to be the most efficient model used to date. In this paper, we propose the RLCkt model based on an exquisite design deep deterministic policy gradient (DDPG) algorithm [8], which offers several advantages over existing methods. 1) RLCkt offers a solution to the problem of having to start over with training when transitioning to new performance metrics. 2) It achieves a 94x faster performance than traditional genetic algorithms while still maintaining comparable results. Additionally,

RLCkt surpasses the state-of-the-art in both convergence speed and generalization capability.

METHODOLOGY

Problem Definition

The sizing of an analog circuit is to determine a suitable circuit parameters combination to achieve the desired specifications of a given circuit topology.

Formally, we define parameter space as $x \in R^n$ and performance space as $y \in R^m$, where the parameter space is a continuous space in R^n and y is the actual performance limit of the circuit obtained through extensive experiments.

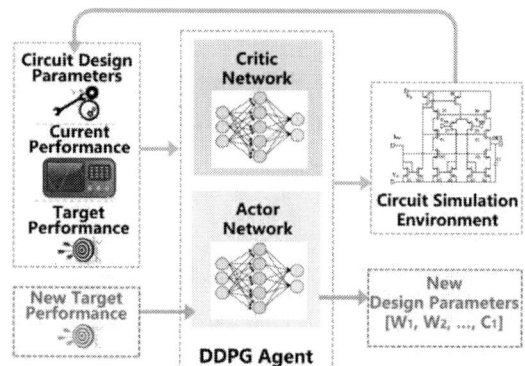

Figure 1: RLCkt method overview

Strategy Generation

Figure 1 depicts the overview of RLCkt. RLCkt using DDPG algorithm with two deep neural networks for the actor and critic respectively. Specifically, the action in RL corresponds to the amount of variation in the circuit design parameters. The reward is determined by the normalized distance between the current performance of the circuit and the target performance. The inputs to the agent network constitute the design parameters, current performances and target performances which are all normalized. In each training step, the DDPG agent updates the design parameters with its outputs. The reward is then assessed by the SPICE simulation results, which is used to guide the update of the actor and critic networks.

979-8-3503-1101-3/23 $31.00 © 2023 IEEE

Training and Deployment

In each training episode, the target performance is randomly sampled from the training dataset constituting M performance points sampled uniformly inside the performance space y. Starting with fixed initial parameters, the agent steps forward according to the output of the actor network. After each step, the actor and critic networks are updated according to the reward. Also, a noise is added to the output action to encourage the exploration of the agent as well as avoid getting trapped in a local optimum.

After training, the actor network is directly deployed on new target performances and predicts the design parameters. Note that the deployment environment can be different from the one used in training.

EXPERIMENTS

Experimental Setup

The circuit simulation environment is a two-level operational amplifier, which schematic is shown in Figure 2(a). The circuit has a total of seven design parameters, including five width parameters of transistors (from W1 to W5), bias current (Ibias) and output capacitance (Cout). The variation range of the above design parameters are shown in Tab. 1. The contest specifies gain and cut-off frequency (Ft) as optimization target. The performance space of the circuit is determined by generating circuits with design parameters randomly sampled from the range above. By connecting the outermost performance points, the performance space is then determined as shown by the red curve in Figure 2(b).

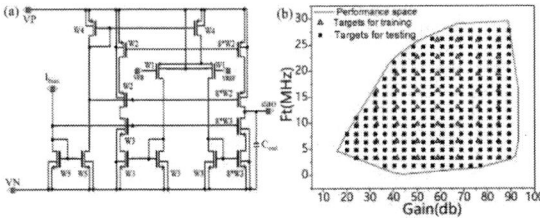

Figure 2: (a) Two-level operational amplifier used in our work. (b) Distribution of training dataset and testing dataset in performance space

TABLE I. The Range of Design Variables

Model	W1-W5	I_{bias}	C_{out}
Upper Limit	1μm	1μA	1pF
Lower Limit	10μm	10μA	10pF

Model Training

The RL agent is then trained with a training dataset constituting 53 performance points sampled uniformly inside the performance space, as shown in Figure 2(b). The training is performed episode-wise. Figure 3(a) shows the evolution of the mean episode reward accumulating a

reward of 50 steps in an episode. The increasing mean episode reward curve indicates that RLCkt has learned continuously to achieve a positive goal-optimized state with the training process moving on. Three thousand episodes of training is used to reach a stably high reward.

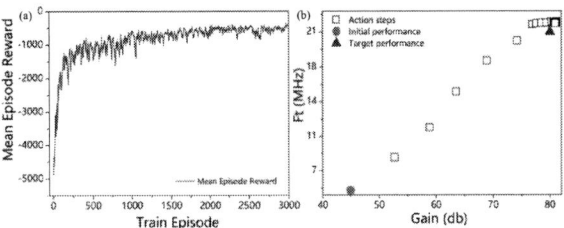

Figure 3: (a) Mean episode reward vs. train episode during model training. (b) Convergence to the target performance point

Experimental Results

After training, the model is tested by 255 performance points sampled from the performance space which has no any overlap with the training dataset, as shown in Figure 2(b). For each of the test target, the target performance is used as input to the trained actor network which outputs the variation of the design parameters. The design parameters are then updated each step, and eventually the model converges to the target performance. The dynamic tracking process of the target performance is shown in Figure 3(b). Figure 4(a) shows the evolution of the relative error with the simulation steps. Considering all 255 test targets, the mean error decreases from over 60% to only 5.1% at only 9 simulation steps. It further reduces to 3.7% at 20 simulation steps. A few outliers are observed which are mainly found at the boundary of the performance space out of the training dataset range, see the distribution of error the performance space in Figure 4(b). In most part of the performance space, the error is well below 7.0%, indicating that the RLCkt model converges very well.

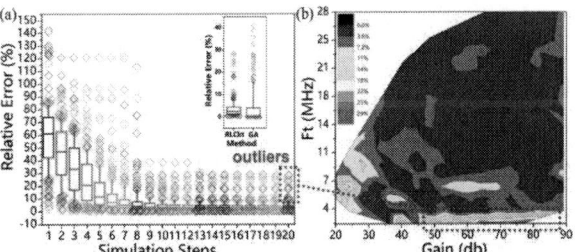

Figure 4: (a) Relative error evolution at different simulation steps. (b) Error distribution across the whole performance space

As explained above, once trained, the RLCkt model converges in only 9 steps to any new performance target in the performance space. This takes only 0.8 seconds in a conventional laptop, and the time for 20 simulations is

1.6s. For comparison, we also perform the optimization using genetic algorithm which has been developed previously [7]. Averagely, the GA takes 153.3 seconds, which is 94 times slower than the RLCkt model when converging to the same performance targets with similar mean error.

Then, one thousand target design specifications the agent has never seen before are randomly generated in the range of the entire performance space specified during training. We deploy RLCkt trained for 3000 episodes on these targets and compare the results with those of AutoCkt [9] and GA [7]. The population size of GA is 50. The numeral comparison is shown in Table II. The target-oriented convergence is shown in Figure 4. Once trained, the RLCkt agent is allowed a strategy length of 20 simulation steps to converge with low mean error and reaches 986 of the 1000 targets across the design space. Moreover, RLCkt doubles the convergence speed more than the state-of-the-art and develops better generalization ability.

TABLE II. Acceleration and Generalization Comparison

Model	Mean Runtime(s)	Acceleration	Generalization
AutoCkt	-	40×	963/1000
GA	153.3	1	970/1000
RLCkt	1.6	94×	986/1000

The RLCkt agent accumulates experience from previous training dataset and memorizes this experience in the modified actor network. However, the GA agent has no memory, and it starts from scratch every time with new target performance. AutoCkt uses discrete actions, which cannot be traded off for speed and accuracy. Correspondingly, RLCkt avoids this problem by switching to a continuous action space. The use of RLCkt model to track new performance target is very fast, but the training of the model is indeed time-consuming, which takes up to 4.5 hours at the same computer. However, the training time of the model can be further reduced by using high performance computer and parallel computing. Then, the trained model can be deployed in other personal computers for high-speed target tracking application.

CONCLUSION

This paper proposes an RL-based framework that designs analog circuit parameters satisfying arbitrary performance targets in design space. Results show that the trained RLCkt runs 94× faster than a traditional genetic algorithm with comparable performance, and the generalization capability is better than the state-of-the-art.

ACKNOWLEDGMENTS

This work was supported by the National Key R&D Program of China (2021YFA1200500), Shanghai Science and Technology Commission "explorer project" (21TS1401300), Natural Science Foundation of Shanghai (23ZR1405900). Guangdong Province Research and Development in Key Fields from Guangdong Greater Bay Area Institute of Integrated Circuit and System(No.2021B0101280002) and Guangzhou City Research and Development Program in Key Field (No.20210302001).

REFERENCES

[1] Budak A F, Bhansali P, Liu B, et al. DNN-Opt: An RL Inspired Optimization for Analog Circuit Sizing using Deep Neural Networks[C]//2021 58th ACM/IEEE Design Automation Conference (DAC). IEEE, 2021: 1219-1224.

[2] J. Clerk Maxwell, A Treatise on Electricity and Magnetism, 3rd ed., vol. 2. Oxford: Clarendon, 1892, pp.68–73.

[3] Sönmez Ö S, Dündar G. Simulation-based analog and RF circuit synthesis using a modified evolutionary strategies algorithm[J]. Integration, 2011, 44(2): 144-154.

[4] Cohen M W, Aga M, Weinberg T. Genetic algorithm software system for analog circuit design[J]. Procedia CIRP, 2015, 36: 17-22.

[5] Lyu W, Xue P, Yang F, et al. An efficient bayesian optimization approach for automated optimization of analog circuits[J]. IEEE Transactions on Circuits and Systems I: Regular Papers, 2017, 65(6): 1954-1967.

[6] Wang H, Wang K, Yang J, et al. GCN-RL circuit designer: Transferable transistor sizing with graph neural networks and reinforcement learning[C]//2020 57th ACM/IEEE Design Automation Conference (DAC). IEEE, 2020: 1-6.

[7] Settaluri K, Haj-Ali A, Huang Q, et al. Autockt: Deep reinforcement learning of analog circuit designs[C]//2020 Design, Automation & Test in Europe Conference & Exhibition (DATE). IEEE, 2020: 490-495.

[8] Y. Hou, L. Liu, Q. Wei, X. Xu and C. Chen, "A novel DDPG method with prioritized experience replay," 2017 IEEE International Conference on Systems, Man, and Cybernetics (SMC), Banff, AB, Canada, 2017, pp. 316-321, doi: 10.1109/SMC.2017.8122622.

[9] Mao W, Wei J H, Wan J . Automatic design of analog integrated circuit based on multi-objective optimization[C]// 2020 IEEE 15th International Conference on Solid-State & Integrated Circuit Technology (ICSICT). IEEE, 2020.

Convolutional Neural Networks on the Edge: A Comparison Between FPGA and GPU

Yichen Wei, Siyi Gong, Hongfei Mei, Longxing Shi and Xinfei Guo*
University of Michigan – Shanghai Jiao Tong University Joint Institute
Shanghai Jiao Tong University, Shanghai, China
{xinfei.guo}@sjtu.edu.cn, *Corresponding author

Abstract—With more computation tasks being pushed towards the edge, it becomes increasingly challenging to decide the right hardware approach for running edge AI tasks. Besides customized AI accelerators, existing hardware platforms such as FPGAs have been employed to support edge computing. Recently, small GPUs have also been developed to run AI tasks that are tailored for edge. While each platform exhibits its own advantages, the shortcomings are also identified. In this work, we present a study where a head to head comparison is conducted by running the same CNN on two popular edge platforms, FPGAs and GPUs. We compare multiple dimensions such as power, inference speed, ease of development and accuracy. The comparison results can provide an initial guidance for edge computing developers who don't have immediate access to both platforms.

I. INTRODUCTION

In the edge computing era, data is stored and processed locally on a resource-constrained tiny device. This offers numerous benefits such as less data movement, faster response time and improved quality of service. It also poses challenges such as power efficiency, accuracy, scalability and more. To efficiently run these compute-intensive tasks such as convoluntional neural network (CNN) on the edge, the hardware needs to be tailored to support both AI and non-AI tasks. GPUs are the "default" platform for running AI tasks due to its highly paralleled SIMD architectures. Recently, embedded GPU platform such as NVIDIA Jetson Nano board was introduced and featured processor that was heavily optimized for edge computing applications, achieving a superior throughput to power ratio [1], [2]. Unlike GPUs, FPGAs are well known for its excellence in accelerating domain-specific tasks such as AI and being adaptive to a variety of workloads [3]. Recent work shows that FPGAs are promising at the edge due to its spatio-temporally parallelized hardware architecture that can be leveraged to accelerate highly parallel tasks such as CNN with good power efficiency. The debate on which platform to choose for edge computing is never ending, thus there is still no clear answer yet. On one hand, FPGAs seem to be able to overcome many challenges that were introduced by GPUs, such as non-deterministic execution model, high power consumption and poor scalability. On another hand, GPUs are more accessible to developers and are capable of delivering massive parallelism and in-situ training capabilities. In this work, we perform a thorough comparison with two popular embedded platforms that are suitable for edge computing. We

report our experiment details, observations and findings in this paper.

II. EDGE PLATFORMS

In this work, we benchmark our implementation with two popular platforms that are optimized for edge computing applications. This section will give a brief introduction of the two platforms.

A. NVIDIA Jetson Nano

Jetson Nano is a small, power-efficient embedded computing board developed by NVIDIA in 2019 intended for embedded IoT, AI and robotics applications on edge devices. The board consists of a Quad-core ARM CPU and a Maxwell GPU with 128 NVIDIA CUDA cores. It's powered by modified Ubuntu and integrated with JetPack SDK, which contains accelerated libraries for deep learning, computer vision, graphics, multimedia, etc [4]. Jetson Nano provides complete support for neural network frameworks like Tensorflow, PyTorch and Caffe, and also supports network optimizer TensorRT [5] to improve inference performance.

B. Xilinx PYNQ-Z2

PYNQ-Z2 is a development platform designed around the Zynq-7000 all programmable system-on-chip (AP SoC) from Xilinx. The Zynq-7000 architecture tightly integrates a dual-core, 650 MHz ARM Cortex-A9 processor with Xilinx 7-series Field Programmable Gate Array (FPGA) logic. This pairing grants the ability to surround a powerful processor with a unique set of software defined peripherals and controllers, tailored by you for the target application [6]. With the help of Xilinx's high-performance Vivado Design Suite and Jupiter notebook and python, embedded programmers can run machine learning (ML) models on FPGA.

To run ML models on FPGA, there are basically the following steps:

- Get information of FPGA architecture (PYNQ-Z2)
- Generate TCU accelerator design (RTL code)
- Synthesize for PYNQ-Z2
- Compile ML model for TCU
- Execute using PYNQ

III. EXPERIMENTAL RESULTS

In this section, we will present our findings in running machine learning models on two aforementioned platforms.

979-8-3503-1101-3/23 $31.00 © 2023 IEEE

TABLE I: Experiment results for various networks running on Jetson Nano board

	CPU usage	GPU usage	ram	swap	voltage	current	avg one inference time	accuracy (cat, dog)
No Task	15%	0%	0.7G	0.0G	5.00V	0.09A		
Resnet-18	35%	95%	1.6G	0.0G	4.92V	0.28A	0.01503s	100.00%, 99.66%
Resnet-50	33%	96%	1.5G	0.0G	4.92V	0.30A	0.03705s	60.96%, 45.78%
Resnet-101	30%	99%	1.9G	0.5G	4.90V	0.30A	0.05709s	53.27%, 56.35%
Resnet-152	25%	99%	1.9G	0.5G	4.91V	0.30A	0.08702s	79.74%, 68.02%
Inception-v4	20%	99%	1.9G	0.0G	4.93V	0.29A	0.10271s	93.90%, 70.21%
AlexNet	35%	99%	1.8G	0.0G	4.93V	0.30A	0.01473	21.93%, 14.63%

TABLE II: Experiment results for runnging YOLOv2 on Jetson Nano board

	CPU usage	GPU usage	ram	swap	voltage	current	avg one inference time	accuracy (cat, dog)
No Task	15%	0%	0.8G	0.0G	5.00V	0.08A		
YOLOv2	34%	43%	1.8G	0.0G	4.92V	0.22A	0.12652s	78%, 82%

A. GPU Results

1) Setup: Jetson nano is set up following the guidelines in jetson-inference, which is a library for inference accelerated with TensorRT provided by NVIDA. We tested several networks in it to measure the performance of GPU. Furthermore, to test the usefulness to hardware accelerator TensorRT, we run inference on MNist dataset [7] both without TensorRT and with TensorRT. Finally, we run inference on YOLOv2 to compare the efficiency on running inference with FPGA. A power meter is employed to measure the real-time power consumption. Fig. 1 shows the experiment setup.

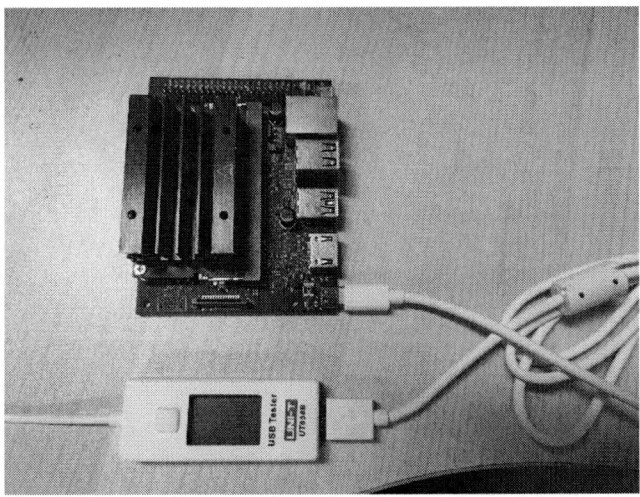

Fig. 1: Jetson Nano Board Setup (connected with a power meter to measure the dynamic power)

2) Implementation Results: Several data is recorded during the inference for various models, the key metrics include CPU usage (average value of 4 cores usage), GPU usage, RAM and swap usage (necessary since Jetson nano 2G version only has 2G RAM) and total power (measured with the USB power estimator shown in Fig. 1). They are summarized in Table I. In almost all models, GPU is fully used and for those large models, some swap is used. The total power is similar when running inference on different models.

Table II summarizes the results when running YOLOv2 model, a popular application for real-time object detection. To make comparison, we also list the baseline where no task runs on the board. It is able to achieve an accuracy of $\sim 80\%$ for recognizing cat or dog cases.

B. FPGA Results

1) Setup: PYNQ-Z2 is set up according to the guide in PYNQ's official documentation, starting by burning the image file to the SD card for the board. Paired with Jupyter Notebook, the board is accessed through a web browser from a PC on the same LAN. After studying the characteristics of existing FPGA-based neural networks accelerators, we finally completed the process of mapping YOLOv2 and Resnet-20 to the FPGA, which is state-of-the-art on standard detection tasks.We use Vivado HLS 2019.2 for accelerator design and Vivado 2019.2 for synthesis. Fig. 2 shows the setup for the FPGA board. To test out the flow, we firstly ran the ResNet-

Fig. 2: FPGA Setup with PYNQ Z-2

20 model. A breakdown of the resource consumption is shown in Fig. 3.

2) Implementation Results: Similar to the GPU evaluation, the same YOLOv2 model was implemented on the FPGA.To make fair comparison, the same dataset was used. The power,

TABLE III: FPGA resource utilization in vivado for YOLOv2

DSP	LUTs	LUTRAM	BRAM	Freq[MHz]
151(168.64%)	34034(63.97%)	6155(35.37%)	87.5(62.5%)	150

TABLE IV: Accuracy and run time summary for the YOLOv2 model

	Accuracy [%]	image preprocess	load image to memory	fpga process	region layer process	post process	total time
cat	84.16	0.74s	0.04s	1.41s	1.39s	1.23s	4.81s
dog	81.75	0.72s	0.06s	1.50s	1.49s	1.30s	5.07s

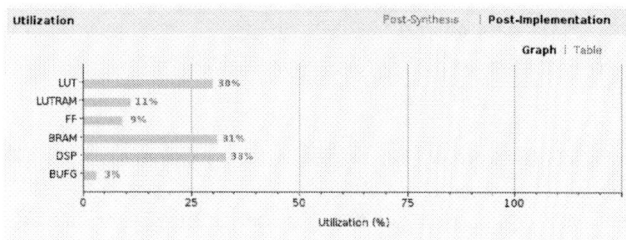

Fig. 3: FPGA resource consumption for Resnet-20

performance and area (LUT count) were recorded from the report when generating bitstream from the vivado while the inference time and accuracy were sampled from the jupyter notebook. Table III and IV summarize the hardware utilization, accuracy and run time for running the YOLOv2 model on the FPGA.

For both GPU and FPGA platforms, similar accuracy has been achieved. This is because the same model yolov2 is used and we run inference with the same pictures. GPUs offer faster inference time compared to FPGA. While FPGA implementation utilizes less hardware resources and thus offers better power efficiency. During the experiments, we also observed that GPU implementation is more user friendly, and the deployment of the model is more straightforward than the FPGA platform.

IV. CONCLUSIONS

In this work, we performed a study by running various CNN models on two popular edge platforms, which are FPGAs and embedded GPUs. We looked into power, inference speed and accuracy aspects. The ongoing work includes generation of more data to understand the differences from the architectural perspectives. The ultimate goal is to develop a methodology to make fair comparisons among different existing hardware platforms for edge computing.

ACKNOWLEDGEMENT

This work was partially supported by a CCF-Tencent Open Fund No. RAGR20220114, a SJTU Explore-X Research Grant No. D6040004/038, a startup fund from UM-SJTU JI and a PRP project No. 20539 from the Shanghai Jiao Tong University. We would like to thank the Xilinx University Program and NVIDIA for their generous hardware donation. Lastly, we also thank NVIDIA developer community and Xilinx developer community for proving all the foundation materials that enabled this study.

REFERENCES

[1] I. Colbert, J. Daly, K. Kreutz-Delgado, and S. Das, "A competitive edge: Can fpgas beat gpus at dcnn inference acceleration in resource-limited edge computing applications?" *arXiv preprint arXiv:2102.00294*, 2021.

[2] L. Pettersson, "Convolutional neural networks on fpga and gpu on the edge: A comparison," 2020.

[3] C. Xu, S. Jiang, G. Luo, G. Sun, N. An, G. Huang, and X. Liu, "The case for fpga-based edge computing," *IEEE Transactions on Mobile Computing*, 2020.

[4] NVIDIA, "Nvidia developer - jetson nano," Website, 2022, https://developer.nvidia.com/embedded/jetson-nano.

[5] H. Vanholder, "Efficient inference with tensorrt," in *GPU Technology Conference*, vol. 1, 2016, p. 2.

[6] Digilent, "Arty z7 reference manual," Website, 2022, https://digilent.com/reference/programmable-logic/arty-z7/reference-manual.

[7] Y. LeCun, L. Bottou, Y. Bengio, and P. Haffner, "Gradient-based learning applied to document recognition," *Proceedings of the IEEE*, vol. 86, no. 11, pp. 2278–2324, 1998.

LOGIC OPTIMIZATION SEQUENCE TUNING BASED ON POLICY SEARCH DEEP REINFORCEMENT LEARNING

Yu Jin*, Haijiao Huang, Wenzhe Ye and Xuebing Zhang

Beijing University of Chemical Technology, Beijing, China

*Corresponding Author's Email: 2014500056@buct.edu.cn

ABSTRACT

Logic optimization involves a set of Boolean transformations , and the different ways of permutation of Boolean transformations affect the circuit performance. The design space composed of these Boolean transformations is exponential. However, heuristic algorithms often need the guidance of human experience, and the result is not ideal for this large-scale search space problem. In this paper, the problem of logic synthesis optimization is mapped to the sequential decision process. We introduce policy search based reinforcement learning, which makes optimization decisions by analyzing data flow generated by synthesis tools, and find the optimal solution for logic synthesis. The evaluation of EPFL benchmark circuit shows that compared with the original design, the nodes are reduced by 19.60%, the depth is reduced by 13.76%. Compared with ABC script resyn2, the number of nodes is reduced by 6.70%, the depth is reduced by 7.76%. In addition, the performance of the scheme is successfully verified by the sequential circuit.

INTRODUCTION

The Boolean network in logic synthesis tool ABC [1] is represented by And-Inverter graph (AIG), and the size of the graph is reduced by a sequence of Boolean transformations on AIG. The complexity of the logic synthesis optimization process lies in the search of the exponential solution space, and every permutation and combination of Boolean transformations will make an impact on the optimization result [2].

Reinforcement learning (RL) is a branch of machine learning that supports large-scale search. The actor-critic framework is used to explore the design space, and the policy and value functions use networks to fit parameters, achieving better results than heuristics and expert scripts [3]. RL algorithm combined with graph convolution neural network is used to explore the design space [4, 5]. Yu et al. took a different approach to map the logic synthesis design flow composition problem to the classification problem [2].

The schemes based on Boolean transformations sequence optimization are inseparable from the process of fitting the policy function by the policy network. The essence of these works [3, 4, 5] formulates the problem of logic synthesis flow into Markov decision process (MDP), and obtains ABC optimization scripts through the exploration of data flow by RL. The graph size is greatly

reduced through ABC's core procedure such as rewrite, resub, and balance. Whether it involves graph networks or adding other elements to the optimization model, the core procedure remains the permutation problem of these Boolean transformations. Moreover, optimizing the Boolean transformations sequence is a black box operation. No matter which RL model is used, the training result completely depends on the statistical information of ABC and the setting of reward. We believe that operating directly on the policy function will also obtain better results.

In view of this, we introduce a new method based on RL, which makes optimization decisions by directly manipulating the data flow generated by the policy function, and analyzing the synthesis tool to find the optimal scheme. Our contribution to this work is as follows:

- We alleviate the scale of the solution space for optimization to a game like environment that RL agents can understand, and feature sets are extracted from AIG features.
- We introduce policy search to logic synthesis procedure to obtain the reward value by minimizing the dual objective function of node number and depth.
- We demonstrate the capabilities of our proposed method in EPFL benchmark suite [6]. We compare our results with the results of resyn2 [5], a heuristic script of ABC. The comparison shows that our scheme obtain better results.
- We also parsed the combination logic in the sequential circuit and optimized it with the proposed scheme. The comparison shows that the better results are obtained by this scheme.

REINFORCEMENT LEARNING

In the framework of logic synthesis, each permutation of these transformations results in a different area and delay, thus leading to exponentially growing searches. So we desire an efficient search algorithm to find the best solution. We set A={a1, a2, ..., an} is defined as the set of transformations in the logic synthesis tool. The length of a sequence of Boolean transformations is defined as m. So the size of the search space is n^m. This exponentially growing spatial search of data is challenging. RL is a method to solve this search problem, so we briefly discuss the background needed to develop our method next.

In reinforcement learning, agents are trained to select actions in an iterative manner to maximize their expected future returns. Policy search is a kind of policy-based reinforcement learning, which parameterizes the policy function. And by iterative training, the parameters are constantly updated in order to find the maximum cumulative expected return as follow:

$$\eta_\theta = E[\sum_{t=0}^{h} R(S_t) \mid \pi_\theta] \qquad (1)$$

Where π_θ is the current parameterized policy function and θ is the parameter. Let τ denote a set of state-action sequences $s_0, a_0 ... s_h, a_h$; $R(\tau)$ represents the return of trajectory τ ; $P(\tau;\theta)$ represents the probability of trajectory τ . The objective function of reinforcement learning can be rewritten as follows:

$$J(\theta) = E(\sum_{t=0} R(s_t, u_t); \pi_\theta) = \sum_{\tau} P(\tau;\theta) R(\tau) \qquad (2)$$

There are many ways to solve optimization problems, among which the most commonly used is the steepest descent method, which is called the policy gradient method . Through mathematical theorem transformation, after sampling N trajectories of the current policy, the empirical average of N trajectories can be used to estimate the calculation of policy gradient as follows:

$$\nabla_\theta J(\pi_\theta) = \frac{1}{N} \sum_{t=0}^{T} \nabla_\theta \ln(\pi_\theta(A_t \mid S_t)) \sum_{t=0}^{T} R_t \qquad (3)$$

IMPLEMENTATION DETAILS AND RESULT ANALYSIS

Implementation details

The process of logic synthesis optimization based on policy search can effectively solve the Boolean transformations sequence tuning problem. In policy search for logic optimization , Fig. 1 shows the basic framework, *.v, *.blif, *.aiger and other formats of file input logic synthesis environment ABC, extract relevant features, such as input pins, output pins, number of nodes and so on, and get the initial state S_t . Under the interaction between action A_t and environment ABC, the policy search agent generate a new state S_{t+1} , and the environment will give an immediate return Rt according to the reward rules. The policy search agent through the full connection layer training, obtain the probability distribution function of 11 actions. Then, according to the probability distribution function, A_t is continuously selected and iterated with the environment for 50 times to get episode return. Later, the parameters θ_{t+1} are updated to proceed to the next episode. Finally, the parameters are constantly updated by iterative training in order to find the

maximum cumulative expected return. In policy search for logic optimization, when the optimized workflow is obtained, and the number of nodes and depth are weighed in the training process, and the results are compared at last. Next, we will discuss the key points of the whole scheme in detail.

1. State representation in RL. In order to model the optimization of logic synthesis as a RL environment, we define the state of the logic synthesis environment as a set of parameters output from the synthesis tool for a given circuit design, input pins, output pins, number of nodes, number of edges, number of levels, number of latches, number of ands. Is used as the feature set of the policy search agent, and the feature set is used as the input of the neural network.

2. Exploration of synthesis flow. Considering the synthesis flow exploration problem, the set of actions is the selected synthesis transformations. Let A be the decision sequence, where A={a1, a2, ..., an}. The decision sequence A is a synthesis flow action include 11 unique transformations, specifically, A={rewrite rewrite-z rewrite-l rewrite-lz refactor refactor-z refactor-l refactor-lz resub resub-z balance}. We use policy gradient ascent to search for the maximum reward of the optimal flow A, and the agent explores the search space of 11 original transformations within the ABC comprehensive framework.

3. Reward Settings. We define a dual objective reward function, which takes into account the variation of design nodes and depth. For a transformation performed on a given AIG state that reduces the number of nodes and the depth, we give a maximum award of two. When the transformation performed increased the number and depth of design nodes, we gave the lowest negative return of Minus two. Between these two extremes, the value and magnitude of the reward was carefully chosen to aid in agency exploration.

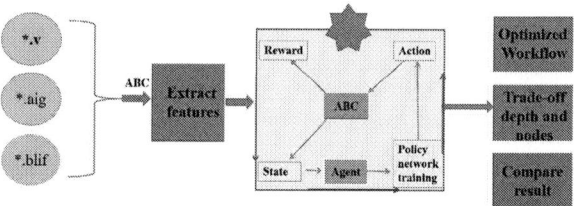

Figure 1 Overall frameworks of policy search for logic optimization

Result analysis

Training is performed on an EPFL benchmarks with 50 episodes，50 iterations per episode. The test circuit is 10 EPFL benchmarks [6] and mapped to the SMIC 55 nm standard cell library. We demonstrate the proposed approach using the open source logic synthesis tool ABC V1.01. We implemented training in Python V3.6.0 and

TABLE I.   DETAILED TEST DATA OF EPFL BENCHMARK

Benchmark(*.v)	Initial Design		ABC scripts resyn2				Policy-Based Method			
	Nodes	Level	Nodes	Imp	Level	Imp	Nodes	Imp	Level	Imp
voter	10324	61	9756	5.50%	57	6.56%	7943	18.58%	58	4.92%
router	240	26	177	26.25%	19	26.92%	139	21.47%	16	38.46%
priority	810	239	676	16.54%	203	15.06%	414	38.76%	117	51.05%
square	16915	248	16623	1.73%	248	0.00%	15793	4.99%	248	0.00%
alu_ctrl	123	10	108	12.20%	8	20.00%	88	18.52%	8	20.00%
i2c	1185	16	1162	1.94%	15	6.25%	1022	12.05%	14	12.50%
mem_ctrl	46602	112	45614	2.12%	110	1.79%	39171	14.13%	101	9.82%
cavlc	668	16	662	0.90%	16	0.00%	622	6.04%	16	0.00%
divisor	40681	4389	40772	-0.22%	4361	0.64%	20760	49.08%	4368	0.48%
adder	1020	256	1019	0.10%	255	0.39%	893	12.37%	255	0.39%
Ave-imp				6.70%		7.76%		19.60%		13.76%

policy search agent neural network using TensorFlow R2.12 training. Experimental parameters were set as: Network size:3 fully connected layers with 20 hidden units each, Optimizer: Adam, learning rate: (α) : 0.01, discount rate (γ) : 0.95. The experimental result is to run Centos7 on Linux version 3.10.0-1160.15.2.el7.x86_64 with an Intel(R) .

The results show that both our scheme and resyn2 script optimization have good improvement effect. Compared with the original design node, the depth of our scheme is reduced by 19.60% and 13.76% on average, while the ABC script resyn2 is only reduced by 6.70% and 7.76% on average, and the policy search scheme compared with resyn2 has an improvement of about 17%. The specific test circuit results are shown in Table I.

In addition, we also analyzed the timing test circuit fiedler-cooley of yosys [7]. The number of nodes in the original design was 218, and the depth was 24. After our scheme training, the number of nodes is 161 and the depth 21, and the average improvement rate is 26.1% and 12.5%. We used design compiler for logic synthesis of this design. The area and time delay are 262.64 square microns and 1.67 nanoseconds, respectively. After the optimization of my own scheme and mapping, the area and delay of 246.68 square microns and 1.55 nanosecond are obtained. Area is 6% better than design compiler. The overall level is comparable to that of design compiler.

CONCLUSION

This paper presents a reinforcement learning framework for Boolean logic optimization. The policy search reinforcement learning scheme is used to optimize the sequence of AIG related Boolean transformation. The spatial solution problem is mapped to a reinforcement learning environment in which the policy search agent iteratively selects the Boolean transformation with the highest expected return. In addition, we use a new objective reward function to guide the exploration process of the agent so that it can balance the number of nodes and depth. By evaluating the benchmark, our proposed method shows superior results to heuristic Boolean transformation sequences. Moreover, it is proved that our scheme can also be used for temporal logic optimization to get better results.

REFERENCES

[1] Mishchenko, A., ABC: A system for sequential synthesis and verification. 2007.

[2] C. Yu, H. Xiao and G. De Micheli, "Developing Synthesis Flows Without Human Knowledge," 2018 55th ACM/ESDA/IEEE Design Automation Conference (DAC), 2018, pp. 1-6, doi: 10.1109/DAC.2018.8465913.A.

[3] A. Hosny, S. Hashemi, M. Shalan and S. Reda, "DRiLLS: Deep Reinforcement Learning for Logic Synthesis," 2020 25th Asia and South Pacific Design Automation Conference (ASP-DAC), 2020, pp. 581-586, doi: 10.1109/ASP-DAC47756.2020.9045559.

[4] W. Haaswijk et al., "Deep Learning for Logic Optimization Algorithms," 2018 IEEE International Symposium on Circuits and Systems (ISCAS), 2018, pp. 1-4, doi: 10.1109/ISCAS.2018.8351885.

[5] K. Zhu, M. Liu, H. Chen, Z. Zhao and D. Z. Pan, "Exploring Logic Optimizations with Reinforcement Learning and Graph Convolutional Network," 2020 ACM/IEEE 2nd Workshop on Machine Learning for CAD (MLCAD), 2020, pp. 145-150, doi: 10.1145/3380446.3430622.

[6] Amarù, L.G., Pierre-Emmanuel, De Micheli, Giovanni, The EPFL Combinational Benchmark Suite. Proceedings of the 24th International Workshop on Logic & Synthesis (IWLS), 2015.

[7] C. Wolf, "Yosys Open Synthesis Suite," 2016.

Agile Full-Chip Sign-Off in the Post-Moore Era

Xiao Dong[1], Songyu Sun[1], Zhengrui Chen[1], Jianyi Yang[2], and Cheng Zhuo[2*]

[1]College of Information Science & Electronic Engineering, Zhejiang University, Hangzhou, China
[2]School of Micro-Nano Electronics, Zhejiang University, Hangzhou, China
*Corresponding Email: czhuo@zju.edu.cn

ABSTRACT

Sign-off is a crucial step in the chip design flow to guarantee the performance and reliability of chips prior to tape-out. However, the ever-growing integration density and voltage scaling in the post-Moore era have made conventional sign-off inaccurate and expensive, presenting challenges such as dynamic parasitics effects, large-scale circuits, and multi-objective coupling. To overcome these challenges, this paper introduces the concept of agile dynamic sign-off, which has significant potential for chip design and validation, combining dynamic modeling, artificial intelligence assisted acceleration, and multi-objective coupling analysis. Two representative works are provided to demonstrate the application of agile dynamic sign-off in system modeling and simulation.

INTRODUCTION

In the VLSI design flow, sign-off is the final approval step before tape-out, which validates the timing, power, and noise of the design to ensure its functionality and completeness [1], [2]. In the post-Moore era, as silicon technology continues to scale down to achieve higher integration density, power supply voltage has been decreased at a much faster pace than the threshold voltage, causing noise margin degradation and hence more accurate validation [3]. On the other hand, the engineering costs of chip validations are growing due to the increasing integration complexity [4]. Thus, low-cost and high-accuracy sign-off is essential for agile chip design in the post-Moore era.

In the post-Moore era, the new technologies, architectures and integration schemes have brought more critical challenges to sign-off, which have not been well addressed in the conventional methodologies:

- *Dynamic parasitics effects*: With the shrinking process geometries, the wire shapes or stack geometries are more complex due to the increasingly complicated design rules, leading to the increased impact of second-order effects caused by coupling capacitance and inductance [5]. Additionally, the coupling capacitance between the power or ground net and signal net causes state-dependent dynamic parasitics effects, where the on-die parasitics change with the input vectors. This increases the complexity of sign-off and makes it challenging to achieve accurate noise and timing validation.
- *Large scale of circuits*: With the ever-increasing number of transistors on a chip and the addition of new functionality, the size of chips has grown so large that sign-off has become a challenging task. Firstly, it is difficult to build a comprehensive and accurate model to capture all the information of a large-scale system. On the other hand, the iterative process of engineering change order (ECO) further increases the time and resource cost of sign-off, given the high demand for accuracy [6]. Moreover, noise validation needs to be executed with a large number of test vectors during sign-off to ensure the performance of the circuit in various application scenarios [7]. Thus, sign-off is notably time-consuming in the post-Moore era.
- *Multi-objective coupling*: In conventional sign-off, noise validation and timing analysis are usually performed separately. Due to the interaction of power supply voltage and switching activities, worst-case noise (or timing) information is referred in timing analysis (or noise validation) [8]. However, it leads to conservative and pessimistic results, resulting in multiple iterations with high cost [9]. Hence, it is crucial to consider how to perform accurate and efficient sign-off with multi-objective coupling.

Due to these challenges, it may take up to several days to weeks to complete a single cycle of full-chip noise or timing sign-off. Moreover, the iterative ECO process based on sign-off results typically requires many more iterations, which can be very expensive in terms of time and resource consumption. As a result, there is an urgent need to develop a novel sign-off methodology that can effectively address the challenges arising in the post-Moore era.

AGILE DYNAMIC SIGN-OFF

The primary objective of sign-off procedures is to validate nominal designs, which operate within a specific set of conditions in order to achieve their intended function. With the continuous scaling of semiconductor technology and increasing integration density, the manufacturing process has become more complex, making precise control increasingly difficult [10]. Thus, process variation has emerged as a significant influence on circuit performance [11]. To ensure that the circuit meets the required specification under varying process conditions, sign-off needs to be executed with multiple process corners [12], [13]. Moreover, due to the numerous

979-8-3503-1101-3/23 $31.00 © 2023 IEEE

possible application scenarios for a circuit, it is necessary to perform noise validation with tens of test vectors during sign-off. This has led to the development of multi-corner and multi-scenario sign-off.

While conventional sign-off is challenged in post-Moore era. To address these challenges, we propose a novel sign-off methodology, known as agile dynamic sign-off, which has great potential to facilitate more accurate and efficient chip design and validation in post-Moore era with the following characteristics:

- *Dynamic modeling*: The agile dynamic sign-off utilizes dynamic models to accurately capture the dynamic behavior of the circuit. This involves building different models that can effectively model the variable system when the on-die parasitics change with input vectors [14]. Unlike conventional static analysis, which ignores the dynamic parasitics effects that lead to state-dependent system, dynamic modeling based sign-off can be more precise and provide more accurate characterizations.

- *Artificial intelligence (AI) assisted acceleration*: Agile dynamic sign-off leverages artificial intelligence techniques to speed up the sign-off process. Machine learning algorithms can be used to identify patterns and relationships in large datasets, reducing the time and cost required for sign-off [15]-[18]. In addition, the use of AI techniques can also improve the accuracy of sign-off results, as it enables the identification of previously unknown correlations and dependencies between system components.

- *Multi-objective coupling analysis*: Agile dynamic sign-off also integrates multi-objective coupling analysis, which allows for a more comprehensive and efficient analysis of the relationship between timing, power and noise in the system. By considering multiple objectives at the same time, it is possible to obtain a more accurate and realistic characterization of the system and make better decisions in terms of trade-offs between performance, power consumption and noise.

With the agile dynamic sign-off, more accurate and efficient full-chip sign-off can be achieved in the post-Moore era.

APPLICATIONS OF AGILE DYNAMIC SIGN-OFF

Some of our previous works have proposed agile dynamic sign-off methodologies in system modeling and simulation, which have demonstrated good performance [14], [16], [19]-[21]. By building dynamic models to capture the dynamic behavior of the system and using AI algorithms to speed up the sign-off process, the proposed approaches can achieve accurate and efficient system analysis. Moreover, the multi-objective coupling analysis

capability of agile dynamic sign-off enables a more comprehensive and integrated analysis of the system, taking into account the interplay between timing, power and noise. Due to space limit, two of representative works about dynamic modeling and AI-assisted acceleration are introduced in this paper to demonstrate the effectiveness of agile dynamic sign-off [14], [16].

Dynamic Modeling

We propose a silicon-validated methodology for power delivery (PD) modeling and simulation on a 32-nm double data-rate (DDR) I/O design. We extract the intrinsic interconnect coupling capacitance as well as power/ground grid resistance and inductance netlist from chip layout and active device capacitance from design SPICE netlist. The intentional decap model is obtained from its netlist and then connected to the on-die resistance grid. The S-parameter of the package and motherboard is extracted and then converted into an equivalent RLGC circuit. With the I/O excitation that is modeled as voltage-controlled current sources, the PD system is constructed as a linear system for simulation. We will mainly discuss the dynamic parasitics effects in the following.

On-die decap consists of intentional decap inserted by designers and intrinsic decap. In tradition, the intrinsic decap is considered to be mainly composed of power and ground (P-G) coupling capacitance and active devices capacitance [22], [23]. While the measurement of on-die capacitance (C_{die}) indicates that only accounting for the intentional decap along with P-G coupling and active device capacitance will lead to underestimation [14]. Thus, we revisit the potential C_{die} and find out two other possible

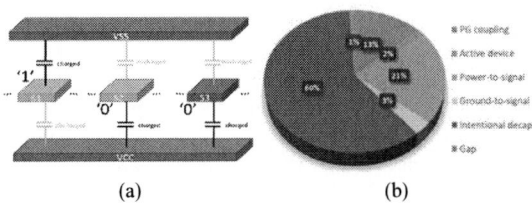

(a)   (b)

Figure 1: (a) Illustration of P/G-S coupling capacitance; (b) Distribution of estimated on-die capacitance [14].

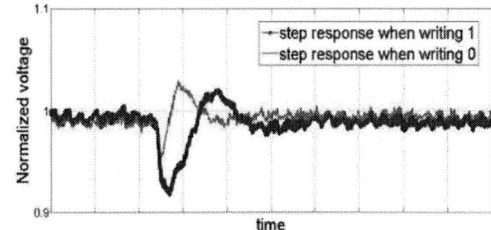

Figure 2: Step response of the DDR PD system when writing 1 and 0 [14].

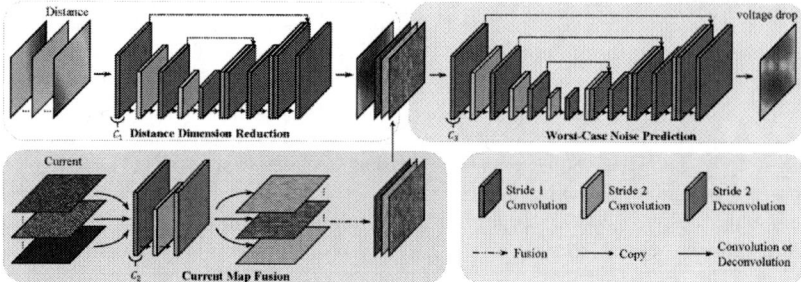

Figure 3: Architecture of the worst-case dynamic PDN noise prediction model [16].

contributors, i.e. well capacitance and power/ground and signal wires (P/G-S) coupling capacitance. The latter is the coupling capacitance between the power (or ground) net and signal net, which depends on the logic value of signal. As shown in Figure 1(a), when signal is set to 1, the capacitance between the signal and vss is charged while the capacitance between the signal and power is ineffective. The reverse holds when signal is set to 0. Figure 1(b) shows the distribution of the estimated C_{die}. The gap between estimation and measurement is reduced to only 1%, which indicates that accounting for the P/G-S coupling capacitance can help improve the accuracy of C_{die} estimation.

Due to the dynamic properties of P/G-S coupling capacitance, the on-die PD system is asymmetry, which means that the system response is different when signal is 1 and 0. Figure 2 shows the step response of a system without intentional decap when the system is writing 1 and 0, which demonstrates the asymmetry in the system. Based on it, we propose a step-response-based approach for waveform reconstruction to accelerate the simulation. Due to the asymmetry of the system, we extract the rise step response and fall step response respectively to accurately model the system. Then, the output waveform can be calculated by linear superposition according to the input vector. If the system needs to be simulated with multiple periodic patterns, the proposed approach can significantly improve the simulation efficiency.

AI-Assisted Acceleration

To resolve the difficulty of large-scale system sign-off, we propose a worst-case dynamic noise prediction framework for power distribution network (PDN) using machine learning. We first partition the PDN into an array of tiles to reduce the input and output dimensions. Temporal compression is also employed to remove the unimportance segments in input test vectors, which do not contribute to the worst-case noise. Then, we extract load current and distance to power bumps as the features, which are easily accessible and provide sufficient information for learning without much training overhead. The load current models the excitation of the system, while the distance to power bumps represents the

capability of supplying currents from the external sources to the tiles. We design a novel convolutional neural network (CNN) architecture to predict the worst-case dynamic noise for each tile. The simulation result from a commercial sign-off tool is taken as ground truth to train the model. Then the worst-case noise can be predicted by the trained model given the test vectors with high efficiency and accuracy.

The proposed CNN architecture is shown in Figure 3, which consists of three subnets for distance feature processing, current map fusion and noise prediction respectively. The dimension of distance feature map is reduced by dimension reduction subnet first since only a few of the power bumps has significant impact on worst-case noise for a particular tile. Then each sampled current map is sent to the fusion subnet separately to learning the critical timing information in the input vector, after which three fused maps are obtained. The dimension-reduced distance map is then concatenated with fused maps and sent to the worst-case noise prediction subnet. The final output is the predicted worst-case noise map. It is noted that the model is fully convolutional so that the noise distribution map of the whole PDN can be predicted with one-time execution, which greatly improve the efficiency.

Figure 4 shows the predicted worst-case noise maps and the corresponding ground truths on three test cases. The noise maps predicted by the proposed framework are

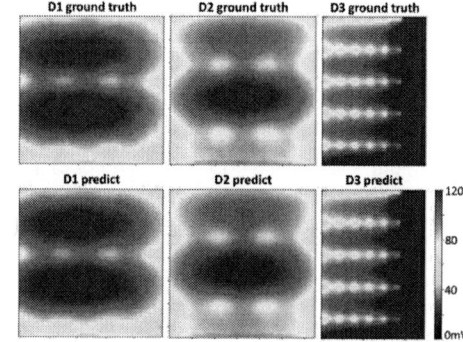

Figure 4: Comparison between the ground truth and predicted worst-case noise map [16].

almost identical to the simulation results. On the other hand, the proposed framework achieves 25-69× speedup compared to the commercial tool. Therefore, the proposed framework can effectively accelerate the noise validation while maintaining high accuracy.

CONCLUSIONS

In conclusion, agile dynamic sign-off is a novel concept that addresses the challenges faced by conventional sign-off in the post-Moore era. With its dynamic modeling, AI-assisted acceleration, and multi-objective coupling analysis capabilities, agile dynamic sign-off is of great help to improve the accuracy, efficiency and comprehensiveness of chip validation. This paper presents two representative works that demonstrate the effectiveness of agile dynamic sign-off in system modeling and simulation, and highlights its potential to support more efficient chip design in the future.

ACKNOWLEDGEMENTS

This work was supported in part by Guangdong Provincial Key R&D program (Grant No. 2021B1101270003), Zhejiang Provincial NSF (Grant No. LD21F040003), NSFC (Grant No. 62034007, 61974133, and 62141404) and SGC Cooperation Project (Grant No. M-0612).

REFERENCES

[1] C. Zhuo, et al., "From layout to system: Early stage power delivery and architecture co-exploration," *IEEE Trans. Comput.-Aided Design Integr. Circuits Syst.*, vol. 38, no. 7, pp. 1291-1304, 2019.

[2] A. Hosny, et al., "Characterizing and optimizing eda flows for the cloud," *IEEE Trans. Comput.-Aided Design Integr. Circuits Syst.*, vol. 41, no. 9, pp. 3040-3051, 2022.

[3] E. Chiprout, "On-die power grids: The missing link," in *Proc. Design Autom. Conf.*, pp. 940-945, 2010.

[4] C.-J. R. Shi, "Mixed-signal system-on-chip verification using a recursively-verifying-modeling (rvm) methodology," in *Proc. Int. Symp. Circuits Syst.*, pp. 1432-1435, 2010.

[5] S. Pant, et al., "Power grid physics and implications for cad," *IEEE Des. Test Comput.*, vol. 24, no. 3, pp. 246-254, 2007.

[6] C. Zhuo, et al., "A fast method to estimate through-bump current for power delivery verification," *IEEE Trans. Comput.-Aided Design Integr. Circuits Syst.*, 2022.

[7] K. Arabi, et al., "Power supply noise in socs: Metrics, management, and measurement," *IEEE Des. Test Comput.*, vol. 24, no. 3, pp. 236-244, 2007.

[8] H. H. Chen, et al., "Power supply noise analysis methodology for deep-submicron vlsi chip design," in *Proc. Design Autom. Conf.*, pp. 638-643, 1997.

[9] A. Devgan, et al., "Block-based static timing analysis with uncertainty," in *Proc. Int. Conf. Comput.-Aided Design*, pp. 607-614, 2003.

[10] D. Blaauw, et al., "Statistical timing analysis: From basic principles to state of the art," *IEEE Trans. Comput.-Aided Design Integr. Circuits Syst.*, vol. 27, no. 4, pp. 589-607, 2008.

[11] V. Mehrotra, et al., "Technology scaling impact of variation on clock skew and interconnect delay," in *Proc. Int. Interconnect Technol. Conf.*, pp. 122-124, 2001.

[12] A. Agarwal, et al., "Statistical timing analysis for intra-die process variations with spatial correlations," in *Proc. Int. Conf. Comput.-Aided Design*, pp. 900-907, 2003.

[13] B. Frankel, et al., "Post-silicon analysis of shielded interconnect delays for useful skew clock design," *IEEE Trans. Electron Devices*, vol. 66, no. 11, pp. 4875-4882, 2019.

[14] C. Zhuo, et al., "Silicon-validated power delivery modeling and analysis on a 32-nm ddr i/o interface," *IEEE Trans. VLSI Syst.*, vol. 23, no. 9, pp. 1760-1771, 2015.

[15] Z. Xie, et al., "Fast ir drop estimation with machine learning," in *Proc. Int. Conf. Comput.-Aided Design*, pp. 1-8, 2020.

[16] X. Dong, et al., "Worst-case dynamic power distribution network noise prediction using convolutional neural network," in *Proc. Design Autom. Conf.*, pp. 1225-1230, 2022.

[17] X. Dong, et al. "Worst-case power integrity prediction using convolutional neural network," *ACM Trans. Des. Automat. Electron. Syst.*, pp. 1-19, 2022.

[18] Y. Chen, et al., "Application of deep learning in back-end simulation: Challenges and opportunities," in *Proc. Asia South Pac. Des. Autom. Conf.*, pp. 641-646, 2022.

[19] C. Zhuo, et al., "A cross-layer approach for early-stage power grid design and optimization," *ACM J. Emerg. Tech. Com.*, vol. 12, no. 3, pp. 1-20, 2015.

[20] J. Chen, et al., "A multi-core chip load model for pdn analysis considering voltage-current-timing interdependency and operation mode transitions," *IEEE Trans. Comp., Packag., Manufact. Technol.*, vol. 9, no. 9, pp. 1669-1679, 2019.

[21] K. Unda, et al., "Cn-sim: A cycle-accurate full system power delivery noise simulator," in *Proc. Asia South Pac. Des. Autom. Conf.*, pp. 554-559, 2017.

[22] R. Panda, et al., "Model and analysis for combined package and on-chip power grid simulation," in *Proc. Int. Symp. Low Power Electron. Design*, pp. 179-184, 2000.

[23] A. V. Mezhiba, et al., "Inductive properties of highperformance power distribution grids," *IEEE Trans. VLSI Syst.*, vol. 10, no. 6, pp. 762-776, 2002.

A 2-D MULTI-DIELECTRIC CAPACITANCE SOLVER BASED ON FLOATING RANDOM WALK METHOD

Jiahao Xu[*], Yibin Zhang[*], Shenghan Gao, Jiecheng Huang, Ming Yang, and Wenjian Yu

Department of Computer Science and Technology, Tsinghua University, Beijing 100084, China

Email: yu-wj@tsinghua.edu.cn

ABSTRACT

We present a 2-D capacitance solver based on floating random walk (FRW) method for handling realistic interconnect structures in ICs. The solver is able to accurately simulate the structures with multiple and conformal dielectrics, and circular conductors in the cross-section of 3-D IC. An approach of using transition circles and space management is proposed to accelerate the computation for structure with circular conductors. A visualization interface is presented to demonstrate the FRW algorithm as well. Experiments with industrial cases show that the presented solver runs much faster than the existing solvers based on finite difference method (FDM), while preserving reliable accuracy with error within 3%.

INTRODUCTION

Capacitance extraction is a fundamental building block for physical design and verification of nanometer integrated circuits (ICs) [1, 2]. It captures the electrostatic effect among IC interconnect wires to express as capacitance components, which together with resistances form the equivalent RC circuits for modeling parasitic effects of interconnects. Capacitance extraction is more difficult than resistance extraction, and developing accurate and efficient capacitance solver is a crucial task for relevant research and EDA tool [1].

The methods for capacitance solver can be classified as domain discretization method [3-7] and floating random walk (FRW) method [8-11]. The former includes finite difference method (FDM), finite element method (FEM) and boundary element method (BEM). They form and solve a linear equation system after domain discretization, and thus are suitable for small structures. The latter is based on Monte Carlo method, so that it scales to large structure. There is no discretization, and the error of FRW method is mainly stochastic error. Therefore, it has more reliable accuracy than the domain discretization method.

In this paper, we present an efficient 2-D capacitance solver based on FRW method, which has reliable accuracy and can be used in different scenarios. Firstly, we present the technique to handle multi-dielectric structures, with multi-layered and conformal dielectrics. Then, an approach is presented to handle circular conductors which are the cross-sections of through-silicon vias (TSVs) in 3-D IC. The strategy using transition circle and space management is proposed to accelerate the computation. Finally, a visualization interface based on OpenCV is implemented for better debugging and demonstration of

the algorithm. Experimental results on industrial cases validate the accuracy the proposed solver and show its runtime speedup over the 2-D FDM based solver [3, 4].

BACKGROUND

A 2-D interconnect structure consists of a number of conductors (usually of rectangle shape) embedded in multiple dielectrics. One conductor is called master conductor, and capacitance solver is used to compute the total and couple capacitances of the master conductor.

The FRW method originates from the formula to calculate the electric potential $\phi(r)$ [8, 9]:

$$\phi(r) = \oint_S P(r, r')\phi(r')dr' \tag{1}$$

where S is a closed surface surrounding point r and $P(r, r')$ is a real-valued function called surface Green's function. $P(r, r')$ can be regarded as a probability density function with respect to r'. Therefore, with Eq. (1) $\phi(r)$ can be calculated with Monte Carlo method, which relies on random samples of r' and its potentials. If for a sample $\phi(r')$ is unknown, one can recursively applies (1) and find a sample point with known potential. This results in a spatial "random walk" path, and with thousands of such paths the electric potential can be well calculated. For the 2-D problem, S is always a square centered at r. So, after discretizing the square edges into small segments, a set of probabilities for the locations of r' can be calculated and stored in advance as Green's function table (GFT). And, with this pre-computed GFT and random walk can be quickly executed.

For capacitance extraction, the Gauss theorem infers that the electric charge of master conductor i satisfies [9]

$$Q_i = \oint_{G_i} F(r)g \oint_S \omega(r, r')P(r, r')\phi(r')dr'dr \tag{2}$$

where G_i is the Gaussian surface enclosing conductor i, $F(r)$ is the dielectric permittivity at r, and $\omega(r, r')$ is

$$\omega(r, r') = -\frac{\nabla_r P(r, r') \cdot \hat{n}(r)}{gP(r, r')} \tag{3}$$

with g satisfying $\oint_{G_i} F(r)gdr = 1$ (meaning $F(r)g$ is a probability density function). So, Eq. (2) can also be computed with a nested Monte Carlo method, i.e. the random walk paths. Note $\omega(r, r')$ is the sample value for the coefficient of electric potential in the linear expression of electric charge, and is thus the sample value of capacitance. Like GFT, $\omega(r, r')$ can also be calculated easily with the pre-computed and stored weight value table (WVT). This is the FRW method for capacitance extraction, with thousands of random-walk paths. Each path is a sampling process which repeatedly constructs the transition square S and randomly chooses r' until the last r' touches a conductor (see Fig. 1).

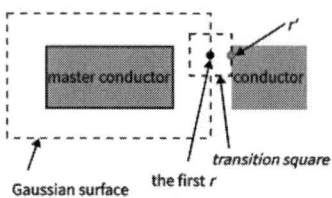

Figure 1: The FRW method for capacitance extraction.

HANDLING MULTI-DIELECTRIC CASES

The 2-D FRW method was proposed in [8], which only considers single-dielectric environment. There is no literature to disclose a 2-D FRW algorithm handling actual structure with multi-layered dielectrics. To resolve it, we borrow the idea from the 3-D FRW algorithm [9]. Similarly, we utilize a transition square with two dielectric layers to make the walk across dielectric interface. For this purpose, we enumerate some possible positions of the interface between two kinds of dielectrics, and some possible permittivity ratios of the two dielectrics. For each configuration we use an FDM method [9] to compute the GFT and WVT (see Fig. 2). This produces a set of GFTs and WVTs to enable handling multi-dielectric structures.

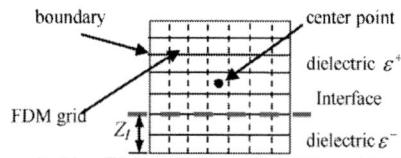

Figure 2: Use FDM to compute GFTs and WVTs [9].

Conformal dielectric is a kind of nonplanar dielectric covering the surface of conductor. The above technique with two-layer transition square can handle it, but leads to large computing time, as if the sample point falls near the vertices of conformal dielectric, the transition square will be very small. To overcome this issue, we use the quadrant transition square in Fig. 3(b) instead, inspired by [10] for 3-D structure. During the FRW process, the average permittivity for each quarter of the transition square is calculated, and then its GFT is analytically obtained.

The 2-D FRW method with above techniques is implemented in C++ and become a program named RW2D. Four interconnect cross-section structures from our industrial partner are used to validate the performance of

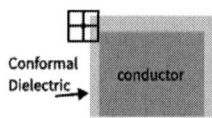 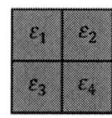

(a) conformal dielectric   (b)quadrant transition square
Figure 3: Handle conformal dielectric with quadrant square.

RW2D. All experiments below are carried out with a laptop with Intel Core i7-10750H CPU @2.60GHz. The four cases all include layered dielectrics and conformal dielectrics. The computational results are listed in Table I. Raphael is a widely used commercial software [1] as golden value, while FDCap2d is the FDM solver developed in [2]. From Table I we see that the technique using quadrant transition square brings **2X to 4X speedup** while preserving accuracy (only total capacitances are listed). The runtime and accuracy of RW2D is comparable to FDCap2d. Because the FRW method is very suitable for parallel computing, we also run it with 8 threads. The results show this brings remarkable speedup, and makes RW2D much faster than FDCap2d.

Table I. Results for structures with conformal dielectrics.

Case	Algorithm	N_{thread}	Capacitance(aF)	Error	Time
1	Raphael[3]	–	189.08	–	–
	FDCap2d[4]	–	187.87	0.64%	0.51s
	RW2D (two-layer)	1	193.32	2.24%	5.06s
		8	193.48	2.33%	1.02s
	RW2D (quadrant)	1	185.22	2.04%	1.29s
		8	184.19	2.59%	**0.27s**
2	Raphael[3]	–	191.19	–	–
	FDCap2d[4]	–	190.06	0.59%	0.64s
	RW2D (two-layer)	1	186.13	2.65%	2.35s
		8	191.50	0.16%	0.49s
	RW2D (quadrant)	1	188.44	1.44%	0.70s
		8	193.18	1.04%	**0.14s**
3	Raphael[3]	–	194.99	–	–
	FDCap2d[4]	–	194.08	0.47%	0.40s
	RW2D (two-layer)	1	197.41	1.24%	0.88s
		8	189.53	2.80%	0.23s
	RW2D (quadrant)	1	196.42	0.73%	0.45s
		8	193.68	0.67%	**0.10s**
4	Raphael[3]	–	200.68	–	–
	FDCap2d[4]	–	198.47	1.10%	0.70s
	RW2D (two-layer)	1	194.45	3.10%	2.48s
		8	200.70	0.01%	0.48s
	RW2D (quadrant)	1	200.29	0.19%	1.35s
		8	199.38	0.65%	**0.30s**

HANDLING CIRCULAR TSV CASES

3-D IC is a new architecture with multiple tiers of silicon dies, where different dies are interconnected with TSVs. The TSV has larger size and the capacitance coupling among them are prominent. The 2-D capacitance solver can be used to extract the top-view cross-section of 3-D IC to extract the TSV capacitances. For this aim, we enhance RW2D to handle this structure of circular TSVs.

The first method is to replace transition square with transition circle (which is also maximized in size). Noting that since the transition circle intersects with any one conductor in no more than one point, the walk should terminate as long as it reaches a position that is near enough to one conductor.

The second method is derived from previous research of 3-D case [11], which still applies transition squares. However, if the maximal size of transition square is restricted by a circular conductor, this method attempts to rotate the transition square letting one of its sides touch the restricting conductor (see Fig. 4(a)). With this strategy, the average number of hops per walk in FRW process is expected to be reduced. However, its overall efficiency may be inferior due to the larger amount of computation.

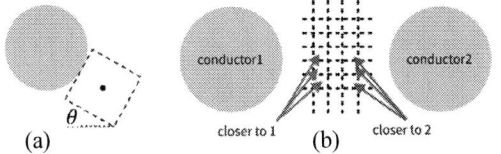

Figure 4: (a)*Rotate square (b) Circle with pre-gridding.*

An optimization technique can be devised for handling structure with a large number of circular conductors. It results in less computing time for finding the nearest conductor. It is similar to the grid-based space management. In the initialization stage, the 2-D space is divided into grid, the attempt is to find out whether all the points in one unit have the same conductor restricting the maximal transition circle (see Fig. 4(b)). If so, the process of enumerating all the conductors can be omitted when the start point of a hop is inside this unit. We combine it with the first method using transition circle.

We generated 40 structures with circular conductors (each includes at most 50 conductors), and tested the above algorithms. The total runtime and average errors are listed in Table II, where "circle1" and "square" label the first two methods respectively, and "circle2" labels the first method plus the grid-based space management. We see that all these methods have good accuracy. The proposed method using transition circle plus grid-based space management accelerates the computation, which is **3.5X faster** than the naïve method derived from [11]. RW2D is over 100X faster than Raphael with less than 1.5% error.

Table II. Results for TSV cases with circular conductors

Case	Algorithm	N_{thread}	Error(avg)	Error(max)	Time
TSV1~ TSV40	Raphael[3]	–	–	–	1318s
	RW2D(circle1)	1	0.43%	1.32%	29.82s
		8	0.46%	1.08%	6.97s
	RW2D(square)	1	0.39%	1.16%	50.15s
		8	0.39%	1.29%	10.66s
	RW2D(circle2)	1	0.43%	1.32%	11.11s
		8	0.46%	1.08%	**3.06s**

VISUALIZATION AND MORE RESULTS

For RW2D, we also developed a visualization interface

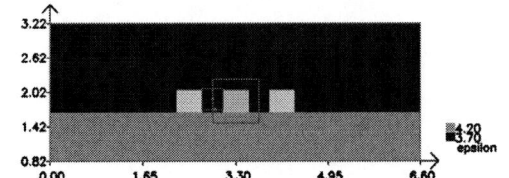

Figure 5: The visualization of 2-D FRW algorithm.

based on OpenCV to display the process of the floating random walk. This makes debugging more convenient (see Fig. 5). More performance evaluation is performed with a dataset "iscas89", which contains 500 structures (each includes 4~5 conductors and no more than 5 dielectric layers). The results are listed in Table III, showing that RW2D runs **5X faster** than FDCap2d, with a little error within 2% on total capacitance. We have also checked that the coupling capacitances in each case is of good accuracy.

Table III. Results for more test cases.

Case	Algorithm	N_{thread}	Error(avg)	Error(max)	Time
"iscas89" dataset	Raphael[3]	–	–	–	255s
	FDCap2d[4]	–	0.37%	1.75%	180s
	RW2D	1	0.48%	1.70%	121.5s
		8	0.36%	1.52%	**33.5s**

CONCLUSIONS

In this work, we have developed a 2-D capacitance solver RW2D based on FRW algorithm. It includes novel techniques to handle multi-dielectric cases with conformal dielectrics, and the structures with circular TSVs. The experiments show that the solver is much faster than the solvers based on FDM while preserving reliable accuracy.

REFERENCES

[1] W. Yu and X. Wang, *Advanced Field-Solver Techniques for RC Extraction of Integrated Circuits*, Springer Inc., Apr. 2014.
[2] W. Gong, W. Yu, Y. Lu, et al., "A parasitic extraction method of VLSI interconnects for pre-route timing analysis," in *Proc. ICCCAS*, Chengdu, China, July 2010, pp. 871-875.
[3] Synopsys Inc., TCAD-Raphael, https://www.synopsys.com /silicon/tcad/interconnect-simulation/raphael.html.
[4] W. Liang and W. Yu, "A 2-D capacitance solver with finite difference method," in *Proc. CSTIC*, Shanghai, China, Jun. 2020.
[5] K. Nabors K and J. White, "FastCap: A multipole accelerated 3-D capacitance extraction program," *IEEE Trans. Computer-Aided Design.*, 10(11): 1447-1459, 1991.
[6] W. Yu, Z. Wang, et al, "Preconditioned multi-zone boundary element analysis for fast 3D electric simulation," *Engineering Analysis with Boundary Elements*, 28(9): 1035-1044, 2004.
[7] W. Yu, Q. Zhang, Z. Ye, and Z. Luo, "Efficient statistical capacitance extraction of nanometer interconnects considering the on-chip line edge roughness," *Microelectronics Reliability*, 52(4): 704-710, 2012.
[8] Y. Le Coz and R. B. Iverson, "A stochastic algorithm for high speed capacitance extraction in integrated circuits," *Solid-State Electron.*, 35(7): 1005-1012, 1992.
[9] W. Yu, H. Zhuang, C. Zhang, G. Hu, and Z. Liu, "RWCap: A floating random walk solver for 3-D capacitance extraction of very-large-scale integration interconnects," *IEEE Trans. Computer-Aided Design.*, 32(3): 353–366, Mar. 2013.
[10] M. Yang and W. Yu, "Floating random walk capacitance solver tackling conformal dielectric with on-the-fly sampling on eight-octant transition cubes," *IEEE Trans. Computer-Aided Design*, 39(12): 4935-4943, 2020.
[11] C. Zhang, W. Yu, Q. Wang, and Y. Shi, "Fast random walk based capacitance extraction for the 3-D IC structures with cylindrical inter-tier-vias," *IEEE Trans. Computer-Aided Design*, 34(12): 1977-1990, 2015

CORRELATION ANALYSIS BETWEEN DEFECT SCANNING AND MACHINE COMPONENTS

Ming Guo[1]

[1] Shanghai Huali Integrated Circuit Corporation, Shanghai 201314, China

ABSTRACT

Based on the correlation between the scanning defect wafer backside rubbing information and the components, the problem components and machine information can be located according to the wafer backside defects. The existing technical methods mainly rely on experience to find the correlation with the scanning components. The defect detection has developed from the traditional manual classification to the method based on machine vision [1]. This method mainly uses the machine learning method to extract the component feature information. Based on the scanning defect feature and the position point of the machine component, and combined with the pixel point feature information, the correlation between the yield of the scanning defect and the correlation between the defect and the scanning machine component is analyzed. The defect feature of the wafer backside image is obtained by extracting the feature of the online scanning defect image, Analyze the correlation between scanning defects and product yield, and locate the problem parts. This method can trace the scanning defects according to the characteristic information of the scanning defects and product yield, and automatically locate the relevant machine parts in time.

INTRODUCTION

In view of the correlation between scanning defects and machine components, the defects on the back of the crystal include surface cracks, scratches, pinholes and other types of defects [2]. The traditional method is to compare the positioning of the sampling frame, determine the defects through the comparison accuracy [3], and fit and compare the defect marks with the sampling least square method [4]. This method uses the machine recognition method to learn the component information. Based on the scanning defect features and the position points and pixel point features of the machine components, the yield of the scanning defects and the correlation between the defects and the scanning machine components are analyzed. Based on the feature information of the scanning defects and the yield of the products, the component traceability analysis of the scanning defects can be carried out to automatically locate the relevant machine components.

COMPONENT MATCHING PROCESS

In order to better match the defect features with the component features, it is necessary to locate the component features based on the position and position information of the component feature points and lock the component defect features; Compare the component features with the wafer backside features to extract the position information of the key points of the component;

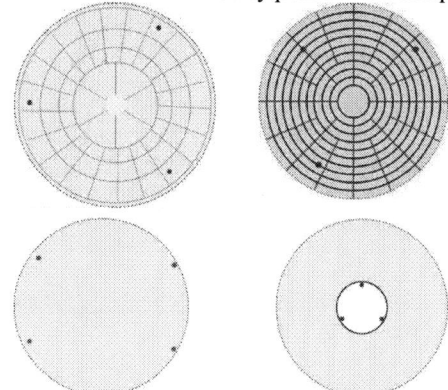

Figure 1: Component Key Markers

Feature recognition of wafer backside defects

This method is mainly used to endow the defect feature pixel Map coordinate system information after preprocessing the scanned wafer backside image, conduct defect superimposition on the Map in combination with the yield of the product, and then compare with the machine part database through the correlation between the yield and the online defect scanning results, so as to automatically locate the machine part with the highest correlation.

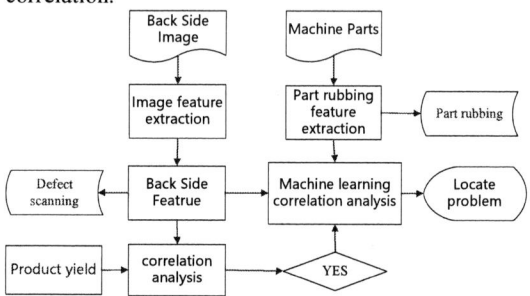

Figure 2: Part to scan defect correlation comparison process

Component feature labeling and yield matching

The characteristics of wafer backside image and

product yield are extracted, the correlation between the two is compared, combined with the component rubbing result information, compared with the component features, the correlation analysis of the three is carried out, and the product yield and scanning defects are automatically located to the problem machine parts through the correlation analysis of component correlation and product yield.

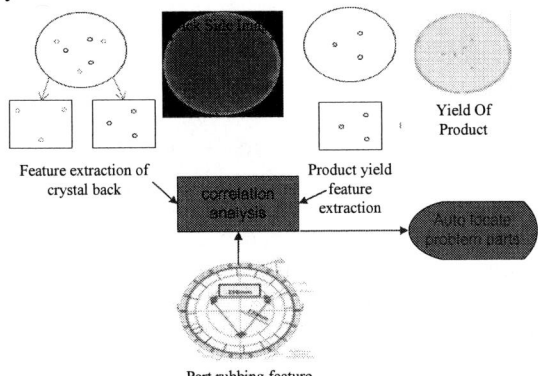

Figure 3: BackSide Image With Yield Map Auto Locate Problems Of Parts

Wafer backside defect feature matching effect

Through the correlation analysis method between scanning defects and machine parts, the image recognition data can be queried by importing images at the front end, which can realize the automatic matching of scanned defect images and defects with product parts, and automatically locate problem machine parts.

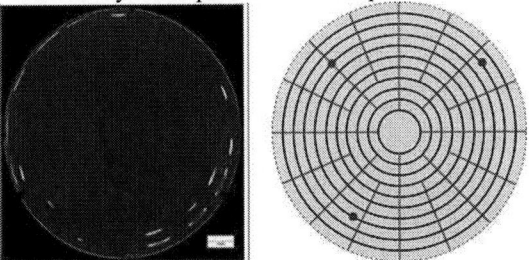

Figure 4: Automatic defect matching part information

CONCLUSION

Combined with the key point feature information of the wafer backside image and the component rubbing information for correlation matching, the correlation between the scanning defect and the expansion of the feature point of the part is analyzed, and the correlation of the relevant part information is automatically analyzed by the imported wafer backside image, and then the problem part machine can be quickly located, combined with the product yield information, the part information that affects the product yield can be quickly tracked, and the technical support is provided for the scanning defect and the

positioning of the problem part, which is conducive to quickly analyzing the correlation between the yield and the problem part.

ACKNOWLEDGEMENTS

Thank you very much for the leadership and colleagues support of my work

REFERENCES

[1] WANG Sainan, MENG Xianjiao, ZHAO Yingjian. Design of Part Defect Detection System Based on Machine Vision[J]. Science and Technology Innovation Herald, 2019, 16(23):2.

[2] SU Junhong, LIU Shengli. Detection and morphology analysis of surface defects of cylindrical high-precision parts[J].Progress in Laser & Optoelectronics,2014,51(4):150:154.)

[3] DU Liuqing, YU Yongwei. Machine vision detection and identification method for magnetic tile surface defects[J].Chinese Journal of Graphics,2014,35(4):590:594.)

[4] LIU Yuanpeng, ZHANG Dinghua, GUI Yuankun, et al. Fitting planar circular curve by least squares method withconstraints[J].Journal of Computer Aided Design&ComputerGraphics,2004,16(10):1382:1385)

Essential Steps to Enable Analyzing Effective Resistance of ESD Paths-PG Routing Network Pruning and Resistance Contribution by Layer

Frank Feng, Abner Huang, Joe Huang, Dawson Chiou, Jeff Byrd, Nicholas Palmer,
Charles McFalls, and Akhil A. Gore

Synopsys Inc. & Synopsys Taiwan Co., Ltd

690 East Middlefield Road Mountain View, CA 94043

hhfeng@synopsys.com, abnercyh@synopsys.com, joehuang@synopsys.com, dawsonc@synopsys.com

ABSTRACT

Electrostatic Discharge (ESD) compliant interconnects are critical to protect internal circuitry. Full chip dynamic simulation can't be a practical solution now, the compliance is analyzed in static points-to-points effective resistance (P2PR) and current density of interconnects along as-designed ESD paths. This paper focuses on P2PR. Performing ESD paths P2PR measurements on full chip are performance challenging, analyzing ESD paths P2PR incompliance are very complex too. This paper introduces two essential steps to make ESD paths P2PR measurement & incompliance analysis feasible on full chip production design flow. They are Power & Ground routing networks pruning in graph theory assisted by manufacturing data, and a reliable method to breakdown lumped P2PR value into resistance contribution by layer for incompliance analysis to help layout fix.

INTRODUCTION

For static P2P effective resistance (P2PR) measurements relevant to interconnect of ESD paths, each P2P path is constituted of source (a group of pins or physical pads where current is injected on), sink (a group of pins or physical pads where branches of current are converged at), and interconnects along as-designed ESD paths. For example, on certain power net, source of a P2P path could be a group of ESD diode finger pins connecting to this power net, sink of the same P2P path could be a group of power clamp MOS finger pins connecting to this power net. The P2PR is generated through a sequence of steps that find P2P source & sink pins cluster on layout extracted netlist, extract resistor network for relevant net, and matrix solver to solve current (I) and voltage (V) for all resistors of entire resistor network associated to targeted P2P paths, then the lumped P2PR is calculated as ($V_{Source} - V_{Sink}$) / I_{Inject}. A full chip design of advanced node can easily contain thousands of ESD paths involving power/ground nets. Because of this fact, a practical and reliable EDA methodology to calculate P2PR for as-design ESD paths in a full chip is nontrivial to build. In concerning of both accuracy and performance, unique reduction in resistance extraction engine, power/ground (P/G) nets pruning, and matrix solving algorithm tuning are all to be optimized to handle huge amounts of polygons and resistors data according to ESD path characteristics. This methodology does not estimate P2PR using shortest path scheme among the layout or resistors network between source and sink, it is executed through an accurate matrix solving focused on ESD paths centric resistors network. These resistors networks are generated by resistance extractor applying on inputted layout polygons data, if polygons belong to P/G nets, then they are usually processed by pruning engine prior to resistance extraction step to remove polygons considered to have minor or no current flowing through when an ESD event occurs. Without prune, resistance extraction on P/G nets will be very difficult to complete in reasonable turnaround for a transistor-level layout database. ESD paths centric P/G nets pruning is essential to full chip ESD paths P2PR measurements. Besides performance concerns, an useful prune heuristic can't cut off ESD paths of interest, and needs to preserve enough amount of polygons to result in good accuracy.

In typical full chip ESD paths P2PR measurements, the so-called "reduction mode" is turned on in resistance extractor. Its function is to reduce size of resistors network at netlisting stage of resistance extraction, and still maintain good P2PR accuracy in comparison to results at no reduction. This is necessary when working on a big design and requiring quick turnaround. However, due to reduction mode is turned on during resistance extraction, the resulted resistors network can't contain exact layer correlation information to the resistors. After matrix solver solved resistors network of P2P paths, the measurement flow can generate accurate, but lumped (or single) P2PR value for each selected P2P path. All P2PR values are then compared against constraints, if exceeding constraint, the corresponding ESD paths are considered as ineffective to protect internal functional circuits from ESD events. To fix such ESD design issue, that means to bring the P2PR down below (or close to) the constraint, some layout change is needed. From layout and electrically conducting points of view, a P2P path (can represent a segment of or completed as-designed ESD path) is consist of a pair of source/sink (group of device pins or physical pads) connected by polygons of multiple interconnecting stacked layers forming path between source and sink. It is difficult to rationalize a lumped P2PR from complex resistors network in-between source and sink, and the layout fix to meet ESD compliance can't be only on some single polygon. User needs to understand what to be focused on, so it is useful to breakdown a single lumped P2PR value into a list of resistance contribution by layer [1]. Such information will let user understand the higher resistance contribution layers and indicate various scenarios of 1. Should placement of source/sink need to be changed 2. More source/sink device fingers to be added into layout 3. Some metal / via layer routings belong to the concerned P2P path are to be changed. P2PR contribution by layer data is essential to rationalize & fix interconnect ESD incompliance.

ESD PATHS CENTRIC P/G PRUNING

Our graph-based pruning [2] is assisted by manufacturing data to apply on polygons of P/G nets, we use the following techniques to achieve the expected results of accuracy and performance.

Layer Order Traversal

Layer-Order-Traversal (LOT) method uses graph traversal to reduce data size. Starting with current sources/sinks points at the device layers, we keep all visited nodes and prune out all unvisited nodes. Layer-Order (manufacturing connection stack order of layers) information is used during traversal: It's only allowed to traverse from lower layer to same or higher layer, but not allowed to traverse from higher layer to lower layer. It implies that polygon paths with little impact to fair calculation of P2PR will not be chosen. In our application, a node represents a polygon, and an edge represents a geometric overlapping of two polygons. The core concept of graph pruning is to identify the nodes which have no impact or ignorable impact to the accuracy of P2PR. We model the pruning problem as the graph traversal problem, i.e., traverse the graph from starting nodes via edges until no more nodes can be visited. By resolving the graph traversal problem, the graph pruning problem can also be resolved by keeping all visited nodes and pruning unvisited nodes.

Our layer-order traversing method modeling information of semiconductor manufacturing process as direction information of graphs. In this paper, we will use typical algorithm breadth-first-search (BFS) as underlying traversing method of LOT for convenience.

In graphs, an edge of two polygons (nodes) means that they are overlapping in geometric viewpoint. The concept of geometric overlapping is commutative. It implies that all edges are not directed. The connectivity of layers and layer-stack are defined by each semiconductor manufacturing process of foundries. We call the information as Layer-Order hereafter. Layer-Order will be collected for a particular semiconductor manufacturing process, and then we maintain a query data structure so that we can convert an undirected graph into a directed graph. The figure I is an example of the LOT result on a network with one source, one sink and 7 layers. The nodes with star symbol are the selected nodes by LOT with BFS, and the nodes without star symbol will be pruned.

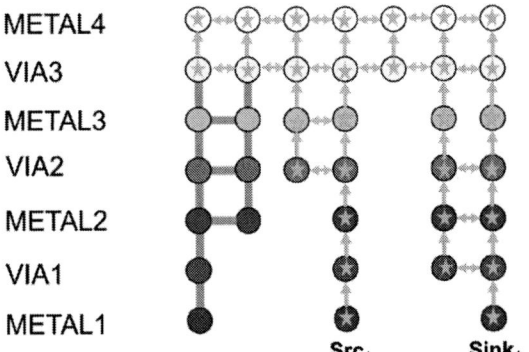

Figure 1: An example of Layer-Order-Traversal from one source, one sink, and 7 layers

Auto Layer Preserving

Auto-Layer-Preserving (ALP) [3] is to further improve pruning runtime and P2PR accuracy, our heuristic chooses the layers we do not want to prune beforehand. In modern VLSI designs, top-most layers are much wider metal lines as in Table I and consist of relatively small portion of polygons of P/G nets. Nevertheless, most of them act as major paths in ESD events, in another word, they significantly influence P2PR of ESD paths. Applying prune on them may likely introduce significant accuracy loss. By using the Layer-Order information and the width of layer via manufacturing information, we can automatically find these wide metal layers which are highly impactive to P2PR accuracy. We keep whole polygons of these layers from being pruned.

TABLE I
Layer Name and Width

Layer name	Width
PAD	3.0 um
METAL15	1.2 um
METAL14	1.2 um
METAL13	0.25 um
METAL12	0.12 um

TABLE II
Prune vs. non-Prune

7 nm Design	Overall P2PR Run Time (hr:min)	Resistors network node count (billion)
No Prune	Can't finish the run	15.98
Prune	23:49	1.35
5nm Design	Overall P2PR Run Time (hr:min)	Resistors network node count (billion)
No Prune	114:47	11.77
Prune	10:59	0.186

Combining LOT and ALP techniques provide an effective way to reduce polygon data size and to make accurate P2PR measurements on transistor-level full chip designs feasible. Test results are shown in Table II. The physical meaning behind the nature of LOT and ALP is that, in the semiconductor manufacturing process of VLSI, top layers play critical role to aggregate the currents from lower layers. For ESD paths, most of nodes from lower layers (or FEOL) of P/G resistors network are irrelevant for calculating equivalent resistance since they are not critical elements of ESD paths except those are near sources and sinks. This insight leads to a simple and fast approach to prune ESD network.

ANALYZE P2PR WITH RESISTANCE CONTRIBUTION BY LAYER

In a production layout design flow, it is only practical to generate data of resistance contribution by layer for P2P paths in violation of constraint, not for all ESD P2P paths of a full chip. To enable breaking down a lump P2P effective resistance to a list of resistance contribution by layer, basically, reduction mode needs to be turned off and via nodes need to be kept during resistance extraction process of a P2P effective resistance measurement. By turning off reduction mode, each resistor presented in the resistors network (in DSPF or SPEF format) is modeled from one individual fractured polygon of a certain layer and has carried layer specific electrical / geometrical properties. It is not an equivalent resistor which may be electrically reduced/modeled from a group of fractured polygons belong to different layer. With non-reduced resistor networks and source/sink pairs of P2P paths for as-designed ESD paths, these data are fed into matrix solver to solve V on each resistor node and I flowing through each resistor. The lumped P2P effective resistance (R_{total}) for each P2P path is still calculated by equation of ($V_{Source} - V_{Sink}$) / I_{Inject}. The resistance contribution for each layer (R_{layer}) is calculated as $R_{layer} = R_{total}$ * $ratio_{layer}$, and the $ratio_{layer}$ is obtained as dividing a collective I/R related electrical property summed from resistors of one layer by a collective I/R related electrical property summed from resistors of all layers for the simulated P2P path.

Below, using commercial EDA tool to demonstrate lumped P2P effective resistance and breakdown into resistance contribution by layer for a few P2P paths. P2P effective resistance checking flow is built in IC Validator PERC of Synopsys. The resistance extraction is executed by industry golden parasitic extraction tool (StarRC). As shown in Fig. 2a, a P2P path is plotted on layout from source (a pad) to sink (a cluster of diode devices/fingers) in fly line, and the P2P path is also displayed by V (voltage value) heatmap map in Red (higher V) → yellow → green → blue (lower V). The lumped P2P effective resistance is 0.5835 ohm, and its breakdown of resistance contribution by layer is shown in Fig. 2b. In the name column, 1st row named as Total is pointing to lumped resistance value. Moving down from 2nd row are displayed by layer name and in the order of contribution percentage from high to low. User can see in the resistance contribution by layer result, resistance contribution is dominated by AP layer contributing to 58.69% and M14 layer contributing to 14.59%. If decide to modify layout to bring lumped P2P effective resistance down for this P2P path, then fix path routing (or placement) directly related to these two layers are first to be considered. For example, bring Source (pad) and Sink (diode devices) closer can directly reduce AP layer routing length.

Figure 2a: P2P path from pad (source) to diodes (sink) plotted in V heatmap

Figure 2b: P2P resistance contribution by layer for P2P path in Fig. 2a

Another example as shown in Fig. 3a, a P2P path is plotted on layout from source (a cluster of diode devices) to sink (a cluster of power clamp nmos devices) in fly line, and the P2P path is also displayed by I (current value) heatmap map in Red (higher I) → yellow → green → blue (lower I). The P2P path in Fig. 2a is on a power net, which is routed all over the design. User can see the higher current path of I heatmap is distributed close to and between source and sink, and rest of area showing in blue in the heatmap indicate there is little current flow through these area. The lumped P2P effective resistance is 2.7089 ohm, and its breakdown of resistance contribution by layer is shown in Fig. 3b. In the name column, 1st row named as Total is pointing to lumped resistance value. Moving down from 2nd row are displayed by layer name and in the order of contribution percentage from high to low. User can see in the resistance contribution by layer result, resistance contribution is dominated by n_odtap_io layer contributing to ~38% and VD_MD_OD_N_IO layer contributing to ~13%. If decide to modify layout to bring lumped P2P effective resistance down for this P2P path, then fix path routing (or placement) directly related to these two layers are first to be considered. Indeed, based on these two layers belonging to diode device (source) layer of this P2P path, it suggests that adding diode device fingers (creating more parallel routing channels to reach source diode pins) is better choice to decrease lumped P2P effective resistance.

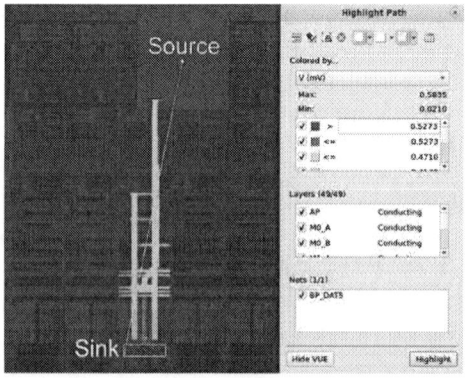

Checking" in *Proc. Taiwan ESD and Reliability Conference*, 2022.

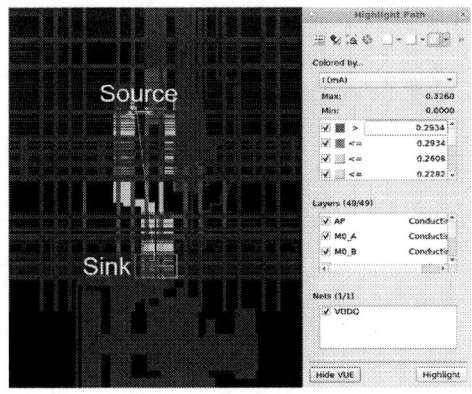

Figure 3a: P2P path from diodes (source) to power clamps (sink) plotted in I heatmap

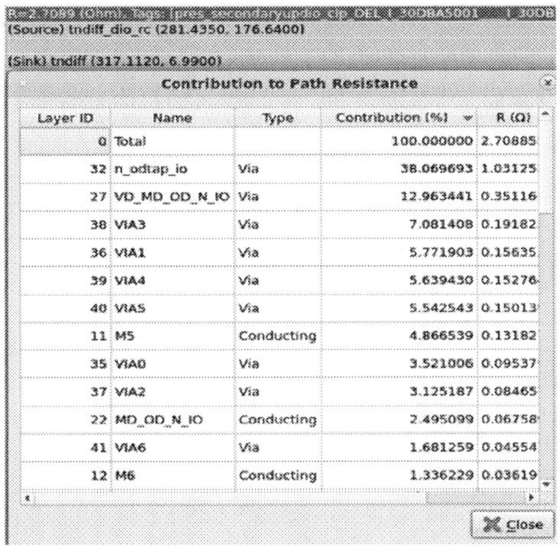

R=2.7089 (Ohm), Tags: [pres secondaryupdio clp DEL 1 30DBA500 1 30DB
(Source) tndiff_dio_rc (281.4350, 176.6400)
(Sink) tndiff (317.1120, 6.9900)

Contribution to Path Resistance

Layer ID	Name	Type	Contribution (%) ▼	R (Ω)
0	Total		100.000000	2.70885
32	n_odtap_io	Via	38.069693	1.03125
27	VD_MD_OD_N_IO	Via	12.963441	0.35116
38	VIA3	Via	7.081408	0.19182
36	VIA1	Via	5.771903	0.15635
39	VIA4	Via	5.639430	0.15276
40	VIA5	Via	5.542543	0.15013
11	M5	Conducting	4.866539	0.13182
35	VIA0	Via	3.521006	0.09537
37	VIA2	Via	3.125187	0.08465
22	MD_OD_N_IO	Conducting	2.495099	0.06758
41	VIA6	Via	1.681259	0.04554
12	M6	Conducting	1.336229	0.03619

Figure 3b: P2P resistance contribution by layer for P2P path in Fig. 3a

SUMMARY

Details of Layer-Order-Traversal & Auto-Layer-Preserving prune heuristic are described, our test data prove its criticality in making full chip P2PR measurements feasible in production chip design flow. Details of obtaining resistance contribution by layers are also described, our examples demonstrate its importance to rationalize P2PR results for guideline of layout fix. These two essential steps are currently implemented in IC Validator PERC for P2P effective resistance checking flow, they are key features to make foundry advanced node tool certification and customer chip tape-out successful for ESD design verification.

REFERENCES

[1] Joe Huang, Judge Lin, Frank Feng, Blanche Wu, Hsin-Fei Weng, and Siao-Lung Hwong, "P2P Resistance Contribution by Layer – An Effective Way of Analyzing ESD Path Interconnect Robustness" in *Proc. Taiwan ESD and Reliability Conference*, 2022.

[2] N. Palmer, C. Lo, A. Huang, D. Chiou, K. Lin, T. Chuang, J. Byrd, C. McFalls, "Power Routing Reduction in Large Scale Design for Electrostatic Discharge Checking" *US patent* No. 63/348,937, Jun 2022.

[3] Abner Huang, Dawson Chiou, "Power Routing Reduction in Large Scale Design for Electrostatic Discharge

AN 18-BIT 2MSPS SAR ADC WITH DOUBLE PASSIVE NOISE-SHAPING CALIBRATION

Xiao-Wei Zhang[1], Jian-Xiong Xi[1], Tao Wang[1], Le-Nian He[1*]

[1]School of Micro-Nano Electronics, Zhejiang University, Hangzhou 310000, China
*Corresponding Author's Email: helenian@zju.edu.cn

ABSTRACT

This paper describes an 18-bit 2MSPS successive approximation register (SAR) analog-to-digital converter (ADC) with double passive noise-shaping calibration. This ADC is intended to have higher accuracy and it only needs minor modifications to conventional architecture, including adding the assistant DACs (ADACs) and integrated capacitors. The ADACs are used to store offset information and mismatch information while integrated capacitors are applied to noise-shaping, combined with the four-port comparator. The proposed SAR ADC with calibration is designed and simulated in a 55-nm CMOS process. The simulated SNDR is improved from 81.24 dB to 111.89 dB.

Keywords: successive approximation register; analog-to-digital converter; assistant DAC; noise-shaping; calibration

INTRODUCTION

In recent years, successive approximation register (SAR) analog-to-digital converter (ADC) has been of great popularity based on the high power efficiency architecture and simple building blocks in nanometer technology for medium-resolution applications. However, the tough requirement on the comparator noise and capacitor mismatch limits the improvement of SAR ADC accuracy. To address the challenges, diverse methods have been explored to reduce the effect of non-ideal factors. Calibration is one of the highly attractive techniques, including foreground calibration and background calibration. In reference [1], a foreground calibration based on deterministic self-calibration and stochastic quantization was proposed to estimate and compensate for the mismatch of DAC. What's more, reference [2] employed a background calibration method for the capacitive DAC and elaborated a way for suppressing the input-referred comparator noise in SAR ADC, which achieves good results. Apart from the development of calibration, there have been emerging efforts to adopt hybrid ADC architectures, which combine the merits of SAR and delta-sigma ADCs. Noise-shaping (NS) SAR ADC provides higher SNDR and reduces the requirement of noise and mismatch while it was first put forward in reference [3]. Furthermore, the SNDR was improved to 90.5dB in reference [4] with the help of 2nd-order MES and aggressive mismatch shaping.

In this work, we introduce a foreground calibration based on the passive noise-shaping technique, which doesn't consume any extra static current and doesn't affect the dynamic range of SAR ADC. Compared with the conventional SAR ADC, assistant DAC and integrated capacitors are added. In order to achieve the goal of noise-shaping, two input ports are added to the comparators. The proposed SAR ADC is designed and simulated in a 55-nm CMOS process.

ARCHITECTURE AND CIRCUITS

The proposed fully-differential 18-bit SAR ADC architecture is illustrated in Fig.1, mainly consisting of 16-bit CDAC arrays, 13-bit ADAC arrays, four-input high-performance comparator, sar logic and calibration logic. CDAC arrays are the main capacitor arrays and ADAC arrays are connected with CDAC arrays through the split capacitors, which are unit capacitors. The bridge-capacitor architecture is used in CDAC arrays to reduce the area and power. However, more non-linearity is introduced. What's more, the offset of the comparator also affects the results of ADC. Thus the calibration is proposed to reduce the effect of non-ideal factors while the noise-shaping technique is used to improve the resolution without expanding the scale of CDAC arrays.

The operation of SAR ADC includes two phases, the calibration phase and the normal work phase. The calibration phase means that the non-ideal factors are detected by ADAC arrays, including offset and mismatch of CDAC arrays. Based on the related information, the compensation value is calculated and stored in the registers, which will be used in the normal work phase. The actions of ADAC arrays make the input of comparators approach continuously during the calibration phase. If the inputs of comparators are connected to the same signal in the beginning, the offset code is the same as the output codes of the comparator after finishing the whole sar logic conversion. Similarly, the mismatch code of every capacitor is achieved in the same way. Assuming that the weight of the target capacitor is equal to the weight of all lower capacitors, the difference between them is the mismatch of the target capacitor. The switching of ADAC arrays provides compensation value and the mismatch code is related to the output of comparators. The mismatch of ADAC arrays themselves doesn't affect the calibration. As a result, the size of the capacitors in ADAC arrays is smaller than that of CDAC

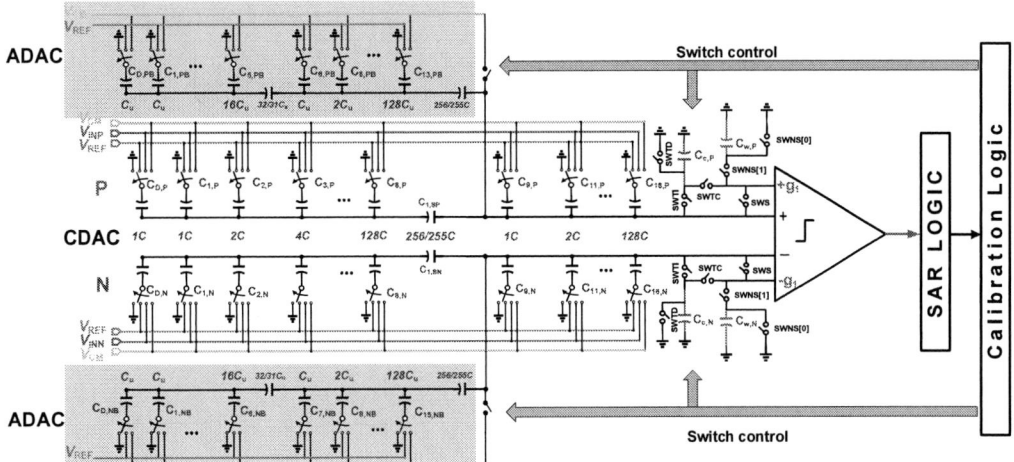

Fig.1: The architecture of the proposed 18-bit SAR ADC.

the 13-bit ADAC arrays could realize a similar effect to 15-bit ADAC arrays because new first-order passive noise-shaping is used, which is shown in Fig.2 (a). During the process of calibration, once the ADAC arrays complete one conversion of the offset detection or mismatch detection, the residual voltage left on the top plates of CDAC arrays is sampled and integrated by the capacitor C_C. After the proportional operation, the residual voltage is added to the comparator with the next sampling signal of the ADAC arrays, which realizes the noise-shaping.

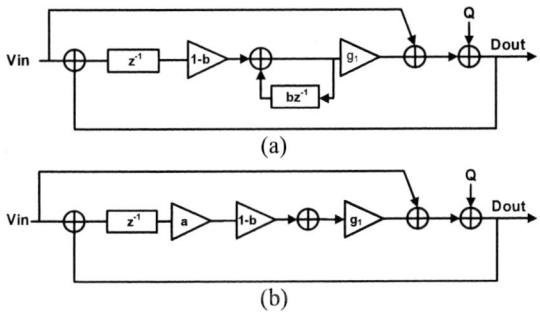

Fig.2: (a) new first-order passive noise-shaping used in the calibration phase (b) first-order passive feedforward noise-shaping method used in the normal work phase.

In the normal work phase, the corresponding corrected codes, including offset corrected codes and mismatch corrected codes, are added to ADAC arrays at the right stage to provide compensation. On the other hand, the residual voltage generated by conversion is sampled by capacitor C_C and it is integrated by C_W. When the next conversion begins, the residual voltage and the input signal are added proportionally through the comparators, which is considered first-order passive feedforward noise shaping, as shown in Fig.2 (b). The capacitor C_C is

multiplexed in the calibration phase and normal work phase. In the actual circuit, the capacitor C_C is split into four capacitors. Four capacitors are connected in parallel for sampling and integration in the calibration phase while any two capacitors are selected for integration in turn during the normal work phase in order to realize dynamic matching and reduce the influence of the capacitor mismatch. As a result, the area utilization of DAC improves. What's more, the noise shaping technique makes the resolution up to 18-bit.

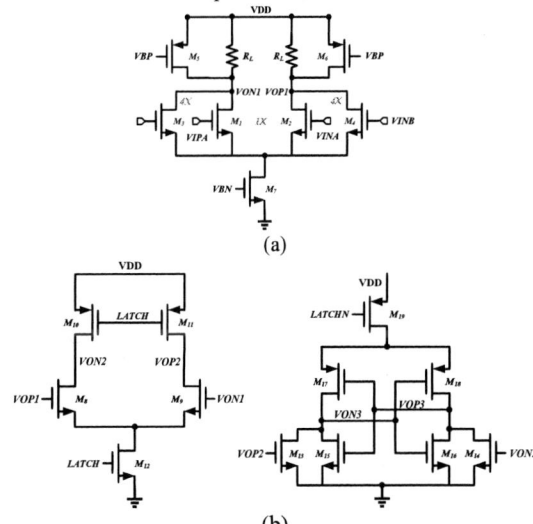

Fig.3: (a) the pre-amplifier and (b) the double-tail dynamic comparator.

The comparator in the proposed SAR ADC includes a pre-amplifier and double-tail dynamic comparator, as depicted in Fig.3. The pre-amplifier is used to amplify the input signal and suppress the noise because the high

979-8-3503-1101-3/23 $31.00 © 2023 IEEE

resolution requires small signal comparison. The transistors M_5 and M_6 are used to provide compensated current, avoiding large current following the load resistors R_L, and it doesn't change the DC gain. In addition, the pre-amplifier has two pairs of input transistors with different sizes, supporting noise shaping.

Double-tail dynamic comparator is based on the separate input and cross-coupled stage, which enables fast operation and wide supply voltage. What's more, the kickback noise of the double-tail dynamic comparator is less than that of strong-arm latch architecture. The combination of amplifier and double-tail architecture achieves the goal of finishing the small signal comparison in a short time.

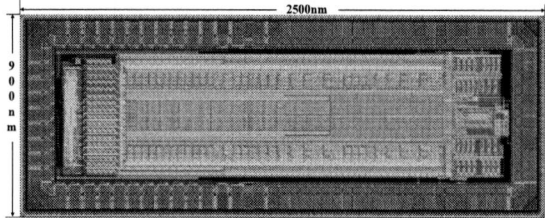

Fig.4: The layout of the proposed SAR ADC

EXPERIMENTAL RESULTS

The proposed SAR ADC is designed and simulated in a 55nm process, as shown in Fig.4. In order to verify the effect of double passive noise shaping calibration, the offset and the mismatch parameters, which are related to the design and the process, are added to the simulated model. The dynamic performances of SAR ADC are depicted in Fig.5, including calibrated results and uncalibrated results. The offset decreases from 3.95mV to 3.54uV and the harmonics caused by mismatch are obviously improved. Apart from the problems of non-ideal factors solved, the 16-bit architecture realizes the resolution of 18-bit. After the calibration, the SNDR increased from 81.24dB to 111.89dB.

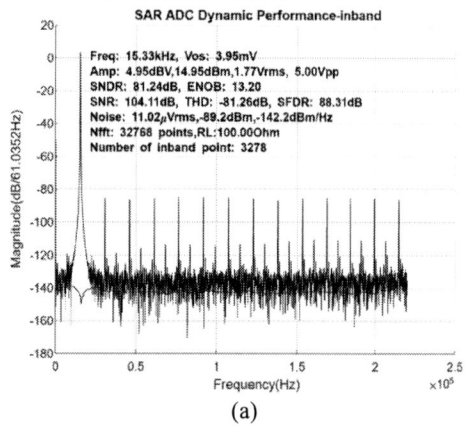

(a)

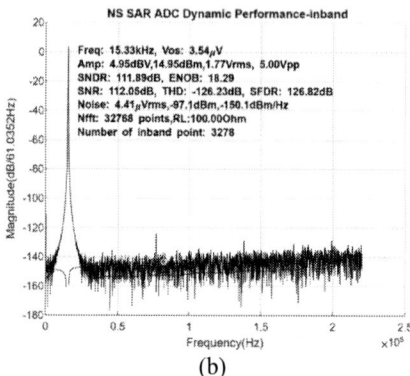

(b)

Fig.5: simulated dynamic performance of the SAR ADC model (a) spectrum after calibration (b) spectrum before calibration.

CONCLUSION

In this paper, an 18-bit 2MSPS SAR ADC with double passive noise-shaping is presented. The effect of offset and mismatch is reduced by the calibration and the noise-shaping is realized by the switched capacitor circuits instead of OTA, which is a simple method and consumes lower power. What's more, the related capacitors can be reused and don't need extra area during both the calibration phase and the normal work phase. The SNDR of ADC after calibration is 111.89 dB, based on the simulated results of the model, which is greatly improved.

REFERENCES

[1] Bagheri, Mojtaba, et al. "A mismatch calibration technique for SAR ADCs based on deterministic self-calibration and stochastic quantization." IEEE Transactions on Circuits and Systems I: Regular Papers 67.9 (2020): 2883-2896.

[2] Liang, Yuhua, et al. "A 14-b 20-MS/s 78.8 dB-SNDR energy-efficient SAR ADC with background mismatch calibration and noise-reduction techniques for portable medical ultrasound systems." IEEE Transactions on Biomedical Circuits and Systems 16.2 (2022): 200-210.

[3] Fredenburg, Jeffrey A., and Michael P. Flynn. "A 90-ms/s 11-mhz-bandwidth 62-db sndr noise-shaping sar adc." IEEE Journal of Solid-State Circuits 47.12 (2012): 2898-2904.

[4] Liu, Jiaxin, et al. "9.3 A 40kHz-BW 90dB-SNDR Noise-Shaping SAR with 4× Passive Gain and 2 nd-Order Mismatch Error Shaping." 2020 IEEE International Solid-State Circuits Conference -(ISSCC). IEEE, 2020.

DESIGN AND SIMULATION OF A PFM-PWM HYBRID CONTROLLER FOR DCDC CONVERTER WITH CLLC TOPOLOGY

Hai Liu[1]*, Lenian He[1], Quan Sun[2], Changyou Men[2]

[1]School of Micro-Nano Electronics, Zhejiang University, Hangzhou 311200, China
[2]Hangzhou Vango Technology Inc., Hangzhou 310052, China
*Corresponding author's email: liuhai@zju.edu.cn

ABSTRACT

This paper aims at providing a system design of high-power bidirectional DCDC converter for electric vehicles application. Based on previous power-stage design, this paper focuses on control stage-design. The control loop consists of PWM loop and PFM loop to achieve high efficiency compared to traditional PFM only; The control mode consists of power-on mode and normal mode to achieve less overshoot when powering on. Functional blocks level, circuit level, and layout level designs are presented to illustrates the structure and operating of control-stage, especially the control core. Simulations of control IC design in SMIC180nm provide basic function and high performance verification.

Keywords—bidirectional DCDC converter, full-bridge CLLC, PFM-PWM hybrid control

INTRODUCTION

Electric Vehicles are becoming more and more popular in the age of resource-shortage and high environmental stress. Present electric vehicles are usually equipped with multi-voltage-level batteries for various on-board applications, for example, 48V for motor driving and 12V for intelligent control system. To achieve high power efficiency, long distance per charge, short charging time, redistribution of electric energy and high reliability for driving safety, high-performance high-power bidirectional DCDC converter between batteries with different voltage levels is required.

To achieve high-power conversion up to 1kW per converter, power-stage topology with soft-switching is required. Full-bridge CLLC topology, as shown in Figure 1, can move the phase of current and voltage of power MOSFETs by its intrinsic resonance to achieve soft-switching. Full-bridge CLLC topology has advantages of high stability and high conversion efficiency while disadvantages of requiring more power devices and high complexity of control[1][2]. Under this application condition, advantages satisfy requirements while disadvantages can be tolerated. Full-bridge CLLC topology is chosen as power-stage topology for this DCDC converter.

Previous researches have shown that the intrinsic frequency of CLLC is determined by resonance components. When operating frequency is a little lower than intrinsic frequency, operating period is a little longer than time required for resonance, which will reserve a piece of time for zero-current and then soft-switching of power MOSFETs, as shown in Figure 2[3]. The controller of the converter is designed to adjusting operating frequency to appropriate range automatically[4].

Traditional control method is PFM, but only PFM can not adjust duty cycle which will result in more power loss. To overcome the shortage, this paper chooses PFM-PWM hybrid control method and proceeds with its design[5]. The controller is designed in and planned to be fabricated in SMIC180nm IC process. The whole design includes functional blocks level, circuit level, and layout level. This paper will mainly introduce circuit-level design and verify the whole design by simulation results.

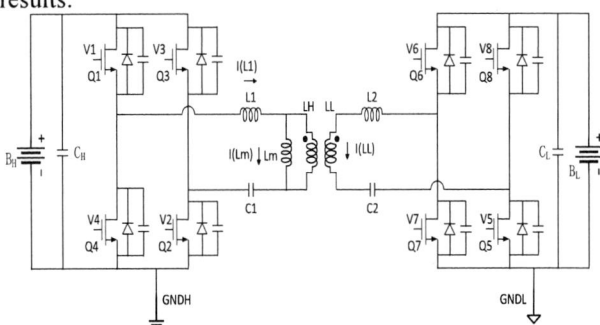

Figure 1: Power-stage full-bridge CLLC topology

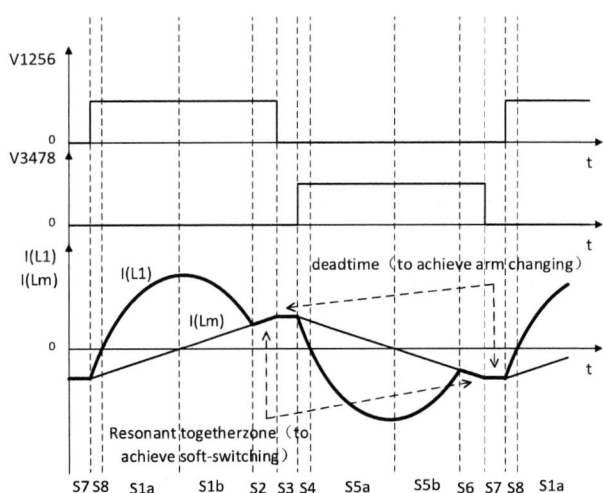

Figure 2: Designed operating waveforms of power-stage

DESIGN OF THE CONTROLLER

Figure 3 indicates main control signal loop with important functional blocks of the controller that is applied to power stage as shown in Figure 1. The controller consists of 7 functional blocks: ① Isolated input ② Signal choosing ③ Control mode choosing ④ Frequency modulation ⑤ Duty modulation ⑥Driving ⑦Isolated output.

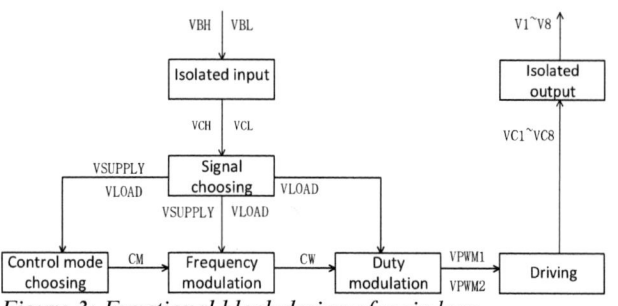

Figure 3: Functional block design of main loop

①Isolated input block changes batteries' voltages VBH and VBL that are relative to GNDH and GNDL from power stage into analog voltages VCH and VCL that are relative to reference ground GNDC of controller.

②Signal choosing block choosing signals from VCH and VCL to VSUPPLY and VLOAD by external direction-indicator signal.

③Control mode block determines the operating mode by the condition of VSUPPLY and VLOAD and outputs operating-mode-indicator signal CM.

④ Frequency modulation block generates frequency-indicator signal VFM by executing VSUPPLY and VLOAD, modulates the frequency of pulse signal VPFM by VFM , and executes edges of VPFM to get trapezoid wave CW.

⑤ Duty modulation block amplifies the error between VLOAD and reference which generates modulation wave MW1 and MW2, and compares MW1 and MW2 with carrier wave CW which generates modulated signals VPWM1 and VPWM2 respectively.

⑥Driving block imposes delays and driving protection on VPWM1 and VPWM2 to get 8 driving signals VC1~VC8.

⑦Isolated output block changes VC1~VC8 that are relative to GNDC into V1~V8 that are relative to source voltage of Q1~Q8.

Figure 4 shows control core design in circuit level. Signal choosing block is achieved by a transmission gate array TGA1. When external direction-indicator signal DR is high, which means the power stage is operating in buck mode, VCH is connected to VSUPPLY and VCL is connected to VLOAD. Control mode block gives control mode indicator signal CM to frequency modulation block. When VSUPPLY is between 2 reference voltages and VSUPPLY enter this window from bottom (rise to enter this window) and VLOAD is lower than another voltage reference, CM is logically high, which means power-on control mode; otherwise, CM is logically low, which means normal control mode. This block is achieved by a schmidt trigger, a comparator, and an "AND" gate. Frequency modulation block executes PFM operation. In normal mode, frequency modulation block firstly amplifies the error of VSUPPLY and its relating reference, which generates VSUPPLYEA; secondly "adds" VSUPPLY and VREF by an analog adder, which generates frequency indicator signal VFM; thirdly modulates VFM to a pulse signal whose frequency is adjusted by VFM by a VCO, which generates VPFM; finally all the straight edges of VPFM will be reshaped into slopes by a analog integer, which generates CW as carrier wave for PWM comparing in next block. In power-on mode, besides operations in normal mode, the adder takes VLOADEA into

account for reducing load side overcharge when powering on. Duty modulation block executes PWM operation. This block takes VLOAD and some references as input, imposes reverse amplification operations to VLOAD, generates modulating wave MW1 and MW2 for PWM comparing, and compares CW with MW1 and MW2 respectively, which finally outputs early driving signals VPWM1 and VPWM2. VPWM1 is the early driving signal for Group1 power MOSFETs (including Q1 Q2 Q5 and Q6) in power stage as shown in Figure 1, and VPWM2 for Group2 (another 4 MOSFETs).

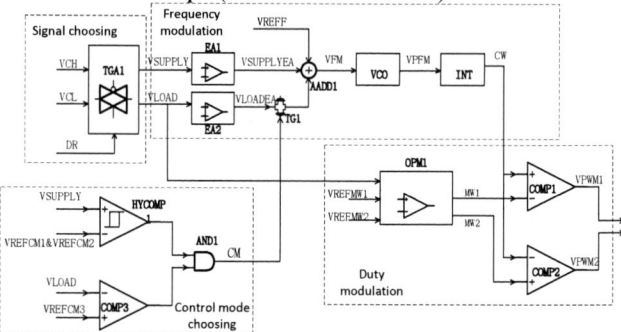

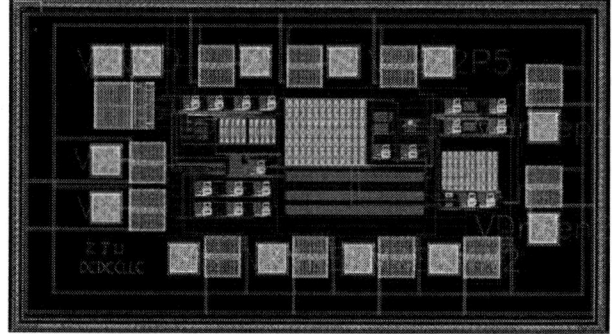

Figure 4: Circuit design of control core

Figure 5 shows the layout design of the control core. The whole layout covers 0.9963mm2 (1350um*738um) and includes 13 pins.

Figure 5: Layout deisgn of the control core

SIMULATION RESULTS AND ANALYSES

Figure 6 shows the simulation results. When Vin varies around expected value 48V, Vout will persist fluctuating in a small range around expected value 12V, which verifies that PFM PWM hybrid control method can achieve the basic function that is stabilizing output voltage for load side. Moreover, it verifies the feasibility of the control method, system structure, and core circuit design. Vout is the output voltage under the whole control core design while Vout_contrast is the output voltage under the control core design without power-on mode. By comparing these 2 waves, it can be found that the whole control design can significantly reduce the overcharge of the load when powering on, which will relieve the stress of the battery in load side. When the power stage is working in the other direction, we can get similar results. When converting power is 1kW and Rdson of power MOSFETs is 30m Ω , the power-stage converting efficiency is 98.2% and 96.3% in step-down and step-up mode respectively.

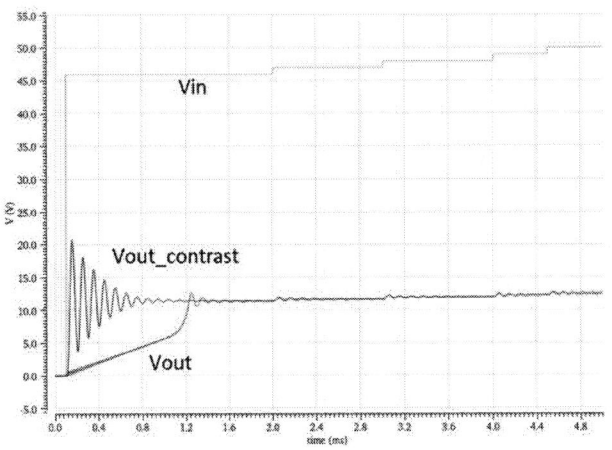

Figure 6: Simulation results

CONCLUSION

The whole system is designed for kW-level power bidirectional DCDC converter applied in electric vehicles, which includes power-stage and control-stage. Full-bridge CLLC and specific operating frequency is designed for power-stage. The whole design described in this paper includes design in 3 levels: functional blocks, circuit, and layout. There are 7 functional blocks illustrating complete signal flow in PWM and PFM control loop; there are 4 blocks of control core with detailed circuit design and operating presented; there is also layout overview of the control core. Simulation results verifies the fundamental functions of voltage converting and output stabilizing and high performances of converting efficiency.

ACKNOWLEDGEMENTS

This work is supported by Collaborative Innovation Project of Manufacturing High Quality Development Industry Chain of Zhejiang Province, No. SGTYHT/20-JS-222.

REFERENCES

[1] P. He and A. Khaligh, "Comprehensive Analyses and Comparison of 1 kW Isolated DC–DC Converters for Bidirectional EV Charging Systems," *IEEE Transactions on Transportation Electrification*, vol. 3, no. 1, pp. 147-156, March 2017, doi: 10.1109/TTE.2016.2630927.

[2] P. He and A. Khaligh, "Design of 1 kW bidirectional half-bridge CLLC converter for electric vehicle charging systems," *2016 IEEE International Conference on Power Electronics, Drives and Energy Systems (PEDES)*, 2016, pp. 1-6, doi: 10.1109/PEDES.2016.7914445.

[3] H. Liu and L. He, "1kW Bidirectional 48V-12V DCDC Converter Design Based on Full Bridge CLLC Topology and FDP Controlling Method for Electric Vehicles Application," *2021 International Conference on Electrical, Electronics and Computing Technology (EECT)*, 2021, pp.201-207 doi:10.1088/1742-6596/1914/1/012026.

[4] H. Liu and L. He, "Analyses and Design of a High Power Bidirectional 48V-12V DCDC Converter System for Electric Vehicle Application," *2021 Industrial Electronics Conference (IECON)*, 2021, pp.1-6 doi:10.1109/IECON48115.2021.9589662.

[5] K. Luo, H. Wang, C. Gou, K. Peng, F. Yang and Y. Hu,

"A Novel PWM/PFM Control Technique for Transient Improvement and High Efficiency Over a Wide Load Range in Buck DC-DC Converter," *2018 IEEE 3rd International Conference on Integrated Circuits and Microsystems (ICICM)*, 2018, pp. 169-173, doi: 10.1109/ICAM.2018.8596546.

DESIGN OF AN 8-CHANNEL 12BITS 1MSPS SAR ADC

*Zhengxue Shi[1], Quan Sun[2], Changyou Men[2] and Lenian He[1, *]*

[1]School of Micro-Nano Electronics, Zhejiang University, Hangzhou 310058, China
[2] Hangzhou Vango Technology Inc., Hangzhou 310052, China
*Corresponding Author's Email: helenian@zju.edu.cn

ABSTRACT

In this paper, an 8-channel 12-bit 1MSPS single-ended SAR ADC (Successive Approximation Register Analog Digital Converter) was implemented in a standard 55nm CMOS technology. Bottom plate sampling and bootstrap switch is adopted to remove distortion in single-ended structure. A segmented DAC array with a unit value bridge capacitor is used to reduce the size of capacitor array and avoid the problem caused by fractional capacitor. A two-stages comparator was used to reduce noise and save power.

The core of this SAR ADC consumes 0.63mW power and the active area is only 0.1mm^2 while achieving 11.42 bits ENOB at 1MSPS sampling rates.

INTRODUCTION

Analog-to-Digital Converters (ADCs) are widely used as interface between analog and digital signals. SAR ADC has attracted more and more attention thanks to its low power consumption, simple structure, increased speed and resolution, and is more friendly to scaling down. SAR ADC can give an accurate reading based upon only one sample of a signal and can minimize latency compared to pipeline and sigma-delta structures, so it is also ideally suited to multi-channel data acquisition systems for industrial control and monitoring systems.

Differential structures are widely used for precision and high conversion rates SAR ADCs due to its immunity to even-order distortions and doubled input-range [1] [2] [3]. However, many natural signals are still measured with respect to the ground in the forms of single-ended signals [4]. Although we can use an additionally single-to-differential converter, it is much more convenient, power saving and area efficient to use a single-ended SAR ADC in single-ended input type applications. However, the trade-offs for this simplicity are the more serious non ideal effects. Therefore, we must solve non ideal effects like even-order distortion and noise compared to differential input to maintain performance.

In this study, we focus on the implementation of a single-ended 12-bit 8-channel SAR ADC. It features high throughput rate of 1MSPS with low power consumption and minimized core area. It also has 8 Single-Ended Inputs with a Channel Sequencer. A consecutive sequence of channels or a selected channel on which the ADC cycles and converts. Moreover, bottom plate sampling and bootstrap switch are used to suppress distortion. A second-stage comparator is used to reduce noise while saving

power. Therefore, we can achieve high resolution with single-ended input structures.

CIRCUIT IMPLEMENTATION
The architecture of the proposed SAR ADC

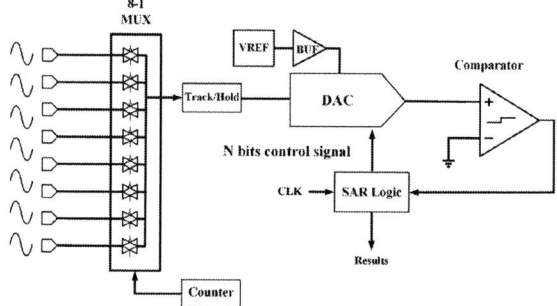

Figure 1: System architecture diagram of the proposed SAR ADC

Fig.1 shows the architecture of the SAR ADC. The ADC is comprised of 8-1 multiplexer, SAR control logic, Track/Hold circuit, capacitive DAC, comparator and internal reference and bias current generator. The 8-1 multiplexer consists of 8 large aspect ratio PMOS-NMOS complementary switches controlled by a programmable channel sequencer. The Track and hold circuit sample and hold the VIN signal selected by the multiplexer. The control logic controls the capacitive DAC to add or subtract fixed amounts of charge according to the comparison result to bring the comparator back into a balanced condition. The control logic also outputs the final quantization results.

Bottom plate sampling

At the instant of transitioning from acquisition to conversion phase, some of charge stored in sampling switch dumped onto sampling capacitor depending on the value of VGS. The variation of VGS will introduce input dependent error voltage and will cause distortion. However, if we use bottom plate sampling method, the charge injection is to first-order independent of VIN, only a dc offset is added.

Fig.2 and Fig.3 show the two phases, acquisition phase and conversion phase for this ADC using bottom plate sampling method. According to Fig.2, during acquisition phase, SW2 is closed and SW1 is in position A. The sampling capacitor acquires the signal on the selected VIN channel. When the ADC starts a conversion,

979-8-3503-1101-3/23 $31.00 © 2023 IEEE

according to Fig.3, SW2 opens, a constant amount of charge is injected and the charge on the sampling capacitors will be frozen afterwards, then SW1 moves to position B, no charge is injected onto sampling capacitor since the charge is frozen. The comparator inputs become unbalanced and comparison result is made. When the SAR decision is made according to the comparison result, the comparator inputs are rebalanced.

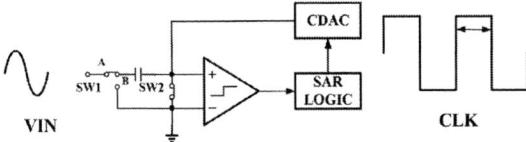

Figure 2: ADC Acquisition Phase

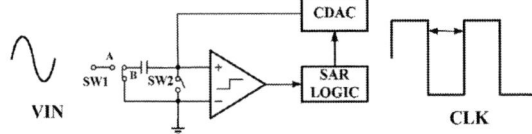

Figure 3: ADC Conversion Phase

Bootstrap switch

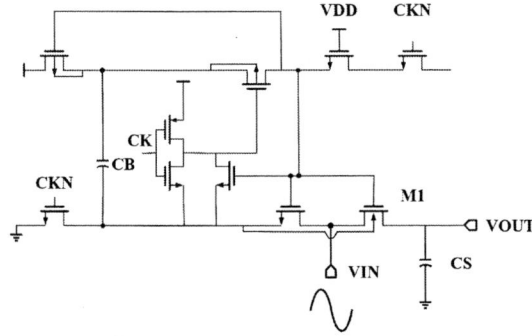

Figure 4: The schematic of proposed bootstrap switch

MOS-Transistors exhibits an input-dependent on resistance due to the variation of gate-source voltage, thereby introducing distortion. While it can be reduced by using a large transistor, however, the charge injection problem will largely degrade the performance. Although differential architectures can eliminate second order harmonic distortion, it cannot be achieved for single-ended inputs in our design. As a result, we need to adopt bootstrap switch to minimize the variation of switch on resistance and charge injection.

The proposed bootstrap switch is shown in Fig.4. The capacitor CB is pre-charged to VDD during tracking period and can bootstrap the gate voltage of the switch transistor M1 during sampling period by VDD, therefore we can keep the device on with a constant VGS equals to VDD. Compared to traditional bootstrap switch in [5], the body of M1 is also tied to the bottom plate of CB and the voltage difference Vbs remains at 0 during sampling phase,

which removes the influence of body effect on linearity and further enhances the performance.

DAC Array

In order to reduce the capacitor array area, a segmented capacitor array is adopted. The DAC array is segmented into two parts, the lower 4 bits or 4 Least Significant Bits (LSB), C8-C11 and the higher 8 bits or 8 Most Significant Bits (MSB), C0-C7. Assuming the unit capacitor value is C_u. The value of C8-C11 is $8C_u$, $4C_u$, $2C_u$, C_u, respectively, and the value of C0-C7 is $128C_u$, $64C_u$, $32C_u$, $16C_u$, $8C_u$, $4C_u$, $2C_u$, C_u, respectively.

The two parts are connected by a bridge capacitor Cs, the value of Cs is calculated as follows:

When the bottom voltage of C7 changes from 0 to VREF, the voltage variation at the top plate, or the comparator's input is:

$$dV_1 = \frac{C_7}{C_{Mt} + C_s // C_{Lt}} \quad (1)$$

C_{Mt} is the total capacitance of the MSB array and C_{Lt} is the total capacitance of the LSB array.

On the other hand, when the bottom voltage of C8 changes from 0 to VREF, the voltage variation at the top plate, or the comparator's input is:

$$dV_2 = \frac{C_8}{C_{Lt}} \frac{C_s // C_{Lt}}{C_{Mt} + C_s // C_{Lt}} \quad (2)$$

The value of Cs should ensure the voltage variation dV_1 equals to $2dV_2$:

$$dV_1 = 2dV_2 \quad (3)$$

Substituting equation (1) and equation (2) into (3), it can be calculated that the value of Cs is exactly one unit capacitor C_u, which can be easily and precisely produced. However, it will introduce 1LSB gain error for this DAC array structures, which will not degrade linearity and cause influence on ENOB performance. As a result, it can be disregarded in our design.

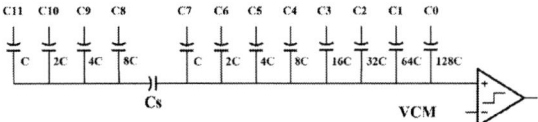

Figure 5: The DAC array structure

Comparator

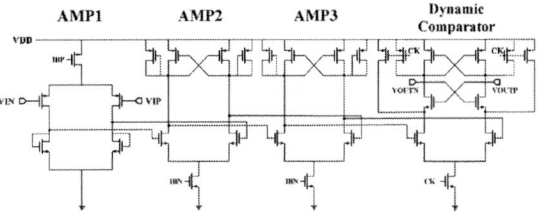

Figure 5: The schematic of comparator

Comparator is one of the dominant noise sources besides the reference buffer, the sampling circuit and the quantization [6]. As illustrated in Fig.5, a two-stage

979-8-3503-1101-3/23 $31.00 © 2023 IEEE 651

comparator is used to lower the noise while consuming low power. The comparator consists of a preamplifier and a regenerative latch.

The preamplifier has three consecutive stages to lower the comparator bandwidth therefore filter out noise without losing settling accuracy. The comparator's input stage needs to detect half an LSB, but the error band of the second and third stages can be relaxed by the gain of the previous stage and therefore the bandwidth can be reduced. The reduction of the bandwidth reduces the power consumption and the total output noise of the comparator.

The regenerative latch is a strong-arm latch which consumes zero static power and directly produces rail-to-rail outputs It can easily meet the speed and precision requirements and is very suitable for our applications.

Flicker noise is suppressed by using a large size PMOS input pair for the first preamplifier stage.

LAYOUT AND SIMULATION RESULTS

This 12-bits ADC is designed and simulated in SMIC 55-nm CMOS process in cadence. Fig.6 shows the layout of the SAR ADC. It can be measured that the overall area is 0.122 mm^2 (230um*530um) including bandgap reference and bias current generator and is only 0.1 mm^2 for SAR ADC core excluding the area of bandgap reference and bias current generator.

TABLE I.   PERFORMANCE COMPARISON

	This work	*[4]*	*[3]*
Supply	2.5/3.3	1.8	1.8
Input type	Single ended	Single ended	Differential
Technology	55nm	180nm	180nm
Speed	1MSPS	10MSPS	0.2MSPS
ENOB (bit)	11.42	9.04	11.59
Power (uW)	632	2290	336
FOM (fJ/step)	230	435	545
Area(mm^2)	0.1	0.207	0.525

Figure 6: Layout of SAR ADC prototype

To verify our SAR ADC prototype, we extracted parasitic and conducted post-simulation in cadence. The result's Fast Fourier Transform (FFT) spectrum shows the effective number of bits (ENOB) is 11.42 bits, signal to noise and distortion ratio (SNDR) is 70.51dB, and spurious free dynamic range (SFDR) is 79.6dB respectively for a 497 kHz input frequency at 1MSPS sampling rate with supply voltage of 2.5V. The average current consumption of the entire circuit is 463uA, while the SAR ADC core circuit consumes 253uA current (excluding the current consumption of bandgap reference and bias current generator). And the power consumption for our SAR ADC core is 632uW, resulting in a Figure of merit (FOM) of 230fJ/conv.

Compared to the recent relevant works listed in TABLE I, this SAR ADC achieves a low FOM and occupies a small area.

CONCLUSION

This paper presents an 8-channel, 12-bit, 1MSPS SAR ADC. This SAR ADC features low-power, small active area and the post-simulation result shows it achieves 11.42 bits of ENOB for a 497 kHz input frequency at 1MSPS sampling rate. The performance is maintained for all channels.

REFERENCES

[1] C. -C. Liu, S. -J. Chang, G. -Y. Huang and Y. -Z. Lin, IEEE Journal of Solid-State Circuits, vol. 45, no. 4, pp. 731-740.

[2] J. Shen, A. Shikata, L. D. Fernando, N. Guthrie, B. Chen, M. Maddox, N. Mascarenhas, R. Kapusta, M. C. W. Coln, "A 16-bit 16-MS/s SAR ADC With On-Chip Calibration in 55-nm CMOS," in IEEE Journal of Solid-State Circuits, vol. 53, no. 4, pp. 1149-1160, April 2018.

[3] X. Zhang, M. Wang, L. Guo and X. Wang, 2017 IEEE 2nd Advanced Information Technology, Electronic and Automation Control Conference (IAEAC), 2017, pp. 2531-2535.

[4] Jisu Son, Young-Chan Jang, IEICE Electronics Express, 2019, Volume 16, Issue 22, Pages 20190597.

[5] B. Razavi, IEEE Solid-State Circuits Magazine, vol. 7, no. 3, pp. 12-15, Summer 2015.

[6] P. Nuzzo, F. De Bernardinis, P. Terreni and G. Van der Plas, IEEE Transactions on Circuits and Systems I: Regular Papers, vol. 55, no. 6, pp. 1441-1454, July 2008.

AUTHOR INDEX

Ai, Juan 464
An, Fubang 566
Aziz, Tariq 58
Bai, Ke 165
Bao, Jie 494
Bao, Yu 223
Bao, Yunjiao 75
Bisheng, Wang 394
Bu, Weihai 84
Byrd, Jeff 640
Cai, Puyang 78
Cai, Shuting 593
Cai, Ying 6
Cao, Jinxiu 105
Cao, Lei 41, 530
Cao, Wei 150, 566
Cao, Yamin 425
Chang, Yanyan 385
Che, Siyuan 196
Chen, C.-Z. 563
Chen, Canny 367, 398
Chen, Chen 186, 188, 221
Chen, Chun-Zhang 557
Chen, Ding 536
Chen, Donglin 596
Chen, Fairy 363
Chen, Fen 507
Chen, Hanyan 370, 380
Chen, Hualun 48, 92, 95, 102
Chen, Hui 539
Chen, Hunglin 90, 401
Chen, Jiaxiang 432
Chen, Jun 482
Chen, Junhao 533
Chen, Kuo-Jung 238
Chen, Lei 447
Chen, Lu 190
Chen, Qilin 464
Chen, Quanhua 105
Chen, Shanshan 90
Chen, Tianyu 385
Chen, Wei 99, 115, 415
Chen, Xing 524
Chen, Xu 404
Chen, Yanye 545
Chen, Yaping 518
Chen, Yekuan 617
Chen, Yiling 617

Chen, Yingxin 9
Chen, Yunbo 238
Chen, Zehua 299
Chen, Zhengrui 630
Chen, Zhongkui 238
Chen, Zijian 179
Chen-Chen 219
Cheng, Jie 328
Cheng, Shiyao 288
Cheng, Shuo 288
Cheng, Tian 215
Cheng, Weichi 295
Cheng, Xingming 497
Cheng, Xinhua 183
Chi, Min-Hwa 118
Chi, Yushan 262
Chiou, Dawson 640
Chu, Zhufei 584, 587
Cong, Dehong 596
Cossins, John 485
Cui, Dexing 343, 346
Cui, Liang 497
Cui, Xingzhi 497
Da, Haixia 536
Dai, Hongwei 464
Dai, Johnny 458
Dan, Wang 277
De Chen, Yu 29, 306
Deng, Dempsey 306
Deng, Hao 269, 271
Ding, Mingzheng 521
Ding, Rongzheng 560
Ding, Shenglan 464
Ding, Yang 81
Dong, Xiao 630
Du, Chongkai 18
Du, Kuiling 542
Du, Yihang 95, 102
Duan, Songhan 92
Duan, Wenting 6
Duan, Wenxu 288
Duan, Xin-Lv 277
Duan, Yangyang 497
Ervin, Joseph 29
Fan, Dandan 497
Fan, Jiaming 337
Fan, Weihai 377
Fang, Chic-Kuo 391

Fang, Jingxu	319
Fang, Jingxun	317, 325
Fang, Lvye	472
Fang, Xiaoli	557
Fang, Yanfen	482
Fang, Zhiqiang	511
Fang, Ziquan	25
Feng, Frank	640
Feng, Lei	485
Feng, Long	3, 303
Feng, Xuewei	32, 58
Feng, Yingxiong	190
Fu, Boyi	71
Fu, Jianyu	521
Fu, Wentao	223
Fu, Zhiyuan	67
Fuh, Yiin-Kuen	256
Fujimori, Toru	142
Gao, Bin	533
Gao, Changcheng	162
Gao, Dawei	408, 421
Gao, Dongliang	246, 303, 309
Gao, Jianfeng	530
Gao, Shenghan	635
Gao, Shuo	328
Gao, Yuanda	333, 337
Geng, Di	277
Geng, Xiao	99
Gong, Siyi	624
Gore, Akhil A.	640
Gu, Jiongjiong	514
Gu, Lin	92, 95, 102
Gu, Yu	35
Guan, Fenglin	183
Guan, Lulu	190
Guan, Tianpeng	243
Guan, Zijun	333, 337
Gui, Youshan	536
Gui, Zhi-Yuan	388
Guo, Chunxiang	190
Guo, Jiannan	569, 572
Guo, Ming	638
Guo, Q. J.	249
Guo, Xinfei	578, 624
Guo, Xuan	581
Guo, Yaozu	527, 614
Han, Baodong	168, 202
Han, Dandan	125
Han, G. C.	249
Han, Genquan	408
Han, Hongyan	84
Han, Muzi	132

Han, Wu-Hao	188
Han, Xuelian	507
Han, Yanchu	530
Hang, Mingguang	183
He, Cunzhe	246, 303, 309
He, Guangzhi	475
He, Lenian	545, 647, 650
He, Le-Nian	644
He, Shi-Qiang	391
He, Xiao	411
Hong, Hong	428
Hou, Jianqiu	172, 179
Hou, Linjie	168
Hou, Ziyang	360
Hu, Jiawei	530
Hu, Juan	494
Hu, Kunmei	587
Hu, Li-Song	186, 212, 219
Hu, Xianghua	475
Hu, Yangyang	38
Hu, Zengwen	179
Hu, Zhenhua	262
Huang, Abner	640
Huang, Haijiao	627
Huang, Jacky	29, 306
Huang, Jian.	176
Huang, Jiecheng	635
Huang, Joe	640
Huang, Jun	193, 196
Huang, Qianqian	67, 71, 84
Huang, Qiao	190
Huang, Ran	425
Huang, Ru	67, 71, 78, 84
Huang, Rutian	99, 115
Huang, Shan	243
Huang, Yaodong	428
Huang, Yifei	620
Humphries, Andrew	485
Ji, Xiaoli	527, 614
Ji, Xiaoxiao	299
Ji, Zhigang	78, 84
Jia, Hanbo	581
Jia, Lili	183
Jia, Shuhuai	12
Jiang, Hao	374
Jiang, Ming-Yu	256
Jiang, Xiaojuan	497
Jiang, Yi	421
Jiang, Zhongwei	198, 200, 206, 209
Jin, Feng	6
Jin, Johnny	428
Jin, Yu	627

Jing, Wang	236
Kang, Bok-Moon	277
Kim, Cheolkyu	428, 458
Kong, Weiran	48
Kou, Xufeng	35
Kuai, Wei	288
Lai, Da-Wei	461
Lan, Jiang	527
Lan, Jun	32, 58
Lau, W. S.	253, 274
Lei, Tong	262
Li, Cheng	29
Li, Fang	183
Li, Fei.	551
Li, Fei	620
Li, Fengyang	238
Li, Gordon	25, 48
Li, Hao	179
Li, Hongyun	507
Li, Hu	319
Li, Jie	32
Li, Jizhou	377
Li, Junfeng	87, 530
Li, Junjie	41, 530
Li, Lanxia	497
Li, Lei	367, 464
Li, Lianlian	61, 530
Li, Qingkun	41
Li, Quanbao.	176
Li, Quanbo	193, 196, 243
Li, Shilin	464
Li, Shipu	1
Li, Tingyu	533
Li, Tomi T.	256
Li, Wei	450
Li, Weimin	333, 337
Li, Xiang	411
Li, Xiaokang	1
Li, Xingquan	593
Li, Xinyi	533
Li, Yang	415, 425, 590
Li, Yanli	132, 147, 153
Li, Yida	32, 58
Li, Yijun	533
Li, Yinglu	530
Li, Yue	231, 536
Li, Yuhui	150
Li, Zeyu	581
Li, Zhi	464
Li, Zhirui	3
Li, Zhixiong	32
Li, Zhi-Yu	388

Li, Ziwen	602
Liang, Xiaofeng	468
Liang, Yidan	461
Liang, Zhiyang	312
Liao, Yiming	527, 614
Lin, Hong-Bo	188
Lin, Longyang	32, 58
Ling, Haiyang	6
Litian, Xu	236
Liu, Botong	92
Liu, Donghua	6, 25
Liu, Enxu	530
Liu, Fredric	25
Liu, Guoping	150
Liu, Hai	647
Liu, Hongjie	497
Liu, Hongmin	162
Liu, Jian	190
Liu, Jianghao	354, 360
Liu, Jianshe	99, 115
Liu, Jianzhong	138
Liu, Junhua	84
Liu, Lianxi	575
Liu, Lifeng	226, 231
Liu, Lihong	514
Liu, Lingge	620
Liu, Lingling	497
Liu, Min	343
Liu, Ming-Xu	277
Liu, Mingying	303
Liu, Shitong	343
Liu, Shuxin	468
Liu, Tiantian	271
Liu, Wenyi	271
Liu, Xianhe	132, 147, 153
Liu, Xiaomeng	269, 271, 288
Liu, Xinqi	35
Liu, Xinyu	581
Liu, Xue	265
Liu, Xuhui	557, 563
Liu, Yan	507
Liu, Yang	530
Liu, Yi	238
Liu, Yi-Chang	186, 219
Liu, Yichang	217
Liu, Yiran	333, 337
Liu, Yuan	105
Liu, Zhao	202
Liu, Zhejun	145
Liu, Zhiwei	78
Liu, Zhongming	122
Lo, Hsiao-Han	256

Lo, Tsz-Chun	499, 503
Lois, Liao Jinzhi	394
Long, Yin	90, 401
Lu, Kuan	468
Lu, Lian	193, 196
Lu, Wei	145
Lu, Xinchun	322
Lu, Xing	432
Lu, Xiuzhen	299
Lu, Xudong	599
Lu, Ye	560
Luan, Tongtong	35
Luan, Zhiwen	179
Luo, Cong	464
Luo, Fu	357
Luo, Jie	168
Luo, Jin	67, 71
Luo, Jun	41
Luo, Yanna	41, 61, 64, 75, 530
Lyu, Haochang	168
Lyu, Pengfei	29
Ma, D.	569
Ma, Dejing	12, 16
Ma, Haolan	527
Ma, Hong	122
Ma, Li	507
Ma, Weiwei	425
Ma, Xing	319
Ma, Yiming	200, 206
Ma, Yinan	231
Ma, Yu	602
Mao, Guiyun	404
McFalls, Charles	640
Mei, Hongfei	624
Mei, Na	411, 418
Mei, Pei	188
Men, Changyou	545, 647, 650
Meng, F. T.	249
Meng, Fanshun	238
Meng, Xiangguo	196
Meng, Yiqi	617
Miao, Qinhua	328
Min, Liu	346
Mou, Jiangtao	165
Mu, Johnny	428
Mu, Tengteng	575
Mu, Xiaodong	95
Mukundhan, Priya	458
Ni, Qiliang	475
Ning, Le	312
Ning, Li	295
Niu, Chuqiao	75

Niu, Xinhuan	354, 357, 360
Ouyang, Keqing	418, 441, 447, 608
Palmer, Nicholas	640
Pan, Feilong	447
Pan, Hongyang	584
Pan, Junjie	193
Pan, Lingran	461
Pan, Quan	557
Peng, Jingsong	81
Peng, Minqiang	441, 447, 608
Peng, Yue	408
Peng, Yufei	539
Qi, Li	186, 217, 219
Qi, Xiangyu	92
Qian, He	533
Qian, Hongtao	472, 490
Qian, Jun	1, 18, 539
Qian, Kai	193
Qian, Weikang	554
Qian, Wensheng	6, 25
Qiao, Jingchao	596
Qin, Meiyan	530
Qin, Xulei	87
Qiu, Chenchen	18, 539
Qiu, Yihang	593
Qu, Haolan	432
Qu, Minghui	360
Qu, Xiaofeng	243
Qu, Yang	450
Que, Yurong	317, 319
Ren, Kailin	38
Ren, Yanming	333
Ren, Ye	84
Ruan, Dapeng	265
Rui, Qiang	238
Rzeznik, Maria	499, 503
Sang, Guanqiao	87
Sang, Qiang-Qiang	186, 219
Shao, Hanyong	67
Shen, Alan	25
Shen, Mei	32, 58
Shen, Yijiang	128
Shen, Yingtao	611
Sheng, Qiang	464
Shi, Feng	317
Shi, Jian	554
Shi, Longxing	624
Shi, W. X.	444
Shi, Xiaoping	288
Shi, Yushu	450
Shi, Zhengxue	650
Simeng, Wei	281

So, Ka-Kiu 499, 503
Song, Kaixing 428
Song, Wan 95
Song, Xin 380, 482
Song, Zhiyu 61, 64, 75
Su, Xinruo 226
Sui, Jin 432
Sui, Zhenchao 231, 246, 309
Sun, Chang 18, 539
Sun, Chao 238
Sun, Chonghui 563, 569, 572
Sun, Daniel 453
Sun, Fangce 109, 112
Sun, Hongbo 168
Sun, Lei 223, 243
Sun, Lifei 29
Sun, Peng 521
Sun, Quan 545, 647, 650
Sun, Songyu 630
Sun, Wen 533
Sun, Xiaoyan 138
Sun, Y. H. 249
Sun, Yao 179
Tan, Enghoe 349, 352
Tan, Wei 497
Tang, April 165
Tang, Gang 542
Tang, Jianshi 533
Tang, Lixin 435, 438, 596
Tang, Yukun 84
Tao, Ouwen 560
Tao, Rushuo 563, 569, 572
Tao, Y. 249
Tian, Cheng 212
Tian, Guoliang 61, 64, 75
Ting, Fai-Lung 499, 503
Tong, Yi 135, 536
Tong, Yuxin 92
Tseng, Xue-Li 256
Tu, Tiancheng 157
Vincent, Benjamin 29
Wan, J. 9
Wan, Jing. 551
Wan, Jing 620
Wan, Ting 374
Wang, Dan 122, 138, 160
Wang, Donghan 200
Wang, Fengjiao 12
Wang, Gui-Lei 277
Wang, Hongdi 312
Wang, Hongzhe 438
Wang, Jiajin 238

Wang, Jing 198, 200, 206, 209, 241
Wang, Jingang 295
Wang, Jingfeng 475
Wang, Jinlei 183
Wang, Jun 226, 231
Wang, Kai 90, 401, 441
Wang, Kaifeng 84
Wang, Ke 527, 614
Wang, Kefeng 299
Wang, Kitty 377
Wang, Kun 415
Wang, Liang 557
Wang, Lingli 566
Wang, Luping 421
Wang, Mengke 605
Wang, Mudan 157
Wang, Ning 22, 81
Wang, Peter J. 256
Wang, Qi 132, 147, 153
Wang, Qifei 200, 206
Wang, Qingpeng 29, 306
Wang, Runsheng 78
Wang, Sen 464
Wang, Shuai 441
Wang, Tao 644
Wang, Tongqing 322
Wang, Weilun 418
Wang, Wenhui 58
Wang, Xinyang 288
Wang, Xue-Hua 212, 215
Wang, Y. H. 249
Wang, Ya 52, 292
Wang, Yansheng 404
Wang, Ye 343, 346
Wang, Yefang 380
Wang, Yong 404
Wang, Yue 138
Wang, Yun 135
Wang, Zhenhui 52, 292
Wang, Zhigao 524
Wang, Zhuangzhuang 95, 102
Wang, Zhuqiu 411, 418
Wang, Zoe 190
Wei, Jianan 265
Wei, Lanying 415
Wei, Qianhui 530
Wei, Shuhua 530
Wei, Wei 138
Wei, Yanzhao 41, 55
Wei, Yayi 125, 135
Wei, Yichen 624
Wei, Zhengying 404, 539

Wen, Boyu 328
Wu, Boping 511, 514
Wu, Cinan 530
Wu, Hanming 557
Wu, Huaqiang 533
Wu, Jun 415
Wu, Pengfei 343, 346
Wu, Qiang 132, 147, 153
Wu, Qiongtao 162
Wu, Tong 363
Wu, Wei-Gang 499, 503
Wu, Wenbo 262
Wu, Xiao-Peng 188
Wu, Xie-Shuai 277
Wu, Xinyu 99, 115
Wu, Xiuheng 581
Wu, Yongqin 84
Wu, Zhuolun 418
Wu, Ziyang 172
Xi, Jian-Xiong 644
Xi, Yue 533
Xi, Zhang 394
Xia, Qing 464
Xia, Yinshui 605
Xian, Wenhao 343, 346
Xiang, Jiaying 385
Xiang, Jin-Juan 277
Xiao, Jiaxing 3, 246, 303
Xiao, Yichen 385
Xiao, Zhiqiang 3, 246, 303, 309
Xiaomin, Li 394
Xie, Hui 9
Xie, Steve 478
Xie, Yuanxiang 288
Xing, Yan 593
Xiong, Shaoyou 122
Xiong, Xiaoming 593
Xu, Chang 388
Xu, Dongyong 128
Xu, Fangyuan 581
Xu, Gaobo 61, 64, 75, 521
Xu, Haoqing 75
Xu, Jiahao 635
Xu, Jiawei 32
Xu, Kaidong 190
Xu, Kangning 190
Xu, Li-Tian 186, 188, 209, 212, 215, 219, 221, 241
Xu, Litian 200, 217
Xu, Renhui 223
Xu, Renren 55
Xu, Ruge 578
Xu, Sheng 614

Xu, Shuaihang 32
Xu, Vina 172
Xu, Weikai 71
Xu, Xingxing 176
Xu, Yong 105
Xu, Yuan 494
Xu, Yun 363, 490
Xu, Zhaozhao 25, 102
Xue, Hui 179
Yan, Feng 527, 614
Yan, Gangping 61, 64, 75
Yan, Haitao 3, 246, 309
Yan, Han 357, 388
Yan, Kevin 453
Yan, Shijia 464
Yan, Sun 281
Yan, Yinan 61, 64, 521
Yang, Chenchen 349, 352
Yang, Cheng 569, 572
Yang, Chenghao 605
Yang, Dan 411, 418
Yang, Fan 548
Yang, Guang 198, 200, 209
Yang, Guan-Hua 277
Yang, Jianyi 630
Yang, Kun 563, 569, 572
Yang, Mengmeng 38
Yang, Ming 635
Yang, Shangbo 61, 64, 75
Yang, Tao 530
Yang, X. L. 249
Yang, Yanbin 408, 421
Yang, Yang 548
Yang, Yi 122
Yang, Yu 325
Yang, Yuhao 196
Yang, Zhaohui 605
Yao, Chun 95, 102
Yao, Jiaxin 41, 55, 530
Yao, Min 507
Yao, Xiang 48
Yao, Xing-Jun 186, 188, 219, 221
Yao, Xingjun 217
Yao, Yu 349, 352
Ye, Pengfei 536
Ye, Tianchun 135
Ye, Wenzhe 627
Yin, Huaxiang 41, 45, 55, 61, 64, 75, 87, 521
Yin, Luqiao 299
Yin, Yin 81
Ying, Wen 398
Younan, Hua 394

Yu, Jinghao 536
Yu, Mingfei 317
Yu, Qinqin 472
Yu, Shaofeng 518, 560
Yu, Shiri 138
Yu, Shirui 157, 160
Yu, Wenjian 635
Yu, Wenjie 333, 337
Yu, Yexiao 122
Yu, Yinsheng 150
Yuan, Ning-Hsiu 256
Yuan, Qin 150
Yuanxiang, Xie 281
Zaheer, Muhammad 58
Zeng, Li 198, 209, 241
Zeng, Xuan 548
Zha, Xiaojing 605
Zha, Zeqi 52, 292
Zhan, Ni 354, 357, 360
Zhang, Alan 188
Zhang, Baoguo 343, 346
Zhang, Chaoran 92
Zhang, Chi 150
Zhang, Fan 45, 490
Zhang, J. W. 444
Zhang, Jian 317, 319, 325
Zhang, Jianhua 38, 299
Zhang, Jian-Kun 186
Zhang, Jia-Yun 188
Zhang, Jing 277
Zhang, Kai 172
Zhang, Kegang 22, 48, 81
Zhang, Lei 325
Zhang, Libin 135
Zhang, Lifei 322
Zhang, Lijuan 340
Zhang, Liqian 620
Zhang, Luoyun 75, 521
Zhang, Nichole 188
Zhang, Qihui 269
Zhang, Qingtian 533
Zhang, Qingzhu 41, 55, 87, 530
Zhang, Rui 408, 421
Zhang, Ruibing 584
Zhang, Ruohan 432
Zhang, Tianfu 16
Zhang, Wenwen 461
Zhang, Xiao-Lei 391
Zhang, Xiao-Wei 644
Zhang, Xin 246, 309
Zhang, Xingdi 401
Zhang, Xinyu 271

Zhang, Xuebing 627
Zhang, Xuechun 295
Zhang, Xuexiang 41
Zhang, Yang 122
Zhang, Yan-Qiu 391
Zhang, Yibin 635
Zhang, Yinchan 357
Zhang, Yintong 25
Zhang, Yu 193, 196, 243, 317, 319, 325
Zhang, Yueyu 138, 160
Zhang, Zhaohao 45, 530
Zhang, Zhijie 312
Zhang, Zihan 217
Zhang, Zi-Han 241
Zhao, Atman 367, 398
Zhao, Chao 168, 277
Zhao, Han 533
Zhao, Hui 157
Zhao, Junfeng 435
Zhao, Lianfu 217
Zhao, T. G. 444
Zhao, Tianxiang 38
Zhao, Xiaotian 578
Zhao, Yage 560
Zhaoa, Hongwen 150
Zheng, Deguang 468
Zheng, Feijun 461
Zheng, Haoqi 16
Zheng, Jiaqi 408
Zheng, Jinfu 165
Zhong, Bo 468
Zhong, Kun 45
Zhong, Yujia 306
Zhong, Yuting 464
Zhou, Bing 32
Zhou, Dylan 25
Zhou, Feichi 32, 58
Zhou, Guohua 441, 447, 608
Zhou, Hexin 176
Zhou, Ke 374
Zhou, Lunmao 608
Zhou, Na 41, 530
Zhou, Pingqiang 590, 602
Zhou, Pingsheng 48, 81
Zhou, Qiguang 160
Zhou, Wei 425
Zhou, Wei-Yu 256
Zhou, Xin 16
Zhou, Xuegong 566
Zhou, Ya 179
Zhou, Yongjie 200
Zhou, Yunyan 494

Zhu, Haiyun 198, 206
Zhu, Hongliang 12, 16
Zhu, Jia 415
Zhu, Lei 333, 337
Zhu, Lu 138
Zhu, Lunan 243
Zhu, Meng-Jiao 241
Zhu, Min 432
Zhu, Peng 223
Zhu, Quanzhou 32, 58
Zhu, Rujun 105
Zhu, Shaojia 317
Zhu, Tianxiang 78
Zhu, Wei 614
Zhu, Xiaolei. 563
Zhu, Xiaolei 569, 572, 617
Zhu, Xiaona 518, 560
Zhu, Yebo 354
Zhu, Yingjie 32
Zhu, Yueqin 374
Zhu, Zheng-Yong 277
Zhu, Zhijun 262
Zhuang, Junjun 404
Zhuo, Cheng 599, 630
Zhuo, Jiaxiang 128
Zihan, Zhang 236
Zou, An 611
Zou, Xinbo 432
Zou, Yida 354, 360
Zuo, Jiaxin. 551
Zuo, Wangge 620